JACARANDA MATHS QUEST
MATHEMATICAL METHODS 11

VCE UNITS 1 AND 2 | THIRD EDITION

JACARANDA MATHS QUEST

MATHEMATICAL METHODS 11

VCE UNITS 1 AND 2 | THIRD EDITION

SUE MICHELL

BEVERLY LANGSFORD WILLING

RAYMOND ROZEN

MARGARET SWALE

jacaranda
A Wiley Brand

Third edition published 2023 by
John Wiley & Sons Australia, Ltd
155 Cremorne Street, Cremorne, Vic 3121

First edition published 2016
Second edition published 2019

Typeset in 10.5/13 pt TimesLTStd

ISBN: 978-1-119-87641-0

The covers of the *Jacaranda Maths Quest VCE Mathematics* series are the work of Victorian artist Lydia Bachimova.

Lydia is an experienced, innovative and creative artist with over 10 years of professional experience, including five years of animation work with Walt Disney Studio in Sydney. She has a passion for hand drawing, painting and graphic design.

Illustrated by diacriTech and Wiley Composition Services

Typeset in India by diacriTech

A catalogue record for this book is available from the National Library of Australia

Printed in Singapore
M121038_080822

Contents

About this resource

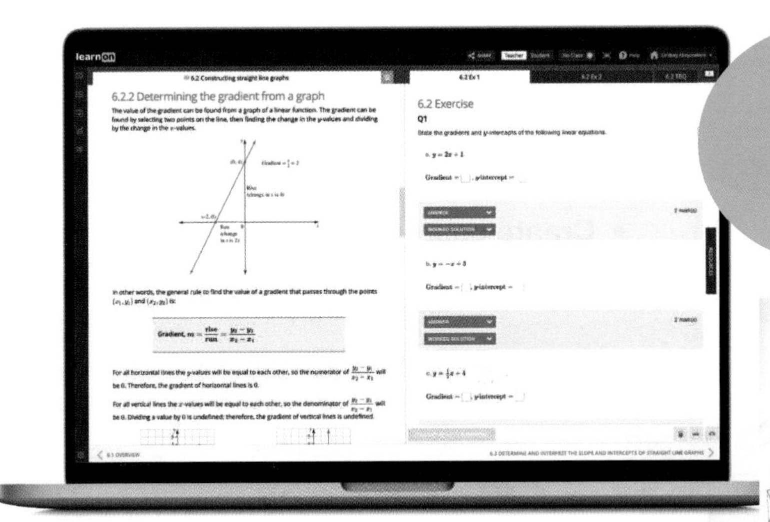

Everything you need for your students to succeed

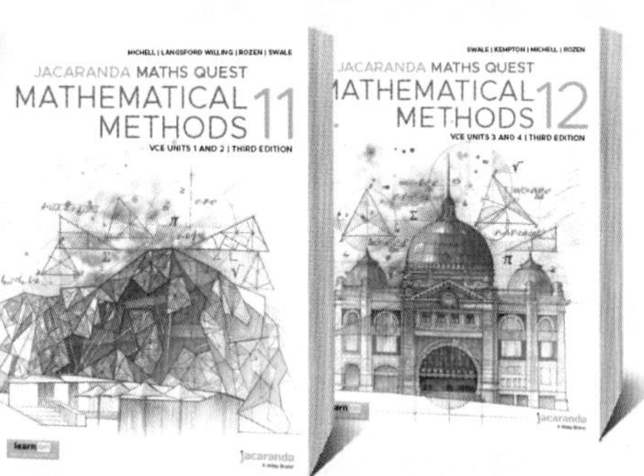

JACARANDA MATHS QUEST

MATHEMATICAL METHODS 11 VCE UNITS 1 AND 2 | THIRD EDITION

Developed by expert Victorian teachers for VCE students

Tried, tested and trusted. The NEW Jacaranda VCE Mathematics series continues to deliver curriculum-aligned material that caters to students of all abilities.

Completely aligned to the VCE Mathematics Study Design

Our expert author team of practising teachers and assessors ensures 100 per cent coverage of the new VCE Mathematics Study Design (2023–2027).

Everything you need for your students to succeed, including:

- NEW! Access targeted question sets including exam-style questions and all relevant past VCAA exam questions since 2013. Ensure assessment preparedness with practice Mathematical investigations.
- NEW! Be confident your students can get unstuck and progress, in class or at home. For every question online they receive immediate feedback and fully worked solutions.
- NEW! Teacher-led videos to unpack concepts, plus VCAA exam questions and exam-style questions to fill learning gaps after COVID-19 disruptions.

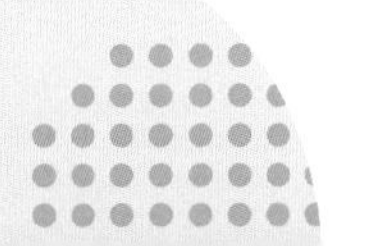

Learn online with Australia's most

Everything you need for each of your lessons in one simple view

- **Trusted, curriculum-aligned theory**
- **Engaging, rich multimedia**
- **All the teacher support resources you need**
- **Deep insights into progress**
- **Immediate feedback for students**
- **Create custom assignments in just a few clicks.**

Practical teaching advice and ideas for each lesson provided in teachON

Each lesson linked to the Key Knowledge (and Key Skills) from the VCE Mathematics Study Design

Reading content and rich media including embedded videos and interactivities

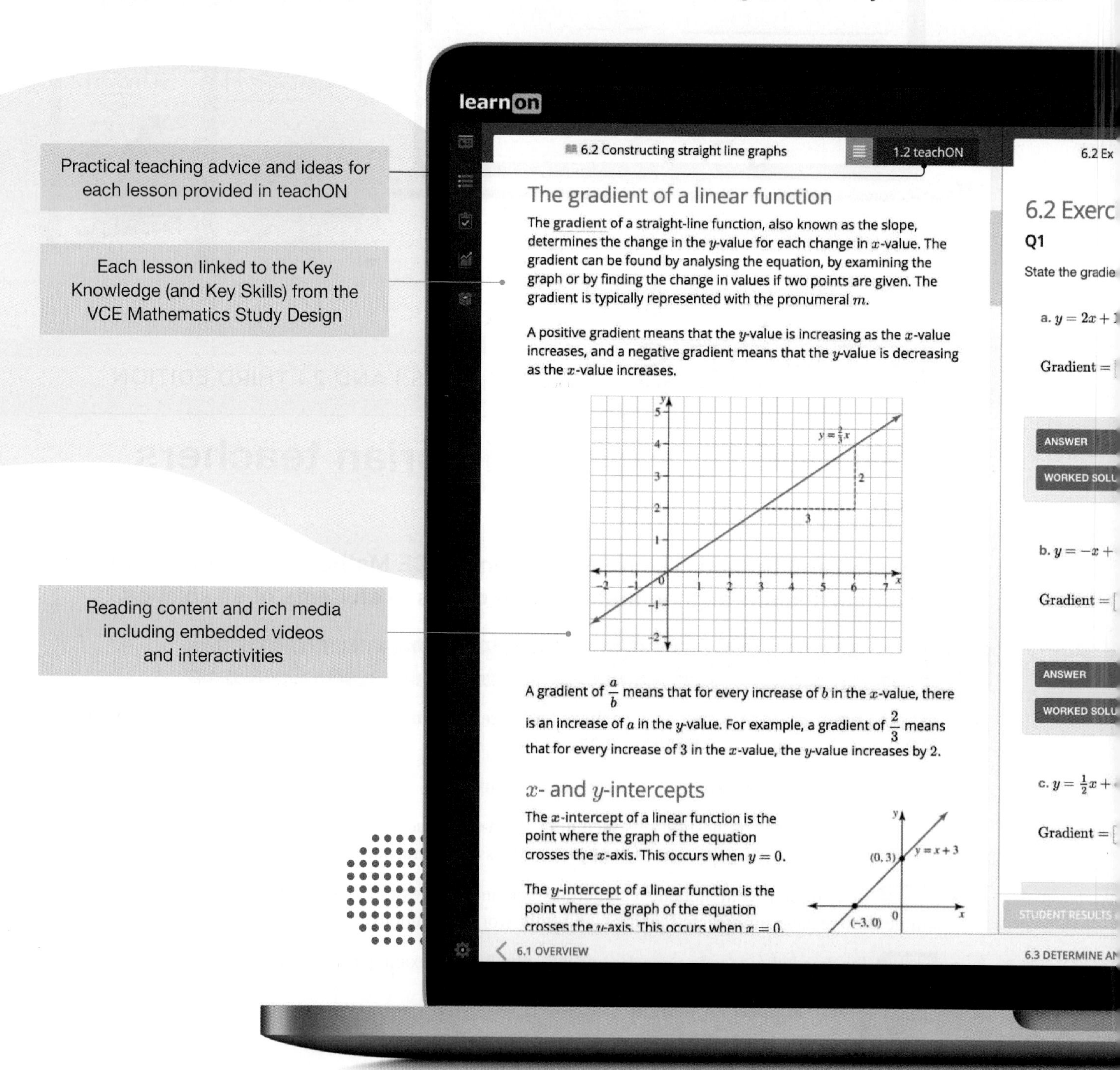

powerful learning tool, learnON

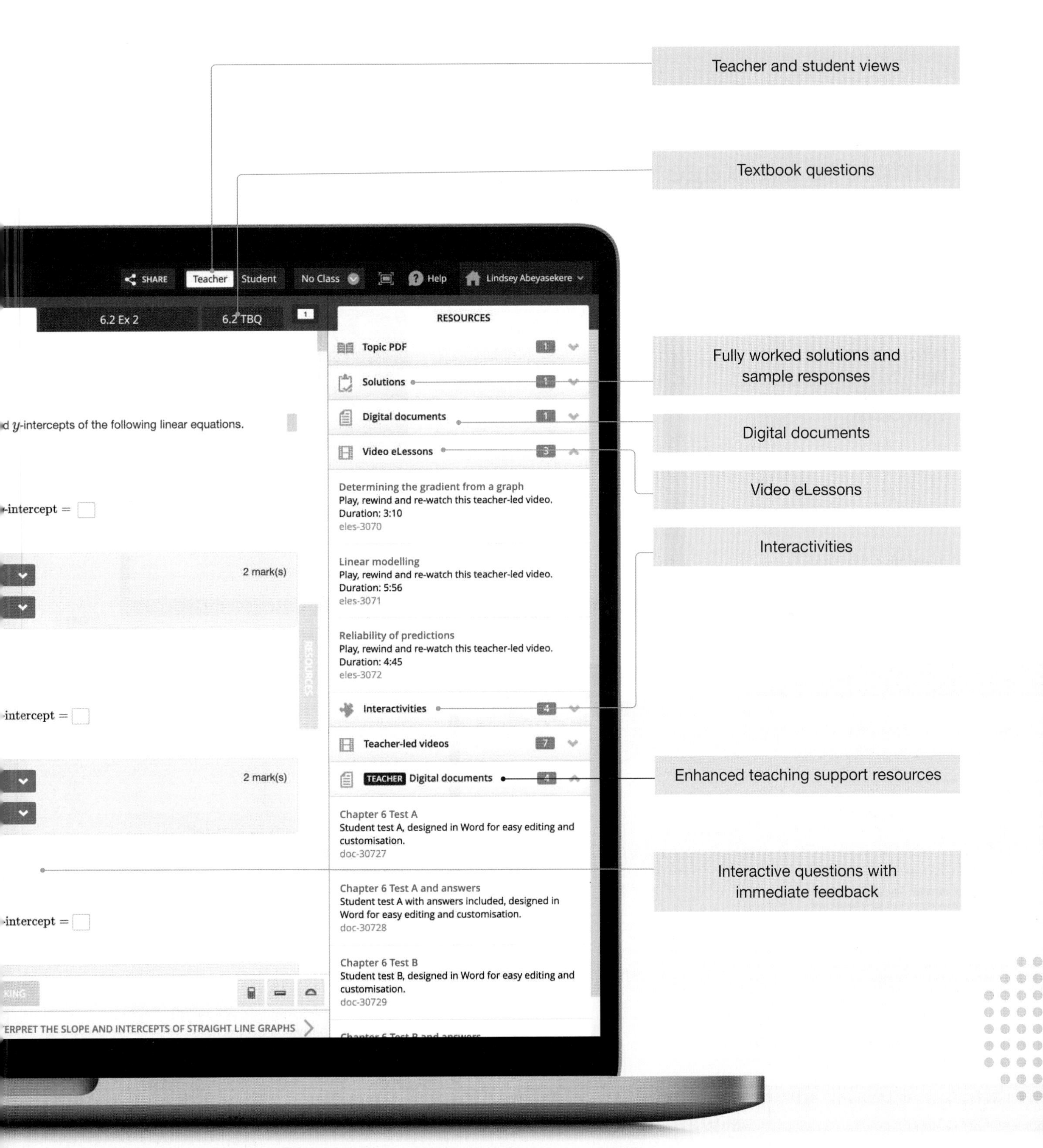

Get the most from your online resources

Online, these new editions are the **complete package**

Trusted Jacaranda theory, plus tools to support teaching and make learning more engaging, personalised and visible.

Each topic is linked to Key Knowledge (and Key Skills) from the VCE Mathematics Study Design.

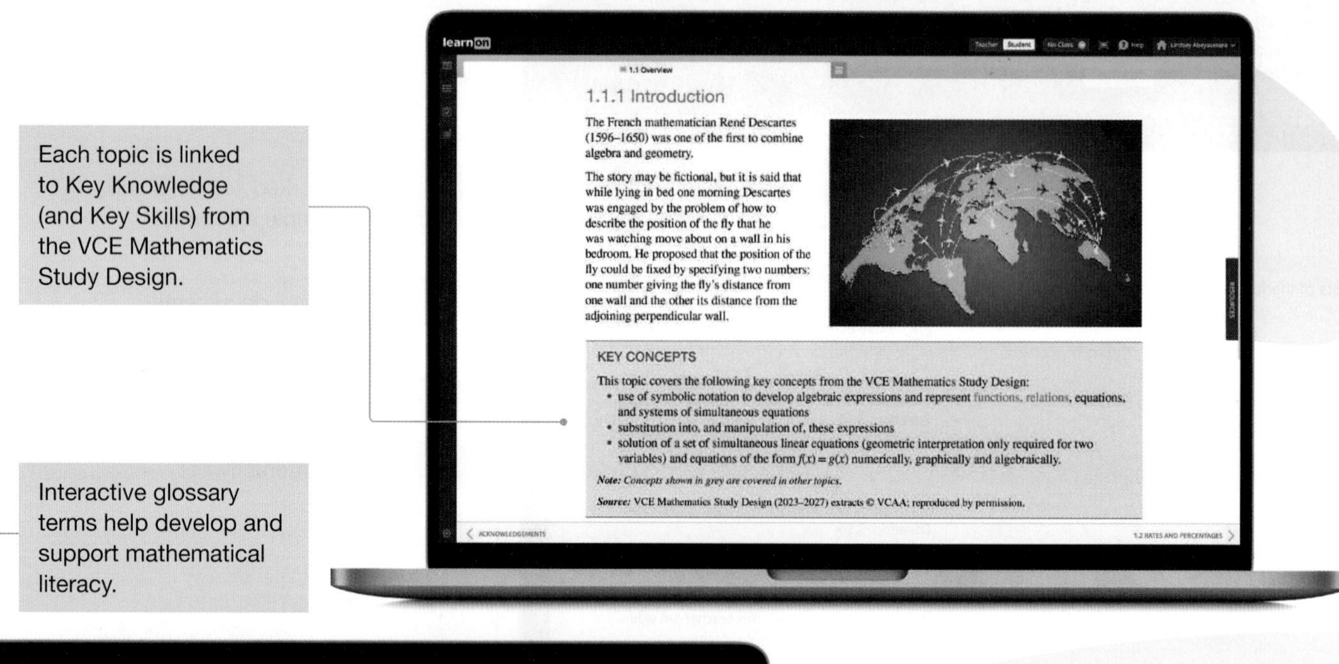

Interactive glossary terms help develop and support mathematical literacy.

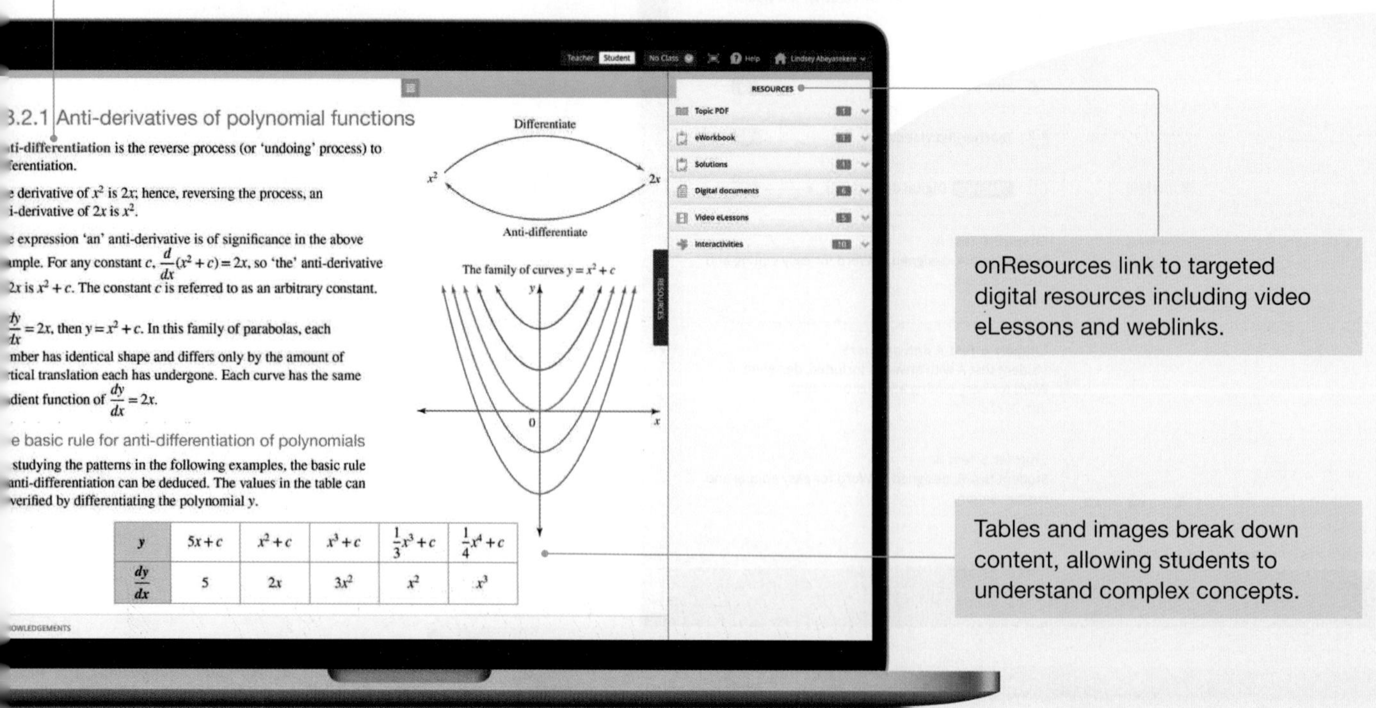

onResources link to targeted digital resources including video eLessons and weblinks.

Tables and images break down content, allowing students to understand complex concepts.

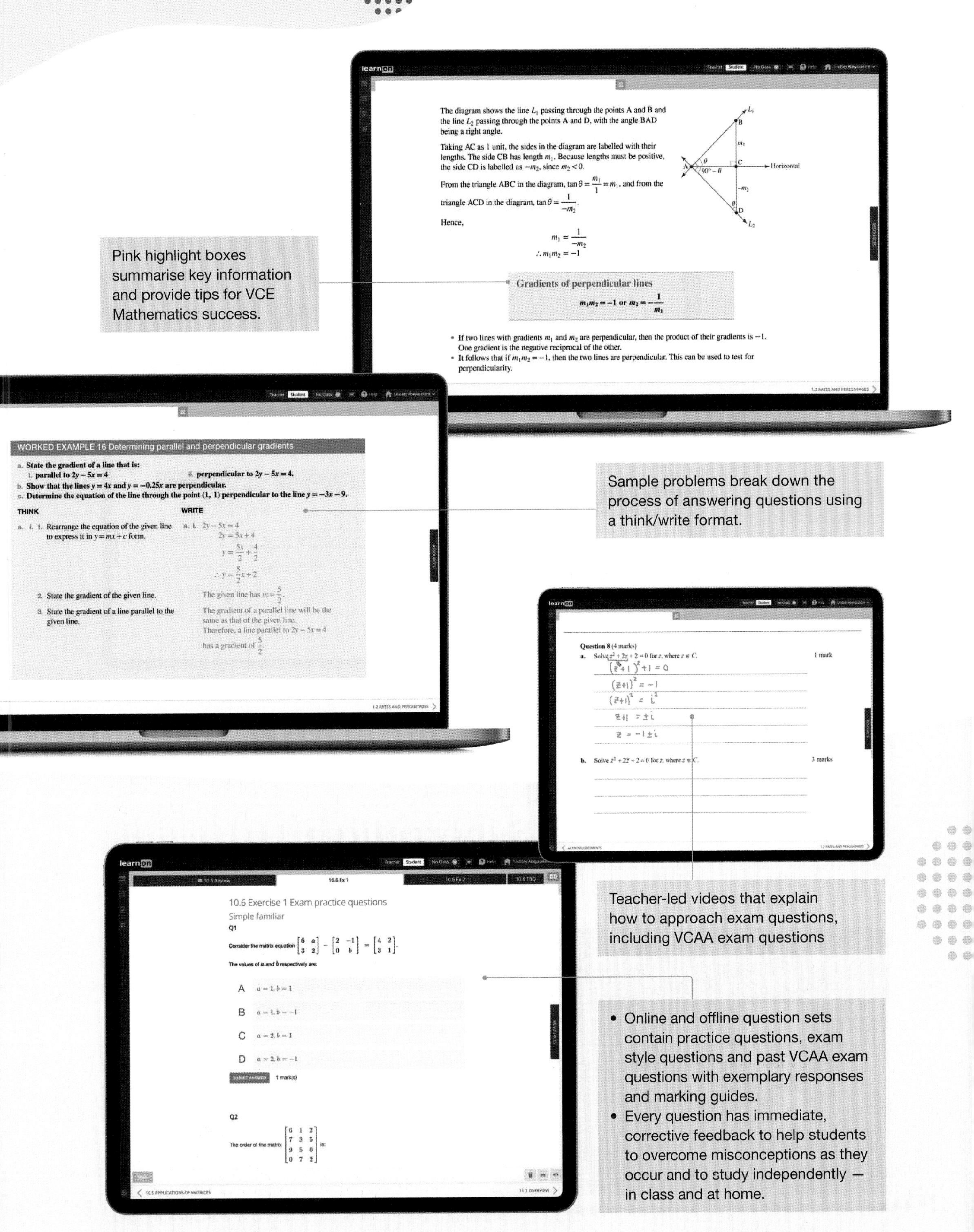
Pink highlight boxes summarise key information and provide tips for VCE Mathematics success.
Gradients of perpendicular lines
WORKED EXAMPLE 16 Determining parallel and perpendicular gradients
Sample problems break down the process of answering questions using a think/write format.
Teacher-led videos that explain how to approach exam questions, including VCAA exam questions
10.6 Exercise 1 Exam practice questions
• Online and offline question sets contain practice questions, exam style questions and past VCAA exam questions with exemplary responses and marking guides.
• Every question has immediate, corrective feedback to help students to overcome misconceptions as they occur and to study independently — in class and at home.

Topic reviews

Topic reviews include online summaries and topic level review exercises that cover multiple concepts. Topic level exam questions are structured just like the exams.

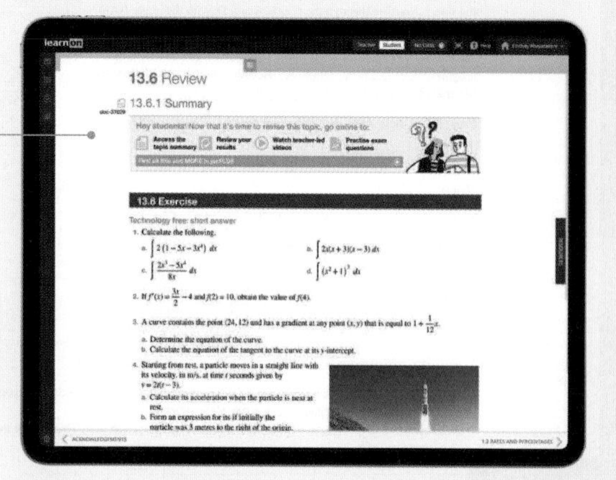

End-of-topic exam questions include relevant past VCE exam questions and are supported by teacher-led videos.

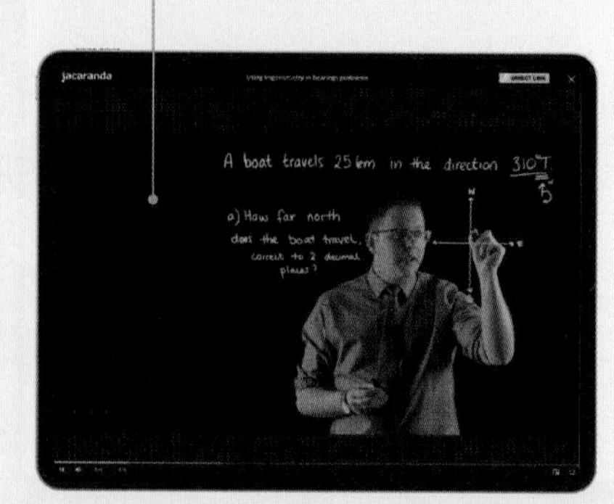

Get exam-ready!

Students can start preparing from lesson one, with exam questions embedded in every lesson — with relevant past VCAA exam questions since 2013.

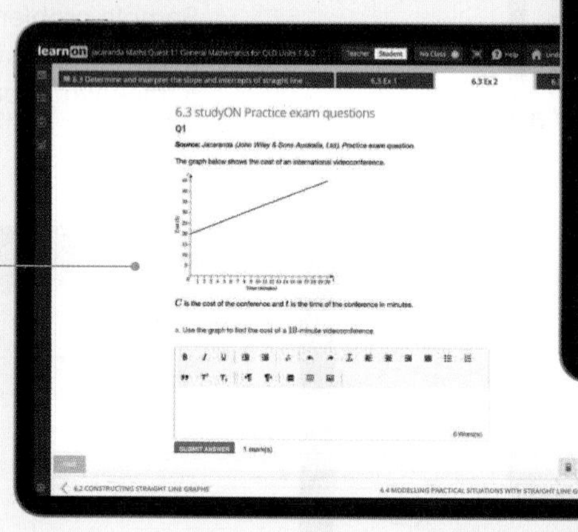

Practice, customisable Mathematical investigations available to build student competence and confidence.

Combine units flexibly with the Jacaranda Supercourse

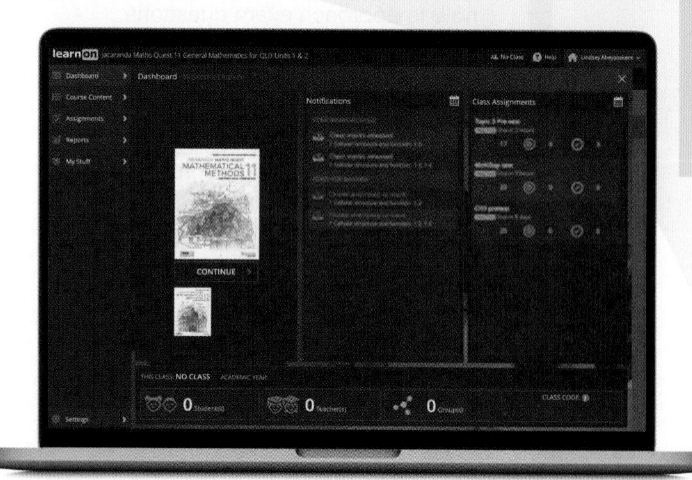

Build the course you've always wanted with the Jacaranda Supercourse. You can combine all Mathematical Methods Units 1 to 4, so students can move backwards and forwards freely. Or Methods and General Units 1 & 2 for when students switch courses. The possibilities are endless!

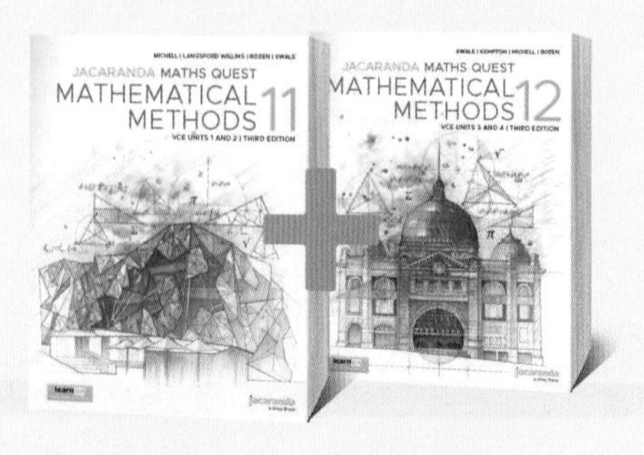

A wealth of teacher resources

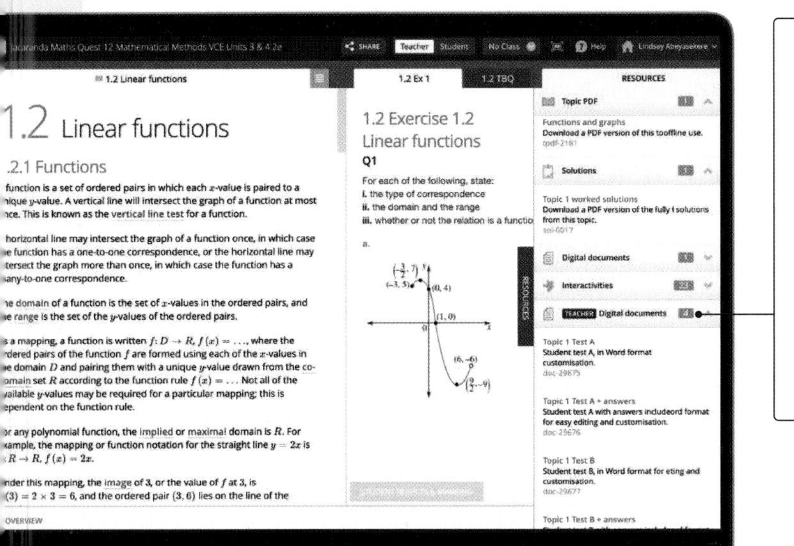

Enhanced teacher support resources, including:

- work programs and curriculum grids
- teaching advice and additional activities
- quarantined topic tests (with solutions)
- quarantined Mathematical investigations (with worked solutions and marking rubrics).

Customise and assign

A testmaker enables you to create custom tests from the complete bank of thousands of questions (including past VCAA exam questions).

Reports and results

Data analytics and instant reports provide data-driven insights into performance across the entire course.

Show students (and their parents or carers) their own assessment data in fine detail. You can filter their results to identify areas of strength and weakness.

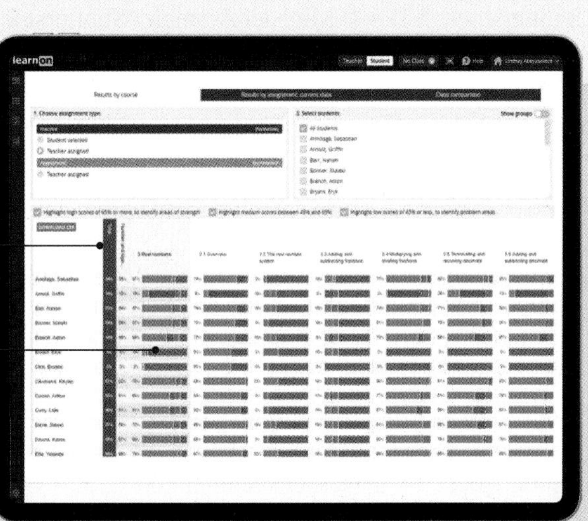

Acknowledgements

The authors and publisher would like to thank the following copyright holders, organisations and individuals for their assistance and for permission to reproduce copyright material in this book.

Selected extracts from the VCE Mathematics Study Design (2023–2027) are copyright Victorian Curriculum and Assessment Authority (VCAA), reproduced by permission. VCE* is a registered trademark of the VCAA. The VCAA does not endorse this product and makes no warranties regarding the correctness and accuracy of its content. To the extent permitted by law, the VCAA excludes all liability for any loss or damage suffered or incurred as a result of accessing, using or relying on the content. Current VCE Study Designs and related content can be accessed directly at www.vcaa.vic.edu.au. Teachers are advised to check the VCAA Bulletin for updates.

Images

• CasioIndiaShop: **72** • © Anton Snarikov / Shutterstock: **441** • Shutterstock / Constantine Panki: **505** • Shutterstock / Dhanu Narenthiran: **456** • Shutterstock / ModernNomads: **455** • Shutterstock / Susan Law Cain: **465** • © afoto6267 / Shutterstock: **734** • © Africa Studio / Shutterstock: **286** • © Alex Staroseltsev / Shutterstock: **792** • © ALEXEY FILATOV/Shutterstock: **522** • © AlohaHawaii / Shutterstock: **312** • © Andrey Shtanko / Shutterstock: **738** • © Ann in the uk / Shutterstock: **235** • © Anteromite / Shutterstock: **132** • © ArtFamily\n / Shutterstock: **12** • © Arthimedes / Shutterstock: **622** • © balein\n / Shutterstock: **27** • © barbaliss / Shutterstock: **510** • © Billion Photos / Shutterstock: **547** • © Bonnie Taylor Barry / Shutterstock: **755** • © Brandon Bourdages / Shutterstock: **330** • © Calin Stan / Shutterstock: **585** • © ChameleonsEye / Shutterstock: **257** • © Cryptographer / Shutterstock: **233** • © D Kvarfordt\n / Shutterstock: **42** • © DaCek / Shutterstock: **129** • © Damian Pankowiec/Shutterstock: **654** • © Daniel Prudek / Shutterstock: **621** • © Davi Sales Batista / Shutterstock: **352** • © Dhoxax / Shutterstock: **330** • © Dirk Ott / Shutterstock: **737** • © Dmitrydesign/Shutterstock: **2** • © DronG / Shutterstock: **620** • © Elena Elisseeva\n / Shutterstock: **23** • © FooTToo / Shutterstock: **547** • © Fotos593 / Shutterstock: **702** • © Fototaras / Shutterstock: **792** • © Fred Fokkelman / Shutterstock: **748** • © FWStudio / Shutterstock: **363** • © Gabe Smith / Shutterstock: **98** • © George Dolgikh / Shutterstock: **620** • © givaga / Shutterstock: **86** • © glebchik / Shutterstock: **234** • © Gorodenkoff / Shutterstock: **793** • © gvictoria / Shutterstock: **557** • © Have a nice day Photo / Shutterstock: **619** • © hddigital / Shutterstock: **313** • © Holger Wulschlaeger / Shutterstock: **663** • © Jaimie Duplass / Shutterstock: **65** • © JASPERIMAGE / Shutterstock: **737** • © jezzerharper / Shutterstock: **656** • © JF19\n / Shutterstock: **14** • © Johan Larson / Shutterstock: **331** • © Joy Rector / Shutterstock: **792** • © JP Phillippe / Shutterstock: **183** • © Kamol Jindamanee / Shutterstock: **110** • © kezza / Shutterstock: **656** • © ktasimar / Shutterstock: **620** • © locrifa / Shutterstock: **55** • © LuFeeTheBear / Shutterstock: **621** • © Maciej Kopaniecki / Shutterstock: **622** • © Mandy Creighton / Shutterstock: **701** • © MarcelClemens / Shutterstock: **595** • © Mattia Mazzucchelli / Shutterstock: **521** • © Michael Zysman / Shutterstock: **687** • © Michal Kowalski / Shutterstock: **163** • © Mirt Alexander / Shutterstock: **258** • © momopixs\n / Shutterstock: **13** • © Monkey Business Images / Shutterstock: **13, 545** • © NASA Images / Shutterstock: **56** • © Natata/Shutterstock: **772** • © Nicholas Rjabow / Shutterstock: **325** • © Nikifor Todorov / Shutterstock: **330** • © Nils Versemann / Shutterstock: **109** • © Pedro Maria / Shutterstock: **620** • © Pete Pahham / Shutterstock: **738** • © photo-denver / Shutterstock: **95** • © Photoseeker / Shutterstock: **164** • © PhotoStockImage / Shutterstock: **165** • © photoyh / Shutterstock: **235** • © Prrrettty / Shutterstock: **748** • © Rafael Lavenere\n / Shutterstock: **43** • © RAGMA IMAGES/Shutterstock: **127** • © Rattiya Thongdumhyu / Shutterstock: **233** • © rodho / Shutterstock: **236** • © roseed abbas / Shutterstock: **68** • © Sarawut Chainawarat / Shutterstock: **736** • © Seregam / Shutterstock: **687** • © Sergey Nivens / Shutterstock: **806** • © shootz photography / Shutterstock: **754** • © Shots Studio / Shutterstock: **13** • © Simon_g / Shutterstock: **621** • © sommart sombutwanitkul / Shutterstock: **653** • © Tamara Kulikova / Shutterstock: **792** • © tatevrika / Shutterstock: **746** • © Thanawan Wisetsin / Shutterstock: **80** • © tobkatrina / Shutterstock: **547** • © TORWAISTUDIO / Shutterstock: **662** • © Valentyn Volkov / Shutterstock: **132** • © Visual Collective / Shutterstock: **1** • © worker / Shutterstock: **232** • © Yaping / Shutterstock: **687** • © Zhukov Oleg / Shutterstock:

131 • © Anan Kaewkhammul / Shutterstock: **434** • © Dimitris Leonidas / Shutterstock: **431** • © Gelpi / Shutterstock: **390** • © Kamira / Shutterstock: **389** • © Leon Rafael / Shutterstock: **410** • © Leonid Andronov / Shutterstock: **400** • © Maxim Godkin / Shutterstock: **433** • © Mega Pixel / Shutterstock: **402** • © Michael D Brown / Shutterstock: **411** • © michaeljung / Shutterstock: **402** • © oliveromg / Shutterstock: **432** • © patpitchaya / Shutterstock: **441** • © Sashkin / Shutterstock: **412** • © Sergey Peterman / Shutterstock: **447** • © Siberian Art / Shutterstock: **411** • © SmartPhotoLab / Shutterstock: **417** • © szefei / Shutterstock: **431** • © Texas Instruments Incorporated: **72** • © Aleks Melnik / Shutterstock: **86** • © drawhunter / Shutterstock: **184** • © EsHanPhot / Shutterstock: **586** • © Jesscr7 / Shutterstock: **301** • © Nearbirds / Shutterstock: **133** • © Toponium / Shutterstock: **85** • © Universal Images Group North America LLC / Alamy Stock Photo: **302** • © Vladi333 / Shutterstock: **771** • © Alaettin YILDIRIM / Shutterstock: **446** • © Lorelyn Medina / Shutterstock: **447** • © Luis Molinero / Shutterstock: **400** • © Ninell / Shutterstock: **444** • © Teguh Mujiono / Shutterstock: **412**

Text

Every effort has been made to trace the ownership of copyright material. Information that will enable the publisher to rectify any error or omission in subsequent reprints will be welcome. In such cases, please contact the Permissions Section of John Wiley & Sons Australia, Ltd.

1 Lines and linear relationships

LEARNING SEQUENCE

Fully worked solutions for this topic are available online.

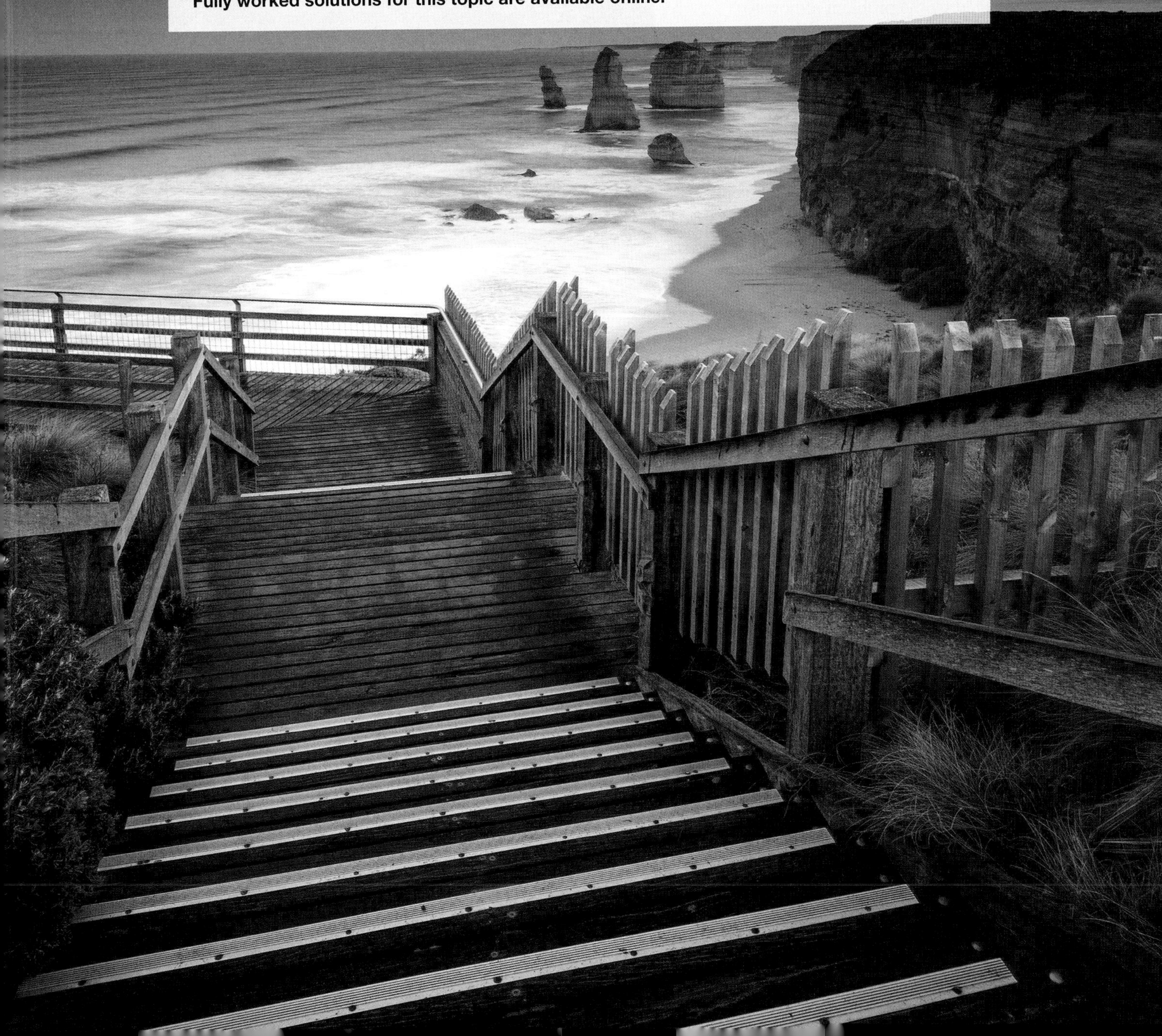

1.1 Overview

Hey students! Bring these pages to life online

Watch videos

Engage with interactivities

Answer questions and check results

Find all this and MORE in jacPLUS

1.1.1 Introduction

The French mathematician René Descartes (1596–1650) was one of the first to combine algebra and geometry.

The story may be fictional, but it is said that while lying in bed one morning Descartes was engaged by the problem of how to describe the position of the fly that he was watching move about on a wall in his bedroom. He proposed that the position of the fly could be fixed by specifying two numbers: one number giving the fly's distance from one wall and the other its distance from the adjoining perpendicular wall. The concept of specifying the position of a point using coordinates had come to Descartes.

Furthermore, he recognised that algebra and geometry could be combined. Graphs, such as lines and circles, could be specified by equations. This branch of mathematics, developed by Descartes, Fermat and others, is called analytic geometry or coordinate geometry.

The idea of using a reference frame was not entirely new, as the ancient Greeks had used such a concept. The astronomer Eudoxus (c.408–335 BCE) devised a complex coordinate system to represent the motion of the sun and moon with the Earth as origin.

Today coordinate systems are used in large-scale real-world applications, such as GPS tracking of vehicles from aircraft to delivery vans. GPS tracks positions relative to the two fixed axes of the Greenwich Prime Meridian and the equator. In the abstract, digital world, the three-dimensional polar and cylindrical coordinate systems are used in all computer animation.

KEY CONCEPTS

This topic covers the following key concepts from the VCE Mathematics Study Design:

- use of symbolic notation to develop algebraic expressions and represent functions, relations, equations, and systems of simultaneous equations
- substitution into, and manipulation of, these expressions
- solution of a set of simultaneous linear equations (geometric interpretation only required for two variables) and equations of the form $f(x) = g(x)$ numerically, graphically and algebraically.

Note: *Concepts shown in grey are covered in other topics.*

1.2 Linear equations and inequations

LEARNING INTENTION

At the end of this subtopic you should be able to:
- re-arrange and solve simple algebraic equations and inequalities by hand.

1.2.1 Linear equations

Throughout this subject, the ability to solve equations and inequations of varying complexity will be required. In this section we revise the underlying skills in solving linear equations, noting that a **linear equation** involves a variable that has an index or power of one. For example, $2x+3=7$ is a linear equation in the variable x, whereas $2x^2+3=7$ is not (it is a quadratic equation).

WORKED EXAMPLE 1 Solving linear equations

Solve the following linear equations for x.

a. $4(3+2x)=22+3x$

b. $\dfrac{2x+3}{5}-\dfrac{1-4x}{7}=x$

THINK

a. Expand the brackets and collect all terms in x together on one side in order to solve the equation.

WRITE

a.
$$\begin{aligned}4(3+2x)&=22+3x\\12+8x&=22+3x\\8x-3x&=22-12\\5x&=10\\x&=\frac{10}{5}\\\therefore x&=2\end{aligned}$$

THINK

b. 1. Express fractions with a common denominator and simplify.

WRITE

b.
$$\begin{aligned}\frac{2x+3}{5}-\frac{1-4x}{7}&=x\\\frac{7(2x+3)-5(1-4x)}{35}&=x\\\frac{14x+21-5+20x}{35}&=x\\\frac{34x+16}{35}&=x\end{aligned}$$

THINK

2. Remove the fraction by multiplying both sides by the common denominator and solve for x.

WRITE

$$\begin{aligned}34x+16&=35x\\16&=35x-34x\\\therefore x&=16\end{aligned}$$

TI | THINK

b. 1. On a Calculator page, press MENU and select:
3. Algebra
1. Solve
Complete the entry line as:
solve

$$\left(\frac{2x+3}{5}-\frac{1-4x}{7}=x,x\right)$$

Then press ENTER.

DISPLAY/WRITE

solve$\left(\frac{2\cdot x+3}{5}-\frac{1-4\cdot x}{7}=x,x\right)$ $x=16$

2. The answer appears on the screen.

$x=-1, y=-2, z=1$

CASIO | THINK

b. 1. On a Main screen, complete the entry line as:
solve

$$\left(\frac{2x+3}{5}-\frac{1-4x}{7}=x,x\right)$$

Then press EXE.

DISPLAY/WRITE

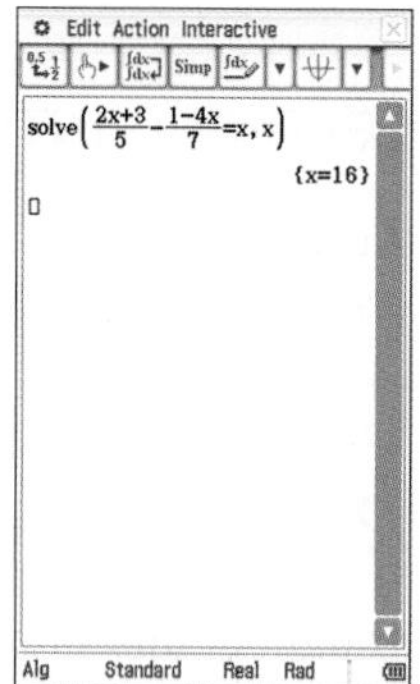

2. The answer appears on the screen.

$x=16$

Literal linear equations

Literal equations contain pronumerals rather than known numbers. The solution to a literal equation in x expresses x as a combination of these pronumerals. Always check to see if answers may be simplified using algebraic skills such as factorisation.

WORKED EXAMPLE 2 Solving literal linear equations

Solve for x: $\frac{x+a}{b}=\frac{b-x}{a}$.

THINK	WRITE
1. Place each fraction on the common denominator and then multiply each side by that term. *Note:* Since there is only one fraction on each side of this equation, a quick way to do this is to 'cross-multiply'.	$\frac{x+a}{b}=\frac{b-x}{a}$ $a(x+a)=b(b-x)$ $ax+a^2=b^2-bx$
2. Collect all the terms in x together and take out x as the common factor.	$ax+bx=b^2-a^2$ $x(a+b)=b^2-a^2$
3. Divide by the coefficient of x to obtain an expression for x.	$\therefore x=\frac{b^2-a^2}{a+b}$
4. Simplify the expression, if possible.	The numerator can be factorised as a difference of two squares. $x=\frac{b^2-a^2}{a+b}$ $=\frac{(b-a)(b+a)}{a+b}$
5. Cancel the common factor to give the solution in its simplest form.	$x=\frac{(b-a)\cancel{(b+a)}}{\cancel{a+b}}$ $\therefore x=b-a$

1.2.2 Linear inequations

An **inequation** contains one of the order or inequality symbols:

$<$ less than
$\leq$ less than or equal to
$>$ greater than
$\geq$ greater than or equal to.

Linear inequations are solved in a similar way to linear equations; however, care must be taken when multiplying or dividing by a negative number. Remember to reverse the order symbol when multiplying or dividing by a negative number.

To illustrate this, consider the inequality statement $-6 < 15$. If this inequality is divided by -3, then the statement must become $2 > -5$, so the order symbol has been reversed.

The solutions to linear inequations are sets of values satisfying an inequality.

When illustrating inequalities on a number line:
- an open circle, o, is used when the end point is not included ($<$ or $>$)
- a closed circle, •, is used when the end point is included ($\leq$ or $\geq$).

WORKED EXAMPLE 3 Solving linear inequations

Calculate the values of x for which $5 - \frac{4x}{5} > 13$ and show this set of values on a number line.

THINK	WRITE
1. Subtract 5 from both sides of the inequation. *Note:* Subtracting a number does not affect the inequality symbol.	$5 - \frac{4x}{5} > 13$ $-\frac{4x}{5} > 13 - 5$ $-\frac{4x}{5} > 8$
2. Multiply both sides by 5. *Note:* Multiplying by a positive number does not affect the inequality symbol.	$-4x > 8 \times 5$ $-4x > 40$
3. Divide both sides by -4. *Note:* Dividing by a negative number does require the symbol to be reversed.	$x < \frac{40}{-4}$ $\therefore x < -10$
4. Illustrate this set of values on a number line.	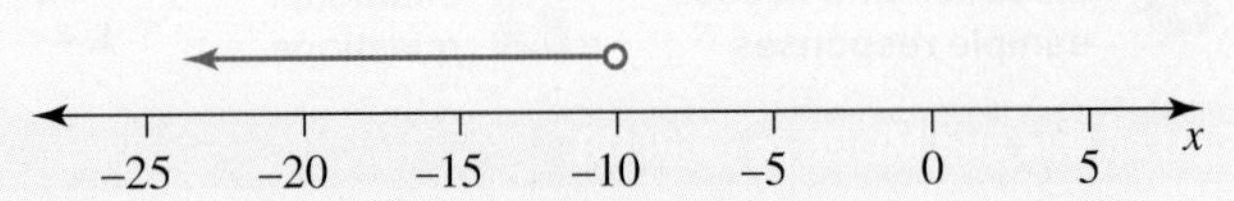 The number line has an open end at $x = -10$ since this value is not included in the set of solutions, $x < -10$.

1.2.3 Solving problems using equations

Algebra enables real-life problems to be expressed as equations.

In setting up equations:
- define the symbols used for the variables, specifying units where appropriate
- ensure any units used are consistent
- express answers in the context of the problem.

WORKED EXAMPLE 4 Solving problems using linear equations

The organisers of an annual student fundraising event for charity know there will be a fixed cost of \$120 plus an estimated cost of 60 cents per student for incidental costs on the day of the fundraiser. The entry fee to the fundraising event is set at \$5.

a. Form an algebraic model for the profit the event can expect to make.

b. Determine the least number of students who must attend the event to avoid the organisers making a loss.

THINK	WRITE
a. 1. Define the variables.	a. Let n = the number of students attending the event. Let P = the profit made in dollars.
2. Form expressions for the cost and the revenue, ensuring units are consistent.	Profit depends on costs and revenues. Revenue (\$) $= 5n$ Costs (\$) $= 120 + 0.60n$
3. Form the expression for the profit to define the algebraic model.	Profit = revenue − costs Hence, $P = 5n - (120 + 0.60n)$ $\therefore P = 4.4n - 120$ gives the linear model for the profit.
b. 1. Impose the condition required to avoid a loss and calculate the consequent restriction on n.	b. A loss is made if $P < 0$. To avoid making a loss, the organisers require $P \geq 0$. $4.4n - 120 \geq 0$ $4.4n \geq 120$ $n \geq \frac{120}{4.4}$ $\therefore n \geq 27.2727...$
2. Express the answer in the context of the question.	To avoid making a loss, at least 28 students need to attend the event.

1.2 Exercise

Students, these questions are even better in jacPLUS

Receive immediate feedback and access sample responses

Access additional questions

Track your results and progress

Find all this and MORE in jacPLUS

Technology free

1. WE1 Solve the following linear equations for x.

a. $3(5x - 1) = 4x - 14$

b. $\frac{4 - x}{3} + \frac{3x - 2}{4} = 5$

2. Solve the following linear equations for x.

a. $2x - 5 = 11$

b. $\frac{x}{2} + 4 = -1$

c. $2x - 4 = 5x + 3$

d. $5x + 3 = -4(1 - x)$

e. $\frac{3 - 2x}{6} = \frac{x + 4}{4}$

f. $\frac{4x - 5}{6} - \frac{x + 4}{12} = 1$

3. Solve the following linear equations for x.

a. $7(2x-3)=5(3+2x)$

b. $\frac{4x}{5}-9=7$

c. $4-2(x-6)=\frac{2x}{3}$

d. $\frac{3x+5}{9}=\frac{4-2x}{5}$

e. $\frac{x+2}{3}+\frac{x}{2}-\frac{x+1}{4}=1$

f. $\frac{7x}{5}-\frac{3x}{10}=2\left(x+\frac{9}{2}\right)$

4. **WE2** Solve for x: $\frac{d-x}{a}=\frac{a-x}{d}$.

5. Solve the following literal equations for x, expressing solutions in simplest form.

a. $ax+b=c$

b. $a(x-b)=bx$

c. $a^2x+a^2=ab+abx$

d. $\frac{x}{a}+\frac{x}{b}=a+b$

e. $\frac{bx-a}{c}=\frac{cx+a}{b}$

f. $\frac{x+a}{b}-2=\frac{x-b}{a}$

6. **WE3** Calculate the values of x for which $7-\frac{3x}{8}\leq -2$ and show this set of values on a number line.

7. Solve the following inequations for x.

a. $4-2x\geq 5$

b. $\frac{x-6}{3}+4<1$

c. $2x-3<4x+1$

d. $-2(x-5)-x>3(x+4)$

e. $1-\frac{4-x}{2}>-1$

f. $\frac{5-x}{2}<-\frac{3x+2}{8}$

8. Solve the following inequations for x.

a. $3x-5\leq -8$

b. $4(x-6)+3(2-2x)<0$

c. $1-\frac{2x}{3}\geq -11$

d. $\frac{5x}{6}-\frac{4-x}{2}>2$

e. $8x+7(1-4x)\leq 7x-3(x+3)$

f. $\frac{2}{3}(x-6)-\frac{3}{2}(x+4)>1+x$

Technology active

9. Solve for x: $\frac{7(x-3)}{8}+\frac{3(2x+5)}{4}=\frac{3x}{2}+1$.

10. Solve for x: $b(x+c)=a(x-c)+2bc$.

11. Solve for x: $4(2+3x)>8-3(2x+1)$.

12. Use CAS technology to solve the following equations.

a. $3(5x-2)+5(3x-2)=8(x-2)$

b. $\frac{2x-1}{5}+\frac{3-2x}{4}<\frac{3}{20}$

13. Solve the following problems by first forming linear equations.

a. The sum of two consecutive even numbers is nine times their difference. Calculate the two numbers.

b. Three is subtracted from a certain number and the result is then multiplied by 4 to give 72. Calculate the number.

c. The sum of three consecutive numbers is the same as the sum of 36 and one-quarter of the smallest number. Calculate the three numbers.

d. The length of a rectangle is 12 cm greater than twice its width. If the perimeter of the rectangle is 48 cm, calculate its length and width.

e. The ratio of the length to the width to the height of a rectangular prism is 2 : 1 : 3, and the sum of the lengths of all its edges is 360 cm. Calculate the height of this rectangular prism.

14. **WE4** Although the organisers of a secondhand book sale are allowed free use of the local Scouts Hall for their fete, they must contribute \$100 towards heating and lighting costs and in addition donate 20 c from the sale of each book to the Scouts Association. The books are intended to be sold for \$2.50 each.

a. Form an algebraic model for the profit the book sale can expect to make.

b. Determine the least number of books that must be sold to ensure the organisers make a profit.

15. A company manufactures a special batch of mobile phone covers. The fixed costs are \$250 and materials cost \$1.80 per cover. The company sells the covers for \$12 each. Determine how many mobile phone covers the company will need to sell to cover all of their costs.

16. The local football club is raising funds for new equipment by setting up a takeaway stall during all of their games. The stall cost the club \$65. They sell drinks for \$4 each, but the club purchases the drinks for \$1.20 each. The cost of the new equipment is \$450. Determine how many drinks the club will need to sell before it has the funds for this new equipment.

1.2 Exam questions

Question 1 (3 marks) TECH-FREE

Solve the literal equation for x, expressing the answer in simplest form.

$$\frac{x-a}{b}-2=\frac{b-x}{a}$$

Question 2 (1 mark) TECH-ACTIVE

MC The solution to the equation $3(x-4)\leq 6$ is

A. $x\geq 6$ B. $x\leq 2$ C. $x\leq -2$ D. $x\geq -2$ E. $x\leq 6$

Question 3 (3 marks) TECH-FREE

Solve for x.

$$\frac{x}{5}-\frac{2x-1}{3}\geq -2$$

More exam questions are available online.

1.3 Systems of simultaneous linear equations

LEARNING INTENTION

At the end of this subtopic you should be able to:

- set up and solve systems of simultaneous linear equations involving up to four unknowns, including by hand for a system of two equations in two unknowns.

To determine the value of one variable, one equation is needed. However, to find the values of two variables, two equations are required. These two equations form a 2×2 **system of equations**. To determine the values of three variables, three equations are required, that is a 3×3 system of equations. This pattern continues for four variables with a 4×4 system of equations.

1.3.1 Systems of 2×2 simultaneous linear equations

The usual algebraic approach for finding the values of two variables, x and y, that satisfy two linear equations simultaneously is by using either a *substitution* method or an *elimination* method. Technology may also be used to solve simultaneous linear equations.

Substitution method

In the substitution method, one equation is used to express one variable in terms of the other, and this expression is substituted in place of that variable in the second equation. This creates an equation with just one variable that can then be solved.

Elimination method

In the elimination method, the two equations are combined in such a way as to eliminate one variable, leaving an equation with just one variable that can then be solved.

WORKED EXAMPLE 5 Solving 2 × 2 simultaneous equations

a. Use the substitution method to solve the following system of simultaneous equations for x and y.

$$\begin{aligned} y &= 3x - 1 \\ x - 3y &= 11 \end{aligned}$$

b. Use the elimination method to solve the following system of simultaneous equations for x and y.

$$\begin{aligned} 8x + 3y &= -13 \\ 5x + 4y &= -6 \end{aligned}$$

THINK	WRITE
a. 1. Write the two equations and number them for ease of reference.	a. $y = 3x - 1$ [1] $x - 3y = 11$ [2]
2. Since the first equation already has y in terms of x, it can be substituted for y in the second equation.	Substitute equation [1] into equation [2]: $x - 3(3x - 1) = 11$
3. Solve for x.	$x - 9x + 3 = 11$ $-8x = 11 - 3$ $-8x = 8$ $\therefore x = -1$
4. Substitute the value of x into one of the original equations and calculate the y-value.	Substitute $x = -1$ into equation [1]: $y = 3(-1) - 1$ $\therefore y = -4$
5. State the answer.	The solution is $x = -1$ and $y = -4$.
6. It is a good idea to check that these values satisfy both equations by substituting the values for x and y into the other equation.	Check: In equation [2], if $x = -1$ and $y = -4$, then: $x - 3y = (-1) - 3 \times (-4)$ $= -1 + 12$ $= 11$ True
b. 1. Write the two equations and number them for ease of reference.	b. $8x + 3y = -13$ [1] $5x + 4y = -6$ [2]
2. Choose whether to eliminate x or y and adjust the coefficient of the chosen variable in each equation. *Note:* Neither equation readily enables one variable to be expressed simply in terms of the other, so the elimination method is the more appropriate method here.	Eliminate y. Multiply equation [1] by 4: $32x + 12y = -52$ [3] Multiply equation [2] by 3: $15x + 12y = -18$ [4]
3. Eliminate y and solve for x. *Note:* 'Same Signs Subtract' can be remembered as the SSS rule.	Equation [3] − equation [4]: $32x - 15x = -52 - (-18)$ $17x = -34$ $\therefore x = -2$

4. Substitute the value of x into one of the original equations and calculate the y-value.
Note: Alternatively, start again but this time eliminate x.

Substitute $x = -2$ into equation [2]:

$$5(-2) + 4y = -6$$
$$-10 + 4y = -6$$
$$4y = -6 + 10$$
$$4y = 4$$
$$\therefore y = 1$$

5. State the answer.

The solution is $x = -2$, $y = 1$.

TI | THINK

b. 1. On a Calculator page, press MENU and select:
3. Algebra
1. Solve
Complete the entry line as:
solve $(8x + 3y = -13$ and $5x + 4y = -6, \{x, y\})$
Then press ENTER.

DISPLAY/WRITE

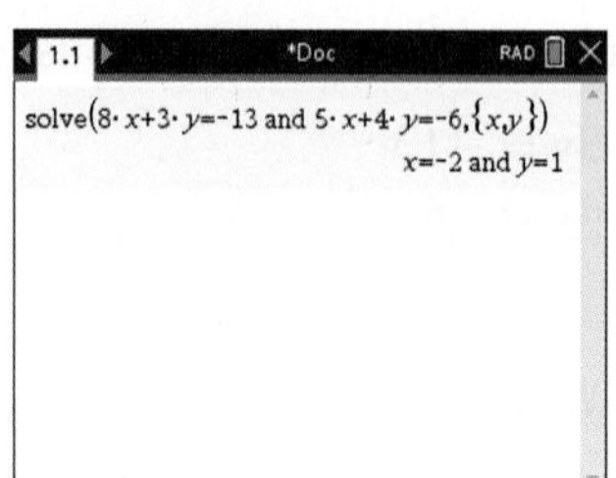

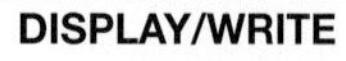

2. The answer appears on the screen.
Note: Alternately, choose the symbol for simultaneous equations on the keyboard.

$x = -2, y = 1$

CASIO | THINK

b. 1. On a Main screen, select:
- Action
- Advanced
- solve

Complete the entry line as:
solve $(\{8x + 3y = -13, 5x + 4y = -6\}, \{x, y\})$
Then press EXE.

DISPLAY/WRITE

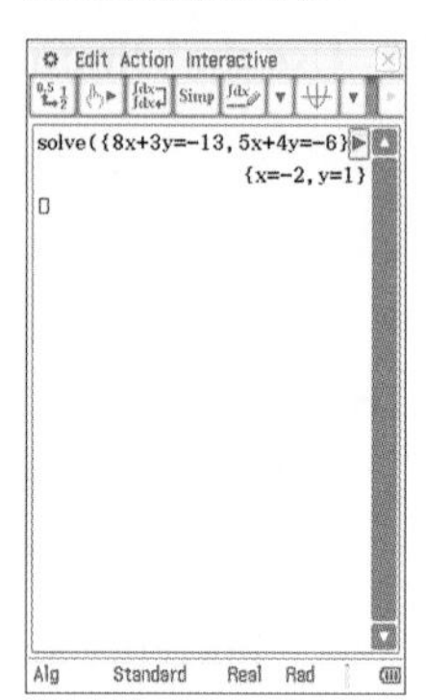

2. The answer appears on the screen.

$x = -2, y = 1$

1.3.2 Problem solving with simultaneous linear equations

The same method used to solve linear problems is used to solve problems that require solving simultaneous linear equations. Remember to always define your variables, taking particular care that all units are consistent.

WORKED EXAMPLE 6 Using simultaneous equations in problem solving

The annual student fundraising event for charity has been scheduled for the next week. If 30 students and 5 staff attend, the revenue gained on entry fees is \$145, whereas if 55 students and 15 staff attend, the revenue is \$312.50. Determine how much the proposed entry fee is for a student and how much it is for a member of the staff.

THINK

a. 1. Define the variables and express the given information using a pair of simultaneous equations.

WRITE

a. Let the entry fees be:
s dollars for students and
t dollars for staff.

$$30s + 5t = 145 \quad [1]$$
$$55s + 15t = 312.5 \quad [2]$$

2. Solve this system of simultaneous equations by elimination.

Eliminate t.
Multiply equation [1] by 3:
$90s + 15t = 435$ [3]
$55s + 15t = 312.5$ [2]
Subtract equation [2] from equation [3]:
$90s - 55s = 435 - 312.5$
$35s = 122.5$
$\therefore s = \frac{122.5}{35}$
$\therefore s = 3.5$
Substitute $s = 3.5$ into equation [1]:
$30(3.5) + 5t = 145$
$105 + 5t = 145$
$5t = 40$
$\therefore t = 8$

3. Write the answer in a sentence.

The student entry fee is \$3.50 and the staff entry fee is \$8.

1.3.3 Systems of simultaneous linear equations with technology

A CAS calculator or other technology is used to solve 3×3 system of equations or 4×4 system of equations.

WORKED EXAMPLE 7 Using technology to solve systems of simultaneous equations

Solve the following 3×3 system of simultaneous equations for x, y and z.

$$3x - 4y + 5z = 10$$
$$2x + y - 3z = -7$$
$$5x + y - 2z = -9$$

THINK

Use your CAS calculator with the steps below to obtain the result.

WRITE

$x = -1, y = -2$ and $z = 1$

TI | THINK

b. 1. On a Calculator page, press MENU and select:
3. Algebra
7. Solve System of Equations
1. Solve System of Equations...
Number of equations: 3
Variables: x, y, z
Press OK.
Enter the equations into the template.
Then press ENTER.

DISPLAY/WRITE

1.1 *Doc RAD
solve$\left(\begin{cases}3\cdot x-4\cdot y+5\cdot z=10\\2\cdot x+y-3\cdot z=-7\\5\cdot x+y-2\cdot z=-9\end{cases}, \{x,y,z\}\right)$
$x=-1$ and $y=-2$ and $z=1$

2. The answer appears on the screen.

$x = -1, y = -2, z = 1$

CASIO | THINK

b. 1. On a Main screen, press [template] to get the template for systems of equations.
Press again to get another row.
Enter the equations into the template and the variables to the right.
Then press EXE.

DISPLAY/WRITE

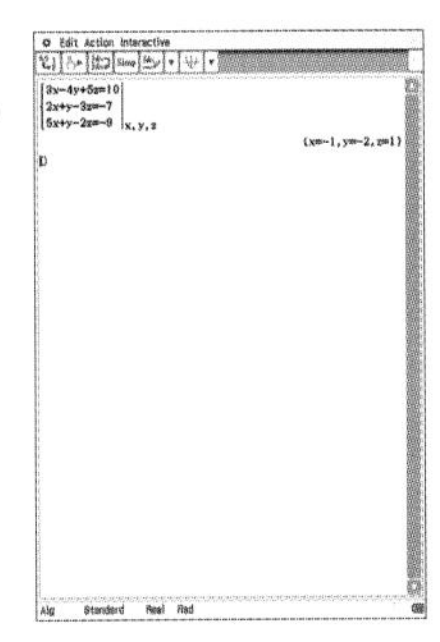

2. The answer appears on the screen.

$x = -1, y = -2, z = 1$

1.3 Exercise

Students, these questions are even better in jacPLUS

Receive immediate feedback and access sample responses

Access additional questions

Track your results and progress

Find all this and MORE in jacPLUS

Technology free

1. WE5 a. Use the substitution method to solve the following system of simultaneous equations for x and y.

$$x = 2y + 5$$
$$4x - 3y = 25$$

b. Use the elimination method to solve the following system of simultaneous equations for x and y.

$$5x + 9y = -38$$
$$-3x + 2y = 8$$

2. Solve each of the following sets of simultaneous equations for x and y.

a. $4x + 3y = 5$, $y = x - 3$

b. $x = 2y - 4$, $x = 1 - 8y$

c. $7x - 3y = 11$, $2x + 3y = 7$

d. $2x + 3y = 10$, $x - y = -5$

e. $2x + 3y = 11$, $3x + 5y = 18$

f. $4x - 3y = -38$, $5x + 2y = -13$

3. Solve the following simultaneous equations for x and y.

a. $y = 5x - 1$, $x + 2y = 9$

b. $x = 5 + \dfrac{y}{2}$, $-4x - 3y = 35$

4. Solve the following simultaneous equations for x and y.

a. $8x + 3y = 8$, $-2x + 11y = \dfrac{35}{6}$

b. $\dfrac{x}{2} + \dfrac{y}{3} = 8$, $\dfrac{x}{3} + \dfrac{y}{2} = 7$

5. Solve the following simultaneous equations for x and y.

a. $2x - y = 7$, $7x - 5y = 42$

b. $ax - by = a$, $bx + ay = b$

6. Solve the following simultaneous equations for x and y.

a. $4x - 3y = 23$, $7x + 4y = 31$

b. $3(x + 2) = 2y$, $7x - 6y = 146$

7. WE6 a. When 4 adults and 5 children attend a pantomime, the total cost of the tickets is \$160, whereas the cost of the tickets for 3 adults and 7 children is \$159. Determine the cost of an adult ticket costs and a child's ticket.

b. A householder's electricity bill consists of a fixed payment together with an amount proportional to the number of units used. When the number of units used was 1428 the total bill was \$235.90, and when the number of units was 2240 the bill was \$353.64.

Let the fixed payment be \$$a$ and the cost per unit be \$$k$.

Determine how much the householder's bill will be if 3050 units are used.

Technology active

8. **WE7** Solve the following 3×3 system of simultaneous equations for x, y and z.

$$\begin{aligned} 5x - 2y + z &= 3 \\ 3x + y + 3z &= 5 \\ 6x + y - 4z &= 62 \end{aligned}$$

9. Solve the following 3×3 system of simultaneous equations for x, y and z.

$$\begin{aligned} z &= 12 - x + 4y \\ z &= 4 + 5x + 3y \\ z &= 5 - 12x - 5y \end{aligned}$$

10. Solve the following 3×3 systems of simultaneous equations for x, y and z.

a. $\begin{aligned} 2x + 3y - z &= 3 \\ 5x + y + z &= 15 \\ 4x - 6y + z &= 6 \end{aligned}$

b. $\begin{aligned} x - 2y + z &= -1 \\ x + 4y + 3z &= 9 \\ x - 7y - z &= -9 \end{aligned}$

c. $\begin{aligned} 2x - y + z &= -19 \\ 3x + y + 9z &= -1 \\ 4x + 3y - 5z &= -5 \end{aligned}$

d. $\begin{aligned} 2x + 3y - 4z &= -29 \\ -5x - 2y + 4z &= 40 \\ 7x + 5y + z &= 21 \end{aligned}$

e. $\begin{aligned} 3x - 2y + z &= 8 \\ 3x + 6y + z &= 32 \\ 3x + 4y - 5z &= 14 \end{aligned}$

f. $\begin{aligned} y &= 3x - 5 \\ \frac{x - z}{2} - y + 10 &= 0 \\ 9x + 2y + z &= 0 \end{aligned}$

In questions 11–13, set up a system of simultaneous equations to use to solve the problems.

11. At a recent art exhibition the total entry cost for a group of 3 adults, 2 concession holders and 3 children came to \$96; 2 adult, 1 concession and 6 child tickets cost \$100; and 1 adult, 4 concession and 1 child ticket cost \$72.

Calculate the cost of an adult ticket, a concession ticket and a child ticket.

12. Agnes, Bjork and Chi are part-time outsource workers for a manufacturing industry. When Agnes works 2 hours, Bjork 3 hours and Chi 4 hours, their combined earnings total \$194. If Agnes works 4 hours, Bjork 2 hours and Chi 3 hours, their total earnings are \$191; and if Agnes works 2 hours, Bjork 5 hours and Chi 2 hours their combined earnings total \$180.

Calculate the hourly rate of pay for each person.

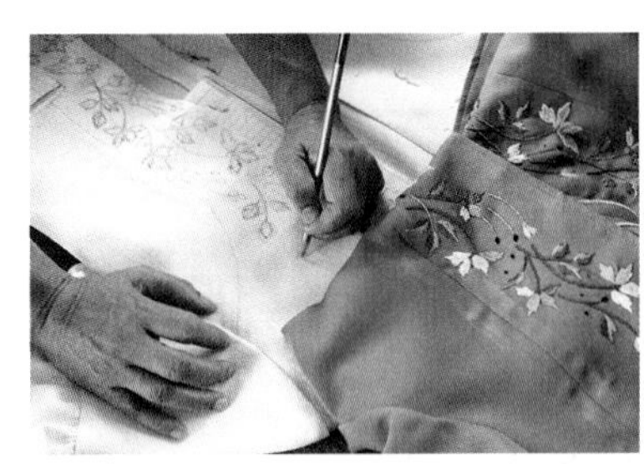

13. A student buys a sandwich at lunchtime from the school canteen for \$4.20 and pays the exact amount using 50-cent coins, 20-cent coins and 10-cent coins. The number of 20-cent coins is the same as half the number of 10-cent coins plus four times the number of 50-cent coins, and the student pays the cashier with 22 coins in total.

Determine how many coins of each type the student used.

14. a. Use CAS technology to obtain the values of x, y and z for the following system of equations.

$$\begin{aligned} 2x+6y+5z&=2\\ 5x-10y-8z&=20.8\\ 7x+4y+10z&=1 \end{aligned}$$

b. Use CAS technology to obtain the values of x, y, z and w for the following 4×4 system of equations.

$$\begin{aligned} x-y+4z-2w&=8\\ 3x+2y-2z+10w&=67\\ 2x+8y+18z+w&=-14\\ 8x-7y-80z+7w&=117 \end{aligned}$$

c. Use CAS technology to obtain the values of x_1, x_2, x_3 and x_4 for the following 4×4 system of equations.

$$\begin{aligned} 4x_1+2x_2+3x_3+6x_4&=-13\\ 12x_1-11x_2-7.5x_3+9x_4&=16.5\\ x_1+18x_3-12x_4&=8\\ -3x_1+12x_2-x_3+10x_4&=-41 \end{aligned}$$

15. A nutritionist at a zoo needs to produce a food compound in which the concentration of fats is 6.8 kg unsaturated fats, 3.1 kg saturated fats and 1.4 kg trans fats. The food compound is formed from three supplements whose concentrations of fats per kg are shown in the following table.

	Unsaturated fat	Saturated fat	Trans fat
Supplement X	6%	3%	1%
Supplement Y	10%	4%	2%
Supplement Z	8%	4%	3%

Calculate how many kilograms of each supplement must be used in order to create the food compound.

1.3 Exam questions

Question 1 (1 mark) TECH-ACTIVE

MC If $3y=-4x-4$ and $2x-\frac{y}{4}=5$, the solution (x, y) equals

A. $(-2,-36)$ B. $(2,-4)$ C. $(-4,52)$ D. $(-1,-28)$ E. $(1,-12)$

Question 2 (1 mark) TECH-ACTIVE

MC If the equations $x-4=4y+8$ and $3x-6=2y+20$ are solved simultaneously, the y-coordinate of their point of intersection would equal

A. 2 B. 14 C. -1 D. -1.4 E. -2

Question 3 (1 mark) TECH-ACTIVE

MC The solution to the simultaneous equations $y=x+2$ and $\frac{1}{4}x+y=0$ is

A. $(1,-1)$ B. $(4,-1)$ C. $(2,0)$ D. $(0,4,2)$ E. $(-1.6, 0.4)$

More exam questions are available online.

1.4 Linear graphs and their equations

LEARNING INTENTION

At the end of this subtopic you should be able to:
- determine the gradient and equation of a straight line
- sketch graphs of linear equations by hand.

Plotting ordered pairs (x, y) that satisfy the linearly related variables x and y creates a straight-line graph. From earlier study in Years 9 and 10, you will be familiar with the linear relationship $y = mx + c$, where m is the gradient of the line and c is the y-intercept.

1.4.1 Gradient of a line

The **gradient** or **slope** measures the steepness of the line as the ratio $\frac{\text{rise}}{\text{run}}$.

If a line contains the two points with coordinates (x_1, y_1) and (x_2, y_2), then the gradient, m, of the line is calculated as follows.

Gradient

$$m = \frac{\text{rise}}{\text{run}} = \frac{y_2 - y_1}{x_2 - x_1}$$

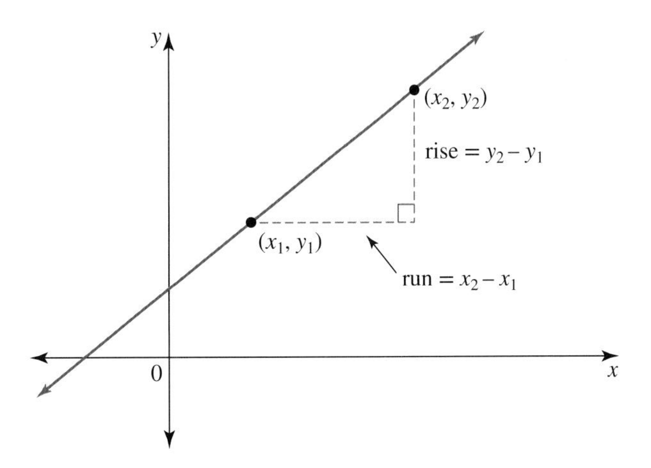

An oblique line has either a positive or a negative gradient.
- If it is **increasing**, as the run increases, the rise increases and $m > 0$.
- If it is **decreasing**, as the run increases, the rise decreases or falls and $m < 0$.

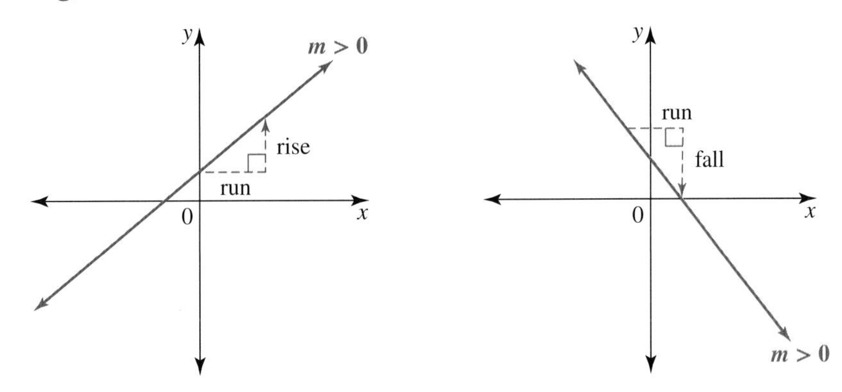

Horizontal lines have a gradient of zero since the rise between any two points is zero. The gradient of a **vertical** line is undefined since the run between any two points is zero.

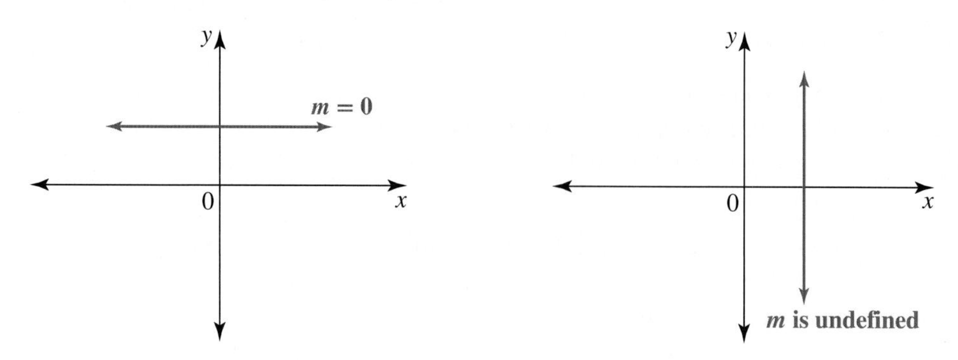

All vertical lines are **parallel**, and all horizontal lines are parallel. For other lines, if they have the same gradient, then they have the same steepness, so they must be parallel to each other.

WORKED EXAMPLE 8 Calculating the gradient between two points

Calculate the gradient of the given line.

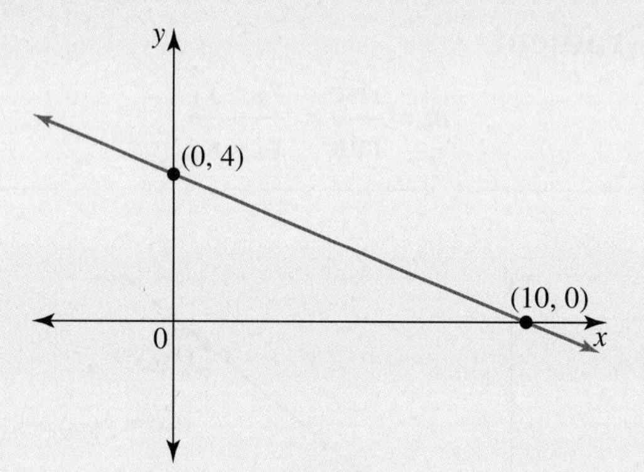

THINK	WRITE
1. Examine the diagram to locate two known points on the line and state their coordinates.	The intercepts with the coordinate axes are shown. Given points: $(0,4)$, $(10,0)$
2. Apply the gradient formula using one point as (x_1, y_1) and the other as (x_2, y_2).	Let $(x_1, y_1) = (0, 4)$ and $(x_2, y_2) = (10, 0)$. $m = \frac{y_2 - y_1}{x_2 - x_1} = \frac{0-4}{10-0} = -\frac{4}{10}$ $\therefore m = -0.4$
3. State the answer.	The gradient of the given line is -0.4.
4. An alternative method would be to calculate the rise and the run from the diagram.	Run = 10 and rise (fall) = -4 $m = \frac{\text{rise}}{\text{run}}$ $\therefore m = \frac{-4}{10} = -0.4$

1.4.2 Sketching straight lines

Although many straight lines may be drawn through a single point, only one line can be drawn through two fixed points. Two points are said to determine the line. The coordinates of every point on a line must satisfy the rule or equation that describes the line.

Although any two points whose coordinates satisfy the equation of a line can be used to sketch it, the two points that are usually chosen are the x- and y-intercepts. Should the line pass through the origin, then the coordinates of the x- and y-intercepts are both (0, 0), which gives one point to use. A second point would then need to be identified. The coordinates of a second point may be obtained by substituting a value for x or y into the rule or equation that describes the line, or alternatively the gradient could be used.

Sketching horizontal and vertical lines

The equation $y = c$, where c is a constant, represents a horizontal line with a y-axis intercept at $(0, c)$.

The equation $x = d$, where d is a constant, represents a vertical line with an x-axis intercept at $(d, 0)$.

WORKED EXAMPLE 9 Sketching straight lines

Sketch the set of points for which:

a. $y = -\dfrac{3x}{2}$ b. $2x - y = 6$ c. $x = 4$

THINK	WRITE
a. 1. Calculate the y-intercept.	a. $y = -\dfrac{3x}{2}$ When $x = 0$, $y = 0$. $\Rightarrow (0, 0)$ is both the x- and the y-intercept. The line must pass through the origin.
2. A second point is needed. Substitute another value for x in the equation of the line.	Point: let $x = 2$. $y = -\dfrac{3 \times (2)}{2}$ $= -3$ $\Rightarrow (2, -3)$ is a point on the line.
3. Plot the two points and sketch the line.	y $y = -\dfrac{3x}{2}$ (0, 0) 0 x (2, –3)
b. 1. Calculate the y-intercept.	b. $2x - y = 6$ y-intercept: let $x = 0$. $2 \times 0 - y = 6$ $-y = 6$ $\therefore y = -6$ $\Rightarrow (0, -6)$ is the y-intercept.

2. Calculate the x-intercept.

x-intercept: let $y=0$.

$$2x-0=6$$
$$2x=6$$
$$\therefore x=3$$

$\Rightarrow (3,0)$ is the x-intercept.

3. Plot the two points and draw a straight line through them. Label the points and the line appropriately.

c. $x=4$ is the equation of a vertical line. Sketch this line.

$x=4$

x-intercept $(4,0)$

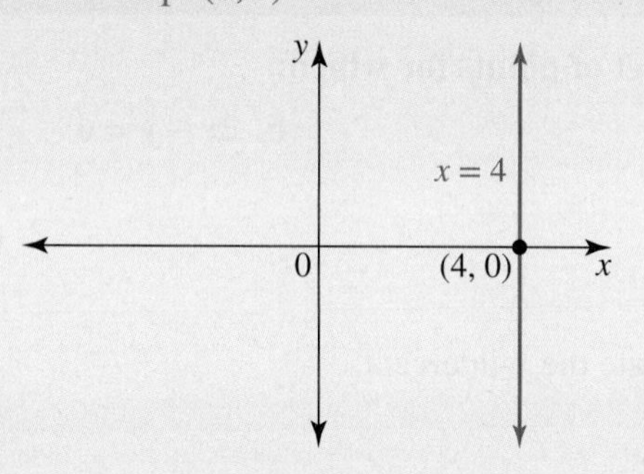

1.4.3 Forming equations of lines

Regardless of whether the line is oblique, horizontal or vertical, two pieces of information are required in order to form its equation. The forms of the equation of an oblique line that are most frequently used are:

- the **point–gradient form**
- the **gradient–y-intercept form**.

Point–gradient form of the equation of a line

Given the gradient m and a point (x_1, y_1) on the line, the equation of the line can be formed as follows.

For any point (x, y) on the line with gradient m:

$$m=\frac{y-y_1}{x-x_1}$$
$$\therefore\ y-y_1=m\left(x-x_1\right)$$

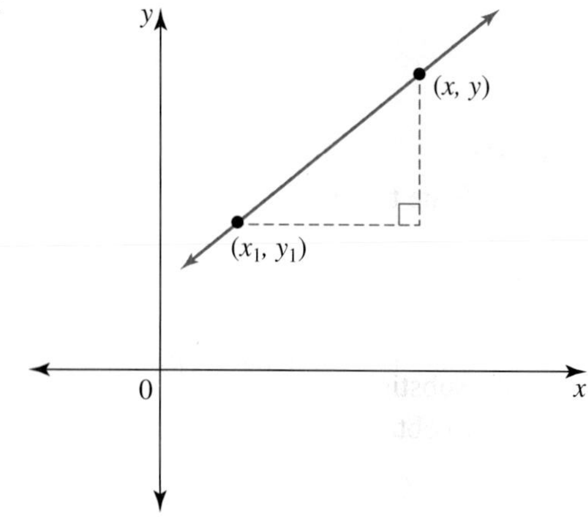

> **Point–gradient form**
>
> **A line with gradient m and passing through the point (x_1, y_1) has the equation:**
>
> $$y - y_1 = m(x - x_1)$$

Gradient–y-intercept form

For a line with gradient m cutting the y-axis at the point $(0, c)$, use $(0, c)$ for (x_1, y_1) in the point–gradient form:

$$\begin{aligned} y - y_1 &= m(x - x_1) \\ \therefore y - c &= m(x - 0) \\ \therefore y &= mx + c \end{aligned}$$

> **Gradient–y-intercept form**
>
> **A line with gradient m and y-intercept c has the equation:**
>
> $$y = mx + c$$

1.4.4 General form of the equation

The general form of the equation of a line can be written as $ax + by + c = 0$. The equation $3x + y - 2 = 0$ can be expressed in equivalent forms that include $y = -3x + 2$ and $y = 2 - 3x$.

WORKED EXAMPLE 10 Determining the equation of a straight line

a. **Form the equation of the line with gradient 4 passing through the point $(3, -7)$.**
b. **Form the equation of the line passing through the points $(5, 9)$ and $(12, 0)$.**
c. **For the line shown, determine its equation.**
d. **Obtain the gradient and the coordinates of the y-intercept of the line with the equation $3x - 8y + 5 = 0$.**

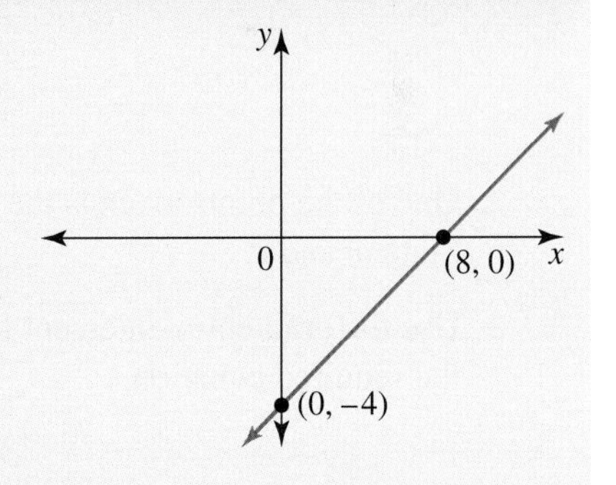

THINK	WRITE
a. 1. State the given information.	a. The gradient and a point are given. $m = 4, (x_1, y_1) = (3, -7)$
2. Write the point–gradient form of the equation.	$y - y_1 = m(x - x_1)$
3. Substitute the given information and simplify to obtain the equation.	$y - (-7) = 4(x - 3)$ $y + 7 = 4x - 12$ $\therefore y = 4x - 19$
b. 1. State the given information.	b. Two points are given. Let $(x_1, y_1) = (5, 9)$ and $(x_2, y_2) = (12, 0)$

2. Use the two points to calculate the gradient.	$m = \frac{0-9}{12-5}$ $= -\frac{9}{7}$
3. Write the point–gradient form of the equation.	$y - y_1 = m(x - x_1)$
4. The point–gradient equation can be used with either of the points. Substitute one of the points and simplify to obtain the equation.	Let $(12, 0)$ be the given point (x_1, y_1) in this equation. $y - 0 = -\frac{9}{7}(x - 12)$ $y = -\frac{9}{7}(x - 12)$
5. The equation could be expressed without fractions. Although this is optional, it looks more elegant.	$7y = -9(x - 12)$ $7y = -9x + 108$ $\therefore 9x + 7y = 108$
6. Had the point $(5, 9)$ been used, the same answer would have been obtained.	Check: $y - 9 = -\frac{9}{7}(x - 5)$ $7(y - 9) = -9(x - 5)$ $7y - 63 = -9x + 45$ $\therefore 9x + 7y = 108$ This is the same equation as before.
c. 1. Calculate the gradient from the graph (or use the coordinates of the y-intercept and the x-intercept points).	**c.** $m = \frac{\text{rise}}{\text{run}}$ $m = \frac{4}{8}$ $\therefore m = \frac{1}{2}$
2. One of the points given is the y-intercept. State m and c.	$m = \frac{1}{2}, c = -4$
3. Use the gradient–y-intercept form to obtain the required equation.	$y = mx + c$ $\therefore y = \frac{1}{2}x - 4$
d. 1. Express the equation in the form $y = mx + c$.	**d.** $3x - 8y + 5 = 0$ $3x + 5 = 8y$ $\therefore y = \frac{3x}{8} + \frac{5}{8}$
2. State m and c.	$m = \frac{3}{8}, c = \frac{5}{8}$
3. Express the answer in the required form.	The gradient is $\frac{3}{8}$ and the y-intercept is $\left(0, \frac{5}{8}\right)$.

Resources

Interactivity Equations from point–gradient and gradient–*y*-intercept forms (int-2551)

1.4 Exercise

Students, these questions are even better in jacPLUS

 Receive immediate feedback and access sample responses

 Access additional questions

 Track your results and progress

Find all this and MORE in jacPLUS

Technology free

1. WE8 Calculate the gradient of the given line.

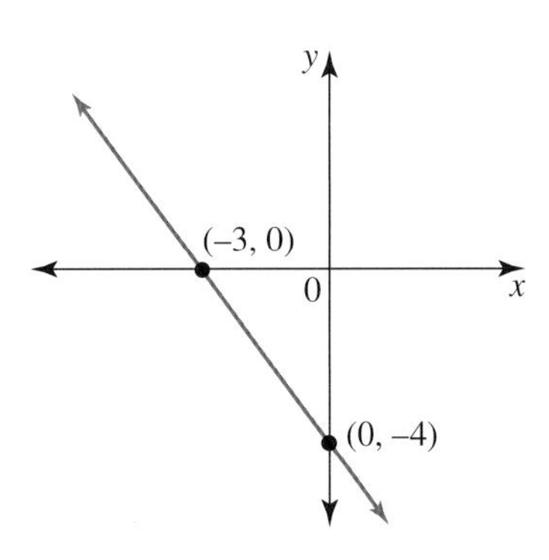

2. a. Determine the gradient of the straight line with x-intercept at $(-2, 0)$ and y-intercept at $(0, 4)$.
   b. Determine the gradient of the line shown.

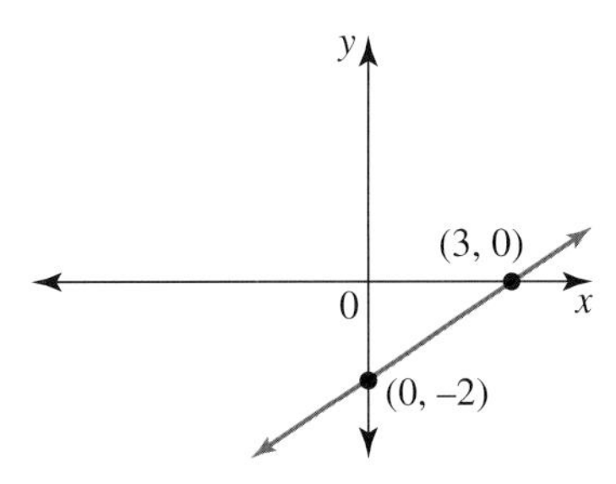

   c. Determine the gradient of the line shown.

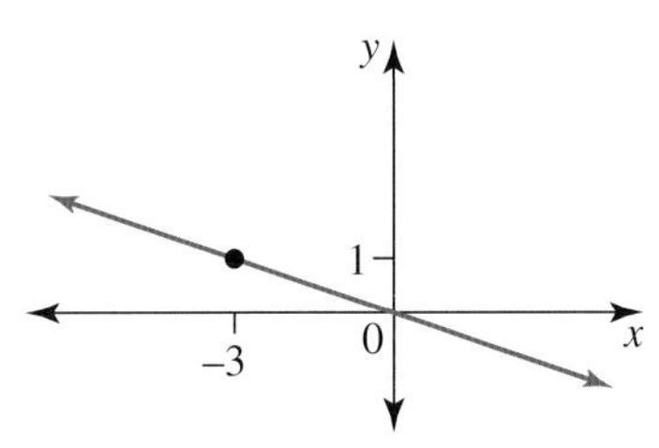

   d. Determine the gradient of the line passing through the points $(7, -2)$ and $(2, 5)$.

3. Calculate the gradients of the lines joining the following points.
   a. $(-3,\ 8)$ and $(-7,\ 18)$
   b. $(0,\ -4)$ and $(12,\ 56)$
   c. $(-2,\ -5)$ and $(10,\ -5)$
   d. $(3,\ -3)$ and $(3,\ 15)$

4. Show that the line passing through the points (a, b) and $(-b, -a)$ is parallel to the line joining the points $(-c, d)$ and $(-d, c)$.

5. **WE9** Sketch the set of points for which:

 a. $y = 4x$ b. $3x + 2y = 6$ c. $y = 2$

6. Sketch the following linear graphs.

 a. $y = 3x - 6$ b. $y = -4x + 1$ c. $y = \frac{x}{2} - 5$

 d. $y + 3x = 8$ e. $y = 4x$ f. $y = -0.5x$

7. Sketch the lines with the following equations.

 a. $y = 3x + 8$ b. $4y - x + 4 = 0$ c. $6x + 5y = 30$

 d. $3y - 5x = 0$ e. $\frac{x}{2} - \frac{3y}{4} = 6$ f. $y = -\frac{6x}{7}$

8. **WE10** a. Form the equation of the line with gradient -2 passing through the point $(-8, 3)$.

 b. Form the equation of the line passing through the points $(4, -1)$ and $(-3, 1)$.

 c. Determine the equation of the line shown.

 d. Obtain the gradient and the coordinates of the y-intercept of the line with equation $6y - 5x - 18 = 0$.

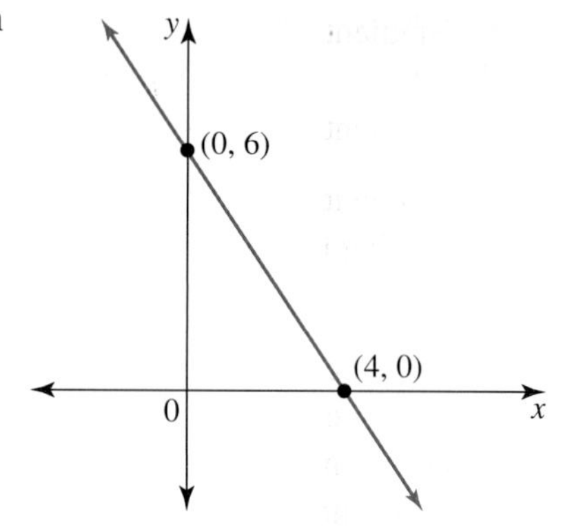

9. Determine the equation of the line parallel to the x-axis that passes through the point $(2, 10)$.

10. a. State the equation of the line with a y-intercept at $(0, 2)$ and a gradient of 5.

 b. Determine the equation of the graph shown.

 c. Determine the equation of the line with a gradient of 3 and passing through the point $(1, 2)$.

 d. Form the equation of the linear graph passing through the origin with a gradient of -5.

 e. Form the equation of the line that passes through the points $(-1, 0)$ and $(3, -2)$.

 f. Determine the equation in the form $ax + by + c = 0$ for the line with gradient $-\frac{3}{4}$ and passing through the point $(-3, 5)$.

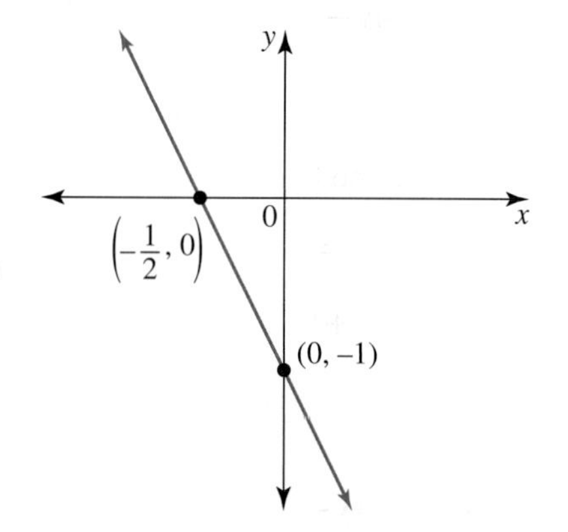

11. a. Form the equations of the given graphs.

 i.

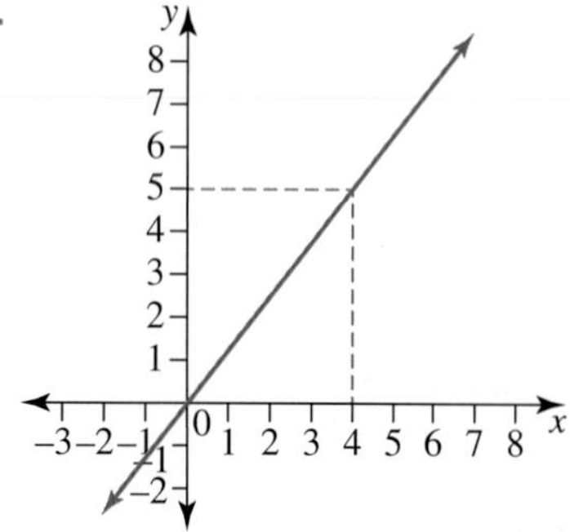

 ii.

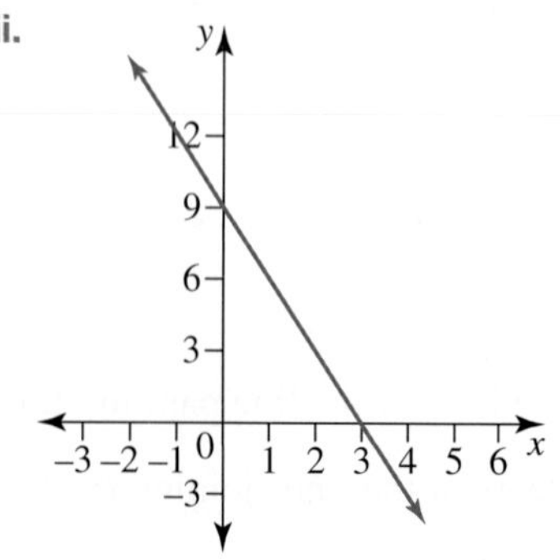

iii.

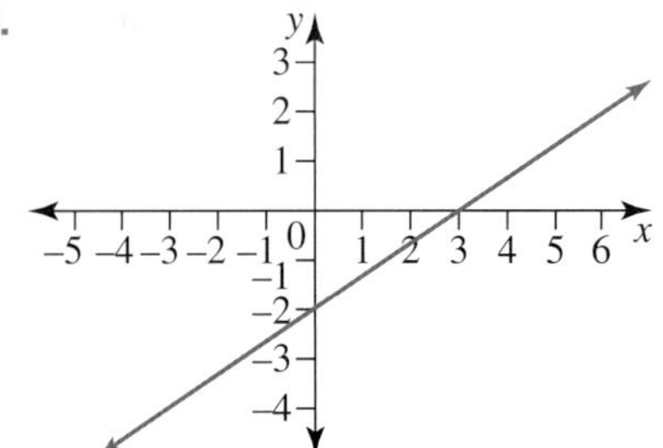

iv.

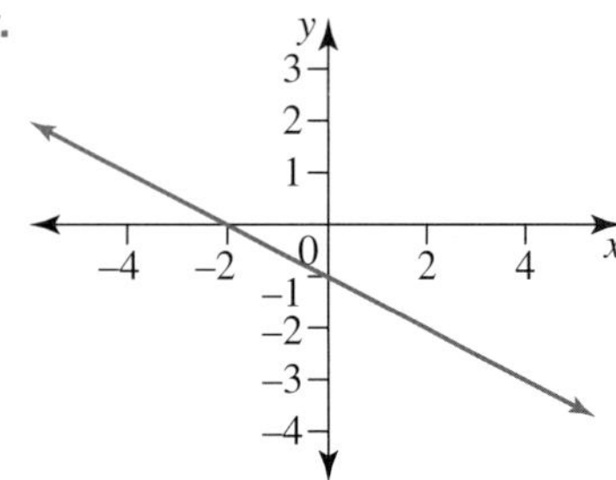

b. A line contains the points $(-12, 8)$ and $(-12, -1)$. Form its equation and sketch its graph.
c. Determine the equation of the line passing through $(10, -8)$ that is parallel to the line $y = 2$ and sketch its graph.

12. Determine the equations of the straight lines that are described by the following information.
    a. Gradient -5, passing through the point $(7, 2)$
    b. Gradient $\frac{2}{3}$, passing through the point $(-4, -6)$
    c. Gradient $-1\frac{3}{4}$, passing through the point $(0, -9)$
    d. Gradient -0.8, passing through the point $(0.5, -0.2)$
    e. Passing through the points $(-1, 8)$ and $(-4, -2)$
    f. Passing through the points $(0, 10)$ and $(10, -10)$

13. a. State the gradient and the coordinates of the y-intercept of the graph that has the equation $y = 2x - 8$.
    b. State the gradient of the line with equation $5x - 3y - 6 = 0$.
    c. Determine the gradient and the coordinates of the y-intercept of the line with equation $4y - 3x = 4$.
    d. Determine which of these lines are parallel.
        i. $3x - 4y - 4 = 0$
        ii. $4y - 3x - 6 = 0$
        iii. $6x - 8y - 6 = 0$
        iv. $2y - 6x - 12 = 0$

14. Determine the gradients and the y-intercepts of the lines with the following equations.
    a. $4x + 5y = 20$
    b. $\frac{2x}{3} - \frac{y}{4} = -5$
    c. $x - 6y + 9 = 0$
    d. $2y - 3 = 0$

Technology active

15. Sketch with CAS technology the lines with the following equations.
    a. $2y - 4x = -11$
    b. $x = 5$
    c. $y = -3$

16. a. Determine the value of a if the point $(2a, 2 - a)$ lies on the line given by $5y = -3x + 4$.
    b. Form the equation of the line containing the points (p, q) and $(-p, -q)$.

17. If the cost, C dollars, of hiring a rowing boat is \$30 plus \$1.50 per hour or part thereof:
    a. determine the rule for C in terms of the hire time in hours, t.
    b. sketch the graph of C versus t and state its gradient.

1.4 Exam questions

Question 1 (1 mark) TECH-ACTIVE

MC The gradient of the line passing through the points $(-3, 5)$ and $(1, -4)$ is

A. $\frac{9}{2}$ **B.** $-\frac{9}{4}$ **C.** -1 **D.** $-\frac{1}{4}$ **E.** $\frac{1}{9}$

Question 2 (1 mark) TECH-ACTIVE

MC State which of the following graphs can represent the equation $-3y + 2x = -6$.

A.

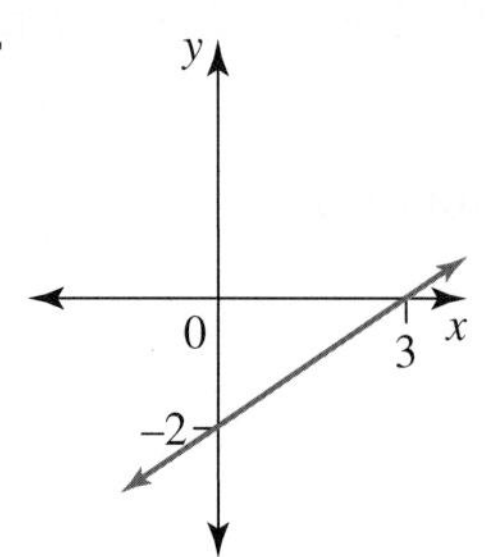

B.

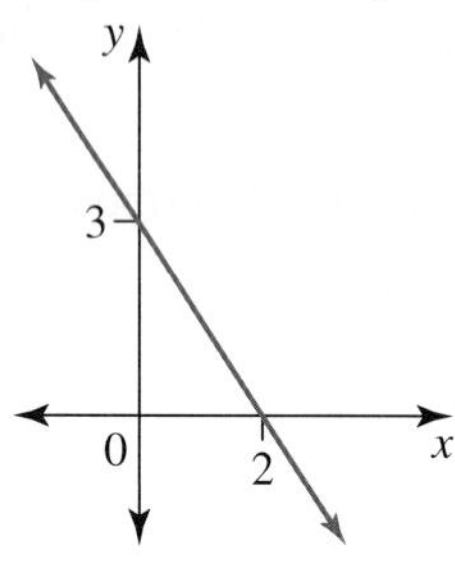

C.

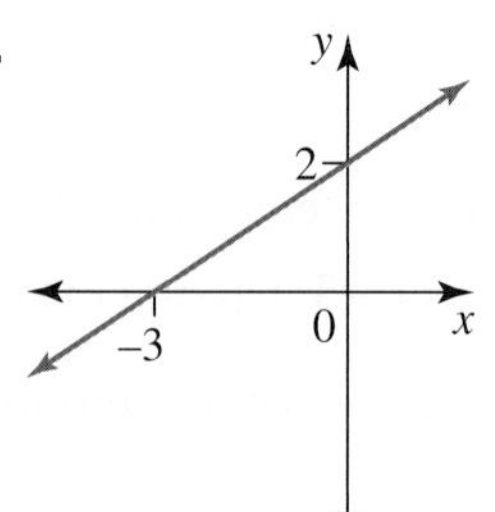

D.

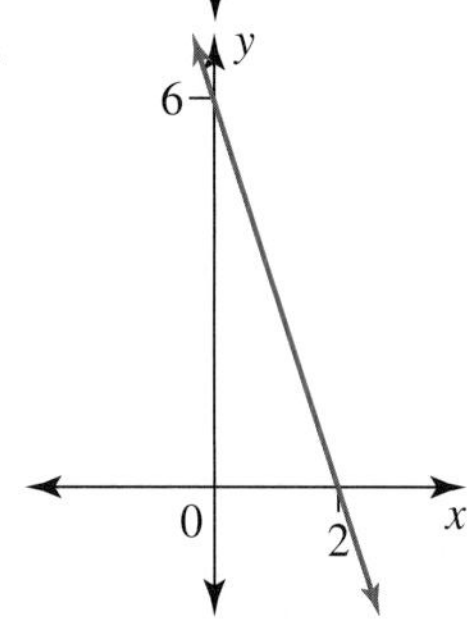

E.

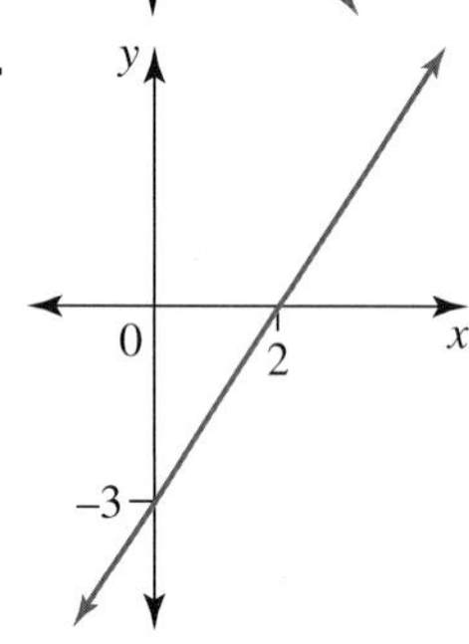

Question 3 (2 marks) TECH-FREE

Form the equation of the line passing through the points $(3, 1)$ and $(-1, 3)$.

More exam questions are available online.

1.5 Intersections of lines and their applications

LEARNING INTENTION

At the end of this subtopic you should be able to:

- set up and solve systems of linear equations by hand.

Graphs allow visual comparisons to be made between models.

For example, if a revenue model and a cost model are graphed together, it is possible to see when a profit is made.

A profit will be made only when the revenue graph lies above the cost graph. At the point of intersection of the two graphs — also known as the break-even point — the revenue equals the costs. Before this point, a loss occurs, since the revenue graph lies below the cost graph.

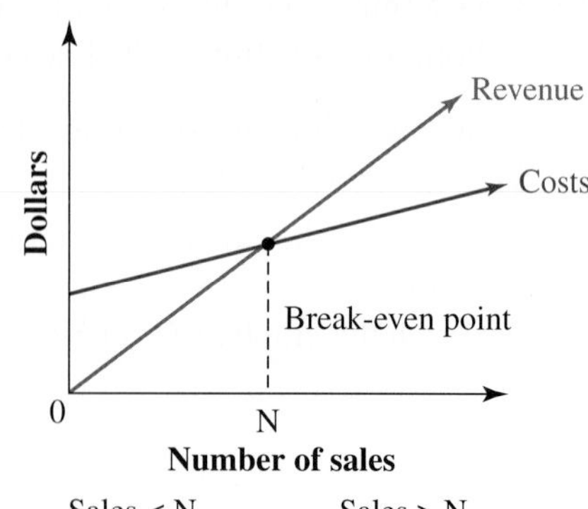

1.5.1 Intersections of lines

Two lines with different gradients intersect at a point. Since this point must lie on both lines, its coordinates are found algebraically using simultaneous equations.

WORKED EXAMPLE 11 Solving problems using a point of intersection

The model for the revenue in dollars, d, from the sale of n items is $d_R = 20n$, and the cost of manufacture of the n items is modelled by $d_C = 500 + 5n$.

a. **Find the coordinates of the point of intersection of the graphs of these two models and sketch the graphs on the same set of axes.**

b. **Obtain the smallest value of n for a profit to be made.**

THINK	WRITE
a. 1. Write the equation at the point of intersection.	a. At the intersection or break-even point: $d_R = d_C$ $20n = 500 + 5n$
2. Solve to find n.	$15n = 500$ $n = \dfrac{500}{15}$ $\therefore n = 33\dfrac{1}{3}$
3. Calculate the d coordinate.	When $n = \dfrac{100}{3}$, $d = 20 \times \dfrac{100}{3}$ $\therefore d = 666\dfrac{2}{3}$
4. State the coordinates of the point of intersection.	The point of intersection is $\left(33\frac{1}{3}, 666\frac{2}{3}\right)$.
5. Both graphs contain the intersection point. Find one other point on each graph.	Points: Let $n = 0$. $\therefore d_R = 0$ and $d_C = 500$ The revenue graph contains the points $(0, 0)$, $\left(33\frac{1}{3}, 666\frac{2}{3}\right)$. The cost graph contains the points $(0, 500)$ and $\left(33\frac{1}{3} \text{ and } 666\frac{2}{3}\right)$.
6. Sketch the graphs.	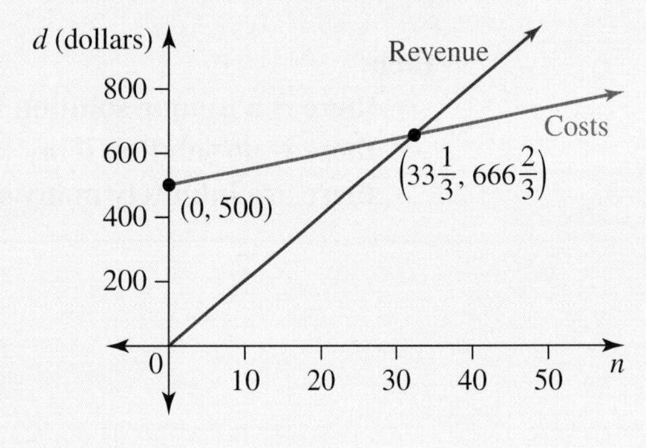

b. 1. State the condition for a profit to be made.	**b.** For a profit, $d_R > d_C$. From the graph, $d_R > d_C$ when $n > 33\frac{1}{3}$.
2. Write the answer in a sentence.	Therefore, at least 34 items need to be sold for a profit to be made.

1.5.2 The number of solutions to systems of 2×2 linear simultaneous equations

Since a linear equation represents a straight line graph, by considering the possible intersections of two lines, three possible outcomes arise when solving a system of simultaneous equations.

Case 1

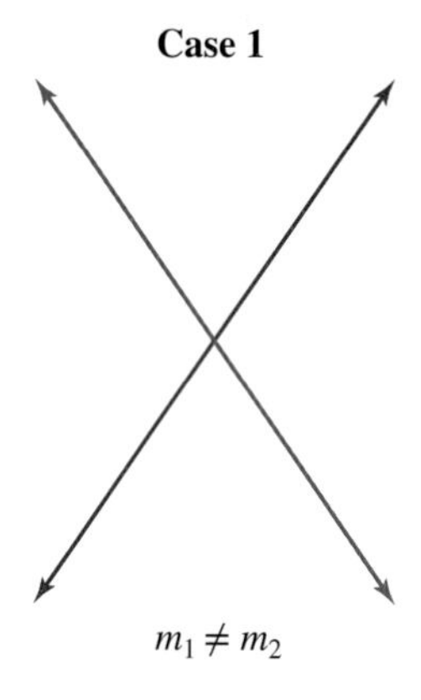

$m_1 \neq m_2$

Case 2

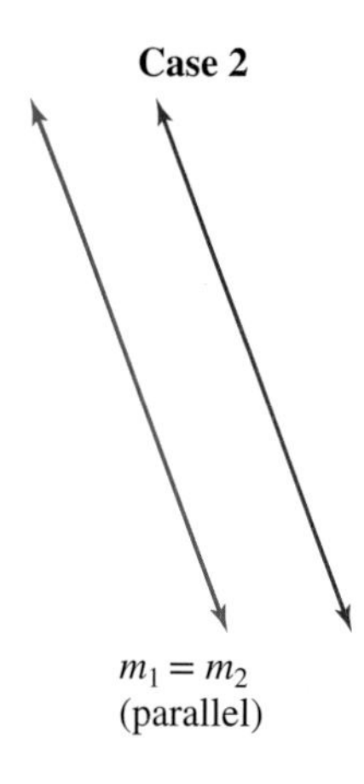

$m_1 = m_2$
(parallel)

Case 3

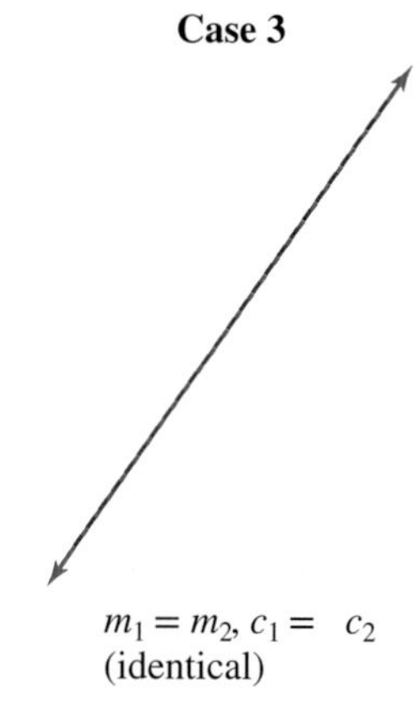

$m_1 = m_2$, $c_1 = c_2$
(identical)

Case 1: Unique solution to the system. The equations represent two lines that intersect at a single point.

Case 2: No solution to the system. The equations represent parallel lines.

Case 3: Infinitely many solutions. The equations are equivalent and represent the same line. Every point on the line is a solution.

Linear simultaneous equations

If a system of equations is rearranged to be in the form

$$y = m_1 x + c_1$$
$$y = m_2 x + c_2$$

then:

- **there is a unique solution if $m_1 \neq m_2$**
- **there is no solution if $m_1 = m_2$ and $c_1 \neq c_2$**
- **there are infinitely many solutions if $m_1 = m_2$ and $c_1 = c_2$.**

WORKED EXAMPLE 12 Determining the number of solutions of a 2×2 system

Determine the value of m so that the system of equations shown below has no solutions.

$$mx - y = 2$$
$$3x + 4y = 12$$

THINK	WRITE
1. Rearrange both equations to the $y = mx + c$ form.	$mx - y = 2 \Rightarrow y = mx - 2$ and $3x + 4y = 12 \Rightarrow y = -\frac{3}{4}x + 3$
2. State the gradients of the lines the equations represent.	The gradients are m and $-\frac{3}{4}$.
3. State the condition for the system of equations to have no solution, and calculate m.	For the system of equations to have no solution, the lines must be parallel but have different y-intercepts. For the lines to be parallel, the two gradients have to be equal. $\therefore m = -\frac{3}{4}$
4. The possibility of the equations being equivalent has to be checked.	Substitute $m = -\frac{3}{4}$ into the $y = mx + c$ forms of the equations. $y = -\frac{3}{4}x - 2$ and $y = -\frac{3}{4}x + 3$ represent parallel lines since they have the same gradients and different y-intercepts.
5. State the answer.	Therefore, if $m = -\frac{3}{4}$, the system will have no solution.

1.5.3 Concurrent lines

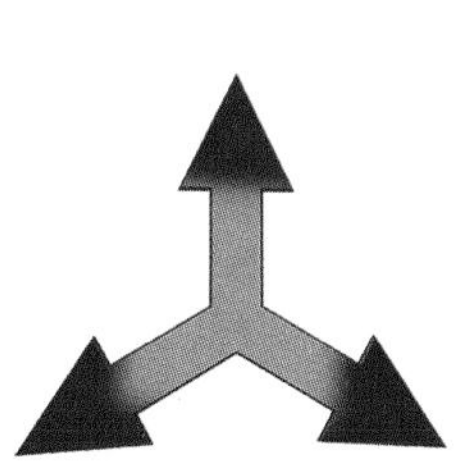

Three or more lines that intersect at a common point are said to be **concurrent**. Their point of intersection is known as the point of concurrency.

To show that three lines are concurrent, the simplest method is to find the point of intersection of two of the lines and then check whether that point lies on the third line.

WORKED EXAMPLE 13 Determining if lines are concurrent

Show that the three lines with equations $5x + 3y = 1$, $4x + 7y = 10$ and $2x - y = -4$ are concurrent.

THINK	WRITE
1. Select a pair of equations to solve simultaneously.	Consider the following equations: $4x + 7y = 10$ [1] $2x - y = -4$ [2]

2. Solve this system of equations.	Eliminate x. Multiply equation [2] by 2 and subtract it from equation [1]. $7y-2\times(-y)=10-2\times(-4)$ $7y+2y=10+8$ $9y=18$ $\therefore y=2$ Substitute $y=2$ into equation [2]. $2x-2=-4$ $2x=-2$ $\therefore x=-1$ Lines [1] and [2] intersect at $(-1,2)$.
3. Test whether the values for x and y satisfy the third equation.	Substitute $x=-1, y=2$ into $5x+3y=1$. $\text{LHS}=5\times(-1)+3\times(2)$ $=1$ $=\text{RHS}$ Therefore, $(-1,2)$ lies on $5x+3y=1$.
4. Write a conclusion.	Since $(-1,2)$ lies on all three lines, the three lines are concurrent. The point $(-1,2)$ is their point of concurrency.

Resources

Interactivity Intersecting, parallel and identical lines (int-2552)

1.5 Exercise

Students, these questions are even better in jacPLUS

Receive immediate feedback and access sample responses

Access additional questions

Track your results and progress

Find all this and MORE in jacPLUS

Technology free

1. Use simultaneous equations to find the coordinates of the point of intersection of the lines with equations $3x-2y=15$ and $x+4y=54$.

2. Determine the coordinates of the point of intersection of each of the following pairs of lines.

   a. $4x-3y=13$ and $2y-6x=-7$

   b. $y=\dfrac{3x}{4}-9$ and $x+5y+7=0$

   c. $y=-5$ and $x=7$

3. A triangle is bounded by the lines with equations $x=3$, $y=6$ and $y=-3x$.

   a. State the coordinates of its vertices.

   b. Calculate its area in square units.

4. WE11 If the model for the revenue in dollars, d, from the sale of n items is $d_R = 25n$, and the cost of manufacture of the n items is modelled by $d_C = 260 + 12n$:
   a. determine the coordinates of the point of intersection of the graphs of these two models and sketch the graphs on the same set of axes
   b. obtain the smallest value of n for a profit to be made.
5. The daily cost of hiring a bicycle from the Pedal On company is $10 plus 75 cents per kilometre, whereas from the Bikes R Gr8 company the cost is a flat rate of $20 with unlimited kilometres.
   a. State the linear equations that model the costs of hiring the bicycles from each company.
   b. On one set of axes, sketch the graphs showing the cost versus the number of kilometres ridden, for each company.
   c. Determine the number of kilometres after which the costs are equal.
   d. Shay wishes to hire a bicycle for the day. Explain how Shay can decide which of the two companies would be cheaper.
6. WE12 Determine the value of m so that the following system of equations has no solutions.

$$\begin{aligned} 2mx + 3y &= 2m \\ 4x + y &= 5 \end{aligned}$$

7. Determine the values of a and b for which the following system of equations will have infinitely many solutions.

$$\begin{aligned} ax + y &= b \\ 3x - 2y &= 4 \end{aligned}$$

8. Determine the value of p for which the lines $2x + 3y = 23$ and $7x + py = 8$ will not intersect.
9. a. Express the lines given by $px + 5y = q$ and $3x - qy = 5q$, $q \neq 0$, in $y = mx + c$ form.
   b. Hence, determine the values of p and q so the system of equations shown will have infinitely many solutions.

$$\begin{aligned} px + 5y &= q \\ 3x - qy &= 5q \end{aligned}$$

   c. State the relationship that must exist between p and q so the lines $px + 5y = q$ and $3x - qy = 5q$ will intersect.

Technology active

10. Use the graphing facility on CAS technology to obtain the point of intersection of the pair of lines $y = \dfrac{17 + 9x}{5}$ and $y = 8 - \dfrac{3x}{2}$, to 2 decimal places.
11. The line passing through the point $(4, -8)$ with gradient -2 intersects the line with gradient 3 and y-intercept 5 at the point Q. Find the coordinates of Q.
12. WE13 Show that the three lines with equations $2x + 3y = 0$, $x - 8y = 19$ and $9x + 5y = 17$ are concurrent.
13. Determine the value of a so that the three lines defined by $x + 4y = 13$, $5x - 4y = 17$ and $-3x + ay = 5$ are concurrent.
14. Show that the following three lines are concurrent and state their point of concurrency: $3x - y + 3 = 0$, $5x + 2y + 16 = 0$ and $9x - 5y + 3 = 0$.
15. Determine the values of d so that the three lines $x + 4y = 9$, $3x - 2y = -1$ and $4x + 3y = d$ are not concurrent.

16. The graph shows cost, C, in dollars, versus distance x, in kilometres, for two different car rental companies A and B.
The cost models for each company are $C = 300 + 0.05x$ and $C = 250 + 0.25x$.

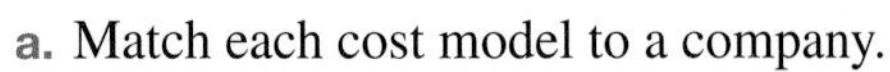

a. Match each cost model to a company.
b. Explain what the gradient of each graph represents.

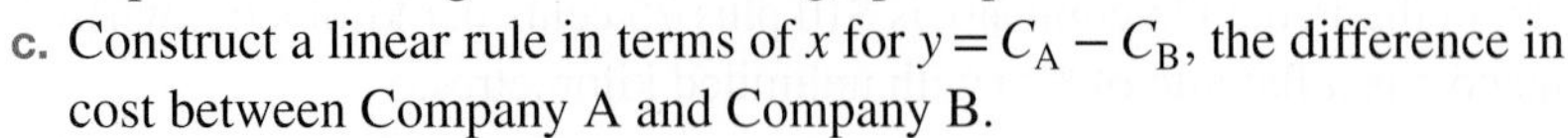

c. Construct a linear rule in terms of x for $y = C_A - C_B$, the difference in cost between Company A and Company B.

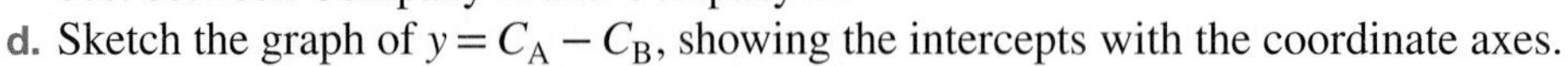

d. Sketch the graph of $y = C_A - C_B$, showing the intercepts with the coordinate axes.
e. Use your graph from part **d** to determine the number of kilometres when:
   i. the costs of each company are the same
   ii. the costs of Company A are cheaper than those of Company B.

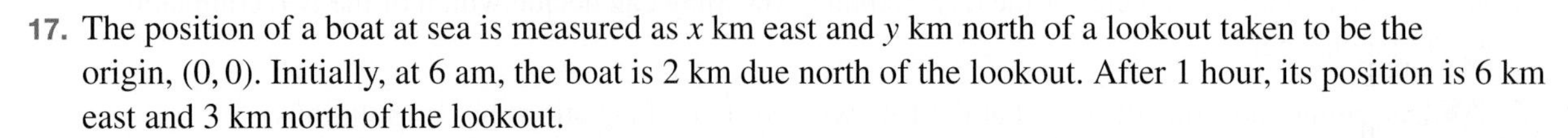

17. The position of a boat at sea is measured as x km east and y km north of a lookout taken to be the origin, $(0, 0)$. Initially, at 6 am, the boat is 2 km due north of the lookout. After 1 hour, its position is 6 km east and 3 km north of the lookout.
   a. Write down the coordinates of the two positions of the boat and, assuming the boat travels in a straight line, form the equation of its path.
   b. The boat continues to sail on this linear path and at some time t hours after 6 am, its distance east of the lookout is $6t$ km. At that time, show that its position north of the lookout is $(t + 2)$ km.
   c. Determine the coordinates of the position of the boat at 9.30 am.
   d. The positions east and north of the lighthouse of a second boat, a large fishing trawler, sailing along a linear path are given by $x = \frac{4t - 1}{3}$ and $y = t$ respectively, where t is the time in hours since 6 am. Find the coordinates of the positions of the trawler at 6 am and 7 am, and hence (or otherwise) find the Cartesian equation of its linear path.
   e. Show that the paths of the boat and the trawler contain a common point, and give the coordinates of this point.
   f. Sketch the paths of the boats on the same axes and explain whether the boat and the trawler collide.
18. At time t, a particle P_1 moving on a straight line has coordinates given by $x = t, y = 3 + 2t$, while at the same time a second particle P_2 moving along another straight line has coordinates given by $x = t + 1, y = 4t - 1$.
   a. Use CAS technology to sketch their paths simultaneously and so determine whether the particles collide.
   b. State the coordinates of the common point on the two paths.

1.5 Exam questions

Question 1 (1 mark) TECH-ACTIVE

MC The simultaneous equations $y = 2x - 1$ and $y = 2x + 1$

A. intersect at $(2, 0)$.
B. intersect at $(-1, 1)$.
C. do not intersect.
D. intersect at $\left(\frac{1}{2}, -\frac{1}{2}\right)$.
E. intersect at $(0, 2)$.

Question 2 (1 mark) TECH-ACTIVE

MC The value of a such that the lines $y = 4(2a + 1)x + 2$ and $y = 2(a - 1)x - \frac{1}{2}$ will not intersect is

A. 1
B. -1
C. 4
D. 2
E. $-\frac{1}{2}$

Question 3 (5 marks) TECH-FREE

Determine the values of a and b for which the system of equation $ax + y = b$ and $x + 3y = 2$ will have infinitely many solutions.

More exam questions are available online.

1.6 Straight lines and gradients

LEARNING INTENTION

At the end of this subtopic you should be able to:
- solve problems involving gradients of parallel and perpendicular lines.

1.6.1 Collinearity

The gradient of a line has several applications. It determines whether lines are parallel or perpendicular, it determines the angle a line makes with the horizontal, and it determines whether three or more points lie on the same line.

Three or more points that lie on the same line are said to be **collinear**.

If $m_{AB} = m_{BC}$, then the line through the points A and B is parallel to the line through the points B and C. Since the point B is common to both AB and BC, the three points A, B and C must lie on the same line.

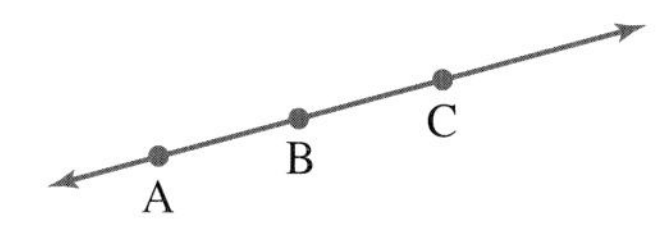

Alternatively, the equation of the line through two of the points can be used to test whether the third point also lies on that line.

WORKED EXAMPLE 14 Determining collinearity of three points

Show that the points A (−5, −3), B (−1, 7) and C (1, 12) are collinear.

THINK	WRITE
1. Select two of the points.	For the points A and B, let (x_1, y_1) be $(-5, -3)$ and (x_2, y_2) be $(-1, 7)$.
2. Calculate the gradient of AB.	$m = \frac{y_2 - y_1}{x_2 - x_1}$ $= \frac{7-(-3)}{-1-(-5)}$ $= \frac{10}{4}$ $\therefore m_{AB} = \frac{5}{2}$
3. Select another pair of points containing a common point with the interval AB.	For the points B and C, let (x_1, y_1) be $(-1, 7)$ and (x_2, y_2) be $(1, 12)$.
4. Calculate the gradient of BC. *Note:* The interval AC could equally as well have been chosen.	$m = \frac{y_2 - y_1}{x_2 - x_1}$ $m = \frac{12-7}{1-(-1)}$ $\therefore m_{BC} = \frac{5}{2}$
5. Compare the gradients to determine collinearity.	Since $m_{AB} = m_{BC}$ and the point B is common, the three points lie on the same line, so they are collinear.

1.6.2 Angle of inclination of a line to the horizontal

If θ is the angle a line makes with the positive direction of the x-axis (or another horizontal line), and m is the gradient of the line, then a relationship between the gradient and this angle can be formed using trigonometry.

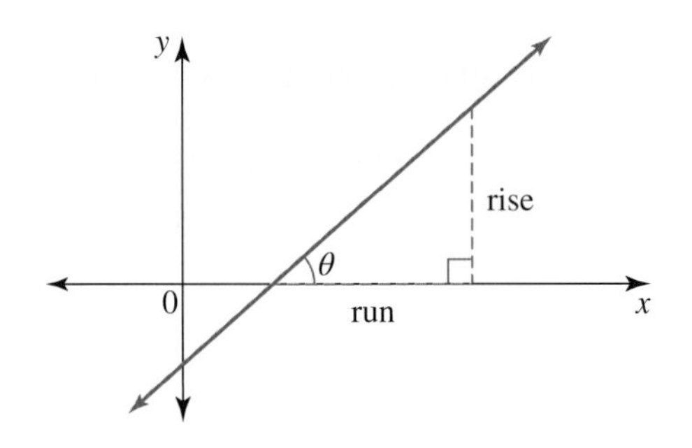

In the right-angled triangle,

$$\tan\,\theta = \frac{\text{opposite}}{\text{adjacent}}$$
$$= \frac{\text{rise}}{\text{run}}$$
$$\therefore \tan\,\theta = m$$

Angle of inclination

The gradient of a line is given by

$$m = \tan\theta$$

where θ is measured in the positive (or anticlockwise) direction from the x-axis.

If the line is vertical, $\theta = 90°$. If the line is horizontal, $\theta = 0°$.

For oblique lines, the angle is either acute ($0° < \theta < 90°$) or obtuse ($90° < \theta < 180°$), and a calculator is used to find θ.

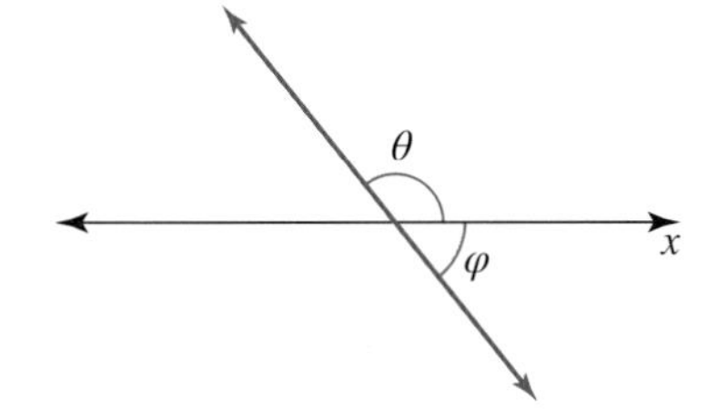

When the gradient, m, is negative, the calculator gives a negative angle for $\tan^{-1} m$. That is, the angle has been measured in the negative or clockwise direction from the x-axis. The angle of inclination, θ, is measured in the anticlockwise direction. Thus, $\theta = 180 - \psi$.

Positive and negative gradients

If $m > 0$, then θ is an acute angle, so $\theta = \tan^{-1}(m)$.

If $m < 0$, then θ is an obtuse angle, so $\theta = 180° - \tan^{-1}(|m|)$.

WORKED EXAMPLE 15 Calculating angles of inclination

a. **Calculate, correct to 2 decimal places, the angle made with the positive direction of the x-axis by the line that passes through the points $(-1, 2)$ and $(2, 8)$.**
b. **Calculate the angle of inclination with the horizontal made by a line that has a gradient of -0.6.**
c. **Obtain the equation of the line that passes through the point $(5, 3)$ at an angle of $45°$ to the horizontal.**

THINK	WRITE
a. 1. Calculate the gradient of the line through the given points.	**a.** Points $(-1, 2)$ and $(2, 8)$ $m = \frac{8-2}{2-(-1)}$ $= \frac{6}{3}$ $= 2$
2. Write down the relationship between the angle and the gradient.	$\tan\theta = m$ $\therefore\ \tan\theta = 2$
3. Find θ correct to 2 decimal places using a calculator.	$\theta = \tan^{-1}(2)$ $\therefore \theta \simeq 63.43°$
4. Answer in context.	Therefore, the required angle is 63.43°.
b. 1. State the gradient.	**b.** The gradient -0.6 is given: $m = -0.6$
2. Determine $\tan^{-1}(m)$.	$\tan^{-1}(-0.6) = -30.96°$
3. Calculate the angle of inclination.	$\theta = 180 - 30.96°$ $\therefore \theta \simeq 149.04°$
4. Answer in context.	Therefore, the required angle is 149° to the nearest degree.
c. 1. State the relationship between the gradient and the angle.	**c.** $m = \tan\theta$
2. Substitute the given angle for θ and calculate the gradient.	$m = \tan(45°)$ $\therefore m = 1$
3. Use the point–gradient form to obtain the equation of the line.	$y - y_1 = m(x - x_1),\ m = 1, (x_1, y_1) = (5, 3)$ $y - 3 = 1(x - 5)$ $\therefore y = x - 2$
4. State the answer.	The equation of the line is $y = x - 2$.

1.6.3 Parallel lines

Parallel lines have the same gradient.

If two lines with gradients m_1 and m_2 are parallel, then $m_1 = m_2$.

Each line would be inclined at the same angle to the horizontal.

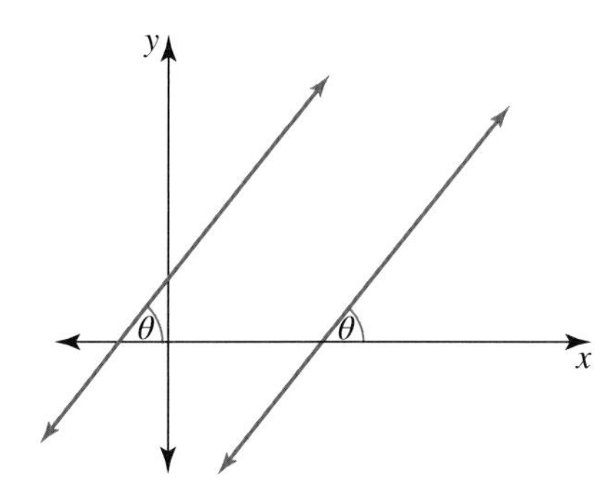

1.6.4 Perpendicular lines

A pair of lines are perpendicular to each other when the angle between them is 90°. For a pair of oblique lines, one must have a positive gradient and the other a negative gradient.

To find the relationship between these gradients, consider two **perpendicular lines** L_1 and L_2, with gradients m_1 and m_2 respectively. Suppose $m_1 > 0$ and $m_2 < 0$, and that L_1 is inclined at an angle θ to the horizontal.

The diagram shows the line L_1 passing through the points A and B and the line L_2 passing through the points A and D, with the angle BAD being a right angle.

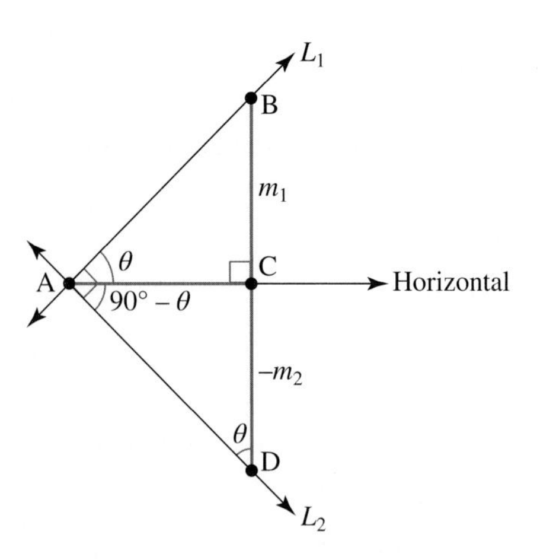

Taking AC as 1 unit, the sides in the diagram are labelled with their lengths. The side CB has length m_1. Because lengths must be positive, the side CD is labelled as $-m_2$, since $m_2 < 0$.

From the triangle ABC in the diagram, $\tan\theta = \frac{m_1}{1} = m_1$, and from the triangle ACD in the diagram, $\tan\theta = \frac{1}{-m_2}$.

Hence,

$$m_1 = \frac{1}{-m_2}$$
$$\therefore m_1 m_2 = -1$$

Gradients of perpendicular lines

$$m_1 m_2 = -1 \text{ or } m_2 = -\frac{1}{m_1}$$

- If two lines with gradients m_1 and m_2 are perpendicular, then the product of their gradients is -1. One gradient is the negative reciprocal of the other.
- It follows that if $m_1 m_2 = -1$, then the two lines are perpendicular. This can be used to test for perpendicularity.

WORKED EXAMPLE 16 Determining parallel and perpendicular gradients

a. State the gradient of a line that is:
i. parallel to $2y - 5x = 4$ **ii. perpendicular to $2y - 5x = 4$.**
b. Show that the lines $y = 4x$ and $y = -0.25x$ are perpendicular.
c. Determine the equation of the line through the point (1, 1) perpendicular to the line $y = -3x - 9$.

THINK	WRITE
a. i. 1. Rearrange the equation of the given line to express it in $y = mx + c$ form.	a. i. $2y - 5x = 4$ $2y = 5x + 4$ $y = \frac{5x}{2} + \frac{4}{2}$ $\therefore y = \frac{5}{2}x + 2$
2. State the gradient of the given line.	The given line has $m = \frac{5}{2}$.
3. State the gradient of a line parallel to the given line.	The gradient of a parallel line will be the same as that of the given line. Therefore, a line parallel to $2y - 5x = 4$ has a gradient of $\frac{5}{2}$.

ii. 1. State the gradient of a perpendicular line to the given line.	ii. The gradient of a perpendicular line will be the negative reciprocal of the gradient of the given line. If $m_1 = \frac{5}{2}$, then $m_2 = -\frac{2}{5}$. Therefore, a line perpendicular to $2y - 5x = 4$ has a gradient of $-\frac{2}{5}$.
b. 1. Write down the gradients of each line.	b. Lines: $y = 4x$ and $y = -0.25x$ Gradients: $m_1 = 4$, $m_2 = -0.25$
2. Test the product of the gradients.	$m_1 m_2 = 4 \times -0.25$ $\therefore m_1 m_2 = -1$ Therefore, the lines are perpendicular.
c. 1. State the gradient of the given line and calculate the gradient of the perpendicular line.	c. For $y = -3x - 9$, its gradient is $m_1 = -3$. The perpendicular line has gradient $m_2 = -\frac{1}{m_1}$. Therefore, the perpendicular line has gradient $\frac{1}{3}$.
2. Use the point–gradient form to obtain the equation of the required line.	$m = \frac{1}{3}$ and $(x_1, y_1) = (1, 1)$ $y - y_1 = m(x - x_1)$ $y - 1 = \frac{1}{3}(x - 1)$ $3(y - 1) = 1(x - 1)$ $3y - 3 = x - 1$ $3y = x + 2$ $y = \frac{1}{3}x + \frac{2}{3}$
3. State the answer.	The required line has equation $y = \frac{1}{3}x + \frac{2}{3}$.

1.6 Exercise

Students, these questions are even better in jacPLUS

Receive immediate feedback and access sample responses

Access additional questions

Track your results and progress

Find all this and MORE in jacPLUS

Technology free

1. WE14 Show that the points A $(-3, -12)$, B $(0, 3)$ and C $(4, 23)$ are collinear.
2. Determine whether the points A $(-4, 13)$, B $(7, -9)$ and C $(12, -19)$ are collinear.
3. Find the value of b if the three points $(3, b)$, $(4, 2b)$ and $(8, 5 - b)$ are collinear.

4. For the points P $(-6, -8)$, Q $(6, 4)$ and R $(-32, 34)$, find the equation of the line through P and Q, and hence determine if the three points are collinear.

5. Explain whether or not the points A $(-15, -95)$, B $(12, 40)$ and C $(20, 75)$ may be joined to form a triangle.

6. **WE16** a. State the gradient of a line that is:

   i. parallel to $3y - 6x = 1$ ii. perpendicular to $3y - 6x = 1$.

   b. Show that the lines $y = x$ and $y = -x$ are perpendicular.
   c. Determine the equation of the line through the point $(1, 1)$ perpendicular to the line $y = 5x + 10$.

7. a. Determine the gradient of the line parallel to the line $x - 2y = 6$.
   b. Determine the gradient of the line parallel to the line $3y - 4x + 2 = 0$.
   c. Determine the gradient of the line perpendicular to $y = 3x - 4$.
   d. Determine the gradient of the line perpendicular to $4y - 2x = 8$.

8. a. Determine the gradient of the line perpendicular to $5x + 4y - 1 = 0$.
   b. Determine the gradient of the line that is perpendicular to the line connecting the two points $(-3, 5)$ and $(2, -7)$.
   c. Determine the gradient of the line that is perpendicular to the line connecting the two points $(2, 4)$ and $(7, -1)$.
   d. Show that the lines $y = 0.2x$ and $y = -5x$ are perpendicular.

9. Determine, in the form $ax + by = c$, the equation of the line that:

   a. passes through the point $(0, 6)$ and is parallel to the line $7y - 5x = 0$
   b. passes through the point $\left(-2, \frac{4}{5}\right)$ and is parallel to the line $3y + 4x = 2$
   c. passes through the point $\left(-\frac{3}{4}, 1\right)$ and is perpendicular to the line $2x - 3y + 7 = 0$
   d. passes through the point $(0, 0)$ and is perpendicular to the line $3x - y = 2$.

10. Find the coordinates of the x-intercept of the line that passes through the point $(8, -2)$ and is parallel to the line $2y - 4x = 7$.

11. **WE15** a. Calculate, correct to 2 decimal places, the angle made with the positive direction of the x-axis by the line that passes through the points $(1, -8)$ and $(5, -2)$.
   b. Calculate the angle of inclination with the horizontal made by a line that has a gradient of -2.
   c. Obtain the equation of the line that passes through the point $(2, 7)$ at an angle of $135°$ to the horizontal.

12. a. Determine the gradient of the line that passes through the point $(1, 2)$ at an angle of $40°$ to the horizontal.
   b. A line passes through the x-axis inclined at an angle of $145°$ with the positive direction of the x-axis. Calculate the gradient of this line.

13. a. Calculate the angle of inclination with the horizontal made by the line that has a gradient of 0.5.
   b. Calculate the angle of inclination with the horizontal made by the line that has a gradient of -0.5.

14. Calculate the magnitudes of the angles the following lines make with the positive direction of the x-axis, expressing your answers correct to 2 decimal places where appropriate.

   a.

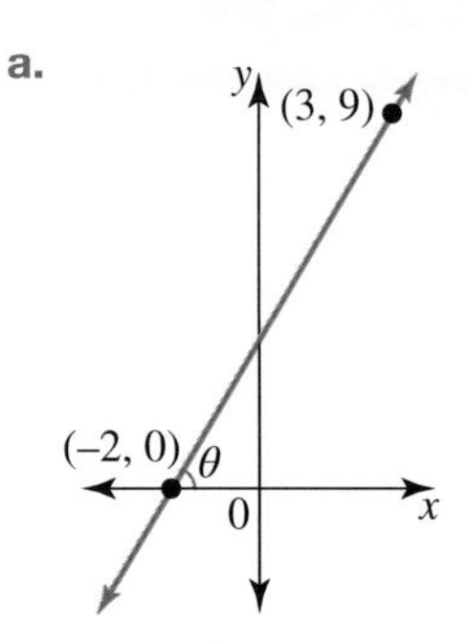

b. The line that cuts the x-axis at $x=4$ and the y-axis at $y=3$
c. The line that is parallel to the y-axis and passes through the point $(1, 5)$
d. The line with gradient -7

15. Calculate the angle of inclination with the horizontal made by each of the lines whose gradients are 5 and 4 respectively, and hence find the magnitude of the acute angle between these two lines.

16. Determine the equation of the line that passes through the point $(-6, 12)$ and makes an angle of $\tan^{-1}(1.5)$ with the horizontal.

17. A line L cuts the x-axis at the point A where $x=4$, and is inclined at an angle of 123.69° to the positive direction of the x-axis.

   a. Form the equation of the line L, specifying its gradient to 1 decimal place.
   b. Form the equation of a second line, K, that passes through the same point A at right angles to the line L.
   c. State the distance between the y-intercepts of K and L.

18. a. Find the value of a so that the line $ax-7y=8$ is:

   i. parallel to the line $3y+6x=7$
   ii. perpendicular to the line $3y+6x=7$.

   b. Find the value of c if the line through the points $(2c, -c)$ and $(c, -c-2)$ makes an angle of 45° with the horizontal.
   c. Find the value of d so the line containing the points $(d+1, d-1)$ and $(4, 8)$ is:

   i. parallel to the line that cuts the x-axis at $x=7$ and the y-axis at $y=-2$
   ii. parallel to the x-axis
   iii. perpendicular to the x-axis.

   d. The angle between the two lines with gradients -1.25 and 0.8 respectively has a magnitude of $\alpha°$. Calculate the value of α.

19. Given the points P$(-2, -3)$, Q$(2, 5)$, R$(6, 9)$ and S$(2, 1)$, show that PQRS is a parallelogram. State whether PQRS is a rectangle.

20. Determine the equation of the line that passes through the point of intersection of the lines $2x-3y=18$ and $5x+y=11$, and is perpendicular to the line $y=8$.

1.6 Exam questions

Question 1 (1 mark) TECH-ACTIVE

MC The equation of the line that would pass through the points shown on the graph is

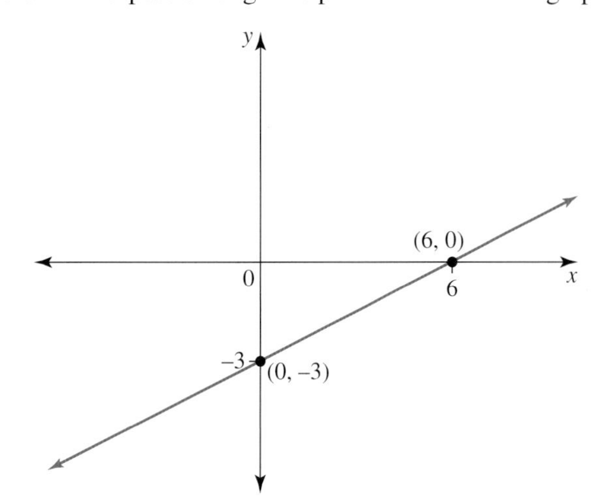

A. $y=\frac{x}{2}-3$ B. $y=\frac{x}{2}-3$ C. $y=2x-3$ D. $y=-2x-3$ E. $y=-\frac{x}{2}+3$

Question 2 (4 marks) TECH-FREE

a. Line L_1 has equation $y = 3x$ and line L_2 has the equation $y = 3 - x$. Sketch the graphs for each line on the same axes, labelling them clearly. **(2 marks)**

b. Write down the equation of the line with the same gradient as L_1 and the same x-intercept as L_2. **(2 marks)**

Question 3 (1 mark) TECH-ACTIVE

MC To the nearest degree, the line with equation $6x + 10y - 3 = 0$ is inclined to the positive direction of the x-axis at an angle of:

A. 99° **B.** 17° **C.** 31° **D.** 149° **E.** 81°

More exam questions are available online.

1.7 Bisection and lengths of line segments

LEARNING INTENTION

At the end of this subtopic you should be able to:

- calculate the distance between two points located on a Cartesian plane
- calculate the midpoint and gradient of a line segment (interval) on the Cartesian plane

1.7.1 Line segments

Sections of lines with two end points that have finite lengths are called **line segments**. The notation AB is used to name the line segment with end points A and B.

1.7.2 The coordinates of the midpoint of a line segment

The point of bisection of a line segment is its **midpoint**. This point is equidistant from the end points of the interval.

Let the end points of the line segment be A (x_1, y_1) and B (x_2, y_2).

Let the midpoint of AB be the point M $(\bar{x}, \bar{y})$, where $\bar{x}$ is the mean of the x-values for A and B and $\bar{y}$ is the mean of the y-values for A and B. Since M bisects AB, AM = MB and the triangles ACM and MDB are congruent (identical).

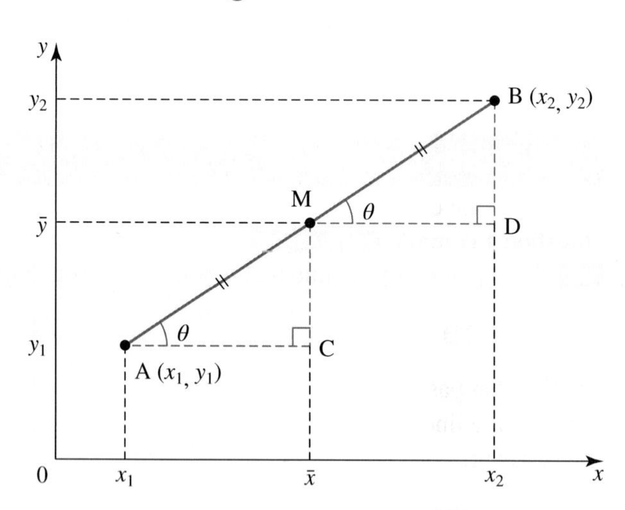

Equating the 'runs':

$$
\begin{aligned}
AC &= MD \\
\bar{x} - x_1 &= x_2 - \bar{x} \\
2\bar{x} &= x_1 + x_2 \\
\bar{x} &= \frac{x_1 + x_2}{2}
\end{aligned}
$$

and equating the 'rises':

$$\begin{aligned} \text{CM} &= \text{DB} \\ \bar{y} - y_1 &= y_2 - \bar{y} \\ 2\bar{y} &= y_1 + y_2 \\ \bar{y} &= \frac{y_1 + y_2}{2} \end{aligned}$$

Hence, the coordinates of the midpoint of a line segment are found by averaging the coordinates of the end points.

Coordinates of the midpoint

The coordinates of the midpoint of a line segment with end points (x_1, y_1) and (x_2, y_2) are:

$$\textbf{midpoint} = \left(\frac{x_1 + x_2}{2}, \frac{y_1 + y_2}{2}\right).$$

WORKED EXAMPLE 17 Calculating midpoints

Calculate the coordinates of the midpoint of the line segment joining the points (−3, 5) and (7, −8).

THINK	WRITE
1. Average the x- and y-coordinates of the end points.	$\bar{x} = \frac{x_1 + x_2}{2} = \frac{(-3) + 7}{2} = \frac{4}{2} = 2$ $\bar{y} = \frac{y_1 + y_2}{2} = \frac{5 + (-8)}{2} = \frac{-3}{2} = -1.5$
2. Write the coordinates as an ordered pair.	Therefore, the midpoint is (2, −1.5).

1.7.3 The perpendicular bisector of a line segment

The line that passes through the midpoint of a line segment at right angles to the line segment is called the **perpendicular bisector** of the line segment.

To find the equation of the perpendicular bisector:

- its gradient is $-\frac{1}{m_{AB}}$, since it is perpendicular to the line segment AB
- the midpoint of the line segment AB also lies on the perpendicular bisector.

Perpendicular bisector of AB

WORKED EXAMPLE 18 Determing the equation of a perpendicular bisector

Determine the equation of the perpendicular bisector of the line segment joining the points A (6, 3) and B (−8, 5).

THINK	WRITE
1. Calculate the gradient of the line segment.	A (6, 3) and B (−8, 5) $m = \frac{5-3}{-8-6}$ $= \frac{2}{-14}$ $= -\frac{1}{7}$
2. Obtain the gradient of a line perpendicular to the line segment.	Since $m_{AB} = -\frac{1}{7}$, the gradient of a line perpendicular to AB is $m_{\perp} = 7$.
3. Calculate the coordinates of the midpoint of the line segment.	$\overline{x} = \frac{6+(-8)}{2} = \frac{-2}{2} = -1$ $\overline{y} = \frac{3+5}{2} = \frac{8}{2} = 4$ The midpoint is (−1, 4). Point (−1, 4); gradient $m = 7$
4. Form the equation of the perpendicular bisector using the point–gradient equation.	$y - y_1 = m(x - x_1)$ $y - 4 = 7(x + 1)$ $y - 4 = 7x + 7$ $\therefore y = 7x + 11$
5. State the answer.	The equation of the perpendicular bisector is $y = 7x + 11$.

1.7.4 The length of a line segment

The length of a line segment is the distance between its end points. For a line segment AB with end points A (x_1, y_1) and B (x_2, y_2), the run $x_2 - x_1$ measures the distance between x_1 and x_2, and the rise $y_2 - y_1$ measures the distance between y_1 and y_2.

Using Pythagoras' theorem, $(AB)^2 = (x_2 - x_1)^2 + (y_2 - y_1)^2$.

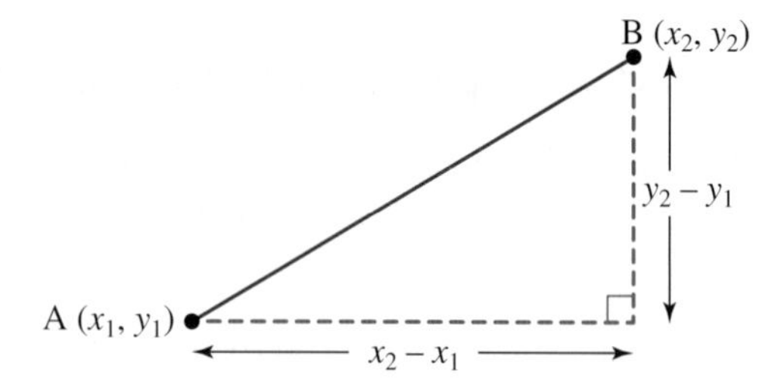

The length of a line segment

$$AB = \sqrt{(x_2 - x_1)^2 + (y_2 - y_1)^2}$$

This could be expressed as the formula for the distance between two points:

$d_{AB} = \sqrt{(x_2 - x_1)^2 + (y_2 - y_1)^2}$, where d_{AB} represents the distance between the points A and B.

WORKED EXAMPLE 19 Calculating the distance between two points

Calculate the length of the line segment joining the points A (−2, −5) and B (1, 3).

THINK	WRITE
1. Write the distance formula.	$d_{AB} = \sqrt{(x_2 - x_1)^2 + (y_2 - y_1)^2}$
2. Substitute the coordinates of the two points. *Note:* It does not matter which point is labelled (x_1, y_1) and which (x_2, y_2).	A (−2, −5) and B (1, 3) Let A be (x_1, y_1) and B be (x_2, y_2). $d_{AB} = \sqrt{(1-(-2))^2 + (3-(-5))^2}$ $= \sqrt{(3)^2 + (8)^2}$ $= \sqrt{9 + 64}$ $= \sqrt{73}$
3. State the answer. By choice, both the exact surd value and its approximate value to 2 decimal places have been given.	Therefore, the length of AB is $\sqrt{73} \simeq 8.54$ units.

Resources

Interactivity Midpoint of a line segment and the perpendicular bisector (int-2553)

1.7 Exercise

Students, these questions are even better in jacPLUS

Receive immediate feedback and access sample responses

Access additional questions

Track your results and progress

Find all this and MORE in jacPLUS

Technology free

1. WE17 Calculate the coordinates of the midpoint of the line segment joining the points (12, 5) and (−9, −1).

2. Determine the coordinates of the midpoint of the line segment joining the points:
   a. (−2, 8) and (12, −2)
   b. (1, 0) and (−5, 4).

3. Determine the coordinates of the midpoint of the line segment joining the points:
   a. (7, 3) and (−4, 2)
   b. (24, 12) and (16, 12).

4. MC M is the midpoint of the line segment AN. Given that M has coordinates (5, 6) and A is the point (3, 7), select the correct coordinates for point N.
   A. (4, 6.5)
   B. (4, −2)
   C. (7, 5)
   D. (1, 8)
   E. (−7, −5)

5. If the midpoint of PQ has coordinates (3, 0) and Q is the point (−10, 10), find the coordinates of point P.

6. Determine the equation of the line that has a gradient of −3 and passes through the midpoint of the segment joining (5, −4) and (1, 0).

7. Given the points A $(3, 0)$, B $(9, 4)$, C $(5, 6)$ and D $(-1, 2)$, show that AC and BD bisect each other.

8. **WE18** Determine the equation of the perpendicular bisector of the line segment joining the points A $(-4, 4)$ and B $(-3, 10)$.

9. Determine the equation of the perpendicular bisector of the line segment joining the points A $(-6, 0)$ and B $(2, 4)$.

10. Determine the equation of the perpendicular bisector of the line segment joining the points A $(1, 2)$ and B $(-3, 5)$.

11. Given that the line $ax + by = c$ is the perpendicular bisector of the line segment CD where C is the point $(-2, -5)$ and D is the point $(2, 5)$, find the smallest non-negative values possible for the integers a, b and c.

12. **WE19** Calculate the length of the line segment joining the points $(6, -8)$ and $(-4, -5)$.

13. Calculate the distance between the points $(10, -3)$ and $(-2, 6)$.

14. Find the exact distance between the points:
    a. $(-3, 2)$ and $(3, -4)$.
    b. $(-1, -5)$ and $(5, -1)$.

15. Triangle ABC has vertices A $(-2, 0)$, B $(2, 3)$ and C $(3, 0)$. Determine which of its sides has the shortest length.

Technology active

16. Calculate the distance between the point $(3, 10)$ and the midpoint of the line segment AB where A is the point $(-1, 1)$ and B is the point $(6, -1)$. Give the answer correct to 2 decimal places.

17. If the distance between the points $(p, 8)$ and $(0, -4)$ is 13 units, find two possible values for p.

18. Given the points A $(-7, 2)$ and B $(-13, 10)$, obtain:
    a. the distance between the points A and B
    b. the coordinates of the midpoint of the line segment AB
    c. the equation of the perpendicular bisector of AB
    d. the coordinates of the point where the perpendicular bisector meets the line $3x + 4y = 24$.

19. Triangle CDE has vertices C $(-8, 5)$, D $(2, 4)$ and E $(0.4, 0.8)$.
    a. Calculate its perimeter to the nearest whole number.
    b. Show that the magnitude of angle CED is $90°$.
    c. Find the coordinates of M, the midpoint of its hypotenuse.
    d. Show that M is equidistant from each of the vertices of the triangle.

20. A circle has its centre at $(4, 8)$ and one end of the diameter at $(-2, -2)$.
    a. Specify the coordinates of the other end of the diameter.
    b. Calculate the area of the circle as a multiple of π.

21. Two friends planning to spend some time bushwalking argue over which one of them should carry a rather heavy rucksack containing food and first aid items. Neither is keen, so they agree to each throw a small coin towards the base of a tree, and the person whose coin lands the greater distance from the tree will have to carry the rucksack. Taking the tree as the origin and the distances in centimetres east and north of the origin as (x, y) coordinates, Anna's coin lands at $(-2.3, 1.5)$ and Liam's coin lands at $(1.7, 2.1)$. Determine who carries the rucksack. Support your answer with a mathematical argument.

22. The diagram shows a main highway through a country town. The section of this highway running between a petrol station at P and a restaurant at R can be considered a straight line. Relative to a fixed origin, the coordinates of the petrol station and restaurant are P (3, 7) and R (5, 3) respectively. Distances are measured in kilometres.

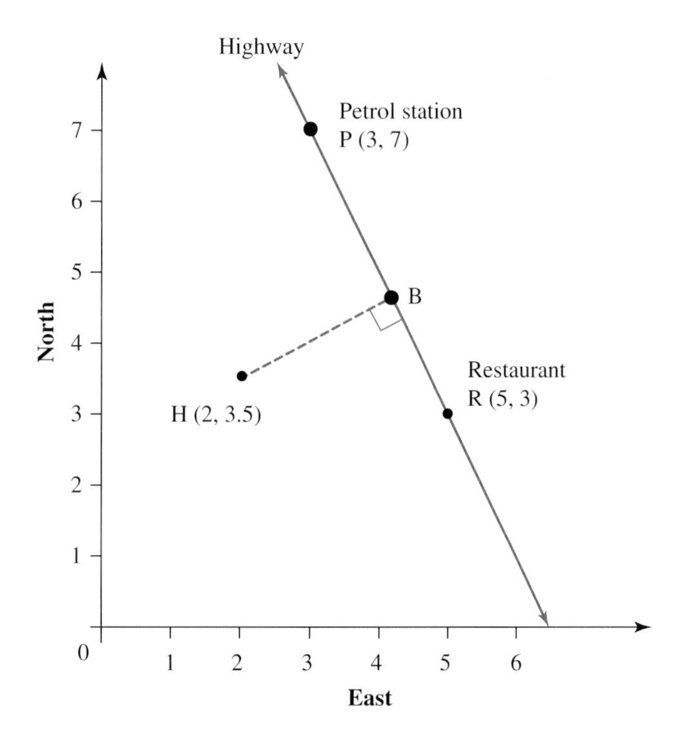

a. Calculate how far apart the petrol station and restaurant are. Give your answer correct to 1 decimal place.
b. Form the equation of the straight line PR.

Ada is late for their waitressing job at the restaurant and is still at home at the point H (2, 3.5). There is no direct route to the restaurant from Ada's home, but there is a bicycle track that goes straight to the nearest point B on the highway. Ada decides to ride to point B and then to travel along the highway from B to the restaurant.

c. Form the equation of the line through H perpendicular to PR.
d. Hence, find the coordinates of B, the closest point on the highway from Ada's home.
e. If Ada's average speed is 10 km/h, find how long, to the nearest minute, it takes Ada to reach the restaurant from home.

23. a. Using CAS technology, construct the perpendicular bisectors of each of the three sides of the triangle. Write down what you notice about the bisectors. Repeat this procedure using other triangles. State whether your observation appears to hold for these triangles.
b. For the triangle formed by joining the points $O(0,0), A(6,0), B(4,4)$, find the point of intersection of the perpendicular bisectors of each side. Check your answer algebraically.

24. a. Using CAS technology, construct the line segments joining each vertex to the midpoint of the opposite side (these are called medians). Write down what you notice about the line segments. Repeat this procedure using other triangles. State whether your observation appears to hold for these triangles.
b. For the triangle formed by joining the points $O(0,0), A(6,0)$ and $B(4,4)$, find the point of intersection of the medians drawn to each side. Check your answer algebraically.

1.7 Exam questions

Question 1 (1 mark) TECH-ACTIVE

MC The distance between the points A and B on the graph shown is

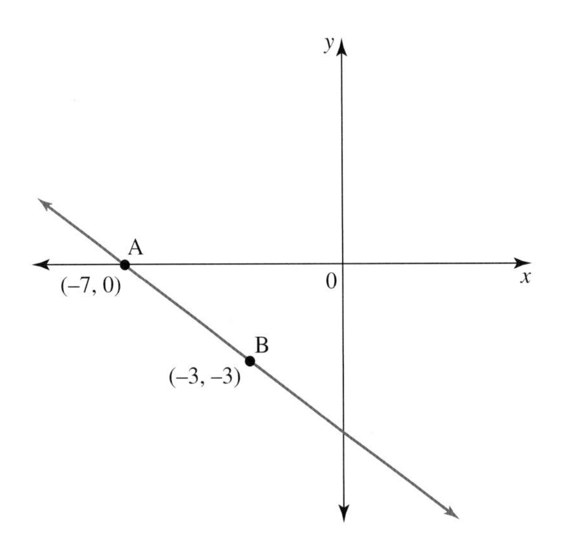

A. 2 B. 13 C. 7 D. 5 E. 25

Question 2 (1 mark) TECH-ACTIVE

MC The midpoint of interval joining $P(-5,-3)$ and $Q(2,-1)$ is

A. $\left(-\frac{7}{2},-2\right)$ B. $\left(-\frac{7}{2},-1\right)$ C. $(-7,-4)$ D. $(-3,-4)$ E. $\left(-\frac{3}{2},-2\right)$

Question 3 (4 marks) TECH-FREE

For the points $A(1,-5)$ and $B(4,-2)$, find

a. the gradient of the line joining A and B **(1 mark)**
b. the equation of the line that is perpendicular to AB and passes through A. **(3 marks)**

More exam questions are available online.

1.8 Review

1.8.1 Summary

doc-37016

Hey students! Now that it's time to revise this topic, go online to:

 Access the topic summary

 Review your results

 Watch teacher-led videos

 Practise exam questions

Find all this and MORE in jacPLUS 

1.8 Exercise

Technology free: short answer

1. Solve the following equations for x.

   a. $3(5x-2)+5(3x-2)=8(x-2)$

   b. $\frac{2x-1}{5}+\frac{3-2x}{4}=\frac{3}{20}$

   c. $ax+3c=3a+cx$

   d. $\frac{5x-b}{2b}-\frac{2x}{b}=2$

2. Solve the following system of equations for x and y.

$$2x+y=6$$
$$5x-2y=24$$

3. Solve the following inequations for x.

   a. $3-2x\leq 1$

   b. $\frac{2x}{3}-5>2+3x$

4. Sketch, labelling any axis intercepts with their coordinates, the line with the equation $3x-4y=24$.

5. Find the equation of the line:

   a. through the point $(-5,8)$ parallel to the line $2x-7y=2$
   b. through the point $(4,0)$ perpendicular to the line $2x-7y=2$
   c. through the two points $(9,2)$ and $(6,-7)$.

6. Determine the equation of the line through the points $(-4,6)$ and $(2,-2)$.

Technology active: multiple choice

7. **MC** If $4(1-3x)+2(3+x)>5$, then:

   A. $x>2$
   B. $x<2$
   C. $x<-2$
   D. $x<\frac{1}{2}$
   E. $x>\frac{1}{2}$

8. **MC** The solutions to the simultaneous equations $7x-2y=11$ and $3x+y=1$ are:

   A. $x=9, y=-26$
   B. $x=5, y=11$
   C. $x=-1, y=2$
   D. $x=1, y=-2$
   E. $x=1, y=4$

9. **MC** A calculator purchased for \$200 depreciates each year by an amount of \$25. The value, V, of the calculator after t years is:

   A. $V=25t$
   B. $V=-25t$
   C. $V=200-25t$
   D. $V=200+25t$
   E. $V=25t-200$

10. **MC** The equation of the graph shown is:

A. $x - 2y = 6$
B. $2y + x = 6$
C. $6x - 3y = 1$
D. $y = -\frac{1}{2}x - 3$
E. $y = 2x - 3$

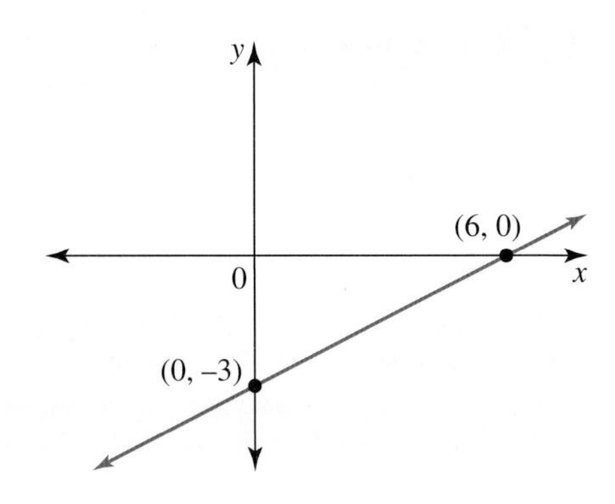

11. **MC** For the points $(-1, -2)$, $(4, 3)$ and $(9, b)$ to be collinear, b would equal:

A. -3 B. -2 C. -1 D. 8 E. 10

12. **MC** The angle of inclination to the positive direction of the x-axis made by the line with equation $\frac{x}{3} - \frac{y}{2} = 1$ is closest to:

A. 26.6° B. 33.7° C. 56.3° D. 123.7° E. 146.3°

13. **MC** The equation of the line through $(9, 5)$ parallel to $y = -2$ is:

A. $y = -2x + 13$ B. $2y = x + 1$ C. $x = 9$ D. $y = 9$ E. $y = 5$

14. **MC** The midpoint of a line segment AB is $(3, -5)$. If A is the point $(13, 11)$, then the coordinates of B are:

A. $(8, 3)$ B. $(-23, -27)$ C. $(23, 27)$ D. $(-7, -21)$ E. $(7, 21)$

15. **MC** The value of a such that there would be no point of intersection between the two lines $ay + 3x = 4$ and $2y + 4x = 3$ is:

A. 2 B. 1.5 C. $\frac{8}{3}$ D. -0.5 E. -2

16. **MC** The distance between the points $(-3, 5)$ and $(-6, 12)$ is:

A. 4 B. $\sqrt{40}$ C. $\sqrt{58}$ D. $\sqrt{370}$ E. $\sqrt{388}$

Technology active: extended response

17. After school, Tenzin rides their bike along a straight path from school to a golf range. Tenzin travels one-third of the way at an average speed of 20 km/h and the remainder of the way at 10 km/h.

a. Find the distance Tenzin travels from school to the golf range if the journey takes 45 minutes.

At the golf range Tenzin is trying to improve their putting. A set of Cartesian coordinates can be imagined to be marked so that Tenzin will be aiming to hit the ball at the point T$(1.5, 4)$ into the hole at the origin $(0, 0)$.

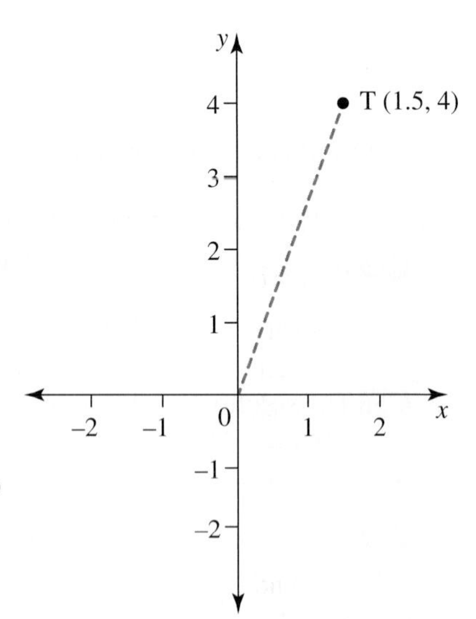

b. If all units are measured in metres, state how far Tenzin's golf ball is from the hole. Give your answer correct to 2 decimal places.
c. State the equation of the straight-line path Tenzin's golf ball needs to travel along for the ball to reach the hole.
d. In fact, Tenzin hits the golf ball along the path with equation $6x - 2y = 1$.
   i. Sketch this path.
   ii. Determine the equation of the line through $(0, 0)$ perpendicular to this path and hence find, to the nearest centimetre, the closest distance of the golf ball's path to the hole at $(0, 0)$.

18. A small fireworks rocket travels along a path that can be considered to be a straight line. On a Cartesian set of x- and y-axes where the units are in metres, the x-coordinates give the horizontal distance the rocket travels and the y-coordinates give the height of the rocket above the ground.
The fireworks rocket is launched from a point S (0, 1) at an angle of $\theta°$ with the horizontal. The fireworks explode on reaching a point E, which is at a height of 10 metres above the ground.

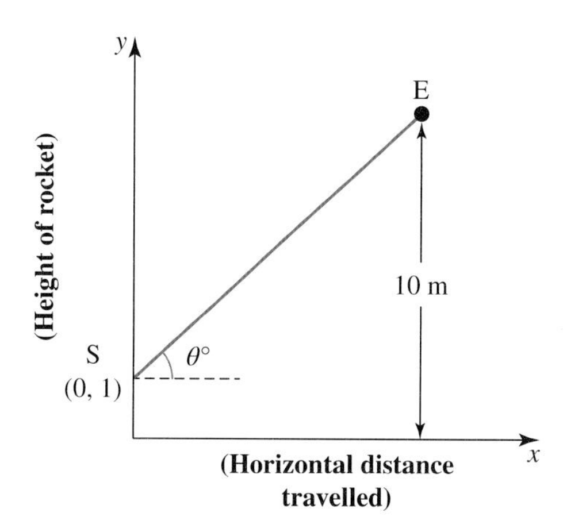

a. i. The first rocket is launched at an angle of 45° to the horizontal. Find the equation of its path and the coordinates of point E.
   ii. After the explosion, part of the debris travels from point E along a line perpendicular to the rocket's path. Find the equation of this path and work out how far horizontally from E the debris reaches the ground.
b. The angle at which the rockets are launched from S (0, 1) is varied and the fireworks explode at E (k, 10), $k > 0$. Show that the equation of the paths of all possible rockets is given by $9x - ky + k = 0$.
c. i. Let r be the horizontal distance from E at which the debris travelling on a line perpendicular to the path SE reaches the ground. Form an expression for r in terms of k.
   ii. It is desirable for the debris not to be too widely scattered. Find the possible values for k such that $4 \leq r \leq 6$.

19. Rain water is collected in a water tank. On 1 April the tank contained 1000 litres of water. Ten days later it contained 1250 litres. Assume the amount of water increases uniformly.
a. Find the rate of increase in litres per day.
b. Form the linear relationship between the volume V litres of water in the tank t days after 1 April.
c. State how much water the linear model predicts should have been in the tank on 30 March.
d. The tank needs replacing and quotes are obtained from two companies. The Latasi company charges $500 for materials plus $26 per hour of construction. The Natano company charges $600 for materials plus $18 per hour of construction.
   i. Form linear models for the costs for each company, defining the symbols used.
   ii. Determine when the costs are the same for each company and calculate this cost.
   iii. Sketch both cost models on the same axes.
   iv. It is estimated that the construction should take approximately 8 hours of work. Determine which company could do the job at a cheaper cost.

20. Three friends visit their local supermarket to buy snacks for a party. One friend buys 3 bags of chips, 1 bag of lollies and 2 boxes of biscuits. The second friend buys 2 bags of chips, 2 bags of lollies and 4 boxes of biscuits. The third friend buys 1 bag of chips, 3 bags of lollies and 3 boxes of biscuits. The three friends spent a total of $24.10, $33.40 and $29.50 respectively.
Determine:
a. the cost of one bag of chips
b. the cost of one bag of lollies
c. the cost of one box of biscuits.

1.8 Exam questions

Question 1 (2 marks) TECH-ACTIVE

Sam is a very good student, and his latest test results show that the sum of his marks for Mathematical Methods, Chemistry and Physics is 256. Mathematical Methods is his best subject, and the difference between his Mathematical Methods and Chemistry marks is 8. His Chemistry mark was 1 mark higher than his Physics mark.

Determine his marks for each subject.

Question 2 (1 mark) TECH-ACTIVE

MC The gradient, m, and y-intercept, c, of the line with equation $3x + y - 5 = 0$ are

A. $m = 3, c = 5$
B. $m = 3, c = -5$
C. $m = -5, c = 3$
D. $m = -3, c = 5$
E. $m = -3, c = -5$

Question 3 (4 marks) TECH-FREE

Find the value of a such that the system of equations has no solution.

$$2ax - 3y = 2a$$
$$8x - 10 = 2y$$

Question 4 (1 mark) TECH-ACTIVE

MC The angle that the line shown makes with the positive direction of the x-axis is nearest to

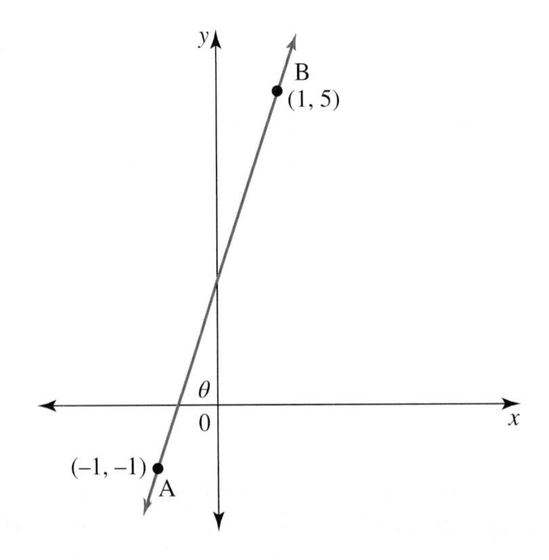

A. 37°
B. 53°
C. 72°
D. 76°
E. 0.05°

Question 5 (1 mark) TECH-ACTIVE

MC The equation of the line passing through the point (3, −2) and perpendicular to the line with equation $2x - 3y - 1 = 0$ is

A. $2x - 3y = 0$
B. $3x + 2y - 5 = 0$
C. $3x - 2y - 13 = 0$
D. $2x + 3y - 1 = 0$
E. $2x - 3y - 12 = 0$

More exam questions are available online.

Answers

Topic 1 Lines and linear relationships

1.2 Linear equations and inequations

1.2 Exercise

1. a. $x = -1$ b. $x = 10$
2. a. $x = 8$ b. $x = -10$ c. $x = -\frac{7}{3}$
   d. $x = -7$ e. $x = -\frac{6}{7}$ f. $x = \frac{26}{7}$
3. a. $x = 9$ b. $x = 20$ c. $x = 6$
   d. $x = \frac{1}{3}$ e. $x = 1$ f. $x = -10$
4. $x = a + d$
5. a. $x = \frac{c-b}{a}$ b. $x = \frac{ab}{a-b}$ c. $x = -1$
   d. $x = ab$ e. $x = \frac{a}{b-c}$ f. $x = b - a$
6. $x \geq 24$

   23 24 25 26 27 28 29 x

7. a. $x \leq -\frac{1}{2}$ b. $x < -3$ c. $x > -2$
   d. $x < -\frac{1}{3}$ e. $x > 0$ f. $x > 22$
8. a. $x \leq -1$ b. $x > -9$ c. $x \leq 18$
   d. $x > 3$ e. $x \geq \frac{2}{3}$ f. $x < -6$
9. $x = -\frac{1}{7}$
10. $x = c$
11. $x > -\frac{1}{6}$
12. a. $x = 0$ b. $x > 4$
13. a. 8 and 10
    b. 21
    c. 12, 13 and 14
    d. Length is 20 cm; width is 4 cm.
    e. Height is 45 cm.
14. a. The profit for the sale of n books is $P = 2.3n - 100$.
    b. 44 books
15. 25 phone covers
16. 184 drinks

1.2 Exam questions

Note: Mark allocations are available with the fully worked solutions online.

1. $x = a + b$
2. E
3. $x \leq 5$

1.3 Systems of simultaneous linear equations

1.3 Exercise

1. a. $x = 7, y = 1$ b. $x = -4, y = -2$
2. a. $x = 2, y = -1$ b. $x = -3, y = \frac{1}{2}$
   c. $x = 2, y = 1$ d. $x = -1, y = 4$
   e. $x = 1, y = 3$ f. $x = -5, y = 6$
3. a. $x = 1, y = 4$ b. $x = -\frac{1}{2}, y = -11$
4. a. $x = \frac{3}{4}, y = \frac{2}{3}$ b. $x = 12, y = 6$
5. a. $x = -\frac{7}{3}, y = -\frac{35}{3}$ b. $x = 1, y = 0$
6. a. $x = 5, y = -1$ b. $x = -82, y = -120$
7. a. An adult ticket costs \$25; a child ticket costs \$12.
   b. \$471.09
8. $x = 5, y = 8, z = -6$
9. $x = 1, y = -2, z = 3$
10. a. $x = 2, y = 1, z = 4$ b. $x = 4, y = 2, z = -1$
    c. $x = -6, y = 8, z = 1$ d. $x = -2, y = 5, z = 10$
    e. $x = 4, y = 3, z = 2$ f. $x = -2, y = -11, z = 40$
11. Adult ticket \$14; concession ticket \$12; child ticket \$10
12. Agnes \$20 per hour; Bjork \$18 per hour; Chi \$25 per hour
13. Two 50-cent coins, twelve 20-cent coins and eight 10-cent coins
14. a. $x = 3, y = 1.5, z = -2.6$
    b. $x = 10,\ y = -6,\ z = 0.5,\ w = 5$
    c. $x_1 = -2,\ x_2 = -4,\ x_3 = \frac{2}{3},\ x_4 = \frac{1}{6}$
15. The food compound requires 50 kg of supplement X, 30 kg of supplement Y and 10 kg of supplement Z.

1.3 Exam questions

Note: Mark allocations are available with the fully worked solutions online.

1. B
2. C
3. E

1.4 Linear graphs and their equations

1.4 Exercise

1. $m = -\frac{4}{3}$
2. a. 2 b. $\frac{2}{3}$
   c. $-\frac{1}{3}$ d. $-\frac{7}{5}$
3. a. $m = -2.5$ b. $m = 5$
   c. $m = 0$ d. Undefined
4. Show both gradients equal 1; sample responses can be found in the worked solutions in the online resources.

5. a.

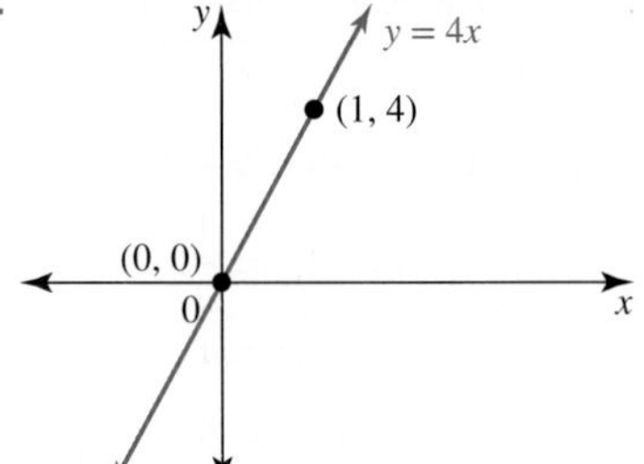

b.

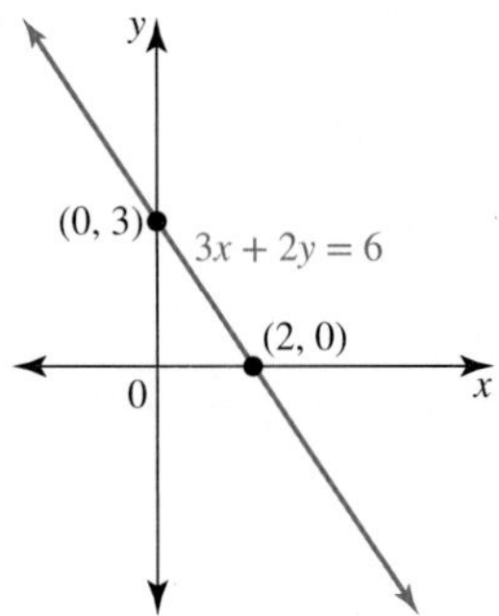

c.

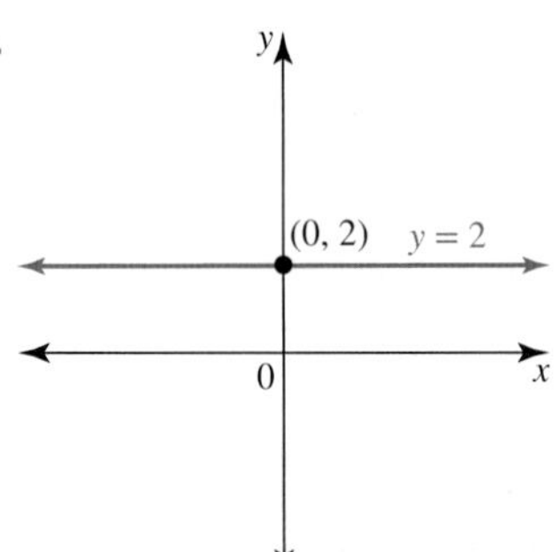

6. a.

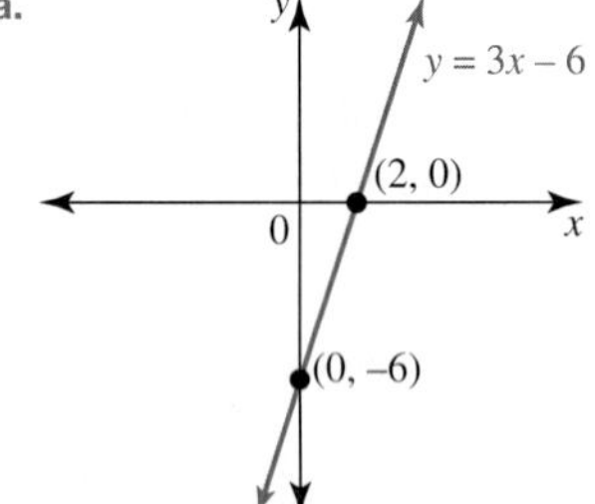

b.

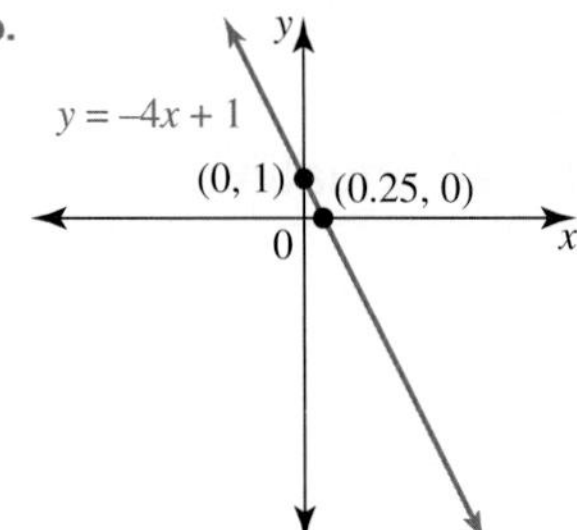

c.

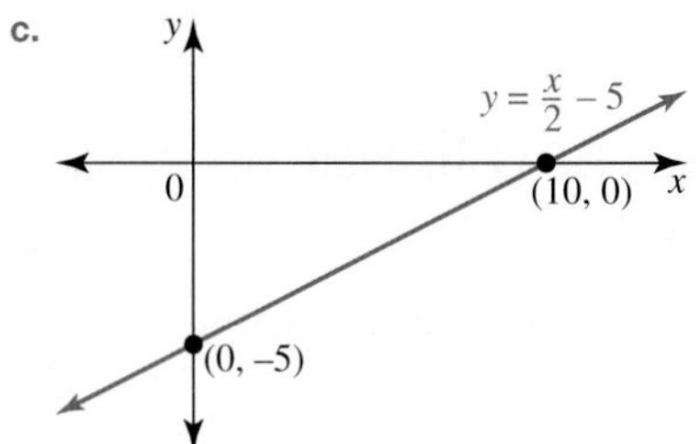

d.

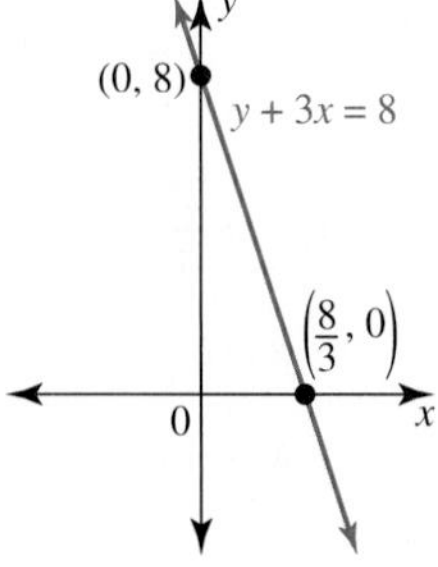

e.

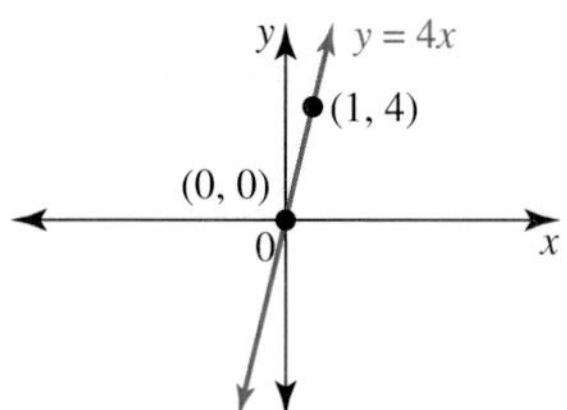

f.

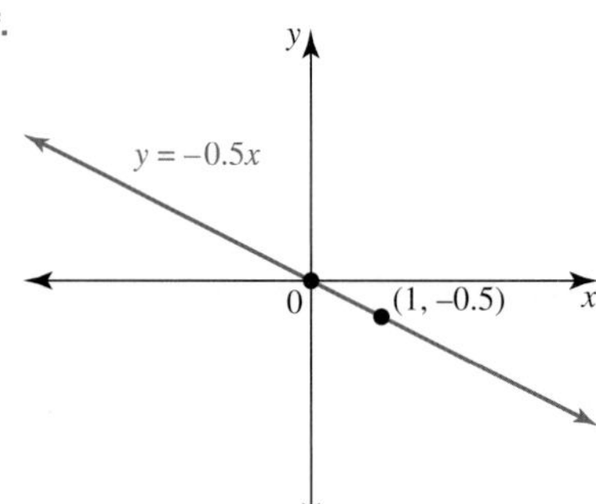

7. a.

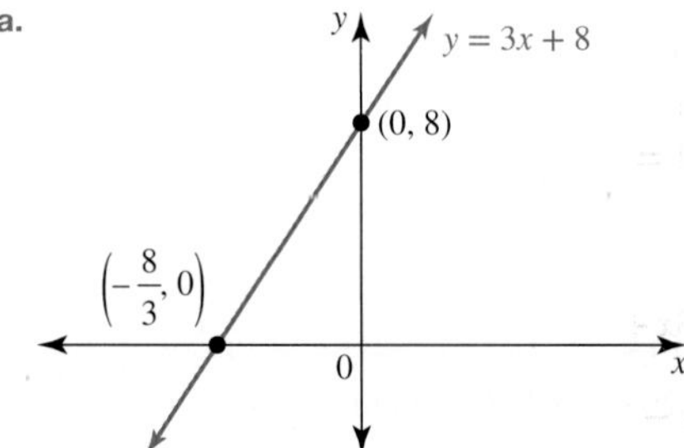

b.

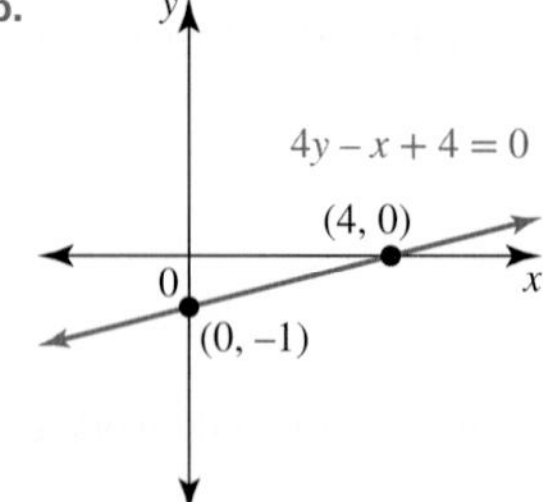

c.

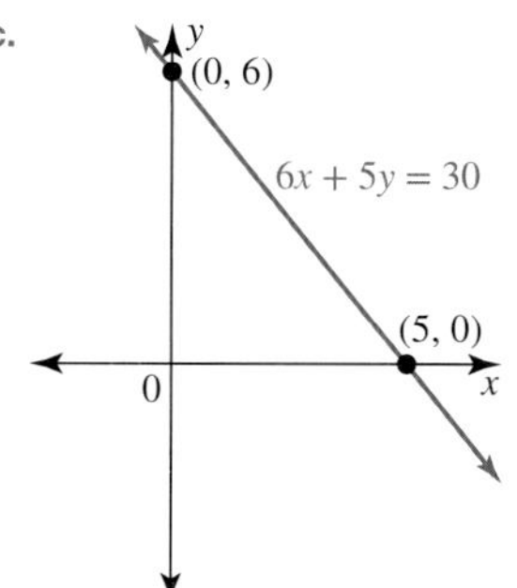

d.

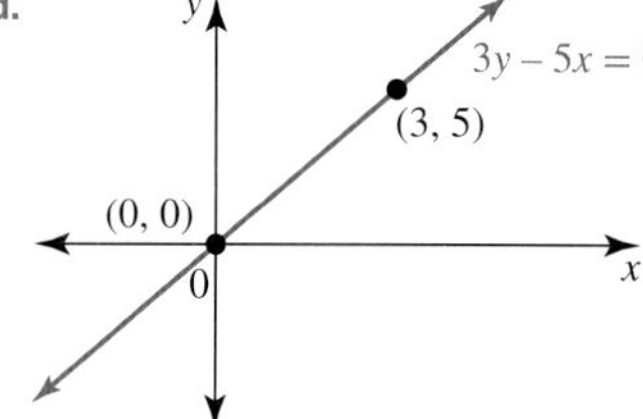

e.

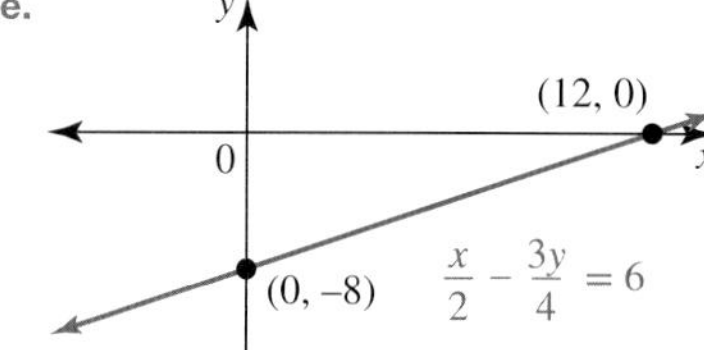

f.

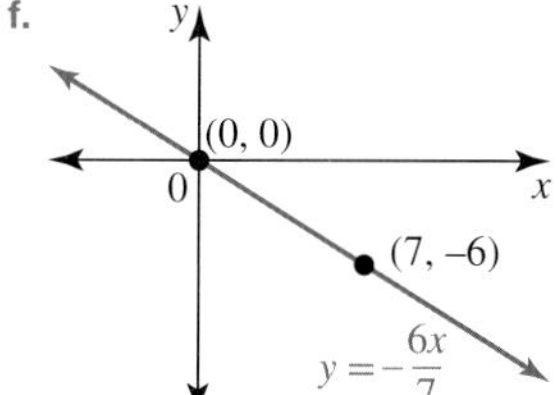

8\. a. $y = -2x - 13$ b. $7y + 2x = 1$

c. $y = -1.5x + 6$ d. $m = \frac{5}{6}; (0, 3)$

9\. $y = 10$

10\. a. $y = 5x + 2$ b. $y = -2x - 1$

c. $y = 3x - 1$ d. $y = -5x$

e. $x + 2y + 1 = 0$ f. $3x + 4y - 11 = 0$

11\. a. i. $y = \frac{5x}{4}$

ii. $y = -3x + 9$

iii. $y = \frac{2}{3}x - 2$

iv. $y = -\frac{1}{2}x - 1$

b. $x = -12$

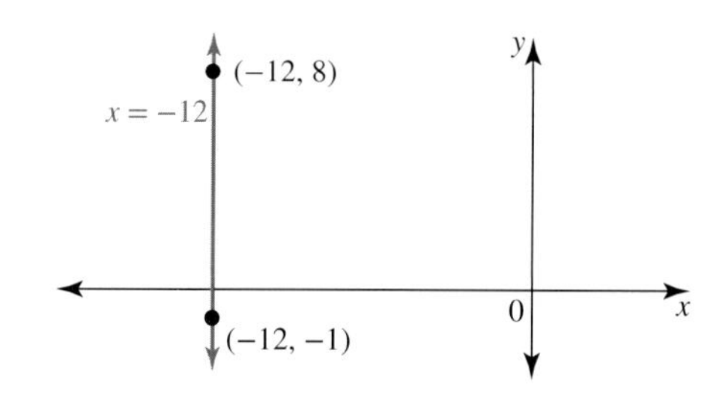

c. $y = -8$

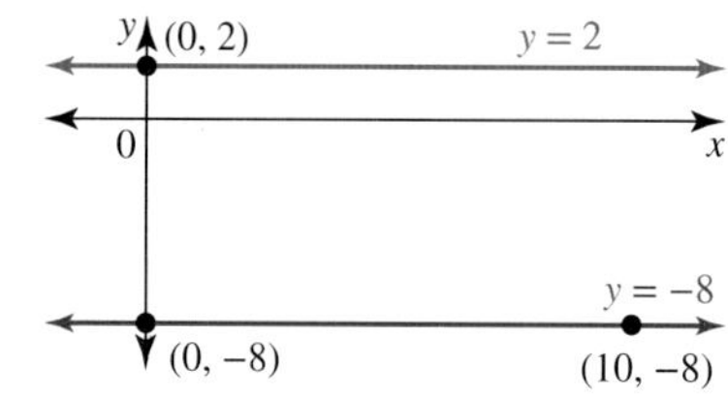

12\. a. $y = -5x + 37$ b. $3y - 2x + 10 = 0$

c. $4y + 7x + 36 = 0$ d. $y = -0.8x + 0.2$

e. $3y - 10x = 34$ f. $y = -2x + 10$

13\. a. $m = 2, \ (0, -8)$ b. $\frac{5}{3}$

c. $m = \frac{3}{4}, \ (0, 1)$ d. i, ii and iii are parallel.

14\. a. $m = -\frac{4}{5}, c = 4$ b. $m = \frac{8}{3}, c = 20$

c. $m = \frac{1}{6}, c = \frac{3}{2}$ d. $m = 0, c = \frac{3}{2}$

15\. a.

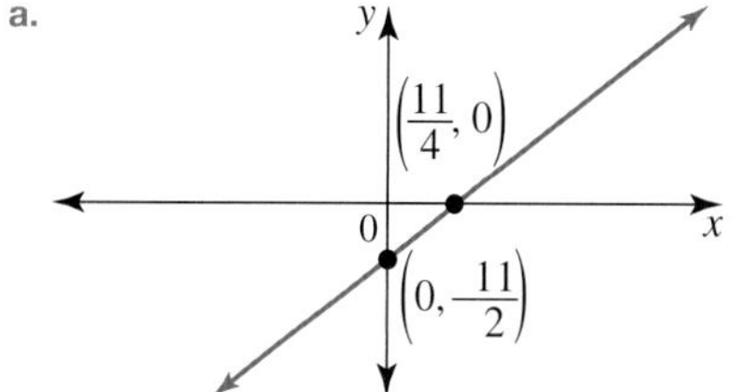

b.

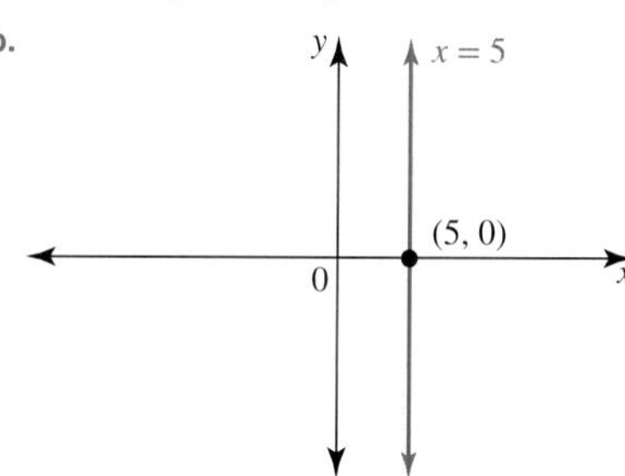

c.

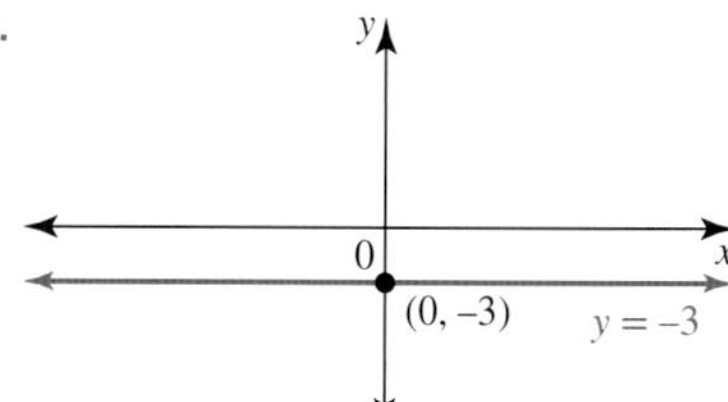

16\. a. $a = -6$ b. $y = \frac{q}{p}x$

17\. a. $C = 30 + 1.5t$

b.

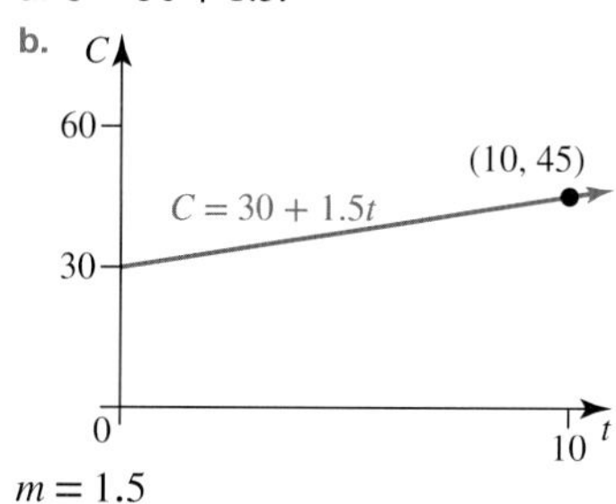

$m = 1.5$

1.4 Exam questions

Note: Mark allocations are available with the fully worked solutions online.

1. B
2. C
3. $y = -\frac{1}{2}x + \frac{5}{2}$

1.5 Intersections of lines and their applications

1.5 Exercise

1. $(12, 10.5)$
2. a. $(-0.5, -5)$ b. $(8, -3)$ c. $(7, -5)$
3. a. $(-2, 6), (3, 6), (3, -9)$
   b. 37.5 square units
4. a. $(20, 500)$

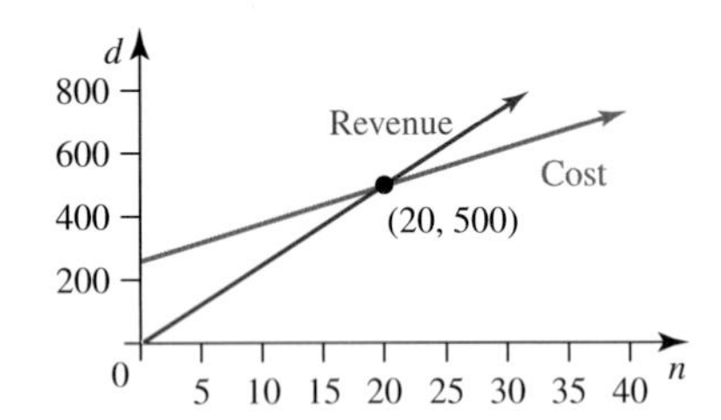

   b. At least 21 items
5. a. $C_1 = 10 + 0.75x$, $C_2 = 20$ where C is the cost and x the distance
   b.

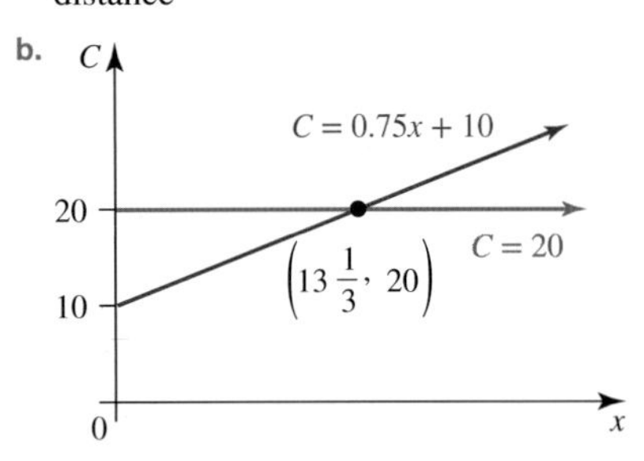

   c. $13\frac{1}{3}$ km
   d. If the distance is less than $13\frac{1}{3}$ km, Pedal On is cheaper; if the distance exceeds $13\frac{1}{3}$ km, Bikes R Gr8 is cheaper.
6. $m = 6$
7. $a = -1.5, b = -2$
8. $p = 10.5$
9. a. $y = -\frac{px}{5} + \frac{q}{5}, y = \frac{3x}{q} - 5$
   b. $p = 0.6, q = -25$
   c. $pq \neq -15$
10. $(1.39, 5.91)$
11. $(-1, 2)$
12. Sample responses can be found in the worked solutions in the online resources. Point of concurrency at $(3, -2)$
13. $a = 10$
14. $(-2, -3)$
15. Any real number except for $d = 10$
16. a. $C_A = 300 + 0.05x$, $C_B = 250 + 0.25x$
    b. Cost per kilometre of travel
    c. $y = 50 - 0.2x$
    d.

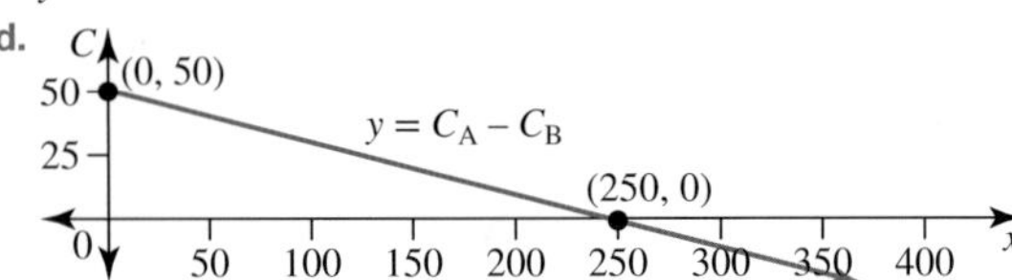

    e. i. 250 km
       ii. More than 250 km
17. a. 6 am $(0, 2)$; 7 am $(6, 3)$; $y = \frac{x}{6} + 2$
    b. Sample responses can be found in the worked solutions in the online resources.
    c. $(21, 5.5)$
    d. 6 am $\left(-\frac{1}{3}, 0\right)$; 7 am $(1, 1)$; $y = \frac{3x}{4} + \frac{1}{4}$
    e. $(3, 2.5)$
    f. There is no collision, since the boat is at the common point at 6.30 am and the trawler is there at 8.30 am.

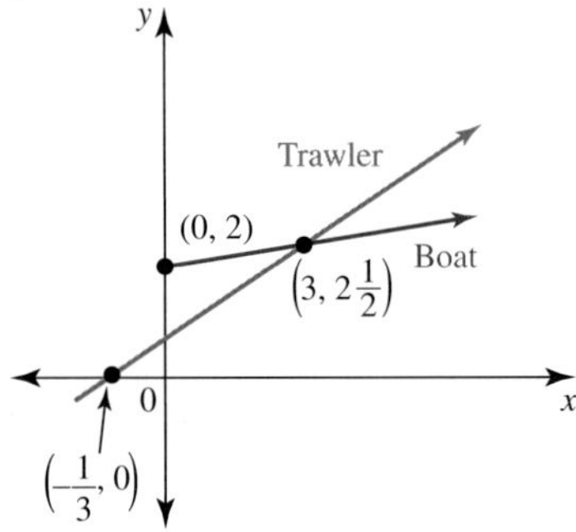

18. a. No collision b. $(4, 11)$

1.5 Exam questions

Note: Mark allocations are available with the fully worked solutions online.

1. C
2. B
3. $a = \frac{1}{3}, b = \frac{2}{3}$

1.6 Straight lines and gradients

1.6 Exercise

1. Sample responses can be found in the worked solutions in the online resources.
2. Collinear
3. $b = \frac{5}{7}$
4. $y = x - 2$; not collinear
5. The points are not collinear, so a triangle can be formed.
6. a. i. $m = 2$ ii. $m = -0.5$
   b. $m_1 m_2 = -1$
   c. $5y + x = 6$
7. a. $\frac{1}{2}$ b. $\frac{4}{3}$ c. $-\frac{1}{3}$ d. -2

8. a. $\frac{4}{5}$ b. $\frac{5}{12}$
c. 1 d. $0.2 \times -5 = -1$

9. a. $-5x + 7y = 42$ b. $20x + 15y = -28$
c. $12x + 8y = -1$ d. $x + 3y = 0$

10. $(9, 0)$

11. a. $56.31°$ b. $116.57°$ c. $y = -x + 9$

12. a. 0.839 b. -0.7

13. a. $26.6°$ b. $153.4°$

14. a. $60.95°$ b. $143.13°$ c. $90°$
d. $98.13°$

15. $78.69°, 75.96°, 2.73°$

16. $3x - 2y = -42$

17. a. $y = -1.5x + 6$ b. $2x - 3y = 8$
c. $\frac{26}{3}$ units

18. a. i. $a = -14$ ii. $a = 3.5$
b. $c = 2$
c. i. $d = 11.4$ ii. $d = 9$ iii. $d = 3$
d. $\alpha = 90$

19. $m_{PQ} = m_{SR} = 2, m_{PS} = m_{QR} = 1$
PQRS is a parallelogram as opposite sides are parallel. Adjacent sides are not perpendicular, so PQRS is not a rectangle.

20. $x = 3$

1.6 Exam questions

Note: Mark allocations are available with the fully worked solutions online.

1. B

2. a.

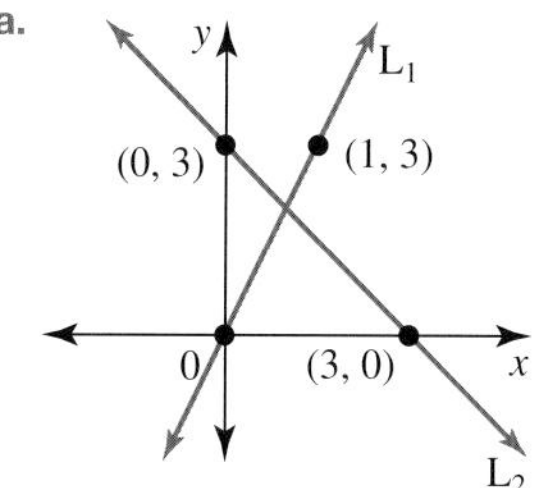

b. $y = 3x - 9$

3. D

1.7 Bisection and lengths of line segments

1.7 Exercise

1. $(1.5, 2)$

2. a. $(5, 3)$ b. $(-2, 2)$

3. a. $(1.5, 2.5)$ b. $(20, 12)$

4. C

5. $P(16, -10)$

6. $y = -3x + 7$

7. $(4, 3)$ is midpoint of both AC and BD.

8. $2x + 12y = 77$

9. $y = -2x - 2$

10. $6y - 8x - 29 = 0$

11. $a = 2, b = 5, c = 0$

12. $\sqrt{109} \approx 10.44$

13. 15

14. a. $\sqrt{72} = 6\sqrt{2}$ b. $\sqrt{52} = 2\sqrt{13}$

15. BC

16. 10.01

17. $p = \pm 5$

18. a. 10 units b. $(-10, 6)$
c. $4y - 3x = 54$ d. $\left(-5, 9\frac{3}{4}\right)$

19. a. 23 units
b. Sample responses can be found in the worked solutions in the online resources.
c. $(-3, 4.5)$
d. $ME = MC = MD = \sqrt{25.25}$

20. a. $(10, 18)$ b. 136π square units

21. Anna

22. a. 4.5 km b. $y = -2x + 13$
c. $y = 0.5x + 2.5$ d. $(4.2, 4.6)$
e. 26 minutes

23. a. Perpendicular bisectors are concurrent.
b. $(3, 1)$

24. a. Medians are concurrent.
b. $\left(\frac{10}{3}, \frac{4}{3}\right)$

1.7 Exam questions

Note: Mark allocations are available with the fully worked solutions online.

1. D

2. E

3. a. 1
b. Equation of the line perpendicular to AB and passing through A is $y + x + 4 = 0$.

1.8 Review

1.8 Exercise

Technology free: short answer

1. a. $x = 0$ b. $x = 4$
c. $x = 3$ d. $x = 5b$

2. $x = 4, y = -2$

3. a. $x \geq 1$ b. $x < -3$

4.

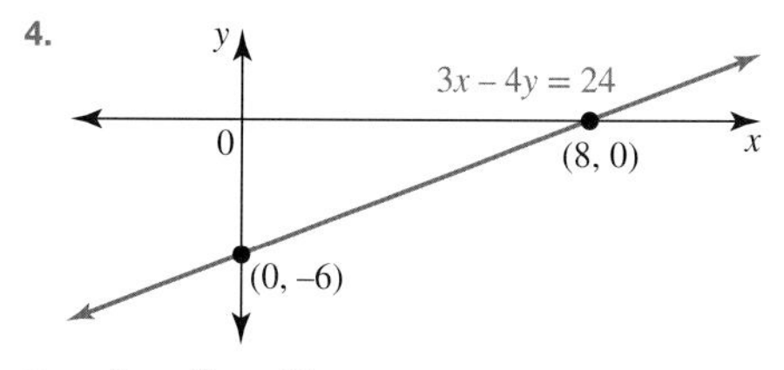

5. a. $7y - 2x = 66$
b. $2y + 7x = 28$
c. $y = 3x - 25$

6. $y = -\frac{4}{3}x + \frac{2}{3}$

Technology active: multiple choice

7. D 8. D 9. C 10. A 11. D
12. B 13. E 14. D 15. B 16. C

Technology active: extended response

17. a. 9 km b. 4.27 m c. $3y - 8x = 0$

d. i.

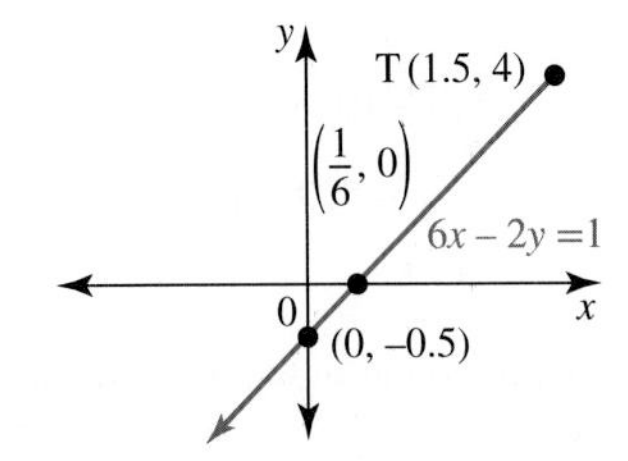

ii. $y = -\frac{x}{3}$; 16 cm

18. a. i. $y = x + 1$; E(9, 10)

ii. $y = -x + 19$; 10 m

b. Sample responses can be found in the worked solutions in the online resources.

c. i. $r = \frac{90}{k}$

ii. $15 \leq k \leq 22.5$

19. a. 25 litres/day

b. t is the independent variable; $V = 1000 + 25t$

c. 950 litres

d. i. $C_L = 500 + 26t$; $C_N = 600 + 18t$; cost is C dollars, construction time is t hours.

ii. 12.5 hours; $825

iii.

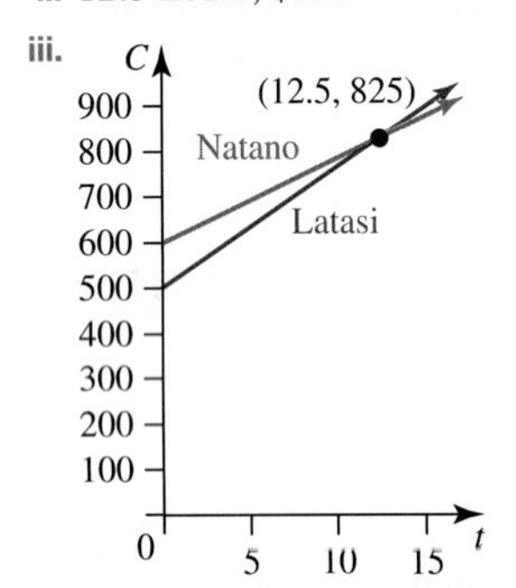

iv. Latasi Company is cheaper.

20. a. $3.70

b. $4.20

c. $4.40

1.8 Exam questions

Note: Mark allocations are available with the fully worked solutions online.

1. Mathematical Methods: 91; Chemistry: 83; Physics: 82
2. D
3. $a = -6$
4. C
5. B

2 Algebraic foundations

LEARNING SEQUENCE

Fully worked solutions for this topic are available online.

2.1 Overview

Hey students! Bring these pages to life online

Watch videos

Engage with interactivities

Answer questions and check results

Find all this and MORE in jacPLUS

2.1.1 Introduction

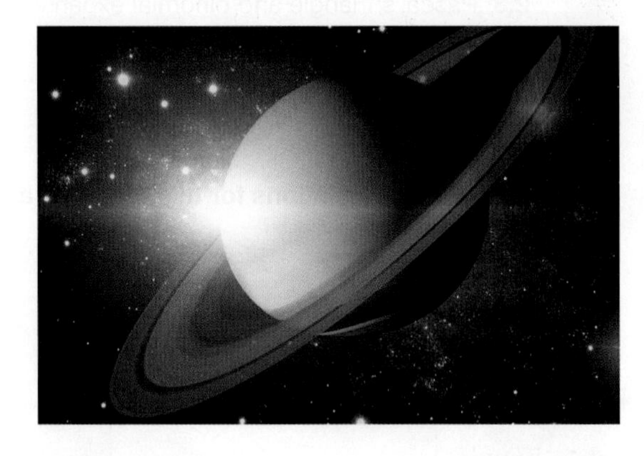

In 2017 the Cassini spacecraft disintegrated in Saturn's atmosphere, having successfully completed a 7-year mission to observe that planet and its moons. In all, it completed 22 orbits of Saturn, sending extraordinary images and information back to Earth. In an earlier mission, the Voyager 1 spacecraft passed Pluto in 1990, and in 2004 it left our solar system. On board Voyager 1 was a time capsule carrying information about Earth and our achievements for, should they exist, any intelligent alien life forms it may reach.

None of this would have been possible without mathematics. Mathematics and physics are essential to the launch and success of all space missions. It has been argued that should there ever be communication between Earthlings and intelligent extra-terrestrials, communication will be through mathematics. This is because mathematics is universal. It is universally true that Pythagoras' theorem, $a^2 + b^2 = c^2$, holds in every country on Earth and, by extension, in any galaxy in the universe.

Expressing the rule for Pythagoras' theorem in symbols gives an example of the succinct nature of algebra. It is the language of mathematics — the unifying thread that has evolved through the ages — that underlies its different branches. Elementary algebra, as studied at school, seeks to establish the fundamentals vital for confident and automatic use of this language. There are other algebras, higher-order ones such as group theory, rings and fields, that have developed only since the 19th century. Despite its abstract nature, research in these fields is highly valued today for its applications in cryptography, encryption and other aspects of internet security.

Two modern-day mathematicians include Cheryl Praeger, an Australian who works in group theory and combinatorics, and the late Maryam Mirzakhani, so far the only female recipient of the Fields medal — the highest award in Mathematics — for her work in the dynamics and geometry of complex surfaces.

KEY CONCEPTS

This topic covers the following key concepts from the VCE Mathematics Study Design:

- use of symbolic notation to develop algebraic expressions and represent functions, relations, equations, and systems of simultaneous equations
- substitution into, and manipulation of, these expressions
- recognition of equivalent expressions and simplification of algebraic expressions involving different forms of polynomial and power functions, the use of distributive and exponent laws applied to these functions, and manipulation from one form of expression to an equivalent form.

Note: *Concepts shown in grey are covered in other topics.*

2.2 Algebraic skills

LEARNING INTENTION

At the end of this subtopic you should be able to:
- expand and factorise linear and simple quadratic expressions with integer coefficients by hand
- factorise the sum or difference of two cubes
- simplify algebraic fractions.

This topic covers some of the algebraic skills required for the foundation to learning and understanding of Mathematical Methods. Some basic algebraic techniques will be revised and some new techniques will be introduced.

2.2.1 Review of expansion and factorisation

Expansion

The **distributive law** is fundamental in expanding to remove brackets.

Distributive law

$$a(b+c)=ab+ac$$

Some simple expansions include:

$$(a+b)(c+d)=ac+ad+bc+bd$$
$$(a+b)^2=a^2+2ab+b^2$$
$$(a-b)^2=a^2-2ab+b^2$$
$$(a+b)(a-b)=a^2-b^2$$

WORKED EXAMPLE 1 Expanding and simplifying an expression

Expand $2(4x-3)^2-(x-2)(x+2)+(x+5)(2x-1)$ and state the coefficient of the x term.

THINK	WRITE
1. Expand each pair of brackets. *Note:* The first term contains a perfect square, the second a difference of two squares and the third a quadratic trinomial.	$2(4x-3)^2-(x-2)(x+2)+(x+5)(2x-1)$ $=2(16x^2-24x+9)-(x^2-4)+(2x^2-x+10x-5)$
2. Expand fully, taking care with signs.	$=32x^2-48x+18-x^2+4+2x^2+9x-5$
3. Collect like terms together.	$=33x^2-39x+17$
4. State the answer. *Note:* Read the question again to ensure the answer given is as requested.	The expansion gives $33x^2-39x+17$ and the coefficient of x is -39.

TI | THINK	DISPLAY/WRITE	CASIO | THINK	DISPLAY/WRITE
1. On a Calculator page, press MENU and select: 3. Algebra 3. Expand Complete the entry line as: expand $(2\times(4x-3)^2-(x-2)(x+2)+(x+5)(2x-1))$ Then press ENTER.	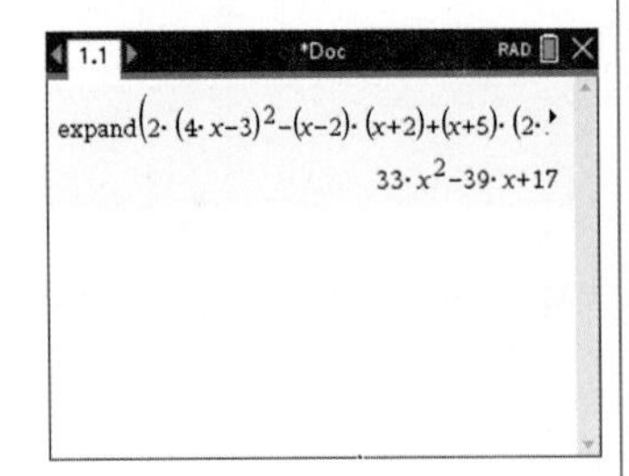	1. On a Main screen, complete the entry line as: expand $(2\times(4x-3)^2-(x-2)(x+2)+(x+5)(2x-1))$ Then press EXE.	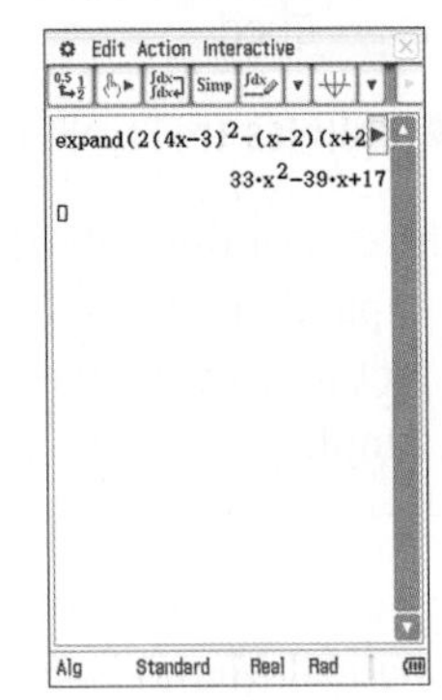
2. The answer appears on the screen.	The expansion gives $33x^2-39x+17$.	2. The answer appears on the screen.	The expansion gives $33x^2-39x+17$.

Factorisation

Some simple **factors** include:

- common factors
- the difference of two **perfect squares**
- perfect squares and factors of other quadratic trinomials.

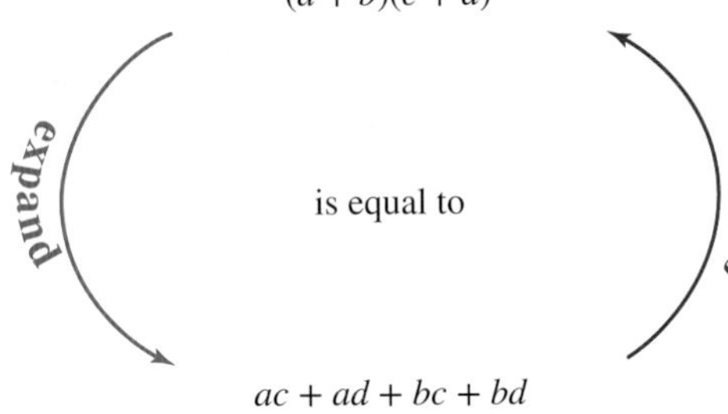

A systematic approach to factorising is displayed in the following diagram.

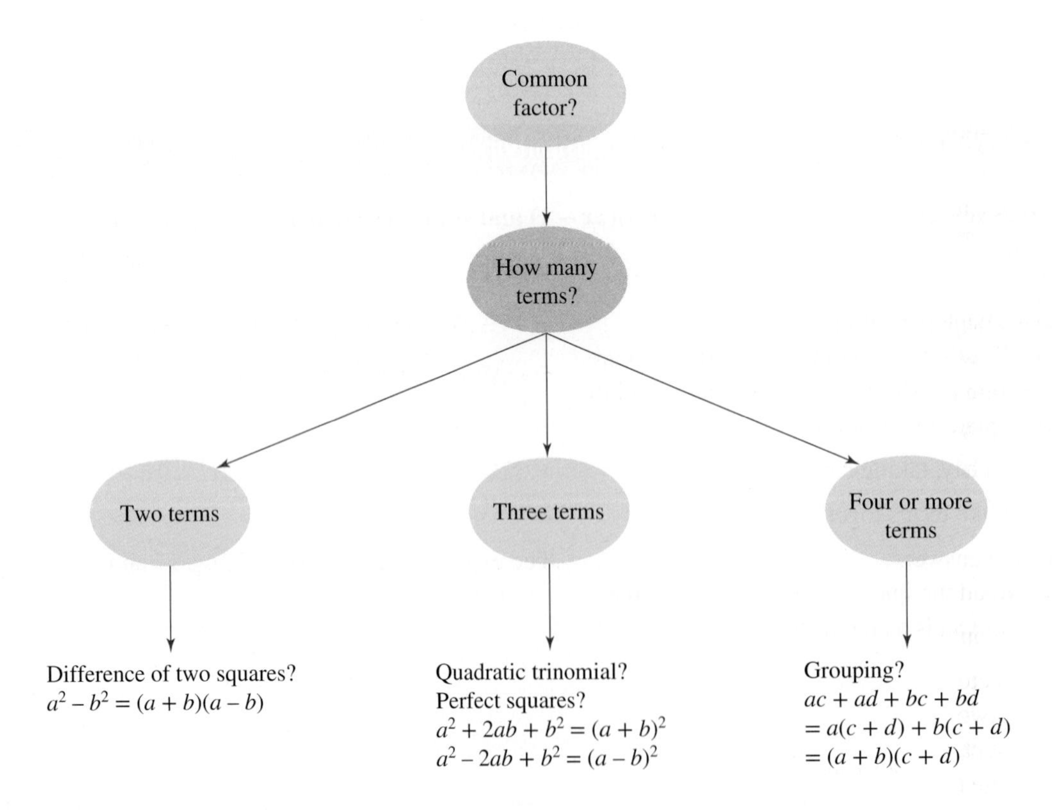

Grouping terms is commonly referred to as grouping '2 and 2' and grouping '3 and 1', depending on the number of terms grouped together.

For example, as the first three terms of $a^2+2ab+b^2-c^2$ are a perfect square, grouping '3 and 1' would create a difference of two squares expression, allowing the whole expression to be factorised.

$$\begin{aligned}a^2+2ab+b^2-c^2 &= (a^2+2ab+b^2)-c^2\\ &= (a+b)^2-c^2\\ &= (a+b-c)(a+b+c)\end{aligned}$$

WORKED EXAMPLE 2 Factorising algebraic expressions

Factorise:

a. $3x^2-12y^2$
b. x^2+7x-8
c. $3x^2-17x+10$
d. $2x^3+5x^2y-12y^2x$
e. $4y^2-x^2+10x-25$
f. $7(x+1)^2-8(x+1)+1$ **using the substitution** $a=(x+1)$**.**

THINK	WRITE
a. 1. Take out the common factor.	a. $3x^2-12y^2=3\left(x^2-4y^2\right)$
2. Recognise the difference of two squares.	$=3\left(x^2-(2y)^2\right)$
3. Factorise using the difference of two squares, $a^2-b^2=(a-b)(a+b)$.	$=3(x-2y)(x+2y)$
b. 1. Recognise the trinomial ax^2+bx+c.	b. x^2+7x-8, where $a=1$, $b=7$ and $c=-8$
2. Find two numbers that multiply to the outside product, ac, and add to the middle, b.	$ac=-8, b=7$ The numbers are -1 and 8.
3. Split the middle term of the given equation.	$x^2+7x-8=x^2-1x+8x-8$
4. Pair and factorise.	$=x^2-1x+8x-8$ $=x(x-1)+8(x-1)$
5. Take out the common factor. *Note:* The middle term may be split as $7x=-1x+8x$ or $7x=8x-1x$.	$=(x-1)(x+8)$
c. 1. Recognise the trinomial ax^2+bx+c.	c. $3x^2-17x+10$, where $a=3$, $b=-17$ and $c=10$
2. Find two numbers that multiply to the outside product, ac, and add to the middle, b.	$ac=30, b=-17$ The numbers are -15 and -2.
3. Split the middle term of the given equation.	$3x^2-17x+10=3x^2-2x-15x+10$
4. Pair and factorise.	$=3x^2-2x-15x+10$ $=x(3x-2)-5(3x-2)$
5. Take out the common factor.	$=(3x-2)(x-5)$
d. 1. Take out the common factor.	d. $2x^3+5x^2y-12xy^2=x\left(2x^2+5xy-12y^2\right)$
2. To factorise the expression in the brackets, find two numbers that multiply to the outside product $(2x-12=-24)$ and add to the middle $(-3+8=5)$.	The numbers are 8 and -3.

3. Split the middle term of the expression in brackets. $x\left(2x^2+5xy-12y^2\right)=x\left(2x^2-3xy+8xy-12y^2\right)$

4. Pair and factorise.
$=x\left(2x^2-3xy+8xy-12y^2\right)$
$=x\left(x\left(2x-3y\right)+4y\left(2x-3y\right)\right)$

5. Take out the common factor and write the three factors. $=x\left(2x-3y\right)\left(x+4y\right)$

e. 1. The last three terms of the expression can be grouped together to form a perfect square. **e.** $4y^2-x^2+10x-25=4y^2-\left(x^2-10x+25\right)$

2. Use the grouping '3 and 1' technique to create a difference of two squares.
$=4y^2-\left(x-5\right)^2$
$=\left(2y\right)^2-\left(x-5\right)^2$

3. Factorise the difference of two squares. $=\left[2y-\left(x-5\right)\right]\left[2y+\left(x-5\right)\right]$

4. Remove the inner brackets to obtain the answer. $=\left(2y-x+5\right)\left(2y+x-5\right)$

f. 1. Substitute $a=(x+1)$ to form a quadratic trinomial in a. **f.** $7(x+1)^2-8(x+1)+1$
$=7a^2-8a+1$ where $a=(x+1)$

2. Factorise the trinomial using any suitable method. $=(7a-1)(a-1)$

3. Substitute $(x+1)$ back in place of a. $=(7(x+1)-1)((x+1)-1)$

4. Remove the inner brackets and simplify to obtain the answer.
$=(7x+7-1)(x+1-1)$
$=(7x+6)(x)$
$=x(7x+6)$

TI | THINK	DISPLAY/WRITE	CASIO | THINK	DISPLAY/WRITE
b. 1. On a Calculator page, press MENU and select: 3. Algebra 2. Factor Complete the entry line as: factor $(4y^2-x^2+10x-25)$ Then press ENTER.	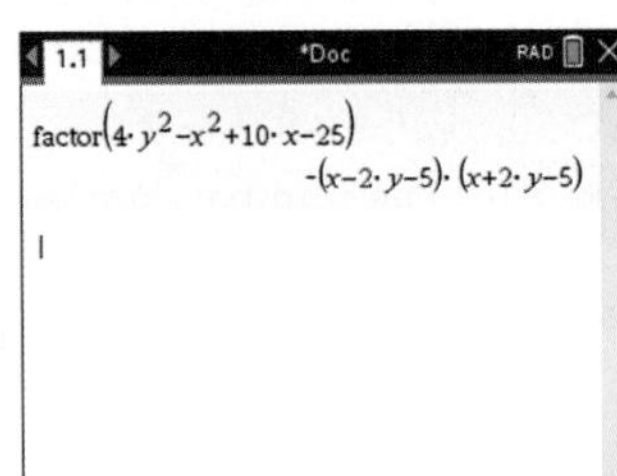	**b.** 1. On a Main screen, complete the entry line as: factor $(4y^2-x^2+10x-25)$ Then press EXE.	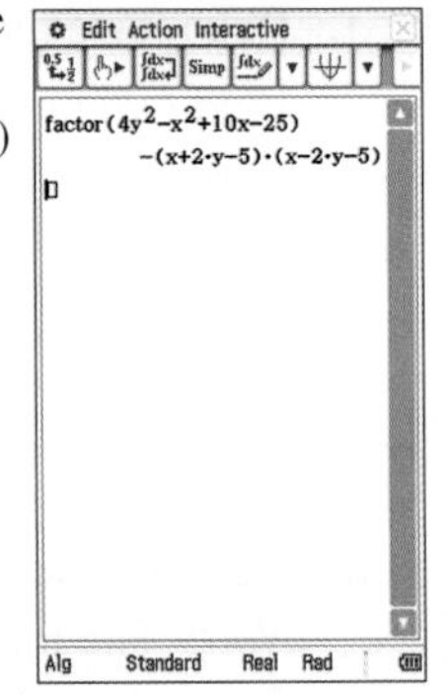
2. The answer appears on the screen.	$-(x-2y-5)(x+2y-5)$	2. The answer appears on the screen.	$-(x+2y-5)$ $(x-2y-5)$

2.2.2 Factorising sums and differences of two perfect cubes

Check the following by hand or by using CAS technology.

Expanding $(a+b)(a^2-ab+b^2)$ gives a^3+b^3, the sum of two cubes.
Expanding $(a-b)(a^2+ab+b^2)$ gives a^3-b^3, the difference of two cubes.

Hence, the factors of the sum and difference of two cubes are as follows.

Sum and difference of two perfect cubes

$$a^3 + b^3 = (a + b)(a^2 - ab + b^2)$$
$$a^3 - b^3 = (a - b)(a^2 + ab + b^2)$$

WORKED EXAMPLE 3 Factorising perfect cubes

Factorise:

a. $x^3 - 27$

b. $2x^3 + 16$.

THINK	WRITE
a. 1. Express $x^3 - 27$ as a difference of two cubes.	a. $x^3 - 27 = x^3 - 3^3$
2. Apply the factorisation rule for the difference of two cubes.	Using $a^3 - b^3 = (a - b)(a^2 + ab + b^2)$ with $a = x, b = 3,$ $x^3 - 3^3 = (x - 3)(x^2 + 3x + 3^2)$
3. State the answer.	$\therefore x^3 - 27 = (x - 3)(x^2 + 3x + 9)$
b. 1. Take out the common factor.	b. $2x^3 + 16 = 2(x^3 + 8)$
2. Express $x^3 + 8$ as a sum of two cubes.	$= 2(x^3 + 2^3)$
3. Apply the factorisation rule for the sum of two cubes.	Using $a^3 + b^3 = (a + b)(a^2 - ab + b^2)$ with $a = x, b = 2,$ $x^3 + 2^3 = (x + 2)(x^2 - 2x + 2^2)$ $2(x^3 + 2^3) = 2(x + 2)(x^2 - 2x + 2^2)$
4. State the answer.	$\therefore 2x^3 + 16 = 2(x + 2)(x^2 - 2x + 4)$

2.2.3 Algebraic fractions

The same methods used to simplify, add, subtract, multiply or divide arithmetic fractions are used to simplify algebraic fractions.

Simplifying algebraic fractions

An algebraic fraction can be simplified by cancelling any common factor between its numerator and its denominator. For example:

$$\frac{ab + ac}{ad} = \frac{\cancel{a}(b + c)}{\cancel{a}d}$$
$$= \frac{b + c}{d}$$

Multiplication and division of algebraic fractions

For the product of algebraic fractions, once numerators and denominators have been factorised, any common factors can then be cancelled. The remaining numerator terms are usually left in factors, as are any remaining denominator terms. For example:

$$\frac{\cancel{a}(b + c)}{\cancel{a}\cancel{d}} \times \frac{\cancel{d}(a + c)}{b} = \frac{(b + c)(a + c)}{b}$$

Note that b is not a common factor of the numerator so it cannot be cancelled with the b in the denominator. As in arithmetic, to divide by an algebraic fraction, multiply by its reciprocal.

$$\frac{a}{b} \div \frac{c}{d} = \frac{a}{b} \times \frac{d}{c}$$

WORKED EXAMPLE 4 Simplifying algebraic fractions

Simplify:

a. $\dfrac{x^2 - 2x}{x^2 - 5x + 6}$

b. $\dfrac{x^4 - 1}{x - 3} \div \dfrac{1 + x^2}{3 - x}$

THINK	WRITE
a. 1. Factorise both the numerator and the denominator. *Note:* The numerator has a common factor; the denominator is a quadratic trinomial.	a. $\dfrac{x^2 - 2x}{x^2 - 5x + 6} = \dfrac{x(x-2)}{(x-3)(x-2)}$
2. Cancel the common factor in the numerator and denominator.	$= \dfrac{x\cancel{(x-2)}}{(x-3)\cancel{(x-2)}}$
3. Write the fraction in its simplest form.	$= \dfrac{x}{x-3}$ No further cancellation is possible.
b. 1. Change the division into multiplication by replacing the divisor by its reciprocal.	b. $\dfrac{x^4 - 1}{x - 3} \div \dfrac{1 + x^2}{3 - x} = \dfrac{x^4 - 1}{x - 3} \times \dfrac{3 - x}{1 + x^2}$
2. Factorise where possible. *Note:* The aim is to create common factors of both the numerator and denominator. For this reason, write $(3 - x)$ as $-(x - 3)$.	Since $x^4 - 1 = (x^2)^2 - 1^2$ $= (x^2 - 1)(x^2 + 1)$ then: $\dfrac{x^4 - 1}{x - 3} \times \dfrac{3 - x}{1 + x^2}$ $= \dfrac{(x^2 - 1)(x^2 + 1)}{x - 3} \times \dfrac{-(x-3)}{1 + x^2}$ $= \dfrac{(x^2 - 1)\cancel{(x^2 + 1)}}{\cancel{x - 3}} \times \dfrac{-\cancel{(x-3)}}{\cancel{1 + x^2}}$
3. Cancel the two sets of common factors of the numerator and denominator.	$= \dfrac{(x^2 - 1)}{1} \times \dfrac{-1}{1}$
4. Multiply the remaining terms in the numerator together and the remaining terms in the denominator together.	$= \dfrac{-(x^2 - 1)}{1}$
5. State the answer.	$= -(x^2 - 1)$ $= 1 - x^2$

TI | THINK	DISPLAY/WRITE	CASIO | THINK	DISPLAY/WRITE
b. 1. On a Calculator page, complete the entry line as: $\frac{x^4-1}{x-3}/\frac{1+x^2}{3-x}$ Then press ENTER.	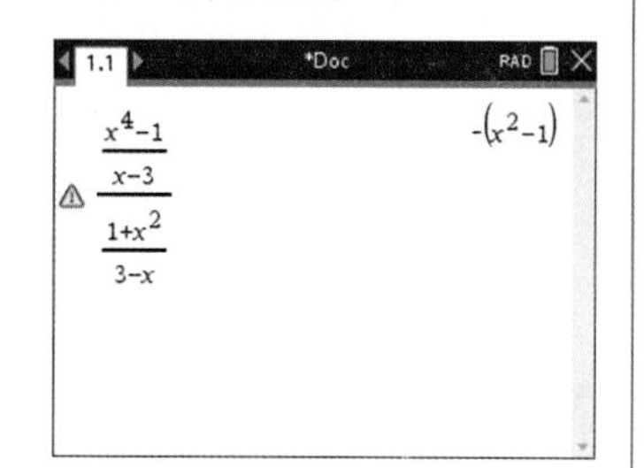	**b. 1.** On a Main screen, complete the entry line as: $\frac{x^4-1}{x-3}/\frac{1+x^2}{3-x}$ Press EXE, then press SIMP to simplify the answer.	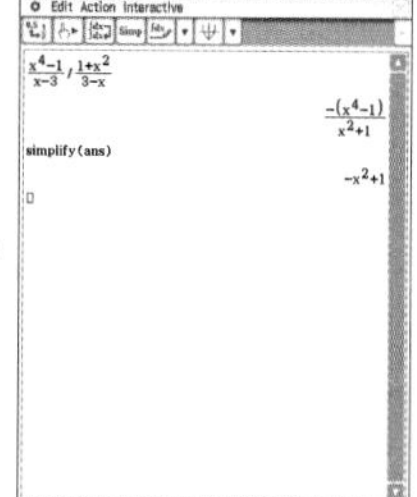
2. The answer appears on the screen.	$-(x^2-1)$	**2.** The answer appears on the screen.	$-x^2+1$

Addition and subtraction of algebraic fractions

Factorisation and expansion techniques are often required when adding or subtracting algebraic fractions.

- Denominators should be factorised in order to select the lowest common denominator.
- Express each fraction with this lowest common denominator.
- Simplify by expanding the terms in the numerator and collect any like terms together.

WORKED EXAMPLE 5 Adding and subtracting algebraic fractions

Simplify $\frac{2}{3x+3}-\frac{1}{x-2}+\frac{x}{x^2-x-2}$.

THINK	WRITE
1. Factorise each denominator.	$\frac{2}{3x+3}-\frac{1}{x-2}+\frac{x}{x^2-x-2}$ $=\frac{2}{3(x+1)}-\frac{1}{(x-2)}+\frac{x}{(x+1)(x-2)}$
2. Select the lowest common denominator and express each fraction with this as its denominator.	$=\frac{2\times(x-2)}{3(x+1)(x-2)}-\frac{1\times 3(x+1)}{3(x+1)(x-2)}+\frac{x\times 3}{3(x+1)(x-2)}$
3. Combine the fractions into one fraction.	$=\frac{2(x-2)-3(x+1)+3x}{3(x+1)(x-2)}$
4. Expand the terms in the numerator. *Note:* It is not necessary to expand the denominator terms.	$=\frac{2x-4-3x-3+3x}{3(x+1)(x-2)}$
5. Collect like terms in the numerator and state the answer. *Note:* Since there are no common factors between the numerator and the denominator, the fraction is in its simplest form.	$=\frac{2x-7}{3(x+1)(x-2)}$

2.2 Exercise

Students, these questions are even better in jacPLUS

Receive immediate feedback and access sample responses

Access additional questions

Track your results and progress

Find all this and MORE in jacPLUS

Technology free

1. Expand the following and simplify where appropriate.

a. $4m(m-2)+3m$
b. $5(m^2-3m+2)-(m+2)$
c. $(x-3)(x+5)$
d. $(3m-2)(5m-4)$
e. $(4-3x)(4+3x)$
f. $(2x-5)^2$

2. Expand and simplify the following.

a. $2(x-5)(x+5)-3(2x-3)$
b. $3-2(x+5)(3x-2)$
c. $(3-2x)(x-5)-(x+3)(x+4)$
d. $3(2x-1)(2x+1)+(x-5)^2$

3. **WE1** Expand $3(2x+1)^2+(7x+11)(7x-11)-(3x+4)(2x-1)$ and state the coefficient of the x term.

4. Expand each of the following expressions.

a. $(2x+3)^2$
b. $4a(b-3a)(b+3a)$
c. $10-(c+2)(4c-5)$
d. $(5-7y)^2$
e. $(3m^3+4n)(3m^3-4n)$
f. $(x+1)^3$

5. Expand and simplify the following, and state the coefficient of the x term.

a. $2(2x-3)(x-2)+(x+5)(2x-1)$
b. $(2+3x)(4-6x-5x^2)-(x-6)(x+6)$
c. $(4x+7)(4x-7)(1-x)$
d. $(x+1-2y)(x+1+2y)+(x-1)^2$
e. $(3-2x)(2x+9)-3(5x-1)(4-x)$
f. $x^2+x-4(x^2+x-4)$

6. Factorise the following.

a. x^2-36
b. $4-25a^2$
c. $9m^2-1$
d. $4a^2-64$
e. $2m^2-98x^2$
f. $1-9(1-m)^2$

7. Factorise the following.

a. $x^2-9x+18$
b. x^2-6x+9
c. $x^2+7x-60$
d. $4x^2+4x-15$
e. $4x^2-20x+25$
f. $8x^2-48xy+72y^2$

8. **WE2** Factorise:

a. $5x^2-45y^2$
b. $x^2-9x-10$
c. $8x^2-14x-15$
d. $4x^3-8x^2y-12xy^2$
e. $9y^2-x^2-8x-16$
f. $4(x-3)^2-3(x-3)-22$ using the substitution $a=x-3$.

9. Factorise:

a. $49-168x+144x^2$
b. $2(x-1)^2+13(x-1)+20$
c. $40(x+2)^2-18(x+2)-7$
d. $144x^2-36y^2$
e. $3a^2x+9ax-a-3$
f. $16x^2+8x+1-y^2$

10. Fully factorise the following.

a. $x^3 + 2x^2 - 25x - 50$
b. $100p^3 - 81pq^2$
c. $4n^2 + 4n + 1 - 4p^2$
d. $49(m + 2n)^2 - 81(2m - n)^2$
e. $13(a - 1) + 52(1 - a)^3$
f. $a^2 - b^2 - a + b + (a + b - 1)^2$

11. Use a substitution method to factorise the following.

a. $(x + 5)^2 + (x + 5) - 56$
b. $2(x + 3)^2 - 7(x + 3) - 9$
c. $70(x + y)^2 - y(x + y) - 6y^2$
d. $x^4 - 8x^2 - 9$
e. $9(p - q)^2 + 12(p^2 - q^2) + 4(p + q)^2$
f. $a^2\left(a + \frac{1}{a}\right)^2 - 4a^2\left(a + \frac{1}{a}\right) + 4a^2$

12. WE3 Factorise:

a. $x^3 - 125$
b. $3 + 3x^3$

13. Factorise the following.

a. $x^3 - 8$
b. $x^3 + 1000$
c. $1 - x^3$
d. $27x^3 + 64y^3$
e. $x^4 - 125x$
f. $(x - 1)^3 + 216$

14. Fully factorise the following.

a. $xy^3 - 27x$
b. $-x^3 - 216$
c. $3 - 81x^3$
d. $32x^3 + 4m^3$
e. $27m^3 + 64n^3$
f. $250x^3 - 128m^3$

15. Fully factorise the following.

a. $24x^3 - 81y^3$
b. $8x^4y^4 + xy$
c. $125(x + 2)^3 + 64(x - 5)^3$
d. $2(x - y)^3 - 54(2x + y)^3$
e. $a^5 - a^3b^2 + a^2b^3 - b^5$
f. $x^6 - y^6$

16. WE4 Simplify:

a. $\dfrac{x^2 + 4x}{x^2 + 2x - 8}$
b. $\dfrac{x^4 - 64}{5 - x} \div \dfrac{x^2 + 8}{x - 5}$

17. Simplify the following algebraic fractions.

a. $\dfrac{x}{(x + 1)(x - 2)} \times \dfrac{x - 2}{3x}$
b. $\dfrac{5}{x(3x + 1)} \div \dfrac{10}{x(x + 3)}$
c. $\dfrac{x^2 + 5x + 6}{4x + 8}$
d. $\dfrac{16 - 9x^2}{8 + 6x} \times \dfrac{2x + 10}{3x - 4}$
e. $\dfrac{4x}{3x^2 + 5x + 2} \div \dfrac{18x^2 - 6x}{9x^2 - 1}$
f. $\dfrac{2x^2 - 3x - 5}{2x^2 - 11x + 15} \times \dfrac{3x^2 - 5x - 12}{3x^2 + 7x + 4}$

18. Simplify the following.

a. $\dfrac{3x^2 - 7x - 20}{25 - 9x^2}$
b. $\dfrac{x^3 + 4x^2 - 9x - 36}{x^2 + x - 12}$
c. $\dfrac{(x + h)^3 - x^3}{h}$
d. $\dfrac{2x^2}{9x^3 + 3x^2} \times \dfrac{1 - 9x^2}{18x^2 - 12x + 2}$
e. $\dfrac{m^3 - 2m^2n}{m^3 + n^3} \div \dfrac{m^2 - 4n^2}{m^2 + 3mn + 2n^2}$
f. $\dfrac{1 - x^3}{1 + x^3} \times \dfrac{1 - x^2}{1 + x^2} \div \dfrac{1 + x + x^2}{1 - x + x^2}$

19. WE5 Simplify $\dfrac{6}{5x - 25} + \dfrac{1}{x - 1} - \dfrac{2x}{x^2 - 6x + 5}$.

20. Express each of the following as a single algebraic fraction.

a. $\frac{2x}{3}+\frac{5x}{4}$

b. $\frac{x}{7}-\frac{3x}{2}$

c. $\frac{4}{x-3}+\frac{5}{x+5}$

d. $\frac{x}{3x-1}-\frac{5}{1-2x}$

e. $\frac{x-3}{2x+1}-\frac{x+2}{x-1}$

f. $\frac{3}{x^2-9}-\frac{1}{x+3}+\frac{5}{x-3}$

Technology active

21. Simplify the following expressions.

a. $\frac{4}{x^2+1}+\frac{4}{x-x^2}$

b. $\frac{4}{x^2-4}-\frac{3}{x+2}+\frac{5}{x-2}$

c. $\frac{5}{x+6}+\frac{4}{5-x}+\frac{3}{x^2+x-30}$

d. $\frac{1}{4y^2-36y+81}+\frac{2}{4y^2-81}-\frac{1}{2y^2-9y}$

22. a. Expand $(2+3x)(x+6)(3x-2)(6-x)$.

b. Factorise $x^2-6x+9-xy+3y$.

c. Factorise $2y^4+2y(x-y)^3$.

23. Expand the following.

a. $(g+12+h)^2$

b. $(2p+7q)^2(7q-2p)$

c. $(x+10)(5+2x)(10-x)(2x-5)$

24. Simplify the following.

a. $\frac{x^3-125}{x^2-25}\times\frac{5}{x^3+5x^2+25x}$

b. $\left(\frac{4}{x+1}-\frac{3}{(x+1)^2}\right)\div\frac{16x^2-1}{x^2+2x+1}$

c. $\frac{1}{p-q}-\frac{p}{p^2-q^2}-\frac{q^3}{p^4-q^4}$

d. $(a+6b)\div\left(\frac{7}{a^2-3ab+2b^2}-\frac{5}{a^2-ab-2b^2}\right)$

25. Using CAS technology:

a. expand $(x+5)(2-x)(3x+7)$

b. factorise $27(x-2)^3+64(x+2)^3$

c. simplify $\frac{3}{x-1}+\frac{8}{x+8}$.

2.2 Exam questions

Question 1 (1 mark) TECH-ACTIVE

MC The coefficient of x in the expansion of $x(x-2)(x+1)+(x-3)^2$ is

A. 12 **B.** 8 **C.** 4 **D.** −4 **E.** −8

Question 2 (1 mark) TECH-ACTIVE

MC The factors of $9(2x-1)^2-9$ are

A. $(1, 0)$ **B.** $2x-1, 9$ **C.** $\left(-\frac{1}{2}, 9\right)$ **D.** $36x, x-1$ **E.** $36x^2-36x$

Question 3 (1 mark) TECH-FREE

Factorise $8(x+3)^2+24(x+3)+16$.

More exam questions are available online.

2.3 Pascal's triangle and binomial expansions

LEARNING INTENTION

At the end of this subtopic you should be able to:
- expand and simplify perfect cubes
- use binomial coefficients from Pascal's triangle to expand higher powers of $(a + b)$.

2.3.1 Expansions of perfect cubes

The perfect square $(a+b)^2$ may be expanded quickly by the rule $(a+b)^2 = a^2 + 2ab + b^2$. The **perfect cube** $(a+b)^3$ can also be expanded by a rule. This rule is derived by expressing $(a+b)^3$ as the product of repeated factors and expanding.

$$\begin{aligned}(a+b)^3 &= (a+b)(a+b)(a+b)\\ &= (a+b)(a+b)^2\\ &= (a+b)(a^2+2ab+b^2)\\ &= a^3+2a^2b+ab^2+ba^2+2ab^2+b^3\\ &= a^3+3a^2b+3ab^2+b^3\end{aligned}$$

Therefore, the rules for expanding a perfect cube are as follows.

Perfect cubes

$$(a+b)^3 = a^3+3a^2b+3ab^2+b^3$$
$$(a-b)^3 = a^3-3a^2b+3ab^2-b^3$$

Features of the rule for expanding perfect cubes

- The powers of the first term, a, decrease as the powers of the second term, b, increase.
- The coefficients of each term in the expansion of $(a+b)^3$ are 1, 3, 3, 1.
- The coefficients of each term in the expansion of $(a-b)^3$ are 1, −3, 3, −1.
- The signs alternate − + − in the expansion of $(a-b)^3$.

WORKED EXAMPLE 6 Expanding a perfect cube

Expand $(2x-5)^3$.

THINK	WRITE
1. Use the rule for expanding a perfect cube.	$(2x-5)^3$ Using $(a-b)^3 = a^3-3a^2b+3ab^2-b^3$, let $2x = a$ and $5 = b$. $(2x-5)^3 = (2x)^3-3(2x)^2(5)+3(2x)(5)^2-(5)^3$
2. Simplify each term.	$= 8x^3-3\times4x^2\times5+3\times2x\times25-125$ $= 8x^3-60x^2+150x-125$
3. State the answer.	$\therefore (2x-5)^3 = 8x^3-60x^2+150x-125$

2.3.2 Pascal's triangle

Pascal's triangle contains many fascinating patterns. Each row from row 1 onwards begins and ends with '1'. Each other number along a row is formed by adding the two terms to its left and right from the preceding row.

Row 0					1				
Row 1				1		1			
Row 2			1		2		1		
Row 3		1		3		3		1	
Row 4	1		4		6		4		1

The numbers in each row are called **binomial coefficients**.

The numbers 1, 2, 1 in row 2 are the coefficients of the terms in the expansion of $(a+b)^2$.

$$(a+b)^2 = 1a^2 + 2ab + 1b^2$$

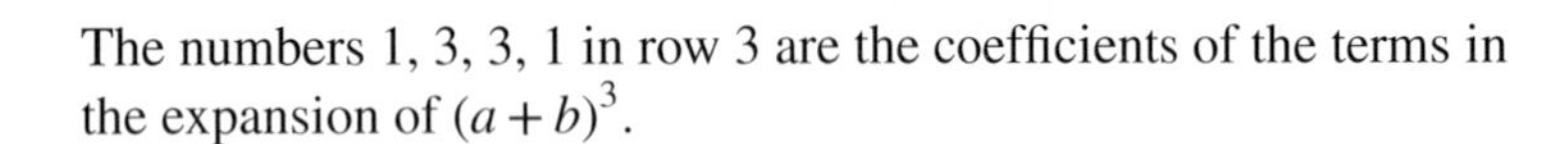

The numbers 1, 3, 3, 1 in row 3 are the coefficients of the terms in the expansion of $(a+b)^3$.

$$(a+b)^3 = 1a^3 + 3a^2b + 3ab^2 + 1b^3$$

Each row of Pascal's triangle contains the coefficients in the expansion of a power of $(a+b)$.

To expand $(a+b)^4$ we would use the binomial coefficients 1, 4, 6, 4, 1 from row 4 to obtain:

$$\begin{aligned}(a+b)^4 &= 1a^4 + 4a^3b + 6a^2b^2 + 4ab^3 + 1b^4 \\ &= a^4 + 4a^3b + 6a^2b^2 + 4ab^3 + b^4\end{aligned}$$

Notice that the powers of a decrease by 1 as the powers of b increase by 1, with the sum of the powers of a and b always totalling 4 for each term in the expansion of $(a+b)^4$.

For the expansion of $(a-b)^4$ the signs would alternate:

$$(a-b)^4 = a^4 - 4a^3b + 6a^2b^2 - 4ab^3 + b^4$$

By extending Pascal's triangle, higher powers of such binomial expressions can be expanded.

WORKED EXAMPLE 7 Using Pascal's triangle in binomial expansions

Form the rule for the expansion of $(a-b)^5$ and hence expand $(2x-1)^5$.

THINK	WRITE
1. Choose the row in Pascal's triangle that contains the required binomial coefficients.	For $(a-b)^5$, the power of the binomial is 5. Therefore, the binomial coefficients are in row 5. The binomial coefficients are: $1, 5, 10, 10, 5, 1$.
2. Write down the required binomial expansion.	Alternate the signs: $(a-b)^5 = a^5 - 5a^4b + 10a^3b^2 - 10a^2b^3 + 5ab^4 - b^5$

3. State the values to substitute in place of a and b.	To expand $(2x-1)^5$, $a=2x, b=1$.
4. Write down the expansion.	$(2x)^5-5(2x)^4(1)+10(2x)^3(1)^2-10(2x)^2(1)^3+5(2x)(1)^4-(1)^5$
5. Evaluate the coefficients and state the answer.	$=32x^5-5\times16x^4+10\times8x^3-10\times4x^2+10x-1$ $=32x^5-80x^4+80x^3-40x^2+10x-1$ $\therefore (2x-1)^5=32x^5-80x^4+80x^3-40x^2+10x-1$

Resources

Interactivity Pascal's triangle and binomial coefficients (int-2554)

2.3 Exercise

Students, these questions are even better in jacPLUS

Receive immediate feedback and access sample responses

Access additional questions

Track your results and progress

Find all this and MORE in jacPLUS

Technology free

1. **WE6** Expand $(3x-2)^3$.

2. Expand the following and simplify where appropriate.
   a. $(x-3)^3$
   b. $(2x-1)^3$
   c. $(x+4)^3$

3. Expand the following.
   a. $(3x+1)^3$
   b. $(1-2x)^3$
   c. $(5x+2y)^3$

4. **MC** Select the correct statement(s).
   A. $(x+2)^3=x^3+6x^2+12x+8$
   B. $(x+2)^3=x^3+2^3$
   C. $(x+2)^3=(x+2)(x^2-2x+4)$
   D. $(x+2)^3=(x+2)(x^2+2x+4)$
   E. $(x+2)^3=x^3+3x^2+3x+8$

5. Expand $\left(a+2b^2\right)^3$ and give the coefficient of a^2b^2.

6. Copy and complete the following table by making use of Pascal's triangle.

Binomial power	Expansion	Number of terms in the expansion	Sum of indices in each term
$(x+a)^2$			
$(x+a)^3$			
$(x+a)^4$			
$(x+a)^5$			

7. **WE7** Form the rule for the expansion of $(a-b)^6$ and hence expand $(2x-1)^6$.

8. Expand $(3x+2y)^4$.

Technology active

9. Expand the following using the binomial coefficients from Pascal's triangle.
   a. $(x+4)^5$
   b. $(x-4)^5$
   c. $(xy+2)^5$
10. Expand the following using the binomial coefficients from Pascal's triangle.
    a. $(3x-5y)^4$
    b. $(3-x^2)^4$
11. Expand and simplify $(1+x)^6-(1-x)^6$ using the binomial coefficients from Pascal's triangle.
12. Find the coefficient of x^2 in the following expressions.
    a. $(x+1)^3-3x(x+2)^2$
    b. $3x^2(x+5)(x-5)+4(5x-3)^3$
    c. $(x-1)(x+2)(x-3)-(x-1)^3$
    d. $(2x^2-3)^3+2(4-x^2)^3$
13. Expand and simplify $[(x-1)+y]^4$.
14. Find the term independent of x in the expansion of $\left(\frac{x}{2}+\frac{2}{x}\right)^6$.
15. If the coefficient of x^2y^2 in the expansion of $(x+ay)^4$ is 3 times the coefficient of x^2y^3 in the expansion of $(ax^2-y)^4$, find the value of a.
16. Find the coefficient of x in the expansion of $(1+2x)(1-x)^5$.
17. a. Expand $(1+x)^4$.
    b. Using a suitably chosen value for x, evaluate 1.1^4 using the expansion in part **a**.
18. Expand $(x+1)^5-(x+1)^4$ and hence show that $(x+1)^5-(x+1)^4=x(x+1)^4$.
19. Prove $(x+1)^{n+1}-(x+1)^n=x(x+1)^n$.
20. A section of Pascal's triangle is shown. Determine the values of a, b and c.

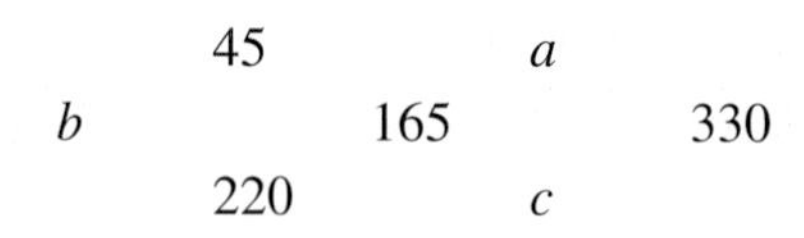

2.3 Exam questions

Question 1 (1 mark) TECH-ACTIVE

MC The coefficient of x^3 in the expansion of $(x+2)^5$ is

A. 120
B. 40
C. 1
D. 8
E. 20

Question 2 (1 mark) TECH-ACTIVE

MC The coefficient of the term independent of in the expansion of $\left(x-\frac{2}{x}\right)^4$ is

A. 4
B. −6
C. 6
D. −24
E. 24

Question 3 (1 mark) TECH-FREE

Expand $(x-2)^4-(x-2)^3$ and hence show that the coefficient of x is −44.

More exam questions are available online.

2.4 The binomial theorem

LEARNING INTENTION

At the end of this subtopic you should be able to:

- evaluate expressions using factorial notation
- calculate combinations
- expand using the binomial theorem
- determine particular terms using the binomial theorem.

Note: The binomial theorem is not part of the Study Design but is included here to enhance understanding. Pascal's triangle is useful for expanding small powers of binomial terms. However, to obtain the coefficients required for expansions of higher powers, the triangle needs to be extensively extended. The binomial theorem provides the way around this limitation by providing a rule for the expansion of $(x + y)^n$. Before this theorem can be presented, some notation needs to be introduced.

2.4.1 Factorial notation

In this and later topics, calculations such as $7 \times 6 \times 5 \times 4 \times 3 \times 2 \times 1$ will be encountered. Such expressions can be written in shorthand as 7! and are read as '7 factorial'. There is a factorial key on most calculators.

Definition

$n! = n \times (n-1) \times (n-2) \times ... \times 3 \times 2 \times 1$ for any natural number n.

It is also necessary to define $0! = 1$.

7! is equal to 5040. It can also be expressed in terms of other factorials such as:

$$\begin{aligned} 7! &= 7 \times (6 \times 5 \times 4 \times 3 \times 2 \times 1) \\ &= 7 \times 6! \end{aligned} \quad \text{or} \quad \begin{aligned} 7! &= 7 \times 6 \times (5 \times 4 \times 3 \times 2 \times 1) \\ &= 7 \times 6 \times 5! \end{aligned}$$

This is useful when working with fractions containing factorials. For example:

$$\begin{aligned} \frac{7!}{6!} &= \frac{7 \times \cancel{6!}}{\cancel{6!}} \\ &= 7 \end{aligned} \quad \text{or} \quad \begin{aligned} \frac{\cancel{5!}}{7!} &= \frac{\cancel{5!}}{7 \times 6 \times \cancel{5!}} \\ &= \frac{1}{42} \end{aligned}$$

By writing the larger factorial in terms of the smaller factorial, the fractions were simplified.

Factorial notation is just an abbreviation, so factorials cannot be combined arithmetically. For example, $3! - 2! \neq 1!$. This is verified by evaluating $3! - 2!$.

$$\begin{aligned} 3! - 2! &= 3 \times 2 \times 1 - 2 \times 1 \\ &= 6 - 2 \\ &= 4 \\ &\neq 1 \end{aligned}$$

WORKED EXAMPLE 8 Evaluating an expression with factorial notation

Evaluate $5! - 3! + \frac{50!}{49!}$.

THINK	WRITE
1. Expand the two smaller factorials.	$5! - 3! + \frac{50!}{49!}$ $= 5 \times 4 \times 3 \times 2 \times 1 - 3 \times 2 \times 1 + \frac{50!}{49!}$
2. To simplify the fraction, write the larger factorial in terms of the smaller factorial.	$= 5 \times 4 \times 3 \times 2 \times 1 - 3 \times 2 \times 1 + \frac{50 \times 49!}{49!}$
3. Calculate the answer.	$= 120 - 6 + \frac{50 \times \cancel{49!}}{\cancel{49!}}$ $= 120 - 6 + 50$ $= 164$

TI | THINK

1. On a Calculator page, complete the entry line as:
$5! - 3! + \frac{50!}{49!}$
Then press ENTER.
Note: The factorial symbol is located in CTRL Menu, Symbols OR
   - Menu
   - Probability
   - factorial

DISPLAY/WRITE

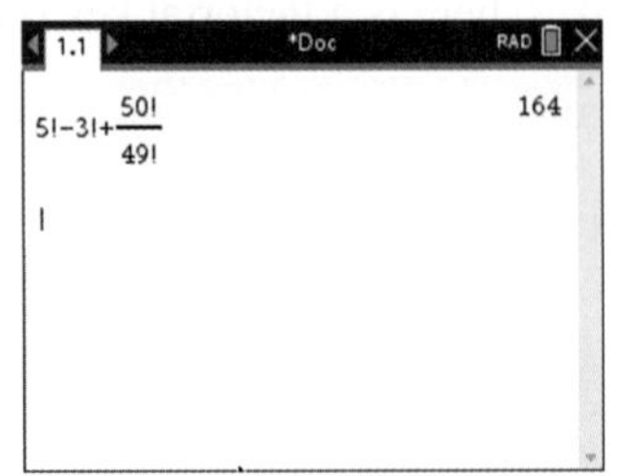

2. The answer appears on the screen.

164

CASIO | THINK

1. On a Main screen, complete the entry line as:
$5! - 3! + \frac{50!}{49!}$
Then press EXE.
Note: The factorial symbol is located in the Advanced menu of the keyboard.

DISPLAY/WRITE

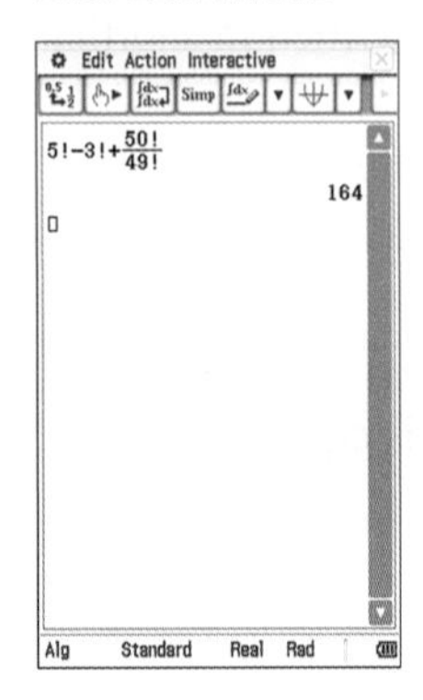

2. The answer appears on the screen.

164

Resources

Interactivity The binomial theorem (int-2555)

2.4.2 Formula for binomial coefficients

Each of the terms in the rows of Pascal's triangle can be expressed using factorial notation. For example, row 3 contains the coefficients 1, 3, 3, 1.

These can be written as $\frac{3!}{0! \times 3!}, \frac{3!}{1! \times 2!}, \frac{3!}{2! \times 1!}, \frac{3!}{3! \times 0!}$.

(Remember that 0! was defined to equal 1.)

The coefficients in row 5 (1, 5, 10, 10, 5, 1) can be written as:

$$\frac{5!}{0! \times 5!}, \frac{5!}{1! \times 4!}, \frac{5!}{2! \times 3!}, \frac{5!}{3! \times 2!}, \frac{5!}{4! \times 1!}, \frac{5!}{5! \times 0!}$$

The third term of row 4 would equal $\frac{4!}{2! \times 2!}$ and so on.

The $(r+1)$th term of row n would equal $\dfrac{n!}{r! \times (n-r)!}$.

This is normally written using the notations ${}^nC_r$ or $\binom{n}{r}$.

These expressions for the binomial coefficients are referred to as **combinatoric coefficients**. They occur frequently in other branches of mathematics, including probability theory.

Combinatoric coefficients

$$\binom{n}{r} = \frac{n!}{r!\,(n-r)!} = {}^nC_r$$

where $r \leq n$ and r and n are non-negative whole numbers.

2.4.3 Pascal's triangle with combinatoric coefficients

Pascal's triangle can now be expressed using this notation.

Row 0					$\binom{0}{0}$				
Row 1				$\binom{1}{0}$		$\binom{1}{1}$			
Row 2			$\binom{2}{0}$		$\binom{2}{1}$		$\binom{2}{2}$		
Row 3		$\binom{3}{0}$		$\binom{3}{1}$		$\binom{3}{2}$		$\binom{3}{3}$	
Row 4	$\binom{4}{0}$		$\binom{4}{1}$		$\binom{4}{2}$		$\binom{4}{3}$		$\binom{4}{4}$

Binomial expansions can be expressed using this notation for each of the binomial coefficients.

The expansion $(a+b)^3 = \binom{3}{0} a^3 + \binom{3}{1} a^2b + \binom{3}{2} ab^2 + \binom{3}{3} b^3$.

Note the following patterns:

- $\binom{n}{0} = 1 = \binom{n}{n}$ (the start and end of each row of Pascal's triangle)
- $\binom{n}{1} = n = \binom{n}{n-1}$ (the second from the start and the second from the end of each row)
- $\binom{n}{r} = \binom{n}{n-r}$.

WORKED EXAMPLE 9 Evaluating combinations

Evaluate $\binom{8}{3}$.

THINK	WRITE
1. Apply the formula.	$\binom{n}{r} = \frac{n!}{r!\,(n-r)!}$ Let $n = 8$ and $r = 3$. $\binom{8}{3} = \frac{8!}{3!\,(8-3)!}$ $= \frac{8!}{3!\,5!}$
2. Write the largest factorial in terms of the next largest factorial and simplify.	$= \frac{8 \times 7 \times 6 \times \cancel{5!}}{3!\,\cancel{5!}}$ $= \frac{8 \times 7 \times 6}{3!}$
3. Calculate the answer.	$= \frac{8 \times 7 \times \cancel{6}}{\cancel{3 \times 2} \times 1}$ $= 8 \times 7$ $= 56$

2.4.4 The binomial theorem

The binomial coefficients in row n of Pascal's triangle can be expressed as $\binom{n}{0}, \binom{n}{1}, \binom{n}{2}, \ldots \binom{n}{n}$, and hence the expansion of $(x+y)^n$ can be formed.

The binomial theorem

The binomial theorem gives the rule for the expansion of $(x+y)^n$ as:

$$(x+y)^n = x^n + \binom{n}{1} x^{n-1}y + \binom{n}{2} x^{n-2}y^2 + \ldots + \binom{n}{r} x^{n-r}y^r + \ldots + y^n$$

since $\binom{n}{0} = 1 = \binom{n}{n}$.

Features of the binomial theorem formula for the expansion of $(x+y)^n$

- There are $(n+1)$ terms.
- In each successive term, the powers of x decrease by 1 as the powers of y increase by 1.
- For each term, the powers of x and y add up to n.
- For the expansion of $(x-y)^n$, the signs of each term alternate $+ - + - + \ldots$ with every even term assigned the $-$ sign and every odd term assigned the $+$ sign.

WORKED EXAMPLE 10 Expanding using the binomial theorem

Use the binomial theorem to expand $(3x+2)^4$.

THINK	WRITE
1. Write out the expansion using the binomial theorem. *Note:* There should be 5 terms in the expansion.	$(3x+2)^4$ $= (3x)^4 + \binom{4}{1}(3x)^3(2) + \binom{4}{2}(3x)^2(2)^2 + \binom{4}{3}(3x)(2)^3 + (2)^4$
2. Evaluate the binomial coefficients.	$= (3x)^4 + 4 \times (3x)^3(2) + 6 \times (3x)^2(2)^2 + 4 \times (3x)(2)^3 + (2)^4$
3. Complete the calculations and state the answer.	$= 81x^4 + 4 \times 27x^3 \times 2 + 6 \times 9x^2 \times 4 + 4 \times 3x \times 8 + 16$ $\therefore (3x+2)^4 = 81x^4 + 216x^3 + 216x^2 + 96x + 16$

2.4.5 Using the binomial theorem

The binomial theorem is very useful for expanding $(x+y)^n$. However, for powers $n \geq 7$ the calculations can become quite tedious. If a particular term is of interest, forming an expression for the general term of the expansion is an easier option.

The general term of the binomial theorem

Consider the terms of the expansion:

$$(x+y)^n = x^n + \binom{n}{1}x^{n-1}y + \binom{n}{2}x^{n-2}y^2 + \ldots + \binom{n}{r}x^{n-r}y^r + \ldots + y^n$$

Term 1: $t_1 = \binom{n}{0}x^n y^0$

Term 2: $t_2 = \binom{n}{1}x^{n-1}y^1$

Term 3: $t_3 = \binom{n}{2}x^{n-2}y^2$

Following the pattern gives:

Term $(r+1)$: $t_{r+1} = \binom{n}{r}x^{n-r}y^r$

The general term of the binomial theorem

For the expansion of $(x+y)^n$, the general term is $t_{r+1} = \binom{n}{r}x^{n-r}y^r$.

For the expansion of $(x-y)^n$, the general term can be expressed as $t_{r+1} = \binom{n}{r}x^{n-r}(-y)^r$.

The general term formula enables a particular term to be evaluated without the need to carry out the full expansion. As there are $(n+1)$ terms in the expansion, if the middle term is sought, there will be two middle terms if n is odd and one middle term if n is even.

WORKED EXAMPLE 11 Determining a term in the binomial expansion

Determine the fifth term in the expansion of $\left(\frac{x}{2}-\frac{y}{3}\right)^9$.

THINK	WRITE
1. State the general term formula of the expansion.	$\left(\frac{x}{2}-\frac{y}{3}\right)^9$ The $(r+1)$th term is $t_{r+1}=\binom{n}{r}\left(\frac{x}{2}\right)^{n-r}\left(-\frac{y}{3}\right)^r$. Since the power of the binomial is 9, $n=9$. $\therefore t_{r+1}=\binom{9}{r}\left(\frac{x}{2}\right)^{9-r}\left(-\frac{y}{3}\right)^r$
2. Choose the value of r for the required term.	For the fifth term, t_5: $r+1=5$ $r=4$ $t_5=\binom{9}{4}\left(\frac{x}{2}\right)^{9-4}\left(-\frac{y}{3}\right)^4$ $=\binom{9}{4}\left(\frac{x}{2}\right)^5\left(-\frac{y}{3}\right)^4$
3. Evaluate to obtain the required term.	$=126\times\frac{x^5}{32}\times\frac{y^4}{81}$ $=\frac{7x^5y^4}{144}$

2.4.6 Identifying a term in the binomial expansion

The general term can also be used to determine which term has a specified property, such as the term independent of x or the term containing a particular power of x.

WORKED EXAMPLE 12 Determining a particular term

Identify which term in the expansion of $(8-3x^2)^{12}$ would contain x^8 and express the coefficient of x^8 as a product of its prime factors.

THINK	WRITE
1. Write down the general term for this expansion.	$(8-3x^2)^{12}$ The general term: $t_{r+1}=\binom{12}{r}(8)^{12-r}(-3x^2)^r$
2. Rearrange the expression for the general term by grouping the numerical parts together and the algebraic parts together.	$t_{r+1}=\binom{12}{r}(8)^{12-r}(-3)^r(x^2)^r$ $=\binom{12}{r}(8)^{12-r}(-3)^r x^{2r}$

3. Find the value of r required to form the given power of x.

For x^8, $2r = 8$, so $r = 4$.

4. Identify which term is required.

Hence, it is the fifth term that contains x^8.

5. Obtain an expression for this term.

With $r = 4$,

$$t_5 = \binom{12}{4}(8)^{12-4}(-3)^4 x^8$$
$$= \binom{12}{4}(8)^8(-3)^4 x^8$$

6. State the required coefficient.

The coefficient of x^8 is $\binom{12}{4}(8)^8(-3)^4$.

7. Express the coefficient in terms of prime numbers.

$$\binom{12}{4}(8)^8(-3)^4 = \frac{12\times 11\times 10\times 9}{4\times 3\times 2\times 1}\times (2^3)^8 \times 3^4$$
$$= 11\times 5\times 9\times 2^{24}\times 3^4$$
$$= 11\times 5\times 3^2\times 2^{24}\times 3^4$$
$$= 11\times 5\times 3^6\times 2^{24}$$

8. State the answer.

Therefore, the coefficient of x^8 is $11\times 5\times 3^6\times 2^{24}$.

2.4 Exercise

Students, these questions are even better in jacPLUS

Receive immediate feedback and access sample responses

Access additional questions

Track your results and progress

Find all this and MORE in jacPLUS

Technology free

1. Evaluate the following.
   a. $3!$ b. $4!$ c. $5!$ d. $2!$ e. $0!\times 1!$ f. $\frac{7!}{6!}$
2. Rewrite the following using factorial notation.
   a. $7\times 6\times 5\times 4\times 3\times 2\times 1$
   b. $8\times 7\times 6\times 5\times 4\times 3\times 2\times 1$
   c. $8\times 7!$
   d. $9\times 8!$
3. **WE8** Evaluate $6! + 4! - \frac{10!}{9!}$.
4. Simplify $\frac{n!}{(n-2)!}$.
5. Evaluate the following.
   a. $6!$
   b. $4! + 2!$
   c. $7\times 6\times 5!$
   d. $\frac{6!}{3!}$

6. Evaluate the following.

a. $\frac{26!}{24!}$

b. $\frac{42!}{43!}$

c. $\frac{49!}{50!} \div \frac{69!}{70!}$

d. $\frac{11!+10!}{11!-10!}$

7. **WE9** Evaluate $\binom{7}{4}$.

8. Determine an algebraic expression for $\binom{n}{2}$ and use this to evaluate $\binom{21}{2}$.

9. Evaluate the following.

a. $\binom{5}{2}$

b. $\binom{5}{3}$

c. $\binom{12}{12}$

d. ${}^{20}C_3$

e. $\binom{7}{0}$

f. $\binom{13}{10}$

10. Simplify the following.

a. $\binom{n}{3}$

b. $\binom{n}{n-3}$

c. $\binom{n+3}{n}$

d. $\binom{2n+1}{2n-1}$

e. $\binom{n}{2}+\binom{n}{3}$

f. $\binom{n+1}{3}$

11. Simplify the following.

a. $(n+1)\times n!$

b. $(n-1)(n-2)(n-3)!$

c. $\frac{n!}{(n-3)!}$

d. $\frac{(n-1)!}{(n+1)!}$

e. $\frac{(n-1)!}{n!}-\frac{(n+1)!}{(n+2)!}$

f. $\frac{n^3-n^2-2n}{(n+1)!}\times\frac{(n-2)!}{n-2}$

Technology active

12. **WE10** Use the binomial theorem to expand $(2x+3)^5$.

13. Use the binomial theorem to expand $(x-2)^7$.

14. Expand the following.

a. $(x+1)^5$

b. $(2-x)^5$

c. $(2x+3y)^6$

d. $\left(\frac{x}{2}+2\right)^7$

e. $\left(x-\frac{1}{x}\right)^8$

f. $(x^2+1)^{10}$

15. **WE11** Determine the fourth term in the expansion of $\left(\frac{x}{3}-\frac{y}{2}\right)^7$.

16. Determine the middle term in the expansion of $\left(x^2+\frac{y}{2}\right)^{10}$.

17. **WE12** Identify which term in the expansion of $(4+3x^3)^8$ would contain x^{15} and express the coefficient of x^{15} as a product of its prime factors.

18. Determine the term independent of x in the expansion of $\left(x+\frac{2}{x}\right)^6$.

For questions 19 and 20, use either expansion or the general term formula.

19. Obtain each of the following terms.
    a. The fourth term in the expansion of $(5x+2)^6$
    b. The third term in the expansion of $(3x^2-1)^6$
    c. The middle term(s) in the expansion of $(x+2y)^7$

20. a. Obtain the coefficient of x^6 in the expansion of $(1-2x^2)^9$.
    b. Express the coefficient of x^5 in the expansion of $(3+4x)^{11}$ as a product of its prime factors.
    c. Determine the term independent of x in the expansion of $\left(x^2+\frac{1}{x^3}\right)^{10}$.

21. Evaluate the following using CAS technology.
    a. $15!$
    b. $\binom{15}{10}$

22. a. Solve for n: $\binom{n}{2}=1770$.
    b. Solve for r: $\binom{12}{r}=220$.

2.4 Exam questions

Question 1 (1 mark) TECH-ACTIVE

MC The value of $\frac{4!+3!-2!}{4!-3!+2!}$ is

A. 1
B. -1
C. 20
D. $\frac{7}{5}$
E. $-\frac{7}{5}$

Question 2 (1 mark) TECH-ACTIVE

MC The value of $3\times\binom{5}{3}+4\times\binom{4}{2}$ is

A. 13
B. 16
C. 21
D. 54
E. 78

Question 3 (4 marks) TECH-ACTIVE

Identify which term in the expansion of $(3-4x^3)^7$ would contain x^{12} and express the coefficient of x^{12} as a product of its prime factors.

More exam questions are available online.

2.5 Sets of real numbers

LEARNING INTENTION

At the end of this subtopic you should be able to:

- classify numbers as elements of the real number system
- use interval notation and number lines to describe sets.

The concept of numbers in counting and the introduction of symbols for numbers marked the beginning of a major intellectual development in the minds of humans. Every civilisation appears to have developed a system for counting using written or spoken symbols for numbers. Over time, technologies were devised to assist in counting and computational techniques, and from these counting machines the computer was developed.

Over the course of history, different categories of numbers have been defined that collectively form the real number system. Real numbers are all the numbers that are positive, zero or negative. Before further describing and classifying the real number system, a review of some mathematical notation is given.

Resources

Interactivity Sets (int-2556)

2.5.1 Set notation

A **set** is a collection of objects, these objects being referred to as the **elements** of the set.

For example, if the **universal set** is $\varepsilon = \{1, 2, 3, 4, 5, 6, 7, 8, 9, 10\}$, then sets could include set $A = \{1, 2, 3, 4, 5\}$, set $B = \{1, 3, 5\}$ and set $C = \{2, 4, 6, 8, 10\}$.

The table below gives the common symbols used in set notation.

Symbol	Example	Meaning
$\in$	$2 \in A$	2 is an **element** of set A, or belongs to set A.
$\notin$	$2 \notin B$	2 is **not** an element of set B, or does not belong to set B.
$\subset$	$B \subset A$	Set B is a **subset** of set A, or every element of set B is an element of set A.
$\not\subset$	$A \not\subset B$	Set A is **not** a subset of set B.
$\cup$	$A \cup C$	The **union** of sets A and C; contains the elements that are either in A or in C or in both, giving $A \cup C = \{1, 2, 3, 4, 5, 6, 8, 10\}$
$\cap$	$A \cap C$	The **intersection** of sets A and C; contains the elements that are in both A and C, giving $A \cap C = \{2, 4\}$
$\backslash$	$A \backslash C$	The **exclusion** symbol; the elements of set A that are not in set C, giving $A \backslash C = \{1, 3, 5\}$
$\varnothing$	$B \cap C = \varnothing$	The empty set, or **disjoint** sets, since sets B and C do not have any common elements.
$'$	$A \cap A' = \varnothing$	The **complement** of A; the set of elements **not** in A, with the symbol A', giving $A' = \{6, 7, 8, 9, 10\}$

2.5.2 Classification of numbers

Although counting numbers are sufficient to solve equations such as $2+x=3$, they are not sufficient to solve, for example, $3+x=2$, where negative numbers are needed, or $3x=2$, where fractions are needed.

The table below shows the classification of numbers in our real number system.

$N=\{1, 2, 3, 4, ...\}$	The **set of natural numbers**, that is the positive whole numbers or counting numbers
$Z=\{...-2, -1, 0, 1, 2 ...\}$	The **set of integers**, that is all positive and negative whole numbers and the number zero
$Q=\left\{...\frac{-9}{8}, 0.75, 0.\bar{3}, 5, ..\right\}$	The **set of rational numbers**, all numbers that can be expressed in the form $\frac{p}{q}, q\neq 0$ and $p, q \in Z$, that is all fractions, integers and finite and recurring decimals.
$I=\left\{...-\sqrt{3}, .., \pi, ..\sqrt{2}, ..\right\}$	The **set of irrational numbers** where $Q \cap I = \varnothing$ or $I=R\backslash Q$, that is all real numbers that are not rational. Some irrational numbers, such as π and e, are called **transcendental** numbers.
$R=Q\cup I$	The **set of real numbers**, that is the union of the set of rational numbers and the set of irrational numbers

The relationship between the sets, $N \subset Z \subset Q \subset R$, can be illustrated as shown.

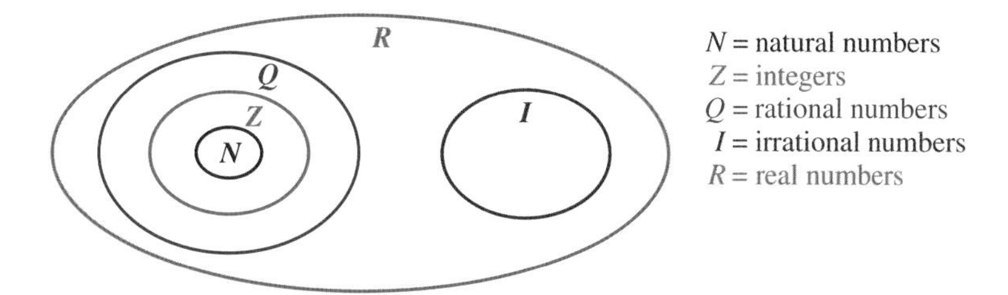

The set of all real numbers forms a number line continuum on which all of the positive, zero or negative numbers are placed. Hence, $R=R^- \cup \{0\} \cup R^+$.

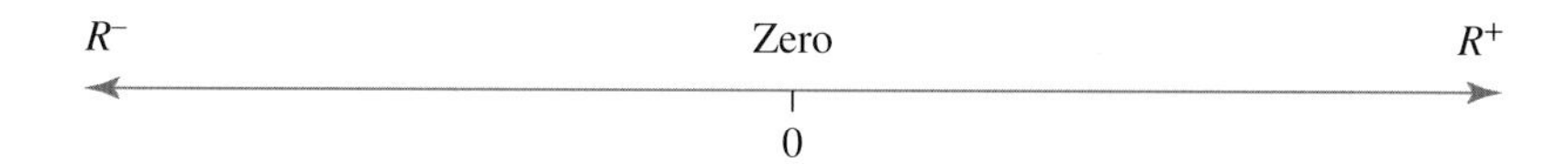

The sets of the real number system can also be viewed as the following hierarchy.

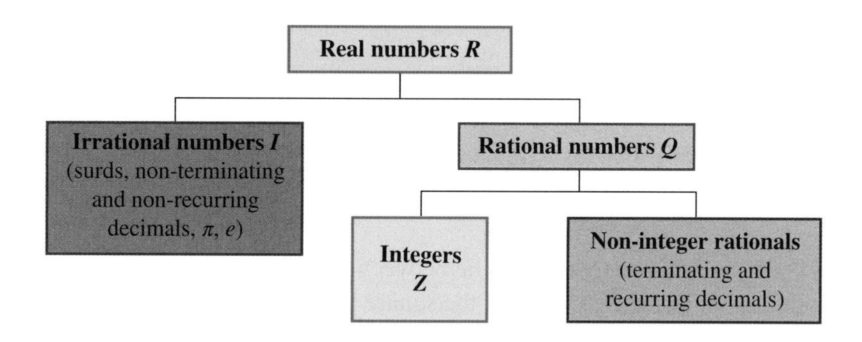

Expressions and symbols that do not represent real numbers

It is important to recognise that the following are not numbers.

- The symbol for infinity ∞ may suggest this is a number, but that is not so. We can speak of numbers getting larger and larger and approaching infinity, but infinity is a concept, not an actual number.
- Any expression of the form $\frac{a}{0}$ does not represent a number, since division by zero is not possible. If $a = 0$, the expression $\frac{0}{0}$ is said to be **indeterminate**. It is not defined as a number.

This is beyond the Mathematical Methods course, but there are numbers that are not elements of the set of real numbers. For example, the square roots of negative numbers, such as $\sqrt{-1}$, are unreal, but these square roots are numbers. They belong to the set of complex numbers. These numbers are very important in higher levels of mathematics.

WORKED EXAMPLE 13 Classifying numbers as elements of the real number system

a. Classify each of the following numbers as an element of a subset of the real numbers.

i. $-\frac{3}{5}$ ii. $\sqrt{7}$ iii. $6 - 2 \times 3$ iv. $\sqrt{9}$

b. Identify which of the following are correct statements.

i. $5 \in Z$ ii. $Z \subset N$ iii. $R^- \cup R^+ = R$

THINK	WRITE
a. i. Fractions are rational numbers.	a. i. $-\frac{3}{5} \in Q$
ii. Surds are irrational numbers.	ii. $\sqrt{7} \in I$
iii. Evaluate the number using the correct order of operations.	iii. $6 - 2 \times 3 = 6 - 6$ $= 0$ $\therefore (6 - 2 \times 3) \in Z$
iv. Evaluate the square root.	iv. $\sqrt{9} = 3$ $\therefore \sqrt{9} \in Z$
b. i. Z is the set of integers.	b. i. $5 \in Z$ is a correct statement since 5 is an integer.
ii. N is the set of natural or counting numbers.	ii. $Z \subset N$ is incorrect since $N \subset Z$.
iii. This is the union of R^-, the set of negative real numbers, and R^+, the set of positive real numbers.	iii. $R^- \cup R^+ = R$ is incorrect since R includes the number zero, which is neither positive nor negative.

2.5.3 Interval notation

Interval notation provides an alternative and often convenient way of describing certain sets of numbers.

Closed interval

$[a, b] = \{x : a \le x \le b\}$ is the set of real numbers that lie between a and b, including the end points, a and b. The inclusion of the end points is indicated by the use of the square brackets [].

This is illustrated on a number line using closed circles at the end points.

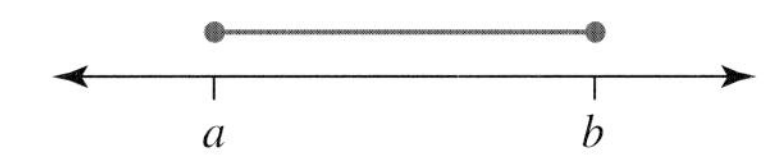

Open interval

$(a, b) = \{x : a < x < b\}$ is the set of real numbers that lie between a and b, not including the end points, a and b. The exclusion of the end points is indicated by the use of the round brackets ().

This is illustrated on a number line using open circles at the end points.

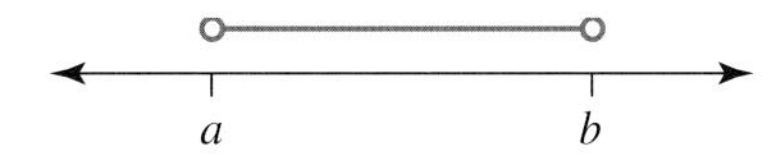

Half-open intervals

Half-open intervals have only one end point included.

$$[a, b) = \{x : a \leq x < b\}$$

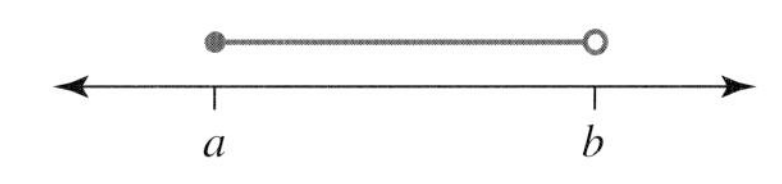

$$(a, b] = \{x : a < x \leq b\}$$

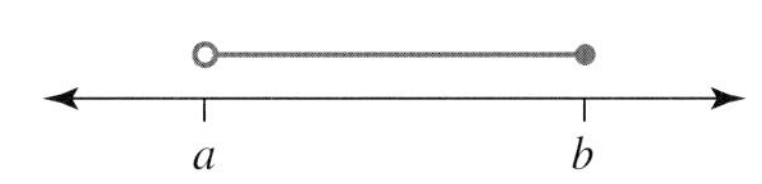

Interval notation can be used for infinite intervals using the symbol for infinity with an open end.

For example, the set of real numbers, R, is the same as the interval $(-\infty, \infty)$.

WORKED EXAMPLE 14 Illustrating a set on the number line

a. Illustrate each of the following sets on a number line and express each set in alternative notation.

i. $(-2, 2]$ **ii.** $\{x : x \geq 1\}$ **iii.** $\{1, 2, 3, 4\}$

b. Use interval notation to describe the sets of numbers shown on the following number lines.

i.

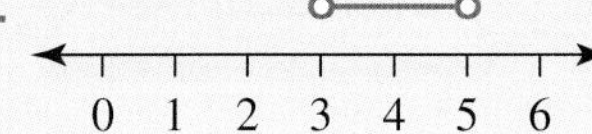

ii.

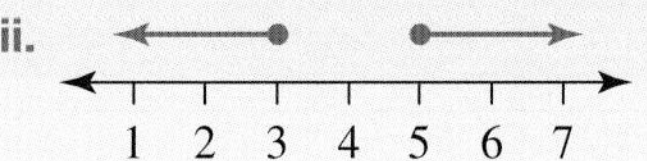

THINK	WRITE
a. i. 1. Describe the given interval. *Note:* The round bracket indicates 'not included' and the square bracket indicates 'included'.	**a. i.** $(-2, 2]$ is the interval representing the set of numbers between -2 and 2, closed at 2, open at -2. 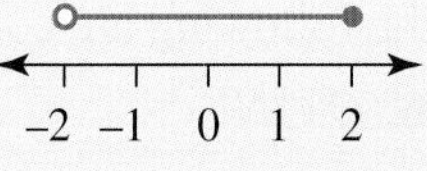
2. Write the set in alternative notation.	An alternative notation for the set is $(-2, 2] = \{x : -2 < x \leq 2\}$.

ii. 1. Describe the given set.	**ii.** $\{x : x \ge 1\}$ is the set of all numbers greater than or equal to 1. This is an infinite interval that has no right-hand end point.
	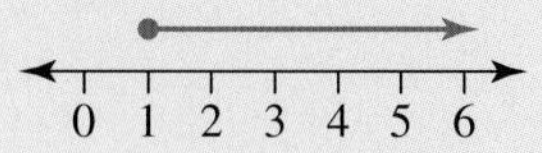
2. Write the set in alternative notation.	An alternative notation is $\{x : x \ge 1\} = [1, \infty)$.
iii. 1. Describe the given set. *Note:* This set does not contain all numbers between the beginning and end of an interval.	**iii.** $\{1, 2, 3, 4\}$ is a set of discrete elements. 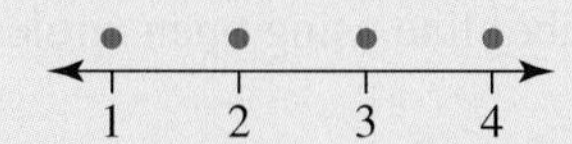
2. Write the set in alternative notation.	Alternative notations could be $\{1, 2, 3, 4\} = \{x : 1 \le x \le 4, x \in N\}$ or $\{1, 2, 3, 4\} = [1, 4] \cap N$.
b. i. Describe the set using interval notation with appropriate brackets.	**b. i.** The set of numbers lie between 3 and 5 with both end points excluded. The set is described as $(3, 5)$.
ii. 1. Describe the set as the union of the two disjoint intervals.	**ii.** The left branch is $(-\infty, 3]$ and the right branch is $[5, \infty)$. The set of numbers is the union of these two. It can be described as $(-\infty, 3] \cup [5, \infty)$.
2. Describe the same set by considering the interval that has been excluded from R.	Alternatively, the diagram can be interpreted as showing what remains after the set $(3, 5)$ is excluded from the set R. An alternative description is $R \backslash (3, 5)$.

2.5 Exercise

Students, these questions are even better in jacPLUS

Receive immediate feedback and access sample responses

Access additional questions

Track your results and progress

Find all this and MORE in jacPLUS

Technology free

1. WE13a Classify each of the following numbers as an element of a subset of the real numbers.

 a. $\dfrac{6}{11}$ b. $\sqrt{27}$ c. $(6-2) \times 3$ d. $\sqrt{0.25}$

2. WE13b Identify which of the following are correct statements.

 a. $17 \in N$ b. $Q \subset N$ c. $Q \cup I = R$

3. Classify each of the following numbers as a subset of the real numbers.

 a. $\dfrac{12}{\sqrt{9}}$ b. 3π c. $\dfrac{\sqrt{25}}{2}$ d. $-\sqrt{4}$

4. Determine the value(s) of x for which $\dfrac{x-5}{(x+1)(x-3)}$ would be undefined.

5. **MC** Identify which of the following does not represent a real number.

A. $\sqrt[3]{-4}$ B. $\sqrt{0}$ C. $\dfrac{6\times2-3\times4}{7}$ D. $5\pi^2+3$ E. $\dfrac{(8-4)\times2}{8-4\times2}$

6. Explain why each of the following statements is false and then rewrite it as a correct statement.

a. $\sqrt{16+25}\in Q$ b. $\left(\dfrac{4}{9}-1\right)\in Z$ c. $R^+=\{x: x\geq0\}$ d. $\sqrt{2.25}\in I$

7. Determine any values of x for which the following would be undefined.

a. $\dfrac{1}{x+5}$ b. $\dfrac{x+2}{x-2}$ c. $\dfrac{x+8}{(2x+3)(5-x)}$ d. $\dfrac{4}{x^2-4x}$

8. State whether the following are true or false.

a. $R^-\subset R$ b. $N\subset R^+$ c. $Z\cup N=R$ d. $Q\cap Z=Z$ e. $I\cup Z=R\backslash Q$ f. $Z\backslash N=Z^-$

9. Select the irrational numbers from the following set of numbers.

$$\left\{\sqrt{11}, \frac{2}{11}, 11^{11}, 11\pi, \sqrt{121}, 2^{\pi}\right\}$$

10. **WE14a** Illustrate each of the following sets on a number line and express each set in alternative notation.

a. $[-2, 2)$
b. $\{x: x<-1\}$
c. $\{-2, -1, 0, 1, 2\}$

11. **WE14b** Use interval notation to describe the sets of numbers shown on the following number lines.

a. 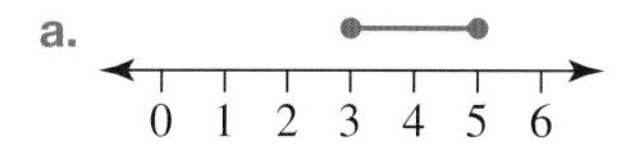

b. 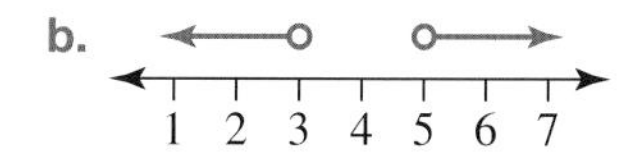

12. Use interval notation to describe the intervals shown on the following number lines.

a. 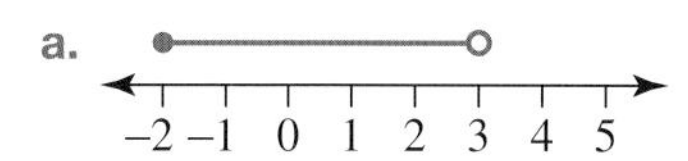

b. 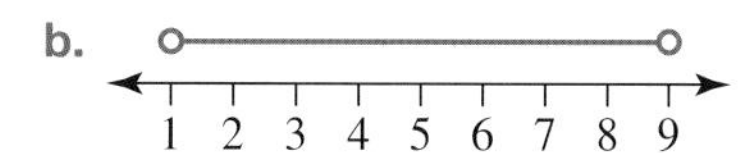

c. 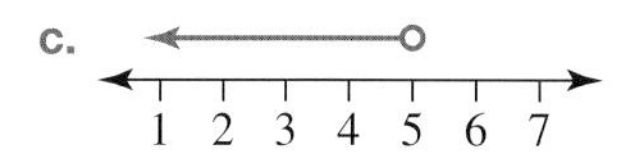

d.

13. Describe the intervals shown below using interval notation.

a. 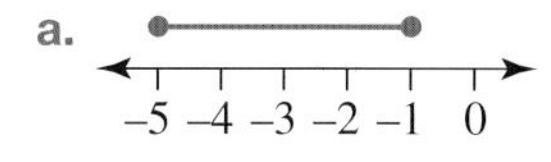

b.

c.

d.

14. Write $R\backslash\{x: 1<x\leq4\}$ as the union of two sets expressed in interval notation.

15. Express the following in interval notation.

a. $\{x: 4<x\leq8\}$ b. $\{x: x>-3\}$ c. $\{x: x\leq0\}$ d. $\{x: -2\leq x\leq0\}$

16. Show the following intervals on a number line.

a. $[-5, 5)$
b. $(4, \infty)$
c. $[-3, 7]$
d. $(-3, 7]$
e. $(-\infty, 3]$
f. $(-\infty, \infty)$

17. Illustrate the following on a number line.

a. $R \setminus [-2, 2]$
b. $(-\infty, \sqrt{2}) \cup (\sqrt{2}, \infty)$
c. $[-4, 2) \cap (0, 4)$
d. $[-4, 2) \cup (0, 4)$
e. $\{-1, 0, 1\}$
f. $R \setminus \{0\}$

18. Use an alternative form of notation to describe the following sets.

a. $\{x : 2 < x < 6, x \in Z\}$
b. $R \setminus (-1, 5]$
c. R^-
d. $(-\infty, -4) \cup [2, \infty)$

Technology active

19. Determine which of the following are rational and which are irrational numbers.

a. $\sqrt{7\,225}$
b. $\sqrt{75\,600}$
c. $0.234\,234\,234\ldots$

20. The ancient Egyptians devised the formula $A = \frac{64}{81}d^2$ for calculating the area, A, of a circle of diameter d. Use this formula to find a rational approximation for π and evaluate it to 9 decimal places. State whether it is a better approximation than $\frac{22}{7}$.

2.5 Exam questions

Question 1 (1 mark) TECH-ACTIVE

MC Identify which of the following represents a rational number.

A. $\sqrt{4+1}$
B. $\left(\sqrt{3+\sqrt{5}}\right)^2$
C. $\frac{2}{3-\sqrt{9}}$
D. $\sqrt{9}+\sqrt{16}$
E. $\sqrt{9-9\times 4}$

Question 2 (1 mark) TECH-FREE

Determine the values of x for which $\frac{x-2}{x(x-1)(x+3)}$ would be undefined. State the reason.

Question 3 (1 mark) TECH-ACTIVE

MC The set of numbers described by $R \setminus [-2, 5)$ is the same as

A. $\{x : -2 \le x \le 5\}$
B. $\{x : x < -2\} \cup \{x : x \ge 5\}$
C. $\{x : x \le -2\} \cup \{x : x > 5\}$
D. $\{x : -5 \le x \le 2\}$
E. $\{x : x < 5\} \cup \{x : x \ge -2\}$

More exam questions are available online.

2.6 Surds

LEARNING INTENTION

At the end of this subtopic you should be able to:
- simplify and order surds
- expand and simplify expressions involving surds
- rationalise denominators with surds.

A **surd** is an nth root, $\sqrt[n]{x}$. Surds are **irrational numbers**; they cannot be expressed in the quotient form $\frac{p}{q}$. Any decimal value obtained from a calculator is just an approximation.

All surds have **radical** signs, but not all numbers with radical signs are surds. For surds, the roots cannot be evaluated exactly. Hence, $\sqrt{26}$ is a surd. $\sqrt{25}$ is not a surd, since 25 is a perfect square; $\sqrt{25} = 5$, which is rational.

2.6.1 Ordering surds

Surds are real numbers and therefore have positions on the number line. To estimate the position of $\sqrt{6}$, we can place it between two rational numbers by placing 6 between its closest perfect squares.

$$
\begin{aligned}
4 &< 6 < 9 \\
\sqrt{4} &< \sqrt{6} < \sqrt{9} \\
\therefore 2 &< \sqrt{6} < 3
\end{aligned}
$$

So $\sqrt{6}$ lies between 2 and 3, closer to 2, since 6 lies closer to 4 than to 9. Checking with a calculator, $\sqrt{6} \simeq 2.4495$. Note that the symbol $\sqrt{\ }$ always gives a positive number, so the negative surd $-\sqrt{6}$ would lie on the number line between -3 and -2 at the approximate position -2.4495.

To order the sizes of two surds such as $3\sqrt{5}$ and $5\sqrt{3}$, express each as an entire surd.

$$
\begin{aligned}
3\sqrt{5} &= \sqrt{9} \times \sqrt{5} \\
&= \sqrt{9 \times 5} \\
&= \sqrt{45}
\end{aligned}
\qquad \text{and} \qquad
\begin{aligned}
5\sqrt{3} &= \sqrt{25} \times \sqrt{3} \\
&= \sqrt{25 \times 3} \\
&= \sqrt{75}
\end{aligned}
$$

Since $\sqrt{45} < \sqrt{75}$, it follows that $3\sqrt{5} < 5\sqrt{3}$.

2.6.2 Surds in simplest form

Surds are said to be in simplest form when the number under the square root sign contains no perfect square factors. This means that $3\sqrt{5}$ is the simplest form of $\sqrt{45}$ and $5\sqrt{3}$ is the simplest form of $\sqrt{75}$.

If the radical sign is a cube root, then the simplest form has no perfect cube factors under the cube root.

To express $\sqrt{128}$ in its simplest form, find perfect square factors of 128.

$$
\begin{aligned}
\sqrt{128} &= \sqrt{64 \times 2} \\
&= \sqrt{64} \times \sqrt{2} \\
&= 8\sqrt{2} \\
\therefore \sqrt{128} &= 8\sqrt{2}
\end{aligned}
$$

WORKED EXAMPLE 15 Simplifying and ordering surds

a. Express $\left\{6\sqrt{2}, 4\sqrt{3}, 2\sqrt{5}, 7\right\}$ with its elements in increasing order.

b. Express the following in simplest form.

i. $\sqrt{56}$

ii. $2\sqrt{252a^2b}$ assuming $a > 0$

THINK	WRITE
a. 1. Express each number entirely as a square root.	a. $\left\{6\sqrt{2}, 4\sqrt{3}, 2\sqrt{5}, 7\right\}$ $6\sqrt{2} = \sqrt{36} \times \sqrt{2} = \sqrt{72}$, $4\sqrt{3} = \sqrt{16} \times \sqrt{3} = \sqrt{48}$, $2\sqrt{5} = \sqrt{4} \times \sqrt{5} = \sqrt{20}$ and $7 = \sqrt{49}$.
2. Order the terms.	$\sqrt{20} < \sqrt{48} < \sqrt{49} < \sqrt{72}$ In increasing order, the set of numbers is $\left\{2\sqrt{5}, 4\sqrt{3}, 7, 6\sqrt{2}\right\}$.
b. i. Find a perfect square factor of the number under the square root sign.	b. i. $\sqrt{56} = \sqrt{4 \times 14}$ $= \sqrt{4} \times \sqrt{14}$ $= 2\sqrt{14}$
ii. 1. Find any perfect square factors of the number under the square root sign.	ii. $2\sqrt{252a^2b} = 2\sqrt{4 \times 9 \times 7a^2b}$
2. Express the square root terms as products and simplify where possible.	$= 2 \times \sqrt{4} \times \sqrt{9} \times \sqrt{7} \times \sqrt{a^2} \times \sqrt{b}$ $= 2 \times 2 \times 3 \times \sqrt{7} \times a \times \sqrt{b}$ $= 12a\sqrt{7b}$
3. Try to find the largest perfect square factor for greater efficiency.	Alternatively, recognising that 252 is 36×7, $2\sqrt{252a^2b} = 2\sqrt{36 \times 7a^2b}$ $= 2 \times \sqrt{36} \times \sqrt{7} \times \sqrt{a^2} \times \sqrt{b}$ $= 2 \times 6 \times \sqrt{7} \times a \times \sqrt{b}$ $= 12a\sqrt{7b}$

2.6.3 Operations with surds

As surds are real numbers, they obey the usual laws for addition and subtraction of like terms and the laws of multiplication and division.

Addition and subtraction with surds

$$a\sqrt{c} + b\sqrt{c} = (a + b)\sqrt{c}$$
$$a\sqrt{c} - b\sqrt{c} = (a - b)\sqrt{c}$$

Expressions such as $\sqrt{2} + \sqrt{3}$ cannot be expressed in any simpler form as $\sqrt{2}$ and $\sqrt{3}$ are 'unlike' surds. Like surds have the same number under the square root sign. Expressing surds in simplest form enables any like surds to be recognised.

Multiplication and division with surds

$$\sqrt{c} \times \sqrt{d} = \sqrt{cd}$$
$$a\sqrt{c} \times b\sqrt{d} = ab\sqrt{cd}$$
$$\frac{\sqrt{c}}{\sqrt{d}} = \sqrt{\frac{c}{d}}$$

Note that $\left(\sqrt{c}\right)^2 = c$ because $\left(\sqrt{c}\right)^2 = \sqrt{c} \times \sqrt{c} = \sqrt{c^2} = c$.

WORKED EXAMPLE 16 Simplifying surds

Simplify the following.

a. $3\sqrt{5} + 7\sqrt{2} + 6\sqrt{5} - 3\sqrt{2}$ **b.** $3\sqrt{98} - 4\sqrt{72} + 2\sqrt{125}$ **c.** $4\sqrt{3} \times 6\sqrt{15}$

THINK	WRITE
a. Collect like surds together and simplify.	a. $3\sqrt{5} + 7\sqrt{2} + 6\sqrt{5} - 3\sqrt{2}$ $= 3\sqrt{5} + 6\sqrt{5} + 7\sqrt{2} - 3\sqrt{2}$ $= 9\sqrt{5} + 4\sqrt{2}$
b. 1. Write each surd in simplest form.	b. $3\sqrt{98} - 4\sqrt{72} + 2\sqrt{125}$ $= 3\sqrt{49 \times 2} - 4\sqrt{36 \times 2} + 2\sqrt{25 \times 5}$ $= 3 \times 7\sqrt{2} - 4 \times 6\sqrt{2} + 2 \times 5\sqrt{5}$ $= 21\sqrt{2} - 24\sqrt{2} + 10\sqrt{5}$
2. Collect like surds together.	$= (21\sqrt{2} - 24\sqrt{2}) + 10\sqrt{5}$ $= -3\sqrt{2} + 10\sqrt{5}$
c. 1. Multiply the rational numbers together and multiply the surds together.	c. $4\sqrt{3} \times 6\sqrt{15}$ $= (4 \times 6)\sqrt{(3 \times 15)}$ $= 24\sqrt{45}$
2. Write the surd in its simplest form.	$= 24\sqrt{9 \times 5}$ $= 24 \times 3\sqrt{5}$ $= 72\sqrt{5}$

2.6.4 Expansions

Expansions of brackets containing surds are carried out using the distributive law in the same way as algebraic expansions.

Distributive law with surds

$$\sqrt{a}\left(\sqrt{b} + \sqrt{c}\right) = \sqrt{ab} + \sqrt{ac}$$
$$\left(\sqrt{a} + \sqrt{b}\right)\left(\sqrt{c} + \sqrt{d}\right) = \sqrt{ac} + \sqrt{ad} + \sqrt{bc} + \sqrt{bd}$$

The perfect squares formula for binomial expansions involving surds becomes $\left(\sqrt{a} \pm \sqrt{b}\right)^2 = a \pm 2\sqrt{ab} + b$, as follows.

Perfect squares involving surds

$$\begin{aligned}(\sqrt{a} \pm \sqrt{b})^2 &= (\sqrt{a})^2 \pm 2\sqrt{a}\sqrt{b} + (\sqrt{b})^2 \\ &= a \pm 2\sqrt{ab} + b\end{aligned}$$

The difference of two squares formula becomes $\left(\sqrt{a} + \sqrt{b}\right)\left(\sqrt{a} - \sqrt{b}\right) = a - b$, as follows.

Difference of two squares involving surds

$$\left(\sqrt{a} + \sqrt{b}\right)\left(\sqrt{a} - \sqrt{b}\right) = \left(\sqrt{a}\right)^2 - \left(\sqrt{b}\right)^2 = a - b$$

The binomial theorem can be used to expand higher powers of binomial expressions containing surds.

WORKED EXAMPLE 17 Expanding and simplifying expressions with surds

Expand and simplify the following.

a. $3\sqrt{2}\left(4\sqrt{2} - 5\sqrt{6}\right)$

b. $\left(2\sqrt{3} - 5\sqrt{2}\right)\left(4\sqrt{5} + \sqrt{14}\right)$

c. $\left(3\sqrt{3} + 2\sqrt{5}\right)^2$

d. $\left(\sqrt{7} + 3\sqrt{2}\right)\left(\sqrt{7} - 3\sqrt{2}\right)$

THINK	WRITE
a. Use the distributive law to expand, then simplify each term.	a. $3\sqrt{2}\left(4\sqrt{2} - 5\sqrt{6}\right) = 12\sqrt{4} - 15\sqrt{12}$ $= 12 \times 2 - 15\sqrt{4 \times 3}$ $= 24 - 15 \times 2\sqrt{3}$ $= 24 - 30\sqrt{3}$
b. 1. Expand as for algebraic terms.	b. $\left(2\sqrt{3} - 5\sqrt{2}\right)\left(4\sqrt{5} + \sqrt{14}\right)$ $= 8\sqrt{15} + 2\sqrt{42} - 20\sqrt{10} - 5\sqrt{28}$
2. Simplify where possible.	$= 8\sqrt{15} + 2\sqrt{42} - 20\sqrt{10} - 5\sqrt{4 \times 7}$ $= 8\sqrt{15} + 2\sqrt{42} - 20\sqrt{10} - 5 \times 2\sqrt{7}$ $= 8\sqrt{15} + 2\sqrt{42} - 20\sqrt{10} - 10\sqrt{7}$ There are no like surds, so no further simplification is possible.
c. 1. Use the rule for expanding a perfect square.	c. $\left(3\sqrt{3} + 2\sqrt{5}\right)^2$ $= \left(3\sqrt{3}\right)^2 + 2\left(3\sqrt{3}\right)\left(2\sqrt{5}\right) + \left(2\sqrt{5}\right)^2$
2. Simplify each term, remembering that $(\sqrt{a})^2 = a$, and collect any like terms together.	$= 9 \times 3 + 2 \times 3 \times 2\sqrt{3 \times 5} + 4 \times 5$ $= 27 + 12\sqrt{15} + 20$ $= 47 + 12\sqrt{15}$

d. Use the rule for expanding a difference of two squares.

d. $\left(\sqrt{7}+3\sqrt{2}\right)\left(\sqrt{7}-3\sqrt{2}\right)$

$=\left(\sqrt{7}\right)^2-\left(3\sqrt{2}\right)^2$

$=7-9\times2$

$=7-18$

$=-11$

2.6.5 Rationalising denominators

It is usually desirable to express any fraction whose denominator contains surds as a fraction with a denominator containing only a rational number. This does not necessarily mean the rational denominator form of the fraction is simpler, but it can provide a form that allows for easier manipulation, and it can enable like surds to be recognised in a surdic expression.

The process of obtaining a rational number for the denominator is called rationalising the denominator. There are different methods for rationalising denominators, depending on how many terms there are in the denominator.

Rationalising monomial denominators

Consider $\frac{a}{b\sqrt{c}}$, where $a, b, c \in Q$. This fraction has a monomial denominator, since its denominator contains one term, $b\sqrt{c}$.

In order to rationalise the denominator of this fraction, we use the fact that $\sqrt{c}\times\sqrt{c}=c$, a rational number.

Multiply both the numerator and the denominator by $\sqrt{c}$. As this is equivalent to multiplying by 1, the value of the fraction is not altered.

Rationalising monomial denominators

$$\frac{a}{b\sqrt{c}}=\frac{a}{b\sqrt{c}}\times\frac{\sqrt{c}}{\sqrt{c}}$$
$$=\frac{a\sqrt{c}}{b(\sqrt{c}\times\sqrt{c})}$$
$$=\frac{a\sqrt{c}}{bc}$$

By this process, $\frac{a}{b\sqrt{c}}=\frac{a\sqrt{c}}{bc}$ and the denominator, bc, is now rational.

Once the denominator has been rationalised, it may be possible to simplify the expression if, for example, any common factor exists between the rationals in the numerator and denominator.

Rationalising binomial denominators

$\sqrt{a}+\sqrt{b}$ and $\sqrt{a}-\sqrt{b}$ are called **conjugate surds**. Multiplying a pair of conjugate surds always results in a rational number, since $\left(\sqrt{a}+\sqrt{b}\right)\left(\sqrt{a}-\sqrt{b}\right)=a-b$. This fact is used to rationalise binomial denominators.

Consider $\dfrac{1}{\sqrt{a}+\sqrt{b}}$, where $a, b \in Q$. This fraction has a binomial denominator since its denominator is the addition of two terms.

To rationalise the denominator, multiply both the numerator and the denominator by $\sqrt{a}-\sqrt{b}$, the conjugate of the surd in the denominator. This is equivalent to multiplying by 1, so the value of the fraction is unaltered; however, it creates a difference of two squares on the denominator.

Rationalising binomial denominators

$$\frac{1}{\sqrt{a}+\sqrt{b}} = \frac{1}{\sqrt{a}+\sqrt{b}} \times \frac{\sqrt{a}-\sqrt{b}}{\sqrt{a}-\sqrt{b}}$$
$$= \frac{\sqrt{a}-\sqrt{b}}{(\sqrt{a}+\sqrt{b})(\sqrt{a}-\sqrt{b})}$$
$$= \frac{\sqrt{a}-\sqrt{b}}{a-b}$$

By this process we have $\dfrac{1}{\sqrt{a}+\sqrt{b}} = \dfrac{\sqrt{a}-\sqrt{b}}{a-b}$, where the denominator, $a-b$, is a rational number.

WORKED EXAMPLE 18 Rationalising denominators

a. Express each of the following with a rational denominator.

i. $\dfrac{12}{5\sqrt{3}}$ **ii.** $\dfrac{5\sqrt{3}-3\sqrt{10}}{4\sqrt{5}}$

b. Simplify $5\sqrt{2}-\dfrac{4}{\sqrt{2}}+3\sqrt{18}$.

c. Express $\dfrac{6}{5\sqrt{3}-3\sqrt{2}}$ **with a rational denominator.**

d. Given $p = 3\sqrt{2}-1$, **calculate** $\dfrac{1}{p^2-1}$, **expressing the answer with a rational denominator.**

THINK	WRITE
a. i. 1. The denominator is monomial. Multiply both numerator and denominator by the surd part of the monomial term.	**a. i.** $\dfrac{12}{5\sqrt{3}} = \dfrac{12}{5\sqrt{3}} \times \dfrac{\sqrt{3}}{\sqrt{3}}$
2. Multiply the numerator terms together and multiply the denominator terms together.	$= \dfrac{12\sqrt{3}}{5\times 3}$ $= \dfrac{12\sqrt{3}}{15}$
3. Cancel the common factor between the numerator and denominator.	$= \dfrac{{}^{4}\cancel{12}\sqrt{3}}{\cancel{15}_{5}}$ $= \dfrac{4\sqrt{3}}{5}$

ii. 1. The denominator is monomial. Multiply both numerator and denominator by the surd part of the monomial term.

ii. $\frac{5\sqrt{3}-3\sqrt{10}}{4\sqrt{5}} = \frac{\left(5\sqrt{3}-3\sqrt{10}\right)}{4\sqrt{5}} \times \frac{\sqrt{5}}{\sqrt{5}}$

$= \frac{\sqrt{5}\left(5\sqrt{3}-3\sqrt{10}\right)}{4\times 5}$

$= \frac{5\sqrt{15}-3\sqrt{50}}{20}$

2. Simplify the surds, where possible.

$= \frac{5\sqrt{15}-3\times 5\sqrt{2}}{20}$

$= \frac{5\sqrt{15}-15\sqrt{2}}{20}$

3. Take out the common factor in the numerator since it can be cancelled as a factor of the denominator.

$= \frac{{}^{1}\cancel{5}\left(\sqrt{15}-3\sqrt{2}\right)}{\cancel{20}_{4}}$

$= \frac{\sqrt{15}-3\sqrt{2}}{4}$

b. Rationalise any denominators containing surds and simplify all terms in order to identify any like surds that can be collected together.

b. $5\sqrt{2}-\frac{4}{\sqrt{2}}+3\sqrt{18}$

$=5\sqrt{2}-\frac{4}{\sqrt{2}}\times\frac{\sqrt{2}}{\sqrt{2}}+3\times 3\sqrt{2}$

$=5\sqrt{2}-\frac{{}^{2}\cancel{4}\sqrt{2}}{{}_{1}\cancel{2}}+9\sqrt{2}$

$=5\sqrt{2}-2\sqrt{2}+9\sqrt{2}$

$=12\sqrt{2}$

c. 1. The denominator is binomial. Multiply both numerator and denominator by the conjugate of the binomial surd contained in the denominator.

c. The conjugate of $5\sqrt{3}-3\sqrt{2}$ is $5\sqrt{3}+3\sqrt{2}$.

$\frac{6}{5\sqrt{3}-3\sqrt{2}} = \frac{6}{5\sqrt{3}-3\sqrt{2}} \times \frac{5\sqrt{3}+3\sqrt{2}}{5\sqrt{3}+3\sqrt{2}}$

$= \frac{6\left(5\sqrt{3}+3\sqrt{2}\right)}{\left(5\sqrt{3}-3\sqrt{2}\right)\left(5\sqrt{3}+3\sqrt{2}\right)}$

2. Expand the difference of two squares in the denominator.
Note: This expansion should always result in a rational number.

$= \frac{6\left(5\sqrt{3}+3\sqrt{2}\right)}{\left(5\sqrt{3}\right)^2-\left(3\sqrt{2}\right)^2}$

$= \frac{6\left(5\sqrt{3}+3\sqrt{2}\right)}{25\times 3-9\times 2}$

$= \frac{6\left(5\sqrt{3}+3\sqrt{2}\right)}{57}$

3. Cancel the common factor between the numerator and the denominator.
Note: The numerator could be expanded but there is no further simplification to gain by doing so.

$$= \frac{{}^{2}\cancel{6}\left(5\sqrt{3}+3\sqrt{2}\right)}{\cancel{57}_{19}}$$

$$= \frac{2\left(5\sqrt{3}+3\sqrt{2}\right)}{19} \text{ or } \frac{10\sqrt{3}+6\sqrt{2}}{19}$$

d. 1. Substitute the given value and simplify.

d. Given $p = 3\sqrt{2} - 1$,

$$\frac{1}{p^2-1} = \frac{1}{\left(3\sqrt{2}-1\right)^2 - 1}$$

$$= \frac{1}{\left(9\times 2 - 6\sqrt{2}+1\right) - 1}$$

$$= \frac{1}{18-6\sqrt{2}}$$

2. Factorise the denominator so that the binomial surd is simpler.

$$= \frac{1}{6\left(3-\sqrt{2}\right)}$$

3. Multiply numerator and denominator by the conjugate of the binomial surd contained in the denominator.

$$= \frac{1}{6\left(3-\sqrt{2}\right)} \times \frac{3+\sqrt{2}}{3+\sqrt{2}}$$

$$= \frac{3+\sqrt{2}}{6\left(3-\sqrt{2}\right)\left(3+\sqrt{2}\right)}$$

4. Expand the difference of two squares and simplify.

$$= \frac{3+\sqrt{2}}{6\left((3)^2 - \left(\sqrt{2}\right)^2\right)}$$

$$= \frac{3+\sqrt{2}}{6\,(9-2)}$$

$$= \frac{3+\sqrt{2}}{6\times 7}$$

$$= \frac{3+\sqrt{2}}{42}$$

TI | THINK	DISPLAY/WRITE	CASIO | THINK	DISPLAY/WRITE
c. 1. On a Calculator page, complete the entry line as: $\frac{6}{5\sqrt{3}-3\sqrt{2}}$ Then press ENTER.	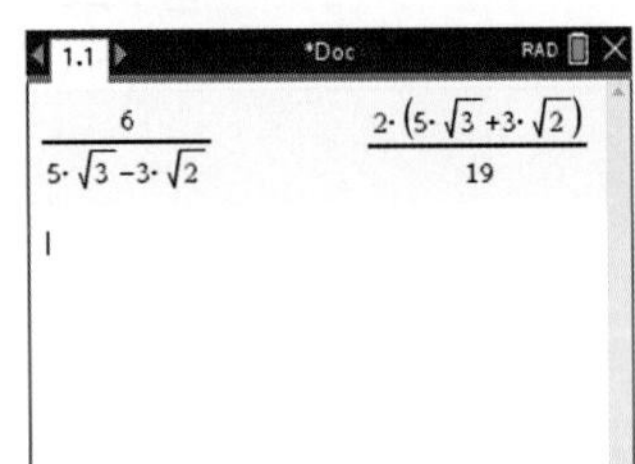	**c.** 1. On a Main screen, complete the entry line as: simplify $\left(\frac{6}{5\sqrt{3}-3\sqrt{2}}\right)$ Then press EXE.	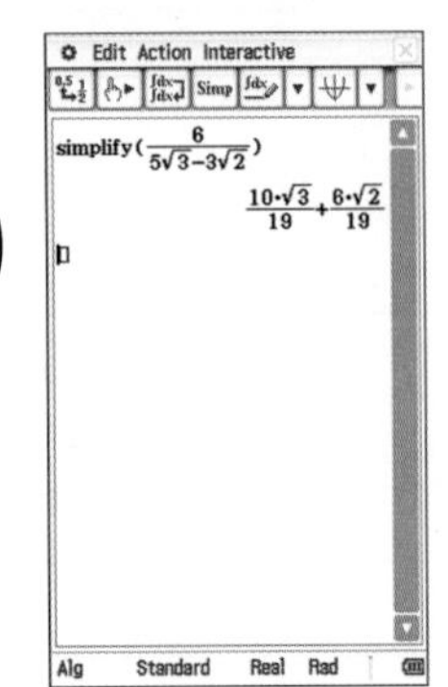
2. The answer appears on the screen.	$\frac{2\left(5\sqrt{3}+3\sqrt{2}\right)}{19}$	2. The answer appears on the screen.	$\frac{10\sqrt{3}}{19}+\frac{6\sqrt{2}}{19}$

2.6 Exercise

Students, these questions are even better in jacPLUS

Receive immediate feedback and access sample responses

Access additional questions

Track your results and progress

Find all this and MORE in jacPLUS

Technology free

1. **WE15** a. Express $\left\{3\sqrt{3}, 4\sqrt{5}, 5\sqrt{2}, 5\right\}$ with its elements in increasing order.
   b. Express the following in simplest form.
      i. $\sqrt{84}$
      ii. $2\sqrt{108ab^2}$ assuming $b > 0$

2. Select the surds from the following set of real numbers.
   $$\left\{\sqrt{8}, \sqrt{900}, \sqrt{\frac{4}{9}}, \sqrt{1.44}, \sqrt{10^3}, \pi, \sqrt[3]{27}, \sqrt[3]{36}\right\}$$

SCHOOL

SPEED LIMIT

$\sqrt{900}$

3. Express the following as entire surds.
   a. $4\sqrt{5}$
   b. $2\sqrt[3]{6}$
   c. $\dfrac{9\sqrt{7}}{4}$
   d. $\dfrac{3}{\sqrt{3}}$
   e. $ab\sqrt{c}$
   f. $m\sqrt[3]{n}$

4. Simplify each of the following.
   a. $\sqrt{32}$
   b. $2\sqrt{44}$
   c. $\sqrt{52}$
   d. $3\sqrt{80}$
   e. $\sqrt{45}$
   f. $\dfrac{\sqrt{99}}{18}$

5. Express each of the following in simplified form.
   a. $\sqrt{75}$
   b. $5\sqrt{48}$
   c. $\sqrt{2000}$
   d. $3\sqrt{288}$
   e. $2\sqrt{72}$
   f. $\sqrt[3]{54}$

6. **WE16** Simplify the following.
   a. $\sqrt{5} - 4\sqrt{7} - 7\sqrt{5} + 3\sqrt{7}$
   b. $3\sqrt{48} - 4\sqrt{27} + 3\sqrt{32}$
   c. $3\sqrt{5} \times 7\sqrt{15}$

7. Simplify the following by collecting like terms together.
   a. $5\sqrt{11} - 2\sqrt{3} - 9\sqrt{3} + 4\sqrt{11}$
   b. $5\sqrt{2} + \sqrt{3} + 4\sqrt{27} - \frac{1}{2}\sqrt{72}$
   c. $4\sqrt{3} + 2\sqrt{8} - 7\sqrt{2} + 5\sqrt{48}$
   d. $2\sqrt{3} - 3\sqrt{20} + 4\sqrt{12} - \sqrt{5}$
   e. $2\sqrt{12} - \sqrt{125} - \sqrt{50} + 2\sqrt{180}$
   f. $\sqrt{36} - \sqrt{108} + 2\sqrt{75} - 4\sqrt{300}$

8. Simplify the following.
   a. $3\sqrt{7} + 8\sqrt{3} + 12\sqrt{7} - 9\sqrt{3}$
   b. $10\sqrt{2} - 12\sqrt{6} + 4\sqrt{6} - 8\sqrt{2}$
   c. $3\sqrt{50} - \sqrt{18}$
   d. $8\sqrt{45} + 2\sqrt{125}$
   e. $\sqrt{6} + 7\sqrt{5} + 4\sqrt{24} - 8\sqrt{20}$
   f. $2\sqrt{12} - 7\sqrt{243} + \frac{1}{2}\sqrt{8} - \frac{2}{3}\sqrt{162}$

9. Simplify the following.

a. $4\sqrt{6}\times\sqrt{21}$

b. $-4\sqrt{27}\times-\sqrt{28}$

c. $4\sqrt{5}\left(\sqrt{5}+\sqrt{10}\right)$

d. $3\sqrt{7}\left(2\sqrt{7}-3\sqrt{14}\right)$

e. $\left(\sqrt{5}-2\right)\left(4-\sqrt{5}\right)$

f. $\left(4\sqrt{3}-2\right)\left(2\sqrt{5}+3\sqrt{7}\right)$

10. Carry out the following operations and express answers in simplest form.

a. $4\sqrt{5}\times2\sqrt{7}$

b. $-10\sqrt{6}\times-8\sqrt{10}$

c. $3\sqrt{8}\times2\sqrt{5}$

d. $\sqrt{18}\times\sqrt{72}$

e. $\dfrac{4\sqrt{27}\times\sqrt{147}}{2\sqrt{3}}$

f. $5\sqrt{2}\times\sqrt{3}\times4\sqrt{5}\times\dfrac{\sqrt{6}}{6}+3\sqrt{2}\times7\sqrt{10}$

11. Expand and simplify the following.

a. $\left(\sqrt{7}+\sqrt{3}\right)\left(\sqrt{7}-\sqrt{3}\right)$

b. $\left(\sqrt{2}-8\right)\left(\sqrt{2}+8\right)$

c. $\left(\sqrt{7}-2\right)^2$

d. $\left(\sqrt{11}+\sqrt{2}\right)^2$

e. $\left(4-2\sqrt{5}\right)\left(4+2\sqrt{5}\right)$

f. $\left(3\sqrt{5}+2\sqrt{3}\right)^2$

12. **WE17** Expand and simplify the following.

a. $2\sqrt{3}\left(4\sqrt{15}+5\sqrt{3}\right)$

b. $\left(\sqrt{3}-8\sqrt{2}\right)\left(5\sqrt{5}-2\sqrt{21}\right)$

c. $\left(4\sqrt{3}-5\sqrt{2}\right)^2$

d. $\left(3\sqrt{5}-2\sqrt{11}\right)\left(3\sqrt{5}+2\sqrt{11}\right)$

13. Expand and simplify the following.

a. $\sqrt{2}\left(3\sqrt{5}-7\sqrt{6}\right)$

b. $5\sqrt{3}\left(7-3\sqrt{3}+2\sqrt{6}\right)$

c. $2\sqrt{10}-3\sqrt{6}\left(3\sqrt{15}+2\sqrt{6}\right)$

d. $\left(2\sqrt{3}+\sqrt{5}\right)\left(3\sqrt{2}+4\sqrt{7}\right)$

14. Expand the following.

a. $\left(2\sqrt{2}+3\right)^2$

b. $\left(3\sqrt{6}-2\sqrt{3}\right)^2$

c. $\left(\sqrt{7}-\sqrt{5}\right)^2$

d. $\left(2\sqrt{5}+\sqrt{3}\right)\left(2\sqrt{5}-\sqrt{3}\right)$

e. $\left(10\sqrt{2}-3\sqrt{5}\right)\left(10\sqrt{2}+3\sqrt{5}\right)$

f. $\left(\sqrt{3}+\sqrt{2}+1\right)\left(\sqrt{3}+\sqrt{2}-1\right)$

15. Rationalise the denominator and simplify where appropriate.

a. $\dfrac{\sqrt{5}}{\sqrt{3}}$

b. $\dfrac{2\sqrt{3}}{3\sqrt{2}}$

c. $\dfrac{\sqrt{2}+3\sqrt{5}}{\sqrt{6}}$

d. $\dfrac{3\sqrt{5}-5\sqrt{2}}{2\sqrt{10}}$

e. $\dfrac{3}{\sqrt{5}-\sqrt{2}}$

f. $\dfrac{\sqrt{5}}{2\sqrt{2}+3}$

16. Express the following in simplest form with rational denominators.

a. $\dfrac{3\sqrt{2}}{4\sqrt{3}}$

b. $\dfrac{\sqrt{5}+\sqrt{2}}{\sqrt{2}}$

c. $\dfrac{\sqrt{12}-3\sqrt{2}}{2\sqrt{18}}$

d. $\dfrac{1}{\sqrt{6}+\sqrt{2}}$

e. $\dfrac{2\sqrt{10}+1}{5-\sqrt{10}}$

f. $\dfrac{3\sqrt{3}+2\sqrt{2}}{\sqrt{3}+\sqrt{2}}$

Technology active

17. a. **WE18** Express each of the following with a rational denominator.

i. $\dfrac{6}{7\sqrt{2}}$

ii. $\dfrac{3\sqrt{5}+7\sqrt{15}}{2\sqrt{3}}$

b. Simplify $14\sqrt{6}+\dfrac{12}{\sqrt{6}}-5\sqrt{24}$.

c. Express $\dfrac{10}{4\sqrt{3}+3\sqrt{2}}$ with a rational denominator.

d. Given $p=4\sqrt{3}+1$, calculate $\dfrac{1}{p^2-1}$, expressing the answer with a rational denominator.

18. Express $\sqrt[3]{384}$ in simplest form.

19. Simplify $\dfrac{1}{2}\sqrt{12}-\dfrac{1}{5}\sqrt{80}+\dfrac{\sqrt{10}}{\sqrt{2}}+\sqrt{243}+5$.

20. a. Expand $\left(\sqrt{2}+1\right)^3$.

b. If $\left(\sqrt{2}+\sqrt{6}\right)^2-2\sqrt{3}\left(\sqrt{2}+\sqrt{6}\right)\left(\sqrt{2}-\sqrt{6}\right)=a+b\sqrt{3}$, $a, b \in N$, find the values of a and b.

21. a. Simplify $\dfrac{2\sqrt{3}-1}{\sqrt{3}+1}-\dfrac{\sqrt{3}}{\sqrt{3}+2}$ by first rationalising each denominator.

b. Show that $\dfrac{\sqrt{3}-1}{\sqrt{3}+1}+\dfrac{\sqrt{3}+1}{\sqrt{3}-1}$ is rational by first placing each fraction on a common denominator.

22. Express each of the following as a single fraction in simplest form.

a. $4\sqrt{5}-2\sqrt{6}+\dfrac{3}{\sqrt{6}}-\dfrac{10}{3\sqrt{5}}$

b. $\sqrt{2}\left(2\sqrt{10}+9\sqrt{8}\right)-\dfrac{\sqrt{5}}{\sqrt{5}-2}$

c. $\dfrac{3}{2\sqrt{3}\left(\sqrt{2}+\sqrt{3}\right)^2}+\dfrac{2}{\sqrt{3}}$

d. $\dfrac{2\sqrt{3}-\sqrt{2}}{2\sqrt{3}+\sqrt{2}}+\dfrac{2\sqrt{3}+\sqrt{2}}{2\sqrt{3}-\sqrt{2}}$

e. $\dfrac{\sqrt{2}}{16-4\sqrt{7}}-\dfrac{\sqrt{14}+2\sqrt{2}}{9\sqrt{7}}$

f. $\dfrac{\left(2-\sqrt{3}\right)^2}{2+\sqrt{3}}+\dfrac{2\sqrt{3}}{4-3\sqrt{2}}$

23. a. If $x=2\sqrt{3}-\sqrt{10}$, calculate the value of the following.

i. $x+\dfrac{1}{x}$

ii. $x^2-4\sqrt{3}x$

b. If $y=\dfrac{\sqrt{7}+2}{\sqrt{7}-2}$, calculate the value of the following.

i. $y-\dfrac{1}{y}$

ii. $\dfrac{1}{y^2-1}$

c. Determine the values of m and n for which each of the following is a correct statement.

i. $\dfrac{1}{\sqrt{7}+\sqrt{3}}-\dfrac{1}{\sqrt{7}-\sqrt{3}}=m\sqrt{7}+n\sqrt{3}$

ii. $\left(2+\sqrt{3}\right)^4-\dfrac{7\sqrt{3}}{\left(2+\sqrt{3}\right)^2}=m+\sqrt{n}$

d. The real numbers x_1 and x_2 are a pair of conjugates. If $x_1 = \dfrac{-b+\sqrt{b^2-4ac}}{2a}$:

i. state x_2
ii. calculate the sum $x_1 + x_2$
iii. calculate the product x_1x_2.

24. A triangle has vertices at the points A$\left(\sqrt{2}, -1\right)$, B$\left(\sqrt{5}, \sqrt{10}\right)$ and C$\left(\sqrt{10}, \sqrt{5}\right)$.

a. Calculate the lengths of each side of the triangle in simplest surd form.
b. Calculate from the surd form the length of the longest side to 1 decimal place.

25. A rectangular lawn has dimensions $\left(\sqrt{6}+\sqrt{3}+1\right)$ m by $\left(\sqrt{3}+2\right)$ m. Hew agrees to mow the lawn for the householder.

a. Calculate the exact area of the lawn.
b. The householder received change of \$23.35 from \$50. Calculate the cost per square metre that Hew charged for mowing the lawn.

2.6 Exam questions

Question 1 (1 mark) TECH-ACTIVE

MC Simplify $5\sqrt{6}\times 3\sqrt{2}-\sqrt{3}\left(4\sqrt{3}+2\right)$.

A. $28\sqrt{3}-12$
B. $8\sqrt{12}-12-\sqrt{6}$
C. $32\sqrt{3}-12$
D. $16\sqrt{6}-12$
E. $9\sqrt{3}+5\sqrt{2}$

Question 2 (1 mark) TECH-FREE

Expand and simplify $\left(3\sqrt{2}-2\sqrt{3}\right)\left(3\sqrt{2}+2\sqrt{3}\right)$.

Question 3 (1 mark) TECH-ACTIVE

MC $\dfrac{\sqrt{2}-\sqrt{3}}{\sqrt{2}+\sqrt{3}}$ is equal to

A. $5+2\sqrt{6}$
B. $-5-2\sqrt{6}$
C. -1
D. $5-2\sqrt{6}$
E. $2\sqrt{6}-5$

More exam questions are available online.

2.7 Review

2.7.1 Summary

doc-37017

Hey students! Now that it's time to revise this topic, go online to:

Access the topic summary

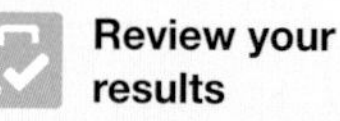
Review your results

Watch teacher-led videos

Practise exam questions

Find all this and MORE in jacPLUS

2.7 Exercise

Technology free: short answer

1. Expand and simplify where possible.
   a. $(5x+2y)^2-3(1+2y)(1-2y)$
   b. $(2x-3)(x^2-6x+2)-(5-x)(5+x)$
   c. $\left(2x+\sqrt{3}\right)^3$
   d. $(2x-5)^4$
   e. $\left(1-\frac{2x}{3}\right)^5$
   f. $(x+1-y)^2$

2. Factorise the following.
   a. $72x^2+41xy-45y^2$
   b. $x^2y^3-36x^2y$
   c. $4y^2-x^2+18x-81$
   d. $64x^3+1$
   e. $(x+2)^3-8(x-1)^3$
   f. $2x^3+6x^2y+6xy^2+2y^3$

3. a. Evaluate $\binom{10}{6}$.
   b. i. Write down row 5 of Pascal's triangle.
      ii. Hence, or otherwise, expand $(1+2xy)^5$.
   c. Given row 6 of Pascal's triangle is 1 6 15 20 15 6 1, calculate the middle term(s) in the expansion of $(3-2x)^6$.

4. Simplify the following.
   a. $4\sqrt{3}+3\sqrt{8}-2\sqrt{72}-\sqrt{75}$
   b. $\left(\sqrt{7}-\sqrt{3}\right)\left(\sqrt{7}+\sqrt{3}\right)$
   c. $\frac{5}{\sqrt{50}}+\sqrt{108}\div\sqrt{3}$
   d. $\frac{11}{2\sqrt{3}-3\sqrt{5}}$
   e. $4\sqrt{5}-\frac{2}{3\sqrt{5}}-\frac{\sqrt{125}}{2}$
   f. $\frac{3}{\sqrt{5}-\sqrt{2}}-\frac{3}{\sqrt{5}+\sqrt{2}}$

5. Simplify the following.
   a. $\frac{x^2+7x+12}{(x+1)^2-9}$
   b. $\frac{4}{x^2-9}+\frac{3}{x^2-6x+9}$
   c. $\frac{5x}{x^3+27}\div\frac{5x^2-15x}{9-x^2}$
   d. $\frac{2}{3\sqrt{5}}+\frac{4\sqrt{5}}{3\sqrt{5}-1}$
   e. $\frac{3}{\sqrt{5}-\sqrt{2}}\times\frac{\sqrt{10}}{\sqrt{5}-\sqrt{2}}$
   f. $\frac{3\sqrt{2}}{\sqrt{5}+2\sqrt{3}}-\frac{2\sqrt{5}+\sqrt{6}}{2\sqrt{2}-1}$

6. Arrange the elements of the following sets of numbers in decreasing order.

a. $\left\{2\sqrt{6}, 4\sqrt{3}, 3\sqrt{5}, 5\sqrt{2}, 2\sqrt{10}\right\}$

b. $\left\{7!, (4!)^2, \dfrac{16!}{14!}, 4! \times 5!\right\}$

c. Given $\sqrt{10} \approx 3.162$, calculate an approximate decimal value for:

i. $\dfrac{1}{\sqrt{10}}$

ii. $\dfrac{\left(\sqrt{5}-\sqrt{2}\right)^2}{3-\sqrt{10}}$

d. Given $x = 1 - 2\sqrt{5}$, calculate the value of x^2 and hence deduce $\sqrt{21 - 4\sqrt{5}}$.

Technology active: multiple choice

7. **MC** The factors of $2(x+1)^2 + 9(x+1) - 5$ are:

A. $(2x+1)(x+6)$
B. $(2x-1)(x+5)$
C. $(2x+3)(x-4)$
D. $(2x+1)(x-5)$
E. $4x^2 + 13x + 6$

8. **MC** The expanded form of $(1-x)^3$ is:

A. $1 - x^3$
B. $1 + x + x^2 - 3x^3$
C. $1 - 3x - 3x^2 - x^3$
D. $1 - 3x + 3x^2 - x^3$
E. $-x^3 - 2x^2 + 2x + 1$

9. **MC** The factorised form of $24x^3 - 81y^3$ is:

A. $3(2x+3y)^3$
B. $(2x-3y)^3$
C. $3(2x-3y)(4x^2+6xy+9y^2)$
D. $3(2x-3y)(4x^2+12xy+9y^2)$
E. $(8x+27y)(8x^2-2xy+27y^2)$

10. **MC** $\dfrac{(n+1)!+n!}{(n+1)!-n!}$ is equal to:

A. $\dfrac{n+2}{n}$
B. -1
C. 1
D. 2
E. $n!$

11. **MC** The solution to the equation $\dbinom{n}{1} = 10$ is:

A. $n = 2$
B. $n = 5$
C. $n = 9$
D. $n = 10$
E. $n = 11$

12. **MC** Select the correct statement about the expansion of $(2x-3)^5$.

A. The coefficient of x^4 would be 5.
B. The coefficient of x^4 would be -5.
C. The constant term would be 243.
D. The third term would be $\dbinom{5}{4}(2x)^3(3)^2$.
E. There are 6 terms in the expansion.

13. **MC** Select the incorrect statement.

A. $I = R \setminus Q$
B. $Z^+ = N$
C. $0! \notin N$
D. $\dfrac{17+3}{5} \in N$
E. $Z \setminus N = Z^- \cup \{0\}$

14. **MC** Select the expression that represents a rational number.

A. $\sqrt{4+9}$
B. $\sqrt{11-4\times 9}$
C. $\dfrac{1}{\sqrt{4}+\sqrt{9}}$
D. $\sqrt[3]{\dfrac{4}{9}}$
E. $\dfrac{4}{\sqrt{9}-3}$

15. **MC** If $a = 18$, then the value of $\frac{6}{\sqrt{a}} + \frac{\sqrt{a}}{\sqrt{6}}$ is:

A. $6\sqrt{2}$ B. $\sqrt{2} + \sqrt{3}$ C. $9 + \sqrt{3}$ D. $\sqrt{6}$ E. $\sqrt{2} + 3\sqrt{6}$

16. **MC** The section of the number line illustrated could be described as:

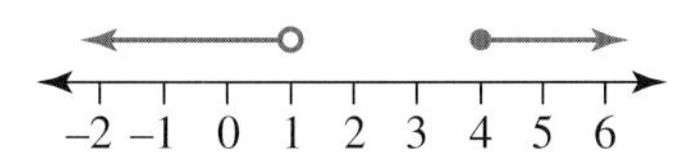

A. $(-\infty, 1] \cup (4, \infty)$ B. $(-\infty, 1) \cap [4, \infty)$ C. $R\backslash(1, 4]$
D. $R\backslash[1, 4)$ E. $R \cap (1, 4)$

Technology active: extended response

17. a. Solve the following system of simultaneous equations to obtain the values of x and y in simplest form with rational denominators.

$$5\sqrt{3}x - y = 12$$
$$2\sqrt{3}x + 3y = 15$$

b. Express the solution set for $\{x : x - \sqrt{2}x < \sqrt{2}\}$ in interval notation.

c. Simplify $\frac{a^3 - 1}{a^3 + 1} - \frac{a^3 + 1}{a^3 - 1} + \frac{3a^3}{a^6 - 1}$.

d. i. Expand and simplify $(a + b + c)(a^2 + b^2 + c^2 - ab - bc - ca)$.
ii. Hence, show that if $a + b + c = 0$, then $a^3 + b^3 + c^3 = 3abc$.
iii. Hence, calculate $102^3 - 100^3 - 2^3$.

18. a. State a rational number that lies between $\frac{2}{3}$ and $\frac{5}{6}$.

b. i. If a and b are any two real numbers, show that it is possible to find another real number that lies between them.
ii. If a and b are integers, determine whether it is always possible to find another integer that lies between them.

c. By expanding $\left(1 + \sqrt{3}\right)^6 + \left(1 - \sqrt{3}\right)^6$, show it is rational.

d. Given m is a prime number, find the values of m and n for which $\frac{n}{\sqrt{m} - \sqrt{2}} + \frac{n}{\sqrt{m} + \sqrt{2}} = 8\sqrt{3}$.

19. Two friends arrange to meet in a park in order to do some walking as part of their exercise regime.

a. On Monday they walk around a square of edge $80\sqrt{2}$ metres. It takes them 20 minutes to complete a circuit. Find their average walking speed in km/h.

b. On Tuesday they walk around the circumference of the circle that the square encloses.
If they walk on Tuesday at an average speed of $\sqrt{8}$ km/h, obtain an exact expression for the time, in minutes, that it takes them to do two circuits of the circular path.

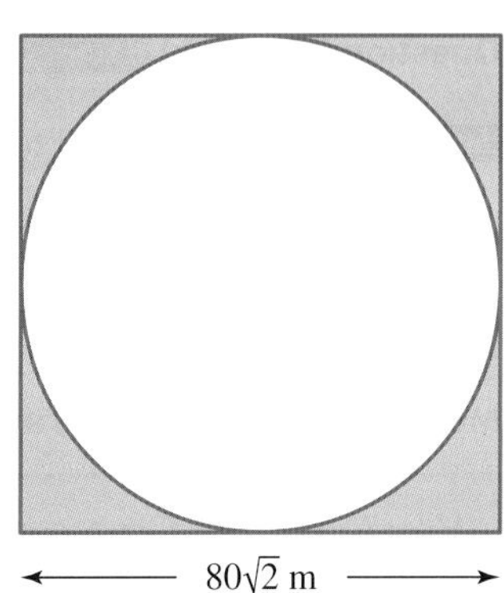

c. On Wednesday the two friends cut across the corners of the square park to create a regular octagonal route to walk along. Calculate the total distance they walk around the perimeter of this path in simplest surd form.

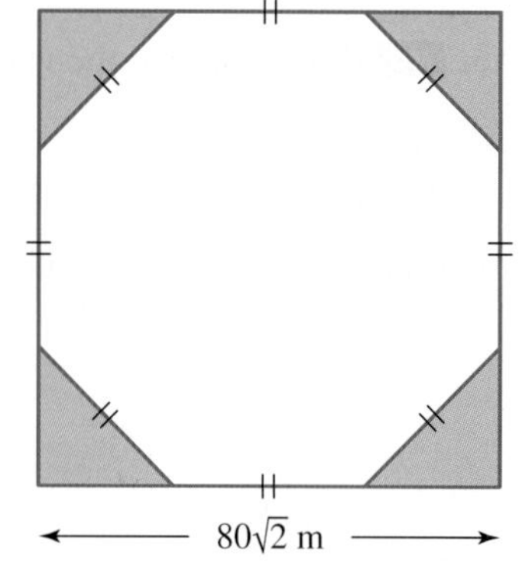

d. On Thursday the friends go to another area, which is bounded by a triangle, PQR. The midpoints of the sides PQ, PR and RQ of the triangle are $\left(\sqrt{2}, \sqrt{3}\right)$, $\left(\sqrt{10} - \sqrt{2}, \sqrt{5}\right)$ and $\left(\sqrt{2}, \sqrt{3} - \sqrt{2}\right)$, respectively (measured in kilometres).

   i. Determine the coordinates of P, Q and R.
   ii. Find how far it is for them to walk directly from P to R.

20. a. Calculate the coefficient of x^3 in the expansion of $(1 - 2x + 3x^2)^5$.
    b. Consider the expansion of $(x + 1)^{17}$.
       i. Show that the ratio of the coefficients of the $(r + 1)$th term to the $(r + 2)$th term is $\dfrac{r+1}{17-r}$.
       ii. Obtain the two terms for which this ratio is 2 : 1.
       iii. Determine the other values of n, $n < 17$, for which the ratio of the coefficients of a pair of consecutive terms in the expansion of $(x + 1)^n$ would also be 2 : 1.

2.7 Exam questions

Question 1 (1 mark) TECH-ACTIVE

MC $\dfrac{(a-b)^3}{a^3 - b^3}$ simplifies to

A. $\dfrac{a-b}{a+b}$ B. $\dfrac{(a-b)^2}{(a+b)^2}$ C. $\dfrac{(a-b)^2}{a^2+ab+b^2}$ D. $-\dfrac{1}{ab}$ E. 1

Question 2 (1 mark) TECH-FREE

Simplify $\dfrac{p^2 - 9q^2}{(p-3q)^2} \div \dfrac{(p+3q)^2}{4p - 12q}$.

Question 3 (1 mark) TECH-ACTIVE

MC Select the sets to which π and $\sqrt{169}$ respectively belong.

A. R and N B. Q and N C. R and Z D. R and Q E. Q and Z

Question 4 (2 marks) TECH-FREE

Illustrate the interval $(-3, 1]$ on a number line and write the set in alternative notation.

Question 5 (1 mark) TECH-ACTIVE

MC $\dfrac{5\sqrt{6} - 2\sqrt{2}}{\sqrt{8}}$ is equal to

A. $\dfrac{5\sqrt{6}-2}{4}$ B. $\dfrac{5\sqrt{3}-2}{2}$ C. $\dfrac{5\sqrt{12}-4}{8}$ D. $5\sqrt{3}$ E. $\dfrac{3\sqrt{8}}{8}$

More exam questions are available online.

Answers

Topic 2 Algebraic foundations

2.2 Algebraic skills

2.2 Exercise

1. a. $4m^2 - 5m$ b. $5m^2 - 16m + 8$
   c. $x^2 + 2x - 15$ d. $15m^2 - 22m + 8$
   e. $16 - 9x^2$ f. $4x^2 - 20x + 25$
2. a. $2x^2 - 6x - 41$ b. $23 - 6x^2 - 26x$
   c. $6x - 3x^2 - 27$ d. $13x^2 - 10x + 22$
3. The expansion gives $55x^2 + 7x - 114$. The coefficient of x is 7.
4. a. $4x^2 + 12x + 9$ b. $4ab^2 - 36a^3$
   c. $20 - 3c - 4c^2$ d. $25 - 70y + 49y^2$
   e. $9m^6 - 16n^2$ f. $x^3 + 3x^2 + 3x + 1$
5. a. $6x^2 - 5x + 7; -5$
   b. $44 - 29x^2 - 15x^3; 0$
   c. $-16x^3 + 16x^2 + 49x - 49; 49$
   d. $2x^2 - 4y^2 + 2; 0$
   e. $39 - 75x + 11x^2; -75$
   f. $-3x^2 - 3x + 16; -3$
6. a. $(x - 6)(x + 6)$ b. $(2 - 5a)(2 + 5a)$
   c. $(3m - 1)(3m + 1)$ d. $4(a - 4)(a + 4)$
   e. $2(m - 7x)(m + 7x)$ f. $(3m - 2)(4 - 3m)$
7. a. $(x - 3)(x - 6)$
   b. $(x - 3)(x - 3) = (x - 3)^2$
   c. $(x - 5)(x + 12)$
   d. $(2x - 3)(2x + 5)$
   e. $(2x - 5)(2x - 5) = (2x - 5)^2$
   f. $8(x - 3y)(x - 3y) = 8(x - 3y)^2$
8. a. $5\left(x - 3y\right)\left(x + 3y\right)$
   b. $(x - 10)(x + 1)$
   c. $(2x - 5)(4x + 3)$
   d. $4x\left(x + y\right)\left(x - 3y\right)$
   e. $\left(3y - x - 4\right)\left(3y + x + 4\right)$
   f. $(x - 1)(4x - 23)$
9. a. $(7 - 12x)^2$ b. $(2x + 3)(x + 3)$
   c. $(4x + 9)(10x + 13)$ d. $36(2x + y)(2x - y)$
   e. $(3ax - 1)(a + 3)$ f. $(4x + 1 - y)(4x + 1 + y)$
10. a. $(x + 2)(x - 5)(x + 5)$
    b. $p(10p - 9q)(10p + 9q)$
    c. $(2n + 1 - 2p)(2n + 1 + 2p)$
    d. $5(n + 5m)(23n - 11m)$
    e. $13(a - 1)(3 - 2a)(2a - 1)$
    f. $(a + b - 1)(2a - 1)$
11. a. $(x + 13)(x - 2)$ b. $(2x - 3)(x + 4)$
    c. $(7x + 9y)(10x + 7y)$ d. $(x - 3)(x + 3)(x^2 + 1)$
    e. $(5p - q)^2$ f. $(a - 1)^4$
12. a. $(x - 5)(x^2 + 5x + 25)$ b. $3(1 + x)(1 - x + x^2)$
13. a. $(x - 2)(x^2 + 2x + 4)$
    b. $(x + 10)(x^2 - 10x + 100)$
    c. $(1 - x)(1 + x + x^2)$
    d. $(3x + 4y)(9x^2 - 12xy + 16y^2)$
    e. $x(x - 5)(x^2 + 5x + 25)$
    f. $(x + 5)(x^2 - 8x + 43)$
14. a. $x(y - 3)\left(y^2 + 3y + 9\right)$
    b. $-(x + 6)\left(x^2 - 6x + 36\right)$
    c. $3(1 - 3x)\left(1 + 3x + 9x^2\right)$
    d. $4(2x + m)(4x^2 - 2mx + m^2)$
    e. $(3m + 4n)\left(9m^2 - 12mn + 16n^2\right)$
    f. $2(5x - 4m)\left(25x^2 + 20mx + 16m^2\right)$
15. a. $3(2x - 3y)(4x^2 + 6xy + 9y^2)$
    b. $xy(2xy + 1)(4x^2y^2 - 2xy + 1)$
    c. $7(9x - 10)(3x^2 + 100)$
    d. $-2(5x + 4y)(43x^2 + 31xy + 7y^2)$
    e. $(a - b)(a + b)^2(a^2 - ab + b^2)$
    f. $(x - y)(x + y)(x^2 + xy + y^2)(x^2 - xy + y^2)$
16. a. $\dfrac{x}{x - 2}$ b. $8 - x^2$
17. a. $\dfrac{1}{3(x + 1)}$ b. $\dfrac{x + 3}{2(3x + 1)}$
    c. $\dfrac{x + 3}{4}$ d. $-x - 5$
    e. $\dfrac{2(3x + 1)}{3(3x + 2)(x + 1)}$ f. 1
18. a. $\dfrac{x - 4}{5 - 3x}$ b. $x + 3$
    c. $3x^2 + 3xh + h^2$ d. $\dfrac{1}{3(1 - 3x)}$
    e. $\dfrac{m^2}{m^2 - mn + n^2}$ f. $\dfrac{(1 - x)^2}{1 + x^2}$
19. $\dfrac{x - 31}{5(x - 5)(x - 1)}$
20. a. $\dfrac{23x}{12}$ b. $-\dfrac{19x}{14}$
    c. $\dfrac{9x + 5}{(x - 3)(x + 5)}$ d. $\dfrac{-2x^2 - 14x + 5}{(3x - 1)(1 - 2x)}$
    e. $\dfrac{-x^2 - 9x + 1}{(2x + 1)(x - 1)}$ f. $\dfrac{4x + 21}{(x^2 - 9)}$
21. a. $\dfrac{4(x + 1)}{x(1 - x)(x^2 + 1)}$ b. $\dfrac{2x + 20}{x^2 - 4}$
    c. $\dfrac{x - 46}{(x + 6)(x - 5)}$ d. $\dfrac{2y^2 - 9y + 81}{y(2y - 9)^2(2y + 9)}$
22. a. $-9x^4 + 328x^2 - 144$ b. $(x - 3)(x - 3 - y)$
    c. $2xy(x^2 - 3xy + 3y^2)$
23. a. $g^2 + h^2 + 2gh + 24g + 24h + 144$
    b. $343q^3 + 98q^2p - 28qp^2 - 8p^3$
    c. $-4x^4 + 425x^2 - 2500$
24. a. $\dfrac{5}{x(x + 5)}$ b. $\dfrac{1}{4x - 1}$
    c. $\dfrac{p^2q}{p^4 - q^4}$ d. $\dfrac{1}{2}(a - 2b)(a - b)(a + b)$

25. a. $-3x^3 - 16x^2 + 9x + 70$
b. $(13x^2 + 28x + 148)(7x + 2)$
c. $\frac{11x + 16}{(x + 8)(x - 1)}$

2.2 Exam questions

Note: Mark allocations are available with the fully worked solutions online.

1. E
2. D
3. $8(x + 5)(x + 4)$

2.3 Pascal's triangle and binomial expansions

2.3 Exercise

1. $27x^3 - 54x^2 + 36x - 8$
2. a. $x^3 - 9x^2 + 27x - 27$
b. $8x^3 - 12x^2 + 6x - 1$
c. $x^3 + 12x^2 + 48x + 64$
3. a. $27x^3 + 27x^2 + 9x + 1$
b. $1 - 6x + 12x^2 - 8x^3$
c. $125x^3 + 150x^2y + 60xy^2 + 8y^3$
4. A
5. $a^3 + 6a^2b^2 + 12ab^4 + 8b^2$
The coefficient of a^2b^2 is 6.
6. See the table at the bottom of the page.*
7. $(a - b)^6 = a^6 - 6a^5b + 15a^4b^2 - 20a^3b^3 + 15a^2b^4 - 6ab^5 + b^6$
$(2x - 1)^6 = 64x^6 - 192x^5 + 240x^4 - 160x^3 + 60x^2 - 12x + 1$
8. $81x^4 + 216x^3y + 216x^2y^2 + 96xy^3 + 16y^4$
9. a. $x^5 + 20x^4 + 160x^3 + 640x^2 + 1280x + 1024$
b. $x^5 - 20x^4 + 160x^3 - 640x^2 + 1280x - 1024$
c. $x^5y^5 + 10x^4y^4 + 40x^3y^3 + 80x^2y^2 + 80xy + 32$
10. a. $81x^4 - 540x^3y + 1350x^2y^2 - 1500xy^3 + 625y^4$
b. $81 - 108x^2 + 54x^4 - 12x^6 + x^8$
11. $12x + 40x^3 + 12x^5$
12. a. -9 b. -975 c. 1 d. -42
13. $x^4 + y^4 - 4x^3 - 4y^3 + 6x^2 + 6y^2 - 4x - 4y + 4x^3y + 4xy^3 - 12x^2y - 12xy^2 + 6x^2y^2 + 12xy + 1$
14. 20
15. $a = -2$
16. -3
17. a. $1 + 4x + 6x^2 + 4x^3 + x^4$
b. Let $x = 0.1$; $1.1^4 = 1.4641$.
18. $(x + 1)^5 - (x + 1)^4 = x^5 + 4x^4 + 6x^3 + 4x^2 + x$; sample responses can be found in the worked solutions in the online resources.
19. Sample responses can be found in the worked solutions in the online resources.
20. $a = 120, b = 55, c = 495$

2.3 Exam questions

Note: Mark allocations are available with the fully worked solutions online.

1. B
2. E
3. $x^4 - 9x^3 + 30x^2 - 44x + 24$
The coefficient of $x = -44$

2.4 The binomial theorem

2.4 Exercise

1. a. 6 b. 24 c. 120 d. 2 e. 1 f. 7
2. a. 7! b. 8! c. 8! d. 9!
3. 734
4. $n(n - 1)$
5. a. 720 b. 26 c. 5040 d. 120
6. a. 650 b. $\frac{1}{43}$ c. $\frac{7}{5}$ d. $\frac{6}{5}$
7. 35
8. $\binom{n}{2} = \frac{n(n-1)}{2}$ and $\binom{21}{2} = 210$
9. a. 10 b. 10 c. 1 d. 1140 e. 1 f. 286
10. a. $\frac{n(n-1)(n-2)}{6}$ b. $\frac{n(n-1)(n-2)}{6}$
c. $\frac{(n+3)(n+2)(n+1)}{6}$ d. $n(2n + 1)$
e. $\frac{n(n-1)(n+1)}{6}$ f. $\frac{n(n-1)(n+1)}{6}$
11. a. $(n + 1)!$ b. $(n - 1)!$ c. $n(n - 1)(n - 2)$
d. $\frac{1}{n(n+1)}$ e. $\frac{2}{n(n+2)}$ f. $\frac{1}{n-1}$
12. $32x^5 + 240x^4 + 720x^3 + 1080x^2 + 810x + 243$
13. $x^7 - 14x^6 + 84x^5 - 280x^4 + 560x^3 - 672x^2 + 448x - 128$
14. a. $x^5 + 5x^4 + 10x^3 + 10x^2 + 5x + 1$
b. $32 - 80x + 80x^2 - 40x^3 + 10x^4 - x^5$

*6.

Binomial power	Expansion	Number of terms in the expansion	Sum of indices in each term
$(x + a)^2$	$x^2 + 2xa + a^2$	3	2
$(x + a)^3$	$x^3 + 3x^2a + 3xa^2 + a^3$	4	3
$(x + a)^4$	$x^4 + 4x^3a + 6x^2a^2 + 4xa^3 + a^4$	5	4
$(x + a)^5$	$x^5 + 5x^4a + 10x^3a^2 + 10x^2a^3 + 5xa^4 + a^5$	6	5

c. $64x^6 + 576x^5y + 2160x^4y^2 + 4320x^3y^3 + 4860x^2y^4 + 2916xy^5 + 729y^6$

d. $\frac{x^7}{128} + \frac{7x^6}{32} + \frac{21x^5}{8} + \frac{35x^4}{2} + 70x^3 + 168x^2 + 224x + 128$

e. $x^8 - 8x^6 + 28x^4 - 56x^2 + 70 - \frac{56}{x^2} + \frac{28}{x^4} - \frac{8}{x^6} + \frac{1}{x^8}$

f. $x^{20} + 10x^{18} + 45x^{16} + 120x^{14} + 210x^{12} + 252x^{10} + 210x^8 + 120x^6 + 45x^4 + 10x^2 + 1$

15. $-\frac{35}{648}x^4y^3$

16. $\frac{63x^{10}y^5}{8}$

17. Sixth term; the coefficient of x^{15} is $2^9 \times 3^5 \times 7$.

18. $t_4 = 160$

19. a. $t_4 = 20\,000x^3$

b. $t_3 = 1215x^8$

c. $t_4 = 280x^4y^3, t_5 = 560x^3y^4$

20. a. -672 b. $11 \times 7 \times 3^7 \times 2^{11}$

c. 210

21. a. 1 307 674 368 000 b. 3003

22. a. $n = 60$ b. $r = 3$ or $r = 9$

2.4 Exam questions

Note: Mark allocations are available with the fully worked solutions online.

1. D
2. D
3. Fourth term; the coefficient of x^{12} is $7 \times 5 \times 3^3 \times 2^8$.

2.5 Sets of real numbers

2.5 Exercise

1. a. $\frac{6}{11} \in Q$

b. $\sqrt{27} \in I$

c. $12 \in N$ or $12 \in Z$ or $12 \in Q$

d. $0.5 \in Q$

2. a. True b. False c. True

3. a. N b. I c. Q d. Z

4. $x = -1$ or $x = 3$

5. E, since $\frac{8}{0}$ is not defined

6. a. $\sqrt{41} \in I$ b. $-\frac{5}{9} \in Q$

c. $R^+ = \{x : x > 0\}$ d. $\sqrt{2.25} \in Q$

7. a. -5 b. 2 c. $-\frac{3}{2}, 5$ d. 0, 4

8. a. True b. True c. False
d. True e. False f. False

9. $\sqrt{11}, 11\pi, 2^\pi$

10. a. $\{x : -2 \le x < 2\}$

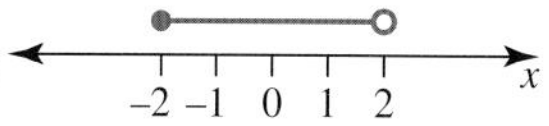
x
−2 −1 0 1 2

b. $(-\infty, -1)$

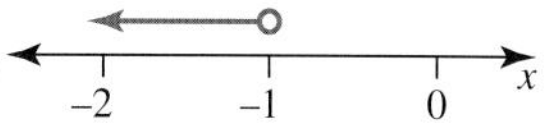
x
−2
−1
0

c. $Z \cap [-2, 2]$ or $\{x : -2 \le x \le 2, x \in Z\}$

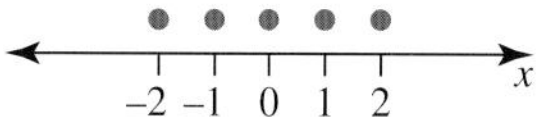
x
−2 −1 0 1 2

11. a. $[3, 5]$

b. $R\backslash[3, 5]$ or $(-\infty, 3) \cup (5, \infty)$

12. a. $[-2, 3)$ b. $(1, 9)$ c. $(-\infty, 5)$ d. $(0, 4]$

13. a. $[-5, -1]$

b. $(-2, \infty)$

c. $[-3, -2) \cup (2, 3]$

d. $(-\infty, 2) \cup (4, \infty), R\backslash[2, 4]$

14. $(-\infty, 1] \cup (4, \infty)$

15. a. $(4, 8]$ b. $(-3, \infty)$ c. $(-\infty, 0]$ d. $[-2, 0]$

16. a.

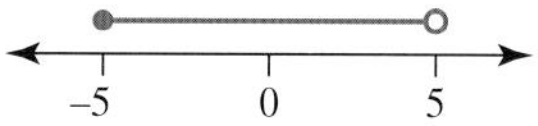
−5
0
5

b.

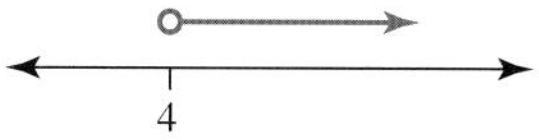
4

c.

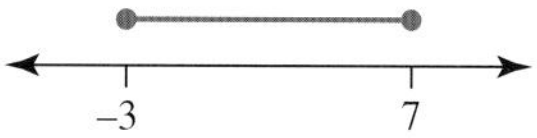
−3
7

d.

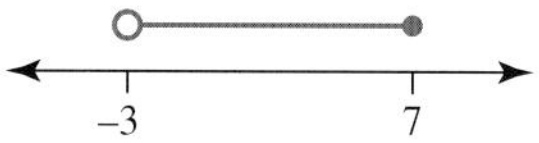
−3
7

e.

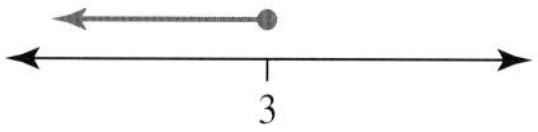
3

f.

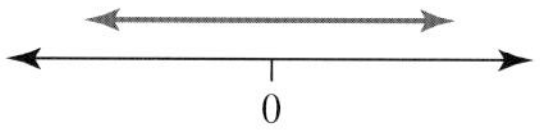
0

17. a.

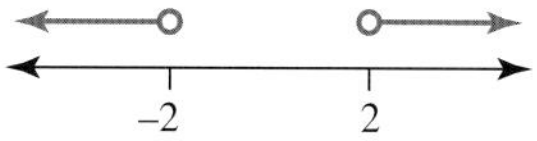
−2
2

b.

−√2
√2

c.

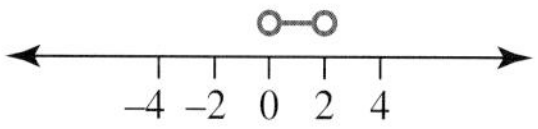
−4 −2 0 2 4

d.

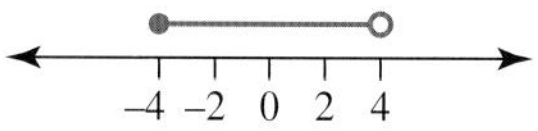
−4 −2 0 2 4

e.

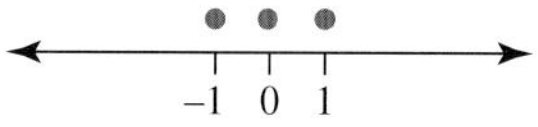
−1 0 1

f.

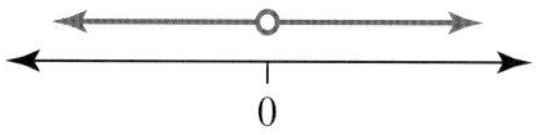
0

18. a. $\{3, 4, 5\}$ b. $(-\infty, -1] \cup (5, \infty)$
c. $(-\infty, 0)$ d. $R\backslash[-4, 2)$

19. a. Rational b. Irrational c. Rational

20. $\pi \approx \frac{256}{81}$; $\frac{256}{81} = 3.160\,493\,827$ to 9 decimal places.
$\frac{22}{7}$ is a better approximation.

2.5 Exam questions

Note: Mark allocations are available with the fully worked solutions online.

1. D
2. $\frac{x-2}{x(x-1)(x+3)}$ is undefined if $x = 0, 1, -3$.
3. B

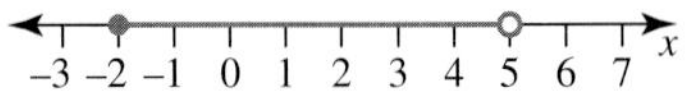

2.6 Surds

2.6 Exercise

1. a. $\{5, 3\sqrt{3}, 5\sqrt{2}, 4\sqrt{5}\}$
   b. i. $2\sqrt{21}$
   ii. $12b\sqrt{3a}$
2. $\sqrt{8}$, $\sqrt{10^3}$, $\sqrt[3]{36}$ are surds.
3. a. $\sqrt{80}$ b. $\sqrt[3]{48}$ c. $\sqrt{\frac{567}{16}}$
   d. $\sqrt{3}$ e. $\sqrt{a^2b^2c}$ f. $\sqrt[3]{m^3n}$
4. a. $4\sqrt{2}$ b. $4\sqrt{11}$ c. $2\sqrt{13}$
   d. $12\sqrt{5}$ e. $3\sqrt{5}$ f. $\frac{\sqrt{11}}{6}$
5. a. $5\sqrt{3}$ b. $20\sqrt{3}$ c. $20\sqrt{5}$
   d. $36\sqrt{2}$ e. $12\sqrt{2}$ f. $3\sqrt[3]{2}$
6. a. $-6\sqrt{5}-\sqrt{7}$ b. $12\sqrt{2}$ c. $105\sqrt{3}$
7. a. $9\sqrt{11}-11\sqrt{3}$ b. $2\sqrt{2}+13\sqrt{3}$
   c. $24\sqrt{3}-3\sqrt{2}$ d. $10\sqrt{3}-7\sqrt{5}$
   e. $4\sqrt{3}+7\sqrt{5}-5\sqrt{2}$ f. $6-36\sqrt{3}$
8. a. $15\sqrt{7}-\sqrt{3}$ b. $2\sqrt{2}-8\sqrt{6}$ c. $12\sqrt{2}$
   d. $34\sqrt{5}$ e. $9\sqrt{6}-9\sqrt{5}$ f. $-59\sqrt{3}-5\sqrt{2}$
9. a. $12\sqrt{14}$
   b. $24\sqrt{21}$
   c. $20+20\sqrt{2}$
   d. $42-63\sqrt{2}$
   e. $6\sqrt{5}-13$
   f. $8\sqrt{15}+12\sqrt{21}-4\sqrt{5}-6\sqrt{7}$
10. a. $8\sqrt{35}$ b. $160\sqrt{15}$ c. $12\sqrt{10}$
    d. 36 e. $42\sqrt{3}$ f. $62\sqrt{5}$
11. a. 4 b. −62 c. $11-4\sqrt{7}$
    d. $13+2\sqrt{22}$ e. −4 f. $57+12\sqrt{15}$
12. a. $24\sqrt{5}+30$
    b. $5\sqrt{15}-6\sqrt{7}-40\sqrt{10}+16\sqrt{42}$
    c. $98-40\sqrt{6}$
    d. 1
13. a. $3\sqrt{10}-14\sqrt{3}$
    b. $35\sqrt{3}+30\sqrt{2}-45$
    c. $-25\sqrt{10}-36$
    d. $6\sqrt{6}+8\sqrt{21}+3\sqrt{10}+4\sqrt{35}$
14. a. $17+12\sqrt{2}$ b. $66-36\sqrt{2}$ c. $12-2\sqrt{35}$
    d. 17 e. 155 f. $4+2\sqrt{6}$
15. a. $\frac{\sqrt{15}}{3}$ b. $\frac{\sqrt{6}}{3}$ c. $\frac{2\sqrt{3}+3\sqrt{30}}{6}$
    d. $\frac{3\sqrt{2}-2\sqrt{5}}{4}$ e. $\sqrt{5}+\sqrt{2}$ f. $3\sqrt{5}-2\sqrt{10}$
16. a. $\frac{\sqrt{6}}{4}$ b. $\frac{\sqrt{10}+2}{2}$ c. $\frac{\sqrt{6}-3}{6}$
    d. $\frac{\sqrt{6}-\sqrt{2}}{4}$ e. $\frac{25+11\sqrt{10}}{15}$ f. $5-\sqrt{6}$
17. a. i. $\frac{3\sqrt{2}}{7}$ ii. $\frac{\sqrt{15}+7\sqrt{5}}{2}$
    b. $6\sqrt{6}$
    c. $\frac{4\sqrt{3}-3\sqrt{2}}{3}$
    d. $\frac{6-\sqrt{3}}{264}$
18. $4\sqrt[3]{6}$
19. $10\sqrt{3}+\frac{\sqrt{5}}{5}+5$
20. a. $7+5\sqrt{2}$ b. $a=8, b=12$
21. a. $\frac{13-7\sqrt{3}}{2}$ b. 4, which is rational
22. a. $\frac{20\sqrt{5}-9\sqrt{6}}{6}$ b. $2\sqrt{5}+31$
    c. $\frac{19\sqrt{3}-18\sqrt{2}}{6}$ d. $\frac{14}{5}$
    e. $-\frac{\sqrt{14}}{252}$ f. $26-19\sqrt{3}-3\sqrt{6}$
23. a. i. $\frac{6\sqrt{3}-\sqrt{10}}{2}$ ii. −2
    b. i. $\frac{8\sqrt{7}}{3}$ ii. $\frac{11\sqrt{7}-28}{56}$
    c. i. $m=0, n=-\frac{1}{2}$ ii. $m=181, n=147$
    d. i. $x_2 = \frac{-b-\sqrt{b^2-4ac}}{2a}$
    ii. $x_1+x_2 = \frac{-b}{a}$
    iii. $x_1x_2 = \frac{c}{a}$
24. a. $AB = 3\sqrt{2}$, $BC = \sqrt{30-20\sqrt{2}}$ or $2\sqrt{5}-\sqrt{10}$, $AC = \sqrt{18-2\sqrt{5}}$
    b. $AB \approx 4.2$ units
25. a. $2\sqrt{6}+3\sqrt{3}+3\sqrt{2}+5\ \text{m}^2$
    b. $1.38 per square metre

2.6 Exam questions

Note: Mark allocations are available with the fully worked solutions online.

1. A
2. 6
3. E

2.7 Review

2.7 Exercise

Technology free: short answer

1. a. $25x^2 + 20xy + 16y^2 - 3$
   b. $2x^3 - 14x^2 + 22x - 31$
   c. $8x^3 + 12\sqrt{3}x^2 + 18x + 3\sqrt{3}$
   d. $16x^4 - 160x^3 + 600x^2 - 1000x + 625$
   e. $1 - \frac{10x}{3} + \frac{40x^2}{9} - \frac{80x^3}{27} + \frac{80x^4}{81} - \frac{32x^5}{243}$
   f. $x^2 + 2x + 1 - 2xy - 2y + y^2$
2. a. $(9x - 5y)(8x + 9y)$
   b. $x^2y(y - 6)(y + 6)$
   c. $(2y - x + 9)(2y + x - 9)$
   d. $(4x + 1)(16x^2 - 4x + 1)$
   e. $(4 - x)(7x^2 - 2x + 4)$
   f. $2(x + y)^3$
3. a. 210
   b. i. 1 5 10 10 5 1
   ii. $1 + 10xy + 40x^2y^2 + 80x^3y^3 + 80x^4y^4 + 32x^5y^5$
   c. $t_4 = -4320x^3$
4. a. $-\sqrt{3} - 6\sqrt{2}$
   b. 4
   c. $\frac{\sqrt{2}}{2} + 6$
   d. $-\frac{2\sqrt{3} + 3\sqrt{5}}{3}$
   e. $\frac{41\sqrt{5}}{30}$
   f. $2\sqrt{2}$
5. a. $\frac{x+3}{x-2}$
   b. $\frac{7x-3}{(x+3)(x-3)^2}$
   c. $\frac{-1}{x^2 - 3x + 9}$
   d. $\frac{37\sqrt{5} + 225}{165}$
   e. $\frac{7\sqrt{10} + 20}{3}$
   f. $\frac{5\sqrt{6} - 4\sqrt{3} - 2\sqrt{5} - 7\sqrt{10}}{7}$
6. a. $\left\{5\sqrt{2}, 4\sqrt{3}, 3\sqrt{5}, 2\sqrt{10}, 2\sqrt{6}\right\}$
   b. $\left\{7!, 4! \times 5!, (4!)^2, \frac{16!}{14!}\right\}$
   c. i. 0.3162
   ii. −4.162
   d. $21 - 4\sqrt{5}, 2\sqrt{5} - 1$

Technology active: multiple choice

7. A
8. D
9. C
10. A
11. D
12. E
13. C
14. C
15. B
16. D

Technology active: extended response

17. a. $x = \sqrt{3}, y = 3$
    b. $\left(-2 - \sqrt{2}, \infty\right)$
    c. $\frac{a^3}{1 - a^6}$
    d. i. $a^3 + b^3 + c^3 - 3abc$
    ii. Sample responses can be found in the worked solutions in the online resources.
    iii. 61 200
18. a. Many answers are possible, such as $\frac{3}{4}$.
    b. i. Suppose $a < b$.
    $\therefore a + a < a + b$ and $a + b < b + b$
    $\therefore 2a < a + b < 2b$
    $\therefore a < \frac{a+b}{2} < b$
    Therefore, the real number $\frac{a+b}{2}$ lies between a and b, showing it is always possible to find another real number between the two.
    ii. No
    c. 416
    d. $m = 3$; $n = 4$
19. a. $0.96\sqrt{2}$ km/h
    b. 4.8π minutes
    c. $640\left(2 - \sqrt{2}\right)$ metres
    d. i. $P\left(\sqrt{10} - \sqrt{2}, \sqrt{5} + \sqrt{2}\right)$,
    $Q\left(3\sqrt{2} - \sqrt{10}, 2\sqrt{3} - \sqrt{5} - \sqrt{2}\right)$,
    $R\left(\sqrt{10} - \sqrt{2}, \sqrt{5} - \sqrt{2}\right)$
    ii. $2\sqrt{2}$ km
20. a. −200
    b. i. Sample responses can be found in the worked solutions in the online resources.
    ii. $t_{12} = \binom{17}{11}x^6$ and $t_{13} = \binom{17}{12}x^5$
    iii. $n = 2, 5, 8, 11, 14$

2.7 Exam questions

Note: Mark allocations are available with the fully worked solutions online.

1. C
2. $\frac{4}{p + 3q}$
3. C
4. $\{x : -3 < x \leq 1\}$

   −3 −2 −1 0 1 2 x
5. B

3 Quadratic relationships

LEARNING SEQUENCE

Fully worked solutions for this topic are available online.

3.1 Overview

Hey students! Bring these pages to life online

Watch videos

Engage with interactivities

Answer questions and check results

Find all this and MORE in jacPLUS

3.1.1 Introduction

Watching water cascade from a fountain or a ball bouncing along a pathway may be the first encounter a young child has with a parabola. The arches of water rise and fall along parabolas; the ball follows a series of parabolas, each with a maximum turning point lower than that of the previous one. It was Galileo who showed that the path of a projectile is parabolic.

In later life, that child may be lucky enough to catch a glimpse in the night sky of the return of Halley's Comet as its nearly parabolic path passes closest to Earth in 2061. Isaac Newton, having studied Halley's Comet in 1680, initially theorised that the path of the comet was consistent with that of a parabola. This was later corrected to be an ellipse, the same shape as orbits of the planets around the Sun. Nevertheless, the paths of many other comets are parabolic. The graph of a parabola is not closed; the 'arms' of the parabola go on forever, so a comet travelling on a parabolic path would never return unless other factors affected its path. Halley's Comet does return about every 75 years.

The parabola is one of the more sturdy and powerful shapes and is therefore often favoured by engineers and architects. Many bridges have parabolic arches, including Sydney Harbour Bridge, the pedestrian bridge over the Yarra River at Southgate, San Francisco's Golden Gate Bridge and Glasgow's Millennium Bridge.

KEY CONCEPTS

This topic covers the following key concepts from the VCE Mathematics Study Design:

- graphs of power functions $f(x) = x^n$ for $n \in \left\{-2, -1, \frac{1}{3}, \frac{1}{2}, 1, 2, 3, 4\right\}$, and transformations of these graphs to the form $y = a(x+b)^n + c$ where $a, b, c \in R$ and $a \neq 0$
- graphs of polynomial functions of low degree, and interpretation of key features of these graphs
- recognition of equivalent expressions and simplification of algebraic expressions involving different forms of polynomial and power functions, the use of distributive and exponent laws applied to these functions, and manipulation from one form of expression to an equivalent form
- use of parameters to represent families of functions and determination of rules of simple functions and relations from given information
- solution of polynomial equations of low degree, numerically, graphically and algebraically, including numerical approximation of roots of simple polynomial functions using the bisection method algorithm.

Note: *Concepts shown in grey are covered in other topics.*

3.2 Quadratic equations with rational roots

LEARNING INTENTION

At the end of this subtopic you should be able to:
- solve quadratic equations using the Null Factor Law (including perfect squares).

Expressions such as $2x^2 + 3x + 4$ and $x^2 - 9$ are in quadratic form. Quadratic expansions always have an x^2 term as the highest power of x. An expression such as $x^3 + 2x^2 + 3x + 4$ is not quadratic; it is called a cubic due to the presence of the x^3 term.

3.2.1 Quadratic equations and the Null Factor Law

The general quadratic equation can be written as $ax^2 + bx + c = 0$, where a, b, c are real constants and $a \neq 0$. If the quadratic expression on the left-hand side of this equation can be factorised, the solutions to the quadratic equation may be obtained using the **Null Factor Law**.

Null Factor Law

The Null Factor Law states that, for any a and b, if the product $ab = 0$, then $a = 0$ or $b = 0$ or both $a = 0$ and $b = 0$.

Applying the Null Factor Law to a quadratic equation expressed in the factorised form as:

$$\begin{aligned} (x-\alpha)(x-\beta) &= 0 \\ \text{then } (x-\alpha) &= 0 \text{ or } (x-\beta) = 0 \\ \therefore x &= \alpha \text{ or } x = \beta \end{aligned}$$

Roots, zeros and factors

The solutions of an equation are also called the **roots** of the equation or the **zeros** of the expression. (This terminology applies to all algebraic equations, not just quadratic equations.) The quadratic equation $(x-1)(x-2) = 0$ has roots $x = 1$ and $x = 2$. These solutions are the zeros of the quadratic expression $(x-1)(x-2)$, since substituting either of $x = 1$ or $x = 2$ in the quadratic expression $(x-1)(x-2)$ makes the expression equal zero.

As a converse of the Null Factor Law, it follows that if the roots of a quadratic equation, or the zeros of a quadratic, are $x = \alpha$ and $x = \beta$, then $(x-\alpha)$ and $(x-\beta)$ are linear factors of the quadratic. The quadratic would be of the form $(x-\alpha)(x-\beta)$ or any multiple of this form, $a(x-\alpha)(x-\beta)$.

WORKED EXAMPLE 1 Using the Null Factor Law to solve quadratic equations

a. Solve the equation $5x^2 - 18x = 8$.
b. Given that $x = 2$ and $x = -2$ are zeros of a quadratic, form its linear factors and expand the product of these factors.

THINK	WRITE
a. 1. Rearrange the given equation to make one side of the equation equal zero.	a. $5x^2 - 18x = 8$ Rearrange: $5x^2 - 18x - 8 = 0$
2. Factorise the quadratic trinomial.	$(5x+2)(x-4) = 0$
3. Apply the Null Factor Law.	$5x + 2 = 0$ or $x - 4 = 0$

4. Solve these linear equations for x.	$5x = -2$ or $x = 4$ $x = -\frac{2}{5}$ or $x = 4$
b. 1. Use the converse of the Null Factor Law.	**b.** Since $x = 2$ is a zero, then $(x - 2)$ is a linear factor, and since $x = -2$ is a zero, then $(x - (-2)) = (x + 2)$ is a linear factor. Therefore, the quadratic has the linear factors $(x - 2)$ and $(x + 2)$.
2. Expand the product of the two linear factors.	The product $= (x - 2)(x + 2)$. Expanding, $(x - 2)(x + 2) = x^2 - 4$ The quadratic has the form $x^2 - 4$ or any multiple of this form, $a(x^2 - 4)$.

 Resources

 **Interactivity** Roots, zeros and factors (int-2557)

3.2.2 Using the perfect square form of a quadratic

As an alternative to solving a quadratic equation by using the Null Factor Law, if the quadratic is a perfect square, solutions to the equation can be found by taking square roots of both sides of the equation. A simple illustration is as follows.

Square root method		**Null Factor Law method**
$x^2 = 9$ $x = \pm\sqrt{9}$ $= \pm 3$	or	$x^2 = 9$ $x^2 - 9 = 0$ $(x - 3)(x + 3) = 0$ $x = \pm 3$

If the square root method is used, both the positive and negative square roots must be considered.

WORKED EXAMPLE 2 Solving quadratics using perfect squares

Solve the equation $(2x + 3)^2 - 25 = 0$.

THINK	WRITE
1. Rearrange so that each side of the given equation contains a perfect square.	$(2x + 3)^2 - 25 = 0$ $(2x + 3)^2 = 25$
2. Take the square roots of both sides.	$2x + 3 = \pm 5$
3. Separate the linear equations and solve.	$2x + 3 = 5$ or $2x + 3 = -5$ $2x = 2$ $2x = -8$ $x = 1$ or $x = -4$

4. An alternative method uses the Null Factor Law.

Alternatively:

$$(2x+3)^2 - 25 = 0$$

Factorise:

$$((2x+3)-5)((2x+3)+5) = 0$$
$$(2x-2)(2x+8) = 0$$
$$2x = 2 \quad \text{or} \quad 2x = -8$$
$$\therefore x = 1 \quad \text{or} \quad x = -4$$

 Resources

 **Interactivity** Perfect square form of a quadratic (int-2558)

3.2.3 Equations that reduce to quadratic form

Substitution techniques can be applied to reduce equations such as $ax^4 + bx^2 + c = 0$ to quadratic form.

The equation $ax^4 + bx^2 + c = 0$ can be expressed in the form $a(x^2)^2 + bx^2 + c = 0$. Letting $u = x^2$, this becomes $au^2 + bu + c = 0$, a quadratic equation in variable u.

By solving the quadratic equation for u, then substituting back x^2 for u, any possible solutions for x can be obtained. Since x^2 cannot be negative, it would be necessary to reject negative u values, since $x^2 = u$ would have no real solutions.

WORKED EXAMPLE 3 Solving equations by substitution

Solve the equation $4x^4 - 35x^2 - 9 = 0$.

THINK	WRITE
1. Use an appropriate substitution to reduce the given equation to quadratic form.	$4x^4 - 35x^2 - 9 = 0$ Let $u = x^2$. $4u^2 - 35u - 9 = 0$
2. Solve for u by factorising and applying the Null Factor Law.	$(4u+1)(u-9) = 0$ $\therefore u = -\frac{1}{4}$ or $u = 9$
3. Substitute back, replacing u with x^2.	$x^2 = -\frac{1}{4}$ or $x^2 = 9$
4. Since x^2 cannot be negative, any negative value of u needs to be rejected.	Reject $x^2 = -\frac{1}{4}$, since there are no real solutions.
5. Solve the remaining equation for x.	$x^2 = 9$ $x = \pm\sqrt{9}$ $x = \pm 3$

TI | THINK

1. On a Calculator page, press MENU and select:
3. Algebra
1. Solve
Complete the entry line as:
solve
$(4x^4 - 35x^2 - 9 = 0, x)$
Then press ENTER.

DISPLAY/WRITE

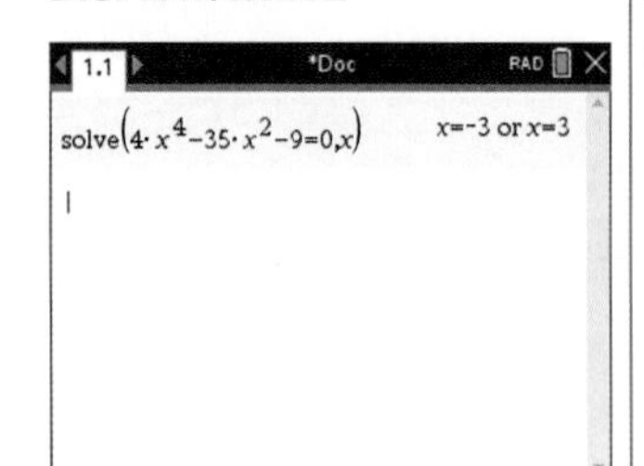

2. The answer appears on the screen.

$x = -3,\ x = 3$

CASIO | THINK

1. On a Main screen, complete the entry line as:
solve
$(4x^4 - 35x^2 - 9 = 0, x)$
Then press EXE.

DISPLAY/WRITE

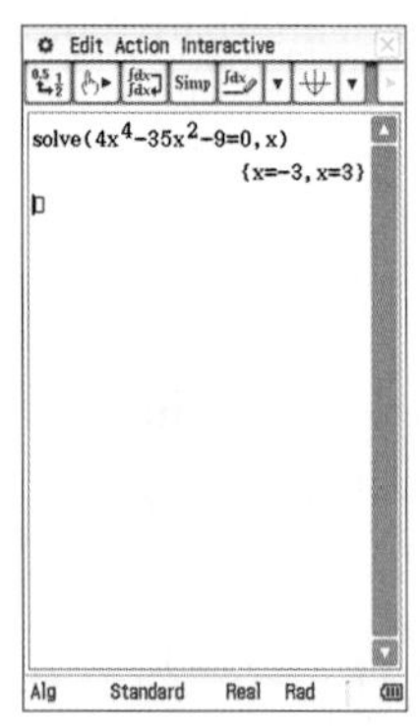

2. The answer appears on the screen.

$x = -3,\ x = 3$

3.2 Exercise

Technology free

1. Solve each of the given equations using the Null Factor Law.
   a. $3x(5 - x) = 0$
   b. $(3 - x)(7x - 1) = 0$
   c. $(x + 8)^2 = 0$
   d. $2(x + 4)(6 + x) = 0$
2. Use the Null Factor Law to solve the following quadratic equations for x.
   a. $(3x - 4)(2x + 1) = 0$
   b. $x^2 - 7x + 12 = 0$
   c. $8x^2 + 26x + 21 = 0$
   d. $10x^2 = 2x$
   e. $12x^2 + 40x - 32 = 0$
   f. $\frac{1}{2}x^2 - 5x = 0$
3. **WE1** a. Solve the equation $10x^2 + 23x = 21$.
   b. Given that $x = -5$ and $x = 0$ are zeros of a quadratic, form its linear factors and expand the product of these factors.
4. Solve the following quadratic equations.
   a. $6x^2 + 5x + 1 = 0$
   b. $12x^2 - 7x = 10$
   c. $49 = 14x - x^2$
   d. $5x + 25 - 30x^2 = 0$
5. **WE2** Solve the equation $(5x - 1)^2 - 16 = 0$.
6. Solve the following quadratic equations for x.
   a. $(x + 2)^2 = 9$
   b. $(x - 1)^2 - 25 = 0$
   c. $(x - 7)^2 + 4 = 0$
   d. $(2x + 11)^2 = 81$
   e. $(7 - x)^2 = 0$
   f. $8 - \frac{1}{2}(x - 4)^2 = 0$
7. Obtain the solutions to the following equations.
   a. $x^2 = 121$
   b. $9x^2 = 16$
   c. $(x - 5)^2 = 1$
   d. $(5 - 2x)^2 - 49 = 0$
   e. $2(3x - 1)^2 - 8 = 0$
   f. $(x^2 + 1)^2 = 100$

8. WE3 Solve the equation $9x^4 + 17x^2 - 2 = 0$.

9. Determine the roots (solutions) of the following equations

a. $18(x-3)^2 + 9(x-3) - 2 = 0$

b. $5(x+2)^2 + 23(x+2) + 12 = 0$

c. $x + 6 + \frac{8}{x} = 0$

d. $2x + \frac{3}{x} = 7$

10. Use a substitution technique to solve the following equations.

a. $(3x+4)^2 + 9(3x+4) - 10 = 0$

b. $2(1+2x)^2 + 9(1+2x) = 18$

c. $x^4 - 29x^2 + 100 = 0$

d. $2x^4 = 31x^2 + 16$

11. Express each of the following equations in quadratic form and hence solve the equations for x.

a. $x(x-7) = 8$

b. $4x(3x-16) = 3(4x-33)$

c. $(x+4)^2 + 2x = 0$

d. $(2x+5)(2x-5) + 25 = 2x$

12. Express each of the following equations in quadratic form and hence solve the equations for x.

a. $2 - 3x = \frac{1}{3x}$

b. $\frac{4x+5}{x+125} = \frac{5}{x}$

c. $7x - \frac{2}{x} + \frac{11}{5} = 0$

d. $\frac{12}{x+1} - \frac{14}{x-2} = 19$

13. Obtain the solutions to the following equations.

a. $x^4 = 81$

b. $(9x^2 - 16)^2 = 20(9x^2 - 16)$

14. Solve the following equations.

a. $\left(x - \frac{2}{x}\right)^2 - 2\left(x - \frac{2}{x}\right) + 1 = 0$

b. $2\left(1 + \frac{3}{x}\right)^2 + 5\left(1 + \frac{3}{x}\right) + 3 = 0$

c. $\left(x + \frac{1}{x}\right)^2 - 4\left(x + \frac{1}{x}\right) + 4 = 0.$

Technology active

15. Solve the equation $(px+q)^2 = r^2$ for x in terms of p, q and r, $r > 0$.

16. Express the value of x in terms of the positive real numbers a and b.

a. $(x-2b)(x+3a) = 0$

b. $2x^2 - 13ax + 15a^2 = 0$

c. $(x-a-b)^2 = 4b^2$

d. $(x+a)^2 - 3b(x+a) + 2b^2 = 0$

17. Consider the quadratic equation $(x-\alpha)(x-\beta) = 0$.

a. If the roots of the equation are $x = 1$ and $x = 7$, form the equation.
b. If the roots of the equation are $x = -5$ and $x = 4$, form the equation.
c. If the roots of the equation are $x = 0$ and $x = 10$, form the equation.
d. If the only root of the equation is $x = 2$, form the equation.

18. If the zeros of the quadratic expression $4x^2 + bx + c$ are $x = -4$ and $x = \frac{3}{4}$, find the values of the integer constants b and c.

In questions 19 and 20, use CAS technology to solve the equations.

19. Solve the following.

a. $60x^2 + 113x - 63 = 0$
b. $4x(x-7) + 8(x-3)^2 = x - 26$

20. a. Find the roots of the equation $32x^2 - 96x + 72 = 0$.
b. Solve $44 + 44x^2 = 250x$.

3.2 Exam questions

Question 1 (1 mark) TECH-ACTIVE

MC The factorised form of $12a^2 - a - 20$ is

A. $(4a-5)(3a-4)$
B. $(4a+5)(3a-4)$
C. $(4a-5)(3a+4)$
D. $(12a+5)(a-4)$
E. $(12a-4)(a+5)$

Question 2 (1 mark) TECH-ACTIVE

MC The solution(s) to $3x^2 - 243 = 0$ is/are

A. -81
B. $+81$
C. 9
D. ± 9
E. -9

Question 3 (3 marks) TECH-FREE

Solve the equation for a.

$2(a-1)^2 - 7(a-1) + 6 = 0$

More exam questions are available online.

3.3 Quadratics over *R*

LEARNING INTENTION

At the end of this subtopic you should be able to:

- solve quadratic equations using the quadratic formula
- use the discriminant to determine the type and number of solutions of a quadratic equation
- express $ax^2 + bx + c$ in completed square form where $a, b, c \in Z$ and $a \neq 0$, by hand.

When $x^2 - 4$ is expressed as $(x-2)(x+2)$, it has been factorised over Q, as both of the zeros are rational numbers. However, over Q, the quadratic expression $x^2 - 3$ cannot be factorised into linear factors. Surds need to be permitted for such an expression to be factorised.

3.3.1 Factorisation over *R*

The quadratic $x^2 - 3$ can be expressed as the difference of two squares, $x^2 - 3 = x^2 - \left(\sqrt{3}\right)^2$, using surds. This can be factorised over R because it allows the factors to contain surds.

$$\begin{aligned} x^2 - 3 &= x^2 - \left(\sqrt{3}\right)^2 \\ &= \left(x - \sqrt{3}\right)\left(x + \sqrt{3}\right) \end{aligned}$$

If a quadratic can be expressed as the difference of two squares, then it can be factorised over R. To express a quadratic trinomial as a difference of two squares, a technique called '**completing the square**' is used.

'Completing the square' technique

Expressions of the form $x^2 \pm px + \left(\frac{p}{2}\right)^2 = \left(x \pm \frac{p}{2}\right)^2$ are perfect squares; for example, $x^2 + 4x + 4 = (x+2)^2$.

To illustrate the 'completing the square' technique, consider the quadratic trinomial $x^2 + 4x + 1$.

If 4 is added to the first two terms $x^2 + 4x$, then this will form a perfect square $x^2 + 4x + 4$. However, 4 must also be subtracted in order not to alter the value of the expression.

$$x^2 + 4x + 1 = x^2 + 4x + 4 - 4 + 1$$

Grouping the first three terms together to form the perfect square and evaluating the last two terms,

$$= \left(x^2 + 4x + 4\right) - 4 + 1$$
$$= (x+2)^2 - 3$$

By writing this difference of two squares form using surds, factors over R can be found.

$$= (x+2)^2 - \left(\sqrt{3}\right)^2$$
$$= \left(x+2-\sqrt{3}\right)\left(x+2+\sqrt{3}\right)$$
$$\text{Thus, } x^2 + 4x + 1 = \left(x+2-\sqrt{3}\right)\left(x+2+\sqrt{3}\right)$$

Completing the square is the method used to factorise **monic** quadratics over R. A monic quadratic is one for which the coefficient of x^2 equals 1.

For a monic quadratic, to complete the square, add and then subtract the square of half the coefficient of x. This squaring will always produce a positive number regardless of the sign of the coefficient of x.

Completing the square

$$x^2 \pm px = \left[x^2 \pm px + \left(\frac{p}{2}\right)^2\right] - \left(\frac{p}{2}\right)^2$$
$$= \left(x \pm \frac{p}{2}\right)^2 - \left(\frac{p}{2}\right)^2$$

To complete the square on $ax^2 + bx + c$, the quadratic should first be written as $a\left(x^2 + \frac{bx}{a} + \frac{c}{a}\right)$ and the technique applied to the monic quadratic in the bracket. The common factor a is carried down through all the steps in the working.

WORKED EXAMPLE 4 Completing the square

a. Complete the square in the following.

i. $x^2 + 8x - 2$ **ii.** $3x^2 - 12x + 1$

b. Factorise the following over R.

i. $x^2 - 14x - 3$ **ii.** $2x^2 + 7x + 4$ **iii.** $4x^2 - 11$

THINK	WRITE
a. i. 1. Add and subtract the square of half the coefficient of x.	a. i. $x^2 + 8x - 2$ $= x^2 + 8x + 4^2 - 4^2 - 2$
2. Group the first three terms together to form a perfect square and evaluate the last two terms.	$= \left(x^2 + 8x + 16\right) - 16 - 2$ $= (x+4)^2 - 18$
ii. 1. Create a monic quadratic by taking the coefficient x^2 of out as a common factor. It may create fractions.	ii. $3x^2 - 12x + 1$ $= 3\left(x^2 - 4x + \frac{1}{3}\right)$

2. Add and subtract the square of half the coefficient of x.

$$= 3\left(x^2 - 4x + 2^2 - 2^2 + \frac{1}{3}\right)$$

$$= 3\left(\left(x^2 - 4x + 4\right) - 4 + \frac{1}{3}\right)$$

3. Group the first three terms together to form a perfect square and evaluate the last two terms.

$$= 3\left((x-2)^2 - \frac{11}{3}\right)$$

or

$$= 3(x-2)^2 - 11$$

b. i. 1. Add and subtract the square of half the coefficient of x.
Note: The negative sign of the coefficient of x becomes positive when squared.

b. i. $x^2 - 14x - 3$

$$= x^2 - 14x + 7^2 - 7^2 - 3$$

2. Group the first three terms together to form a perfect square and evaluate the last two terms.

$$= \left(x^2 - 14x + 49\right) - 49 - 3$$

$$= (x-7)^2 - 52$$

3. Factorise the difference of two squares expression.

$$= (x-7)^2 - \left(\sqrt{52}\right)^2$$

$$= \left(x - 7 - \sqrt{52}\right)\left(x - 7 + \sqrt{52}\right)$$

4. Express any surds in their simplest form.

$$= \left(x - 7 - 2\sqrt{13}\right)\left(x - 7 + 2\sqrt{13}\right)$$

5. State the answer.

Therefore, $x^2 - 14x - 3$

$$= \left(x - 7 - 2\sqrt{13}\right)\left(x - 7 + 2\sqrt{13}\right)$$

ii. 1. First create a monic quadratic by taking the coefficient of x^2 out as a common factor. This may create fractions.

ii. $2x^2 + 7x + 4$

$$= 2\left(x^2 + \frac{7}{2}x + 2\right)$$

2. Add and subtract the square of half the coefficient of x for the monic quadratic expression.

$$= 2\left(x^2 + \frac{7}{2}x + \left(\frac{7}{4}\right)^2 - \left(\frac{7}{4}\right)^2 + 2\right)$$

3. Within the bracket, group the first three terms together and evaluate the remaining terms.

$$= 2\left[\left(x^2 + \frac{7}{2}x + \frac{49}{16}\right) - \frac{49}{16} + 2\right]$$

$$= 2\left[\left(x + \frac{7}{4}\right)^2 - \frac{49}{16} + 2\right]$$

$$= 2\left[\left(x + \frac{7}{4}\right)^2 - \frac{49}{16} + \frac{32}{16}\right]$$

$$= 2\left[\left(x + \frac{7}{4}\right)^2 - \frac{17}{16}\right]$$

4. Factorise the difference of two squares that has been formed.

$$= 2\left[\left(x + \frac{7}{4}\right) - \sqrt{\frac{17}{16}}\right]\left[\left(x + \frac{7}{4}\right) + \sqrt{\frac{17}{16}}\right]$$

$$= 2\left(x + \frac{7}{4} - \frac{\sqrt{17}}{4}\right)\left(x + \frac{7}{4} + \frac{\sqrt{17}}{4}\right)$$

5. State the answer.

$$2x^2 + 7x + 4$$
$$= 2\left(x + \frac{7}{4} - \frac{\sqrt{17}}{4}\right)\left(x + \frac{7}{4} + \frac{\sqrt{17}}{4}\right)$$
$$= 2\left(x + \frac{7 - \sqrt{17}}{4}\right)\left(x + \frac{7 + \sqrt{17}}{4}\right)$$

iii. The quadratic is a difference of two squares. Factorise it.

iii. $4x^2 - 11$
$$= (2x)^2 - \left(\sqrt{11}\right)^2$$
$$= \left(2x - \sqrt{11}\right)\left(2x + \sqrt{11}\right)$$

TI | THINK	DISPLAY/WRITE	CASIO | THINK	DISPLAY/WRITE
a. ii. 1. On a Calculator page, press MENU and select: 3. Algebra 5. Complete the square. Complete the entry line as: $(3x^2 - 12x + 1,\ x)$ Then press ENTER.	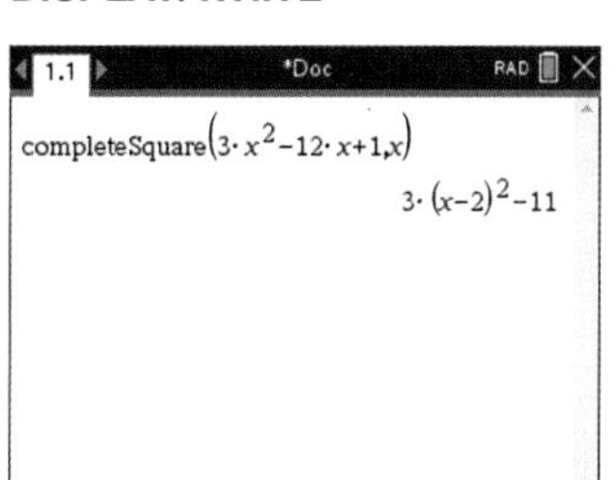	a. ii. 1. On a Conics page, complete the entry line as: $y = 3x^2 - 12x + 1$ Select: • Fit • Fit into Conics Form • $y = A(x - H)^2 + k$	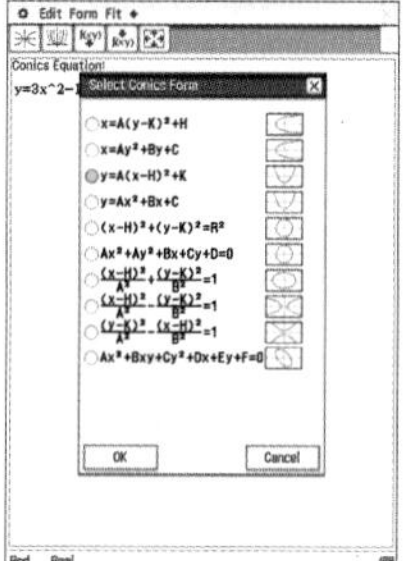 
2. The answer appears on the screen.	$3(x - 2)^2 - 11$	2. The answer appears on the screen.	$3(x - 2)^2 - 11$
b. i. 1. On a Calculator page, press MENU and select: 3. Algebra 2. Factor Complete the entry line as: factor $(x^2 - 14x - 3, x)$ Then press ENTER.	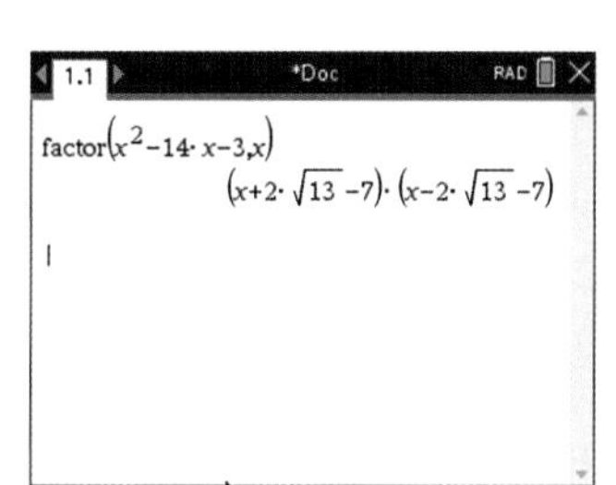	b. i. 1. On a Main screen, select: • Action • Transformation • Factor • rFactor Complete the entry line as: rFactor $(x^2 - 14x - 3)$ Then press EXE. *Note*: If the factors are irrational, rFactor must be used to find the factors.	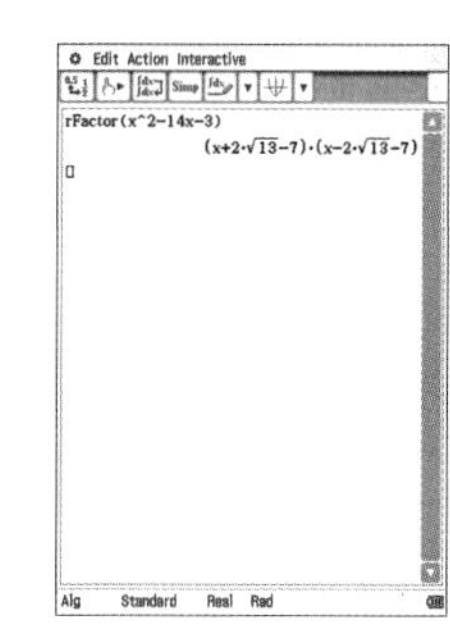
2. The answer appears on the screen.	$\left(x + 2\sqrt{13} - 7\right)\left(x - 2\sqrt{13} - 7\right)$	2. The answer appears on the screen.	$\left(x + 2\sqrt{13} - 7\right)$ $\left(x - 2\sqrt{13} - 7\right)$

on Resources

Interactivity Completing the square (int-2559)

3.3.2 Quadratic equations with real roots

The choices of method to consider for solving the quadratic equation $ax^2 + bx + c = 0$ are:

- factorise over Q and use the Null Factor Law
- factorise over R by completing the square and use the Null Factor Law
- use the formula $x = \dfrac{-b \pm \sqrt{b^2 - 4ac}}{2a}$.

The quadratic formula

The **quadratic formula** is used for solving quadratic equations and is obtained by completing the square on the left-hand side of the equation $ax^2 + bx + c = 0$.

Using completing the square, it can be shown that

$$
\begin{aligned}
ax^2 + bx + c &= a\left(\left(x + \frac{b}{2a}\right)^2 - \frac{b^2 - 4ac}{4a^2}\right) \\
ax^2 + bx + c &= 0 \\
a\left(\left(x + \frac{b}{2a}\right)^2 - \frac{b^2 - 4ac}{4a^2}\right) &= 0 \\
\left(x + \frac{b}{2a}\right)^2 - \frac{b^2 - 4ac}{4a^2} &= 0 \\
\left(x + \frac{b}{2a}\right)^2 &= \frac{b^2 - 4ac}{4a^2} \\
x + \frac{b}{2a} &= \pm\sqrt{\frac{b^2 - 4ac}{4a^2}} \\
x &= -\frac{b}{2a} \pm \frac{\sqrt{b^2 - 4ac}}{2a} \\
&= \frac{-b \pm \sqrt{b^2 - 4ac}}{2a}
\end{aligned}
$$

The quadratic formula

The solutions of the quadratic equation $ax^2 + bx + c = 0$ are

$$x = \frac{-b \pm \sqrt{b^2 - 4ac}}{2a}$$

Using the quadratic formula is less tedious than completing the square. Although the formula can also be used to solve a quadratic equation that factorises over Q, factorisation is usually simpler, making it the preferred method.

WORKED EXAMPLE 5 Solving using the quadratic formula

Use the quadratic formula to solve the equation $x(9-5x)=3$.

THINK	WRITE
1. The given equation needs to be expressed in the general quadratic form, $ax^2+bx+c=0$.	$x(9-5x)=3$ $9x-5x^2=3$ $5x^2-9x+3=0$
2. State the values of a, b and c.	$a=5, b=-9, c=3$
3. State the formula for solving a quadratic equation.	$x=\dfrac{-b\pm\sqrt{b^2-4ac}}{2a}$
4. Substitute the a, b, c values and evaluate.	$=\dfrac{-(-9)\pm\sqrt{(-9)^2-4(5)(3)}}{2\times(5)}$ $=\dfrac{9\pm\sqrt{81-60}}{10}$ $=\dfrac{9\pm\sqrt{21}}{10}$
5. Express the roots in simplest surd form and state the answers. *Note:* If the question asked for answers correct to 2 decimal places, you would use your calculator to find approximate answers of $x\simeq0.44$ and $x\simeq1.36$. Otherwise, do *not* approximate answers.	The solutions are $x=\dfrac{9-\sqrt{21}}{10}$, $x=\dfrac{9+\sqrt{21}}{10}$.

on Resources

 **Interactivity** The quadratic formula (int-2561)

The discriminant

The term b^2-4ac in the quadratic formula is called the **discriminant**. It is denoted by the Greek letter delta, Δ.

The discriminant

$$\boldsymbol{\Delta = b^2-4ac}$$

in the quadratic equation ax^2+bx+c.

WORKED EXAMPLE 6 Calculating the discriminant

For each of the following quadratics, calculate the discriminant.

a. $2x^2 + 15x + 13$ b. $5x^2 - 6x + 9$ c. $-3x^2 + 3x + 8$

THINK	WRITE
a. 1. State the values of a, b and c needed to calculate the discriminant.	a. $2x^2 + 15x + 13, a = 2, b = 15, c = 13$
2. State the formula for the discriminant.	$\Delta = b^2 - 4ac$
3. Substitute the values of a, b and c.	$\therefore \Delta = (15)^2 - 4(2)(13)$ $= 225 - 104$ $= 121$
b. 1. State the values of a, b and c, and calculate the discriminant.	b. $5x^2 - 6x + 9, a = 5, b = -6, c = 9$ $\Delta = b^2 - 4ac$ $= (-6)^2 - 4(5)(9)$ $= 36 - 180$ $= -144$
c. 1. State the values of a, b and c, and calculate the discriminant.	c. $-3x^2 + 3x + 8, a = -3, b = 3, c = 8$ $\Delta = b^2 - 4ac$ $= (3)^2 - 4(-3)(8)$ $= 9 + 96$ $= 105$

Resources

Interactivity The discriminant (int-2560)

3.3.3 The role of the discriminant in quadratic equations

The type of factors determines the type of solutions to an equation, so it is no surprise that the discriminant determines the number and type of solutions as well as the number and type of factors.

The formula for the solution to the quadratic equation $ax^2 + bx + c = 0$ can be expressed as $x = \dfrac{-b \pm \sqrt{\Delta}}{2a}$, where the discriminant $\Delta = b^2 - 4ac$.

The role of the discriminant

- **If $\Delta < 0$, there are no real solutions to the equation.**
- **If $\Delta = 0$, there is one real solution (or two equal solutions) to the equation.**
- **If $\Delta > 0$, there are two distinct real solutions to the equation.**

For $a, b, c \in Q$, the results are as follows:

- If $\Delta > 0$ and is a perfect square, the roots are **rational**, the factors are rational, and the quadratic factorises over Q.
- If $\Delta > 0$ and not a perfect square, the roots are **irrational**, the factors contain surds, and the quadratic factorises over R.
- If $\Delta = 0$, the quadratic factorises to a perfect square giving one distinct real solution.
- If $\Delta < 0$, the quadratic has no solutions and cannot be factorised over R.

WORKED EXAMPLE 7 Type and number of roots from discriminants

a. Use the discriminant to determine the number and type of roots to the equation:

i. $15x^2 + 8x - 5 = 0$ **ii. $3x^2 + 8 = 4x$**

b. Find the values of k so the equation $x^2 + kx - k + 8 = 0$ will have one real solution and check the answer.

THINK	WRITE
a. i. 1. Identify the values of a, b and c from the general form $ax^2 + bx + c = 0$.	a. $15x^2 + 8 - 5 = 0, a = 15, b = 8\ c = -5$
2. State the formula for the discriminant.	$\Delta = b^2 - 4ac$
3. Substitute the values of a, b and c, and evaluate.	$= (8)^2 - 4(15)(-5)$ $= 64 + 300$ $= 364$
4. Interpret the result.	Since the discriminant is positive but not a perfect square, the equation has two irrational roots.
ii. 1. Rewrite in the quadratic form $ax^2 + bx + c = 0$.	$3x^2 + 8 = 4x$ $3x^2 - 4x + 8 = 0$
2. Identify the values of a, b and c.	$a = 3, b = -4, c = 8$
3. State the formula for the discriminant, substitute and evaluate.	$\Delta = b^2 - 4ac$ $= (-4)^2 - 4 \times 3 \times 8$ $= 16 - 96$ $= -80$
4. Interpret the result.	Since the discriminant is negative, the equation has no real roots.
b. 1. Express the equation in general form and identify the values of a, b and c.	b. $x^2 + kx - k + 8 = 0$ $\therefore x^2 + kx + (-k + 8) = 0$ $a = 1, b = k, c = (-k + 8)$
2. Substitute the values of a, b and c, and obtain an algebraic expression for the discriminant.	$\Delta = b^2 - 4ac$ $= (k)^2 - 4(1)(-k + 8)$ $= k^2 + 4k - 32$
3. State the condition on the discriminant for the equation to have one solution.	For one solution, $\Delta = 0$.
4. Solve for k.	$k^2 + 4k - 32 = 0$ $(k + 8)(k - 4) = 0$ $k = -8, k = 4$
5. Check the solutions of the equation for each value of k.	If $k = -8$, the original equation becomes: $x^2 - 8x + 16 = 0$ $(x - 4)^2 = 0$ $\therefore x = 4$ This equation has one solution. If $k = 4$, the original equation becomes: $x^2 - 4x + 4 = 0$ $(x + 2)^2 = 0$ $\therefore x = -2$ This equation has one solution.

3.3.4 Quadratic equations with rational and irrational coefficients

From the converse of the Null Factor Law, if the roots of a quadratic equation are $x=\alpha$ and $x=\beta$, then the factorised form of the quadratic equation would be $a(x-\alpha)(x-\beta)=0, a\in R$. The roots could be rational or irrational. If both roots are rational, expanding the factors will always give a quadratic with rational coefficients. However, if the roots are irrational, not all of the coefficients of the quadratic may be rational.

For example, if the roots of a quadratic equation are $x=\sqrt{2}$ and $x=3\sqrt{2}$, then the factorised form of the equation is $\left(x-\sqrt{2}\right)\left(x-3\sqrt{2}\right)=0$. This expands to $x^2-4\sqrt{2}x+6=0$, where the coefficient of x is irrational.

However, if the roots were a pair of conjugate surds such as $x=\sqrt{2}$ and $x=-\sqrt{2}$, then the equation is $\left(x-\sqrt{2}\right)\left(x+\sqrt{2}\right)=0$. This expands to $x^2-2=0$, where all coefficients are rational.

The formula gives the roots of the general equation $ax^2+bx+c=0, a, b, c\in Q$ as

$$x_1=\frac{-b-\sqrt{\Delta}}{2a} \text{ and } x_2=\frac{-b+\sqrt{\Delta}}{2a}.$$

If $\Delta>0$ but not a perfect square, these roots are irrational, and x_1 and x_2 are a pair of conjugate surds.

- If a quadratic equation with rational coefficients has irrational roots, the roots must occur in conjugate surd pairs.
- The conjugate surd pairs are of the form $m-n\sqrt{p}$ and $m+n\sqrt{p}$ with $m, n, p\in Q$ and $p>0$.

WORKED EXAMPLE 8 Solutions involving conjugate pairs

a. Use completing the square to solve the quadratic equation $x^2+8\sqrt{2}x-17=0$. State whether the solutions are a pair of conjugate surds.

b. One root of the quadratic equation with rational coefficients $x^2+bx+c=0$, $b, c\in Q$ is $x=6+\sqrt{3}$. State the other root and calculate the values of b and c.

THINK	WRITE
a. 1. Add and also subtract the square of half the coefficient of x.	a. $x^2+8\sqrt{2}x-17=0$ $x^2+8\sqrt{2}x+\left(4\sqrt{2}\right)^2-\left(4\sqrt{2}\right)^2-17=0$
2. Group the first three terms of the left-hand side of the equation together to form a perfect square and evaluate the last two terms.	$x^2+8\sqrt{2}x+\left(4\sqrt{2}\right)^2-32-17=0$ $\left(x+4\sqrt{2}\right)^2-49=0$
3. Rearrange so that each side of the equation contains a perfect square.	$\left(x+4\sqrt{2}\right)^2=49$
4. Take the square roots of both sides.	$x+4\sqrt{2}=\pm\sqrt{49}$ $x+4\sqrt{2}=\pm7$
5. Solve for x.	$x=\pm7-4\sqrt{2}$ $\therefore x=7-4\sqrt{2}, x=-7-4\sqrt{2}$

6. State whether or not the solutions are conjugate surds.	The solutions are not a pair of conjugate surds. The conjugate of $7-4\sqrt{2}$ is $7+4\sqrt{2}$, not $-7-4\sqrt{2}$.
b. 1. State the conjugate surd root.	**b.** $x^2+bx+c=0, b, c \in Q$ Since the equation has rational coefficients, the roots occur in conjugate surd pairs. Therefore, the other root is $x=6-\sqrt{3}$.
2. Form the two linear factors of the equation in factorised form.	$x=6-\sqrt{3} \Rightarrow \left(x-\left(6-\sqrt{3}\right)\right)$ is a factor and $x=6+\sqrt{3} \Rightarrow \left(x-\left(6+\sqrt{3}\right)\right)$ is a factor.
3. Write the equation.	The equation is: $(x-(6+\sqrt{3}))(x-(6-\sqrt{3}))=0$
4. Expand as a difference of two squares.	$\therefore ((x-6)+\sqrt{3})((x-6)-\sqrt{3})=0$ $(x-6)^2-(\sqrt{3})^2=0$
5. Express in the form $x^2+bx+c=0$.	$x^2-12x+36-3=0$ $x^2-12x+33=0$
6. State the values of b and c.	$b=-12, c=33$

3.3.5 Equations of the form $\sqrt{x}=ax+b$

Using the substitution $u=\sqrt{x}$, equations of the form $\sqrt{x}=ax+b$ could be reduced to a quadratic equation in u, where $u=au^2+b$. Any negative solution for u would need to be rejected as $\sqrt{x}\geq 0$. Alternatively, the quadratic form could be obtained by squaring both sides of the equation $\sqrt{x}=ax+b$ to form $x=(ax+b)^2$, with no substitution required.

However, since the same quadratic equation would be obtained by squaring $\sqrt{x}=-(ax+b)$, the squaring process may produce extraneous 'solutions' — ones that do not satisfy the original equation. It is always necessary to verify the solutions by testing whether they satisfy the original equation.

WORKED EXAMPLE 9 Solving quadratics by squaring

Solve the equation $3+2\sqrt{x}=x$ for x.

THINK	WRITE
1. Isolate the surd term on one side of the given equation.	$3+2\sqrt{x}=x$
2. Square both sides to remove the surd term.	$2\sqrt{x}=x-3$ $(2\sqrt{x})^2=(x-3)^2$ $4x=(x-3)^2$ $4x=x^2-6x+9$

3. Expand to form the quadratic equation and solve this equation.	$x^2-10x+9=0$ $(x-9)(x-1)=0$ $\therefore x=9 \quad \text{or} \quad x=1$
4. Check whether the solution $x=9$ is valid using the original equation.	Substitute $x=9$ into $3+2\sqrt{x}=x$. LHS $=3+2\sqrt{9}$ RHS $=x$ $=3+2\sqrt{9}$ $\quad=9$ $=9$ $\quad=$ LHS Therefore, accept $x=9$ as a solution.
5. Check whether the solution $x=1$ is valid using the original equation.	Substitute $x=1$ in $3+2\sqrt{x}=x$. LHS $=3+2\sqrt{x}$ RHS $=x$ $=3+2\sqrt{1}$ $\quad=1$ $=5$ $\quad\neq$ LHS Therefore, reject $x=1$ as a solution.
6. State the answer.	The answer is $x=9$.

TI | THINK	DISPLAY/WRITE	CASIO | THINK	DISPLAY/WRITE
1. On a Calculator page, press MENU and select: 3. Algebra 1. Solve Complete the entry line as: solve $(3+2\sqrt{x}=x,x)$ Then press ENTER.	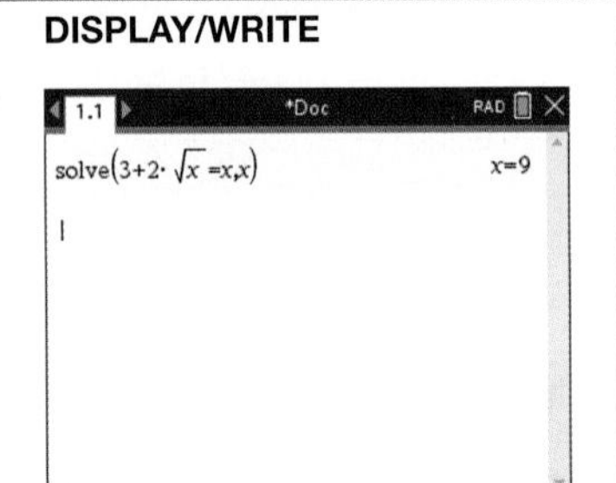	1. On a Main screen, complete the entry line as: solve $(3+2\sqrt{x}=x,x)$ Then press EXE.	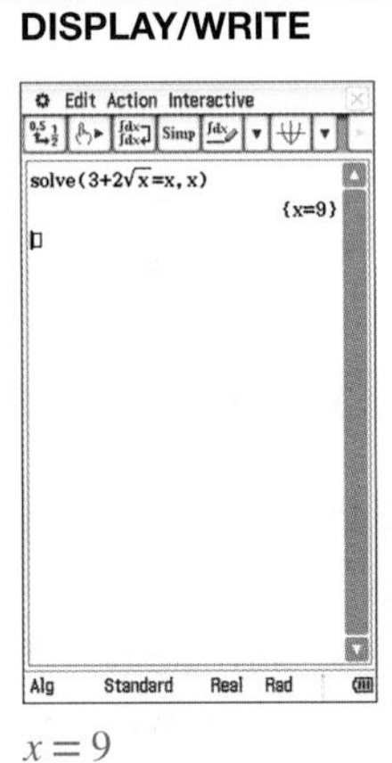
2. The answer appears on the screen.	$x=9$	2. The answer appears on the screen.	$x=9$

3.3 Exercise

Students, these questions are even better in jacPLUS

Receive immediate feedback and access sample responses

Access additional questions

Track your results and progress

Find all this and MORE in jacPLUS

Technology free

1. Complete the following statements about perfect squares.

a. $x^2+10x+\ldots=(x+\ldots)^2$ b. $x^2-7x+\ldots=(x-\ldots)^2$ c. $x^2+x+\ldots=(x+\ldots)^2$

2. WE4 Factorise the following over R.

a. $x^2-10x-7$ b. $3x^2+7x+3$ c. $5x^2-9$

3. Use the 'completing the square' method to factorise $3x^2 - 8x + 5$ and check the answer by using another method of factorisation.

4. Factorise the following where possible.

a. $3(x - 8)^2 - 6$ b. $(x - 7)^2 + 9$

5. Use the 'completing the square' method to factorise, where possible, the following over R.

a. $x^2 - 6x + 7$ b. $x^2 + 4x - 3$ c. $x^2 - 2x + 6$
d. $2x^2 + 5x - 2$ e. $-x^2 + 8x - 8$ f. $3x^2 + 4x - 6$

6. Factorise the following over R, where possible.

a. $x^2 - 12$ b. $x^2 - 12x + 4$ c. $x^2 + 9x - 3$
d. $2x^2 + 5x + 1$ e. $3x^2 + 4x + 3$ f. $1 + 40x - 5x^2$

7. **WE5** Use the quadratic formula to solve the equation $(2x + 1)(x + 5) - 1 = 0$.

8. Use the quadratic formula $x = \dfrac{-b \pm \sqrt{b^2 - 4ac}}{2a}$ to solve the following equations, expressing solutions in simplest surd form.

a. $3x^2 - 5x + 1 = 0$ b. $-5x^2 + x + 5 = 0$
c. $2x^2 + 3x + 4 = 0$ d. $x(x + 6) = 8$

9. Solve the following quadratic equations, giving the answers in simplest exact form.

a. $9x^2 - 3x - 4 = 0$ b. $5x(4 - x) = 12$
c. $(x - 10)^2 = 20$ d. $x^2 + 6x - 3 = 0$
e. $6x = x^2$ f. $8x^2 - 22x + 12 = 0$

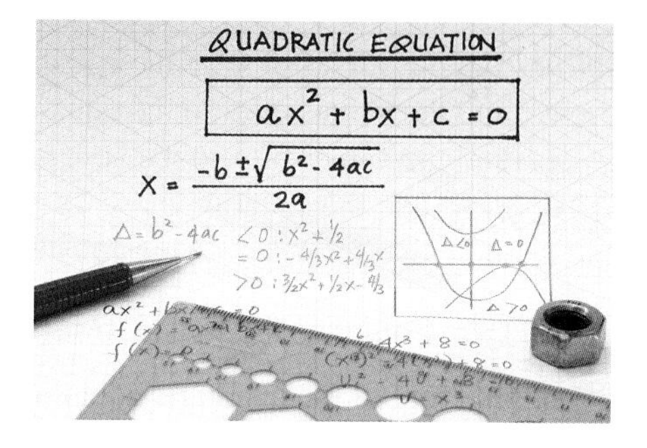

10. **WE6** For each of the following quadratics, calculate the discriminant. Hence, state the number and type of factors, and whether the 'completing the square' method would be needed to obtain the factors.

a. $4x^2 + 5x + 10$ b. $169x^2 - 78x + 9$
c. $-3x^2 + 11x - 10$ d. $\frac{1}{3}x^2 - \frac{8}{3}x + 2$

11. For each of the following, calculate the discriminant, and hence state the number and type of linear factors.

a. $5x^2 + 9x - 2$ b. $12x^2 - 3x + 1$
c. $121x^2 + 110x + 25$ d. $x^2 + 10x + 23$

12. a. Calculate the discriminant for the equation $3x^2 - 4x + 1 = 0$.
b. Use the result of a to determine the number and nature of the roots of the equation $3x^2 - 4x + 1 = 0$.
In parts c to f, use the discriminant to determine the number and type of solutions to the given equation.
c. $-x^2 - 4x + 3 = 0$
d. $2x^2 - 20x + 50 = 0$
e. $x^2 + 4x + 7 = 0$
f. $1 = x^2 + 5x$

13. Without actually solving the equations, determine the number and the nature of the roots of the following equations.

a. $-5x^2 - 8x + 9 = 0$ b. $4x^2 + 3x - 7 = 0$ c. $4x^2 + x + 2 = 0$
d. $28x - 4 - 49x^2 = 0$ e. $4x^2 + 25 = 0$ f. $3\sqrt{2}x^2 + 5x + \sqrt{2} = 0$

Technology active

14. Solve the equation $3(2x+1)^4 - 16(2x+1)^2 - 35 = 0$ for $x \in R$.

15. Solve the equation $\sqrt{2}x^2 + 4\sqrt{3}x - 8\sqrt{2} = 0$, expressing solutions in simplest surd form.

16. Use an appropriate substitution to reduce the following equations to quadratic form and hence obtain all solutions over R.

a. $(x^2-3)^2 - 4(x^2-3) + 4 = 0$
b. $5x^4 - 39x^2 - 8 = 0$
c. $x^2(x^2-12) + 11 = 0$
d. $\left(x+\frac{1}{x}\right)^2 + 2\left(x+\frac{1}{x}\right) - 3 = 0$
e. $(x^2-7x-8)^2 = 3(x^2-7x-8)$
f. $3\left(x^2+\frac{1}{x^2}\right) + 2\left(x+\frac{1}{x}\right) - 2 = 0$ given that $x^2+\frac{1}{x^2} = \left(x+\frac{1}{x}\right)^2 - 2$

17. WE7 a. Use the discriminant to determine the number and type of roots to the equation $0.2x^2 - 2.5x + 10 = 0$.
b. Find the values of k so the equation $kx^2 - (k+3)x + k = 0$ will have one real solution.

18. Show that the equation $mx^2 + (m-4)x = 4$ will always have real roots for any real value of m.

19. a. Find the values of m so the equation $x^2 + (m+2)x - m + 5 = 0$ has one root.
b. Find the values of m so the equation $(m+2)x^2 - 2mx + 4 = 0$ has one root.
c. Find the values of p so the equation $3x^2 + 4x - 2(p-1) = 0$ has no real roots.
d. Show that the equation $kx^2 - 4x - k = 0$ always has two solutions for $k \in R\backslash\{0\}$.
e. Show that for $p, q \in Q$, the equation $px^2 + (p+q)x + q = 0$ always has rational roots.

20. WE8 a. Complete the square to solve the equation $x^2 - 20\sqrt{5}x + 100 = 0$.
b. One root of the quadratic equation with rational coefficients $x^2 + bx + c = 0, b, c \in Q$ is $x = 1 - \sqrt{2}$. State the other root and calculate the values of b and c.

21. Form linear factors from the following information and expand the product of these factors to obtain a quadratic expression.

a. The zeros of a quadratic are $x = \sqrt{2}$ and $x = -\sqrt{2}$.
b. The zeros of a quadratic are $x = -4 + \sqrt{2}$ and $x = -4 - \sqrt{2}$.

22. a. Solve the following for x, expressing solutions in simplest surd form with rational denominators.

i. $x^2 + 6\sqrt{2}x + 18 = 0$
ii. $2\sqrt{5}x^2 - 3\sqrt{10}x + \sqrt{5} = 0$

b. One root of the quadratic equation with rational coefficients, $x^2 + bx + c = 0$, $b, c, \in Q$ is $x = \frac{-1+\sqrt{5}}{2}$.

i. State the other root.
ii. Form the equation and calculate the values of b and c.

23. WE9 a. Solve the equation $4x - 3\sqrt{x} = 1$ for x.
b. Use the substitution $u = \sqrt{x}$ to solve $4x - 3\sqrt{x} = 1$.

24. Solve the equation $2\sqrt{x} = 8 - x$ by the following two methods:

a. squaring both sides of the equation
b. using a suitable substitution to reduce to quadratic form.

25. Solve $1 + \sqrt{x+1} = 2x$.

3.3 Exam questions

Question 1 (1 mark) TECH-ACTIVE

MC The discriminant of $3x^2 + 5x - 1$ is equal to

A. 37 **B.** 13 **C.** 15 **D.** -15 **E.** 45

Question 2 (1 mark) TECH-ACTIVE

MC The solutions to the equation $2x^2 - 7x + 1 = 0$ are

A. $\frac{7 \pm \sqrt{57}}{4}$ **B.** $\frac{-7 \pm \sqrt{41}}{2}$ **C.** $\frac{2 \pm \sqrt{41}}{2}$ **D.** $\frac{2 \pm \sqrt{41}}{4}$ **E.** $\frac{7 \pm \sqrt{41}}{4}$

Question 3 (4 marks) TECH-FREE

Determine the value(s) of m for which the quadratic $mx^2 + (m+3)x = -3$ will have real roots.

More exam questions are available online.

3.4 Applications of quadratic equations

LEARNING INTENTION

At the end of this subtopic you should be able to:
- solve real world problems using quadratic equations.

Quadratic equations may occur in problem solving and when setting up mathematical models of real-life situations.

3.4.1 Solving problems using quadratic equations

Similar methods to solving problems with linear relationships are used when solving quadratic equations.

In setting up equations:
- define the symbols used for the variables, specifying units where appropriate
- ensure any units used are consistent
- check whether mathematical solutions are feasible in the context of the problem
- express answers in the context of the problem.

WORKED EXAMPLE 10 Using quadratic equations in problem solving

The owner of a gift shop imported a certain number of paperweights for \$900. They were pleased when all except 4 were sold for \$10 more than each paperweight had cost to import. From the sale of the paperweights, the gift shop owner received a total of \$1400. Find how many paperweights were imported.

THINK	WRITE
1. Define the key variable.	Let x be the number of paperweights imported.
2. Find an expression for the cost of importing each paperweight.	The total cost of importing x paperweights is \$900. Therefore, the cost of each paperweight is $\left(\frac{900}{x}\right)$ dollars.
3. Find an expression for the selling price of each paperweight and identify how many are sold.	The number of paperweights sold is $(x-4)$, and each is sold for $\left(\frac{900}{x}+10\right)$ dollars.
4. Create the equation showing how the sales revenue of \$1400 is formed.	$\left(\frac{900}{x}+10\right)\times(x-4)=1400$
5. Now the equation has been formulated, solve it.	Expand: $900-\frac{3600}{x}+10x-40=1400$ $-\frac{3600}{x}+10x=540$ $-3600+10x^2=540x$ $10x^2-540x-3600=0$ $x^2-54x-360=0$ $(x-60)(x+6)=0$ $\therefore x=60, x=-6$
6. Check the feasibility of the mathematical solutions.	Reject $x=-6$, since x must be a positive whole number.
7. Write the answer in context.	Therefore, 60 paperweights were imported by the gift shop owner.

3.4.2 Quadratically related variables

The formula for the area, A, of a circle in terms of its radius, r, is $A=\pi r^2$. This is of the form $A=kr^2$, as π is a constant. The area varies directly as the square of its radius with the constant of proportionality $k=\pi$. This is a quadratic relationship between A and r.

r	0	1	2	3
A	0	π	4π	9π

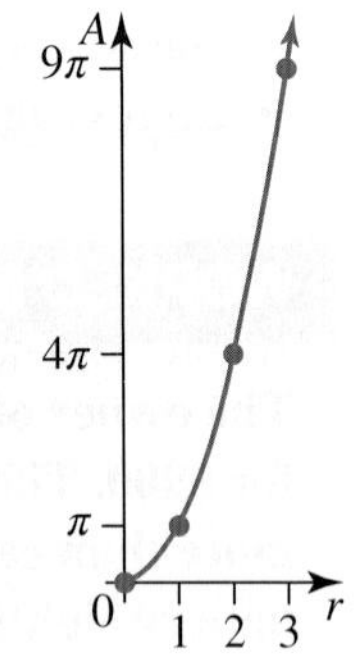

Plotting the graph of A against r gives a curve that is part of a **parabola**.

For any variables x and y, if y is directly proportional to x^2, then $y=kx^2$, where k is the constant of proportionality.

Other quadratically related variables would include, for example, those where y was the sum of two parts, one part of which was constant and the other part of which was in direct proportion to x^2 so that $y=c+kx^2$.

WORKED EXAMPLE 11 Solving problems with quadratic equations

The volume of a cone of fixed height is directly proportional to the square of the radius of its base. When the radius is 3 cm, the volume is 30π cm^3. Calculate the radius when the volume is 480π cm^3.

THINK	WRITE
1. Write the variation equation, defining the symbols used.	$V = kr^2$, where V is the volume of a cone of fixed height and radius r. k is the constant of proportionality.
2. Use the given information to find k.	$r = 3, V = 30\pi \Rightarrow 30\pi = 9k$ $\therefore k = \frac{30\pi}{9}$ $= \frac{10\pi}{3}$
3. Write the rule connecting V and r.	$V = \frac{10\pi}{3} r^2$
4. Substitute $V = 480\pi$ and find r.	$480\pi = \frac{10\pi}{3} r^2$ $10\pi r^2 = 480\pi \times 3$ $r^2 = \frac{480\pi \times 3}{10\pi}$ $r^2 = 144$ $r = \pm\sqrt{144}$ $r = \pm 12$
5. Check the feasibility of the mathematical solutions.	Reject $r = -12$, since r must be positive. $\therefore r = 12$
6. Write the answer in context.	The radius of the cone is 12 cm.

3.4 Exercise

Students, these questions are even better in jacPLUS

Receive immediate feedback and access sample responses

Access additional questions

Track your results and progress

Find all this and MORE in jacPLUS

Technology active

1. WE10 The owner of a fish shop bought x kilograms of salmon for \$400 from the wholesale market. At the end of the day, all except for 2 kg of the fish were sold at a price per kg that was \$10 more than what the owner paid at the market. From the sale of the fish, a total of \$540 was made.
   Calculate how many kilograms of salmon the fish-shop owner bought at the market.

2. **WE11** The surface area of a sphere is directly proportional to the square of its radius. When the radius is 5 cm, the area is 100π cm^2. Calculate the radius when the area is 360π cm^2.

3. The cost of hiring a chainsaw is \$10 plus an amount that is proportional to the square of the number of hours for which the chainsaw is hired. If it costs \$32.50 to hire the chainsaw for 3 hours, find, to the nearest half hour, the length of time for which the chainsaw was hired if the cost of hire was \$60.

4. The area of an equilateral triangle varies directly as the square of its side length. A triangle of side length $2\sqrt{3}$ cm has an area of $3\sqrt{3}$ cm^2. Calculate the side length if the area is $12\sqrt{3}$ cm^2.

5. The cost of producing x hundred litres of olive oil is $20 + 5x$ dollars. If the revenue from the sale of x hundred litres of the oil is $1.5x^2$ dollars, calculate, to the nearest litre, the number of litres that must be sold to make a profit of \$800.

6. The product of two consecutive even natural numbers is 440. Determine the numbers.

7. The sum of the squares of two consecutive natural numbers plus the square of their sums is 662. Determine the numbers.

8. The ratio of the height of a triangle to the length of its base is $\sqrt{2} : 1$. If the area of the triangle is $\sqrt{32}$ cm^2, calculate the length of its base and its height.

9. The hypotenuse of a right angled-triangle is $(3x + 3)$ cm and the other two sides are $3x$ cm and $(x - 3)$ cm. Determine the value of x and calculate the perimeter of the triangle.

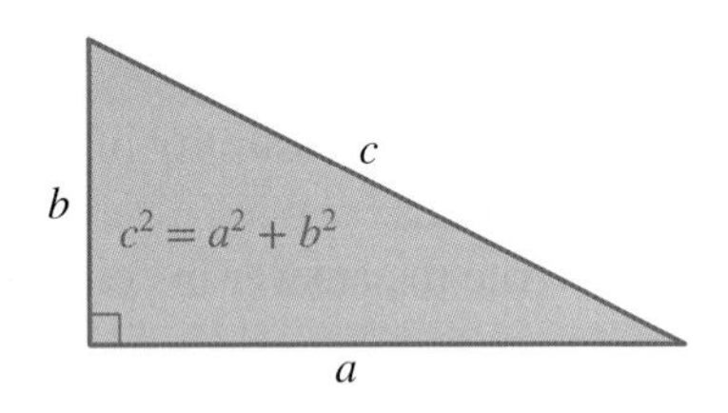

10. A photograph, 17 cm by 13 cm, is placed in a rectangular frame. The border around the photograph is of uniform width and has an area of 260 cm^2, measured to the nearest cm^2. Calculate the dimensions of the frame measured to the nearest cm.

11. A gardener has 16 metres of edging to place around three sides of a rectangular garden bed, the fourth side of which is bounded by the backyard fence.

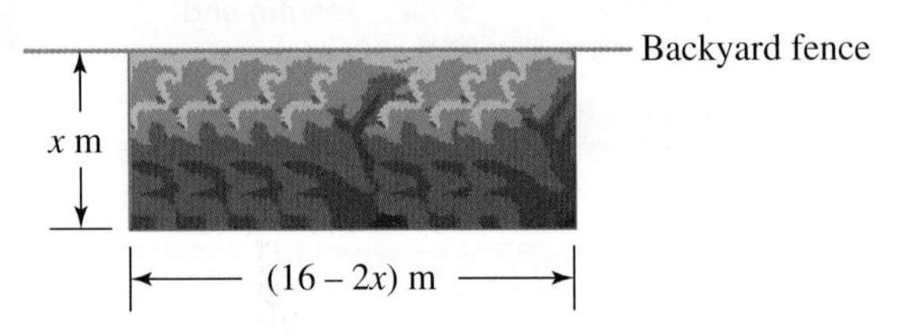

a. If the width of the garden bed is x metres, explain why its length is $(16 - 2x)$ metres.
b. If the area of the rectangular garden is k square metres, show that $2x^2 - 16x + k = 0$.
c. Calculate the discriminant and hence determine the values of k for which this equation will have:
   i. no solutions
   ii. one solution
   iii. two solutions.

d. Calculate the largest possible area of the garden bed and its dimensions in this case.

e. The gardener decides the area of the garden bed is to be 15 square metres. Given that the gardener would also prefer to use as much of the backyard fence as possible as a boundary to the garden bed, calculate the dimensions of the rectangle in this case, correct to 1 decimal place.

12. A young collector of fantasy cards buys a parcel of the cards in a lucky dip at a fete for \$10 but finds on opening the parcel that only two are of interest. Keeping those two cards aside, the collector decides to resell the remaining cards to an unsuspecting friend for \$1 per card more than the original cost, thereby making a nice profit of \$6. Calculate how many cards the collector's friend received.

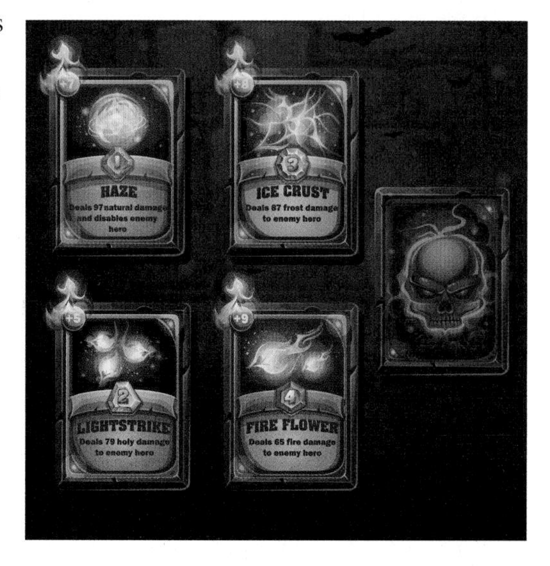

13. The formula for the total surface area A of a cone of base radius r and slant height l is $A = \pi r^2 + \pi rl$.

a. Calculate, correct to 3 decimal places, the radius of the base of a cone with slant height 5 metres and total surface area 20 m^2.

b. For any cone that has a surface area of 20 m^2, determine r in terms of l and use this expression to check the answer to question part **a**.

14. The distance a particle falls from rest is in direct proportion to the square of the time of fall. Determine the effect on the distance fallen if the time of fall is doubled.

15. The number of calories of heat produced in a wire in a given time varies as the square of the voltage. Determine the effect on the number of calories of heat produced if the voltage is reduced by 20%.

3.4 Exam questions

Question 1 (1 mark) TECH-ACTIVE

MC A farmer wants to fence off a small section of natural vegetation alongside a river. He has 100 metres of fencing materials to make the three sides of the rectangular section abutting the river. If the length of the side parallel to the river is x metres, the area of the section can be written as

A. $50x - \dfrac{x^2}{2}$ **B.** $50x + \dfrac{x^2}{2}$ **C.** $100x - x^2$ **D.** $100x + x^2$ **E.** $50x + 2x^2$

Question 2 (1 mark) TECH-ACTIVE

MC A right-angled triangle has a hypotenuse of $(5x + 4)$ cm and two sides $5x$ cm and $(x - 4)$ cm, respectively. The value of x is

A. 16 **B.** 25 **C.** 32 **D.** 8 **E.** 48

Question 3 (4 marks) TECH-ACTIVE

The product of two numbers is −483. If the difference between the two numbers is 44, determine the smaller of the two numbers.

More exam questions are available online.

3.5 Graphs of quadratic polynomials

LEARNING INTENTION

At the end of this subtopic you should be able to:
- sketch quadratic polynomial functions in various forms
- identify and describe simple transformations of quadratic polynomial functions.

A quadratic **polynomial** is an algebraic expression of the form $ax^2 + bx + c$, where each power of the variable x is a positive whole number, with the highest power of x being 2. It is called a second-degree polynomial, whereas a linear polynomial of the form $ax + b$ is a first-degree polynomial since the highest power of x is 1.

The graph of a quadratic polynomial is called a parabola.

3.5.1 The graph of $y = x^2$ and its transformations

The simplest parabola has the equation $y = x^2$.

Key features of the graph of $y = x^2$:
- It is symmetrical about the y-axis.
- The **axis of symmetry** has the equation $x = 0$.
- The graph is **concave up** (opens upwards).
- It has a minimum turning point, or **vertex**, at the point $(0, 0)$.

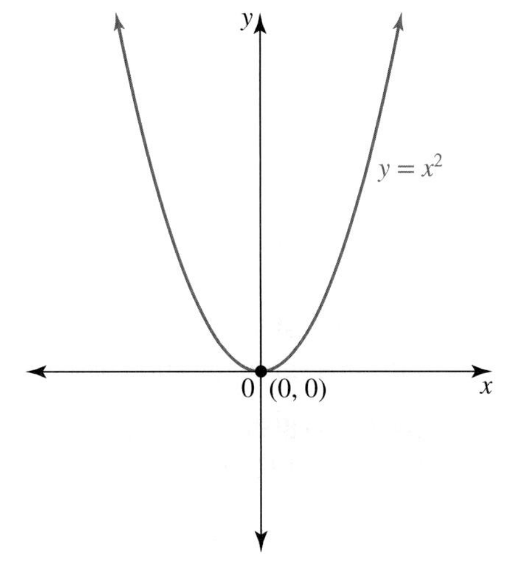

Making the graph wider or narrower

The graphs of $y = ax^2$ for $a = \frac{1}{3}$, 1 and 3 are drawn on the same set of axes.

Comparison of the graphs of $y = x^2$, $y = 3x^2$ and $y = \frac{1}{3}x^2$ shows that the graph of $y = ax^2$ will be:
- narrower than the graph of $y = x^2$ if $a > 1$
- wider than the graph of $y = x^2$ if $0 < a < 1$.

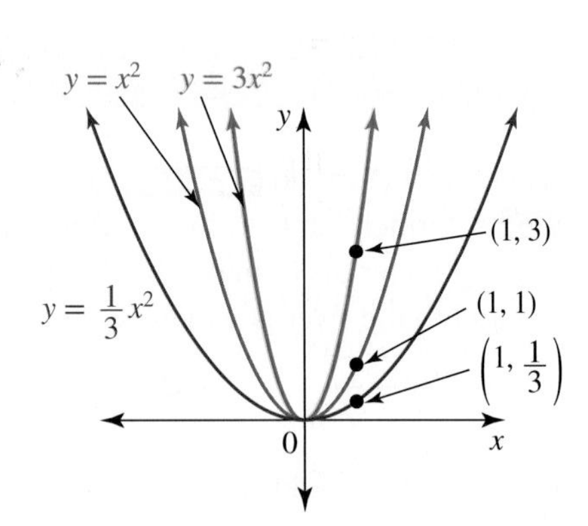

The coefficient of x^2, a, is called the **dilation factor**. It measures the amount of stretching or compression from the x-axis.

For $y = ax^2$, the graph of $y = x^2$ has been dilated by a factor of a from the x-axis or by a factor of a parallel to the y-axis.

Translating the graph up or down

The graphs of $y = x^2 + k$ for $k = -2, 0$ and 2 are drawn on the same set of axes.

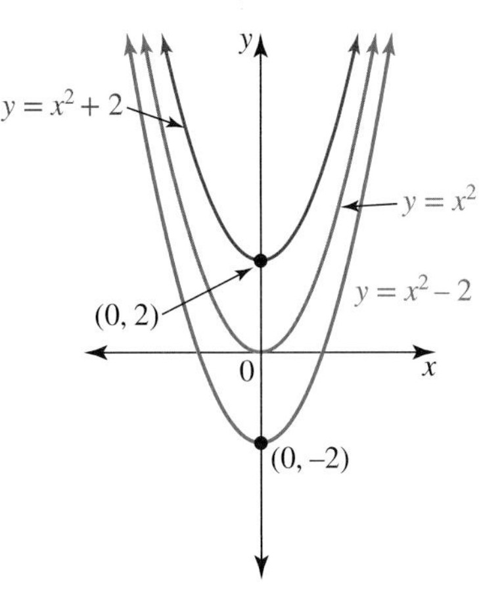

Comparison of the graphs of $y = x^2$, $y = x^2 + 2$ and $y = x^2 - 2$ shows that the graph of $y = x^2 + k$ will:

- have a turning point at $(0, k)$
- move the graph of $y = x^2$ vertically upwards by k units if $k > 0$
- move the graph of $y = x^2$ vertically downwards by k units if $k < 0$.

The value of k gives the **vertical translation**. For the graph of $y = x^2 + k$, the graph of $y = x^2$ has been translated by k units from the x-axis.

Translating the graph left or right

The graphs of $y = (x - h)^2$ for $h = -2, 1$ and 4 are drawn on the same set of axes.

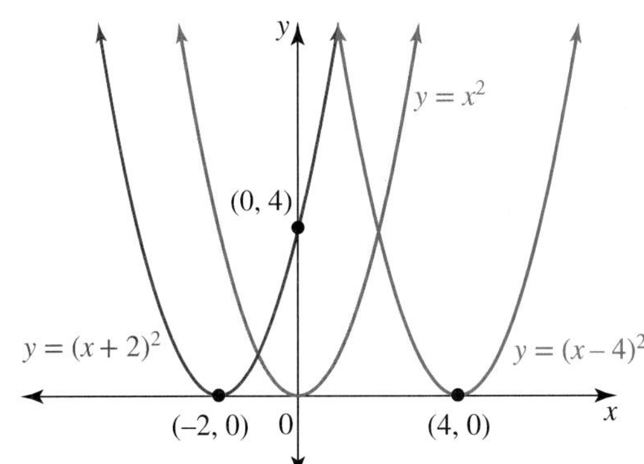

Comparison of the graphs of $y = x^2$, $y = (x + 2)^2$ and $y = (x - 4)^2$ shows that the graph of $y = (x - h)^2$ will:

- have a turning point at $(h, 0)$
- move the graph of $y = x^2$ horizontally to the right by h units if $h > 0$
- move the graph of $y = x^2$ horizontally to the left by h units if $h < 0$.

The value of h gives the **horizontal translation**. For the graph of $y = (x - h)^2$, the graph of $y = x^2$ has been translated by h units from the y-axis.

Reflecting the graph in the x-axis

The graph of $y = -x^2$ is obtained by reflecting the graph of $y = x^2$ in the x-axis.

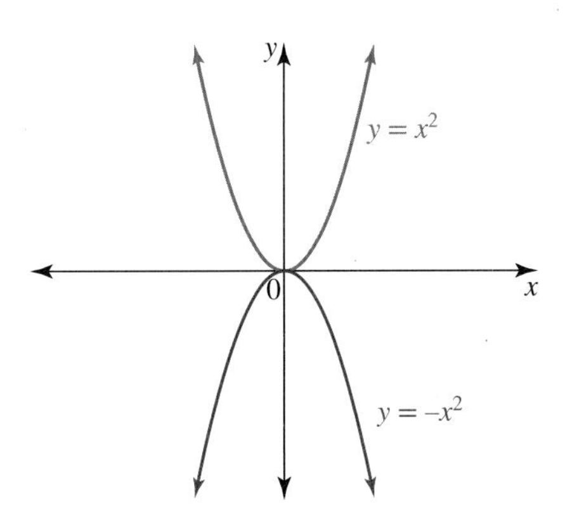

Key features of the graph of $y = -x^2$:

- It is symmetrical about the y-axis.
- The axis of symmetry has the equation $x = 0$.
- The graph is **concave down** (opens downwards).
- It has a maximum turning point, or vertex, at the point $(0, 0)$.

A negative coefficient of x^2 indicates the graph of a parabola has been **reflected** in the x-axis.

WORKED EXAMPLE 12 Matching parabolas to equations

Match the graphs of the parabolas A, B and C with the following equations.

a. $y = -x^2 + 3$ b. $y = -3x^2$ c. $y = (x - 3)^2$

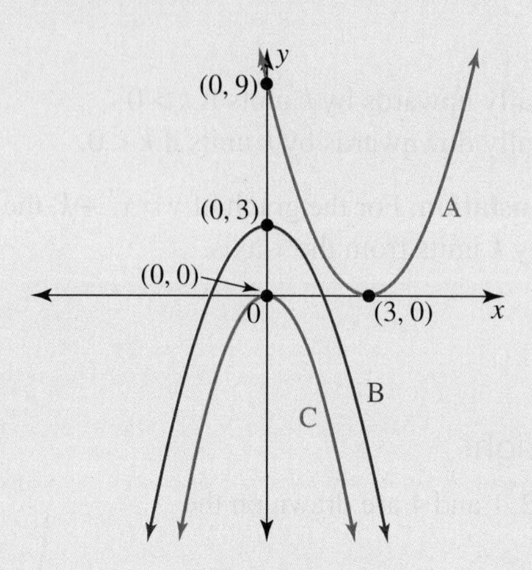

THINK	WRITE
1. Compare graph A with the basic graph $y = x^2$ to identify the transformations.	Graph A opens upwards and has been moved horizontally to the right. Graph A matches with equation **c**, $y = (x - 3)^2$.
2. Compare graph B with the basic graph $y = x^2$ to identify the transformations.	Graph B opens downwards and has been moved vertically upwards. Graph B matches with equation **a**, $y = -x^2 + 3$.
3. Check graph C for transformations.	Graph C opens downwards. It is narrower than both graphs A and B. Graph C matches with equation **b**, $y = -3x^2$.

Resources

Interactivity Quadratic functions (int-2562)

3.5.2 Sketching parabolas from their equations

The key points required when sketching a parabola are:

- the turning point, or vertex
- the y-intercept
- any x-intercepts.

The axis of symmetry is also a key feature of the graph.

The equation of a parabola allows this information to be obtained but in different ways, depending on the form of the equation.

We shall consider three forms for the equation of a parabola:

- general form
- turning point form
- x-intercept form.

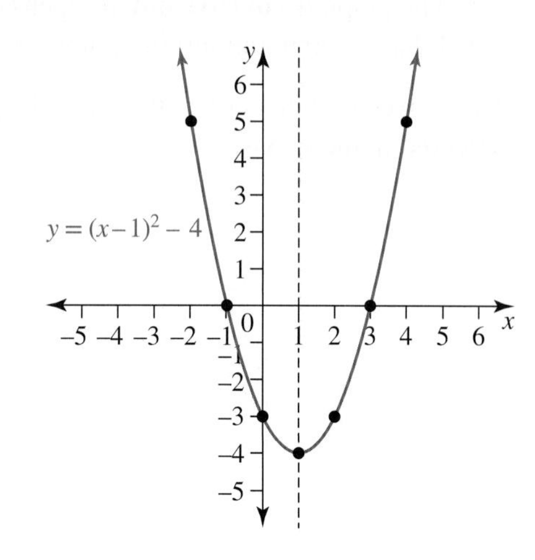

The general or polynomial form, $y=ax^2+bx+c$

If $a>0$, then the parabola is concave up and has a minimum turning point.

If $a<0$, then the parabola is concave down and has a maximum turning point.

The dilation factor $a, a>0$, determines the width of the parabola. The dilation factor is always a positive number (so it could be expressed as $|a|$).

The methods to determine the key features of the graph are as follows.

- Substitute $x=0$ to obtain the y-intercept (the y-intercept is obvious from the equation).
- Substitute $y=0$ and solve the quadratic equation $ax^2+bx+c=0$ to obtain the x-intercepts. There may be 0, 1 or 2 x-intercepts, as determined by the discriminant.
- The equation of the axis of symmetry is $x=-\frac{b}{2a}$.

  This is because the formula for solving $ax^2+bx+c=0$ gives $x=-\frac{b}{2a}\pm\frac{\sqrt{b^2-4ac}}{2a}$ as the x-intercepts, and these are symmetrical about their midpoint $x=-\frac{b}{2a}$.
- The turning point lies on the axis of symmetry, so its x-coordinate is $x=-\frac{b}{2a}$. Substitute this value into the parabola's equation to calculate the y-coordinate of the turning point.

Axis of symmetry of a parabola

$$x=-\frac{b}{2a}$$

WORKED EXAMPLE 13 Sketching parabolas given in general form

Sketch the graph of $y=\frac{1}{2}x^2-x-4$ and label the key points with their coordinates.

THINK	WRITE
1. Write down the y-intercept.	$y=\frac{1}{2}x^2-x-4$ y-intercept: if $x=0$, then $y=-4 \Rightarrow (0,-4)$
2. Obtain any x-intercepts.	x-intercepts: let $y=0$. $\frac{1}{2}x^2-x-4=0$ $x^2-2x-8=0$ $(x+2)(x-4)=0$ $\therefore x=-2,4$ $\Rightarrow (-2,0),(4,0)$
3. Find the equation of the axis of symmetry.	Axis of symmetry formula: $x=-\frac{b}{2a}$ $a=\frac{1}{2}, b=-1$ $x=-\frac{-1}{\left(2\times\frac{1}{2}\right)}$ $=1$

4. Find the coordinates of the turning point.	Turning point: when $x = 1$, $y = \frac{1}{2} - 1 - 4$ $= -4\frac{1}{2}$ $\Rightarrow \left(1, -4\frac{1}{2}\right)$ is the turning point.
5. Identify the type of turning point.	Since $a > 0$, the turning point is a minimum turning point.
6. Sketch the graph using the information obtained in the previous steps. Label the key points with their coordinates.	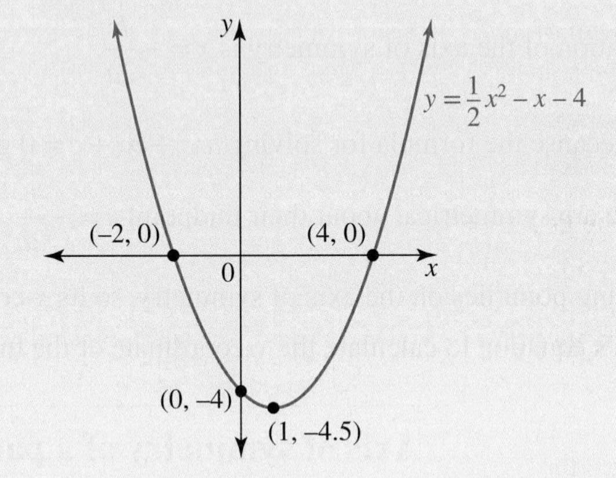

TI | THINK	DISPLAY/WRITE	CASIO | THINK	DISPLAY/WRITE
1. On a Graphs page, complete the entry line as: $f1(x) = \frac{1}{2}x^2 - x - 4$ Then press ENTER to view the graph.		1. On a Graphs & Table screen, complete the entry line as: $y1 = \frac{1}{2}x^2 - x - 4$ Tap the Graph icon to view the graph.	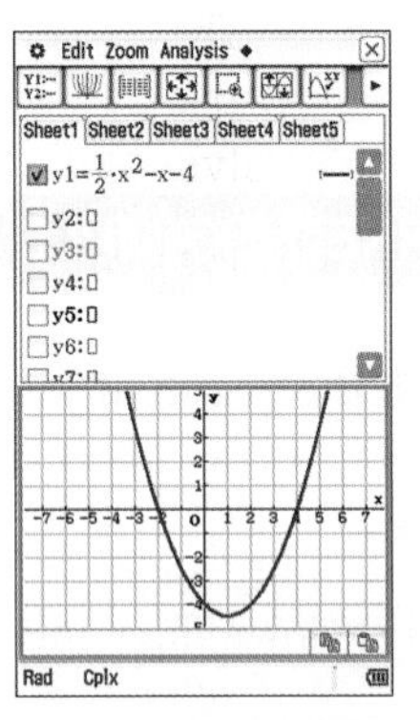
2. To view the key points, select: • MENU • 6. Analyze Graph • 1. Zero or • 2. Minimum or • 3. Maximum Follow the prompts to show the key points.	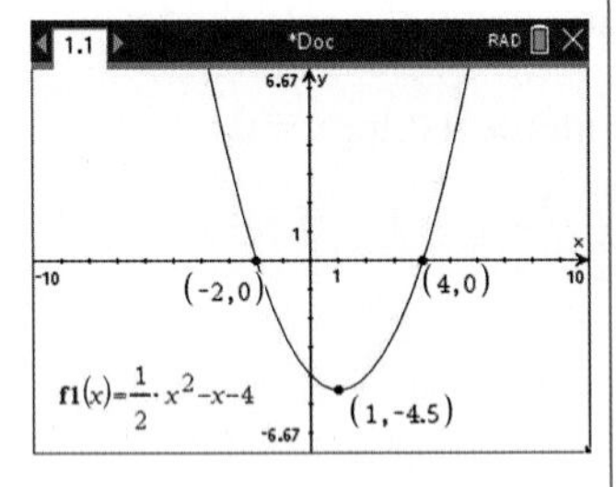	2. Tap $y = 0$ or the MIN or the MAX icons to view the key points.	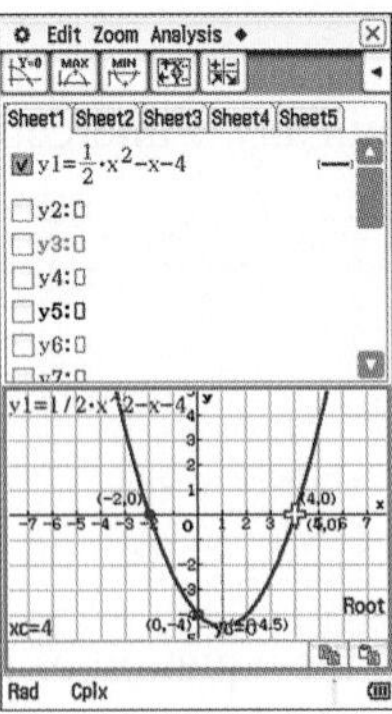

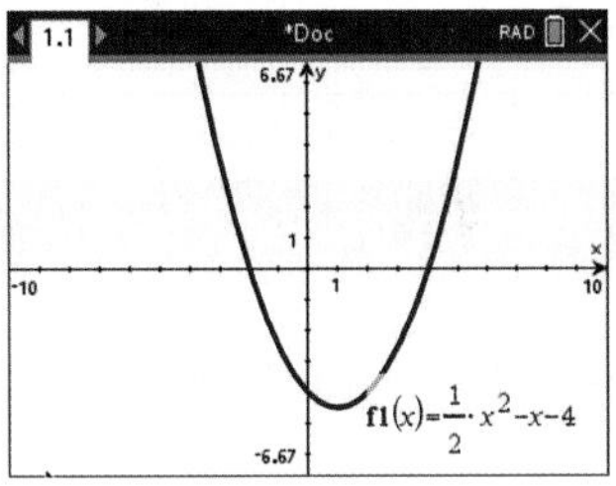

Turning point form, $y = a(x-h)^2 + k$

Since h represents the horizontal translation and k the vertical translation, this form of the equation readily provides the coordinates of the turning point.

- The turning point has coordinates (h, k).
  If $a > 0$, the turning point is a minimum, and if $a < 0$, it is a maximum. Depending on the nature of the turning point, the y-coordinate of the turning point gives the minimum or maximum value of the quadratic.
- Find the y-intercept by substituting $x = 0$.
- Find the x-intercepts by substituting $y = 0$ and solving the equation $a(x-h)^2 + k = 0$. However, before attempting to find x-intercepts, consider the type of turning point and its y-coordinate, as this will indicate whether there are any x-intercepts.

The general form of the equation of a parabola can be converted to turning point form by the use of the 'completing the square' technique. By expanding the turning point form, the general form would be obtained.

WORKED EXAMPLE 14 Sketching parabolas given in turning point form

a. Sketch the graph of $y = -2(x+1)^2 + 8$ and label the key points with their coordinates.

b. i. Express $y = 3x^2 - 12x + 18$ in the form $y = a(x-h)^2 + k$ and hence state the coordinates of its vertex.

ii. Sketch its graph.

THINK	WRITE
a. 1. Obtain the coordinates and the type of turning point from the given equation. *Note:* The x-coordinate of the turning point could also be obtained by letting $(x+1) = 0$ and solving this for x.	a. $y = -2(x+1)^2 + 8$ $\therefore y = -2(x-(-1))^2 + 8$ Maximum turning point at $(-1, 8)$
2. Calculate the y-intercept.	Let $x = 0$. $\therefore y = -2(1)^2 + 8$ $= 6$ $\Rightarrow (0, 6)$
3. Calculate any x-intercepts. *Note:* The graph is concave down with a maximum y-value of 8, so there will be x-intercepts.	x-intercepts: let $y = 0$. $0 = -2(x+1)^2 + 8$ $2(x+1)^2 = 8$ $(x+1)^2 = 4$ $(x+1) = \pm\sqrt{4}$ $x = \pm 2 - 1$ $x = -3, 1$ $\Rightarrow (-3, 0), (1, 0)$
4. Sketch the graph, remembering to label the key points with their coordinates.	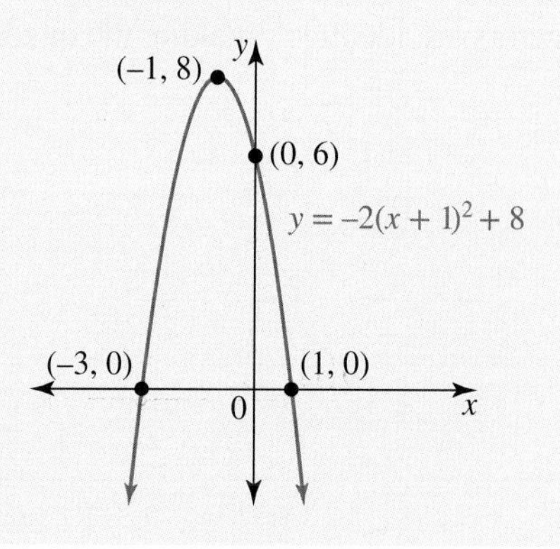

b. i. 1. Apply the 'completing the square' technique to the general form of the equation.	**b.** $y = 3x^2 - 12x + 18$ $= 3(x^2 - 4x + 6)$ $= 3[(x^2 - 4x + (2)^2) - (2)^2 + 6]$ $= 3[(x-2)^2 + 2]$
2. Expand to obtain the form $y = a(x-h)^2 + k$.	$= 3(x-2)^2 + 6$ $\therefore y = 3(x-2)^2 + 6$
3. State the coordinates of the vertex (turning point).	The vertex is $(2, 6)$.
ii. 1. Obtain the y-intercept from the general form.	The y-intercept is $(0, 18)$.
2. Will the graph have x-intercepts?	Since the graph is concave up with a minimum y-value of 6, there are no x-intercepts.
3. Sketch the graph.	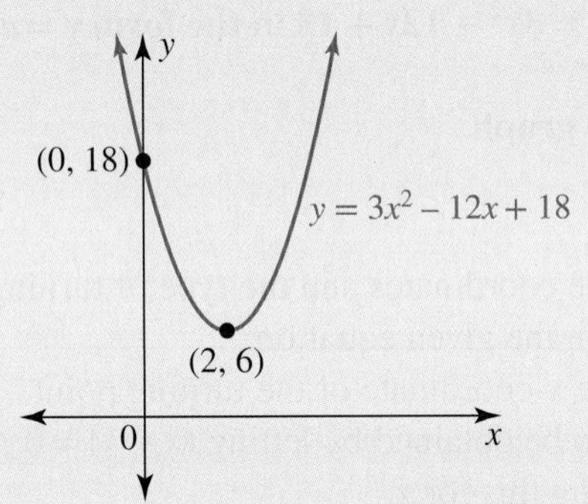

Factorised or x-intercept form, $y = a(x - x_1)(x - x_2)$

This form of the equation readily provides the x-intercepts.

- The x-intercepts occur at $x = x_1$ and $x = x_2$.
- The axis of symmetry lies halfway between the x-intercepts and its equation, $x = \frac{x_1 + x_2}{2}$, gives the x-coordinate of the turning point.
- The turning point is obtained by substituting $x = \frac{x_1 + x_2}{2}$ into the equation of the parabola and calculating the y-coordinate.
- The y-intercept is obtained by substituting $x = 0$.

If the linear factors are distinct, the graph cuts through the x-axis at each x-intercept.

If the linear factors are identical, making the quadratic a perfect square, the graph touches the x-axis at its turning point.

WORKED EXAMPLE 15 Sketching parabolas given in factorised form

Sketch the graph of $y = -\frac{1}{2}(x + 5)(x - 1)$.

THINK	WRITE
1. Identify the x-intercepts.	$y = -\frac{1}{2}(x + 5)(x - 1)$ x-intercepts: let $y = 0$. $\frac{1}{2}(x + 5)(x - 1) = 0$ $x + 5 = 0 \quad \text{or } x - 1 = 0$ $x = -5 \text{ or} \quad x = 1$ The x-intercepts are $(-5, 0), (1, 0)$.
2. Calculate the equation of the axis of symmetry.	The axis of symmetry has the equation $x = \frac{-5 + 1}{2}$ $\therefore x = -2$
3. Obtain the coordinates of the turning point.	Turning point: substitute $x = -2$ into the equation. $y = -\frac{1}{2}(x + 5)(x - 1)$ $= -\frac{1}{2}(3)(-3)$ $= \frac{9}{2}$ The turning point is $\left(-2, \frac{9}{2}\right)$.
4. Calculate the y-intercept.	$y = -\frac{1}{2}(x + 5)(x - 1)$ y-intercept: let $x = 0$. $y = -\frac{1}{2}(5)(-1)$ $= \frac{5}{2}$ The y-intercept is $\left(0, \frac{5}{2}\right)$.
5. Sketch the graph.	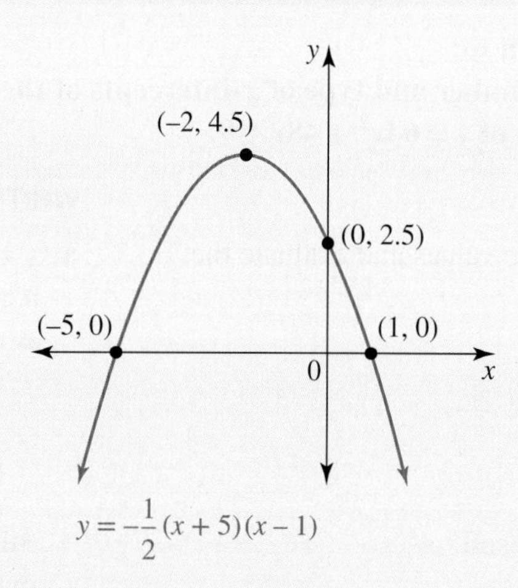

3.5.3 The discriminant and the x-intercepts

The zeros of the quadratic expression $ax^2 + bx + c$, the roots of the quadratic equation $ax^2 + bx + c = 0$ and the x-intercepts of the graph of a parabola with rule $y = ax^2 + bx + c$ all have the same x-values. The discriminant determines the type and number of these values.

- If $\Delta > 0$, there are two x-intercepts. The graph *cuts* through the x-axis at two different places.
- If $\Delta = 0$, there is one x-intercept. The graph *touches* the x-axis at its turning point.
- If $\Delta < 0$, there are no x-intercepts. The graph does not intersect the x-axis and lies entirely above or entirely below the x-axis, depending on its concavity.

If $a > 0$ and $\Delta < 0$, the graph lies entirely above the x-axis and every point on it has a positive y-coordinate. $ax^2 + bx + c$ is called **positive definite** in this case.

If $a < 0$ and $\Delta < 0$, the graph lies entirely below the x-axis and every point on it has a negative y-coordinate. $ax^2 + bx + c$ is called **negative definite** in this case.

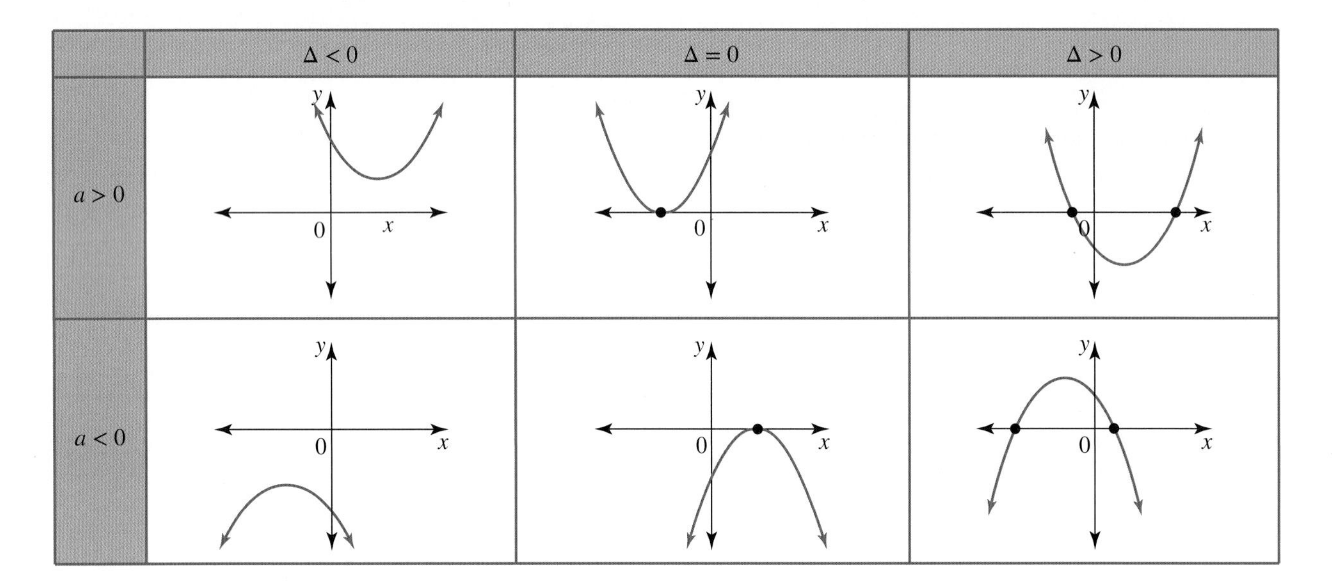

When $\Delta \geq 0$ and for $a, b, c \in Q$, the x-intercepts are rational if Δ is a perfect square and irrational if Δ is not a perfect square.

WORKED EXAMPLE 16 Using the discriminant to help sketching parabolas

Use the discriminant to:

a. determine the number and type of x-intercepts of the graph defined by $y = 64x^2 + 48x + 9$

b. sketch the graph of $y = 64x^2 + 48x + 9$.

THINK	WRITE
a. 1. State the a, b, c values and evaluate the discriminant.	**a.** $y = 64x^2 + 48x + 9$ $a = 64, b = 48, c = 9$ $\Delta = b^2 - 4ac$ $= (48)^2 - 4(64)(9)$ $= 2304 - 2304$ $= 0$
2. Interpret the result.	Since the discriminant is zero, the graph has one rational x-intercept.

b. 1. Interpret the implication of a zero discriminant for the factors.

b. The quadratic must be a perfect square.

$y = 64x^2 + 48x + 9$

$= (8x + 3)^2$

2. Identify the key points.

x-intercept: let $y = 0$.

$(8x + 3)^2 = 0$

$x = -\frac{3}{8}$

Therefore, $\left(-\frac{3}{8}, 0\right)$ is both the x-intercept and the turning point.

y-intercept: let $x = 0$ in $y = 64x^2 + 48x + 9$.

$\therefore y = 9$

Therefore, $(0, 9)$ is the y-intercept.

3. Sketch the graph.

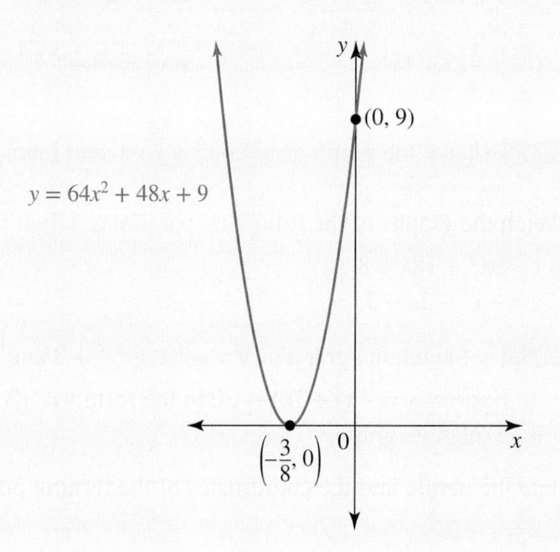

3.5 Exercise

Students, these questions are even better in jacPLUS

Receive immediate feedback and access sample responses

Access additional questions

Track your results and progress

Find all this and MORE in jacPLUS

Technology free

1. Sketch the following parabolas on the same set of axes.

a. $y = 2x^2$ **b.** $y = -2x^2$ **c.** $y = 0.5x^2$

d. $y = -0.5x^2$ **e.** $y = (2x)^2$ **f.** $y = \left(-\frac{x}{2}\right)^2$

2. **WE12** Match the graphs of the parabolas A, B and C with the following equations.

i. $y = x^2 - 2$ ii. $y = -2x^2$
iii. $y = -(x + 2)^2$

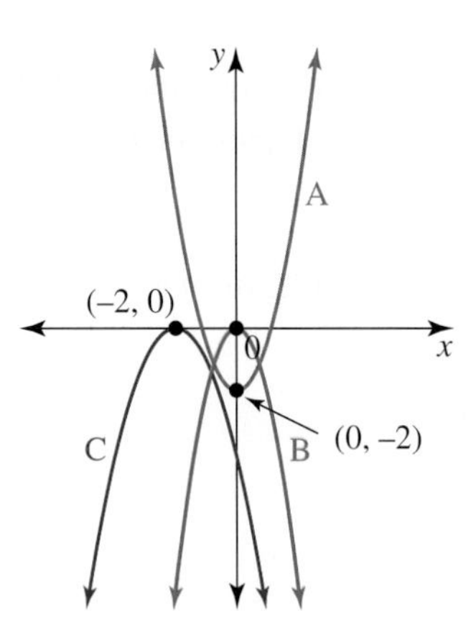

3. State the coordinates of the turning points of the parabolas with the following equations.

a. $y = x^2 + 8$ b. $y = x^2 - 8$
c. $y = 1 - 5x^2$ d. $y = (x - 8)^2$
e. $y = (x + 8)^2$ f. $y = -\frac{1}{2}(x + 12)^2$

4. Sketch the following graphs, showing the turning point and any intercepts with the coordinate axes.

a. $y = \frac{1}{2}x^2 + 2$ b. $y = 2x^2 + 4$ c. $y = (x - 2)^2$
d. $y = -\frac{1}{4}(x + 1)^2$ e. $y = x^2 - 4$ f. $y = -x^2 + 2$

5. **WE13** Sketch the graph of $y = \frac{1}{3}x^2 + x - 6$ and label the key points with their coordinates.

6. Sketch the graphs of the following parabolas, labelling their key points with their coordinates.

a. $y = 9x^2 + 18x + 8$ b. $y = -x^2 + 7x - 10$
c. $y = -x^2 - 2x - 3$ d. $y = x^2 - 4x + 2$

7. **WE14** a. Sketch the graph of $y = -2(x + 1)^2 + 8$ and label the key points with their coordinates.

b. i. Express $y = -x^2 + 10x - 30$ in the form $y = a(x - h)^2 + k$ and hence state the coordinates of its vertex.
ii. Sketch its graph.

8. State the nature and the coordinates of the turning point for each of the following parabolas.

a. $y = 4 - 3x^2$ b. $y = (4 - 3x)^2$

9. **MC** Select which of the following is the equation of a parabola with a turning point at $(-5, 2)$.

A. $y = -5x^2 + 2$ B. $y = 2 - (x - 5)^2$ C. $y = (x + 2)^2 - 5$
D. $y = -(x + 5)^2 + 2$ E. $y = (x + 5)^2 - 2$

10. a. Sketch the graph of $y = (x + 4)^2 - 1$.
b. State the turning point of $y = 3 - (x + 4)^2$ and sketch its graph.
c. i. Use the 'completing the square' technique to express $x^2 + 6x + 12$ in the form $(x + h)^2 + k$.
ii. Hence, state the coordinates of the turning point of $y = x^2 + 6x + 12$ and sketch its graph.
d. State the turning point of $y = -(2x + 5)^2$ and sketch its graph.

11. **WE15** Sketch the graph of $y = 2x(4 - x)$.

12. Sketch the graph of $y = (2 + x)^2$.

13. Sketch the following graphs, showing all intercepts with the coordinate axes and the turning point.

a. $y = (x + 1)(x - 3)$ b. $y = (x - 5)(2x + 1)$

Technology active

For each of the parabolas in questions 14 to 18:

i. give the coordinates of the turning point

ii. give the coordinates of the y-intercept

iii. give the coordinates of any x-intercepts

iv. sketch the graph.

14. a. $y = x^2 - 9$ b. $y = (x - 9)^2$ c. $y = 6 - 3x^2$ d. $y = -3(x + 1)^2$

15. a. $y = x^2 + 6x - 7$ b. $y = 3x^2 - 6x - 7$ c. $y = 5 + 4x - 3x^2$ d. $y = 2x^2 - x - 4$

16. a. $y = (x - 5)^2 + 2$ b. $y = 2(x + 1)^2 - 2$ c. $y = -2(x - 3)^2 - 6$ d. $y = -(x - 4)^2 + 1$

17. a. $y = -2(1 + x)(2 - x)$ b. $y = (2x + 1)(2 - 3x)$ c. $y = 0.8x(10x - 27)$ d. $y = (3x + 1)^2$

18. a. $y = \frac{1}{4}(1 - 2x)^2$ b. $y = -0.25(1 + 2x)^2$ c. $y = -2x^2 + 3x - 4$ d. $y = 10 - 2x^2 + 8x$

19. a. i. Express $2x^2 - 12x + 9$ in the form $a(x + b)^2 + c$.
    ii. Hence, state the coordinates of the turning point of the graph of $y = 2x^2 - 12x + 9$.
    iii. State the minimum value of the polynomial $2x^2 - 12x + 9$.

    b. i. Express $-x^2 - 18x + 5$ in the form $a(x + b)^2 + c$.
    ii. Hence, state the coordinates of the turning point of the graph of $y = -x^2 - 18x + 5$.
    iii. State the maximum value of the polynomial $-x^2 - 18x + 5$.

20. **WE16** Use the discriminant to:

    a. determine the number and type of x-intercepts of the graph defined by $y = 42x - 18x^2$
    b. sketch the graph of $y = 42x - 18x^2$.

21. Show that $7x^2 - 4x + 9$ is positive definite.

22. Use the discriminant to determine the number and type of intercepts each of the following graphs makes with the x-axis.

    a. $y = 9x^2 + 17x - 12$
    b. $y = -5x^2 + 20x - 21$
    c. $y = -3x^2 - 30x - 75$
    d. $y = 0.02x^2 + 0.5x + 2$

23. Find the values of k for which the graph of $y = 5x^2 + 10x - k$ has:

    a. one x-intercept b. two x-intercepts c. no x-intercepts.

24. a. Find the values of m for which $mx^2 - 2x + 4$ is positive definite.

    b. i. Show that there is no real value of p for which $px^2 + 3x - 9$ is positive definite.
    ii. If $p = 3$, determine the equation of the axis of symmetry of the graph of $y = px^2 + 3x - 9$.

    c. i. Find the values of t for which the turning point of $y = 2x^2 - 3tx + 12$ lies on the x-axis.
    ii. Find the values of t for which the equation of the axis of symmetry of $y = 2x^2 - 3tx + 12$ is $x = 3t^2$.

25. Use CAS technology to give the exact coordinates of:

    i. the turning point
    ii. any x-intercepts for the parabolas defined by:

    a. $y = 12x^2 - 18x + 5$ b. $y = -8x^2 + 9x + 12$.

3.5 Exam questions

Question 1 (1 mark) TECH-ACTIVE

MC The graph of $y=(x-2)^2+3$ is the graph of $y=x^2$

A. translated 2 units to the right and 3 down.
B. translated 2 units to the right and 3 up.
C. translated 2 units to the left and 3 up.
D. translated 2 units to the left and 3 down.
E. translated 3 units to the right and 2 down.

Question 2 (1 mark) TECH-ACTIVE

MC The turning point of the graph of the quadratic function $f(x)=x^2-6x+7$ is

A. $(-3, 2)$
B. $(-3, -2)$
C. $(3, -2)$
D. $(3, 7)$
E. $(3, -7)$

Question 3 (5 marks) TECH-ACTIVE

Sketch the graph of $y=2x^2+x-3$, labelling all key points.

More exam questions are available online.

3.6 Determining the rule of a quadratic polynomial from a graph

LEARNING INTENTION

At the end of this subtopic you should be able to:
- identify quadratics and their key features from graphs.

Whether the equation of the graph of a quadratic polynomial is expressed in $y=ax^2+bx+c$ form, $y=a(x-h)^2+k$ form or $y=a(x-x_1)(x-x_2)$ form, each equation contains three unknowns. Hence, three pieces of information are needed to fully determine the equation. This means that exactly one parabola can be drawn through three non-collinear points.

3.6.1 Using different forms of the parabola

If the information given includes the turning point or the intercepts with the axes, one form of the equation may be preferable over another.

Identifying quadratics from a graph

As a guide:
- **if the turning point is given, use the $y=a(x-h)^2+k$ form**
- **if the x-intercepts are given, use the $y=a(x-x_1)(x-x_2)$ form**
- **if 3 points on the graph are given, use the $y=ax^2+bx+c$ form.**

WORKED EXAMPLE 17 Determining the equation of a parabola

Determine the rules for the following parabolas.

a.

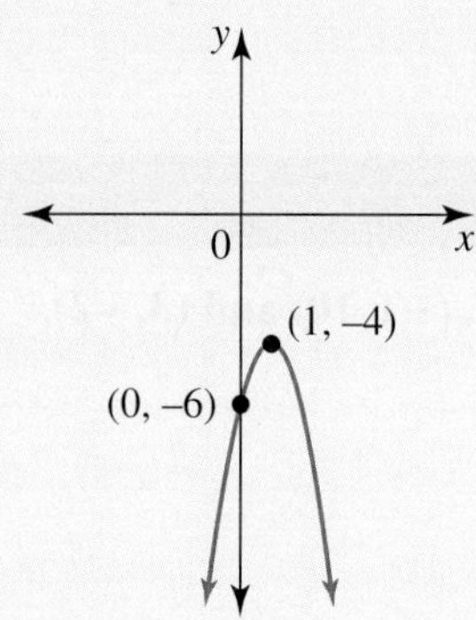

b.

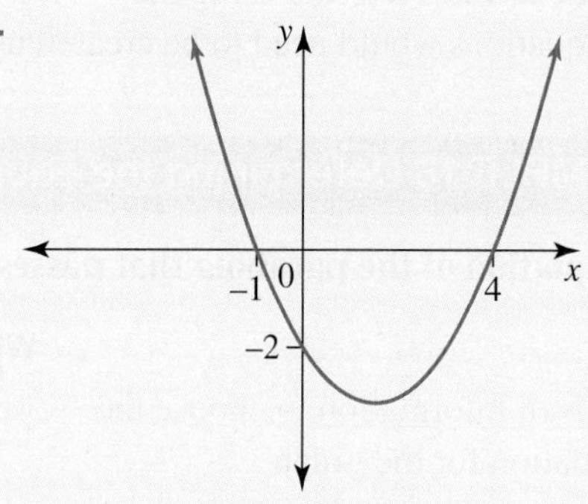

THINK	WRITE
a. 1. Consider the given information to choose the form of the equation for the graph.	**a.** Let the equation be $y = a(x - h)^2 + k$. Turning point $(1, -4)$ $\therefore y = a(x - 1)^2 - 4$
2. Determine the value of a.	Substitute the given point, $(0, -6)$. $-6 = a(0 - 1)^2 - 4$ $-6 = a - 4$ $\therefore a = -2$
3. Is the sign of a appropriate?	Check: the graph is concave down, so $a < 0$.
4. Write the rule for the graph. *Note:* Check if the question specifies whether the rule needs to be expanded into general form.	The equation of the parabola is $y = -2(x - 1)^2 - 4$.
b. 1. Consider the given information to choose the form of the equation for the graph.	**b.** Let the equation be $y = a(x - x_1)(x - x_2)$. Given $x_1 = -1, x_2 = 4$ $\therefore\ y = a(x + 1)(x - 4)$
2. Determine the value of a.	Substitute the third given point, $(0, -2)$ $-2 = a(0 + 1)(0 - 4)$ $-2 = a(1)(-4)$ $-2 = -4a$ $a = \frac{-2}{-4}$ $= \frac{1}{2}$
3. Is the sign of a appropriate?	Check: the graph is concave up, so $a > 0$.
4. Write the rule for the graph.	The equation is $y = \frac{1}{2}(x + 1)(x - 4)$.

3.6.2 Using simultaneous equations

In Worked example 17b, three points were available, but because two of them were key points, the x-intercepts, we chose to form the rule using the $y = a(x - x_1)(x - x_2)$ form. If the points were not key points, then simultaneous equations would need to be created using the coordinates given.

WORKED EXAMPLE 18 Determining the equation given 3 points

Determine the equation of the parabola that passes through the points (1, −4), (−1, 10) and (3, −2).

THINK	WRITE
1. Consider the given information to choose the form of the equation for the graph.	Let $y = ax^2 + bx + c$.
2. Substitute the first point to form an equation in a, b and c.	First point: $(1, -4) \Rightarrow -4 = a(1)^2 + b(1) + c$ $\therefore -4 = a + b + c$ [1]
3. Substitute the second point to form a second equation in a, b and c.	Second point: $(-1, 10) \Rightarrow 10 = a(-1)^2 + b(-1) + c$ $\therefore 10 = a - b + c$ [2]
4. Substitute the third point to form a third equation in a, b and c.	Third point: $(3, -2) \Rightarrow -2 = a(3)^2 + b(3) + c$ $\therefore -2 = 9a + 3b + c$ [3]
5. Write the equations as a system of 3×3 simultaneous equations.	$-4 = a + b + c$ [1] $10 = a - b + c$ [2] $-2 = 9a + 3b + c$ [3]
6. Solve the system of simultaneous equations using CAS technology.	The equation of the parabola is $y = 2x^2 - 7x + 1$.

TI | THINK	DISPLAY/WRITE
1. On a Calculator page, press MENU and select: 3. Algebra 7. Solve system of equations Number of equations: 3 Variables: a, b, c Then press OK. Complete the entry line as: solve $(-4 = a \times 1^2 + b \times 1 + c$ and $10 = a \times (-1)^2 + b \times -1 + c$ and $-2 = a \times 3^2 + b \times 3 + c, \{a, b, c\})$ Then press ENTER.	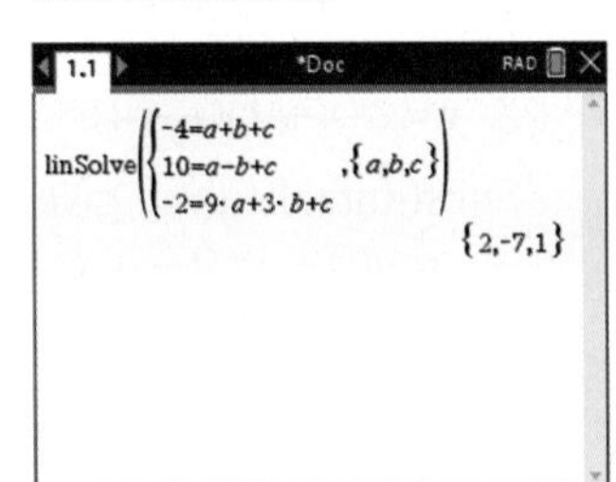
2. Substitute the values in the general quadratic equation.	$y = 2x^2 - 7x + 1$

CASIO | THINK	DISPLAY/WRITE
1. On a Main screen, complete the entry line as: solve $(\{-4 = a \times 1^2 + b \times 1 + c, 10 = a \times (-1)^2 + b \times -1 + c, -2 = a \times 3^2 + b \times 3 + c\}\{a, b, c\})$ Then press EXE.	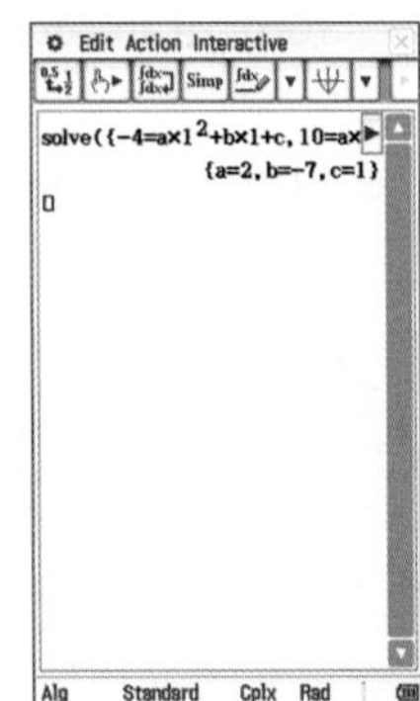
2. Substitute the values in the general quadratic equation.	$y = 2x^2 - 7x + 1$

3.6 Exercise

Students, these questions are even better in jacPLUS

Receive immediate feedback and access sample responses

Access additional questions

Track your results and progress

Find all this and MORE in jacPLUS

Technology free

1. a. A parabola with equation $y = x^2 + c$ passes through the point $(1, 5)$. Determine the value of c and state the equation of the parabola.
   b. A parabola with equation $y = ax^2$ passes through the point $(6, -2)$. Calculate the value of a and state the equation of the parabola.
   c. A parabola with equation $y = a(x - 2)^2$ passes through the point $(0, -12)$. Calculate the value of a and state the equation of the parabola.

2. The parabola $y = x^2 + bx$ passes through the point $(-3, 3)$.
   a. Calculate b and form the parabola's equation.
   b. Find the x-intercepts of the graph of the parabola.

3. a. Form the equation of the parabola $y = (x - a)^2 + c$ given it has a turning point at $(4, -8)$.
   b. Form the equation of the parabola $y = -2(x - h)^2 + k$ given it has a turning point at $(-1, 3)$.

4. a. Express the equation of the parabola $y = A(x - h)^2 + k$ in terms of A, given it has a vertex at $(5, 12)$.
   b. If the parabola in part **a** also has a y-intercept at $(0, 7)$, calculate the value of A and state the parabola's equation.

5. a. State the two linear factors of the equation of the parabola whose x-intercepts occur at $x = 3$ and at $x = 8$, and form a possible equation for this parabola.
   b. The x-intercepts of a parabola occur at $x = -11$ and $x = 2$. Form a possible equation for this parabola.

6.  WE17 Determine the rules for the following parabolas.

a.

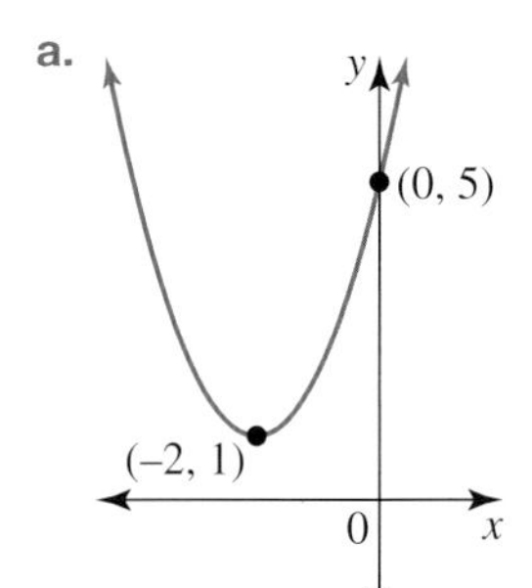

b.

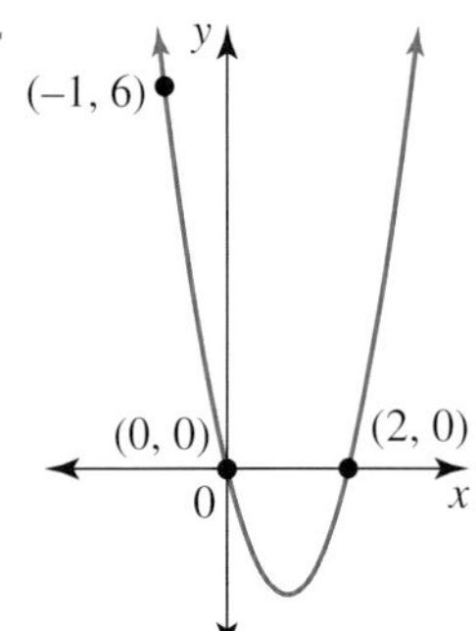

7. Form the rule for the following parabola. Express the rule in expanded polynomial form.

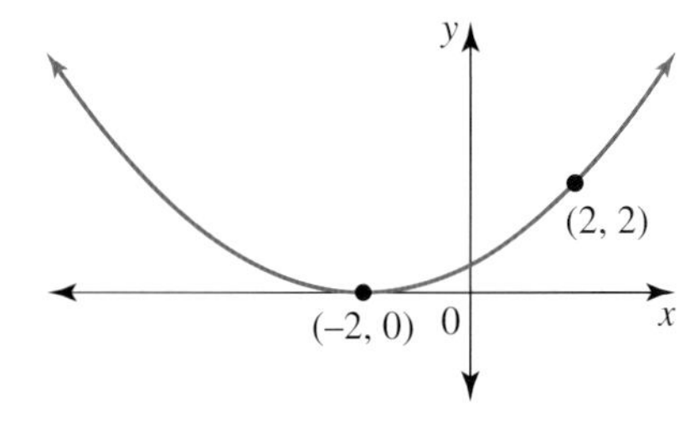

8. Determine the equation of each of the parabolas shown in the diagrams.

a.

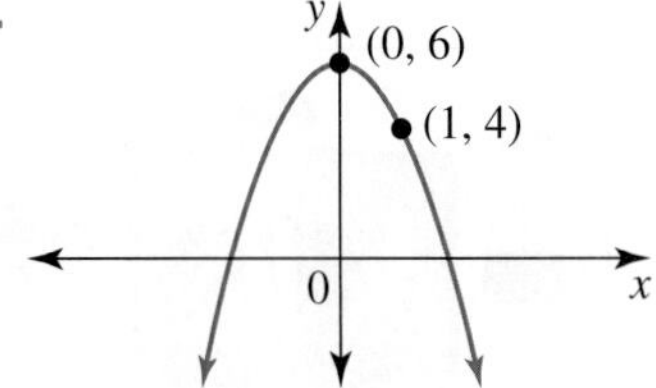

b.

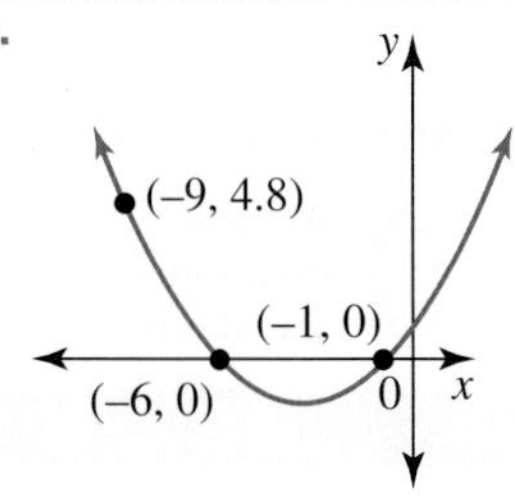

9. For the graph of the parabola shown:

a. determine its rule

b. calculate the length between the intercepts of the graph and the x-axis.

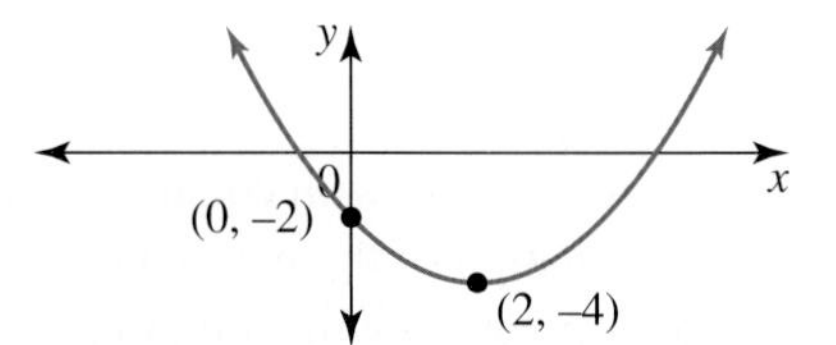

10. a. A parabola has the same shape as $y = 3x^2$, but its vertex is at $(-1, -5)$. Write its equation in $y = ax^2 + bx + c$ form.

b. The parabola with equation $y = (x - 3)^2$ is translated 8 units to the left. Write the equation of its image.

11. For each of the following graphs, two possible equations are given. Select the correct equation for each graph.

a.

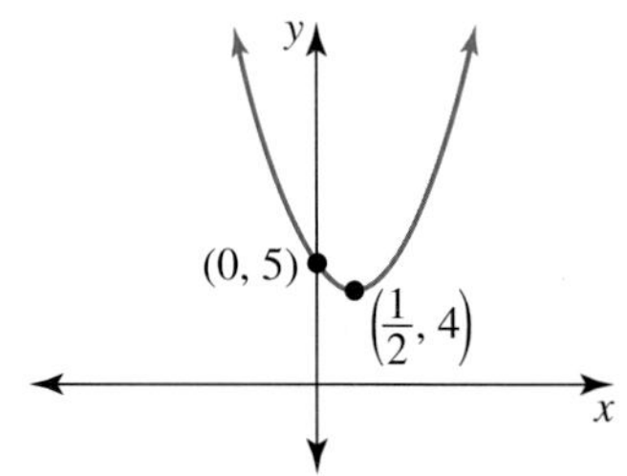

Equation A: $y = (2x - 1)^2 + 4$ Equation B: $y = \frac{1}{4}\left(x - \frac{1}{2}\right)^2 + 4$.

b.

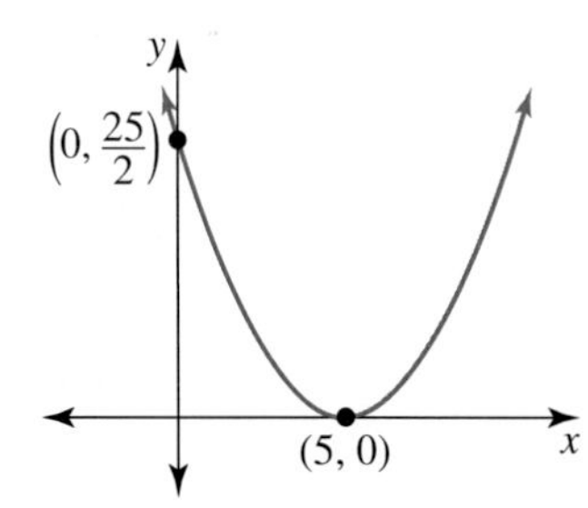

Equation A: $y = \frac{1}{2}(5 - x)^2$ Equation B: $y = 2(x - 5)^2$.

c.

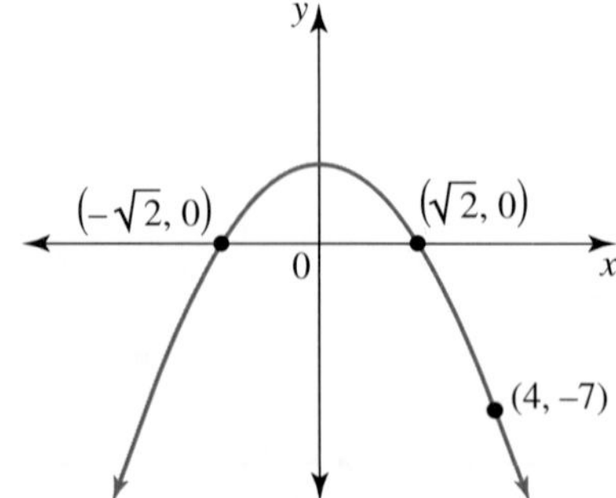

Equation A: $y = -\frac{1}{2}(x^2 - 2)$ Equation B: $y = -\frac{7}{18}(x^2 + 2)$.

Technology active

12. WE18 Determine the equation of the parabola that passes through the points $(-1, -7)$, $(2, -10)$ and $(4, -32)$.

13. Determine the equation of the parabola that passes through the points $(0, -2)$, $(-1, 0)$ and $(4, 0)$.

14. A parabola contains the three points A $(-1, 10)$, B $(1, 0)$ and C $(2, 4)$.
   a. Determine its equation.
   b. Determine the coordinates of its intercepts with the coordinate axes.
   c. Determine the coordinates of its vertex.
   d. Sketch the graph, showing the points A, B and C.

15. a. Give equations for three possible members of the family of parabolas that have x-intercepts of $(-3, 0)$ and $(5, 0)$.
   b. One member of this family of parabolas has a y-intercept of $(0, 45)$. Determine its equation and its vertex.

16. The axis of symmetry of a parabola has the equation $x = 4$. If the points $(0, 6)$ and $(6, 0)$ lie on the parabola, form its equation, expressing it in $y = ax^2 + bx + c$ form.

17. A parabola has the equation $y = (ax + b)(x + c)$. When $x = 5$, its graph cuts the x-axis, and when $y = -10$, the graph cuts the y-axis.
   a. Show that $y = ax^2 + (2 - 5a)x - 10$.
   b. Express the discriminant in terms of a.
   c. If the discriminant is equal to 4, determine the equation of the parabola and the coordinates of its other x-intercept.

18. a. The graph of a parabola touches the x-axis at $x = -4$ and passes through the point $(2, 9)$. Determine its equation.
   b. A second parabola touches the x-axis at $x = p$ and passes through the points $(2, 9)$ and $(0, 36)$. Show there are two possible values for p. For each possible value, form the equation of the parabola, and sketch the parabolas on the same axes.

3.6 Exam questions

Question 1 (1 mark) TECH-ACTIVE

MC The equation of the parabola shown could be

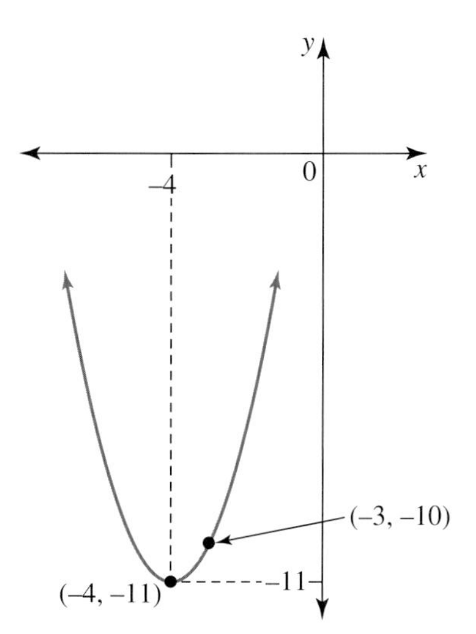

A. $y = x^2 + 8x + 5$ **B.** $y = x^2 - 8x - 5$ **C.** $y = x^2 + 8x + 16$
D. $y = x^2 - 8x + 16$ **E.** $y = x^2 - 8x + 5$

Question 2 (2 marks) TECH-ACTIVE

Determine the equation of the parabola in the form $y = ax^2 + bx + c$ that has a y-intercept of -12 and x-intercepts of 4 and -3.

Question 3 (1 mark) TECH-ACTIVE

MC A parabola has a turning point of (2, 3) and a y-intercept of 1. The equation of this parabola could be

A. $y = (x+2)^2 - 3$ **B.** $y = (x+2)^2 + 3$ **C.** $y = \frac{1}{2}(x+2)^2 + 3$

D. $y = -\frac{1}{2}(x-2)^2 + 3$ **E.** $y = (x-2)^2 + 3$

More exam questions are available online.

3.7 Quadratic inequations

LEARNING INTENTION

At the end of this subtopic you should be able to:

- solve simple algebraic inequalities by hand
- use the discriminant to determine whether a straight line intersects a parabola.

If $ab > 0$, this could mean $a > 0$ and $b > 0$, or it could mean $a < 0$ and $b < 0$.

Solving a quadratic inequation involving the product of factors is not as straightforward as solving a linear inequation. To assist in the solution of a quadratic inequation, either the graph or its **sign diagram** is a useful reference.

3.7.1 Sign diagrams of quadratics

A **sign diagram** is like a 'squashed' graph with only the x-axis showing. The sign diagram indicates the values of x where the graph of a quadratic polynomial is above, on or below the x-axis. It shows the x-values for which $ax^2 + bx + c > 0$, the x-values for which $ax^2 + bx + c = 0$, and the x-values for which $ax^2 + bx + c < 0$. A graph of the quadratic shows the same information and could be used, but usually the sign diagram is simpler to draw when solving a quadratic inequation. Unlike a graph, scaling and turning points that do not lie on the x-axis are not important in a sign diagram.

The three types of graphs with either 2, 1 or 0 x-intercepts are shown together with their matching sign diagrams for concave up and concave down parabolas.

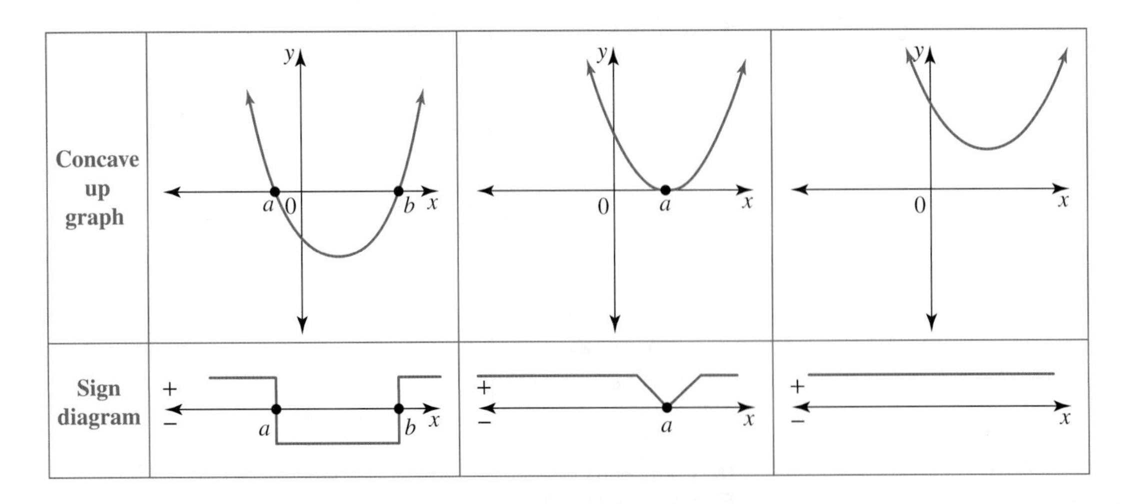

Concave down graph	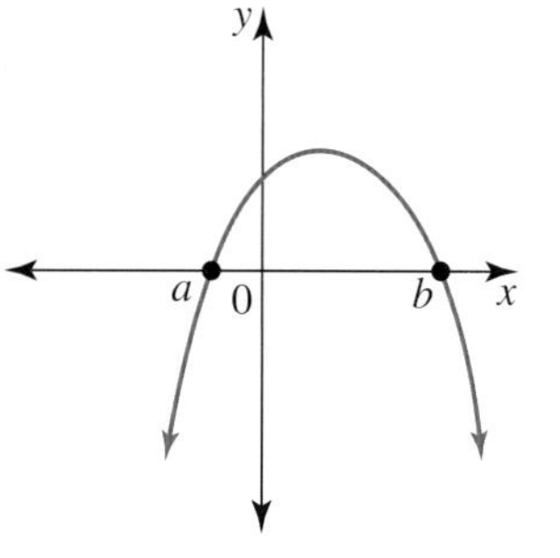	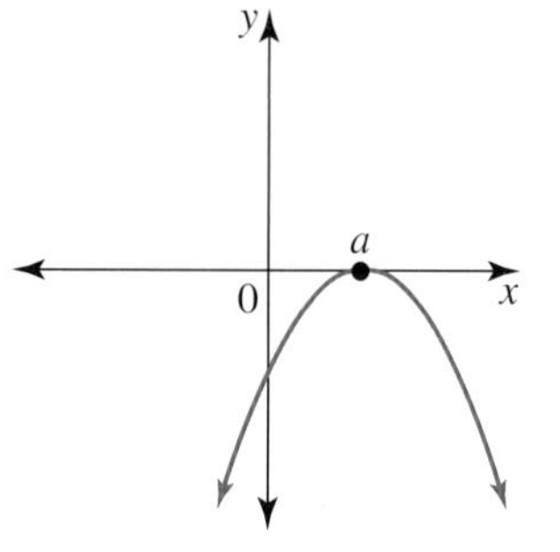	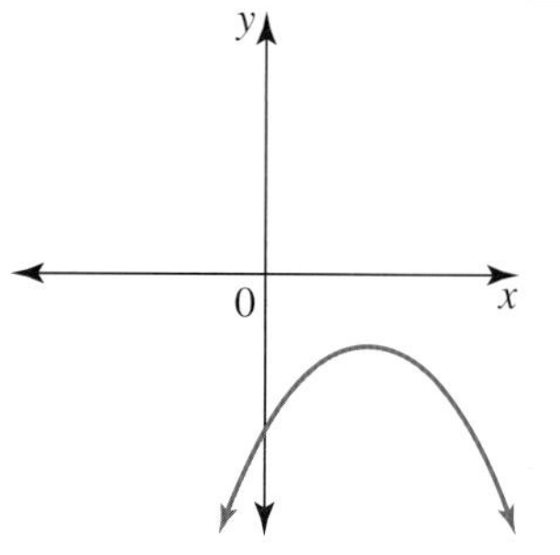
Sign diagram	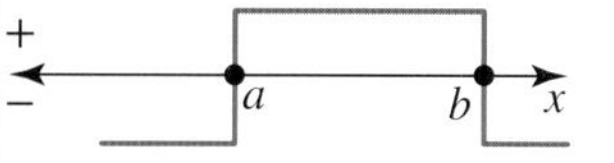	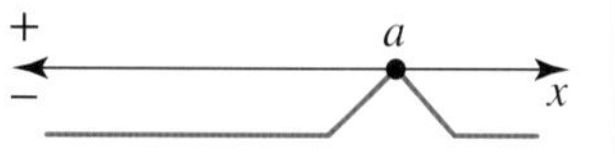	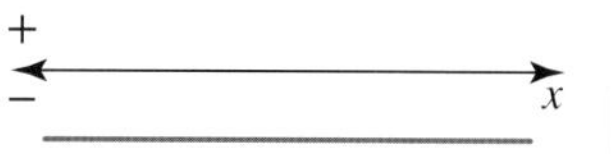

Sign diagrams of quadratics

To draw a sign diagram of $ax^2 + bx + c$:

- **find the zeros of the quadratic expression by solving $ax^2 + bx + c = 0$**
- **start the sign diagram below the axis if $a < 0$ and above the axis if $a > 0$**
- **either touch or cut through the axis at each zero depending whether the zero is a repeated one or not.**

Repeated zeros are said to have **multiplicity** 2, whereas non-repeated ones have multiplicity 1.

WORKED EXAMPLE 19 Drawing a sign diagram

Draw the sign diagram of $(4 - x)(2x - 3)$.

THINK	WRITE
1. Find the zeros of the quadratic.	Zeros occur when $(4 - x)(2x - 3) = 0$. $(4 - x) = 0$ or $(2x - 3) = 0$ $x = 4$ or $x = 1.5$
2. Draw the x-axis and mark the zeros in the correct order.	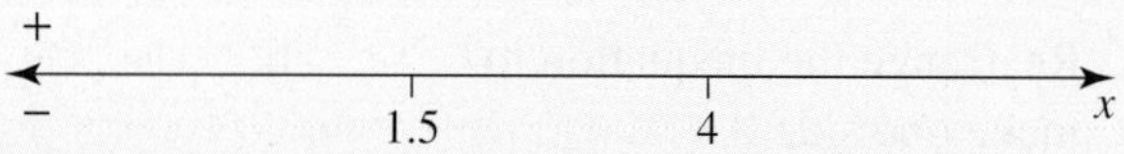
3. Consider the coefficient of x^2 to determine the concavity.	Multiplying the x terms from each bracket of $(4 - x)(2x - 3)$ gives $-2x^2$. Therefore, the graph is concave down.
4. Draw the sign diagram.	The sign diagram starts below the x-axis and cuts the axis at each zero. 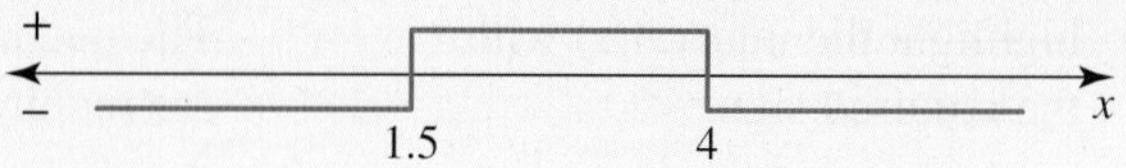

3.7.2 Solving quadratic inequations

To solve a quadratic inequation:

- rearrange the terms in the inequation, if necessary, so that one side of the inequation is 0 (similar to solving a quadratic equation)
- calculate the zeros of the quadratic inequation and draw its sign diagram
- read from the sign diagram the set of values of x that satisfy the inequation.

WORKED EXAMPLE 20 Using sign diagrams to solve quadratic inequations

a. Solve the quadratic inequation $(4-x)(2x-3)>0$ using the sign diagram from Worked example 19 and check the solution using a selected value for x.

b. Solve $\{x : x^2 \geq 3x+10\}$.

THINK / **WRITE**

a. 1. Copy the sign diagram from Worked example 19.

a.

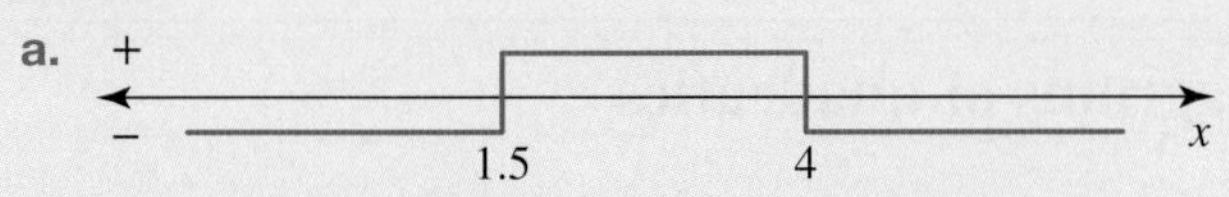

2. Highlight the part of the sign diagram that shows the values required by the inequation.

$(4-x)(2x-3)>0$

Positive values of the quadratic lie above the x-axis.

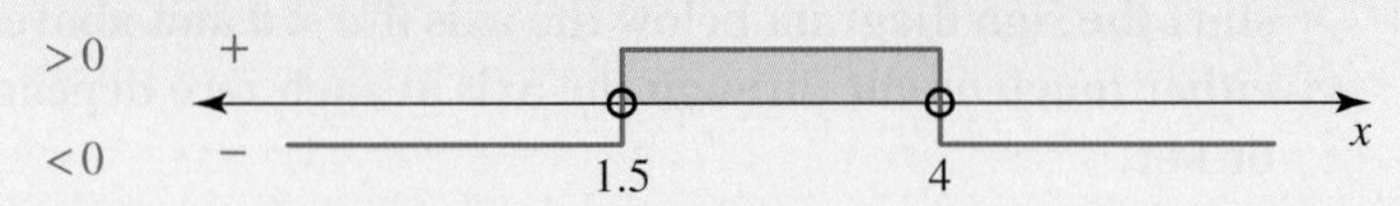

3. State the interval in which the required solutions lie.
Note: The interval is open because the original inequation has an open inequality sign.

Therefore, the solution is $1.5<x<4$.

4. Choose an x-value that lies in the solution interval and check it satisfies the inequation.

Let $x=2$.
Substitute in the LHS of $(4-x)(2x-3)>0$.

$$\begin{aligned}(4-2)(2\times2-3)&=2\times1\\&=2\\&>0\end{aligned}$$

Therefore, $x=2$ is in the solution set.

b. 1. Rearrange the inequation to make one side 0.

b.

$$\begin{aligned}x^2&\geq 3x+10\\x^2-3x-10&\geq 0\end{aligned}$$

2. Calculate the zeros of the quadratic.

Let $x^2-3x-10=0$.

$$\begin{aligned}(x-5)(x+2)&=0\\x&=5, \text{ or } x=-2\end{aligned}$$

3. Draw the sign diagram and highlight the interval(s) with the required sign.

$x^2-3x-10\geq 0$

The quadratic is concave up, so values above or on the x-axis are required.

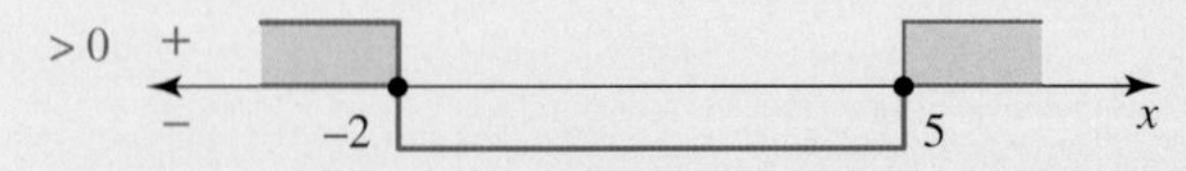

4. State the intervals required.

$\therefore (x-5)(x+2)\geq 0$ when $x\leq -2$ or $x\geq 5$

5. State the answer in set notation.

The answer is $\{x : x\leq -2\}\cup\{x : x\geq 5\}$.

3.7.3 Intersections of lines and parabolas

The possible number of points of intersection between a straight line and a parabola will be either 0, 1 or 2.

- If there is no point of intersection, the line makes no contact with the parabola.

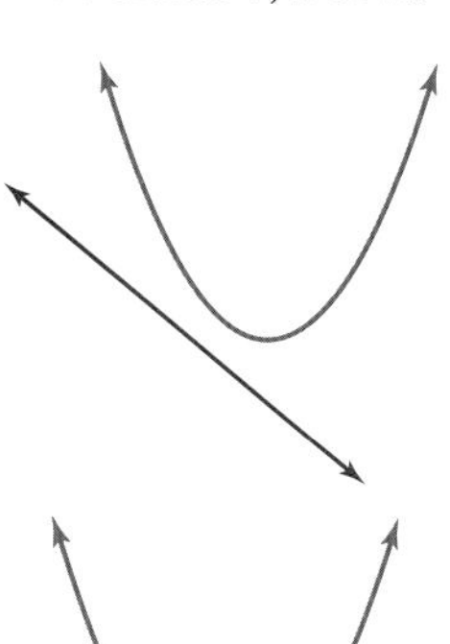

- If there is 1 point of intersection, a non-vertical line is a **tangent line** to the parabola, touching the parabola at that one point of contact.

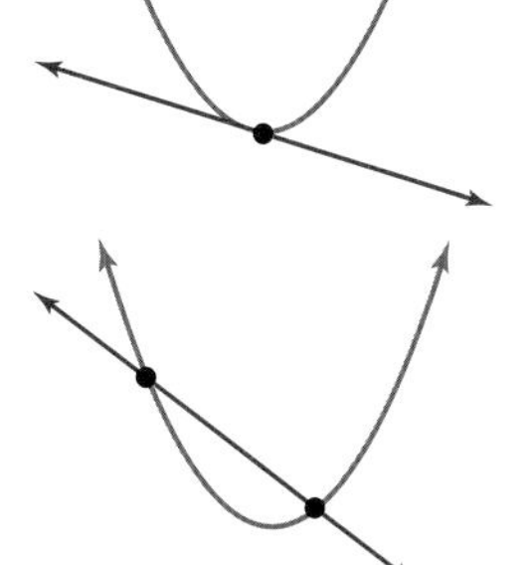

- If there are 2 points of intersection, the line cuts through the parabola at these points.

WORKED EXAMPLE 21 Determining points of intersections

a. Calculate the coordinates of the points of intersection of the parabola $y = x^2 - 3x - 4$ and the line $y - x = 1$.

b. Find how many points of intersection there are between the graphs of $y = 2x - 5$ and $y = 2x^2 + 5x + 6$.

THINK	WRITE
a. 1. Set up the simultaneous equations.	a. $y = x^2 - 3x - 4 \quad [1]$ $y - x = 1 \quad [2]$
2. Substitute from the linear equation into the quadratic equation.	From equation [2], $y = x + 1$. Substitute this into equation [1]. $x + 1 = x^2 - 3x - 4$ $x^2 - 4x - 5 = 0$
3. Solve the newly created quadratic equation for the x-coordinates of the points of intersection of the line and parabola.	$x^2 - 4x - 5 = 0$ $(x + 1)(x - 5) = 0$ $x = -1$ or $x = 5$
4. Find the matching y-coordinates using the simpler linear equation.	In equation [2]: when $x = -1, y = 0$ when $x = 5, y = 6$.
5. State the coordinates of the points of intersection.	The points of intersection are $(-1, 0)$ and $(5, 6)$.

THINK	WRITE
b. 1. Set up the simultaneous equations.	**b.** $y = 2x - 5$ [1] $y = 2x^2 + 5x + 6$ [2]
2. Substitute the linear equation into the quadratic equation. Rearrange the resulting equation into polynomial form.	Substitute equation [1] in equation [2]. $2x - 5 = 2x^2 + 5x + 6$ $2x^2 + 3x + 11 = 0$
3. The discriminant of this quadratic equation determines the number of solutions.	$\Delta = b^2 - 4ac, a = 2, b = 3, c = 11$ $= (3)^2 - 4(2)(11)$ $= -79$ $\therefore \Delta < 0$ There are no points of intersection between the two graphs.

TI | THINK — **DISPLAY/WRITE**

a. 1. On a Graphs page, complete the entry lines as:
$f1(x) = x^2 - 3x - 4$
$f2(x) = x + 1$
Press ENTER after each entry to view the graphs.

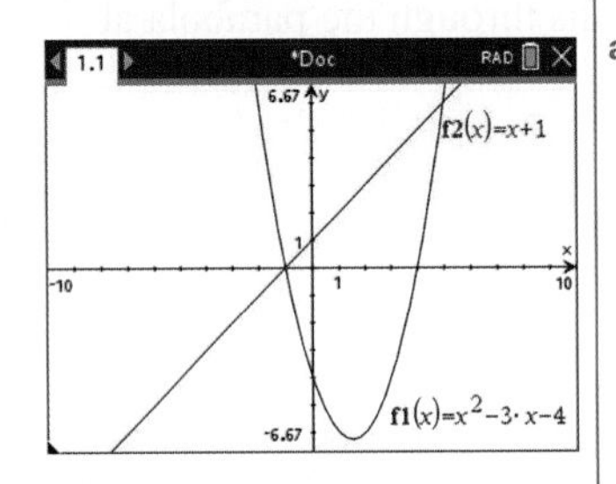

2. To view the point of intersections, select:
- MENU
- 6. Analyze Graph.
- 4. Intersection

Follow the prompts to show the key points.

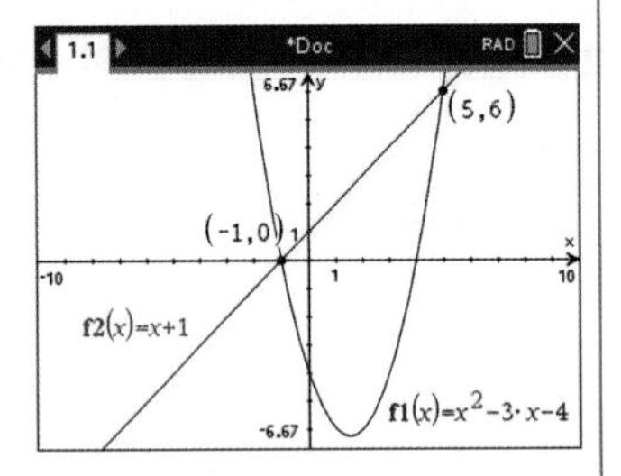

3. State the points of intersection.

$(-1, 0)$ and $(5, 6)$

CASIO | THINK — **DISPLAY/WRITE**

a. 1. On a Graphs & Table screen, complete the entry lines as:
$y1 = x^2 - 3x - 4$
$y2 = x + 1$
Tap the Graph icon to view the graphs.

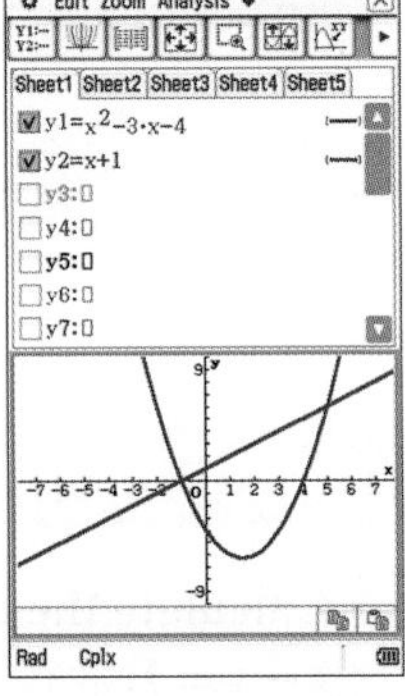

2. To locate the points of intersection, select:
- Analysis
- Trace

Trace around the graph to locate the points of intersection.

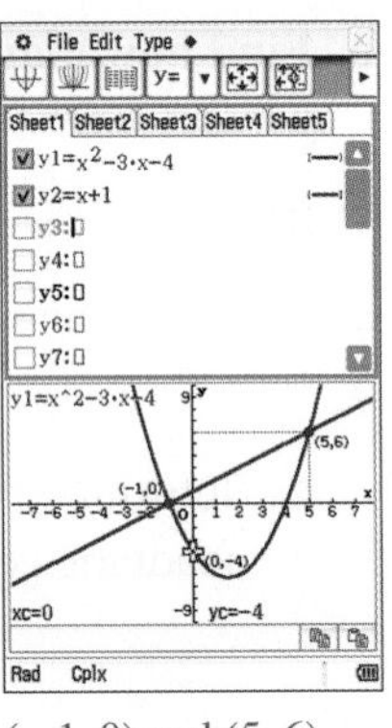

3. State the points of intersection.

$(-1, 0)$ and $(5, 6)$

3.7.4 Quadratic inequations in discriminant analysis

The need to solve a quadratic inequation as part of the analysis of a problem can occur in a number of situations, an example of which arises when a discriminant is itself a quadratic polynomial in some variable.

WORKED EXAMPLE 22 Using the discriminant for points of intersection

Find the values of m for which there will be at least one intersection between the line $y = mx + 5$ and the parabola $y = x^2 - 8x + 14$.

THINK	WRITE
1. Set up the simultaneous equations and form the quadratic equation from which any solutions will be generated.	$y = mx + 5$ [1] $y = x^2 - 8x + 14$ [2] Substitute from equation [1] into equation [2]. $mx + 5 = x^2 - 8x + 14$ $x^2 - 8x - mx + 9 = 0$ $x^2 - (8 + m)x + 9 = 0$
2. Obtain an algebraic expression for the discriminant of this equation.	$\Delta = b^2 - 4ac, a = 1, b = -(8 + m), c = 9$ $= (-(8 + m))^2 - 4(1)(9)$ $= (8 + m)^2 - 36$
3. For at least one intersection, $\Delta \geq 0$. Impose this condition on the discriminant to set up a quadratic inequation.	$\Delta \geq 0$ for one or two intersections. $\therefore\ (8 + m)^2 - 36 \geq 0$
4. Solve this inequation for m by finding the zeros and using a sign diagram of the discriminant.	$((8 + m) - 6)((8 + m) + 6) \geq 0$ $(2 + m)(14 + m) \geq 0$ The zeros are $m = -2, m = -14$. The sign diagram of $(2 + m)(14 + m)$ will be that of a concave up quadratic. 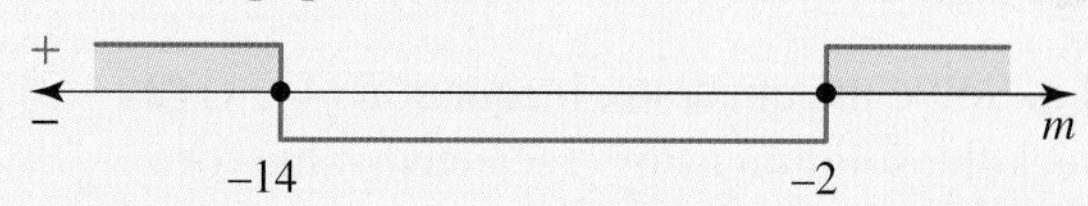 $\therefore m \leq -14$ or $m \geq -2$
5. State the answer.	If $m \leq -14$ or $m \geq -2$, there will be at least one intersection between the line and the parabola.

3.7 Exercise

Students, these questions are even better in jacPLUS

Receive immediate feedback and access sample responses

Access additional questions

Track your results and progress

Find all this and MORE in jacPLUS

Technology free

1. a. Draw the sign diagram of $(x+5)(x-5)$.
   b. Hence, state the values of x for which $(x+5)(x-5) \geq 0$.

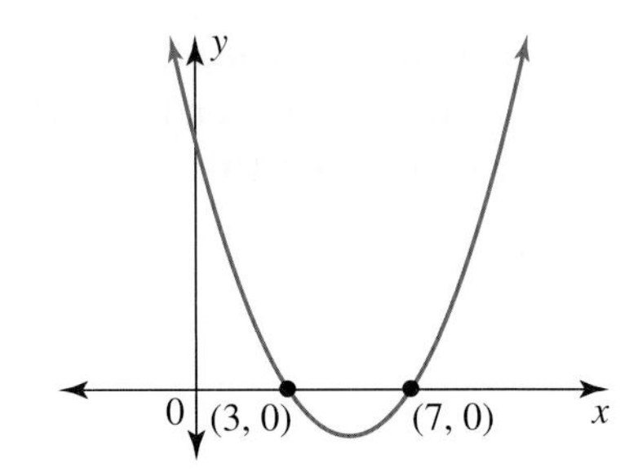

2. The diagram shows part of the graph of $y=(x-3)(x-7)$.
   a. Draw the sign diagram for this graph.
   b. Sate the values of x for which $(x-3)(x-7) \leq 0$.

3. The sign diagram of $x(3-x)$ is shown. Use the sign diagram to state:
   a. $\{x : x(3-x) < 0\}$
   b. $\{x : x(3-x) > 0\}$.

4. a. Draw the sign diagram of $(x-6)^2$.
   b. Hence, state the set of values of x for which $(x-6)^2 > 0$.

5. Consider the quadratic polynomial expression $9+3x-2x^2$.
   a. Factorise $9+3x-2x^2$ into its two linear factors.
   b. State the zeros of $9+3x-2x^2$ and draw its sign diagram.
   c. Find the values of x for which $9+3x-2x^2 \geq 0$.

6. **WE19** Draw the sign diagram of $(x+3)(x-4)$.

7. Draw the sign diagram of $81x^2-18x+1$.

8. **WE20** a. Solve the quadratic inequation $(4-x)(2x-3) \leq 0$ using the sign diagram from Worked example 19 and check the solution using a selected value for x.
   b. Calculate $\{x : 6x^2 < x+2\}$.

9. Solve $6x < x^2+9$.

10. Solve the following quadratic inequations.
    a. $x^2+8x-48 \leq 0$
    b. $-x^2+3x+4 \leq 0$
    c. $3(3-x) < 2x^2$
    d. $(x+5)^2 < 9$
    e. $9x < x^2$
    f. $5(x-2) \geq 4(x-2)^2$

11. Calculate the following sets.
    a. $\{x : 36-12x+x^2 > 0\}$
    b. $\{x : 6x^2-12x+6 \leq 0\}$
    c. $\{x : -8x^2+2x < 0\}$
    d. $\{x : x(1+10x) \leq 21\}$

12. **WE21** a. Calculate the coordinates of the points of intersection of the parabola $y=x^2+3x-10$ and the line $y+x=2$.
    b. Find how many points of intersection there are between the graphs of $y=6x+1$ and $y=-x^2+9x-5$.

13. Solve each of the following pairs of simultaneous equations.
    a. $y=5x+2$
       $y=x^2-4$
    b. $4x+y=3$
       $y=x^2+3x-5$
    c. $2y+x-4=0$
       $y=(x-3)^2+4$
    d. $\dfrac{x}{3}+\dfrac{y}{5}=1$
       $x^2-y+5=0$

14. Obtain the coordinates of the point(s) of intersection of:
 a. the line $y = 2x + 5$ and the parabola $y = -5x^2 + 10x + 2$
 b. the line $y = -5x - 13$ and the parabola $y = 2x^2 + 3x - 5$
 c. the line $y = 10$ and the parabola $y = (5 - x)(6 + x)$
 d. the line $19x - y = 46$ and the parabola $y = 3x^2 - 5x + 2$.

Technology active

15. Use a discriminant to determine the number of intersections of:
 a. the line $y = 4 - 2x$ and the parabola $y = 3x^2 + 8$
 b. the line $y = 2x + 1$ and the parabola $y = -x^2 - x + 2$
 c. the line $y = 0$ and the parabola $y = -2x^2 + 3x - 2$.

16. Show that the line $y = 4x$ is a tangent to the parabola $y = x^2 + 4$ and sketch the line and parabola on the same diagram, labelling the coordinates of the point of contact.

17. WE22 Find the values of m for which there will be at least one intersection between the line $y = mx - 7$ and the parabola $y = 3x^2 + 6x + 5$.

18. Determine the values of k for which there will be no intersection between $y = kx + 9$ and $y = x^2 + 14$.

19. Determine the values of k so that $y = (k - 2)x + k$ and $y = x^2 - 5x$ will have:
 a. no intersections
 b. one point of intersection
 c. two points of intersection.

20. a. Determine the values of p for which the equation $px^2 - 2px + 4 = 0$ will have real roots.
 b. Determine the real values of t for which the line $y = tx + 1$ will not intersect the parabola $y = 2x^2 + 5x + 11$.
 c. Determine the values of n for which the line $y = x$ will be a tangent to the parabola $y = 9x^2 + nx + 1$.

21. Consider the line $2y - 3x = 6$ and the parabola $y = x^2$.
 a. Calculate the coordinates of their points of intersection, correct to 2 decimal places.
 b. Sketch the line and the parabola on the same diagram.
 c. Use inequations to describe the region enclosed between the two graphs.
 d. Calculate the y-intercept of the line parallel to $2y - 3x = 6$ that is a tangent to the parabola $y = x^2$.

22. Use CAS technology to determine the solutions to:
 a. $19 - 3x - 5x^2 < 0$
 b. $6x^2 + 15x \leq 10$.

23. a. Sketch the parabolas $y = (x + 2)^2$ and $y = 4 - x^2$ on the one diagram and hence determine their points of intersection.
 b. i. Show that the parabolas $y = (x + 2)^2$ and $y = k - x^2$ have one point of intersection if $k = 2$.
  ii. Sketch $y = (x + 2)^2$ and $y = 2 - x^2$ on the one diagram, labelling their common point C with its coordinates.
  iii. The line $y = ax + b$ is the tangent to both curves at point C. Find its equation.

24. a. The equation $x^2 - 5x + 4 = 0$ gives the x-coordinates of the points of intersection of the parabola $y = x^2$ and a straight line. State the equation of this line.
 b. The equation $3x^2 + 9x - 2 = 0$ gives the x-coordinates of the points of intersection of the parabola and the straight line $y = 3x + 1$. State the equation of the parabola.

25. a. Using CAS technology, draw on the one diagram the graphs of $y = 2x^2 - 10x$ with the family of lines $y = -4x + a$, for $a = -6, -4, -2, 0$, and determine the points of intersection for each of these values of a.
 b. Use CAS technology to solve the equation $2x^2 - 10x = -4x + a$ to obtain x in terms of a.
 c. Hence, obtain the value of a for which $y = -4x + a$ is a tangent to $y = 2x^2 - 10x$ and give the coordinates of the point of contact in this case.

3.7 Exam questions

Question 1 (1 mark) TECH-ACTIVE

MC The solution to the inequality $(4x-3)(2x+1)>0$ is

A. $\{x : x>1\}$
B. $\{x : -0.75<x<0.5\}$
C. $\{x : -0.5<x<0.75\}$
D. $\{x : x<0.75\} \cup \{x : x>0.5\}$
E. $\{x : x<-0.5\} \cup \{x : x>0.75\}$

Question 2 (1 mark) TECH-ACTIVE

MC The solution to the inequality $-2x^2-x+3 \geq 0$ is

A. $x \in \left(-\infty, -\frac{3}{2}\right) \cup (1, \infty)$
B. $x \in \left(-\frac{3}{2}, 1\right)$
C. $x \in \left(-\infty, -\frac{3}{2}\right] \cup [1, \infty)$
D. $x \in \left(-1, \frac{3}{2}\right)$
E. $x \in \left[-1, \frac{3}{2}\right]$

Question 3 (3 marks) TECH-FREE

Determine the values of m for which the line $y=mx+24$ will be a tangent to the parabola $x^2+4x+33$.

More exam questions are available online.

3.8 Quadratic models and applications

LEARNING INTENTION

At the end of this subtopic you should be able to:
- solve problems using quadratic models.

Quadratic polynomials can be used to model a number of situations such as the motion of a falling object and the time of flight of a projectile. They can be used to model the shape of physical objects such as bridges, and they can also occur in economic models of cost and revenue.

3.8.1 Maximum and minimum values

The greatest or least value of the quadratic model is often of interest.

A quadratic reaches its maximum or minimum value at its turning point. The y-coordinate of the turning point represents the maximum or minimum value, depending on the nature of the turning point.

$a<0$	The quadratic is concave down, giving $a(x-h)^2+k \leq k$, so the maximum value of the quadratic is k.	(h, k)
$a>0$	The quadratic is concave up, giving $a(x-h)^2+k \geq k$, so the minimum value of the quadratic is k.	(h, k)

WORKED EXAMPLE 23 Solving problems involving quadratic models

A stone is thrown vertically into the air so that its height h metres above the ground after t seconds is given by $h = 1.5 + 5t - 0.5t^2$.

a. **Find the greatest height that the stone reaches.**
b. **Find how many seconds the stone takes to reach its greatest height.**
c. **Find the times at which the stone is 6 metres above the ground. State why this occurs two times.**
d. **Sketch the graph and give the time to return to the ground to 1 decimal place.**

THINK	WRITE
a. 1. The turning point is required. Calculate the coordinates of the turning point and state its type.	a. $h = 1.5 + 5t - 0.5t^2$ $a = -0.5, b = 5, c = 1.5$ Turning point: The axis of symmetry has equation $t = -\frac{b}{2a}$. $t = -\frac{5}{2 \times (-0.5)}$ $= 5$ When $t = 5$, $h = 1.5 + 5(5) - 0.5(5)^2$ $= 14$ The turning point is $(5, 14)$. This is a maximum turning point as $a < 0$.
2. State the answer. *Note:* The turning point is in the form (t, h) as t is the independent variable and h the dependent variable. The greatest height is the h-coordinate.	Therefore, the greatest height the stone reaches is 14 metres above the ground.
b. The required time is the t-coordinate of the turning point.	b. The stone reaches its greatest height after 5 seconds.
c. 1. Substitute the given height and solve for t.	c. $h = 1.5 + 5t - 0.5t^2$ When $h = 6$, $6 = 1.5 + 5t - 0.5t^2$. $0.5t^2 - 5t + 4.5 = 0$ $t^2 - 10t + 9 = 0$ $(t-1)(t-9) = 0$ $\therefore t = 1$ or $t = 9$
2. Interpret the answer.	Therefore, the first time the stone is 6 metres above the ground is 1 second after it has been thrown into the air and is rising upwards. It is again 6 metres above the ground after 9 seconds when it is falling down.

d. 1. Calculate the time the stone returns to the ground.

d. The stone returns to the ground when $h = 0$.

$$0 = 1.5 + 5t - 0.5t^2$$
$$t^2 - 10t - 3 = 0$$
$$t^2 - 10t = 3$$
$$t^2 - 10t + 25 = 3 + 25$$
$$(t-5)^2 = 28$$
$$t = 5 \pm \sqrt{28}$$
$$t \simeq 10.3 \text{ (reject negative value)}$$

The stone reaches the ground after 10.3 seconds.

2. Sketch the graph from its initial height to when the stone hits the ground. Label the axes appropriately.

When $t = 0, h = 1.5$, so the stone is thrown from a height of 1.5 metres.
Initial point: (0, 1.5)
Maximum turning point: (5, 14)
End point: (10.3, 0)

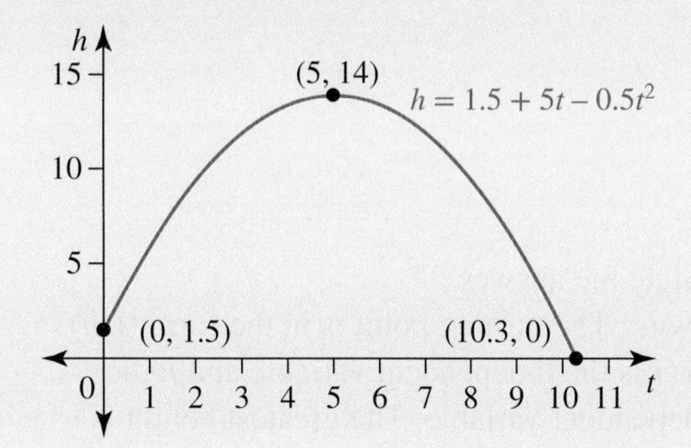

Resources

Interactivity Projectiles (int-2563)

3.8 Exercise

Students, these questions are even better in jacPLUS

Receive immediate feedback and access sample responses

Access additional questions

Track your results and progress

Find all this and MORE in jacPLUS

Technology active

1. A gardener has 30 metres of edging to enclose a rectangular area using the back fence as one edge.
   a. Show the area function is $A = 30x - 2x^2$, where A square metres is the area of the garden bed of width x metres.
   b. Calculate the dimensions of the garden bed for the maximum area that can be enclosed.
   c. Calculate the maximum area.

2. **WE23** A missile is fired vertically into the air from the top of a cliff so that its height h metres above the ground after t seconds is given by $h = 100 + 38t - \frac{19}{12}t^2$.

   a. Find the greatest height the missile reaches.
   b. Find how many seconds the missile takes to reach its greatest height.
   c. Sketch the graph and give the time to return to the ground to 1 decimal place.

3. A child throws a ball vertically upwards so that after t seconds its height h metres above the ground is given by $10h = 16t + 4 - 9t^2$.

   a. Calculate how long the ball takes to reach the ground.
   b. Determine whether the ball will strike the foliage overhanging from a tree if the foliage is 1.6 metres vertically above the ball's point of projection.
   c. Calculate the greatest height the ball could reach.

4. In a game of volleyball, a player serves a 'sky-ball' serve from the back of a playing court of length 18 metres. The path of the ball can be considered to be part of the parabola $y = 1.2 + 2.2x - 0.2x^2$ where x (in metres) is the horizontal distance travelled by the ball from where it was hit and y (in metres) is the vertical height the ball reaches.

   a. Use the 'completing the square' technique to express the equation in the form $y = a(x - b)^2 + c$.
   b. Calculate how high the volleyball reaches.
   c. The net is 2.43 metres high and is placed in the centre of the playing court. Show that the ball clears the net and calculate by how much.

5. Georgie has a large rectangular garden area with dimensions l metres by w metres, which they wish to divide into three sections to grow different vegetables. Georgie plans to put a watering system along the perimeter of each section. This will require a total of 120 metres of hosing.

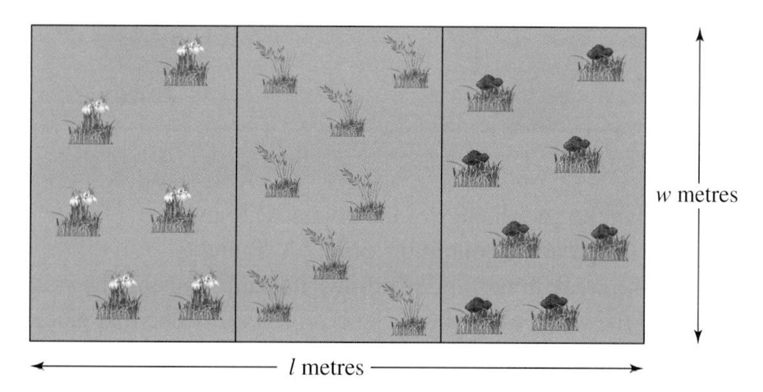

   a. Show the total area of the three sections, A m^2, is given by $A = 60w - 2w^2$ and hence calculate the dimensions when the total area is a maximum.
   b. Using the maximum total area, Georgie decides the areas of the three sections should be in the ratio 1 : 2 : 3. Find the total length of hosing for the watering system that is now required.

6. The number of bacteria cells in a slowly growing culture at time t hours after 8.00 am is given by $N = 100 + 46t + 2t^2$.

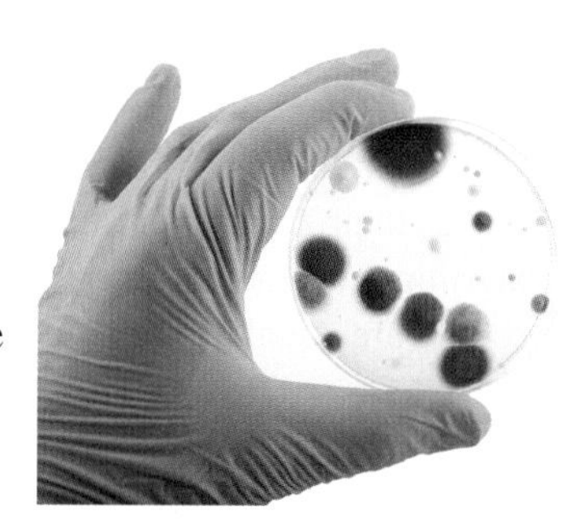

   a. Calculate how long it takes for the initial number of bacteria cells to double.
   b. Calculate how many cells are present at 1.00 pm.
   c. At 1.00 pm a virus is introduced that initially starts to destroy the bacteria cells so that t hours after 1.00 pm the number of cells is given by $N = 380 - 180t + 30t^2$. Find the minimum number that the population of bacteria cells reaches and the time at which this number is reached.

7. Let $z = 5x^2 + 4xy + 6y^2$. Given $x + y = 2$, find the minimum value of z and the values of x and y for which z is minimum.

8. A piece of wire of length 20 cm is cut into two sections, and each is used to form a square. The sum of the areas of these two squares is S cm^2.
   a. If one square has a side length of 4 cm, calculate the value of S.
   b. If one square has a side length of x cm, express S in terms of x and hence determine how the wire should be cut for the sum of the areas to be a minimum.

9. The cost, C dollars, of manufacturing n dining tables is the sum of three parts. One part represents the fixed overhead costs of \$20, another represents the cost of raw materials and is directly proportional to n, and the third part represents the labour costs, which are directly proportional to the square of n.
   a. If 5 tables cost \$195 to manufacture and 8 tables cost \$420 to manufacture, determine the relationship between C and n.
   b. Determine the maximum number of dining tables that can be manufactured if costs are not to exceed \$1000.

10. The arch of a bridge over a small creek is parabolic in shape with its feet evenly spaced from the ends of the bridge. Relative to the coordinate axes, the points A, B and C lie on the parabola.

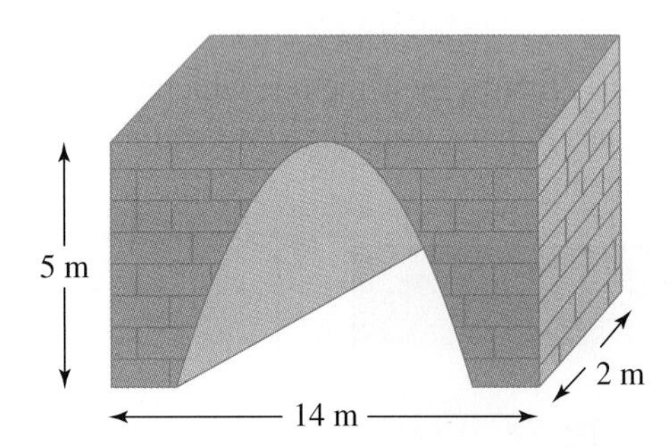

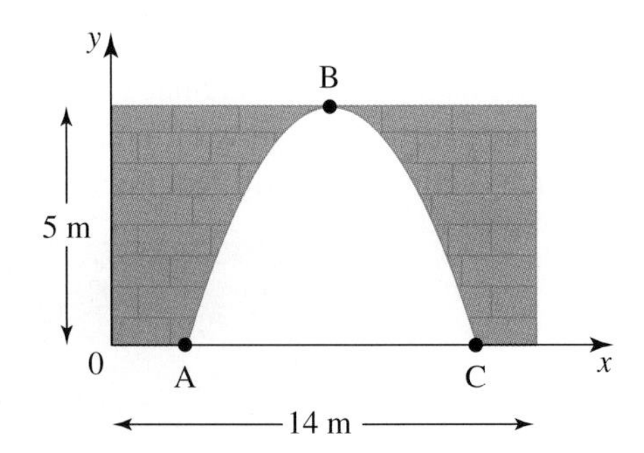

   a. If AC = 8 metres, write down the coordinates of the points A, B and C.
   b. Determine the equation of the parabola containing points A, B and C.
   c. Following heavy rainfall, the creek floods and overflows its bank, causing the water level to reach 1.5 metres above AC. Find the width of the water level, correct to 1 decimal place.

11. The daily cost, C dollars, of producing x kg of plant fertiliser for use in market gardens is $C = 15 + 10x$. The manufacturer decides that the fertiliser will be sold for v dollars per kg where $v = 50 - x$.
    Determine an expression for the daily profit in terms of x and hence determine the price per kilogram that should be charged for maximum daily profit.

12. a. If the sum of two numbers is 16, determine the numbers for which:
       i. their product is greatest
       ii. the sum of their squares is least.
    b. If the sum of two non-zero numbers is k:
       i. express their greatest product in terms of k
       ii. determine whether there are any values of k for which the sum of the squares of the numbers and their product are equal. If so, state the values; if not, explain why.

13. Meteorology records for the heights of tides above mean sea level in Tuvalu predict the tide levels shown in the following table.

Time of day	Height of tide (in metres)
10:15 am	1.05
4:21 pm	3.26
10:30 pm	0.94

a. Use CAS technology to find the equation of a quadratic model that fits these three data points in the form $h = at^2 + bt + c$, where h is the height in metres of the tide t hours after midnight. Express the coefficients to 2 decimal places.

b. Calculate the greatest height of the tide above sea level and the time of day it is predicted to occur.

14. A piece of wire of length 20 cm is cut into two sections. One section is used to create a square and the other section a circle. The sum of the areas of the square and the circle is S cm^2.

a. If the square has a side length x cm, express S in terms of x.

b. Graph the S–x relationship and hence calculate the lengths of the two sections of the wire for S to be a minimum. Give the answer to 1 decimal place.

c. Use the graph to find the value of x for which S is a maximum.

3.8 Exam questions

Question 1 (1 mark) TECH-ACTIVE

MC Jill has 80 metres of wire to enclose a rectangular-shaped pen for her new alpaca pair. If she can use the back of her house as one side, the maximum area, in square metres, that she can create for her new animals is

A. 400
B. 800
C. 1000
D. 1200
E. 1600

Question 2 (1 mark) TECH-ACTIVE

MC An artist decides to include a 40 cm length rod in her latest sculpture. She wants to make two squares from the rod and so cuts it into two parts. If x cm is the side length of one of the squares, the equation for the sum of the areas of the two squares is

A. $A = 2x^2 - 20x + 100$
B. $A = 2x^2 - 20x + 40$
C. $A = x^2 - 20x + 100$
D. $A = x^2 - 20x + 40$
E. $A = 4x^2 - 20x + 40$

Question 3 (3 marks) TECH-ACTIVE

Nick is standing on a 20-metre cliff overlooking a beach. Nick throws a ball vertically up into the air. Ruby is lying on the beach below watching and later calculated that the motion of the ball can be described as $h = 1.5 + 4t - 0.2t^2$, where h is the height in metres above the cliff from where Nick threw the ball and t is the time in seconds.

Find the greatest height the ball reached above Ruby and how many seconds it took to reach that height.

More exam questions are available online.

3.9 Review

3.9.1 Summary

doc-37018

Hey students! Now that it's time to revise this topic, go online to:

 **Access the topic summary**  **Review your results**  **Watch teacher-led videos**  **Practise exam questions**

Find all this and MORE in jacPLUS

3.9 Exercise

Technology free: short answer

1. Solve for x.

   a. $x^2 - 4x - 21 = 0$ b. $10x^2 + 37x + 7 = 0$ c. $(x^2 + 4)^2 - 7(x^2 + 4) - 8 = 0$

   d. $2x^2 = 3x(x - 2) + 1$ e. $x = \dfrac{12}{x - 2} - 2$ f. $3 + \sqrt{x} = 2x$

2. Solve the following quadratic inequations.

   a. $2x^2 - 5x - 3 > 0$ b. $10 - x^2 \geq 0$ c. $20x^2 + 20x + 5 \geq 0$

3. Sketch the graphs of the following, showing all key points.

   a. $y = 2(x - 3)(x + 1)$ b. $y = 1 - (x + 2)^2$ c. $y = x^2 + x + 9$

4. Factorise over R.

   a. $-x^2 + 20x + 24$ b. $4x^2 - 2x - 9$

5. a. Calculate the discriminant and hence state the number and type of solutions to $5x^2 + 8x - 2 = 0$.
   b. Find the values of k for which the equation $kx^2 - 4x(k + 2) + 36 = 0$ has no real roots.

6. a. Use an algebraic method to find the coordinates of the points of intersection of the parabola $y = x^2 + 2x$ and the line $y = x + 2$.
   b. Sketch $y = x^2 + 2x$ and $y = x + 2$ on the same set of axes.
   c. Give an algebraic description of the region enclosed by the parabola $y = x^2 + 2x$ and the line $y = x + 2$.
   d. i. Find the value of k for which the line $y = x + k$ is a tangent to the parabola $y = x^2 + 2x$.
      ii. Sketch this tangent on the diagram drawn in part b, identifying the point of contact with the parabola.

Technology active: multiple choice

7. **MC** The solutions of the equation $(x - 2)(x + 1) = 4$ are:

   A. $x = 2, x = -1$
   B. $x = 6, x = -1$
   C. $x = -6, x = 1$
   D. $x = 3, x = -2$
   E. $x = -3, x = 2$

8. **MC** The values of x for which $-5x^2 + 8x + 3 = 0$ are closest to:

   A. $-0.6, -1$
   B. $0.6, -1$
   C. $-0.3, 1.9$
   D. $0.3, -1.9$
   E. $-0.2, -3$

9. **MC** For the graph of the parabola $y=ax^2+bx+c$ shown, with $\Delta=b^2-4ac$, the restrictions on a and Δ are:

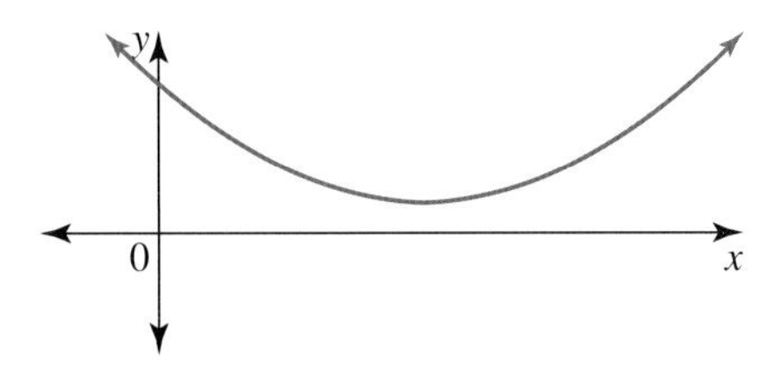

A. $a>0$ and $\Delta>0$ **B.** $a>0$ and $\Delta<0$ **C.** $a<0$ and $\Delta<0$
D. $a<0$ and $\Delta>0$ **E.** $a>0$ and $\Delta=0$

10. **MC** The parabola with equation $y=x^2$ is translated so that its image has its vertex at $(-4,3)$. The equation of the image is:

A. $y=(x-4)^2+3$ **B.** $y=(x-3)^2+4$ **C.** $y=(x+4)^2+3$
D. $y=(x+3)^2-4$ **E.** $y=-4x^2+3$

11. **MC** If x^2+4x-6 is expressed in the form $(x+b)^2+c$, the values of b and c will be:

A. $b=2, c=-10$ **B.** $b=-2, c=-10$ **C.** $b=4, c=-2$
D. $b=-4, c=-2$ **E.** $b=2, c=-8$

12. **MC** The equation of the parabola shown is:

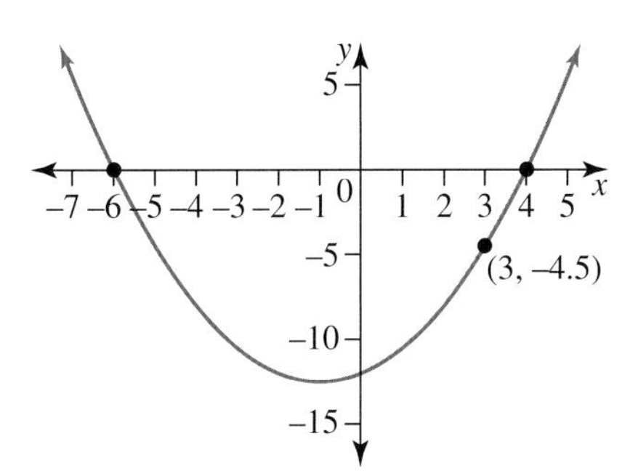

A. $y=x^2+2x-24$ **B.** $y=0.5x^2+x-12$ **C.** $y=x^2-2x-24$
D. $y=0.5x^2-x-12$ **E.** $y=(x+1)^2-12$

13. **MC** The solution set of $\{x : x^2<4x\}$ is:

A. $\{x : x<4\}$ **B.** $\{x : -4<x<0\}$ **C.** $\{x : 0<x<4\}$
D. $\{x : x<0\}\cup\{x : x>4\}$ **E.** $\{x : x<-4\}\cup\{x : x>0\}$

14. **MC** A quadratic graph touches the x-axis at $x=-6$ and cuts the y-axis at $y=-10$. Its equation is:

A. $y=(x+6)^2-10$ **B.** $y=(x+6)(x+10)$ **C.** $y=\frac{5}{18}x^2-10$
D. $y=\frac{5}{18}(x+6)^2$ **E.** $y=-\frac{5}{18}(x+6)^2$

15. **MC** The x-coordinates of the points of intersection of the parabola $y=3x^2-10x+2$ with the line $2x-y=1$ can be determined from the equation:

A. $3x^2-10x+2=0$ **B.** $3x^2-12x+3=0$ **C.** $x^2-6x+1=0$
D. $3x^2-8x+1=0$ **E.** $(2x-1)^2=0$

16. **MC** The maximum value of $4-2x-x^2$ is:

A. 5 **B.** 4 **C.** 3 **D.** 1 **E.** -1

Technology active: extended response

17. At a winter skiing championship, two competitors, one from Japan and the other from Canada, compete for the gold medal in one of the jump events.
Each competitor leaves the ski run at point S and travels through the air, landing back on the ground at some point G. The winner will be the competitor who covers the greater horizontal distance OG.
The Canadian skier jumps first and her height (h metres) above the ground is described by $h = -\frac{1}{35}(x^2 - 60x - 700)$, where x metres is the horizontal distance travelled.

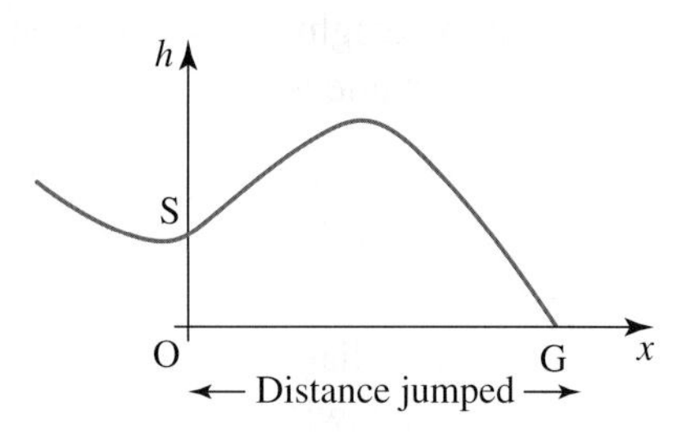

   a. Show that the point S is 20 metres above O.
   b. Calculate how far the Canadian skier jumps.

   The Japanese skier jumps next. She reaches a maximum height of 35 metres above the ground after a horizontal distance of 30 metres has been covered.

   c. Assuming the path is a parabola, form the equation for h in terms of x which describes this competitor's path.
   d. Decide which competitor receives the gold medal. Your decision should be supported with appropriate mathematical reasoning.

18. The diagram shows the arch of a bridge where the shape of the curve, OAB, is a parabola. OB is the horizontal road level. Taking O as the origin, the equation of the curve OAB is $y = 2.5x - 0.3125x^2$. All measurements are in metres.

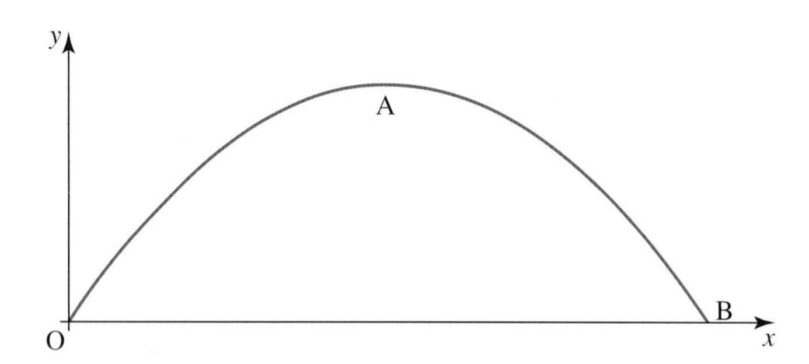

   a. Calculate the length of OB, the span of the bridge.
   b. Calculate the height above the road of point A, the highest point on the curve.
   c. A car towing a caravan needs to drive under the bridge. The caravan is 5 metres wide and has a height of 2 metres. Only one single lane of traffic can pass under the bridge. Explain clearly, using mathematical analysis, whether the caravan can be towed under this bridge.

   To avoid accidents, the bridge engineers decide to place height and width limits. Only vehicles whose height and width fit into the greatest allowable dimensions are permitted to travel under the bridge.

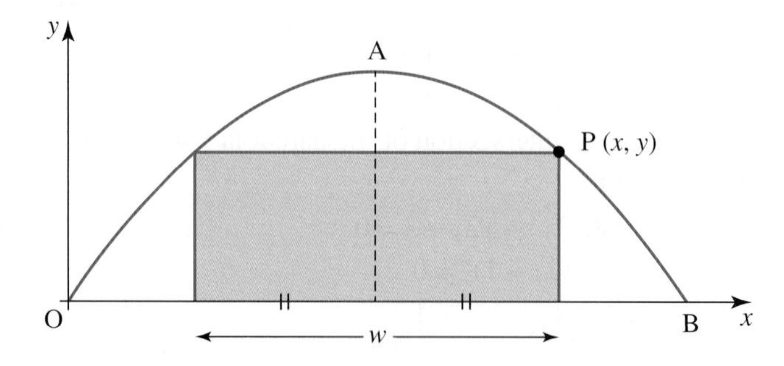

   P (x, y) lies on the curve and is a corner of the rectangle formed by the height and width restrictions.

d. Express the width, w, of the rectangle in terms of x.
e. If the height restriction is 3.2 m, calculate the x-coordinate of P.
f. Determine whether the caravan would be permitted to be towed under the bridge under these restrictions.

19. ABCD is a rectangle of length one more unit than its width. Point F lies on AB and divides AB in the ratio $x : 1$ so that AF is x units in length and FB is 1 unit in length. Point G lies on DC and divides DC in the same ratio, $x : 1$.
a. Draw a diagram showing this information.
b. State the width of rectangle ABCD.
c. If the area of the square AFGD is one more square unit than the area of the rectangle FBCG, show that $x^2 - x - 1 = 0$.
d. Hence, find the value of x in simplest surd form.
e. The value found for x is called the golden ratio and usually given the symbol ϕ. Calculate $\frac{1}{\phi}$ and give its relationship to the other root of the equation $x^2 - x - 1 = 0$.
f. Show $\frac{1}{\phi} = \phi - 1$ and explain this relationship using the equation $x^2 - x - 1 = 0$.

20. Ignoring air resistance, the path of a cricket ball hit by a batsman can be considered to travel on a parabolic path which starts at the point $(0, 0)$ where the ball is struck by the batsman.
Let x metres measure the horizontal distance of the ball from the batsman in the direction the ball travels, and y metres measure the vertical height above the ground that the ball reaches.

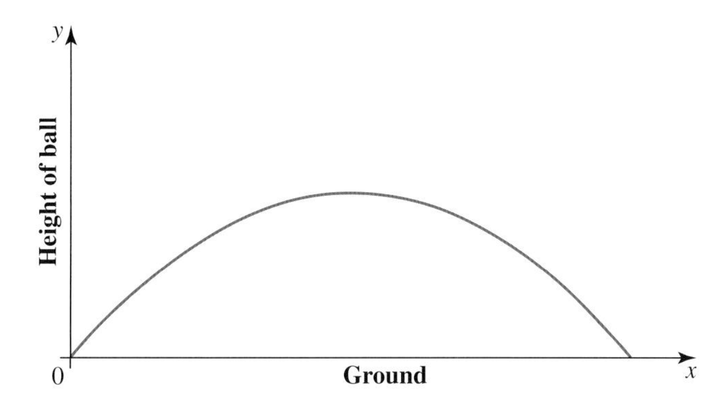

A batsman hits a cricket ball towards a fielder who is 65 metres away. The ball is struck with a horizontal speed of 28 m/s, which is assumed to remain constant throughout the flight of the ball. On its way, the ball reaches a maximum height of 4.9 metres after 1 second.
a. Calculate the coordinates of the turning point of the quadratic path of the ball.
b. Form the equation of the path of the ball.

The fielder starts running forward at the instant the ball is hit and catches it at a height of 1.3 metres above the ground.
c. Calculate the time it takes the fielder to reach the ball.
d. Hence, obtain the uniform speed in m/s at which the fielder runs in order to catch the ball.

3.9 Exam questions

Question 1 (2 marks) TECH-FREE

Solve for x.

$$\left(x+\frac{1}{x}\right)^2+4\left(x+\frac{1}{x}\right)+4=0$$

Question 2 (1 mark) TECH-ACTIVE

MC If the equation $ax^2+bx+c=0$ has two unequal solutions, then b^2-4ac is

A. ≥ 0 **B.** >0 **C.** $=0$ **D.** <0 **E.** ≤ 0

Question 3 (1 mark) TECH-ACTIVE

MC The graph of $-4x^2-x-3$ has

A. no x-intercepts.
B. one x-intercept.
C. two x-intercepts.
D. three x-intercepts.
E. one x-intercept and one y-intercept.

Question 4 (1 mark) TECH-ACTIVE

MC The equation of the graph shown may be

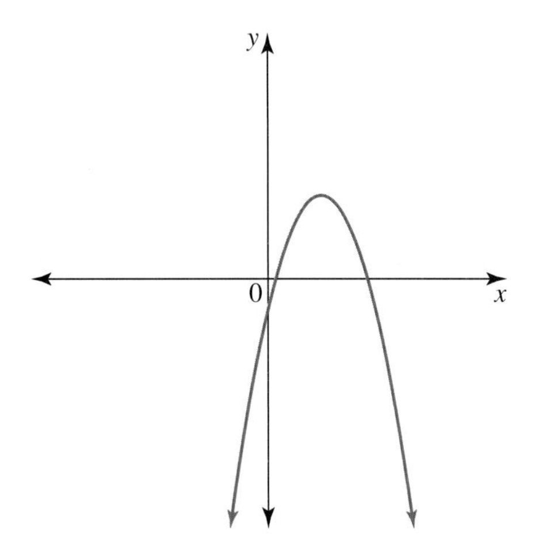

A. $y=-(x+2)^2-3$ **B.** $y=(x+2)^2+3$ **C.** $y=(x-2)^2+3$
D. $y=-(x-2)^2+3$ **E.** $y=-(x+2)^2+3$

Question 5 (3 marks) TECH-FREE

Determine the values of k for which there will be at least one intersection between the line $y=kx+4$ and the parabola $y=x^2-8x+12$.

More exam questions are available online.

Answers

Topic 3 Quadratic relationships

3.2 Quadratic equations with rational roots

3.2 Exercise

1. a. $0, 5$ b. $\frac{1}{7}, 3$ c. -8 d. $-6, -4$

2. a. $x = \frac{4}{3}, -\frac{1}{2}$ b. $x = 4, 3$ c. $x = -\frac{3}{2}, x = -\frac{7}{4}$
 d. $x = 0, \frac{1}{5}$ e. $x = \frac{2}{3}, -4$ f. $x = 0, 10$

3. a. $x = -3, \frac{7}{10}$ b. $(x+5)x = x^2 + 5x$

4. a. $-\frac{1}{2}, -\frac{1}{3}$ b. $-\frac{2}{3}, \frac{5}{4}$
 c. 7 d. $-\frac{5}{6}, 1$

5. $x = -\frac{3}{5}, 1$

6. a. $x = -5, 1$ b. $x = -4, 6$ c. No real solutions
 d. $x = -10, -1$ e. $x = 7$ f. $x = 0, 8$

7. a. ± 11 b. $\pm \frac{4}{3}$ c. $4, 6$
 d. $-1, 6$ e. $-\frac{1}{3}, 1$ f. ± 3

8. $x = \pm \frac{1}{3}$

9. a. $x = \frac{19}{6}, \frac{7}{3}$ b. $x = -6, -\frac{13}{5}$
 c. $x = -2, -4$ d. $x = \frac{1}{2}, 3$

10. a. $-\frac{14}{3}, -1$ b. $-\frac{7}{2}, \frac{1}{4}$
 c. $\pm 2, \pm 5$ d. ± 4

11. a. $-1, 8$ b. $\frac{11}{6}, \frac{9}{2}$ c. $-8, -2$ d. $0, \frac{1}{2}$

12. a. $\frac{1}{3}$ b. $\pm \frac{25}{2}$ c. $-\frac{5}{7}, \frac{2}{5}$ d. $0, \frac{17}{19}$

13. a. ± 3 b. $\pm \frac{4}{3}, \pm 2$

14. a. $-1, 2$ b. $-\frac{3}{2}, -\frac{6}{5}$ c. $x = 1$

15. $x = \frac{r-q}{p}, x = -\frac{(r+q)}{p}$

16. a. $x = -3a, x = 2b$ b. $x = \frac{3a}{2}, x = 5a$
 c. $x = a - b, x = a + 3b$ d. $x = b - a, x = 2b - a$

17. a. $(x-1)(x-7) = 0$ b. $(x+5)(x-4) = 0$
 c. $x(x-10) = 0$ d. $(x-2)^2 = 0$

18. $b = 13, c = -12$

19. a. $x = -\frac{7}{3}, x = \frac{9}{20}$ b. $x = \frac{7}{4}, x = \frac{14}{3}$

20. a. $x = \frac{3}{2}$ b. $\frac{2}{11}, \frac{11}{2}$

3.2 Exam questions

Note: Mark allocations are available with the fully worked solutions online.

1. B
2. D
3. $a = \frac{5}{2}, a = 3$

3.3 Quadratics over *R*

3.3 Exercise

1. a. $x^2 + 10x + 25 = (x+5)^2$
 b. $x^2 - 7x + \frac{49}{4} = \left(x - \frac{7}{2}\right)^2$
 c. $x^2 + x + \frac{1}{4} = \left(x + \frac{1}{2}\right)^2$

2. a. $\left(x - 5 - 4\sqrt{2}\right)\left(x - 5 + 4\sqrt{2}\right)$
 b. $3\left(x + \frac{7 - \sqrt{13}}{6}\right)\left(x + \frac{7 + \sqrt{13}}{6}\right)$
 c. $\left(\sqrt{5}x - 3\right)\left(\sqrt{5}x + 3\right)$

3. $(x-1)(3x-5)$

4. a. $3\left(x - 8 - \sqrt{2}\right)\left(x - 8 + \sqrt{2}\right)$
 b. No linear factors over R

5. a. $\left(x - 3 - \sqrt{2}\right)\left(x - 3 + \sqrt{2}\right)$
 b. $\left(x + 2 - \sqrt{7}\right)\left(x + 2 + \sqrt{7}\right)$
 c. No linear factors over R
 d. $2\left(x + \frac{5}{4} - \frac{\sqrt{41}}{4}\right)\left(x + \frac{5}{4} + \frac{\sqrt{41}}{4}\right)$
 e. $-\left(x - 4 - 2\sqrt{2}\right)\left(x - 4 + 2\sqrt{2}\right)$
 f. $3\left(x + \frac{2}{3} - \frac{\sqrt{22}}{3}\right)\left(x + \frac{2}{3} + \frac{\sqrt{22}}{3}\right)$

6. a. $\left(x - 2\sqrt{3}\right)\left(x + 2\sqrt{3}\right)$
 b. $\left(x - 6 - 4\sqrt{2}\right)\left(x - 6 + 4\sqrt{2}\right)$
 c. $\left(x + \frac{9 - \sqrt{93}}{2}\right)\left(x + \frac{9 + \sqrt{93}}{2}\right)$
 d. $2\left(x + \frac{5 - \sqrt{17}}{4}\right)\left(x + \frac{5 + \sqrt{17}}{4}\right)$
 e. $3\left(\left(x + \frac{2}{3}\right)^2 + \frac{5}{9}\right)$, no linear factors over R

f. $-5\left(x-4-\frac{9\sqrt{5}}{5}\right)\left(x-4+\frac{9\sqrt{5}}{5}\right)$

7. $x=\frac{-11\pm\sqrt{89}}{4}$

8. a. $x=\frac{5\pm\sqrt{13}}{6}$ b. $x=\frac{1\pm\sqrt{101}}{10}$

c. No real solutions d. $-3\pm\sqrt{17}$

9. a. $\frac{1\pm\sqrt{17}}{6}$ b. $\frac{10\pm2\sqrt{10}}{5}$ c. $10\pm2\sqrt{5}$

d. $-3\pm2\sqrt{3}$ e. $x=0,6$ f. $x=\frac{3}{4},2$

10. a. $\Delta=-135$, no real factors

b. $\Delta=0$, one repeated rational factor

c. $\Delta=1$, two rational factors

d. $\Delta=\frac{40}{9}$, two real factors; completing the square is needed to obtain the factors.

11. a. $\Delta=121$, two rational factors

b. $\Delta=-39$, no real factors

c. $\Delta=0$, one repeated rational factor

d. $\Delta=8$, two irrational factors

12. a. 4

b. Two rational roots

c. Two irrational solutions

d. One rational solution

e. No real solutions

f. Two irrational solutions

13. a. Two irrational roots

b. Two rational roots

c. No real roots

d. One rational root

e. No real roots

f. Two irrational roots

14. $x=\frac{-1\pm\sqrt{7}}{2}$

15. $x=-\sqrt{6}\pm\sqrt{14}$

16. a. $\pm\sqrt{5}$ b. $\pm2\sqrt{2}$

c. $\pm\sqrt{11},\pm1$ d. $\frac{-3\pm\sqrt{5}}{2}$

e. $-1,8,\frac{7\pm\sqrt{93}}{2}$ f. -1

17. a. There are no real roots. b. $k=-1,k=3$

18. $\Delta=(m+4)^2\Rightarrow\Delta\geq0$

19. a. $m=-4\pm4\sqrt{2}$ b. $m=2\pm2\sqrt{3}$

c. $p<\frac{1}{3}$ d. $\Delta>0$

e. Δ is a perfect square.

20. a. $x=10\sqrt{5}\pm20$

b. $x=1+\sqrt{2}, b=-2, c=-1$

21. a. $\left(x-\sqrt{2}\right)\left(x+\sqrt{2}\right)=x^2-2$

b. $\left(x+4-\sqrt{2}\right)\left(x+4+\sqrt{2}\right)=x^2+8x+14$

22. a. i. $x=-3\sqrt{2}$ ii. $x=\frac{3\sqrt{2}\pm\sqrt{10}}{4}$

b. i. $x=\frac{-1-\sqrt{5}}{2}$

ii. $x^2+x-1=0, b=1, c=-1$

23. a. $x=1$ b. $x=1$

24. a. 4 b. 4

25. $x=\frac{5}{4}$

3.3 Exam questions

Note: Mark allocations are available with the fully worked solutions online.

1. A 2. E 3. $\Delta=(m-3)^2$

3.4 Applications of quadratic equations

3.4 Exercise

1. 20 kg
2. $3\sqrt{10}$ cm
3. $4\frac{1}{2}$ hours
4. $4\sqrt{3}$ cm
5. 2511 litres
6. 20 and 22
7. 10 and 11
8. Base $2\sqrt{2}$ cm; height 4 cm
9. $x=24$; perimeter = 168 cm
10. Length 24 cm; width 20 cm
11. a. Sample responses can be found in the worked solutions in the online resources.

b. Sample responses can be found in the worked solutions in the online resources.

c. $\Delta=256-8k$

i. $k>32$ ii. $k=32$ iii. $0<k<32$

d. 32 m^2; width 4 m, length 8 m

e. Width 1.1 metres and length 13.8 metres

12. 8 cards
13. a. 1.052 metres

b. $r=\frac{-\left(\pi l-\sqrt{\pi^2l^2+80\pi}\right)}{2\pi}$

14. The distance is quadrupled.
15. Heat is reduced by 36%.

3.4 Exam questions

Note: Mark allocations are available with the fully worked solutions online.

1. A
2. E
3. The smaller number is −23.

3.5 Graphs of quadratic equations

3.5 Exercise

1. 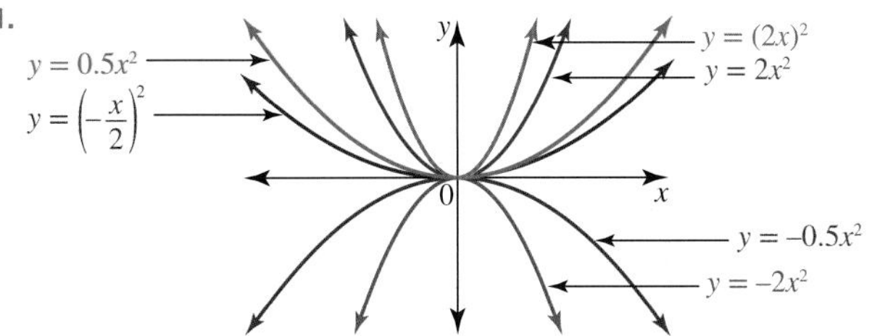

2. A. i. $y = x^2 - 2$
   B. ii. $y = -2x^2$
   C. iii. $y = -(x+2)^2$

3. a. $(0, 8)$ b. $(0, -8)$ c. $(0, 1)$
   d. $(8, 0)$ e. $(-8, 0)$ f. $(-12, 0)$

4. a. 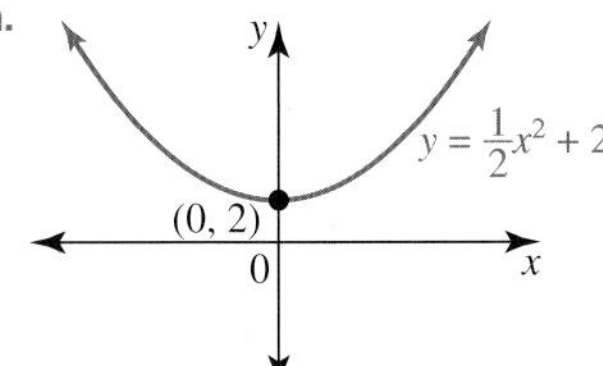

b. 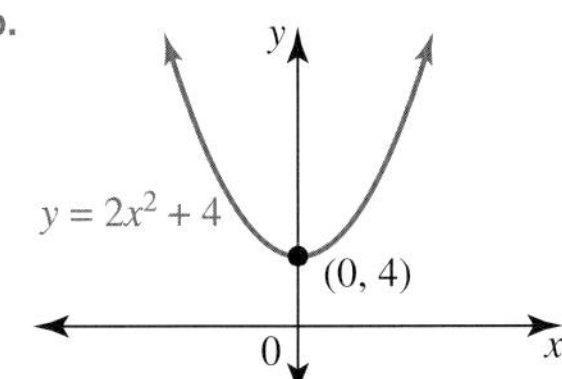

c. 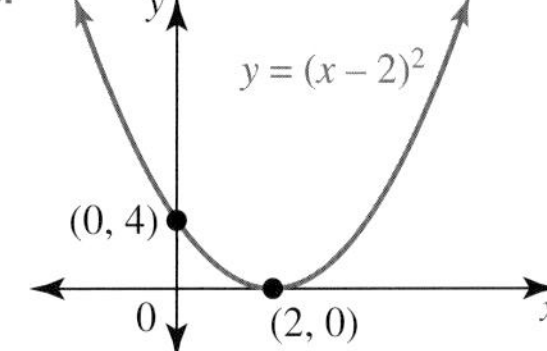

d. 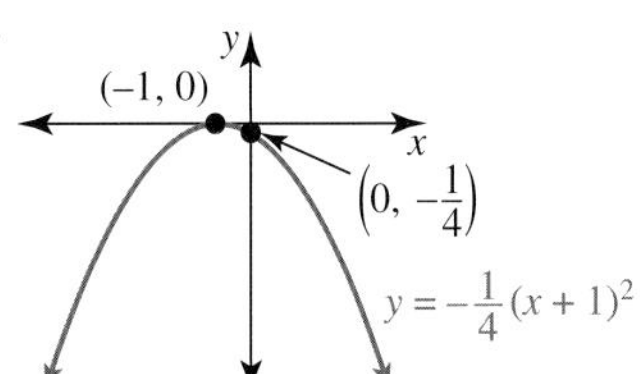

e. 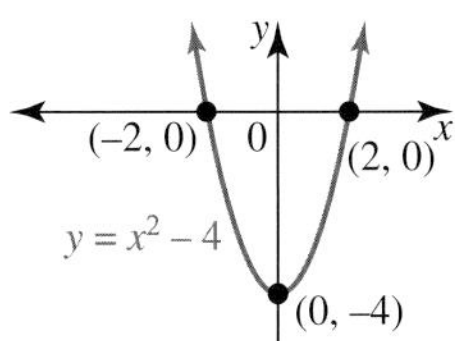

f. 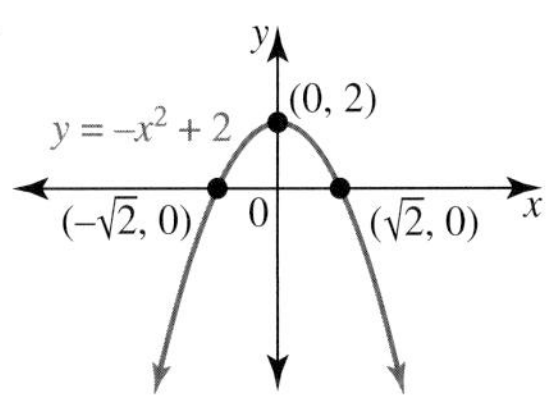

5. Axis intercepts $(-6, 0)$, $(3, 0)$, $(0, -6)$; minimum turning point $(-1.5, -6.75)$

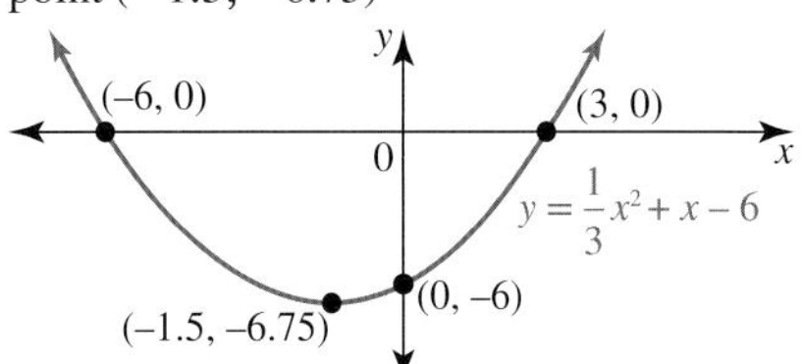

6. a. 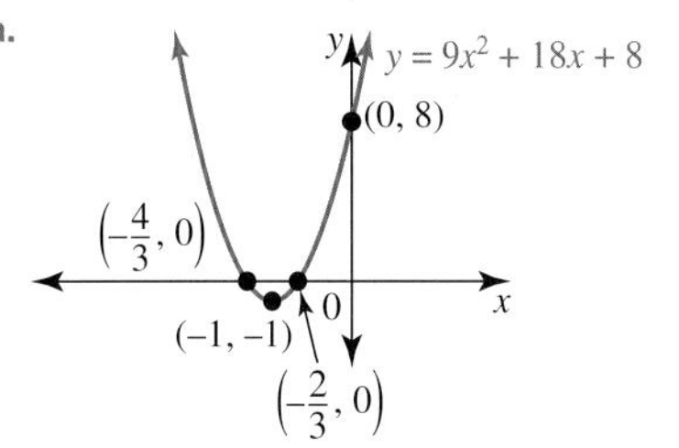

b. 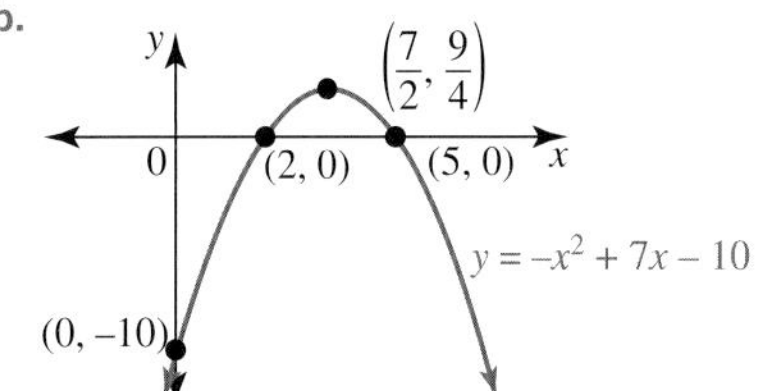

c. 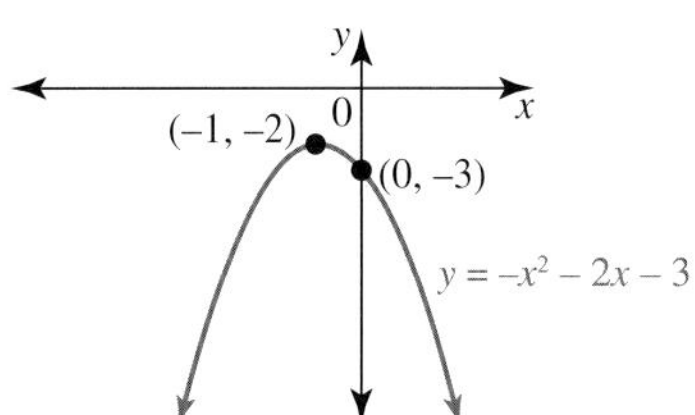

d. 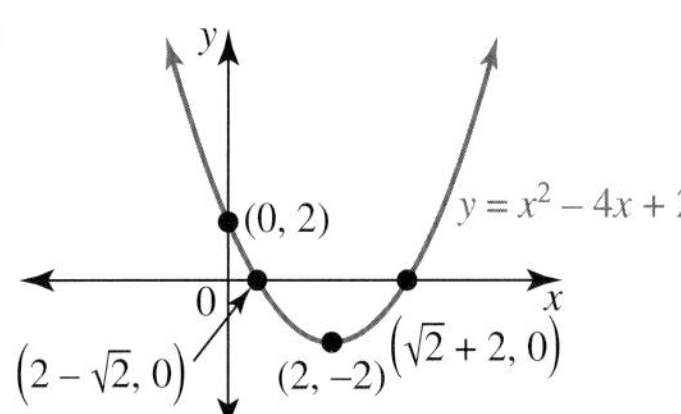

7. a. Maximum turning point $(-1, 8)$; axis intercepts $(0, 6)$, $(-3, 0)$, $(1, 0)$

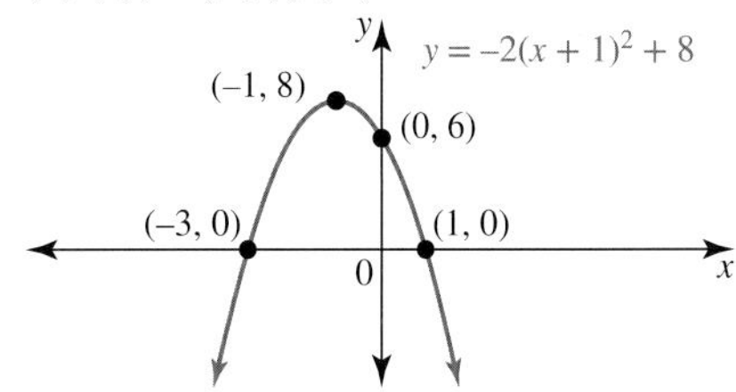

b. i. $y = -(x-5)^2 - 5$
   The vertex is $(5, -5)$, a maximum turning point
   ii. y-intercept $(0, -30)$, no x-intercepts

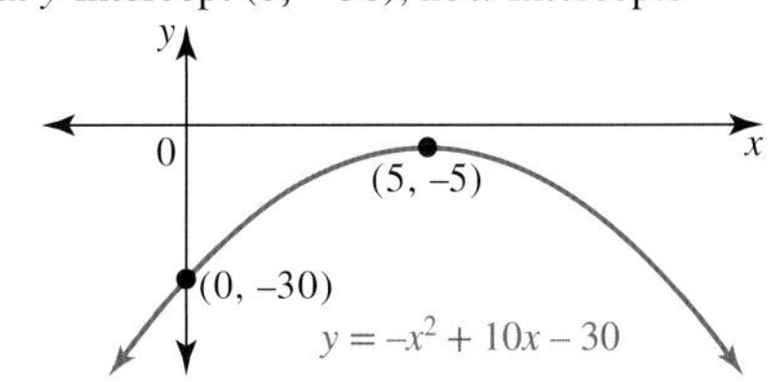

8. a. Maximum turning point at $(0, 4)$

b. Minimum turning point at $\left(\frac{4}{3}, 0\right)$

9. D

10. a.

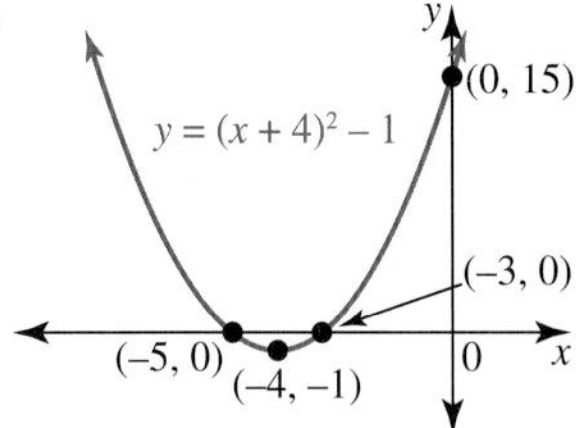

b. Turning point $(-4, 3)$

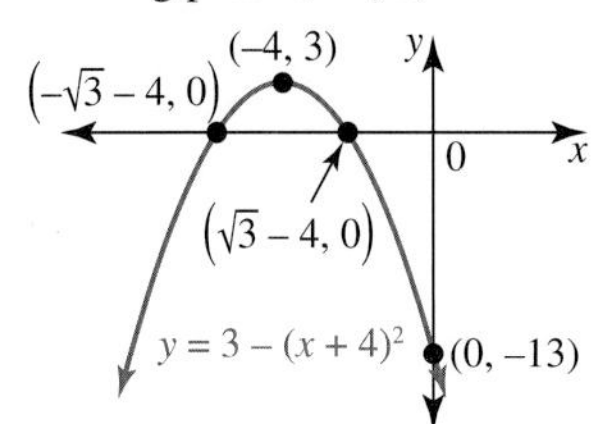

c. i. $x^2 + 6x + 12 = (x + 3)^2 + 3$

ii. Turning point $(-3, 3)$

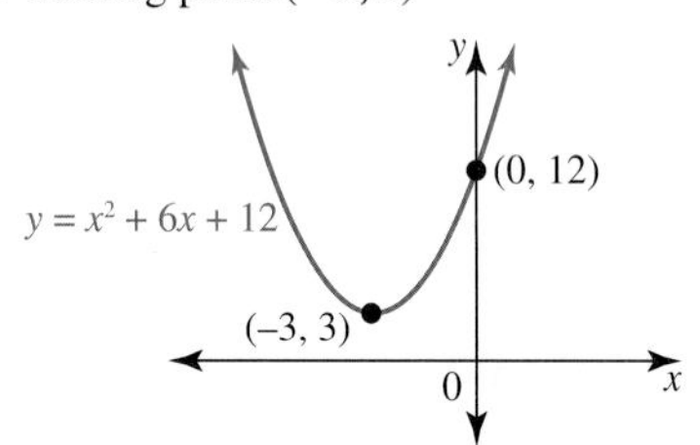

d. Turning point $\left(-\frac{5}{2}, 0\right)$

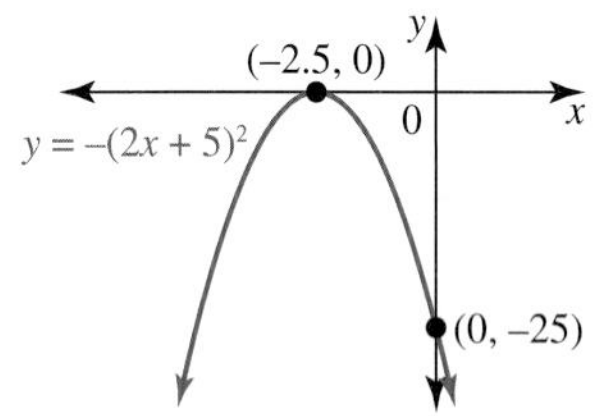

11. Axis intercepts $(0, 0)$, $(4, 0)$; maximum turning point $(2, 8)$

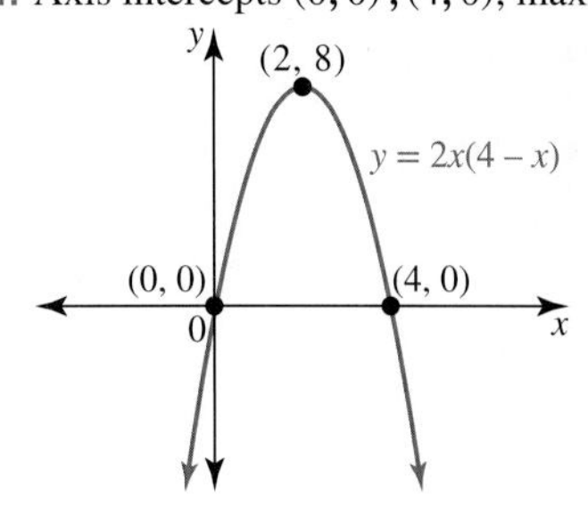

12. Minimum turning point and x-intercept $(-2, 0)$, y-intercept $(0, 4)$

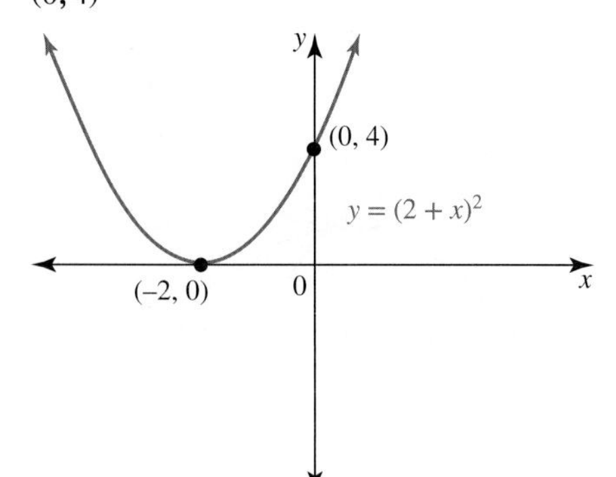

13. a.

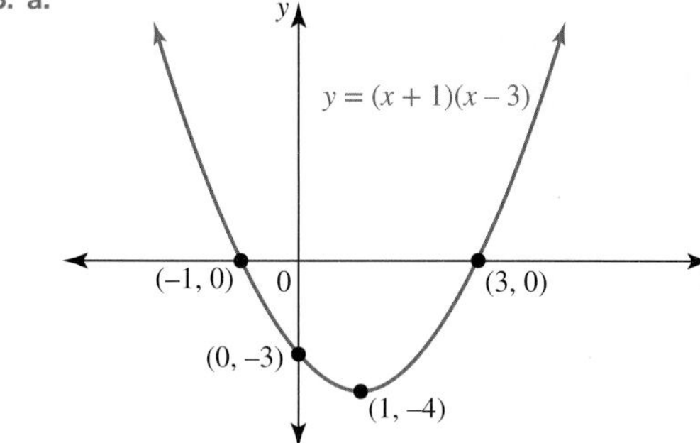

b.

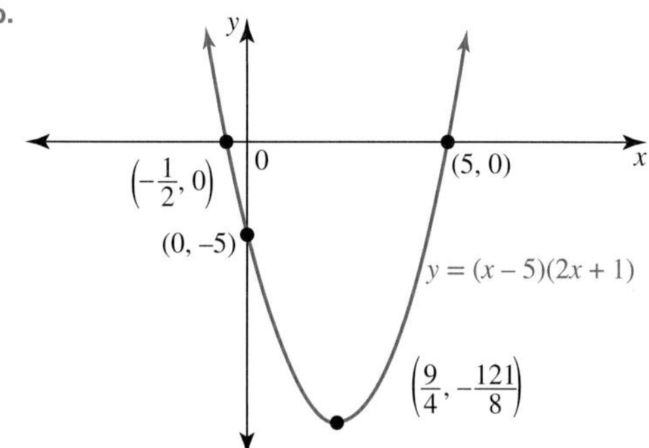

14.

	Turning point	y-intercept	x-intercepts
a.	$(0, -9)$	$(0, -9)$	$(\pm 3, 0)$
b.	$(9, 0)$	$(0, 81)$	$(9, 0)$
c.	$(0, 6)$	$(0, 6)$	$\left(\pm\sqrt{2}, 0\right)$
d.	$(-1, 0)$	$(0, -3)$	$(-1, 0)$

a.

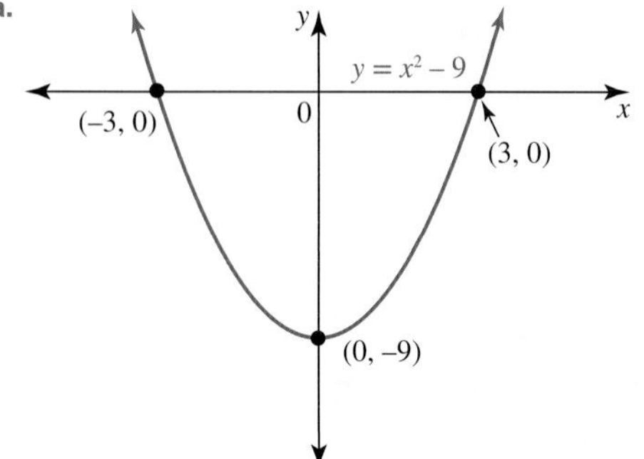

b.

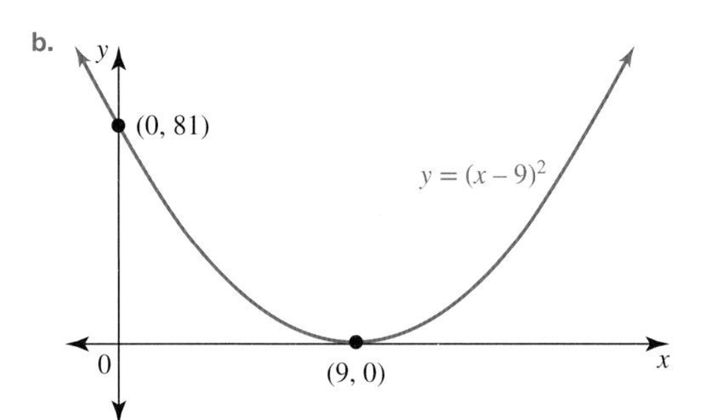

c.

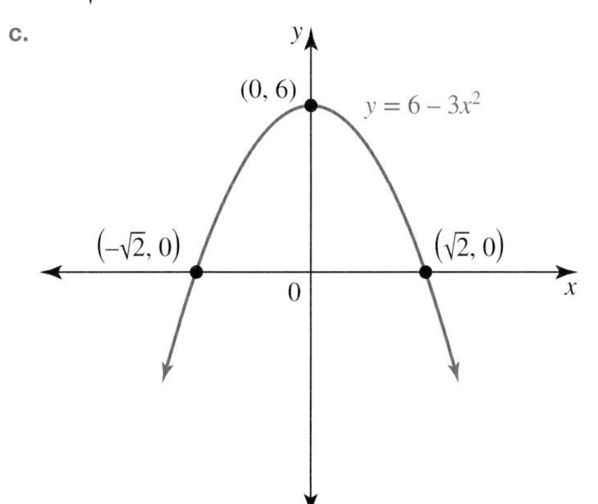

d.

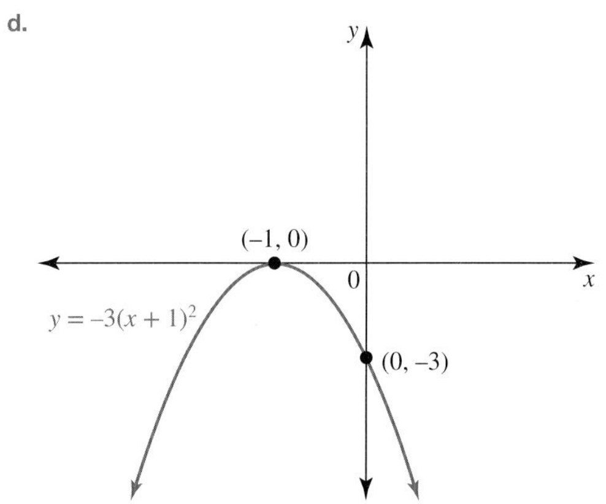

15.

	Turning point	y-intercept	x-intercepts
a.	$(-3,-16)$	$(0,-7)$	$(-7,0),(1,0)$
b.	$(1,-10)$	$(0,-7)$	$\left(1\pm\dfrac{\sqrt{30}}{3},0\right)$
c.	$\left(\dfrac{2}{3},\dfrac{19}{3}\right)$	$(0,5)$	$\left(\dfrac{2\pm\sqrt{19}}{3},0\right)$
d.	$\left(\dfrac{1}{4},-\dfrac{33}{8}\right)$	$(0,-4)$	$\left(\dfrac{1\pm\sqrt{33}}{4},0\right)$

a.

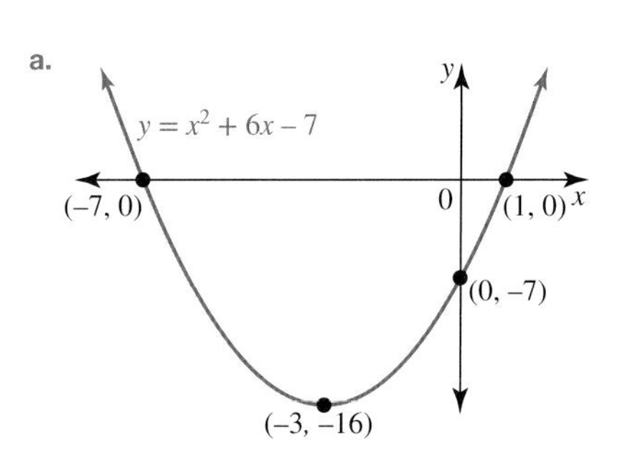

b.

$\left(1-\dfrac{\sqrt{30}}{3},0\right)$ $\left(1+\dfrac{\sqrt{30}}{3},0\right)$ $(0,-7)$ $y=3x^2-6x-7$ $(1,-10)$

c.

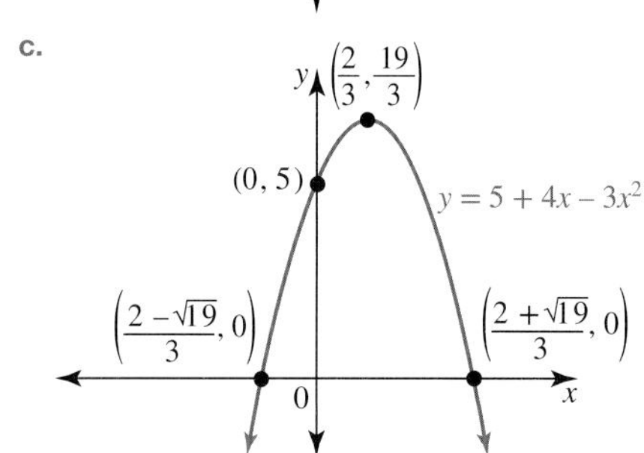

d.

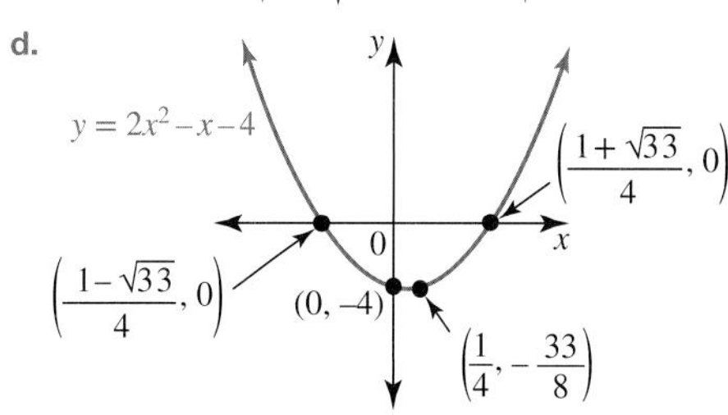

16.

	Turning point	y-intercept	x-intercepts
a.	$(5,2)$	$(0,27)$	none
b.	$(-1,-2)$	$(0,0)$	$(0,0),(-2,0)$
c.	$(3,-6)$	$(0,-24)$	none
d.	$(4,1)$	$(0,-15)$	$(3,0),(5,0)$

a.

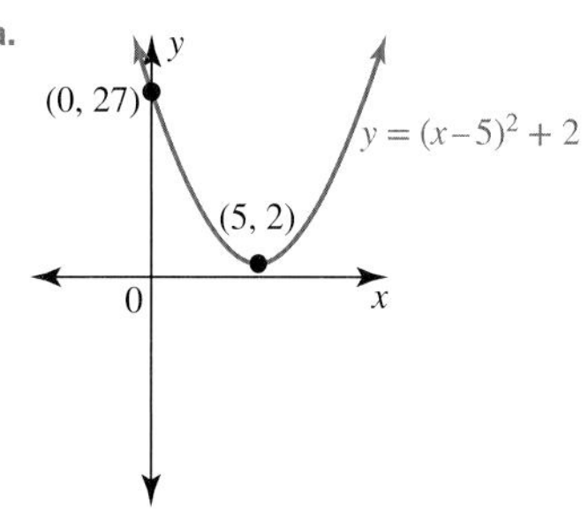

b.

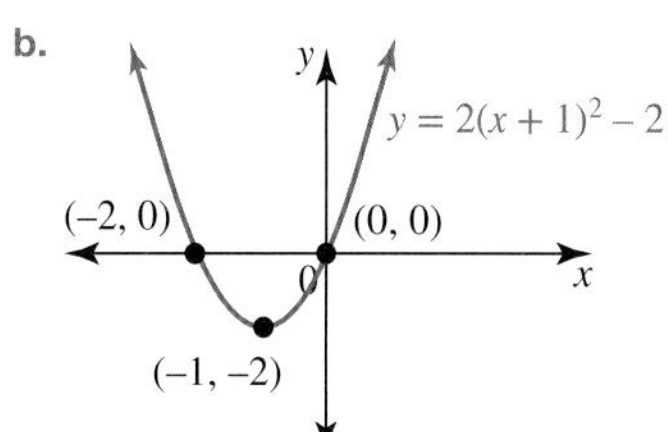

c.

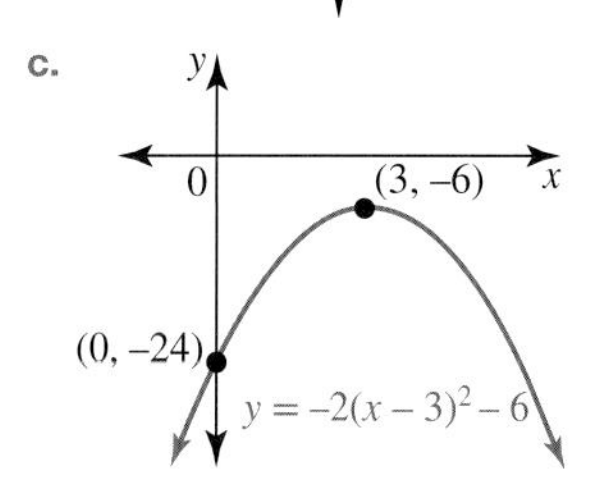

d.

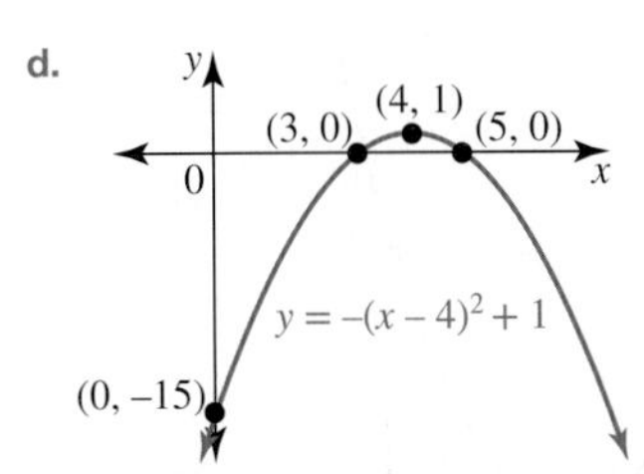

17.

	Turning point	y-intercept	x-intercepts
a.	$(0.5, -4.5)$	$(0, -4)$	$(-1, 0), (2, 0)$
b.	$\left(\frac{1}{12}, \frac{49}{24}\right)$	$(0, 2)$	$\left(-\frac{1}{2}, 0\right), \left(\frac{2}{3}, 0\right)$
c.	$(1.35, -14.58)$	$(0, 0)$	$(0, 0), (2.7, 0)$
d.	$\left(-\frac{1}{3}, 0\right)$	$(0, 1)$	$\left(-\frac{1}{3}, 0\right)$

a.

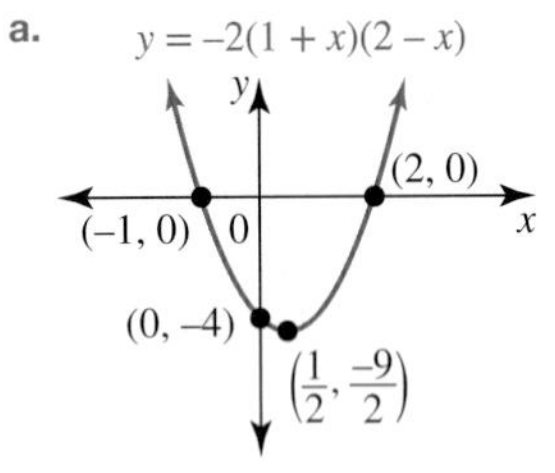

b.

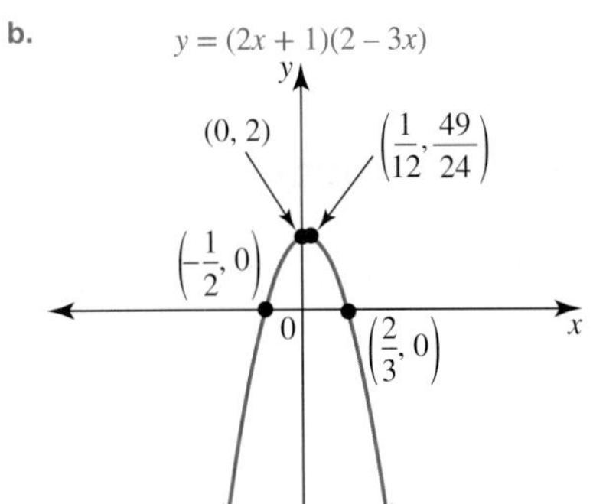

c.

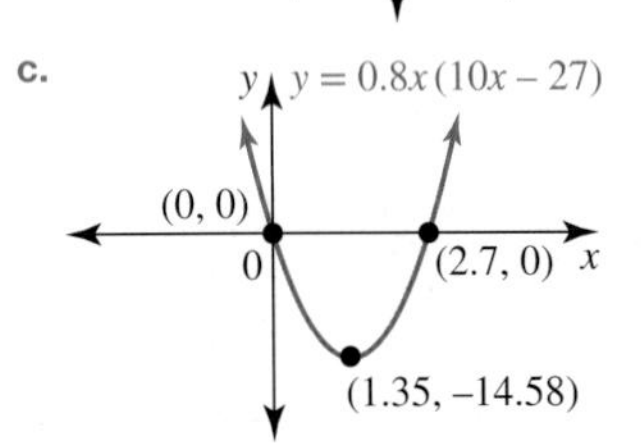

d.

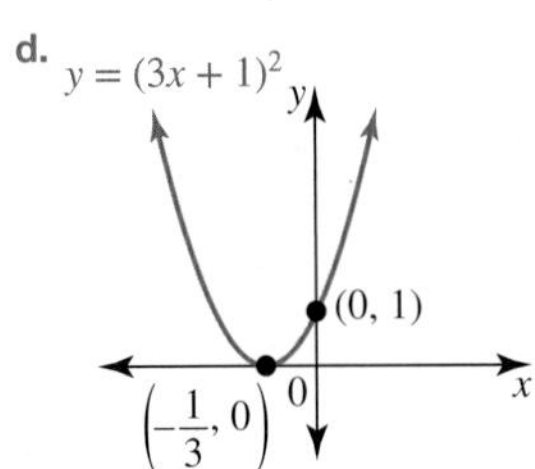

18.

	Turning point	y-intercept	x-intercepts
a.	$\left(\frac{1}{2}, 0\right)$	$\left(0, \frac{1}{4}\right)$	$\left(\frac{1}{2}, 0\right)$
b.	$\left(-\frac{1}{2}, 0\right)$	$\left(0, -\frac{1}{4}\right)$	$\left(-\frac{1}{2}, 0\right)$
c.	$\left(\frac{3}{4}, -\frac{23}{8}\right)$	$(0, -4)$	none
d.	$(2, 18)$	$(0, 10)$	$(-1, 0), (5, 0)$

a.

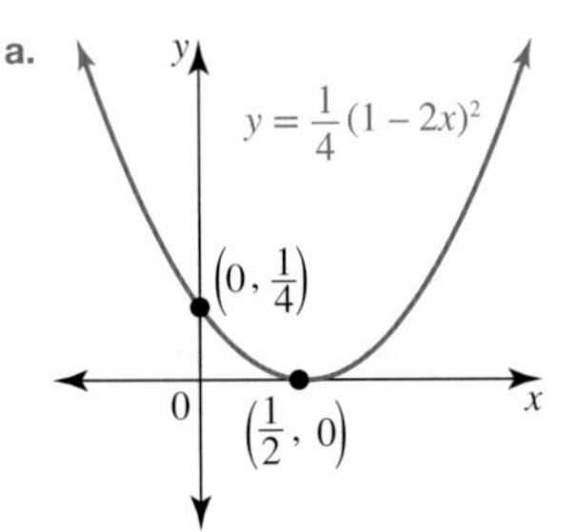

b.

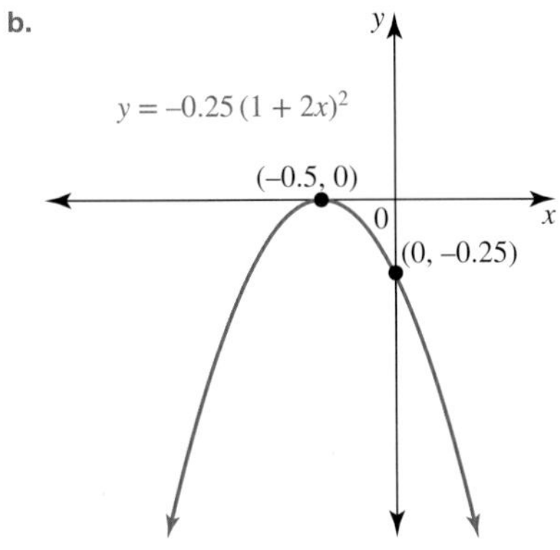

c.

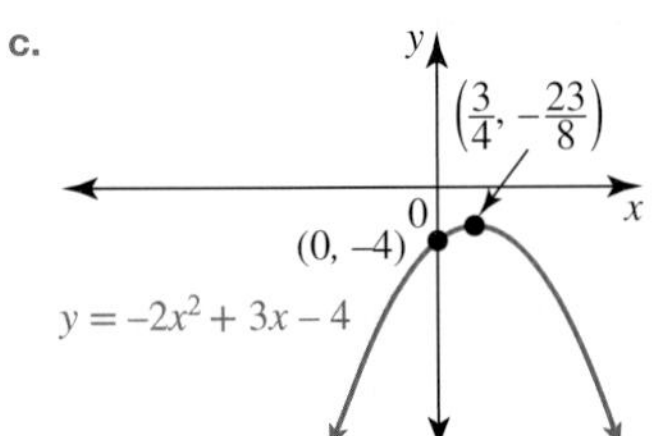

d.

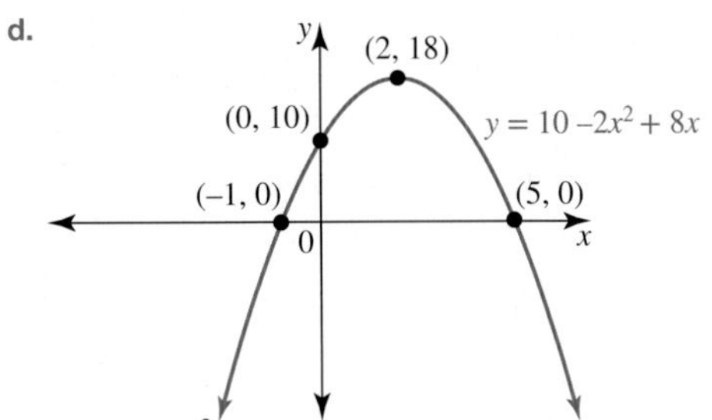

19. a. i. $2(x-3)^2 - 9$

ii. $(3, -9)$

iii. -9

b. i. $-(x+9)^2 + 86$

ii. $(-9, 86)$

iii. 86

20. a. Two rational x-intercepts

b. Axis intercepts $(0, 0)$, $\left(\frac{7}{3}, 0\right)$;

maximum turning point $\left(\frac{7}{6}, \frac{49}{2}\right)$

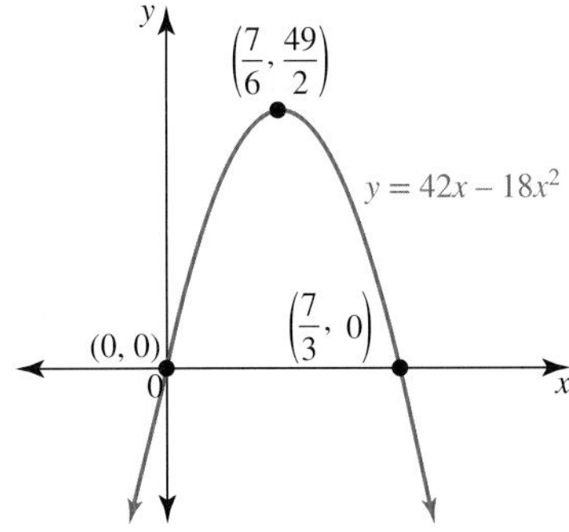

21. $\Delta < 0$ and $a > 0$
22. a. Two irrational x-intercepts
    b. No x-intercepts
    c. One rational x-intercept
    d. Two rational x-intercepts
23. a. $k = -5$ b. $k > -5$ c. $k < -5$
24. a. $m > \frac{1}{4}$
    b. i. Sample responses can be found in the worked solutions in the online resources.
       ii. $x = -\frac{1}{2}$
    c. i. $t = \pm \frac{4\sqrt{6}}{3}$ ii. $t = 0, 0.25$
25. a. i. $\left(\frac{3}{4}, -\frac{7}{4}\right)$ ii. $\left(\frac{9 \pm \sqrt{21}}{12}, 0\right)$
    b. i. $\left(\frac{9}{16}, \frac{465}{32}\right)$ ii. $\left(\frac{9 \pm \sqrt{465}}{16}, 0\right)$

3.5 Exam questions

Note: Mark allocations are available with the fully worked solutions online.

1. B
2. C
3. Turning point $= \left(-\frac{1}{4}, -3\frac{1}{8}\right)$,
   intercepts $= \left(-\frac{3}{2}, 0\right), (1, 0), (0, -3)$

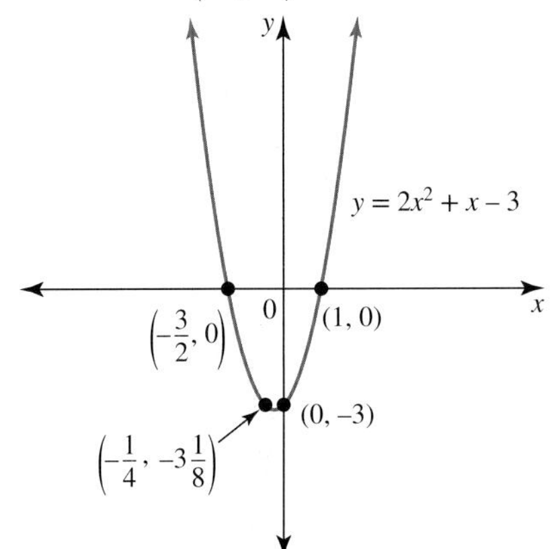

3.6 Determining the rule of a quadratic polynomial from a graph

3.6 Exercise

1. a. $c = 4, y = x^2 + 4.$ b. $a = -\frac{1}{18}, y = -\frac{1}{18}x^2$
   c. $a = -3, y = -3(x - 2)^2$
2. a. $b = 2, y = x^2 + 2x$ b. $(-2, 0), (0, 0).$
3. a. $y = (x - 4)^2 - 8$ b. $y = -2(x + 1)^2 + 3$
4. a. $y = A(x - 5)^2 + 12$
   b. $A = -\frac{1}{5}, y = -\frac{1}{5}(x - 5)^2 + 12$
5. a. $x - 3, x - 8, y = (x - 3)(x - 8)$
   b. $y = (x + 11)(x - 2)$
6. a. $y = (x + 2)^2 + 1$ b. $y = 2x(x - 2)$
7. $y = \frac{1}{8}x^2 + \frac{1}{2}x + \frac{1}{2}$
8. a. $y = -2x^2 + 6$ b. $y = 0.2(x + 6)(x + 1)$
9. a. $y = \frac{1}{2}(x - 2)^2 - 4$ b. $4\sqrt{2}$ units
10. a. $y = 3(x + 1)^2 - 5 = 3x^2 + 6x - 2$
    b. $y = (x + 5)^2$
11. a. A b. A c. A
12. $y = -2x^2 + x - 4$
13. $y = \frac{1}{2}x^2 - \frac{3}{2}x - 2$, which is the same as
    $y = \frac{1}{2}(x + 1)(x - 4)$
14. a. $y = 3x^2 - 5x + 2$
    b. $\left(\frac{2}{3}, 0\right), (1, 0), (0, 2)$
    c. $\left(\frac{5}{6}, -\frac{1}{12}\right)$
    d.

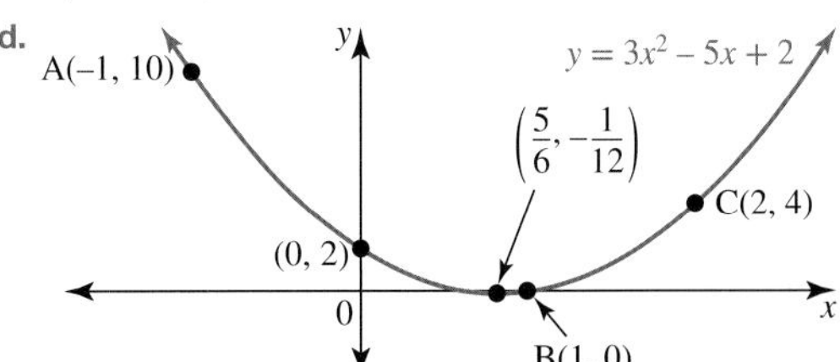

15. a. Answers will vary but should be of the form $y = a(x + 3)(x - 5)$.
    b. $y = -3(x + 3)(x - 5)$, vertex $(1, 48)$
16. $y = \frac{1}{2}x^2 - 4x + 6$
17. a. Sample responses can be found in the worked solutions in the online resources.
    b. $\Delta = (2 + 5a)^2$
    c. $y = -\frac{4}{5}x^2 + 6x - 10, \left(\frac{5}{2}, 0\right)$

18. a. $y = \frac{1}{4}(x+4)^2$

b. $p = \frac{4}{3} \Rightarrow y = \frac{9}{4}(3x-4)^2$

$p = 4 \Rightarrow y = \frac{9}{4}(x-4)^2$

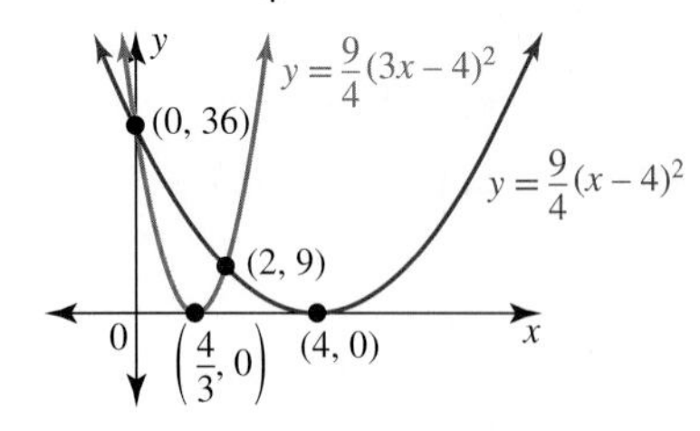

3.6 Exam questions

Note: Mark allocations are available with the fully worked solutions online.

1. A
2. $y = x^2 - x - 12$
3. D

3.7 Quadratic inequations

3.7 Exercise

1. a.

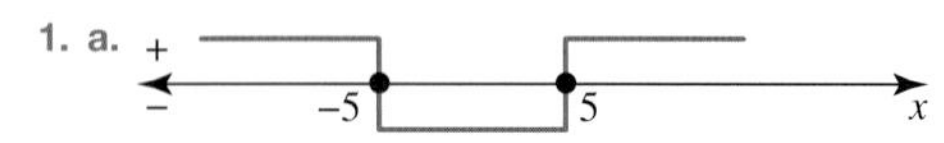

b. $x \leq -5$ or $x \geq 5$

2. a.

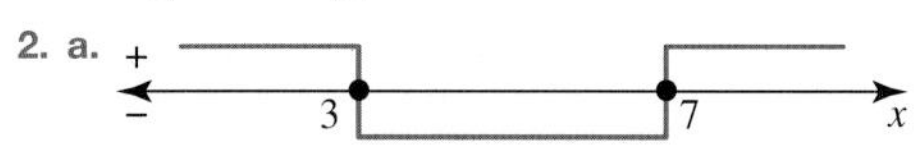

b. $3 \leq x \leq 7$

3. a. $\{x : x < 0\} \cup \{x : x > 3\}$ b. $\{x : 0 < x < 3\}$

4. a.

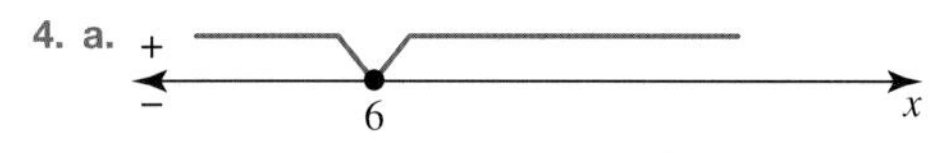

b. $R \setminus \{6\}$

5. a. $(3-x)(3+2x)$

b. $x = -\frac{3}{2}, x = 3,$

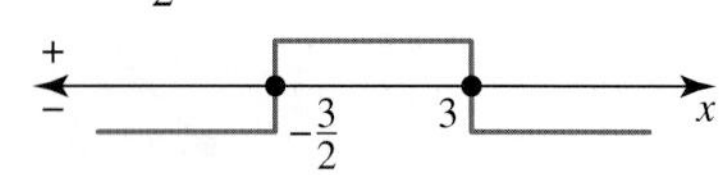

c. $-\frac{3}{2} \leq x \leq 3$

6. Zeros $x = -3, x = 4$, concave up sign diagram

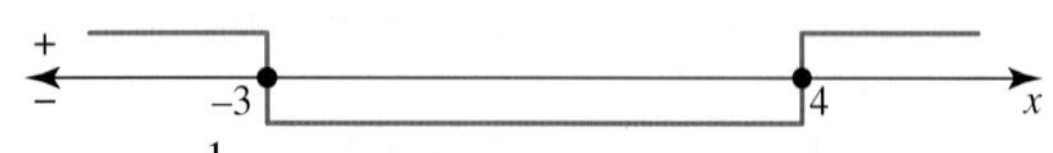

7. Zero $x = \frac{1}{9}$, multiplicity 2, concave up sign diagram touching at the zero required

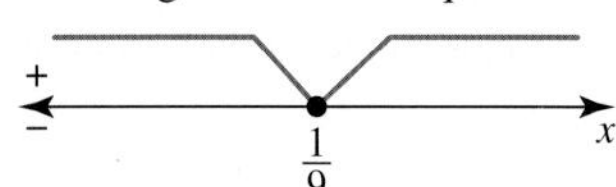

8. a. $x \leq \frac{3}{2}$ or $x \geq 4$ b. $\left\{x : -\frac{1}{2} < x < \frac{2}{3}\right\}$

9. $R \setminus \{3\}$

10. a. $-12 \leq x \leq 4$ b. $x \leq -1$ or $x \geq 4$

c. $x < -3$ or $x > \frac{3}{2}$ d. $-8 < x < -2$

e. $x < 0$ or $x > 9$ f. $2 \leq x \leq \frac{13}{4}$

11. a. $R \setminus \{6\}$ b. $\{1\}$

c. $\{x : x < 0\} \cup \left\{x : x > \frac{1}{4}\right\}$ d. $\left\{x : -\frac{3}{2} \leq x \leq \frac{7}{5}\right\}$

12. a. $(-6, 8)$ and $(2, 0)$ b. No intersections

13. a. $(6, 32)$ or $(-1, -3)$ b. $(-8, 35)$ or $(1, -1)$

c. No solution d. $(0, 5)$ or $\left(-\frac{5}{3}, \frac{70}{9}\right)$

14. a. $\left(\frac{3}{5}, \frac{31}{5}\right), (1, 7)$ b. $(-2, -3)$

c. $(-5, 10), (4, 10)$ d. $(4, 30)$

15. a. No intersections

b. Two intersections

c. No intersections

16. $(2, 8)$

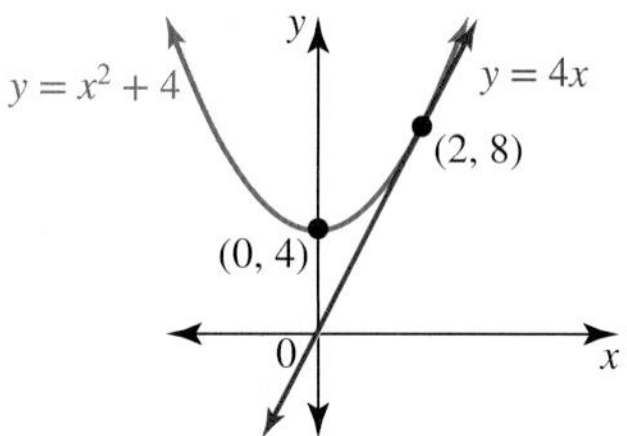

17. $m \leq -6$ or $m \geq 18$

18. $-2\sqrt{5} < k < 2\sqrt{5}$

19. a. $-9 < k < -1$ b. $k = -9, k = -1$

c. $k < -9$ or $k > -1$

20. a. $p < 0$ or $p \geq 4$ b. $5 - 4\sqrt{5} < t < 5 + 4\sqrt{5}$

c. $n = -5, n = 7$

21. a. $(-1.14, 1.29), (2.64, 6.96)$

b.

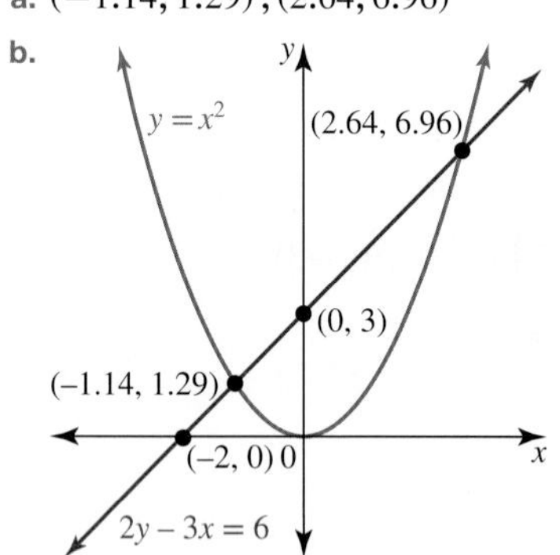

c. $2y - 3x \leq 6$ and $y \geq x^2$

d. $\left(0, -\frac{9}{16}\right)$

22. a. $x < -\frac{1}{10}\left(\sqrt{389} + 3\right)$ or $x > \frac{1}{10}\left(\sqrt{389} - 3\right)$

b. $-\frac{1}{12}\left(\sqrt{465} + 15\right) \leq x \leq \frac{1}{12}\left(\sqrt{465} - 15\right)$

23. a. $(-2, 0), (0, 4)$

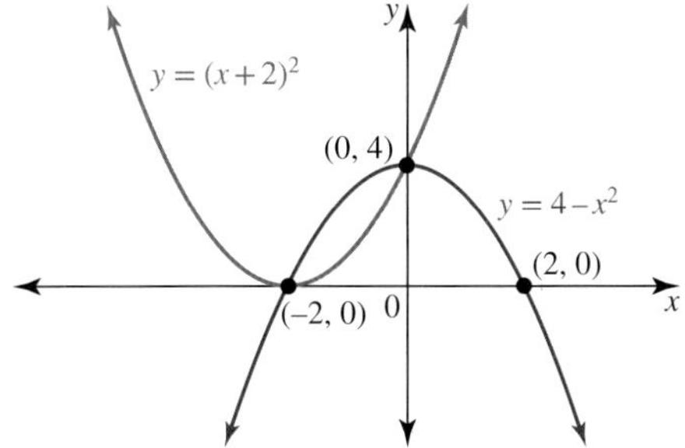

b. i. Sample responses can be found in the worked solutions in the online resources.

ii. $C(-1, 1)$

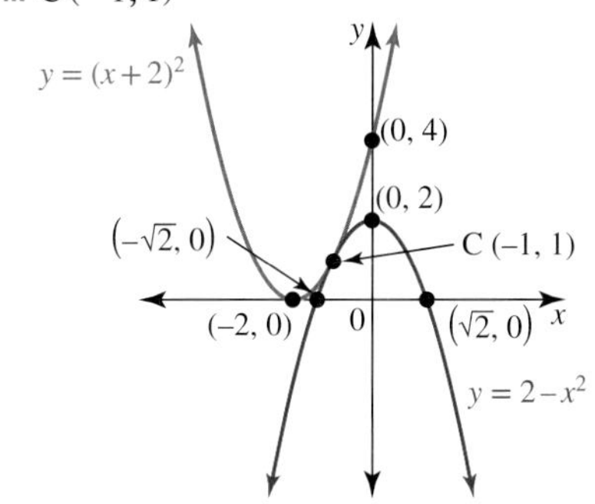

iii. $y = 2x + 3$

24. a. $y = 5x - 4$

b. $y = \frac{5}{3} - x^2$; other answers possible

25. a. Use CAS technology to sketch the graph. $a = -6$, no intersections; $a = -4$, points $(1, -8), (2, -12)$;

$a = -2$, points $(0.38, -3.53), (2.62, -12.47)$;

$a = 0$, points $(0, 0), (3, -12)$

b. $x = -\frac{1}{2}\left(\sqrt{2a+9} - 3\right)$ or $x = \frac{1}{2}\left(\sqrt{2a+9} + 3\right)$

c. $a = -4.5$, point $(1.5, -10.5)$

3.7 Exam questions

Note: Mark allocations are available with the fully worked solutions online.

1. E
2. E
3. The line $y = mx + 24$ will be a tangent to the parabola $x^2 + 4x + 33$ for $m = -2$ and $m = 10$.

3.8 Quadratic models and applications

3.8 Exercise

1. a. Sample responses can be found in the worked solutions in the online resources.

b. Width 7.5 metres, length 15 metres

c. 112.5 square metres

2. a. 328 metres

b. 12 seconds

c. The missile reaches the ground after 26.4 seconds.

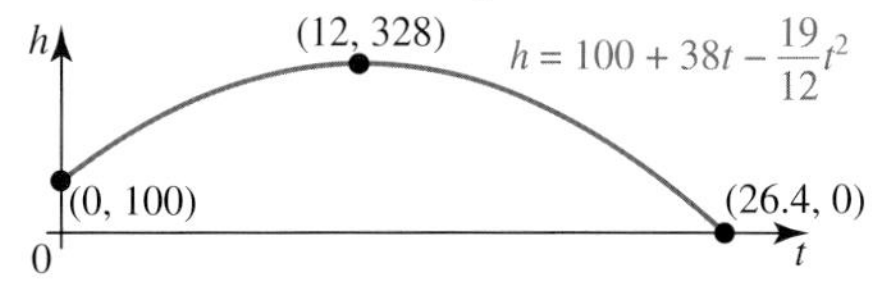

3. a. 2 seconds

b. The ball does not strike the foliage.

c. $\frac{10}{9}$ metres

4. a. $y = -0.2(x - 5.5)^2 + 7.25$

b. 7.25 metres

c. 2.37 metres

5. a. Sample responses can be found in the worked solutions in the online resources; length 30 metres, width 15 metres.

b. 120 metres

6. a. 2 hours

b. 380

c. 110 bacteria cells at 4 pm

7. $z = \frac{104}{7}$ when $x = \frac{8}{7}, y = \frac{6}{7}$

8. a. 17

b. $S = 2x^2 - 10x + 25$, two pieces of 10 cm

9. a. $C = 20 + 10n + 5n^2$ b. 13

10. a. $A(3, 0), B(7, 5), C(11, 0)$

b. $y = -\frac{5}{16}(x - 7)^2 + 5$

c. 6.7 metres

11. Profit $= -x^2 + 40x - 15$, \$30 per kg

12. a. i. Both numbers are 8.

ii. Both numbers are 8.

b. i. $\frac{k^2}{4}$ ii. No values possible

13. a. $h = -0.06t^2 + 1.97t - 12.78$

b. 3.39 metres above mean sea level at 4.25 pm

14. a. $S = x^2 + \frac{(10 - 2x)^2}{\pi}$ b. 11.2 cm and 8.8 cm

c. $x = 0$

3.8 Exam questions

Note: Mark allocations are available with the fully worked solutions online.

1. B
2. A
3. The ball's height above Ruby was 41.5 metres and it took 10 seconds to reach this height.

3.9 Review

3.9 Exercise

Technology free: short answer

1. a. $x = -3, 7$ b. $x = -\frac{1}{5}, -\frac{7}{2}$ c. $x = \pm 2$

d. $x = 3 \pm 2\sqrt{2}$ e. $x = \pm 4$ f. $x = \frac{9}{4}$

2. a. $x < -\frac{1}{2}$ or $x > 3$ b. $-\sqrt{10} \le x \le \sqrt{10}$

c. $x \in R$

3. a. x-intercepts $(3, 0), (-1, 0)$; y-intercept $(0, -6)$; turning point $(1, -8)$

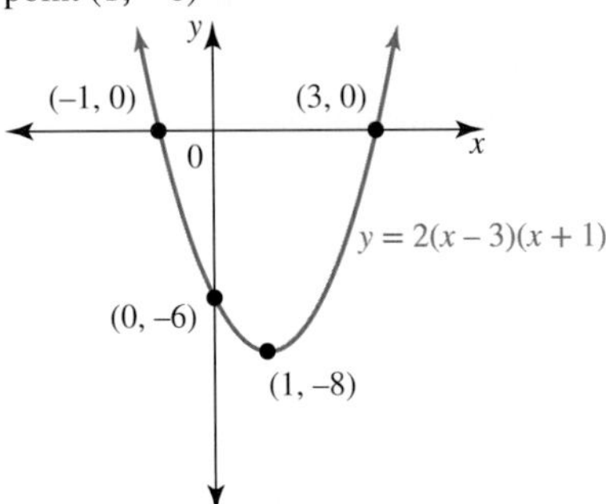

b. x-intercepts $(-3, 0), (-1, 0)$; y-intercept $(0, -3)$; turning point $(-2, 1)$

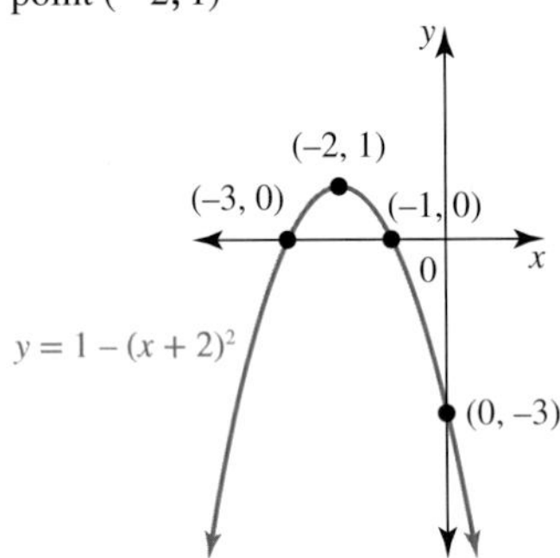

c. No x-intercepts; y-intercept $(0, 9)$; turning point $(-0.5, 8.75)$

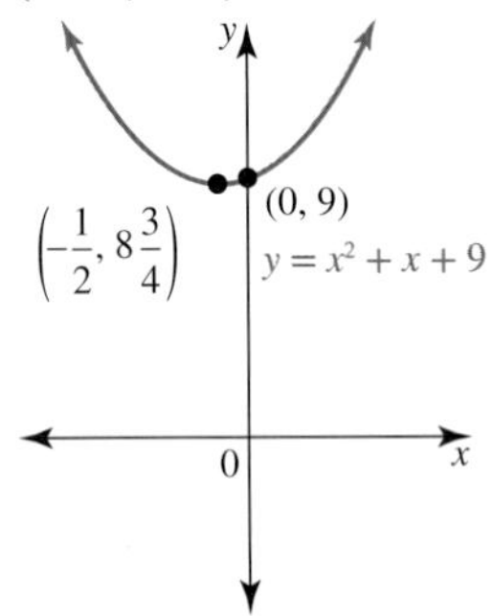

4. a. $-\left(x - 10 - 2\sqrt{31}\right)\left(x - 10 + 2\sqrt{31}\right)$

b. $4\left(x - \frac{1+\sqrt{37}}{4}\right)\left(x - \frac{1-\sqrt{37}}{4}\right)$

5. a. $\Delta = 104$, two irrational solutions.

b. $1 < k < 4$

6. a. $(-2, 0), (1, 3)$

b. Parabola: x-intercepts $(-2, 0), (0, 0)$; turning point $(-1, -1)$

Line through $(-2, 0), (0, 2)$. Both graphs meet at $(-2, 0), (1, 3)$.

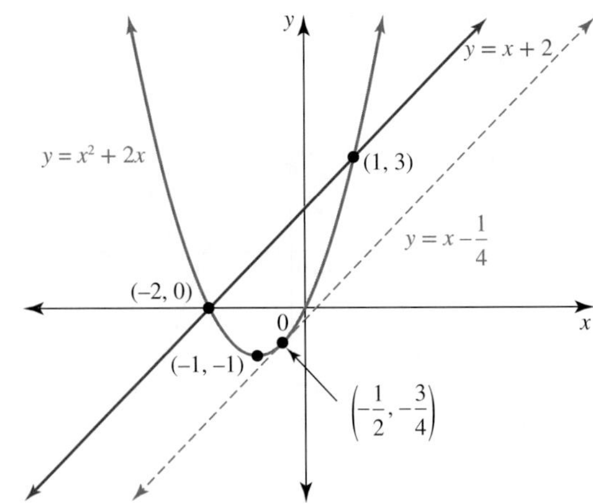

c. $y \le x + 2$ and $y \ge x^2 + 2x$

d. i. $k = -\frac{1}{4}$

ii. The tangent is added to the graph in part b. Point of contact: $\left(-\frac{1}{2}, -\frac{3}{4}\right)$.

Technology active: multiple choice

7. D
8. C
9. B
10. C
11. A
12. B
13. C
14. E
15. B
16. A

Technology active: extended response

17. a. $h = -\frac{1}{35}(x^2 - 60x - 700)$

When $x = 0$, $h = -\frac{1}{35}(-700)$.

$\therefore h = 20$ and the point S is $(0, 20)$.

Therefore, S is 20 metres above O.

b. 70 metres

c. $h = -\frac{1}{60}(x - 30)^2 + 35$

d. The Japanese competitor wins.

18. a. 8 metres

b. 5 metres

c. The caravan can be towed under the bridge.

d. $8 - 2x$

e. 6.4

f. No

19. a.

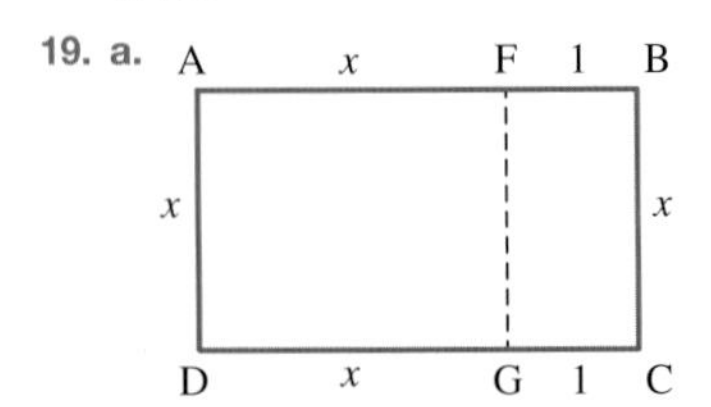

b. x units

c. Area measure of rectangle FBCG is $1 \times x = x$.

$$\therefore x^2 = x + 1$$

$$\therefore x^2 - x - 1 = 0$$

d. $\dfrac{1 + \sqrt{5}}{2}$

e. $\dfrac{1}{\phi} = -\dfrac{1 - \sqrt{5}}{2}$

f. $$\begin{aligned}\phi - 1 &= \frac{1 + \sqrt{5}}{2} - 1 \\ &= \frac{1 + \sqrt{5} - 2}{2} \\ &= \frac{\sqrt{5} - 1}{2}\end{aligned}$$

and:

$$\begin{aligned}\frac{1}{\phi} &= -\frac{1 - \sqrt{5}}{2} \\ &= \frac{\sqrt{5} - 1}{2} \\ &= \phi - 1\end{aligned}$$

As $x = \phi$ is a root of $x^2 - x - 1 = 0$,

$$\begin{aligned}\phi^2 - \phi - 1 &= 0 \\ \phi(\phi - 1) &= 1 \\ \phi - 1 &= \frac{1}{\phi}\end{aligned}$$

20. a. $(28, 4.9)$

b. $y = -\dfrac{1}{160}(x - 28)^2 + 4.9$

c. $\dfrac{13}{7}$ seconds

d. 7 m/s

3.9 Exam questions

Note: Mark allocations are available with the fully worked solutions online.

1. $x = -1$
2. B
3. A
4. D
5. The zeros are $-8 + 4\sqrt{2}, -8 - 4\sqrt{2}$.

   $k \in \left(-\infty, -8 - 4\sqrt{2}\right] \cup \left[-8 + 4\sqrt{2}, \infty\right)$

4 Cubic polynomials

LEARNING SEQUENCE

Fully worked solutions for this topic are available online.

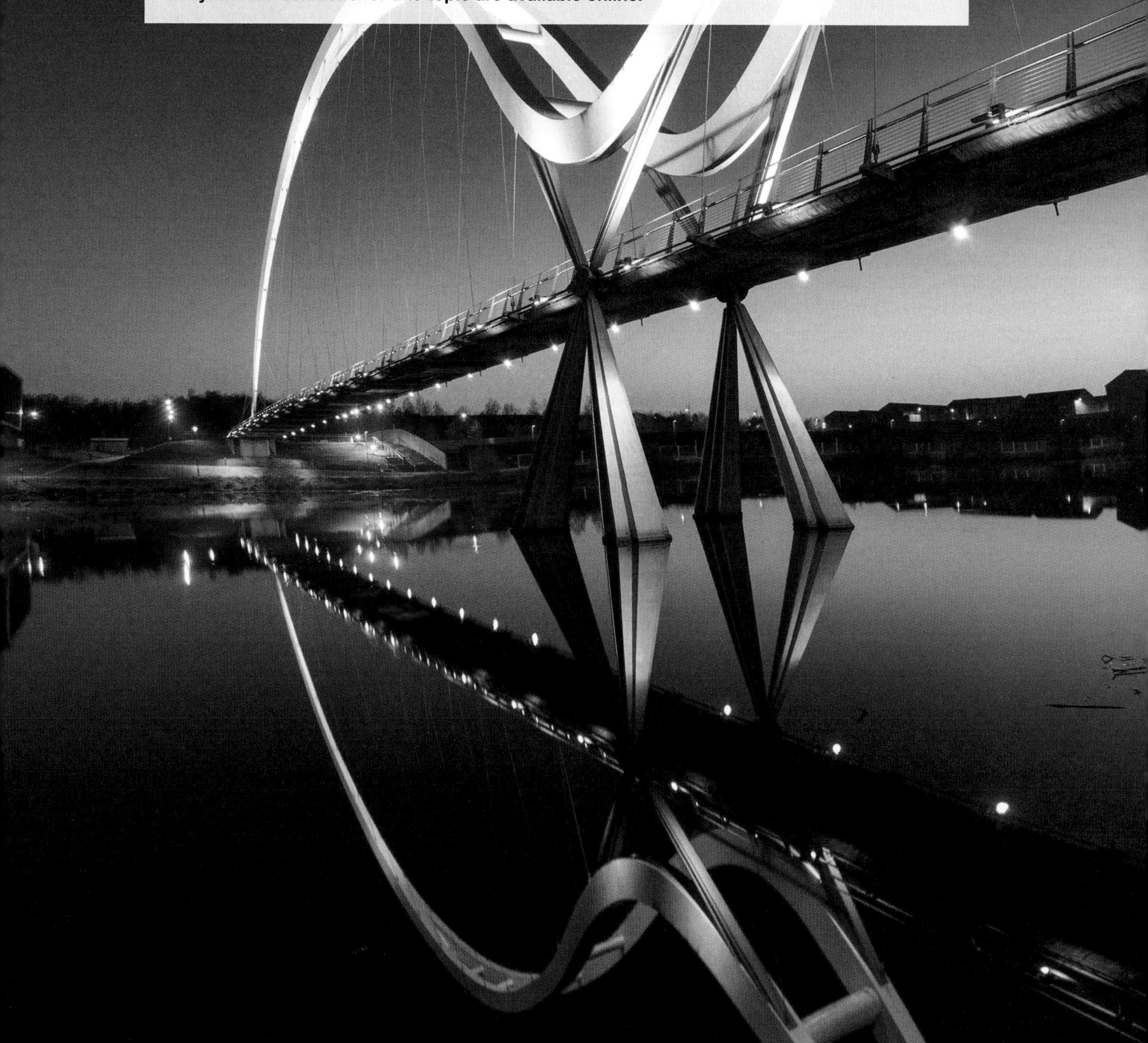

4.1 Overview

Hey students! Bring these pages to life online

 Watch videos

 Engage with interactivities

 Answer questions and check results

Find all this and MORE in jacPLUS

4.1.1 Introduction

Mathematicians are driven by curiosity and a desire to generalise. One historical example of this is: 'Since quadratic equations can be solved by a formula, is the same true of cubic equations?'

The Babylonians could solve quadratic equations, and by the 9th century the general formula had essentially been formulated. Not until the 16th century was the general solution of the cubic equation found by two Italian mathematicians, Gerolamo Cardano and Niccolo Fontana; the latter was also known as Tartaglia due to a speech defect (*tartagliare* means 'to stutter').

Tartaglia devised a method to solve cubic equations of a particular form, but he refused to publish his method. Eventually, he was prevailed upon to share his method with Cardano, on the condition that Cardano kept it secret. Knowledge of Tartaglia's work enabled Cardano to complete the general method for solving cubics, and he published this in 1545. Although he acknowledged his debt to Tartaglia, Tartaglia was outraged at the broken promise and a long feud ensued between them. Not all mathematical communities are cooperative and harmonious.

Cardano's method contained square roots of negative numbers; for this he is remembered as the first person to use what are now called complex numbers. It took another 300 years before complex numbers were accepted and their usefulness recognised, not only in higher mathematics, but in practical areas such as electrical engineering. Once more, mathematical research had led to a major discovery, the significance of which could not have been predicted.

KEY CONCEPTS

This topic covers the following key concepts from the VCE Mathematics Study Design:

- graphs of power functions $f(x) = x^n$ for $n \in \left\{-2, -1, \frac{1}{3}, \frac{1}{2}, 1, 2, 3, 4\right\}$, and transformations of these graphs to the form $y = a(x+b)^n + c$ where $a, b, c \in R$ and $a \neq 0$
- graphs of polynomial functions of low degree, and interpretation of key features of these graphs
- the connection between the roots of a polynomial function, its factors and the horizontal axis intercepts of its graph, including the remainder, factor and rational root theorems
- solution of polynomial equations of low degree, numerically, graphically and algebraically, including numerical approximation of roots of simple polynomial functions using the bisection method algorithm.

Note: *Concepts shown in grey are covered in other topics.*

Source: VCE Mathematics Study Design (2023–2027) extracts © VCAA; reproduced by permission.

4.2 Polynomials

LEARNING INTENTION

At the end of this subtopic you should be able to:
- classify and evaluate polynomials
- express a cubic polynomial $p(x)$, with integer coefficients, in the form $p(x) = (x - a)\, q(x) + r$ and determine $\frac{p(x)}{x - a}$ by hand.

A **polynomial** is an algebraic expression in which the power of the variable is a positive whole number. For example, $3x^2 + 5x - 1$ is a quadratic polynomial in the variable x, but $\frac{3}{x^2} + 5x - 1$, that is $3x^{-2} + 5x - 1$, is not a polynomial because of the $3x^{-2}$ term. Note that the coefficients of x can be positive or negative integers, and rational or irrational real numbers.

4.2.1 Classification of polynomials

- The **degree of a polynomial** is the highest power of the variable.
  For example, linear polynomials have degree 1, quadratic polynomials have degree 2 and cubic polynomials have degree 3.
- The **leading term** is the term containing the highest power of the variable.
- If the coefficient of the leading term is 1, then the polynomial is said to be **monic**.
- The **constant term** is the term that does not contain the variable.

General form of a polynomial

A polynomial of degree n has the form $a_n x^n + a_{n-1} x^{n-1} + \ldots + a_1 x + a_0$, where $n \in N$ and the coefficients $a_n, a_{n-1}, \ldots a_1, a_0 \in R$. The leading term is $a_n x^n$ and the constant term is a_0.

WORKED EXAMPLE 1 Classifying polynomials

Select the polynomials from the following list of algebraic expressions and for these polynomials, state the degree, the coefficient of the leading term, the constant term and the type of coefficients.

A. $5x^3 + 2x^2 - 3x + 4$ **B.** $5x - x^3 + \frac{x^4}{2}$ **C.** $4x^5 + 2x^2 + 7x^{-3} + 8$

THINK	WRITE
1. Check the powers of the variable x in each algebraic expression.	A and B are polynomials since all the powers of x are positive integers. C is not a polynomial due to the $7x^{-3}$ term.
2. For polynomial A, state the degree, the coefficient of the leading term and the constant term.	Polynomial A: the leading term of $5x^3 + 2x^2 - 3x + 4$ is $5x^3$. Therefore, the degree is 3 and the coefficient of the leading term is 5. The constant term is 4.
3. Classify the coefficients of polynomial A as elements of a subset of R.	The coefficients in polynomial A are integers. Therefore, A is a polynomial over Z.

4. For polynomial B, state the degree, the coefficient of the leading term and the constant term.	Polynomial B: the leading term of $5x - x^3 + \frac{x^4}{2}$ is $\frac{x^4}{2}$. Therefore, the degree is 4 and the coefficient of the leading term is $\frac{1}{2}$. The constant term is 0.
5. Classify the coefficients of polynomial B as elements of a subset of R.	The coefficients in polynomial B are rational numbers. Therefore, B is a polynomial over Q.

Polynomial notation

- A polynomial in variable x is often referred to as $p(x)$.
- The value of the polynomial $p(x)$ when $x = a$ is written as $p(a)$.
- $p(a)$ is evaluated by substituting a in place of x in the $p(x)$ expression.

WORKED EXAMPLE 2 Evaluating polynomials

a. If $p(x) = 5x^3 + 2x^2 - 3x + 4$, calculate $p(-1)$.
b. If $p(x) = ax^2 - 2x + 7$ and $p(4) = 31$, obtain the value of a.

THINK	WRITE
a. Substitute -1 in place of x and evaluate.	a. $p(x) = 5x^3 + 2x^2 - 3x + 4$ $p(-1) = 5(-1)^3 + 2(-1)^2 - 3(-1) + 4$ $= -5 + 2 + 3 + 4$ $= 4$
b. 1. Find an expression for $p(4)$ by substituting 4 in place of x, and then simplify.	b. $p(x) = ax^2 - 2x + 7$ $p(4) = a(4)^2 - 2(4) + 7$ $= 16a - 1$
2. Equate the expression for $p(4)$ with 31.	$p(4) = 31$ $\Rightarrow 16a - 1 = 31$
3. Solve for a.	$16a = 32$ $a = 2$

4.2.2 Identity of polynomials

If two polynomials are **identically equal**, then the coefficients of like terms are equal. **Equating coefficients** means that if $ax^2 + bx + c \equiv 2x^2 + 5x + 7$, then $a = 2$, $b = 5$ and $c = 7$. The identically equal symbol '$\equiv$' means the equality holds for all values of x. For convenience, however, we shall replace this symbol with the equality symbol '$=$' in working steps.

WORKED EXAMPLE 3 Determining coefficients

Calculate the values of a, b and c so that $x(x - 7) = a(x - 1)^2 + b(x - 1) + c$.

THINK	WRITE
1. Expand each bracket and express both sides of the equality in expanded polynomial form.	$x(x - 7) = a(x - 1)^2 + b(x - 1) + c$ $\therefore x^2 - 7x = a(x^2 - 2x + 1) + bx - b + c$ $\therefore x^2 - 7x = ax^2 + (-2a + b)x + (a - b + c)$

2. Equate the coefficients of like terms.

Equate the coefficients.

$$x^2: 1 = a \quad [1]$$
$$x: -7 = -2a + b \quad [2]$$
$$\text{Constant}: 0 = a - b + c \quad [3]$$

3. Solve the system of simultaneous equations.

Since $a = 1$, substitute $a = 1$ into equation [2].

$$-7 = -2(1) + b$$
$$b = -5$$

Substitute $a = 1$ and $b = -5$ into equation [3].

$$0 = 1 - (-5) + c$$
$$c = -6$$

4. State the answer.

$\therefore a = 1, b = -5, c = -6$

4.2.3 Operations on polynomials

The addition, subtraction or multiplication of two or more polynomials results in another polynomial. For example, if $p(x) = x^2$ and $q(x) = x^3 + x^2 - 1$, then $p(x) + q(x) = x^3 + 2x^2 - 1$, a polynomial of degree 3; $p(x) - q(x) = -x^3 + 1$, a polynomial of degree 3; and $p(x)q(x) = x^5 + x^4 - x^2$, a polynomial of degree 5.

WORKED EXAMPLE 4 Operations with polynomials

Given $p(x) = 3x^3 + 4x^2 + 2x + m$ and $q(x) = 2x^2 + kx - 5$, find the values of m and k for which $2p(x) - 3q(x) = 6x^3 + 2x^2 + 25x - 25$.

THINK

WRITE

1. Form a polynomial expression for $2p(x) - 3q(x)$ by collecting like terms together.

$$2p(x) - 3q(x)$$
$$= 2(3x^3 + 4x^2 + 2x + m) - 3(2x^2 + kx - 5)$$
$$= 6x^3 + 2x^2 + (4 - 3k)x + (2m + 15)$$

2. Equate the two expressions for $2p(x) - 3q(x)$.

Hence, $6x^3 + 2x^2 + (4 - 3k)x + (2m + 15)$
$$= 6x^3 + 2x^2 + 25x - 25$$

3. Calculate the values of m and k.

Equate the coefficients of x.

$$4 - 3k = 25$$
$$k = -7$$

Equate the constant terms.

$$2m + 15 = -25$$
$$m = -20$$

4. State the answer.

Therefore, $m = -20, k = -7$.

4.2.4 Division of polynomials

There are several methods for performing the **division of polynomials**, and CAS technology computes the division readily. Here, two 'by hand' methods will be shown.

The inspection method for division

The division of one polynomial by another polynomial of equal or lesser degree can be carried out by expressing the numerator in terms of the denominator.

To divide $(x+3)$ by $(x-1)$, or to find $\frac{x+3}{x-1}$, write the numerator $x+3$ as $(x-1)+1+3=(x-1)+4$.

$$\frac{x+3}{x-1}=\frac{(x-1)+4}{x-1}$$

This expression can then be split into the sum of **partial fractions** as:

$$\begin{aligned}\frac{x+3}{x-1}&=\frac{(x-1)+4}{x-1}\\&=\frac{x-1}{x-1}+\frac{4}{x-1}\\&=1+\frac{4}{x-1}\end{aligned}$$

The inspection method for division of polynomials

The division is in the form: $\frac{\textbf{dividend}}{\textbf{divisor}}=\textbf{quotient}+\frac{\textbf{remainder}}{\textbf{divisor}}$

In the language of division, when the dividend $(x+3)$ is divided by the divisor $(x-1)$, it gives a quotient of 1 and a remainder of 4.

Note that from the division statement $\frac{x+3}{x-1}=1+\frac{4}{x-1}$, we can write $x+3=1\times(x-1)+4$.

This is similar to the division of integers. For example, 7 divided by 2 gives a quotient of 3 and a remainder of 1.

$$\begin{aligned}\frac{7}{2}&=3+\frac{1}{2}\\\therefore 7&=3\times2+1\end{aligned}$$

This inspection process of division can be extended, with practice, to division involving non-linear polynomials. It could be used to show that $\frac{x^2+4x+1}{x-1}=\frac{x(x-1)+5(x-1)+6}{x-1}$ and therefore $\frac{x^2+4x+1}{x-1}=x+5+\frac{6}{x-1}$. This result can be verified by checking that $x^2+4x+1=(x+5)(x-1)+6$.

WORKED EXAMPLE 5 Determining the quotient and remainder

a. Calculate the quotient and the remainder when $(x+7)$ is divided by $(x+5)$.

b. Use the inspection method to find $\frac{3x-4}{x+2}$.

THINK	WRITE
a. 1. Write the division of the two polynomials as a fraction.	a. $\frac{x+7}{x+5}=\frac{(x+5)-5+7}{x+5}$
2. Write the numerator in terms of the denominator.	$=\frac{(x+5)+2}{x+5}$
3. Split into partial fractions.	$=\frac{(x+5)}{x+5}+\frac{2}{x+5}$

4. Simplify.	$= 1 + \dfrac{2}{x+5}$
5. State the answer.	The quotient is 1 and the remainder is 2.
b. 1. Express the numerator in terms of the denominator.	b. The denominator is $(x+2)$. Since $3(x+2) = 3x+6$, the numerator is $3x-4 = 3(x+2) - 6 - 4$ $\therefore 3x-4 = 3(x+2) - 10$
2. Split the given fraction into its partial fractions.	$\dfrac{3x-4}{x+2} = \dfrac{3(x+2)-10}{x+2}$ $= \dfrac{3(x+2)}{(x+2)} - \dfrac{10}{x+2}$
3. Simplify and state the answer.	$= 3 - \dfrac{10}{x+2}$ $\therefore \dfrac{3x-4}{x+2} = 3 - \dfrac{10}{x+2}$

TI | THINK / **DISPLAY/WRITE**

a. 1. On a Calculator page, press MENU, then select:
3: Algebra
8: Polynomial Tools
5: Quotient of Polynomial
Complete the entry line as:
polyQuotient $(x+7,\ x+5)$
then press ENTER.

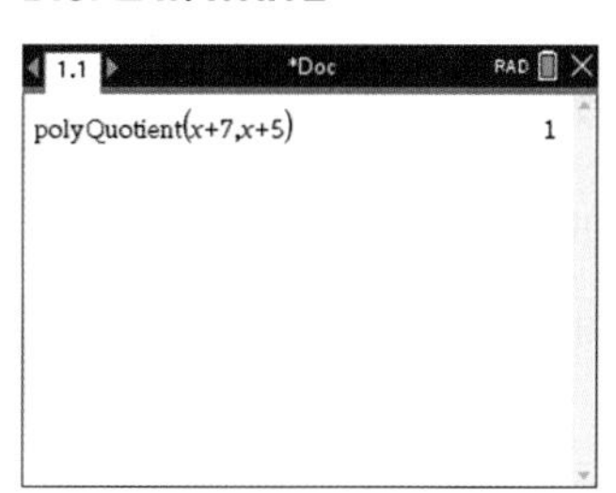

2. Press MENU, then select:
3: Algebra
8: Polynomial Tools
4: Remainder of Polynomial
Complete the entry line as:
polyRemainder
$(x+7,\ x+5)$
then press ENTER.

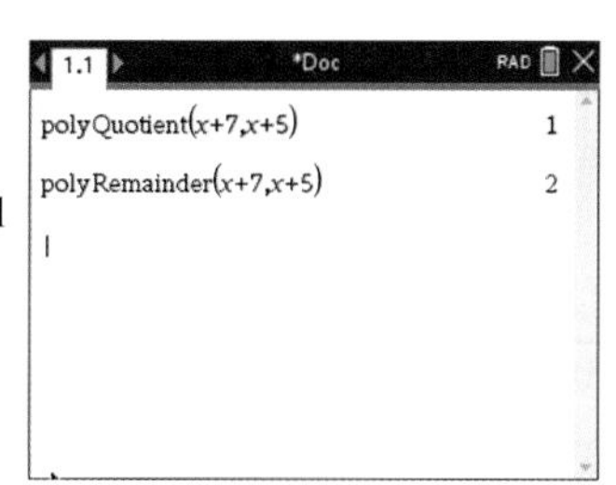

3. The answers appear on the screen. — The quotient is 1 and the remainder is 2.

4. Alternatively, press MENU, then select:
3: Algebra
9: Fraction Tools
1: Proper Fraction
Complete the entry line as:
propFrac $\left(\dfrac{x+7}{x+5}\right)$
then press ENTER.

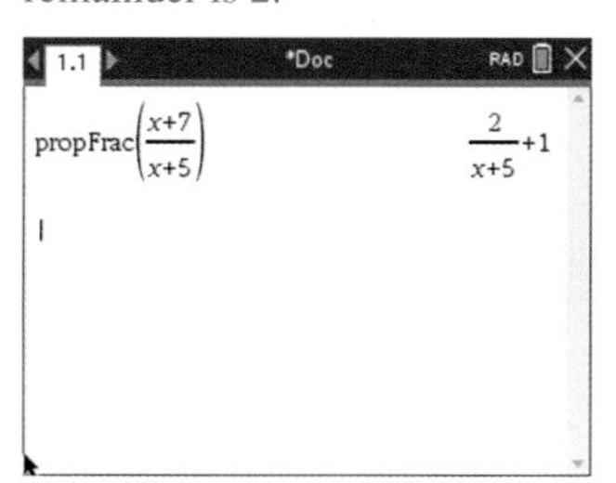

5. The answer appears on the screen. — The quotient is 1 and the remainder is 2.

CASIO | THINK / **DISPLAY/WRITE**

a. 1. On a Main screen, select:
- Interactive
- Transformation
- Fraction
- prop Frac

Complete the entry line as:
$(x+7)/(x+5)$
then select OK.

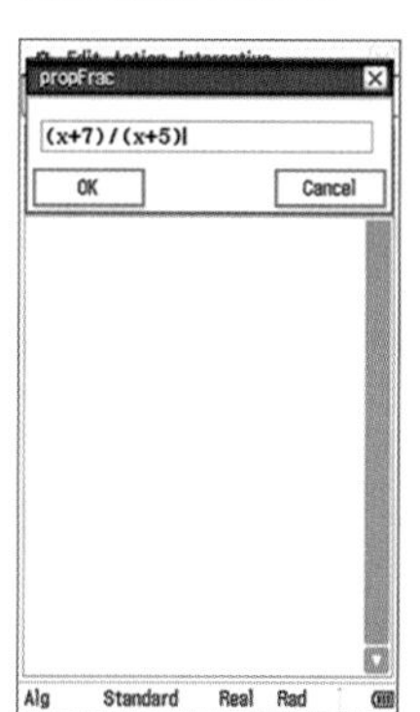

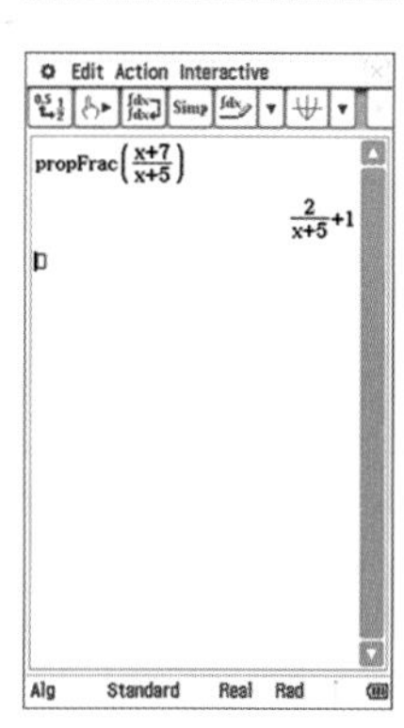

2. The answers appear on the screen. — The quotient is 1 and the remainder is 2.

Algorithm for long division of polynomials

The inspection method of division is very efficient, particularly when the division involves only linear polynomials. However, it is also possible to use the **long division algorithm** to divide polynomials.

Long division of polynomials

The steps in the long division algorithm are:

1. **Divide the leading term of the divisor into the leading term of the dividend.**
2. **Multiply the divisor by this quotient.**
3. **Subtract the product from the dividend to form a remainder of lower degree.**
4. **Repeat this process until the degree of the remainder is lower than that of the divisor.**

To illustrate this process, consider (x^2+4x+1) divided by $(x-1)$. This is written as:

$$x-1\overline{)x^2+4x+1}$$

Step 1. The leading term of the divisor $(x-1)$ is x; the leading term of the dividend (x^2+4x+1) is x^2.

Dividing x into x^2, we get $\frac{x^2}{x}=x$. We write this quotient x on top of the long division symbol.

$$\begin{array}{r} x \\ x-1\overline{)x^2+4x+1} \end{array}$$

Step 2. The divisor $(x-1)$ is multiplied by the quotient x to give $x(x-1)=x^2-x$. This product is written underneath the terms of (x^2+4x+1); like terms are placed in the same columns.

$$\begin{array}{r} x \\ x-1\overline{)x^2+4x+1} \\ x^2-x \quad\; \end{array}$$

Step 3. x^2-x is subtracted from $\left(x^2+4x+1\right)$. This cancels out the x^2 leading term to give $x^2+4x+1-(x^2-x)=5x+1$.

$$\begin{array}{r} x \\ x-1\overline{)x^2+4x+1} \\ -(x^2-x) \quad\; \\ \hline 5x+1 \end{array}$$

The division statement, so far, would be $\frac{x^2+4x+1}{x-1}=x+\frac{5x+1}{x-1}$. This is incomplete since the remainder $(5x+1)$ is not of a smaller degree than the divisor $(x-1)$. The steps in the algorithm must be repeated with the same divisor $(x-1)$ but with $(5x+1)$ as the new dividend.

Continue the process.

Step 4. Divide the leading term of the divisor $(x-1)$ into the leading term of $(5x+1)$; this gives $\frac{5x}{x}=5$.

Write this as $+5$ on the top of the long division symbol.

$$\begin{array}{r} x+5 \\ x-1\overline{)x^2+4x+1} \\ -(x^2-x) \quad\; \\ \hline 5x+1 \end{array}$$

Step 5. Multiply $(x-1)$ by 5 and write the result underneath the terms of $(5x+1)$.

$$\begin{array}{r l} & x+5 \\ x-1 \,\big) & \overline{x^2+4x+1} \\ & -(x^2-x) \\ \hline & 5x+1 \\ & 5x-5 \end{array}$$

Step 6. Subtract $(5x-5)$ from $(5x+1)$.

$$\begin{array}{r l} & x+5 \leftarrow \text{Quotient} \\ x-1 \,\big) & \overline{x^2+4x+1} \\ & -(x^2-x) \\ \hline & 5x+1 \\ & -(5x-5) \\ \hline & 6 \leftarrow \text{Remainder} \end{array}$$

The remainder is of lower degree than the divisor, so no further division is possible and we have reached the end of the process.

Thus, $\dfrac{x^2+4x+1}{x-1} = x+5+\dfrac{6}{x-1}$.

This method can be chosen instead of the inspection method or if the inspection method becomes harder to use.

WORKED EXAMPLE 6 Using long division to find a quotient and a remainder

a. **Given $p(x) = 4x^3 + 6x^2 - 5x + 9$, use the long division method to divide $p(x)$ by $(x+3)$ and state the quotient and the remainder.**

b. **Use the long division method to calculate the remainder when $(9x^3 + 5x)$ is divided by $(5+3x)$.**

THINK	WRITE
a. 1. Set up the long division.	a. $x+3\,\overline{)4x^3+6x^2-5x+9}$
2. The first stage of the division is to divide the leading term of the divisor into the leading term of the dividend.	$\begin{array}{r l} & 4x^2 \\ x+3\,\big) & \overline{4x^3+6x^2-5x+9} \end{array}$
3. The second stage of the division is to multiply the result of the first stage by the divisor. Write this product, placing like terms in the same columns.	$\begin{array}{r l} & 4x^2 \\ x+3\,\big) & \overline{4x^3+6x^2-5x+9} \\ & 4x^3+12x^2 \end{array}$
4. The third stage of the division is to subtract the result of the second stage from the dividend. This will yield an expression of lower degree than the original dividend.	$\begin{array}{r l} & 4x^2 \\ x+3\,\big) & \overline{4x^3+6x^2-5x+9} \\ & -(4x^3+12x^2) \\ \hline & -6x^2-5x+9 \end{array}$
5. The algorithm needs to be repeated. Divide the leading term of the divisor into the leading term of the newly formed dividend.	$\begin{array}{r l} & 4x^2-6x \\ x+3\,\big) & \overline{4x^3+6x^2-5x+9} \\ & -(4x^3+12x^2) \\ \hline & -6x^2-5x+9 \end{array}$

6. Multiply the result by the divisor and write this product keeping like terms in the same columns.

$$\begin{array}{r} 4x^2 - 6x \\ x+3\,\overline{)\,4x^3 + 6x^2 - 5x + 9} \\ -\,(4x^3 + 12x^2) \\ \hline -\,6x^2 - 5x + 9 \\ -\,6x^2 - 18x \end{array}$$

7. Subtract to yield an expression of lower degree.
Note: The degree of the expression obtained is still not less than the degree of the divisor, so the algorithm will need to be repeated again.

$$\begin{array}{r} 4x^2 - 6x \\ x+3\,\overline{)\,4x^3 + 6x^2 - 5x + 9} \\ -\,(4x^3 + 12x^2) \\ \hline -\,6x^2 - 5x + 9 \\ -\,(-\,6x^2 - 18x) \\ \hline 13x + 9 \end{array}$$

8. Divide the leading term of the divisor into the dividend obtained in the previous step.

$$\begin{array}{r} 4x^2 - 6x + 13 \\ x+3\,\overline{)\,4x^3 + 6x^2 - 5x + 9} \\ -\,(4x^3 + 12x^2) \\ \hline -\,6x^2 - 5x + 9 \\ -\,(-\,6x^2 - 18x) \\ \hline 13x + 9 \end{array}$$

9. Multiply the result by the divisor and write this product, keeping like terms in the same columns.

$$\begin{array}{r} 4x^2 - 6x + 13 \\ x+3\,\overline{)\,4x^3 + 6x^2 - 5x + 9} \\ (4x^3 + 12x^2) \\ \hline -\,6x^2 - 5x + 9 \\ -\,(-\,6x^2 - 18x) \\ \hline 13x + 9 \\ 13x + 39 \end{array}$$

10. Subtract to yield an expression of lower degree.
Note: The term reached is a constant, so its degree is less than that of the divisor. The division is complete.

$$\begin{array}{r} 4x^2 - 6x + 13 \\ x+3\,\overline{)\,4x^3 + 6x^2 - 5x + 9} \\ -\,(4x^3 + 12x^2) \\ \hline -\,6x^2 - 5x + 9 \\ -\,(-\,6x^2 - 18x) \\ \hline 13x + 9 \\ -(13x + 39) \\ \hline -\,30 \\ \hline \end{array}$$

11. State the answer.

$$\frac{4x^3 + 6x^2 - 5x + 9}{x+3} = 4x^2 - 6x + 13 - \frac{30}{x+3}$$

The quotient is $4x^2 - 6x + 13$ and the remainder is -30.

b. 1. Set up the division, expressing both the divisor and the dividend in decreasing powers of x. This creates the columns for like terms.

b. $9x^3 + 5x = 9x^3 + 0x^2 + 5x + 0$
$5 + 3x = 3x + 5$

$$3x + 5\,\overline{)\,9x^3 + 0x^2 + 5x + 0}$$

2. Divide the leading term of the divisor into the leading term of the dividend, multiply this result by the divisor and then subtract this product from the dividend.

$$\begin{array}{r|l} & 3x^2 \\ 3x+5 & 9x^3+0x^2+5x+0 \\ & -(9x^3+15x^2) \\ \hline & -15x^2+5x+0 \end{array}$$

3. Repeat the three steps of the algorithm using the dividend created by the first application of the algorithm.

$$\begin{array}{r|l} & 3x^2-5x \\ 3x+5 & 9x^3+0x^2+5x+0 \\ & -(9x^3+15x^2) \\ \hline & -15x^2+5x+0 \\ & -(-15x^2-25x) \\ \hline & 30x+0 \end{array}$$

4. Repeat the algorithm using the dividend created by the second application of the algorithm.

$$\begin{array}{r|l} & 3x^2-5x+10 \\ 3x+5 & 9x^3+0x^2+5x+0 \\ & -(9x^3+15x^2) \\ \hline & -15x^2+5x+0 \\ & -(-15x^2-25x) \\ \hline & 30x+0 \\ & -(30x+50) \\ \hline & -50 \end{array}$$

5. State the answer.

$$\frac{9x^3+5x}{3x+5} = 3x^2-5x+10+\frac{-50}{3x+5}$$
$$= 3x^2-5x+10-\frac{50}{(3x+5)}$$

The remainder is -50.

Resources

Interactivity Long division of polynomials (int-2564)

4.2 Exercise

Students, these questions are even better in jacPLUS

Receive immediate feedback and access sample responses

Access additional questions

Track your results and progress

Find all this and MORE in jacPLUS

Technology free

1. **WE1** Select the polynomials from the following list of algebraic expressions and state their degree, the coefficient of the leading term, the constant term and the type of coefficients.

a. $30x+4x^5-2x^3+12$

b. $\frac{3x^2}{5}-\frac{2}{x}+1$

c. $5.6+4x-0.2x^2$

2. For those of the following that are polynomials, state their degree. For those that are not polynomials, state a reason why not.

a. $7x^4 + 3x^2 + 5$

b. $9 - \frac{5}{2}x - 4x^2 + x^3$

c. $-9x^3 + 7x^2 + 11\sqrt{x} - \sqrt{5}$

d. $\frac{6}{x^2} + 6x^2 + \frac{x}{2} - \frac{2}{x}$

3. Consider the following list of algebraic expressions.

A. $3x^5 + 7x^4 - \frac{x^3}{6} + x^2 - 8x + 12$

B. $9 - 5x^4 + 7x^2 - \sqrt{5}x + x^3$

C. $\sqrt{4x^5} - \sqrt{5}x^3 + \sqrt{3}x - 1$

D. $2x^2\left(4x - 9x^2\right)$

E. $\frac{x^6}{10} - \frac{2x^8}{7} + \frac{5}{3x^2} - \frac{7x}{5} + \frac{4}{9}$

F. $\left(4x^2 + 3 + 7x^3\right)^2$

a. Select the polynomials from the list and for each of these polynomials state:

i. the degree

ii. the type of coefficients

iii. the leading term

iv. the constant term.

b. Give a reason why each of the remaining expressions is not a polynomial.

4. Write down a monic polynomial over R in the variable y for which the degree is 7, the coefficient of the y^2 term is $-\sqrt{2}$, the constant term is 4 and the polynomial contains four terms.

5. a. If $p(x) = -x^3 + 2x^2 + 5x - 1$, calculate $p(1)$.

b. If $p(x) = 2x^3 - 4x^2 + 3x - 7$, calculate $P(-2)$.

c. For $p(x) = 3x^3 - x^2 + 5$, find $p(3)$ and $p(-x)$.

d. For $p(x) = x^3 + 4x^2 - 2x + 5$, find $p(-1)$ and $p(2a)$.

6. Given $p(x) = 2x^3 + 3x^2 + x - 6$, evaluate the following.

a. $p(3)$

b. $p(-2)$

c. $p(1)$

d. $p(0)$

e. $p\left(-\frac{1}{2}\right)$

f. $p(0.1)$

7. If $p(x) = x^2 - 7x + 2$, obtain expressions for the following.

a. $p(a) - p(-a)$

b. $p(1 + h)$

c. $p(x + h) - p(x)$

8. WE2 a. If $p(x) = 7x^3 - 8x^2 - 4x - 1$, calculate $p(2)$.

b. If $p(x) = 2x^2 + kx + 12$ and $p(-3) = 0$, find k.

9. a. If $p(x) = ax^2 + 9x + 2$ and $p(1) = 3$, calculate the value of a.

b. Given $p(x) = -5x^2 + bx - 18$, calculate the value of b if $p(3) = 0$.

c. Given $p(x) = -2x^3 + 3x^2 + kx - 10$, calculate the value of k if $p(-1) = -7$.

d. If $p(x) = x^3 - 6x^2 + 9x + m$ and $p(0) = 2p(1)$, determine the value of m.

e. If $p(x) = -2x^3 + 9x + m$ and $p(1) = 2p(-1)$, determine m.

f. If $q(x) = -x^2 + bx + c$ and $q(0) = 5$ and $q(5) = 0$, obtain the values of b and c.

10. WE3 Calculate the values of a, b and c so that $(2x + 1)(x - 5) \equiv a(x + 1)^2 + b(x + 1) + c$.

11. a. Determine the values of a and b so that $x^2 + 10x + 6 \equiv x(x + a) + b$.

b. Express $8x - 6$ in the form $ax + b(x + 3)$.

c. Express the polynomial $6x^2 + 19x - 20$ in the form $(ax + b)(x + 4)$.

d. Determine the values of a, b and c so that $x^2 - 8x = a + b(x + 1) + c(x + 1)^2$ for all values of x.

12. Express $(x+2)^3$ in the form $px^2(x+1)+qx(x+2)+r(x+3)+t$.

13. a. If $3x^2+4x-7 \equiv a(x+1)^2+b(x+1)+c$, calculate a, b and c.
    b. If $x^3+mx^2+nx+p \equiv (x-2)(x+3)(x-4)$, calculate m, n and p.
    c. If $x^2-14x+8 \equiv a(x-b)^2+c$, calculate a, b and c and hence express $x^2-14x+8$ in the form $a(x-b)^2+c$.
    d. Express $4x^3+2x^2-7x+1$ in the form $ax^2(x+1)+bx(x+1)+c(x+1)+d$.

14. **WE4** Given $p(x)=4x^3-ax^2+8$ and $q(x)=3x^2+bx-7$, determine the values of a and b for which $p(x)+2q(x)=4x^3+x^2-8x-6$.

15. a. If $p(x)=2x^2-7x-11$ and $q(x)=3x^4+2x^2+1$, determine, expressing the terms in descending powers of x:
    i. $q(x)-p(x)$ ii. $3p(x)+2q(x)$ iii. $p(x)q(x)$
    b. If $p(x)$ is a polynomial of degree m and $q(x)$ is a polynomial of degree n where $m>n$, state the degree of:
    i. $p(x)+q(x)$ ii. $p(x)-q(x)$ iii. $p(x)q(x)$

16. **WE5** a. Calculate the quotient and the remainder when $(x-12)$ is divided by $(x+3)$.
    b. Use the inspection method to find $\dfrac{4x+7}{2x+1}$.

17. Carry out the following divisions and specify the remainder in each case.
    a. $\dfrac{x+5}{x+1}$ b. $\dfrac{2x-3}{x+4}$
    c. $\dfrac{4x+11}{4x+1}$ d. $\dfrac{6x+13}{2x-3}$

18. **WE6** a. Given $p(x)=2x^3-5x^2+8x+6$, divide $p(x)$ by $(x-2)$ and state the quotient and the remainder.
    b. Use the long division method to calculate the remainder when (x^3+10) is divided by $(1-x)$.

19. Carry out the following divisions and specify the quotient and the remainder
    a. $(x+7)$ is divided by $(x-2)$
    b. $(8x+5)$ is divided by $(2x+1)$
    c. $(x^2+6x-17)$ is divided by $(x-1)$
    d. $(2x^2-8x+3)$ is divided by $(x+2)$
    e. (x^3+2x^2-3x+5) is divided by $(x-3)$
    f. (x^3-8x^2+9x-2) is divided by $(x-1)$.

20. a. Divide x^3-x^2+3x-5 by $x-2$ and state the quotient and remainder.
    b. Divide $3x^3-x^2+6x-5$ by $3x-1$ and state the quotient and remainder.
    c. Divide $6x^3-3x^2+x+1$ by $2x-7$ and state the quotient and remainder.
    d. Divide $6x^3-5x^2+x+3$ by $2x+3$ and state the quotient and remainder.

Technology active

21. Calculate the values of a, b and c if $3x^2-6x+5 \equiv ax(x+2)+b(x+2)+c$, and hence express $\dfrac{3x^2-6x+5}{x+2}$ in the form $q(x)+\dfrac{r}{x+2}$ where $q(x)$ is a polynomial and $r \in R$.

22. If $p(x)=x^4+3x^2-7x+2$ and $q(x)=x^3+x+1$, expand the product $p(x)q(x)$ and state its degree.

23. Obtain the quotient and remainder when $(x^4-3x^3+6x^2-7x+3)$ is divided by $(x-1)^2$.

24. Perform the following divisions.

a. $(8x^3 + 6x^2 - 5x + 15)$ divided by $(1 + 2x)$

b. $(4x^3 + x + 5)$ divided by $(2x - 3)$

c. $(x^3 + 6x^2 + 6x - 12) \div (x + 6)$

d. $(2 + x^3) \div (x + 1)$

e. $\dfrac{x^4 + x^3 - x^2 + 2x + 5}{x^2 - 1}$

f. $\dfrac{x(7 - 2x^2)}{(x + 2)(x - 3)}$

25. a. Determine the values of a, b, c and d if $p(x) = x^3 - 3x^2 + cx - 2$, $q(x) = ax^3 + bx^2 + 3x - 2a$ and $2p(x) - q(x) = 5(x^3 - x^2 + x + d)$.

b. i. Express $4x^4 + 12x^3 + 13x^2 + 6x + 1$ in the form $(ax^2 + bx + c)^2$ where $a > 0$.

ii. Hence, state a square root of $4x^4 + 12x^3 + 13x^2 + 6x + 1$.

26. $p(x) = x^4 + kx^2 + n^2$, $q(x) = x^2 + mx + n$ and the product $p(x)q(x) = x^6 - 5x^5 - 7x^4 + 65x^3 - 42x^2 - 180x + 216$.

a. Calculate k, m and n.

b. Obtain the linear factors of $p(x)q(x) = x^6 - 5x^5 - 7x^4 + 65x^3 - 42x^2 - 180x + 216$.

27. a. Use CAS technology to divide $(4x^3 - 7x^2 + 5x + 2)$ by $(2x + 3)$.

b. State the remainder and the quotient.

c. Evaluate the dividend if $x = -\dfrac{3}{2}$.

d. Evaluate the divisor if $x = -\dfrac{3}{2}$.

28. a. Define $p(x) = 3x^3 + 6x^2 - 8x - 10$ and $q(x) = -x^3 + ax - 6$.

b. Evaluate $p(-4) + p(3) - p\left(\dfrac{2}{3}\right)$.

c. Give an algebraic expression for $p(2n) + 24q(n)$.

d. Obtain the value of a so that $q(-2) = -16$.

4.2 Exam questions

Question 1 (1 mark) TECH-FREE

Given $p(x) = x^3 - 2x^2 - 4x + 2$, use the long division method to divide $p(x)$ by $(x - 1)$, and state the quotient and the remainder.

Question 2 (1 mark) TECH-ACTIVE

MC Select the expression that is not a polynomial.

A. $x^4 - x^3 + \dfrac{3x}{7} - 1$

B. $5x^4 + 3x^3$

C. $-3x^3 + x^2 - 2\sqrt{x}$

D. $4x$

E. $3x^2 + \dfrac{x}{\sqrt{5}}$

Question 3 (1 mark) TECH-ACTIVE

MC The degree, coefficient of the leading term and constant term of the polynomial with equation $5x + 6 - 3x^2 - x^3 - 5$ are, in order

A. $2, -3, 6$

B. $3, -1, -5$

C. $1, 5, 6$

D. $3, -1, 1$

E. $3, 1, -5$

More exam questions are available online.

4.3 The remainder and factor theorems

LEARNING INTENTION

At the end of this subtopic you should be able to:
- use the remainder and factor theorems to find linear factors of cubic polynomials
- use the rational root theorem to factorise cubic polynomials
- use the Null Factor Law to solve cubic equations.

The remainder obtained when dividing $p(x)$ by the linear divisor $(x-a)$ is of interest because if the remainder is zero, then the divisor must be a linear factor of the polynomial. To pursue this interest we need to be able to calculate the remainder quickly without the need to do a lengthy division.

4.3.1 The remainder theorem

The actual division, as we know, will result in a quotient and a remainder. This is expressed in the division statement $\frac{p(x)}{x-a} = \text{quotient} + \frac{\text{remainder}}{x-a}$.

Since $(x-a)$ is linear, the remainder will be some constant term independent of x.

From the division statement it follows that:

$$p(x) = (x-a) \times \text{quotient} + \text{remainder}$$

If we let $x = a$, this makes $(x-a)$ equal to zero and the statement becomes:

$$p(a) = 0 \times \text{quotient} + \text{remainder}$$

Therefore:

$$p(a) = \text{remainder}$$

This result is known as the **remainder theorem**.

Remainder theorem

If a polynomial $p(x)$ is divided by $(x-a)$, the remainder is $p(a)$.

Note that:
- if $p(x)$ is divided by $(x+a)$, the remainder is $p(-a)$, since replacing x by $-a$ would make the $(x+a)$ term equal zero
- if $p(x)$ is divided by $(ax+b)$, the remainder is $p\left(-\frac{b}{a}\right)$, since replacing x by $-\frac{b}{a}$ would make the $(ax+b)$ term equal zero.

WORKED EXAMPLE 7 Calculating remainders

Determine the remainder when $p(x) = x^3 - 3x^2 - 2x + 9$ is divided by:

a. $x - 2$

b. $2x + 1$

THINK	WRITE
a. 1. What value of x will make the divisor zero?	a. $(x - 2) = 0 \Rightarrow x = 2$
2. Write an expression for the remainder.	$p(x) = x^3 - 3x^2 - 2x + 9$ The remainder is $p(2)$.
3. Evaluate to obtain the remainder.	$p(2) = (2)^3 - 3(2)^2 - 2(2) + 9$ $= 1$ The remainder is 1.
b. 1. Find the value of x that makes the divisor zero.	b. $(2x + 1) = 0 \Rightarrow x = -\frac{1}{2}$
2. Write an expression for the remainder and evaluate it.	The remainder is $p\left(-\frac{1}{2}\right)$. $p\left(-\frac{1}{2}\right) = \left(-\frac{1}{2}\right)^3 - 3\left(-\frac{1}{2}\right)^2 - 2\left(-\frac{1}{2}\right) + 9$ $= -\frac{1}{8} - \frac{3}{4} + 1 + 9$ $= 9\frac{1}{8}$ The remainder is $9\frac{1}{8}$.

TI | THINK	DISPLAY/WRITE
a. 1. On a Calculator page, press MENU, then select: 3: Algebra 8: Polynomial Tools 4: Remainder of Polynomial Complete the entry line as: polyRemainder $(x^3 - 3x^2 - 2x + 9, x - 2)$ then press ENTER.	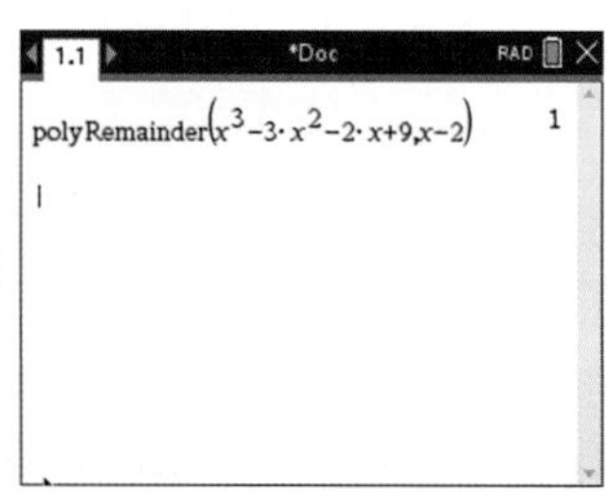
2. The answer appears on the screen.	The remainder is 1.

CASIO | THINK	DISPLAY/WRITE
a. 1. On a Main screen, select: • Interactive • Transformation • Fraction • propFrac Complete the entry line as: $(x^3 - 3x^2 - 2x + 9)/(x - 2)$ then select OK.	 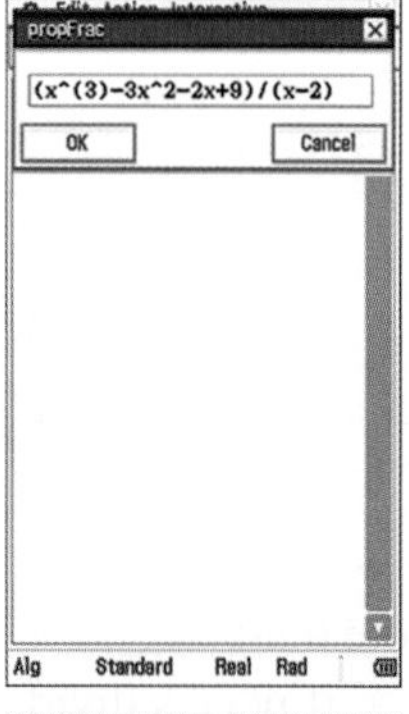 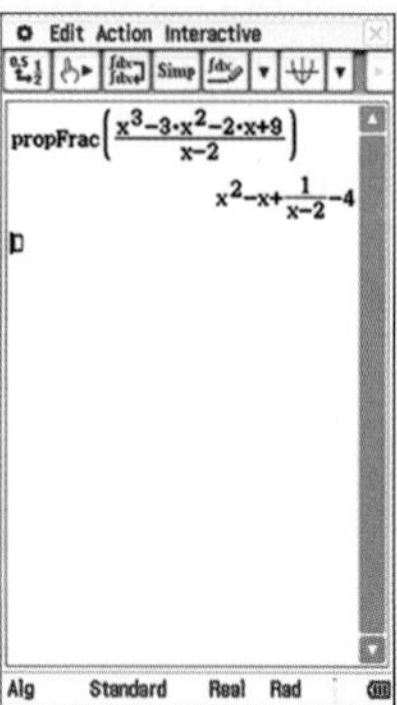
2. The answer appears on the screen.	The remainder is 1.

on Resources

Interactivity The remainder and factor theorems (int-2565)

4.3.2 The factor theorem

We know 4 is a factor of 12 because it divides 12 exactly, leaving no remainder. Similarly, if the division of a polynomial $p(x)$ by $(x-a)$ leaves no remainder, then the divisor $(x-a)$ must be a factor of the polynomial $p(x)$.

$$p(x) = (x-a) \times \text{quotient} + \text{remainder}$$

If the remainder is zero, then

$$p(x) = (x-a) \times \text{quotient}.$$

Therefore,

$$(x-a) \text{ is a factor of } p(x).$$

This is known as the **factor theorem**.

Factor theorem

If $p(x)$ is a polynomial and $p(a) = 0$, then $(x-a)$ is a factor of $p(x)$.

Conversely, if $(x-a)$ is a factor of a polynomial $p(x)$, then $p(a) = 0$, and a is a **zero** of the polynomial.

It also follows from the remainder theorem that if $p\left(-\frac{b}{a}\right) = 0$, then $(ax+b)$ is a factor of $p(x)$ and $-\frac{b}{a}$ is a zero of the polynomial.

WORKED EXAMPLE 8 Using the factor theorem

a. Show that $(x+3)$ is a factor of $q(x) = 4x^4 + 4x^3 - 25x^2 - x + 6$.

b. Determine the polynomial $p(x) = ax^3 + bx + 2$ that leaves a remainder of -9 when divided by $(x-1)$ and is exactly divisible by $(x+2)$.

THINK	WRITE
a. 1. State how the remainder can be calculated when $q(x)$ is divided by the given linear expression.	**a.** $q(x) = 4x^4 + 4x^3 - 25x^2 - x + 6$ When $q(x)$ is divided by $(x+3)$, the remainder equals $q(-3)$.
2. Evaluate the remainder.	$q(-3) = 4(-3)^4 + 4(-3)^3 - 25(-3)^2 - (-3) + 6$ $= 324 - 108 - 225 + 3 + 6$ $= 0$
3. It is important to explain in the answer why the given linear expression is a factor.	Since $q(-3) = 0$, $(x+3)$ is a factor of $q(x)$.
b. 1. Express the given information in terms of the remainders.	**b.** $p(x) = ax^3 + bx + 2$ Dividing by $(x-1)$ leaves a remainder of -9. $\Rightarrow p(1) = -9$ Dividing by leaves a remainder of 0. $\Rightarrow p(-2) = 0$.

2. Set up a pair of simultaneous equations in a and b.	$p(1) = a + b + 2$ $a + b + 2 = -9$ $\therefore a + b = -11$ [1] $p(-2) = -8a + -2b + 2$ $-8a + -2b + 2 = 0$ $\therefore 4a + b = 1$ [2]
3. Solve the simultaneous equations.	$a + b = -11$ [1] $4a + b = 1$ [2] Equation [2] – equation [1]: $3a = 12$ $a = 4$ Substitute $a = 4$ into equation [1]. $4 + b = -11$ $b = -15$
4. Write the answer.	$\therefore p(x) = 4x^3 - 15x + 2$

4.3.3 Factorising polynomials

When factorising a cubic or higher-degree polynomial, the first step should be to check if any of the standard methods for factorising can be used. In particular, look for a common factor, then look to see if a grouping technique can produce either a common linear factor or a difference of two squares. If the standard techniques do not work, then the remainder and factor theorems can be used to factorise, since the zeros of a polynomial enable linear factors to be formed.

Cubic polynomials may have up to three zeros and therefore up to three linear factors.

For example, if the cubic polynomial $p(x)$ has three zeros:

$$x = 1,\ x = 2 \text{ and } x = -4$$

then

$$p(1) = 0,\ \ P(2) = 0 \text{ and } P(-4) = 0$$

and

$$(x - 1), (x - 2) \text{ and } (x + 4) \text{ are its three linear factors.}$$

Integer zeros of a polynomial may be found through a trial-and-error process where factors of the polynomial's constant term are tested systematically. For the polynomial $p(x) = x^3 + x^2 - 10x + 8$, the constant term is 8, so the possibilities to test are $1, -1,\ 2, -2,\ 4, -4,\ 8$ and -8. This is a special case of what is known as the **rational root theorem**.

Rational root theorem

The rational solutions to the polynomial equation

$$a_n x^n + a_{n-1} x^{n-1} + \ldots + a_2 x^2 + a_1 x + a_0 = 0,$$

where the coefficients are integers and a_n and a_0 are non-zero, will have solutions $x = \frac{p}{q}$ (in simplest form), where p is a factor of a_0 and q is a factor of a_n.

In practice, not all of the zeros need to be, nor necessarily can be, found through trial and error. For a cubic polynomial it is sufficient to find one zero by trial and error and form its corresponding linear factor using the factor theorem. Dividing this linear factor into the cubic polynomial gives a quadratic quotient and zero remainder, so the quadratic quotient is also a factor. The standard techniques for factorising quadratics can then be applied.

Factorising polynomials by equating coefficients

For the division step, long division could be used; however, it is more efficient to use a division method based on **equating coefficients**. With practice, this can usually be done by inspection.

To illustrate, consider $p(x) = x^3 + x^2 - 10x + 8$.

$$\text{Since } p(1) = 0, x = 1 \text{ is a zero and is a factor.}$$

$$\text{So, } p(x) = (x-1)(ax^2 + bx + c).$$

The coefficients of the quadratic factor are found by equating coefficients of like terms.

$$x^3 + x^2 - 10x + 8 = (x-1)(ax^2 + bx + c)$$

$$\text{For } x^3\text{: } 1 = a \Rightarrow \ a = 1$$

$$\text{For constants: } 8 = -c \Rightarrow c = -8$$

$$\text{giving } x^3 + x^2 - 10x + 8 = (x-1)(x^2 + bx - 8).$$

The term in x^2 can be formed by the product of the x term in the first bracket with the x term in the second bracket, and also by the product of the constant term in the first bracket with the x^2 term in the second bracket.

$$x^3 + x^2 - 10x + 8 = (x-1)(x^2 + bx - 8)$$

$$\text{Equating coefficients of } x^2: \quad 1 = b - 1$$
$$\therefore b = 2$$

If preferred, the coefficients of x could be equated or used as a check.

It follows that:

$$\begin{aligned} p(x) &= (x-1)\left(x^2 + 2x - 8\right) \\ &= (x-1)(x-2)(x+4) \end{aligned}$$

WORKED EXAMPLE 9 Factorising cubic polynomials

a. Factorise $p(x) = x^3 - 2x^2 - 5x + 6$.
b. Given that $(x+1)$ and $(5-2x)$ are factors of $p(x) = -4x^3 + 4x^2 + 13x + 5$, completely factorise $p(x)$.

THINK	WRITE
a. 1. The polynomial does not factorise by a grouping technique, so a zero needs to be found. The factors of the constant term are potential zeros.	a. $p(x) = x^3 - 2x^2 - 5x + 6$ The factors of 6 are ±1, ±2, ±3 and ±6.

2. Use the remainder theorem to test systematically until a zero is obtained. Then use the factor theorem to state the corresponding linear factor.	$p(1) = 1 - 2 - 5 + 6$ $= 0$ $\therefore (x - 1)$ is a factor.
3. Express the polynomial in terms of a product of the linear factor and a general quadratic factor.	$\therefore x^3 - 2x^2 - 5x + 6 = (x - 1)\left(ax^2 + bx + c\right)$
4. State the values of a and c.	For the coefficient of x^3 to be 1, $a = 1$. For the constant term to be 6, $c = -6$. $\therefore x^3 - 2x^2 - 5x + 6 = (x - 1)\left(x^2 + bx - 6\right)$
5. Calculate the value of b.	Equating the coefficients of x^2 gives: $x^3 - 2x^2 - 5x + 6 = (x - 1)(x^2 + bx - 6)$ $-2 = b - 1$ $b = -1$ $\therefore x^3 - 2x^2 - 5x + 6 = (x - 1)\left(x^2 - x - 6\right)$
6. Factorise the quadratic factor so the polynomial is fully factorised into its linear factors.	Hence, $p(x) = x^3 - 2x^2 - 5x + 6$ $= (x - 1)\left(x^2 - x - 6\right)$ $= (x - 1)(x - 3)(x + 2)$
b. 1. Multiply the two given linear factors to form the quadratic factor.	**b.** $p(x) = -4x^3 + 4x^2 + 13x + 5$ Since $(x + 1)$ and $(5 - 2x)$ are factors, $(x + 1)(5 - 2x) = -2x^2 + 3x + 5$ is a quadratic factor.
2. Express the polynomial as a product of the quadratic factor and a general linear factor.	The remaining factor is linear. $\therefore p(x) = (x + 1)(5 - 2x)(ax + b)$ $= \left(-2x^2 + 3x + 5\right)(ax + b)$
3. Find a and b.	$-4x^3 + 4x^2 + 13x + 5 = \left(-2x^2 + 3x + 5\right)(ax + b)$ Equating coefficients of x^3 gives: $-4 = -2a$ $\therefore a = 2$ Equating constants gives: $5 = 5b$ $\therefore b = 1$
4. State the answer.	$-4x^3 + 4x^2 + 13x + 5 = \left(-2x^2 + 3x + 5\right)(2x + 1)$ $= (x + 1)(5 - 2x)(2x + 1)$ $\therefore p(x) = (x + 1)(5 - 2x)(2x + 1)$

TI | THINK	DISPLAY/WRITE
a. 1. On a Calculator page, press MENU, then select: 3: Algebra 2: Factor Complete the entry line as: factor $\left(x^3 - 2x^2 - 5x + 6\right)$ then press ENTER.	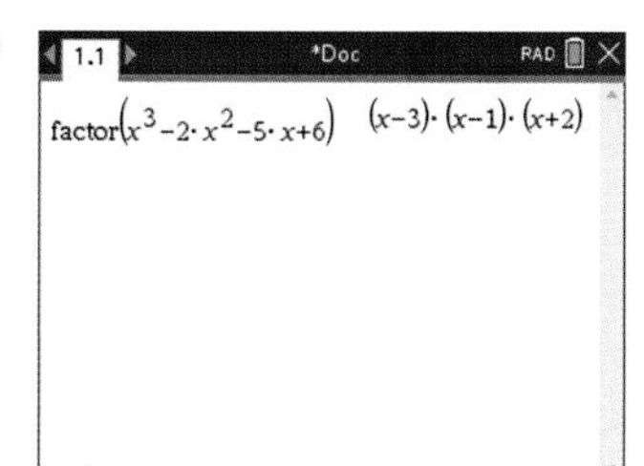
2. The answer appears on the screen.	$x^3 + 2x^2 - 5x + 6$ $= (x - 3)(x - 1)(x + 2)$

CASIO | THINK	DISPLAY/WRITE
a. 1. On a Main screen, select: • Interactive • Transformation • factor • factor Complete the entry line as: $x^3 - 2x^2 - 5x + 6$ then select OK.	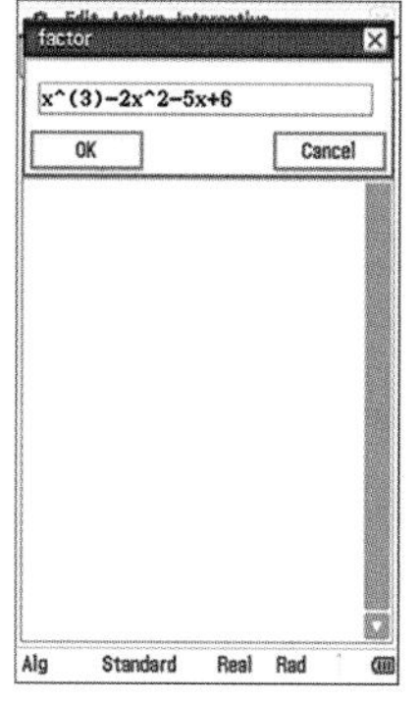 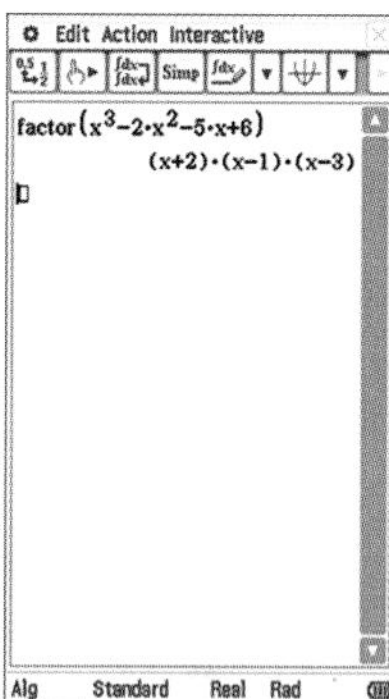
2. The answer appears on the screen.	$x^3 + 2x^2 - 5x + 6$ $= (x - 1)(x + 2)(x - 3)$

4.3.4 Polynomial equations

If a polynomial is expressed in factorised form, then the polynomial equation can be solved using the Null Factor Law.

$$(x-a)(x-b)(x-c)=0$$
$$\therefore (x-a)=0,\ (x-b)=0,\ (x-c)=0$$
$$\therefore x=a,\ x=b \text{ or } x=c$$

$x = a$, $x = b$ and $x = c$ are called the roots or the solutions to the equation $P(x) = 0$.
The factor theorem may be required to express the polynomial in factorised form.

WORKED EXAMPLE 10 Solving cubic equations

Solve the equation $3x^3 + 4x^2 = 17x + 6$.

THINK	WRITE
1. Rearrange the equation so one side is zero.	$3x^3 + 4x^2 = 17x + 6$ $3x^3 + 4x^2 - 17x - 6 = 0$
2. Since the polynomial does not factorise by grouping techniques, use the remainder theorem to find a zero and the factor theorem to form the corresponding linear factor. *Note:* It is simpler to test for integer zeros first.	Let $p(x) = 3x^3 + 4x^2 - 17x - 6$. Test factors of the constant term: $p(1) \neq 0$ $p(-1) \neq 0$ $p(2) = 3(2)^3 + 4(2)^2 - 17(2) - 6$ $= 24 + 16 - 34 - 6$ $= 0$ Therefore, $(x - 2)$ is a factor.

3. Express the polynomial as a product of the linear factor and a general quadratic factor.	$3x^3+4x^2-17x-6=(x-2)\,(ax^2+bx+c)$
4. Find and substitute the values of a and c.	$\therefore\ 3x^3+4x^2-17x-6=(x-2)\,(3x^2+bx+3)$
5. Calculate b.	Equate the coefficients of x^2: $4=b-6$ $b=10$
6. Completely factorise the polynomial.	$3x^3+4x^2-17x-6=(x-2)(3x^2+10x+3)$ $=(x-2)(3x+1)(x+3)$
7. Solve the equation.	The equation $3x^3+4x^2-17x-6=0$ becomes: $(x-2)(3x+1)(x+3)=0$ $x-2=0,\ 3x+1=0,\ x+3=0$ $x=2,\ x=-\frac{1}{3},\ x=-3$

TI | THINK	DISPLAY/WRITE	CASIO | THINK	DISPLAY/WRITE
1. On a Calculator page, press MENU, then select: 3: Algebra 1: Solve Complete the entry line as: solve $(3x^3+4x^2=17x+6,x)$ then press ENTER.	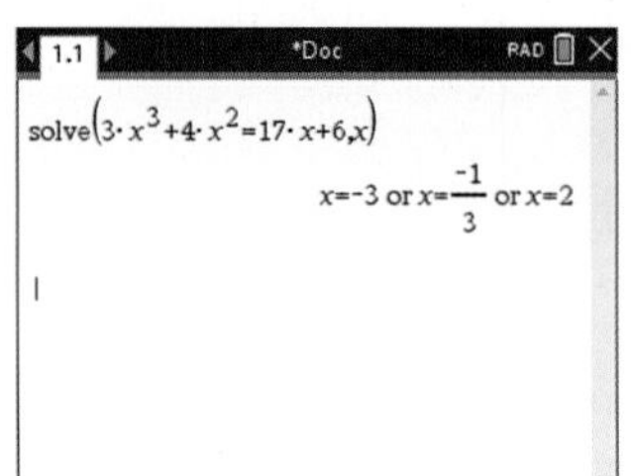	**1.** On a Main screen, complete the entry line as: solve $(3x^3+4x^2=17x+6,x)$ then press EXE.	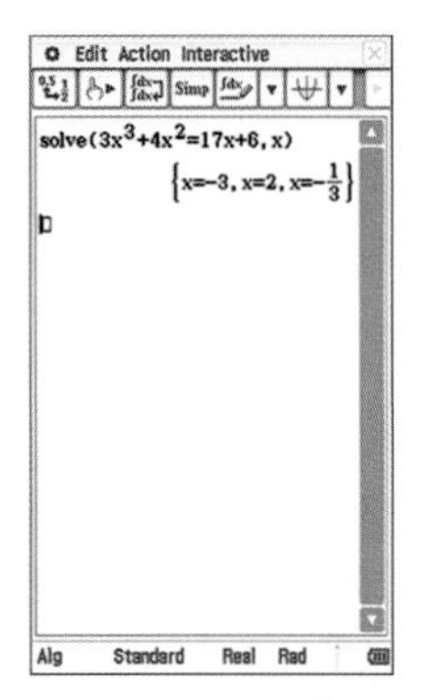
2. The answer appears on the screen.	$x=-3$ or $x=-\frac{1}{3}$ or $x=2$	**2.** The answer appears on the screen.	$x=-3$ or $x=-\frac{1}{3}$ or $x=2$

4.3 Exercise

Technology free

1. **a.** Without actual division, calculate the remainder when $3x^2+8x-5$ is divided by $(x-1)$.
   b. Without dividing, calculate the remainder when $-x^3+7x^2+2x-12$ is divided by $(x+1)$.

2. **WE7** Determine the remainder when $p(x)=x^3+4x^2-3x+5$ is divided by:
   a. $x+2$
   b. $2x-1$.

3. **MC** Select the correct statement for the remainder when $p(x)$ is divided by $(2x+9)$.

A. The remainder is $p(9)$.
B. The remainder is $p(-9)$.
C. The remainder is $p\left(-\frac{9}{2}\right)$.
D. The remainder is $p\left(\frac{9}{2}\right)$.
E. The remainder is $\frac{p(-9)}{2x+9}$.

4. Calculate the remainder without actual division when:

a. x^3-4x^2-5x+3 is divided by $(x-1)$
b. $6x^3+7x^2+x+2$ is divided by $(x+1)$
c. $-2x^3+2x^2-x-1$ is divided by $(x-4)$
d. x^3+x^2+x-10 is divided by $(2x+1)$
e. $27x^3-9x^2-9x+2$ is divided by $(3x-2)$
f. $4x^4-5x^3+2x^2-7x+8$ is divided by $(x-2)$.

5. a. When ax^2-4x-9 is divided by $(x-3)$, the remainder is 15. Determine the value of a.
b. When x^3+x^2+kx+5 is divided by $(x+2)$, the remainder is -5. Determine the value of k.
c. If x^3-kx^2+4x+8 leaves a remainder of 29 when it is divided by $(x-3)$, determine the value of k.

6. **WE8** a. Show that $(x-2)$ is a factor of $q(x)=4x^4+4x^3-25x^2-x+6$.
b. Determine the polynomial $p(x)=3x^3+ax^2+bx-2$ that leaves a remainder of -22 when divided by $(x+1)$ and is exactly divisible by $(x-1)$.

7. a. When $p(x)=x^3-2x^2+ax+7$ is divided by $(x+2)$, the remainder is 11. Determine the value of a.
b. If $p(x)=4-x^2+5x^3-bx^4$ is exactly divisible by $(x-1)$, determine the value of b.
c. If $2x^3+cx^2+5x+8$ has a remainder of 6 when divided by $(2x-1)$, determine the value of c.
d. Given that each of x^3+3x^2-4x+d and x^4-9x^2-7 have the same remainder when divided by $(x+3)$, determine the value of d.

8. Given $(2x+a)$ is a factor of $12x^2-4x+a$, obtain the value(s) of a.

9. a. Calculate the values of a and b for which $q(x)=ax^3+4x^2+bx+1$ leaves a remainder of 39 when divided by $(x-2)$, given $(x+1)$ is a factor of $q(x)$.
b. Dividing $p(x)=\frac{1}{3}x^3+mx^2+nx+2$ by either $(x-3)$ or $(x+3)$ results in the same remainder. If that remainder is three times the remainder left when $p(x)$ is divided by $(x-1)$, determine the values of m and n.

10. a. Show that $x+4$ is a factor of $3x^3+11x^2-6x-8$.
b. Show that $x-5$ is a factor of $-x^3+6x^2+x-30$.
c. Show that $2x-1$ is a factor of $6x^3+7x^2-9x+2$.
d. Show that $x-1$ is not a factor of $2x^3+13x^2+5x-6$.

11. a. Given $x+3$ is a factor of $x^3-13x+a$, determine the value of a.
b. Given $2x-5$ is a factor of $4x^3+kx^2-9x+10$, determine the value of k.

12. a. Given $(x-4)$ is a factor of $p(x)=x^3-x^2-10x-8$, fully factorise $p(x)$.
b. Given $(x+12)$ is a factor of $p(x)=3x^3+40x^2+49x+12$, fully factorise $p(x)$.
c. Given $(5x+1)$ is a factor of $p(x)=20x^3+44x^2+23x+3$, fully factorise $p(x)$.
d. Given $(4x-3)$ is a factor of $p(x)=-16x^3+12x^2+100x-75$, fully factorise $p(x)$.
e. Given $(3x-5)$ is a factor of $p(x)=9x^3-75x^2+175x-125$, fully factorise $p(x)$.
f. Given $(8x-11)$ and $(x-3)$ are factors of $p(x)=-8x^3+59x^2-138x+99$, fully factorise $p(x)$.

13. **WE9** a. Factorise $p(x)=x^3+3x^2-13x-15$.
b. Given that $(x+1)$ and $(3x+2)$ are factors of $p(x)=12x^3+41x^2+43x+14$, completely factorise $p(x)$.

14. Factorise the following:

a. x^3+5x^2+2x-8
b. $x^3+10x^2+31x+30$
c. $2x^3-13x^2+13x+10$
d. $-18x^3+9x^2+23x-4$
e. x^3-7x+6
f. $x^3+x^2-49x-49$

15. **WE10** Solve the equation $6x^3 + 13x^2 = 2 - x$.

16. Solve the following equations for x.

a. $(x+4)(x-3)(x+5) = 0$
b. $2(x-7)(3x+5)(x-9) = 0$
c. $x^3 - 13x^2 + 34x + 48 = 0$
d. $2x^3 + 7x^2 = 9$
e. $3x^2(3x+1) = 4(2x+1)$
f. $2x^4 + 3x^3 - 8x^2 - 12x = 0$.

17. Solve the following equations for x.

a. $(2x-1)(3x+4)(x+1) = 0$
b. $2x^3 - x^2 - 6x + 3 = 0$
c. $8 - (x-5)^3 = 0$
d. $-x^3 + 2x^2 + 13x + 10 = 0$
e. $x^3 + 3x^2 + 2x + 6 = 0$
f. $6x^3 - 11x^2 - 3x + 2 = 0$

Technology active

18. Given the zeros of the polynomial $p(x) = 12x^3 + 8x^2 - 3x - 2$ are not integers, use the rational root theorem to calculate one zero and hence determine the three linear factors of the polynomial.

19. a. A monic polynomial of degree 3 in x has zeros of 5, 9 and −2. Express this polynomial in:
    i. factorised form
    ii. expanded form.

b. A polynomial of degree 3 has a leading term with coefficient −2 and zeros of −4, −1 and $\frac{1}{2}$. Express this polynomial in:
    i. factorised form
    ii. expanded form.

20. a. The polynomial $24x^3 + 34x^2 + x - 5$ has three zeros, none of which are integers. Calculate the three zeros and express the polynomial as the product of its three linear factors.

b. The polynomial $p(x) = 8x^3 + mx^2 + 13x + 5$ has a zero of $\frac{5}{2}$.
    i. State a linear factor of the polynomial.
    ii. Calculate the value of m.
    iii. Fully factorise the polynomial.

21. a. i. Factorise the polynomials $p(x) = x^3 - 12x^2 + 48x - 64$ and $q(x) = x^3 - 64$.
    ii. Hence, show that $\dfrac{p(x)}{q(x)} = 1 - \dfrac{12x}{x^2 + 4x + 16}$.

b. A cubic polynomial $p(x) = x^3 + bx^2 + cx + d$ has integer coefficients and $p(0) = 9$. Two of its linear factors are $\left(x - \sqrt{3}\right)$ and $\left(x + \sqrt{3}\right)$. Calculate the third linear factor and obtain the values of b, c and d.

22. a. Show that $(x-2)$ is a factor of $p(x) = x^3 + 6x^2 - 7x - 18$ and hence fully factorise $p(x)$ over R.

b. Show that $(3x-1)$ is the only real linear factor of $3x^3 + 5x^2 + 10x - 4$.

c. Show that $\left(2x^2 - 11x + 5\right)$ is a factor of $2x^3 - 21x^2 + 60x - 25$ and hence calculate the roots of the equation $2x^3 - 21x^2 + 60x - 25 = 0$.

23. a. If $\left(x^2 - 4\right)$ divides $p(x) = 5x^3 + kx^2 - 20x - 36$ exactly, fully factorise $p(x)$ and hence obtain the value of k.

b. If $x = a$ is a solution to the equation $ax^2 - 5ax + 4(2a-1) = 0$, calculate possible values for a.

c. The polynomials $p(x) = x^3 + ax^2 + bx - 3$ and $q(x) = x^3 + bx^2 + 3ax - 9$ have a common factor of $(x+a)$. Calculate a and b and fully factorise each polynomial.

d. $(x+a)^2$ is a repeated linear factor of the polynomial $p(x) = x^3 + px^2 + 15x + a^2$. Show there are two possible polynomials satisfying this information, and for each polynomial, calculate the values of x that give the roots of the equation $x^3 + px^2 + 15x + a^2 = 0$.

24. Specify the remainder when $\left(9 + 19x - 2x^2 - 7x^3\right)$ is divided by $\left(x - \sqrt{2} + 1\right)$.

25. Solve the equation $10x^3 - 5x^2 + 21x + 12 = 0$ expressing the values of x to 4 decimal places.

4.3 Exam questions

Question 1 (1 mark) TECH-ACTIVE

MC When $2x^3 - 7x^2 + x - 2$ is divided by $(x - 1)$, the remainder is

A. -12 **B.** -6 **C.** 8 **D.** 6 **E.** -8

Question 2 (1 mark) TECH-ACTIVE

MC The cubic polynomial $p(x) = x^3 - kx + 3$ is exactly divisible by $(x + 3)$. The value of must be

A. 8 **B.** -8 **C.** 3 **D.** -3 **E.** 5

Question 3 (2 marks) TECH-FREE

Determine the polynomial $p(x) = ax^3 + bx + 20$ that is exactly divisible by $(x - 1)$ and, when divided by $(x - 2)$, leaves a remainder of -14.

More exam questions are available online.

4.4 Graphs of cubic polynomials

LEARNING INTENTION

At the end of this subtopic you should be able to:

- sketch cubic polynomial functions by hand including cases where an x-axis intercept is a touch point or a stationary point of inflection
- identify the effect of transformations of the plane, dilation, reflection in axes, translation and simple combinations of these transformations, on the graphs of cubic functions.

The graph of the general cubic polynomial has an equation of the form $y = ax^3 + bx^2 + cx + d$, where a, b, c and d are real constants and $a \neq 0$. Since a cubic polynomial may have up to three linear factors, its graph may have up to three x-intercepts. The shape of its graph is affected by the number of x-intercepts.

4.4.1 The graph of $y = x^3$ and transformations

The graph of the simplest cubic polynomial has the equation $y = x^3$.

Key features of the graph of $y = x^3$:

- $(0, 0)$ is a **stationary point of inflection**.
- The shape of the graph changes from concave down to concave up at the stationary point of inflection.
- There is only one x-intercept.
- As the values of x become very large positive, the behaviour of the graph shows its y-values become increasingly large positive also. This means that as $x \to \infty$, $y \to \infty$. This is read as 'as x approaches infinity, y approaches infinity'.
- As the values of x become very large negative, the behaviour of the graph shows its y-values become increasingly large negative. This means that as $x \to -\infty$, $y \to -\infty$.
- The graph starts from below the x-axis and increases as x increases.

y
$y = x^3$
(0, 0)
0
x

Once the basic shape is known, the graph can be dilated, reflected and translated in much the same way as the parabola $y = x^2$.

Dilation

The graph of $y = 4x^3$ will be narrower than the graph of $y = x^3$ due to the dilation factor of 4 from the x-axis.

Similarly, the graph of $y = \frac{1}{4}x^3$ will be wider than the graph of $y = x^3$ due to the dilation factor of $\frac{1}{4}$ from the x-axis.

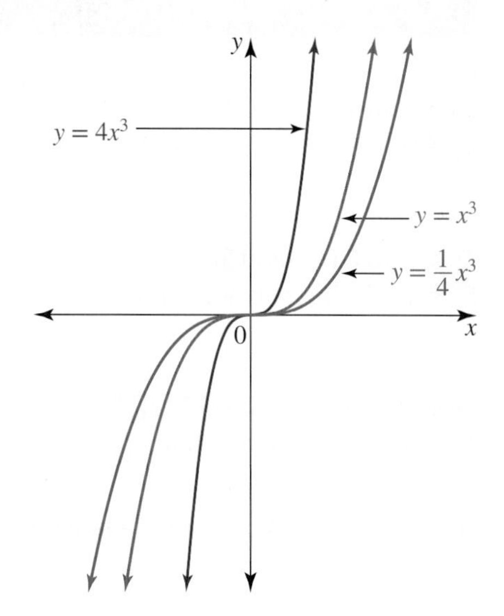

Reflection

The graph of $y = -x^3$ is the reflection of the graph of $y = x^3$ in the x-axis.

For the graph of $y = -x^3$ note that:
- as $x \to \infty$, $y \to -\infty$ and as $x \to -\infty$, $y \to \infty$
- the graph starts from above the x-axis and decreases as x increases
- at $(0, 0)$, the stationary point of inflection, the graph changes from concave up to concave down.

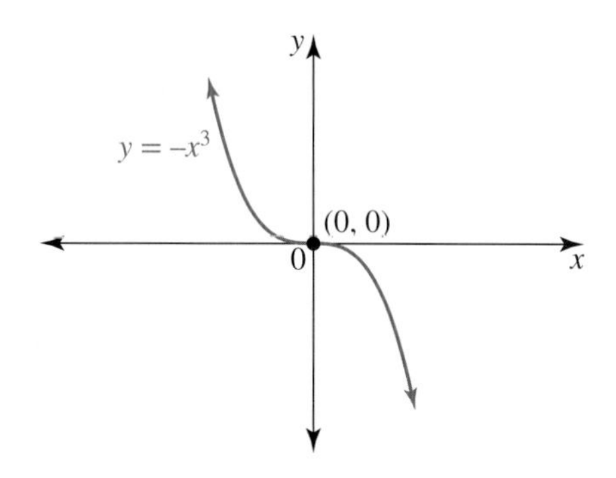

Translation

The graph of $y = x^3 + 4$ is obtained when the graph of $y = x^3$ is translated vertically upwards by 4 units. The stationary point of inflection is at the point $(0, 4)$.

The graph of $y = (x + 4)^3$ is obtained when the graph of $y = x^3$ is translated horizontally 4 units to the left. The stationary point of inflection is at the point $(-4, 0)$.

The transformations from the basic parabola $y = x^2$ are recognisable from the equation $y = a(x - h)^2 + k$, and the equation of the graph of $y = x^3$ can be transformed to a similar form.

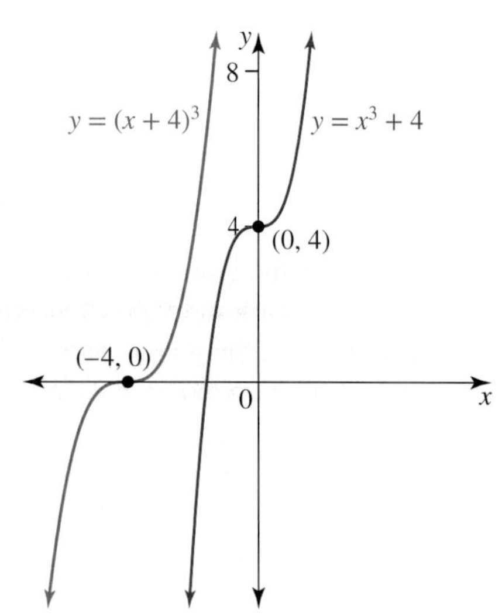

The graph of $y = a(x - h)^3 + k$

The key features of the graph of $y = a(x - h)^3 + k$ are:
- a stationary point of inflection at (h, k)
- a change of concavity at the stationary point of inflection
- if $a > 0$, the graph starts below the x-axis and increases, like $y = x^3$
- if $a < 0$, the graph starts above the x-axis and decreases, like $y = -x^3$
- the one x-intercept is found by solving $a(x - h)^3 + k = 0$
- the y-intercept is found by substituting $x = 0$.

WORKED EXAMPLE 11 Sketching cubic polynomials

Sketch the graphs of the following polynomials.

a. $y = (x+1)^3 + 8$

b. $y = 6 - \frac{1}{2}(x-2)^3$

THINK

WRITE

a. 1. State the point of inflection.

a. $y = (x+1)^3 + 8$
The point of inflection is $(-1, 8)$.

2. Calculate the y-intercept.

y-intercept: let $x = 0$.
$y = (1)^3 + 8$
$= 9$
$\Rightarrow (0, 9)$

3. Calculate the x-intercept.

x-intercept: let $y = 0$.
$(x+1)^3 + 8 = 0$
$(x+1)^3 = -8$
Take the cube root of both sides:
$x + 1 = \sqrt[3]{-8}$
$x + 1 = -2$
$x = -3$
$\Rightarrow (-3, 0)$

4. Sketch the graph. Label the key points and ensure the graph changes concavity at the point of inflection.

The coefficient of x^3 is positive, so the graph starts below the x-axis and increases.

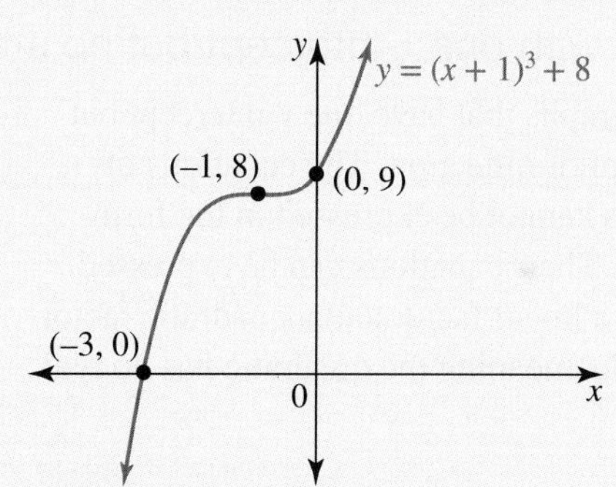

b. 1. Rearrange the equation to the $y = a(x-h)^3 + k$ form and state the point of inflection.

b. $y = 6 - \frac{1}{2}(x-2)^3$
$= -\frac{1}{2}(x-2)^3 + 6$
Point of inflection: $(2, 6)$

2. Calculate the y-intercept.

y-intercept: let $x = 0$.
$y = -\frac{1}{2}(-2)^3 + 6$
$= 10$
$\Rightarrow (0, 10)$

3. Calculate the x-intercept.
Note: A decimal approximation helps locate the point.

x-intercept: let $y = 0$.

$$-\frac{1}{2}(x-2)^3 + 6 = 0$$
$$\frac{1}{2}(x-2)^3 = 6$$
$$(x-2)^3 = 12$$
$$x-2 = \sqrt[3]{12}$$
$$x = 2+\sqrt[3]{12}$$
$$\Rightarrow (2+\sqrt[3]{12}, 0) \approx (4.3, 0)$$

4. Sketch the graph, showing all key features.

$a < 0$, so the graph starts above the x-axis and decreases.

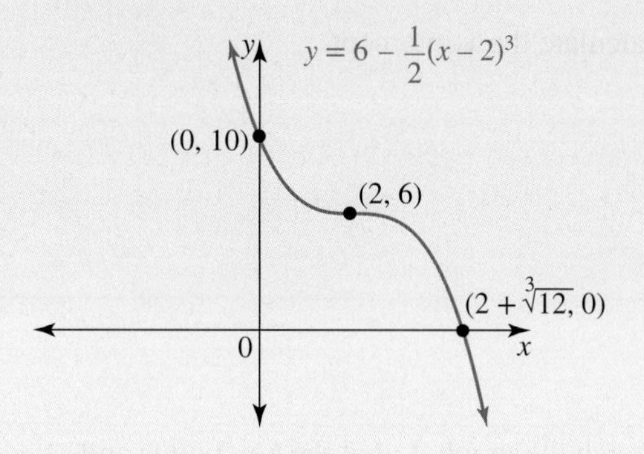

Cubic graphs with one x-intercept but no stationary point of inflection

There are cubic graphs that have one x-intercept but no stationary point of inflection. The equations of such cubic graphs cannot be expressed in the form $y = a(x-h)^3 + k$. Their equations can be expressed as the product of a linear factor and a quadratic factor that is **irreducible**, meaning the quadratic has no real factors.

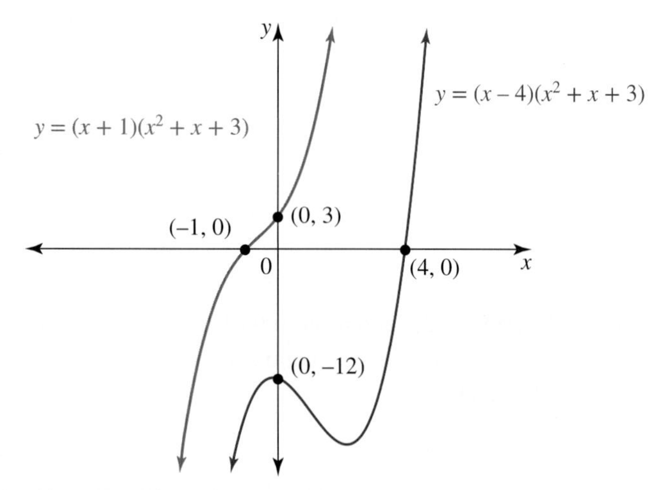

Technology is often required to sketch such graphs. Two examples, $y = (x+1)\left(x^2+x+3\right)$ and $y = (x-4)\left(x^2+x+3\right)$, are shown in the diagram. Each has a linear factor, and the discriminant of the quadratic factor x^2+x+3 is negative; this means it cannot be further factorised over R.

Both graphs maintain the **long-term behaviour** exhibited by all cubics with a positive leading-term coefficient; that is, as $x \to \infty$, $y \to \infty$ and as $x \to -\infty$, $y \to -\infty$.

Every cubic polynomial must have at least one linear factor in order to maintain this long-term behaviour.

 Resources

 **Interactivity** Cubic polynomials (int-2566)

4.4.2 Cubic graphs with three x-intercepts

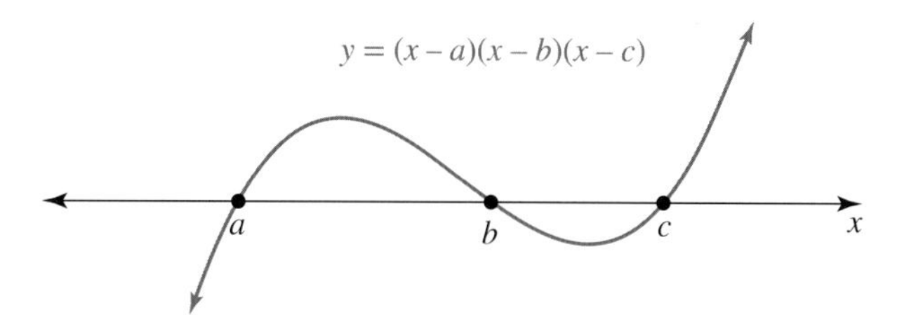

For the graph of a cubic polynomial to have three x-intercepts, the polynomial must have three distinct linear factors. This is the case when the cubic polynomial expressed as the product of a linear factor and a quadratic factor is such that the quadratic factor has two distinct linear factors.

This means that the graph of a monic cubic with an equation of the form $y = (x - a)(x - b)(x - c)$, where $a, b, c \in R$ and $a < b < c$, will have the shape of the graph shown.

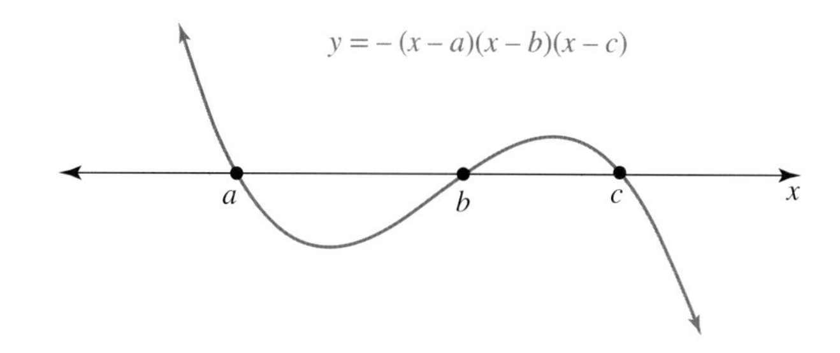

If the graph is reflected in the x-axis, its equation is of the form $y = -(x - a)(x - b)(x - c)$, and the shape of its graph satisfies the long-term behaviour that as $x \to \infty$, $y \to -\infty$, and as $x \to -\infty$, $y \to \infty$.

It is important to note the graph is not a quadratic, so the maximum and minimum turning points do not lie halfway between the x-intercepts. In a later chapter we will learn how to locate these points without using technology.

To sketch the graph, it is usually sufficient to identify the x- and y-intercepts and to ensure the shape of the graph satisfies the long-term behaviour requirement determined by the sign of the leading term.

WORKED EXAMPLE 12 Sketching cubics given in factorised form

Sketch the following without attempting to locate turning points.

a. $y = (x - 1)(x - 3)(x + 5)$
b. $y = (x + 1)(2x - 5)(6 - x)$

THINK	WRITE
a. 1. Calculate the x-intercepts.	a. $y = (x - 1)(x - 3)(x + 5)$ x-intercepts: let $y = 0$. $(x - 1)(x - 3)(x + 5) = 0$ $x = 1,\ x = 3,\ x = -5$ $\Rightarrow (-5, 0), (1, 0), (3, 0)$ are the x-intercepts.
2. Calculate the y-intercept.	y-intercept: let $x = 0$. $y = (-1)(-3)(5)$ $= 15$ $\Rightarrow (0,\ 15)$ is the y-intercept.
3. Determine the shape of the graph.	Multiplying together the terms in x from each bracket gives x^3, so its coefficient is positive. The shape is of a positive cubic.

4. Sketch the graph.

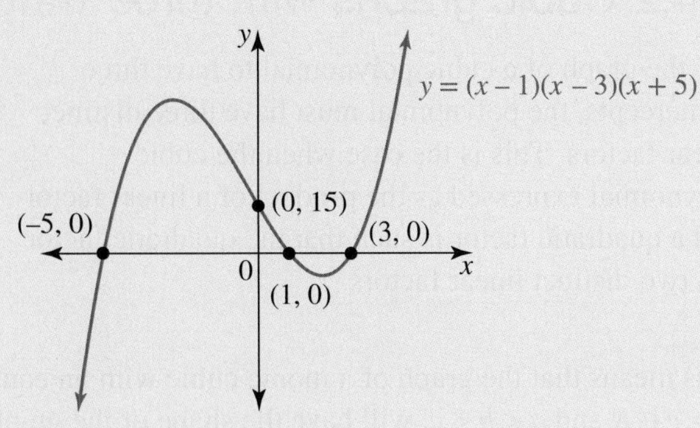

b. 1. Calculate the x-intercepts.

b. $y = (x+1)(2x-5)(6-x)$

x-intercepts: let $y = 0$.

$(x+1)(2x-5)(6-x) = 0$

$x+1 = 0, \quad 2x-5 = 0, \quad 6-x = 0$

$x = -1, \quad x = 2.5, \quad x = 6$

$\Rightarrow (-1, 0), (2.5, 0), (6, 0)$ are the x-intercepts.

2. Calculate the y-intercept.

y-intercept: let $x = 0$.

$y = (1)(-5)(6)$

$= -30$

$\Rightarrow (0, -30)$ is the y-intercept.

3. Determine the shape of the graph.

Multiplying the terms in x from each bracket gives $(x) \times (2x) \times (-x) = -2x^3$, so the shape is of a negative cubic.

4. Sketch the graph.

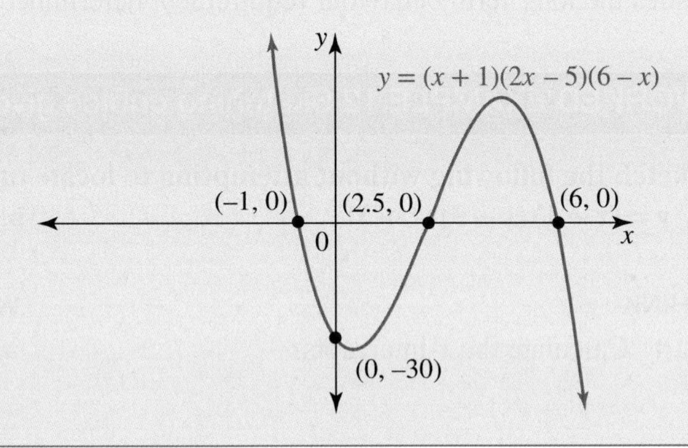

4.4.3 Cubic graphs with two x-intercepts

If a cubic has two x-intercepts, one at $x = a$ and one at $x = b$, then in order to satisfy the long-term behaviour required of any cubic, the graph either touches the x-axis at $x = a$ and turns, or it touches the x-axis at $x = b$ and turns. One of the x-intercepts must be a turning point.

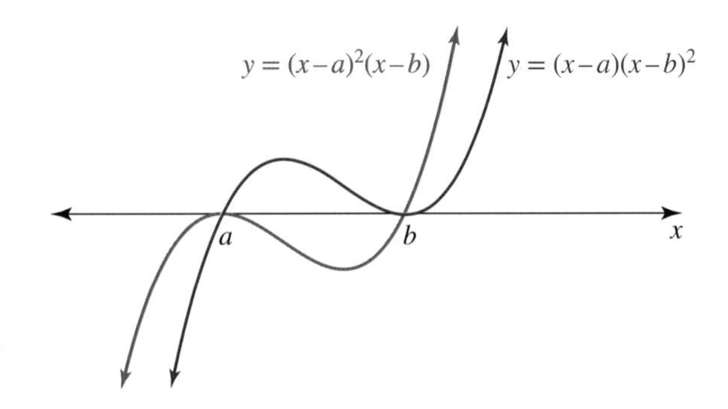

Thinking of the cubic polynomial as the product of a linear and a quadratic factor, for its graph to have two instead of three x-intercepts, the quadratic factor must have two identical factors. Either the factors of the cubic are $(x-a)(x-a)(x-b) = (x-a)^2(x-b)$, or the factors are $(x-a)(x-b)(x-b) = (x-a)(x-b)^2$. The repeated factor identifies the x-intercept that is the turning point. The repeated factor is said to be of multiplicity 2 and the single factor of multiplicity 1.

The graph of a cubic polynomial with equation of the form $y = (x - a)^2 (x - b)$ has a turning point on the x-axis at $(a, 0)$ and a second x-intercept at $(b, 0)$. The graph is said to *touch* the x-axis at $x = a$ and *cut* it at $x = b$.

Although the turning point on the x-axis must be identified when sketching the graph, there will be a second turning point that cannot yet be located without technology.

Note that a cubic graph whose equation has a repeated factor of multiplicity 3, such as $y = (x - h)^3$, would have only one x-intercept, as this is a special case of $y = a(x - h)^3 + k$ with $k = 0$. The graph would cut the x-axis at its stationary point of inflection $(h, 0)$.

WORKED EXAMPLE 13 Sketching cubics with repeated roots

Sketch the graphs of the following polynomials.

a. $y = \frac{1}{4}(x - 2)^2 (x + 5)$ **b.** $y = -2(x + 1)(x + 4)^2$

THINK	WRITE
a. 1. Calculate the x-intercepts and interpret the multiplicity of each factor.	**a.** $y = \frac{1}{4}(x - 2)^2 (x + 5)$ x-intercepts: let $y = 0$. $\frac{1}{4}(x - 2)^2 (x + 5) = 0$ $\therefore x = 2$ (touch), $x = -5$ (cut) x-intercept at $(-5, 0)$ and turning-point x-intercept at $(2, 0)$
2. Calculate the y-intercept.	y-intercept: let $x = 0$. $y = \frac{1}{4}(-2)^2 (5)$ $= 5$ $\Rightarrow (0, 5)$
3. Sketch the graph.	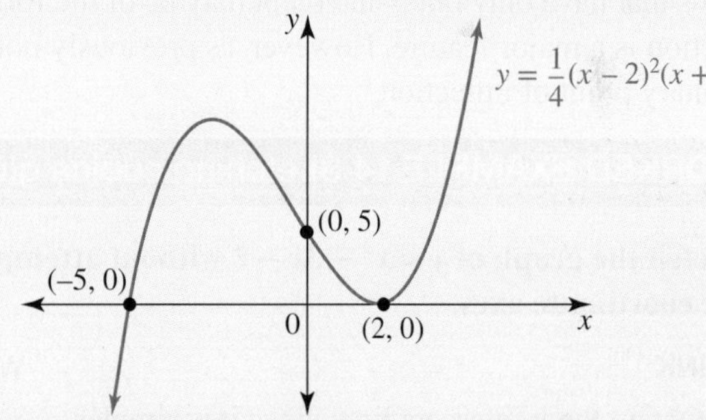
b. 1. Calculate the x-intercepts and interpret the multiplicity of each factor.	**b.** $y = -2(x + 1)(x + 4)^2$ x-intercepts: let $y = 0$. $-2(x + 1)(x + 4)^2 = 0$ $(x + 1)(x + 4)^2 = 0$ $\therefore x = -1$ (cut), $x = -4$ (touch) x-intercept at $(-1, 0)$ and turning-point x-intercept at $(-4, 0)$
2. Calculate the y-intercept.	y-intercept: let $x = 0$. $y = -2(1)(4)^2$ $= -32$ y-intercept at $(0, -32)$

3. Sketch the graph.

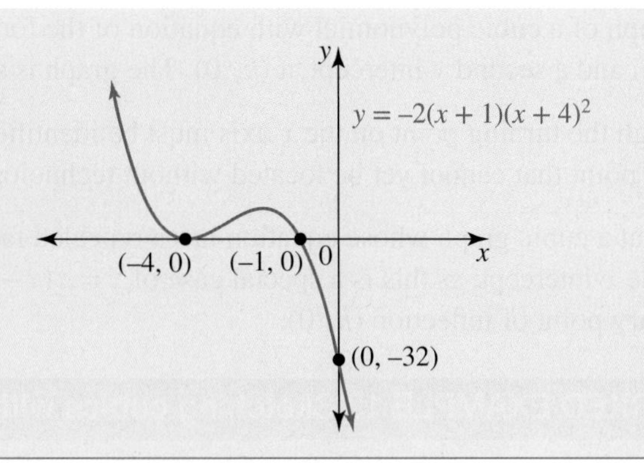

4.4.4 Cubic graphs in the general form $y = ax^3 + bx^2 + cx + d$

If the cubic polynomial with equation $y = ax^3 + bx^2 + cx + d$ can be factorised, then the shape of its graph and its key features can be determined. Standard factorisation techniques such as grouping terms together may be sufficient, or the factor theorem may be required in order to obtain the factors.

The sign of a, the coefficient of x^3, determines the long-term behaviour the graph exhibits.

$$\text{For } a > 0, \text{ as } x \to +\infty, \ y \to +\infty \text{ and as } x \to -\infty, \ y \to -\infty.$$

$$\text{For } a < 0, \text{ as } x \to +\infty, \ y \to -\infty \text{ and as } x \to -\infty, \ y \to +\infty.$$

The value of d determines the y-intercept.

The factors determine the x-intercepts, and the multiplicity of each factor will determine how the graph intersects the x-axis.

Every cubic graph must have at least one x-intercept, and hence the polynomial must have at least one linear factor. Considering a cubic as the product of a linear and a quadratic factor, it is the quadratic factor that determines whether there is more than one x-intercept.

Graphs that have only one x-intercept may be of the form $y = a(x - h)^3 + k$, where the stationary point of inflection is a major feature. However, as previously noted, not all graphs with only one x-intercept have a stationary point of inflection.

WORKED EXAMPLE 14 Sketching cubics given in general form

Sketch the graph of $y = x^3 - 3x - 2$ without attempting to obtain any turning points that do not lie on the coordinate axes.

THINK	WRITE
1. Obtain the y-intercept first, since it is simpler to obtain from the expanded form.	$y = x^3 - 3x - 2$ y-intercept: $(0, -2)$
2. Factorisation will be needed in order to obtain the x-intercepts.	x-intercepts: let $y = 0$. $x^3 - 3x - 2 = 0$
3. The polynomial does not factorise by grouping, so the factor theorem needs to be used.	Let $p(x) = x^3 - 3x - 2$. $p(1) \neq 0$ $p(-1) = -1 + 3 - 2 = 0$ $\therefore (x + 1)$ is a factor. $x^3 - 3x - 2 = (x + 1)(x^2 + bx - 2)$ $= (x + 1)(x^2 - x - 2)$ $= (x + 1)(x - 2)(x + 1)$ $= (x + 1)^2(x - 2)$

$\therefore x^3 - 3x - 2 = 0$

$\Rightarrow (x+1)^2(x-2) = 0$

$\therefore x = -1,\ x = 2$

4. What is the nature of these x-intercepts?

$y = p(x) = (x+1)^2(x-2)$

$x = -1$ (touch) and $x = 2$ (cut)

Turning point at $(-1, 0)$

5. Sketch the graph.

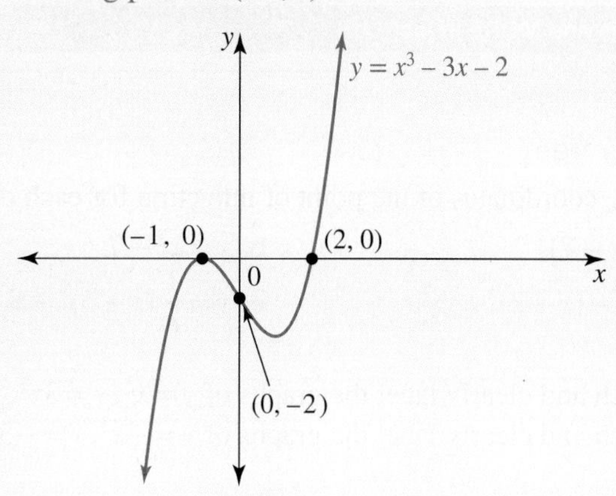

TI | THINK	DISPLAY/WRITE	CASIO | THINK	DISPLAY/WRITE
1. On a Graphs page, complete the entry line for function 1 as: $f1(x) = x^3 - 3x - 2$ then press ENTER.	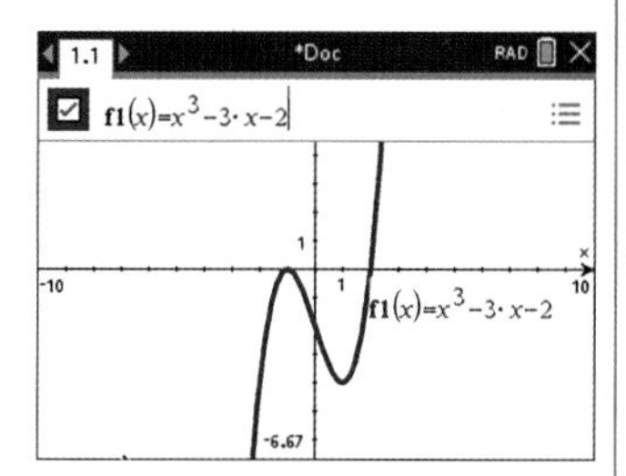	1. On a Graph & Table screen, complete the entry line for $y1$ as: $y1 = x^3 - 3x - 2$ then press EXE. Tap the graph icon to draw the graphs.	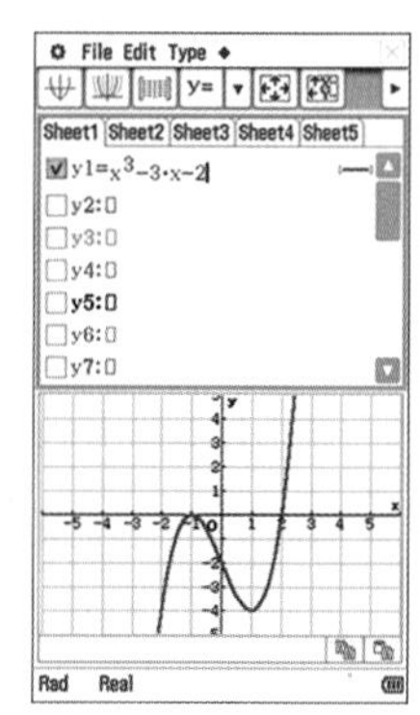
2. To find the x-intercepts, press MENU, then select: 6: Analyze Graph 1: Zero Move the cursor to the left of the x-intercept when prompted for the lower bound and press ENTER. Move the cursor to the right of the x-intercept when prompted for the upper bound and press ENTER. Repeat this process to find the other x-intercept.	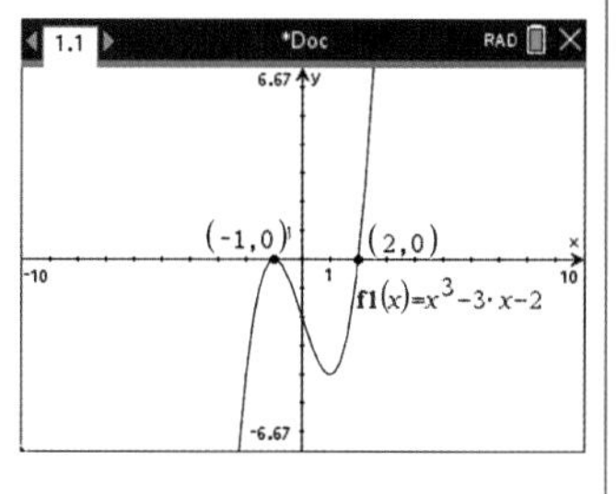	2. To find the x-intercepts, select: • Analysis • G-Solve • Root With the cursor on the first x-intercept, press EXE. Use the left or right arrow to move the cursor to the other x-intercept and press EXE.	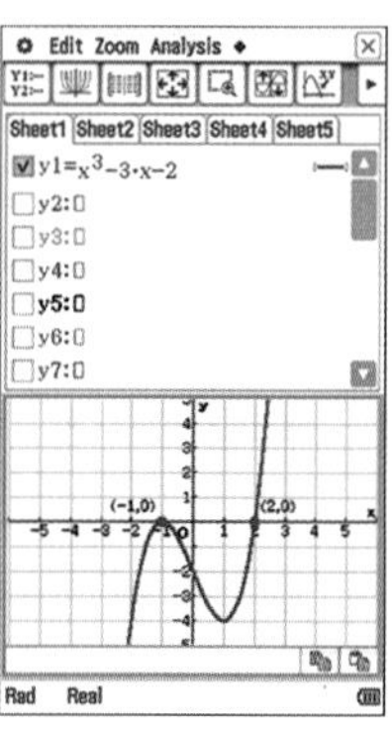
3. To find the y-intercept, press MENU, then select: 5 : Trace 1 : Graph Trace Type 0, then press ENTER twice.	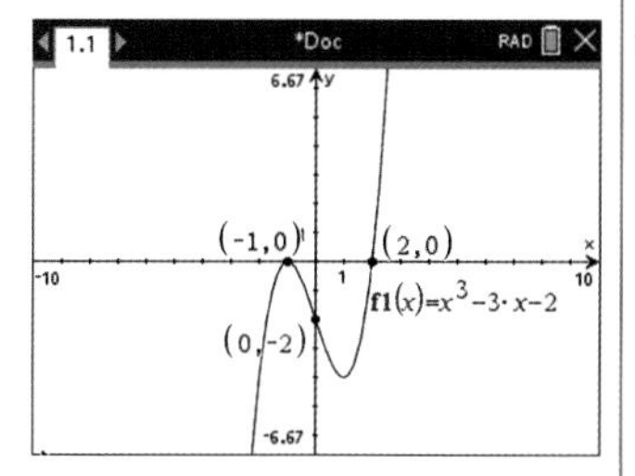	3. To find the y-intercept, select: • Analysis • Trace Type '0', select OK, then press EXE.	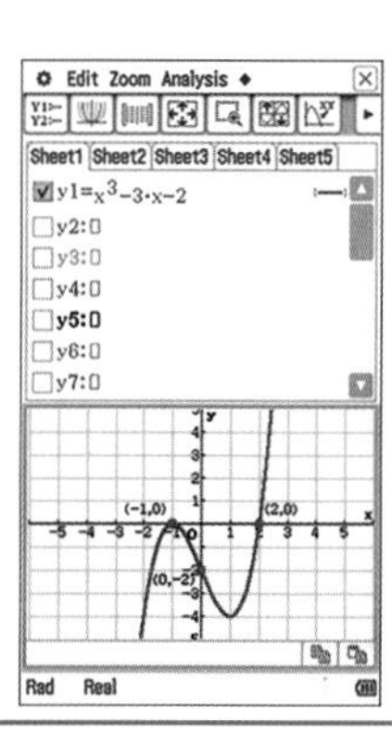

4.4 Exercise

Students, these questions are even better in jacPLUS

Receive immediate feedback and access sample responses

Access additional questions

Track your results and progress

Find all this and MORE in jacPLUS

Technology free

1. State the coordinates of the point of inflection for each of the following.

   a. $y = (x - 7)^3$
   b. $y = x^3 - 7$
   c. $y = -7x^3$
   d. $y = 2 - (x - 2)^3$
   e. $y = \frac{1}{6}(x + 5)^3 - 8$
   f. $y = -\frac{1}{2}(2x - 1)^3 + 5$

2. a. Sketch and clearly label the graphs of $y = x^3, y = 3x^3, y = x^3 + 3$ and $y = (x + 3)^3$ on the one set of axes.
   b. Sketch and clearly label the graphs of $y = -x^3, y = -3x^3, y = -x^3 + 3$ and $y = -(x + 3)^3$ on the one set of axes.

3. WE11 Sketch the graphs of the following polynomials.

   a. $y = (x - 1)^3 - 8$
   b. $y = 1 - \frac{1}{36}(x + 6)^3$

4. a. Sketch the graph $y = -x^3 + 1$. Include all important features.
   b. Sketch the graph $y = 2(3x - 2)^3$. Include all important features.
   c. Sketch the graph $y = 2(x + 3)^3 - 16$. Include all important features.
   d. Sketch the graph of $y = (3 - x)^3 + 1$. Include all important features.

5. Sketch the graphs of the following, identifying all key points.

   a. $y = (x + 4)^3 - 27$
   b. $y = 2(x - 1)^3 + 10$
   c. $y = 27 + 2(x - 3)^3$
   d. $y = 16 - 2(x + 2)^3$
   e. $y = -\frac{3}{4}(3x + 4)^3$
   f. $y = 9 + \frac{x^3}{3}$

6. State the coordinates of the point of inflection and sketch the graph of the following.

   a. $y = \left(\frac{x}{2} - 3\right)^3$
   b. $y = 2x^3 - 2$

7. WE12 Sketch the following without attempting to locate turning points.

   a. $y = (x + 1)(x + 6)(x - 4)$
   b. $y = (x - 4)(2x + 1)(6 - x)$

8. Sketch the graphs of the following without attempting to locate any turning points that do not lie on the coordinate axes.

   a. $y = (x - 2)(x + 1)(x + 4)$
   b. $y = -0.5x(x + 8)(x - 5)$
   c. $y = (x + 3)(x - 1)(4 - x)$
   d. $y = \frac{1}{4}(2 - x)(6 - x)(4 + x)$

9. Sketch $y = 3x\left(x^2 - 4\right)$.

10. WE13 Sketch the graphs of the following polynomials.

    a. $y = \frac{1}{9}(x - 3)^2(x + 6)$
    b. $y = -2(x - 1)(x + 2)^2$

11. Sketch the graphs of the following without attempting to locate any turning points that do not lie on the coordinate axes.

a. $y = -(x+4)^2(x-2)$
b. $y = 2(x+3)(x-3)^2$
c. $y = (x+3)^2(4-x)$
d. $y = \frac{1}{4}(2-x)^2(x-12)$
e. $y = 3x(2x+3)^2$
f. $y = -0.25x^2(2-5x)$

12. Sketch the graphs of the following, showing any intercepts with the coordinate axes and any stationary point of inflection that do not lie on the coordinate axes.

a. $y = (x+3)^3$
b. $y = (x+3)^2(2x-1)$
c. $y = (x+3)(2x-1)(5-x)$
d. $2(y-1) = (1-2x)^3$
e. $4y = x(4x-1)^2$
f. $y = -\frac{1}{2}(2-3x)(3x+2)(3x-2)$

13. **WE14** Sketch the graph of $y = x^3 - 3x^2 - 10x + 24$ without attempting to obtain any turning points that do not lie on the coordinate axes.

14. a. Determine the x- and y-intercepts of the cubic graph $y = -x^3 - 3x^2 + 16x + 48$. Hence, sketch the graph.
b. Factorise $2x^3 + x^2 - 13x + 6$ and sketch the graph of $y = 2x^3 + x^2 - 13x + 6$, showing all intercepts with the coordinate axes.
c. Determine the x- and y-intercepts of the cubic graph $y = x^3 + 5x^2 - x - 5$. Hence, sketch the graph.
d. Factorise $-x^3 - 5x^2 - 3x + 9$ and sketch the graph of $y = -x^3 - 5x^2 - 3x + 9$, showing all intercepts with the coordinate axes.

15. Factorise, if possible, and then sketch the graphs of the cubic polynomials with equations given by:

a. $y = 9x^2 - 2x^3$
b. $y = 9x^3 - 4x$
c. $y = -9x^3 - 9x^2 + 9x + 9$
d. $y = 9x\left(x^2 + 4x + 3\right)$
e. $y = 9x^3 + 27x^2 + 27x + 9$
f. $y = 9x^2 - 3x^3 + x - 3$

Technology active

16. Sketch the graphs of the following without attempting to locate any turning points that do not lie on the coordinate axes.

a. $y = 2x^3 - 3x^2 - 17x - 12$
b. $y = 6 - 55x + 57x^2 - 8x^3$
c. $y = 6x^3 - 13x^2 - 59x - 18$
d. $y = x^3 - 17x + 4$
e. $y = -5x^3 - 7x^2 + 10x + 14$
f. $y = -\frac{1}{2}x^3 + 14x - 24$

17. a. Sketch the graph of $y = -x^3 + 3x^2 + 10x - 30$ without attempting to obtain any turning points that do not lie on the coordinate axes.
b. Determine the coordinates of the stationary point of inflection of the graph with equation $y = x^3 + 3x^2 + 3x + 2$ and sketch the graph.

18. Consider $p(x) = 30x^3 + kx^2 + 1$.

a. Given $(3x - 1)$ is a factor, find the value of k.
b. Hence, express $p(x)$ as the product of its linear factors.
c. State the values of x for which $p(x) = 0$.
d. Sketch the graph of $y = p(x)$.
e. Determine whether the point $(-1, -40)$ lies on the graph of $y = p(x)$. Justify your answer.

19. Consider $y = x^3 - 5x^2 + 11x - 7$.

a. Show that the graph of $y = x^3 - 5x^2 + 11x - 7$ has only one x-intercept.
b. Show that $y = x^3 - 5x^2 + 11x - 7$ cannot be expressed in the form $y = a(x-b)^3 + c$.
c. Describe the behaviour of the graph as $x \to \infty$.
d. Given the graph of $y = x^3 - 5x^2 + 11x - 7$ has no turning points, draw a sketch of the graph.

20. a. Sketch the graphs of the following, locating turning points and intercepts.

i. $y = x^3 + 4x^2 - 44x - 96$
ii. $y = 10x^3 - 20x^2 - 10x - 19$
iii. $y = 9x^3 - 70x^2 + 25x + 500$

b. Show that the turning points are not placed symmetrically in the interval between the adjoining x-intercepts.

4.4 Exam questions

Question 1 (1 mark) TECH-ACTIVE

MC The equation of the graph shown, where is a positive constant, could be

A. $y = a(x+2)(x-4)^2$
B. $y = a(x+2)^2(x-4)$
C. $y = -a(x+2)(x-4)^2$
D. $y = a(x+4)^2(x-2)$
E. $y = -a(x+4)(x-2)^2$

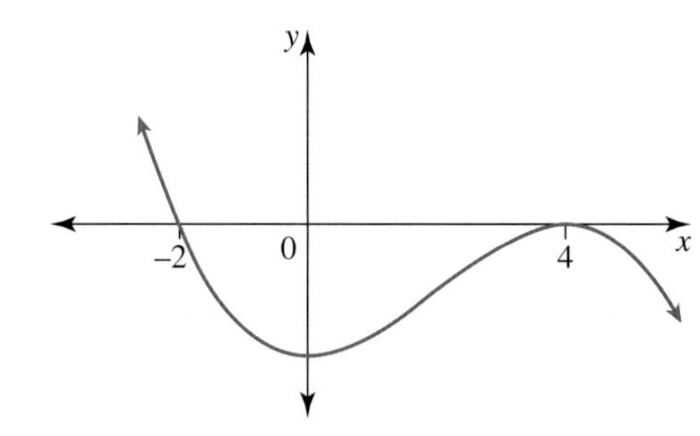

Question 2 (3 marks) TECH-FREE

Sketch the graph of $y = (x+4)(x+1)(x-3)$. (Do not attempt to find the turning points.)

Question 3 (1 mark) TECH-ACTIVE

MC For the function $y = -2(3-x)^3 - 1$, determine which of the following statements is false.

A. The function has been translated 1 unit downwards.
B. The y-intercept is −55.
C. The function has been translated 3 units to the right.
D. The function has been dilated by a factor of 2.
E. The function is a negative cubic.

More exam questions are available online.

4.5 Equations of cubic polynomials

LEARNING INTENTION

At the end of this subtopic you should be able to:

- identify key features of a cubic polynomial from a graph and determine its equation
- solve and graph cubic inequations
- determine points of intersection with other linear and quadratic graphs.

The equation $y = ax^3 + bx^2 + cx + d$ contains four unknown coefficients that need to be specified, so four pieces of information are required to determine the equation of a cubic graph.

4.5.1 Cubic equations

Depending on the information given, one form of the cubic equation may be preferable over another to determine the equation from either a graph or key points on the graph.

Cubic polynomial equations

As a guide:

- **if there is a stationary point of inflection given, use the $y = a(x-h)^3 + k$ form**
- **if the x-intercepts are given, use the $y = a(x-x_1)(x-x_2)(x-x_3)$ form, or the repeated factor form $y = a(x-x_1)^2(x-x_2)$ if there is a turning point at one of the x-intercepts**
- **if four points on the graph are given, use the $y = ax^3 + bx^2 + cx + d$ form.**

WORKED EXAMPLE 15 Determining the cubic equation for a given graph

Determine the equation of each of the following graphs.

a. **The graph of a cubic polynomial that has a stationary point of inflection at the point $(-7, 4)$ and an x-intercept at $(1, 0)$**

b.

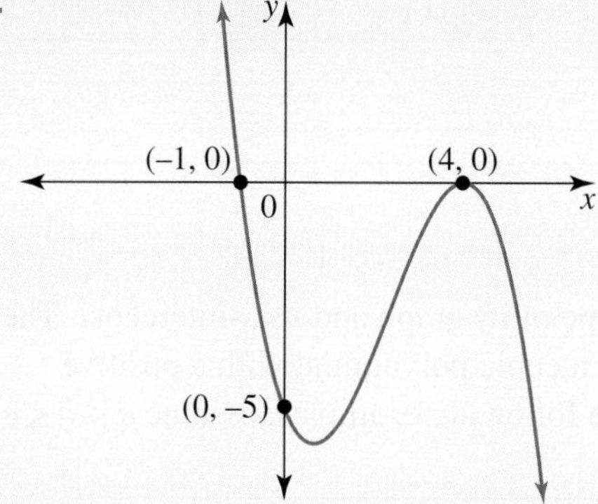

c.

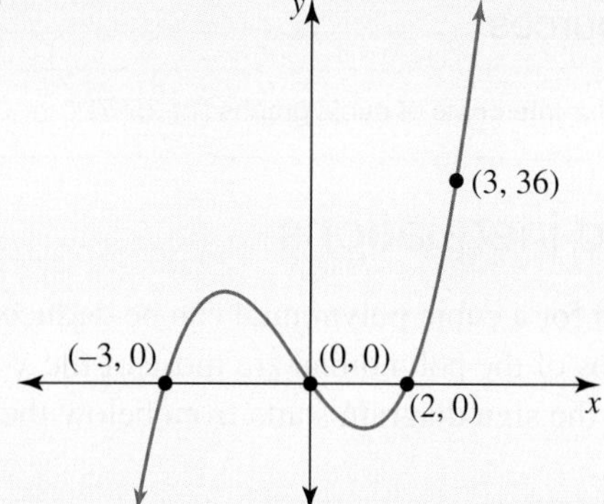

THINK

a. 1. Consider the given information and choose the form of the equation to be used.

WRITE

a. A stationary point of inflection is given.
Let $y = a(x-h)^3 + k$.
The point of inflection is $(-7, 4)$.
$\therefore y = a(x+7)^3 + 4$

2. Calculate the value of a.
Note: The coordinates of the given points show the y-values decrease as the x-values increase, so a negative value for a is expected.

Substitute the given x-intercept point $(1, 0)$.

$$0 = a(8)^3 + 3$$
$$(8)^3 a = -4$$
$$a = \frac{-4}{8 \times 64}$$
$$a = -\frac{1}{128}$$

3. Write the equation of the graph.

The equation is $y = -\frac{1}{128}(x+7)^3 + 4$.

b. 1. Consider the given information and choose the form of the equation to be used.

b. Two x-intercepts are given.
One shows a turning point at $x = 4$ and the other a cut at $x = -1$.
Let the equation be $y = a(x+1)(x-4)^2$.

2. Calculate the value of a.

Substitute the given y-intercept point $(0, -5)$.

$$-5 = a(1)(-4)^2$$
$$-5 = a(16)$$
$$a = -\frac{5}{16}$$

3. Write the equation of the graph.	The equation is $y = -\frac{5}{16}(x+1)(x-4)^2$.
c. 1. Consider the given information and choose the form of the equation to be used.	**c.** Three x-intercepts are given. Let the equation be $y = a(x+3)(x-0)(x-2)$ $\therefore y = ax(x+3)(x-2)$
2. Calculate the value of a.	Substitute the given point (3, 36). $36 = a(3)(6)(1)$ $36 = 18a$ $a = 2$
3. Write the equation of the graph.	The equation is $y = 2x(x+3)(x-2)$.

Resources

Interactivity *x*-intercepts of cubic graphs (int-2567)

4.5.2 Cubic inequations

The sign diagram for a cubic polynomial can be deduced from the shape of its graph and its x-intercepts. The values of the zeros of the polynomial are those of the x-intercepts. For a cubic polynomial with a positive coefficient of x^3, the sign diagram starts from below the x-axis. For the following examples, assume $a < b < c$.

One zero $(x-a)^3$ or $(x-a) \times$ irreducible quadratic factor	+ − a x
Two zeros $(x-a)^2(x-b)$	+ − a b x
Three zeros $(x-a)(x-b)(x-c)$	+ − a b c x

At a zero of multiplicity 2, the sign diagram touches the x-axis. For a zero of odd multiplicity, either multiplicity 1 or multiplicity 3, the sign diagram cuts the x-axis. This 'cut and touch' nature applies if the coefficient of x^3 is negative; however, the sign diagram would start from above the x-axis in that case.

Solving cubic inequations

To solve a cubic inequation, use the following steps:

- **Rearrange the terms in the inequation, if necessary, so that one side of the inequation is 0.**
- **Factorise the cubic expression and calculate its zeros.**
- **Draw the sign diagram, or the graph, of the cubic.**
- **Read from the sign diagram the set of values of x that satisfy the inequation.**

An exception applies to inequations of forms such as $a(x-h)^3 + k > 0$. These inequations are solved in a similar way to solving linear inequations without the need for a sign diagram or a graph. Note the similarity between the sign diagram of the cubic polynomial with one zero and the sign diagram of a linear polynomial.

WORKED EXAMPLE 16 Solving cubic inequations

Solve the following inequations.

a. $(x+2)(x-1)(x-5)>0$ **b.** $\{x : 4x^2 \le x^3\}$ **c.** $(x-2)^3-1>0$

THINK

a. 1. Read the zeros from each factor.

WRITE

a. $(x+2)(x-1)(x-5)>0$

Zeros: $x=-2, x=1, x=5$

2. Consider the leading-term coefficient to draw the sign diagram. This is a positive cubic.

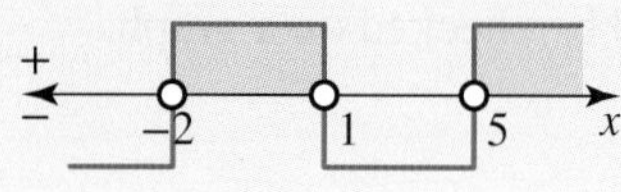

3. State the intervals in which the required solutions lie.

$-2<x<1$ or $x>5$

4. Alternatively, the solution could be obtained from the graph of the cubic.

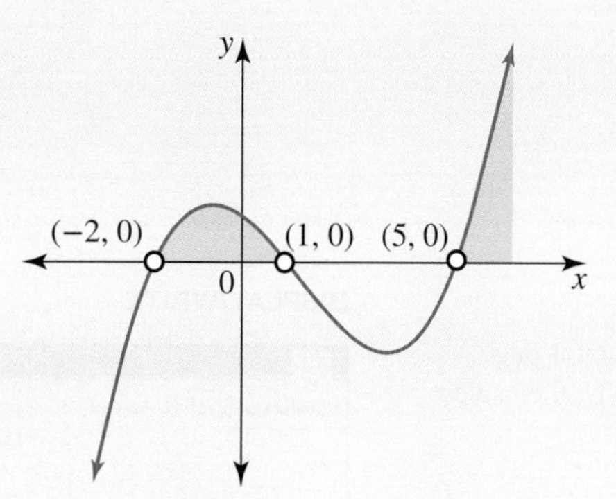

b. 1. Rearrange so one side is 0.

b. $4x^2 \le x^3$

$4x^2 - x^3 \le 0$

$\therefore x^2(4-x) \le 0$

2. Factorise to calculate the zeros.

Let $x^2(4-x)=0$.

$x^2=0$ or $4-x=0$

$x=0$ or $x=4$

Zeros: $x=0$ (multiplicity 2), $x=4$

3. Consider the leading-term coefficient to draw the sign diagram. $4x^2 - x^3$ is a negative cubic.

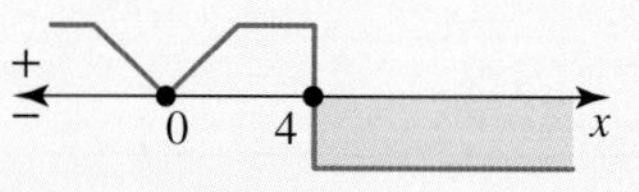

4. State the answer from the sign diagram using set notation, since the question is in set notation.

$\{x : x \ge 4\} \cup \{0\}$

5. Alternatively, use a graph to obtain the solution.

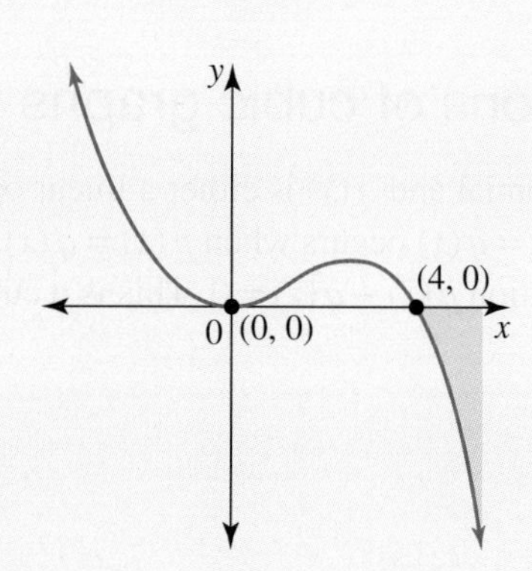

c. 1. Solve for x.	**c.** $(x-2)^3-1>0$ $(x-2)^3>1$ $(x-2)>\sqrt[3]{1}$ $x-2>1$ $x>3$
2. The solution could be checked on a graph.	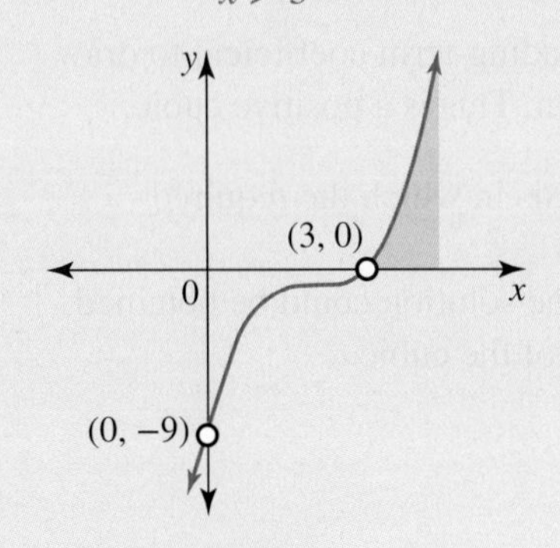

TI | THINK	**DISPLAY/WRITE**	**CASIO | THINK**	**DISPLAY/WRITE**
a. 1. On a Calculator page, press MENU, then select 3: Algebra 1: Solve Complete the entry line as: solve $((x+2)(x-1)(x-5)>0,\ x)$ then press ENTER.	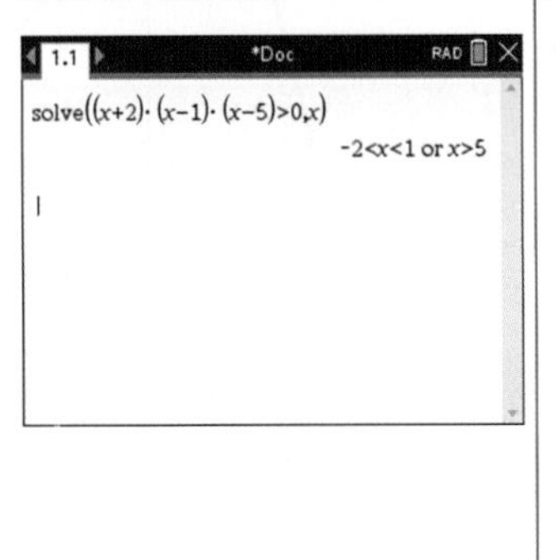	**a. 1.** On a Main screen, complete the entry line as: solve $((x+2)(x-1)(x-5)>0,\ x)$ then press EXE.	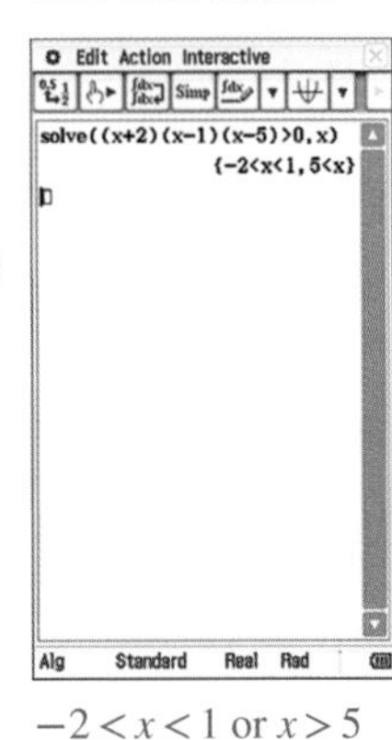
2. The answer appears on the screen.	$-2<x<1$ or $x>5$	**2.** The answer appears on the screen.	$-2<x<1$ or $x>5$

 Resources

 **Interactivity** Cubic inequations (int-2568)

4.5.3 Intersections of cubic graphs with linear and quadratic graphs

If $p(x)$ is a cubic polynomial and $q(x)$ is either a linear or a quadratic polynomial, then the intersection of the graphs of $y=p(x)$ and $y=q(x)$ occurs when $p(x)=q(x)$. Hence, the x-coordinates of the points of intersection are the roots of the equation $p(x)-q(x)=0$. This is a cubic equation since $p(x)-q(x)$ is a polynomial of degree 3.

WORKED EXAMPLE 17 Determining points of intersection

Sketch the graphs of $y = x(x - 1)(x + 1)$ and $y = x$, and calculate the coordinates of the points of intersection.

THINK

1. Sketch the graphs.

WRITE

$y = x(x - 1)(x + 1)$

This is a positive cubic.

x-intercepts: let $y = 0$.

$x(x - 1)(x + 1) = 0$

$x = 0,\ x = \pm 1$

$(-1, 0), (0, 0), (1, 0)$ are the three x-intercepts.

The y-intercept is $(0, 0)$.

Line: $y = x$ passes through $(0, 0)$ and $(1, 1)$.

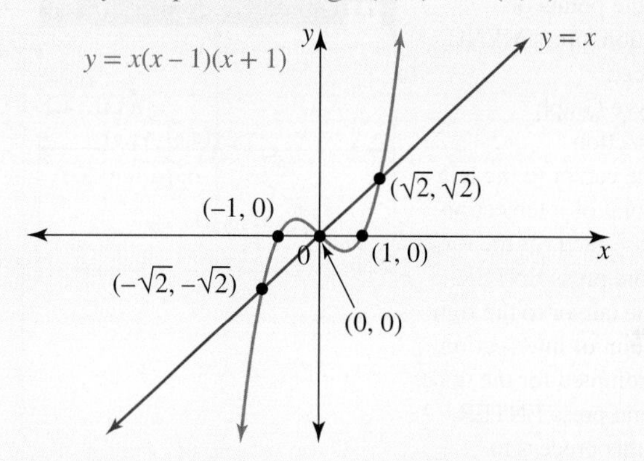

2. Calculate the coordinates of the points of intersection.

At intersection:

$$x(x - 1)(x + 1) = x$$
$$x\left(x^2 - 1\right) - x = 0$$
$$x^3 - 2x = 0$$
$$x\left(x^2 - 2\right) = 0$$
$$x = 0,\ x^2 = 2$$
$$x = 0,\ x = \pm\sqrt{2}$$

Substituting these x-values in the equation of the line $y = x$, the points of intersection are $(0, 0), \left(\sqrt{2}, \sqrt{2}\right), \left(-\sqrt{2}, -\sqrt{2}\right)$.

TI | THINK

1. On a Graphs page, complete the entry line for function 1 as:
$f1(x) = x \cdot (x - 1) \cdot (x + 1)$
and the entry line for function 2 as:
$f2(x) = x$
then press ENTER.

DISPLAY/WRITE

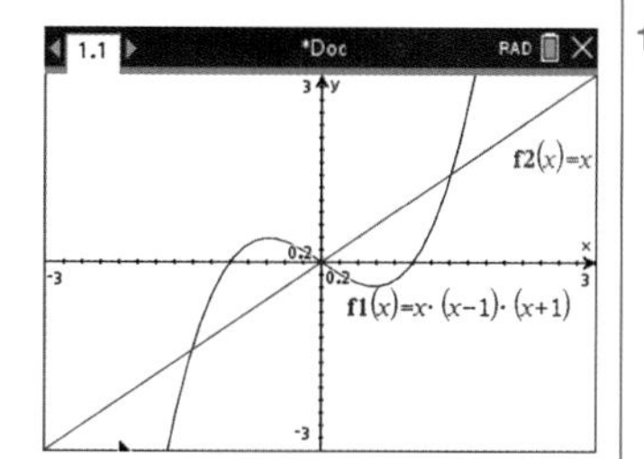

CASIO | THINK

1. On a Graph & Table screen, complete the entry line for $y1$ as:
$y1 = x \cdot (x - 1) \cdot (x + 1)$
then press EXE.
Complete the entry line for $y2$ as:
$y2 = x$
then press EXE.
Tap the graph icon to draw the graphs.
Note: Make sure to use the multiplication operator between x and the brackets in $y1$.

DISPLAY/WRITE

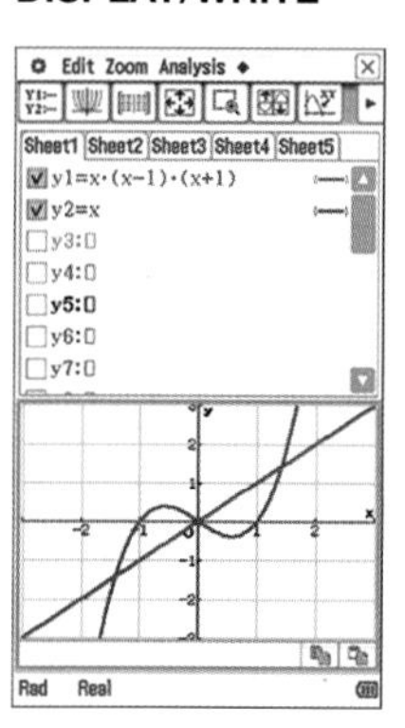

2. To find the x-intercepts, press MENU, then select:
6: Analyze Graph
1: Zero
Click on the blue graph. Move the cursor to the left of the x-intercept when prompted for the lower bound and press ENTER. Move the cursor to the right of the x-intercept when prompted for the upper bound and press ENTER. Repeat this process to find the other x-intercepts.

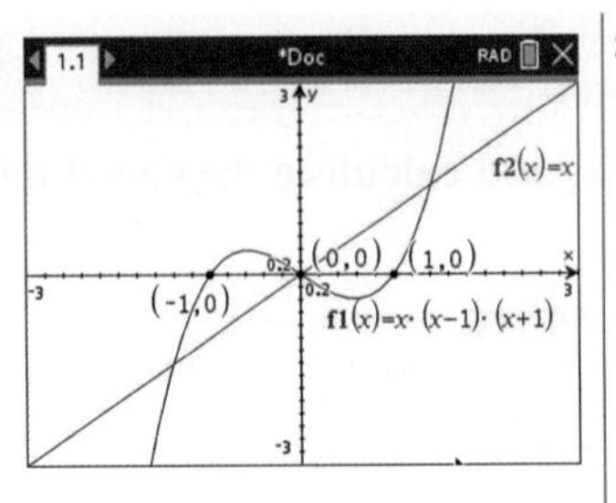

3. To find the points of intersection, press MENU, then select:
6: Analyze Graph
4: Intersection
Move the cursor to the left of the point of intersection when prompted for the lower bound and press ENTER. Move the cursor to the right of the point of intersection when prompted for the upper bound and press ENTER. Repeat this process to find the other points of intersection.

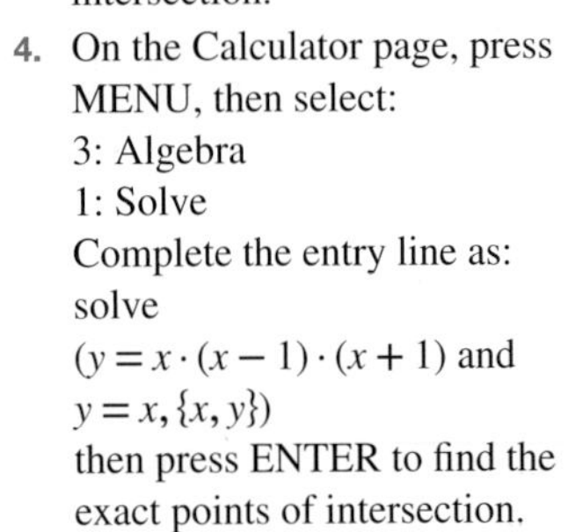

4. On the Calculator page, press MENU, then select:
3: Algebra
1: Solve
Complete the entry line as:
solve
$(y = x \cdot (x-1) \cdot (x+1)$ and $y = x, \{x, y\})$
then press ENTER to find the exact points of intersection.

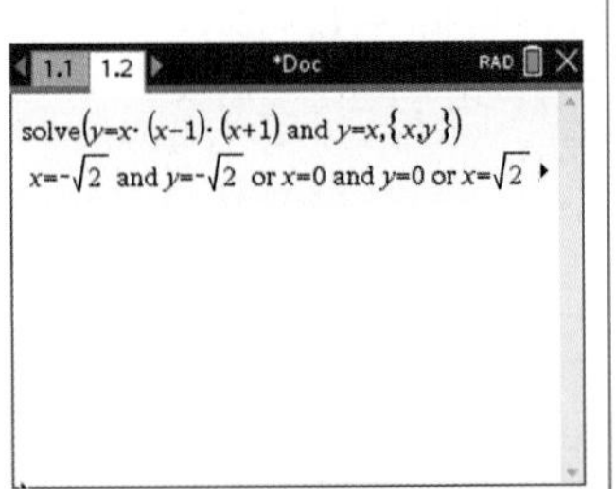

5. Answer the question.

The points of intersection are $(0, 0)$, $\left(\sqrt{2}, \sqrt{2}\right)$ and $\left(-\sqrt{2}, -\sqrt{2}\right)$.

2. To find the x-intercepts, select:
- Analysis
- G-Solve
- Root

Press EXE to select the first curve. With the cursor on the first x-intercept, press EXE. Use the left or right arrow to move the cursor to the other x-intercept and press EXE.

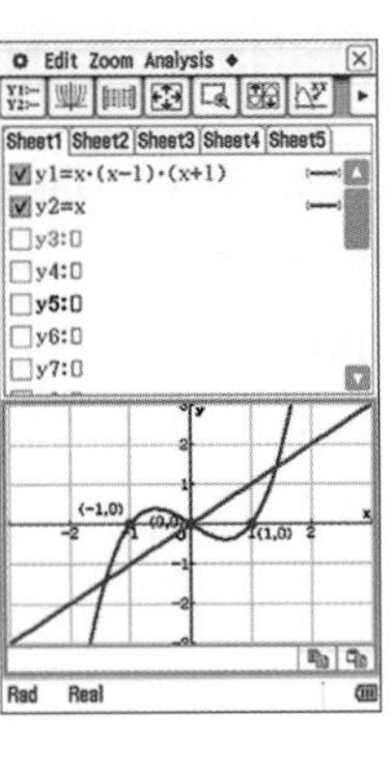

3. To find the points of intersection, select:
- Analysis
- G-Solve
- Intersection

With the cursor on the first point of intersection, press EXE. Use the left or right arrow to move the cursor to the other points of intersection and press EXE.

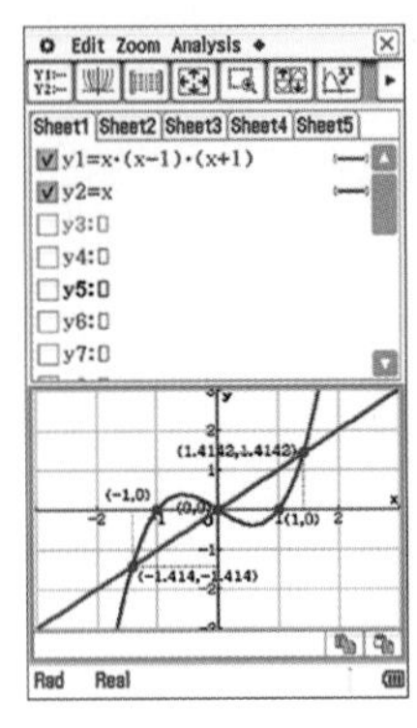

4. On a Main screen, complete the entry line as:
solve $(\{y = x \times (x-1) \times (x+1), y = x\}, \{x, y\})$
then press EXE to find the exact points of intersection.
Note: Make sure to use the multiplication operator between x and the brackets.

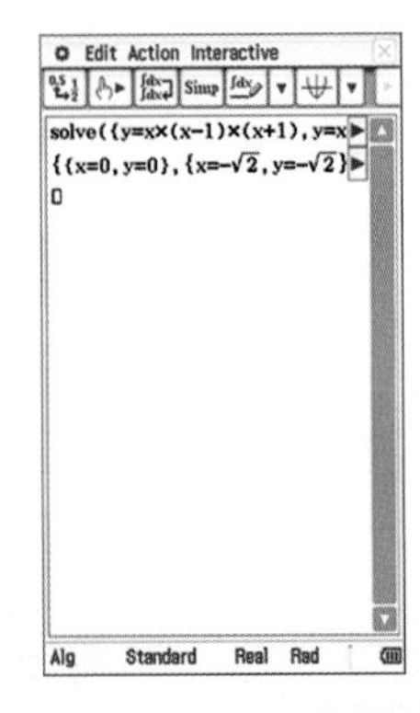

5. Answer the question.

The points of intersection are $(0, 0)$, $\left(\sqrt{2}, \sqrt{2}\right)$ and $\left(-\sqrt{2}, -\sqrt{2}\right)$.

4.5 Exercise

Students, these questions are even better in jacPLUS

Receive immediate feedback and access sample responses

Access additional questions

Track your results and progress

Find all this and MORE in jacPLUS

Technology free

1. WE15 Determine the equation of each of the following graphs.

   a. The graph of a cubic polynomial that has a stationary point of inflection at the point $(3, -7)$ and an x-intercept at $(10, 0)$

   b. 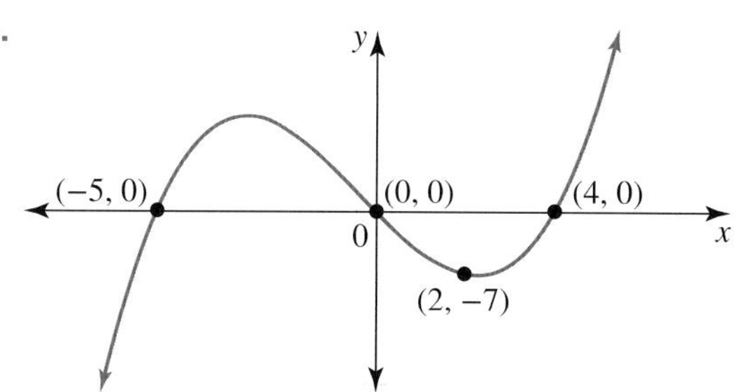

   c. 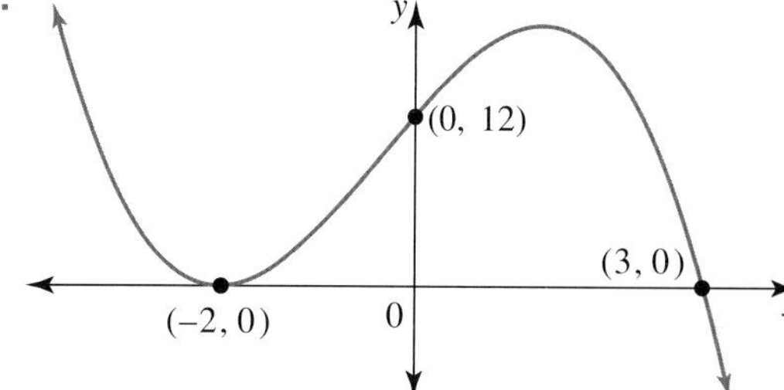

2. a. The graph of a cubic polynomial of the form $y = a(x-h)^3 + k$ has a stationary point of inflection at $(3, 9)$ and passes through the origin. Form the equation of the graph.

   b. The graph of the form $y = a(x-h)^3 + k$ has a stationary point of inflection at $(-2, 2)$ and a y-intercept at $(0, 10)$. Determine the equation.

   c. The graph of the form $y = a(x-h)^3 + k$ has a stationary point of inflection at $(0, 4)$ and passes through the x-axis at $\left(\sqrt[3]{2}, 0\right)$. Determine the equation.

   d. The graph of $y = x^3$ is translated 5 units to the left and 4 units upwards. State its equation after these translations take place.

   e. After the graph $y = x^3$ has been reflected about the x-axis, translated 2 units to the right and translated downwards 1 unit, state what its equation would become.

   f. The graph shown has a stationary point of inflection at $(3, -1)$. Determine its equation.

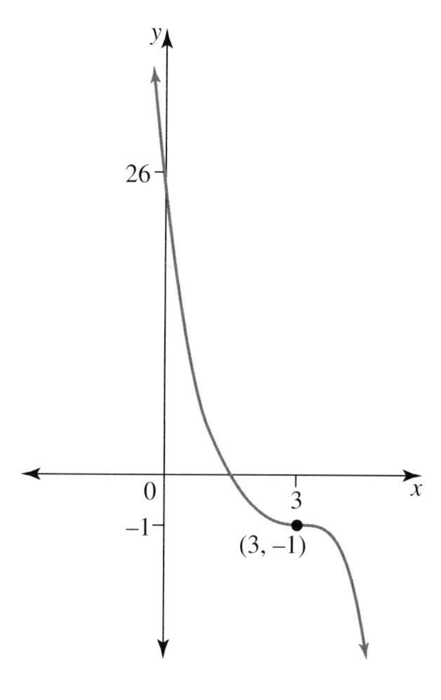

3. Form the equations of the graphs shown.

   a. 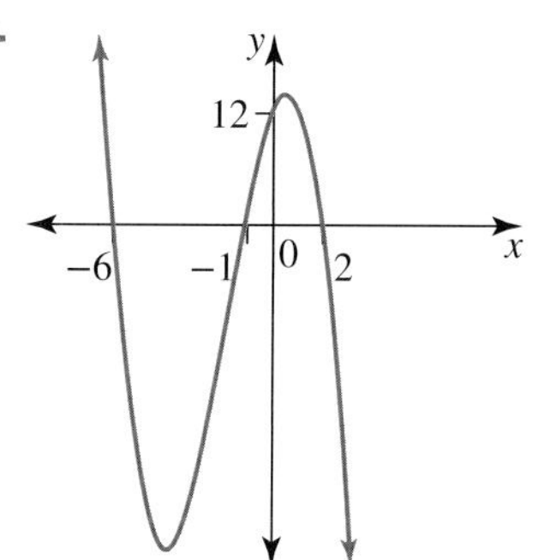

   b. 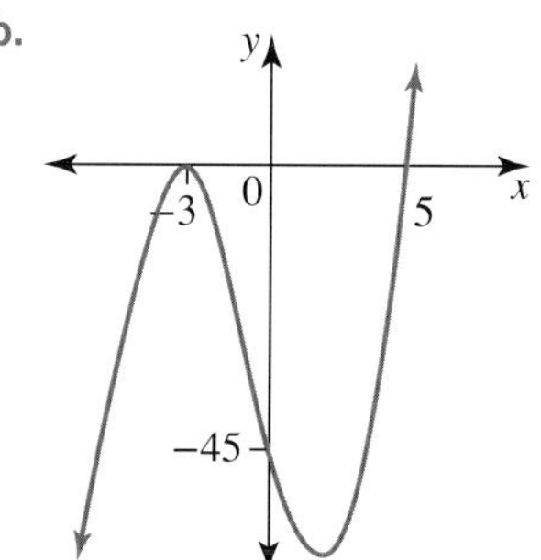

c. Determine the equation that represents the graph shown in the diagram.

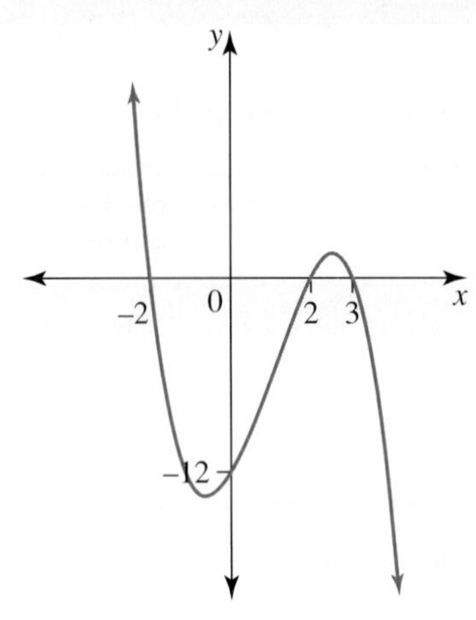

d. Determine the equation of the cubic graph shown and express the equation in polynomial form.

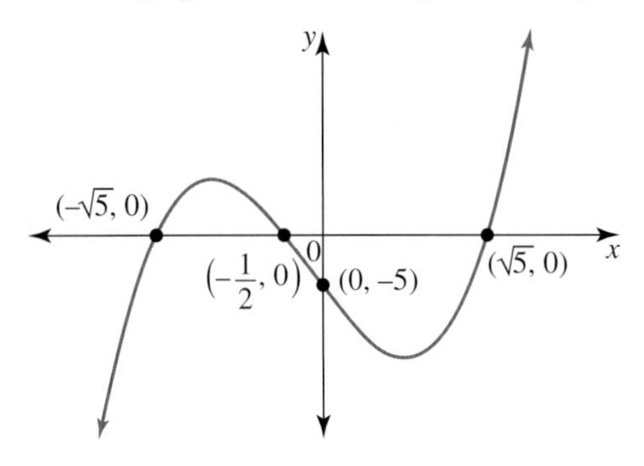

e. Form the equation of the cubic graph that touches the x-axis at the origin, cuts the x-axis at $(2, 0)$ and passes through the point $(-1, 12)$.

f. A cubic graph has x-intercepts at $(-5, 0)$, $\left(\frac{1}{2}, 0\right)$ and $(8, 0)$. The graph crosses the y-axis at $y = 10$. Determine the equation of the graph.

4. Determine the equation for each of the following graphs of cubic polynomials.

a.

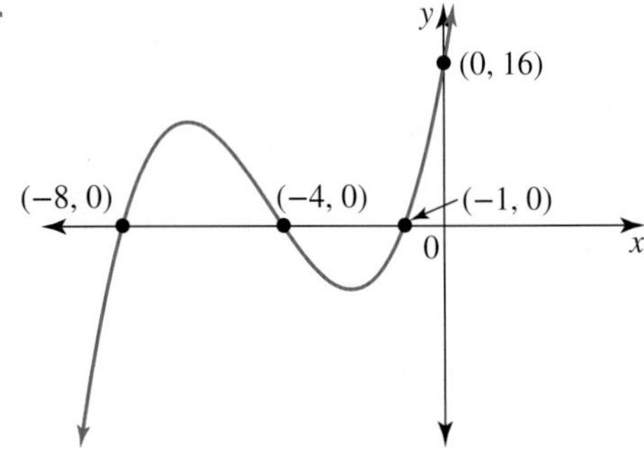

b.

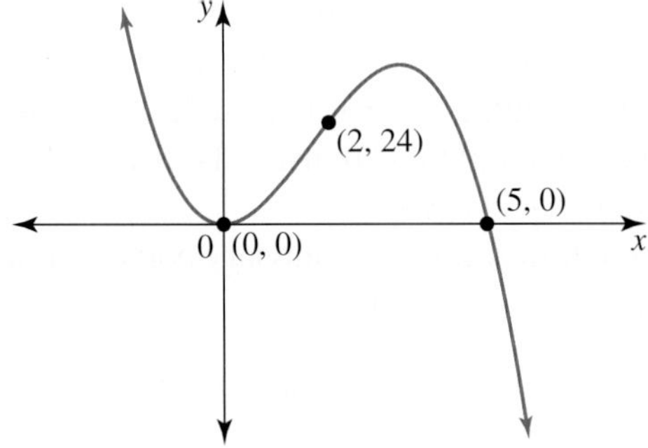

c.

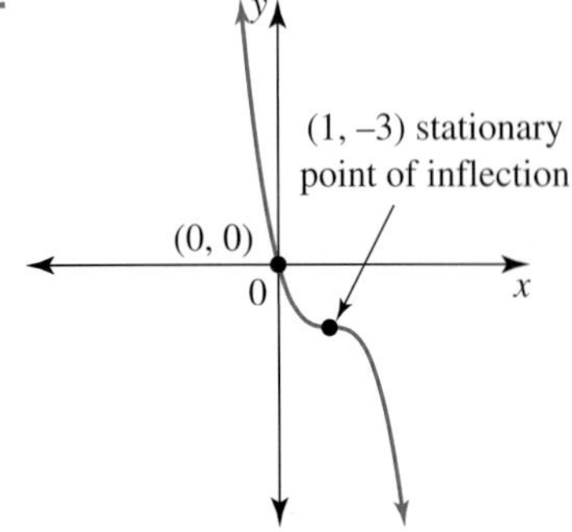

d.

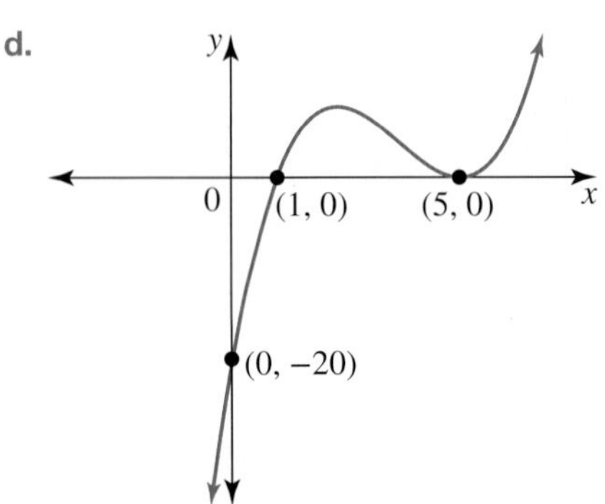

5. a. Give the equation of the graph that has the same shape as $y = -2x^3$ and a point of inflection at $(-6, -7)$.
   b. Calculate the y-intercept of the graph that is created by translating the graph of $y = x^3$ two units to the right and four units down.
   c. A cubic graph has a stationary point of inflection at $(-5, 2)$ and a y-intercept of $(0, -23)$. Calculate its exact x-intercept.
   d. A curve has the equation $y = ax^3 + b$ and contains the points $(1, 3)$ and $(-2, 39)$. Calculate the coordinates of its stationary point of inflection.

6. The equation of the graph shown is $y = (5x + 2)(3x - 1)(1 + 8x)$.

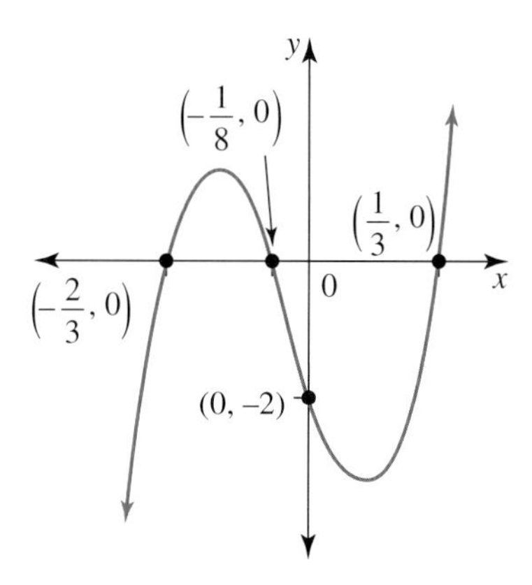

   a. Use the graph to solve the inequation $(5x + 2)(3x - 1)(1 + 8x) \geq 0$.
   b. Draw the sign diagram of the graph.
   c. Use the sign diagram to solve the inequation $(5x + 2)(3x - 1)(1 + 8x) < 0$.

7. **WE16** Solve the following inequations.
   a. $(x - 2)^2 (6 - x) > 0$
   b. $\{x : 4x \leq x^3\}$
   c. $2(x + 4)^3 - 16 < 0$

8. Solve the following cubic inequations.
   a. $(x - 2)(x + 1)(x + 9) \geq 0$
   b. $x^2 - 5x^3 < 0$
   c. $8(x - 2)^3 - 1 > 0$
   d. $x^3 + x \leq 2x^2$
   e. $5x^3 + 6x^2 - 20x - 24 < 0$
   f. $2(x + 1) - 8(x + 1)^3 < 0$

9. Calculate $\{x : 3x^3 + 7 > 7x^2 + 3x\}$.

10. The graph of $y = p(x)$ shown is the reflection of a monic cubic polynomial in the x-axis. The graph touches the x-axis at $x = a$, $a < 0$ and cuts it at $x = b$, $b > 0$.

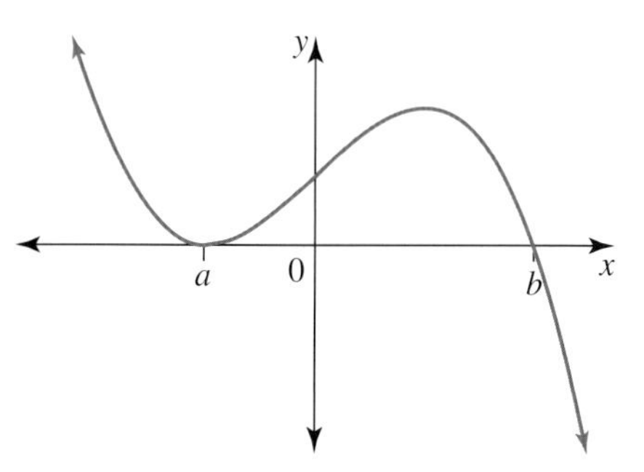

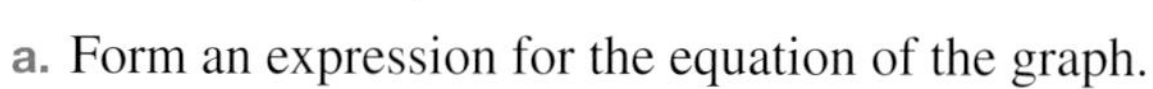

   a. Form an expression for the equation of the graph.
   b. Use the graph to find $\{x : p(x) \geq 0\}$.
   c. Determine how far horizontally to the left the graph would need to be moved so that both of its x-intercepts are negative.
   d. Determine how far horizontally to the right the graph would need to be moved so that both of its x-intercepts are positive.

11. **WE17** Sketch the graphs of $y = (x + 2)(x - 1)^2$ and $y = -3x$, and calculate the coordinates of the points of intersection.

12. Calculate the coordinates of the points of intersection of $y = 4 - x^2$ and $y = 4x - x^3$, and then sketch the graphs on the same axes.

13. a. Determine the coordinates of the points of intersection of the following.
    i. $y = 2x^3$ and $y = x^2$
    ii. $y = 2x^3$ and $y = x - 1$
  b. Illustrate the answers to part **a** with a graph.
  c. Solve the inequation $2x^3 - x^2 \leq 0$ algebraically and explain how you could use your graph from part **b** to solve this inequation.

Technology active

14. a. The number of solutions to the equation $x^3 + 2x - 5 = 0$ can be found by determining the number of intersections of the graphs of $y = x^3$ and a straight line. Determine the equation of this line and how many solutions $x^3 + 2x - 5 = 0$ has.
  b. Use a graph of a cubic and a linear polynomial to determine the number of solutions to the equation $x^3 + 3x^2 - 4x = 0$.
  c. Use a graph of a cubic and a quadratic polynomial to determine the number of solutions to the equation $x^3 + 3x^2 - 4x = 0$.
  d. Solve the equation $x^3 + 3x^2 - 4x = 0$.

15. a. Show the line $y = 3x + 2$ is a tangent to the curve $y = x^3$ at the point $(-1, -1)$.
  b. Determine the coordinates of the point where the line cuts the curve.
  c. Sketch the curve and its tangent on the same axes.
  d. Investigate for what values of m will the line $y = mx + 2$ have one, two or three intersections with the curve $y = x^3$.

16. Use simultaneous equations to determine the equation of the cubic graph containing the points $(0, 3), (1, 4), (-1, 8), (-2, 7)$.

17. A graph of a cubic polynomial with equation $y = x^3 + ax^2 + bx + 9$ has a turning point at $(3, 0)$.
  a. State the factor of the equation with greatest multiplicity.
  b. Determine the other x-intercept.
  c. Calculate the values of a and b.

18. The graph with equation $y = (x + a)^3 + b$ passes through the three points $(0, 0), (1, 7), (2, 26)$.
  a. Use this information to determine the values of a and b.
  b. Determine points of intersection of the graph with the line $y = x$.
  c. Sketch both graphs in part **b** on the same axes.
  d. Hence, with the aid of the graphs determine, $\{x : x^3 + 3x^2 + 2x > 0\}$.

19. The graph of a polynomial of degree 3 cuts the x-axis at $x = 1$ and at $x = 2$. It cuts the y-axis at $y = 12$.
  a. Explain why this is insufficient information to completely determine the equation of the polynomial.
  b. Show that this information identifies a family of cubic polynomials with equation $y = ax^3 + (6 - 3a)x^2 + (2a - 18)x + 12$.
  c. On the same graph, sketch the two curves in the family for which $a = 1$ and $a = -1$.
  d. Determine the equation of the curve for which the coefficient of x^2 is 15. Specify the x-intercepts and sketch this curve.

20. a. Write down, to 2 decimal places, the coordinates of the points of intersection of $y = (x + 1)^3$ and $y = 4x + 3$.
  b. Form the cubic equation $ax^3 + bx^2 + cx + d = 0$ for which the x-coordinates of the points of intersection obtained in part a are the solution.
  c. State what feature of the graph of $y = ax^3 + bx^2 + cx + d$ the x-coordinates of these points of intersection would be.

4.5 Exam questions

Question 1 (1 mark) TECH-ACTIVE

MC The equation of the graph shown could be

A. $y = (x+1)^3 - 5$
B. $y = 2(x+1)^3 - 5$
C. $y = \frac{1}{2}(x+1)^3 - 5$
D. $y = (x-1)^3 - 5$
E. $y = (x+1)^3 - 3$

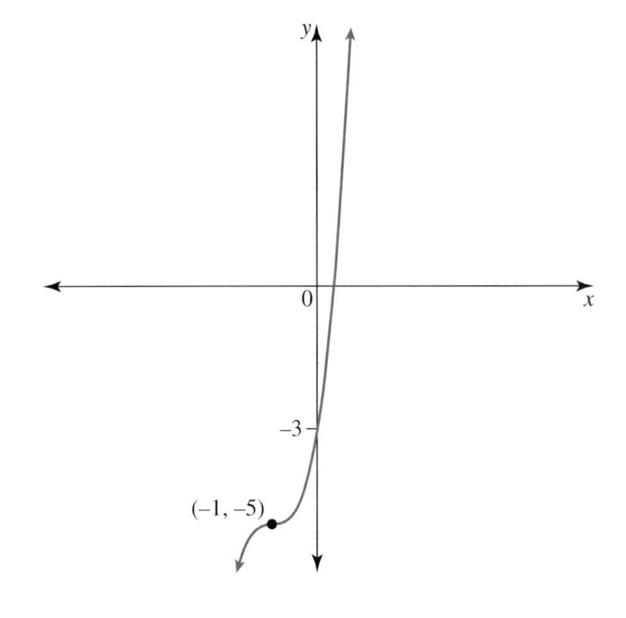

Question 2 (1 mark) TECH-ACTIVE

MC The equation of the cubic function that has the sign diagram shown below is

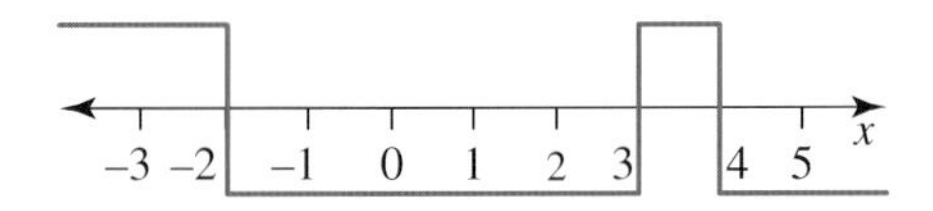

A. $y = (x+2)(x-3)(x-4)$
B. $y = (x-2)(x+3)(x+4)$
C. $y = (x+2)^2(x-3)(x+4)$
D. $y = -(x+2)(x-3)(x-4)$
E. $y = (x+2)^2(x-3)(x+4)$

Question 3 (3 marks) TECH-FREE

Determine the equation of the graph below in both of the forms $y = a(x-b)(x-c)(x-d)$ and $y = ax^3 + bx^2 + cx + d$.

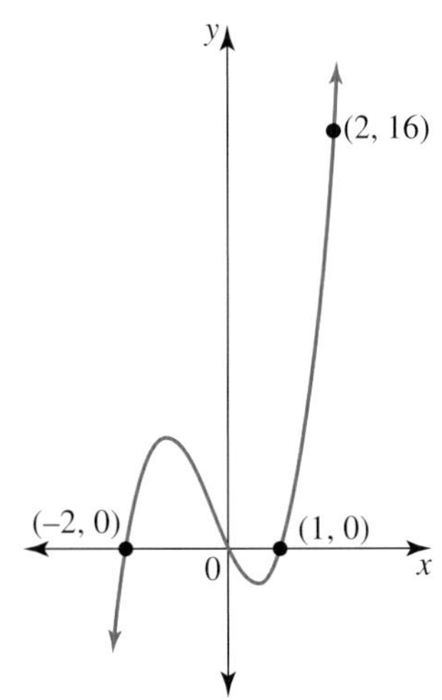

More exam questions are available online.

4.6 Cubic models and applications

LEARNING INTENTION

At the end of this subtopic you should be able to:
- use cubic polynomials as models to solve real-world problems.

Practical situations that use cubic polynomials as models are likely to require a restriction of the possible values the variable may take. This is called a **domain** restriction. The domain is the set of possible values of the variable that the polynomial may take. We shall look more closely at domains in later chapters.

4.6.1 Maximum and minimum values and cubic models

Similar methods to modelling with linear and quadratic relationships are used when using cubic polynomials, with the polynomial model expressed in terms of one variable.

In setting up equations:
- define the symbols used for the variables, specifying units where appropriate
- ensure any units used are consistent and lengths of objects are positive
- determine the links between the variables
- check whether mathematical solutions are feasible in the context of the problem
- express answers in the context of the problem.

Applications of cubic models where a maximum or minimum value of the model is sought will require identification of turning point coordinates. In a later chapter we will see how this is done. For now, obtaining turning points may require the use of CAS technology.

WORKED EXAMPLE 18 Determining cubic models

A rectangular storage container is designed to have an open top and a square base.

The base has side length x cm and the height of the container is h cm. The sum of its dimensions (the sum of the length, width and height) is 48 cm.

a. Express h in terms of x.
b. Show that the volume, V cm^3, of the container is given by $V = 48x^2 - 2x^3$.
c. State any restrictions on the values x can take.
d. Sketch the graph of V against x for appropriate values of x, given its maximum turning point has coordinates $(16, 4096)$.
e. Calculate the dimensions of the container with the greatest possible volume.

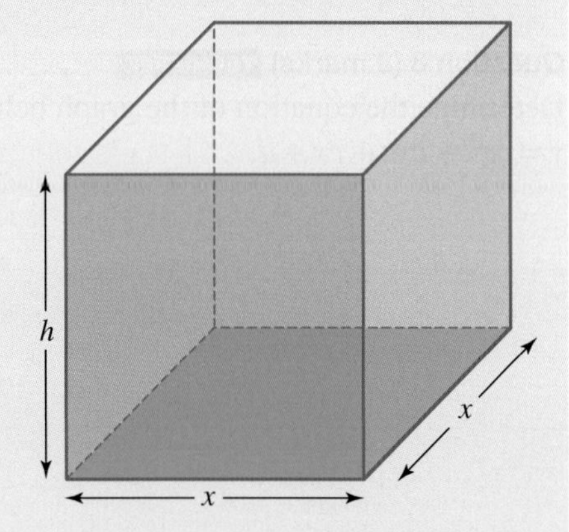

THINK	WRITE
a. Write the given information as an equation connecting the two variables.	a. The sum of the dimensions is 48 cm. $x + x + h = 48$ $h = 48 - 2x$

b. Use the result from part **a** to express the volume in terms of one variable and prove the required statement.

b. The formula for volume of a cuboid is

$V = lwh$

$\therefore V = x^2 h$

Substitute $h = 48 - 2x$.

$V = x^2(48 - 2x)$

$\therefore V = 48x^2 - 2x^3$, as required

c. State the restrictions.
Note: Lengths of objects are positive.

c. Length cannot be negative or zero, so $x > 0$.
Height cannot be negative or zero, so $h > 0$.

$48 - 2x > 0$

$-2x > -48$

$\therefore x < 24$

Hence, the restriction is $0 < x < 24$.

d. Draw the cubic graph but only show the section of the graph for which the restriction applies. Label the axes with the appropriate symbols and label the given turning point.

d. $V = 48x^2 - 2x^3$

$= 2x^2(24 - x)$

x-intercepts: let $V = 0$.

$x^2 = 0$ or $24 - x = 0$

$\therefore x = 0$ (touch), $x = 24$ (cut)

$(0, 0), (24, 0)$ are the x-intercepts.

This is a negative cubic.
Maximum turning point $(16, 4096)$
Draw the section for which $0 < x < 24$.

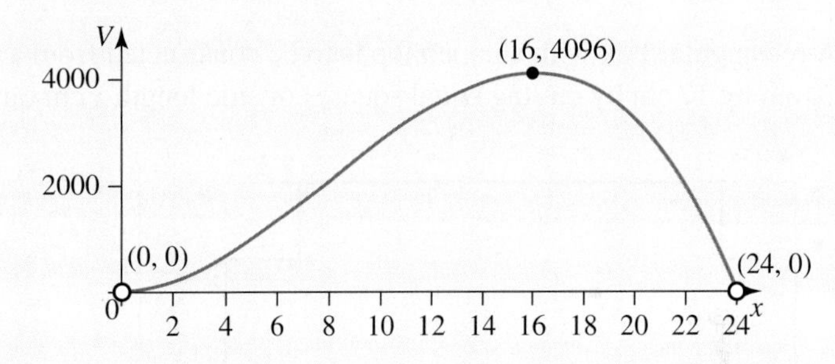

e. 1. Calculate the required dimensions.
Note: The maximum turning point (x, V) gives the maximum value of V and the value of x when this maximum occurs.

2. State the answer.

e. The maximum turning point is $(16, 4096)$. This means the greatest volume is $4096\ \text{cm}^3$. It occurs when $x = 16$.

$\therefore\ h = 48 - 2(16) \Rightarrow h = 16$

Dimensions: length $= 16$ cm, width $= 16$ cm, height $= 16$ cm

The container has the greatest volume when it is a cube of edge 16 cm.

4.6 Exercise

Students, these questions are even better in jacPLUS

 Receive immediate feedback and access sample responses

 Access additional questions

 Track your results and progress

Find all this and MORE in jacPLUS

Technology active

1. **WE18** A rectangular storage container is designed to have an open top and a square base.

   The base has side length x metres and the height of the container is h metres. The total length of its 12 edges is 6 metres.

   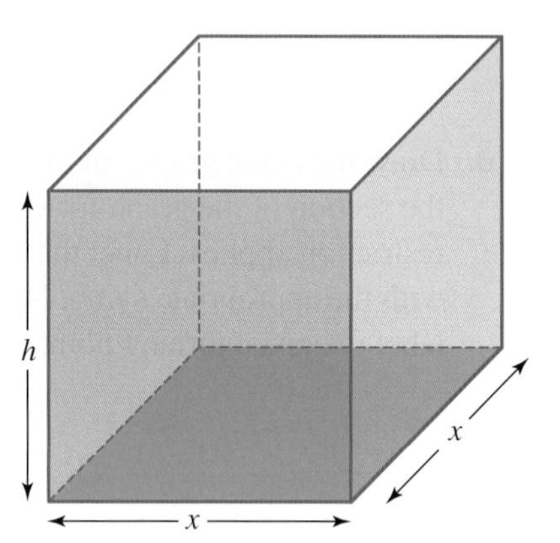

   a. Express h in terms of x.
   b. Show that the volume, $V\text{ m}^3$, of the container is given by $V = 1.5x^2 - 2x^3$.
   c. State any restrictions on the values x can take.
   d. Sketch the graph of V against x for appropriate values of x, given its maximum turning point has coordinates $(0.5, 0.125)$.
   e. Calculate the dimensions of the container with the greatest possible volume.

2. A rectangular box with an open top is to be constructed from a rectangular sheet of cardboard measuring 20 cm by 12 cm by cutting equal squares of side length x cm out of the four corners and folding the flaps up.

   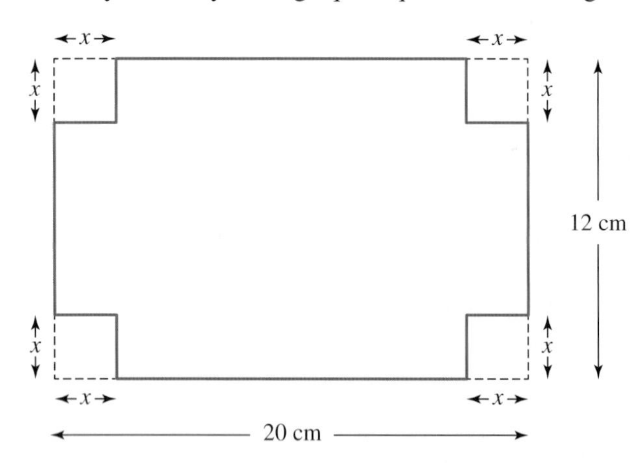

   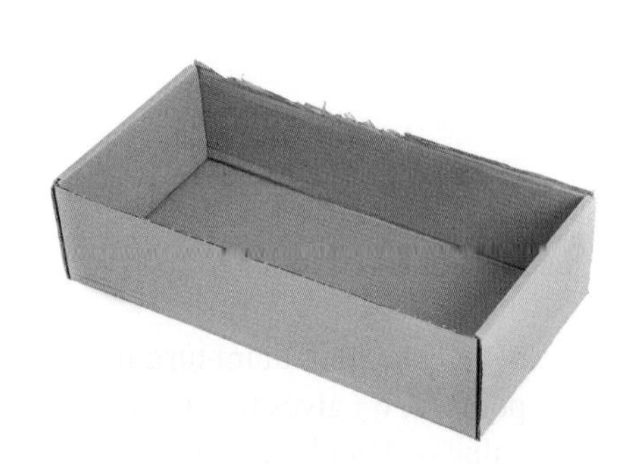

   The box has length l cm, width w cm and volume $V\text{ cm}^3$.

   a. Express l and w in terms of x, and hence express V in terms of x.
   b. State any restrictions on the values of x.
   c. Sketch the graph of V against x for appropriate values of x, given the unrestricted graph would have turning points at $x = 2.43$ and $x = 8.24$.
   d. Calculate the length and width of the box with maximum volume, and give this maximum volume to the nearest whole number.

3. The cost, C dollars, for an artist to produce x sculptures by contract is given by $C = x^3 + 100x + 2000$. Each sculpture is sold for \$500 and as the artist only makes the sculptures by order, every sculpture produced will be paid for. However, too few sales will result in a loss to the artist.
   a. Show the artist makes a loss if only five sculptures are produced and a profit if six sculptures are produced.
   b. Show that the profit, P dollars, from the sale of x sculptures is given by $P = -x^3 + 400x - 2000$.
   c. Explain what will happen to the profit if a large number of sculptures are produced and why this effect occurs.
   d. Calculate the profit (or loss) from the sale of:
      i. 16 sculptures
      ii. 17 sculptures.
   e. Use the above information to sketch the graph of the profit P for $0 \le x \le 20$. Place its intersection with the x-axis between two consecutive integers, but don't attempt to obtain its actual x-intercepts.
   f. State how many sculptures the artist should produce in order to guarantee a profit is made.

4. The number of bacteria in a slow-growing culture at time t hours after 9 am is given by $N = 54 + 23t + t^3$.
   a. Calculate the initial number of bacteria at 9 am.
   b. Calculate how long it takes for the initial number of bacteria to double.
   c. Calculate how many bacteria there are by 1 pm.
   d. Once the number of bacteria reaches 750, the experiment is stopped. Find the time of the day at which this happens.

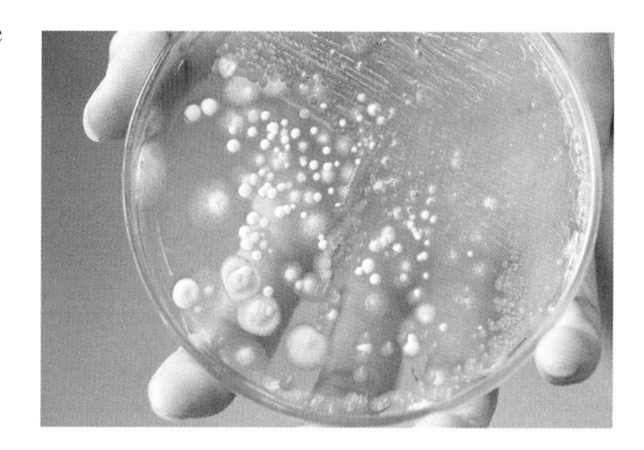

5. Engineers are planning to build an underground tunnel through a city to ease traffic congestion. The cross-section of their plan is bounded by the curve shown.

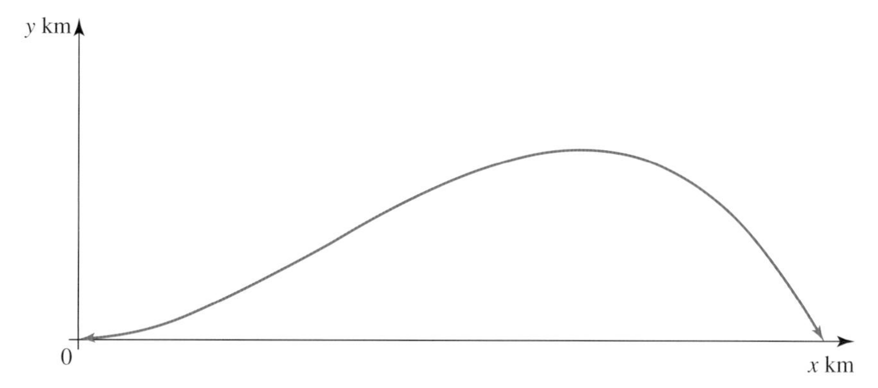

The equation of the bounding curve is $y = ax^2(x - b)$ and all measurements are in kilometres.
It is planned that the greatest breadth of the bounding curve will be 6 km and the greatest height will be 1 km above this level at a point 4 km from the origin.

a. Determine the equation of the bounding curve.
b. If the greatest breadth of the curve was extended to 7 km, determine the greatest height of the curve above this new lowest level.

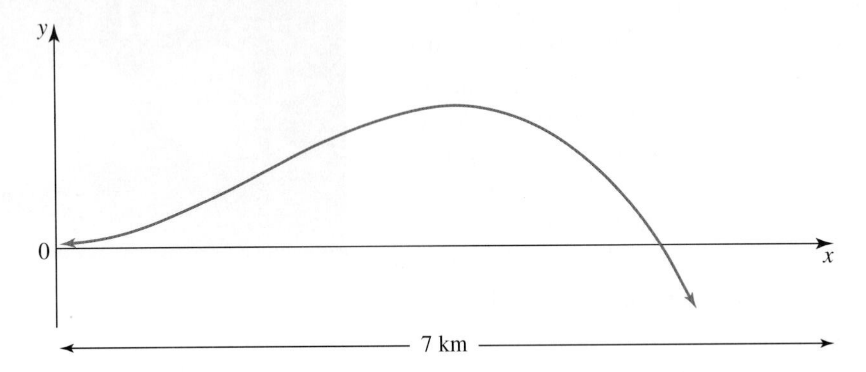

6. Determine the smallest positive integer and the most negative integer for which the difference between the square of 5 more than this number and the cube of 1 more than the number exceeds 22.

7. A tent used by a group of bushwalkers is in the shape of a square-based right pyramid with a slant height of 8 metres.

For the figure shown, let OV, the height of the tent, be h metres and the edge of the square base be $2x$ metres.

a. Use Pythagoras' theorem to express the length of the diagonal of the square base of the tent in terms of x.
b. Use Pythagoras' theorem to show $2x^2 = 64 - h^2$.

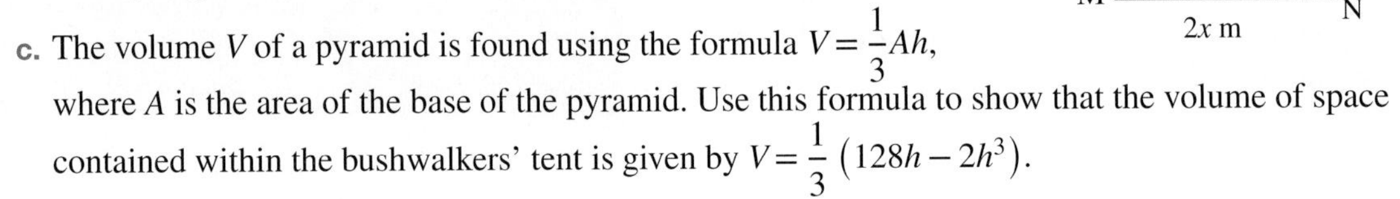

c. The volume V of a pyramid is found using the formula $V = \frac{1}{3}Ah$, where A is the area of the base of the pyramid. Use this formula to show that the volume of space contained within the bushwalkers' tent is given by $V = \frac{1}{3}\left(128h - 2h^3\right)$.
d. i. If the height of the tent is 3 metres, calculate the volume.
ii. Determine the values for the height that this mathematical model allows.
e. Use CAS technology to sketch the graph of $V = \frac{1}{3}\left(128h - 2h^3\right)$ and state the height for which the volume is greatest, correct to 2 decimal places.
f. Show that the greatest volume occurs when the height is half the length of the base.

8. A cylindrical storage container is designed so that it is open at the top and has a surface area of 400π cm^2. Its height is h cm and its radius is r cm.

a. Show that $h = \frac{400 - r^2}{2r}$.
b. Show that the volume, V cm^3, that the container can hold is given by $V = 200\pi r - \frac{1}{2}\pi r^3$.
c. State any restrictions on the values r can take.
d. Sketch the graph of V against r for appropriate values of r.
e. Determine the radius and height of the container if the volume is 396π cm^3.
f. Use CAS technology to obtain the maximum volume and calculate the corresponding dimensions of the cylindrical storage container, expressed to 1 decimal place.

9. A new playground slide for children is to be constructed at a local park. At the foot of the slide the children climb a vertical ladder to reach the start of the slide. The slide must start at a height of 2.1 metres above the ground and end at a point 0.1 metres above the ground and 4 metres horizontally from its foot. A model for the slide is

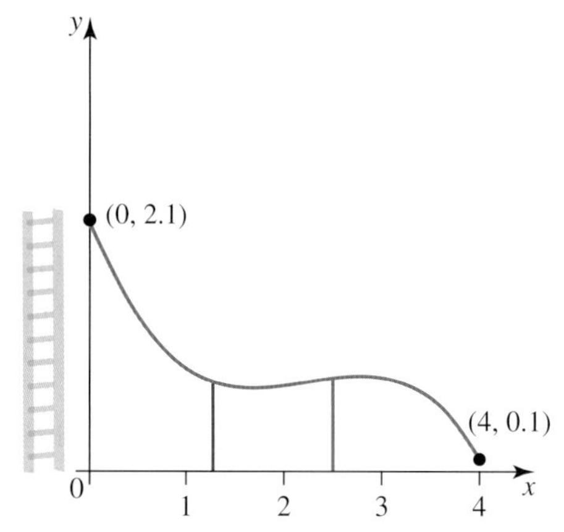

$$h = ax^3 + bx^2 + cx + d$$

where h metres is the height of the slide above ground level at a horizontal distance of x metres from its foot. The foot is at the origin.

The ladder supports the slide at one end, and the slide also requires two vertical struts as support. One strut of length 1 metre is placed at a point 1.25 metres horizontally from the foot of the slide and the other is placed at a point 1.5 metres horizontally from the end of the slide and is of length 1.1 metres.

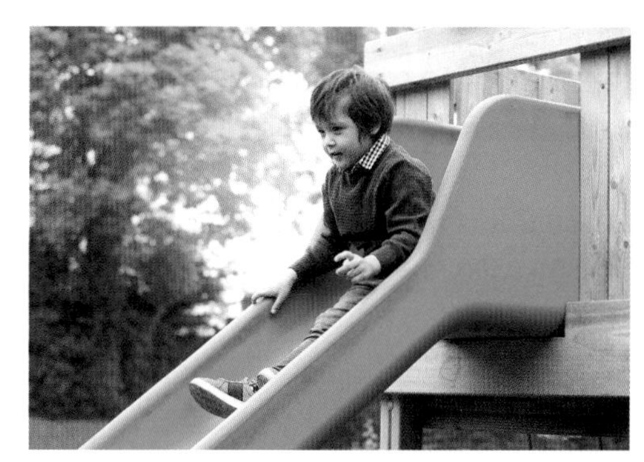

a. Give the coordinates of 4 points that lie on the cubic graph of the slide.
b. State the value of d in the equation of the slide.
c. Form a system of three simultaneous equations, the solutions to which give the coefficients a, b, c in the equation of the slide.
d. Use technology to obtain the equation of the slide.

10. Since 1994, the world record times for the men's 100-m sprint can be roughly approximated by the cubic model $T(t) = -0.000\,05\,(t-6)^3 + 9.85$, where T is the time in seconds and t is the number of years since 1994.

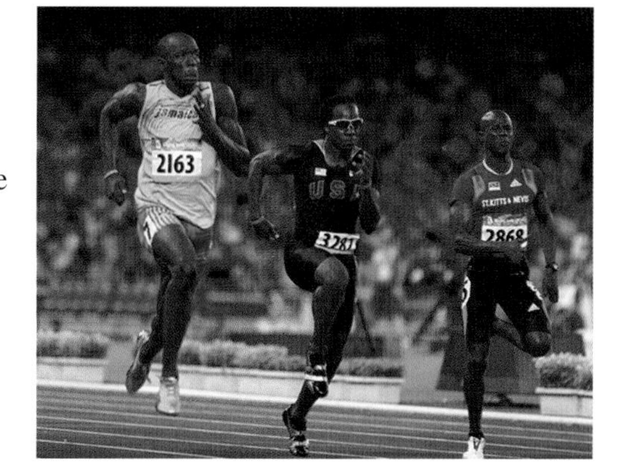

a. In 1997 the world record was 9.86 seconds and in 2014 the record was 9.72 seconds. Compare these times with those predicted by the cubic model.
b. Sketch the graph of T versus t from 1994 to 2014.
c. State what the model predicts for 2022 and explain whether the model is likely to be a good predictor beyond 2022.

11. A rectangle is inscribed under the parabola $y = 9 - (x-3)^2$ so that two of its corners lie on the parabola and the other two lie on the x-axis at equal distances from the intercepts the parabola makes with the x-axis.

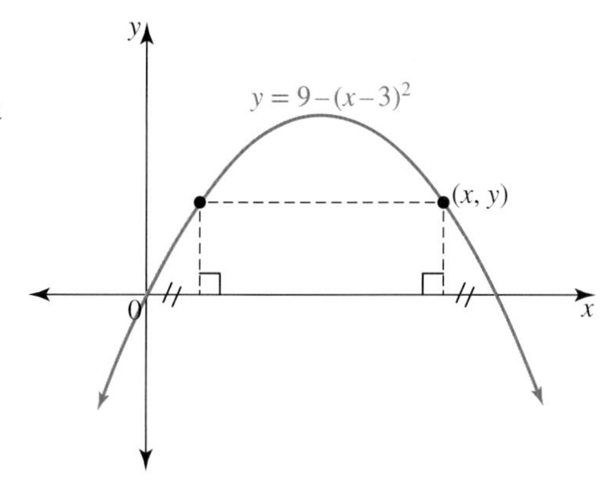

a. Calculate the x-intercepts of the parabola.
b. Express the length and width of the rectangle in terms of x.
c. Hence, show that the area of the rectangle is given by $A = -2x^3 + 18x^2 - 36x$.
d. State the values of x for which this is a valid model of the area.
e. Calculate the value(s) of x for which $A = 16$.
f. Use CAS technology to calculate, to 3 decimal places, the length and width of the rectangle that has the greatest area.

12. A pathway through the countryside passes through five scenic points. Relative to a fixed origin, four of these points have coordinates A $(-3, 0)$, B $\left(-\sqrt{3}, -12\sqrt{3}\right)$, C $\left(\sqrt{3}, 12\sqrt{3}\right)$, D $(3, 0)$; the fifth scenic point is the origin, O $(0, 0)$. The two-dimensional shape of the path is a cubic polynomial.

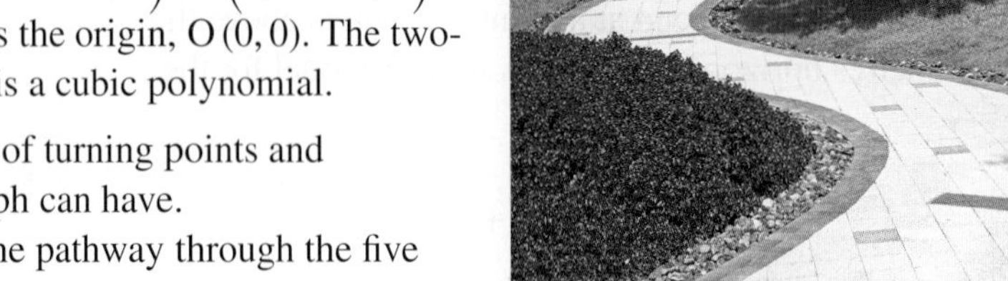

   a. State the maximum number of turning points and x-intercepts that a cubic graph can have.
   b. Determine the equation of the pathway through the five scenic points.
   c. Sketch the path, given that points B and C are turning points of the cubic polynomial graph.
   d. It is proposed that another pathway be created to link B and C by a direct route. Show that if a straight-line path connecting B and C is created, it will pass through O, and give the equation of this line.
   e. An alternative plan is to link B and C by a cubic path that has a stationary point of inflection at O. Determine the equation of this path.

4.6 Exam questions

Question 1 (1 mark) TECH-ACTIVE

MC An expression for the volume of the cylinder shown, with radius $2x$ and length $3x$, would be

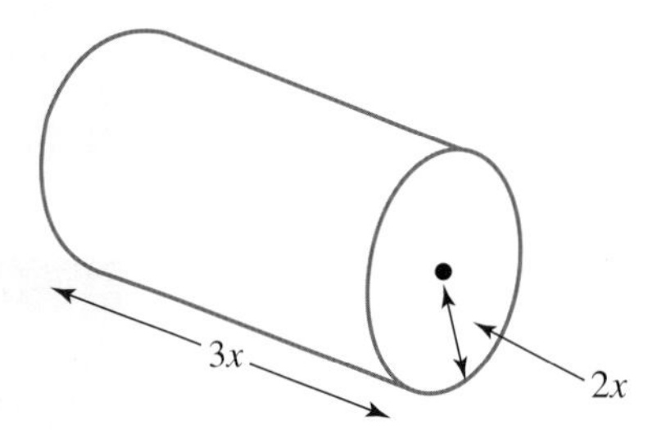

A. $4\pi x^3$ B. $12\pi x^3$ C. $4\pi x^2 + 3x$ D. $3r^2 + \pi r^3$ E. $6\pi x^3$

Question 2 (1 mark) TECH-ACTIVE

MC A rectangular shaped box with a square base has a volume of $60\,\text{cm}^3$. The base of the box has a width 2 cm larger than the height. If x is the height of the box, an expression for the volume is

A. $x^3 + 2x^2 + 2x - 60 = 0$
B. $60 - 2x^3 - 2x^2 - 2x = 0$
C. $x^3 + 4x^2 + 4x - 60 = 0$
D. $x^3 + 4x^2 + 4x + 60 = 0$
E. $60 - x^3 + 4x^2 + 4x = 0$

Question 3 (1 mark) TECH-ACTIVE

MC The volume of a container is given by $V = 2.7x^2 - 3x^3$. The restrictions on the values that x can take are

A. $x > 0.9$
B. $x > 0$
C. $0 \leq x \leq 0.9$
D. $0 < x < 0.9$
E. $x < 0.9$

More exam questions are available online.

4.7 Review

4.7.1 Summary

doc-37019

Hey students! Now that it's time to revise this topic, go online to:

Access the topic summary

Review your results

Watch teacher-led videos

Practise exam questions

Find all this and MORE in jacPLUS

4.7 Exercise

Technology free: short answer

1. Factorise $p(x) = x^3 + 5x^2 + 3x - 9$ into linear factors.
2. The polynomial $p(x) = x^3 - ax^2 + bx - 3$ leaves a remainder of 2 when it is divided by $(x - 1)$ and a remainder of -4 when it is divided by $(x + 1)$. Calculate the values of a and b.
3. Divide $(2x^3 - 3x^2 + x - 1)$ by $(x + 2)$ and state the quotient and the remainder.
4. Sketch the following graphs.
   a. $y = 8 - (x + 3)^3$
   b. $y = -2(4 - x)^2(5 + x)$
   c. $y = (8x - 3)^3$
   d. $y = 2x^3 - x$
5. Sketch the graph of $y = -x^3 + 6x^2 - 11x + 6$ and hence, or otherwise, solve the inequation $x^3 - 6x^2 + 11x - 6 < 0$.
6. Calculate the coordinates of the points of intersection of $y = 4x - x^3$ and $y = -2x$.

Technology active: multiple choice

7. **MC** State which of the following expressions is a polynomial.
   A. $x^3 + 4x - 7x^{-1} + x^{-3}$
   B. $\left(2\sqrt{x} + 7\right)^3$
   C. $\dfrac{4x - 7x^5}{x^3}$
   D. $2^{3x} + 2^{2x} + 2^x + 2$
   E. $\left(\sqrt{3} - 5x\right)^3$
8. **MC** If $x^3 - 2x^2 - 3x + 10 \equiv (x + 2)(ax^2 + bx + c)$, then the values of a, b and c are, respectively:
   A. $1, -2, 5$
   B. $1, 0, 5$
   C. $-2, -3, 10$
   D. $1, -4, 5$
   E. $-2, -4, 10$
9. **MC** When the polynomial $p(x)$ is divided by $(3x + 6)$, the remainder is:
   A. $p(-6)$
   B. $p(-3) + p(-2)$
   C. $\frac{1}{3}p(-2)$
   D. $3p(-2)$
   E. $p(-2)$
10. **MC** If $p(x) = 3 + kx - 5x^2 + 2x^3$ and $p(-1) = 8$, then k is equal to:
    A. 0
    B. 4
    C. -4
    D. 12
    E. -12

11. **MC** A possible equation for the curve shown is:

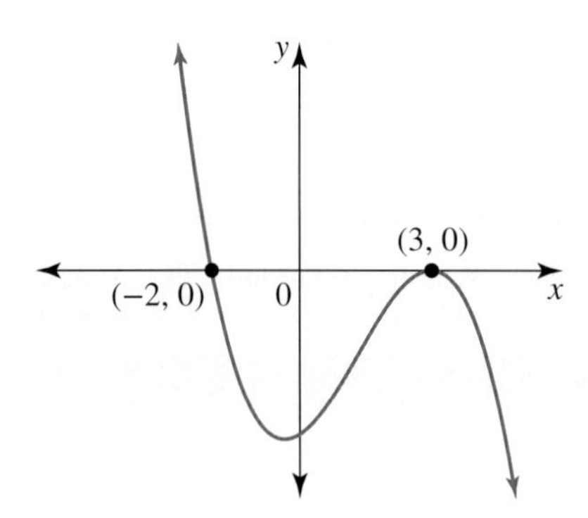

A. $y=(x-2)(x+3)^2$
B. $y=(2-x)(x-3)^2$
C. $y=-(x+2)(x-3)^2$
D. $y=-(x+2)^2(x-3)$
E. $y=(x+2)(x-3)^2$

12. **MC** The graph with the equation $y=2x^3$ is translated 2 units horizontally to the right and 3 units vertically down. The equation of the graph becomes:

A. $y=2(x-2)^3-3$
B. $y=2(x+2)^3-3$
C. $y=2(x-2)^3+3$
D. $y=2(x+2)^3+3$
E. $y=(2x-2)^3-3$

13. **MC** $\dfrac{3-4x}{x+1}$ is equal to:

A. $-4+\dfrac{7}{x+1}$
B. $1-\dfrac{4}{x+1}$
C. $-4+\dfrac{1}{x+1}$
D. $7-\dfrac{4}{3-4x}$
E. $-1+\dfrac{4}{3-4x}$

14. **MC** The solution to the inequation $(x+4)^3>-1$ is:

A. $x>-1$
B. $x>-4$
C. $x>-5$
D. $x<-3$
E. $-7<x<-5$ or $x>1$

15. **MC** The solutions to the equation $2x^3=14x^2+16x$ are:

A. $x=-1, x=8$
B. $x=-1, x=2, x=8$
C. $x=-8, x=2, x=7$
D. $x=-8, x=0, x=7$
E. $x=-1, x=0, x=8$

16. **MC** The graph shown cuts the x-axis at $x=a, x=b$ and $x=4.5$.

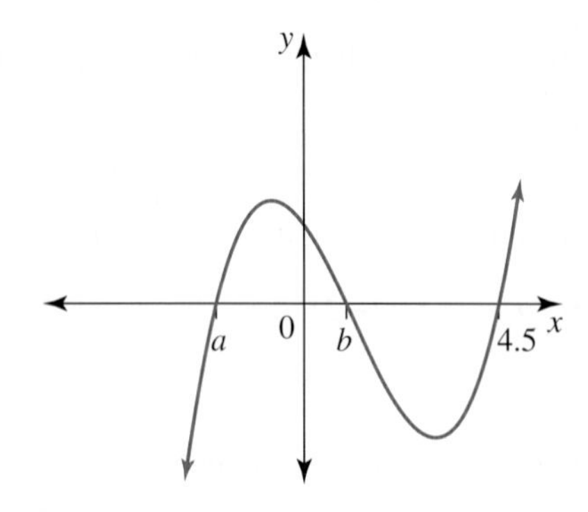

A possible equation for the graph is:

A. $y=(x+a)(x-b)(x-4.5)$
B. $y=-(x-a)(x+b)(x-4.5)$
C. $y=(x-a)(x-b)(2x-9)$
D. $y=(x+a)(x-b)(2x-9)$
E. $y=-(a-x)(x+b)(x+4.5)$

Technology active: extended response

17. Consider the cubic polynomial $p(x) = 8x^3 - 34x^2 + 33x - 9$.
    a. Show that $(x - 3)$ is a factor of $p(x)$.
    b. Hence, completely factorise $p(x)$.
    c. The graph of the polynomial $y = p(x) = 8x^3 - 34x^2 + 33x - 9$ has turning points at $(0.62,\ 0.3)$ and $(2.2, -15.8)$. Sketch the graph, labelling all key points with their coordinates.
    d. Specify $\{x : p(x) \geq 0\}$.
    e. Calculate $\{x : p(x) = -9\}$.
    f. Determine the values of k for which the line $y = k$ will intersect the graph of $y = p(x)$ in:
       i. 3 places
       ii. 2 places
       iii. 1 place.

18. The revenue ($) from the sale of x thousand items is given by $r(x) = 6(2x^2 + 10x + 3)$ and the manufacturing cost ($) of x thousand items is $c(x) = x(6x^2 - x + 1)$.
    a. State the degree of $r(x)$ and of $c(x)$.
    b. Calculate the revenue and the cost if 1000 items are sold and explain whether a profit is made.
    c. Show that the profit ($) from the sale of x thousand items is given by $p(x) = -6x^3 + 13x^2 + 59x + 18$.
    d. Given the graph of $y = -6x^3 + 13x^2 + 59x + 18$ cuts the x-axis at $x = -2$, sketch the graph of $y = p(x)$ for appropriate values of x.
    e. If a loss occurs when the number of items manufactured is d, state the smallest value of d.

19. Relative to a reference point O, two towns A and B are located at the points $(1, 20)$ and $(5, 12)$ respectively. A freeway passing through A and B can be considered to be a straight line.
    a. Determine the equation of the line modelling the freeway.

    Before the freeway was built, the road between A and B followed a scenic route modelled by the equation $y = a(2x - 1)(x - 6)(x + b)$ for $0 \leq x \leq 8$.

    b. Using the fact this road goes through towns A and B, show that $a = \frac{2}{3}$ and $b = -7$.
    c. Determine the coordinates of the end points where the scenic route starts and finishes.
    d. On the same diagram, sketch the scenic route and the freeway. Any end points and intercepts with the axes should be given, and the positions of the points A and B should be marked on your graph.
    e. The freeway meets the scenic route at three places. Calculate the coordinates of these three points.
    f. Find which of the three points found in part **e** is closest to the reference point O.

20. The slant height of a right conical tent has a length of 13 metres. For the figure shown, O is the centre of the circular base of radius r metres. OV, the height of the tent, is h metres.

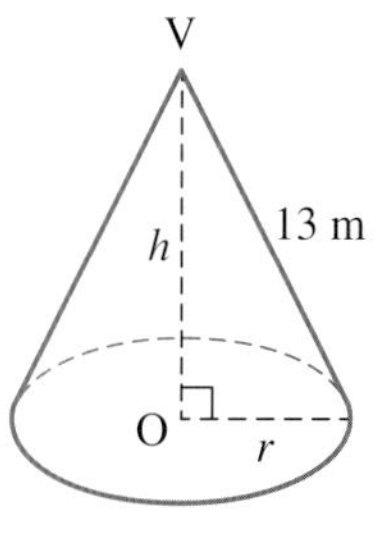

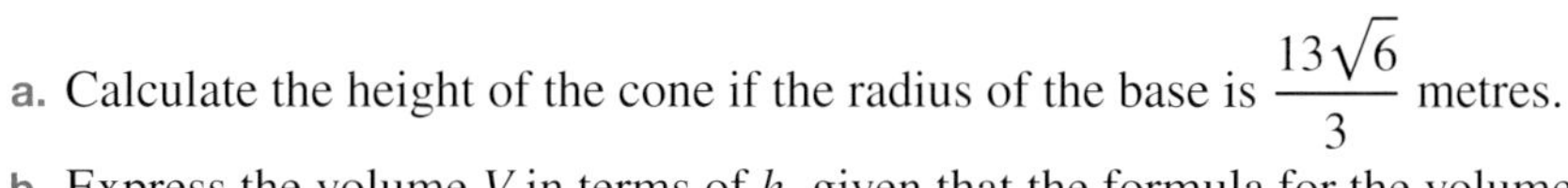

    a. Calculate the height of the cone if the radius of the base is $\frac{13\sqrt{6}}{3}$ metres.

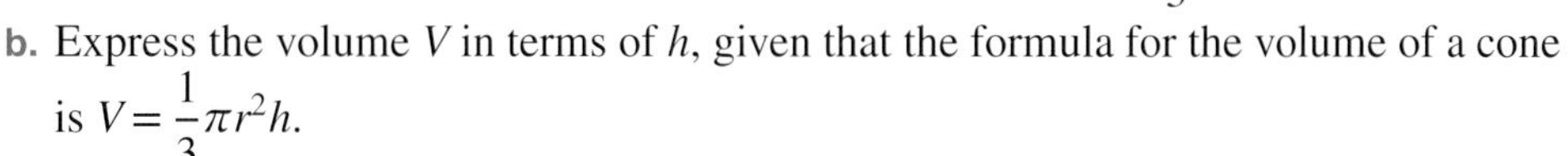

    b. Express the volume V in terms of h, given that the formula for the volume of a cone is $V = \frac{1}{3}\pi r^2 h$.
    c. State any restrictions on the values h can take, and sketch the graph of V against h for these restrictions.
    d. Express the volume as multiples of π for $h = 7,\ h = 8,\ h = 9$, and hence obtain the integer a so that the greatest volume occurs when $a < h < a + 1$.
    e. i. Using the midpoint of the interval $[a, a + 1]$ as an estimate for h, calculate r.
       ii. Use the estimates for h and r to calculate an approximate value for the maximum volume, to the nearest whole number.
    f. i. It can be shown that the greatest volume occurs when $r = \sqrt{2}h$. Use this information to calculate the height and radius that give the greatest volume.
       ii. Specify the greatest volume to the nearest whole number and compare this value with the approximate value obtained in part **e**.

4.7 Exam questions

Question 1 (3 marks) TECH-FREE

Calculate the values of a, b and c for which $4x^2 - 3x + 1 = ax(x-2) + b(x-2) + c$, and hence express $\frac{4x^2 - 3x + 1}{x-2}$ in the form $p(x) + \frac{k}{x-2}$, where $p(x)$ is a polynomial and $k \in R$.

Question 2 (4 marks) TECH-FREE

Solve the equation $6x^3 - 17x^2 = 5x - 6$.

Question 3 (4 marks) TECH-ACTIVE

Consider $p(x) = 3x^3 + kx^2 + 4$.

a. Given that $(3x+2)$ is a factor of $p(x)$, find k and hence express $p(x)$ as a product of linear factors. **(2 marks)**

b. Sketch $p(x)$. **(2 marks)**

Question 4 (4 marks) TECH-FREE

Find the equation of the cubic function $y = ax^3 + bx^2 + cx + d$ given that $a = 1$, $d = 36$ and the curve passes through the points $(-1, 42)$ and $(1, 20)$.

Question 5 (4 marks) TECH-ACTIVE

A triangular box consists of sides x mm in length and height y mm. If the length, width and height added together equal 18 mm, find an expression for the volume and determine the restrictions on the values that x can take. Sketch a graph of x versus volume.

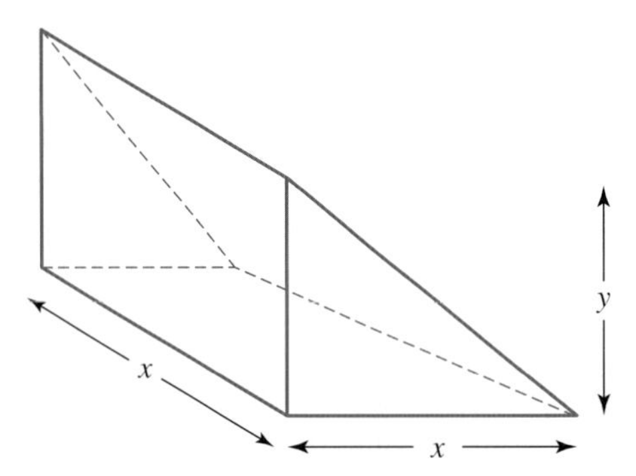

More exam questions are available online.

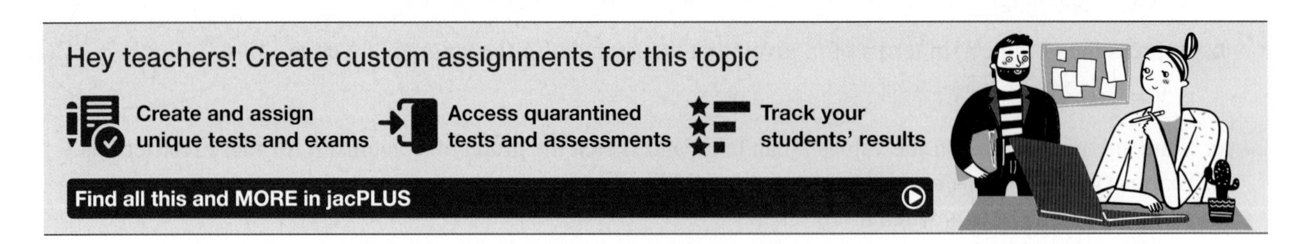

Answers

Topic 4 Cubic polynomials

4.2 Polynomials

4.2 Exercise

1. A: Degree 5; leading coefficient 4; constant term 12; coefficients $\in Z$
   C: Degree 2; leading coefficient -0.2; constant term 5.6; coefficients $\in Q$
2. a. Polynomial of degree 4
   b. Polynomial of degree 3
   c. Not a polynomial due to the $\sqrt{x}$ term.
   d. Not a polynomial due to the $\frac{6}{x^2}$ term and the $\frac{2}{x}$ term.
3. a. A, B, D, F are polynomials.

	Degree	Type of coefficient	Leading term	Constant term
A	5	Q	$3x^5$	12
B	4	R	$-5x^4$	9
D	4	Z	$-18x^4$	0
F	6	N	$49x^6$	9

   b. C is not a polynomial due to the $\sqrt{4x^5} = 2x^{\frac{5}{2}}$ term.
   E is not a polynomial due to the $\frac{5}{3x^2} = \frac{5}{3}x^{-2}$ term.
4. $y^7 + 2y^5 - \sqrt{2}y^2 + 4$ (answers will vary)
5. a. $p(1) = 5$
   b. $p(-2) = -45$
   c. $p(3) = 77, p(-x) = -3x^3 - x^2 + 5$
   d. $p(-1) = 10, p(2a) = 8a^3 + 16a^2 - 4a + 5.$
6. a. 78 b. -12 c. 0
   d. -6 e. -6 f. -5.868
7. a. $-14a$ b. $h^2 - 5h - 4$ c. $2xh + h^2 - 7h$
8. a. 15 b. 10
9. a. $a = -8$ b. $b = 21$ c. $k = 2$
   d. $m = -8$ e. $m = 21$ f. $b = 4; c = 5$
10. $a = 2; b = -13; c = 6$
11. a. $a = 10, b = 6$
    b. $8x - 6 = 10x - 2(x + 3)$
    c. $6x^2 + 19x - 20 = (6x - 5)(x + 4)$
    d. $a = 9, b = -10, c = 1$
12. $(x + 2)^3 = x^2(x + 1) + 5x(x + 2) + 2(x + 3) + 2$
13. a. $a = 3; b = -2; c = -8$
    b. $m = -3; n = -10; p = 24$
    c. $a = 1; b = 7; c = -41; x^2 - 14x + 8 = (x - 7)^2 - 41$
    d. $4x^3 + 2x^2 - 7x + 1 = 4x^2(x + 1) - 2x(x + 1) - 5(x + 1) + 6$
14. $a = 5; b = -4$
15. a. i. $3x^4 + 7x + 12$
    ii. $6x^4 + 10x^2 - 21x - 31$
    iii. $6x^6 - 21x^5 - 29x^4 - 14x^3 - 20x^2 - 7x - 11$
    b. i. m ii. m iii. $m + n$
16. a. $\frac{x - 12}{x + 3} = 1 - \frac{15}{x + 3}$; quotient is 1; remainder is -15.
    b. $\frac{4x + 7}{2x + 1} = 2 + \frac{5}{2x + 1}$
17. a. $\frac{x + 5}{x + 1} = 1 + \frac{4}{x + 1}$, remainder 4.
    b. $\frac{2x - 3}{x + 4} = 2 - \frac{11}{x + 4}$, remainder -11.
    c. $\frac{4x + 11}{4x + 1} = 1 + \frac{10}{4x + 1}$, remainder 10.
    d. $\frac{6x + 13}{2x - 3} = 3 + \frac{22}{2x - 3}$, remainder 22.
18. a. $\frac{2x^3 - 5x^2 + 8x + 6}{x - 2} = 2x^2 - x + 6 + \frac{18}{x - 2}$; quotient is $2x^2 - x + 6$; remainder is 18.
    b. $\frac{x^3 + 10}{1 - x} = -x^2 - x - 1 + \frac{11}{1 - x}$; remainder is 11.
19.

	Quotient	Remainder
a.	1	9
b.	4	1
c.	$x + 7$	-10
d.	$2x - 12$	27
e.	$x^2 + 5x + 12$	41
f.	$x^2 - 7x + 2$	0

20. a. The quotient is $x^2 + x + 5$.
    The remainder is 5.
    b. The quotient is $x^2 + 2$.
    The remainder is -3.
    c. The quotient is $3x^2 + 9x + 32$.
    The remainder is 225.
    d. The quotient is $3x^2 - 7x + 11$.
    The remainder is -30.
21. $a = 3, b = -12, c = 29$; $\frac{3x^2 - 6x + 5}{x + 2} = 3x - 12 + \frac{29}{x + 2}$; $q(x) = 3x - 12, r = 29$
22. $x^7 + 4x^5 - 6x^4 + 5x^3 - 4x^2 - 5x + 2$; degree 7
23. The quotient is $x^2 - x + 3$; the remainder is 0.
24. a. $4x^2 + x - 3 + \frac{18}{1 + 2x}$ b. $2x^2 + 3x + 5 + \frac{20}{2x - 3}$
    c. $x^2 + 6 - \frac{48}{x + 6}$ d. $x^2 - x + 1 + \frac{1}{x + 1}$
    e. $x^2 + x + \frac{3x + 5}{x^2 - 1}$ f. $-2x - 2 - \frac{7x + 12}{(x + 2)(x - 3)}$
25. a. $a = -3; b = -1; c = 4; d = -2$
    b. i. $4x^4 + 12x^3 + 13x^2 + 6x + 1 = (2x^2 + 3x + 1)^2$
    ii. $2x^2 + 3x + 1$
26. a. $k = -13; m = -5; n = 6$
    b. $(x - 3)^2 (x - 2)^2 (x + 3)(x + 2)$

27. a. $2x^2 - \frac{13}{2}x - \frac{139}{4(2x+3)} + \frac{49}{4}$

b. The remainder is $-\frac{139}{4}$; the quotient is $2x^2 - \frac{13}{2}x + \frac{49}{4}$.

c. The dividend equals $-\frac{139}{4}$.

d. The divisor equals 0.

28. a. Define using CAS technology.

b. $\frac{349}{9}$

c. $24n^2 + 24an - 16n - 154$

d. $a = 9$

4.2 Exam questions

Note: Mark allocations are available with the fully worked solutions online.

1. Quotient $= x^2 - x - 5$, remainder $= -3$
2. C
3. D

4.3 The remainder and factor theorems

4.3 Exercise

1. a. 6 b. -6

2. a. The remainder is 19.

b. The remainder is $\frac{37}{8}$.

3. C

4. a. -5 b. 2 c. -101

d. $-10\frac{3}{8}$ e. 0 f. 26

5. a. $a = 4$ b. $k = 3$ c. $k = 2$

6. a. Sample responses can be found in the worked solutions in the online resources.

b. $a = -9$; $b = 8$; $p(x) = 3x^3 - 9x^2 + 8x - 2$

7. a. -10 b. 8 c. -19 d. -19

8. $a = 0$; $a = -1$

9. a. $a = 2$; $b = 3$ b. $m = -\frac{2}{3}$; $n = -3$

10. a. Show $p(-4) = 0$. b. Show $p(5) = 0$.

c. Show $p\left(\frac{1}{2}\right) = 0$. d. $p(1) = 14$, so $p(1) \neq 0$.

11. a. $a = -12$ b. $k = -8$

12. a. $(x-4)(x+1)(x+2)$

b. $(x+12)(3x+1)(x+1)$

c. $(5x+1)(2x+3)(2x+1)$

d. $(4x-3)(5-2x)(5+2x)$

e. $(3x-5)^2(x-5)$

f. $-(x-3)^2(8x-11)$

13. a. $p(-1) = 0 \Rightarrow (x+1)$ is a factor; $p(x) = (x+1)(x+5)(x-3)$

b. $p(x) = (x+1)(3x+2)(4x+7)$

14. a. $(x-1)(x+2)(x+4)$ b. $(x+2)(x+3)(x+5)$

c. $(x-2)(2x+1)(x-5)$ d. $(x+1)(3x-4)(1-6x)$

e. $(x-1)(x+3)(x-2)$ f. $(x+1)(x-7)(x+7)$

15. $x = -2, \frac{1}{3}, -\frac{1}{2}$

16. a. $x = -4, 3, -5$ b. $x = 7, -\frac{5}{3}, 9$

c. $x = -1, 6, 8$ d. $x = 1, -\frac{3}{2}, -3$

e. $x = 1, -\frac{2}{3}$

f. $x = 0, x = -\frac{3}{2}, x = 2, x = -2$

17. a. $x = -\frac{4}{3}, -1, \frac{1}{2}$ b. $x = \frac{1}{2}, \pm\sqrt{3}$

c. $x = 7$ d. $x = -2, -1, 5$

e. $x = -3$ f. $x = -\frac{1}{2}, \frac{1}{3}, 2$

18. The linear factors are $(2x-1)$, $(2x+1)$ and $(3x+2)$.

19. a. i. $(x-5)(x-9)(x+2)$

ii. $x^3 - 12x^2 + 17x + 90$

b. i. $(x+4)(x+1)(1-2x)$

ii. $-2x^3 - 9x^2 - 3x + 4$

20. a. $(2x+1)(3x-1)(4x+5)$

b. i. $(2x-5)$

ii. $m = -26$

iii. $(2x-5)(4x+1)(x-1)$

21. a. i. $p(x) = (x-4)^3$; $q(x) = (x-4)\left(x^2+4x+16\right)$

ii. Sample responses can be found in the worked solutions in the online resources.

b. The third factor is $(x-3)$; $b = c = -3$; $d = 9$.

22. a. $p(x) = (x-2)\left(x+4+\sqrt{7}\right)\left(x+4-\sqrt{7}\right)$

b. Sample responses can be found in the worked solutions in the online resources.

c. $(2x-1)(x-5)^2$; the equation has roots $x = \frac{1}{2}, x = 5$

23. a. $(x-2)(x+2)(5x+9)$; $k = 9$

b. $a = 1$; $a = 2$

c. $a = -3$; $b = 1$; $p(x) = (x-3)\left(x^2+1\right)$, $q(x) = (x-3)(x+3)(x+1)$

d. $x^3 + 7x^2 + 15x + 9$: $x = -3$, $x = -1$; $x^3 - 9x^2 + 15x + 25$: $x = 5$, $x = -1$

24. $-12\sqrt{2} + 33$

25. -0.4696

4.3 Exam questions

Note: Mark allocations are available with the fully worked solutions online.

1. B
2. A
3. $\therefore P(x) = x^3 - 21x + 20$

4.4 Graphs of cubic polynomials

4.4 Exercise

1. a. $(7, 0)$ b. $(0, -7)$ c. $(0, 0)$
 d. $(2, 2)$ e. $(-5, -8)$ f. $\left(\frac{1}{2}, 5\right)$

2. a.

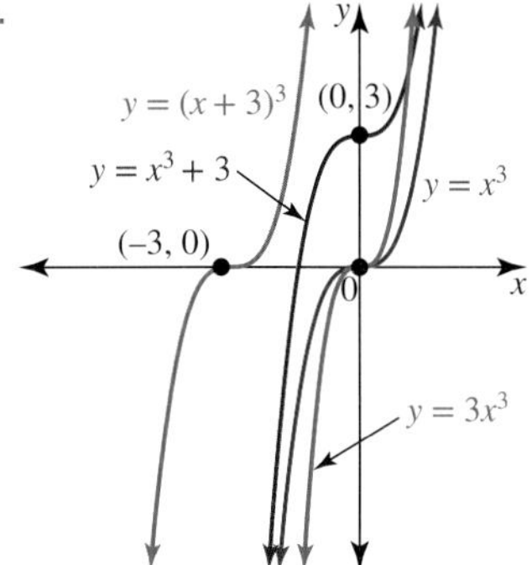

b.

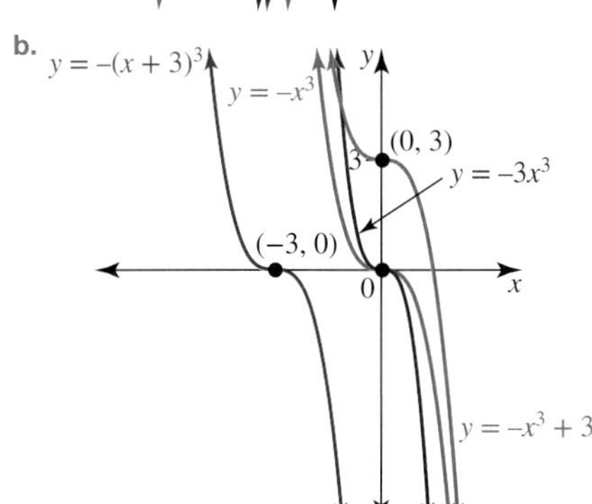

3.

	Inflection point	y-intercept	x-intercept
a.	$(1, -8)$	$(0, -9)$	$(3, 0)$
b.	$(-6, 1)$	$(0, -5)$	$\left(\sqrt[3]{36} - 6, 0\right)$

a.

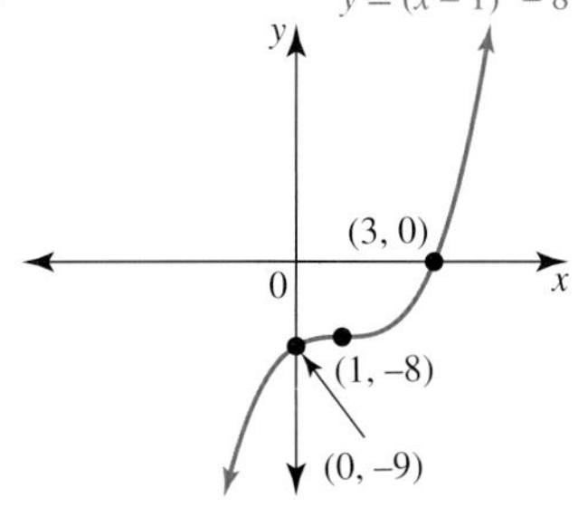

b.

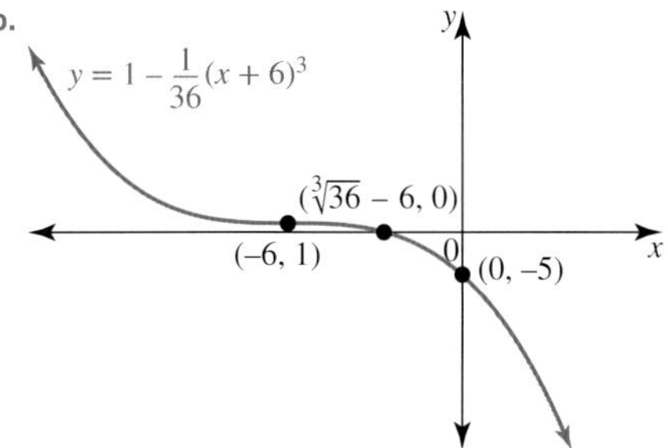

4. a.

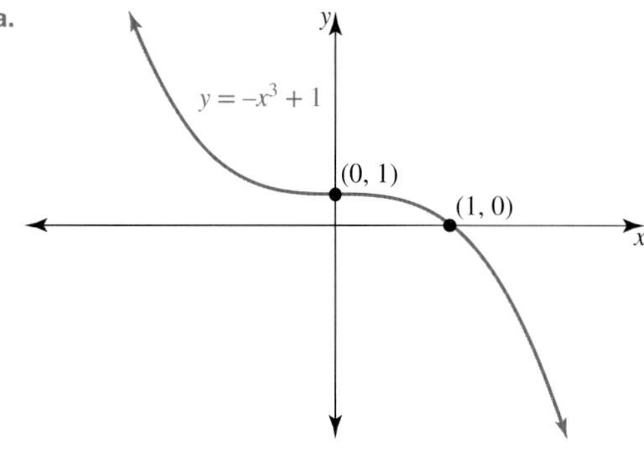

b.

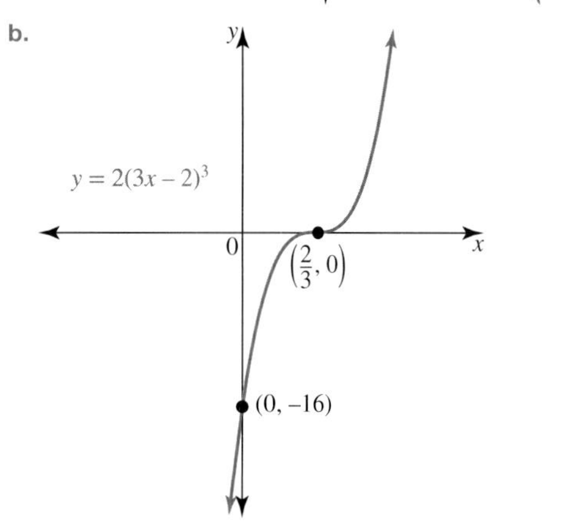

c.

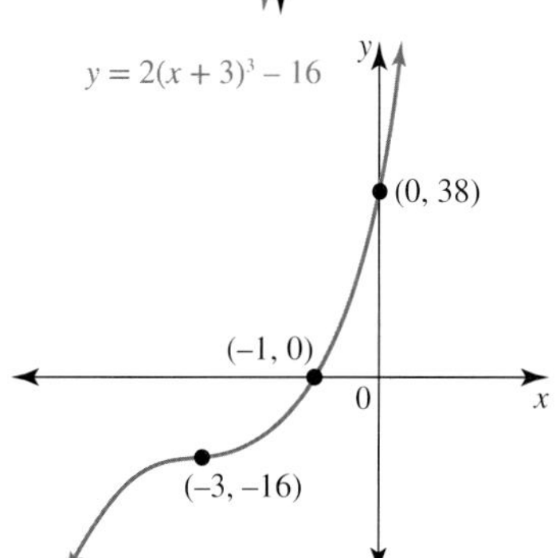

d.

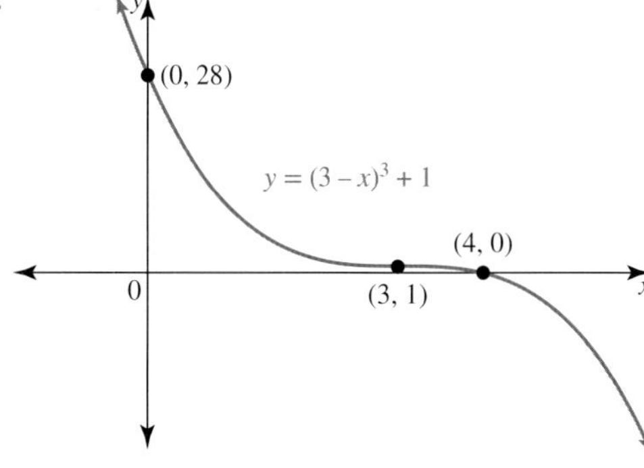

5.

	Inflection point	y-intercept	x-intercept
a.	$(-4, -27)$	$(0, 37)$	$(-1, 0)$
b.	$(1, 10)$	$(0, 8)$	$\left(1-\sqrt[3]{5}, 0\right)$
c.	$(3, 27)$	$(0, -27)$	$\left(3-\dfrac{3}{\sqrt[3]{2}}, 0\right)$
d.	$(-2, 16)$	$(0, 0)$	$(0, 0)$
e.	$\left(-\dfrac{4}{3}, 0\right)$	$(0, -48)$	$\left(-\dfrac{4}{3}, 0\right)$
f.	$(0, 9)$	$(0, 9)$	$(-3, 0)$

a.

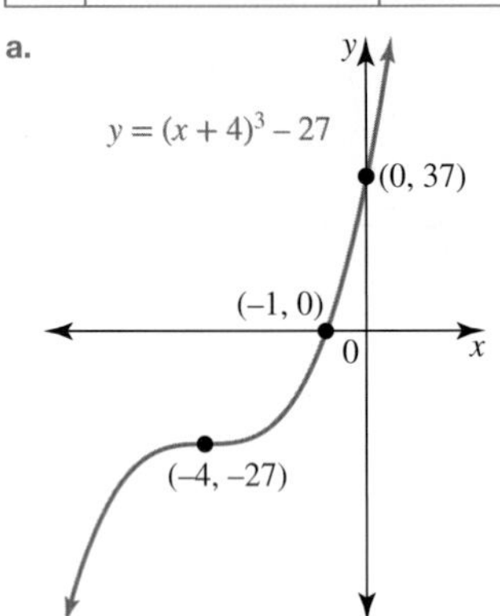

b.

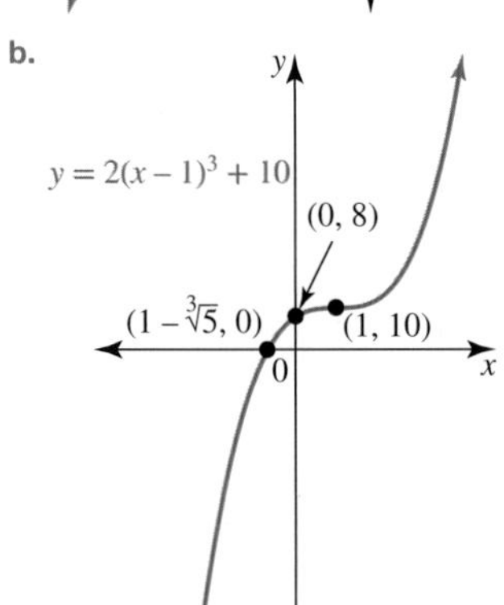

c.

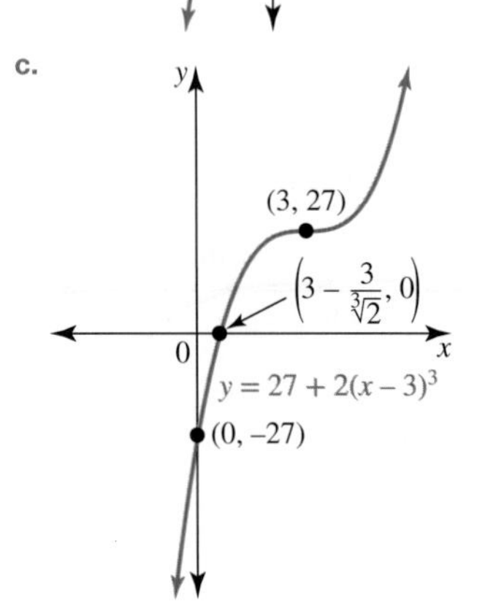

d.

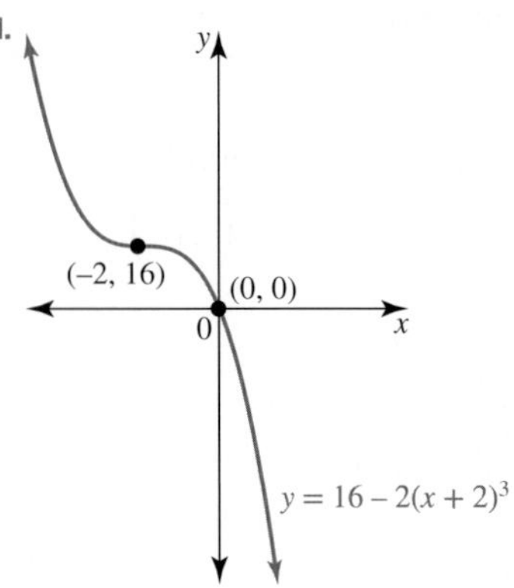

e.

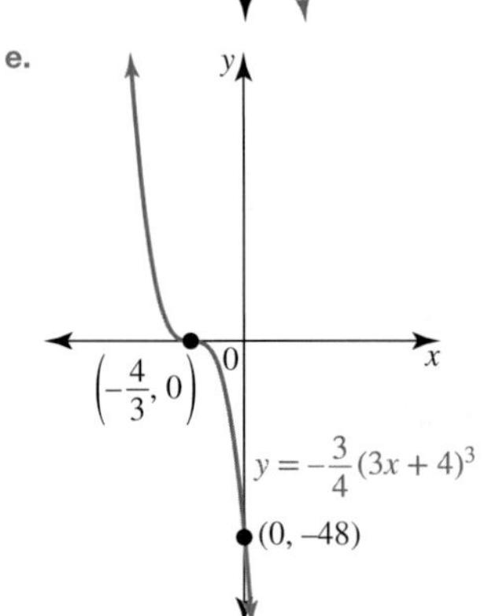

f.

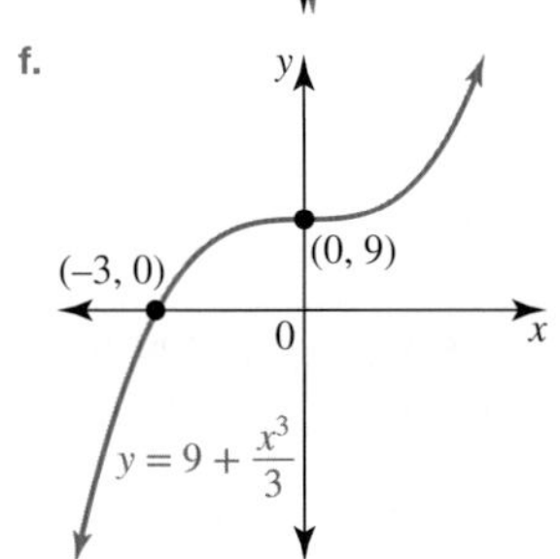

6.

	Inflection point	y-intercept	x-intercept
a.	$(6, 0)$	$(0, -27)$	$(6, 0)$
b.	$(0, -2)$	$(0, -2)$	$(1, 0)$

a.

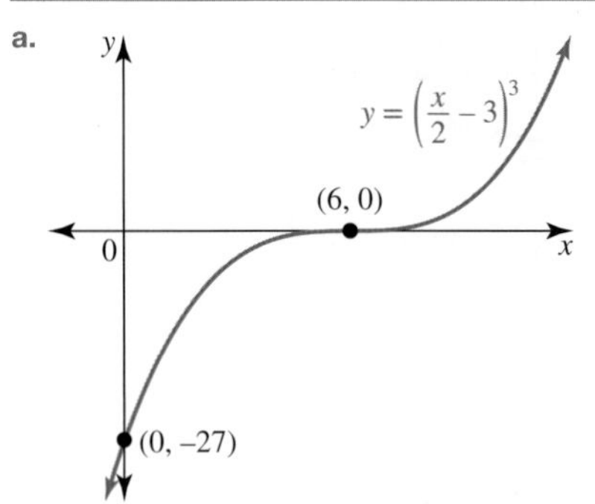

b.

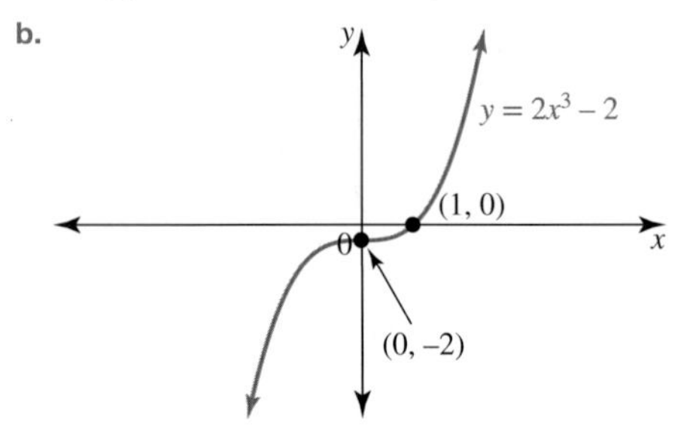

7.

	y-intercept	x-intercepts
a.	$(0, -24)$	$(-6, 0), (-1, 0), (4, 0)$
b.	$(0, -24)$	$\left(-\frac{1}{2}, 0\right), (4, 0), (6, 0)$

a.

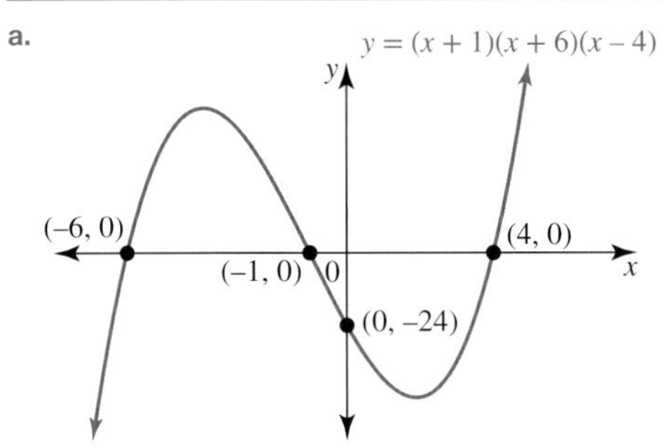

b.

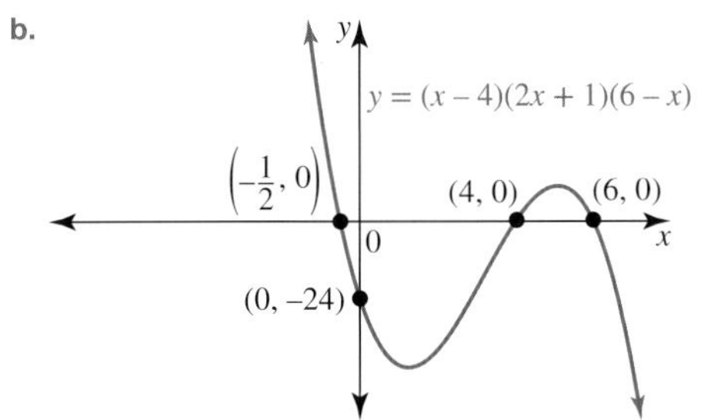

8.

	y-intercept	x-intercepts
a.	$(0, -8)$	$(-4, 0), (-1, 0), (2, 0)$
b.	$(0, 0)$	$(-8, 0), (0, 0), (5, 0)$
c.	$(0, -12)$	$(-3, 0), (1, 0), (4, 0)$
d.	$(0, 12)$	$(-4, 0), (2, 0), (6, 0)$

a.

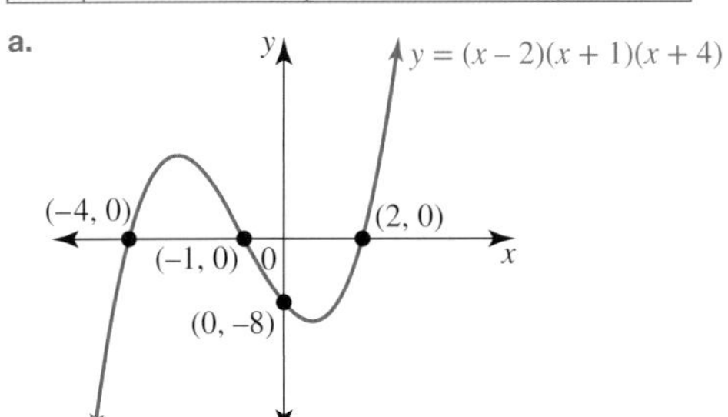

b.

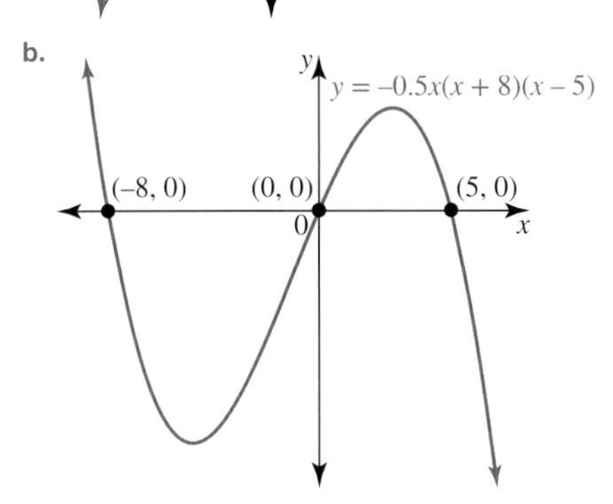

c.

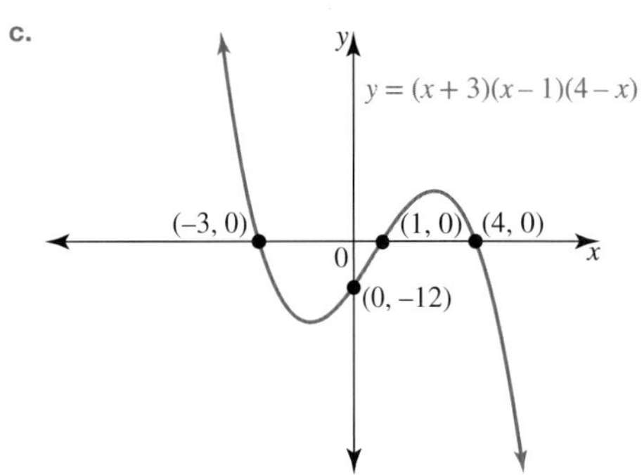

d.

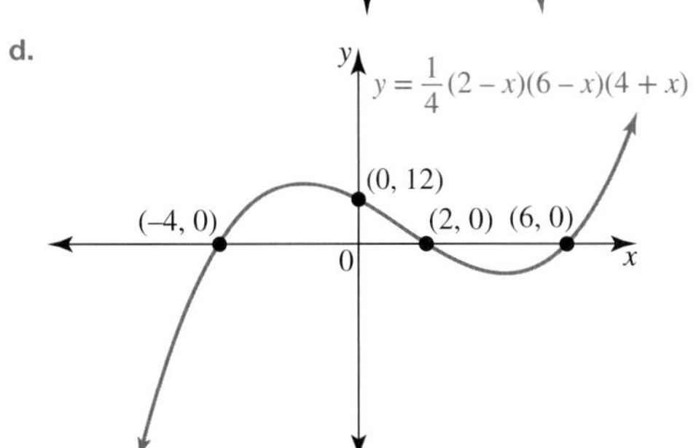

9.

y-intercept	x-intercepts
$(0, 0)$	$(-2, 0), (0, 0), (2, 0)$

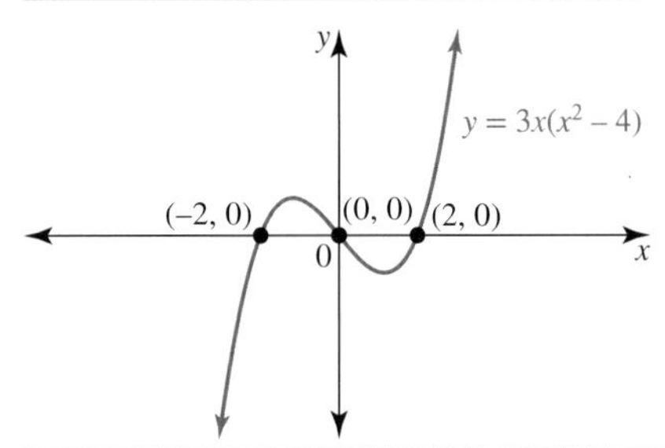

10.

	y-intercept	x-intercepts
a.	$(0, 6)$	$(-6, 0)$ and $(3, 0)$, which is a turning point
b.	$(0, 8)$	$(-2, 0)$, which is a turning point, and $(1, 0)$

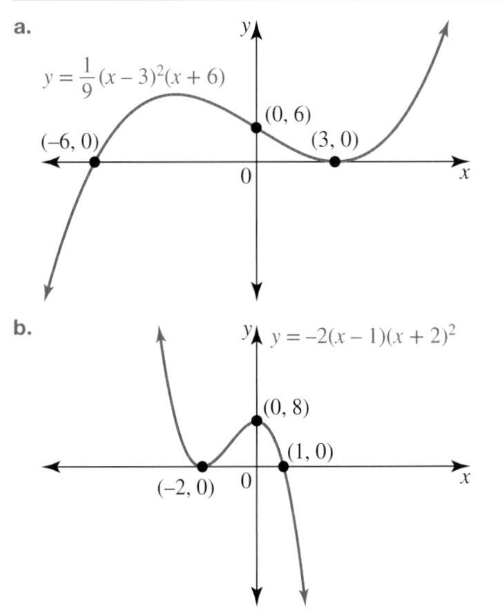

11.

	y-intercept	x-intercepts
a.	$(0, 32)$	$(-4, 0)$ is a turning point; $(2, 0)$ is a cut.
b.	$(0, 54)$	$(3, 0)$ is a turning point; $(-3, 0)$ is a cut.
c.	$(0, 36)$	$(-3, 0)$ is a turning point; $(4, 0)$ is a cut.
d.	$(0, -12)$	$(2, 0)$ is a turning point; $(12, 0)$ is a cut.
e.	$(0, 0)$	$\left(-\frac{3}{2}, 0\right)$ is a turning point; $(0, 0)$ is a cut.
f.	$(0, 0)$	$(0, 0)$ is a turning point; $(0.4, 0)$ is a cut.

a.

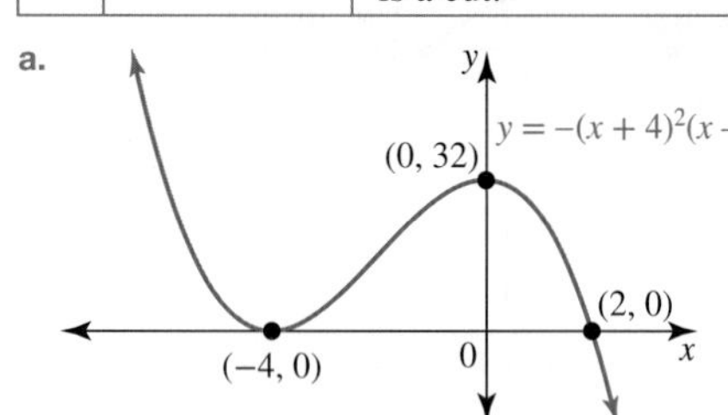

b.

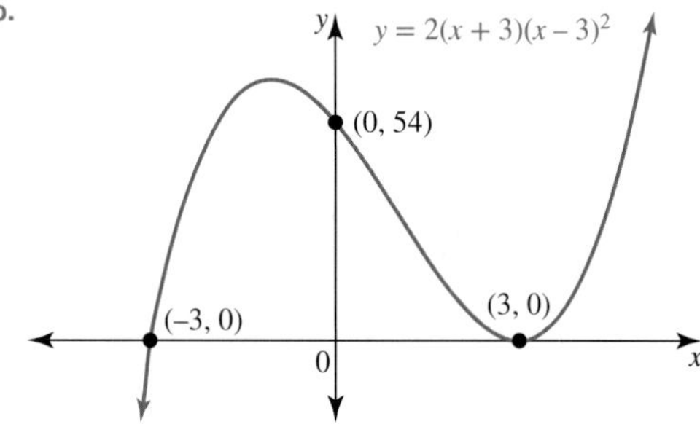

c.

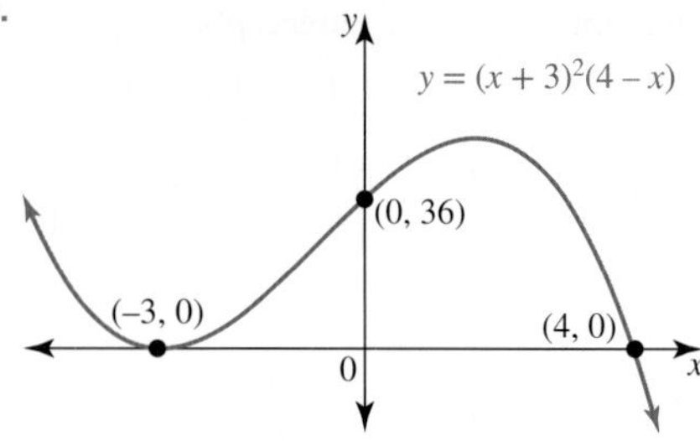

d.

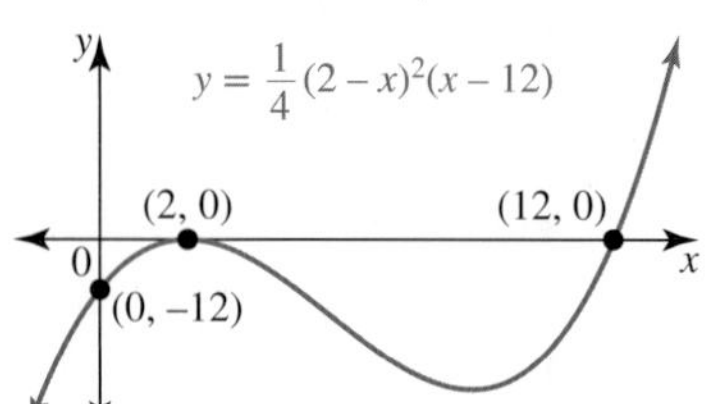

e.

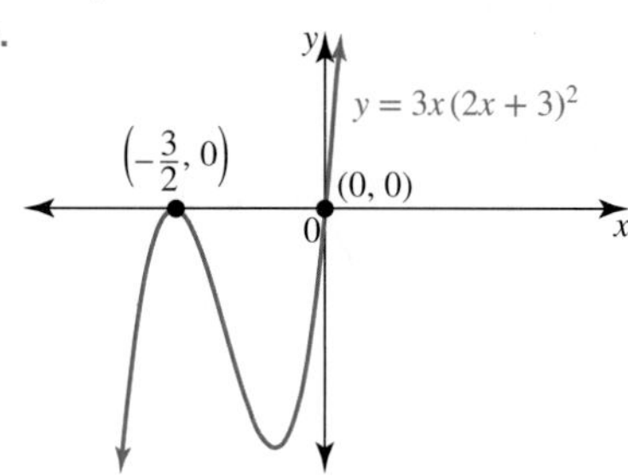

f.

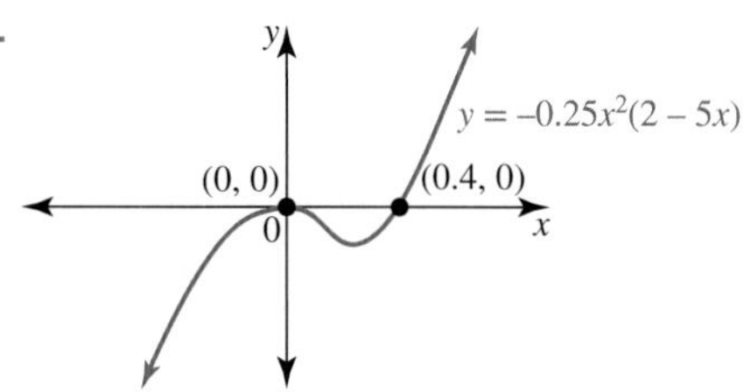

12. See the table at the bottom of the page.*

a.

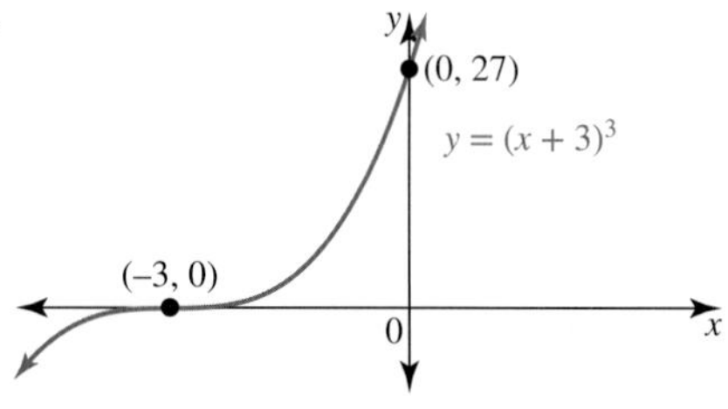

*12.

	Stationary point of inflection	y-intercept	x-intercepts
a.	$(-3, 0)$	$(0, 27)$	$(-3, 0)$
b.	None	$(0, -9)$	$(-3, 0)$ is a turning point; $\left(\frac{1}{2}, 0\right)$ is a cut.
c.	None	$(0, -15)$	$(-3, 0)$, $\left(\frac{1}{2}, 0\right)$, $(5, 0)$
d.	$\left(\frac{1}{2}, 1\right)$	$\left(0, \frac{3}{2}\right)$	$\left(\frac{1 + \sqrt[3]{2}}{2}, 0\right)$ or approx. $(1.1, 0)$
e.	None	$(0, 0)$	$\left(\frac{1}{4}, 0\right)$ is a turning point; $(0, 0)$ is a cut.
f.	None	$(0, 4)$	$\left(\frac{2}{3}, 0\right)$ is a turning point; $\left(-\frac{2}{3}, 0\right)$ is a cut.

b.

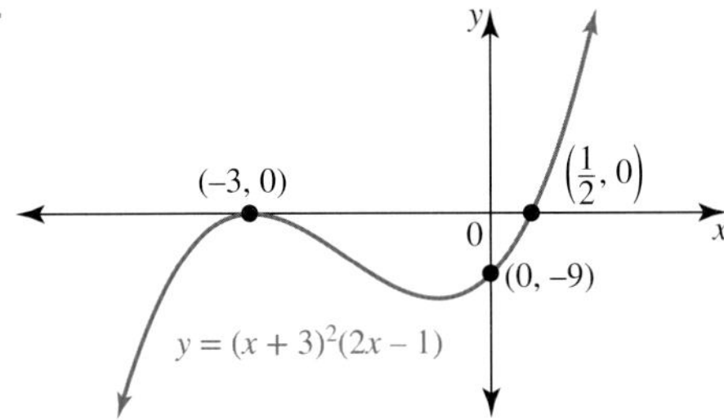

c.

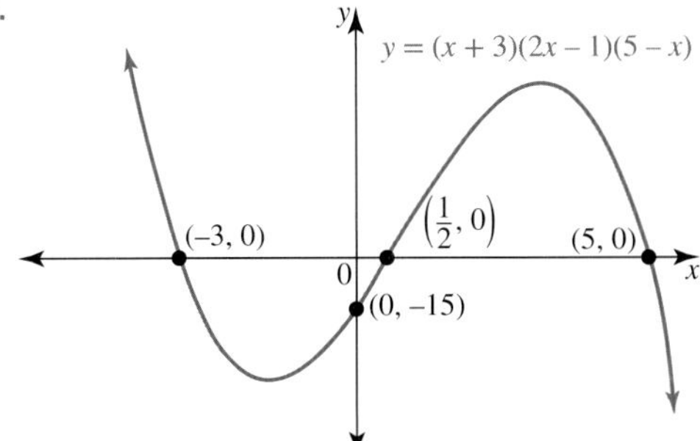

d.

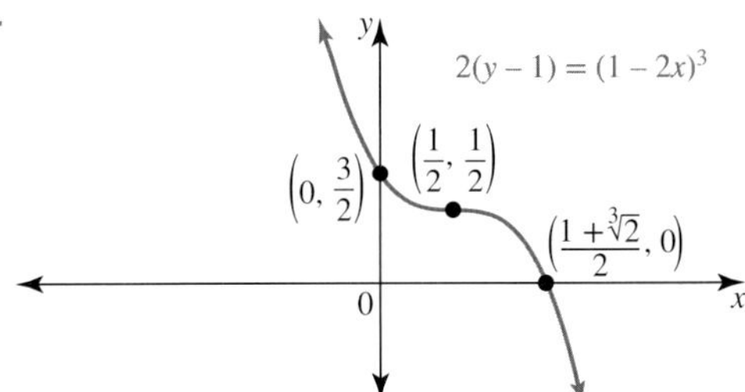

e.

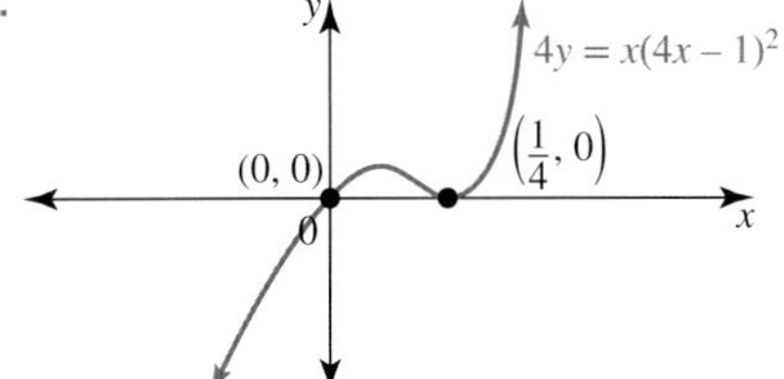

f.

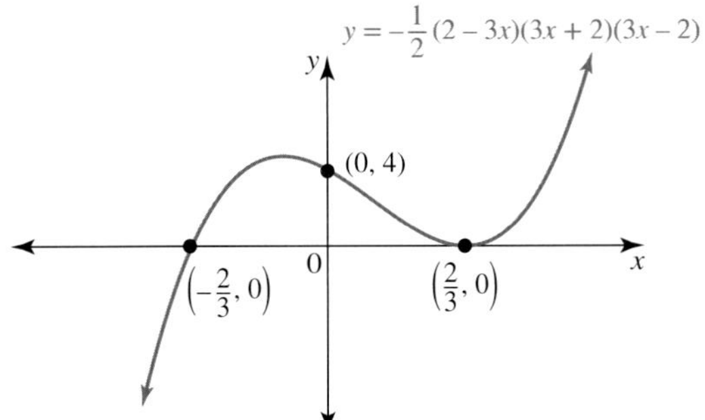

13. $x^3 - 3x^2 - 10x + 24 = (x - 2)(x - 4)(x + 3)$

y-intercept	*x*-intercepts
(0, 24)	(−3, 0), (2, 0), (4, 0)

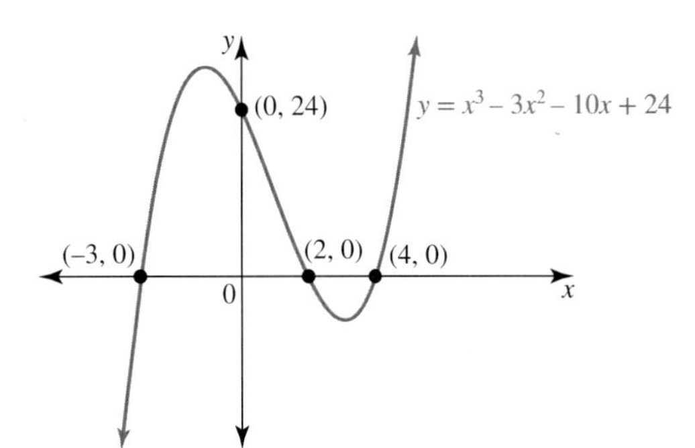

14. a.

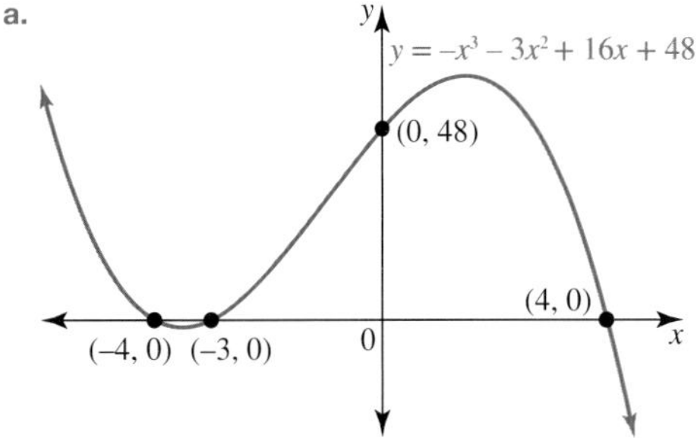

b. $(x - 2)(2x - 1)(x + 3)$

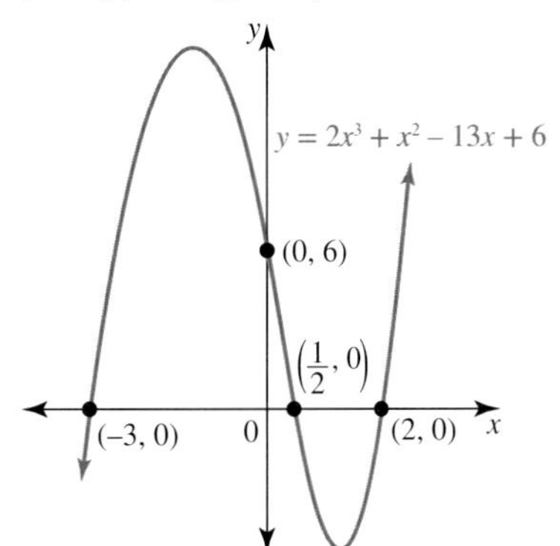

c.

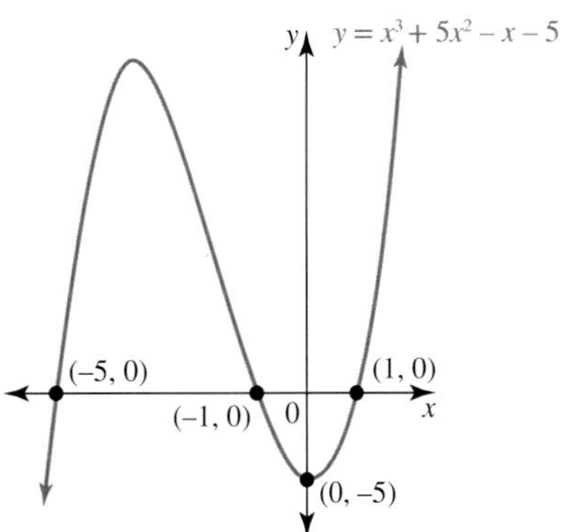

d. $-(x - 1)(x + 3)^2$

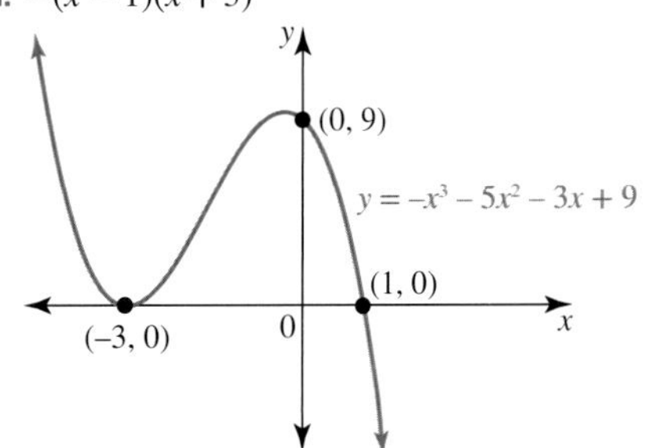

15. See the table at the bottom of the page.*

a.

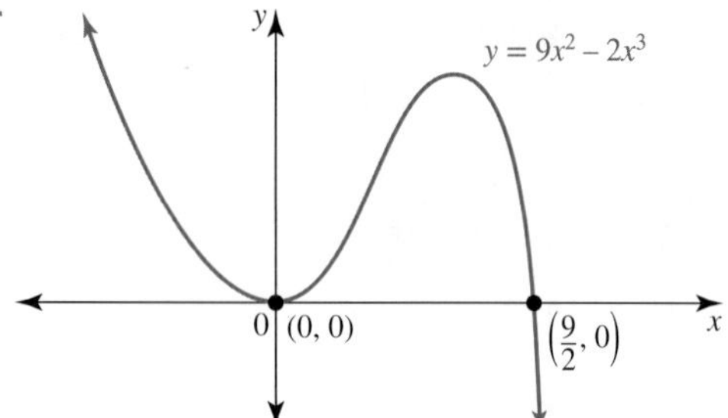

b.

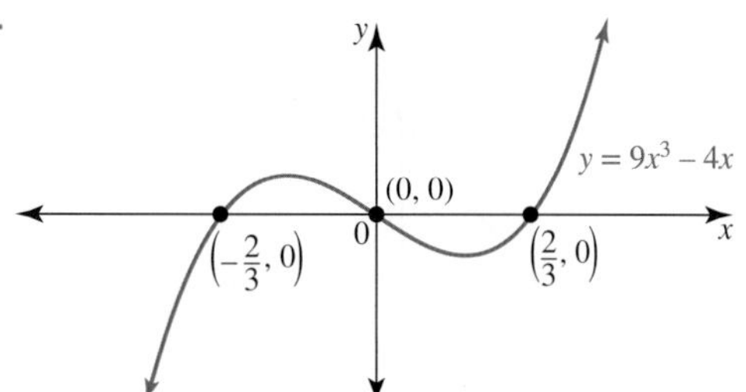

c.

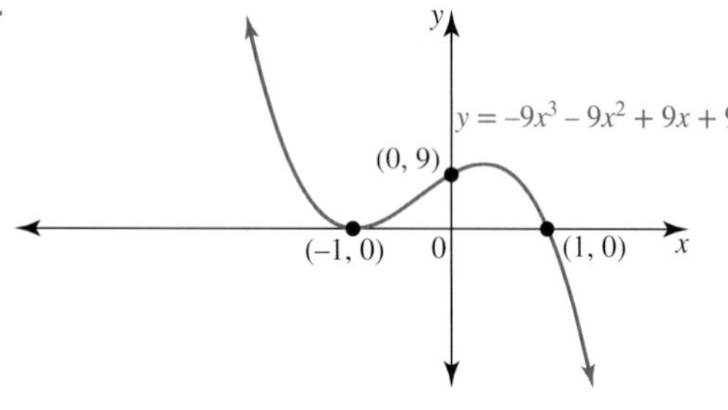

d.

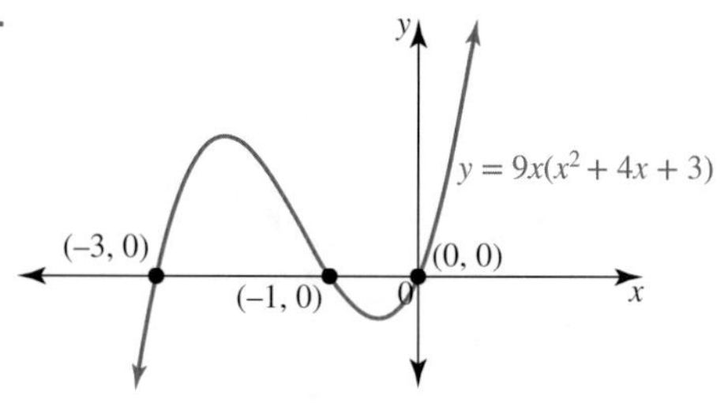

e.

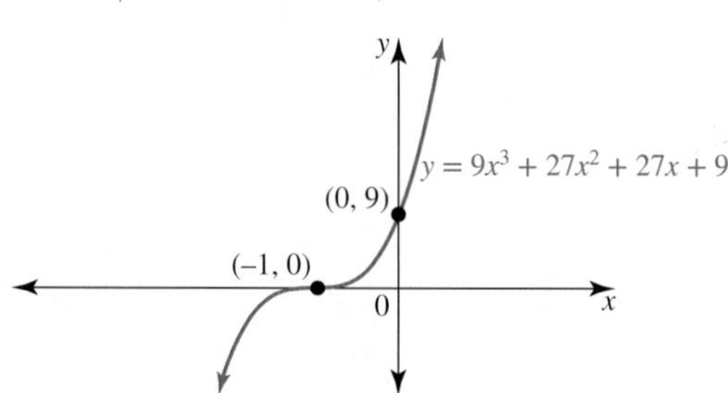

f.

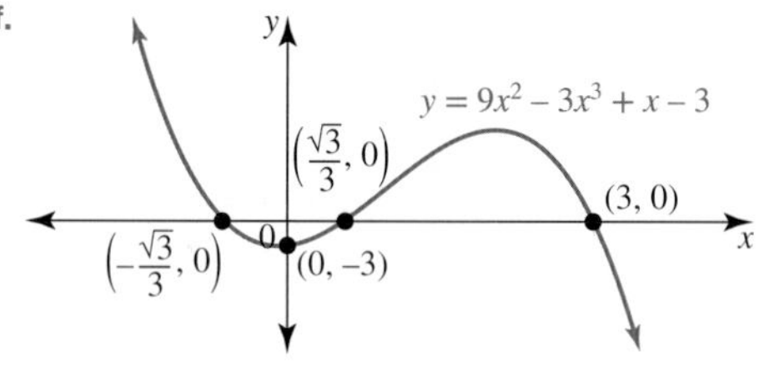

*15.

	Factorised form	Stationary point of inflection	y-intercept	x-intercepts
a.	$y = x^2(9 - 2x)$	None	$(0, 0)$	$(0, 0)$ is a turning point; $\left(\frac{9}{2}, 0\right)$ is a cut.
b.	$y = x(3x - 2)(3x + 2)$	None	$(0, 0)$	$\left(-\frac{2}{3}, 0\right)$, $(0, 0)$ $\left(\frac{2}{3}, 0\right)$
c.	$y = -9(x + 1)^2(x - 1)$	None	$(0, 9)$	$(-1, 0)$ is a turning point; $(1, 0)$ is a cut.
d.	$y = 9x(x + 1)(x + 3)$	None	$(0, 0)$	$(-3, 0), (-1, 0), (0, 0)$
e.	$y = 9(x + 1)^3$	$(-1, 0)$	$(0, 9)$	$(-1, 0)$
f.	$y = -3(x - 3)\left(x - \frac{\sqrt{3}}{3}\right)\left(x + \frac{\sqrt{3}}{3}\right)$	None	$(0, -3)$	$\left(-\frac{\sqrt{3}}{3}, 0\right), \left(\frac{\sqrt{3}}{3}, 0\right), (3, 0)$

16. See the table at the bottom of the page.*

a.

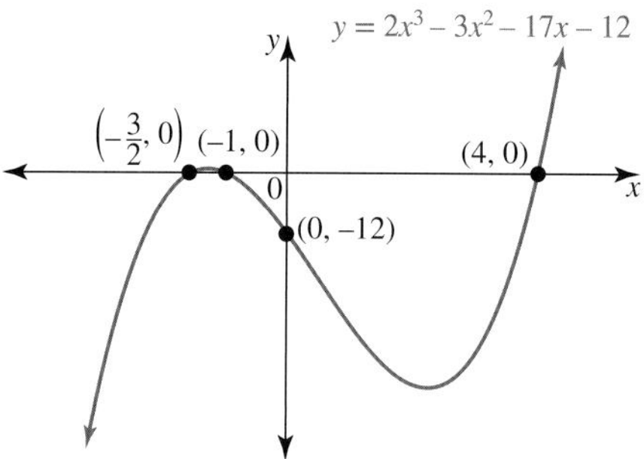

b.

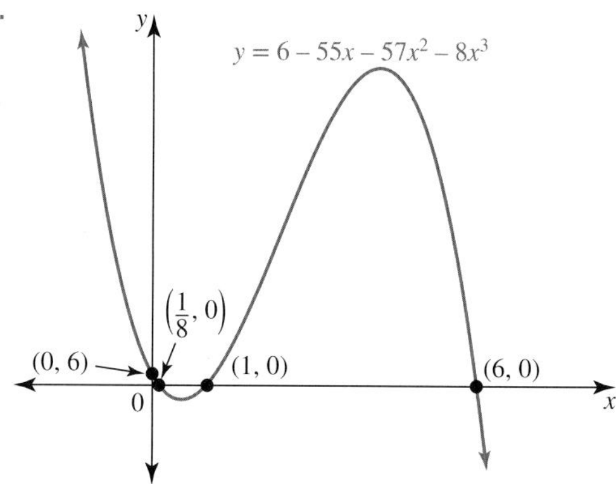

c.

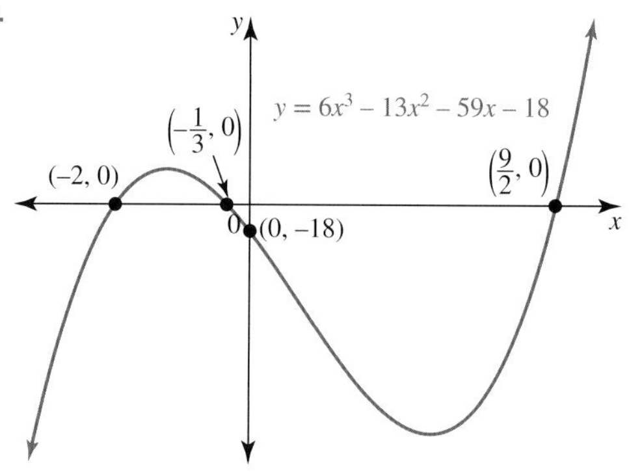

d.

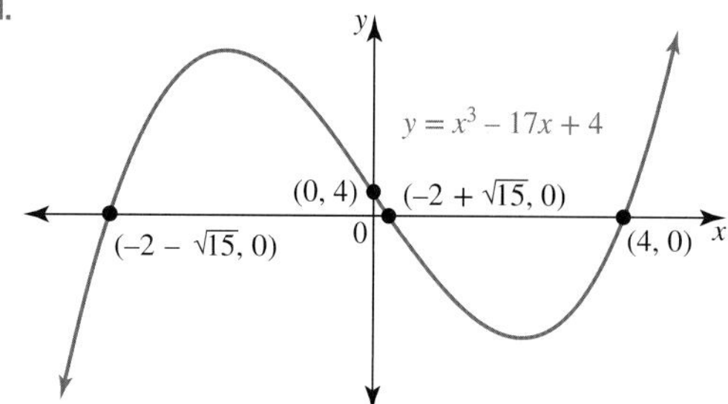

e.

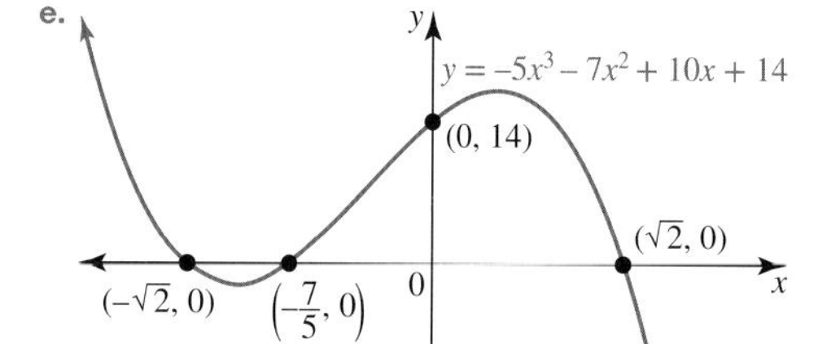

f.

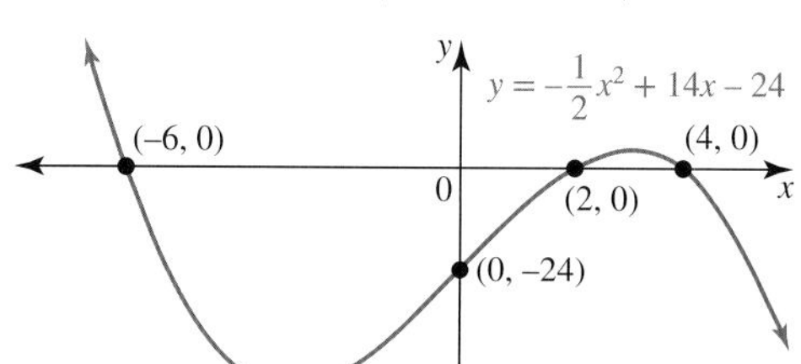

*16.

	Factorised form	*y*-intercept	*x*-intercepts
a.	$y = (x + 1)(2x + 3)(x - 4)$	$(0, -12)$	$\left(-\frac{3}{2}, 0\right), (-1, 0), (4, 0)$
b.	$y = -(x - 1)(8x - 1)(x - 6)$	$(0, 6)$	$\left(\frac{1}{8}, 0\right), (1, 0), (6, 0)$
c.	$y = (x + 2)(3x + 1)(2x - 9)$	$(0, -18)$	$(-2, 0), \left(-\frac{1}{3}, 0\right), \left(\frac{9}{2}, 0\right)$
d.	$y = (x - 4)(x + 2 - \sqrt{5})(x + 2 + \sqrt{5})$	$(0, 4)$	$(-2 - \sqrt{5}, 0), (-2 + \sqrt{5}, 0), (4, 0)$
e.	$y = (5x + 7)(\sqrt{2} - x)(\sqrt{2} + x)$	$(0, 14)$	$(-\sqrt{2}, 0), \left(-\frac{7}{5}, 0\right), (\sqrt{2}, 0)$
f.	$y = -\frac{1}{2}(x - 2)(x + 6)(x - 4)$	$(0, -24)$	$(-6, 0), (2, 0), (4, 0)$

17. a. $-x^3 + 3x^2 + 10x - 30$
$= -(x-3)\left(x-\sqrt{10}\right)\left(x+\sqrt{10}\right)$

y-intercept	*x*-intercepts
$(0, -30)$	$\left(-\sqrt{10}, 0\right), (3, 0), \left(\sqrt{10}, 0\right)$

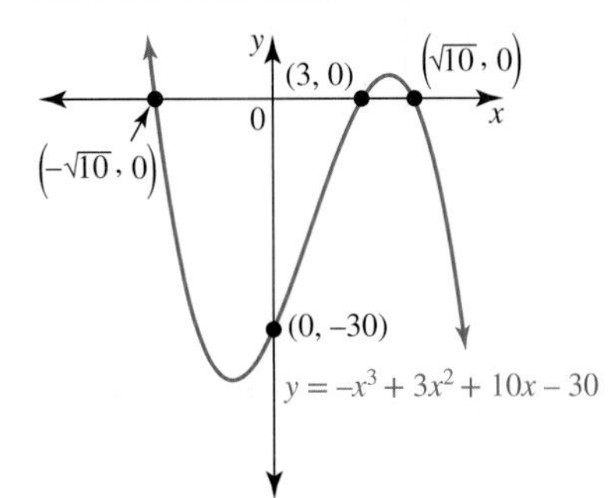

b. Stationary point of inflection $(-1, 1)$; y-intercept $(0, 2)$; x-intercept $(-2, 0)$

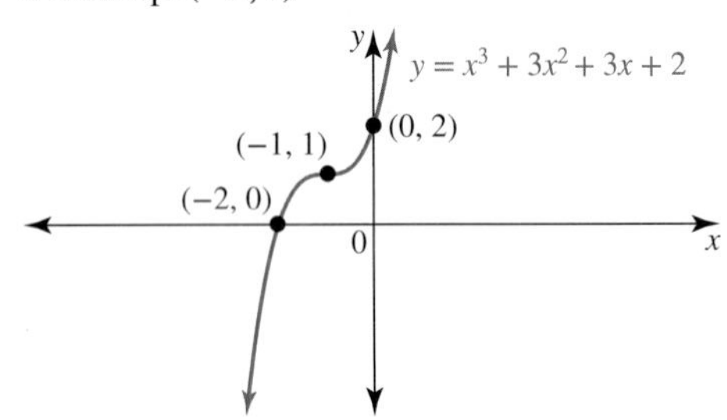

18. a. $k = -19$

b. $P(x) = (3x-1)(5x+1)(2x-1)$

c. $x = -\frac{1}{5}, \frac{1}{3}, \frac{1}{2}$

d. y-intercept $(0, 1)$; x-intercepts

$\left(-\frac{1}{5}, 0\right), \left(\frac{1}{3}, 0\right), \left(\frac{1}{2}, 0\right)$

e. No

19. a. Sample responses can be found in the worked solutions in the online resources.

b. Sample responses can be found in the worked solutions in the online resources

c. $y \to \infty$

d.

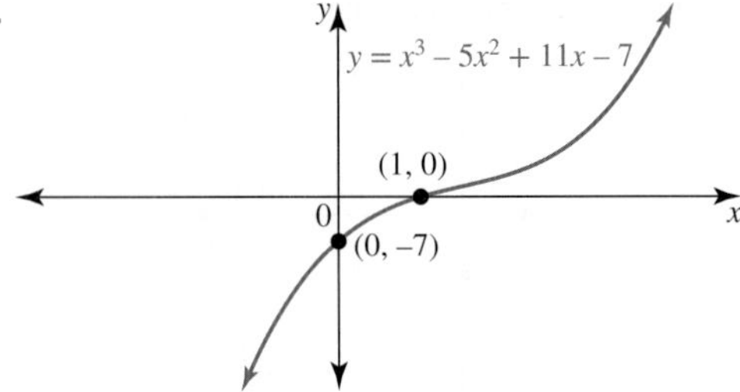

20. a. i. Maximum turning point $(-5.4, 100.8)$; minimum turning point $(2.7, -166.0)$; y-intercept $(0, -96)$; x-intercepts $(-8, 0), (-2, 0), (6, 0)$

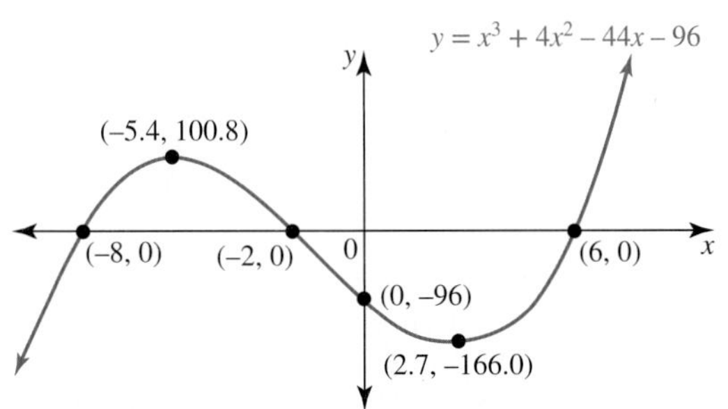

ii. Maximum turning point $(-0.2, -17.9)$; minimum turning point $(1.5, -45.3)$; y-intercept $(0, -19)$; x-intercepts $(2.6, 0)$

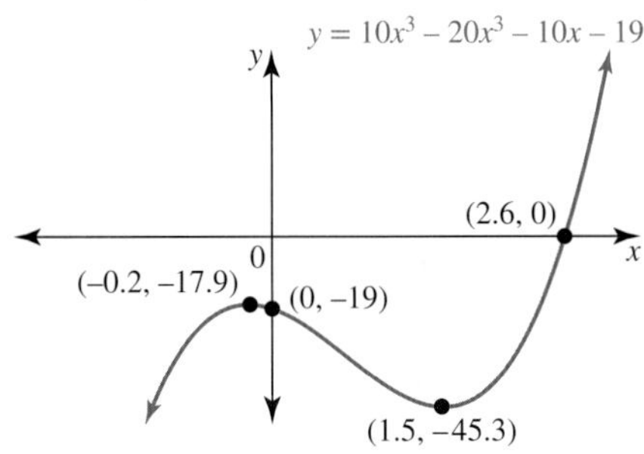

iii. Maximum turning point $(0.2, 502.3)$; minimum turning point $(5, 0)$; y-intercept $(0, 500)$; x-intercepts $(-2.2, 0), (5, 0)$

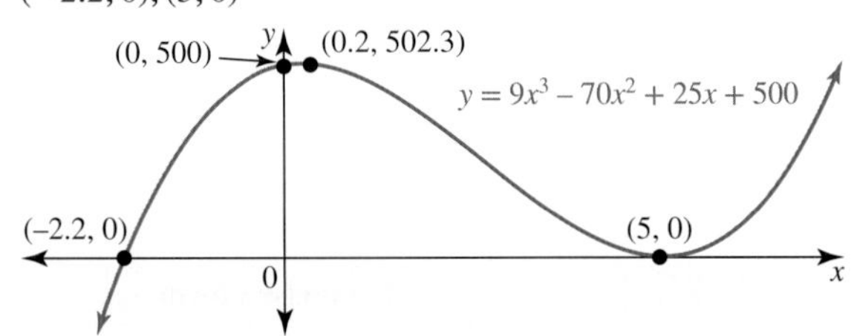

b. Sample responses can be found in the worked solutions in the online resources.

4.4 Exam questions

Note: Mark allocations are available with the fully worked solutions online.

1. C
2. 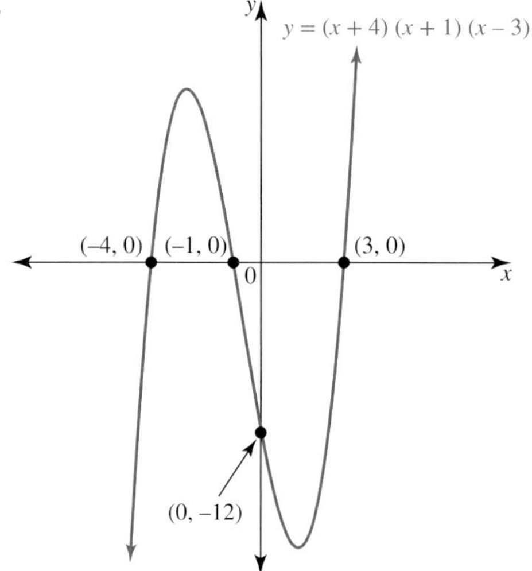

3. E

4.5 Equations of cubic polynomials

4.5 Exercise

1. a. $y = \frac{1}{49}(x-3)^3 - 7$ b. $y = \frac{1}{4}x(x+5)(x-4)$
   c. $y = -(x+2)^2(x-3)$
2. a. $y = \frac{1}{3}(x-3)^3 + 9$ b. $y = (x+2)^3 + 2$
   c. $y = -2x^3 + 4$ d. $y = (x+5)^3 + 4$
   e. $y = -(x-2)^3 - 1$ f. $y = -(x-3)^3 - 1$
3. a. $y = -(x+6)(x+1)(x-2)$
   b. $y = (x+3)^2(x-5)$
   c. $y = -(x+2)(x-2)(x-3)$
   d. $y = (2x+1)\left(x+\sqrt{5}\right)\left(x-\sqrt{5}\right) = 2x^3 + x^2 - 10x - 5$
   e. $y = -4x^2(x-2)$
   f. $y = \frac{1}{4}(2x-1)(x+5)(x-8)$
4. a. $y = \frac{1}{2}(x+8)(x+4)(x+1)$
   b. $y = -2x^2(x-5)$
   c. $y = -3(x-1)^3 - 3$
   d. $y = \frac{4}{5}(x-1)(x-5)^2$
5. a. $y = -2(x+6)^3 - 7$ b. $(0, -12)$
   c. $\left(\sqrt[3]{10} - 5, 0\right)$ d. $(0, 7)$
6. a. $-\frac{2}{5} \leq x \leq -\frac{1}{8}$ or $x \geq \frac{1}{3}$
   b. + – $-\frac{2}{5}$ $-\frac{1}{8}$ $\frac{1}{3}$ x
   c. $x < -\frac{2}{5}$ or $-\frac{1}{8} < x < \frac{1}{3}$
7. a. $x < 6,\ x \neq 2$
   b. $\{x : -2 \leq x \leq 0\} \cup \{x : x \geq 2\}$
   c. $x < -2$
8. a. $-9 \leq x \leq -1$ or $x \geq 2$ b. $x > \frac{1}{5}$
   c. $x > 2.5$ d. $x \leq 0$ or $x = 1$
   e. $x < -2$ or $-\frac{6}{5} < x < 2$ f. $-\frac{3}{2} < x < -1$ or $x > -\frac{1}{2}$
9. $\{x : -1 < x < 1\} \cup \left\{x : x > \frac{7}{3}\right\}$
10. a. $y = -(x-a)^2(x-b)$
    b. $\{x : x \leq b\}$
    c. More than b units to the left
    d. More than $-a$ units to the right
11. The point of intersection is $\left(-\sqrt[3]{2}, 3\sqrt[3]{2}\right)$.
    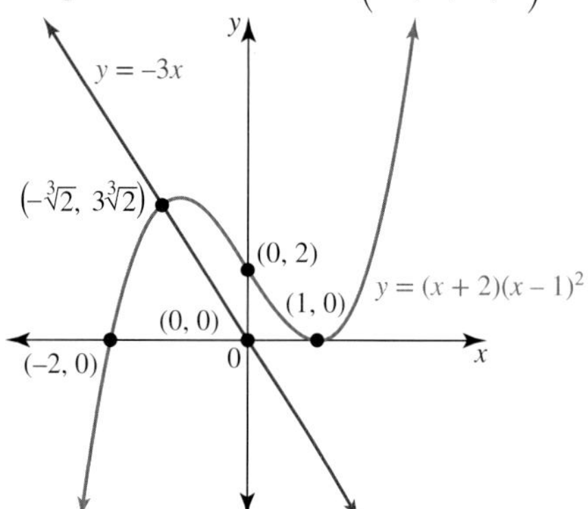

12. The points of intersection are $(1, 3)$, $(2, 0)$, $(-2, 0)$.
    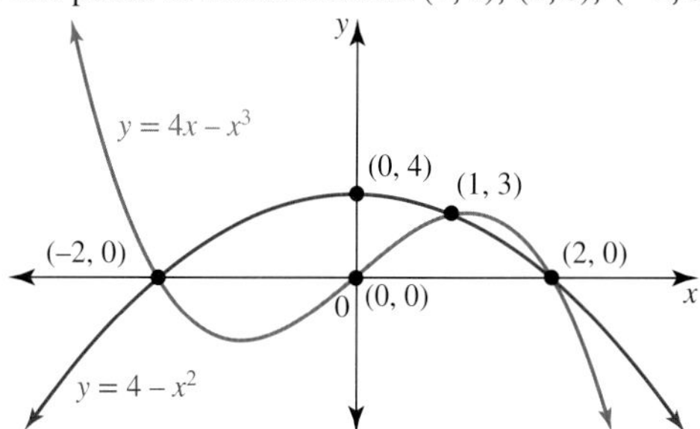

13. a. i. $(0, 0)$, $\left(\frac{1}{2}, \frac{1}{4}\right)$
       ii. $(-1, -2)$
    b. 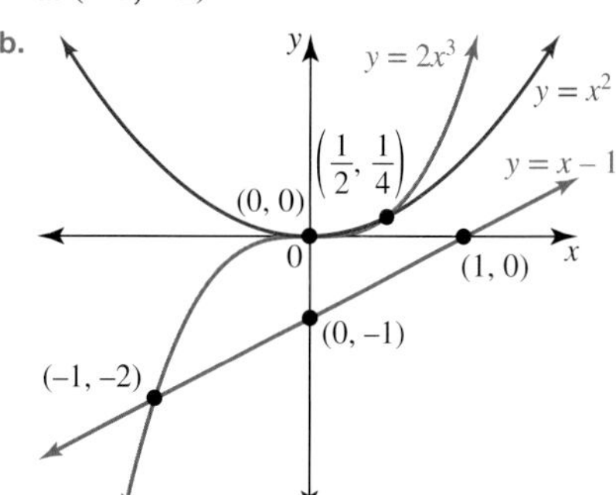

    c. $x \leq 0.5$; sample responses can be found in the worked solutions in the online resources.

14. a. $y = -2x + 5$; 1 solution

b. $y = x^3 + 3x^2$, $y = 4x$; 3 solutions

c. There are 3 solutions. One method is to use $y = x^3$, $y = -3x^2 + 4x$.

d. $x = -4, 0, 1$

15. a. Sample responses can be found in the worked solutions in the online resources.

b. $(2, 8)$

c.

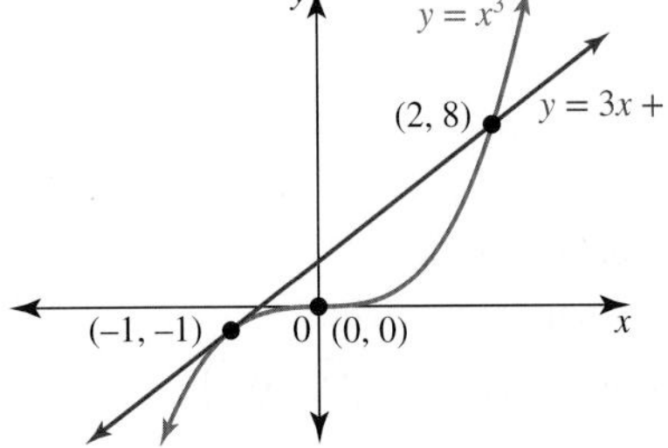

d. One intersection if $m < 3$; two intersections if $m = 3$; three intersections if $m > 3$

16. $y = 2x^3 + 3x^2 - 4x + 3$

17. a. $(x - 3)^2$

b. $(-1, 0)$

c. $a = -5$, $b = 3$

18. a. $a = 1; b = -1$

b. $(-2, -2)$, $(-1, -1)$, $(0, 0)$

c.

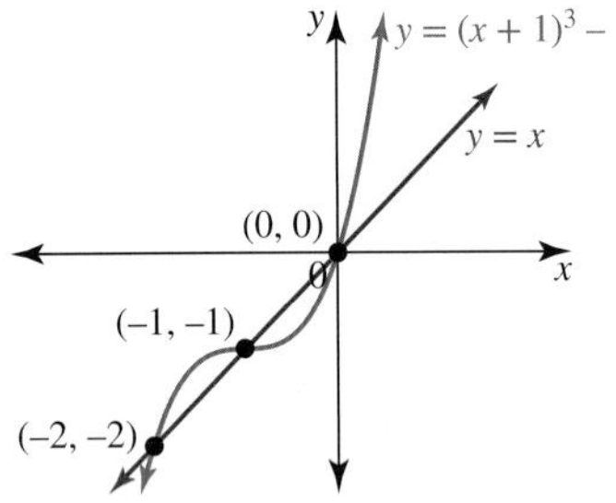

d. $\{x : -2 < x < -1\} \cup \{x : x > 0\}$

19. a. Fewer than four pieces of information are given.

b. Sample responses can be found in the worked solutions in the online resources.

c.

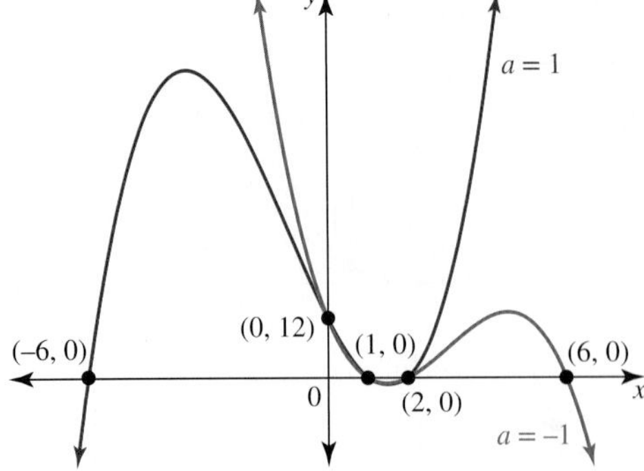

If $a = 1$, the graph contains the points $(1, 0)$, $(2, 0)$, $(0, 12)$, $(-6, 0)$; if $a = -1$, the points are $(1, 0)$, $(2, 0)$, $(0, 12)$, $(6, 0)$.

d. The equation is $y = -3x^3 + 15x^2 - 24x + 12$ or $y = -3(x - 1)(x - 2)^2$; the x-intercepts are $(1, 0)$, $(2, 0)$; $(2, 0)$ is a maximum turning point.

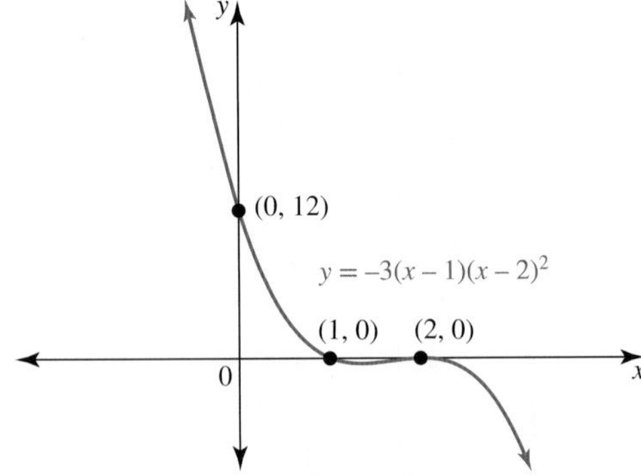

20. a. $(-3.11, -9.46)$, $(-0.75, 0.02)$, $(0.86, 6.44)$

b. $x^3 + 3x^2 - x - 2 = 0$

c. x-intercepts

4.5 Exam questions

Note: Mark allocations are available with the fully worked solutions online.

1. B

2. D

3. $y = 2x^3 + 2x^2 - 4x$ or $y = 2x(x + 2)(x - 1)$

4.6 Cubic models and applications

4.6 Exercise

Sample responses are provided for proofs in the worked solutions in your online resources.

1. a. $h = \dfrac{3 - 4x}{2}$

b. A sample proof can be found in the worked solutions in your online resources.

c. $0 < x < \dfrac{3}{4}$

d.

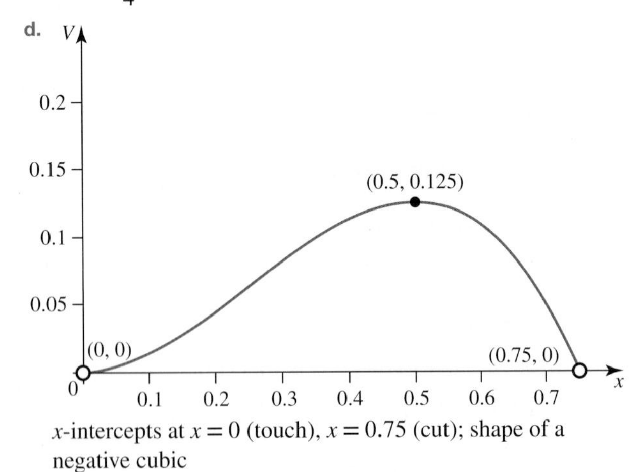

x-intercepts at $x = 0$ (touch), $x = 0.75$ (cut); shape of a negative cubic

e. Cube of edge 0.5 m

2. a. $l = 20 - 2x$; $w = 12 - 2x$; $V = (20 - 2x)(12 - 2x)x$

b. $0 < x < 6$

c. 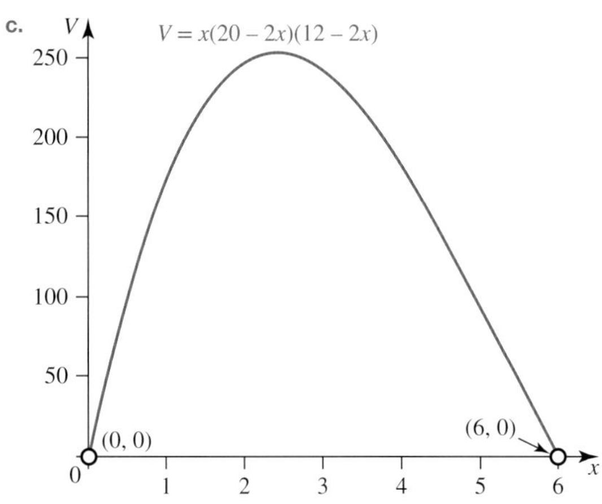

x-intercepts at $x = 10$, $x = 6$, $x = 0$, but since $0 < x < 6$, the graph won't reach $x = 10$; shape of a positive cubic.

d. Length 15.14 cm; width 7.14 cm; height 2.43 cm; greatest volume 263 cm^3

3. a. Loss of \$125; profit of \$184

b. A sample proof can be found in the worked solutions in your online resources.

c. Too many and the costs outweigh the revenue from the sales. A negative cubic tends to $-\infty$ as x becomes very large.

d. i. Profit \$304

ii. Loss \$113

e. See the graph at the bottom of the page.*
The x-intercepts lie between 5 and 6 and between 16 and 17.

f. Between 6 and 16

4. a. 54 b. 2 hours c. 210 d. 5 pm

5. a. $y = -\frac{1}{32}x^2(x-6)$ b. $\frac{81}{32}$ km

6. $-4, 1$

7. a. $2\sqrt{2x}$

b. A sample proof can be found in the worked solutions in your online resources.

c. A sample proof can be found in the worked solutions in your online resources.

d. i. 110 m^3

ii. Mathematically, $0 < h < 8$

e. 4.62 m
See the graph at the bottom of the page.*

f. A sample proof can be found in the worked solutions in your online resources.

8. a. A sample proof can be found in the worked solutions in your online resources.

b. A sample proof can be found in the worked solutions in your online resources.

c. $0 < r < 20$

d. 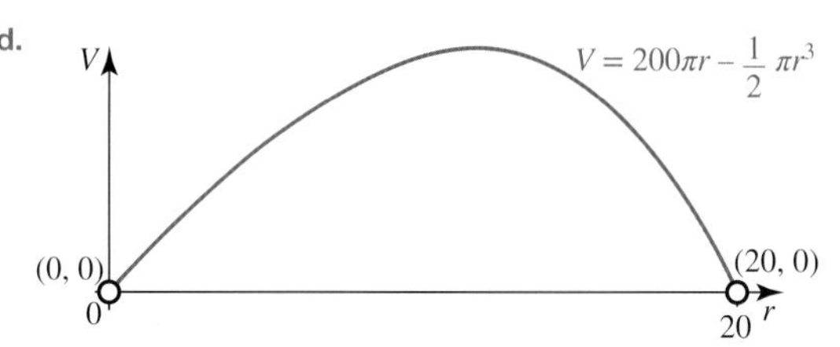

e. Radius 2 cm, height 99 cm, or radius 18.9 cm, height 1.1 cm

f. Greatest volume of 4836.8 cm^3; base radius 11.5 cm; height 11.5 cm

*3 e.
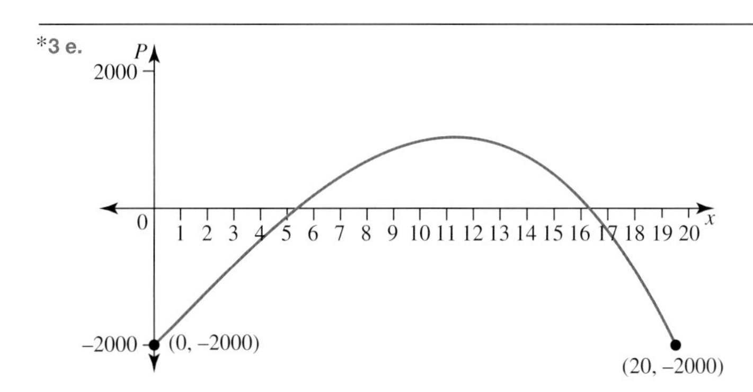

*7 e.
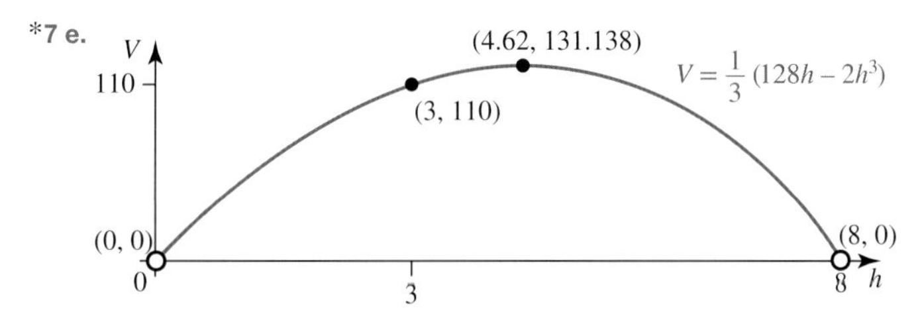

9. a. $(0, 21)$, $(1.25, 1)$, $(2.5, 1.1)$, $(4, 0.1)$

b. $d = 2.1$

c. $125a + 100b + 80c = -70.4$
$125a + 50b + 20c = -8$
$64a + 16b + 4c = -2$

d. $y = -0.164x^3 + x^2 - 1.872x + 2.1$

10. a. $T(3) = 9.85$, $T(20) = 9.71$

b.

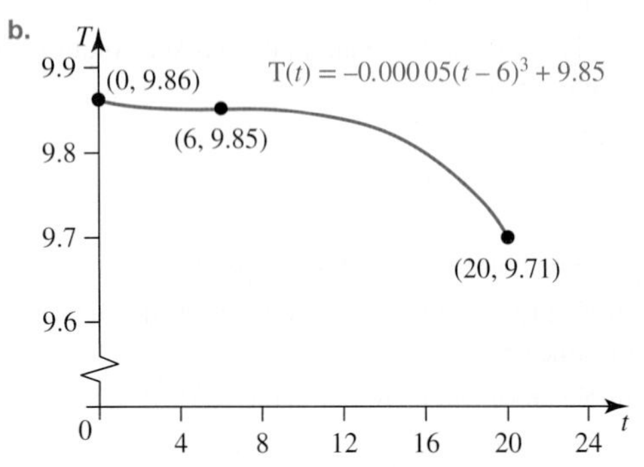

c. $T(28) = 9.32$; unlikely but not totally impossible. The model is probably not a good predictor.

11. a. $x = 0$, $x = 6$

b. Length $2x - 6$; width $6x - x^2$

c. A sample proof can be found in the worked solutions in your online resources.

d. $3 < x < 6$

e. $x = 4$, $x = \dfrac{5 + \sqrt{33}}{2}$

f. Length 3.464 units; width 6.000 units

12. a. 3 x-intercepts; 2 turning points

b. $y = -2x\left(x^2 - 9\right)$

c.

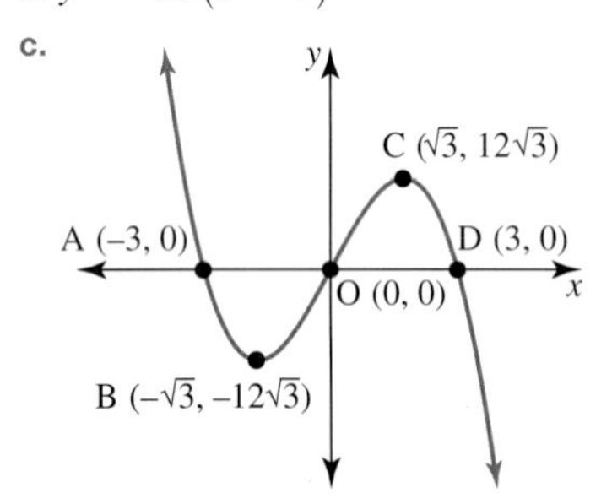

d. $y = 12x$

e. $y = 4x^3$

4.6 Exam questions

Note: Mark allocations are available with the fully worked solutions online.

1. B
2. C
3. D

4.7 Review

4.7 Exercise

Technology free: short answer

1. $(x - 1)(x + 3)^2$
2. $a = -2$, $b = 2$
3. Quotient $2x^2 - 7x + 15$; remainder -31
4. a. Stationary point of inflection $(-3, 8)$; y-intercept $(0, -19)$; x-intercept $(-1, 0)$

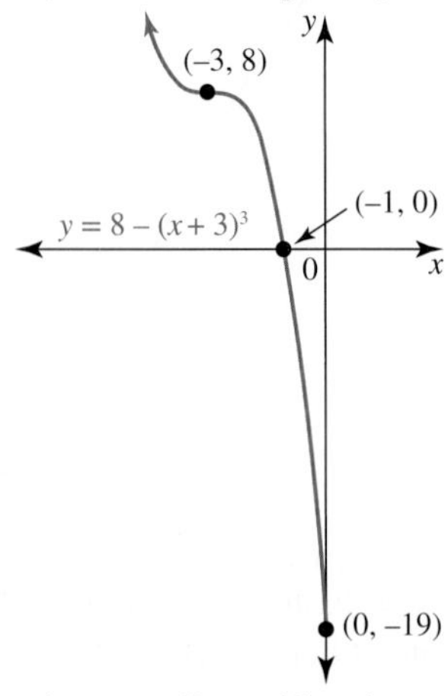

b. y-intercept $(0, -160)$; x-intercepts $(-5, 0)$, $(4, 0)$ (turning point)

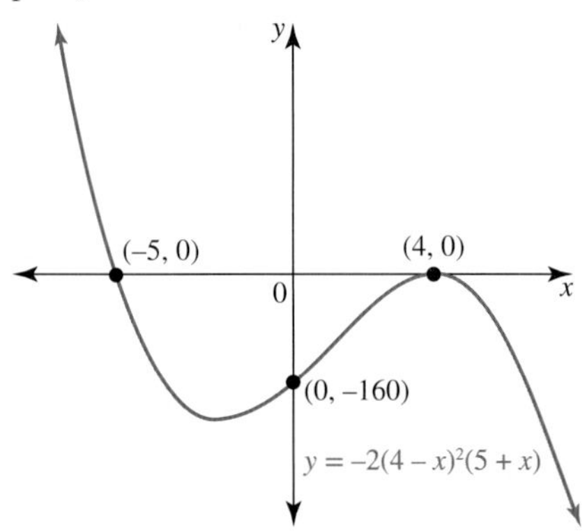

c. Stationary point of inflection $\left(\dfrac{3}{8}, 0\right)$; y-intercept $(0, -27)$

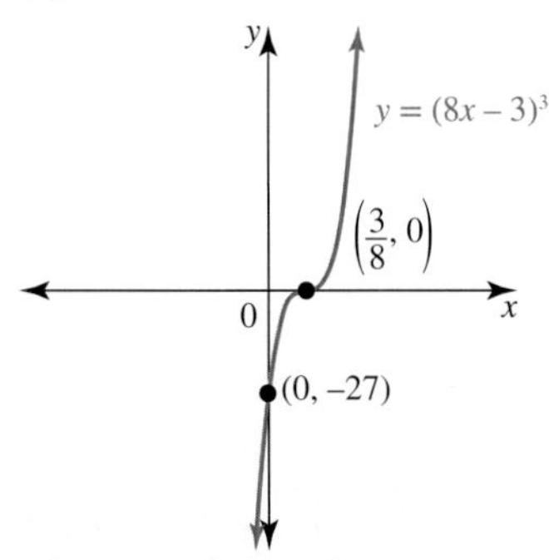

d. y-intercept $(0, 0)$; x-intercepts $\left(-\dfrac{\sqrt{2}}{2}, 0\right), (0, 0), \left(\dfrac{\sqrt{2}}{2}, 0\right)$

5. y-intercept $(0, 6)$; x-intercepts at $x = 1$, $x = 2$, $x = 3$

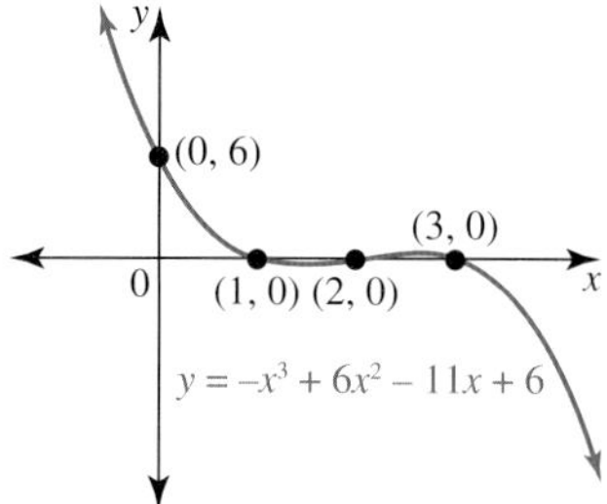

$x < 1$ or $2 < x < 3$

6. $(0, 0)$, $\left(\sqrt{6}, -2\sqrt{6}\right)$, $\left(-\sqrt{6}, 2\sqrt{6}\right)$

Technology active: multiple choice

7. E
8. D
9. E
10. E
11. C
12. A
13. A
14. C
15. E
16. C

Technology active: extended response

17. a. $p(3) = 8(3)^3 - 34(3)^2 + 33(3) - 9$
$= 216 - 306 + 99 - 9$
$= 0$
Since $p(3) = 0$, $(x - 3)$ is a factor.

b. $(x - 3)(4x - 3)(2x - 1)$

c.

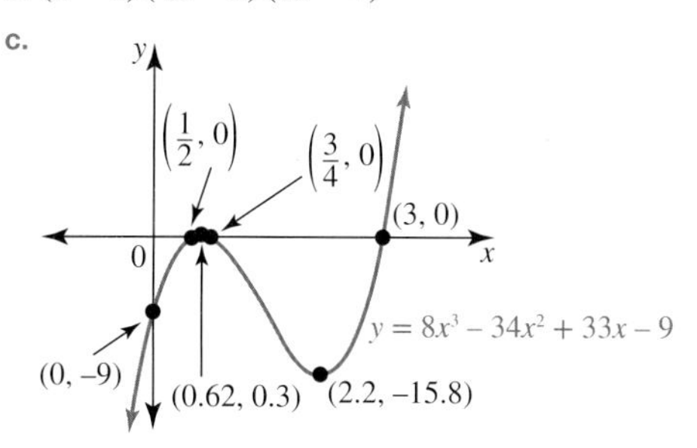

d. $\left\{x : \frac{1}{2} \le x \le \frac{3}{4}\right\} \cup \{x : x \ge 3\}$

e. $\left\{0. \frac{3}{2}, \frac{11}{4}\right\}$

f. i. $-15.8 < k < 0.3$
ii. $k = 0.3$, $k = -15.8$
iii. $k < -15.8$ or $k > 0.3$

18. a. R has degree 2; C has degree 3.

b. Revenue \$90; cost \$6; profit \$84

c. The profit is revenue minus cost, $R - C$.
$\therefore\ p(x) = r(x) - c(x)$
$= 6\left(2x^2 + 10x + 3\right) - x\left(6x^2 - x + 1\right)$
$= 12x^2 + 0x + 18 - 6x^3 + x^2 - x$
$\therefore\ p(x) = -6x^3 + 13x^2 + 59x + 18$

d. Restriction $x \ge 0$; x-intercept $(4.5, 0)$

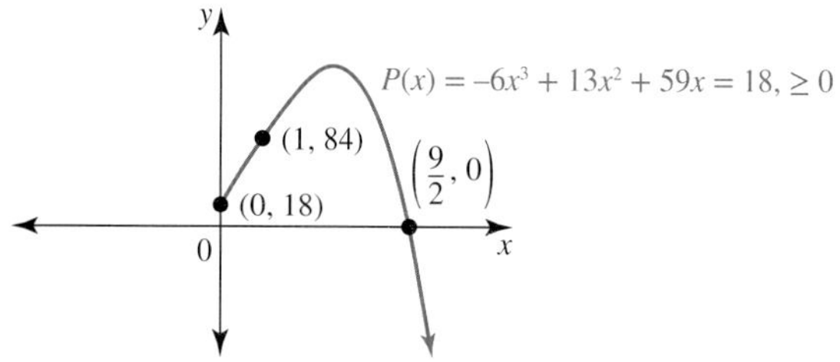

e. $d = 4501$

19. a. $y = -2x + 22$

b. $y = a(2x - 1)(x - 6)(x + b)$, $0 \le x \le 8$
Substitute point A $(1, 20)$.
$20 = a(2 - 1)(1 - 6)(1 + b)$
$20 = -5a(1 + b)$
$a(1 + b) = -4$.... [1]
Substitute point B $(5, 12)$.
$12 = a(10 - 1)(5 - 6)(5 + b)$
$12 = -9a(5 + b)$
$3a(5 + b) = -4$.... (2)
Divide equation [2] by equation [1].
$\frac{3a(5 + b)}{a(1 + b)} = \frac{-4}{-4}$
$\frac{3(5 + b)}{1 + b} = 1,\ a \ne 0$
$15 + 3b = 1 + b$
$2b = -14$
$b = -7$
Substitute $b = -7$ into equation [1].
$a(1 - 7) = -4$
$-6a = -4$
$a = \frac{4}{6}$
$= \frac{2}{3}$

c. End points $(0, -28)$, $(8, 20)$

d.

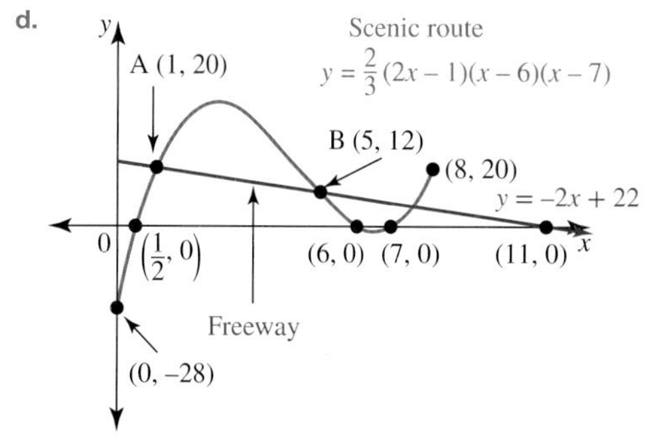

e. $(1, 20)$, $(5, 12)$ and $\left(\frac{15}{2}, 7\right)$

f. $\left(\frac{15}{2}, 7\right)$

20. a. $\frac{13\sqrt{3}}{3}$ metres

b. $V = \frac{1}{3}\pi h\left(169 - h^2\right)$

c. $0 < h < 13$

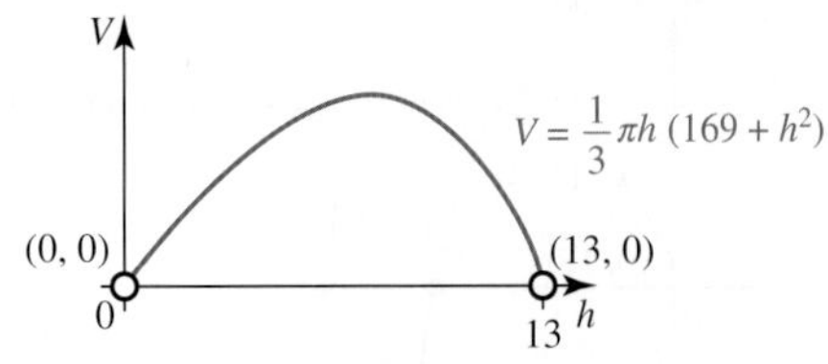

d. $V(7) = 280\pi = V(8)$, $V(9) = 264\pi$, $a = 7$

e. **i.** $h = 7.5$, $r = 10.62$

ii. 886 m^3

f. **i.** $h = \dfrac{13}{\sqrt{3}}$, $r = \dfrac{13\sqrt{6}}{3}$

ii. 886 m^3

4.7 Exam questions

Note: Mark allocations are available with the fully worked solutions online.

1. $\dfrac{4x^2 - 3x + 1}{x - 2} = 4x + 5 + \dfrac{11}{x - 2}$
2. $x = 3, \dfrac{1}{2}, -\dfrac{2}{3}$
3. **a.** $P(x) = (3x + 2)(x - 1)(x - 2)$, $k = -7$

   b.

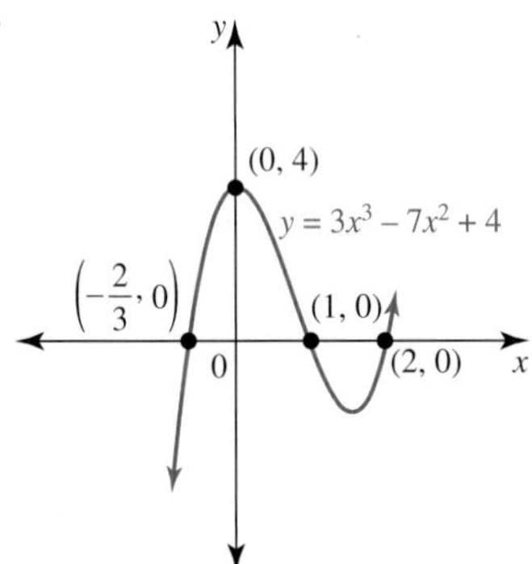

4. The equation is $y = x^3 - 5x^2 - 12x + 36$.
5. $V = 9x^2 - x^3$, $0 < x < 9$

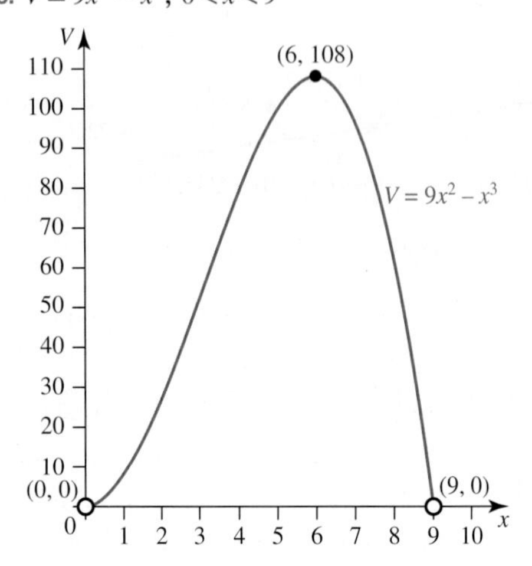

5 Quartic polynomials

LEARNING SEQUENCE

Fully worked solutions for this topic are available online.

5.1 Overview

Hey students! Bring these pages to life online

Watch videos

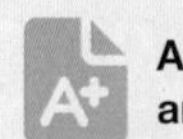

Engage with interactivities

Answer questions and check results

Find all this and MORE in jacPLUS

5.1.1 Introduction

When the Tartaglia–Cardano general method for solving cubic equations was published in 1545, the general method for solving quartic equations was included too. This was the work of Luigi Ferrari, then a student of Cardano's.

With general methods now obtained for degrees 1 to 4, attention turned to higher order polynomial equations, in particular the degree 5 quintic. Over a span of three centuries several mathematicians thought they were close to a general method, only to suffer failure. In the 19th century two young mathematicians finally solved the problem by proving, to everyone's surprise including theirs, that there was no general method.

In 1828 the Norwegian Niels Abel was the first to prove that quintics had no general solution. Shortly afterwards, Évariste Galois from France went further and proved that no general formula was possible for solving any polynomial equation other than those of degrees 1 to 4.

This did not mean higher degree equations could not be solved. Numerical methods of solution were widely used, including the method of bisection. However, it meant there was no general formula or method for any degree beyond 4.

Tragically, the two men had short lives. Abel died of tuberculosis aged 26; Galois was killed in a duel in 1832, aged only 20.

In 1637 Pierre de Fermat also scribbled in a margin, complaining it was too small for his proof of what became known as Fermat's Last Theorem. No-one could solve the problem until finally, in 1995, the British mathematician Andrew Wiles did. Wiles used Galois Theory in his proof, and he received the Abel Prize along with many other accolades.

KEY CONCEPTS

This topic covers the following key concepts from the VCE Mathematics Study Design:

- qualitative interpretation of features of graphs of functions, including those of real data not explicitly represented by a rule, with approximate location of any intercepts, stationary points and points of inflection
- graphs of power functions $f(x) = x^n$ for $n \in \left\{-2, -1, \frac{1}{3}, \frac{1}{2}, 1, 2, 3, 4\right\}$, and transformations of these graphs to the form $y = a(x+b)^n + c$ where $a, b, c \in R$ and $a \neq 0$
- graphs of polynomial functions of low degree, and interpretation of key features of these graphs
- use of parameters to represent families of functions and determination of rules of simple functions and relations from given information
- solution of polynomial equations of low degree, numerically, graphically and algebraically, including numerical approximation of roots of simple polynomial functions using the bisection method algorithm.

Note: Concepts shown in grey are covered in other topics.

Source: VCE Mathematics Study Design (2023–2027) extracts © VCAA; reproduced by permission.

5.2 Quartic polynomials

LEARNING INTENTION

At the end of this subtopic you should be able to:
- sketch graphs of quartic polynomials, identifying the key features
- solve quartic equations and inequations from factored form.

A quartic polynomial is a polynomial of degree 4 and is of the form $P(x) = ax^4 + bx^3 + cx^2 + dx + e$, where $a \neq 0$ and $a, b, c, d, e \in R$.

5.2.1 Graphs of quartic polynomials of the form $y = a(x - h)^4 + k$

The simplest quartic polynomial graph has the equation $y = x^4$. As both negative and positive numbers raised to an even power, in this case 4, will be positive, the long-term behaviour of the graph of $y = x^4$ must be that as $x \to -\infty$ or as $x \to \infty$, then $y \to \infty$.

The graph of $y = x^4$ is similar to that of the parabola $y = x^2$. Both graphs are concave up with a minimum turning point at $(0, 0)$, and both contain the points $(-1, 1)$ and $(1, 1)$. However, for the intervals where $x < -1$ and $x > 1$, the graph of $y = x^4$ lies above the parabola. This is because $x^4 > x^2$ for these intervals. Likewise, the graph of $y = x^4$ lies below that of the parabola for the intervals $-1 < x < 0$ and $0 < x < 1$, since $x^4 < x^2$ for these intervals.

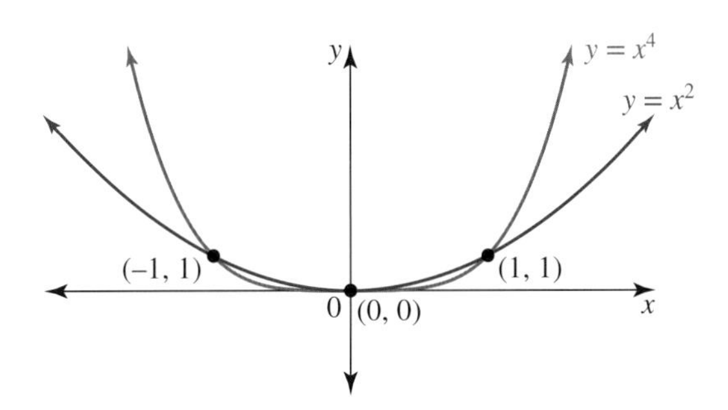

Despite these differences, the two graphs are of sufficient similarity to enable us to obtain the key features of graphs of quartic polynomials of the form $y = a(x - h)^4 + k$ in much the same manner as for quadratics of the form $y = a(x - h)^2 + k$.

Dilated by factor a from the x-axis, a horizontal translation of h units and a vertical translation of k units, the graph of $y = x^4$ is transformed to that of $y = a(x - h)^4 + k$.

The graph of $y = a(x - h)^4 + k$

The key features of the graph of $y = a(x - h)^4 + k$ are:
- **a turning point with coordinates (h, k)**
- **if $a > 0$, the turning point is a minimum**
- **if $a < 0$, the turning point is a maximum**
- **an axis of symmetry with equation $x = h$**
- **zero, one or two x-intercepts, solutions of $a(x - h)^4 + k = 0$**
- **a y-intercept that can be found by substituting $x = 0$.**

WORKED EXAMPLE 1 Sketching quartics in turning point form

Sketch the graphs of the following quartic polynomials.

a. $y = \frac{1}{4}(x+3)^4 - 4$ **b.** $y = -(3x-1)^4 - 7$

THINK	WRITE
a. 1. State the coordinates and type of turning point.	**a.** $y = \frac{1}{4}(x+3)^4 - 4$ The turning point is $(-3, -4)$. As $a = \frac{1}{4}, a > 0$, so the turning point is a minimum.
2. Calculate the y-intercept.	y-intercept: let $x = 0$. $y = \frac{1}{4}(3)^4 - 4$ $= \frac{81}{4} - \frac{16}{4}$ $= \frac{65}{4}$ y-intercept: $\left(0, \frac{65}{4}\right)$
3. Determine whether there will be any x-intercepts.	As the y-coordinate of the minimum turning point is negative, the concave up graph must pass through the x-axis.
4. Calculate the x-intercepts. *Note*: $\pm$ is needed in taking the fourth root of each side.	$\frac{1}{4}(x+3)^4 - 4 = 0$ $\frac{1}{4}(x+3)^4 = 4$ $\therefore (x+3)^4 = 16$ Take the fourth root of both sides. $(x+3) = \pm\sqrt[4]{16}$ $(x+3) = \pm 2$ $\therefore x = -5$ or $x = -1$ x-intercepts: $(-5, 0)$ and $(-1, 0)$
5. Sketch the graph.	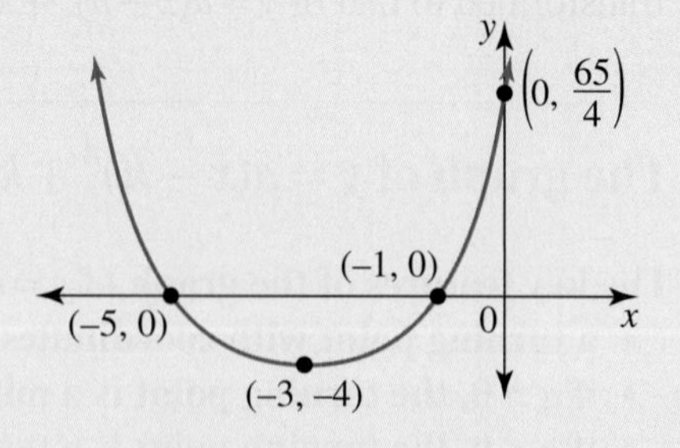
b. 1. Express the equation in the form $y = a(x-h)^4 + k$.	**b.** $y = -(3x-1)^4 - 7$ $= -\left(3\left(x - \frac{1}{3}\right)\right)^4 - 7$ $= -81\left(x - \frac{1}{3}\right)^4 - 7$

2. State the coordinates of the turning point and its type.	The graph has a maximum turning point at $\left(\frac{1}{3}, -7\right)$.
3. Calculate the y-intercept.	y-intercept: let $x = 0$ in the original form. $y = -(3x-1)^4 - 7$ $= -(-1)^4 - 7$ $= -(1) - 7$ $= -8$ y-intercept: $(0, -8)$
4. Determine whether there will be any x-intercepts.	As the y-coordinate of the maximum turning point is negative, the concave down graph will not pass through the x-axis.
5. Sketch the graph.	The graph is symmetric about its axis of symmetry, $x = \frac{1}{3}$.

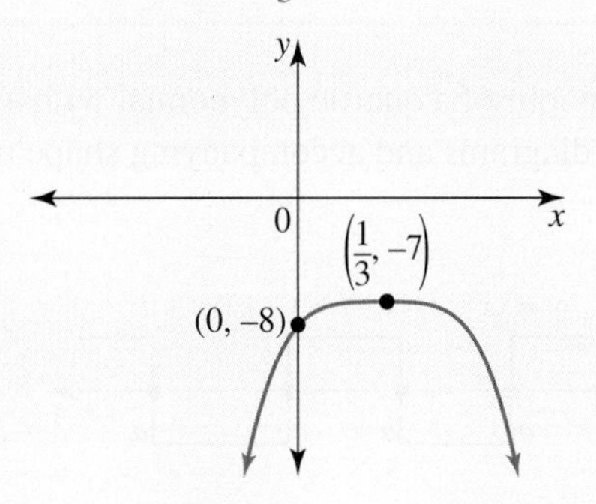

TI | THINK	DISPLAY/WRITE	CASIO | THINK	DISPLAY/WRITE
a. 1. On a Graphs page, complete the entry line as: $f1(x) = \frac{1}{4}(x+3)^4 - 4$ then press ENTER to view the graph. To view the key points on the graph, select: MENU 6: Analyze Graph 1: Zero and MENU 6: Analyze Graph 2: Minimum	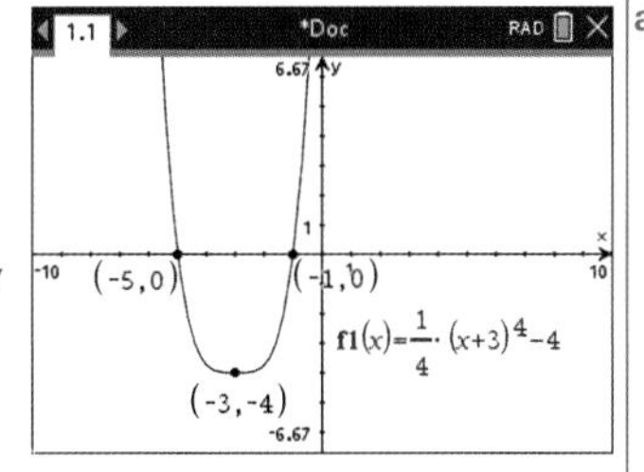	**a. 1.** On a Graphs & Table screen, complete the entry line as: $y1 = \frac{1}{4}(x+3)^4 - 4$ Tap the Graph icon to view the graph. Select the $y = 0$, MIN and x y icons to view the key points on the graph.	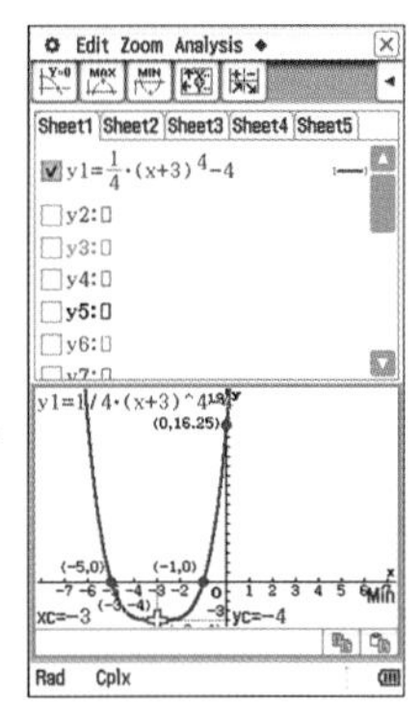

Resources

Interactivity Polynomials of higher degree (int-2569)

5.2.2 Quartic polynomials that can be expressed as the product of linear factors

Not all quartic polynomials have linear factors. However, the graphs of those that can be expressed as the product of linear factors can be readily sketched by analysing these factors.

Quartic polynomial factors

A quartic polynomial may have up to 4 linear factors since it is of fourth degree. The possible combinations of these linear factors are:

- **four distinct linear factors, $y = (x - a)(x - b)(x - c)(x - d)$**
- **one repeated linear factor, $y = (x - a)^2(x - b)(x - c)$**
- **two repeated linear factors, $y = (x - a)^2(x - b)^2$**
- **one factor of multiplicity 3, $y = (x - a)^3(x - b)$**
- **one factor of multiplicity 4, $y = (x - a)^4$.**

The last case, in which the graph has a minimum turning point at $(a, 0)$, has already been considered.

The long-term behaviour of a quartic polynomial with a positive coefficient of x^4 is as $x \to \pm\infty, y \to \infty$. Therefore, the sign diagrams and accompanying shape of the graphs must be of the form shown in the diagrams.

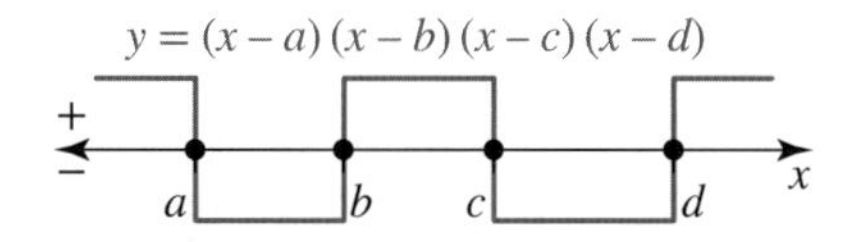

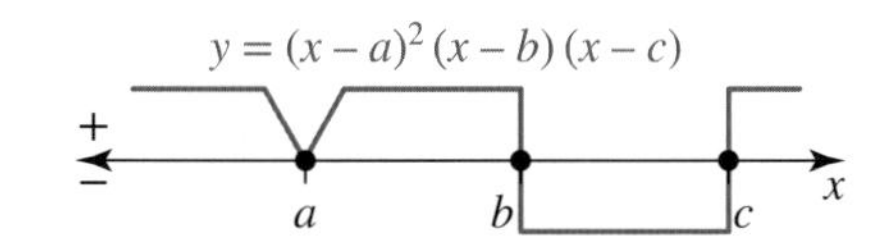

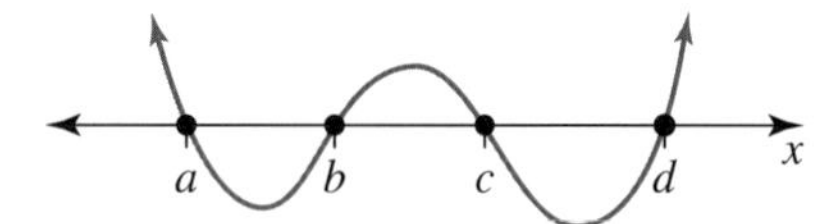

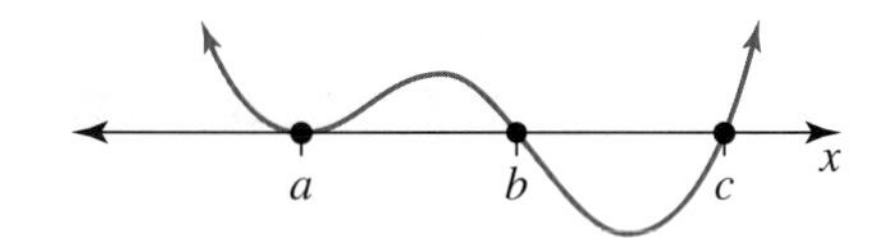

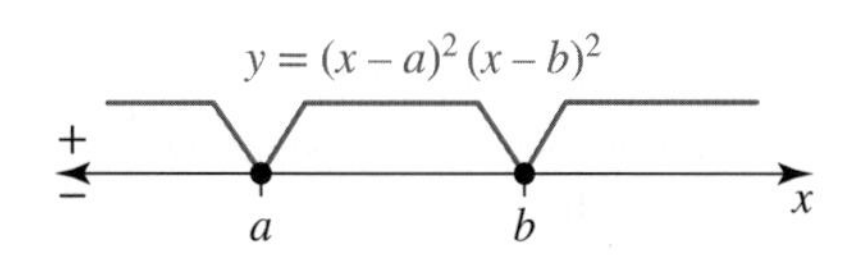

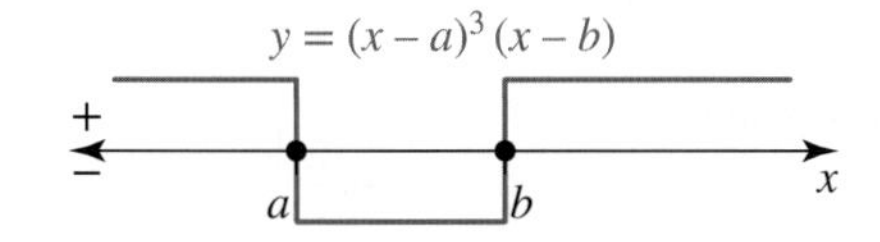

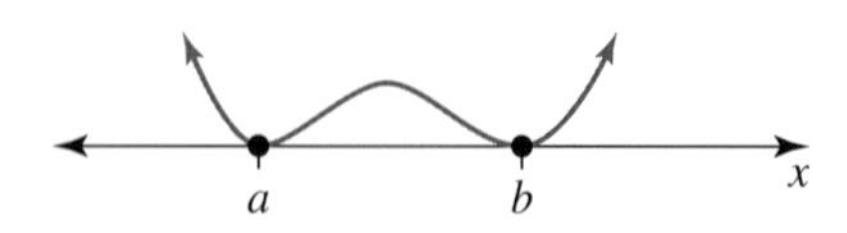

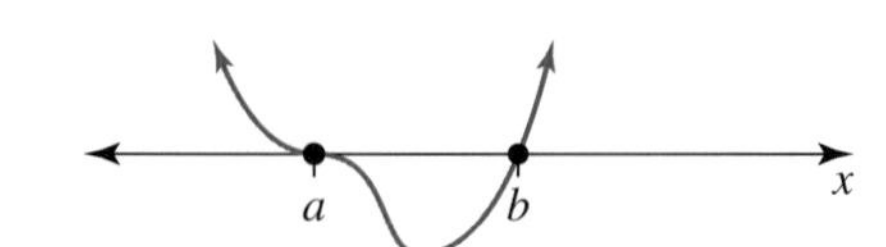

For a negative coefficient of x^4, as $x \to \pm\infty, y \to -\infty$, so the sign diagrams and graphs are inverted.

Quartic graphs

The x-intercept at $x = a$ from:

- **a single factor, $(x - a)$, cuts the x-axis**
- **a repeated factor, $(x - a)^2$, is a turning point on the x-axis**
- **a factor of multiplicity three, $(x - a)^3$, is a stationary point of inflection.**

WORKED EXAMPLE 2 Sketching quartics in factorised form

Sketch the graph of $y = (x+2)(2-x)^3$.

THINK	WRITE
1. Calculate the x-intercepts.	$y=(x+2)(2-x)^3$ x-intercepts: let $y=0$. $(x+2)(2-x)^3=0$ $\therefore (x+2)=0$ or $(2-x)^3=0$ $\therefore x=-2$ or $x=2$ x-intercepts: $(-2,0)$ and $(2,0)$
2. Interpret the nature of the graph at each x-intercept.	Due to the multiplicity of each factor, at $x=-2$ the graph cuts the x-axis, and at $x=2$ its saddle-cuts the x-axis. The point $(2,0)$ is a stationary point of inflection.
3. Calculate the y-intercept.	y-intercept: let $x=0$. $y=(2)(2)^3$ $=16$ y-intercept: $(0,16)$
4. Determine the sign of the coefficient of the leading term and identify the long-term behaviour of the graph.	The leading term is $(x)(-x)^3=-x^4$. The coefficient of the leading term is negative, so as $x\to\pm\infty$, $y\to-\infty$. This means the sketch of the graph must start and finish below the x-axis.
5. Sketch the graph.	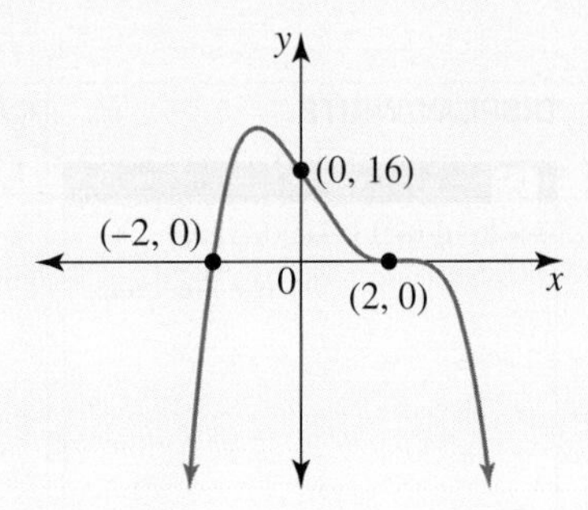

5.2.3 Equations and inequations

Factorisation techniques may enable a quartic polynomial $p(x)$ given in its general form to be rewritten as the product of linear factors.

Quartic equations and inequations given as linear factors

For polynomials given in the form of the product of linear factors, the graph of $y=p(x)$ can be readily sketched and the equation $p(x)=0$ solved using the Null Factor Law. A sign diagram or a graph will help in solving inequations.

WORKED EXAMPLE 3 Solving quartic inequalities

If $p(x) = (x-1)(x+1)(3x+1)(x-2)$, solve the inequation $p(x) \leq 0$.

THINK	WRITE
1. State the zeros of the polynomial.	$p(x) = (x-1)(x+1)(3x+1)(x-2)$ The zeros of the polynomial are: $x = 1, x = -1, x = -\frac{1}{3}, x = 2$
2. Draw the sign diagram.	The leading term has a positive coefficient, so the sign diagram is:

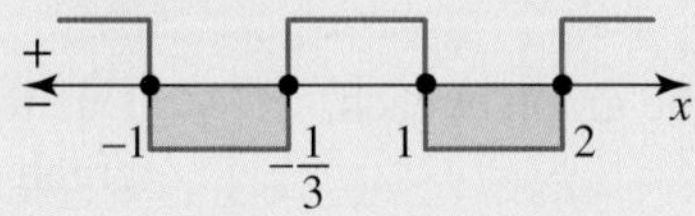

Alternatively, sketch the graph.

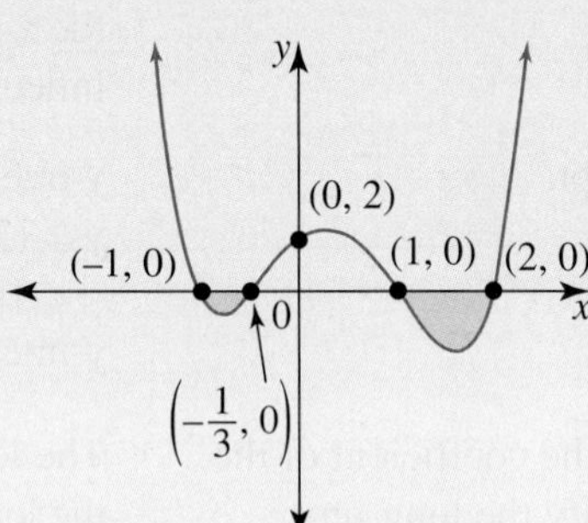

THINK	WRITE
3. State the solution to the inequation.	$p(x) \leq 0$ $\therefore -1 \leq x \leq -\frac{1}{3}$ or $1 \leq x \leq 2$

TI | THINK

1. On a Calculator page, press MENU and select:
   3. Algebra
   1. Solve
   Complete the entry line as:
   solve $(3x^4 - 5x^3 - 5x^2 + 5x + 2 \leq 0, x)$
   then press ENTER.

DISPLAY/WRITE

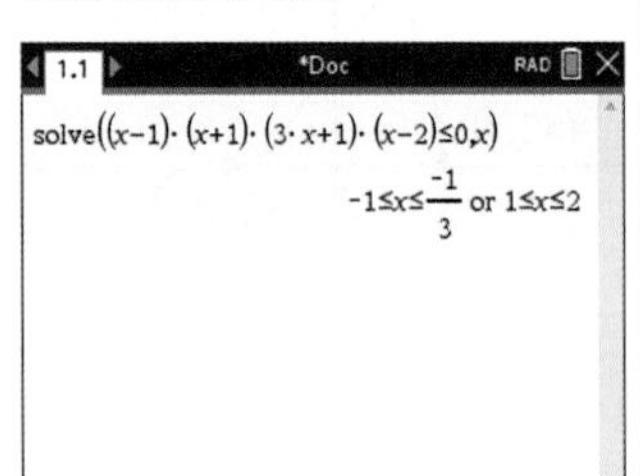

2. The answer appears on the screen.

$-1 \leq x \leq -\frac{1}{3}$ or $1 \leq x \leq 2$

CASIO | THINK

1. On a Main screen, complete the entry line as:
   solve $(3x^4 - 5x^3 - 5x^2 + 5x + 2 \leq 0, x)$
   then press EXE.

DISPLAY/WRITE

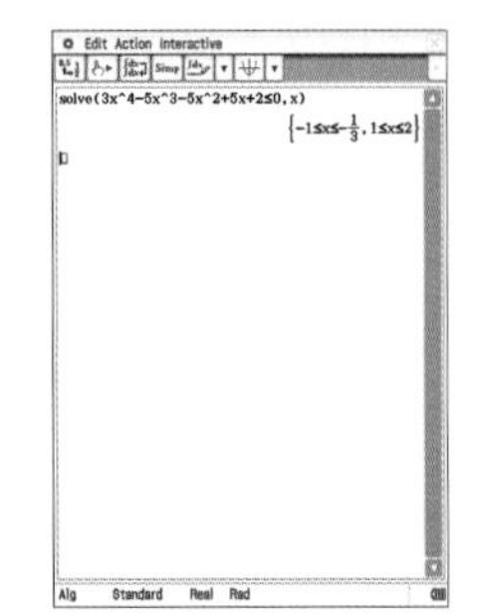

2. The answer appears on the screen.

$-1 \leq x \leq -\frac{1}{3}$ or $1 \leq x \leq 2$

Quartic equations and inequations from factored form

For equations given in the form of the product of factors, the solutions are still found using the Null Factor Law. The factors may be quadratic factors. The quadratic factors may or may not have zeros. Again, either a sign diagram or a sketch of the graph will help in solving inequations.

WORKED EXAMPLE 4 Solving quartic inequations with quadratic factors

For the following quartic polynomials, solve the inequation $p(x) \geq 0$. Give your answer to 2 decimal places where appropriate.

a. $p(x) = (x + 2)(x - 3)\left(x^2 + 4\right)$

b. $p(x) = (x + 2)(3 - x)\left(x^2 + 2x - 4\right)$

c. $p(x) = \left(x^2 + 2x - 4\right)\left(x^2 - 4\right)$

THINK	WRITE
a. 1. State the zeros of the polynomial. *Note:* $x^2 + 4 > 0$ for all x.	$p(x) = (x + 2)(x - 3)\left(x^2 + 4\right)$ The zeros are $x = -2$ or $x = 3$ as $x^2 + 4 \neq 0$.
2. Draw a sign diagram.	The leading term has a positive coefficient, so the sign diagram is:

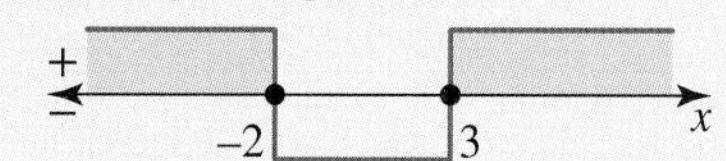

Alternatively, sketch the graph.

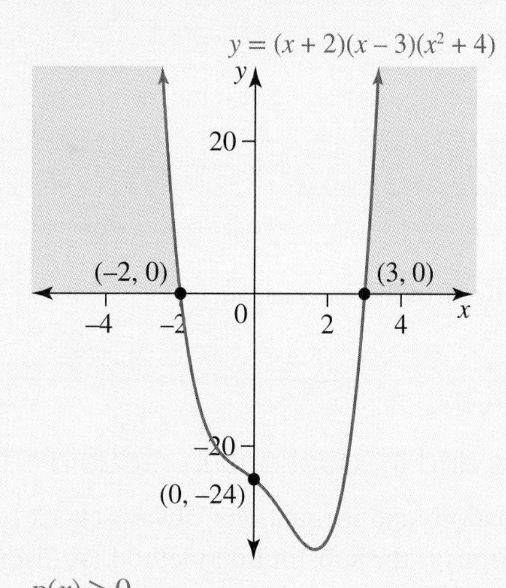

THINK	WRITE
3. State the solutions to the inequation.	$p(x) \geq 0$ $x \leq -2$ or $x \geq 3$
b. 1. State the zeros of the polynomial. Use the quadratic formula to find the zeros of $\left(x^2 + 2x - 4\right) = 0$.	$p(x) = (x + 2)(3 - x)\left(x^2 + 2x - 4\right)$ The zeros are $x = -2, 3, -3.24$ or 1.24.
2. Draw a sign diagram.	The leading term has a negative coefficient, so the sign diagram is:

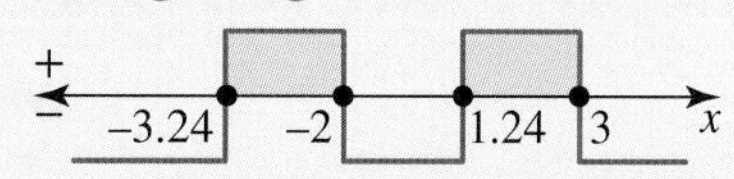

Alternatively, sketch the graph.

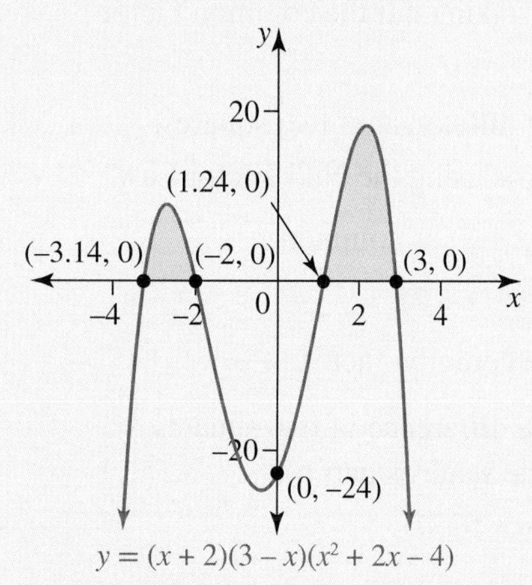

3. State the solutions to the inequation. | $p(x) \geq 0$
$-3.24 \leq x \leq -2$ or $1.24 \leq x \leq 3$

c. 1. State the zeros of the polynomial.
Use the quadratic formula to find the zeros of $\left(x^2+2x-4\right)=0$ and the difference of two squares for $\left(x^2-4\right)=0$.

$p(x)=\left(x^2+2x-4\right)\left(x^2-4\right)$
The zeros are $x=-2, 2, -3.24$ or 1.24.

2. Draw a sign diagram.

The leading term has a positive coefficient, so the sign diagram is:

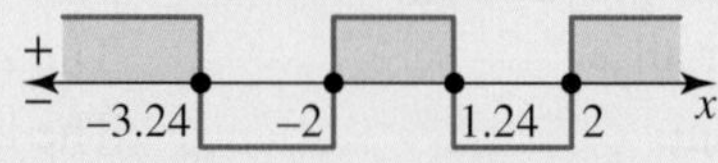

Alternatively, sketch the graph.

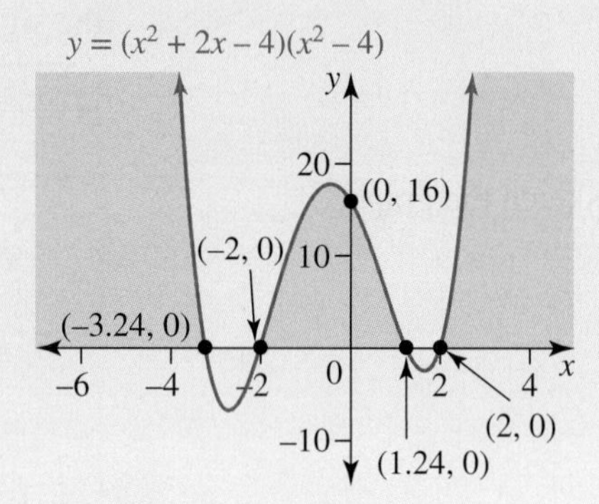

3. State the solutions to the inequation.

$p(x) \geq 0$
$x \leq -3.24$ or $-2 \leq x \leq 1.24$ or $x \geq 2$

Quartic equations and inequations in reducible form

In solving quartic equations and inequations, always check for common factors. Another method to solve quartic equations and inequations is the substitution method, as discussed in Topic 3. The following example illustrates using similar methods in solving quartic equations and inequations.

WORKED EXAMPLE 5 Solving quartic equations and inequations by substitution

a. For the following quartic polynomials, solve the equation $p(x)=0$.
i. $p(x)=x^4-16x^2$ **ii. $p(x)=(x+2)^4-16(x+2)^2$**

b. Solve the inequation $(x-1)^4-10(x-1)^2+9>0$.

THINK	WRITE
a. i. 1. Factorise by taking out the common factor of x^2.	$x^4-16x^2=0$ $x^2\left(x^2-16\right)=0$
2. Factorise by difference of two squares.	$x^2(x-4)(x+4)=0$
3. State the zeros using the Null Factor Law.	$x=0, -4$ or 4
ii. 1. Recognise $(x+2)$ as common.	$(x+2)^4-16(x+2)^2=0$
2. Substitute $a=(x+2)$.	$a^4-16a^2=0$
3. Take out the common factor.	$a^2\left(a^2-16\right)=0$
4. Factorise the difference of two squares (note now the same as part **a**).	$a^2(a-4)(a+4)=0$
5. State the zeros for a.	$a=0, -4$ or 4

6. Substitute $a = (x+2)$. $\quad x+2 = 0, -4$ or 4

7. State the solutions of $p(x) = 0$. $\quad x = -2, -6$ or 2

b. 1. Recognise the quadratic.

$$(x-1)^4 - 10(x-1)^2 + 9 > 0$$

$$\left[(x-1)^2\right]^2 - 10\left[(x-2)^2\right] + 9 > 0$$

2. Substitute $a = (x-1)^2$ to give a quadratic in a.

$$a^2 - 10a + 9 > 0$$

3. Factorise the quadratic in a.

$$(a-9)(a-1) > 0$$

4. Substitute $a = (x-1)^2$.

$$\left((x-1)^2 - 9\right)\left((x-1)^2 - 1\right) > 0$$

5. Factorise each quadratic by the difference of two squares.

$$(x-1-3)(x-1+3)(x-1-1)(x-1+1) > 0$$

6. Simplify.

$$(x-4)(x+2)(x-2)(x) > 0$$

7. State the zeros. $\quad x = 4, -2, 2, 0$

8. Draw the sign diagram. $\quad$ The leading term has a positive coefficient, so the sign diagram is:

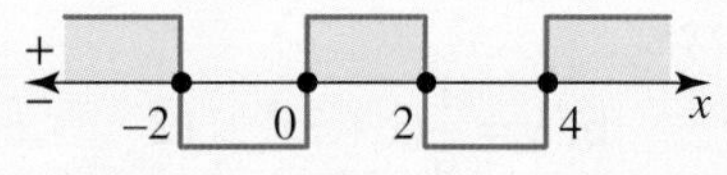

Alternatively, sketch the graph.

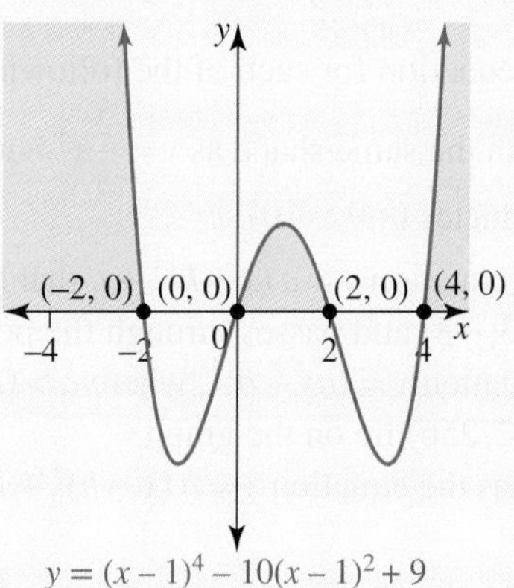

9. State the solutions to the inequation. $\quad p(x) = 0$

$$x < -2 \text{ or } 0 < x < 2 \text{ or } x > 4$$

5.2 Exercise

Technology free

1. a. On the same set of axes, sketch the graphs of $y = x^4$, $y = 2x^4$ and $y = \frac{1}{2}x^4$. Label the points for which $x = -1, 0$ and 1.

   b. On the same set of axes, sketch the graphs of $y = x^4$, $y = -x^4$, $y = -2x^4$ and $y = (-2x)^4$. Label the points for which $x = -1, 0$ and 1.

c. On the same set of axes, sketch the graphs of $y=x^4$, $y=-(x+1)^4$ and $y=(1-x)^4$. Label the points for which $x=-1, 0$ and 1.

d. On the same set of axes, sketch the graphs of $y=x^4$, $y=x^4+2$ and $y=-x^4-1$. Label the points for which $x=-1, 0$ and 1.

2. **WE1** Sketch the graphs of the following quartic polynomials.

a. $y=(x-2)^4-1$

b. $y=-(2x+1)^4$

3. a. State the coordinates and nature of the turning point of the graph of $y=\frac{1}{8}(x+2)^4-2$, and sketch the graph.

b. After the graph of $y=x^4$ has been reflected about the x-axis, translated 1 unit to the right and translated downwards 1 unit:

i. state the coordinates and nature of its turning point.

ii. state its equation and sketch the transformed graph.

c. The equation of the graph of a quartic polynomial is of the form $y=a(x-h)^4+k$. Determine the equation given there is a y-intercept at $(0, 64)$ and a turning point at $(4, 0)$.

4. Sketch the following graphs, identifying the coordinates of the turning point and any point of intersection with the coordinate axes.

a. $y=(x-1)^4-16$

b. $y=\frac{1}{9}(x+3)^4+12$

c. $y=250-0.4(x+5)^4$

d. $y=-\left(6(x-2)^4+11\right)$

5. Determine a possible equation for each of the following.

a. A quartic graph with the same shape as $y=\frac{2}{3}x^4$ but whose turning point has the coordinates $(-9, -10)$.

b. The curve with the equation $y=a(x+b)^4+c$ that has a minimum turning point at $(-3, -8)$ and passes through the point $(-4, -2)$.

c. A curve has the equation $y=(ax+b)^4$, where $a>0$ and $b<0$. The points $(0, 16)$ and $(2, 256)$ lie on the graph.

d. The graph shown has the equation $y=a(x-h)^4+k$.

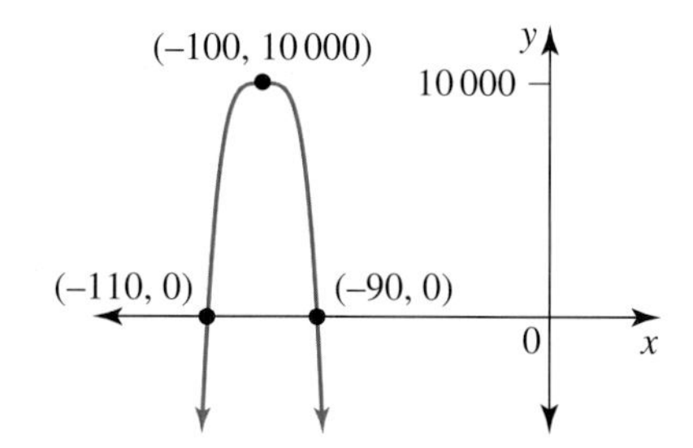

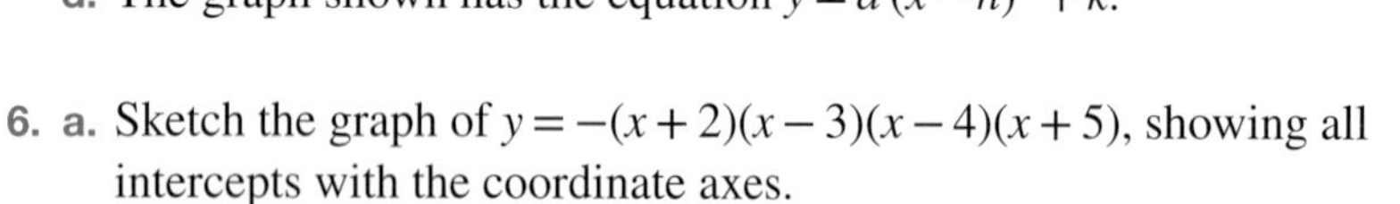

6. a. Sketch the graph of $y=-(x+2)(x-3)(x-4)(x+5)$, showing all intercepts with the coordinate axes.

b. The graph of a quartic polynomial with three x-intercepts is shown.

i. For each of the three x-intercepts, state the corresponding factor in the equation of the graph.

ii. Write the form of the equation.

iii. The graph cuts the y-axis at $(0, -6)$. Determine the equation of the graph.

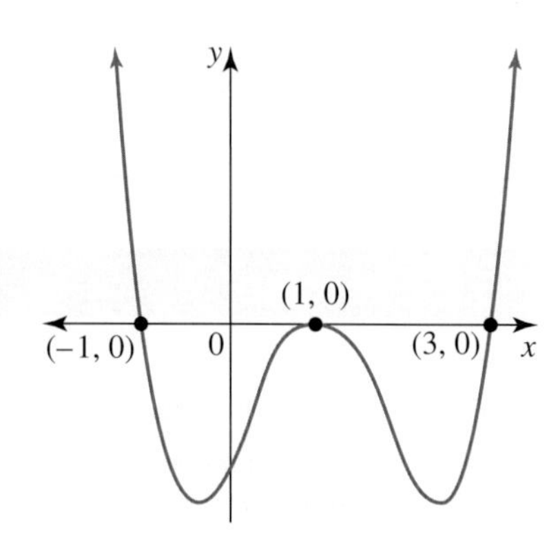

7. **WE2** Sketch the graph of $y=(x+2)^2(2-x)^2$.

8. Give a suitable equation for the graph of the quartic polynomial shown.

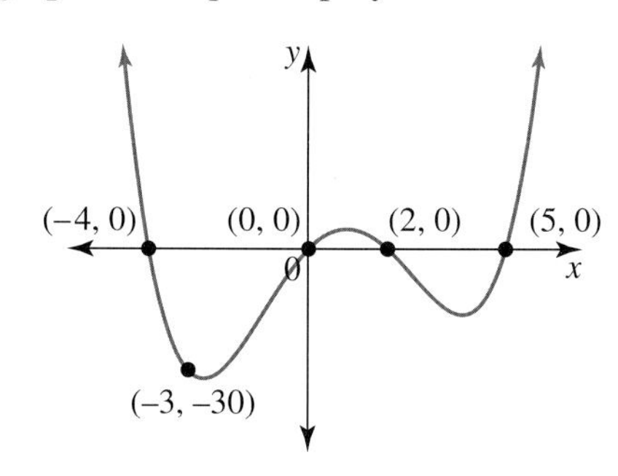

9. Sketch the following quartic polynomials without attempting to locate any turning points that do not lie on the coordinate axes.

a. $y = (x + 8)(x + 3)(x - 4)(x - 10)$

b. $y = -\frac{1}{100}(x + 3)(x - 2)(2x - 15)(3x - 10)$

c. $y = -2(x + 7)(x - 1)^2(2x - 5)$

d. $y = \frac{2}{3}x^2(4x - 15)^2$

e. $y = 3(1 + x)^3(4 - x)$

f. $y = (3x + 10)(3x - 10)^3$

10. For each of the following quartic graphs, form a possible equation.

a.

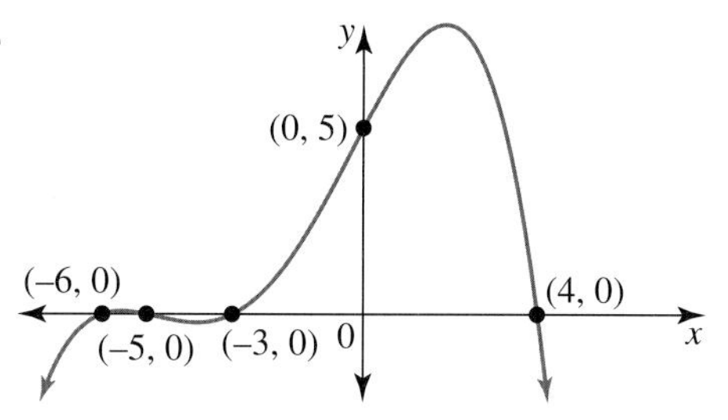

b.

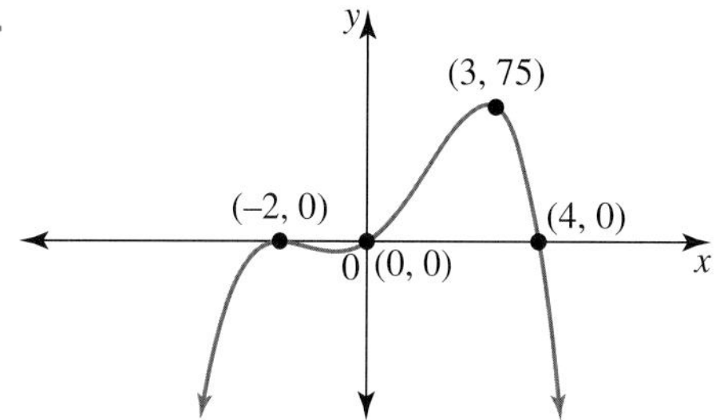

c.

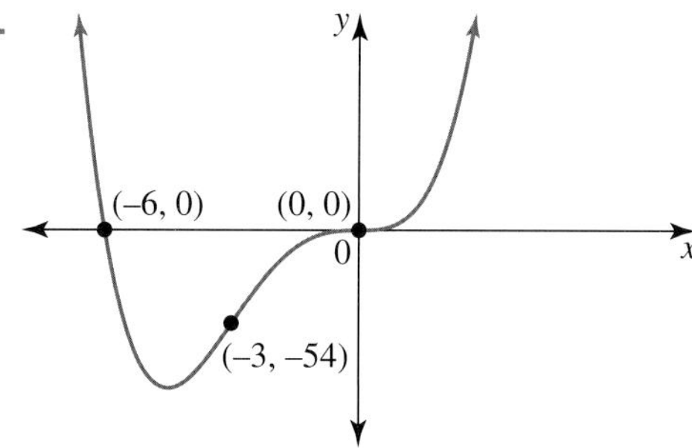

d.

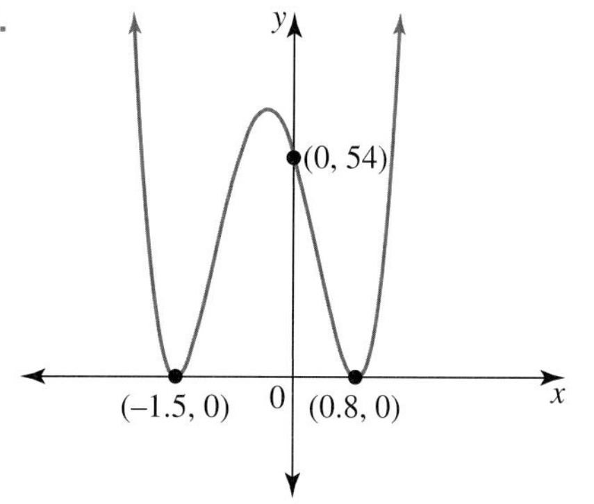

11. If $p(x) = 3x^4 + ax^3 + 2x^2 - 5x + 12$ and $p(-2) = 14$, determine the value of a.

12. Solve the following quartic equations for x.

a. $x(3x + 1)(x - 5)(4x + 3) = 0$

b. $(x - 2)^2(5x + 3)(4 - x) = 0$

c. $(x + 2)^2(2x - 1)^2 = 0$

d. $(x + 4)(2 - x)^3 = 0$

13. **WE3** If $p(x) = (x + 2)(x - 3)(x + 5)(2x + 3)$, solve the inequation $p(x) \geq 0$.

14. For the following quartic polynomials, solve the inequation $p(x) < 0$.

a. $p(x) = x^2(x + 4)(x - 5)$

b. $p(x) = (x - 4)(x - 1)^3$

c. $p(x) = x(x - 2)(4 - x)(2x + 5)$

d. $p(x) = (x + 3)^2(3 - x)^2$

15. **WE4** For the following quartic polynomials, solve the inequation $p(x) \geq 0$. Give your answers to 2 decimal places where appropriate.

a. $p(x) = (x + 5)(x - 2)\left(x^2 + 1\right)$

b. $p(x) = (x - 4)(6 - x)\left(x^2 - 4x - 6\right)$

c. $p(x) = \left(x^2 + 3x - 5\right)\left(x^2 - 9\right)$

16. Solve the following inequations, giving your answers to 2 decimal places where appropriate.

a. $(x-7)(x+2)\left(x^2+1\right)>0$

b. $(x+3)(5-x)\left(x^2-x-7\right)\leq 0$

c. $\left(3x^2-8x+2\right)\left(4x^2-25\right)>0$

d. $\left(x^2-2x+8\right)\left(2x^2-3x+5\right)>0$

17. Solve the equation $(x+2)^4-13(x+2)^2-48=0$.

18. WE5a Solve the following quartic equations for x.

a. $8x^4+72x^3=0$

b. $4x^4=16x^2$

c. $(x+3)^4+2(x+3)^2-15=0$

d. $9(x-1)^4-49(x-1)^2=0$

19. WE5b Solve the following quartic inequations for x.

a. $(x+3)(2x-1)(4-x)(2x-11)<0$

b. $300x^4+200x^3+28x^2\leq 0$

c. $x^3(x+1)-8(x+1)>0$

d. $20(2x-1)^4-8(1-2x)^3\geq 0$

20. A graph with the equation $y=a(x-b)^4+c$ has a maximum turning point at $(-2,4)$ and cuts the y-axis at $y=0$.

a. Determine its equation.

b. Sketch the graph and so determine $\{x: a(x-b)^4+c>0\}$.

Technology active

21. The curve with equation $y=ax^4+k$ passes through the points $(-1,1)$ and $\left(\frac{1}{2},\frac{3}{8}\right)$.

a. Determine the values of a and k.

b. State the coordinates of the turning point and its nature.

c. Give the equation of the axis of symmetry.

d. Sketch the curve.

22. The graph of $y=a(x+b)^4+c$ passes through the points $(-2,3)$ and $(4,3)$.

a. State the equation of its axis of symmetry.

b. Given the greatest y-value the graph reaches is 10, state the coordinates of the turning point of the graph.

c. Determine the equation of the graph.

d. Calculate the coordinates of the point of intersection with the y-axis.

e. Calculate the exact value(s) of any intercepts the graph makes with the x-axis.

f. Sketch the graph.

23. a. Factorise $-x^4+18x^2-81$ by substituting $u=x^2$.

b. Sketch the graph of $y=-x^4+18x^2-81$.

c. Use the graph to obtain $\left\{x: -x^4+18x^2-81>0\right\}$.

d. State the solution set for $\left\{x: x^4-18x^2+81>0\right\}$.

24. The graphs of $y=x^4$ and $y=2x^3$ intersect at the origin and at a point P.

a. Calculate the coordinates of the point P.

b. The parabola $y=ax^2$ and the straight line $y=mx$ pass through the origin and the point P. Determine the values of a and m.

c. Using the values obtained for a and m in part b, sketch the graphs of $y=x^4$, $y=2x^3$, $y=ax^2$ and $y=mx$ on the same set of axes.

d. i. Determine the points at which the graphs of $y=nx^3$ and $y=x^4$ intersect.

ii. If each of the four curves $y=x^4$, $y=nx^3$, $y=ax^2$ and $y=mx$ intersect at the same two points, express a and m in terms of n.

25. Use CAS technology to sketch the graphs of the following, locating turning points and axis intercepts. Give answers to 2 decimal places where appropriate.

a. $y=x^4-x^3-12x^2-4x+4$

b. $y=x^4-7x-8$.

5.2 Exam questions

Question 1 (1 mark) TECH-ACTIVE

MC Which of the following options could **not** be the graph of $y = ax^4 + bx^3 + cx^2 + dx + e$ with $a > 0$?

A.

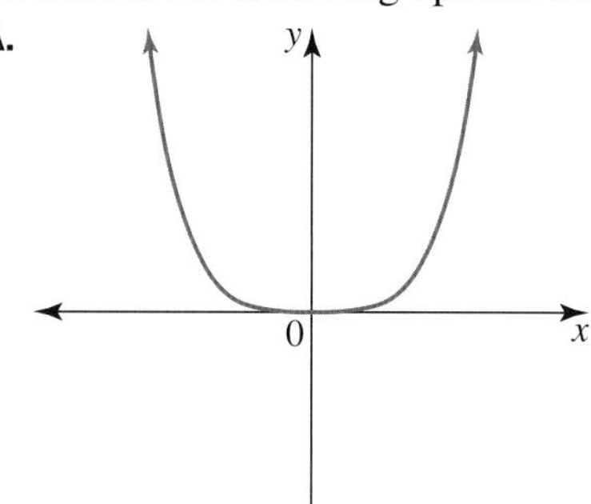

B.

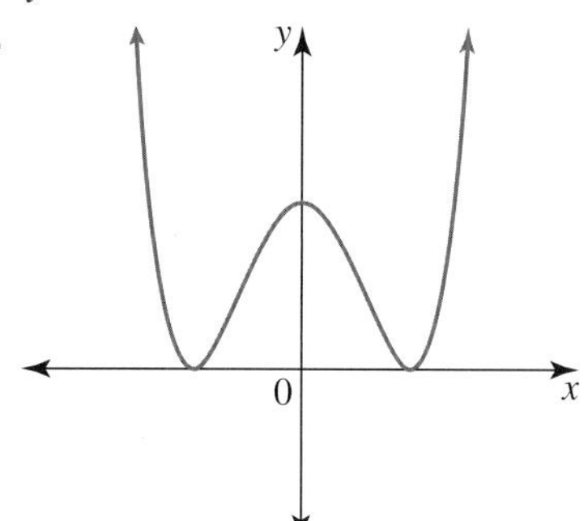

C.

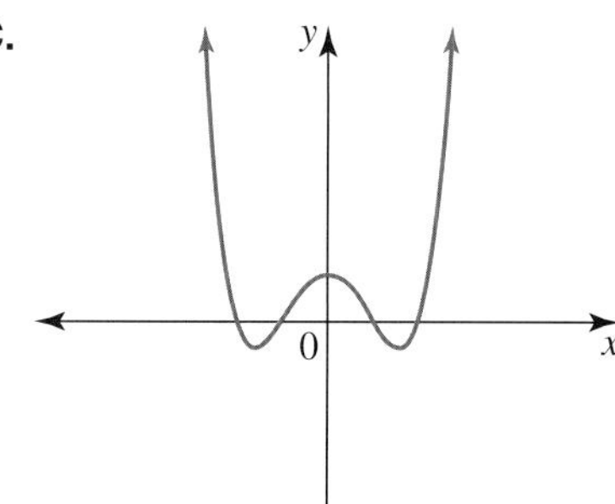

D.

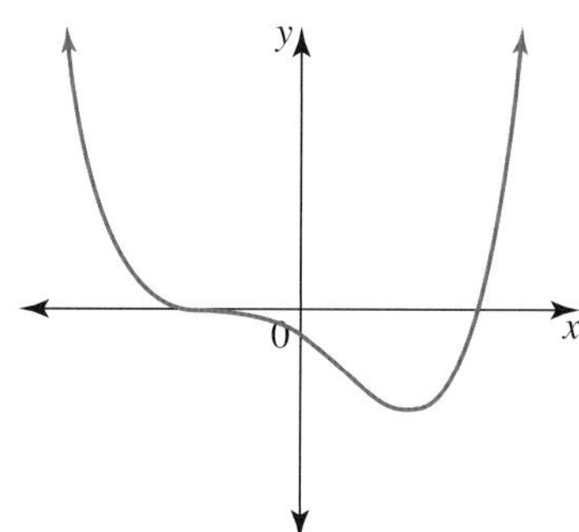

E.

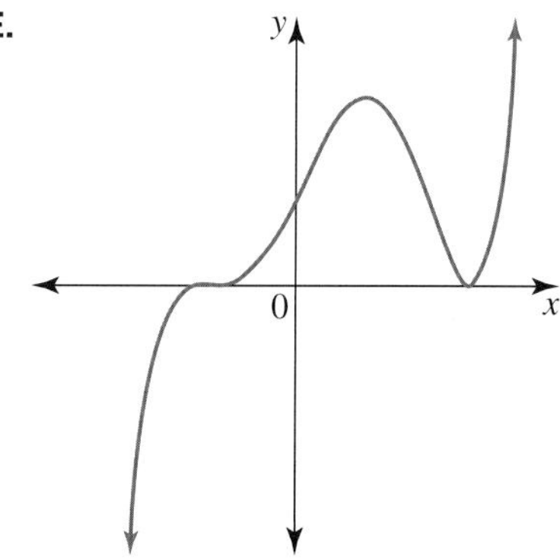

Question 2 (1 mark) TECH-ACTIVE

MC A possible equation for the graph shown, given that $a, b, c > 0$, would be

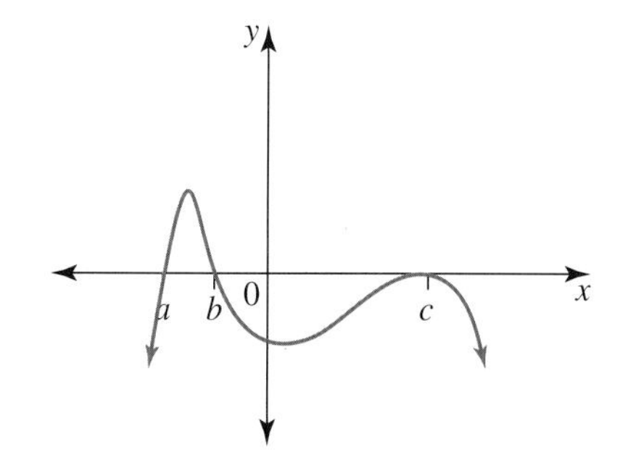

A. $f(x) = -(x-a)(x-b)(x-c)^2$
B. $f(x) = (x-a)(x-b)(x-c)^2$
C. $f(x) = (x-a)(x-b)(x+c)^2$
D. $f(x) = (x+a)(x+b)(x-c)^2$
E. $f(x) = -(x+a)(x+b)(x-c)^2$

Question 3 (4 marks) TECH-FREE

The equation of a quartic graph is $y = -(x+1)^4 + 16$.

Sketch the graph, showing all key points.

More exam questions are available online.

5.3 Families of polynomials

LEARNING INTENTION

At the end of this subtopic you should be able to:
- sketch a family of curves and identify their similarities and differences.

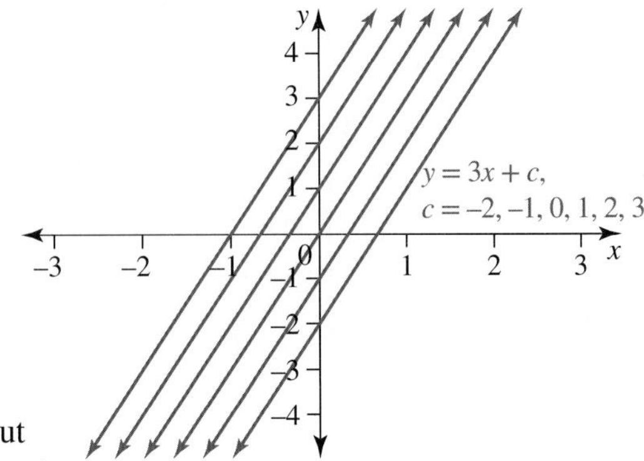

Of all the polynomials, the simplest is the linear polynomial. In some ways it is the exception, because its graph is a straight line, whereas the graphs of all other polynomials are curves. Nevertheless, the graphs of linear polynomials and other polynomials of the same degree display similarities. Sets of polynomials that share a common feature or features may be considered a family. Often these families are described by a common equation containing one or more constants that can be varied in value. Such a varying constant is called a **parameter**.

An example is the set of linear polynomials with a gradient of 3 but with differing y-intercepts. This is the family of parallel lines defined by the equation $y = 3x + c, c \in R$, some members of which are shown in the diagram. This set of lines is generated by allowing the parameter c to take the values $-2, -1, 0, 1, 2$ and 3. As could be anticipated, these values of c are the y-intercepts of each line.

5.3.1 Families of polynomials in turning point form

Polynomials given in the form $p(x) = a(x - h)^n + k, n \in \{2, 3, 4\}$, all have common features. The point (h, k) is a maximum or minimum turning point if $n =$ 2 or 4, and a stationary point of inflection if $n = 3$.

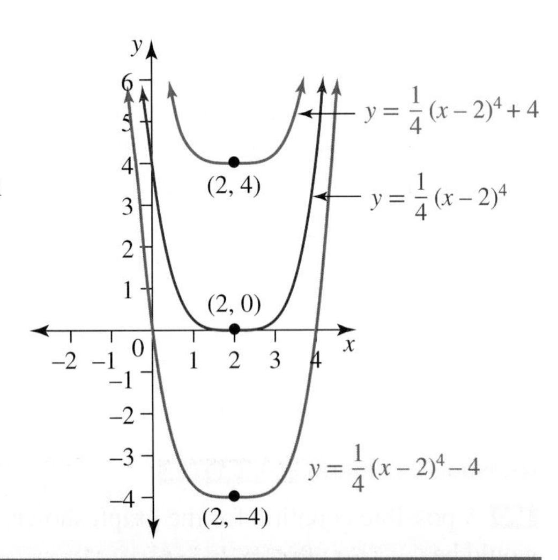

An example of a family of quartic polynomials is shown. This is the family defined by the equation $y = \frac{1}{4}(x - 2)^4 + k, k \in R$. In this family, the turning point, $(2, k)$, is a minimum turning point. The set of polynomials is generated by allowing k to take the values of $-4, 0$ and 4.

WORKED EXAMPLE 6 Sketching families of polynomials

Consider the family of polynomials $p(x) = a(x - h)^4$.

a. i. **Describe the family of polynomials if $a > 0$.**

ii. **On the same set of axes, sketch the family of polynomials if $a = \frac{1}{8}$ and $h = -4, -2, 0, 2, 4$.**

b. **Discuss the differences and similarities in the family of $p(x)$ if $a = -\frac{1}{8}$.**

THINK	WRITE
a. 1. Discuss the turning point, including whether it is a maximum or a minimum turning point.	The polynomial has a turning point at $(h, 0)$ and the x-intercept is at $x = h$. Since $a > 0$, the turning point is a minimum.

2. Sketch the curves and label each carefully.

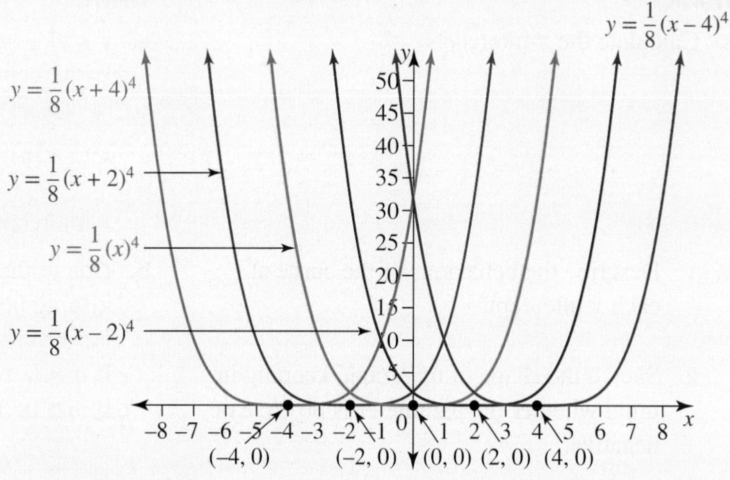

b. 1. Discuss the differences.

If $a = -\frac{1}{8}$ and not $a = \frac{1}{8}$, the quartic curves will be the same shape but reflected in the x-axis. This gives maximum turning points at $(h, 0)$ instead of minimum turning points.

2. Discuss the similarities.

The similarities are the same shape as the dilation factor has not altered and the x-intercepts are still the turning points at $x = h$.

5.3.2 Other families of polynomials

A family of polynomials can be general with parameters. Varying the parameters give different shapes. A point or points is necessary to determine the equation of a member of the family.

For example, the quartic polynomial $p(x) = x^3(x + m)$ defines a family with a stationary point of inflection at $x = 0$ and a root at the point $x = -m$.

If $m > 0$, the root is negative, so it is on the left of the origin.

If $m < 0$, the root is positive, so it is on the right of the origin.

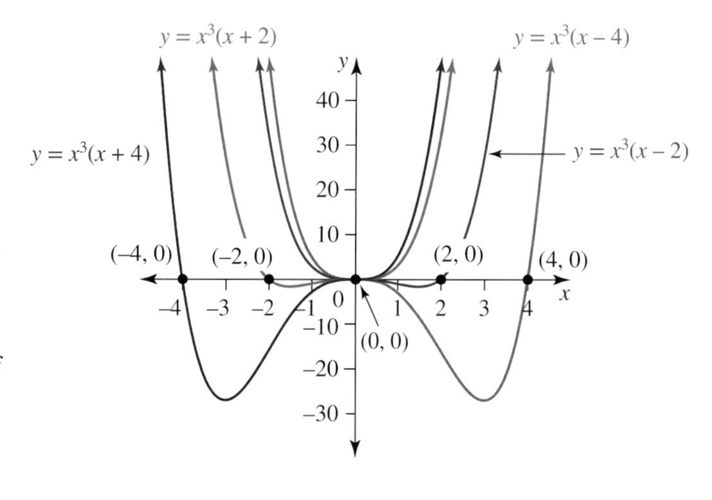

The graphs show the family generated when $m = -4, -2, 2$ and 4.

WORKED EXAMPLE 7 Families of cubic polynomials

Consider the family of cubic polynomials defined by the equation $y = x^3 + mx^2$, where the parameter m is a real non-zero constant.

a. Calculate the x-intercepts and express them in terms of m where appropriate.

b. Draw a sketch of the shape of the curve for positive and negative values of m, and comment on the behaviour of the graph at the origin in each case.

c. Determine the equation of the member of the family that contains the point (7, 49).

THINK	WRITE
a. Calculate the x-intercepts.	**a.** $y = x^3 + mx^2$ x-intercepts: let $y = 0$. $x^3 + mx^2 = 0$ $x^2(x + m) = 0$ $\therefore x = 0$ or $x = -m$ x-intercepts: $(0, 0)$ and $(-m, 0)$
b. 1. Describe the behaviour of the curve at each x-intercept.	**b.** Due to the multiplicity of the factor there is a turning point at $(0, 0)$. At $(-m, 0)$ the graph cuts the x-axis.
2. Sketch the shape of the graph, keeping in mind whether the parameter is positive or negative.	If $m < 0$, then $(-m, 0)$ lies to the right of the origin. If $m > 0$, then $(-m, 0)$ lies to the left of the origin. 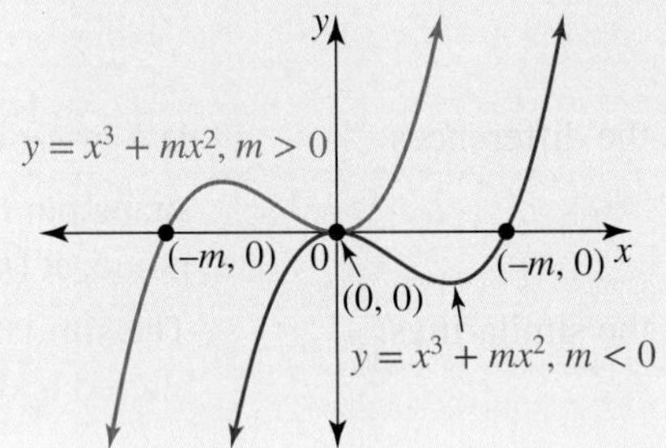
3. Comment on the behaviour of the graph at the origin.	The graph has a maximum turning point at the origin if m is negative. If m is positive, there is a minimum turning point at the origin.
c. 1. Use the given point to determine the value of the parameter.	**c.** $y = x^3 + mx^2$ Substitute the point $(7, 49)$. $49 = 7^3 + m \times 7^2$ $49 = 49(7 + m)$ $1 = 7 + m$ $\therefore m = -6$
2. State the equation of the required curve.	If $m = -6$, then the member of the family that passes through the point $(7, 49)$ has the equation $y = x^3 - 6x^2$.

5.3 Exercise

Technology free

1. On the same set of axes, sketch the graphs of $y = x^4$, $y = x^4 + 16$ and $y = x^4 - 16$.

2. On the same set of axes, sketch the graphs of $y = x^4$, $y = 2x^4$ and $y = \frac{1}{2}x^4$.

3. a. On the same set of axes, sketch the graphs of $y = x^3$ and $y = x$, labelling the points of intersection with their coordinates.
   b. Hence, state $\{x: x^3 \le x\}$.
4. **WE6** Consider the family of polynomials $p(x) = a(x-h)^3$.
   a. i. Describe the family of polynomials if $a > 0$.
      ii. On the same set of axes, sketch the family of polynomials if $a = \frac{1}{2}$ and $h = -4, -2, 0, 2, 4$.
   b. Discuss the differences and similarities in the family of $p(x)$ if $a = -\frac{1}{2}$.
5. Consider the family of polynomials $p(x) = (x-3)^4 + c$.
   a. Describe the family of polynomials if:
      i. $c > 0$ ii. $c = 0$ iii. $c < 0$
   b. On the same set of axes, sketch the family of polynomials if $c = -1, 0, 1$.
6. The equation $y = a(x-b)^4 + c$ represents a family of quartic polynomials. One of the graphs has a minimum turning point at $(-1, -12)$ and passes through the origin.
   a. Determine its equation.
   b. State the equation of its axis of symmetry.
   c. Determine the coordinates of its other x-intercept.
7. Consider the family of polynomials given by the equation $p(x) = a(2-x)(2+x)^3$.
   a. On the same set of axes, sketch the graphs when $a = 3$ and $a = -3$.
   b. Discuss the differences and similarities of the graphs drawn in part a.
8. Lucy was interrupted as she was drawing the graph of a polynomial. A section of her incomplete graph is shown in the diagram.

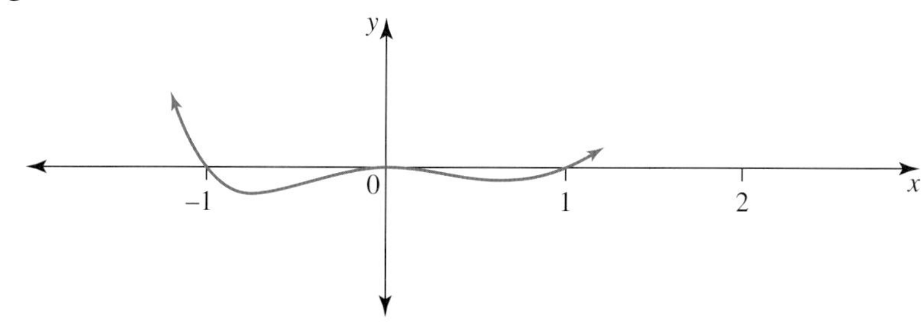

   a. Form the equation of a monic quartic polynomial whose graph would have the features that are shown.
   b. i. State the family of quartic polynomials with these features.
      ii. Determine the equation of the polynomial if it passes through the point $(-2, 6)$.
9. A family of straight lines for which $y = ax$ all pass through the origin.
   a. On the same diagram, sketch the lines for which $a = 1, a = 2$ and $a = -1$.
   b. Determine the value of a for which one of the lines will pass through the point $(-4, -10)$.
10. Consider the family of parabolas for which $y = -x^2 + k$ where $k \in R$.
   a. The point $(-5, 30)$ lies on one of the curves in this family. Give the equation of this parabola.
   b. Express the coordinates of the turning point in terms of k and state its nature.
   c. Determine the values of k for which the parabolas lie below the x-axis.

11. **WE7** Consider the family of quartic polynomials defined by the equation $y = x^4 - mx^3$, where the parameter m is a real non-zero constant.

a. Calculate the x-intercepts and express them in terms of m where appropriate.
b. Draw a sketch of the shape of the curve for positive and negative values of m, and comment on the behaviour of the graph at the origin in each case.
c. Determine the equation of the member of the family that contains the point $(-1, -16)$.

12. a. Using a parameter, form the equation of the family of lines which pass through the point $(2, 3)$.
b. Use the equation from part **a** to find which line in this family also passes through the origin.

13. a. Determine the coordinates of the point that is common to the family of parabolas defined by the equation $y = ax^2 + bx, a \neq 0$.
b. Express the equation of those parabolas, in the form $y = ax^2 + bx, x \neq 0$, that pass through the point $(2, 6)$ in terms of the one parameter, a, only.

14. Calculate the x-intercepts of the graphs of the family of polynomials defined by the equation $y = mx^2 - x^4, m > 0$, and draw a sketch of the shape of the graph.

Technology active

15. The curve belonging to the family of polynomials for which $y = a(x + b)^3 + c$ has a stationary point of inflection at $(-1, 7)$ and passes through the point $(-2, -20)$. Determine the equation of the curve.

16. The graph of a monic polynomial of degree 4 has a turning point at $(-2, 0)$, and one of its other x-intercepts occurs at $(4, 0)$. Its y-intercept is $(0, 48)$. Determine a possible equation for the polynomial and identify the coordinates of its other x-intercept.

17. Consider the family of quadratic polynomials defined by $y = a(x - 3)^2 + 5 - 4a, a \in R \setminus \{0\}$.

a. Show that every member of this family passes through the point $(1, 5)$.
b. Determine the value(s) of a for which the turning point of the parabola will lie on the x-axis.
c. Determine the value(s) of a for which the parabola will have no x-intercepts.

18. Consider the two families of polynomials for which $y = k$ and $y = x^2 + bx + 10$, where k and b are real constants.

a. Describe a feature of each family that is shared by all members of that family.
b. Sketch the graphs of $y = x^2 + bx + 10$ for $b = -7, 0$ and 7.
c. Determine the values of k for which members of the family $y = k$ will intersect $y = x^2 + 7x + 10$:

i. once only ii. twice iii. never.

d. If $k = 7$, determine the values of b for which $y = k$ will intersect $y = x^2 + bx + 10$:

i. once only ii. twice iii. never.

19. Consider the family of cubic polynomials for which $y = ax^3 + (3 - 2a)x^2 + (3a + 1)x - 4 - 2a$, where $a \in R \setminus \{0\}$.

a. Show that the point $(1, 0)$ is common to all this family.
b. For the member of the family that passes through the origin, form its equation and sketch its graph.
c. A member of the family passes through the point $(-1, -10)$. Show that its graph has exactly one x-intercept.
d. By calculating the coordinates of the point of intersection of the polynomials with equations $y = ax^3 + (3 - 2a)x^2 + (3a + 1)x - 4 - 2a$ and $y = (a - 1)x^3 - 2ax^2 + (3a - 2)x - 2a - 5$, show that for all values of a the point of intersection will always lie on a vertical line.

20. Use CAS technology to sketch the family of quartic polynomials for which $y = x^4 + ax^2 + 4, a \in R$, for $a = -6, -4, -2, 0, 2, 4, 6$, and determine the values of a for which the polynomial equation $x^4 + ax^2 + 4 = 0$ will have:

a. four roots b. two roots c. no real roots.

5.3 Exam questions

Question 1 (1 mark) TECH-ACTIVE

MC For the family of graphs $y = ax^3$, where $a > 0$, select the statement that is **not** true.

A. As $x \to \pm\infty, y \to \pm\infty$.
B. Each graph passes through $(-1, -a)$.
C. Each graph has a stationary point of inflection at $(0, 0)$.
D. All graphs are essentially similar shapes.
E. The greater the value of a, the wider the graph.

Question 2 (1 mark) TECH-ACTIVE

MC For the family of graphs of $y = x^n$, where $n \in N$ and $n = 2$ or 4, select the statement that is **not** true.

A. As $x \to \pm\infty, y \to \infty$.
B. Each graph passes through the point $(1, 1)$.
C. Each graph has a stationary point of inflection at $(0, 0)$.
D. All graphs are essentially similar shapes.
E. The larger the power, the narrower the graph.

Question 3 (3 marks) TECH-FREE

Find x- and y-intercepts and any turning points/points of inflection, and sketch the graph of $y = (x - 3)^4 - 1$.

More exam questions are available online.

5.4 Numerical approximation of roots of polynomial equations

LEARNING INTENTION

At the end of this subtopic you should be able to:

- use the bisection method to estimate roots of a polynomial equation
- estimate solutions and turning points using the intersections of two graphs.

For any polynomial $p(x)$, the values of the x-axis intercepts of the graph of $y = p(x)$ are the roots of the polynomial equation $p(x) = 0$. These roots can always be obtained if the polynomial is linear or quadratic, or if the polynomial can be expressed as a product of linear factors. However, there are many polynomial equations that cannot be solved by an algebraic method. In such cases, if an approximate value of a root can be estimated, then this value can be improved upon by a method of **iteration**. An iterative procedure is one that is repeated by using the values obtained from one stage to calculate the value of the next stage and so on.

5.4.1 Existence of roots

Existence of roots

For a polynomial $p(x)$, if $p(a)$ and $p(b)$ are of opposite signs, then there is at least one value of $x \in (a, b)$ where $p(x) = 0$.

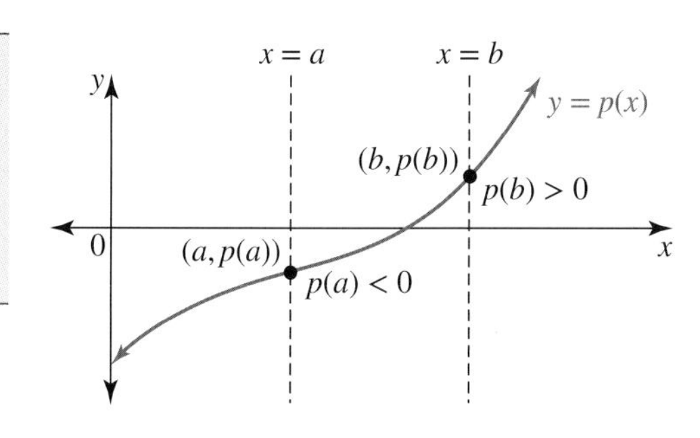

For example, in the diagram shown, $p(a) < 0$ and $p(b) > 0$. The graph cuts the x-axis at a point for which $a < x < b$.

This means that the equation $p(x) = 0$ has a root that lies in the interval (a, b). This gives an estimate of the root. Often the values of a and b are integers, and these may be found through trial and error.

Alternatively, if the polynomial graph has been sketched, it may be possible to obtain their values from the graph. Ideally, the values of a and b are not too far apart in order to avoid, if possible, there being more than one x-intercept of the graph, or one root of the polynomial equation, that lies between them.

5.4.2 The bisection method

Either of the values of a and b for which $p(a)$ and $p(b)$ are of opposite sign provides an estimate for one of the roots of the equation $p(x) = 0$. The **bisection method** is a procedure for improving this estimate by halving the interval in which the root is known to lie.

Let c be the midpoint of the interval $[a, b]$, so $c = \frac{1}{2}(a + b)$.

The value $x = c$ becomes an estimate of the root.

By testing the sign of $p(c)$, it can be determined whether the root lies in $(a, c]$ or in $[c, b)$.

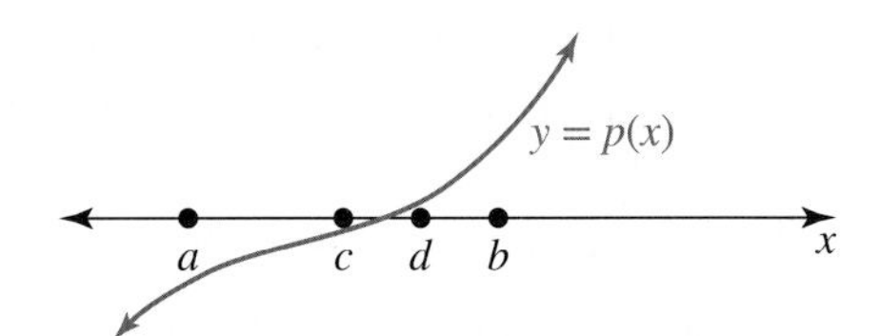

In the diagram shown, $p(a) < 0$ and $p(c) < 0$, so the root does not lie between a and c. It lies between c and b, since $p(c) < 0$ and $p(b) > 0$.

The midpoint d of the interval $[c, b]$ can then be calculated. The value of d may be an acceptable approximation to the root. If not, the accuracy of the approximation can be further improved by testing which of $[c, d]$ and $[d, b]$ contains the root and then halving that interval, and so on. The use of some form of technology helps considerably with the calculations, as it can take many iterations to achieve an estimate that has a high degree of accuracy.

Any other roots of the polynomial equation may be estimated by the same method once an interval in which each root lies has been established.

WORKED EXAMPLE 8 Using the bisection method to estimate roots

Consider the cubic polynomial $p(x) = x^3 - 3x^2 + 7x - 4$.

a. Show that the equation $x^3 - 3x^2 + 7x - 4 = 0$ has a root that lies between $x = 0$ and $x = 1$.

b. State a first estimate of the root.

c. Carry out two iterations of the method of bisection by hand to obtain two further estimates of this root.

d. Continue the iteration using a calculator until the error in using this estimate as the root of the equation is less than 0.05.

THINK	WRITE
a. 1. Calculate the value of the polynomial at each of the given values.	a. $p(x) = x^3 - 3x^2 + 7x - 4$ $p(0) = -4$ $p(1) = 1 - 3 + 7 - 4$ $= 1$
2. Interpret the values obtained.	As $p(0) < 0$ and $p(1) > 0$, there is a value of x that lies between $x = 0$ and $x = 1$ where $p(x) = 0$. Hence, the equation $x^3 - 3x^2 + 7x - 4 = 0$ has a root that lies between $x = 0$ and $x = 1$.
b. By comparing the values calculated, select the one that is closer to the root.	b. $p(0) = -4$ and $p(1) = 1$, so the root is closer to $x = 1$ than to $x = 0$. A first estimate of the root of the equation is $x = 1$.

c. 1. Calculate the midpoint of the first interval.

c. The midpoint of the interval between $x = 0$ and $x = 1$ is $x = \frac{1}{2}$.

2. State the second estimate.

$x = 0.5$ is a second estimate.

3. Determine in which half of the interval the root lies.

$$p\left(\frac{1}{2}\right) = \frac{1}{8} - \frac{3}{4} + \frac{7}{2} - 4$$
$$= -\frac{9}{8}$$

As $p\left(\frac{1}{2}\right) < 0$ and $p(1) > 0$, the root lies between $x = \frac{1}{2}$ and $x = 1$.

4. Calculate the midpoint of the second interval.

The midpoint of the interval between $x = \frac{1}{2}$ and $x = 1$ is:

$$x = \frac{1}{2}\left(\frac{1}{2} + 1\right)$$
$$= \frac{3}{4}$$
$$= 0.75$$

5. State the third estimate.

$x = 0.75$ is a third estimate.

d. 1. Continue the calculations using a calculator.

d. $p(0.75) = -0.015\,625$

As $-0.05 < p(0.75) < 0.05$, this estimate provides a solution to the equation with an error that is less than 0.05.

2. State the estimate of the root.
Note: The value is still an estimate, not the exact value.

Hence, $x = 0.75$ is a good estimate of the root of the equation that lies between $x = 0$ and $x = 1$.

5.4.3 Using the intersections of two graphs to estimate solutions to equations

Consider the quadratic equation $x^2 + 2x - 6 = 0$. Although it can be solved algebraically to give $x = \pm\sqrt{7} - 1$, we shall use it to illustrate another non-algebraic method for solving equations. If the equation is rearranged to $x^2 = -2x + 6$, then any solutions to the equation are the x-coordinates of any points of intersection of the parabola $y = x^2$ and the straight line $y = -2x + 6$.

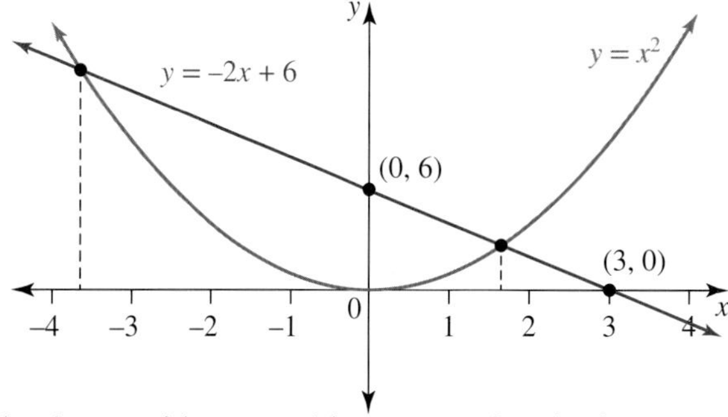

Both of these polynomial graphs are relatively simple graphs to draw. The line can be drawn accurately using its intercepts with the coordinate axes, and the parabola can be drawn with reasonable accuracy by plotting some points that lie on it. The diagram of their graphs shows that there are two points of intersection and hence that the equation $x^2 + 2x - 6 = 0$ has two roots.

Estimates of the roots can be read from the graph. One root is approximately $x = -3.6$ and the other is approximately $x = 1.6$. (This agrees with the actual solutions, which to 3 decimal places are $x = -3.646$ and $x = 1.646$.)

Alternatively, we can confidently say that one root lies in the interval $[-4, -3]$ and the other in the interval $[1, 2]$. By applying the method of bisection, the roots could be obtained to a greater accuracy than that of the estimates that were read from the graph.

Solving polynomials graphically to estimate roots

To use the graphical method to solve the polynomial equation $h(x) = 0$, follow these steps:

- **Rearrange the equation into the form $p(x) = q(x)$ where each of the polynomials $p(x)$ and $q(x)$ have graphs that can be drawn quite simply and accurately.**
- **Sketch the graphs of $y = p(x)$ and $y = q(x)$ with care.**
- **The number of intersections determines the number of solutions to the equation $h(x) = 0$.**
- **The x-coordinates of the points of intersection are the solutions to the equation.**
- **Estimate these x-coordinates by reading off the graph.**
- **Alternatively, an interval in which the x-coordinates lie can be determined from the graph and the method of bisection applied to improve the approximation.**

WORKED EXAMPLE 9 Estimating roots graphically

Use a graphical method to estimate any solutions to the equation $x^4 - 2x - 12 = 0$.

THINK	WRITE
1. Rearrange the equation so that it is expressed in terms of two familiar polynomials.	$x^4 - 2x - 12 = 0$ $x^4 = 2x + 12$
2. State the equations of the two polynomial graphs to be drawn.	The solutions to the equation are the x-coordinates of the points of intersection of the graphs of $y = x^4$ and $y = 2x + 12$.
3. Determine any information that will assist you to sketch each graph with some accuracy.	The straight line $y = 2x + 12$ has a y-intercept at $(0, 12)$ and an x-intercept at $(-6, 0)$. The quartic graph $y = x^4$ has a minimum turning point at $(0, 0)$ and contains the points $(\pm 1, 1)$ and $(\pm 2, 16)$.
4. Carefully sketch each graph on the same set of axes.	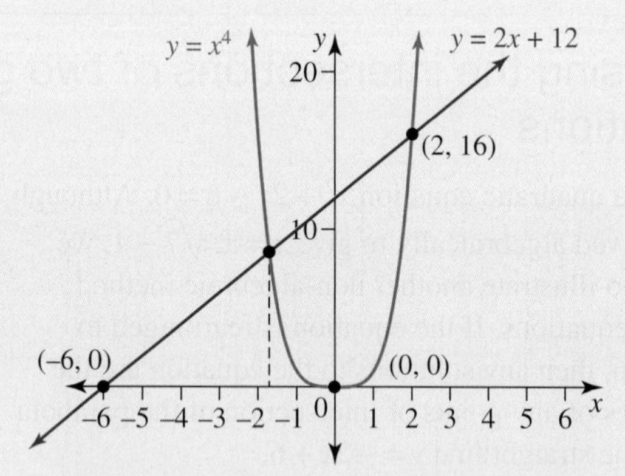
5. State the number of solutions to the original equation given.	As there are two points of intersection, the equation $x^4 - 2x - 12 = 0$ has two solutions.
6. Use the graph to obtain the solutions.	From the graph it is clear that one point of intersection is at $(2, 16)$, so $x = 2$ is an exact solution of the equation. An estimate of the x-coordinate of the other point of intersection is approximately -1.7, so $x = -1.7$ is an approximate solution to the equation.

5.4.4 Estimating coordinates of turning points

If the linear factors of a polynomial are known, sketching the graph of the polynomial is a relatively easy task to undertake. Turning points, other than those that lie on the x-axis, have largely been ignored, or, at best, approximated. In later topics, calculus will be introduced, which will allow for the identification of the exact location of turning points. For now we will address this unfinished aspect of our graph sketching by considering a numerical method of systematic trial and error to locate the approximate position of a turning point.

For any polynomial with zeros at $x = a$ and $x = b$, its graph will have at least one turning point between $x = a$ and $x = b$.

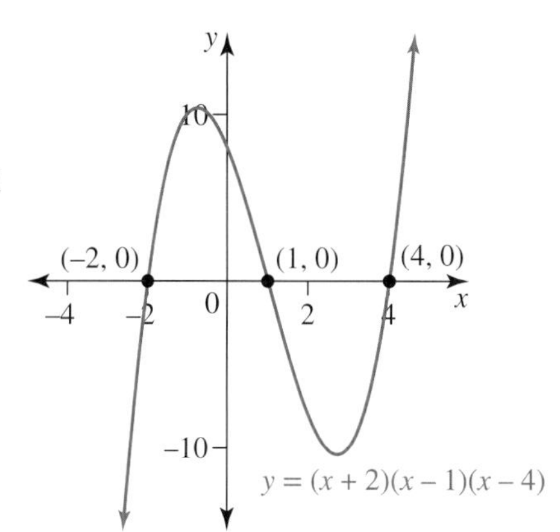

To illustrate, consider the graph of $y = (x + 2)(x - 1)(x - 4)$. The factors show there are three x-intercepts: one at $x = -2$, a second one at $x = 1$ and a third at $x = 4$.

There would be a turning point between $x = -2$ and $x = 1$, and a second turning point between $x = 1$ and $x = 4$. The first turning point must be a maximum and the second one must be a minimum, since it is a positive cubic polynomial.

The interval in which the x-coordinate of the maximum turning point lies can be narrowed using a table of values.

x	−2	−1.5	−1	−0.5	0	0.5	1
y	0	6.875	10	10.125	8	4.375	0

As a first approximation, the maximum turning point lies near the point $(-0.5, 10.125)$. Zooming in further around $x = -0.5$ gives greater accuracy.

x	−0.75	−0.5	−0.25
y	10.390 625	10.125	9.2989

An improved estimate is that the maximum turning point lies near the point $(-0.75, 10.39)$. The process could be continued by zooming in around $x = -0.75$ if greater accuracy is desired.

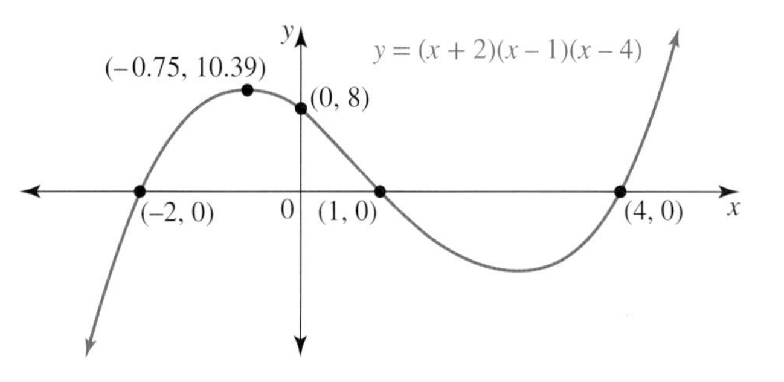

An approximate position of the minimum turning point could be estimated by the same numerical method of systematic trial and error.

The shape of this positive cubic with a y-intercept at $(0, 8)$ could then be sketched.

5.4.5 An alternative approach

For any polynomial $p(x)$, if $p(a) = p(b)$, then its graph will have at least one turning point between $x = a$ and $x = b$. This means for the cubic polynomial shown in the previous diagram, the maximum turning point must lie between the x-values for which $y = 8$ (the y-intercept value).

Substitute $y = 8$ into $y = (x+2)(x-1)(x-4)$:

$$x^3 - 3x^2 - 6x + 8 = 8$$
$$x^3 - 3x^2 - 6x = 0$$
$$x\left(x^2 - 3x - 6\right) = 0$$
$$x = 0 \text{ or } x^2 - 3x - 6 = 0$$

As the cubic graph must have a maximum turning point, the quadratic equation must have a solution. Solving it gives the negative solution as $x = -1.37$. Rather than test values between $x = -2$ and $x = 1$ as we have previously, the starting interval for testing values could be narrowed to between $x = -1.37$ and $x = 0$.

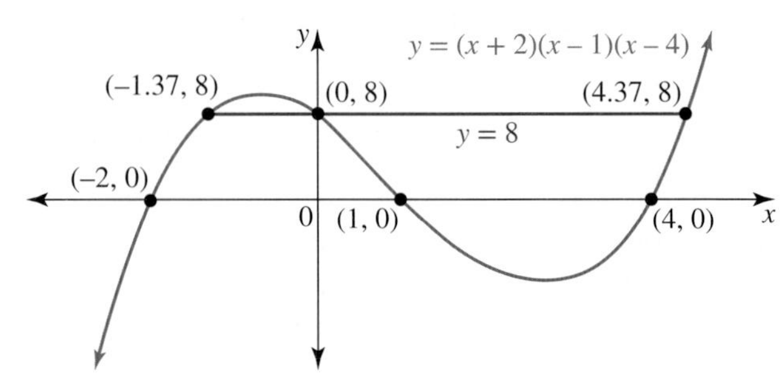

The positive solution $x = 4.37$ indicates that the minimum turning point lies between $x = 0$ and $x = 4.37$. In this case the interval between the two positive x-intercepts provides a narrower and therefore better interval to zoom into.

WORKED EXAMPLE 10 Estimating turning points

a. State an interval in which the x-coordinate of the minimum turning point on the graph of $y = x(x-2)(x+3)$ must lie.

b. Use a numerical method to zoom in on this interval and hence estimate the position of the minimum turning point of the graph. Give the x-coordinate correct to 1 decimal place.

THINK

a. 1. State the values of the x-intercepts in increasing order.

WRITE

a. $y = x(x-2)(x+3)$

The x-intercepts occur when $x = 0, x = 2, -3$.

In increasing order they are $x = -3, x = 0, x = 2$.

2. Determine the pair of x-intercepts between which the required turning point lies.

The graph is a positive cubic, so the first turning point is a maximum and the second is a minimum. The minimum turning point must lie between $x = 0$ and $x = 2$.

b. 1. Construct a table of values that zooms in on the interval containing the required turning point.

b. Values of the polynomial calculated over the interval $[0, 2]$ are tabulated.

x	0	0.5	1	1.5	2
y	0	−2.625	−4	−3.375	0

2. State an estimate of the position of the turning point.

The turning point is near $(1, -4)$.

3. Zoom in further to obtain a second estimate.

Zooming in around $x = 1$ gives the following table of values.

x	0.9	1	1.1	1.2
y	−3.861	−4	−4.059	−4.032

4. State the approximate position.

The minimum turning point is approximately $(1.1, -4.059)$.

5.4 Exercise

Students, these questions are even better in jacPLUS

Receive immediate feedback and access sample responses

Access additional questions

Track your results and progress

Find all this and MORE in jacPLUS

Technology free

1. **MC** For a polynomial equation $p(x) = 0$, it is known that a solution to this equation lies in the interval $x \in (b, c)$. Determine which one of the following statements supports this conclusion.
   A. $p(b) > 0, p(c) > 0$
   B. $p(b) < 0, p(c) < 0$
   C. $p(b) > 0, p(c) < 0$
   D. $p(b) < 0, p(c) > 0$
   E. None of these

2. The equation $x^3 + 7x - 14 = 0$ is known to have exactly one solution.
   a. Show that this solution does not lie between $x = -2$ and $x = -1$.
   b. Show that the solution does lie between $x = 1$ and $x = 2$.
   c. Use the midpoint of the interval $[1, 2]$ to deduce a narrower interval $[a, b]$ in which the root of the equation lies.

3. Let $p(x) = 3x^2 - 3x - 1$. The equation $p(x) = 0$ has two solutions, one negative and one positive.
   a. Evaluate $p(-2)$ and $p(0)$, and hence explain why the negative solution to the equation lies in the interval $[-2, 0]$.
   b. Carry out the method of bisection twice to narrower the interval in which the negative solution lies.
   c. Using your answer from part b, state an estimate of the negative solution to the equation.

4. For each of the following polynomials, show that there is a zero of each in the interval $[a, b]$.
   a. $p(x) = x^2 - 12x + 1, a = 10, b = 12$
   b. $p(x) = -2x^3 + 8x + 3, a = -2, b = -1$
   c. $p(x) = x^4 + 9x^3 - 2x + 1, a = -2, b = 1$

Technology active

5. The following polynomial equations are formed using the polynomials in question 4. Use the method of bisection to obtain two narrower intervals in which the root lies and hence give an estimate of the root which lies in the interval $[a, b]$.
   a. $x^2 - 12x + 1 = 0, a = 10, b = 12$
   b. $-2x^3 + 8x + 3 = 0, a = -2, b = -1$
   c. $x^4 + 9x^3 - 2x + 1 = 0, a = -2, b = 1$

6. **WE8** Consider the cubic polynomial $p(x) = x^3 + 3x^2 - 7x - 4$.
   a. Show that the equation $x^3 + 3x^2 - 7x - 4 = 0$ has a root that lies between $x = 1$ and $x = 2$.
   b. State a first estimate of the root.
   c. Carry out two iterations of the method of bisection by hand to obtain two further estimates of this root.
   d. Continue the iteration using a calculator until the error in using this estimate as the root of the equation is less than 0.05.

7. The quadratic equation $5x^2 - 26x + 24 = 0$ has a root in the interval for which $1 \le x \le 2$.
   a. Use the method of bisection to obtain this root correct to 1 decimal place.
   b. Determine the other root of this equation.
   c. Comment on the efficiency of the method of bisection.

8. Consider the polynomial defined by the rule $y = x^4 - 3$.

a. Complete the table of values for the polynomial.

x	-2	-1	0	1	2
y					

b. Hence, state an interval in which $\sqrt[4]{3}$ lies.

c. Use the method of bisection to show that $\sqrt[4]{3} = 1.32$ to 2 decimal places.

9. The graph of $y = x^4 - 2x - 12$ has two x-intercepts.

a. Construct a table of values for this polynomial rule for $x = -3, -2, -1, 0, 1, 2, 3$.

b. Hence, state an exact solution to the equation $x^4 - 2x - 12 = 0$.

c. State an interval within which the other root of the equation lies, and use the method of bisection to obtain an estimate of this root correct to 1 decimal place.

10. Consider the polynomial equation $p(x) = x^3 + 5x - 2 = 0$.

a. Determine an interval $[a, b]$, $a, b \in Z$, in which there is a root of this equation.

b. Use the method of bisection to obtain this root with an error less than 0.1.

c. State the equations of two graphs, the intersections of which would give the number of solutions to the equation.

d. Sketch the two graphs and hence state the number of solutions to the equation $p(x) = x^3 + 5x - 2 = 0$. State whether the diagram supports the answer obtained in part **b**.

11. **WE9** Use a graphical method to estimate any solutions to the equation $x^4 + 3x - 4 = 0$.

12. Use a graphical method to estimate any solutions to the equation $x^3 - 6x + 4 = 0$.

13. The graph shows that the line $y = 3x - 2$ is a tangent to the curve $y = x^3$ at a point A and that the line intersects the curve again at a point B.

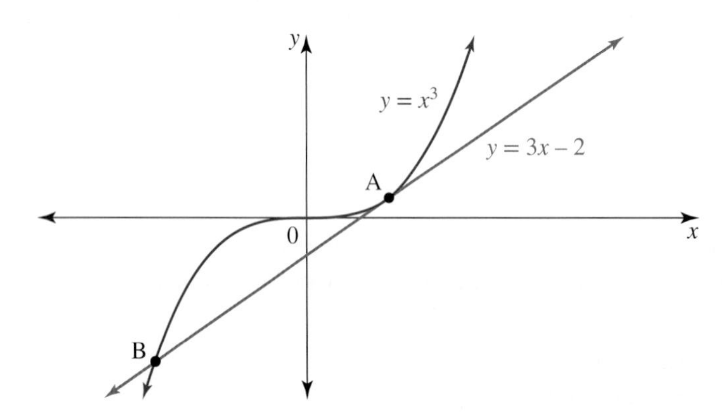

a. Form the polynomial equation $p(x) = 0$ for which the x-coordinates of the points A and B are solutions.

b. Describe the number and multiplicity of the linear factors of the polynomial specified in part **a**.

c. Use an algebraic method to calculate the exact roots of the polynomial equation specified in part **a** and hence state the coordinates of the points A and B.

d. Using the graph, state how many solutions there are to the equation $x^3 - 3x + 1 = 0$.

14. WE10 a. State an interval in which the x-coordinate of the maximum turning point on the graph of $y = -x(x+2)(x-3)$ must lie.

b. Use a numerical method to zoom in on this interval and hence estimate the position of the maximum turning point of the graph. Give the x-coordinate correct to 1 decimal place.

15. Use a systematic numerical trial and error process to estimate the positions of the following turning points. Express x-coordinates correct to 1 decimal place.

a. The maximum turning point of $y = (x+4)(x-2)(x-6)$
b. The minimum turning point of $y = x\,(2x+5)(2x+1)$
c. The maximum and minimum turning points of $y = x^2 - x^4$

16. Consider the cubic polynomial $y = 2x^3 - x^2 - 15x + 9$.

a. State the y-intercept.
b. Determine which other points on the graph have the same y-coordinate as the y-intercept.
c. Determine the two x-values between which the maximum turning point lies.
d. Use a numerical method to zoom in on this interval and hence estimate the position of the maximum turning point of the graph. Give the x-coordinate correct to 1 decimal place.

17. For the following polynomials, $p(0) = d$. Solve the equations $p(x) = d$ and hence state intervals in which the turning points of the graphs of $y = p(x)$ lie.

a. $p(x) = x^3 - 3x^2 - 4x + 9$
b. $p(x) = x^3 - 12x + 18$
c. $p(x) = -2x^3 + 10x^2 - 8x + 1$
d. $p(x) = x^3 + x^2 + 7$

18. The weekly profit y, in tens of dollars, from the sale of $10x$ containers of whey protein sold by a health food business is given by $y = -x^3 + 7x^2 - 3x - 4, x \geq 0$.
The graph of the profit is shown in the diagram.

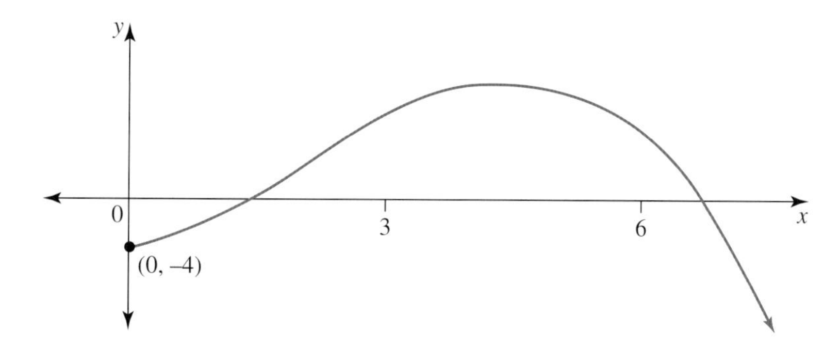

a. Show that the business first started to make a profit when the number of containers sold was between 10 and 20.
b. Use the method of bisection to construct two further intervals for the value of x required for the business to first start making a profit.
c. Hence, use a systematic numerical trial and error process to calculate the number of containers that need to be sold for a profit to be made.
d. Use the graph to state an interval in which the greatest profit lies.
e. Use a systematic numerical trial and error process to estimate the number of containers that need to be sold for greatest profit, and state the greatest profit to the nearest dollar.
f. As the containers are large, storage costs can lower profits. State an estimate from the graph of the number of containers beyond which no profit is made, and improve upon this value with a method of your choice.

19. Consider the cubic polynomial $y = -x^3 - 4$ and the family of lines $y = ax$.

a. Using CAS technology, sketch $y = -x^3 - 4$ and the family of lines $y = ax$, for $a \in [-6, 3], a \in Z$.
b. Determine the largest integer value of a for which the equation $x^3 + ax + 4 = 0$ has three roots.
c. If $a > 0$, state how many solutions there are to $x^3 + ax + 4 = 0$.
d. Determine the root(s) of the equation $x^3 + x + 4 = 0$ using the method of bisection. Express your answer correct to 2 decimal places.

20. A rectangular sheet of cardboard measures 18 cm by 14 cm. Four equal squares of side length x cm are cut from the corners, and the sides are folded to form an open rectangular box in which to place some clothing.

a. Express the volume of the box in terms of x.
b. State an interval within which lies the value of x for which the volume is greatest.
c. Use the spreadsheet option on a CAS calculator to systematically test values in order to determine, to 3 decimal places, the side length of the square needed for the volume of the box to be greatest.

5.4 Exam questions

Question 1 (1 mark) TECH-ACTIVE

MC For the graph $y = x(x-4)(x+5)$, a maximum turning point occurs between

A. $x = -5$ and $x = 0$.
B. $x = 0$ and $x = 4$.
C. $x = 4$ and $x = 20$.
D. $x = 0$ and $x = 5$.
E. $x = 0$ and $x = -4$.

Question 2 (3 marks) TECH-FREE

Consider the cubic polynomial $p(x) = x^3 - 2x^2 + 8x - 5$.

a. Show that the equation $x^3 - 2x^2 + 8x - 5$ has a root that lies between $x = 0$ and $x = 1$. **(1 mark)**
b. State a first estimate of the root and then use the method of bisection to calculate a more accurate estimate. **(2 marks)**

Question 3 (1 mark) TECH-ACTIVE

MC The graph of $y = (x+2)\left(x^2 - 5x + 4\right)$ has

A. one maximum turning point.
B. one minimum turning point.
C. one maximum and one minimum turning point.
D. one maximum and two minimum turning points.
E. two maximum and one minimum turning point.

More exam questions are available online.

5.5 Review

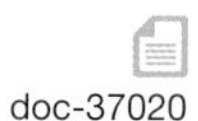

5.5.1 Summary

doc-37020

Hey students! Now that it's time to revise this topic, go online to:

Access the topic summary

Review your results

Watch teacher-led videos

Practise exam questions

Find all this and MORE in jacPLUS

5.5 Exercise

Technology free: short answer

1. Sketch the following graphs.

   a. $y = 16 - (x+3)^4$

   b. $y = \frac{1}{18}(x+2)(x-3)^3$

   c. $y = -x^4 + 8x^2 - 16$

2. Solve the following inequations.

   a. $(x+4)(x+1)^2(x-3) \geq 0$

   b. $(x-5)^3(3x+7) < 0$

3. a. Sketch the graphs of $y = 1 - x$ and $y = x^3$ on the same set of axes and hence state the smallest integers between which the root of the equation $1 - x - x^3 = 0$ lies.

   b. Form a second interval in which the root of the equation $1 - x - x^3 = 0$ lies, using the method of bisection.

4. a. Sketch the graph of $y = 2(x-1)^4 - 2$ and hence solve $2(x-1)^4 - 2 < 0$.

   b. Sketch the graph of $y = x(x-2)^2(x-6)$ and hence solve $x(x-2)^2(x-6) \geq 0$.

5. a. State the family of quartic polynomials, $p(x)$, that intersects the x-axis at $x = 3$ and has a stationary point of inflection at $(-2, 0)$.

   b. Sketch the graph of $y = p(x)$ if the curve passes through the point $(2, 32)$.

6. A family of curves have the common equation $y = 2x^3 + mx^2 - 4x + 5$, where m is a real constant.

   a. One of the curves passes through the point $(-2, 9)$. Calculate the value of m and hence give the equation of this curve.

   b. The family of curves intersect the graph of $y = 2x^3 + 5$ at two points. Calculate the coordinates of these two points in terms of m.

Technology active: multiple choice

7. **MC** Select the correct statement about the graph of $y = \frac{1}{2}(x+6)^4 - 3$.

   A. There is a maximum turning point at $(6, -3)$.

   B. There is a minimum turning point at $(6, -3)$.

   C. There is a stationary point of inflection at $(-6, -3)$.

   D. There is a maximum turning point at $(-3, 3)$.

   E. There is a minimum turning point at $(-6, -3)$.

8. **MC** A possible equation for the quartic graph shown could be:

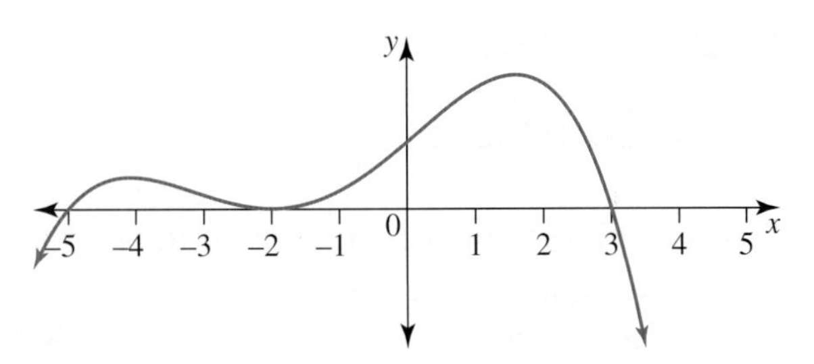

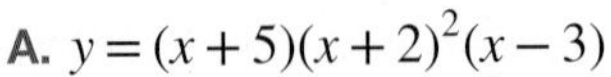

A. $y = (x+5)(x+2)^2(x-3)$
B. $y = -(x-5)(x-2)^2(x+3)$
C. $y = (x+5)(x+2)^2(3-x)$
D. $y = -(x+4)^2(x-2)^2$
E. $y = 5(x+2)^4 + 3$

9. **MC** The solutions to the equation $9x^2 - 81x^4 = 0$ are:

A. $x = -3, x = 3$
B. $x = -3, x = 0, x = 3$
C. $x = -\sqrt{3}, x = 0, x = \sqrt{3}$
D. $x = -\sqrt[4]{3}, x = \sqrt[4]{3}$
E. $x = -\frac{1}{3}, x = 0, x = \frac{1}{3}$

10. **MC** The equation $(5-x)^4 + 25 = 0$ has exactly:

A. 0 solutions
B. 1 solution
C. 2 solutions
D. 3 solutions
E. 4 solutions

11. **MC** The curve defined by $y = a(x+b)^4 + c$ has a turning point at $(0, -7)$ and passes through the point $(-1, -10)$. The sum $a + b + c$ is equal to:

A. −4
B. −6
C. −7
D. −10
E. −12

12. **MC** The graphs of $y = x^4$ and $y = x^3$ intersect at the point(s):

A. $(0, 0)$ only
B. $(1, 1)$ only
C. $(0, 0)$ and $(1, 1)$ only
D. $(-1, 1), (0, 0)$ and $(1, 1)$ only
E. $(-1, -1), (0, 0)$ and $(1, -1)$ only.

13. **MC** For the graph shown, state its degree and the number of x-intercepts.

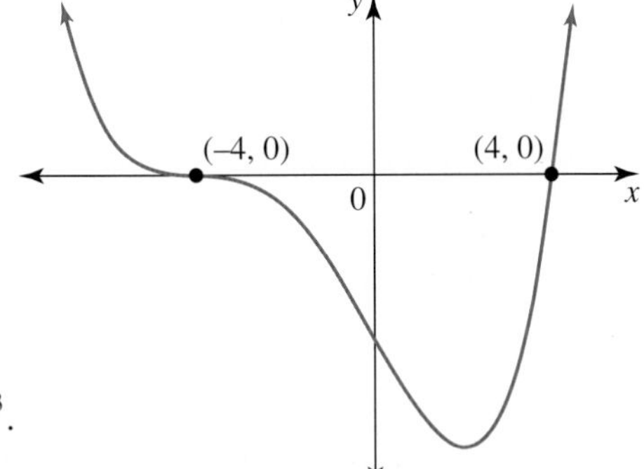

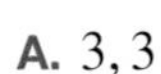

A. 3, 3
B. 4, 2
C. 4, 3
D. 3, 4
E. 2, 4

14. **MC** Select the correct statement about the graph of $y = 0.1(2x+5)^3$.

A. There is a maximum turning point at $(-5, 0)$.
B. There is a minimum turning point at $(-5, 0)$.
C. There is a stationary point of inflection at $(-5, 0)$.
D. There is a stationary point of inflection at $(0, 0)$ and a minimum turning point at $(-2.5, 0)$.
E. There is a stationary point of inflection at $(-2.5, 0)$ and no turning points.

15. **MC** The equation $6x^3 - 7x + 5 = 0$ has only one solution. This solution lies in the interval for which:

A. $-3 \le x \le -2$
B. $-2 \le x \le -1$
C. $-1 \le x \le 0$
D. $0 \le x \le 1$
E. $1 \le x \le 2$

16. **MC** The x-coordinates of the points of intersection of the graphs of $y = \frac{1}{2}x^3$ and $y = 2 - x^2$ are the solutions to the equation:

A. $\frac{1}{2}x^3 - x^2 + 2 = 0$
B. $x^3 + 2x^2 - 4 = 0$
C. $x^3 - 2x^2 - 4 = 0$
D. $x^3 + 2x^2 + 4 = 0$
E. $x^3 + x^2 - 2 = 0$

Technology active: extended response

17. The cross-section of a structure with equation $y = ax^2 + b$ is shown in the diagram.
A small tunnel is to be built through the structure. Its cross-section is the shaded rectangle PQOR and the point $p\,(x, y)$ lies on the curve $y = ax^2 + b$.
Units are in metres.

a. Calculate the values of a and b and state the equation of the cross-section curve.

b. Express the area, A, of the rectangular cross-section of the tunnel in terms of x.

c. The area of the cross-section is 12 square metres. Show that either $x = 1$ or $x^2 + x - 3 = 0$.

d. i. Show that a root of the equation $x^2 + x - 3 = 0$ is found in the interval $x \in [1.25, 1.5]$.

ii. Calculate an estimate, β, of this root using two iterations of the method of bisection.

e. Determine whether the height of the tunnel is less for $x = 1$ or $x = \beta$.

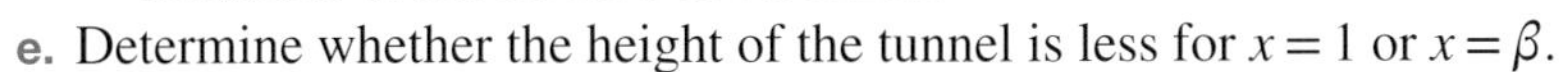

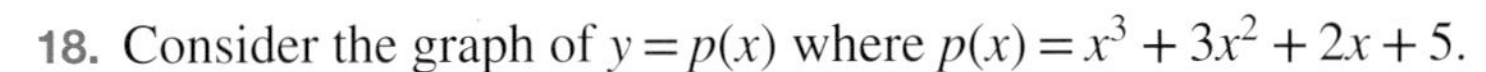

18. Consider the graph of $y = p(x)$ where $p(x) = x^3 + 3x^2 + 2x + 5$.

a. Solve the equation $p(x) = 5$ and hence state the intervals between which the turning points of the graph would lie.

b. Use a systematic numerical method to estimate the coordinates of the turning points to 2 decimal places.

c. Explain why the graph can have only one x-intercept.

d. Locate an interval in which the x-intercept lies. Use the method of bisection to generate three narrower intervals in which the x-intercept lies.

e. State an estimate of the x-intercept to 1 decimal place.

f. Use the information gathered to sketch the graph.

19. A family of curves is defined by the rule $y = (a - 2)x^3 - x^2 + (2 - a)x + 1$ where a is a real constant.

a. i. Determine the value of a for which the rule will represent a parabola.

ii. Use this value of a to form the rule for the parabola, and state the nature and coordinates of its turning point.

b. Form the rule of the curve with $a = 3$ and locate its intercepts with the coordinate axes.

c. Form the rule of the curve with $a = 1$ and locate its intercepts with the coordinate axes.

d. Show that every member of this family of curves has the same y-intercept and shares two x-intercepts with every other curve. Express the possible third x-intercept in terms of a.

e. Calculate the value of a for which the curve has three x-intercepts, one of which is the point $\left(\frac{1}{2}, 0\right)$.

20. A quartic polynomial is defined by the rule $y = x^4 + ax^3 + bx^2 + cx + d$ where $a, b, c, d \in R$. The line $y = -2x$ is a tangent to the graph of this polynomial, touching it at the point $(-3, 6)$. The line also cuts the graph at the origin and at the point $(-6, 12)$, as shown in the diagram.

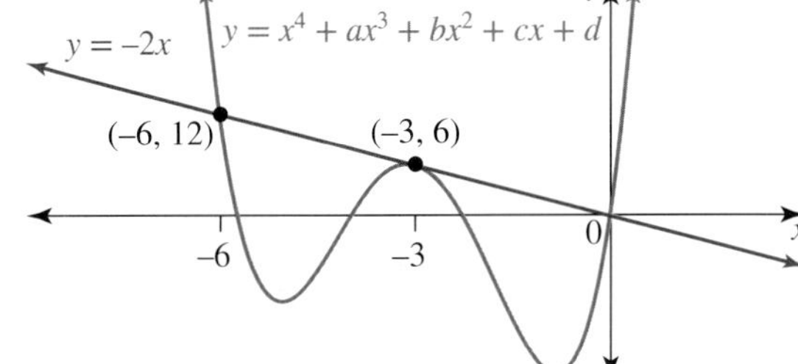

a. State the value of d.

b. Form the equation $p(x) = 0$ for which the x-coordinates of the points of intersection of the two graphs $y = x^4 + ax^3 + bx^2 + cx + d$ and $y = -2x$ are the solutions.

c. Use the information shown in the diagram to write down the factors of the equation $p(x) = 0$ and hence calculate the values of a, b and c.

d. i. State the rule for the quartic polynomial $y = x^4 + ax^3 + bx^2 + cx + d$ shown in the diagram and show that its graph has an x-intercept at $x = -4$.

ii. Calculate the exact values of its other x-intercepts.

5.5 Exam questions

Question 1 (2 marks) TECH-FREE

Given that $(x+1)$, $(x-3)$, $(x+2)$ and $(x-1)$ are factors of the polynomial $x^4 - x^3 - 7x^2 + x + 6$, solve the inequation $x^4 - x^3 - 7x^2 + x + 6 < 0$.

Question 2 (1 mark) TECH-ACTIVE

MC The family of graphs $y = a(x-h)^n + k$, where n is an even positive integer, has

A. one stationary point of inflection.
B. two stationary points of inflection.
C. one turning point.
D. two turning points.
E. one stationary point of inflection and one turning point.

Question 3 (4 marks) TECH-ACTIVE

State the degree of the graph shown and write the equation for the polynomial, given that the point (2, 18) lies on the graph.

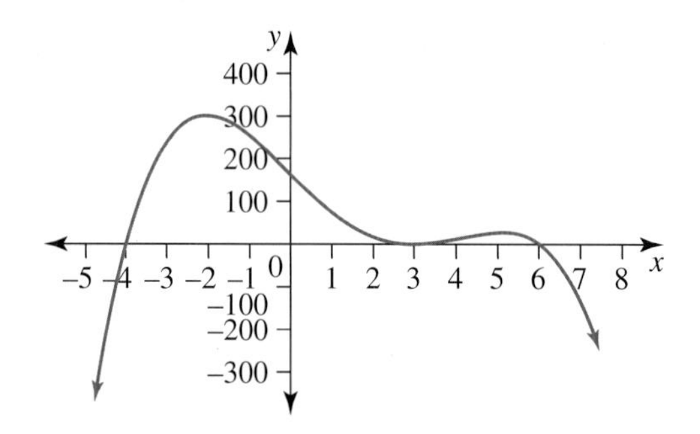

Question 4 (3 marks) TECH-FREE

State the y-intercept for the polynomial $y = 2x^3 - x^2 - 10x + 7$ and the other points on the graph that have the same y-coordinate as the y-intercept. Between which two of these points does the minimum turning point lie?

Question 5 (1 mark) TECH-ACTIVE

MC The Cartesian plane consists of four quadrants as shown.

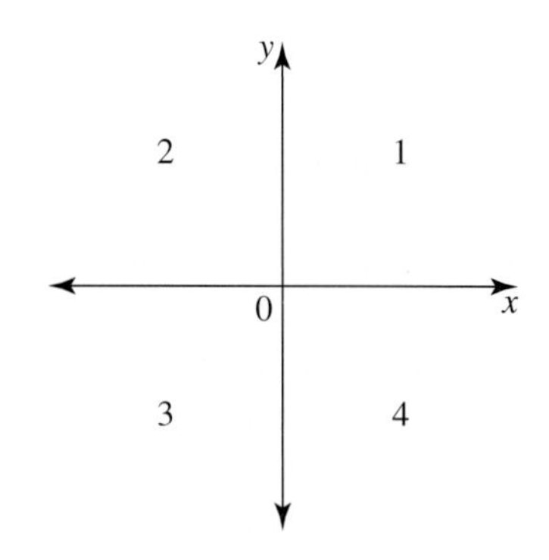

The extremities (or extreme ends) of the graph $y = -2x(x-3)(x+5)^2$ as $x \to \pm\infty$ lie in quadrants

A. 1 and 2
B. 1 and 3
C. 1 and 4
D. 3 and 4
E. 2 and 4

More exam questions are available online.

Answers

Topic 5 Quartic polynomials

5.2 Quartic polynomials

5.2 Exercise

1. a.

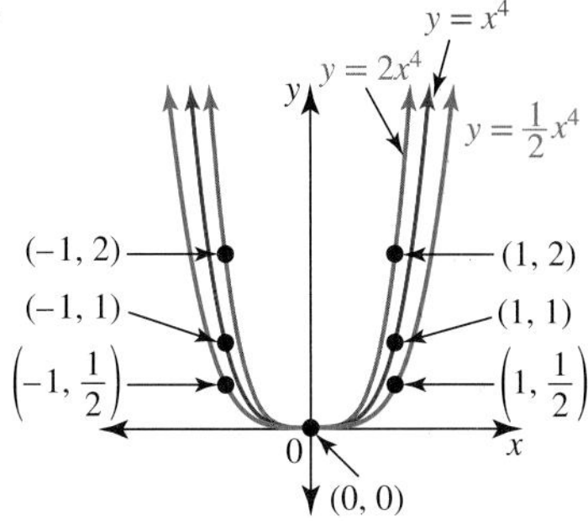

b.

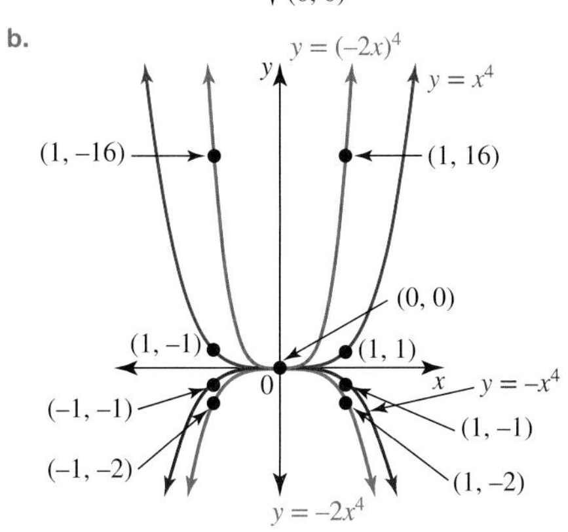

c.

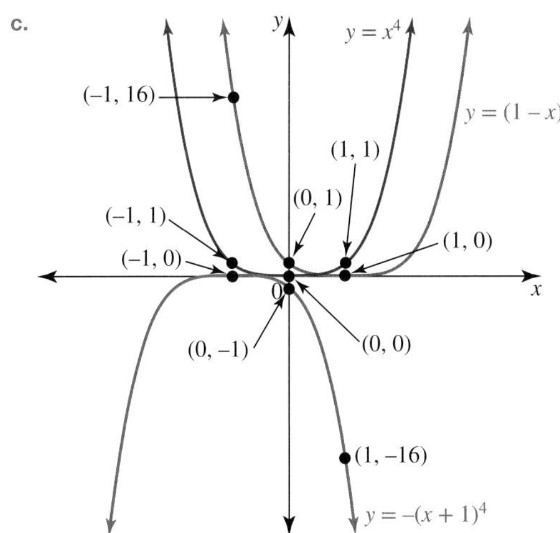

d.

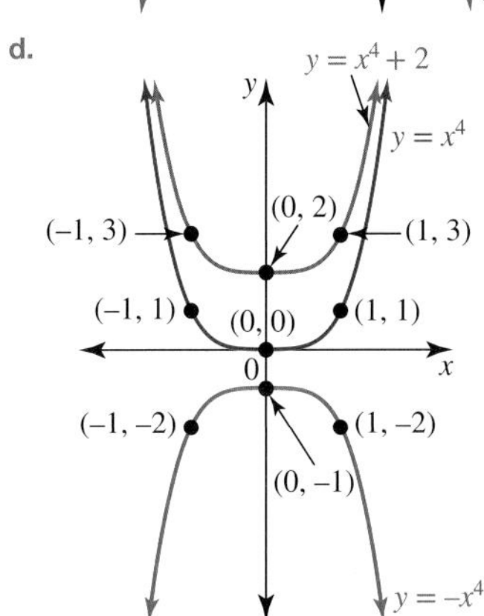

2. a.

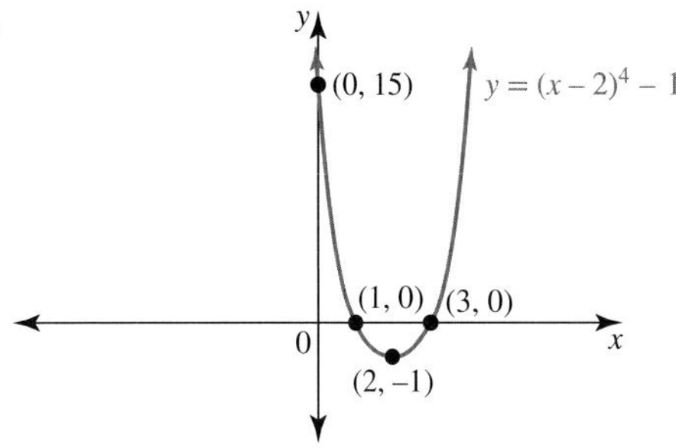

Minimum turning point $(2, -1)$; y-intercept $(0, 15)$; x-intercepts $(1, 0), (3, 0)$

b.

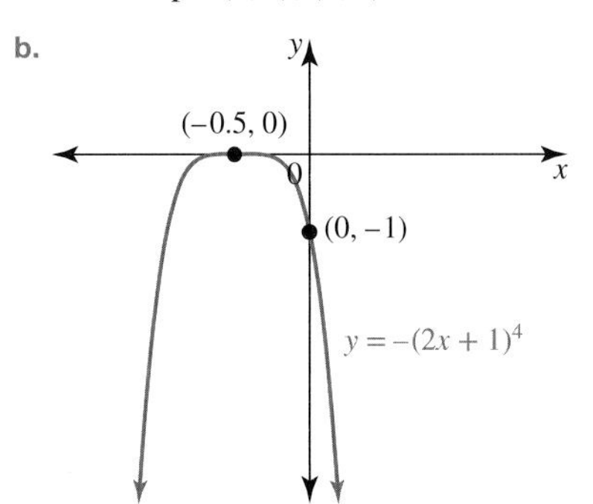

x-intercept and maximum turning point $\left(-\frac{1}{2}, 0\right)$; y-intercept $(0, -1)$

3. a. The turning point is $(-2, -2)$.

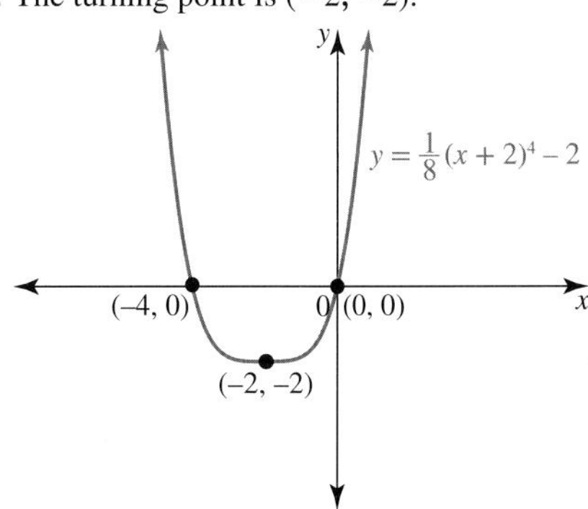

b. i. Maximum turning point $(1, -1)$

ii. $y = -(x - 1)^4 - 1$

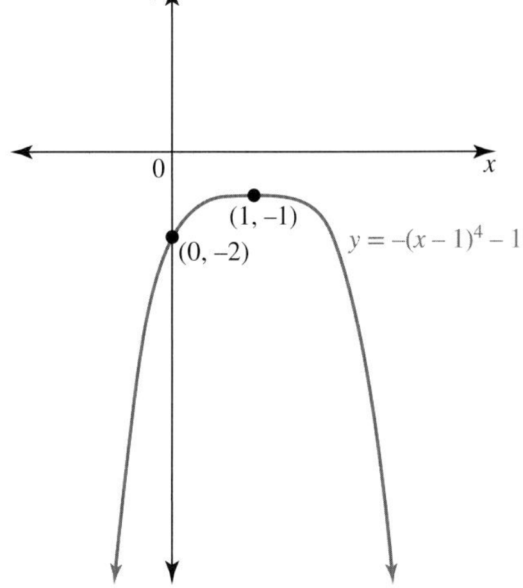

c. $y = \frac{1}{4}(x-4)^4$

4.

	Turning point	y-intercept	x-intercepts
a.	$(1, -16)$ minimum	$(0, -15)$	$(-1, 0), (3, 0)$
b.	$(-3, 12)$ minimum	$(0, 21)$	none
c.	$(-5, 250)$ maximum	$(0, 0)$	$(-10, 0), (0, 0)$
d.	$(2, -11)$ maximum	$(0, -107)$	none

a.

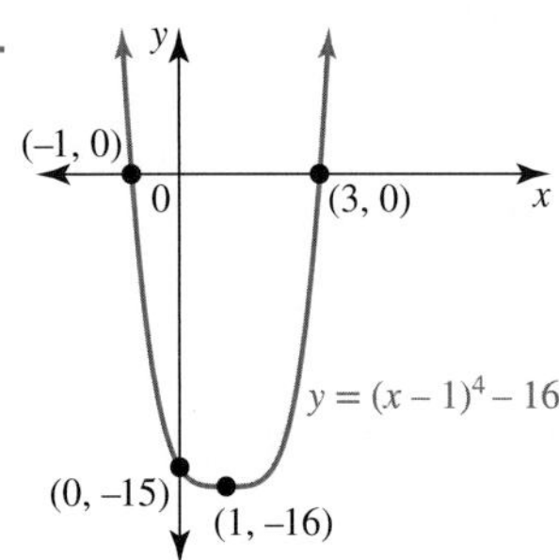

b.

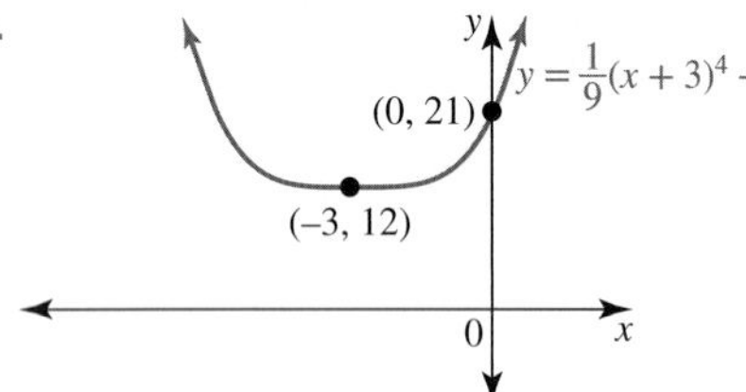

c.

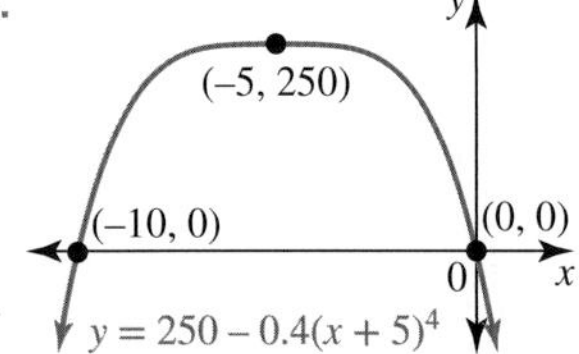

d.

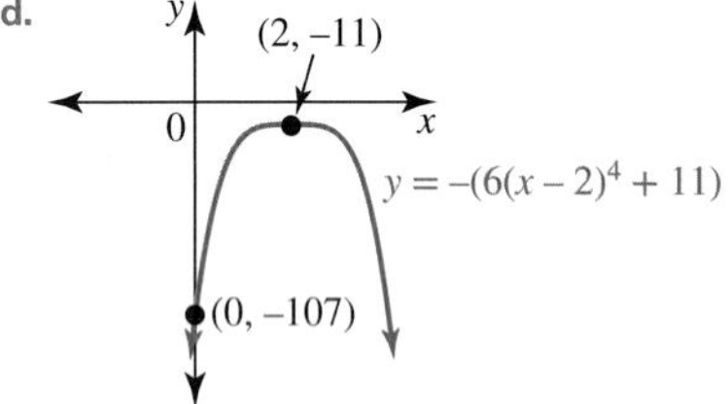

5. a. $y = \frac{2}{3}(x+9)^4 - 10$ b. $y = 6(x+3)^4 - 8$

c. $y = (3x-2)^4$ d. $y = -(x+100)^4 + 10\,000$

6. a.

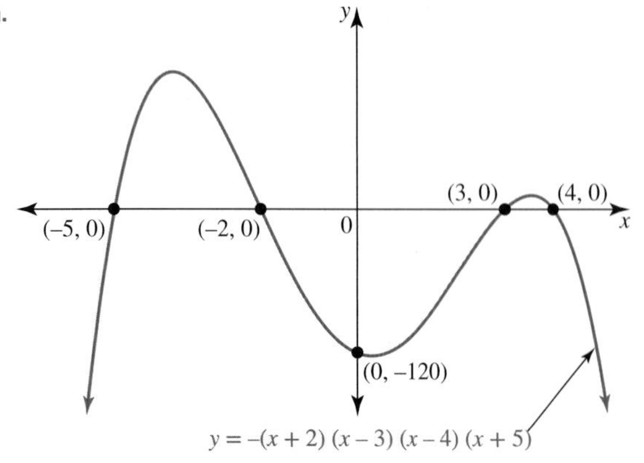

b. i. Cut at $x = -1 \Rightarrow (x+1)$ is a factor, touch at $x = 1 \Rightarrow (x-1)^2$ is a factor, cut at $x = 3 \Rightarrow (x-3)$ is a factor.

ii. $y = a(x+1)(x-1)^2(x-3)$

iii. $y = 2(x+1)(x-1)^2(x-3)$

7.

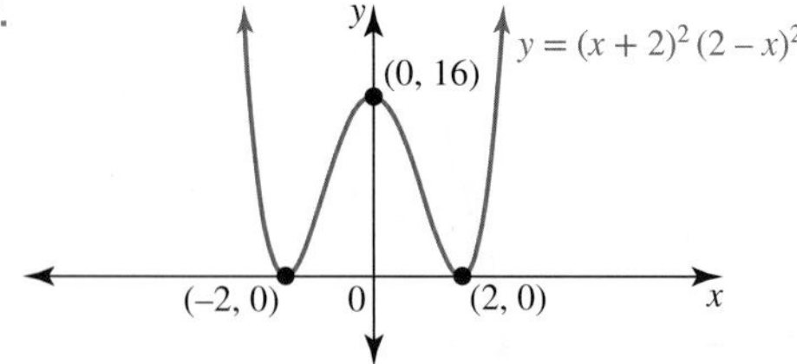

x-intercepts $(-2, 0)$ and $(2, 0)$ are turning points; y-intercept $(0, 16)$

8. $y = \frac{1}{4}x(x+4)(x-2)(x-5)$

9. a. x-intercepts $(-8, 0), (-3, 0), (4, 0), (10, 0)$; y-intercept $(0, 960)$

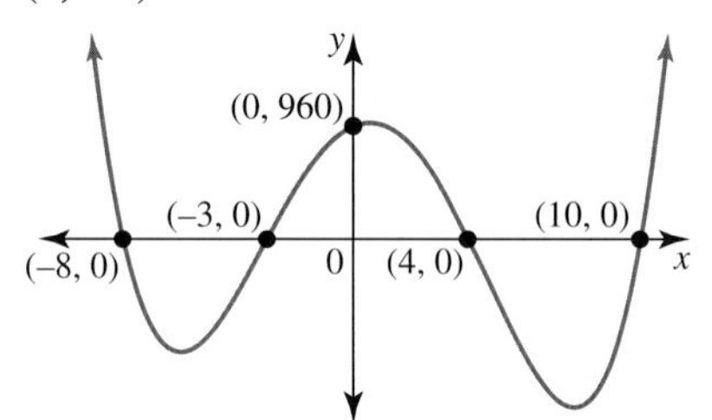

b. x-intercepts $(-3, 0), (2, 0), \left(\frac{15}{2}, 0\right), \left(\frac{10}{3}, 0\right)$; y-intercept $(0, 9)$

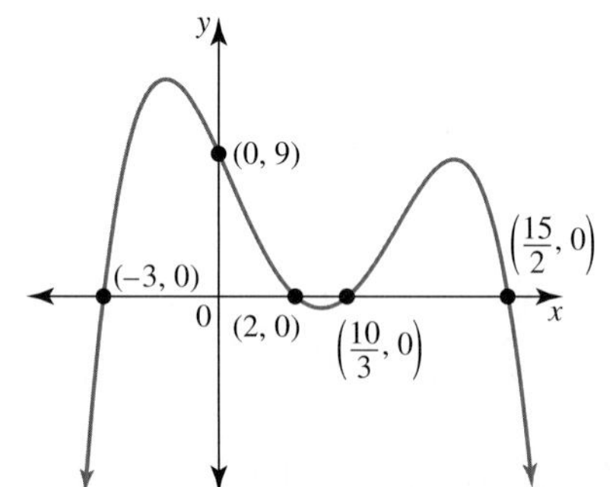

c. x-intercepts $(-7, 0), (2.5, 0)$; turning point $(1, 0)$; y-intercept $(0, 70)$

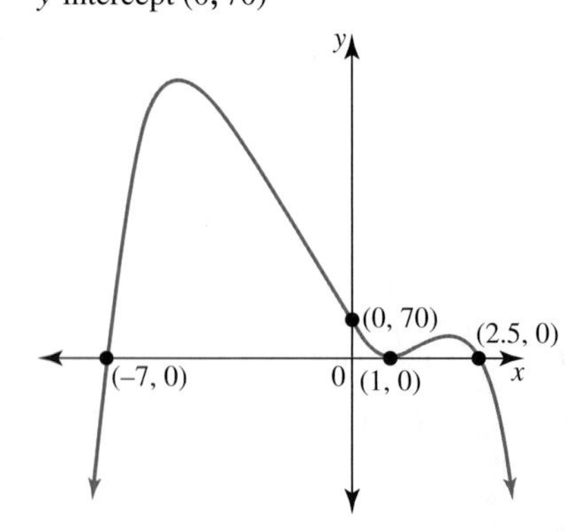

d. x-intercepts and turning points $(0, 0)$, $\left(\frac{15}{4}, 0\right)$

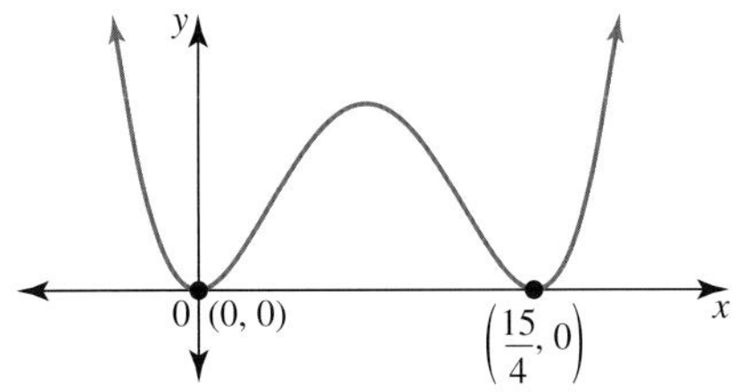

e. x-intercept and stationary point of inflection $(-1, 0)$; x-intercept $(4, 0)$; y-intercept $(0, 12)$

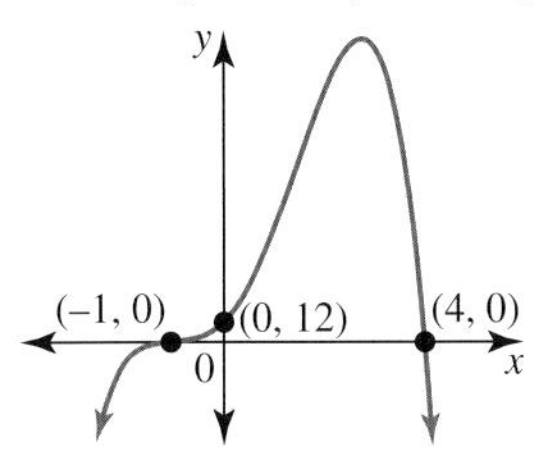

f. x-intercept and stationary point of inflection $\left(\frac{10}{3}, 0\right)$; x-intercept $\left(-\frac{10}{3}, 0\right)$; y-intercept $(0, -10\,000)$

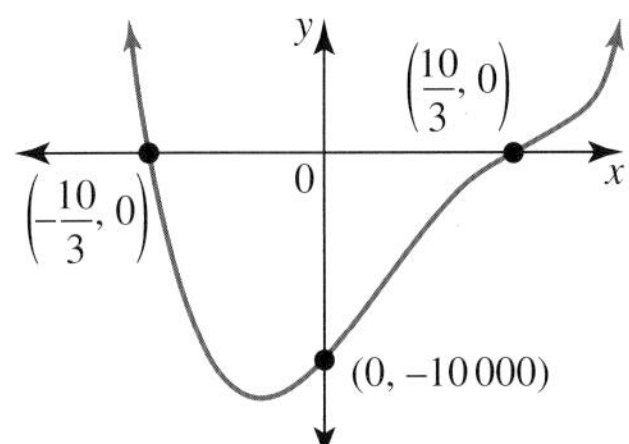

10. a. $y = -\frac{1}{72}(x+6)(x+5)(x+3)(x-4)$

b. $y = -x(x-4)(x+2)^2$

c. $y = \frac{2}{3}x^3(x+6)$

d. $y = \frac{3}{8}(2x+3)^2(5x-4)^2$

11. $a = 8$

12. a. $x = 0, 5, -\frac{1}{3}, -\frac{3}{4}$ b. $x = 2, 4, -\frac{3}{5}$

c. $x = -2, \frac{1}{2}$ d. $x = -4, 2$

13. $x \le -5$ or $-2 \le x \le -\frac{3}{2}$ or $x \ge 3$

14. a. $-4 < x < 0$ or $0 < x < 5$

b. $1 < x < 4$

c. $x < -\frac{5}{2}$ or $0 < x < 2$ or $x > 4$

d. No real solutions

15. a. $x \le -5$ or $x \ge 2$

b. $-1.16 \le x \le 4$ or $5.16 \le x \le 6$

c. $x \le -4.19$ or $-3 \le x \le 1.19$ or $x \ge 3$

16. a. $x < -2$ or $x > 7$

b. $x \le -3$ or $-2.19 \le x \le 3.19$ or $x \ge 5$

c. $x < -2.5$ or $0.28 < x < 2.39$ or $x > 2.5$

d. $x \in R$

17. $x = 2$ or -6

18. a. $x = 0, x = -9$ b. $x = 0, 2, -2$

c. $x = -3 \pm \sqrt{3}$ d. $x = 1, \frac{10}{3}$ or $-\frac{4}{3}$

19. a. $x < -3$ or $0.5 < x < 4$ or > 5.5

b. $-\frac{7}{15} \le x \le -\frac{1}{5}$ or $x = 0$

c. $x < -1$ or $x > 2$

d. $x \le \frac{3}{10}$ or $x \ge \frac{1}{2}$

20. a. $y = -\frac{1}{4}(x+2)^4 + 4$

b.

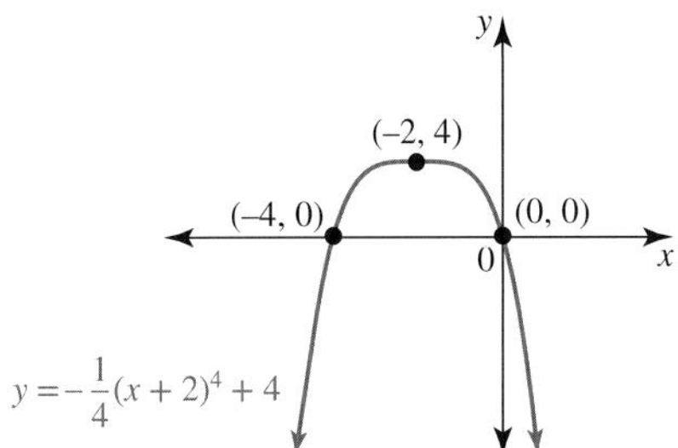

x-intercepts $(-4, 0)$, $(0, 0)$;

$\left\{x: -\frac{1}{4}(x+2)^4 + 4 > 0\right\} = (x: -4 < x < 0)$

21. a. $a = \frac{2}{3}, k = \frac{1}{3}$

b. Minimum turning point $\left(0, \frac{1}{3}\right)$

c. $x = 0$

d.

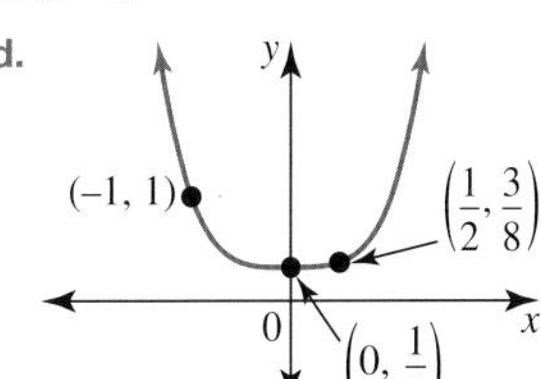

22. a. $x = 1$

b. $(1, 10)$

c. $y = -\frac{7}{81}(x-1)^4 + 10$

d. $\left(0, \frac{803}{81}\right)$

e. $\left(1 - \sqrt[4]{\frac{810}{7}}, 0\right), \left(1 + \sqrt[4]{\frac{810}{7}}, 0\right)$

f.

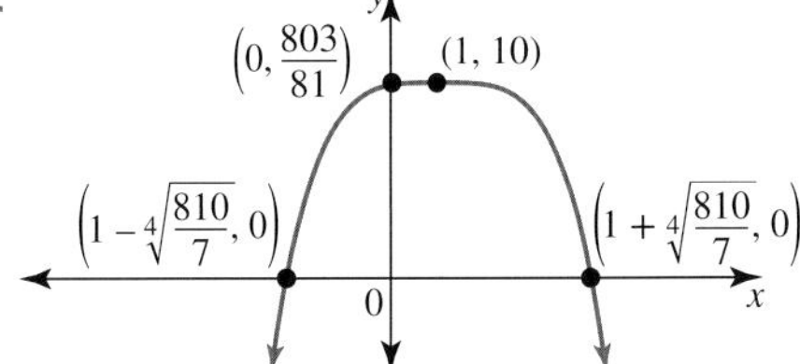

23. a. $-(x-3)^2(x+3)^2$

b. 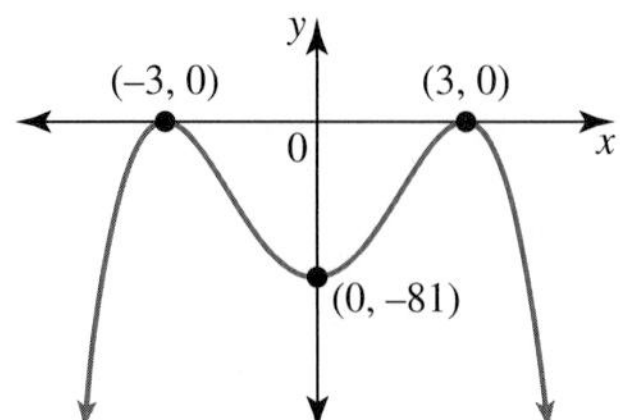

x-intercepts and maximum turning points $(-3, 0), (3, 0)$;
y-intercept and minimum turning point $(0, -81)$

c. $\{x : -x^4 + 18x^2 - 81 > 0\} = \varnothing$

d. $R \setminus \{-3, 3\}$

24. a. $(2, 16)$

b. $a = 4, m = 8$

c. 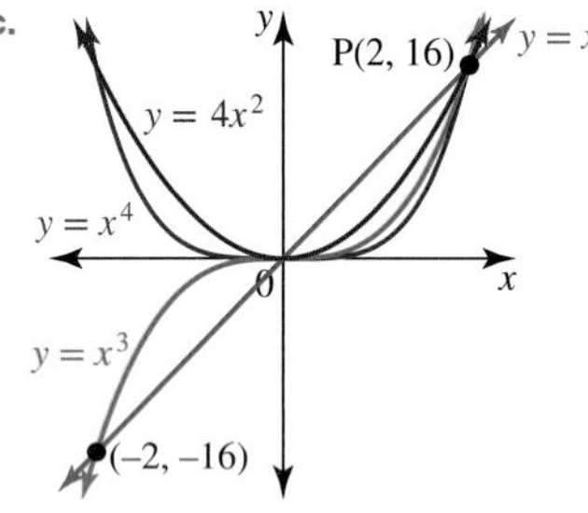

d. i. $(0, 0)$ and $\left(n, n^4\right)$

ii. $a = n^2, m = n^3$

25. a. 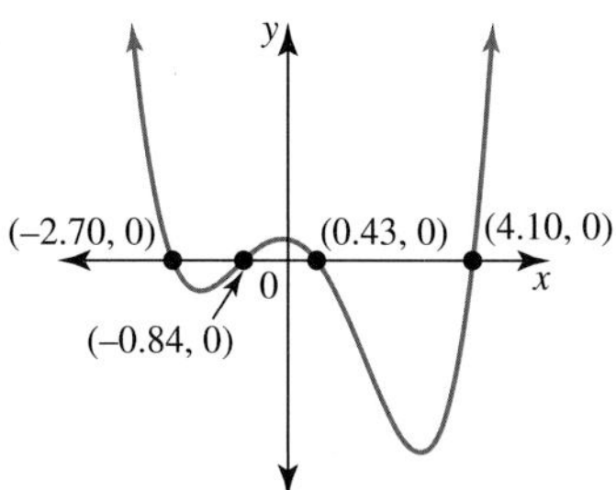

x-intercepts $(-2.70, 0), (-0.84, 0), (0.43, 0), (4.10, 0)$;
minimum turning points $(-2, -12), (2.92, -62.19)$;
maximum turning point $(-0.17, 4.34)$

b. 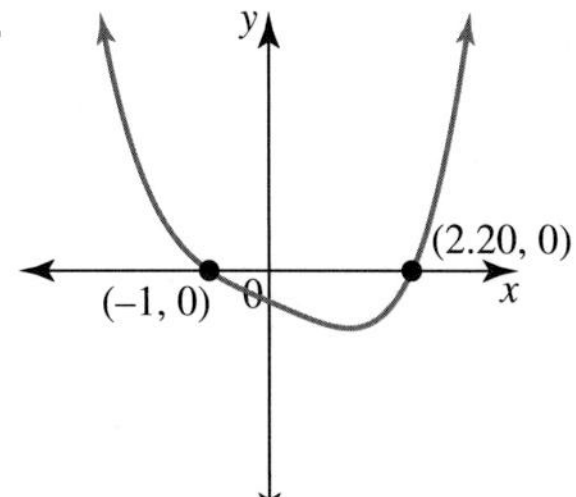

Minimum turning point $(1.21, -14.33)$; x-intercepts $(-1, 0), (2.20, 0)$

5.2 Exam questions

Note: Mark allocations are available with the fully worked solutions online.

1. E
2. A

3. 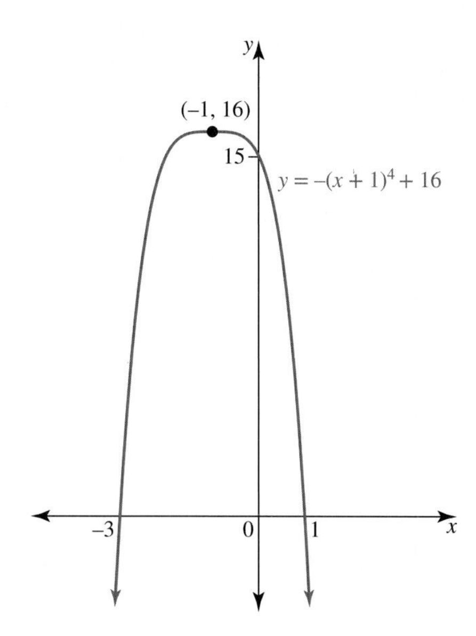

5.3 Families of polynomials

5.3 Exercise

1. 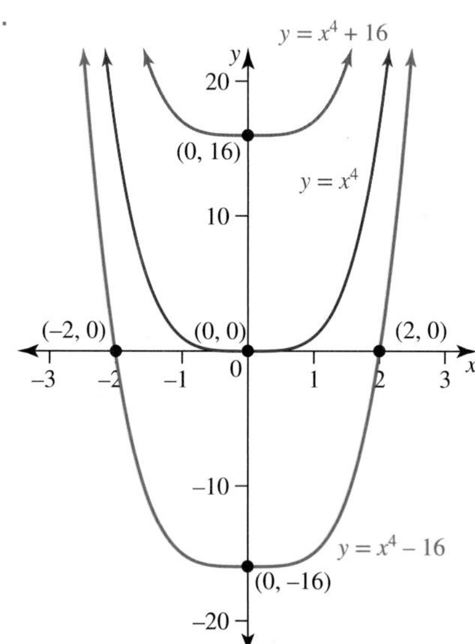

2. 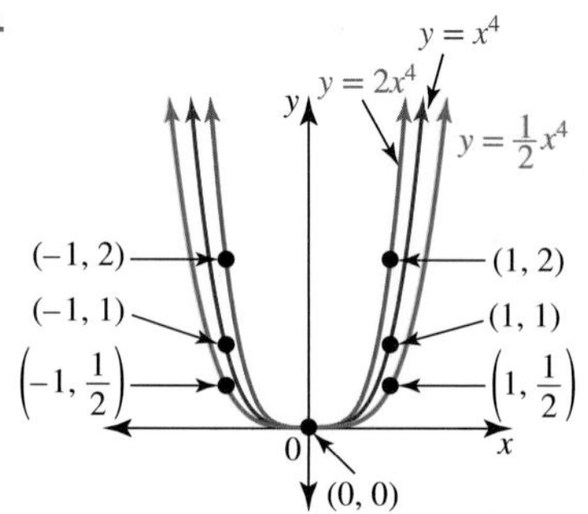

3. a.

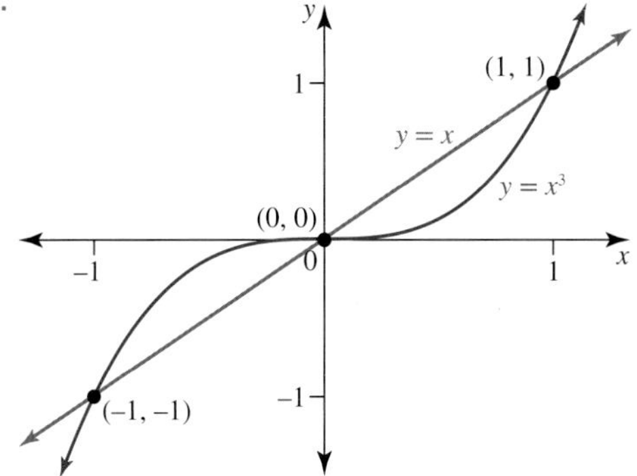

b. $x \leq -1$ or $0 \leq x \leq 1$

4. a. $p(x) = a(x - h)^3$

i. If $a > 0$, the cubic polynomials have a stationary point of inflection at $(h, 0)$.
As $x \to +\infty, y \to +\infty$
and as
$x \to -\infty, y \to -\infty$

ii.

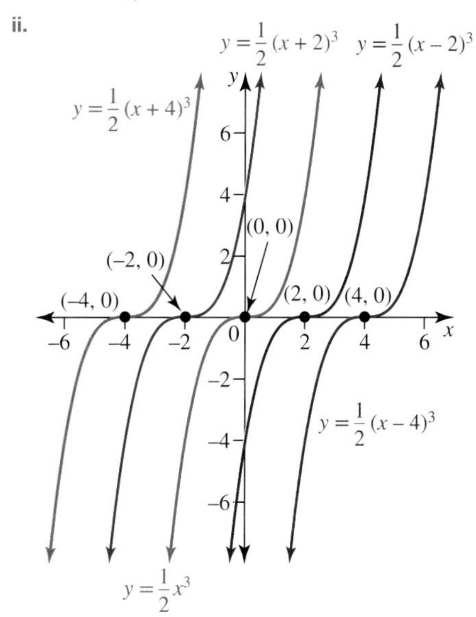

b. If $a = -\frac{1}{2}$, the cubic polynomials would still have points of inflection at $(h, 0)$ but would be reflected in the x-axis, meaning that
as $x \to +\infty, y \to -\infty$
and as
$x \to -\infty, y \to +\infty$

5. a. The minimum turning point is at $(3, c)$.

i. $c > 0$
The minimum turning point is above the x-axis, so the curve is always positive with no x-axis intercepts.

ii. $c = 0$
The minimum turning point is on the x-axis, so the curve touches the x-axis at $x = 3$.

iii. $c < 0$
The minimum turning point is below the x-axis, so the curve has two x-axis intercepts.

b.

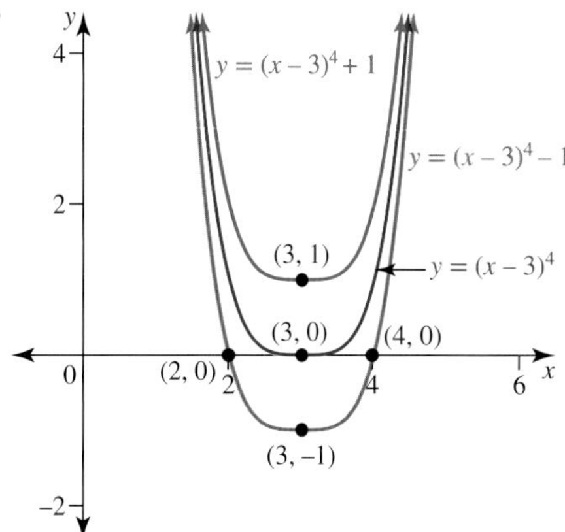

6. a. $y = 12(x + 1)^4 - 12$ b. $x = -1$
c. $(-2, 0)$

7. a.

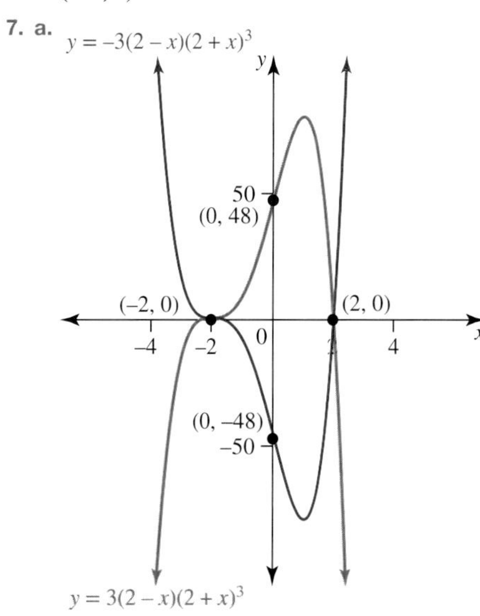

b. The curves are the same shape with x-intercepts at $(2, 0)$ and $(-2, 0)$
One curve is the reflection of the other over the x-axis, so the maximum turning point of one curve is a minimum turning point of the other curve.

8. a. $y = x^2(x + 1)(x - 1)$

b. i. $y = ax^2(x + 1)(x - 1)$ ii. $y = \frac{1}{2}x^2(x + 1)(x - 1)$

9. a.

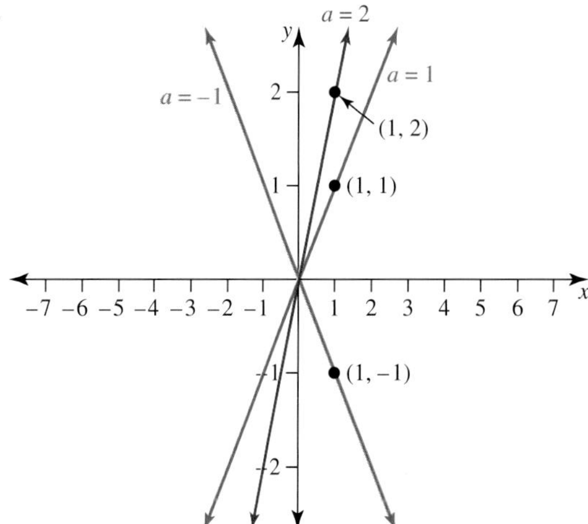

b. $a = 2.5$

10 a. $y = -x^2 + 55$

b. Maximum turning point $(0, k)$

c. $k < 0$

11. a. $(0, 0), (m, 0)$

b.

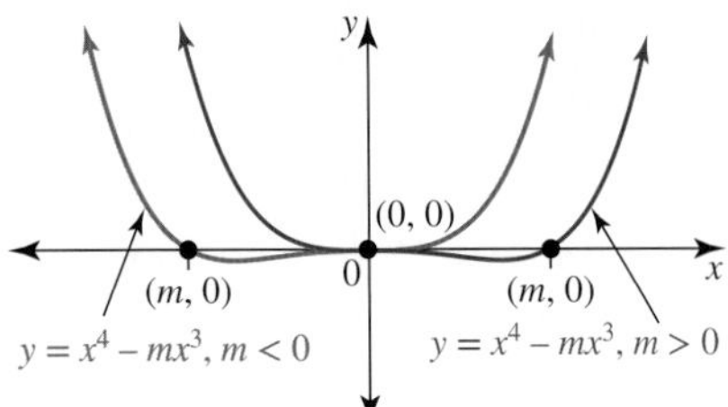

There is a stationary point of inflection at the origin.

c. $y = x^4 + 17x^3$

12. a. $y = mx - 2m + 3$ b. $y = \frac{3}{2}x$

13. a. The origin

b. $y = ax^2 + (3 - 2a)x, a \neq 0.$

14. $\left(-\sqrt{m}, 0\right), \left(\sqrt{m}, 0\right)$ and $(0, 0)$, which is also a turning point

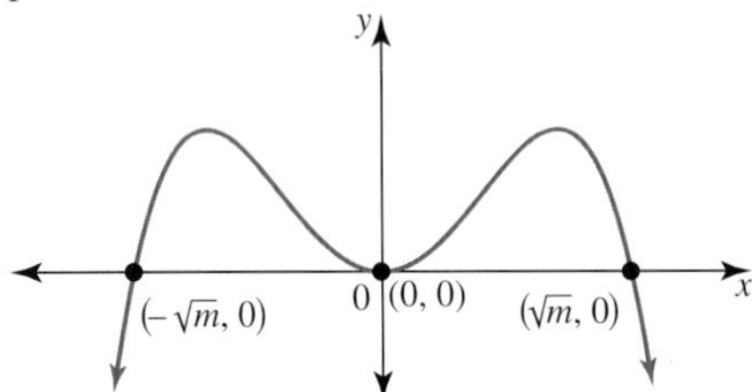

15. $y = 27(x + 1)^3 + 7$

16. $y = (x + 2)^2(x - 4)(x - 3), (3, 0)$

17. a. Sample responses can be found in the worked solutions in the online resources.

b. $a = \frac{5}{4}$

c. $0 < a < \frac{5}{4}$

18. a. The set of horizontal lines and the set of concave up parabolas with y-intercept $(0, 10)$

b.

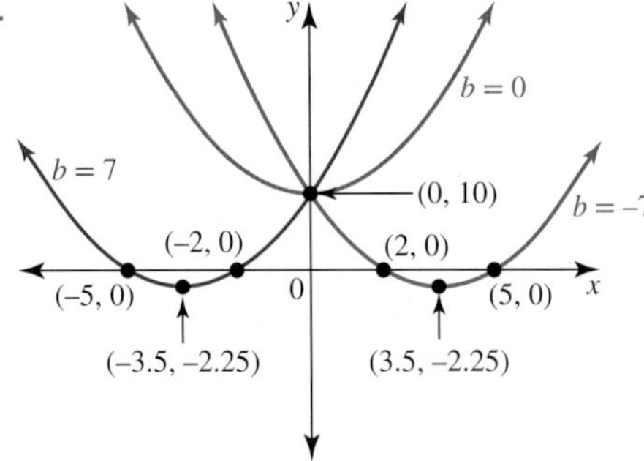

c. i. $k = -2.25$ ii. $k > -2.25$ iii. $k < -2.25$

d. i. $b = \pm 2\sqrt{3}$

ii. $b < -2\sqrt{3}$ or $b > 2\sqrt{3}$

iii. $-2\sqrt{3} < b < 2\sqrt{3}$

19. a. Sample responses can be found in the worked solutions in the online resources.

b. $a = -2$; $y = -2x^3 + 7x^2 - 5x$; x-intercepts at origin, $(1, 0)$ and $\left(\frac{5}{2}, 0\right)$

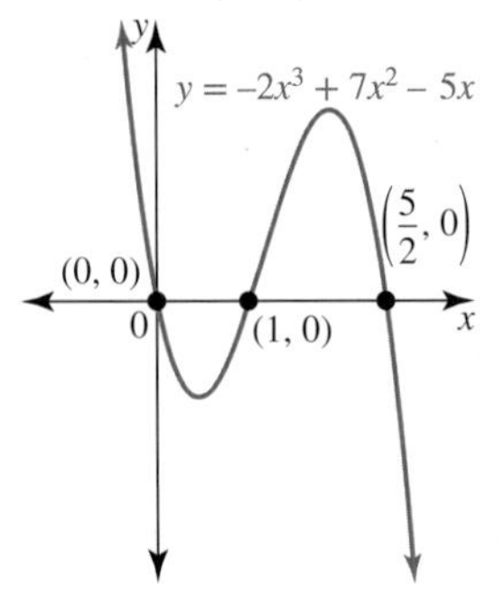

c. $a = 1$; $y = x^3 + x^2 + 4x - 6$; sample responses can be found in the worked solutions in the online resources.

d. The point of intersection $(-1, -8a - 2)$ lies on the vertical line $x = -1$.

20.

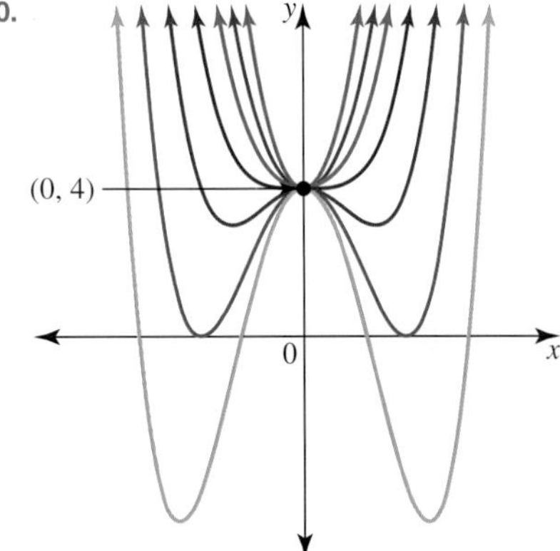

a. $a < -4$

b. $a = -4$

c. $a > -4$

5.3 Exam questions

Note: Mark allocations are available with the fully worked solutions online.

1. E

2. C

3. x-intercepts at $(2, 0), (4, 0)$
   y-intercepts at $(0, 80)$

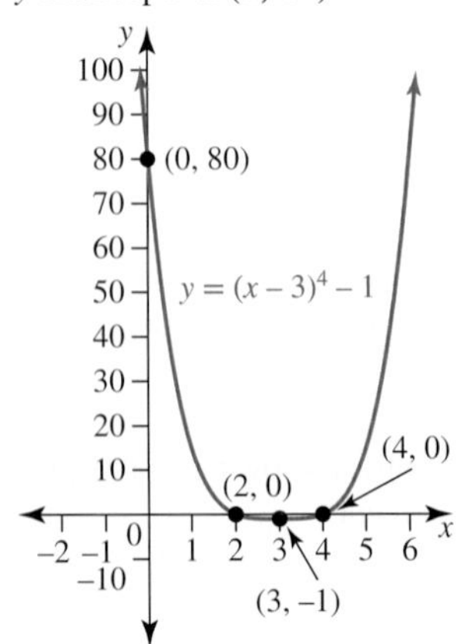

5.4 Numerical approximation of roots of polynomial equations

5.4 Exercise

1. C
2. a. Both $p(-2) < 0$ and $p(-1) < 0$
   b. $p(1) < 0$ and $p(2) > 0$
   c. $\left[\frac{3}{2}, 2\right]$
3. a. $p(-2) = 17 > 0$ and $p(0) = -1 < 0$
   b. $[-1, 0]$, $[-0.5, 0]$
   c. $x = -0.25$
4. a. $p(10) = -19, p(12) = 1$
   b. $p(-2) = 3, p(-1) = -3$
   c. $p(-2) = -51, p(1) = 9$
5. a. $[11, 12]$, $[11.5, 12]$; $x = 11.75$
   b. $[-2, -1.5]$, $[-2, -1.75]$; $x = -1.875$
   c. $[-2, -0.5]$, $[-1.25, -0.5]$; $x = -0.875$
6. a. $p(1) < 0, p(2) > 0$
   b. $x = 2$
   c. $x = 1.5$; $x = 1.75$
   d. $x = 1.875$
7. a. $x = 1.2$
   b. $x = 4$
   c. The method of bisection very slowly converges towards the solution.
8. a.

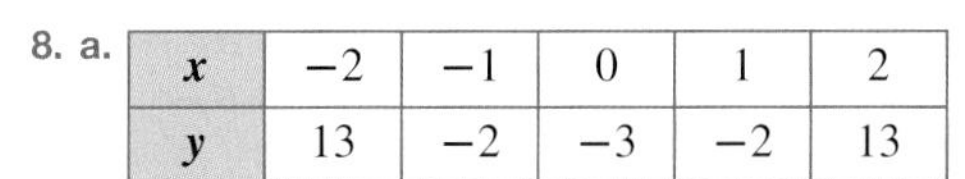

x	-2	-1	0	1	2
y	13	-2	-3	-2	13

   b. $[1, 2]$
   c. Sample responses can be found in the worked solutions in the online resources.
9. a.

x	-3	-2	-1	0	1	2	3
y	75	8	-9	-12	-13	0	63

   b. $x = 2$
   c. $[-2, -1]$; $x = -1.7$
10. a. $[0, 1]$
   b. $x = 0.375$
   c. $y = x^3$ and $y = -5x + 2$ (or other)
   d.

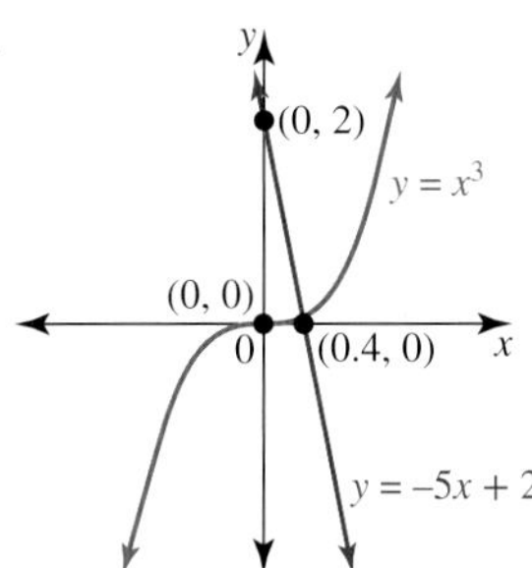

One root close to $x = 0.375$

11.

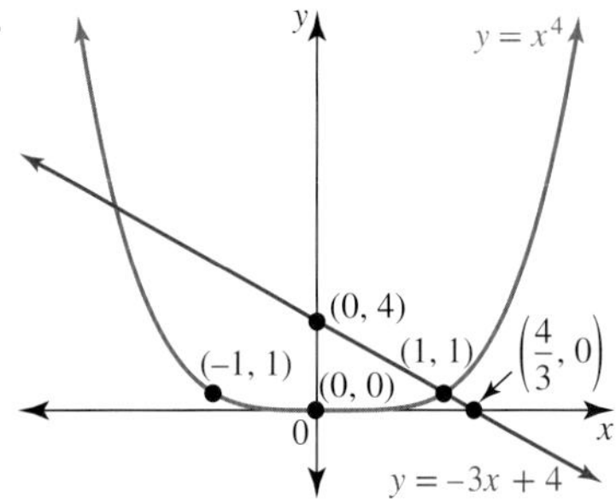

$x = -1.75$ (estimate); $x = 1$ (exact)

12.

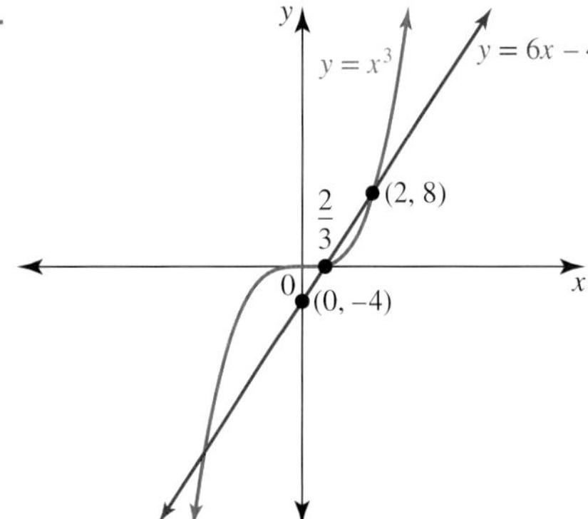

Exact solution $x = 2$; approximate solutions $x = -2.7$ and $x = 0.7$

13. a. $x^3 - 3x + 2 = 0$
   b. Two factors, one of multiplicity 2, one of multiplicity 1
   c. $x = -2, x = 1$, A$(1, 1)$, B$(-2, -8)$
   d. Three solutions
14. a. $x \in [0, 3]$
   b. $(1.8, 8.208)$
15. a. $(-1.6, 65.664)$
   b. $(-0.2, -0.552)$
   c. Maximum turning points approximately $(-0.7, 0.2499)$ and $(0.7, 0.2499)$; minimum turning point exactly $(0, 0)$
16. a. $(0, 9)$
   b. $\left(-\frac{5}{2}, 9\right)$ and $(3, 9)$
   c. Between $x = -\frac{5}{2}$ and $x = 0$
   d. $(-1.4, 22.552)$
17. a. $x \in [-1, 0]$ and $x \in [0, 4]$
   b. $x \in \left[-2\sqrt{3}, 0\right]$ and $x \in \left[0, 2\sqrt{3}\right]$
   c. $x \in [0, 1]$ and $x \in [1, 4]$
   d. $x \in [-1, 0]$ and at the point $(0, 7)$
18. a. $y = 0$ for an $x \in [1, 2]$
   b. $x \in [1, 1.5]$; $x \in [1, 1.25]$
   c. 12 containers
   d. $x \in [3, 6]$
   e. 44 containers; \$331
   f. 65 or more containers

19. a.

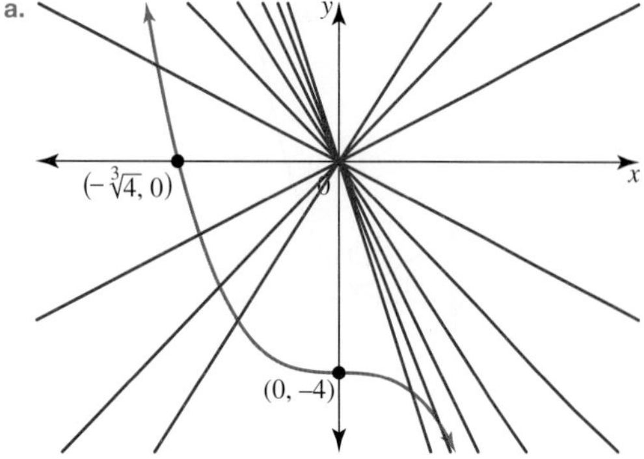

b. $a = -5$

c. One

d. $x = -1.38$

20. a. $V = x(18 - 2x)(14 - 2x)$

b. Between $x = 0$ and $x = 7$

c. 2.605 cm

5.4 Exam questions

Note: Mark allocations are available with the fully worked solutions online.

1. A

2. a. $p(0) < 0,\ p(1) > 0 \Rightarrow$ The root is between $x = 0$ and $x = 1$.

b. The root is closer to $p\left(\frac{1}{2}\right)$, so a more accurate estimate of the root is $x = \frac{1}{2}$.

3. C

5.5 Review

5.5 Exercise

Technology free: short answer

1. a. Maximum turning point $(-3, 16)$; y-intercept $(0, -65)$; x-intercepts $(-5, 0), (-1, 0)$

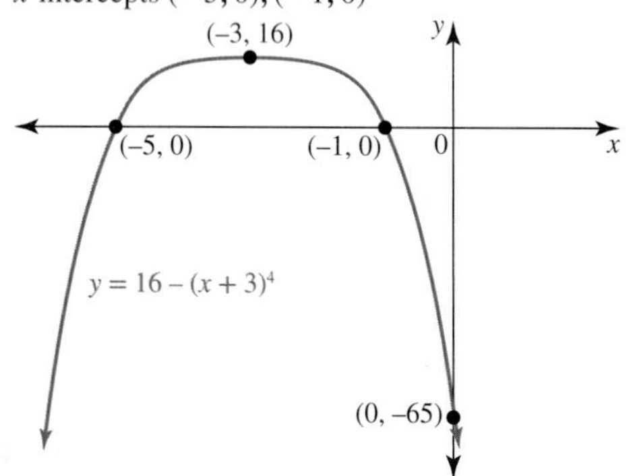

b. y-intercept $(0, -3)$; x-intercepts $(-2, 0)$ cut, $(3, 0)$ saddle cut

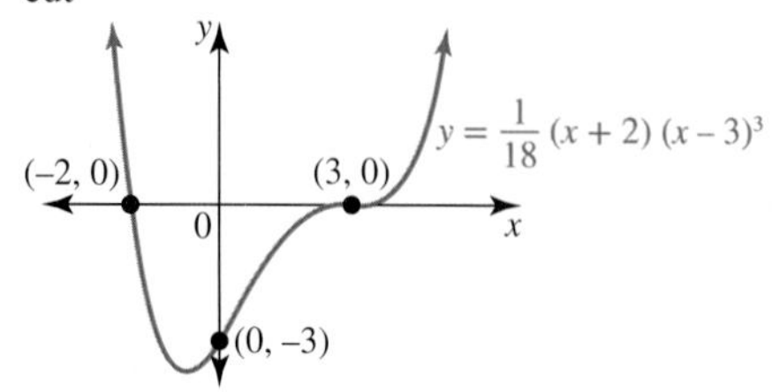

c. y-intercept and turning point $(0, -16)$; x-intercepts and turning points $(-2, 0)$ and $(2, 0)$

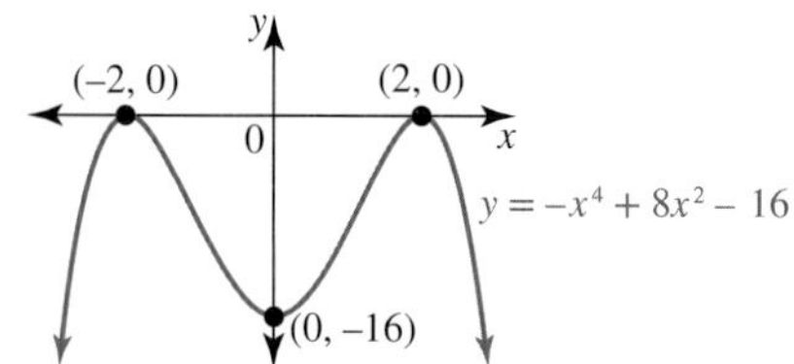

2. a. $x \le -4$ or $x = -1$ or $x \ge 3$

b. $-\frac{7}{3} < x < 5$

3. a. $y = 1 - x$

$y = x^3$

[0, 1]

b. [0.5, 1]

4. a.

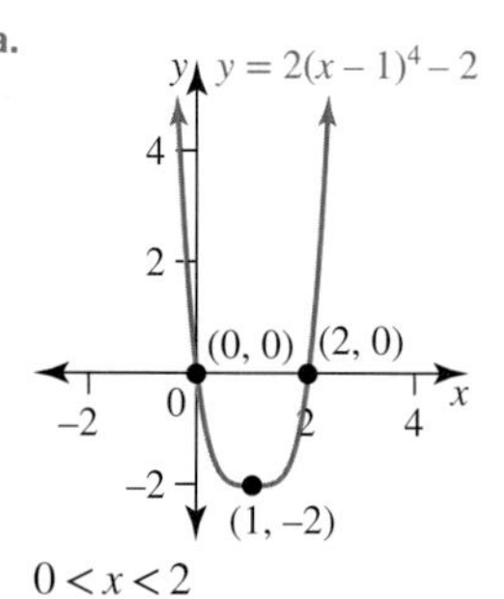

$0 < x < 2$

b.

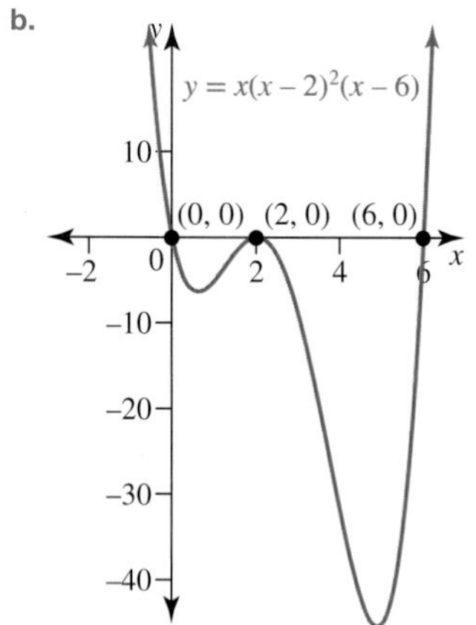

$x \le 0$ or $x = 2$ or $x \ge 6$

5. a. $p(x) = a(x-3)(x+2)^3$

b. $y = -\frac{1}{2}(x-3)(x+2)^3$

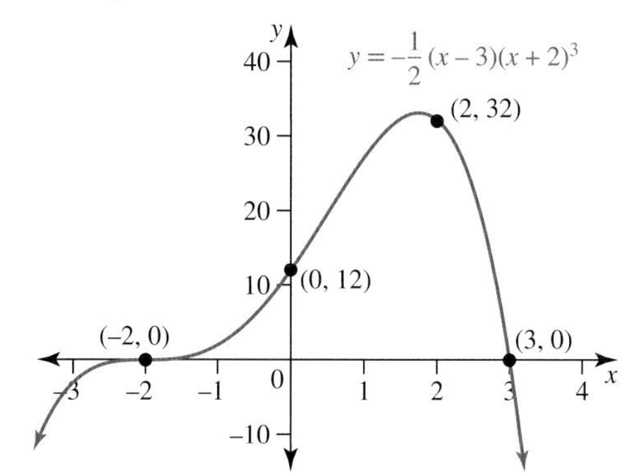

6. a. $m = 3, y = 2x^3 + 3x^2 - 4x + 5$

b. $(0, 5)$ and $\left(\frac{4}{m}, \frac{128}{m^3} + 5\right), m \neq 0$

Technology active: multiple choice

7. E	8. C	9. E	10. A	11. D
12. C	13. B	14. E	15. B	16. B

Technology active: extended response

17. a. $a = -4, b = 16$ and curve: $y = -4x^2 + 16$

b. $A = 16x - 4x^3$

c. Sample responses can be found in the worked solutions.

d. i. Sample responses can be found in the worked solutions.

ii. $\beta = 1.3125$

e. $x = \beta$

18. a. $x = -2, x = -1, x = 0$; maximum turning point $x \in [-2, -1]$; minimum turning point $x \in [-1, 0]$

b. $(-1.58, 5.38)$ and $(-0.42, 4.62)$

c. Both turning points lie above the x-axis.

d. $[-3, -2], [-3, -2.5], [-3, -2.75], [-3, -2.875]$ (answers may vary)

e. $(-2.9, 0)$

f.

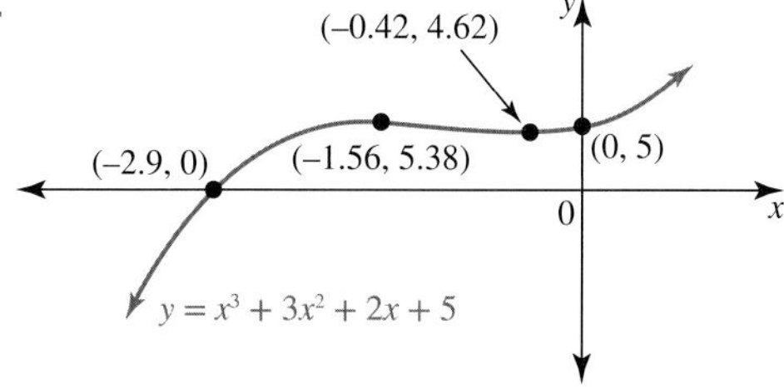

19. a. i. $a = 2$

ii. $y = -x^2 + 1$; maximum turning point $(0, 1)$

b. $y = x^3 - x^2 - x + 1$; $(0, 1), (-1, 0), (1, 0)$

c. $y = -x^3 - x^2 + x + 1$; $(0, 1), (-1, 0), (1, 0)$

d. Sample responses can be found in the worked solutions in the online resources.

e. $a = 4$

20. a. $d = 0$

b. $x^4 + ax^3 + bx^2 + (c+2)x = 0$

c. $a = 12, b = 45, c = 52$

d. i. The rule for the quartic polynomial $y = x^4 + ax^3 + bx^2 + cx + d$ shown in the diagram is $y = x^4 + 12x^3 + 45x^2 + 52x$.

Let $x = -4$.

$$\begin{aligned} y &= (-4)^4 + 12(-4)^3 + 45(-4)^2 + 52(-4) \\ &= 256 - 768 + 720 - 208 \\ &= 976 - 976 \\ &= 0 \end{aligned}$$

There is an x-intercept at $x = -4$.

ii. $(0, 0), \left(-4 - \sqrt{3}, 0\right)$ and $\left(-4 + \sqrt{3}, 0\right)$

5.5 Exam questions

Note: Mark allocations are available with the fully worked solutions online.

1. $\{x: -2 < x < -1\} \cup \{x: 1 < x < 3\}$

2. C

3. $y = -\frac{3}{4}(x+4)(x-3)^2(x-6)$, degree 4

4. $y = 7$ when $x = -2, 0, 2.5$

The graph is a positive odd cubic with two turning points. The second turning point is a minimum turning point and lies somewhere between (0, 7) and (2.5, 7).

5. D

6 Functions and relations

LEARNING SEQUENCE

Fully worked solutions for this topic are available online.

6.1 Overview

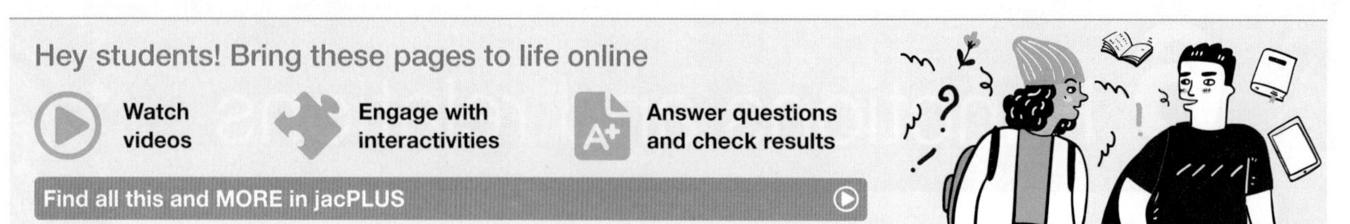

6.1.1 Introduction

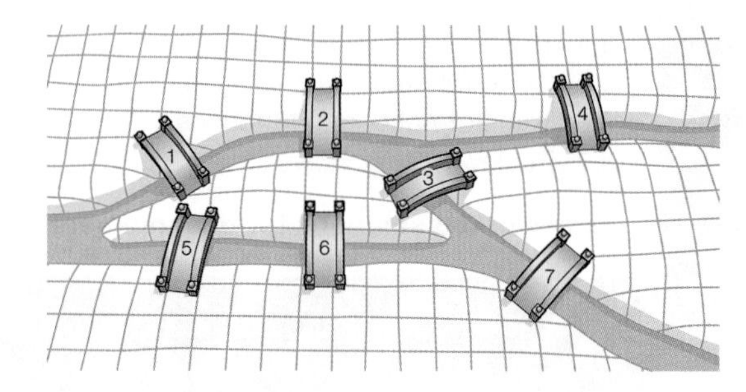

The concept of a function is fundamental in mathematical analysis. The reference to, and use of, the notation $y = f(x)$ is commonplace. The 18th-century mathematician Leonhard Euler is credited with first using the function notation $f(x)$, and that of inverse function notation $f^{-1}(x)$.

Euler devised much of the notation that we use today, including the symbol π for the ratio of the circumference of a circle to its diameter, the symbol Σ for summation, $\sin x$ and $\cos x$, $\log x$ and e^x, i as the symbol for the square root of -1, the convention of naming the angles of a triangle as A, B and C and the sides opposite these angles as a, b and c respectively, and much more.

Although Euler was born and educated in Basel, Switzerland, he spent much of his professional life in St Petersburg in Russia except for a 25-year period in Berlin under the patronage of Frederick the Great. In 1741 he was invited to return to St Petersburg by Catherine the Great and he remained there for the rest of his life. He suffered from poor eyesight from early adulthood, and shortly after returning to St Petersburg he lost his sight completely. This did not deter him; of the work he produced, almost half was completed when he was totally blind. He had a remarkable capacity to carry his calculations in his mind, and his sons helped to produce the books containing his discoveries.

One famous problem that Euler was able to solve was the Seven Bridges of Königsberg problem. Although only two of the original seven bridges remain, due to heavy bombardment in the Second World War, the legacy of the city (now Kaliningrad) is that Euler's solution led to a new branch of mathematics known as graph theory.

Euler's works and name live on in mathematics today.

KEY CONCEPTS

This topic covers the following key concepts from the VCE Mathematics Study Design:

- functions and function notation, domain, co-domain and range, representation of a function by rule, graph and table, inverse functions and their graphs
- graphs of power functions $f(x) = x^n$ for $n \in \left\{-2, -1, \frac{1}{3}, \frac{1}{2}, 1, 2, 3, 4\right\}$, and transformations of these graphs to the form $y = a(x+b)^n + c$ where $a, b, c \in R$ and $a \neq 0$
- use of symbolic notation to develop algebraic expressions and represent functions, relations, equations, and systems of simultaneous equations
- transformations of the plane and application to basic functions and relations by simple combinations of dilations (students should be familiar with both 'parallel to an axis' and 'from an axis' descriptions), reflections in an axis and translations (matrix representation may be used but is not required).

Note: *Concepts shown in grey are covered in other topics.*

6.2 Functions and relations

LEARNING INTENTION

At the end of this subtopic you should be able to:
- classify a function or relation by the type of correspondence between ordered pairs
- identify the domain, co-domain and range of a function
- evaluate functions using function notation.

The mathematical language, notation and concepts that will be introduced in this topic will continue to be used throughout all subsequent topics. Within this topic, concepts will be illustrated using the now-familiar polynomial relationships. They will also be applied to some new relationships.

6.2.1 Relations and functions, domain and range

Special terminology and mathematical language relating to this topic are listed below.
- A **relation** is any set of ordered pairs (x, y).
- The **domain** is the set of all x-values of the ordered pairs.
- The **range** is the set of all y-values of the ordered pairs.
- A **function** is a set of ordered pairs in which every x-value is paired to a unique y-value.
- A **restricted domain** occurs when restrictions are placed on values of variables in some practical situations and models. A restricted domain usually affects the range.

Consider the following examples of sets of ordered pairs.

$A = \{(-2, 4), (1, 5), (3, 4)\}$	$B = \{(x, y) : y = 2x\}$	$C = \{(x, y) : y \le 2x\}$
This is a set of ordered pairs.	This set describes a linear equation.	This set describes a linear inequation.
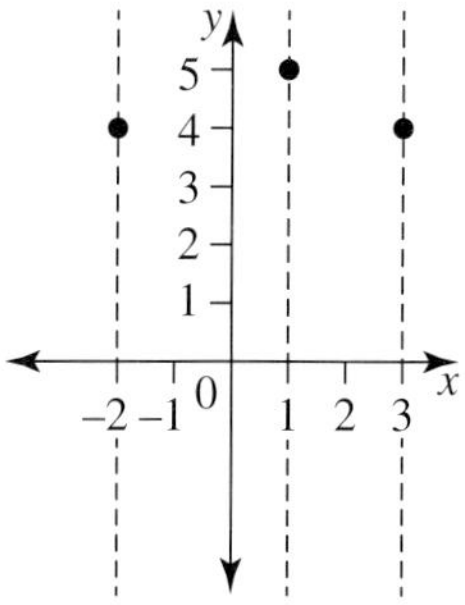	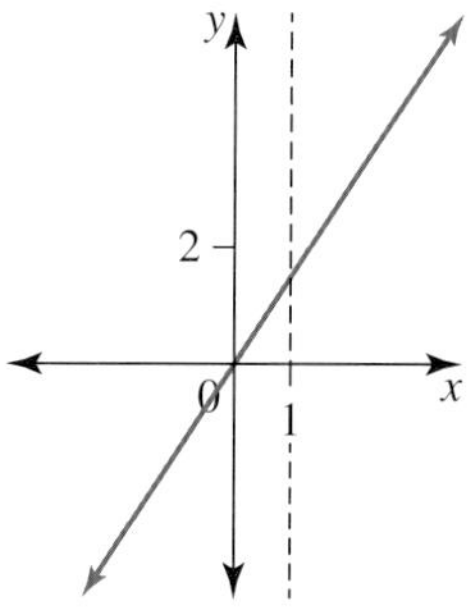	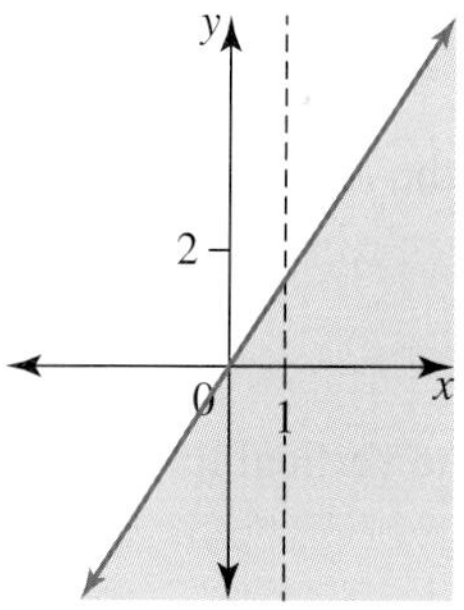
The domain is $x \in \{-2, 1, 3\}$.	The domain is $x \in R$.	The domain is $x \in R$.
The range is $y \in \{4, 5\}$.	The range is $y \in R$.	The range is $y \in R$.
The relation A is a function as every x-value has a unique y-value.	The relation B is a function as every x-value has a unique y-value.	The relation C is not a function as every x-value does not have a unique y-value. $(1, 0), (1, -1)$ and $(1, -2)$ are ordered pairs in the region with the same x-value.

These examples illustrate that functions can be recognised from their graphs by dotting a vertical line. If the vertical line cuts the graph exactly once, the graph is a function.

WORKED EXAMPLE 1 Stating domain and range, and whether a relation is a function

For each of the following, state the domain and range, and whether the relation is a function or not.

a. $\{(1, 4), (2, 0), (2, 3), (5, -1)\}$

b.

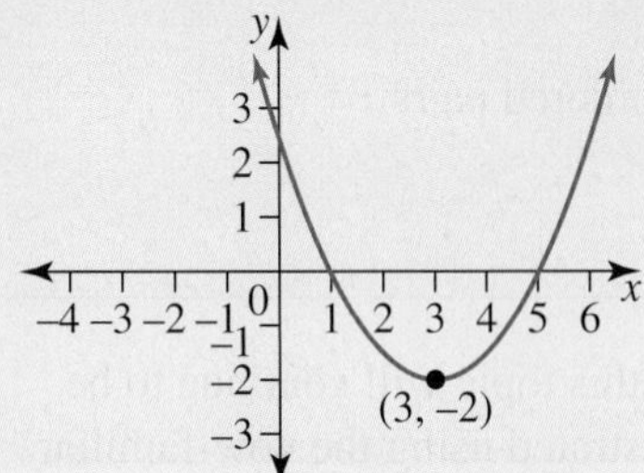

c.

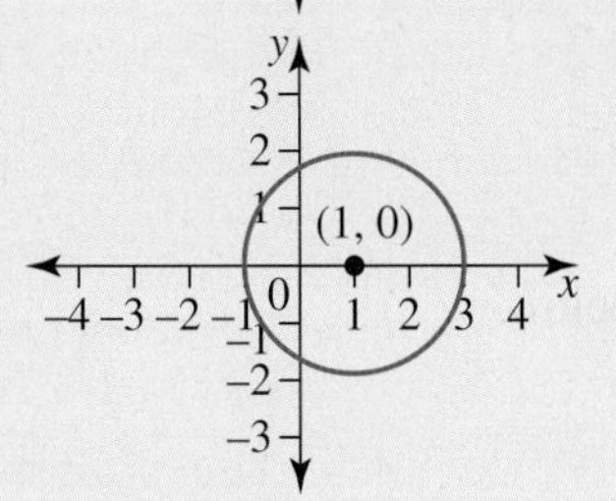

d. $\{(x, y) : y = 4 - x^3\}$

THINK	WRITE
a. 1. State the domain.	**a.** $\{(1, 4), (2, 0), (2, 3), (5, -1)\}$ The domain is the set of x-values: $\{1, 2, 5\}$.
2. State the range.	The range is the set of y-values: $\{-1, 0, 3, 4\}$.
3. Are there any ordered pairs that have the same x-coordinate?	The relation is not a function, since there are two different points with the same x-coordinate: $(2, 0)$ and $(2, 3)$.
b. 1. State the domain.	**b.** Reading from left to right horizontally in the direction of the x-axis, the graph uses every possible x-value.The domain is $(-\infty, \infty)$ or R.
2. State the range.	Reading from bottom to top vertically in the direction of the y-axis, the graph's y-values start at -2 and increase from there. The range is $[-2, \infty)$ or $\{y : y \geq -2\}$.
3. Use the vertical line test.	This is a function, since any vertical line cuts the graph exactly once.
c. 1. State the domain and range.	**c.** The domain is $[-1, 3]$; the range is $[-2, 2]$.
2. Use the vertical line test.	This is not a function, as a vertical line can cut the graph more than once. 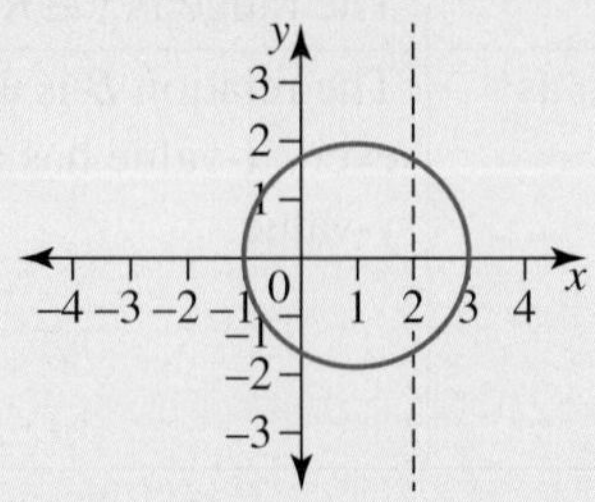
d. 1. State the domain.	**d.** $y = 4 - x^3$ This is the equation of a polynomial, so its domain is R.

2. State the range.

It is the equation of a cubic polynomial with a negative coefficient of its leading term, so as $x \to \pm\infty$, $y \to \mp\infty$. The range is R.

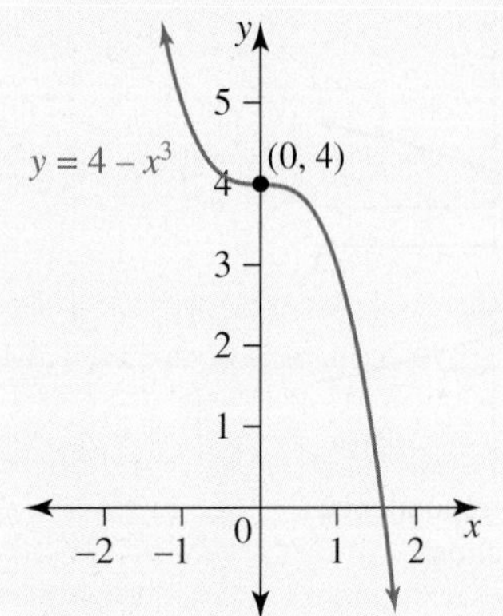

3. Is the relation a function?

This is a function, because all polynomial relations are functions.

Resources

Interactivity Vertical and horizontal line tests (int-2570)

6.2.2 Type of correspondence

Functions and relations can be classified according to the **correspondence** between the coordinates of their ordered pairs. There are four possible types:

- **one-to-one correspondence** where each x-value is paired to a unique y-value, such as $\{(1,2),(3,4)\}$
- **many-to-one correspondence** where more than one x-value may be paired to the same y-value, such as $\{(1,2),(3,2)\}$
- **one-to-many correspondence** where each x-value may be paired to more than one y-value, such as $\{(1,2),(1,4)\}$
- **many-to-many correspondence** where more than one x-value may be paired to more than one y-value, such as $\{(1,2),(1,4),(3,4)\}$.

Here 'many' means more than one.

The graph of a function is recognised by the vertical line test and its type of correspondence is determined by the **horizontal line test**.

A function has a:

- one-to-one correspondence if a horizontal line cuts its graph exactly once
- many-to-one correspondence if a horizontal line cuts its graph more than once.

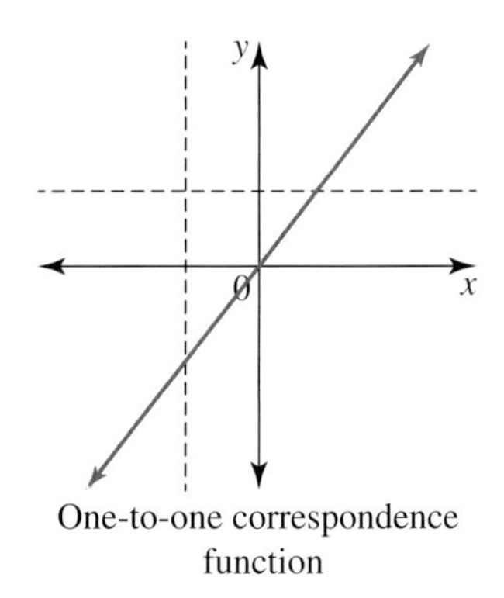

One-to-one correspondence function

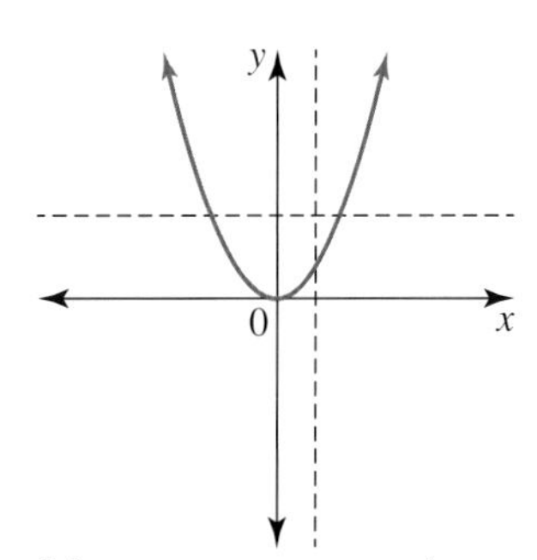

Many-to-one correspondence function

A relation that is not a function has a:

- one-to-many correspondence if a horizontal line cuts its graph exactly once
- many-to-many correspondence if a horizontal line cuts its graph more than once.

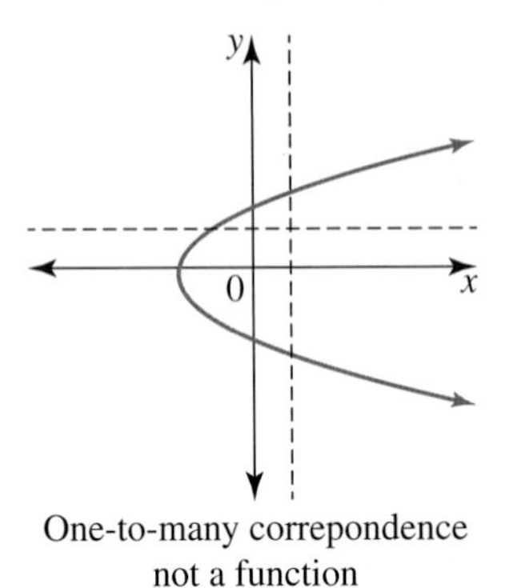

One-to-many correpondence
not a function

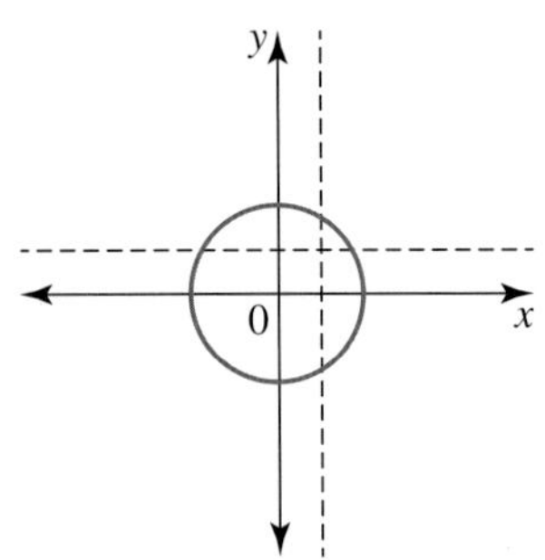

Many-to-many correspondence
not a function

Graphically, the type of correspondence is determined by the number of intersections of a horizontal line (one or many) to the number of intersections of a vertical line (one or many) with the graph. Functions have either a one-to-one or many-to-one correspondence, since their graphs must pass the vertical line test.

WORKED EXAMPLE 2 Identifying types of functions and relations

Identify the type of correspondence and state whether each relation is a function or not.

a. $(x, y) : y = (x+3)(x-1)(x-6)$ **b.** $\{(1,3), (2,4), (1,5)\}$

THINK	WRITE
a. 1. Draw the graph.	**a.** $y = (x+3)(x-1)(x-6)$ x-intercepts: $(-3,0), (1,0), (6,0)$ y-intercept: $(0, 18)$ The graph is a positive cubic.
	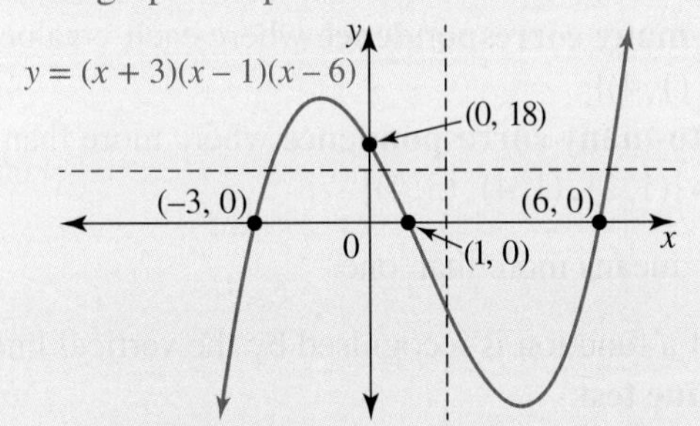
2. Use the horizontal line test and the vertical line test to determine the type of correspondence.	A horizontal line cuts the graph in more than one place. A vertical line cuts the graph exactly once. This is a many-to-one correspondence.
3. State whether the relation is a function.	$y = (x+3)(x-1)(x-6)$ is the equation of a polynomial function with a many-to-one correspondence.
b. 1. Look to see if there are points with the same x- or y-coordinates.	**b.** $\{(1,3), (2,4), (1,5)\}$ $x = 1$ is paired to both $y = 3$ and $y = 5$. The relation has a one-to-many correspondence. It is not a function.

2. Alternatively, or as a check, plot the points and use the horizontal and vertical line tests.

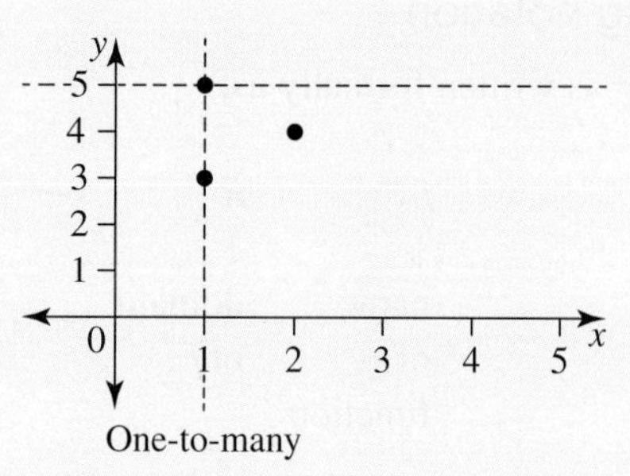

A horizontal line cuts the graph exactly once. A vertical line can cut the graph in more than one place. This is a one-to-many correspondence.

6.2.3 Function notation

The rule for a function such as $y = x^2$ will often be written as $f(x) = x^2$. This is read as 'f of x equals x^2'. We shall also refer to a function as $y = f(x)$, particularly when graphing a function as the set of ordered pairs (x, y) with x as the independent or explanatory variable and y as the dependent or response variable.

$f(x)$ is called the **image** of x under the function **mapping**, which means that if, for example, $x = 2$, then $f(2)$ is the y-value that $x = 2$ is paired with (mapped to), according to the function rule. We have used $p(x)$ notation for polynomial functions.

For $f(x) = x^2$, $f(2) = 2^2 = 4$. The image of 2 under the mapping f is 4; the ordered pair $(2, 4)$ lies on the graph of $y = f(x)$; 2 is mapped to 4 under f: these are all variations of the mathematical language that could be used for this function.

The ordered pairs that form the function with rule $f(x) = x^2$ could be illustrated on a mapping diagram. The mapping diagram shown uses two number lines, one for the x-values and one for the y-values, but there are varied ways to show mapping diagrams.

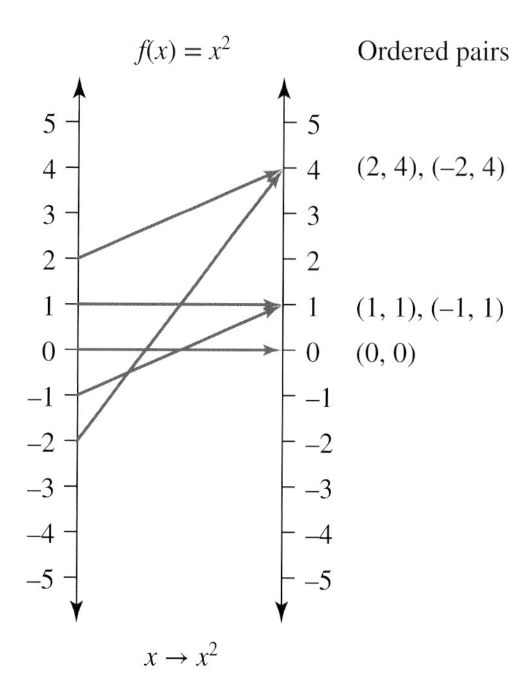

Under the mapping, every x-value in the domain is mapped to its square, $x \to x^2$. The range is the set of the images, or corresponding y-values, of each x-value in the domain. For this example, the polynomial function has a domain of R and a range of $[0, \infty)$, since squared numbers are not negative. Not all of the real numbers on the y number line are elements of the range in this example. The set of all the available y-values, whether used in the mapping or not, is called the **co-domain**. Only the set of those y-values that are used for the mapping form the range. For this example, the co-domain is R and the range is a subset of the co-domain, since $[0, \infty) \subset R$.

The mapping diagram also illustrates the many-to-one correspondence of the function defined by $y = x^2$.

Formal mapping notation

The mapping $x \rightarrow x^2$ is written formally as:

f:	R	$\rightarrow$	R,	$f(x)=x^2$
$\downarrow$	$\downarrow$		$\downarrow$	$\downarrow$
name of function	domain of f		co-domain	rule for, or equation of, f

The domain of the function must always be specified when writing functions formally.

We will always use R as the co-domain. Mappings will be written as $f: D \rightarrow R$, where D is the domain. Usually a graph of the function is required in order to determine its range.

Although f is the commonly used symbol for a function, other symbols may be used.

WORKED EXAMPLE 3 Evaluating and using function notation

Consider $f: R \rightarrow R, f(x)=a+bx$, where $f(1)=4$ and $f(-1)=6$.

a. **Calculate the values of a and b and state the function rule.**

b. **Evaluate $f(0)$.**

c. **Calculate the value of x for which $f(x)=0$.**

d. **Find the image of −5.**

e. **Write the mapping for a function g that has the same rule as f but a domain restricted to R^+.**

THINK	WRITE
a. 1. Use the given information to set up a system of simultaneous equations.	a. $f(x)=a+bx$ $f(1)=4 \Rightarrow 4=a+b\times 1$ $\therefore a+b=4$ [1] $f[-1]=6 \Rightarrow 6=a+b\times -1$ $\therefore a-b=6$ [2]
2. Solve the system of simultaneous equations to obtain the values of a and b.	Equation [1] + equation [2]: $2a=10$ $a=5$ Substitute $a=5$ into equation [1]: $\therefore b=-1$
3. State the answer.	$a=5, b=-1$ $f(x)=5-x$
b. Substitute the given value of x.	b. $f(x)=5-x$ $f(0)=5-0$ $=5$
c. Substitute the rule for $f(x)$ and solve the equation for x.	c. $f(x)=0$ $5-x=0$ $\therefore x=5$
d. Write the expression for the image and then evaluate it.	d. The image of −5 is $f(-5)$. $f(x)=5-x$ $f(-5)=5-(-5)$ $=10$ The image is 10.
e. Change the name of the function and change the domain.	e. $g: R^+ \rightarrow R, g(x)=5-x$

6.2 Exercise

Students, these questions are even better in jacPLUS

Receive immediate feedback and access sample responses

Access additional questions

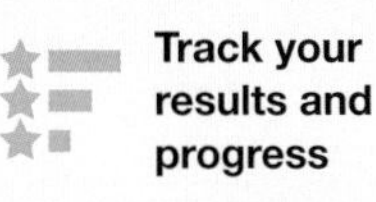
Track your results and progress

Find all this and MORE in jacPLUS

Technology free

1. a. State the domain and range of $\{(-4, 7), (0, 10), (3, 5)\}$.
   b. Express $\{x : -10 \leq x < 4\}$ in interval notation.
   c. Use interval notation to describe the domain and range of the graph shown.

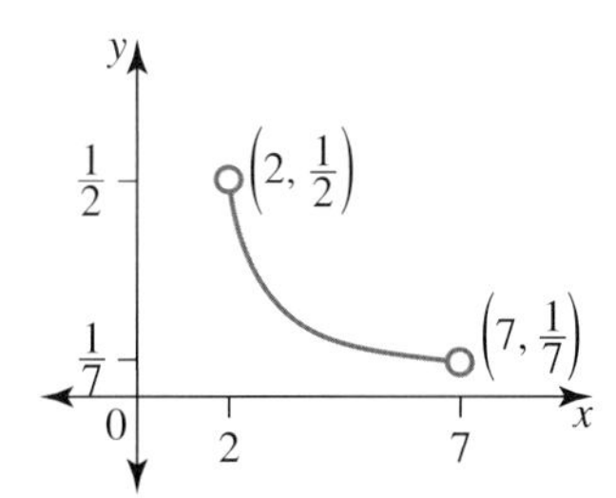

   d. State the domain and range of the given graph.

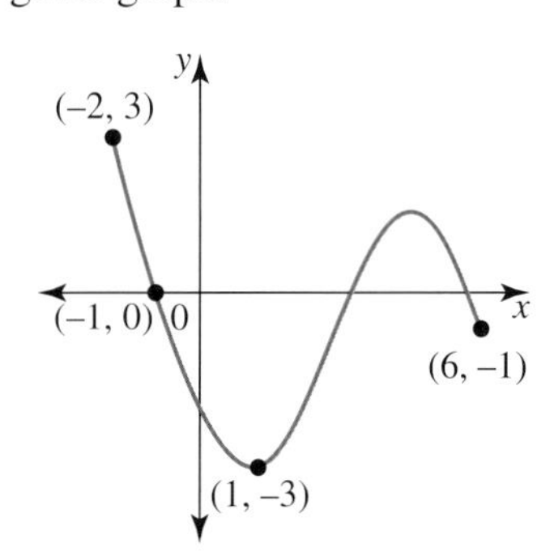

2. Calculate the coordinates of any end points of the following and state the domain and range in interval notation.
   a. $y = 2x, -1 < x \leq 3$
   b. $y = \frac{2}{3}x, x > 3$
   c. $y = -3x, -5 \leq x \leq 5$
   d. $y = x^2, x \in R$
   e. $y = x^2, x \in R^+$
   f. $y = x^2, x \in (-2, 2)$
3. **WE1** For each of the following, state the domain and range, and whether the relation is a function or not.
   a. $\{(4, 4), (3, 0), (2, 3), (0, -1)\}$
   b.

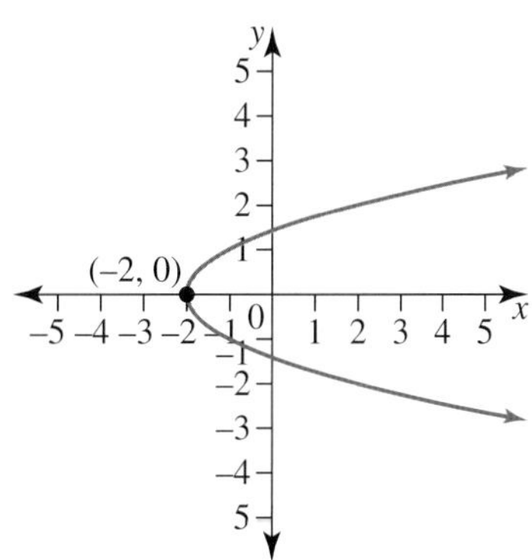

c.

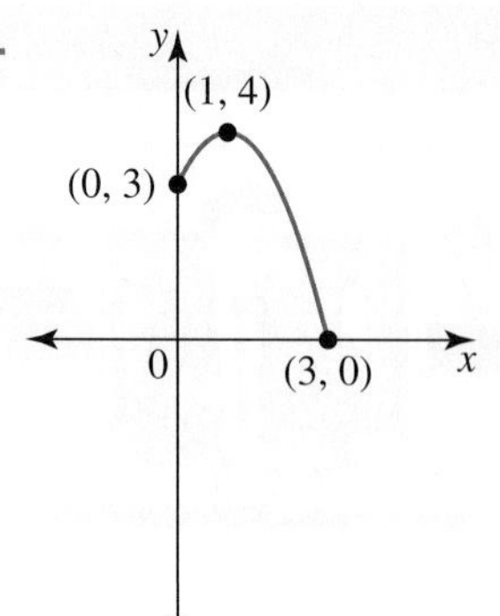

d. $\{(x, y) : y = 4 - x^2\}$

4. Sketch the graph of $y = -4x$, $x \in [-1, 3)$, and state its domain and range.

5. Consider the relation $\{(x.y) : y = 2x - 1, -2 < x \leq 6\}$.
   a. Sketch the graph of the relation.
   b. State the domain and range.
   c. Explain, with a reason, whether or not this relation is a function.

6. Consider the relation $\{(x.y) : y = (x - 2)^3 + 4, -2 \leq x \leq 4\}$.
   a. Sketch the graph of the relation.
   b. State the domain and range.
   c. Explain, with a reason, whether or not this relation is a function.

7. Consider the relation $\{(x.y) : y \geq -1\}$.
   a. Sketch the graph of the relation.
   b. State the domain and range.
   c. Explain, with a reason, whether or not this relation is a function.

8. **WE2** Identify the type of correspondence and state whether each relation is a function or not.
   a. $\{(x, y) : y = 8\,(x + 1)^3 - 1\}$

   b.

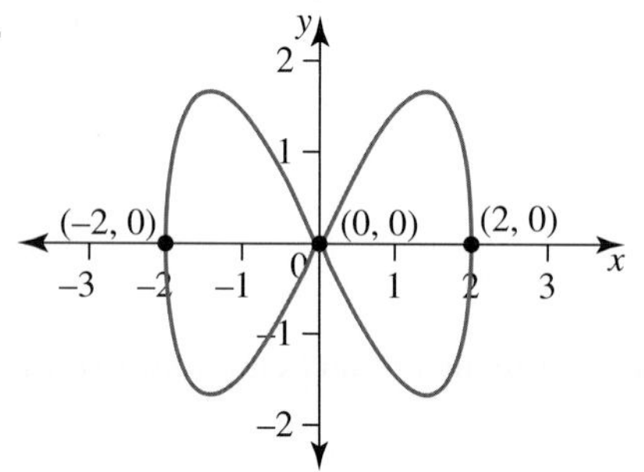

9. For each of the following relations:
   i. state the domain and range
   ii. state the type of correspondence
   iii. identify any of the relations that are not functions.

a.

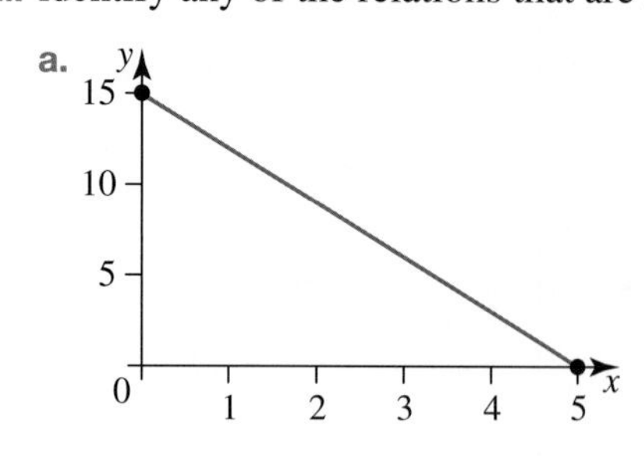

b.

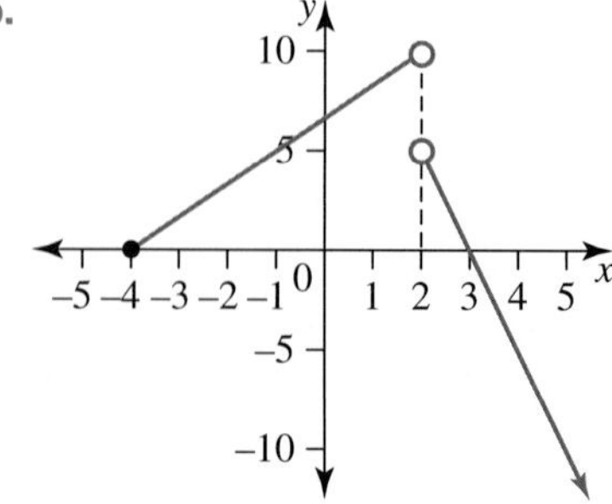

c.

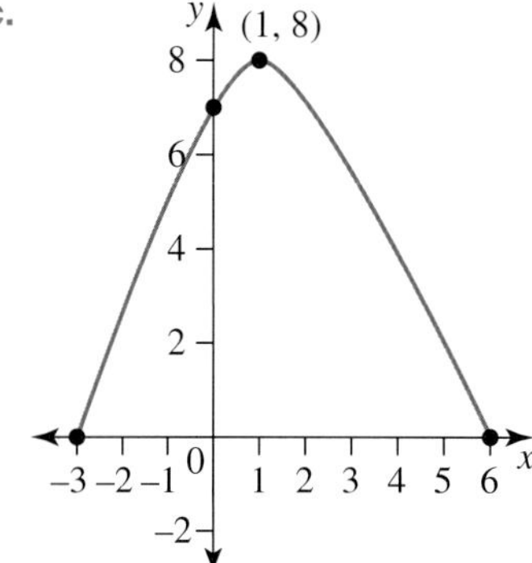

d.

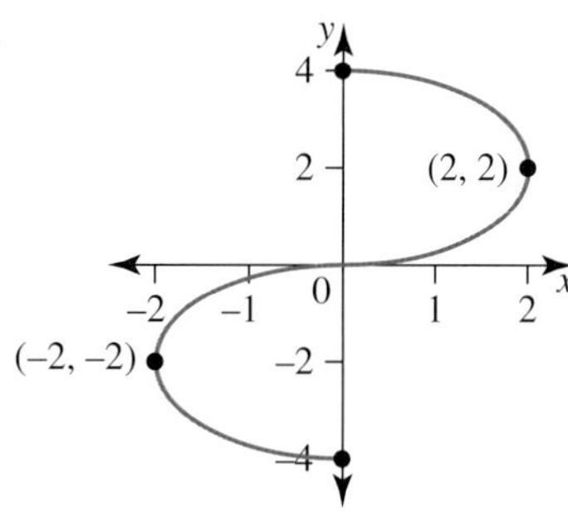

e.

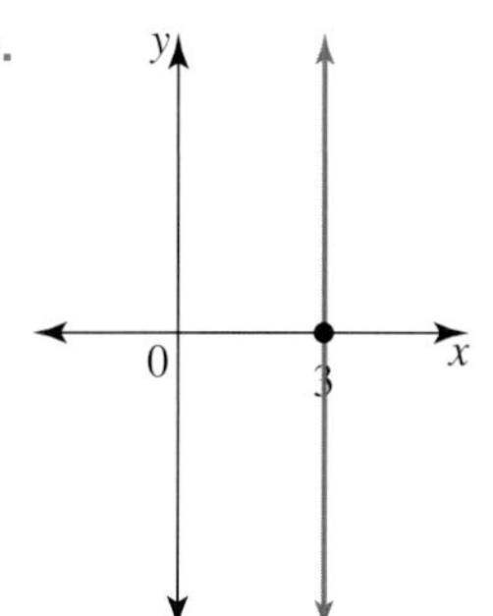

f.

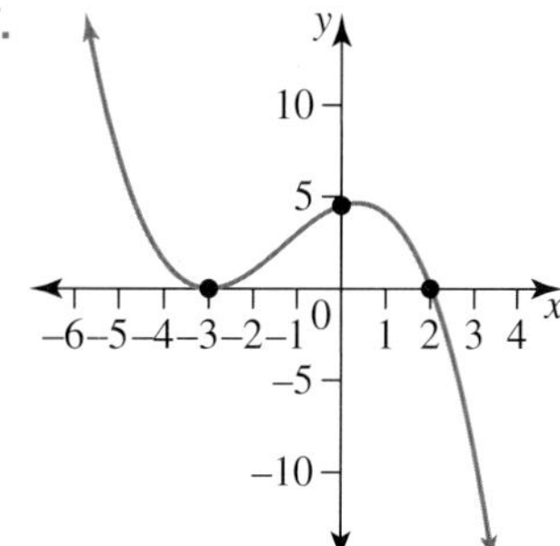

10. For each of the following relations, state the domain, range, type of correspondence and whether the relation is a function.

a. $\{(-11, 2), (-3, 8), (-1, 0), (5, 2)\}$
b. $\{(20, 6), (20, 20), (50, 10), (60, 10)\}$
c. $\{(-14, -7), (0, 0), (0, 2), (14, 7)\}$
d. $\{(x, y) : y = 2(x-16)^3 + 13\}$
e. $\{(x, y) : y = 4 - (x-6)^2\}$
f. $\{(x, y) : y = 3x^2(x-5)^2\}$

11. Sketch each of the functions defined by the following rules and state the domain and range of each.

a. $f(x) = 4x + 2,\ x \in (-1, 1]$
b. $g(x) = 4x(x+2),\ x \geq -2$
c. $h(x) = 4 - x^3,\ x \in R^+$
d. $y = 4,\ x \in R \setminus [-2, 2]$

12. a. Sketch the graph of $y = (x-2)^2$, stating its domain, range and type of correspondence.
b. Restrict the domain of the function defined by $y = (x-2)^2$ so that it will be a one-to-one and increasing function.

13. **WE3** Consider $f : R \to R, f(x) = ax + b$, where $f(2) = 7$ and $f(3) = 9$.
a. Calculate the values of a and b and state the function rule.
b. Evaluate $f(0)$.
c. Calculate the value of x for which $f(x) = 0$.
d. Determine the image of -3.
e. Write the mapping for a function g that has the same rule as f but a domain restricted to $(-\infty, 0]$.

14. Consider $f(x) = 1 - 4x$.
a. Calculate

i. $f(0)$
ii. $f(-5)$
iii. $f\left(\frac{1}{8}\right)$.

b. Determine the value of x for which

i. $f(x) = 0$
ii. $f(x) = -1$
iii. $f(x) = x$.

c. Identify the image of

i. -11
ii. 1.5
iii. $1 - \sqrt{2}$.

d. Determine any value(s) of x for which $f(x) = g(x)$, given $g(x) = x^2 + 5$.

15. If $f(x) = x^2 + 2x - 3$, calculate the following.

a. i. $f(-2)$ ii. $f(9)$

b. i. $f(2a)$ ii. $f(1-a)$

c. $f(x+h) - f(x)$

d. $\{x : f(x) > 0\}$

e. The values of x for which $f(x) = 12$

f. The values of x for which $f(x) = 1 - x$

16. Consider $f: R \to R, f(x) = x^3 - x^2$.

a. Determine the image of 2.

b. Sketch the graph of $y = f(x)$.

c. State the domain and range of the function f.

d. State the type of correspondence.

e. Give a restricted domain so that f is one-to-one and increasing.

f. Calculate $\{x : f(x) = 4\}$.

17. Express $y = x^2 - 6x + 10, 0 \leq x < 7$ in mapping notation and state its domain and range.

18. Given $A = \{(1,2), (2,3), (3,2), (k,3)\}$, state the possible values for k and the type of correspondence, if the following apply.

a. A is not a function.

b. A is a function.

c. Draw a mapping diagram for the relation in part **a** using the chosen k values.

d. Draw a mapping diagram for the function in part **b** using the chosen k values.

19. Select the functions from the following list, express them in function mapping notation and state their ranges.

a. $y = x^2,\ x \in Z^+$

b. $2x + 3y = 6$

c. $y = \pm x$

d. $\{(x, 5),\ x \in R\}$

e. $\{(-1, y),\ y \in R\}$

f. $y = -x^3,\ x \in R^+$

Technology active

20. Consider the functions f and g where $f(x) = a + bx + cx^2$ and $g(x) = f(x-1)$.

a. Given $f(-2) = 0, f(5) = 0$ and $f(2) = 3$, determine the rule for the function f.

b. Express the rule for g as a polynomial in x.

c. Calculate any values of x for which $f(x) = g(x)$.

d. On the same axes, sketch the graphs of $y = f(x)$ and $y = g(x)$, and describe the relationship between the two graphs.

21. a. A toy car moves along a horizontal straight line so that at time t its position x from a child at a fixed origin is given by the function $x(t) = 4 + 5t$. For the interval $t \in [0, 5]$, state the domain and range of this function and calculate how far the toy car travels in this interval.

b. A hat is thrown vertically into the air and at time t seconds its height above the ground is given by the function $h(t) = 10t - 5t^2$. Calculate how long it takes the hat to return to the ground and hence state the domain and range of this function.

c. For part of its growth over a two-week period, the length of a leaf at time t weeks is given by the function $l(t) = 0.5 + 0.2t^3, 0 \leq t \leq 2$.

i. State the domain and determine the range of this function.

ii. Calculate how long it takes for the leaf to reach half the length that it reaches by the end of the time period.

22. Define $f(x) = x^3 + lx^2 + mx + n$. Given $f(3) = -25$, $f(5) = 49$ and $f(7) = 243$, answer the following questions.

a. Calculate the constants l, m and n and hence state the rule for $f(x)$.

b. Determine the image of 1.2.

c. Calculate, correct to 3 decimal places, the value of x such that $f(x) = 20$.

d. Given $x \geq 0$, express the function as a mapping and state its range.

6.2 Exam questions

Question 1 (1 mark) TECH-ACTIVE

MC State which of the following relations is not a function.

A.

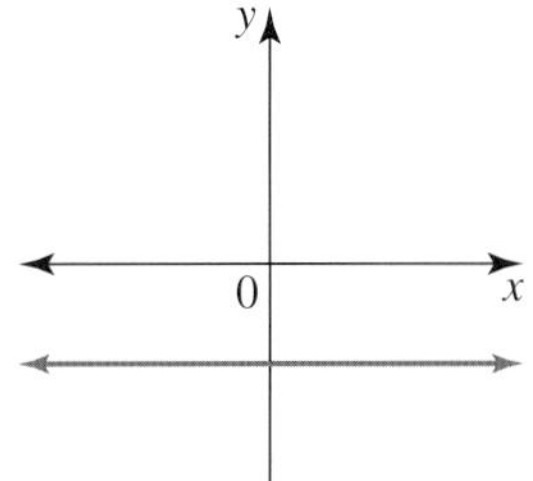

B.

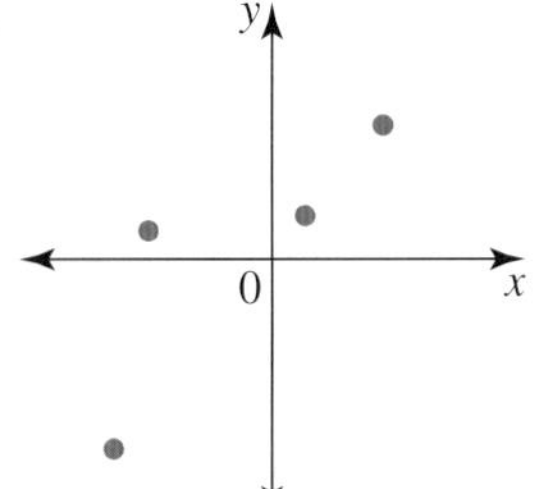

C.

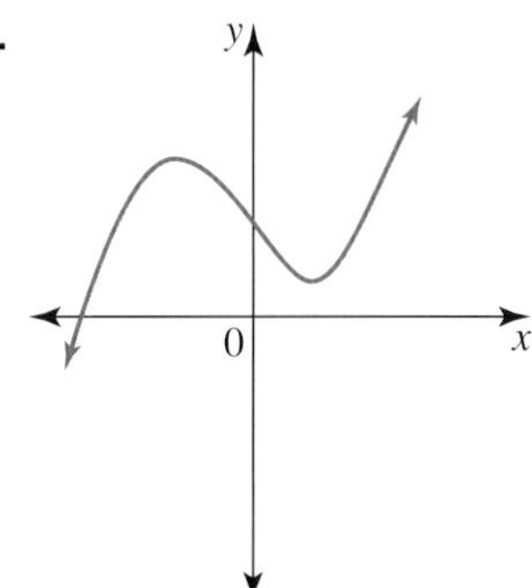

D.

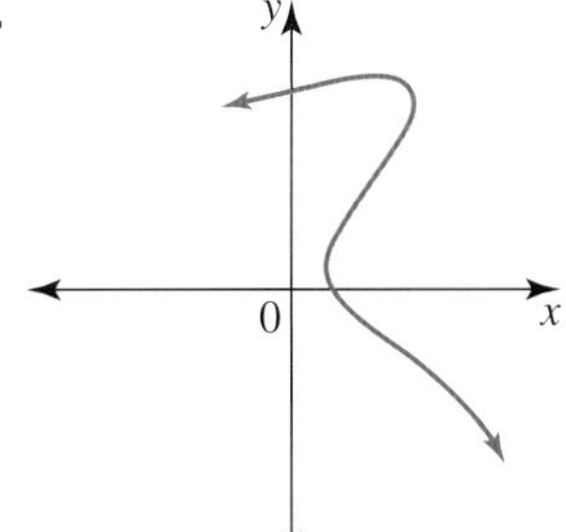

E.

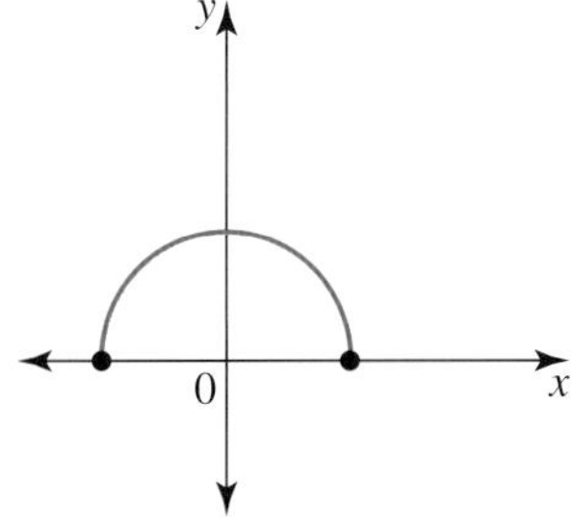

Question 2 (1 mark) TECH-ACTIVE

MC The type of correspondence for a function can be

A. many-to-one and many-to-many.

B. many-to-one and one-to-one.

C. many-to-one and one-to-many.

D. one-to-one and many-to-many.

E. one-to-many and one-to-one.

Question 3 (5 marks) TECH-FREE

If $f: R \rightarrow R, f(x) = 2x^3 - 3x^2$

a. find the image of -1 **(1 mark)**

b. sketch the graph of $y = f(x)$ **(2 marks)**

c. state the domain and range of the function f **(1 mark)**

d. state the type of correspondence of f. **(1 mark)**

More exam questions are available online.

6.3 The rectangular hyperbola and the truncus

LEARNING INTENTION

At the end of this subtopic you should be able to:

- sketch graphs of hyperbolas and truncus, showing all key features, including asymptotes
- form the equation of a graph of a hyperbola or truncus
- determine inversely proportional relationships as an application of hyperbolas
- use hyperbola and truncus graphs in modelling real-life situations.

The family of functions with rules $y = x^n, n \in N$ are the now-familiar polynomial functions. If n is a negative number, however, these functions cannot be polynomials. Here we consider the functions with rule $y = x^n$ where $n = -1$ and $n = -2$.

6.3.1 The graph of $y = \frac{1}{x}$

With $n = -1$, the rule $y = x^{-1}$ can also be written as $y = \frac{1}{x}$. This is the rule for a **rational function** called a **hyperbola**. Two things can be immediately observed from the rule:

- $x = 0$ must be excluded from the domain, since division by zero is not defined.
- $y = 0$ must be excluded from the range, since there is no number whose reciprocal is zero.

The lines $x = 0$ and $y = 0$ are **asymptotes**. An asymptote is a line the graph will approach but never reach. As these two asymptotes $x = 0$ and $y = 0$ are a pair of perpendicular lines, the hyperbola is known as a **rectangular hyperbola**. The asymptotes are a key feature of the graph of a hyperbola.

Completing a table of values can give us a 'feel' for this graph.

x	-10	-4	-2	-1	$-\frac{1}{2}$	$-\frac{1}{4}$	$-\frac{1}{10}$	0	$\frac{1}{10}$	$\frac{1}{4}$	$\frac{1}{2}$	1	2	4	10
y	$-\frac{1}{10}$	$-\frac{1}{4}$	$-\frac{1}{2}$	-1	-2	-4	-10	No value possible	10	4	2	1	$\frac{1}{2}$	$\frac{1}{4}$	$\frac{1}{10}$

The values in the table illustrate that as $x \to \infty, y \to 0$ and as $x \to -\infty, y \to 0$.

The table also illustrates that as $x \to 0$, either $y \to -\infty$ or $y \to \infty$.

These observations describe the asymptotic behaviour of the graph.

The key features of the graph of $y = \frac{1}{x}$ (shown) are as follows:

- The vertical asymptote has the equation $x = 0$ (the y-axis).
- The horizontal asymptote has the equation $y = 0$ (the x-axis).
- The domain is $R \setminus \{0\}$.
- The range is $R \setminus \{0\}$.

- As $x \to \infty, y \to 0$ from above and as $x \to -\infty, y \to 0$ from below. This can be written as: as $x \to \infty, y \to 0^+$ and as $x \to -\infty$, $y \to 0^-$.
- As $x \to 0^-, y \to -\infty$ and as $x \to 0^+, y \to \infty$.
- The graph is of a function with a one-to-one correspondence.
- The graph has two branches separated by the asymptotes.
- As the two branches do not join at $x = 0$, the function is said to be **discontinuous** at $x = 0$.
- The graph lies in **quadrants** 1 and 3 as defined by the asymptotes.

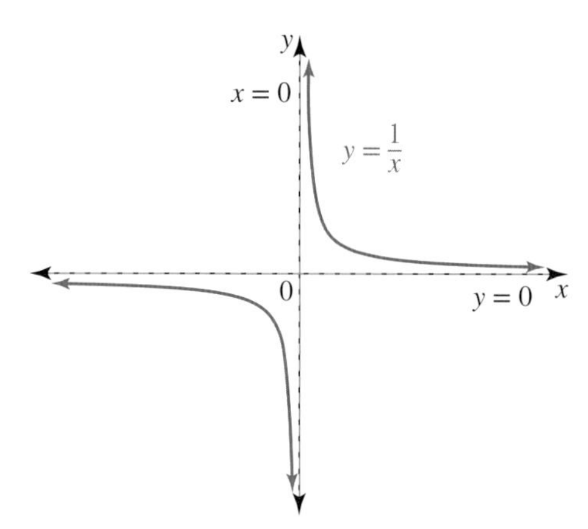

The asymptotes divide the Cartesian plane into four areas or quadrants. The quadrants formed by the asymptotes are numbered 1 to 4 anticlockwise.

With the basic shape of the hyperbola established, transformations of the graph of $y = \frac{1}{x}$ can be studied.

Dilation from the x-axis

The effect of a dilation factor on the graph can be illustrated by comparing $y = \frac{1}{x}$ and $y = \frac{2}{x}$.

For $x = 1$, the point $(1, 1)$ lies on $y = \frac{1}{x}$, whereas the point $(1, 2)$ lies on $y = \frac{2}{x}$. The dilation effect on $y = \frac{2}{x}$ is to move the graph further out from the x-axis. The graph has a dilation factor of 2 from the x-axis.

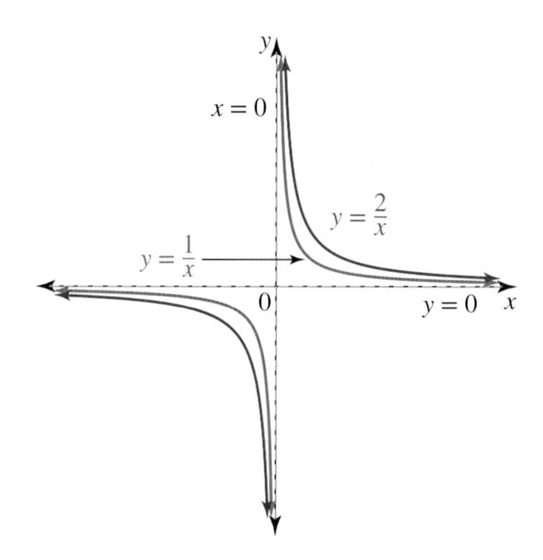

Vertical translation

The graph of $y = \frac{1}{x} + 2$ illustrates the effect of a **vertical translation** of 2 units upwards.

Key features:

- The horizontal asymptote has the equation $y = 2$. This means that as $x \to \pm\infty, y \to 2$.
- The vertical asymptote is unaffected by the vertical translation and remains $x = 0$.
- The domain is $R\backslash\{0\}$.
- The range is $R\backslash\{2\}$.

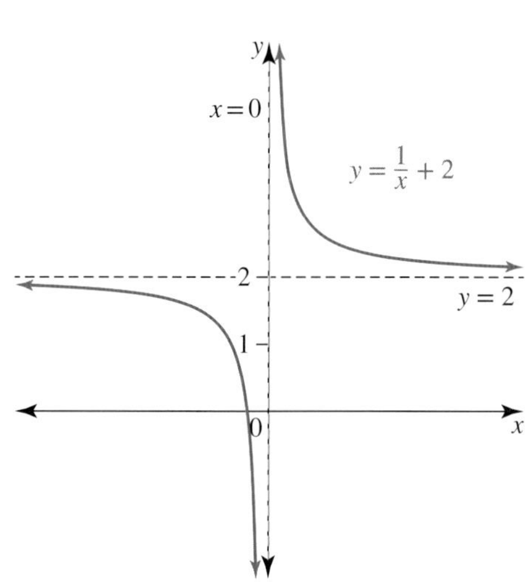

Horizontal translation

For $y = \frac{1}{x-2}$, as the denominator cannot be zero, $x - 2 \neq 0 \Rightarrow x \neq 2$.

The domain must exclude $x = 2$, so the line $x = 2$ is the vertical asymptote.

The graph of $y = \frac{1}{x-2}$ demonstrates the same effect that we have seen with other graphs that are translated 2 units to the right.

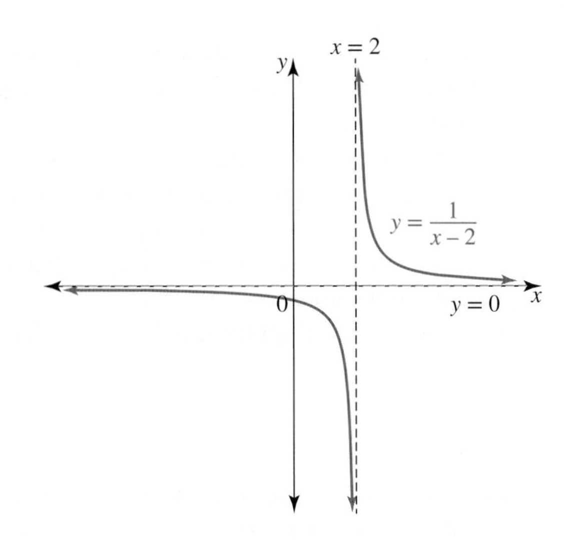

Key features:

- The vertical asymptote has the equation $x = 2$.
- The horizontal asymptote is unaffected by the **horizontal translation** and still has the equation $y = 0$.
- The domain is $R \backslash \{2\}$.
- The range is $R \backslash \{0\}$.

Reflection in the x-axis

The graph of $y = -\frac{1}{x}$ illustrates the effect of inverting the graph by reflecting $y = \frac{1}{x}$ in the x-axis.

The graph of $y = -\frac{1}{x}$ lies in quadrants 2 and 4 as defined by the asymptotes.

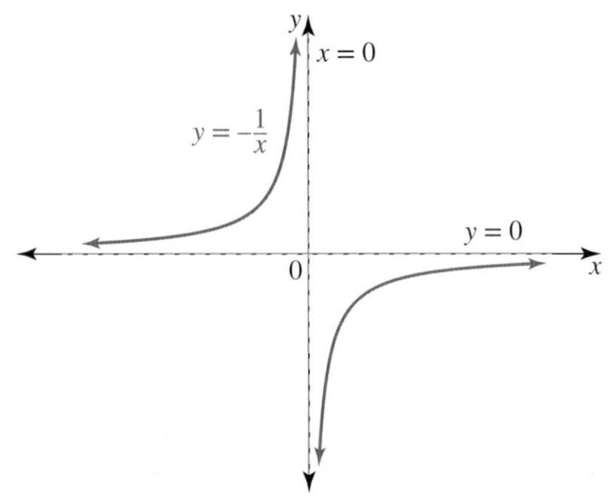

The general equation of a hyperbola

The general equation of the hyperbola

$$y = \frac{a}{x-h} + k$$

The key features of the graph $y = \frac{a}{x-h} + k$ are as follows:

- The vertical asymptote has the equation $x = h$.
- The horizontal asymptote has the equation $y = k$.
- The domain is $R \backslash \{h\}$.
- The range is $R \backslash \{k\}$.
- Asymptotic behaviour: as $x \to \pm\infty, y \to k$ and as $x \to h, y \to \pm\infty$.
- There are two branches to the graph and the graph is discontinuous at $x = h$.
- If $a > 0$, the graph lies in the asymptote-formed quadrants 1 and 3.
- If $a < 0$, the graph lies in the asymptote-formed quadrants 2 and 4.
- $|a|$ gives the dilation factor from the x-axis.

If the equation of the hyperbola is in the form $y = \frac{a}{bx+c} + k$, then the vertical asymptote can be identified by finding the x-value for which the denominator term $bx + c = 0$. The horizontal asymptote is $y = k$ because as $x \to \pm\infty$, $\frac{a}{bx+c} \to 0$ and therefore $y \to k$.

WORKED EXAMPLE 4 Sketching hyperbolas

Sketch the graphs of the following functions, stating the domain and range.

a. $y = \frac{4}{x} - 2$ **b.** $y = \frac{-1}{x-2} + 1$

THINK	WRITE
a. 1. State the equations of the asymptotes.	**a.** $y = \frac{4}{x} - 2$ Vertical asymptote: $x = 0$ Horizontal asymptote: $y = -2$
2. Calculate the coordinates of any axis intercepts.	No y-intercept, since the y-axis is the vertical asymptote x-intercept: let $y = 0$. $0 = \frac{4}{x} - 2$ $\frac{4}{x} = 2$ $4 = 2x$ $x = 2$ $(2, 0)$ is the x-intercept.
3. Anticipating the position of the graph, calculate the coordinates of a point so each branch of the graph will have a known point.	The graph lies in asymptote-formed quadrants 1 and 3. A point to the left of the vertical asymptote is required. Symmetrical point: let $x = -2$. $y = \frac{4}{-2} - 2$ $= -4$ $(-2, -4)$ is a point.
4. Sketch the graph.	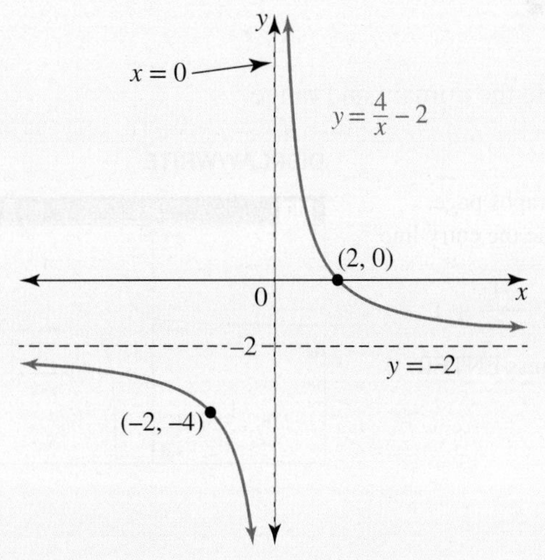
5. State the domain and range.	The domain is $R\backslash\{0\}$; the range is $R\backslash\{-2\}$.

THINK	WRITE
b. 1. State the equations of the asymptotes.	**b.** $y = \frac{-1}{x-2} + 1$ The vertical asymptote occurs when $x - 2 = 0$. Vertical asymptote: $x = 2$ The horizontal asymptote occurs as $x \to \pm\infty$. Horizontal asymptote: $y = 1$
2. Calculate the coordinates of any axis intercepts.	y-intercept: let $x = 0$. $y = \frac{-1}{-2} + 1$ $= \frac{3}{2}$ $\left(0, \frac{3}{2}\right)$ is the y-intercept. x-intercept: let $y = 0$. $\frac{-1}{x-2} + 1 = 0$ $\frac{1}{x-2} = 1$ $x - 2 = 1$ $x = 3$ $(3, 0)$ is the x-intercept.
3. Sketch the graph, ensuring it approaches but does not touch the asymptotes. *Note:* Check the graph has the shape anticipated.	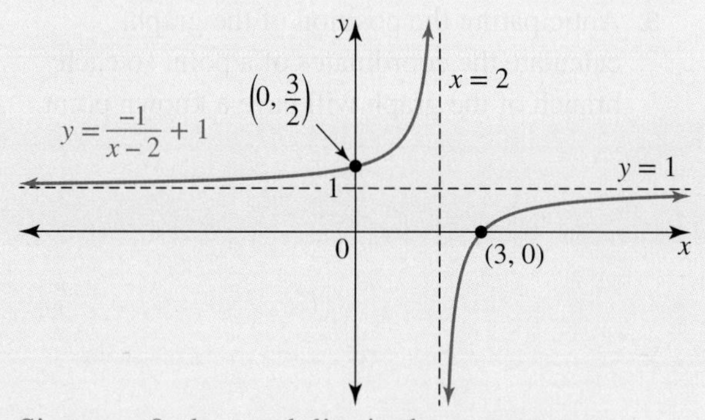 Since $a < 0$, the graph lies in the asymptote-formed quadrants 2 and 4, as expected.
4. State the domain and range.	The domain is $R\backslash\{2\}$; the range is $R\backslash\{1\}$.

TI | THINK	DISPLAY/WRITE	CASIO | THINK	DISPLAY/WRITE
a. 1. On a Graphs page, complete the entry line as: $f1(x) = \frac{-1}{x-2} + 1$ Then press ENTER.	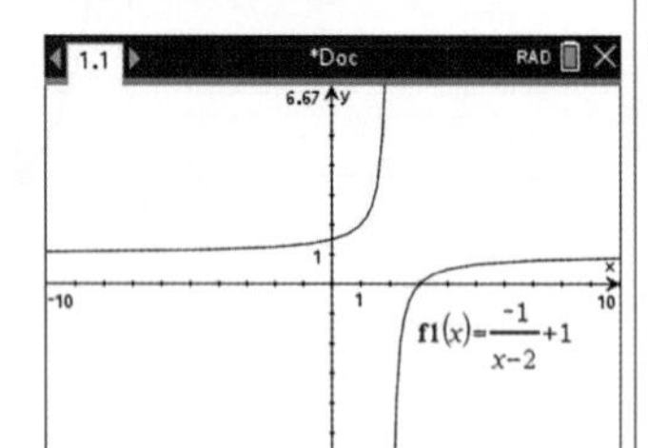	**a. 1.** On a Graphs & Table screen, complete the entry line as: $y1 = \frac{-1}{x-2} + 1$ Then press EXE.	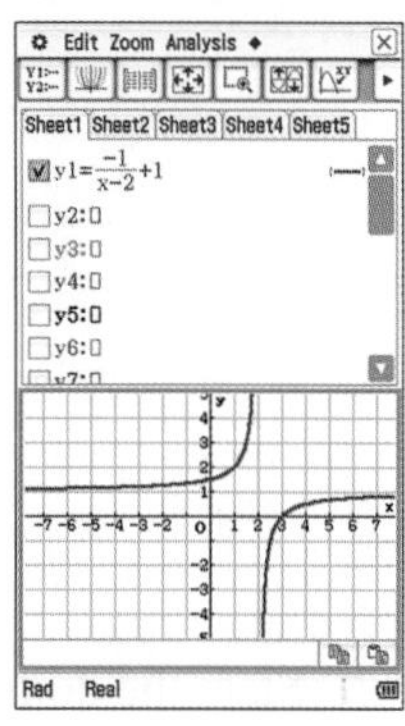
2. State the domain and range.	The domain is $R\backslash\{2\}$. The range is $R\backslash\{1\}$.	**2.** State the domain and range.	The domain is $R\backslash\{2\}$. The range is $R\backslash\{1\}$.

Note: The asymptotes cannot be shown. However, they must be included in your graphs.

Proper rational functions

The equation of the hyperbola $y = \dfrac{a}{x-h} + k$ is expressed in **proper rational function** form. This means the rational term, $\dfrac{a}{x-h}$, has a denominator of a higher degree than that of the numerator. For example, $y = \dfrac{x-1}{x-2}$ is not expressed as a proper rational function: the numerator has the same degree as the denominator. Using division, it can be converted to the proper form $y = \dfrac{1}{x-2} + 1$, which is recognisable as a hyperbola and from which the asymptotes can be obtained.

Resources

Interactivity The rectangular hyperbola (int-2573)

6.3.2 Forming the equation

From the equation $y = \dfrac{a}{x-h} + k$, it can be seen that three pieces of information will be needed to form the equation of a hyperbola. These are usually the equations of the asymptotes and the coordinates of a point on the graph.

WORKED EXAMPLE 5 Determining equations of asymptotes

a. Identify the asymptotes of the hyperbola with equation $y = \dfrac{2x-3}{5-2x}$.

b. Form the equation of the hyperbola shown.

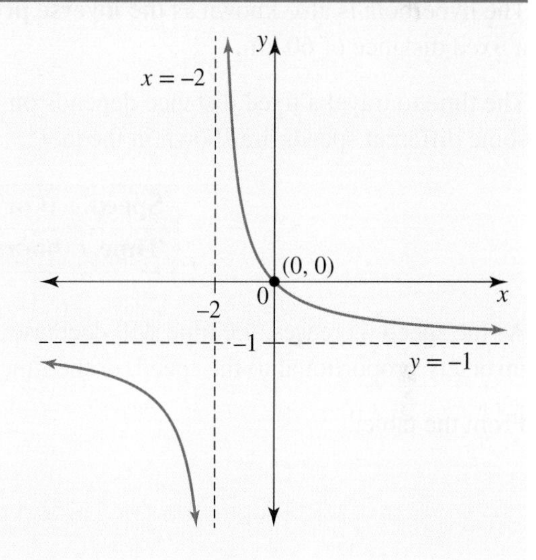

THINK

a. 1. The equation is in improper form, so reduce it to proper form using division.
Note: The long division algorithm could also have been used to reduce the function to proper form.

WRITE

a. $y = \dfrac{2x-3}{5-2x}$

$= \dfrac{-1(5-2x)+2}{5-2x}$

$= -1 + \dfrac{2}{5-2x}$

$y = \dfrac{2}{5-2x} - 1$ is the proper rational function form.

2. State the equations of the asymptotes.

Vertical asymptote when $5 - 2x = 0$

$\therefore x = \dfrac{5}{2}$

The horizontal asymptote is $y = -1$.

b. 1. Recall the general equation of the hyberbola, where the vertical asymptote is $x = h$, and the horizontal asymptote is $y = k$	**b.** The general equation of the graph is $y = \frac{a}{x-h} + k$.
2. Substitute the equations of the asymptotes shown on the graph into the general equation of a hyperbola.	From the graph, the asymptotes have equations $x = -2$ and $y = -1$. Substituting into the general form gives $y = \frac{a}{x+2} - 1$.
3. Use a known point on the graph to determine the remaining unknown constant.	The point $(0, 0)$ lies on the graph. $0 = \frac{a}{2} - 1$ $\therefore 1 = \frac{a}{2}$ $\therefore a = 2$
4. State the equation of the hyperbola.	The equation is $y = \frac{2}{x+2} - 1$.

6.3.3 Inverse proportion

The hyperbola is also known as the **inverse proportion** graph. To illustrate this, consider the time taken to travel a fixed distance of 60 km.

The time to travel a fixed distance depends on the speed of travel. For a distance of 60 km, the times taken for some different speeds are shown in the table.

Speed, v (km/h)	10	15	20	30
Time, t (hours)	6	4	3	2

As the speed increases, the time will decrease; as the speed decreases, the time will increase. The time is inversely proportional to the speed, or the time varies inversely as the speed.

From the table:

$$t \times v = 60$$
$$\therefore t = \frac{60}{v}$$

This is the equation of a hyperbola where v is the independent variable and t the dependent variable.

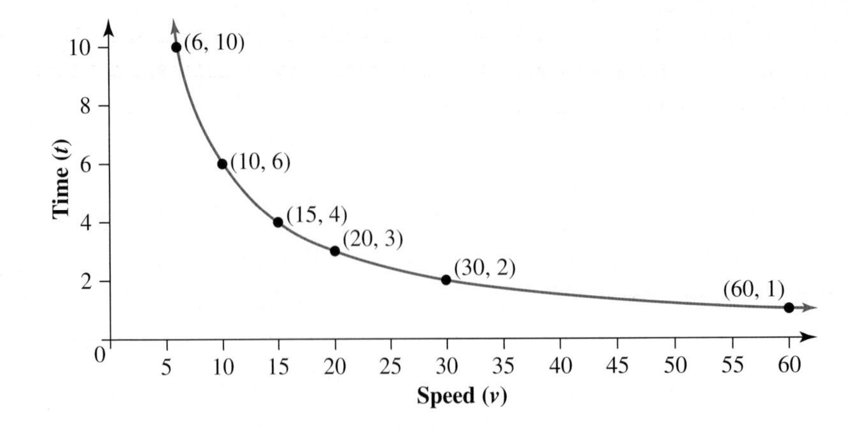

The graph of time versus speed exhibits the asymptotic behaviour typical of a hyperbola. As the speed increases, becoming faster and faster, the time decreases, becoming smaller and smaller. In other words, as $v \to \infty$, $t \to 0$, but t can never reach zero nor can speed reach infinity. Further, as $v \to 0$, $t \to \infty$. Only one branch of the hyperbola is given since neither time nor speed can be negative.

Inverse proportion

In general, the following rules apply.

- **'y is inversely proportional to x' is written as $y \propto \frac{1}{x}$.**
- **If y is inversely proportional to x, then $y = \frac{k}{x}$ where k is the constant of proportionality.**
- **This relationship can also be expressed as $xy = k$, so if the product of two variables is constant, the variables are in inverse proportion.**

WORKED EXAMPLE 6 Determining an inversely proportional relationship

Boyle's Law says that if the temperature of a given mass of gas remains constant, its volume V is inversely proportional to the pressure P.
A container of volume 50 cm^3 is filled with a gas under a pressure of 75 cm of mercury.
a. Determine the relationship between the volume and pressure.
b. The container is connected by a hose to an empty container of volume 100 cm^3. Calculate the pressure in the two containers.

THINK	WRITE
a. 1. Write the rule for the inverse proportion relation.	a. $V \propto \frac{1}{P}$ $\therefore V = \frac{k}{P}$
2. Use the given data to find k and hence the rule.	Substitute $V = 50, P = 75$. $50 = \frac{k}{75}$ $k = 50 \times 75$ $= 3750$ Hence, $V = \frac{3750}{P}$.
b. 1. State the total volume.	b. The two containers are connected and can be thought of as one. Therefore, the combined volume is $100 + 50 = 150$ cm^3.
2. Calculate the pressure for this volume.	$V = \frac{3750}{P}$ When $V = 150$ $150 = \frac{3750}{P}$ $P = \frac{3750}{150}$ $= 25$
3. State the answer.	The gas in the containers is under a pressure of 25 cm of mercury.

6.3.4 The graph of the truncus $y = \frac{1}{x^2}$

With $n = -2$, the rule $y = x^{-2}$ is also written as $y = \frac{1}{x^2}$. The rational function with this rule is called a truncus. Its graph shares similarities with the graph of the hyperbola $y = \frac{1}{x}$, as the rule shows that $x = 0$ and $y = 0$ are vertical and horizontal asymptotes, respectively.

A major difference between the two curves is that for a truncus, whether $x < 0$ or $x > 0$, the y-values must be positive. This means the graph of the truncus $y = \frac{1}{x^2}$ will have two branches, one in quadrant 1 and the other in quadrant 2. The quadrants are formed by the asymptote positions.

The key features of the graph of $y = \frac{1}{x^2}$ (shown) are as follows:

- The vertical asymptote has the equation $x = 0$ (the y-axis).
- The horizontal asymptote has the equation $y = 0$ (the x-axis).
- The domain is $R\backslash\{0\}$.
- The range is R^+.
- As $x \to \pm\infty, y \to 0^+$.
- As $x \to 0, y \to \infty$.
- The graph is of a function with a many-to-one correspondence.
- The graph has two symmetric branches separated by the vertical asymptote.
- The graph lies in quadrants 1 and 2 as defined by the asymptotes.
- The function is discontinuous at $x = 0$.

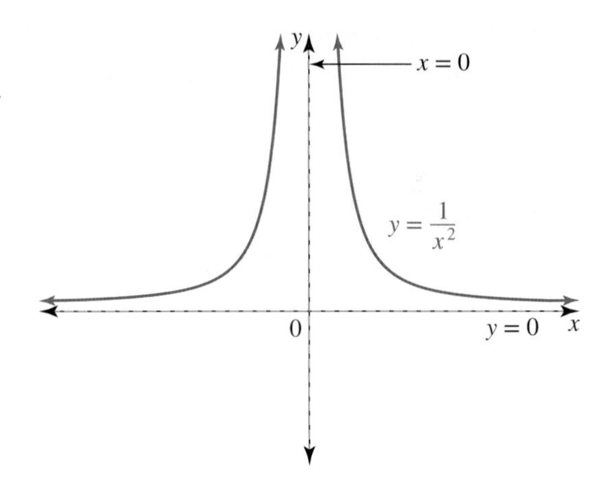

The branches of the truncus $y = \frac{1}{x^2}$ approach the horizontal asymptote more rapidly than those of the hyperbola $y = \frac{1}{x}$ because $\frac{1}{x^2} < \frac{1}{x}$ for $x > 1$. This can be seen when the graphs of the hyperbola and truncus are drawn on the same diagram.

Under a dilation of factor a from the x axis and translations of h units horizontally and k units vertically, $y = \frac{1}{x^2}$ becomes $y = \frac{a}{(x-h)^2} + k$.

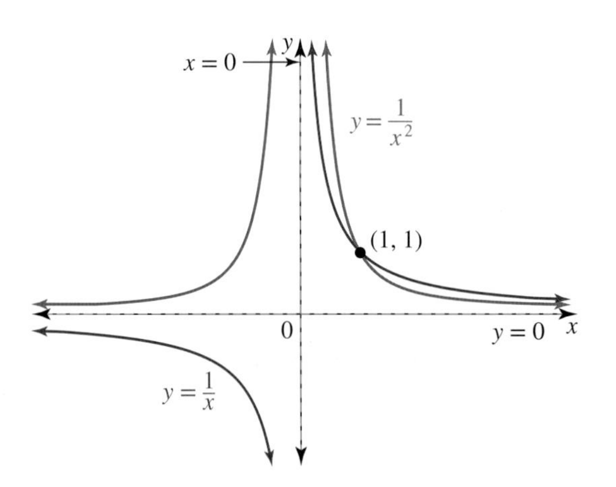

The general equation of the truncus

$$y = \frac{a}{(x-h)^2} + k$$

The key features of the graph of $y = \frac{a}{(x-h)^2} + k$ are as follows:

- The vertical asymptote has the equation $x = h$.
- The horizontal asymptote has the equation $y = k$.
- As $x \to \pm\infty,\ y \to k$.

- If $a > 0$, the graph lies in the asymptote-formed quadrants 1 and 2.
- If $a < 0$, the graph lies in the asymptote-formed quadrants 3 and 4.
- The domain is $R\backslash\{h\}$.
- The function is discontinuous at $x = h$.
- If $a > 0$, the range is (k, ∞).
- If $a < 0$, the range is $(-\infty, k)$.
- $|a|$ gives the dilation factor from the x-axis.

When graphing either a hyperbola or a truncus, each branch should include at least one point.

WORKED EXAMPLE 7 Graphing a truncus

Sketch the graphs of the following functions, stating the domain and range.

a. $y = \dfrac{1}{(x+1)^2} + 2$ **b.** $y = 1 - \dfrac{4}{(x-2)^2}$

THINK / **WRITE**

a. 1. State the equations of the asymptotes.

a. $y = \dfrac{1}{(x+1)^2} + 2$

The vertical asymptote occurs when $(x+1)^2 = 0$.

$x + 1 = 0$

$\therefore x = -1$

Vertical asymptote: $x = -1$

Horizontal asymptote: $y = 2$

2. Calculate the coordinates of any axis intercepts.

y-intercept: let $x = 0$.

$y = \dfrac{1}{(1)^2} + 2$

$= 3$

$(0, 3)$ is the y-intercept.

Since $a > 0$, the graph lies above the horizontal asymptote, $y = 2$.

Hence, there is no x-intercept.

3. Sketch the graph, ensuring that each branch has at least one known point.

Let $x = -2$.

$y = \dfrac{1}{(-1)^2} + 2$

$= 3$

$(-2, 3)$ lies on the graph.

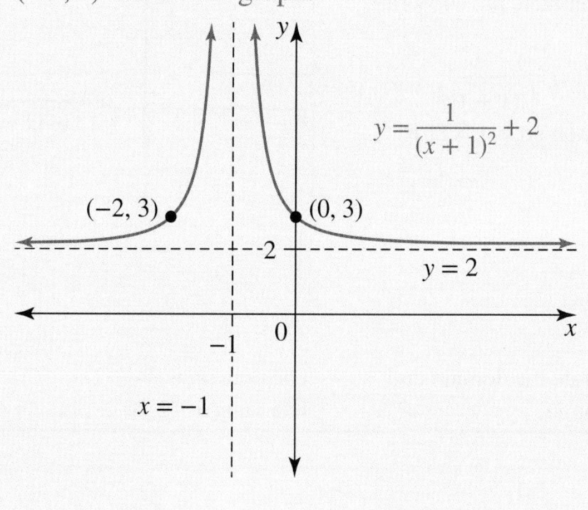

4. State the domain and range. | The domain is $R\backslash\{-1\}$; the range is $(2, \infty)$.

b. 1. State the equations of the asymptotes.

b. $y = 1 - \dfrac{4}{(x-2)^2}$
$= \dfrac{-4}{(x-2)^2} + 1$
Vertical asymptote: $x = 2$
Horizontal asymptote: $y = 1$

2. Calculate the coordinates of any axis intercepts.
Note: The symmetry of the graph about its vertical asymptote may enable key points to be identified.

y-intercept: let $x = 0$.
$y = \dfrac{-4}{(-2)^2} + 1$
$= 0$
The graph passes through the origin $(0, 0)$.
x-intercepts: let $y = 0$.
$\dfrac{-4}{(x-2)^2} + 1 = 0$
$\dfrac{4}{(x-2)^2} = 1$
$(x-2)^2 = 4$
$x - 2 = \pm 2$
$\therefore x = 0, x = 4$
$(4, 0)$ is also an x-intercept.

3. Sketch the graph, ensuring that each branch has at least one known point.

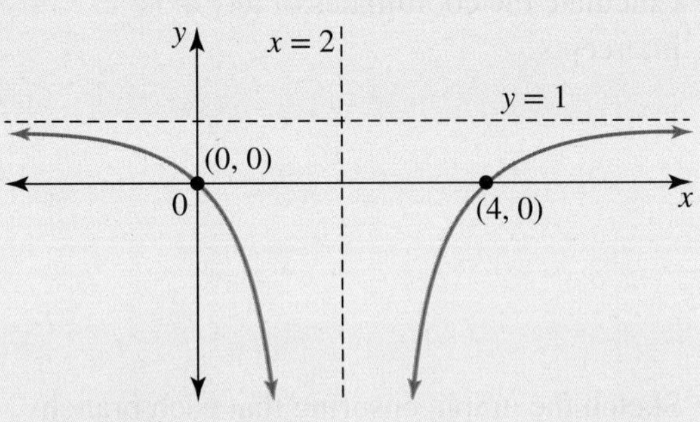

4. State the domain and range.

Domain $R\backslash\{2\}$; range $(-\infty, 1)$

TI | THINK	DISPLAY/WRITE	CASIO | THINK	DISPLAY/WRITE
a. 1. On a Graphs page, complete the entry line as: $f1(x) = \dfrac{1}{(x+1)^2} + 2$ Then press ENTER.	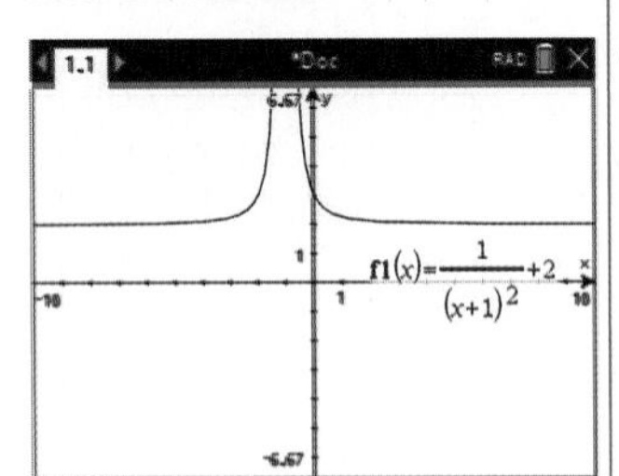	a. 1. On a Graphs & Table screen, complete the entry line as: $f1(x) = \dfrac{1}{(x+1)^2} + 2$ Then press EXE.	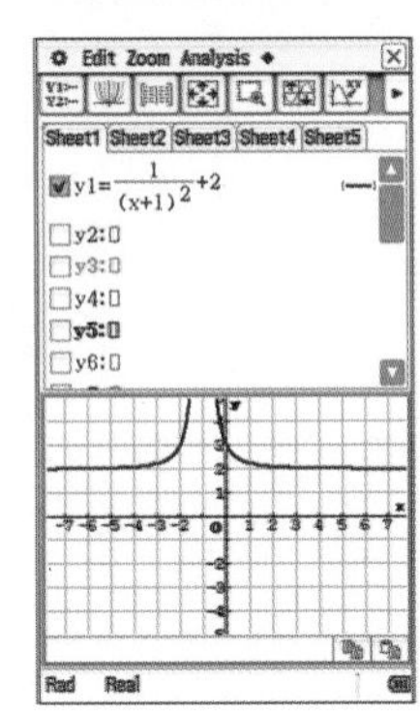
2. State the domain and range.	The domain is $R\backslash\{-1\}$. The range is $(2, \infty)$.	2. State the domain and range.	The domain is $R\backslash\{-1\}$. The range is $(2, \infty)$.

6.3.5 Modelling with the hyperbola and truncus

As for inverse proportionality, practical applications involving hyperbola or truncus models may need domain restrictions. Unlike many polynomial models, neither the hyperbola nor the truncus has maximum and minimum turning points, so the asymptotes are often where the interest will lie. The horizontal asymptote is often of particular interest as it represents the limiting value of the model.

WORKED EXAMPLE 8 Interpreting a model given by a hyperbola

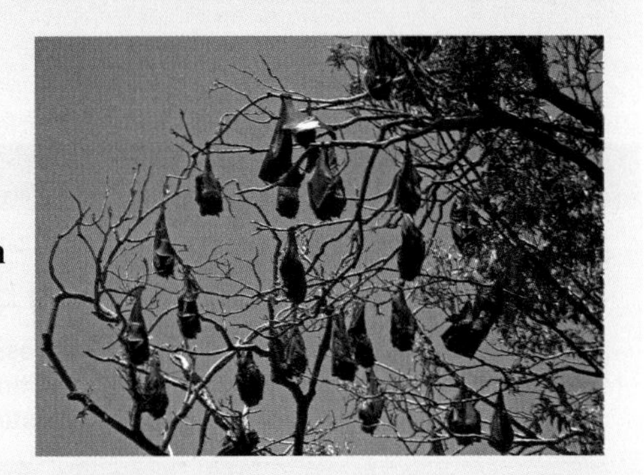

A relocation plan to reduce the number of bats in a public garden is formed, and t months after the plan is introduced the number of bats N in the garden is thought to be modelled by $N = 250 + \dfrac{30}{t+1}$.

a. Calculate how many bats were removed from the garden in the first 9 months of the relocation plan.

b. Sketch the graph of the bat population over time using the given model and state its domain and range.

c. Determine the maximum number of bats that will be relocated according to this model.

THINK

a. Find the number of bats at the start of the plan and the number after 9 months, and calculate the difference.

WRITE

a. $N = 250 + \dfrac{30}{t+1}$

When $t = 0$, $N = 250 + \dfrac{30}{1}$.

Therefore, there were 280 bats when the plan was introduced.

When $t = 9$, $N = 250 + \dfrac{30}{10}$.

Therefore, 9 months later there were 253 bats.

Hence, over the first 9 months, 27 bats were removed.

THINK

b. 1. Identify the asymptotes and other key features that are appropriate for the restriction $t \geq 0$.

WRITE

b. $N = 250 + \dfrac{30}{t+1}, t \geq 0$

Vertical asymptote $t = -1$ (not applicable)

Horizontal asymptote $N = 250$

The initial point is $(0, 280)$.

THINK

2. Sketch the part of the graph of the hyperbola that is applicable and label axes appropriately.

Note: The vertical scale is broken in order to better display the graph.

WRITE

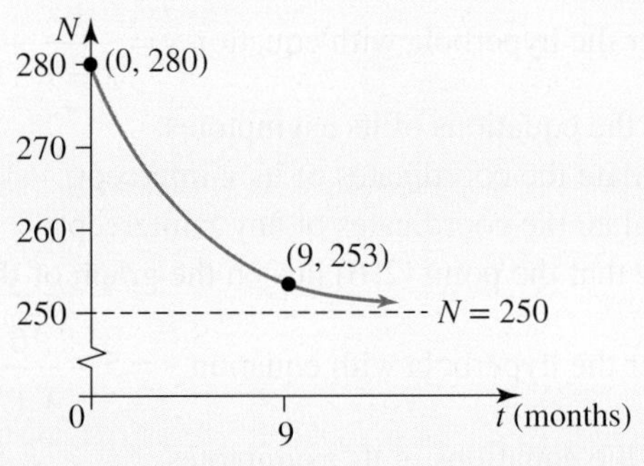

THINK

3. State the domain and range for this model.

WRITE

Domain $\{t : t \geq 0\}$

Range $(250, 280]$

c. 1. Interpret the meaning of the horizontal asymptote.	**c.** The horizontal asymptote shows that as $t \to \infty, N \to 250$. This means $N = 250$ gives the limiting population of the bats.
2. State the answer.	Since the population of bats cannot fall below 250 and there were 280 initially, the maximum number of bats that can be relocated approaches 30.

6.3 Exercise

Students, these questions are even better in jacPLUS

Receive immediate feedback and access sample responses

Access additional questions

Track your results and progress

Find all this and MORE in jacPLUS

Technology free

1. State the equations of the asymptotes of the following hyperbolas.

a. $y = \dfrac{1}{x+5} + 2$ **b.** $y = \dfrac{8}{x} - 3$ **c.** $y = \dfrac{-3}{4x}$ **d.** $y = \dfrac{-3}{14+x} - \dfrac{3}{4}$

2. For each of the following, state:

i. the equations of the asymptotes
ii. the domain
iii. the range.

a. $y = \dfrac{1}{x-6} + 1$ **b.** $y = \dfrac{1}{3+x} - 7$ **c.** $y = 4 + \dfrac{8}{2-x}$ **d.** $y = \dfrac{-5}{4(x-1)} - 3.$

3. **WE4** Sketch the graphs of the following functions, stating the domain and range.

a. $y = \dfrac{2}{x+2} - 1$ **b.** $y = \dfrac{-1}{x-2}$

4. Consider the hyperbola with equation $y = \dfrac{1}{x-1} + 5.$

a. State the equations of its asymptotes.
b. Calculate the coordinates of its y-intercept.
c. Calculate the coordinates of any x-intercept.
d. Show that the point $(2, 6)$ lies on the graph of the hyperbola and sketch the graph.

5. Consider the hyperbola with equation $y = 5 - \dfrac{6}{x+3}.$

a. State the equations of its asymptotes.
b. Calculate the coordinates of its y-intercept.
c. Calculate the coordinates of any x-intercept.
d. Sketch the graph.

6. Sketch the graph of the following functions, stating the domain and range.

a. $y = \dfrac{1}{x+1} - 3$ **b.** $y = 4 - \dfrac{3}{x-3}$ **c.** $y = -\dfrac{5}{3+x}$ **d.** $y = -\left(1 + \dfrac{5}{2-x}\right)$

7. For $f: R \setminus \left\{\frac{1}{2}\right\} \to R,\ f(x) = 4 - \frac{3}{1-2x}$, state the equations of the asymptotes and the asymptote-formed quadrants in which the graph would lie.

8. **WE5** a. Identify the asymptotes of the hyperbola with equation $y = \frac{6x}{3x+2}$.

b. Form the equation of the hyperbola shown.

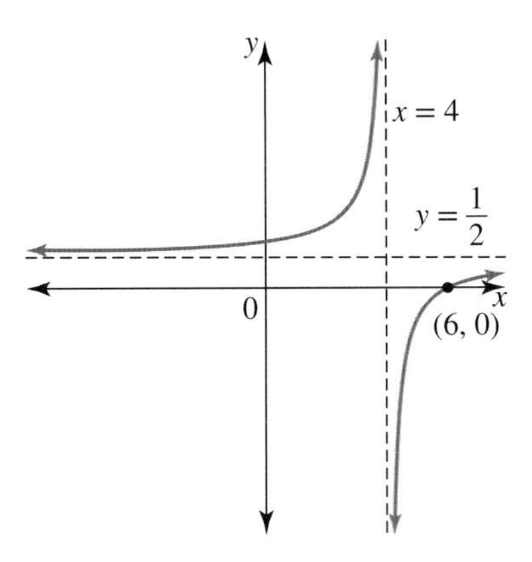

9. a. The equations of the asymptotes of a hyperbola are $x = -4$ and $y = 2$. The graph passes through the point (0, 8). Express its equation in the form $y = \frac{a}{x-h} + k$.

b. The point $(2, -1)$ lies on the graph of a hyperbola. Given the equations of the asymptotes are $x = 0$ and $y = 5$, determine the equation of the hyperbola.

c. Form the equation of the hyperbola shown in the diagram.

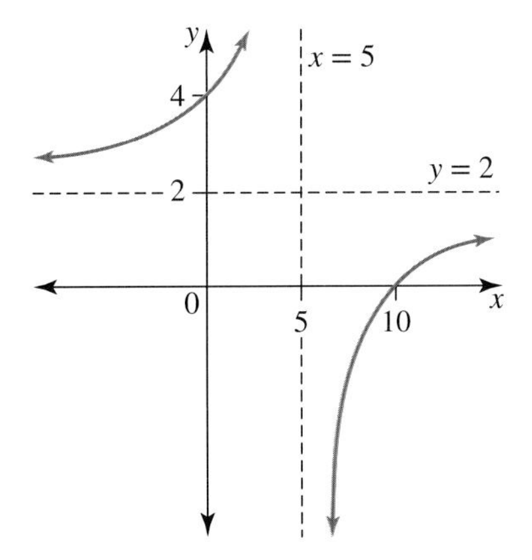

d. Part of a hyperbola is shown in the diagram.

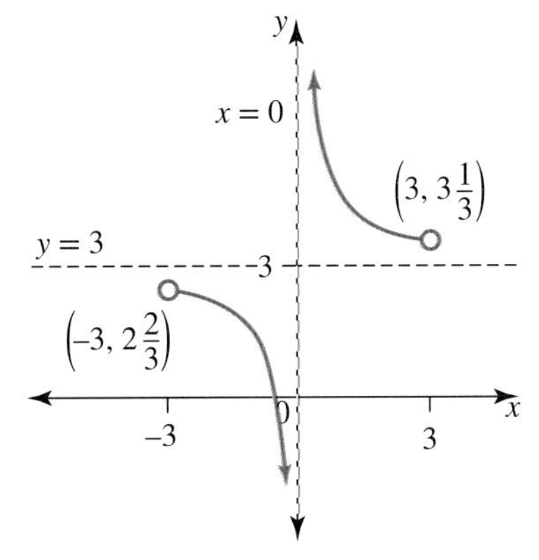

i. State the domain and range of the graph.

ii. Form the equation of the graph.

10. a. If $\frac{11-3x}{4-x}=a-\frac{b}{4-x}$ calculate the values of a and b.

b. Hence, sketch the graph of $y=\frac{11-3x}{4-x}$.

c. Determine the values of x for which $\frac{11-3x}{4-x}>0$.

11. Express in the form $y=\frac{a}{bx+c}+d$ and state the equations of the asymptotes for each of the following.

a. $y=\frac{x}{4x+1}$ b. $(x-4)(y+2)=4$ c. $y=\frac{1+2x}{x}$ d. $2xy+3y+2=0$

12. Determine the domain and range of the hyperbola with equation $xy-4y+1=0$.

13. MC The equations of the asymptotes of a truncus are $x=7$ and $y=0$. State which of the following could be its equation.

A. $y=\frac{1}{x^2-7}$ B. $y=\frac{2}{x-7}$ C. $y=\frac{-3}{(7-x)^2}$ D. $y=\frac{4}{(x+7)^2}$ E. $y=\frac{5}{49+x^2}$

14. The diagram shows the graph of a truncus.

a. State the domain and range of the graph.
b. Form the equation of the graph.

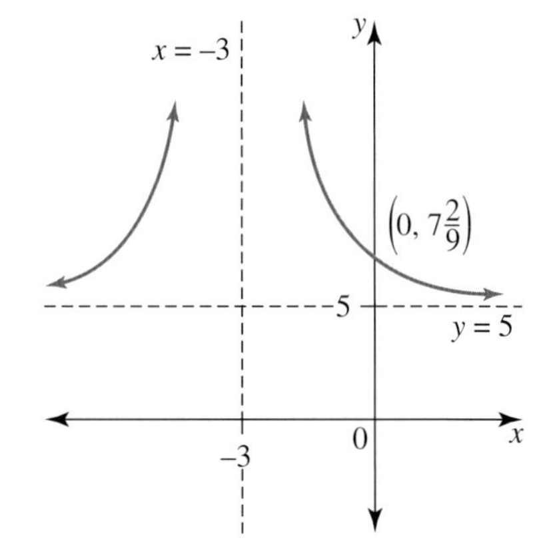

15. WE6 From Ohm's Law, the electrical resistance, R ohms, of a metal conductor in which the voltage is constant is inversely proportional to the current, I amperes.

When the current is 0.6 amperes, the resistance is 400 ohms.

a. Determine the relationship between the resistance and current.
b. Calculate the resistance if the current is increased by 20%. You may choose to use technology to complete your answer.

16. WE7 Sketch the graphs of the following functions, stating the domain and range.

a. $y=\frac{1}{(x-3)^2}-1$ b. $y=\frac{-8}{(x+2)^2}-4$

17. For each of the following:

i. state the equations of the asymptotes
ii. state the domain and range
iii. calculate any intercepts with the coordinate axes
iv. sketch the graph.

a. $y=\frac{4}{3x^2}-1$ b. $y=\frac{-2}{(x-1)^2}+4$.

18. Identify the equations of the asymptotes and sketch the graphs of the following, stating the domain and range.

a. $y = \frac{12}{(x-2)^2} + 5$ **b.** $y = \frac{-24}{(x+2)^2} + 6$ **c.** $y = 7 - \frac{1}{7x^2}$

d. $y = \frac{4}{(2x-1)^2}$ **e.** $y = -2 - \frac{1}{(2-x)^2}$ **f.** $y = \frac{x^2+2}{x^2}$

19. Determine the equations of the asymptotes and state the domain and range of the truncus with equation $y = \frac{3}{2(1-5x)^2}$.

20. The diagrams in parts **a** to **d** are of either a hyperbola or a truncus. Form the equations of each graph.

a.

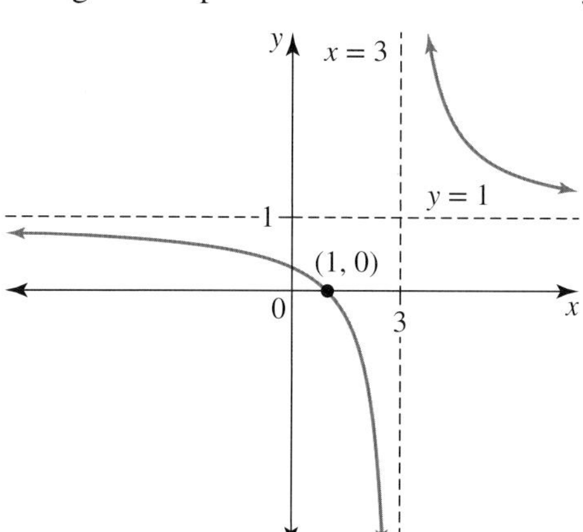

b.

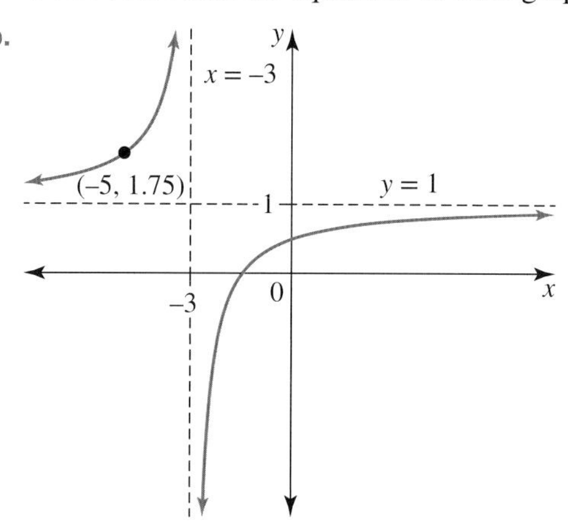

c.

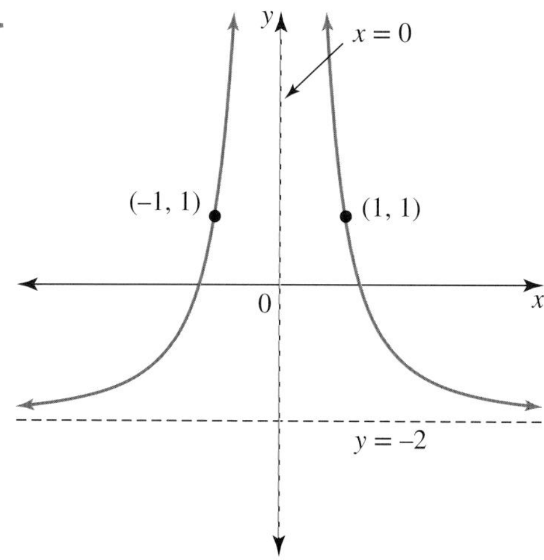

d.

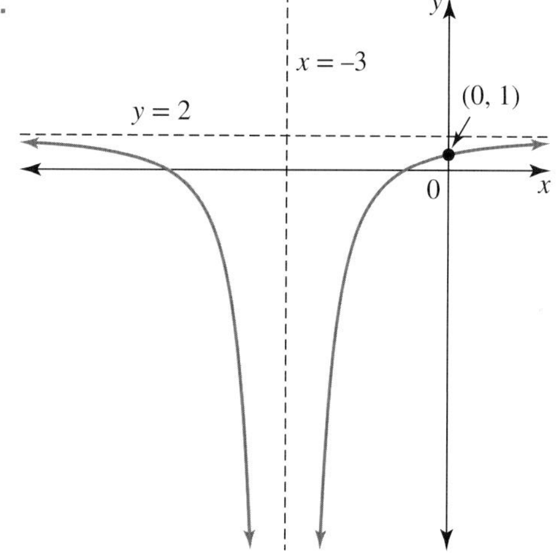

e. The graph of a truncus has the same shape as $y = \frac{4}{x^2}$ but its vertical asymptote has the equation $x = -2$. Give its rule and write the function in mapping notation.

f. A hyperbola is undefined when $x = \frac{1}{4}$. As $x \to -\infty$, its graph approaches the line $y = -\frac{1}{2}$ from below. The graph cuts the x-axis where $x = 1$.

i. Determine the equation of the hyperbola and express it in the form $y = \frac{ax+b}{cx+d}$ where $a, b, c, d \in Z$.

ii. Write the function in mapping notation.

Technology active

21. **WE8** The number of cattle, P, owned by a farmer at a time t years after purchase is modelled by $P = 30 + \dfrac{100}{2+t}$.

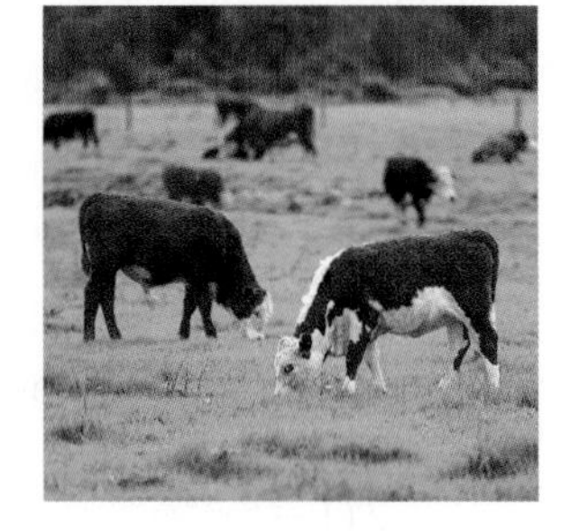

a. Calculate how many cattle the herd is reduced by after the first 2 years.
b. Sketch the graph of the number of cattle over time using the given model and state its domain and range.
c. Determine the minimum number the herd of cattle is expected to reach according to this model.

22. Select the table of data that shows $y \propto \dfrac{1}{x}$ and complete the rule for y in terms of x.

a.

x	0.5	1	2	4		8
y	20.5	11	7	6.5	6.4	

b.

x	0.5	1	2	4		8
y	14.4	7.2	3.6	1.8	6.4	

23. The strength of a radio signal I is given by $I = \dfrac{k}{d^2}$, where d is the distance from the transmitter.

a. Draw a sketch of the shape of the graph of I against d.
b. Describe the effect on the strength of the signal when the distance from the transmitter is doubled.

24. The time t taken to travel a fixed distance of 180 km is given by $t = \dfrac{k}{v}$ where v is the speed of travel.

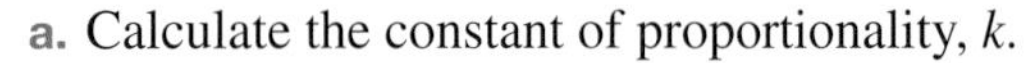

a. Calculate the constant of proportionality, k.
b. Sketch a graph to show the nature of the relationship between the time and the speed.
c. Determine the speed that needs to be maintained if the entire journey is to be completed in $2\frac{1}{4}$ hours.

25. The height, h metres, of a hot air balloon above ground level t minutes after take-off is given by $h = 25 - \dfrac{100}{(t+2)^2}$, $t \geq 0$.

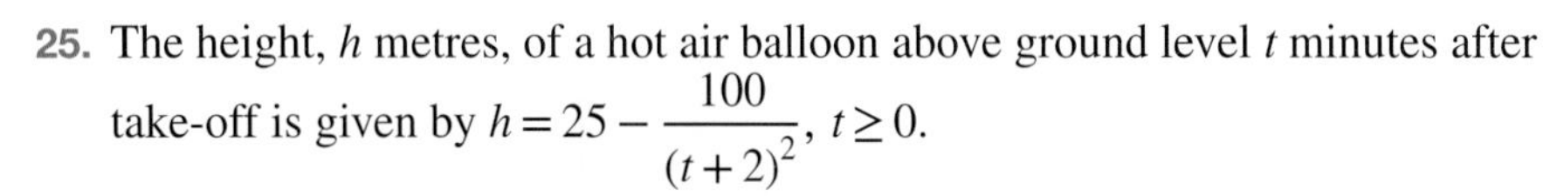

a. Calculate how long, to the nearest second, it takes the balloon to reach an altitude of 12.5 metres above ground level.
b. Determine the balloon's limiting altitude.

26. a. Calculate the coordinates of the point(s) of intersection of $xy = 2$ and $y = \dfrac{x^2}{4}$.

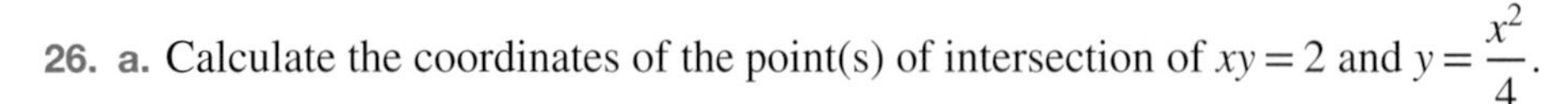

b. On the same axes, sketch the graphs of $xy = 2$ and $y = \dfrac{x^2}{4}$.

c. Calculate the coordinates of the point(s) of intersection of $y = \dfrac{4}{x^2}$ and $y = \dfrac{x^2}{4}$, and sketch the graphs of $y = \dfrac{4}{x^2}$ and $y = \dfrac{x^2}{4}$ on the same set of axes.

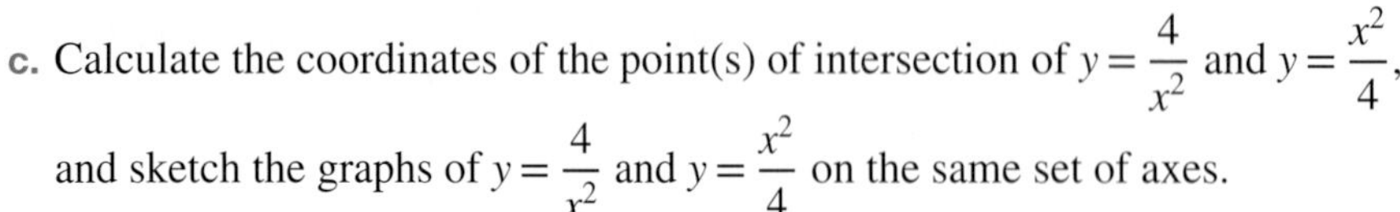

d. Express the coordinates of the point(s) of intersection of $y = \dfrac{a}{x^2}$ and $y = \dfrac{x^2}{a}$ in terms of a, for $a > 0$.

27. In an effort to protect a rare species of stick insect, 20 of the species were captured and relocated to a small island where there were few predators.
After 2 years the population size grew to 240 stick insects.
A model for the size N of the stick insect population after t years on the island is thought to be defined by the function $N: R^+ \cup \{0\} \to R, N(t) = \dfrac{at+b}{t+2}$.

a. Calculate the values of a and b.

b. Determine the length of time, to the nearest month, that it took for the stick insect population to reach 400.

c. Show that $N(t+1) - N(t) = \dfrac{880}{(t+2)(t+3)}$.

d. Hence, or otherwise, find the increase in the stick insect population during the 12th year and compare this with the increase during the 14th year.
Describe what is happening to the growth in population.

e. Determine when the model predicts the number of stick insects will reach 500.

f. Determine how large the stick insect population can grow.

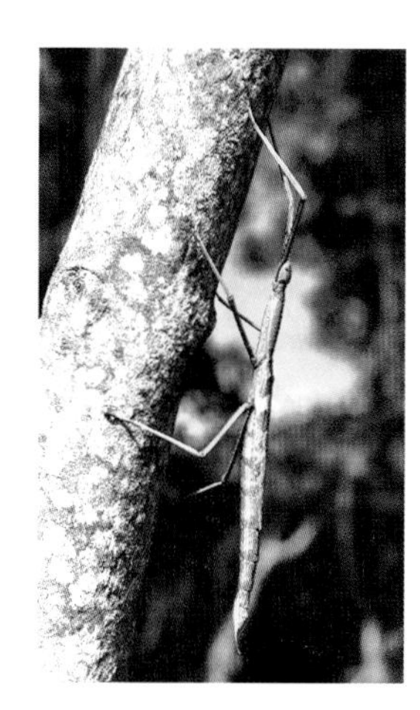

28. Use CAS technology to sketch $y = \dfrac{x+1}{x+2}$ together with its asymptotes, and use the graphing screen to obtain:

a. the number of intersections of $y = x$ with $y = \dfrac{x+1}{x+2}$

b. the values of k for which $y = x + k$ intersects $y = \dfrac{x+1}{x+2}$ once, twice or not at all.

6.3 Exam questions

Question 1 (4 marks) TECH-FREE

a. Sketch $y = \dfrac{1}{(x-2)^2}$, showing axial intercepts and asymptotes. **(2 marks)**

b. State the domain, range and type of correspondence. **(2 marks)**

Question 2 (1 mark) TECH-ACTIVE

MC A possible equation for the graph shown is

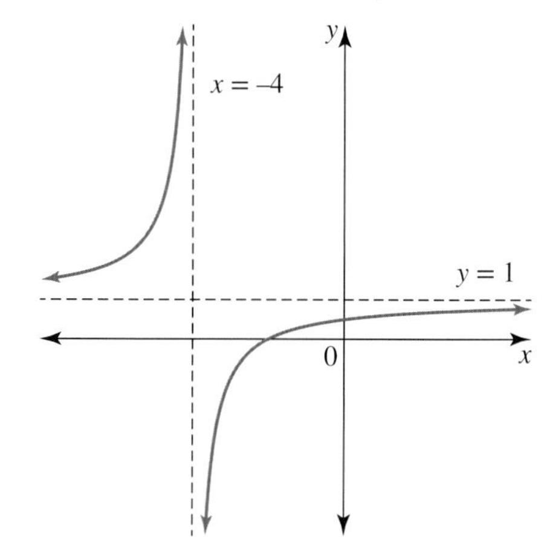

A. $y = \dfrac{2}{x+4} - 1$

B. $y = \dfrac{2}{x+4} + 1$

C. $y = \dfrac{-2}{x-4} + 1$

D. $y = \dfrac{-2}{x+4} + 1$

E. $y = \dfrac{2}{x-1} + 4$

Question 3 (1 mark) TECH-ACTIVE

MC For the graph of the function $y = \dfrac{1}{x} - 3$, state which of the following is *not* true.

A. The function has one-to-one correspondence.

B. The domain is $R\backslash\{0\}$.

C. The graph passes through the point $(1, -2)$.

D. There are asymptotes at $x = 3$, $y = 0$.

E. The range is $R\backslash\{-3\}$.

More exam questions are available online.

6.4 The square root function

LEARNING INTENTION

At the end of this subtopic you should be able to:
- sketch the square root function together with simple transformations
- determine the equation of a given square root graph.

The **square root function** is formed from two branches of the relation $y^2 = x$.

Since $y^2 = x \Rightarrow y = \pm\sqrt{x}$, the upper branch has the equation $y = \sqrt{x}$ and the lower branch has the equation $y = -\sqrt{x}$.

These two functions are drawn on the diagram.

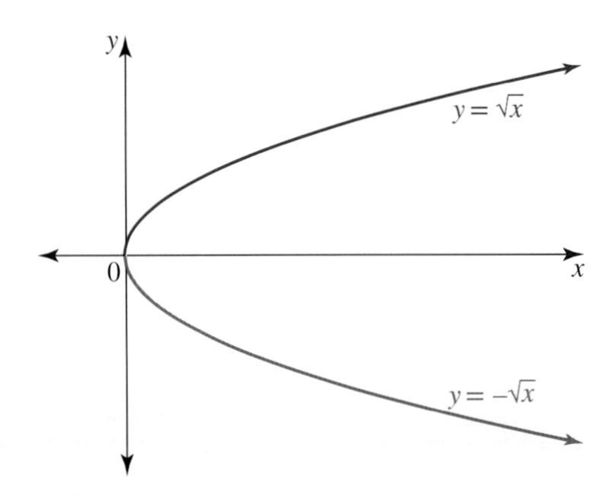

6.4.1 The graph of $y = \sqrt{x}$

The square root function has the rule $y = x^n, n = \frac{1}{2}$. It is not a polynomial function. The square root function is defined by the equation $y = \sqrt{x}$ or $y = x^{\frac{1}{2}}$. The y-values of this function must be such that $y \geq 0$. No term under a square root symbol can be negative, so this function also requires that $x \geq 0$. The graph of the square root function is shown in the diagram.

The key features of the graph of $y = \sqrt{x}$ are as follows:
- The end point is $(0, 0)$.
- As $x \to \infty, y \to \infty$.
- The function is defined only for $x \geq 0$, so the domain is $R^+ \cup \{0\}$ or $[0, \infty)$.
- y-values cannot be negative, so the range is $R^+ \cup \{0\}$ or $[0, \infty)$.
- The graph is of a function with a one-to-one correspondence.
- The shape is the upper half of the sideways parabola $y^2 = x$.

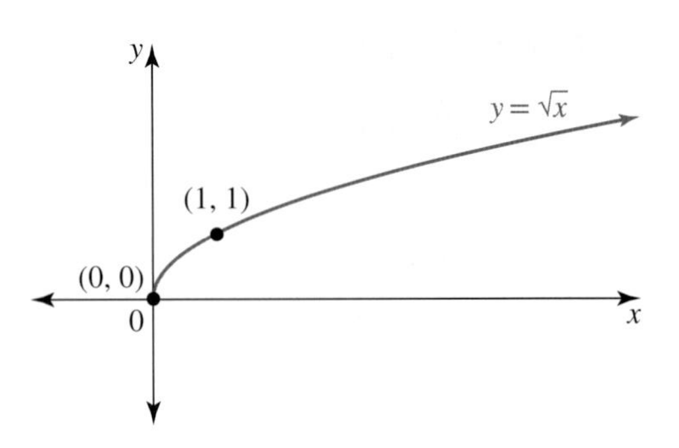

Variations of the basic graph

The term under the square root cannot be negative so $y = \sqrt{-x}$ has a domain requiring $-x \geq 0 \Rightarrow x \in (-\infty, 0]$. This graph can be obtained by reflecting the graph of $y = \sqrt{x}$ in the y-axis. The four variations in position of the basic graph are shown in the diagram.

The diagram could also be interpreted as displaying the graphs of the two relations $y^2 = x$ (on the right of the y-axis) and $y^2 = -x$ (on the left of the y-axis). The end points of the square root functions are the vertices of these sideways parabolas, the point (0, 0).

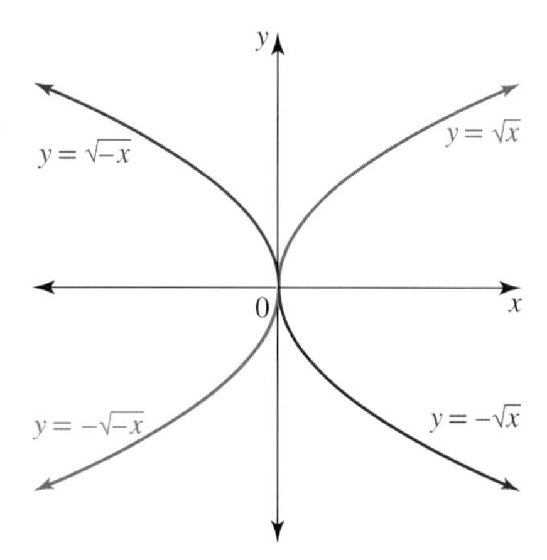

6.4.2 Transformations of the square root function

The general equation of the square root function

$$y = a\sqrt{x - h} + k$$

The key features of the graph of $y = a\sqrt{x - h} + k$ are as follows:

- The end point is (h, k).
- If $a > 0$, the end point is a minimum point.
- If $a < 0$, the end point is a maximum point.
- The domain $[h, \infty)$, since $x - h \geq 0 \rightarrow x \geq h$.
- If $a > 0$, the range is $[k, \infty)$.
- If $a < 0$, the range is $(-\infty, k]$.
- There are either one or no x-intercepts.
- There are either one or no y-intercepts.

The graph of $y = a\sqrt{-(x - h)} + k$ also has its end point at (h, k), but its domain is $(-\infty, h]$.

WORKED EXAMPLE 9 Sketching square root functions

a. **Sketch $y = 2\sqrt{x + 1} - 4$, stating its domain and range. Include any axis intercepts with their coordinates.**

b. **For the function $f(x) = \sqrt{2 - x}$:**
  i. **state its domain**
  ii. **sketch the graph of $y = f(x)$.**

THINK	WRITE
a. 1. State the coordinates of the end point.	a. $y = 2\sqrt{x + 1} - 4$ End point $(-1, -4)$

2. Calculate any intercepts with the coordinate axes.	y-intercept: let $x = 0$. $y = 2\sqrt{1} - 4$ $= -2$ y-intercept $(0, -2)$ x-intercept: let $y = 0$. $2\sqrt{x+1} - 4 = 0$ $\therefore \sqrt{x+1} = 2$ Square both sides: $x + 1 = 4$ $\therefore x = 3$ x-intercept $(3, 0)$
3. Sketch the graph and state its domain and range.	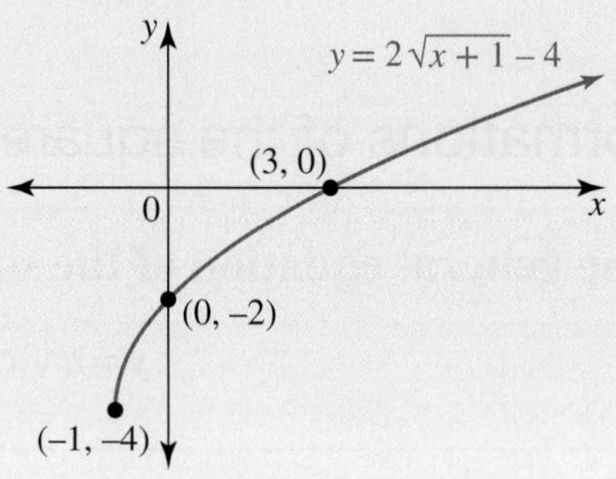 Domain $[-1, \infty)$, range $[-4, \infty)$
b. i. State the requirement on the expression under the square root and then state the domain.	**b. i.** $f(x) = \sqrt{2 - x}$ Domain: require $2 - x \geq 0$. $-x \geq -2$ $\therefore x \leq 2$ Therefore, the domain is $(-\infty, 2]$.
ii. Identify the key points of the graph and sketch.	**ii.** $f(x) = \sqrt{2-x}$ $y = \sqrt{2-x}$ $= \sqrt{-(x-2)}$ The end point $(2, 0)$ is also the x-intercept. y-intercept: let $x = 0$ in $y = \sqrt{2-x}$. $y = \sqrt{2}$ The y-intercept is $(0, \sqrt{2})$. 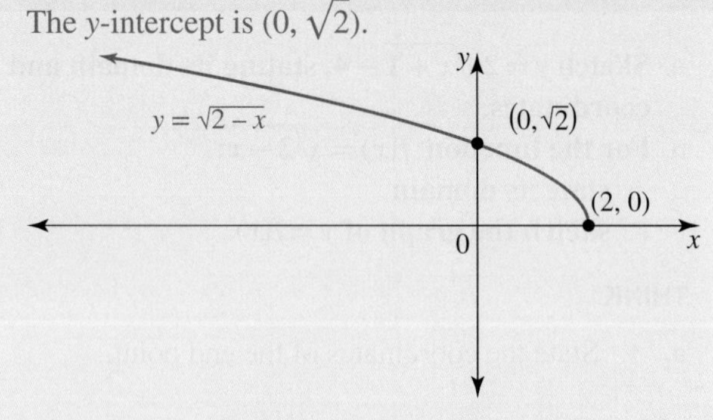

TI | THINK	DISPLAY/WRITE	CASIO | THINK	DISPLAY/WRITE
a. 1. On a Graphs page, complete the entry line as: $f1(x)=2\sqrt{x+1}-4$ Then press ENTER.	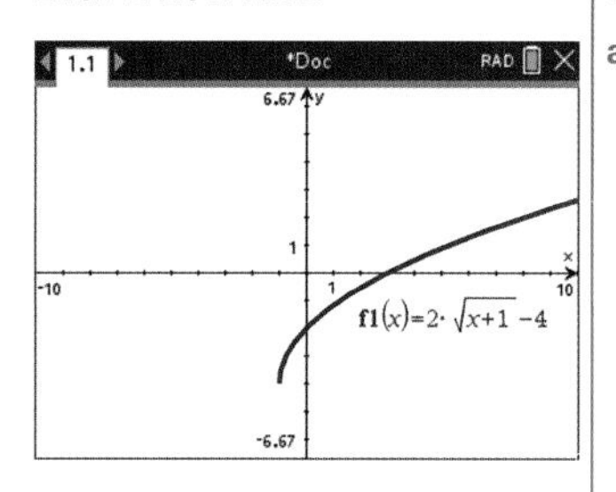	**a. 1.** On a Graphs & Table screen, complete the entry line as: $y1=2\sqrt{x+1}-4$ Then press EXE.	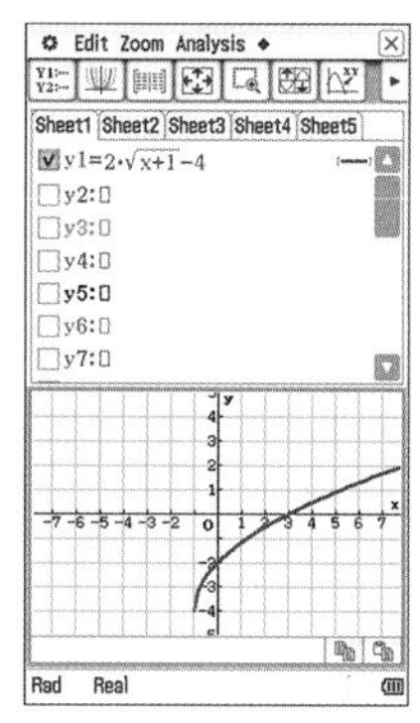
2. State the domain and range.	The domain is $[-1, \infty)$. The range is $[-4, \infty)$.	**2.** State the domain and range.	The domain is $[-1, \infty)$. The range is $[-4, \infty)$.

6.4.3 Determining the equation of a square root function

If a diagram is given, the direction of the graph will indicate whether to use the equation in the form of $y=a\sqrt{x-h}+k$ or of $y=a\sqrt{-(x-h)}+k$. If a diagram is not given, the domain or a rough sketch of the given information may clarify which form of the equation to use.

WORKED EXAMPLE 10 Determining the equation of a square root graph

Form a possible equation for the square root function shown.

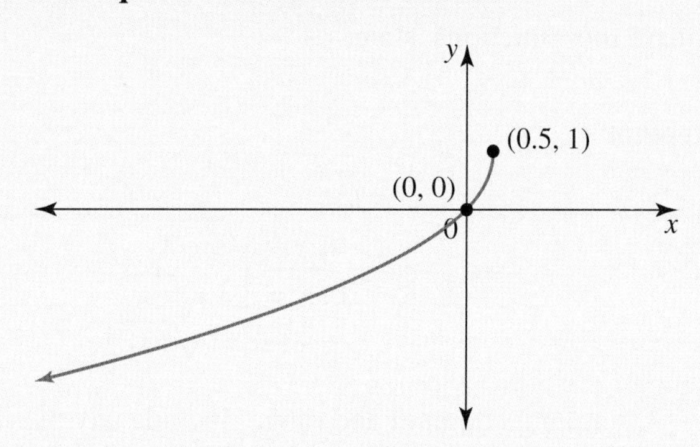

THINK	WRITE
1. Note the direction of the graph to decide which form of the equation to use.	The graph opens to the left with $x \leq 0.5$. Let the equation be $y=a\sqrt{-(x-h)}+k$.
2. Substitute the coordinates of the end point into the equation.	End point (0.5, 1) $y=a\sqrt{-(x-0.5)}+1$ $=a\sqrt{-x+0.5}+1$
3. Use a second point on the graph to determine the value of a.	(0, 0) lies on the graph. $0=a\sqrt{0.5}+1$ $-1=a\times\frac{1}{\sqrt{2}}$ $a=-\sqrt{2}$
4. State the equation of the graph.	The equation is $y=-\sqrt{2}\sqrt{-x+0.5}+1$.

5. Express the equation in a simplified form.	$y=-\sqrt{2}\sqrt{-x+\frac{1}{2}}+1$ $=-\sqrt{2}\sqrt{\frac{-2x+1}{2}}+1$ $=-\sqrt{2}\frac{\sqrt{-2x+1}}{\sqrt{2}}+1$ $=-\sqrt{-2x+1}+1$ Therefore, the equation is $y=1-\sqrt{1-2x}$.

6.4 Exercise

Students, these questions are even better in jacPLUS

Receive immediate feedback and access sample responses

Access additional questions

Track your results and progress

Find all this and MORE in jacPLUS

Technology free

1. For each of the following square root functions, state:
   i. its domain
   ii. the coordinates of its end point
   iii. its range.

   a. $y=\sqrt{x-9}$
   b. $y=\sqrt{4-x}$
   c. $y=-\sqrt{x+3}$
   d. $y=3+\sqrt{-x}$
   e. $y=\sqrt{3x-6}-7$
   f. $y=4-\sqrt{1-2x}$

2. **WE9** a. Sketch $y=\sqrt{x-1}-3$, stating its domain and range. Include any axis intercepts with their coordinates.
   b. For the function $f(x)=-\sqrt{2x+4}$:
      i. state the domain
      ii. sketch the graph of $y=f(x)$.

3. Sketch the graph of each of the following square root functions, showing end points and any intersections with the coordinate axes.
   a. $y=\sqrt{x+4}-1$
   b. $y=3-\sqrt{3x}$
   c. $y=-\sqrt{6-x}$
   d. $y=\sqrt{9-2x}+1$

4. Sketch the following relations and state their domains and ranges.
   a. $y=\sqrt{x+3}-2$
   b. $y=5-\sqrt{5x}$
   c. $y=2\sqrt{9-x}+4$
   d. $y=\sqrt{49-7x}$
   e. $y=2+\sqrt{x+4}$
   f. $y+1+\sqrt{-2x+3}=0$

5. WE10 Form a possible equation for the square root function shown.

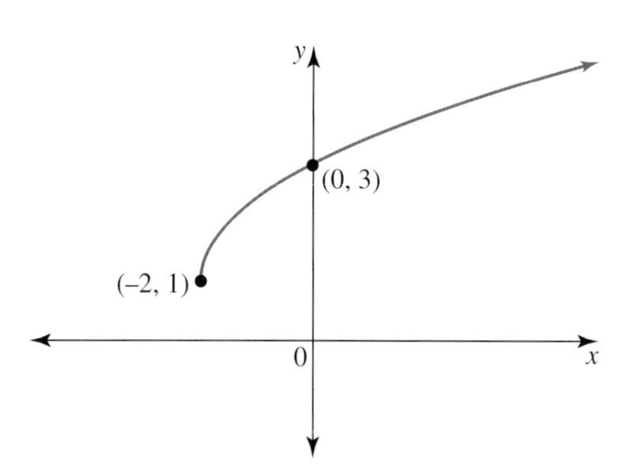

6. Determine the equation for the graph shown, given it represents a square root function with end point $(-2, 2)$.

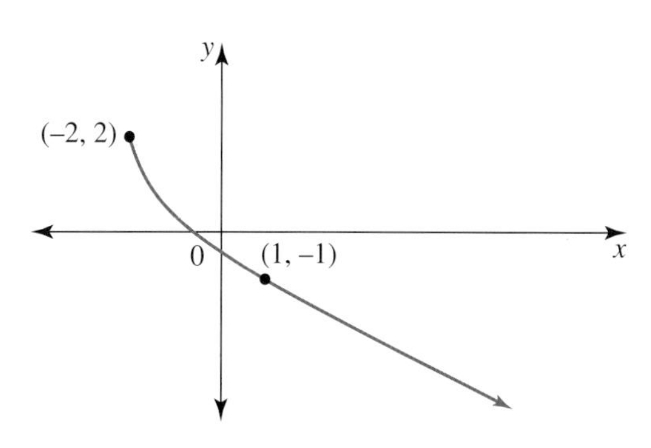

7. Form the equation of the square root function with end point $(4, -1)$ and containing the point $(0, 9)$. Determine the coordinates of the point at which this function cuts the x-axis.

8. Give the equation of the function that has the same shape as $y = \sqrt{-x}$ and an end point with coordinates $(4, -4)$.

9. A square root function graph has an equation in the form $y = a\sqrt{x-h} + k$. If the end point is $(3, 2)$ and the graph passes through the point $(4, 6)$, determine the equation of the graph.

10. Form the equation of the square root function shown in the diagram, expressing it in the form $y = a\sqrt{x-h} + k$.

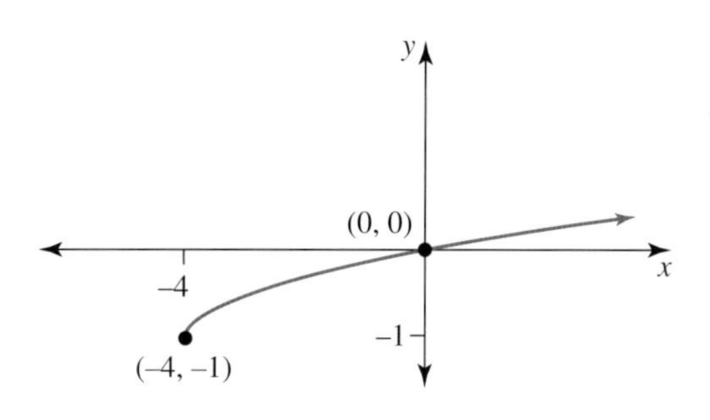

11. The equation of the graph shown has the form $y = \sqrt{a(x-b)} + c$. Determine the values of a, b and c, and state the equation.

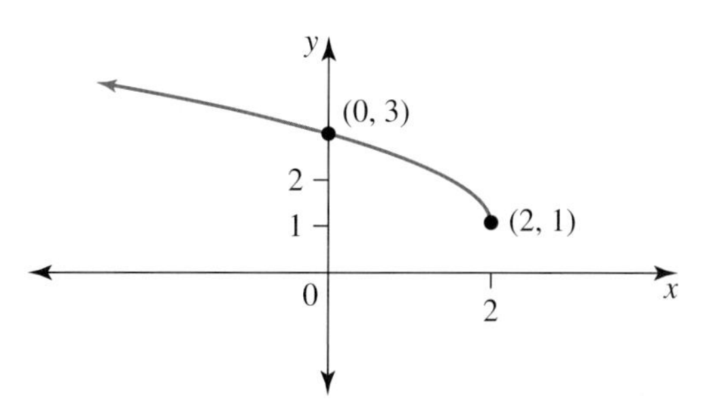

Technology active

12. Consider the function $S = \{(x, y) : y + 2 = 3\sqrt{x-1}\}$.
   a. Find the coordinates of its end point and x-intercept.
   b. Sketch the graph of S.
   c. Form the equation of a square root function that would be related to S.

13. A small object falls to the ground from a vertical height of h metres. The time, seconds, is proportional to the square root of the height. If it takes 2 seconds for the object to reach the ground from a height of 19.6 metres, obtain the rule for t in terms of h.

14. Consider the relation $S = \{(x, y) : (y - 2)^2 = 1 - x\}$.
   a. Express the equation of the relation S with y as the subject.
   b. Define, as mappings, the two functions f and g which together form the relation S.
   c. Sketch, on the same diagram, the graphs of $y = f(x)$ and $y = g(x)$.
   d. Give the image of -8 for each function.

15. The function $y = a - \sqrt{b(x - c)}$ has an end point at $(2, 5)$ and cuts the x-axis at $x = -10.5$. Determine the values of a, b and c, and hence state the equation of the function and its domain and range.

16. A function defined as $f : [0, \infty\} \to R, f(x) = \sqrt{mx} + n, f(1) = 1$ and $f(4) = 4$.
   a. Determine the values of m and n.
   b. Calculate the value of x for which $f(x) = 0$.
   c. Sketch the graph of $y = f(x)$ together with the graph of $y = x$ and state $\{x : f(x) > x\}$.

6.4 Exam questions

Question 1 (1 mark) TECH-ACTIVE

MC The graph of the function $f(x) = 1 + \sqrt{3 - x}$ could be

A.

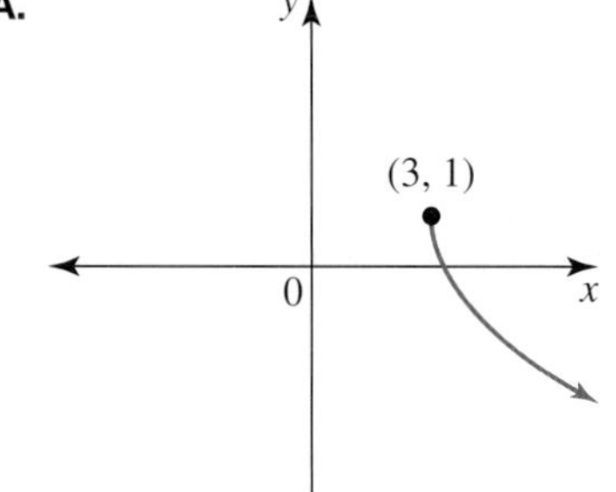

B.

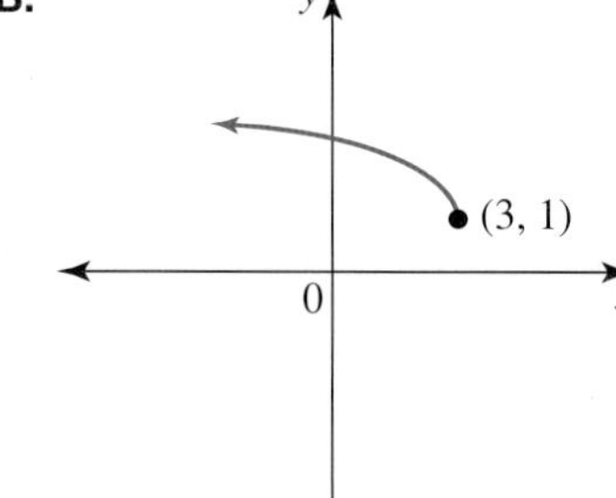

C.

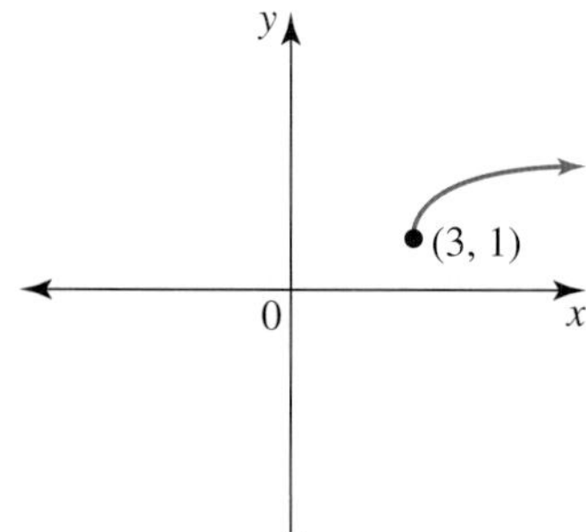

D.

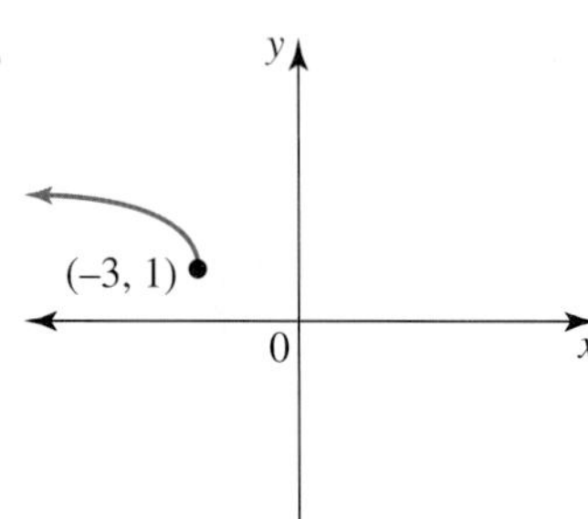

E.

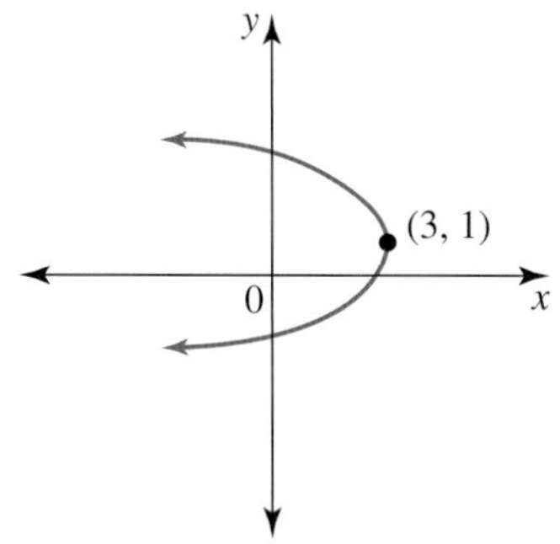

Question 2 (5 marks) TECH-FREE

Sketch $y = 3\sqrt{x + 4} - 3$ and state the domain and range.

Question 3 (1 mark) TECH-ACTIVE

MC The coordinates of the end point of the square root function with equation $y - 2 = \sqrt{3 - x}$ are

A. (2, 3) **B.** (2, −3) **C.** (3, 2) **D.** (−3, 2) **E.** (−3, −2)

More exam questions are available online.

6.5 Other functions and relations

LEARNING INTENTION

At the end of this subtopic you should be able to:
- identify maximal domains of functions
- determine the equations of inverse relations and use the correct notation for inverse functions
- determine the domain and range of inverse functions
- sketch the inverse of a cubic function.

In this section we will use the functions we have learnt about so far to shed insight into and explore some connections between other functions and relations.

6.5.1 Maximal domains

When we write the function for which $f(x)=x^2$, we imply that this function has domain R. This is called the **maximal** or **implied** domain of the function. The knowledge acquired about the domains of polynomial, hyperbola, truncus and square root functions can now be applied to identify the maximal domains of algebraic functions with which we are not familiar.

We know, for example, that the maximal domain of any polynomial function is R; the domain of the square root function $y=\sqrt{x}$ is $R^+ \cup \{0\}$ or $[0, \infty)$, since the expression under the square root symbol cannot be negative; and the domain of the hyperbola $y=\frac{1}{x}$ is $R\backslash\{0\}$, since the denominator cannot be zero.

The maximal domain of any function must exclude:
- any value of x for which the denominator would become zero
- any value of x which would create a negative number or expression under a square root sign.

Hence, to obtain maximal domains, the following conditions would have to be satisfied:

$$y=\frac{g(x)}{f(x)} \Rightarrow f(x)\neq 0$$
$$y=\sqrt{f(x)} \Rightarrow f(x)\geq 0$$
$$y=\frac{g(x)}{\sqrt{f(x)}} \Rightarrow f(x)>0$$

Numerators can be zero. For $y=\frac{g(x)}{f(x)}$, if $g(a)=0$ and $f(a)=2$, for example, the value of $\frac{g(a)}{f(a)}=\frac{0}{2}=0$, which is defined.

At a value of x for which its denominator would be zero, a function will be discontinuous and its graph will have a vertical asymptote. As noted, this value of x is excluded from the function's domain.

Where a function is composed of the sum or difference of two functions, its domain must be the set over which all of its parts are defined.

Maximal domain

The maximal domain of $y=f(x) \pm g(x)$ is $d_f \cap d_g$, where d_f and d_g are the domains of f and g respectively.

WORKED EXAMPLE 11 Identifying maximal domains

Identify the maximal domains of the functions with the following rules.

a. $y = \dfrac{1}{x^2 - 9}$ **b.** $y = \dfrac{x - 4}{x^2 + 9}$ **c.** $y = \dfrac{1}{\sqrt{x^3 - 8}}$

THINK	WRITE
a. 1. Determine any values of x for which the denominator would become zero and exclude these values from the domain.	**a.** $y = \dfrac{1}{x^2 - 9}$ If the denominator $x^2 - 9 = 0$, then: $(x + 3)(x - 3) = 0$ $\therefore x = -3, x = 3$
2. State the maximal domain.	The values $x = -3, x = 3$ must be excluded from the domain. Therefore, the maximal domain is $R\backslash\{\pm 3\}$.
b. 1. Determine any values of x for which the denominator would become zero.	**b.** $y = \dfrac{x - 4}{x^2 + 9}$ If the denominator is zero: $x^2 + 9 = 0$ $x^2 = -9$ There is no real solution. As the denominator $x^2 + 9$ is the sum of two squares, it can never be zero.
2. State the maximal domain.	The denominator cannot be zero and both the numerator and denominator are polynomials, so they are always defined for any x-value. Therefore, the maximal domain of the function is R.
c. 1. State the condition that the expression contained under the square root sign must satisfy.	**c.** $y = \dfrac{1}{\sqrt{x^3 - 8}}$ Since it is under the square root and in the denominator. $x^3 - 8 > 0$.
2. Solve the inequation and state the maximal domain.	$x^3 > 8$ $\therefore x > 2$ The maximal domain is $(2, \infty)$.

6.5.2 Inverse relations and functions

The relation $A = \{(-1, 4), (0, 3), (1, 5)\}$ is formed by the mapping:

$$\begin{aligned} -1 &\rightarrow 4 \\ 0 &\rightarrow 3 \\ 1 &\rightarrow 5 \end{aligned}$$

The **inverse** relation is formed by 'undoing' the mapping where:

$$\begin{aligned} 4 &\rightarrow -1 \\ 3 &\rightarrow 0 \\ 5 &\rightarrow 1 \end{aligned}$$

The inverse of A is the relation $\{(4, -1), (3, 0), (5, 1)\}$.

The x- and y-coordinates of the points in relation A have been interchanged in its inverse. This causes the domains and ranges to be interchanged also.

Domain of $A = \{-1, 0, 1\}$ = range of its inverse

Range of $A = \{3, 4, 5\}$ = domain of its inverse

The equation of the inverse

The effect of the operation 'multiply by 2' is undone by the operation 'divide by 2'; the effect of the operation 'subtract 2' is undone by the operation 'add 2' and so on. These are examples of inverse operations, and we use these 'undoing' operations to solve equations. The 'undoing' operations can also be used to deduce the equation of the inverse of a given relation.

The 'multiply by 2' rule is that of the relation $y = 2x$, so its inverse with the 'divide by 2' rule is $y = \frac{x}{2}$.

Algebraically, interchanging x and y in the equation $y = 2x$ gives $x = 2y$. Rearranging this to make y the subject gives $y = \frac{x}{2}$, the equation of its inverse.

> **The inverse equation**
>
> **To obtain the equation of the inverse:**
> - **interchange x and y in the equation of the original relation or function**
> - **rearrange to make y the subject.**
>
> **Domains and ranges are interchanged between a pair of inverse relations.**

In some cases, the domain will need to be included when stating the equation of the inverse. For example, to find the equation of the inverse of the function $y = \sqrt{x}$, interchanging coordinates gives $x = \sqrt{y}$. Expressing $x = \sqrt{y}$ with y as the subject gives $y = x^2$. This rule is not unexpected, since 'square root' and 'squaring' are inverse operations. However, as the range of the function $y = \sqrt{x}$ is $[0, \infty)$, this must be the domain of its inverse. Hence, the equation of the inverse of $y = \sqrt{x}$ is $y = x^2$, with the restriction that $x \geq 0$.

Graphs of inverse relations

The graphs of the pair of inverses $y = 2x$ and $y = \frac{x}{2}$ show that their graphs are symmetric about the line $y = x$. The coordinates of symmetric points relative to $y = x$ are interchanged on each graph. Both graphs contain the point $(0, 0)$, since interchanging coordinates of this point does not change the point.

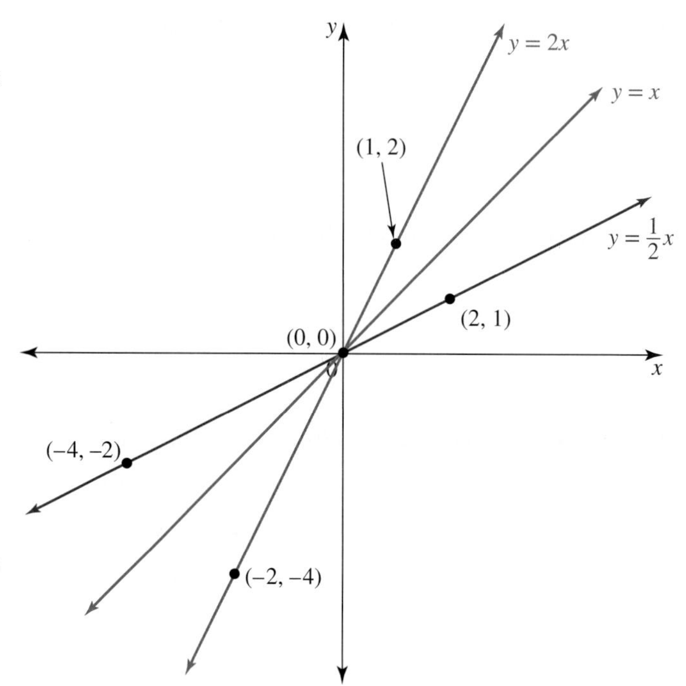

Given the graph of a function or another relation, reflection of this graph in the line $y = x$ will create the graph of its inverse.

The coordinates of known points, such as the axis intercepts, are interchanged by this reflection. If the graphs intersect, they will do so on the line $y = x$ since interchanging the coordinates of any point on $y = x$ would not cause any alteration to the coordinates.

Notation for inverse functions

Neither the original relation nor its inverse have to be functions. However, when both a relation and its inverse are functions, there is a special symbol for the name of the inverse.

Inverse function notation

If the inverse of a function f is itself a function, then the inverse function is denoted by f^{-1}.

The equation of the inverse of the square root function $f(x) = \sqrt{x}$ can be written as $f^{-1}(x) = x^2, x \geq 0$.

In mapping notation, if $f : [0, \infty) \to R, f(x) = \sqrt{x}$, then the inverse function is

$$f^{-1} : [0, \infty) \to R, f^{-1}(x) = x^2.$$

The domain of f^{-1} equals the range of f and the range of f^{-1} equals the domain of f.

Domain and range of $f^{-1}(x)$

Domain of $f^{-1}(x)$ = range of $f(x)$

Range of $f^{-1}(x)$ = domain of $f(x)$

Note that f^{-1} is a function notation and cannot be used for relations that are not functions. Note also that the inverse function f^{-1} and the reciprocal function $\frac{1}{f}$ represent different functions: $f^{-1} \neq \frac{1}{f}$.

WORKED EXAMPLE 12 Determining the domain and range of inverse functions

Consider the linear function $f: [0, 3) \to R, f(x) = 2x - 2$.

a. State the domain and determine the range of f.

b. State the domain and range of f^{-1}, the inverse of f.

c. Form the rule for the inverse and express the inverse function in mapping notation.

d. Sketch the graphs of $y = f(x)$ and $y = f^{-1}(x)$ on the same set of axes.

THINK	WRITE
a. State the domain and use its end points to determine the range.	**a.** $f: [0,3) \to R, f(x) = 2x - 2$ Domain $[0, 3)$ End points: when $x = 0, f(0) = -2$ (closed); when $x = 3, f(3) = 4$ (open). Therefore, the range is $[-2, 4)$.
b. State the domain and range of the inverse.	**b.** $d_{f^{-1}} = r_f$ $= [-2, 4)$ $r_{f^{-1}} = d_f$ $= [0, 3)$ The inverse has domain $[-2, 4)$ and range $[0, 3)$.
c. 1. Interchange x- and y-coordinates to form the equation of the inverse function.	**c.** $f(x) = 2x - 2$ Let $y = 2x - 2$. Function: $y = 2x - 2$ Inverse: $x = 2y - 2$
2. Rearrange the equation to make y the subject.	$x + 2 = 2y$ $\therefore y = \frac{x+2}{2}$ Therefore the rule for the inverse is $f^{-1}(x) = \frac{x+2}{2}$.
3. Express the inverse function as a mapping.	The inverse function is: $f^{-1}: [-2, 4) \to R, f^{-1}(x) = \frac{x+2}{2}$
d. Sketch each line using its end points and include the line $y = x$ as part of the graph. *Note:* The graph of f^{-1} could be deduced from the graph of f by interchanging coordinates of the end points.	**d.** 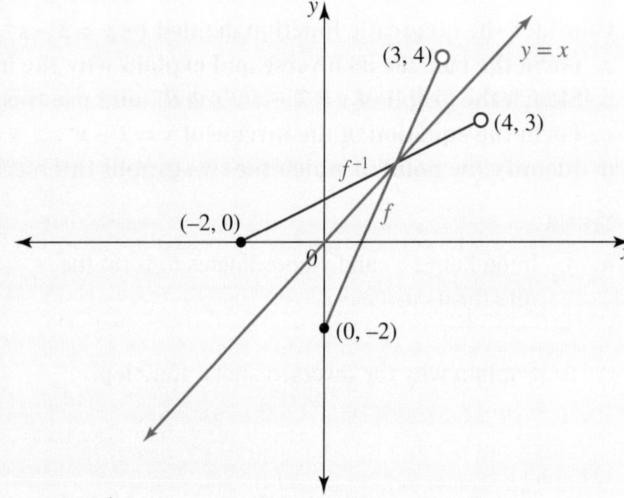 f and f^{-1} intersect on the line $y = x$.

on Resources

Interactivity Inverse functions (int-2575)

6.5.3 Condition for the inverse function f^{-1} to exist

The inverse of the parabola with equation $y = x^2$ can be obtained algebraically by interchanging x- and y-coordinates to give $x = y^2$ as the equation of its inverse.

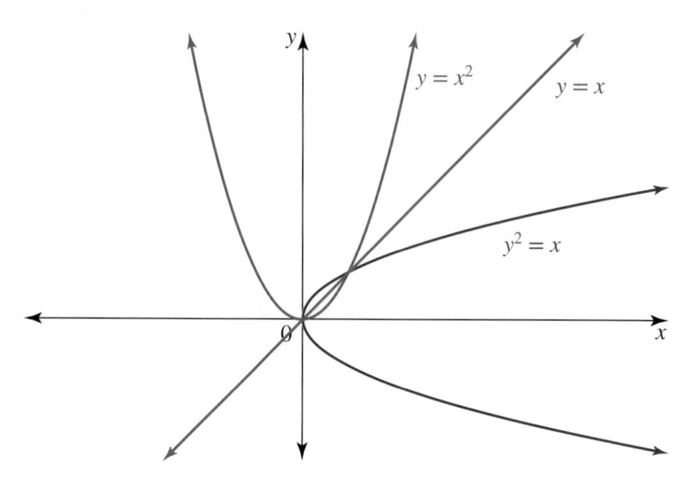

Graphically, the inverse of the parabola is obtained by reflecting the graph of the parabola in the line $y = x$. This not only illustrates that the sideways parabola, $y^2 = x$, is the inverse of the parabola $y = x^2$, but it also illustrates that although the parabola is a function, its inverse is not a function.

The reflection has interchanged the types of correspondence, as well as the domains and ranges. The many-to-one function $y = x^2$ has an inverse with a one-to-many correspondence and therefore its inverse is not a function. This has an important implication for functions.

Inverse function condition

For f^{-1} to exist, f must be a one-to-one function.

To ensure the inverse exists as a function, the domain of the original function may need to be restricted in order to ensure its correspondence is one-to-one.

WORKED EXAMPLE 13 Restricting the domain for inverse functions

Consider the quadratic function defined by $y = 2 - x^2$.

a. Form the rule for its inverse and explain why the inverse is not a function.
b. Sketch the graph of $y = 2 - x^2$, $x \in R^-$ and use this to sketch its inverse on the same diagram.
c. Form the equation of the inverse of $y = 2 - x^2$, $x \in R^-$.
d. Identify the point at which the two graphs intersect.

THINK	WRITE
a. 1. Interchange x- and y-coordinates to form the rule for its inverse.	a. $y = 2 - x^2$ Inverse: $x = 2 - y^2$ $\therefore y^2 = 2 - x$ is the rule for the inverse.
2. Explain why the inverse is not a function.	The quadratic function is many-to-one, so its inverse has a one-to-many correspondence. Therefore, the inverse is not a function.

b. 1. Sketch the graph of the function for the restricted domain.

b. $y = 2 - x^2, x \in R^-$
Domain: R^-
y-intercept: $(0, 2)$ is open since $x \in R^-$.
x-intercept: let $y = 0$.

$$2 - x^2 = 0$$
$$x^2 = 2$$
$$\therefore x = \pm\sqrt{2}$$
$$\Rightarrow x = -\sqrt{2} \text{ since } x \in R^-$$

x-intercept $(-\sqrt{2}, 0)$
Turning point $(0, 2)$

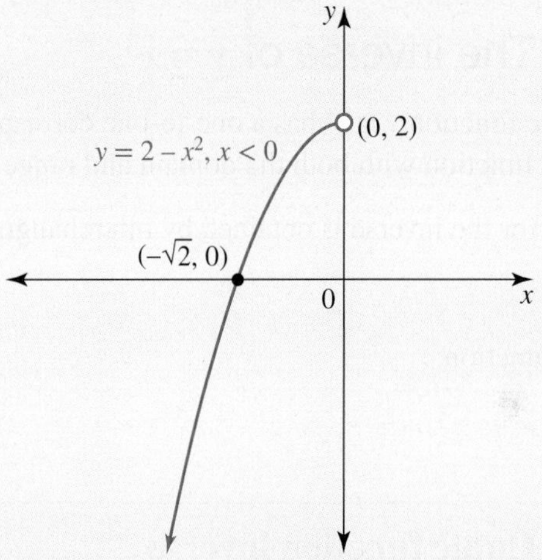

2. Deduce the key features of the inverse and sketch its graph and the line $y = x$ on the same diagram as the graph of the function.

For the inverse, $(2, 0)$ is an open point on the x-axis and $(0, -\sqrt{2})$ is the y-intercept. Its graph is the reflection of the graph of $y = 2 - x^2, x \in R^-$ in the line $y = x$.

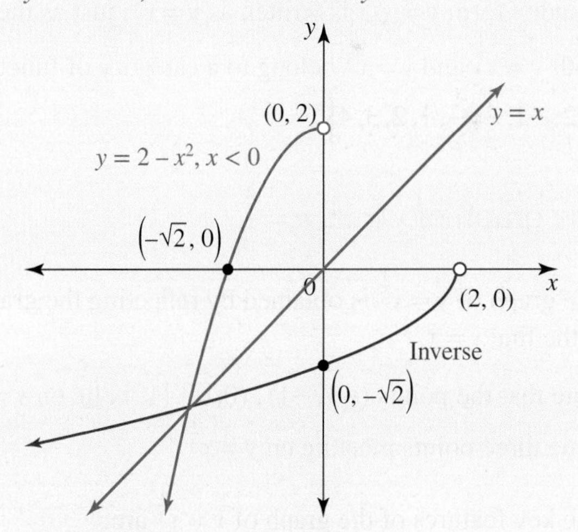

c. Use the range of the inverse to help deduce its equation.

c. From part **a**, the inverse of $y = 2 - x^2$ is:

$$y^2 = 2 - x$$
$$\therefore y = \pm\sqrt{2 - x}$$

The range of the inverse must be R^-, so the branch with the negative square root is required.
Therefore, the equation of the inverse is $y = -\sqrt{2 - x}$.

d. Choose two of the three equations that contain the required point and solve this system of simultaneous equations.

d. The point of intersection lies on $y = x, y = 2 - x^2$ and $y = -\sqrt{2 - x}$.
Using the first two equations, at intersection:
$x = 2 - x^2, x \in R^-$
$x^2 + x - 2 = 0$
$(x + 2)(x - 1) = 0$
$x = -2, x = 1$
Reject $x = 1$ since $x \in R^-$.
Therefore, $(-2, -2)$ is the point of intersection.

6.5.4 The inverse of $y = x^3$

The cubic function $y = x^3$ has a one-to-one correspondence, and both its domain and range are R. Its inverse will also be a function with both the domain and range of R.

The rule for the inverse is obtained by interchanging the x- and y-coordinates.

Function: $y = x^3$

Inverse function: $x = y^3$

$$\therefore y = \sqrt[3]{x}$$

Cubic function inverse

The inverse of the cubic function $y = x^3$ is the cube root function $f : R \to R, f(x) = \sqrt[3]{x}$.

In index form $y = \sqrt[3]{x}$ is written as $y = x^{\frac{1}{3}}$, just as the square root function $y = \sqrt{x}$ can be expressed as $y = x^{\frac{1}{2}}$. Both $y = x^{\frac{1}{3}}$ and $y = x^{\frac{1}{2}}$ belong to a category of functions known as power functions of the form $y = x^n$ where $n \in \left\{-2, -1, \frac{1}{3}, \frac{1}{2}, 1, 2, 3, 4\right\}$.

The graph of $y = x^n$, $n = \frac{1}{3}$

The graph of $y = x^{\frac{1}{3}}$ is obtained by reflecting the graph of $y = x^3$ in the line $y = x$.

Note that the points $(-1, -1)$, $(0, 0)$, $(1, 1)$ lie on $y = x^3$ so the same three points must lie on $y = x^{\frac{1}{3}}$.

The key features of the graph of $y = x^{\frac{1}{3}}$ are:

- point of inflection at $(0, 0)$
- as $x \to \infty, y \to \infty$, and as $x \to -\infty, y \to -\infty$
- domain R
- range R.

Unlike the square root function, the cube root function has domain R, since cube roots of negative numbers can be calculated.

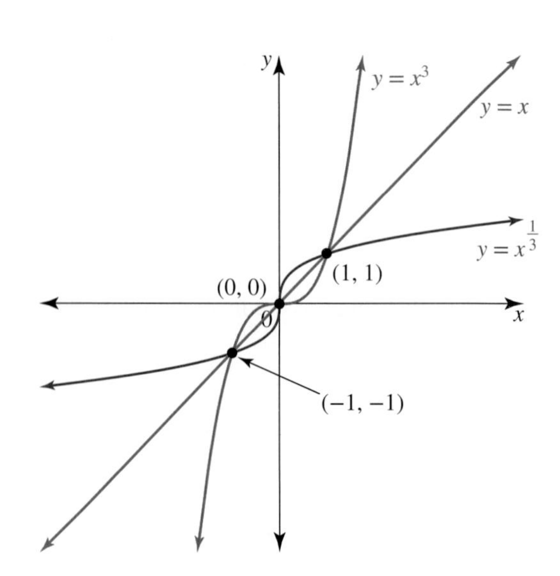

WORKED EXAMPLE 14 Sketching power functions

On the same set of axes sketch the graphs of $y = x^n$ for $n = \frac{1}{2}$ and $n = \frac{1}{3}$. Hence, state $\{x : x^{\frac{1}{3}} > x^{\frac{1}{2}}\}$.

THINK	WRITE
1. State the features of the first graph.	The function $y = x^n$ for $n = \frac{1}{2}$ is the square root function, $y = x^{\frac{1}{2}}$ or $y = \sqrt{x}$. End point: $(0, 0)$ Point: the point $(1, 1)$ lies on the graph. The domain requires $x \geq 0$.
2. State the features of the second graph.	The function $y = x^n$ for $n = \frac{1}{3}$ is the cube root function, $y = x^{\frac{1}{3}}$ or $y = \sqrt[3]{x}$. Point of inflection: $(0, 0)$ Point: the point $(1, 1)$ lies on the graph. The domain allows $x \in R$.
3. Sketch the graphs on the same set of axes.	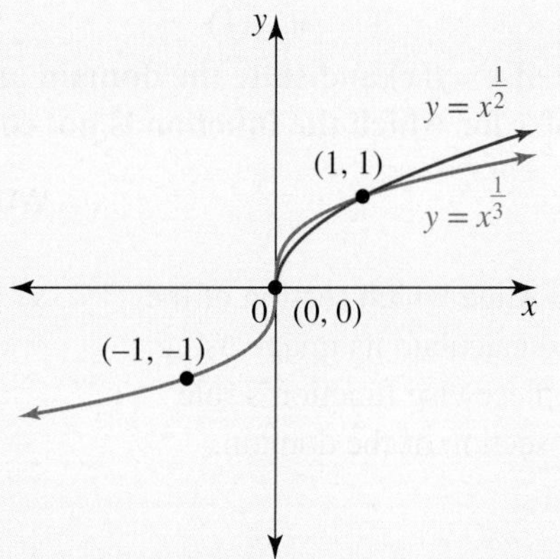
4. State the answer to the inequation.	For $x > 1$, the square root values are larger than the cube root values. For $0 < x < 1$, the cube root values are the larger. $\{x : x^{\frac{1}{3}} > x^{\frac{1}{2}}\}$ is $\{x : 0 < x < 1\}$

6.5.5 Piecewise (hybrid) functions

Note: Although piecewise (hybrid) functions are not part of the Units 1 & 2 course, they are included here in preparation for Units 3 & 4.

A **piecewise function** (also known as a **hybrid function**) is one in which the rule may take a different form over different sections of its domain. An example of a simple piecewise function is one defined by the rule:

$$y = \begin{cases} x, & x \geq 0 \\ -x, & x < 0 \end{cases}$$

Graphing this function would give a line with positive gradient to the right of the y-axis and a line with negative gradient to the left of the y-axis.

This piecewise function is continuous at $x = 0$ since both of its branches join, but that may not be the case for all piecewise functions. If the branches do not join, the function is not continuous for that value of x: it is discontinuous at that point of its domain.

Sketching a piecewise function is like sketching a set of functions with restricted domains all on the same graph. Each branch of the piecewise rule is valid only for part of the domain; if the branches do not join, it is important to indicate which end points are open and which are closed.

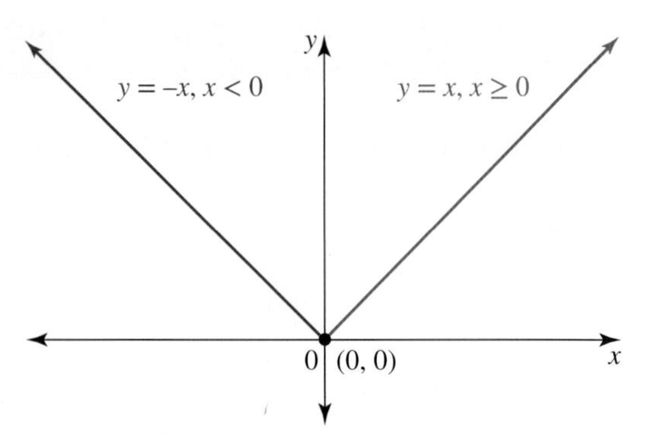

As for any function, each x-value can only be paired to exactly one y-value in a piecewise function. To calculate the corresponding y-value for a given value of x, the choice of which branch of the piecewise rule to use depends on which section of the domain the x-value belongs to.

WORKED EXAMPLE 15 Sketching piecewise functions

Consider the function:

$$f(x) = \begin{cases} x^2, & x < 1 \\ -x, & x \geq 1 \end{cases}$$

a. **Evaluate the following.**
   i. $f(-2)$ ii. $f(1)$ iii. $f(2)$.

b. **Sketch the graph of $y = f(x)$ and state the domain and range.**

c. **State any value of x for which the function is not continuous.**

THINK

a. Decide for each x-value which section of the domain it is in and calculate its image using the branch of the piecewise function's rule applicable to that section of the domain.

WRITE

a. i. $f(x) = \begin{cases} x^2, & x < 1 \\ -x, & x \geq 1 \end{cases}$

$f(-2)$: Since $x = -2$ lies in the domain section $x < 1$, use the rule $f(x) = x^2$.

$f(-2) = (-2)^2$

$\therefore f(-2) = 4$

ii. $f(1)$: Since $x = 1$ lies in the domain section $x \geq 1$, use the rule $f(x) = -x$.

$\therefore f(1) = -1$

iii. $f(2)$: Since $x = 2$ lies in the domain section $x \geq 1$, use the rule $f(x) = -x$.

$\therefore f(2) = -2$

THINK

b. 1. Sketch each branch over its restricted domain to form the graph of the piecewise function.

WRITE

b. Sketch $y = x^2, x < 1$.
Parabola, turning point $(0, 0)$, open end point $(1, 1)$
Sketch $y = -x, x \geq 1$.
Line, closed end point $(1, -1)$
Point $x = 2 \Rightarrow (2, -2)$

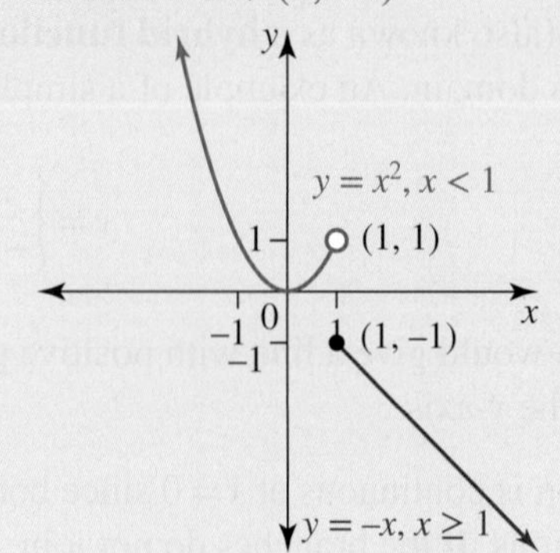

2. State the domain and range.

The domain is R.
The range is $(-\infty, -1] \cup [0, \infty)$.

c. State any value of x where the branches of the graph do not join.

c. The function is not continuous at $x = 1$ because there is a break in the graph.

6.5 Exercise

Technology free

1. **WE11** Identify the maximal domains of the functions with the following rules.

a. $y = \sqrt{3 - 2x}$ b. $y = \dfrac{1}{1 - 2x}$ c. $y = 2x^2 - 3x + 1$

2. Identify the maximal domains of the functions with the following rules.

a. $y = \dfrac{1}{16 - x^2}$ b. $y = \dfrac{2 - x}{x^2 + 3}$ c. $y = \sqrt{x^2 - 4}$

d. $y = \dfrac{1}{\sqrt{4 - x}}$ e. $y = \sqrt{x} + \sqrt{2 - x}$ f. $y = \dfrac{\sqrt{x^2 + 2}}{x^2 + 8}$

3. Give the implied domains of the functions defined by the following rules.

a. $f(x) = \dfrac{1}{x^2 + 5x + 4}$ b. $g(x) = \sqrt{x + 3}$ c. $h(x) = f(x) + g(x)$

4. Consider the set of points $S = \{(2, 9), (3, 10), (4, 11), (5, 12)\}$.
   a. State the domain and range of the set S.
   b. Write down the set of points which form the inverse of the set S and give its domain and range.
   c. Form a rule for the points in set S.
   d. State the rule for the points in the inverse set.

5. **WE12** Consider the linear function $f: (-\infty, 2] \rightarrow R, f(x) = 6 - 3x$.
   a. State the domain and determine the range of f.
   b. State the domain and range of f^{-1}, the inverse of f.
   c. Form the rule for the inverse and express the inverse function in mapping notation.
   d. Sketch the graphs of $y = f(x)$ and $y = f^{-1}(x)$ on the same set of axes.

6. For each of the following functions:
   i. state the domain and range
   ii. form the rule for the inverse
   iii. state the domain and range of the inverse.

a. $y = \dfrac{x + 3}{4}$ b. $y = 2 - \dfrac{5}{x}$

c. $y = 10 - 2x, x \in [-1, 2]$ d. $y = \sqrt{x - 2}$

7. Consider the linear function $f: R \to R, f(x) = 3x - 2$.
   a. Sketch the graph of $y = f(x)$.
   b. Hence, sketch the graph of its inverse.
   c. Determine the coordinates of the point of intersection of the graphs of $y = f(x)$ and its inverse $y = f^{-1}(x)$.
   d. Form the mapping for the inverse function f^{-1}.
8. Determine the rule for the inverse of each of the following.
   a. $4x - 8y = 1$
   b. $y = -\frac{2}{3}x - 4$
   c. $y = 4x^2$
   d. $y = \sqrt{2x + 1}$
9. Consider the hyperbola function with the rule $f(x) = \frac{1}{x - 2}$ defined on its maximal domain.
   a. State the maximal domain and the equations of the asymptotes of the function.
   b. Obtain the rule for its inverse.
   c. State the equations of the asymptotes of the inverse.
   d. For any hyperbola with equation $y = \frac{1}{x - a} + b$, determine the equations of the asymptotes of its inverse. Write down the equation of the inverse.
10. Consider $f: [-2, 4) \to R, f(x) = 4 - 2x$.
    a. State the domain and range of f^{-1}.
    b. Obtain the rule for f^{-1} and write f^{-1} as a mapping.
    c. Sketch the graphs of $y = f(x)$ and $y = f^{-1}(x)$.
    d. Determine the value of x for which $f(x) = f^{-1}(x)$.
11. Consider the function with rule $y = (x + 4)^2 + 2$.
    a. State the maximal domain and explain why the inverse of $y = (x + 4)^2 + 2$ is not a function.
    b. State the domain and range of the inverse of $f: [-4, 0] \to R, f(x) = (x + 4)^2 + 2$ and explain why the inverse is a function.
    c. Form the rule $f^{-1}(x)$.
12. a. Explain why the inverse of $g: R \to R, g(x) = (x - 1)^3 + 2$ will be a function.
    b. Form the inverse function of $g: R \to R, g(x) = (x - 1)^3 + 2$ and state the domain and range of this inverse.
13. a. State the range and the implied domain of $y = 6x - x^2$.
    b. Given $h: (-\infty, k] \to R, h(x) = 6x - x^2$, find the largest value of k so that the inverse of h will be a function.
14. **WE13** Consider the quadratic function $y = (x + 1)^2$ defined on its maximal domain.
    a. Form the rule for its inverse and explain why the inverse is not a function.
    b. Sketch the graph of $y = (x + 1)^2, x \in [-1, \infty)$ and use this to sketch its inverse on the same diagram.
    c. Form the equation of the inverse of $y = (x + 1)^2, x \in [-1, \infty)$.
    d. Identify the coordinates of the point at which the two graphs intersect.
15. Consider the square root function with equation $y = \sqrt{4 - x}$
    a. Sketch the square root function and its inverse on the same set of axes.
    b. State the restrictions that need to be placed on the inverse function.
16. Given $f: [a, \infty) \to R, f(x) = x^2 - 4x + 9$, find the smallest value of a so that f^{-1} exists.
17. Sketch the graphs of the following by reflecting the given relation in the line $y = x$.
    a. $y = 1$ and its inverse
    b. $y = \sqrt{-x}$ and its inverse
    c. $y = \frac{1}{x^2}$ and its inverse
    d. $y = x^3 - 1$ and its inverse

18. **WE14** On the same set of axes sketch the graphs of $y = x^n$ for $n = 1$ and $n = \frac{1}{3}$. Hence, state $\{x : x^{\frac{1}{3}} > x\}$.

19. a. Calculate the coordinates of the points of intersection of the two functions $y = x^n$, $n = -1$ and $n = \frac{1}{3}$.

b. Sketch the graphs of $y = x^n$, $n = -1$ and $n = \frac{1}{3}$ on the same set of axes.

20. **WE15** Consider piecewise function $f(x) = \begin{cases} x^3, & x < 1 \\ 2, & x \geq 1 \end{cases}$.

a. Evaluate the following.

i. $f(-2)$ ii. $f(1)$ iii. $f(2)$

b. Sketch the graph of $y = f(x)$ and state the domain and range.

c. State any value of x for which the function is not continuous.

Technology active

21 Sketch the graph of each of the following piecewise functions and state the domain and range.

a. $y = \begin{cases} 2, & x \leq 0 \\ 1 + x^2, & x > 0 \end{cases}$

b. $y = \begin{cases} x^3, & x < 1 \\ x, & x \geq 1 \end{cases}$

c. $y = \begin{cases} -2x, & x < -1 \\ 2, & -1 \leq x \leq 1 \\ 2x, & x > 1 \end{cases}$

d. $y = \begin{cases} \dfrac{1}{x-1}, & x < 1 \\ \dfrac{1}{2-x}, & x > 2 \end{cases}$

e. $y = \begin{cases} x^{\frac{1}{3}}, & x \geq 0 \\ -x^{-2}, & x < 0 \end{cases}$

f. $y = \begin{cases} x^3, & x < -1 \\ \sqrt[3]{x}, & -1 \leq x \leq 1 \\ 3x, & x > 1 \end{cases}$

22 Consider $f : R \to R$, $f(x) = \begin{cases} -x - 1, & x < -1 \\ \sqrt{2(1-x)}, & -1 \leq x \leq 1 \\ x + 1, & x > 1 \end{cases}$

a. Calculate the values of:

i. $f(0)$ ii. $f(3)$ iii. $f(-2)$ iv. $f(1)$

b. Show the function is not continuous at $x = 1$ or $x = -1$.

c. Sketch the graph of $y = f(x)$ and state the type of correspondence.

23 Form a rule for the graph of each piecewise function.

a.

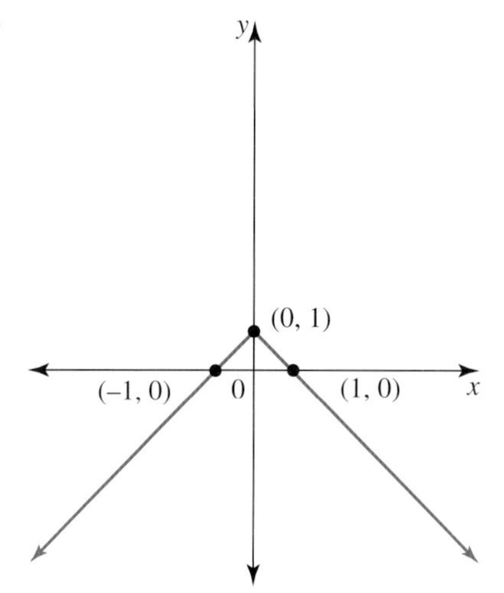

b.

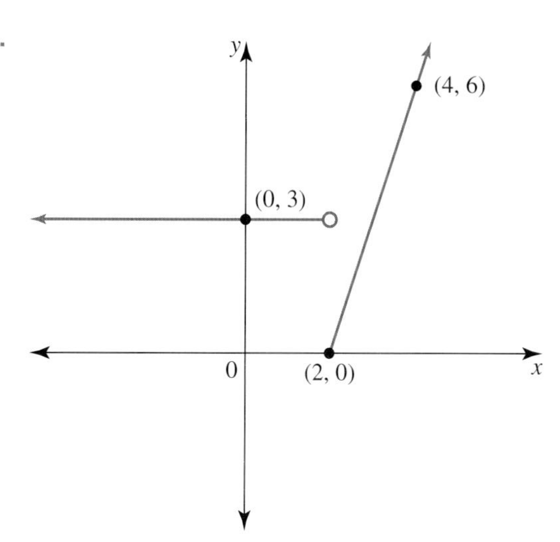

24 a. Consider the function defined by $f(x)=\begin{cases}4x+a, & x<1\\ \dfrac{2}{x}, & 1\le x\le 4\end{cases}$

i. Determine the value of a so the function will be continuous at $x=1$.

ii. Explain whether the function is continuous at $x=0$.

b. Determine the values of a and b so that the function with the rule

$$f(x)=\begin{cases}a, & x\in(-\infty,-3]\\ x+2, & x\in(-3,3)\\ b, & x\in[3,\infty)\end{cases}$$

is continuous for $x\in R$; for these values, sketch the graph of $y=f(x)$.

25 In an effort to reduce the time her children spend in the shower, a mother introduced a penalty scheme with fines to be paid from the children's pocket money according to the following: If someone spends more than 5 minutes in the shower, the fine in dollars is equal to the shower time in minutes; if someone spends up to and including 5 minutes in the shower, there is no fine. If someone chooses not to shower at all, there is a fine of \$2 because that child won't be nice to be near.

Defining appropriate symbols, express the penalty scheme as a mathematical rule in piecewise form and sketch the graph which represents it.

26 Consider the function $g: D\to R, g(x)=x^2+8x-9$.

a. Give two possible domains D for which g^{-1} exists.

b. If the domain of g is chosen to be R^+, form g^{-1} and state its range.

c. i. For this domain R^+, sketch the graphs of $y=g(x)$ and $y=g^{-1}(x)$ on the same set of axes.

ii. Calculate the exact coordinates of any point(s) of intersection of the two graphs.

d. Identify the number that is its own image under the mapping g^{-1}.

27 A function is defined by the rule $f(x)=\begin{cases}\sqrt{x-4}+2, & x>4\\ 2, & 1\le x\le 4.\\ \sqrt{1-x}+2, & x<1\end{cases}$

a. Calculate the values of $f(0)$ and $f(5)$.

b. Sketch $y=f(x)$ and state its domain and range.

c. Determine the values of x for which $f(x)=8$.

d. State whether the inverse of this function will also be a function. Draw a sketch of the graph of the inverse on a new set of axes.

28 a. Use CAS technology to find algebraically the rule for the inverse of $y=x^2+5x-2$.

b. Use CAS technology to draw the graph of $y=x^2+5x-2$ and its inverse and find, to 2 decimal places, the coordinates of the points of intersection of $y=x^2+5x-2$ with its inverse.

6.5 Exam questions

Question 1 (1 mark) TECH-ACTIVE

MC The maximal domain of the function with the rule $f(x)=3+\dfrac{1}{2x-1}$ is

A. $R\backslash\left\{\frac{1}{3}\right\}$ **B.** $R\backslash\left\{\frac{1}{2}\right\}$ **C.** $R\backslash\left\{-\frac{1}{2}\right\}$ **D.** $R\backslash\{1\}$ **E.** $R\backslash\{-1\}$

Question 2 (1 mark) TECH-ACTIVE

MC The inverse function $f^{-1}(x)$ of $f: (2, \infty) \rightarrow R$, where $f(x) = x - 2$, is given by

A. $f^{-1}: [2, \infty) \rightarrow R$ where $f^{-1}(x) = x + 2$.
B. $f^{-1}: R^{+} \rightarrow R$ where $f^{-1}(x) = x - 2$.
C. $f^{-1}: R^{+} \rightarrow R$ where $f^{-1}(x) = x + 2$.
D. $f^{-1}: R \rightarrow R$ where $f^{-1}(x) = x + 2$.
E. $f^{-1}: [2, \infty)$ where $f^{-1}(x) = x - 2$.

Question 3 (1 mark) TECH-ACTIVE

MC The function $f: (-\infty, a] \rightarrow R, f(x) = x^2 + 4x - 5$ has an inverse function if

A. $a < 2$
B. $a = 2$
C. $a \geq 0$
D. $a \leq -2$
E. $a \geq -5$

More exam questions are available online.

6.6 Transformations of functions

LEARNING INTENTION

At the end of this subtopic you should be able to:
- identify and sketch transformations such as dilation, reflection and translation of functions.

Once the basic shape of a function is known, its features can be identified after various **transformations** have been applied to it simply by interpreting the transformed equation of the image. For example, we recognise the equation $y = (x - 1)^2 + 3$ to be a parabola formed by translating the basic parabola $y = x^2$ horizontally 1 unit to the right and vertically 3 units up. In exactly the same way, we recognise $y = \frac{1}{x-1} + 3$ to be a hyperbola formed by translating the basic hyperbola $y = \frac{1}{x}$ horizontally 1 unit to the right and vertically 3 units up.

In this section we extend such interpretation of the transformed equation to consider the image of a general function $y = f(x)$ under a sequence of transformations.

6.6.1 Horizontal and vertical translations of $y = f(x)$

Translations parallel to the x- and y-axes move graphs horizontally to the left or right and vertically up or down, respectively.

Under a **horizontal translation** of h units to the right, the following effect is seen:

$$y = x^2 \quad \rightarrow \quad y = (x - h)^2;$$
$$y = \frac{1}{x} \quad \rightarrow \quad y = \frac{1}{x - h};$$
$$y = \sqrt{x} \quad \rightarrow \quad y = \sqrt{x - h};$$

and so, for any function, $y = f(x) \rightarrow y = f(x - h)$.

Under a **vertical translation** of k units upwards:

$$y = x^2 \quad \rightarrow \quad y = x^2 + k;$$
$$y = \frac{1}{x} \quad \rightarrow \quad y = \frac{1}{x} + k;$$
$$y = \sqrt{x} \quad \rightarrow \quad y = \sqrt{x} + k;$$

and so, for any function, $y = f(x) \rightarrow y = f(x) + k$.

Translation of functions

- **$y=f(x-h)$ is the image of $y=f(x)$ under a horizontal translation of h units to the right.**
- **$y=f(x)+k$ is the image of $y=f(x)$ under a vertical translation of k units upwards.**
- **$y=f(x-h)+k$ is the image of $y=f(x)$ under the combined transformations of h units parallel to the x-axis and k units parallel to the y-axis.**

WORKED EXAMPLE 16 Sketching horizontal transformations

The diagram shows the graph of $y=f(x)$ passing through points $(-2, 0)$ $(0, 2)$ $(3, 1)$.
Sketch the graph of $y = f(x+1)$ using the images of these three points.

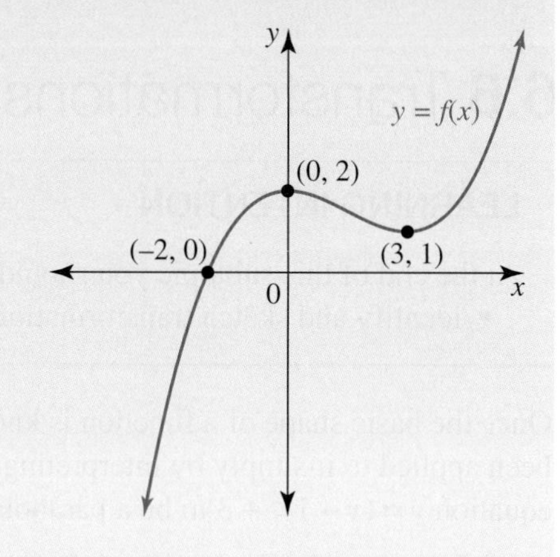

THINK	WRITE
1. Identify the transformation required.	$y=f(x+1)$ This is a horizontal translation 1 unit to the left of the graph of $y=f(x)$.
2. Find the image of each key point.	Under this transformation: $(-2,0) \rightarrow (-3,0)$ $(0,2) \rightarrow (-1,2)$ $(3,1) \rightarrow (2,1)$
3. Sketch the image.	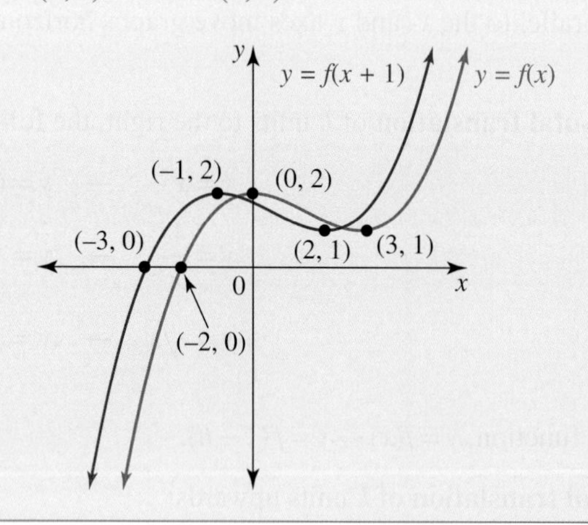

on Resources

Interactivity Transformations of functions (int-2576)

6.6.2 Reflections in the coordinate axes

The point (x, y) becomes $(x, -y)$ when reflected in the x-axis and $(-x, y)$ when reflected in the y-axis.

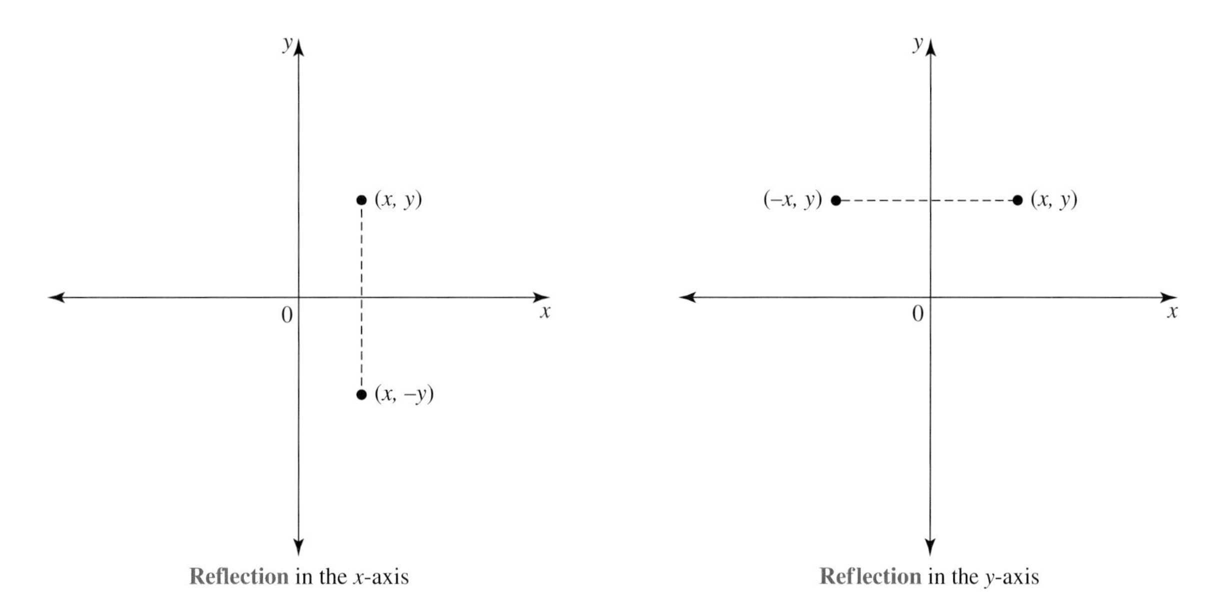

Under a **reflection in the x-axis**, the following effect is seen:

$$y = x^2 \quad \rightarrow \quad y = -x^2$$

$$y = \sqrt{x} \quad \rightarrow \quad y = -\sqrt{x}$$

$$y = \frac{1}{x^2} \quad \rightarrow \quad y = -\frac{1}{x^2}$$

and so for any function $y = f(x) \rightarrow y = -f(x)$.

Under a **reflection in the y-axis**, the following effect is seen:

$$y = x^2 \quad \rightarrow \quad y = (-x)^2 = x^2$$

$$y = \sqrt{x} \quad \rightarrow \quad y = \sqrt{-x}$$

$$y = x^3 \quad \rightarrow \quad y = (-x)^3$$

and so for any function $y = f(x) \rightarrow y = f(-x)$.

Reflection of functions

- **$y = -f(x)$ is the image of $y = f(x)$ under a reflection in the x-axis.**
- **$y = f(-x)$ is the image of $y = f(x)$ under a reflection in the y-axis.**

WORKED EXAMPLE 17 Sketching the reflection of a function

Consider again the graph from Worked example 16.
Sketch the graph of $y = f(-x)$ using the images of the points $(-2, 0), (0, 2), (3, 1)$.

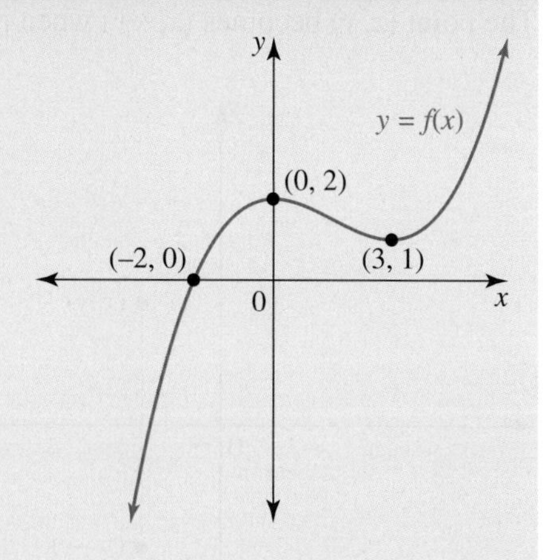

THINK	WRITE
1. Identify the transformation required.	$y = f(-x)$ This is a reflection in the y-axis of the graph of $y = f(x)$.
2. Find the image of each key point.	Under this transformation, $(x, y) \rightarrow (-x, y)$ $(-2, 0) \rightarrow (2, 0)$ $(0, 2) \rightarrow (0, 2)$ $(3, 1) \rightarrow (-3, 1)$
3. Sketch the image.	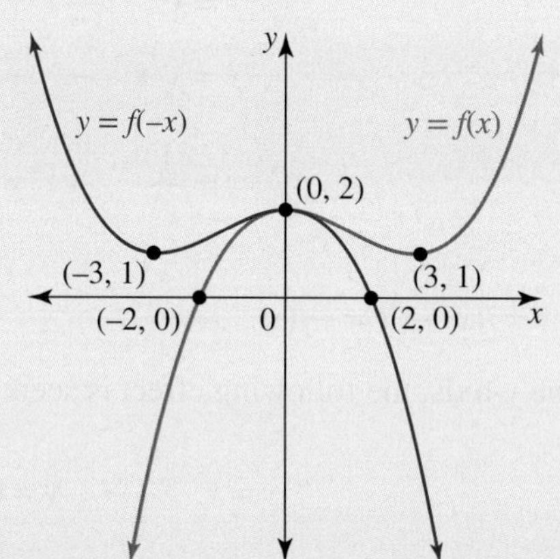

6.6.3 Dilations from the coordinate axes

A dilation from an axis either stretches or compresses a graph from that axis, depending on whether the dilation factor is greater than 1 or between 0 and 1, respectively.

Dilation from the x-axis by factor a

Under a **dilation from the x-axis** by a factor of a, the following effect is seen:

$$y = x^2 \rightarrow y = ax^2$$

$$y = \sqrt{x} \rightarrow y = a\sqrt{x}$$

$$y = \frac{1}{x^2} \rightarrow y = \frac{a}{x^2}$$

and so for any function $y = f(x) \rightarrow y = af(x)$.

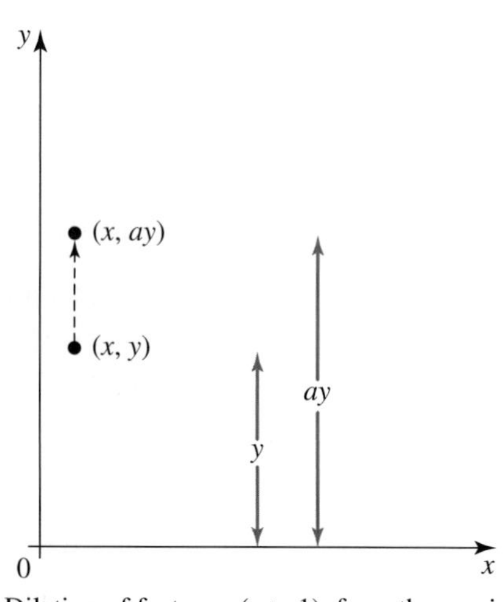

Dilation of factor a, ($a > 1$), from the x-axis

Dilation of functions from the x-axis

- **$y = af(x)$ is the image of $y = f(x)$ under a dilation of factor a from the x-axis, parallel to the y-axis.**

Dilation from the y-axis by factor a

A dilation from the y-axis acts parallel to the x-axis, or in the x-direction.

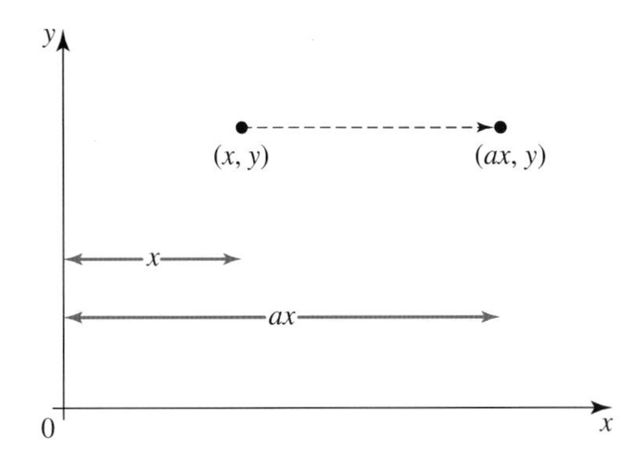

The point $(x, y) \rightarrow (ax, y)$ when dilated by a factor a from the y-axis. To see the effect of this dilation, consider the graph of $y = x(x-2)$ under a dilation of factor 2 from the y-axis. Choosing the key points, under this dilation:

$$\begin{array}{ccc} (0,0) & \rightarrow & (0,0) \\ (1,-1) & \rightarrow & (2,-1) \\ (2,0) & \rightarrow & (4,0) \end{array}$$

and the transformed graph is as shown.

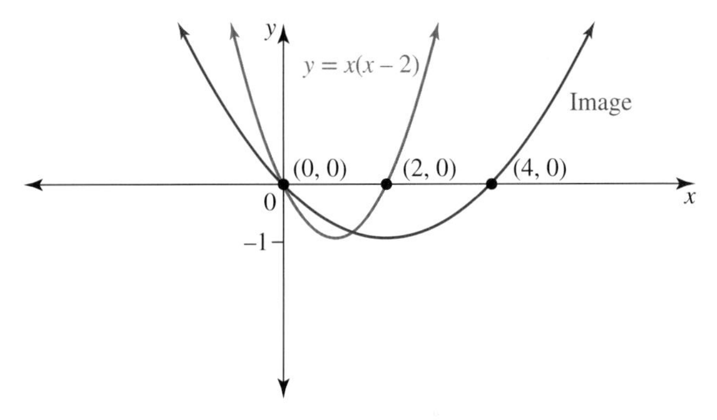

The equation of the image of $y = x(x-2)$ can be found by fitting the points to a quadratic equation. Its equation is

$$y = (0.5x)(0.5x-2) \Rightarrow y = \left(\frac{x}{2}\right)\left(\left(\frac{x}{2}\right)-2\right)$$

Under a **dilation from the y-axis** by a factor of a, the following effect is seen:

$$\begin{array}{ccc} y = x^2 & \rightarrow & y = \left(\dfrac{x}{a}\right)^2 \\ y = \sqrt{x} & \rightarrow & y = \sqrt{\dfrac{x}{a}} \\ y = \dfrac{1}{x^2} & \rightarrow & y = \dfrac{1}{\left(\frac{x}{a}\right)^2} \end{array}$$

and so for any function $y = f(x) \rightarrow y = f\left(\dfrac{x}{a}\right)$.

Dilation of functions from the y-axis

For any function:

- **$y = f(ax)$ is the image of $y = f(x)$ under a dilation of factor $\dfrac{1}{a}$ from the y-axis, parallel to the x-axis.**
- **$y = f\left(\dfrac{x}{a}\right)$ is the image of $y = f(x)$ under a dilation of factor a from the y-axis, parallel to the x-axis.**

Dilation of functions from both axes

- **Under a dilation of factor a from the x-axis (in the y-direction), $y = f(x) \rightarrow y = af(x)$.**
- **Under a dilation of factor a from the y-axis (in the x-direction), $y = f(x) \rightarrow y = f\left(\frac{x}{a}\right)$.**

WORKED EXAMPLE 18 Sketching dilations of a graph

For the graph given in Worked examples 16 and 17, sketch the graph of $y = f(2x)$ using the images of the points $(-2, 0), (0, 2), (3, 1)$.

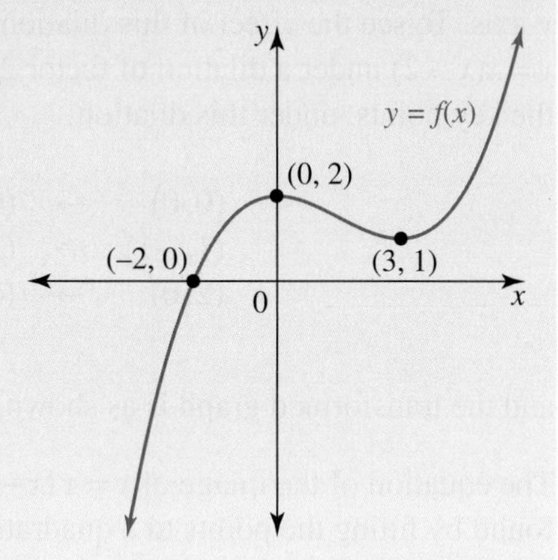

THINK	WRITE
1. Identify the transformation.	$y = f(2x) \Rightarrow y = f\left(\dfrac{x}{\frac{1}{2}}\right)$ The transformation is a dilation from the y-axis of factor $\frac{1}{2}$. This dilation acts in the x-direction.
2. Find the image of each key point.	Under this dilation, $(x, y) \rightarrow \left(\frac{x}{2}, y\right)$. $(-2, 0) \rightarrow (-1, 0)$ $(0, 2) \rightarrow (0, 2)$ $(3, 1) \rightarrow (1.5, 1)$
3. Sketch the image.	

y = f(2x) y = f(x)
y
(0, 2)
(−2, 0) (1.5, 1) (3, 1)
0
x
(−1, 0)

6.6.4 Combinations of transformations

The graph of $y = a(x + b)^n + c$ where $a, b, c \in R$ and $a \neq 0$ is the graph of $y = x^n$ where $n \in \left\{-2, -1, \frac{1}{3}, \frac{1}{2}, 1, 2, 3, 4\right\}$ under a set of simple transformations that are identified as follows.

- a gives the dilation factor $|a|$ from the x-axis or parallel to the y-axis.
- If $a < 0$, there is a reflection in the x-axis.
- b gives the horizontal translation parallel to the x-axis:
  - if $b > 0$, it is a horizontal translation to the left
  - if $b < 0$, it is a horizontal translation to the right.
- c gives the vertical translation parallel to the y-axis:
  - if $c > 0$, it is a vertical translation upwards
  - if $c < 0$, it is a vertical translation downwards.

When applying transformations to $y = f(x)$ to form the graph of $y = af(x + b) + c$, the order of operations can be important, so any dilation or reflection should be applied before any translation.

It is quite possible that more than one order or set of transformations may achieve the same image. For example, $y = 4x^2$ can be considered a dilation of $y = x^2$ by a factor of 4 from the x-axis. Alternatively, as $y = 4x^2 = (2x)^2$, it is also a dilation of $y = x^2$ by a factor of $\frac{1}{2}$ from the y-axis.

WORKED EXAMPLE 19 Describing transformations

a. Describe the transformations applied to the graph of $y = x^3$ to obtain:

i. $y = 4(x - 7)^3$

ii. $y = 6 - 2(x + 5)^3$

b. Describe the transformations applied to the graph of $y = f(x)$ to obtain $y = -3f(x - 8) - 9$.

THINK	WRITE
a. i. Identify the transformations in the correct order of dilation and translation.	a. $y = 4(x - 7)^3$ Dilation of factor 4 from the x-axis, followed by a horizontal translation of 7 units to the right
ii. 1. Express the image equation in the general form.	$y = 6 - 2(x + 5)^3$ $y = -2(x + 5)^3 + 6$
2. Identify the transformations in the correct order of dilation, reflection and translation.	Dilation of factor 2 from the x-axis, followed by a reflection in the x-axis, then a horizontal translation of 5 units to the left and a vertical translation upwards of 6 units
b. Identify the transformations in the correct order of dilation, reflection and translation.	b. $y = -3f(x - 8) - 9$ Dilation of 3 units from the x-axis and a reflection in the x-axis, followed by a horizontal translation of 8 units to the right and a vertical translation downwards of 9 units

WORKED EXAMPLE 20 Equation of an image from given transformations

The graph of $y=\sqrt{x}$ undergoes two transformations in the order given: dilation of factor 5 from the x-axis, followed by a horizontal translation of 6 units to the left.

a. Give the equation of the image.

b. Describe the sequence of transformations that need to be applied to the image to undo the effect of the transformations and revert to the graph of $y=\sqrt{x}$.

THINK	WRITE
a. 1. Consider the general form.	a. $y=a(x+b)^n+c$
2. Dilation of factor 5 from the x-axis, $a=5$	$y=\sqrt{x} \rightarrow y=5\sqrt{x}$
3. Horizontal translation to the left, $b=6$	$\rightarrow y=5\sqrt{(x+6)}$ The image of $y=\sqrt{x}$ under the given set of transformations is $y=5\sqrt{(x+6)}$.
b. 1. Undo the horizontal translation to the left.	b. $y=5\sqrt{(x+6)} \rightarrow y=5\sqrt{x}$
2. Undo the dilation of factor 5 from the x-axis.	$\rightarrow y=\sqrt{x}$
3. State the required transformations in the required order.	Undoing the transformations requires the image to undergo a horizontal translation of 6 units to the right followed by a dilation of factor $\frac{1}{5}$ from the x-axis.

6.6 Exercise

Students, these questions are even better in jacPLUS

Receive immediate feedback and access sample responses

Access additional questions

Track your results and progress

Find all this and MORE in jacPLUS

Technology free

1. Identify the transformations that would be applied to the graph of $y=x^2$ to obtain each of the following graphs.

   a. $y=3x^2$ b. $y=-x^2$ c. $y=x^2+5$ d. $y=(x+5)^2$

2. a. After the graph of $y=x^2$ is translated 3 units upwards and 2 units to the left, state what its equation becomes.

   b. State the equation that represents a transformation of $y=\frac{1}{x}$ by a dilation of factor 2 parallel to the y-axis and a horizontal translation to the right of 5 units.

   c. The graph of $y=\sqrt{x}$ is reflected in the x-axis and dilated by a factor of $\frac{1}{2}$ from the x-axis. State what its equation becomes.

   d. After the graph of $y=x^4$ is moved downwards 4 units and then reflected in the x-axis, state what its equation becomes.

3. Describe the transformations required to change:

a. $y=\sqrt{x}$ into $y=\sqrt{x+4}-5$

b. $y=\dfrac{1}{x}$ into $y=\dfrac{3}{x-2}+4$

c. $y=x^3$ into $y=2-(x-3)^3$

d. $y=\dfrac{1}{x^2}$ into $y=\dfrac{-3}{2(x+1)^2}+4$

e. $y=f(x)$ into $y=f(x+1)+2$

f. $y=f(x)$ into $y=-4f(x)$.

4. **WE16, WE17** For the graph of $y=f(x)$ shown, sketch the following graphs using the images of the points $(-2, 0), (0, 2), (3, 1)$.

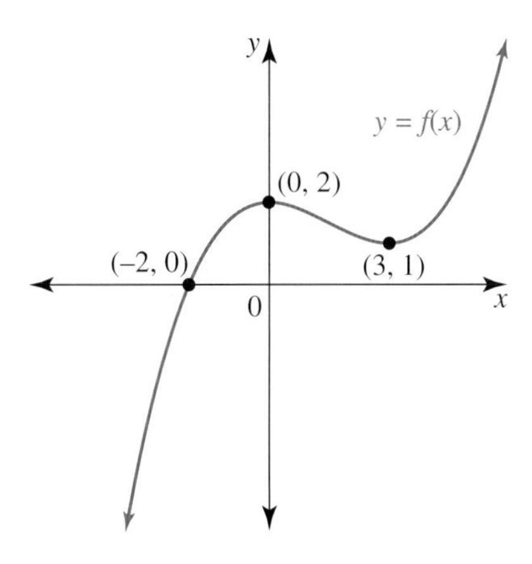

a. $y=f(x)-2$

b. $y=f(x-2)+1$

c. $y=-f(x)$

5. a. The parabola with equation $y=(x-1)^2$ is reflected in the x-axis followed by a vertical translation upwards of 3 units. State the equation of its final image.

b. Obtain the equation of the image if the order of the transformations in part a was reversed. State whether the image is the same as that in part a.

6. Describe the transformations that have been applied to the graph of $y=x^3$ to obtain each of the following graphs.

a. $y=\left(\dfrac{x}{3}\right)^3$

b. $y=8x^3+1$

c. $y=(x-4)^3-4$

d. $y=-2(x+1)^3$

7. **WE18** For the graph shown, sketch the following graphs using the images of the points $(-2, 0), (0, 2), (3, 1)$.

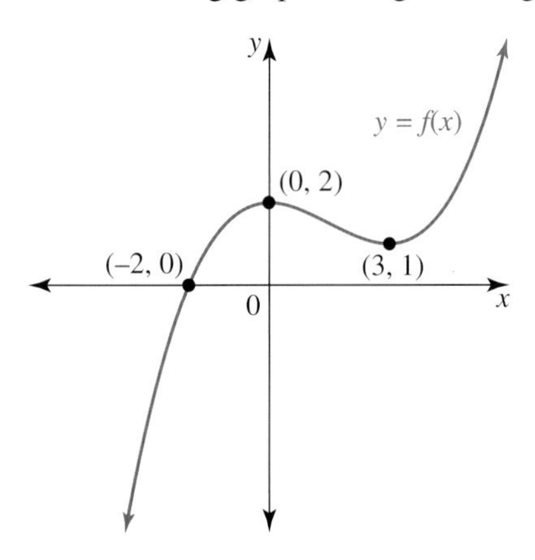

a. $y=f\left(\dfrac{x}{2}\right)$

b. $y=\dfrac{1}{2}f(x)$

8. Give the equations of the images of:

a. $y = \sqrt{x}$

b. $y = x^4$

if their graphs are:

i. dilated by a factor of 2 from the x-axis

ii. dilated by a factor of 2 from the y-axis

iii. reflected in the x-axis and then translated 2 units vertically upwards

iv. translated 2 units vertically upwards and then reflected in the x-axis

v. reflected in the y-axis and then translated 2 units to the right

vi. translated 2 units to the right and then reflected in the x-axis.

9. WE19 a. Describe the transformations applied to the graph of $y = x^4$ to obtain:

i. $y = 5(x+3)^4$

ii. $y = 8 - 7(x-6)^4$

b. Describe the transformations applied to the graph of $y = f(x)$ to obtain $y = 3f(x+4) - 15$.

10. Describe the transformations applied to the graph of $y = \dfrac{1}{x^2}$ to obtain:

a. $y = -\dfrac{1}{x^2} + 2$

b. $y = \dfrac{3}{(x+1)^2} - 5$

11. The graph of $y = f(x)$ is shown.

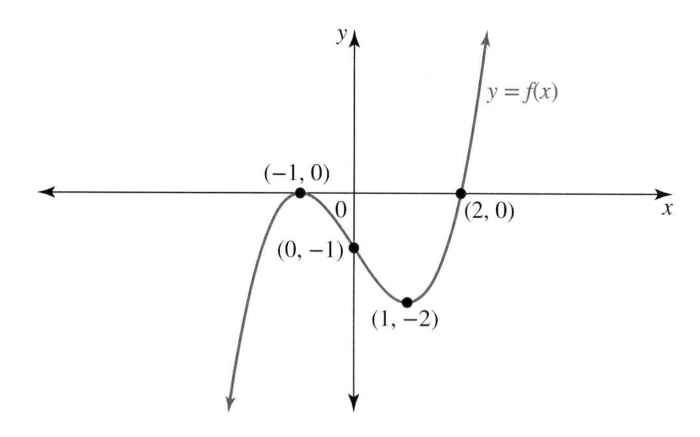

On separate diagrams sketch the graphs of the following.

a. $y = f(x-1)$

b. $y = -f(x)$

c. $y = 2f(x)$

d. $y = f(-x)$

e. $y = f\left(\dfrac{x}{2}\right)$

f. $y = f(x) + 2$

12. a. Describe the transformations applied to $y = f(x)$ if its image is $y = 2f(-x) + 1$.

b. Describe the transformations applied to $y = f(x)$ if its image is $y = -f(3x)$.

13. WE20 a. The graph of $y=\frac{1}{x}$ undergoes two transformations in the order given: dilation of factor 2 from the x-axis, followed by a horizontal translation of 3 units to the left.

Give the equation of its image.

b. Describe the sequence of transformations that need to be applied to the image to undo the effect of the transformations and revert to the graph of $y=\frac{1}{x}$.

14. Give the coordinates of the image of the point $(3, -4)$ if it is:

a. translated 2 units to the left and 4 units down

b. reflected in the y-axis and then reflected in the x-axis

c. dilated by a factor of $\frac{1}{5}$ from the x-axis parallel to the y-axis

d. dilated by a factor of $\frac{1}{5}$ from the y-axis parallel to the x-axis.

Technology active

15. a. i. Give the equation of the image of $y=\frac{1}{x}$ after the two transformations are applied in the order given: dilation of factor 3 from the x-axis, then reflection in the x-axis.

ii. Reverse the order of the transformations and give the equation of the image.

b. i. Give the equation of the image of $y=\frac{1}{x^2}$ after the two transformations are applied in the order given: dilation of factor 3 from the x-axis, then vertical translation 6 units up.

ii. Reverse the order of the transformations and give the equation of the image.

c. If $f(x)=\frac{1}{x^2}$, give the equations of the asymptotes of $y=-2f(x+1)$.

16. Describe the transformations applied to $y=f(x)$ if its image is:

a. $y=2f(x+3)$

b. $y=6f(x-2)+1$

c. $y=-2f(x)+3$

17. Form the equation of the image after the given functions have been subjected to the set of transformations in the order specified.

a. $y=\frac{1}{x^2}$ undergoes a dilation of factor $\frac{1}{3}$ from the x-axis followed by a horizontal translation of 3 units to the left.

b. $y=\frac{1}{x}$ undergoes a reflection in the x-axis followed by a horizontal translation of 1 unit to the right.

c. $y=\sqrt[3]{x}$ undergoes a horizontal translation of 1 unit to the right followed by a dilation of factor 2 from the x-axis.

18. a. The function $g : R \rightarrow R, g(x)=x^2-4$ is reflected in the y-axis. Describe its image.

b. Show that the image of the function $f : R \rightarrow R, f(x)=x^{\frac{1}{3}}$ when it is reflected in the y-axis is the same as when it is reflected in the x-axis.

c. The graph of $y=(x-2)^2+5$ is reflected in both the x-axis and the y-axis. State the nature, and the coordinates, of the turning point of its image.

d. A curve $y=f(x)$ is dilated by a factor of 2 from the x-axis, then vertically translated 1 unit up, then reflected in the x-axis. After these three transformations have been applied, the equation of its image is $y=6(x-2)^3-1$. Determine the equation of $y=f(x)$.

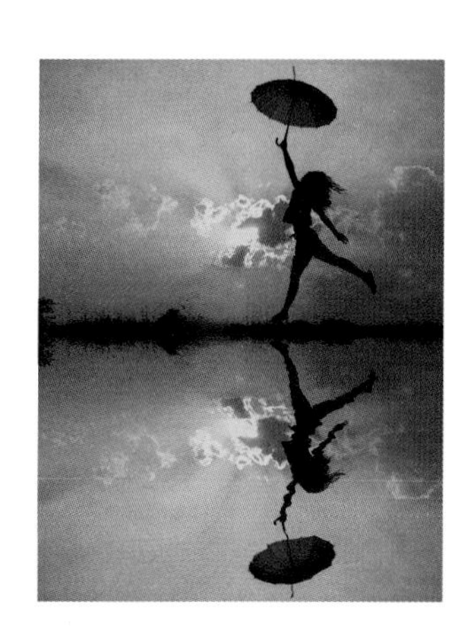

19. The graph of the function $y = g(x)$ is given.

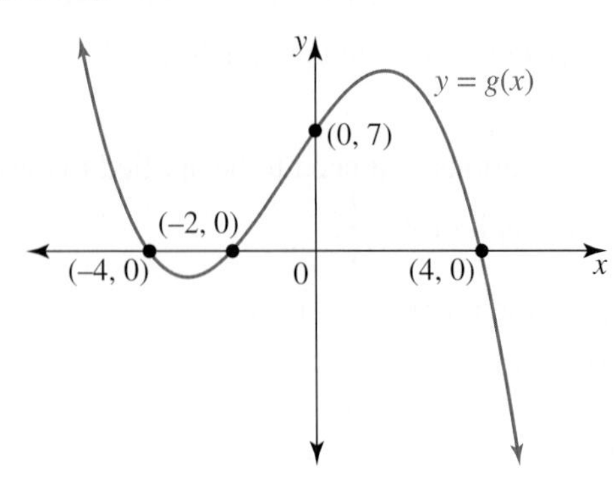

a. Sketch the graph of $y = -g(2x)$.
b. Sketch the graph of $y = g(2 - x)$.
c. Determine the values of h for which all the x-intercepts of the graph of $y = g(x + h)$ will be negative.
d. Give a possible equation for the graph of $y = g(x)$ and hence find an expression for $g(2x)$.

20. Using CAS technology, on the same screen, graph $y_1 = x^2 + 5x - 6$, $y_2 = (2x)^2 + 5(2x) - 6$, and $y_3 = \left(\frac{x}{2}\right)^2 + 5\left(\frac{x}{2}\right) - 6$, and compare the graphs. State which of these graphs are parabolas.

6.6 Exam questions

Question 1 (1 mark) TECH-ACTIVE

MC The image of the point $(-6, 3)$ dilated parallel to the x-axis by a factor of $\frac{1}{3}$ is

A. $(-2, 3)$
B. $(2, 3)$
C. $(-6, 1)$
D. $(-6, -1)$
E. $(-6, -9)$

Question 2 (3 marks) TECH-FREE

Form the image of the graph $y = \sqrt{x}$ dilated by a factor of 2 parallel to the y-axis, reflected in the x-axis and translated 3 units to the left.

Question 3 (3 marks) TECH-FREE

Describe the transformation of $y = f(x)$ required to obtain $y = -3 - 4f(x - 1)$.

More exam questions are available online.

6.7 Review

6.7.1 Summary

doc-37021

Hey students! Now that it's time to revise this topic, go online to:

 Access the topic summary

 Review your results

 Watch teacher-led videos

 Practise exam questions

Find all this and MORE in jacPLUS

6.7 Exercise

Technology active: short answer

1. Sketch each of the following, stating the domain and range.

   a. $y = -\sqrt{4-x}$

   b. $y = 4 + \sqrt[3]{x}$

   c. $y = \dfrac{4}{(4-x)^2} + 4$

   d. $y = \dfrac{6x+5}{3x+1}$

2. Identify the asymptotes of the hyperbola with equation $y = \dfrac{3x+2}{x+3}$.

3. Consider the function defined by $f(x) = \begin{cases} 3, & x < 1 \\ 4 - 2x, & x \geq 1 \end{cases}$

   a. Evaluate $f(0) + f(1) + f(2)$.

   b. Sketch $y = f(x)$.

   c. State the domain and range of the function.

   d. State its type of correspondence.

4. Consider $f : [-2, 4) \to R, f(x) = 1 - \dfrac{x}{2}$.

   a. State the domain and determine the range of f.

   b. Obtain the rule for the inverse function f^{-1}, giving its domain and range.

   c. Sketch f and f^{-1} on the same diagram.

   d. Calculate the coordinates of any point of intersection of the two graphs.

5. a. Form a possible equation for the square root function shown.

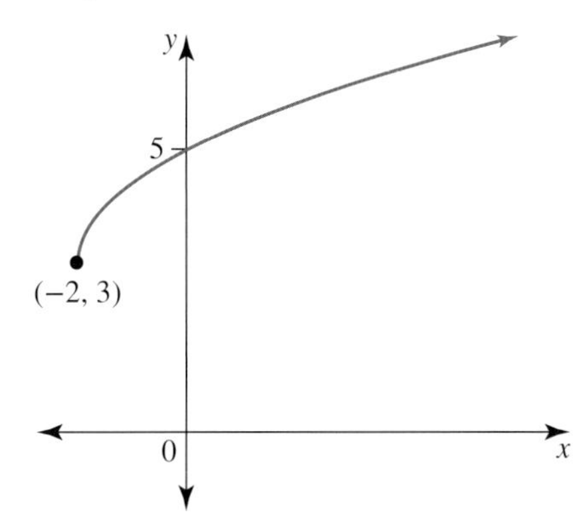

b. Form the rule in hybrid form for the given graph.

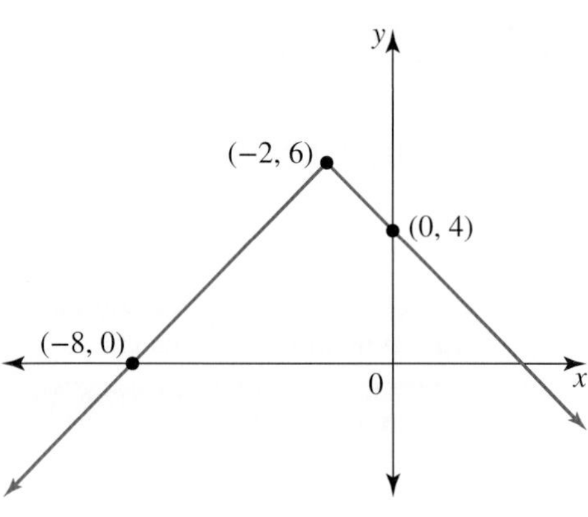

c. Obtain the equation of the hyperbola shown in the diagram.

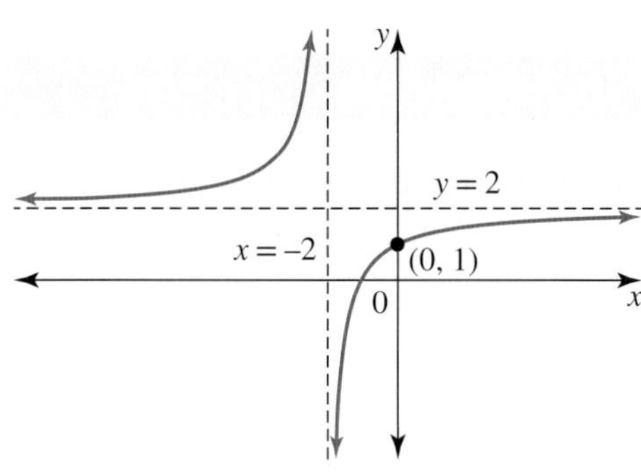

6. a. Determine the rule for the inverse of $y = 1 - 8x^3$.

b. Determine the rule for the inverse of $y = (x + 1)^{\frac{1}{3}} + 5$.

c. i. Obtain the largest value of a so that the inverse of $f : (-\infty, a] \to R, f(x) = (x - 2)^2 + 6$ will be a function.

ii. Use the value obtained for a to determine the inverse function f^{-1}.

Technology active: multiple choice

7. **MC** State which of the following is not a function.

A. $\{(x, y) : y = x^2\}$
B. $\{(x, y) : xy = 1\}$
C. $\{(x, y) : y = \sqrt{-2x}\}$
D. $\{(x, y) : y = \sqrt{2 - x}\}$
E. $\{(x, y) : y = \pm\sqrt{x}\}$

8. **MC** The range of the function $f : [-2, \infty) \to R, f(x) = 3\,[(x + 2)^2 + 5]$ is:

A. $[-2, \infty)$
B. $[5, \infty)$
C. $[6, \infty)$
D. $[15, \infty)$
E. R

9. **MC** The range defined by $f : [-5, 0) \to R$, where $f(x) = x^2 - 16$, is best given by:

A. $[-16, 9)$
B. $[-16, \infty)$
C. $(-16, 9]$
D. $[-16, 9]$
E. $(-16, \infty)$

10. **MC** The equations of the asymptotes of the truncus $y = \dfrac{-3}{(x + 2)^2} + 4$ are:

A. $x = -2, y = 4$
B. $x = 2, y = -4$
C. $x = -4, y = 4$
D. $x = 4, y = -4$
E. $x = -\dfrac{3}{4}, y = 4$

11. **MC** The graph of $y = c - \sqrt{a(x - b)}$ has an end point at $(-2, 5)$ and passes through the point $(6, 1)$. The values of a, b and c are:

A. $a = 9, b = 2, c = -5$
B. $a = -9, b = 2, c = -5$
C. $a = 9, b = 5, c = -2$
D. $a = 2, b = -2, c = 5$
E. $a = -2, b = -2, c = 5$

12. **MC** For the function $f: [-2, 4] \rightarrow R, f(x) = ax + b, f(0) = 1$ and $f(1) = 0$. The image of -2 under the mapping is:

A. -2 B. -1 C. 1 D. 3 E. 4

13. **MC** The image of the point $(4, -3)$ dilated by a factor of $\frac{1}{4}$ from the x-axis is:

A. $(1, -3)$ B. $(-1, -3)$ C. $\left(4, \frac{3}{4}\right)$ D. $\left(-4, \frac{3}{4}\right)$ E. $\left(4, -\frac{3}{4}\right)$

14. **MC** The maximal domain of $y = \frac{x+6}{\sqrt{x-2}}$ is:

A. $R \setminus \{-6\}$ B. $R \setminus \{2\}$ C. $R \setminus \{-6, 2\}$ D. $[-2, \infty)$ E. $(2, \infty)$

15. **MC** The graph of $y = \sqrt{x}$ is translated 4 units upwards and then reflected in the x-axis. The equation of its image is:

A. $y = -\sqrt{x} + 4$ B. $y = \sqrt{-x} + 4$ C. $y = -\sqrt{x} - 4$ D. $y = \sqrt{-x} - 4$ E. $y = -\sqrt{x-4}$

16. **MC** The graph of a function $y = f(x)$ is shown in diagram (i). The graph shown in diagram (ii) is of:

A. $y = f(-2x)$

B. $y = f\left(-\frac{1}{2}x\right)$

C. $y = -2f(x)$

D. $y = -\frac{1}{2}f(x)$

E. $y = f(x+2)$

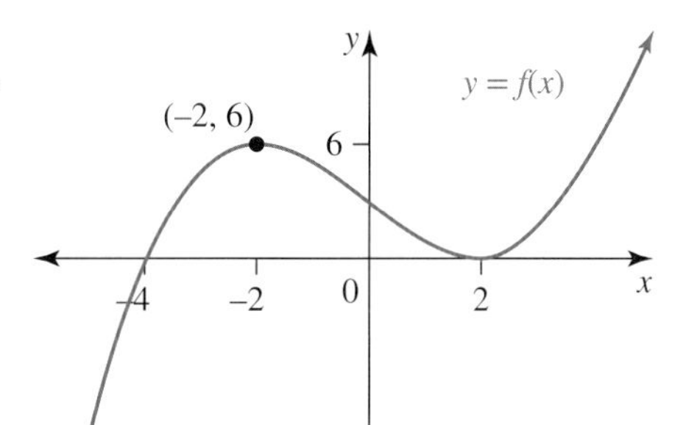

Diagram (i)

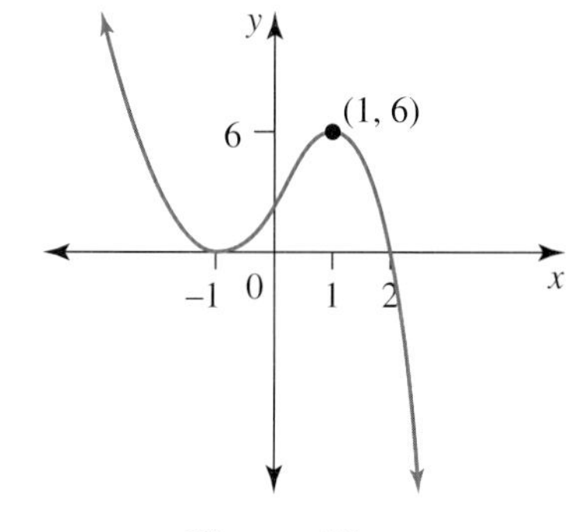

Diagram (ii)

Technology active: extended response

17. Consider the function given by $f(x) = x^2 - 10x + 21$.
   a. Sketch the graph of $y = f(x)$, labelling the turning point and the intersections with the coordinate axes with their coordinates.
   b. Determine the values of k so that the graph of $y = f(x) + k$ will not intersect the x-axis.
   c. Determine the values of h so that the roots of the equation $f(x-h) = 0$ will always be negative.
   d. Sketch the following transformations of the graph of $y = f(x)$, showing the images of its turning point and intersections with the coordinate axes.
      i. $y = f(x+3)$
      ii. $y = f(-x)$
      iii. $y = f(2x)$
      iv. $y = -0.5\,f(x) - 2$
      v. The reflection of $y = f(x)$ in the line $y = x$

18. a. If the intensity, I, of a sound is inversely proportional to the distance, d, from the source of the sound, draw a graph which illustrates this relationship.
   b. If the quantity $y + 2$ is inversely proportional to the square of the quantity $x + 2$, establish the relationship between x and y given $y = -0.5$ when $x = 4$, and hence calculate a so that $y = a$ when $x = a$.
   c. The height, h cm, of a cone of fixed volume varies inversely as the square of its radius, r cm. A set of special cones are to be made with the same volume. One of these special cones has a height of 20 cm has a radius of 3.5 cm.
      i. Determine the height of a special cone if the radius is 7 cm.
      ii. Determine the percentage change in the height of a special cone if the radius has been decreased by 25%. Give your answer correct to 1 decimal place.

19. a. Describe a reflection and a translation with respect to the x- and y-coordinate axes that will result in the curve $y = \dfrac{2}{x-2} - 1$, $x > 2$ being reflected in the line $x = 2$. Give the equation of the image.

b. A small container with height h cm has its longitudinal cross-section in the shape of the curve with the equation $h = \dfrac{2}{(x-2)} - 1$, $2 < x \leq 4$, and its reflection in the line $x = 2$.

Sketch this container and state the diameter of its circular base.

c. If the diameter of the top of the container is 1 cm, calculate its height.

d. If the container is filled with a powder to a height of 1.5 cm, calculate the top circular surface area of the powder.

20. A ray of light comes in along the line $x + y = 2$ above the x-axis and is reflected off the axis so that the angle of departure (the angle of reflection) is equal to the angle of arrival (the angle of incidence).

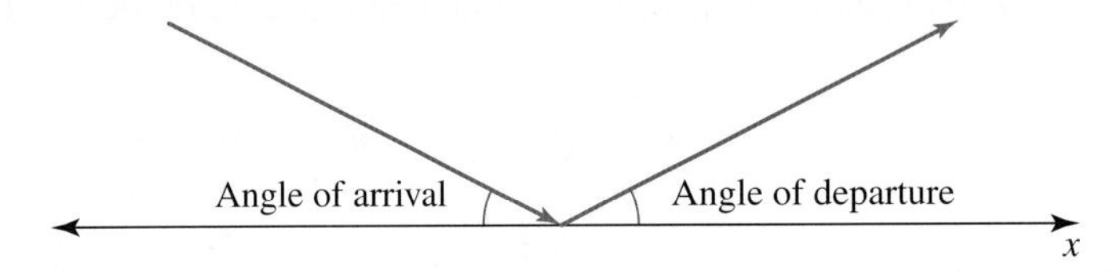

a. i. Calculate the magnitude of the angle of departure. (*Hint:* $m = \tan\theta$.)

ii. Form the equation of the line along which the departing light travels.

iii. Express the path of the incoming and departing rays in terms of a hybrid function.

b. The reflected ray of light strikes the vertical line $x = 4$ and is reflected off this line in the same way so that the angle of departure is equal to the angle of arrival.

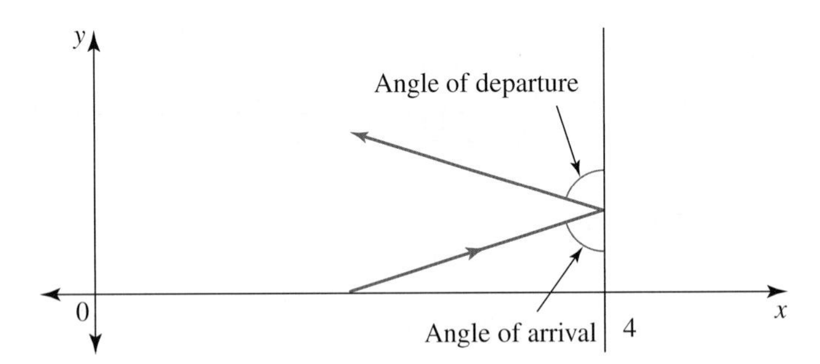

i. Give a reason why this section of the path of the ray of light is not a function.

ii. Form a hybrid rule with x in terms of y, which describes both the incoming and departing paths of the ray of light for this section of its path.

c. The light rays continue to bounce back and forth between the vertical lines $x = 0$ and $x = 4$.

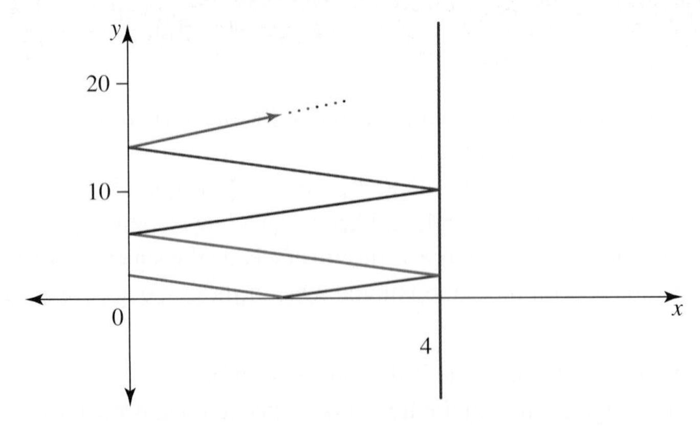

Predict the equation of the line the ray of light will travel along immediately after its fourth reflection on the line $x = 4$.

6.7 Exam questions

Question 1 (3 marks) TECH-FREE

$f(x) = x^2 - 6x$ and $g(x) = f(1 - x)$.

Express the rule for g as a polynomial in x and calculate any values for which $f(x) = g(x)$.

Question 2 (1 mark) TECH-ACTIVE

MC The domain and range respectively of the function $f(x) = 2 - \dfrac{5}{(4-x)^2}$ are given by

A. $R\backslash\{2\}, R\backslash\{4\}$
B. $R\backslash\{4\}, R\backslash\{2\}$
C. $R\backslash\{4\}, R\backslash(-\infty, 2)$
D. $R\backslash\{-4\}, (2, \infty)$
E. $R\backslash\{-4\}, (-\infty, 2)$

Question 3 (5 marks) TECH-FREE

Sketch the graph of $y = -1 + \sqrt{4 + x}$ and its inverse by reflecting the given function in the line $y = x$.

Question 4 (3 marks) TECH-FREE

Consider the function for which $f(x) = \begin{cases} x^2, & \text{if } x < 1 \\ 3 - x, & \text{if } x \geq 1 \end{cases}$.

a. Evaluate $f(-1), f(1)$ and $f(2)$. **(1 mark)**
b. Sketch the graph of $y = f(x)$. State the domain, range and any other values of x for which the function is discontinuous. **(2 marks)**

Question 5 (1 mark) TECH-ACTIVE

MC State which of the following equations best represents the equation of a function $f(x)$ reflected in the y-axis, reflected in the x-axis and translated 2 units parallel to the y-axis in the positive direction.

A. $-f(-x) + 2$
B. $f(-x) + 2$
C. $f(-x) - 2$
D. $-f(-x) - 2$
E. $-f(-x - 2)$

More exam questions are available online.

Answers

Topic 6 Functions and relations

6.2 Functions and relations

6.2 Exercise

1. a. Domain $\{-4, 0, 3\}$, range $\{5, 7, 10\}$
   b. $x \in [-10, 4)$
   c. Domain $(2, 7)$, range $\left(\frac{1}{7}, \frac{1}{2}\right)$
   d. Domain $[-2, 6]$, range $[-3, 3]$
2. a. $(-1, -2)$ is an open end point; $(3, 6)$ is a closed end point. Domain $(-1, 3]$, range $(-2, 6]$
   b. $(3, 2)$ is an open end point. Domain $(3, \infty)$, range $(2, \infty)$
   c. $(-5, 15)$ and $(5, -15)$ are closed end points. Domain $[-5, 5]$, range $[-15, 15]$
   d. No end points. Domain $(-\infty, \infty)$, range $[0, \infty)$
   e. Open end point at $(0, 0)$. Domain $(0, \infty)$, range $(0, \infty)$
   f. $(-2, 4)$ and $(2, 4)$ are open end points. Domain $(-2, 2)$, range $[0, 4)$
3. a. Domain $\{0, 2, 3, 4\}$, range $\{-1, 0, 3, 4\}$; a function
   b. Domain $[-2, \infty)$, range R; not a function
   c. Domain $[0, 3]$, range $[0, 4]$; a function
   d. Maximal domain R, range $(-\infty, 4)$; a function
4.

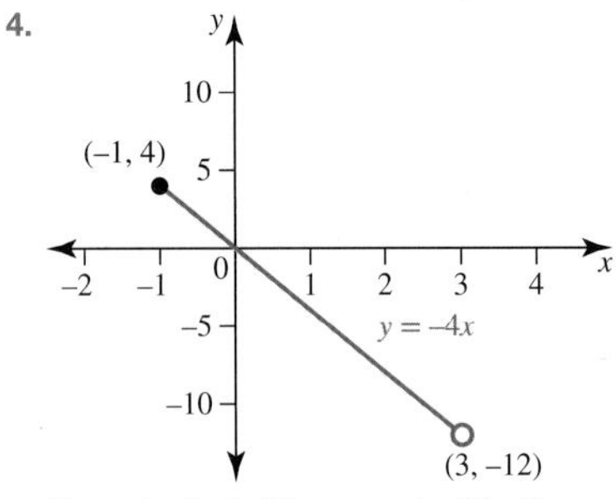

Domain $[-1, 3)$, range $(-12, 4]$

5. a.

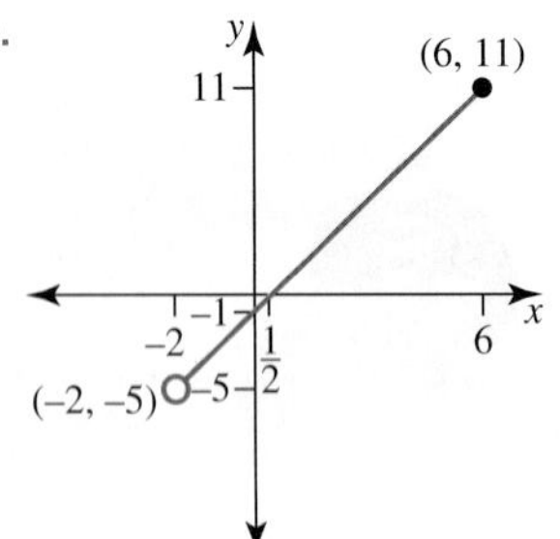

   b. Domain $(-2, 6]$, range $(-5, 11]$
   c. The relation is a function as a vertical line cuts graph exactly once.
6. a.

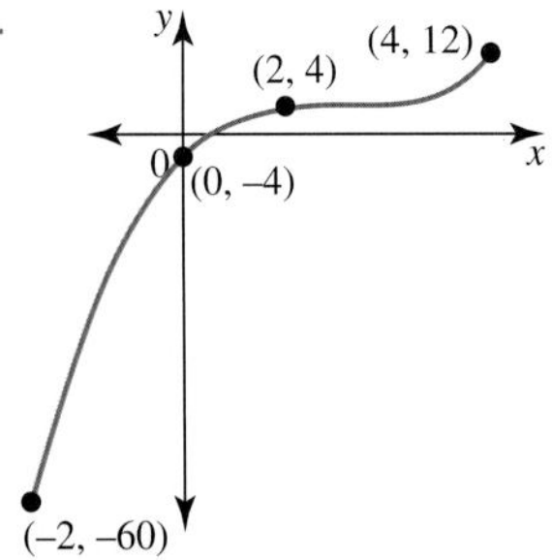

   b. Domain $[-2, 4]$, range $[-60, 12]$
   c. The relation is a function as a vertical line cuts graph exactly once.
7. a.

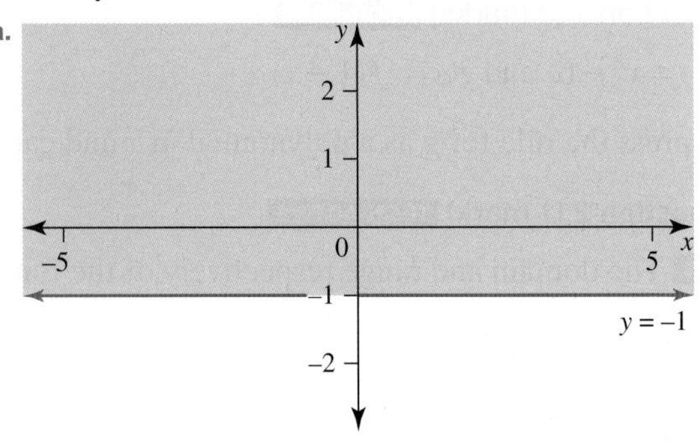

   b. Domain R, range $[-1, \infty)$
   c. The relation is not a function as a vertical line cuts graph at more than one place.
8. a. One-to-one correspondence; a function
   b. Many-to-many correspondence; not a function
9. a. i. Domain $[0, 5]$; range $[0, 15]$
      ii. One-to-one
      iii. Function
   b. i. Domain $[-4, 2) \cup (2, \infty)$; range $(-\infty, 10)$
      ii. Many-to-one
      iii. Function
   c. i. Domain $[-3, 6]$; range $[0, 8]$
      ii. Many-to-one
      iii. Function
   d. i. Domain $[-2, 2]$; range $[-4, 4]$
      ii. One-to-many
      iii. Not a function
   e. i. Domain $\{3\}$; range R
      ii. One-to-many
      iii. Not a function
   f. i. Domain R; range R
      ii. Many-to-one
      iii. Function
10. a. Domain $\{-11, -3, -1, 5\}$; range $\{0, 2, 8\}$; many-to-one; function
    b. Domain $\{20, 50, 60\}$; range $\{6, 10, 20\}$; many-to-many: not a function
    c. Domain $\{-14, 0, 14\}$; range $\{-7, 0, 2, 7\}$; one-to-many: not a function
    d. Domain R; range R; one-to-one; function
    e. Domain R; range $(-\infty, 4]$; many-to-one; function
    f. Domain R; range $[0, \infty)$; many-to-one; function

11. a. Line; end points $(-1, -2)$ (open), $(1, 6)$ (closed); domain $(-1, 1]$; range $(-2, 6]$

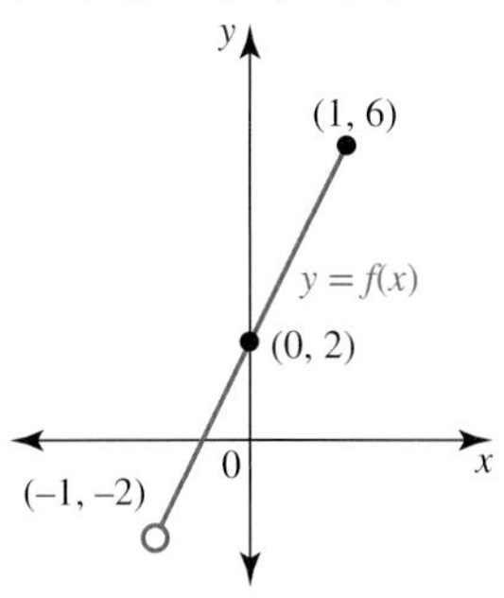

b. Parabola; end point $(-2, 0)$ (closed); minimum turning point $(-1, -4)$; y-intercept $(0, 0)$; domain $[-2, \infty)$; range $[-4, \infty)$

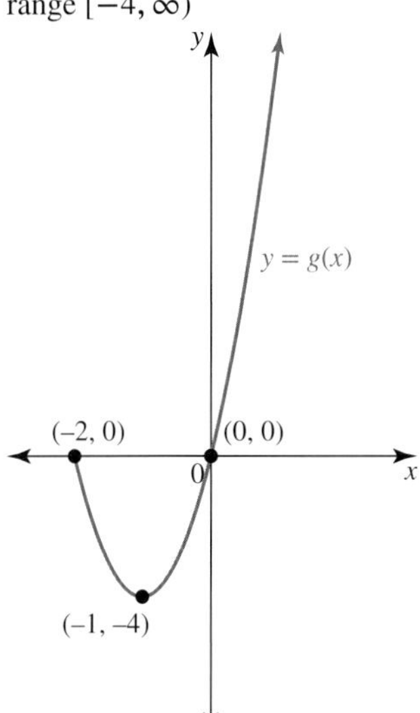

c. Cubic with stationary point of inflection; open end point $(0, 4)$; x-intercept $\left(\sqrt[3]{4}, 0\right)$; domain R^+; range $(-\infty, 4)$

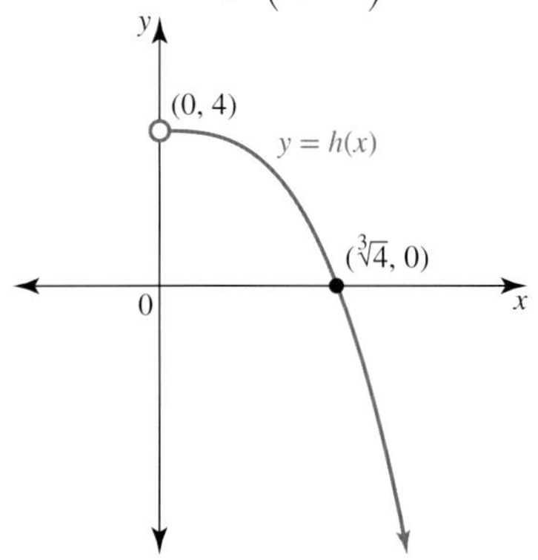

d. Two sections of a horizontal line; open end points at $(\pm 2, 4)$; domain $R \setminus [-2, 2]$; range $\{4\}$

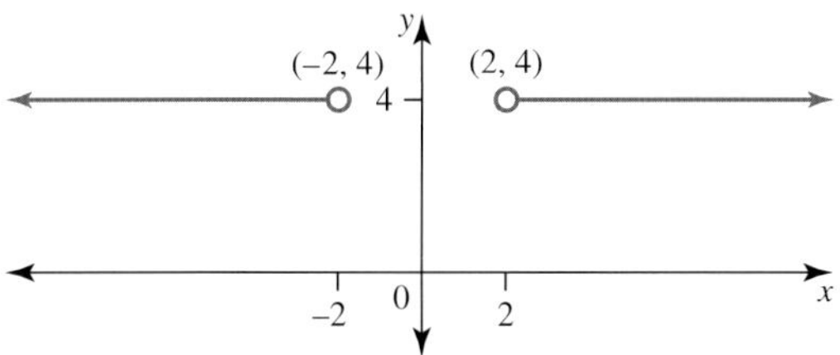

12. a.

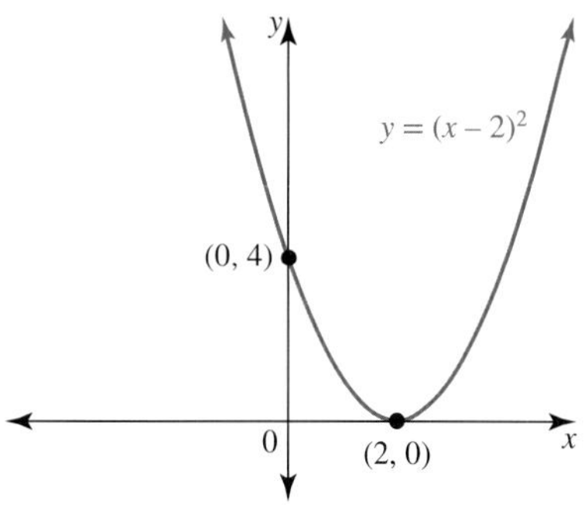

Domain R; range $R^+ \cup \{0\}$; many-to-one correspondence

b. A possible answer is $[2, \infty)$.

13. a. $a = 2, b = 3; f(x) = 2x + 3$

b. $f(0) = 3$

c. $x = -1.5$

d. $f(-3) = -3$

e. $g : (-\infty, 0] \to R, g(x) = 2x + 3$

14. a. i. 1 ii. 21 iii. $\frac{1}{2}$

b. i. $\frac{1}{4}$ ii. $\frac{1}{2}$ iii. $\frac{1}{5}$

c. i. 45 ii. -5 iii. $4\sqrt{2} - 3$

d. $x = -2$

15. a. i. -3 ii. 96

b. i. $f(2a) = 4a^2 + 4a - 3$

ii. $f(1 - a) = a^2 - 4a$

c. $2xh + h^2 + 2h$

d. $\{x : x < -3\} \cup \{x : x > 1\}$

e. $x = -5, x = 3$

f. $x = -4, x = 1$

16. a. 4

b. x-intercepts: $(0, 0)$ at turning point, $(1, 0)$

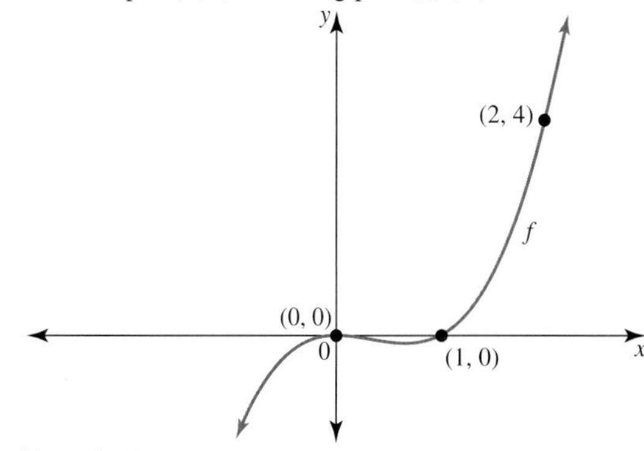

c. Domain R; range R

d. Many-to-one

e. An answer is R^-.

f. $\{2\}$

17. $f : [0, 7) \to R, f(x) = x^2 - 6x + 10$; domain $[0, 7)$; range $[1, 17)$.

18. a. $k = 1$ or $k = 3$; many-to-many

b. $k \in R \setminus \{1, 3\}$; many-to-one

c.

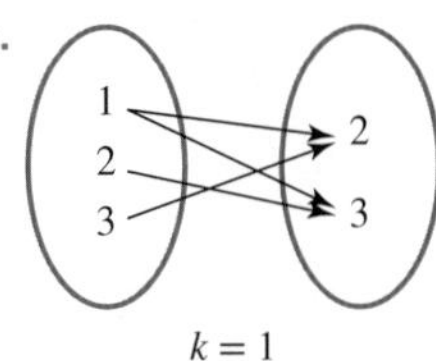

$k = 1$ $k = 3$

d.

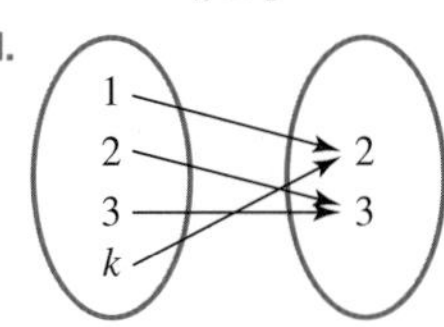

19. a. Function; $f: Z^+ \to R, f(x) = x^2$; domain Z^+; range $\{1, 4, 9, 16, \ldots\}$

b. Function; $f: R \to R, f(x) = \dfrac{6-2x}{3}$; domain R; range R

c. Not a function

d. Function; $f: R \to R, f(x) = 5$; domain R; range $\{5\}$

e. Not a function

f. Function; $f: R^+ \to R,\ f(x) = -x^3$; domain R^+; range R^-

20. a. $f(x) = \dfrac{5}{2} + \dfrac{3}{4}x - \dfrac{1}{4}x^2$

b. $g(x) = \dfrac{3}{2} + \dfrac{5}{4}x - \dfrac{1}{4}x^2$

c. $x = 2$

d.

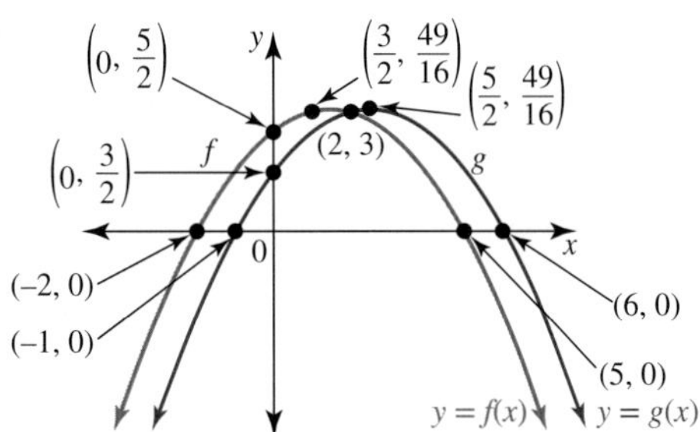

The graphs intersect at (2, 3).

$y = f(x)$: maximum turning point $\left(\dfrac{3}{2}, \dfrac{49}{16}\right)$;

x-intercepts $(-2, 0), (5, 0)$; y-intercept $\left(0, \dfrac{5}{2}\right)$

$y = g(x)$: maximum turning point $\left(\dfrac{5}{2}, \dfrac{49}{16}\right)$;

x-intercepts $(-1, 0), (6, 0)$; y-intercept $\left(0, \dfrac{3}{2}\right)$

$y = g(x)$ has the same shape as $y = f(x)$ but has been translated 1 unit to the right.

21. a. Domain $\{t: 0 \le t \le 5\}$; range $\{x: 4 \le x \le 29\}$; the distance travelled is 25 units.

b. 2 seconds; domain $[0, 2]$; range $[0, 5]$

c. i. Domain $[0, 2]$; range $[0.5, 2.1]$

ii. Approximately 1.4 weeks

22. a. $l = 0; m = -12; n = -16; f(x) = x^3 - 12x - 16$

b. -28.672

c. 4.477

d. $f: [0, \infty) \to R,\ f(x) = x^3 - 12x - 16$; domain $[0, \infty)$; range $[-32, \infty)$

6.2 Exam questions

Note: Mark allocations are available with the fully worked solutions online.

1. D

2. B

3. a. -5

b.

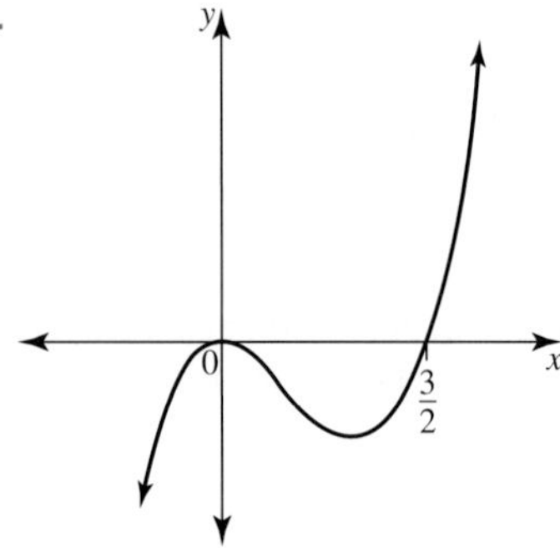

c. Domain $= R$, range $= R$

d. The horizontal line test shows many-to-one correspondence.

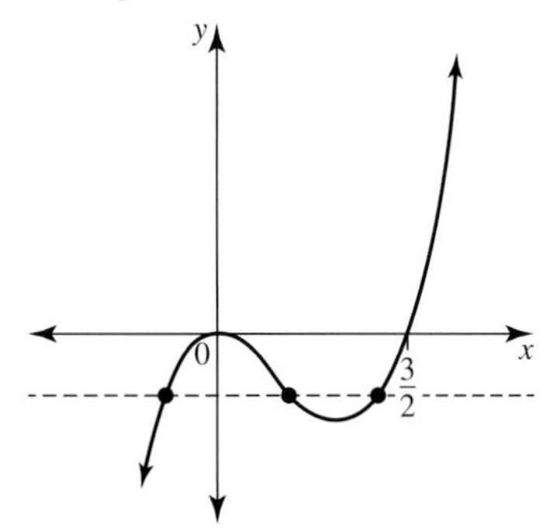

6.3 The rectangular hyperbola and the truncus

6.3 Exercise

1. a. $x = -5, y = 2$

b. $x = 0, y = -3$

c. $x = 0, y = 0$

d. $x = -14, y = -\dfrac{3}{4}$

2. a. i. $x = 6$ and $y = 1$.

ii. $R\backslash\{6\}$

iii. $R\backslash\{1\}$.

b. i. $x = -3$ and $y = -7$.

ii. $R\backslash\{-3\}$

iii. $R\backslash\{-7\}$

c. i. $x = 2$ and $y = 4$.

ii. $R\backslash\{2\}$

iii. $R\backslash\{4\}$

d. i. $x = 1$ and $y = -3$.

ii. $R\backslash\{1\}$

iii. $R\backslash\{-3\}$.

3. a. Vertical asymptote $x = -2$; horizontal asymptote $y = -1$; y-intercept $(0, 0)$; domain $R\backslash\{-2\}$; range $R\backslash\{-1\}$

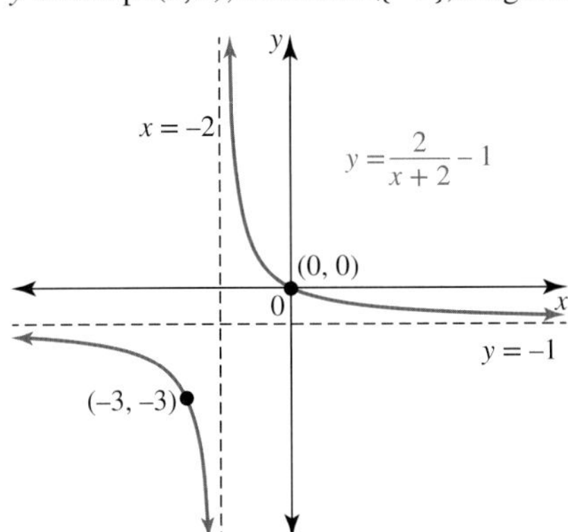

b. Vertical asymptote $x = 2$; horizontal asymptote $y = 0$; y-intercept $\left(0, \dfrac{1}{2}\right)$; domain $R\backslash\{2\}$; range $R\backslash\{0\}$

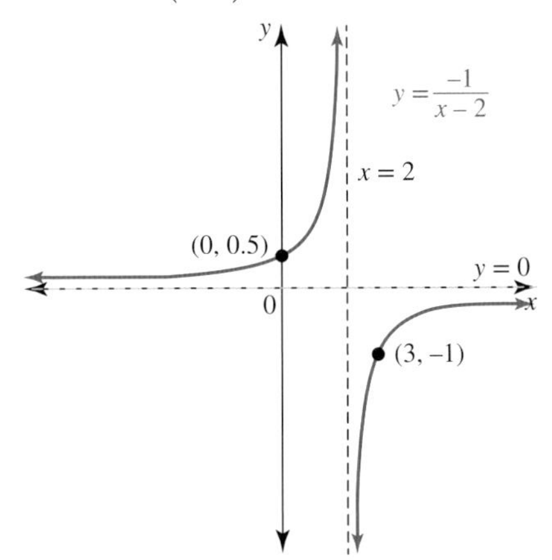

4. a. $x = 1$ and $y = 5$.

b. $(0, 4)$

c. $\left(\dfrac{4}{5}, 0\right)$

d.

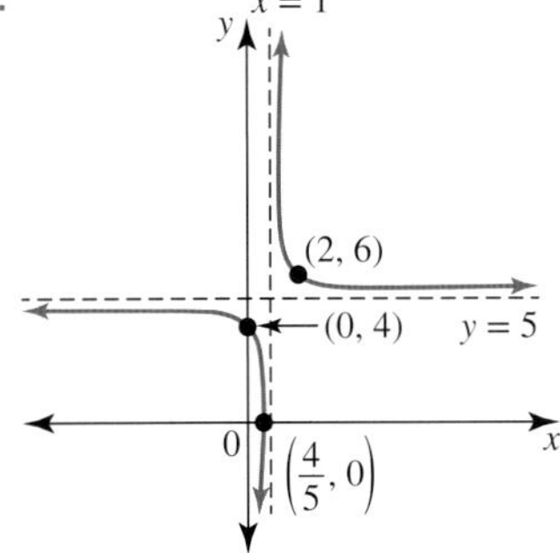

5. a. $x = -3$ and $y = 5$.

b. $(0, 3)$

c. $\left(-\dfrac{9}{5}, 0\right)$

d.

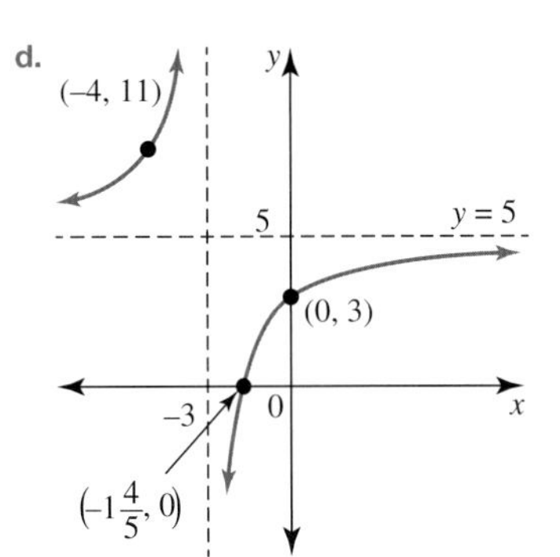

6. a. Domain $R\backslash\{-1\}$; range $R\backslash\{-3\}$

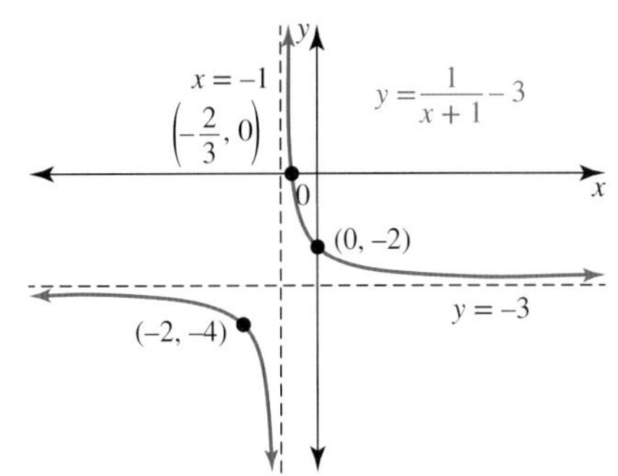

b. Domain $R\backslash\{3\}$; range $R\backslash\{4\}$

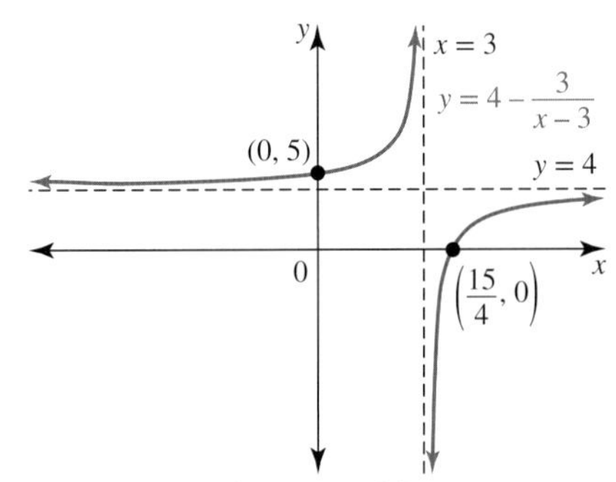

c. Domain $R\backslash\{-3\}$; range $R\backslash\{0\}$

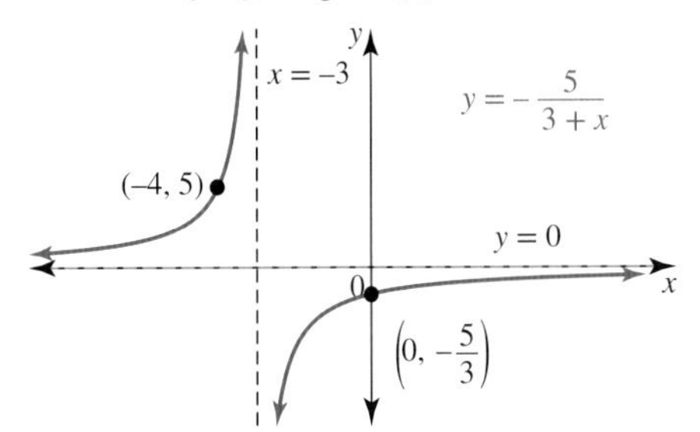

d. Domain $R\backslash\{2\}$; range $R\backslash\{-1\}$

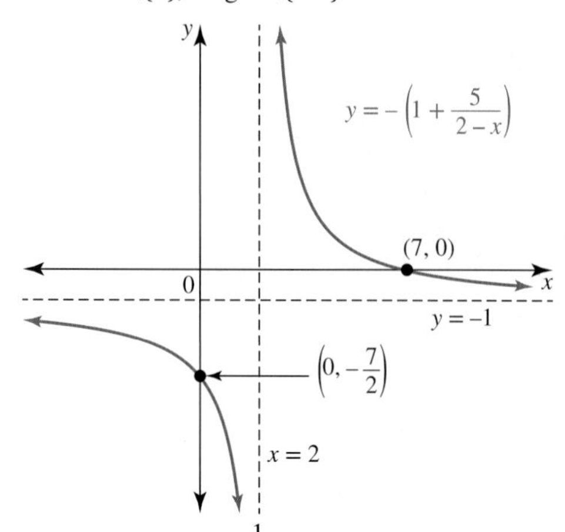

7. Vertical asymptote $x = \frac{1}{2}$; horizontal asymptote $y = 4$; the graph lies in quadrants 1 and 3 (quadrants as defined by the asymptotes).

8. a. Vertical asymptote $x = -\frac{2}{3}$; horizontal asymptote $y = 2$

b. $y = \frac{-1}{x-4} + \frac{1}{2}$

9. a. $y = \frac{24}{x+4} + 2$

b. $y = -\frac{12}{x} + 5$

c. $y = 2 - \frac{10}{x-5}$

d. i. Domain $(-3, 3) \setminus \{0\}$; range $R \setminus \left[\frac{8}{3}, \frac{10}{3}\right]$

ii. $y = \frac{1}{x} + 3, -3 < x < 3, x \neq 0$

10. a. $a = 3; b = 1$

b. Asymptotes $x = 4, y = 3$; y-intercept $\left(0, \frac{11}{4}\right)$; x-intercept $\left(\frac{11}{3}, 0\right)$; point $(5, 4)$

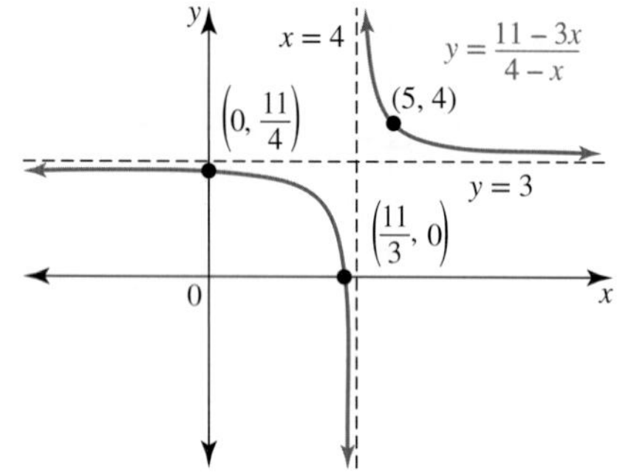

c. $x < \frac{11}{3}$ or $x > 4$

11. a. $y = \frac{-1}{16x+4} + \frac{1}{4}, x = -\frac{1}{4}, y = \frac{1}{4}$

b. $y = \frac{4}{x-4} - 2, x = 4, y = -2$

c. $y = \frac{1}{x} + 2, x = 0, y = 2$

d. $y = \frac{-2}{2x+3}, x = -\frac{3}{2}, y = 0$

12. Domain $R\backslash\{4\}$; range $R\backslash\{0\}$

13. C

14. a. Domain $R\backslash\{-3\}$, range $(5, \infty)$

b. $y = \frac{20}{(x+3)^2} + 5$

15. a. $R = \frac{240}{I}$ b. $333\frac{1}{3}$ ohms

16. a. Domain $R\backslash\{3\}$; range $(-1, \infty)$

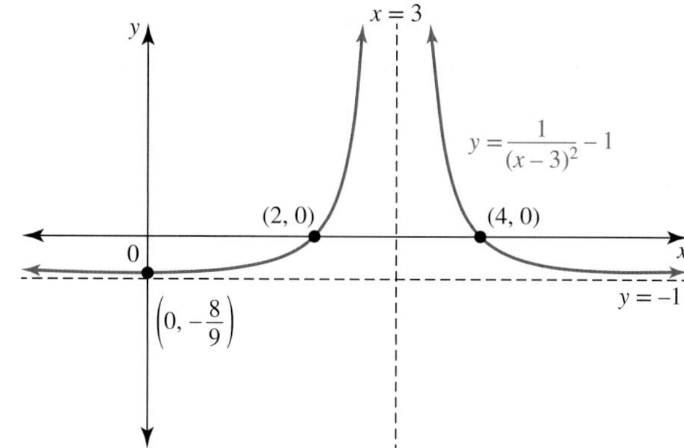

b. Domain $R\backslash\{-2\}$; range $(-\infty, -4)$

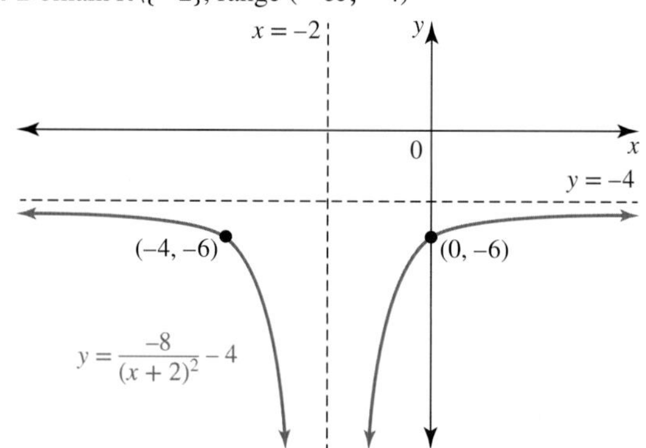

17. a. i. $x = 0$ and $y = -1$

ii. Domain $R\backslash\{0\}$, range $(-1, \infty)$.

iii. $\left(\frac{2\sqrt{3}}{3}, 0\right)$ and $\left(-\frac{2\sqrt{3}}{3}, 0\right)$; no y-intercept

iv.

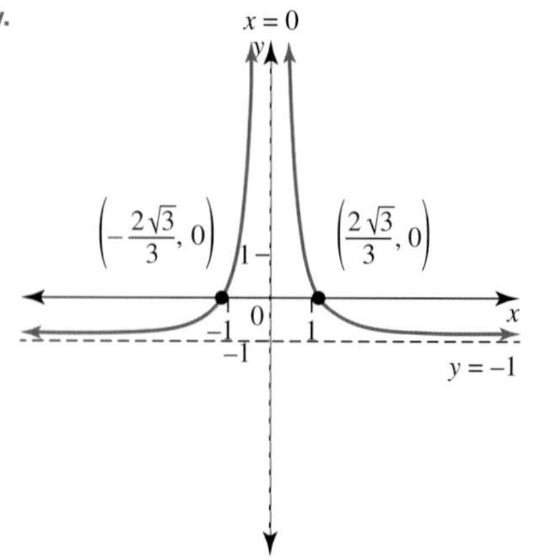

b. i. $x = 1$ and $y = 4$.

ii. Domain $R\backslash\{1\}$, range $(-\infty, 4)$.

iii. $\left(1 - \frac{\sqrt{2}}{2}, 0\right), \left(1 + \frac{\sqrt{2}}{2}, 0\right), (0, 2)$

iv.

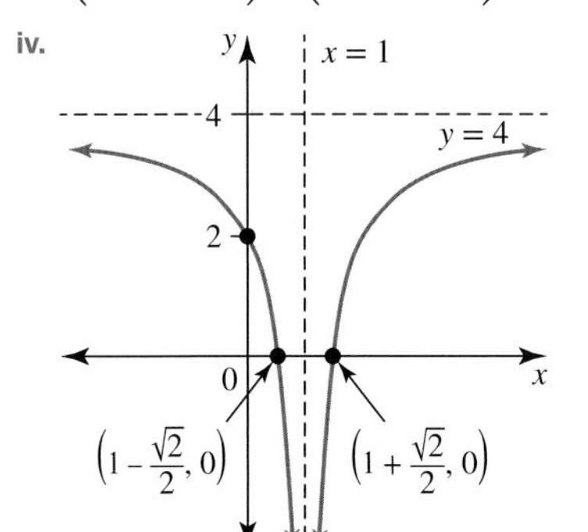

18. a. $x = 2, y = 5$

Domain $R\backslash\{2\}$, range $(5, \infty)$

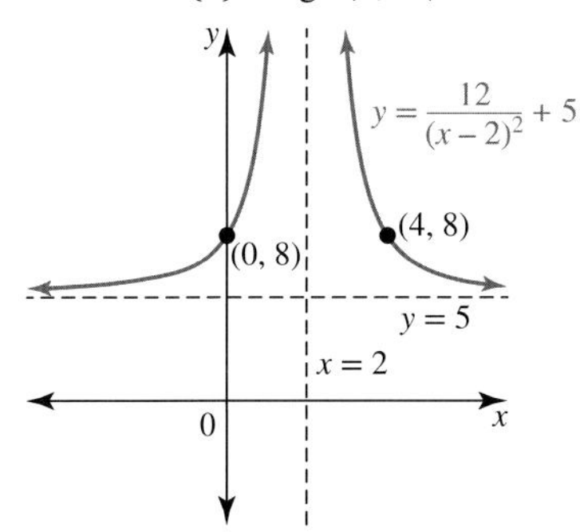

b. $x = -2, y = 6$

Domain $R\backslash\{-2\}$, range $(-\infty, 6)$

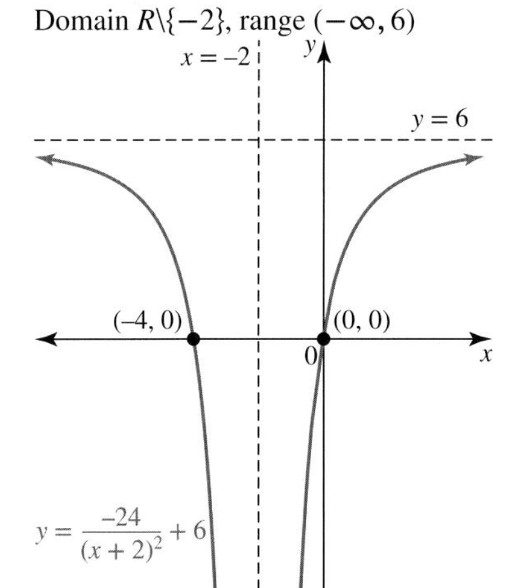

c. $x = 0, y = 7$

Domain $R\backslash\{0\}$, range $(-\infty, 7)$

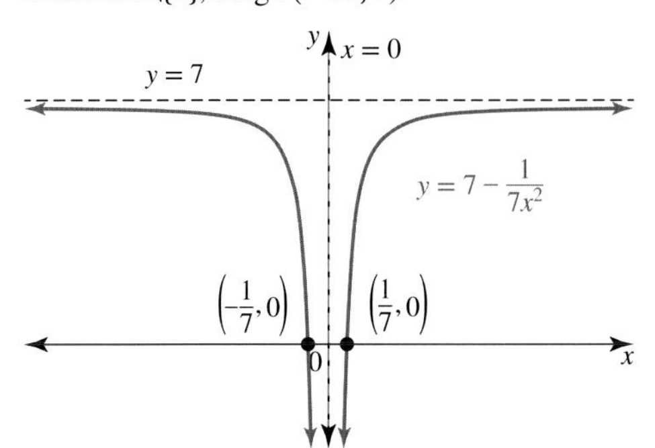

d. $x = \frac{1}{2}, y = 0$

Domain $R\backslash\left\{\frac{1}{2}\right\}$, range $(0, \infty)$

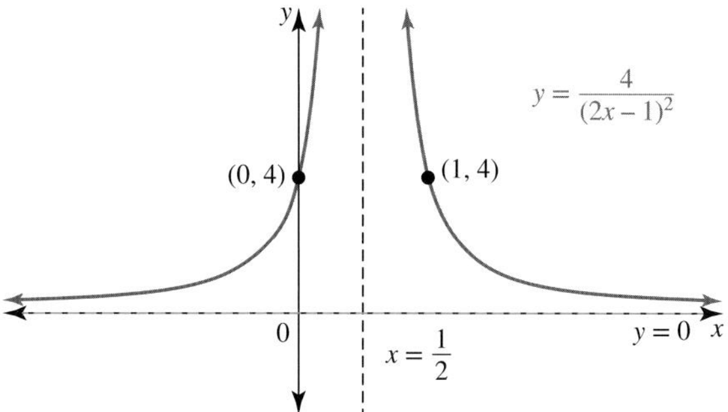

e. $x = 2, y = -2$

Domain $R\backslash\{2\}$, range $(-\infty, -2)$

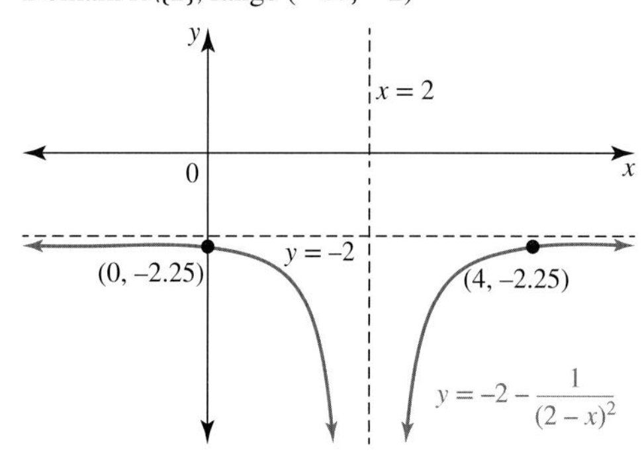

f. $x = 0, y = 1$

Domain $R\backslash\{0\}$, range $(1, -\infty)$

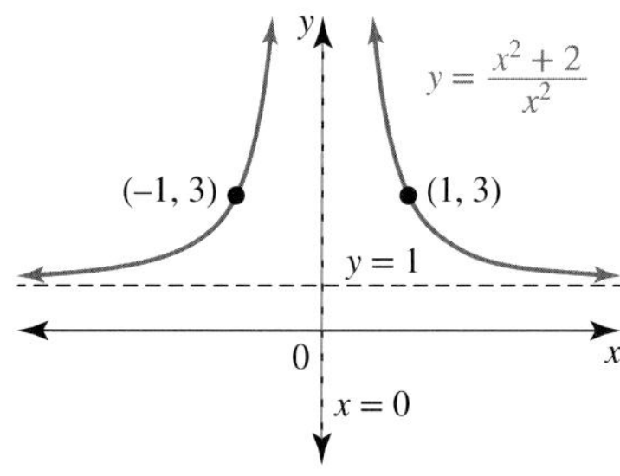

19. Asymptotes $x = \frac{1}{5}, y = 0$; domain $R \setminus \left\{\frac{1}{5}\right\}$; range $(0, \infty)$

20. a. $y = \frac{2}{x-3} + 1$ **b.** $y = \frac{-1.5}{x+3} + 1$

c. $y = \frac{3}{x^2} - 2$ **d.** $y = 2 - \frac{9}{(x+3)^2}$

e. $y = \frac{4}{(x+2)^2}, f: R \setminus \{-2\} \to R, f(x) = \frac{4}{(x+2)^2}$

f. i. $y = \frac{-2x+2}{4x-1}$

ii. $f: R \setminus \left\{\frac{1}{4}\right\} \to R,\ f(x) = \frac{-2x+2}{4x-1}$

21. a. Reduced by 25 cattle

b. Domain $\{t : t \geq 0\}$; range $(30, 80]$

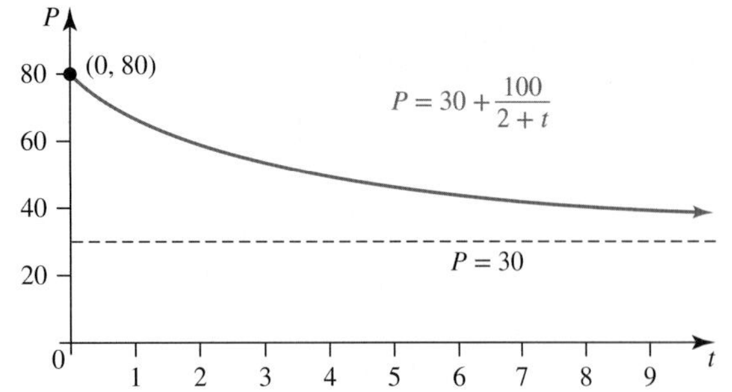

c. The number of cattle will never go below 30.

22. Table b; $y = \dfrac{7.2}{x}$, $y = 6.4, x = 1.125$; $x = 8, y = 0.9$

23. a. One branch of a truncus required

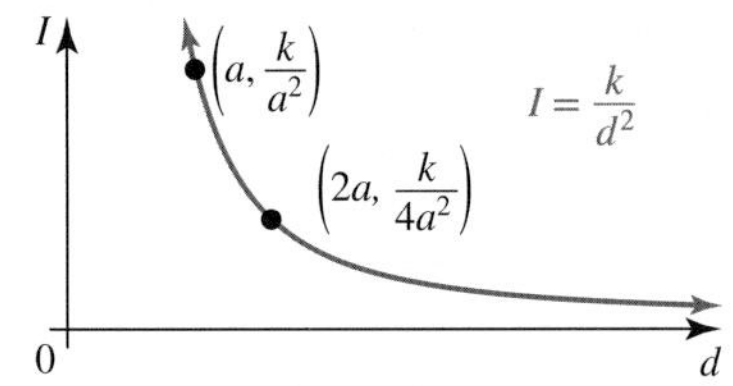

b. Intensity reduced by 75%

24. a. The constant is the distance; $k = 180$.

b.

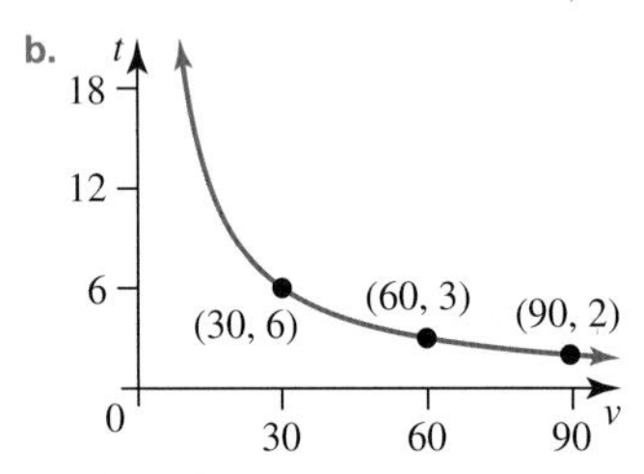

c. 80 km/h

25. a. 50 seconds

b. 25 metres above the ground

26. a. $(2, 1)$

b.

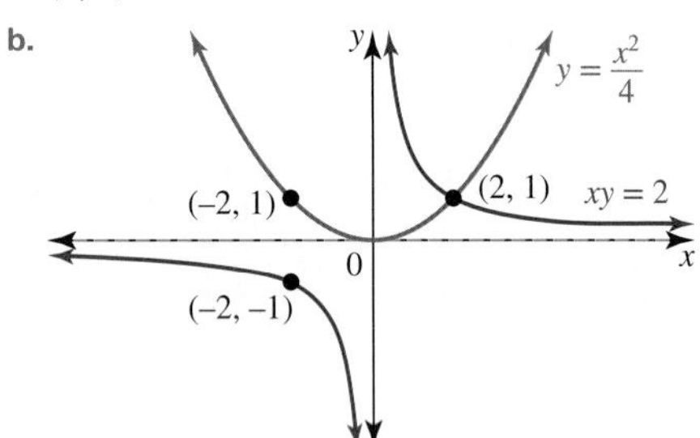

c. $(-2, 1), (2, 1)$

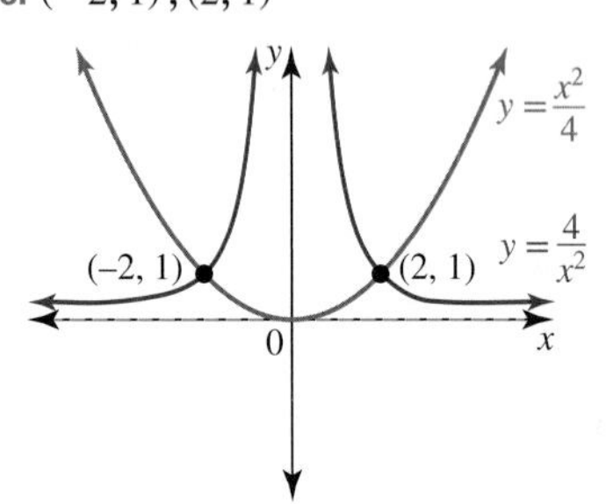

d. $(-\sqrt{a}, 1), (\sqrt{a}, 1)$

27. a. $a = 460$; $b = 40$

b. 12 years 8 months

c. Sample responses can be found in the worked solutions in the online resources.

d. Increases by approximately 4 insects in the 12th year and 3 in the 14th year; growth is slowing.

e. Never reaches 500 insects

f. Cannot be larger than 460

28. a. Two intersections

b. One intersection if $k = 1, 5$; two intersections if $k < 1$ or $k > 5$; no intersections if $1 < k < 5$

6.3 Exam questions

Note: Mark allocations are available with the fully worked solutions online.

1. a.

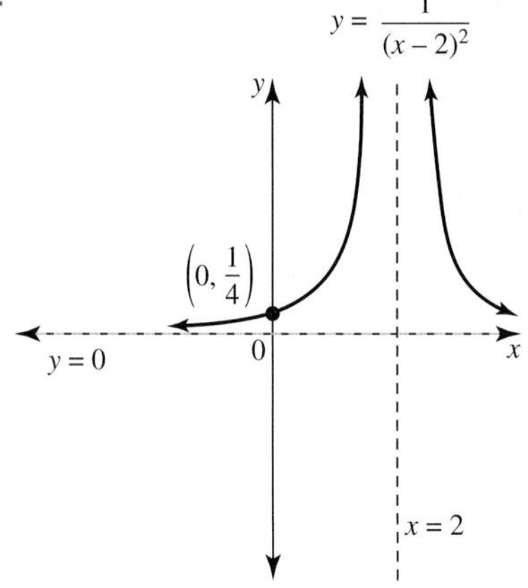

b. Domain $R\backslash\{2\}$, range $(0, \infty)$

A horizontal line test would show many-to-one correspondence.

2. D

3. D

6.4 The square root function

6.4 Exercise

	i.	ii.	iii.
1. a.	$[9, \infty)$	$(9, 0)$	$[0, \infty)$
b.	$(-\infty, 4]$	$(4, 0)$	$[0, \infty)$
c.	$[-3, \infty)$	$(-3, 0)$	$(-\infty, 0]$
d.	$(-\infty, 0]$	$(0, 3)$	$[3, \infty)$
e.	$[2, \infty)$	$(2, -7)$	$[-7, \infty)$

f. i. $\left(-\infty, \dfrac{1}{2}\right]$

ii. $\left(\dfrac{1}{2}, 4\right)$

iii. $(-\infty, 4]$

2. a. Domain $[1, \infty)$; range $[-3, \infty)$

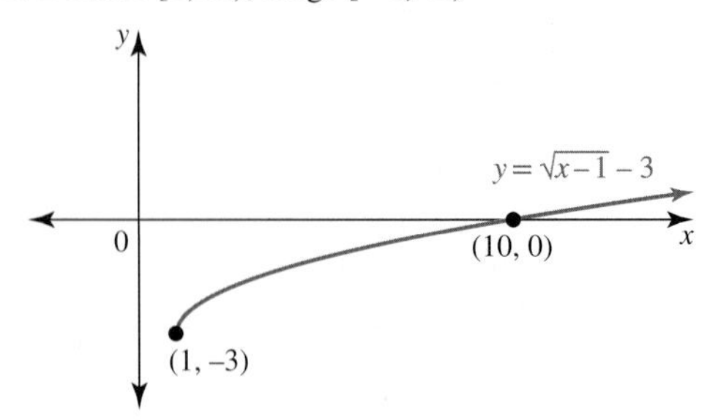

b. i. $[-2, \infty)$

ii.

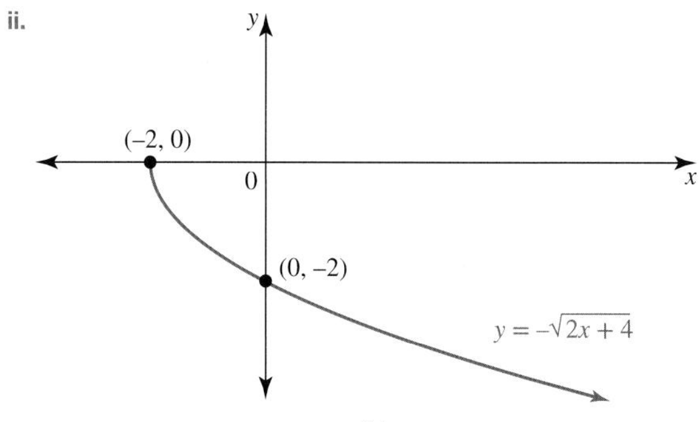

3. a.

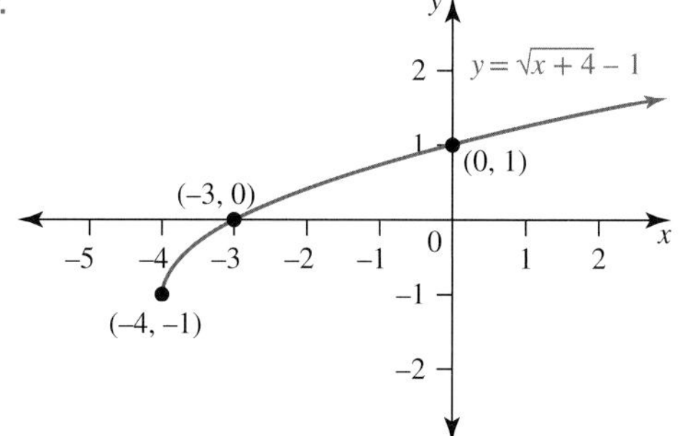

b.

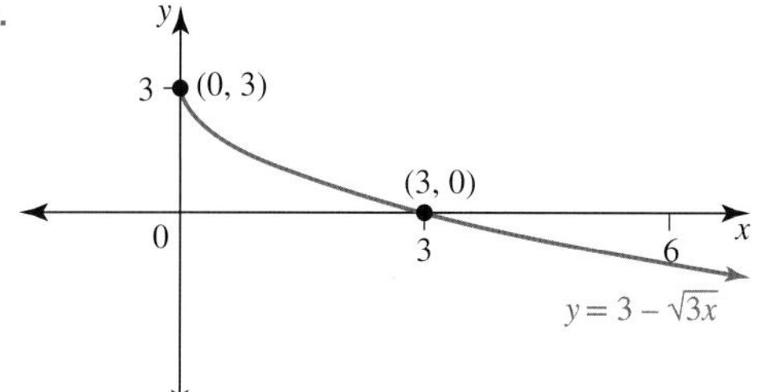

c.

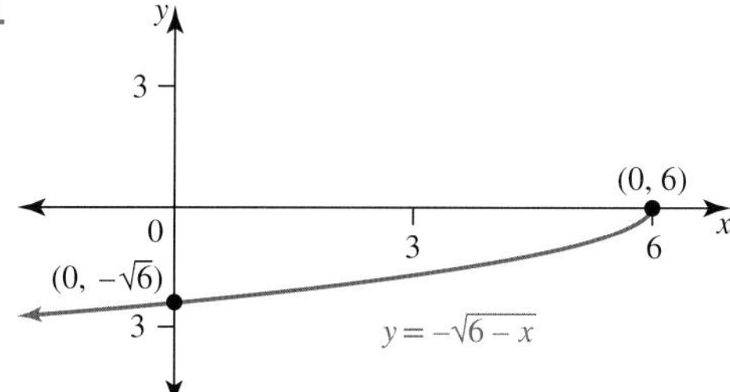

d.

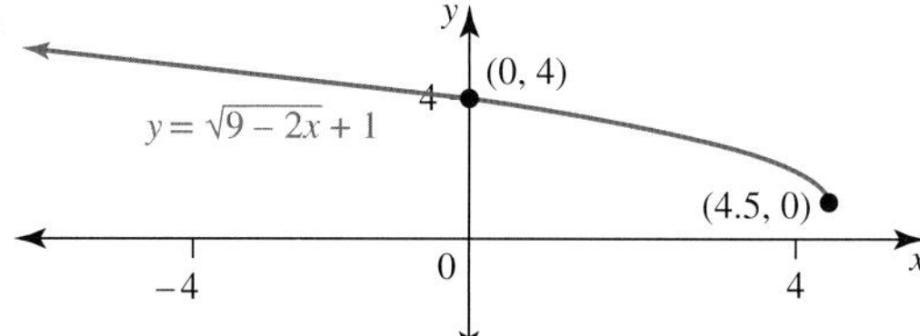

4. a. Domain $[-3, \infty)$; range $[-2, \infty)$

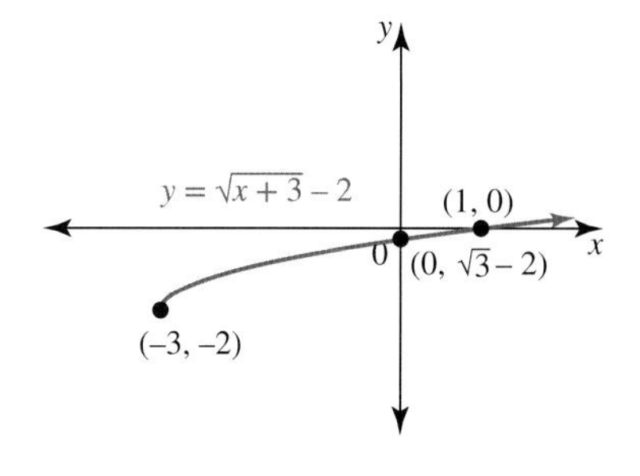

b. Domain $[0, \infty)$; range $(-\infty, 5]$

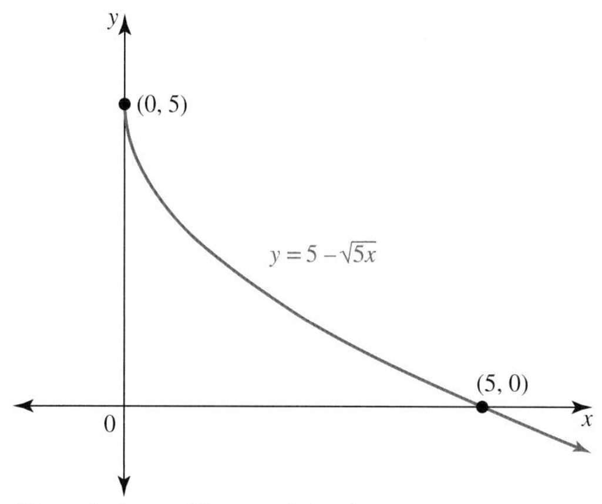

c. Domain $(-\infty, 9]$; range $[4, \infty)$

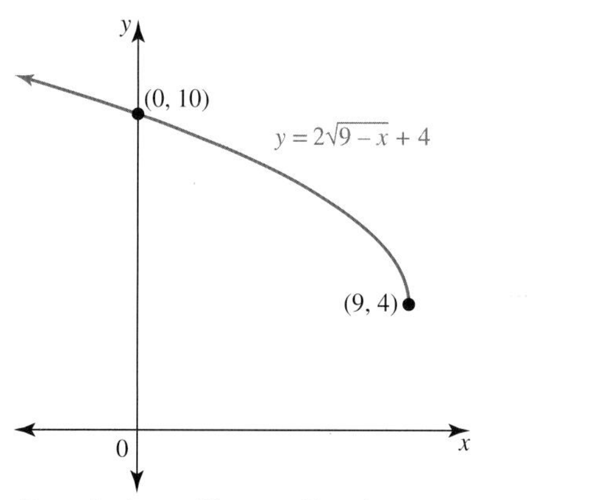

d. Domain $(-\infty, 7]$; range $[0, \infty)$

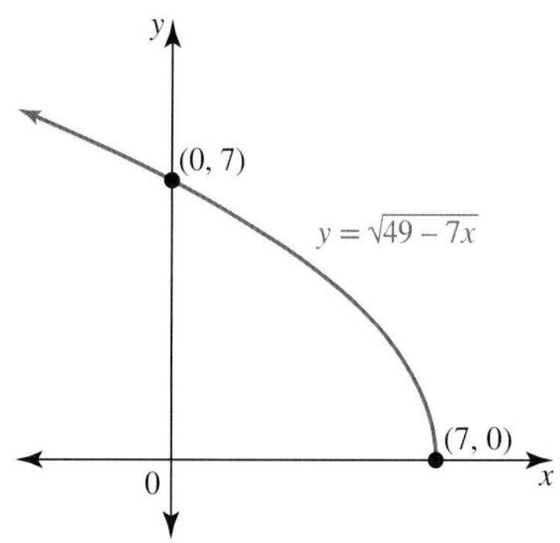

e. Domain $[-4, \infty)$; range $[2, \infty)$

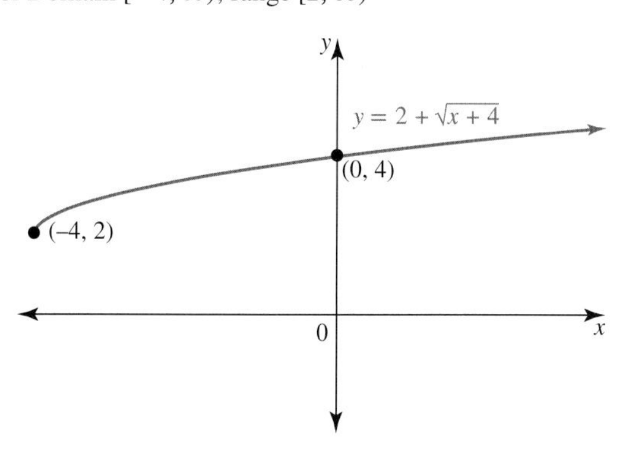

f. Domain $\left(-\infty, \frac{3}{2}\right]$; range $(-\infty, -1]$

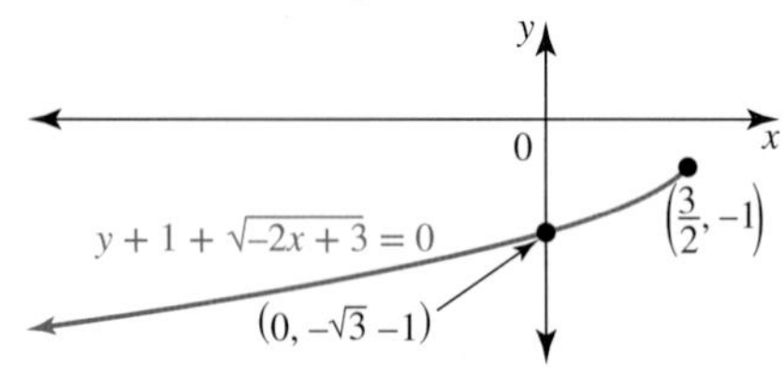

5. $y = \sqrt{2}\sqrt{x+2} + 1 \Rightarrow y = \sqrt{2(x+2)} + 1$
6. $y = -\sqrt{3(x+2)} + 2$
7. $y = 5\sqrt{4-x} - 1; \left(\frac{99}{25}, 0\right)$
8. $y = \sqrt{4-x} - 4$
9. $y = 4\sqrt{x-3} + 2$
10. $y = \frac{1}{2}\sqrt{x+4} - 1$
11. $a = -2, b = 2, c = 1;\ y = \sqrt{-2(x-2)} + 1$ or $y = \sqrt{4-2x} + 1$
12. a. $(1, \ -2), \left(\frac{13}{9}, 0\right)$
    b.

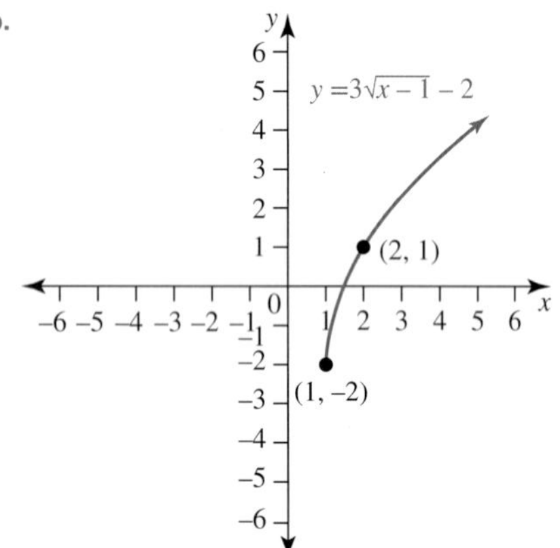

    c. $y = -2 - 3\sqrt{x-1}$ or $y + 2 = -3\sqrt{x-1}$
13. $t = \sqrt{\frac{h}{4.9}}$
14. a. $y = 2 \pm \sqrt{1-x}$
    b. $f : (-\infty, 1] \rightarrow R, f(x) = 2 + \sqrt{1-x}$
    $g : (-\infty, 1] \rightarrow R, g(x) = 2 - \sqrt{1-x}$
    c.

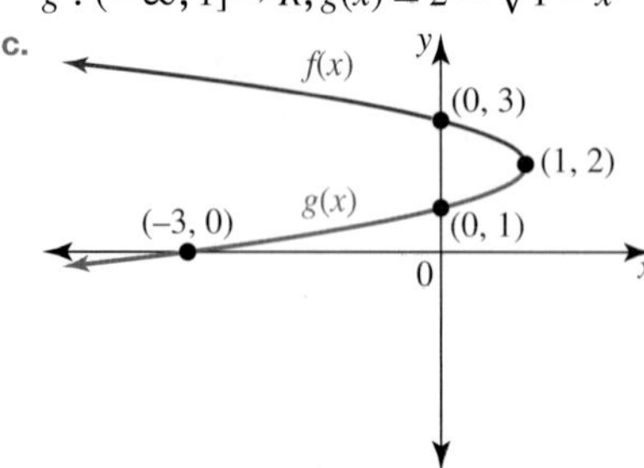

    d. $f(-8) = 5; g(-8) = -1.$
15. $a = 5, b = -2, c = 2$
    Equation: $y = 5 + \sqrt{-2(x-2)}$ or $y = 5 + \sqrt{4-2x}$
    Domain: $(-\infty, \ 2]$
    Range: $(-\infty, 5]$
16. a. $m = 9; n = -2$
    b. $x = \frac{4}{9}$
    c. $\{x : 1 < x < 4\}$

6.4 Exam questions

Note: Mark allocations are available with the fully worked solutions online.

1. B
2.

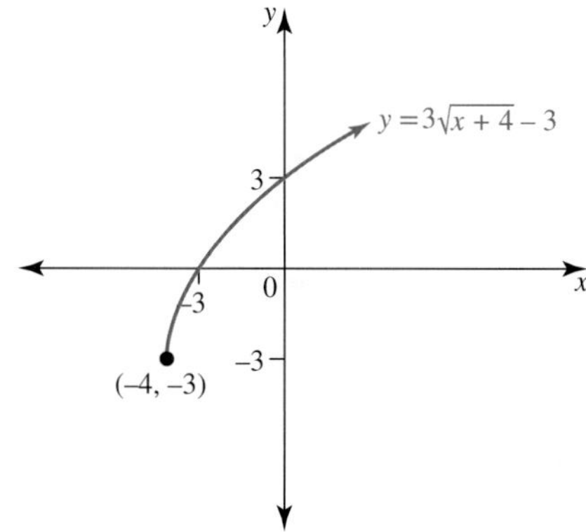

Domain $[-4, \infty)$, range $[-3, \infty)$

3. C

6.5 Other functions and relations

6.5 Exercise

1. a. $x \leq \frac{3}{2}$
   b. $R \setminus \frac{1}{2}$
   c. R
2. a. $R\backslash\{\pm 4\}$
   b. R
   c. $(-1, \infty)$
   d. $R \setminus (-2, 2)$
   e. $R\backslash\{\pm 2\}$
   f. $(-\infty, 4)$
   g. $[0, 2]$
   h. R^+
   i. R
3. a. $R\backslash\{-4, -1\}$
   b. $[-3, \infty)$
   c. $[-3, \infty)\backslash\{-1\}$
4. a. Domain $\{2, 3, 4, 5\}$, range $\{9, 10, 11, 12\}$.
   b. $\{(9, 2), (10, 3), (11, 4), (12, 5)\}$, domain $\{9, 10, 11, 12\}$, range $\{2, 3, 4, 5\}$.
   c. $y = x + 7$
   d. $y = x - 7$
5. a. Domain $(-\infty, 2]$; range $[0, \infty)$
   b. Domain $[0, \infty)$; range $(-\infty, 2]$
   c. $f^{-1} : [0, \infty) \rightarrow R, f^{-1}(x) = \frac{6-x}{3}$

d.

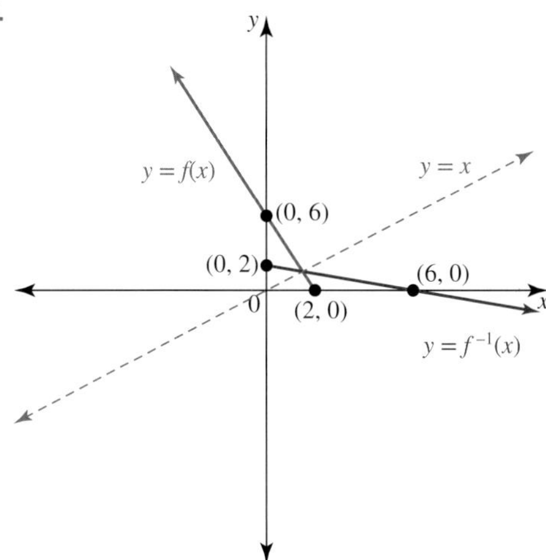

6. a. i. Domain R, range R.
   ii. $y = 4x - 3$
   iii. Domain R, range R.
 b. i. Domain $R\backslash\{0\}$, range $R\backslash\{2\}$.
   ii. $y = -\dfrac{5}{x-2}$
   iii. Domain $R\backslash\{2\}$, range $R\backslash\{0\}$.
 c. i. Domain $[-1, 2]$, range $[6, 12]$.
   ii. $y = 5 - \dfrac{x}{2}, x \in [6, 12]$
   iii. Domain $[6, 12]$, range $[-1, 2]$.
 d. i. Domain $[2, \infty)$, range $[0, \infty)$.
   ii. $y = x^2 + 2, x \geq 0$
   iii. Domain $[0, \infty)$, range $[2, \infty)$

7. a.

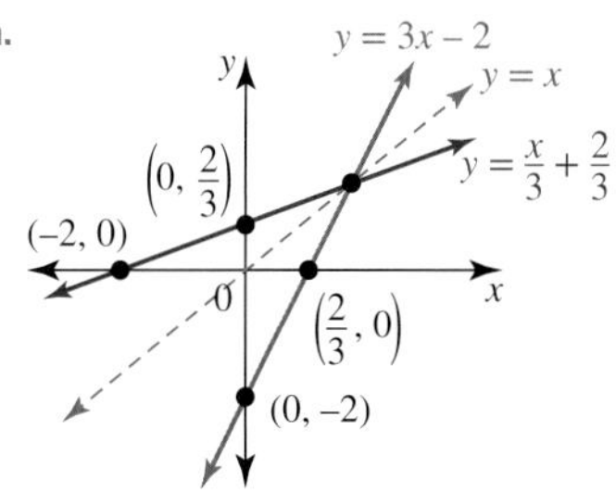

 b. See the diagram in part a.
 c. $(1, 1)$
 d. $f^{-1}: R \to R, f^{-1}(x) = \dfrac{x}{3} + \dfrac{2}{3}$.

8. a. $4y - 8x = 1$
 b. $y = -\dfrac{3x}{2} - 6$
 c. $y^2 = \dfrac{x}{4} \left(\text{or } y = \pm \dfrac{\sqrt{x}}{2}\right)$
 d. $y = \dfrac{x^2 - 1}{2}, x \geq 0$

9. a. Maximal domain $R\backslash\{2\}$; asymptote equations: $x = 2, y = 0$
 b. $f^{-1}(x) = \dfrac{1}{x} + 2$
 c. $x = 0, y = 2$
 d. $x = b; y = a; y = \dfrac{1}{x-b} + a$

10. a. Domain $(-4, 8]$; range $[-2, 4)$
 b. $f^{-1}(x) = \dfrac{4-x}{2}, -4 < x \leq 8;$

 $f^{-1}: (-4, 8] \to R, f^{-1}(x) = \dfrac{4-x}{2}$
 c.

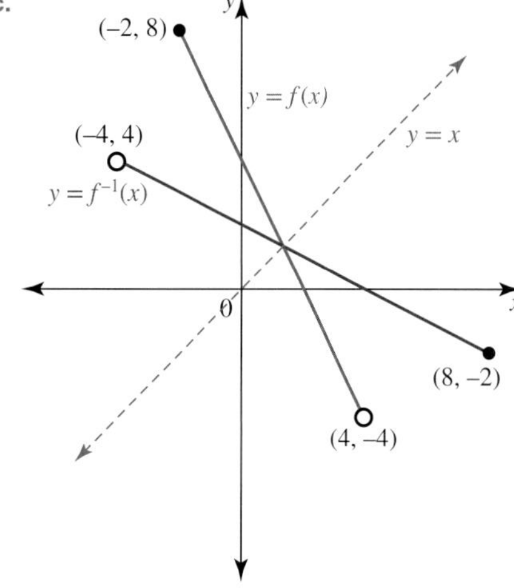

 d. $x = \dfrac{4}{3}$

11. a. R, many-to-one correspondence, not one-to-one
 b. Inverse domain $[2, 18]$, range $[-4, 0]$. f is one-to-one, so the inverse is a function.
 c. $f^{-1}(x) = \sqrt{x-2} - 4$

12. a. g has one-to-one correspondence.
 b. $g^{-1}: R \to R, g^{-1}(x) = \sqrt[3]{x-2} + 1$, domain R, range R.

13. a. Domain R, range $(-\infty, 9]$
 b. $k = 3$

14. a. $(y+1)^2 = x$ or $y = \pm\sqrt{x} - 1$; one-to-many correspondence, so not a function
 b.

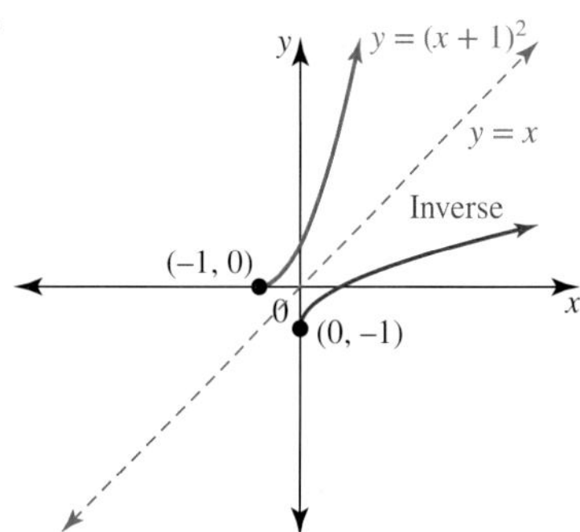

 c. $y = \sqrt{x} - 1$
 d. No intersection

15. a.

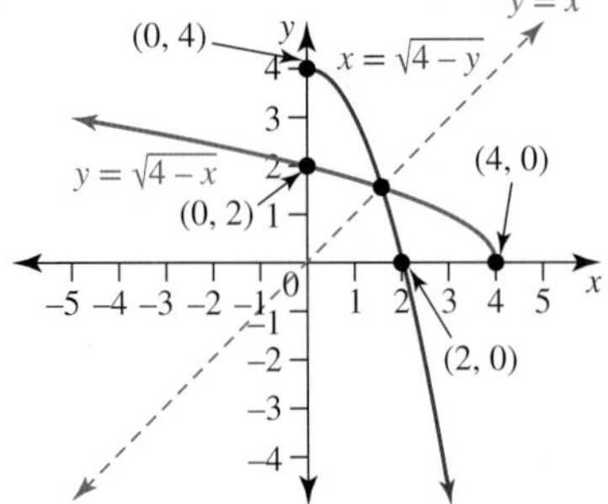

b. Domain $x \geq 0$

16. $a = 2$

17. a.

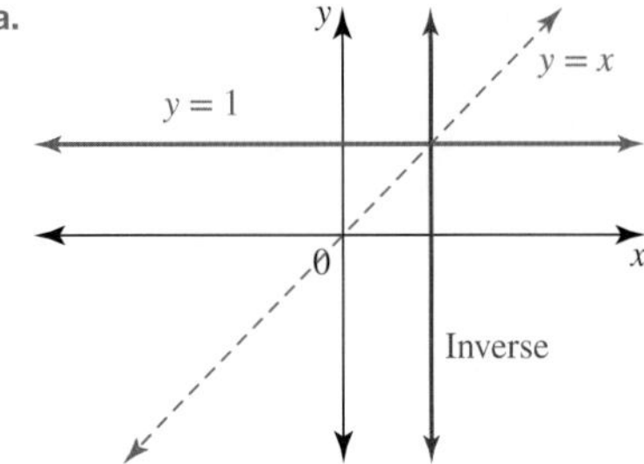

b.

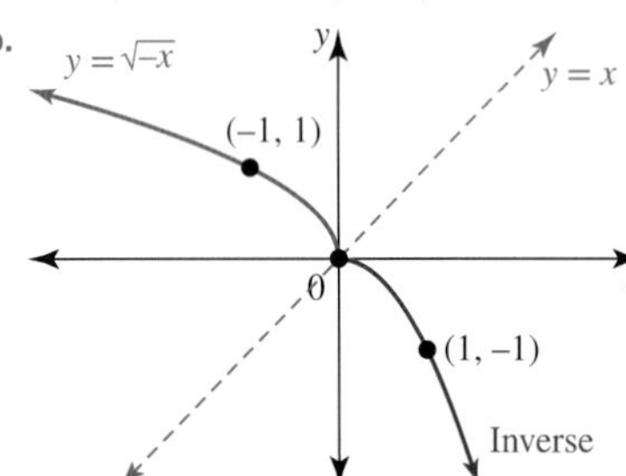

c.

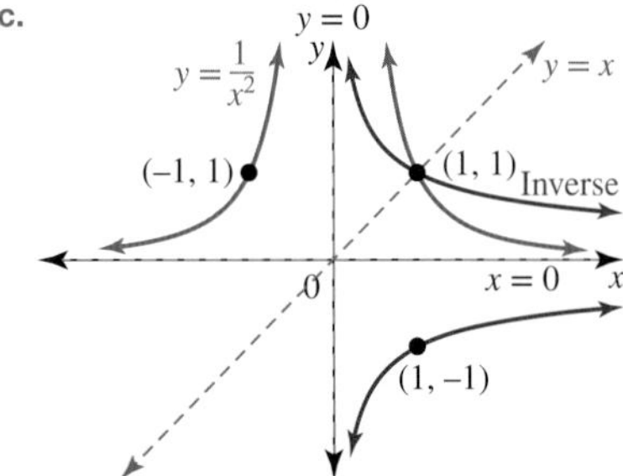

d.

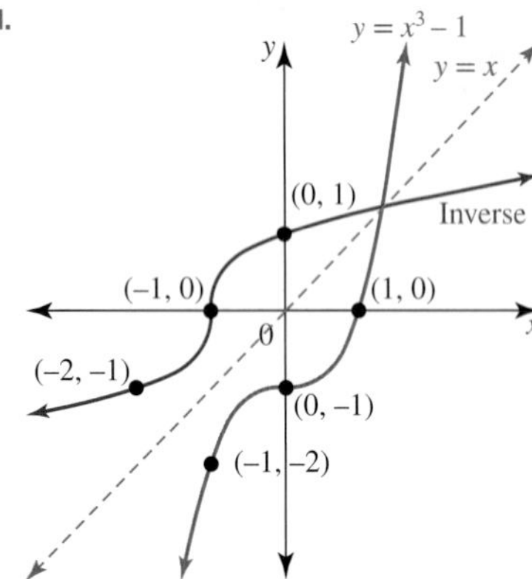

18.

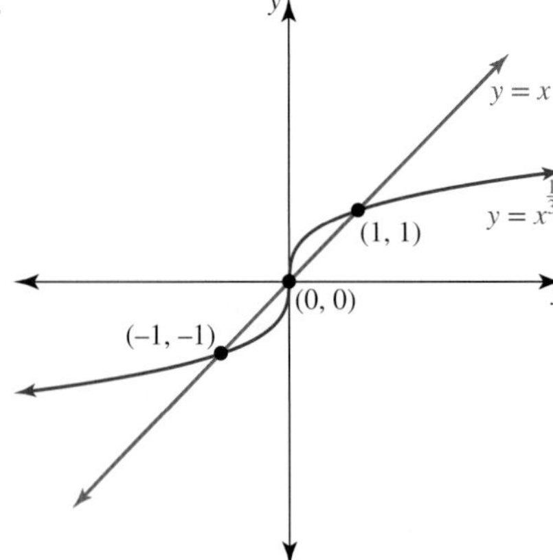

$\{x : x < -1\} \cup \{x : 0 < x < 1\}$

19. a. $(-1, -1), (1, 1)$

b.

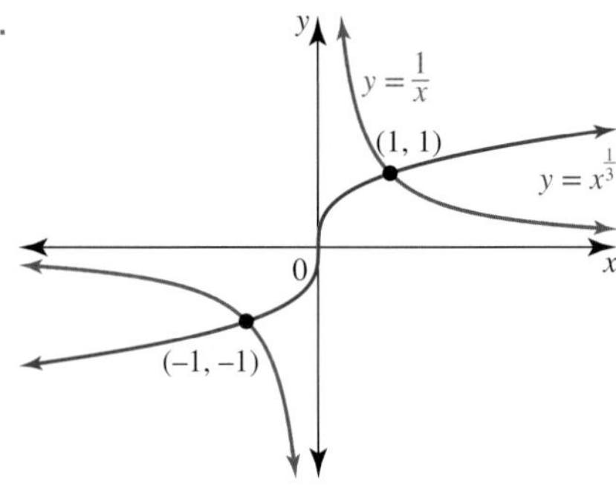

20. a. i. $f(-2) = -8$ ii. $f(1) = 2$ iii. $f(2) = 2$

b. Domain R; range $(-\infty, 1) \cup \{2\}$

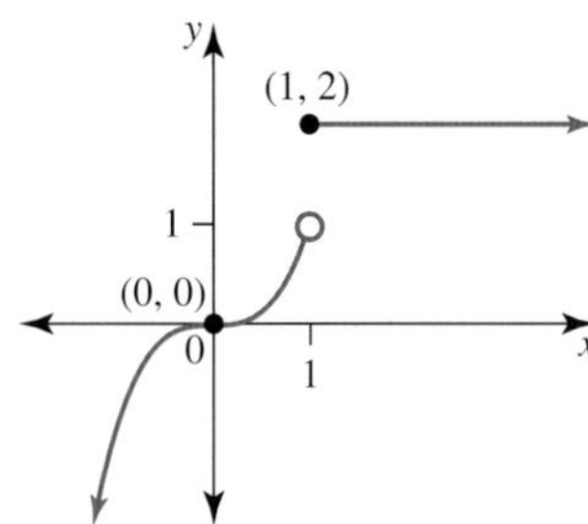

c. Not continuous at $x = 1$

21. a. Domain R; range $(1, \infty)$

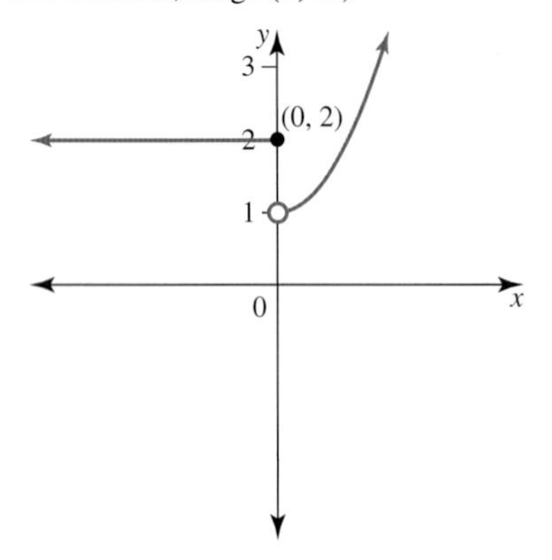

b. Domain R; range R

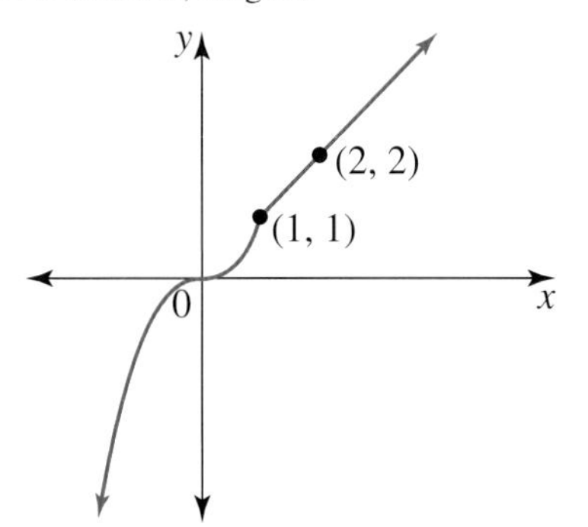

c. Domain R; range $[2, \infty)$

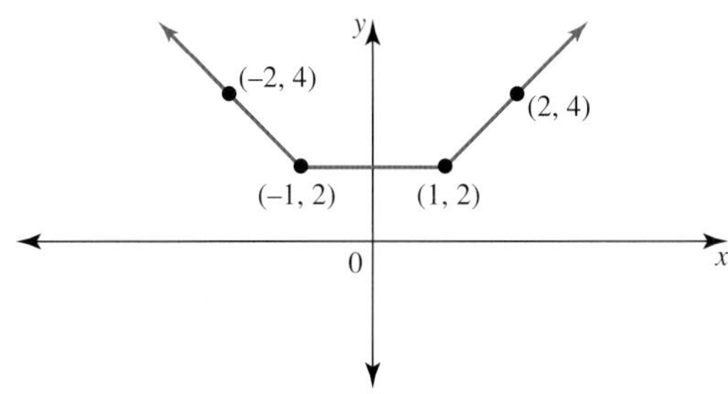

d. Domain $R \setminus [1, 2]$; range R^-

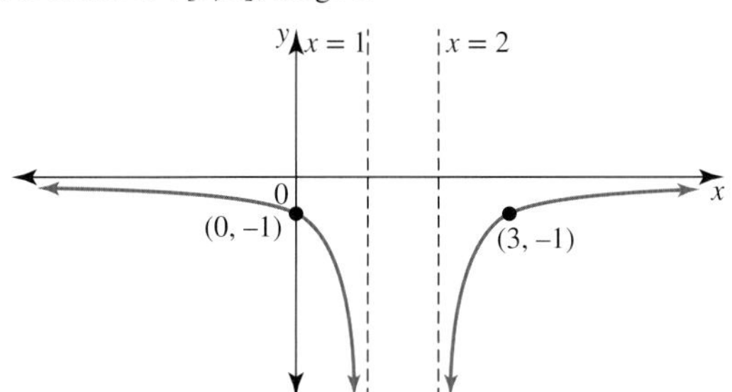

e. Domain R; range R

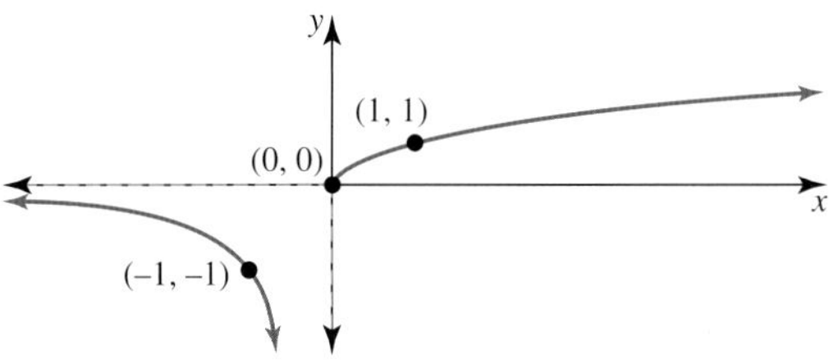

f. Domain R; range $R \setminus (1, 3]$

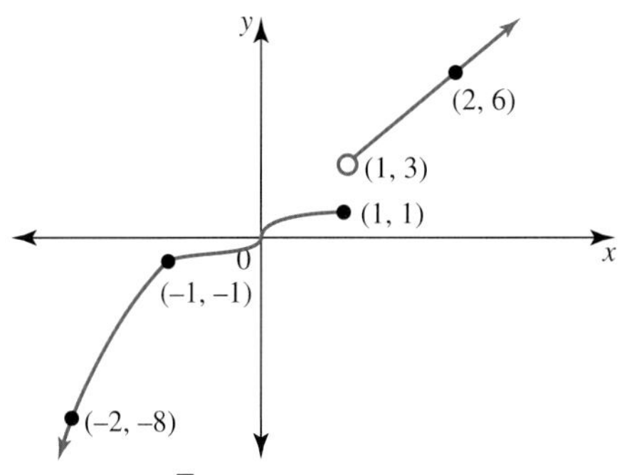

22. a. i. $f(0) = \sqrt{2}$

ii. $f(3) = 4$

iii. $f(-2) = 1$

iv. $f(1) = 0$

b. Sample responses can be found in the worked solutions in the online resources.

c. Many-to-one correspondence

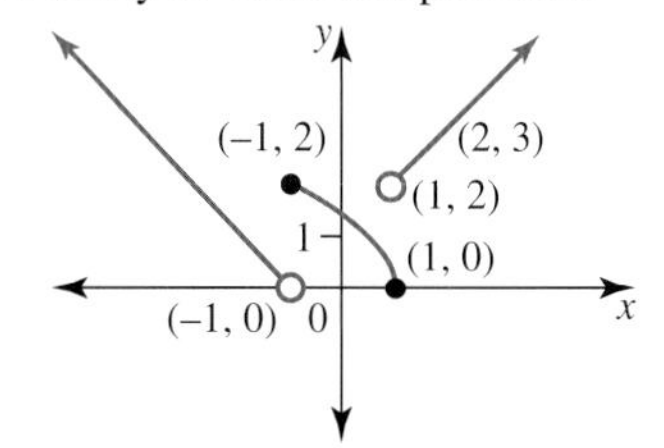

23. a. $y = \begin{cases} x+1, & x \le 0 \\ -x+1, & x > 0 \end{cases}$

b. $y = \begin{cases} 3, & x < 2 \\ 3x-6, & x \ge 2 \end{cases}$

24. a. i. $a = -2$

ii. Continuous

b. $a = -1, b = 5$

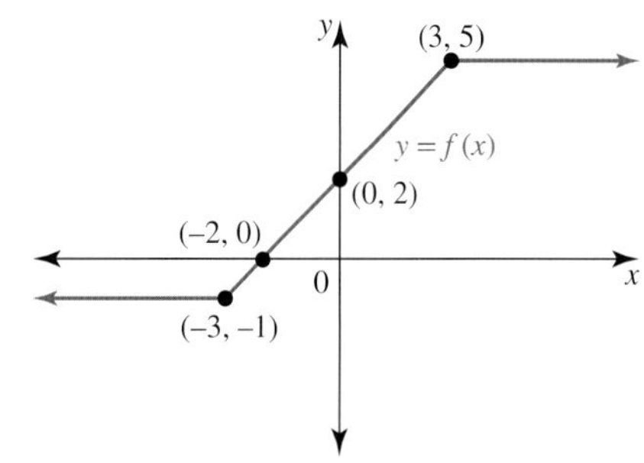

25. With cost C dollars and time t minutes: $C = \begin{cases} 2, & t = 0 \\ 0, & 0 < t \le 5 \\ t, & t > 5 \end{cases}$

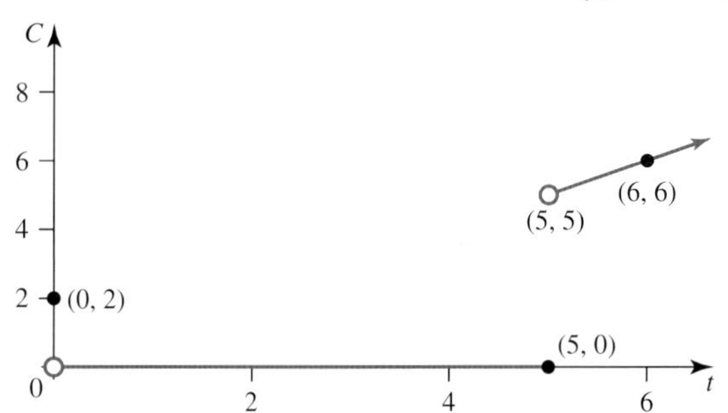

26. a. Either $D = [-4, \infty)$ or $D = (-\infty, -4]$. (Other answers are possible.)

b. $g^{-1} : (-9, \infty) \to R, g^{-1}(x) = \sqrt{x+25} - 4$; range R^+

c. i.

ii. $\left(\dfrac{-7+\sqrt{85}}{2}, \dfrac{-7+\sqrt{85}}{2}\right)$

d. $\dfrac{-7+\sqrt{85}}{2}$

27. a. $f(0) = 3 = f(5)$

b. Domain R; range $[2, \infty)$

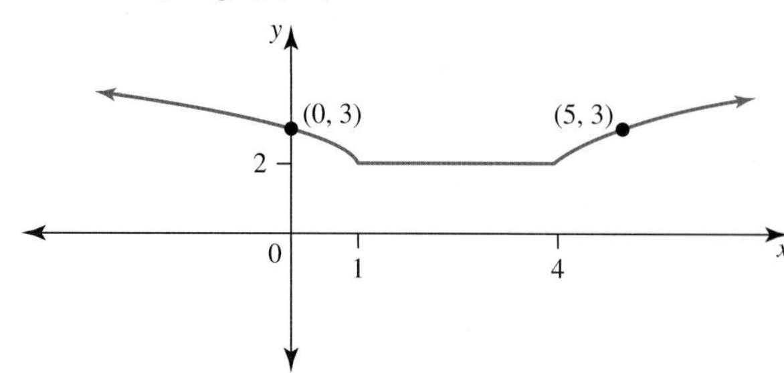

c. $x = -35, x = 40$

d. The inverse is not a function.

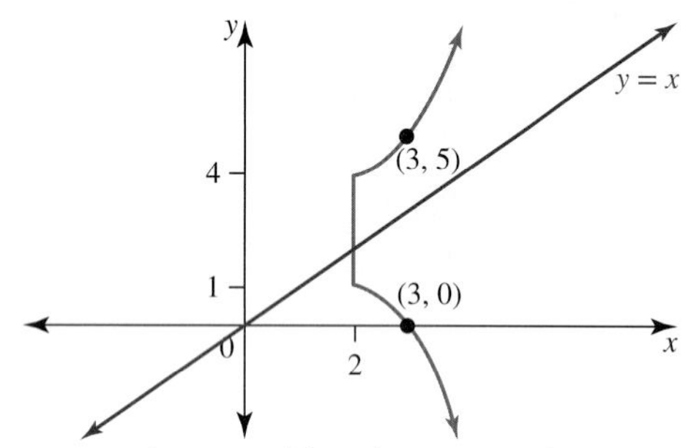

28. a. $y = 0.5\left((4x+33)^{0.5} - 5\right), y = -0.5\left((4x+33)^{0.5} + 5\right)$

b. $(-4.45, -4.45), (0.45, 0.45), (-0.76, -5.24), (-5.24, -0.76)$

6.5 Exam questions

Note: Mark allocations are available with the fully worked solutions online.

1. B
2. C
3. D

6.6 Transformations of functions

6.6 Exercise

1. a. Dilation of factor 3 from the x-axis

b. Reflection in the x-axis

c. Translation 5 units upwards

d. Translation 5 units to the left

2. a. $y = (x+2)^2 + 3$ b. $y = \dfrac{2}{x-5}$

c. $y = -\dfrac{1}{2}\sqrt{x}$ d. $y = 4 - x^4$

3. a. Horizontal translation 4 units left, vertical translation 5 units downwards

b. Dilation of factor 3 from the x-axis (parallel to the y-axis) followed by a horizontal translation 2 units right and vertical translation 4 units upwards

c. Reflection in the x-axis followed by a horizontal translation 3 units right and vertical translation 2 units upwards

d. Reflection in the x-axis and dilation by factor $\dfrac{3}{2}$ from the x-axis (parallel to the y-axis), followed by a horizontal translation 1 unit left and vertical translation 4 units upwards

f. Reflection in the x-axis and dilation of factor 4 from the x-axis (parallel to the y-axis)

4. a.

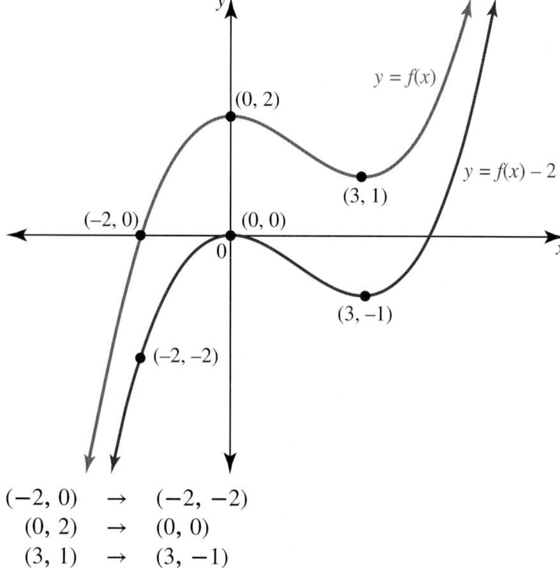

$(-2, 0) \rightarrow (-2, -2)$
$(0, 2) \rightarrow (0, 0)$
$(3, 1) \rightarrow (3, -1)$

b.

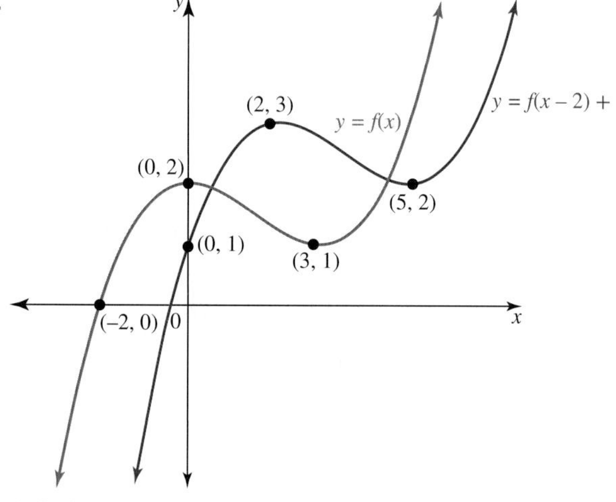

$(-2, 0) \rightarrow (0, 1)$
$(0, 2) \rightarrow (2, 3)$
$(3, 1) \rightarrow (5, 2)$

c.

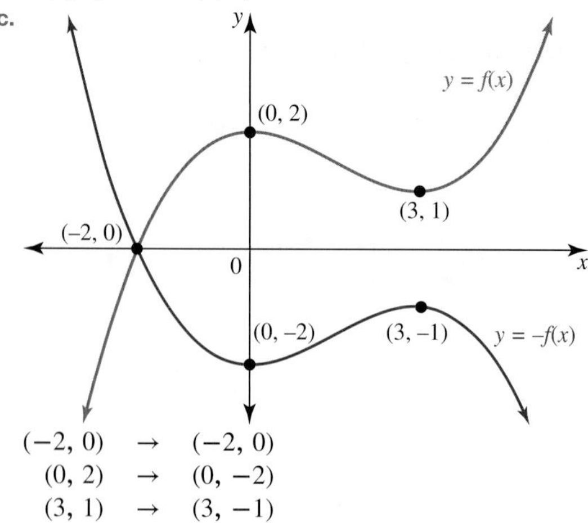

$(-2, 0) \rightarrow (-2, 0)$
$(0, 2) \rightarrow (0, -2)$
$(3, 1) \rightarrow (3, -1)$

5. a. $y = -(x-1)^2 + 3$

b. $y = -(x-1)^2 - 3$; a different image

6. a. Dilation of factor 3 from the y-axis
   b. Dilation of factor 8 from the x-axis and translation 1 unit upwards
   c. Horizontal translation 4 to the right; vertical translation 4 units downwards
   d. Dilation of factor 2 from the x-axis; reflection in the x-axis and translation $\frac{1}{2}$ unit to the left

7. a.

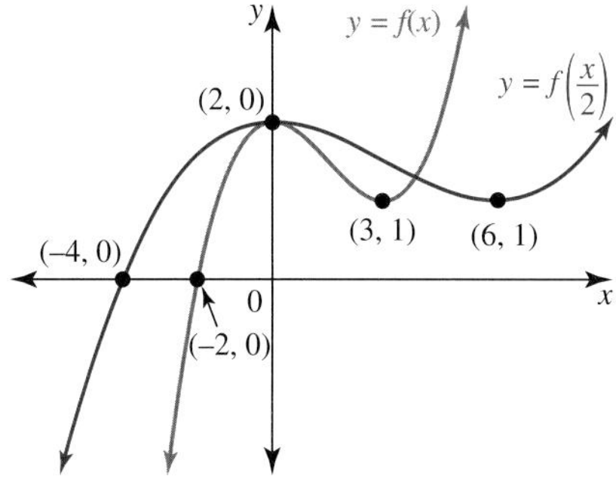

$$(-2, 0) \rightarrow (-4, 0)$$
$$(0, 2) \rightarrow (0, 2)$$
$$(3, 1) \rightarrow (6, 1)$$

   b.

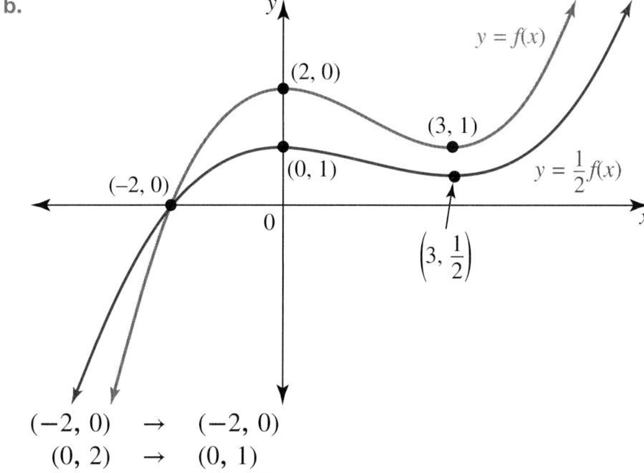

$$(-2, 0) \rightarrow (-2, 0)$$
$$(0, 2) \rightarrow (0, 1)$$
$$(3, 1) \rightarrow \left(3, \frac{1}{2}\right)$$

8. a. i. $y = 2\sqrt{x}$ ii. $y = \sqrt{\frac{x}{2}}$
   iii. $y = -\sqrt{x} + 2$ iv. $y = -\sqrt{x} - 2$
   v. $y = \sqrt{-x}$ vi. $y = -\sqrt{x - 2}$
   b. i. $y = 2x^4$ ii. $y = \frac{1}{16}x^4$
   iii. $y = -x^4 + 2$ iv. $y = -x^4 - 2$
   v. $y = -(x - 2)^4$ vi. $y = -(x - 2)^4$

9. a. i. Dilation of factor 5 from the x-axis (or parallel to the y-axis) followed by a horizontal translation of 3 units to the left
   ii. Dilation of factor 7 from the x-axis (or parallel to the y-axis) and reflection in the x-axis, followed by a horizontal translation of 6 units to the right and a vertical translation upwards of 8 units
   b. Dilation of factor 3 from the x-axis (or parallel to the y-axis) followed by a horizontal translation of 4 units to the left and a vertical translation downwards of 15 units

10. a. Reflection in the x-axis followed by a vertical translation upwards of 2 units
   b. Dilation of factor 3 from the x-axis (or parallel to the y-axis) followed by a horizontal translation of 1 unit to the left and a vertical translation downwards of 5 units

11. a.

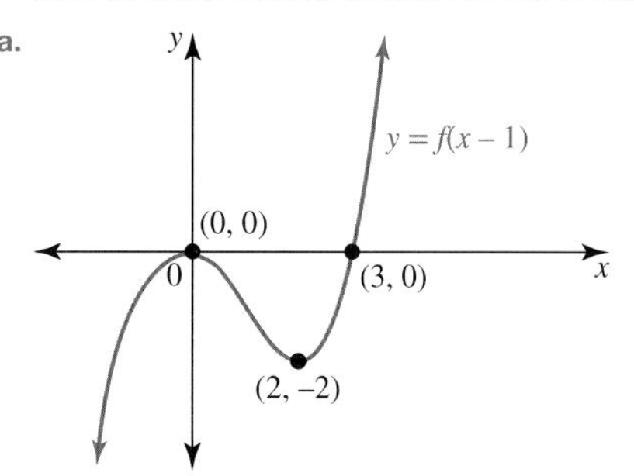

   b.

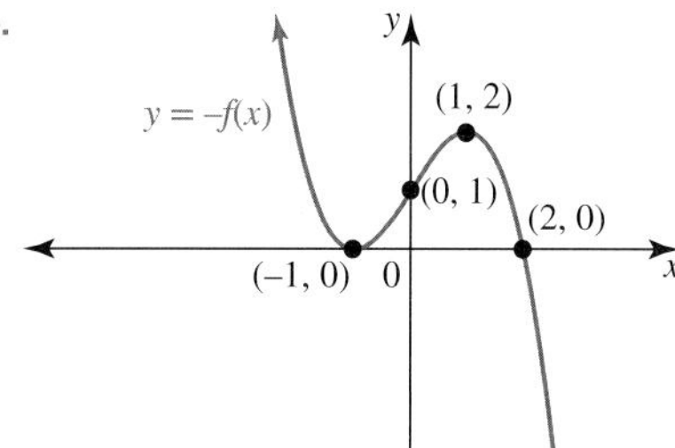

   c.

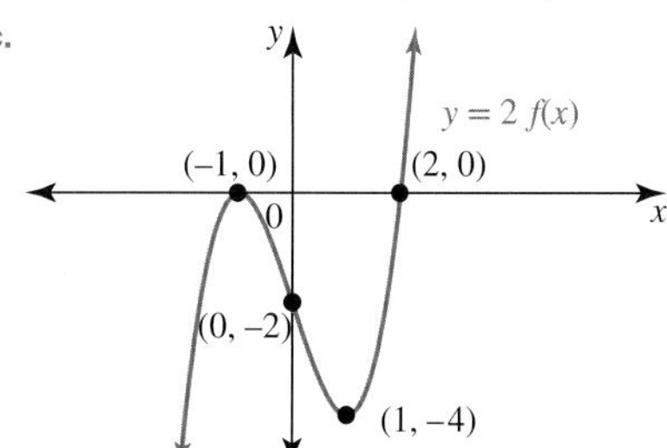

   d.

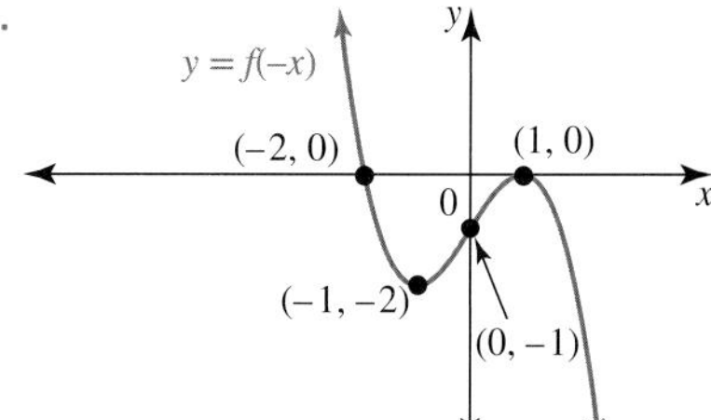

   e.

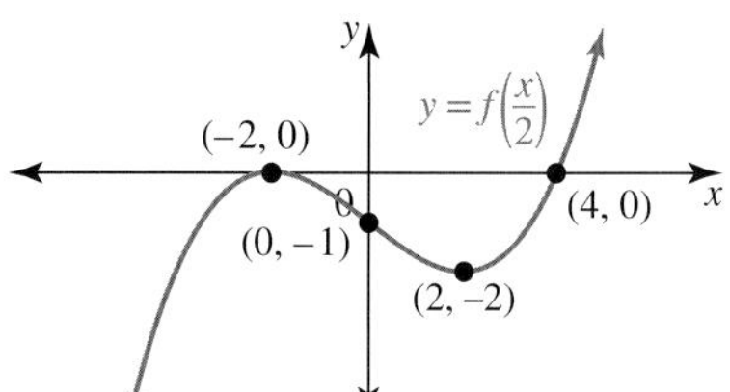

   f.

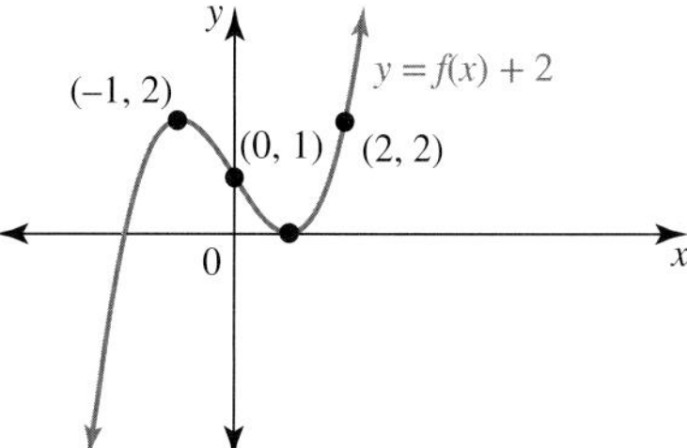

12. a. Dilation of factor 2 from the x-axis (parallel to the y-axis) and reflection in the y-axis followed by a vertical translation of 1 unit upwards

b. Reflection in the x-axis and dilation of factor $\frac{1}{3}$ from the y-axis

13. a. $y = \frac{2}{x+3}$

b. Horizontal translation 3 units to the right followed by dilation of factor $\frac{1}{2}$ from the x-axis

14. a. $(1, -8)$

b. $(-3, 4)$

c. $(3, -0.8)$

d. $(0.6, -4)$

15. a. i. $y = -\frac{3}{x}$

ii. $y = -\frac{3}{x}$

b. i. $y = \frac{3}{x^2} + 6$

ii. $y = \frac{3}{x^2} + 18$

c. $x = -1, y = 0$

16. a. Dilation of factor 2 from the x-axis; translation of 3 units to the left

b. Dilation of factor 6 from the x-axis; translation of 2 units to the right, 1 unit upwards

c. Dilation of factor 2 from the x-axis; reflection in the x-axis; translation of 3 units upwards

17. a. $y = \frac{1}{3(x+3)^2}$

b. $y = \frac{1}{1-x}$

c. $y = 2\sqrt[3]{x-1}$

d. $y = -(x+3)(x-3)(x-7)$

e. $y = \frac{1}{8}x^2(x+4)(x-4)$

18. a. The image is the same parabola, g.

b. The image is $y = -x^{\frac{1}{3}}$ for either reflection.

c. Maximum turning point $(-2, -5)$

d. $f(x) = -3(x-2)^3$

19. a.

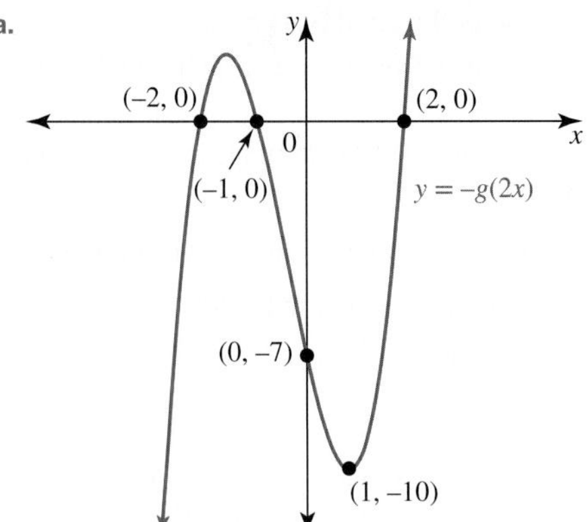

b.

c. $h > 4$

d. $g(x) = -\frac{7}{32}(x+4)(x+2)(x-4)$

$g(2x) = -\frac{7}{4}(x+2)(x+1)(x-2)$

20. y_2 is a dilation of y_1 factor $\frac{1}{2}$ from the y-axis; y_3 is a dilation of y_1 factor 2 from the y-axis. All are parabolas.

6.6 Exam questions

Note: Mark allocations are available with the fully worked solutions online.

1. A

2. $y = -2\sqrt{(x+3)}$

3. Dilation of factor 4 from the x-axis; reflection in the x-axis, followed by a horizontal translation of 1 unit to the right and a vertical translation downwards by 3 units

6.7 Review

6.7 Exercise

Technology free: short answer

1. See the table at the bottom of the page.*

a.

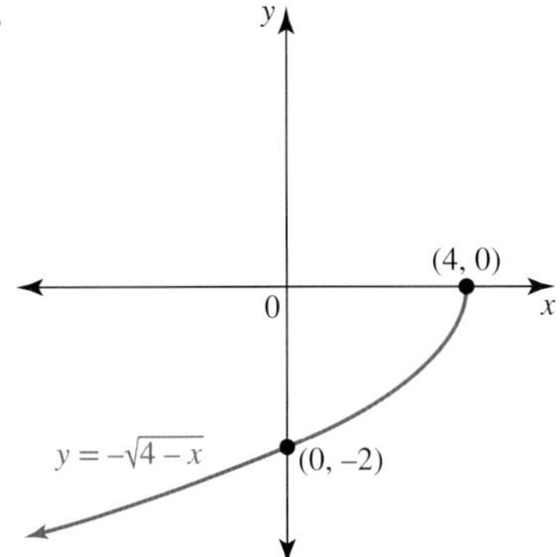

b.

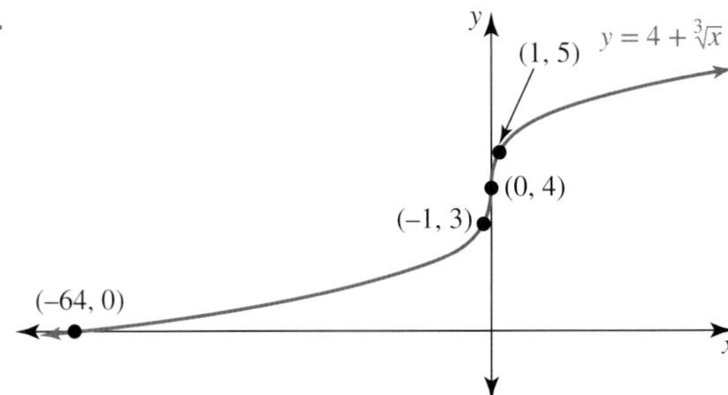

c.

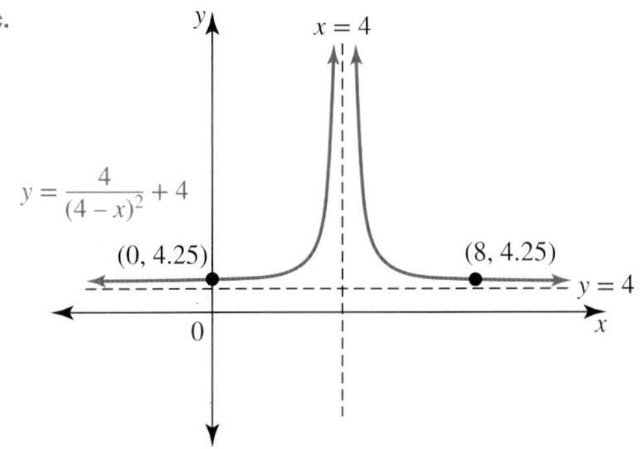

d.

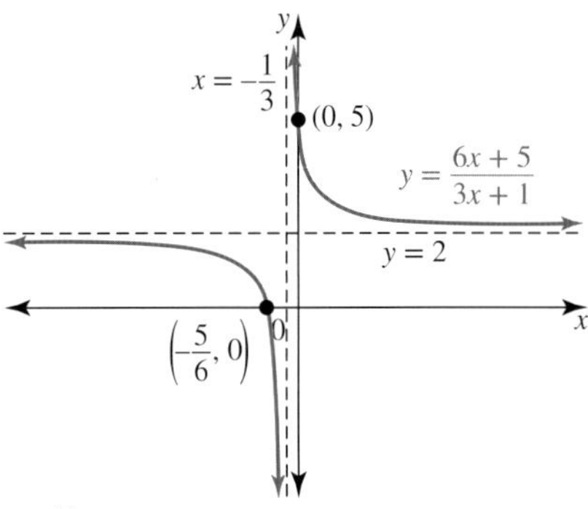

2. $3 - \dfrac{7}{x+3}$; asymptotes $x = -3$, $y = 3$

3. a. $f(0) + f(1) + f(2) = 3 + 2 + 0 = 5$

   b.

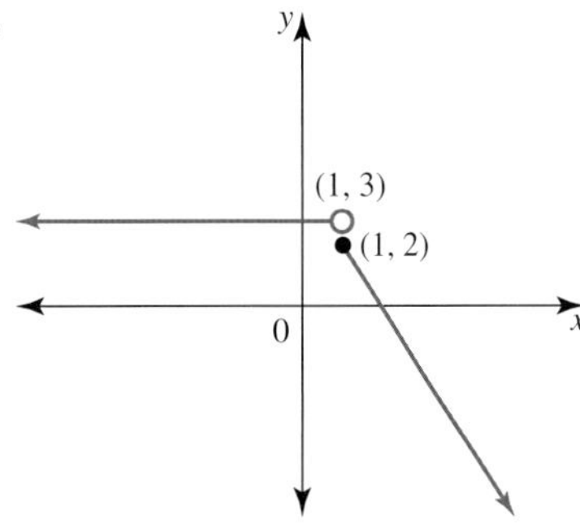

   c. Domain R, range $(-\infty, 2] \cup \{3\}$

   d. Many-to-one correspondence

4. a. Domain $[-2, 4)$, range $(-1, 2]$

   b. $f^{-1}(x) = 2 - 2x$; domain $(-1, 2]$, range $[-2, 4)$

   c.

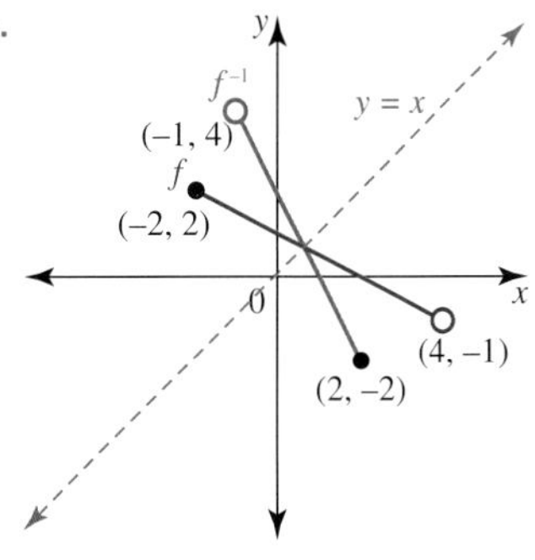

   d. $\left(\dfrac{2}{3}, \dfrac{2}{3}\right)$

*1.

	Shape	x-intercepts	y-intercepts	Domain	Range
a	Square root function, end point $(4, 0)$	$(4, 0)$	$(0, -2)$	$(-\infty, 4]$	$(-\infty, 0]$
b	Cube root function, point of inflection $(0, 4)$	$(-64, 0)$	$(0, 4)$	R	R
c	Truncus, asymptotes $x = 4, y = 4$	None	$\left(0, 4\frac{1}{4}\right)$	$R\backslash\{4\}$	$(4, \infty)$
d	Hyperbola, asymptotes $x = -\frac{1}{3}, y = 2$	$\left(-\frac{5}{6}, 0\right)$	$(0, 5)$	$R \backslash \left\{-\frac{1}{3}\right\}$	$R\backslash\{2\}$

5. a. $y=\sqrt{2x+4}+3$ b. $y=\begin{cases} x+8, x \le -2 \\ 4-x, x>-2 \end{cases}$

c. $y=2-\dfrac{2}{x+2}$

6. a. $y=\dfrac{1}{2}(1-x)^{\frac{1}{3}}$

b. $y=(x-5)^3-1$

c. i. $a=2$

ii. $f^{-1}:[6,\infty)\to R, f^{-1}(x)=2-\sqrt{x-6}$

Technology active: multiple choice

7. E
8. D
9. C
10. A
11. D
12. D
13. E
14. E
15. C
16. A

Technology active: extended response

17. a. Minimum turning point $(5,-4)$, y-intercept $(0,21)$, x-intercept $(3,0),(7,0)$

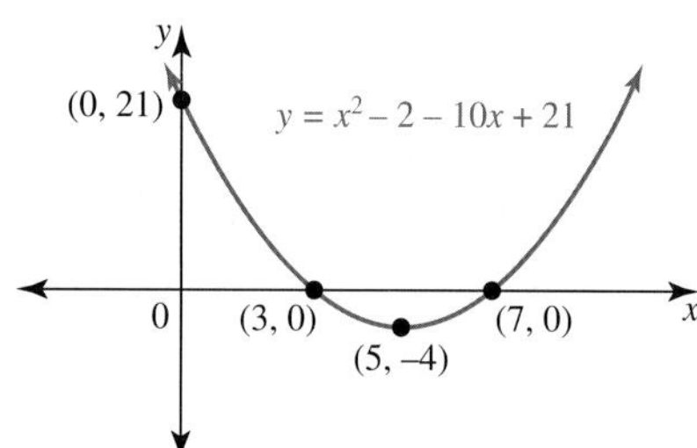

b. $k>4$

c. $h<-7$

d. See the table at the bottom of the page.*

i.

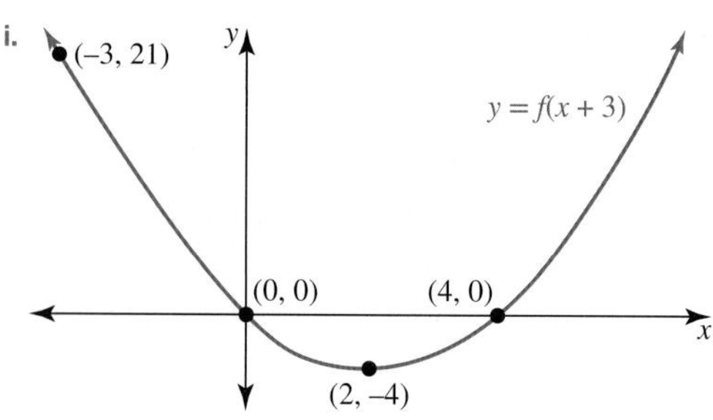

ii.

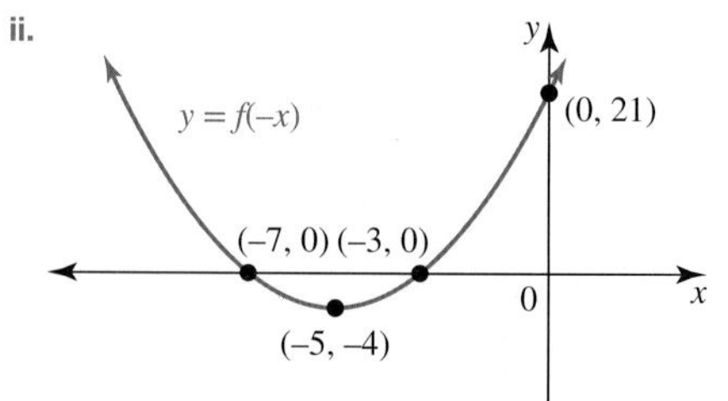

iii.

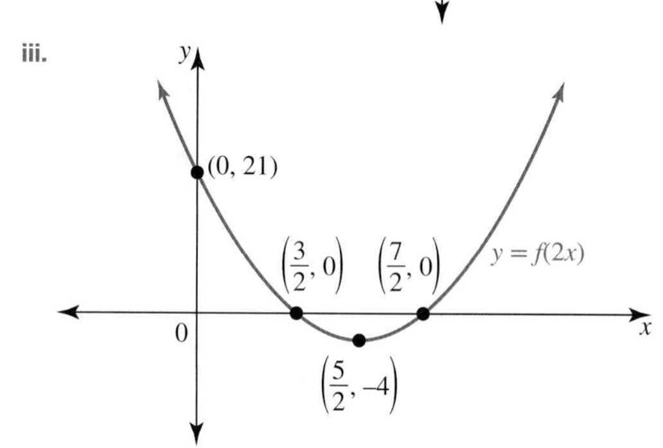

iv.

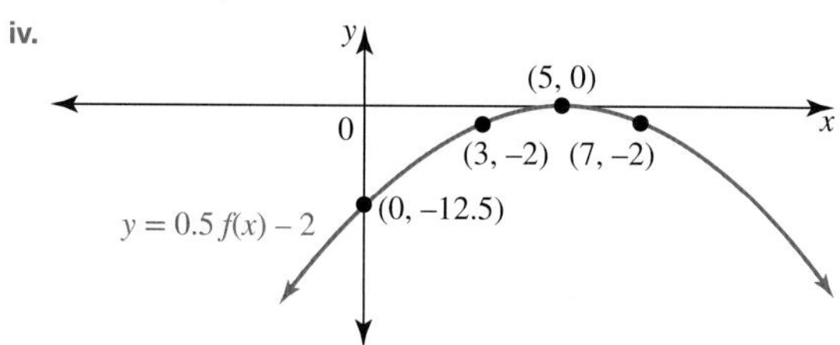

18. a. $I=\dfrac{k}{d}$

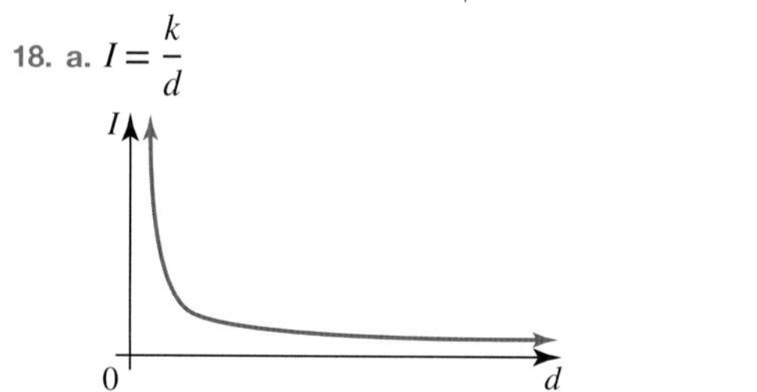

*17. d.

	Transformation	Turning point	x-intercepts	y-intercept	Other
i.	Translation 3 units left	$(2,-4)$	$(0,0),(4,0)$	$(0,0)$	$(-3,21)$
ii.	Reflection in y-axis	$(-5,-4)$	$(-7,0),(-3,0)$	$(0,21)$	
iii.	Dilation of factor $\frac{1}{2}$ from the y-axis	$\left(\frac{5}{2},-4\right)$	$\left(\frac{3}{2},0\right),\left(\frac{7}{2},0\right)$	$(0,21)$	
iv.	Reflection in the x-axis, dilation of factor 0.5 from the x-axis, translation 2 units down	$(5,0)$	$(5,0)$	$(0,-12.5)$	$(3,-2),(7,-2)$

b. $y = \dfrac{54}{(x+2)^2} - 2, a = \sqrt[3]{54} - 2$

c. i. 5 cm　　ii. 77.8%

19. a. Reflect in the y-axis, then translate 4 right; or translate 4 left, then reflect in the y-axis.

$y = \dfrac{-2}{x-2} - 1, x < 2$

b. Diameter 4 cm

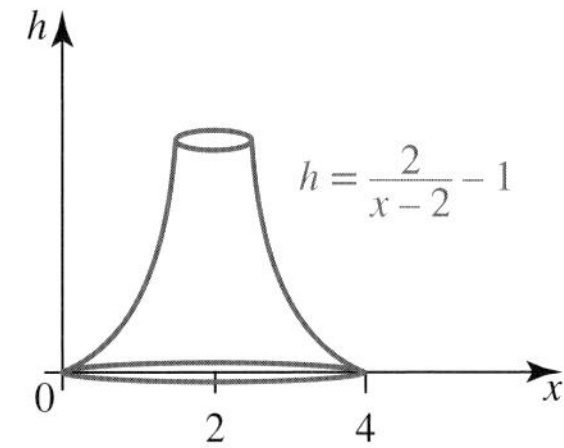

c. 3 cm

d. 0.8 cm, 0.64π cm^2

20. a. i. 45°　　ii. $y = x - 2$

iii. $y = \begin{cases} 2 - x, & x < 2 \\ x - 2, & x \geq 2 \end{cases}$

b. i. Fails the vertical line test

ii. $x = \begin{cases} y + 2, & 0 < y < 2 \\ 6 - y, & y \geq 2 \end{cases}$

c. $y = -x + 30, 0 \leq x \leq 4$

6.7 Exam questions

Note: Mark allocations are available with the fully worked solutions online.

1. $g(x) = x^2 + 4x - 5; \left(\dfrac{1}{2}, -2\dfrac{3}{4}\right)$

2. C

3.

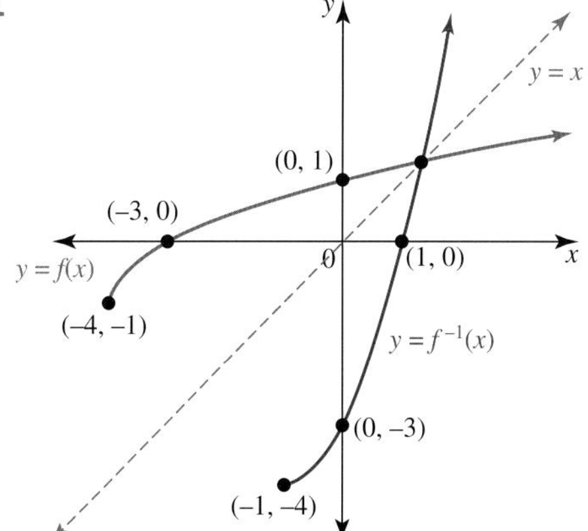

4. a. $f(-1) = 1$

$f(1) = 2$

$f(2) = 1$

b.

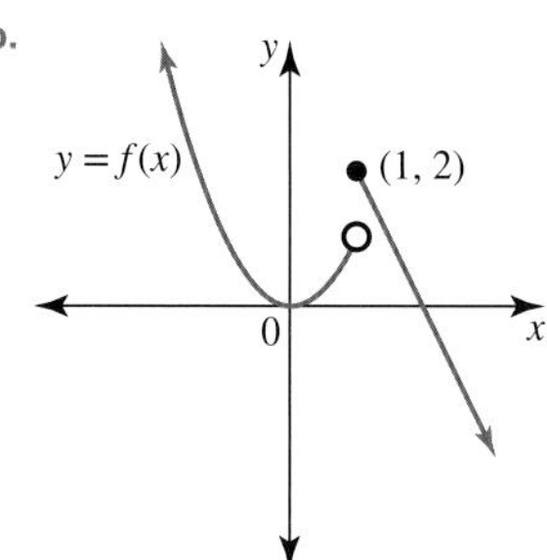

Domain R, range R, discontinuous at $x = 1$

5. A

7 Probability

LEARNING SEQUENCE

Fully worked solutions for this topic are available online.

7.1 Overview

Hey students! Bring these pages to life online

Watch videos

Engage with interactivities

Answer questions and check results

Find all this and MORE in jacPLUS

7.1.1 Introduction

The concept of randomness and random events is a feature of our lives. Unless we are one of a pair of identical twins, the complete set of genes we inherit from our parents is unique, a result of a random selection event at the time of conception.

Randomness is associated with a sense of 'fairness', an unbiasedness — for example, who serves first in a tennis match or which cricket side bats first is determined by the toss of a coin.

Probability theory gives us the means of measuring the likeliness of outcomes of a random event. Although the origins of probability lie in the gambling addictions of Cardano, the 16th-century Italian mathematician, and in the work of the 17th-century French mathematicians Pascal and Fermat, initially on behalf of a gambler (the Chevalier de Méré), probability has far more diverse applications today. Importantly, it is essential for quantum physics, the science of subatomic particles where properties such as position can only be expressed in terms of probabilities.

Probability theory developed substantially throughout the 20th century. As a branch of mathematics and statistics it is widely used in the finance and insurance industries, in market analysis and in the sciences, including actuarial science and biological sciences.

KEY CONCEPTS

This topic covers the following key concepts from the VCE Mathematics Study Design:

- random experiments, sample spaces, outcomes, elementary and compound events, random variables and the distribution of results of experiments
- simulation using simple random generators such as coins, dice, spinners and pseudo-random generators using technology, and the display and interpretation of results, including informal consideration of proportions in samples
- addition and multiplication principles for counting
- combinations including the concept of a selection and computation of ${}^nC_r$ and the application of counting techniques to probability
- probability of elementary and compound events and their representation as lists, grids, Venn diagrams, tables and tree diagrams
- the addition rule for probabilities, $\Pr(A \cup B) = \Pr(A) + \Pr(B) - \Pr(A \cap B)$, and the relation that for mutually exclusive events $\Pr(A \cap B) = 0$, hence $\Pr(A \cup B) = \Pr(A) + \Pr(B)$
- conditional probability in terms of reduced sample space, the relations $\Pr(A \mid B) = \dfrac{\Pr(A \cap B)}{\Pr(B)}$ and $\Pr(A \cap B) = \Pr(A \mid B) \times \Pr(B)$
- the law of total probability for two events $\Pr(A) = \Pr(A \mid B)\Pr(B) + \Pr(A \mid B')\Pr(B')$
- the relations for pairwise independent events A and B, $\Pr(A \mid B) = \Pr(A)$, $\Pr(B \mid A) = \Pr(B)$ and $\Pr(A \cap B) = \Pr(A) \times \Pr(B)$
- simulation to estimate probabilities involving selection with and without replacement.

7.2 Probability review

LEARNING INTENTION

At the end of this subtopic you should be able to:
- display probabilities in various diagrams and tables
- understand complementary events, mutually exclusive events and the addition formula
- calculate and solve probability problems.

The language and the notation used in the theory of probability is that which appears in set theory. Some set notation has already been introduced in earlier topics.

7.2.1 Notation and fundamentals: outcomes, sample spaces and events

Consider the experiment or trial of spinning a wheel that is divided into eight equal sectors, with each sector marked with one of the numbers 1 to 8. If the wheel is unbiased, each of these numbers is equally likely to occur.

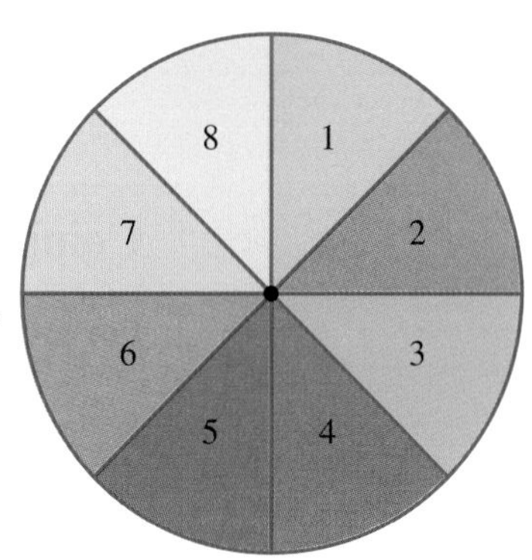

The **outcome** of each trial is one of the eight numbers.

The **sample space**, ξ, is the set of all possible outcomes: $\xi = \{1, 2, 3, 4, 5, 6, 7, 8\}$.

An **event** is a particular set of outcomes that is a subset of the sample space. For example, if M is the event of obtaining a number that is a multiple of 3, then $M = \{3, 6\}$. This set contains two outcomes. This is written in set notation as $n(M) = 2$.

The **probability** of an event is the long-term proportion, or relative frequency, of its occurrence.

For any event A, the probability of its occurrence is $\Pr(A) = \dfrac{n(A)}{n(\xi)}$.

Hence, for the event M:

$$\begin{aligned}\Pr(M) &= \frac{n(M)}{n(\xi)} \\ &= \frac{2}{8} \\ &= \frac{1}{4}\end{aligned}$$

This value does not mean that a multiple of 3 is obtained once in every four spins of the wheel. However, it does mean that after a very large number of spins of the wheel, the proportion of times that a multiple of 3 would be obtained approaches $\frac{1}{4}$.

Probability of an event

For any event A,

$$0 \leq \Pr(A) \leq 1.$$

- If $\Pr(A) = 0$, then it is not possible for A to occur. For example, the chance that the spinner lands on a negative number would be zero.
- If $\Pr(A) = 1$, then the event A is certain to occur. For example, it is 100% certain that the number the spinner lands on will be smaller than 9.

The probability of each outcome $\Pr(1) = \Pr(2) = \Pr(3) = ... = \Pr(8) = \frac{1}{8}$ for this spinning wheel. As each outcome is equally likely to occur, the outcomes are **equiprobable**. In other situations, some outcomes may be more likely than others.

For any sample space, $\Pr(\xi) = \frac{n(\xi)}{n(\xi)} = 1$, and the sum of the probabilities of each of the outcomes in any sample space must total 1.

Complementary events

For the spinner example, the event that the number is not a multiple of 3 is the complement of the event M. The complementary event is written as M' or as $\overline{M}$.

$$\begin{aligned}\Pr(M') &= 1 - \Pr(M)\\ &= 1 - \frac{1}{4}\\ &= \frac{3}{4}\end{aligned}$$

Complementary events

For any complementary events:

$$\mathbf{Pr(A) + Pr(A') = 1}$$

or

$$\mathbf{Pr(A') = 1 - Pr(A)}$$

WORKED EXAMPLE 1 Calculating complementary probabilities

A spinning wheel is divided into eight sectors, each of which is marked with one of the numbers 1 to 8. This wheel is biased so that Pr(8) = 0.3, while the other numbers are equiprobable.

a. Calculate the probability of obtaining the number 4.

b. If A is the event the number obtained is even, calculate Pr(A) and Pr(A').

THINK	WRITE
a. 1. State the complement of obtaining the number 8 and the probability of this.	a. The sample space contains the numbers 1 to 8, so the complement of obtaining 8 is obtaining one of the numbers 1 to 7. As $\Pr(8) = 0.3$, the probability of not obtaining 8 is $1 - 0.3 = 0.7$.
2. Calculate the required probability.	Since each of the numbers 1 to 7 are equiprobable, the probability of obtaining each number is $\frac{0.7}{7} = 0.1$. Hence, $\Pr(4) = 0.1$.

b. 1. Identify the elements of the event.	**b.** $A = \{2, 4, 6, 8\}$
2. Calculate the probability of the event.	$\begin{aligned}\Pr(A) &= \Pr(2 \text{ or } 4 \text{ or } 6 \text{ or } 8)\\ &= \Pr(2) + \Pr(4) + \Pr(6) + \Pr(8)\\ &= 0.1 + 0.1 + 0.1 + 0.3\\ &= 0.6\end{aligned}$
3. State the complementary probability.	$\begin{aligned}\Pr(A') &= 1 - \Pr(A)\\ &= 1 - 0.6\\ &= 0.4\end{aligned}$

7.2.2 Venn diagrams

A **Venn diagram** can be useful for displaying the union and intersection of sets. Such a diagram may be helpful in displaying compound events in probability, as illustrated for the sets or events A and B.

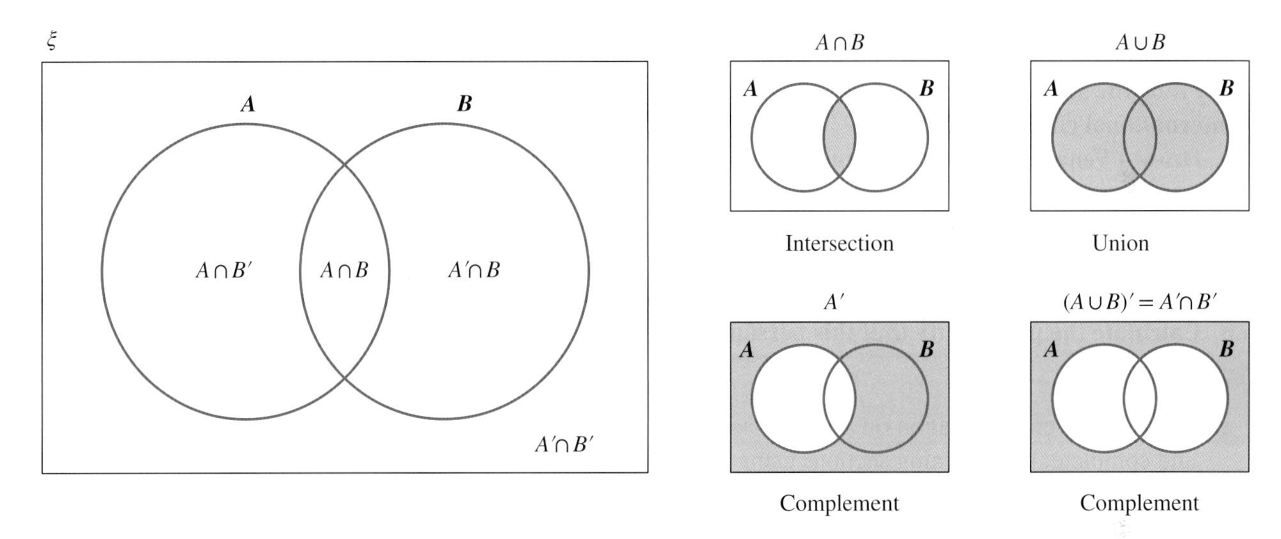

The information shown in the Venn diagram may be the actual outcomes for each event, or it may only show a number that represents the number of outcomes for each event. Alternatively, the Venn diagram may show the probability of each event. The total probability is 1; that is, $\Pr(\xi) = 1$.

The addition formula

The number of elements contained in set A is denoted by $n(A)$.

The Venn diagram illustrates that $n(A \cup B) = n(A) + n(B) - n(A \cap B)$.

Hence, dividing by the number of elements in the sample space gives:

$$\frac{n(A \cup B)}{n(\xi)} = \frac{n(A)}{n(\xi)} + \frac{n(B)}{n(\xi)} - \frac{n(A \cap B)}{n(\xi)}$$

$$\therefore \Pr(A \cup B) = \Pr(A) + \Pr(B) - \Pr(A \cap B)$$

The result is known as the addition formula.

The addition formula

$$\mathbf{Pr(A \cup B) = Pr(A) + Pr(B) - Pr(A \cap B)}$$

If the events A and B are **mutually exclusive**, then they cannot occur simultaneously. For mutually exclusive events, $n(A \cap B) = 0$ and therefore $\Pr(A \cap B) = 0$.

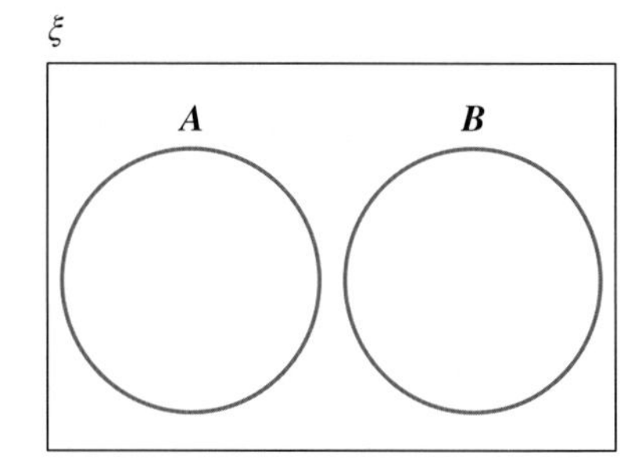

Mutually exclusive events

For mutually exclusive events:

$$\Pr(A \cup B) = \Pr(A) + \Pr(B)$$

WORKED EXAMPLE 2 Solving problems with Venn diagrams

From a survey of 50 people it was found that in the past month 30 people had made a donation to a local charity, 25 had donated to an international charity and 20 had made donations to both local and international charities.

Let L be the set of people donating to a local charity and I the set of people donating to an international charity.

a. **Draw a Venn diagram to illustrate the results of this survey.**

One person from the group is selected at random.

b. **Using appropriate notation, calculate the probability that this person donated to a local charity but not an international one.**

c. **Calculate the probability that this person did not make a donation to either type of charity.**

d. **Calculate the probability that this person donated to at least one of the two types of charity.**

THINK	WRITE
a. Show the given information on a Venn diagram and complete the remaining sections using arithmetic.	a. Given: $n(\xi) = 50$, $n(L) = 30$, $n(I) = 25$ and $n(L \cap I) = 20$ 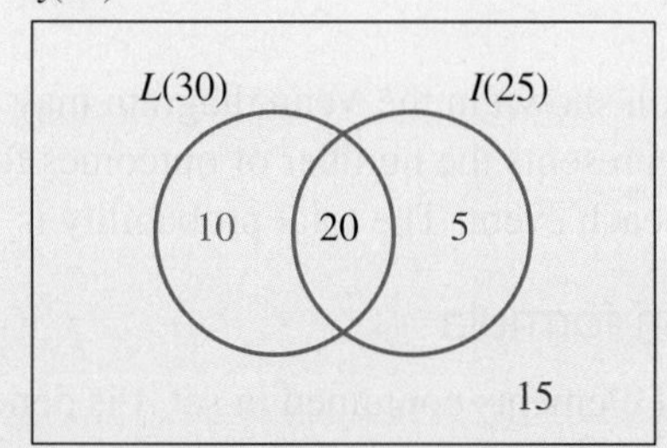
b. 1. State the required probability using set notation.	b. $\Pr(L \cap I') = \dfrac{n(L \cap I')}{n(\xi)}$
2. Identify the value of the numerator from the Venn diagram and calculate the probability.	$\Pr(L \cap I') = \dfrac{10}{50} = \dfrac{1}{5}$
3. Express the answer in context.	The probability that the randomly chosen person donated to a local charity but not an international one is 0.2.

c. 1. State the required probability using set notation.

c. $\Pr(L' \cap I') = \dfrac{n(L' \cap I')}{n(\xi)}$

2. Identify the value of the numerator from the Venn diagram and calculate the probability.

$$\Pr(L \cap I) = \frac{15}{50}$$
$$= \frac{3}{10}$$

3. Express the answer in context.

The probability that the randomly chosen person did not donate is 0.3.

d. 1. State the required probability using set notation.

d. $\Pr(L \cup I) = \dfrac{n(L \cup I)}{n(\xi)}$

2. Identify the value of the numerator from the Venn diagram and calculate the probability.

$$\Pr(L \cup I) = \frac{10 + 20 + 5}{50}$$
$$= \frac{35}{50}$$
$$= \frac{7}{10}$$

3. Express the answer in context.

The probability that the randomly chosen person donated to at least one type of charity is 0.7.

7.2.3 Probability tables

For situations involving two events, a probability table can provide an alternative to a Venn diagram. Consider the Venn diagram shown.

A **probability table** presents any known probabilities of the four compound events $A \cap B$, $A \cap B'$, $A' \cap B$ and $A' \cap B'$ in rows and columns.

	$\boldsymbol{B}$	$\boldsymbol{B'}$	
$\boldsymbol{A}$	$\Pr(A \cap B)$	$\Pr(A \cap B')$	$\mathbf{Pr}(\boldsymbol{A})$
$\boldsymbol{A'}$	$\Pr(A' \cap B)$	$\Pr(A' \cap B')$	$\mathbf{Pr}(\boldsymbol{A'})$
	$\mathbf{Pr}(\boldsymbol{B})$	$\mathbf{Pr}(\boldsymbol{B'})$	$\mathbf{Pr}(\boldsymbol{\xi}) = \mathbf{1}$

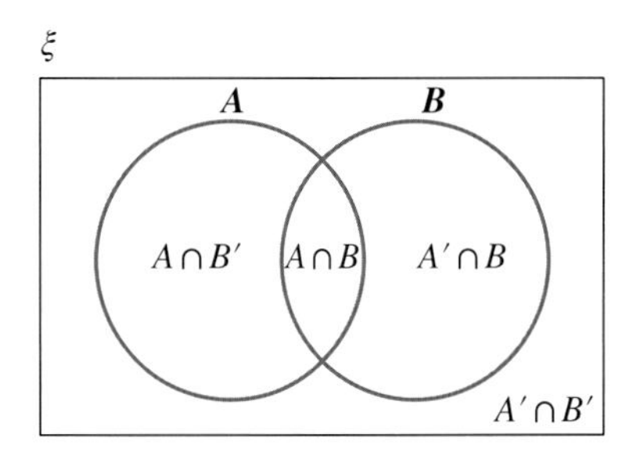

This allows the table to be completed using arithmetic calculations. For example, $\Pr(A) = \Pr(A \cap B) + \Pr(A \cap B')$ and $\Pr(B) = \Pr(A \cap B) + \Pr(A' \cap B)$.

The probabilities of complementary events can be calculated using the formula $\Pr(A') = 1 - \Pr(A)$.

To obtain $\Pr(A \cup B)$, the addition formula $\Pr(A \cup B) = \Pr(A) + \Pr(B) - \Pr(A \cap B)$ can be used.

Probability tables are also known as **Karnaugh maps**.

WORKED EXAMPLE 3 Constructing probability tables

Given $\Pr(A) = 0.4, \Pr(B) = 0.7$ and $\Pr(A \cap B) = 0.2$:

a. **construct a probability table for the events A and B**

b. **calculate $\Pr(A' \cup B)$.**

THINK

a. 1. Enter the given information in a probability table.

WRITE

a. Given: $\Pr(A) = 0.4$, $\Pr(B) = 0.7$, $\Pr(A \cap B) = 0.2$ and also $\Pr(\xi) = 1$

	B	B'	
A	0.2		**0.4**
A'			
	0.7		**1**

2. Add in the complementary probabilities.

$\Pr(A') = 1 - 0.4 = 0.6$ and $\Pr(B') = 1 - 0.7 = 0.3$

	B	B'	
A	0.2		**0.4**
A'			**0.6**
	0.7	**0.3**	**1**

3. Complete the remaining sections using arithmetic.

For the first row, $0.2 + 0.2 = 0.4$.
For the first column, $0.2 + 0.5 = 0.7$.

	B	B'	
A	0.2	0.2	**0.4**
A'	0.5	0.1	**0.6**
	0.7	**0.3**	**1**

b. 1. State the addition formula.

b. $\Pr(A' \cup B) = \Pr(A') + \Pr(B) - \Pr(A' \cap B)$

2. Use the values in the probability table to carry out the calculation.

From the probability table, $\Pr(A' \cap B) = 0.5$.

$\therefore \Pr(A' \cup B) = 0.6 + 0.7 - 0.5$
$= 0.8$

7.2.4 Other ways to illustrate sample spaces

In experiments involving two tosses of a coin or two rolls of a die, the sample space can be illustrated using a simple tree diagram or a lattice diagram.

Simple tree diagram

The outcome of each toss of the coin is either Heads (H) or Tails (T). For two tosses, the outcomes are illustrated by the tree diagram shown.

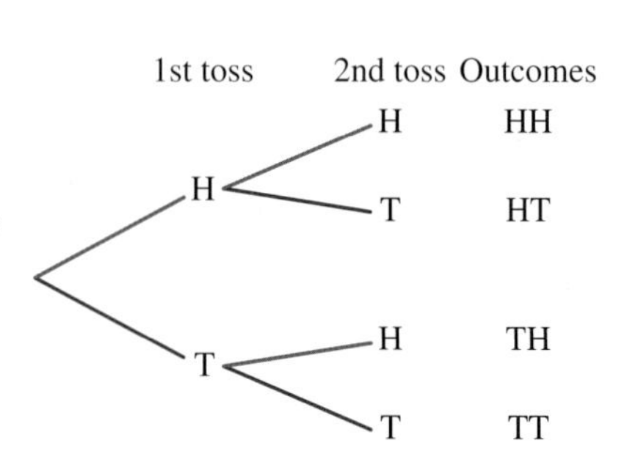

The sample space consists of the four equally likely outcomes HH, HT, TH and TT. This means the probability of obtaining two Heads in two tosses of a coin would be $\frac{1}{4}$.

The tree diagram could be extended to illustrate repeated tosses of the coin.

Lattice diagram

For two rolls of a six-sided die, the outcomes are illustrated graphically using a grid known as a lattice diagram.

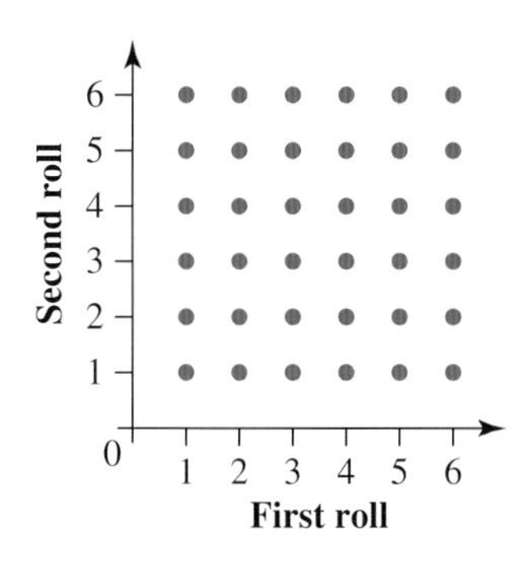

There are 36 points on the grid. These points represent the equally likely elements of the sample space. Each point can be described by a pair of coordinates, with the first coordinate giving the outcome from the first roll of the die and the second coordinate giving the outcome from the second roll of the die. As there is only one point with coordinates $(6, 6)$, this means the probability that both rolls result in a 6 is $\frac{1}{36}$.

Use of physical measurements in infinite sample spaces

When an arrow is fired at an archery target, there are an infinite number of points at which the arrow could land. The sample space can be represented by the area measure of the target, assuming the arrow hits it. The probability of such an arrow landing in a particular section of the target can then be calculated as the ratio of that area to the total area of the target.

WORKED EXAMPLE 4 Drawing a lattice diagram

a. A four-sided tetrahedral die with faces labelled 1, 2, 3, 4 is rolled at the same time a coin is tossed.
  - **i. Draw a lattice diagram to represent the sample space.**
  - **ii. Calculate the probability of obtaining a 1 on the die and a Tail on the coin.**
  - **iii. Calculate the probability of the event of obtaining a number that is at least 3, together with a Head on the coin.**

b. The coin is thrown onto a square sheet of cardboard with 10 cm edges. The coin lands with its centre inside or on the boundary of the cardboard. Given the radius of the circular coin is 1.5 cm, calculate the probability that the coin lands completely inside the area covered by the square piece of cardboard.

THINK	WRITE
a. i. Construct a lattice diagram to illustrate the sample space.	a. i. 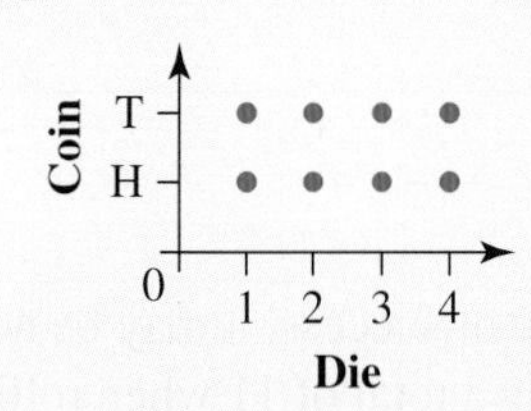
ii. Calculate the required probability.	ii. There are 8 equally likely outcomes in the sample space. Only one of these outcomes is a 1 on the die and a Tail on the coin. The probability of obtaining a 1 on the die and a Tail on the coin is $\frac{1}{8}$.
iii. 1. Identify the number of outcomes that make up the event.	iii. The event of obtaining a number that is at least 3 together with a Head on the coin occurs for the two outcomes 3H and 4H.
2. Calculate the probability.	The probability of obtaining a number that is at least 3 together with a Head is $\frac{2}{8} = \frac{1}{4}$.

b. 1. Draw a diagram to show the area in which the centre of the coin may land.

b. For the coin to land inside the square, its centre must be no less than 1.5 cm from each edge of the square.
The area in which the centre of the coin may land is a square of edge $10 - 2 \times 1.5 = 7$ cm.

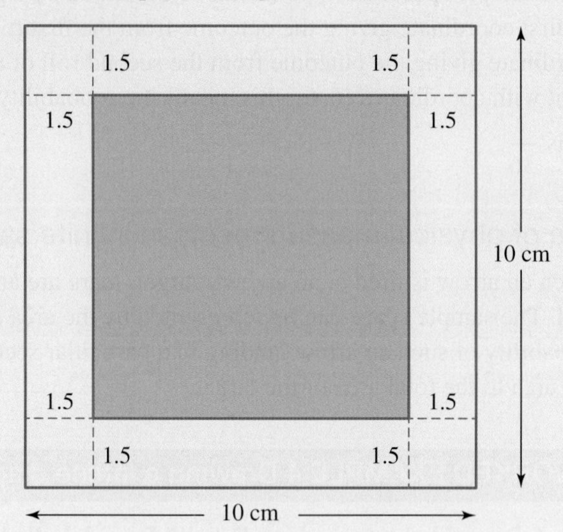

2. Calculate the required probability.

The area that the centre of the coin could land in is the area of the cardboard, which is $10 \times 10 = 100$ cm^2.
The area that the centre of the coin must land in for the coin to be completely inside the area of the cardboard is $7 \times 7 = 49$ cm^2.
Therefore, the probability that the coin lands completely inside the cardboard area is $\frac{49}{100} = 0.49$.

Simulations

To estimate the probability of success, it may be necessary to perform experiments or simulations. For example, the probability of obtaining a total of 11 when rolling two dice can be estimated by repeatedly rolling two dice and counting the number of times a total of 11 appears.

The results of a particular experiment or simulation can only give an estimate of the true probability. The more times the simulation is carried out, the better the estimate of the true probability.

For example, to simulate whether a baby is born male or female, a coin is flipped. If the coin lands Heads up, the baby is a boy. If the coin lands Tails up, the baby is a girl. The coin is flipped 100 times and returns 43 Heads and 57 Tails.

This simulation gives a probability of 0.43 of the baby being a boy and 0.57 for a girl. This closely resembles the theoretical probability of 0.5.

7.2 Exercise

Students, these questions are even better in jacPLUS

Receive immediate feedback and access sample responses

Access additional questions

Track your results and progress

Find all this and MORE in jacPLUS

Technology free

1. A bag contains 5 green, 6 pink, 4 orange and 8 blue counters. A counter is selected at random. Find the probability that the counter is:

   a. green
   b. orange or blue
   c. not blue
   d. black.

2. An unbiased six-sided die is thrown onto a table. State the probability that the number that is uppermost is

   a. 4
   b. not 4
   c. even
   d. smaller than 5
   e. at least 5
   f. greater than 12.

3. One letter from the alphabet, A through to Z, is chosen at random. Calculate the probability that the letter chosen is:

   a. Q
   b. a vowel
   c. either X or Y or Z
   d. not D
   e. either a consonant or a vowel
   f. one of the letters in the word PROBABILITY.

4. Tickets are drawn randomly from a barrel containing 2000 tickets. There are 3 prizes to be won: first, second and third. Josephine has purchased 10 tickets. Calculate the probability that she wins:

   a. the first prize
   b. the second prize but not the first
   c. all three prizes.

5. A card is drawn randomly from a standard pack of 52 cards. Calculate the probability that the card is:

   a. not green
   b. from a red suit
   c. a heart
   d. a 10 or from a red suit
   e. not an ace.

6. **WE1** A spinning wheel is divided into eight sectors, each of which is marked with one of the numbers 1 to 8. This wheel is biased so that $\Pr(6) = \frac{9}{16}$, while the other numbers are equiprobable.

   a. Calculate the probability of obtaining the number 1.
   b. If A is the event that a prime number is obtained, calculate $\Pr(A)$ and $\Pr(A')$.

7. A spinner is divided into 4 sections coloured red, blue, green and yellow. Each section is equally likely to occur. The spinner is spun twice. List all the possible outcomes and hence find the probability of obtaining:

   a. the same colour
   b. a red and a yellow
   c. not a green.

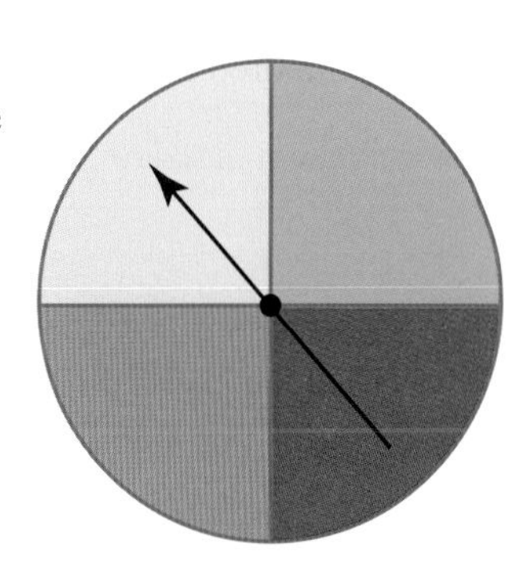

8. A bag contains 20 balls, of which 9 are green and 6 are red. One ball is selected at random.
   a. Calculate the probability that this ball is:
      i. either green or red
      ii. not red
      iii. neither green nor red.
   b. Determine how many additional red balls must be added to the original bag so that the probability that the chosen ball is red is 0.5.
9. A coin is tossed three times.
   a. Draw a simple tree diagram to show the possible outcomes.
   b. Calculate the probability of obtaining at least one Head.
   c. Calculate the probability of obtaining either exactly two Heads or two Tails.
10. A coin is tossed three times. Show the sample space on a tree diagram and hence find the probability of getting:
   a. 2 Heads and 1 Tail
   b. either 3 Heads or 3 Tails
   c. a Head on the first toss of the coin
   d. at least 1 Head
   e. no more than 1 Tail.
11. The 3:38 pm train to the city is late on average 1 day out of 3. Draw a probability tree to show the outcomes on three consecutive days. Hence, find the probability that the bus is:
   a. late on 1 of the days
   b. late on at least 2 days
   c. on time on the last day
   d. on time on all 3 days.

Technology active

12. A survey was carried out to find the type of occupation of 800 adults in a small suburb. There were 128 executives, 180 professionals, 261 trades workers, 178 labourers and 53 unemployed people.
   A person from this group is chosen at random. Calculate the probability that the person chosen is:
   a. a labourer
   b. not employed
   c. not an executive
   d. either a tradesperson or a labourer.

13. Two hundred people applied to do their driving test in October. The results are shown below.

	Passed	Failed
Male	73	26
Female	81	20

   a. Find the probability that a person selected at random has failed the test.
   b. Find the probability that a person selected at random is a female who passed the test.

14. A class of 20 primary school students went on an excursion to a local park. Some of the students carried an umbrella in their backpack, some carried a raincoat in their backpack, and some did not bring either an umbrella or a raincoat. The Venn diagram shown refers to the two sets B: 'students who had an umbrella in their backpack' and C: 'students who had a raincoat in their backpack'.
Use the Venn diagram to answer the following.

ξ (20)

B C

2 3 4

11

a. Calculate how many students had umbrellas.
b. A student's name is chosen at random. Calculate the probability that this student
   i. had an umbrella
   ii. had a raincoat but not an umbrella.
   iii. had neither an umbrella nor a raincoat.
c. Use the addition formula to calculate $\Pr(B' \cup C)$.

15. **WE2** In a group of 42 students, it was found that 30 students studied Mathematical Methods and 15 Geography. Ten of the Geography students did not study Mathematical Methods.
Let M be the set of students studying Mathematical Methods and let G be the set of students studying Geography.
a. Draw a Venn diagram to illustrate this situation.

One student from the group is selected at random.

b. Using appropriate notation, calculate the probability that this student studies Mathematical Methods but not Geography.
c. Calculate the probability that this student studies neither Mathematical Methods nor Geography.
d. Calculate the probability that this student studies only one of Mathematical Methods or Geography.

16. From a set of 18 cards numbered $1, 2, 3, \ldots, 18$, one card is drawn at random.
Let A be the event of obtaining a multiple of 3, B be the event of obtaining a multiple of 4 and let C be the event of obtaining a multiple of 5.
a. List the elements of each event and then illustrate the three events as sets on a Venn diagram.
b. State which events are mutually exclusive.
c. State the value of $\Pr(A)$.
d. Calculate the following.
   i. $\Pr(A \cup C)$
   ii. $\Pr(A \cap B')$
   iii. $\Pr((A \cup B \cup C)')$

17. **WE3** Given $\Pr(A) = 0.65$, $\Pr(B) = 0.5$ and $\Pr(A' \cap B') = 0.2$:
a. construct a probability table for the events A and B
b. calculate $\Pr(B' \cup A)$.

18. For two events A and B it is known that $\Pr(A \cup B) = 0.75$, $\Pr(A') = 0.42$ and $\Pr(B) = 0.55$.
a. Form a probability table for these two events.
b. State $\Pr(A' \cap B')$.
c. Show that $\Pr(A \cup B)' = \Pr(A' \cap B')$.
d. Show that $\Pr(A \cap B) = 1 - \Pr(A' \cup B')$.
e. Draw a Venn diagram for the events A and B.

19. Two unbiased dice are rolled and the larger of the two numbers is noted. If the two dice show the same number, then the sum of the two numbers is recorded. Use a table to show all the possible outcomes.
Hence, find the probability that the result is:

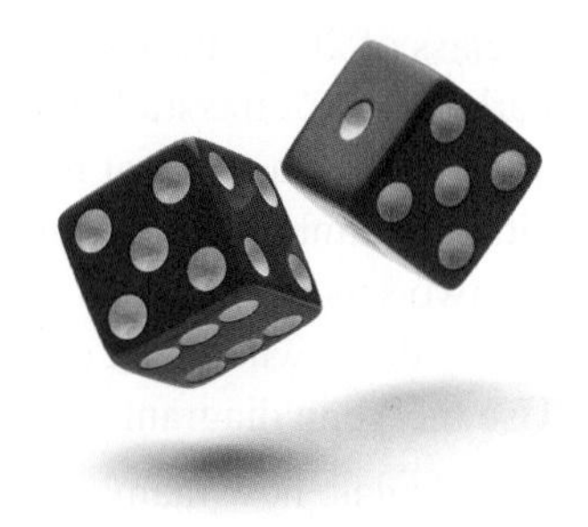

a. 5
b. 10
c. a number greater than 5
d. 7
e. either a two-digit number or a number greater than 6
f. not 9.

20. WE4 a. A six-sided die with faces labelled $1, 2, 3, 4, 5, 6$ is rolled at the same time that a coin is tossed.
i. Draw a lattice diagram to represent the sample space.
ii. Calculate the probability of obtaining a 6 on the die and a Head on the coin.
iii. Calculate the probability of obtaining an even number together with a Tail on the coin.

b. The coin is thrown onto a rectangular sheet of cardboard with dimensions 13 cm by 10 cm. The coin lands with its centre inside or on the boundary of the cardboard. Given the radius of the circular coin is 1.5 cm, calculate the probability that the coin lands completely inside the area covered by the rectangular piece of cardboard.

21. The gender of babies in a set of triplets is simulated by flipping 3 coins. If a coin lands Tails up, the baby is a boy. If a coin lands Heads up, the baby is a girl. In the simulation, the trial is repeated 40 times and the following results show the number of Heads obtained in each trial:

0, 3, 2, 1, 1, 0, 1, 2, 1, 0, 1, 0, 2, 0, 1, 0, 1, 2, 3, 2,
1, 3, 0, 2, 1, 2, 0, 3, 1, 3, 0, 1, 0, 1, 3, 2, 2, 1, 2, 1

a. Calculate the probability that exactly one of the babies in a set of triplets is female.
b. Calculate the probability that more than one of the babies in the set of triplets is female.

22. A die is rolled 20 times. Each roll results in one of the outcomes $\{1, 2, 3, 4, 5, 6\}$.
a. Use the random number generator of a CAS technology to simulate this experiment by generating 20 random integers between 1 and 6.
b. From your simulation, estimate of the chance of obtaining a six.
c. Explain how you could improve your estimate.

23. A sample of 100 first-year university science students were asked if they study physics or chemistry. It was found that 63 study physics, 57 study chemistry and 4 study neither.
A student is then selected at random. Calculate the probability that the student studies:

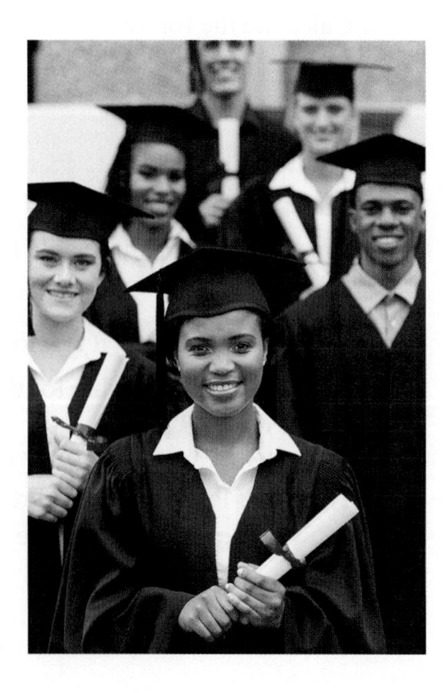

a. either physics or chemistry but not both
b. both physics and chemistry.

In total, there are 1200 first-year university science students.

c. Estimate the number of students who are likely to study both physics and chemistry.

Two students are chosen at random. Given that the same student can be chosen twice, find the probability that:

d. both students study physics and chemistry
e. each student studies just one of the two subjects, physics and chemistry
f. at least one of the two students studies neither physics nor chemistry.

7.2 Exam questions

Question 1 (3 marks) TECH-FREE

a. Draw a simple tree diagram to show the possible outcomes when an unbiased coin is tossed three times. **(1 mark)**

b. Calculate the probability of obtaining Heads on the first two throws and Tails on the third. **(1 mark)**

c. Calculate the probability of obtaining either exactly three Heads or three Tails. **(1 mark)**

Question 2 (1 mark) TECH-ACTIVE

MC Using the following probability table, the incorrect statement is

	B	B'	
A	0.2		0.4
A'		0.2	
	0.6		1

A. $\Pr(A \cap B) = 0.2$
B. $\Pr(A' \cap B) = 0.4$
C. $\Pr(A \cup B) = 0.9$
D. $\Pr(A) + \Pr(B) = 1$
E. $\Pr(A) < \Pr(B)$

Question 3 (3 marks) TECH-FREE

At a school with 55 VCE students, 30 are enrolled in Mathematical methods and 17 are enrolled in Chemistry. The Chemistry teacher noted that, in his class, 5 students were not studying Mathematical Methods.

a. Draw a Venn diagram showing the enrolments. **(1 mark)**

b. The principal randomly selects a student walking down the corridor and asks if they are studying Mathematical Methods but not Chemistry.
Calculate the probability that the student says 'yes'. **(1 mark)**

c. The principal randomly selects a student walking down the corridor and asks if they are studying Mathematical Methods but not Chemistry.
Calculate the probability that the student is not doing Mathematical Methods or Chemistry. **(1 mark)**

More exam questions are available online.

7.3 Conditional probability

LEARNING INTENTION

At the end of this subtopic you should be able to:
- represent conditional probability in correct notation
- calculate conditional probability.

Some given information may reduce the number of elements in a sample space. For example, in two tosses of a coin, if it is known that at least one Head is obtained, then this reduces the sample space from {HH, HT, TH, TT} to {HH, HT, TH}. This affects the probability of obtaining two Heads.

7.3.1 Notation for conditional probability

The probability of obtaining two Heads given at least one Head has occurred is called **conditional probability**. It is written as $\Pr(A \mid B)$, where A is the event of obtaining two Heads and B is the event of at least one Head. The event B is the conditional event known to have occurred. Since event B has occurred, the sample space has been reduced to three elements. This means $\Pr(A \mid B) = \frac{1}{3}$.

If no information is given about what has occurred, the sample space contains four elements and the probability of obtaining two Heads is $\Pr(A) = \frac{1}{4}$.

WORKED EXAMPLE 5 Calculating conditional probabilities

The table shows the results of a survey of 100 people aged between 16 and 29 about their preferred choice of food when eating at a café.

	Vegetarian (V)	Non-vegetarian (V')	Total
Male (M)	18	38	56
Female (F)	25	19	44
Total	43	57	100

One person is selected at random from those surveyed. Use the table to identify the following events and calculate the probabilities.

a. $\Pr(M \cap V)$ **b.** $\Pr(M \mid V)$ **c.** $\Pr(V' \mid F)$ **d.** $\Pr(V)$

THINK	WRITE
a. 1. Describe the event $M \cap V$.	**a.** The event $M \cap V$ is the event of a selected person being both male and vegetarian.
2. Calculate the probability.	$\Pr(M \cap V) = \frac{n(M \cap V)}{n(\xi)}$ $= \frac{18}{100}$ $\therefore \Pr(M \cap V) = 0.18$
b. 1. Describe the conditional probability.	**b.** The event $M \mid V$ is the event of a person being male given that the person is vegetarian.
2. State the number of elements in the reduced sample space.	Since the person is known to be vegetarian, the sample space is reduced to $n(V) = 43$ people.
3. Calculate the required probability.	Of the 43 vegetarians, 18 are male. $\therefore \Pr(M \mid V) = \frac{n(M \cap V)}{n(V)}$ $= \frac{18}{43}$
c. 1. Identify the event.	**c.** The event $V' \mid F$ is the event of a person being non-vegetarian given the person is female.
2. State the number of elements in the reduced sample space.	Since the person is known to be female, the sample space is reduced to $n(F) = 44$ people.
3. Calculate the probability.	Of the 44 females, 19 are non-vegetarian. $\therefore \Pr(V' \mid F) = \frac{n(V' \cap F)}{n(F)}$ $= \frac{19}{44}$

d. 1. State the event.	**d.** The event V is the event the selected person is vegetarian.
2. Calculate the required probability. *Note:* This is not a conditional probability.	$\Pr(V) = \frac{n(V)}{n(\xi)}$ $= \frac{43}{100}$ $\therefore \Pr(V) = 0.43$

The formula for conditional probability

Consider the events A and B:

$$\Pr(A) = \frac{n(A)}{n(\xi)}, \Pr(B) = \frac{n(B)}{n(\xi)} \text{ and } \Pr(A \cap B) = \frac{n(A \cap B)}{n(\xi)}$$

For the conditional probability, $\Pr(A \mid B)$, the sample space is reduced to $n(B)$.

$$\begin{aligned}\Pr(A \mid B) &= \frac{n(A \cap B)}{n(B)} \\ &= n(A \cap B) \div n(B) \\ &= \frac{n(A \cap B)}{n(\xi)} \div \frac{n(B)}{n(\xi)} \\ &= \Pr(A \cap B) \div \Pr(B) \\ &= \frac{\Pr(A \cap B)}{\Pr(B)}\end{aligned}$$

The conditional probability formula

$$\mathbf{Pr(A \mid B) = \frac{Pr(A \cap B)}{Pr(B)}}$$

This formula illustrates that if the events A and B are mutually exclusive so that $\Pr(A \cap B) = 0$, then $\Pr(A \mid B) = 0$. That is, if B occurs, then it is impossible for A to occur.

However, if B is a subset of A so that $\Pr(A \cap B) = \Pr(B)$, then $\Pr(A \mid B) = 1$. That is, if B occurs, then it is certain that A will occur.

Resources

 **Interactivity** Conditional probability and independence (int-6292)

7.3.2 Multiplication of probabilities

Consider the conditional probability formula for $\Pr(B \mid A)$:

$$\Pr(B \mid A) = \frac{\Pr(B \cap A)}{\Pr(A)}$$

Since $B \cap A$ is the same as $A \cap B$, then $\Pr(B \mid A) = \dfrac{\Pr(A \cap B)}{\Pr(A)}$.

Rearranging, the formula for multiplication of probabilities is formed.

Multiplying probabilities

$$\mathbf{Pr(A \cap B) = Pr(A) \times Pr(B \mid A)}$$

For example, the probability of obtaining first an aqua (A) and then a black (B) ball when selecting two balls without replacement from a bag containing 16 balls, 6 of which are aqua and 10 of which are black, would be:
$\Pr(A \cap B) = \Pr(A) \times \Pr(B \mid A)$.

$$= \frac{6}{16} \times \frac{10}{15}$$

The multiplication formula can be extended. For example, the probability of obtaining 3 black balls when selecting 3 balls from a bag containing 16 balls, 10 of which are black, without replacing the 3 selected balls would be $\dfrac{10}{16} \times \dfrac{9}{15} \times \dfrac{8}{14}$.

WORKED EXAMPLE 6 Determining probabilities

a. If $\Pr(A) = 0.6$, $\Pr(A \mid B) = 0.6125$ and $\Pr(B') = 0.2$, calculate $\Pr(A \cap B)$ and $\Pr(B \mid A)$.

b. Three girls each select one ribbon at random, one after the other, from a bag containing 8 green ribbons and 10 red ribbons. Calculate the probability that the first girl selects a green ribbon and both of the other girls select a red ribbon.

THINK	WRITE
a. 1. State the conditional probability formula for $\Pr(A \mid B)$.	a. $\Pr(A \mid B) = \dfrac{\Pr(A \cap B)}{\Pr(B)}$
2. Obtain the value of $\Pr(B)$.	For complementary events: $\Pr(B) = 1 - \Pr(B') = 1 - 0.2 = 0.8$
3. Use the formula to calculate $\Pr(A \cap B)$.	$\Pr(A \mid B) = \dfrac{\Pr(A \cap B)}{\Pr(B)}$ $0.6125 = \dfrac{\Pr(A \cap B)}{0.8}$ $\Pr(A \cap B) = 0.6125 \times 0.8 = 0.49$
4. State the conditional probability formula for $\Pr(B \mid A)$.	$\Pr(B \mid A) = \dfrac{\Pr(B \cap A)}{\Pr(A)}$
5. Calculate the required probability.	$\Pr(B \cap A) = \Pr(A \cap B)$ $\Pr(B \mid A) = \dfrac{\Pr(A \cap B)}{\Pr(A)} = \dfrac{0.49}{0.6} = \dfrac{49}{60}$

b. 1. Define the events.

b. Let G be the event a green ribbon is chosen and R be the event a red ribbon is chosen.

2. Describe the sample space.

There are 18 ribbons in the bag forming the elements of the sample space. Of these 18 ribbons, 8 are green and 10 are red.

3. State the probability that the first ribbon selected is green.

$\Pr(G) = \frac{8}{18}$

4. Calculate the conditional probability that the second ribbon is red by reducing the number of elements in the sample space.

Once a green ribbon has been chosen, there are 7 green and 10 red ribbons remaining, giving a total of 17 ribbons in the bag.

$\therefore \Pr(R \mid G) = \frac{10}{17}$

5. Calculate the conditional probability that the third ribbon is red by reducing the number of elements in the sample space.

Once a green and a red ribbon have been chosen, there are 7 green and 9 red ribbons remaining, giving a total of 16 ribbons in the bag.

$\therefore \Pr(R \mid G \cap R) = \frac{9}{16}$

6. Calculate the required probability.
Note: $\Pr(G \cap R \cap R) = \Pr(G) \times \Pr(R \mid G) \times \Pr(R \mid G \cap R)$

$$\Pr(G \cap R \cap R) = \frac{8}{18} \times \frac{10}{17} \times \frac{9}{16}$$
$$= \frac{5}{34}$$

7.3.3 Probability tree diagrams

The sample space of a two-stage trial where the outcomes of the second stage are dependent on the outcomes of the first stage can be illustrated with a **probability tree diagram**.

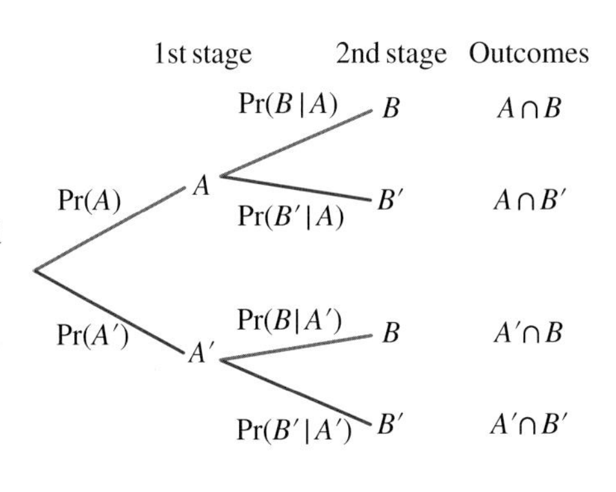

Each branch is labelled with its probability; conditional probabilities are required for the second-stage branches. Calculations are performed according to the addition and multiplication laws of probability.

The formula for multiplication of probabilities is applied by multiplying the probabilities that lie along the respective branches to calculate the probability of an outcome. For example, to obtain the probability of A and B occurring, we need to multiply the probabilities along the branches A to B, since $\Pr(A \cap B) = \Pr(A) \times \Pr(B \mid A)$.

The addition formula for mutually exclusive events is applied by adding the results from separate outcome branches to calculate the union of any of the four outcomes. For example, to obtain the probability that A occurs, add together the results from the two branches where A occurs. This gives $\Pr(A) = \Pr(A)\Pr(B \mid A) + \Pr(A)\Pr(B' \mid A)$.

- Multiply along the branch.
- Add the results from each complete branch.

WORKED EXAMPLE 7 Calculating probabilities

A box of chocolates contains 6 soft-centred and 4 hard-centred chocolates. A chocolate is selected at random and once it has been eaten, a second chocolate is chosen.
Let S_i be the event a soft-centred chocolate is chosen on the ith selection and H_i be the event that a hard-centred chocolate is chosen on the ith selection, $i = 1, 2$.

a. **Deduce the value of $\Pr(H_2 \mid S_1)$.**
b. **Construct a probability tree diagram to illustrate the possible outcomes.**
c. **Calculate the probability that the first chocolate has a hard centre and the second a soft centre.**
d. **Calculate the probability that either both chocolates have soft centres or both have hard centres.**

THINK	WRITE
a. 1. Identify the meaning of $\Pr(H_2 \mid S_1)$.	a. $\Pr(H_2 \mid S_1)$ is the probability that the second chocolate has a hard centre given that the first has a soft centre.
2. State the required probability.	If a soft-centred chocolate has been chosen first, there remain in the box 5 soft- and 4 hard-centred chocolates. $\therefore \Pr(H_2 \mid S_1) = \frac{4}{9}$
b. Construct the two-stage probability tree diagram.	b. 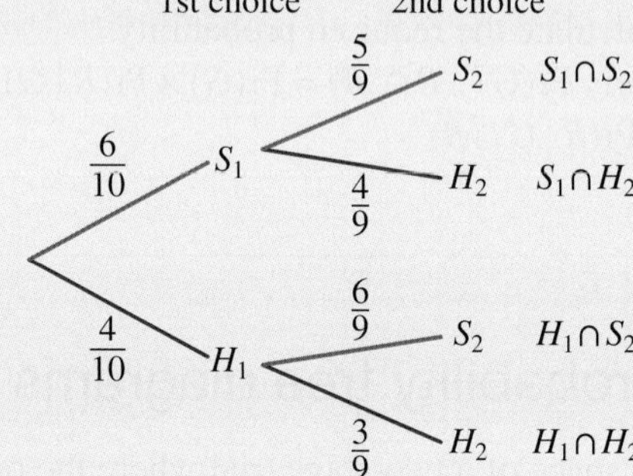
c. Identify the appropriate branch and multiply along it to obtain the required probability. *Note*: The multiplication law for probability is $\Pr(H_1 \cap S_2) = \Pr(H_1) \times \Pr(S_2 \mid H_1)$.	c. The required outcome is $H_1 \cap S_2$. $\Pr(H_1 \cap S_2) = \frac{4}{10} \times \frac{6}{9} = \frac{2}{5} \times \frac{2}{3} = \frac{4}{15}$ The probability is $\frac{4}{15}$.
d. 1. Identify the required outcome.	d. The probability that both chocolates have the same type of centre is $\Pr((S_1 \cap S_2) \cup (H_1 \cap H_2))$.
2. Calculate the probabilities along each relevant branch.	$\Pr(S_1 \cap S_2) = \frac{6}{10} \times \frac{5}{9}$ $\Pr(H_1 \cap H_2) = \frac{4}{10} \times \frac{3}{9}$

3. Use the addition law for mutually exclusive events by adding the probabilities from the separate branches.

The probability that both chocolates have the same type of centre is:

$$\frac{6}{10} \times \frac{5}{9} + \frac{4}{10} \times \frac{3}{9} = \frac{30}{90} + \frac{12}{90}$$
$$= \frac{42}{90}$$
$$= \frac{7}{15}$$

The probability both centres are the same type is $\frac{7}{15}$.

7.3 Exercise

Students, these questions are even better in jacPLUS

Receive immediate feedback and access sample responses

Access additional questions

Track your results and progress

Find all this and MORE in jacPLUS

Technology free

1. A bag contains 3 red (R) balls, 4 purple (P) balls and 2 yellow (Y) balls. One ball is chosen at random and removed from the bag. A second ball is then chosen from the bag.
   a. Given that the first ball chosen is yellow, calculate the probability that the second ball will be red. Write the symbol for this conditional probability.
   b. Given that the first ball chosen is yellow, calculate the probability that the second ball will be yellow. Write the symbol for this conditional probability.
   c. Given that the first ball chosen is red, calculate the probability that the second ball will be purple. Write the symbol for this conditional probability.
   d. If the first ball chosen is not red, calculate the probability that the second ball will be red. Write the symbol for this conditional probability.
2. Of a group of 20 students who study either Art or Biology or both subjects, 13 study Art, 16 study Biology, and 9 study both Art and Biology.
   Let A be the event a student studies Art and B be the event a student studies Biology.
   State whether the following probability statements are true (T) or false (F).

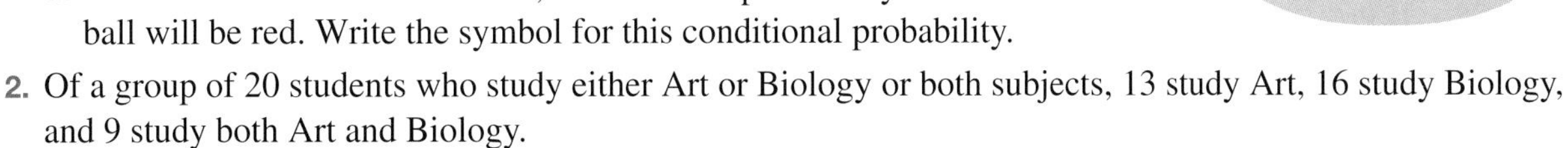

   a. $\Pr(A) = \frac{13}{20}$
   b. $\Pr(A \mid B) = \frac{13}{20}$
   c. $\Pr(A \mid B) = \frac{13}{16}$
   d. $\Pr(A \mid B) = \frac{9}{16}$
   e. $\Pr(B \mid A) = \frac{9}{13}$
   f. $\Pr(B \mid A) = \frac{7}{13}$

3. a. For two events C and D it is known that $\Pr(C \mid D) = \frac{3}{5}$. If $\Pr(D) = \frac{1}{4}$, calculate $\Pr(C \cap D)$.

   b. For two events M and N it is known that $\Pr(N \mid M) = 0.375$ and $\Pr(M \mid N) = 0.6$. Given that $\Pr(M) = 0.8$, calculate $\Pr(N)$.

4. Two unbiased dice are rolled and the sum of the topmost numbers is noted. Given that the sum is an even number, find the probability that the sum is less than 6.

5. Two unbiased dice are rolled. Determine the probability that the sum is greater than 8, given that a 5 appears on the first die.

6. Given $\Pr(A) = 0.7, \Pr(B) = 0.3$ and $\Pr(A \cup B) = 0.8$, determine the following.

   a. $\Pr(A \cap B)$ b. $\Pr(A \mid B)$ c. $\Pr(B \mid A)$ d. $\Pr(A \mid B')$

7. Given $\Pr(A) = 0.6, \Pr(B) = 0.5$ and $\Pr(A \cup B) = 0.8$, determine the following.

   a. $\Pr(A \cap B)$ b. $\Pr(A \mid B)$ c. $\Pr(B \mid A)$ d. $\Pr(A \mid B')$

8. Given $\Pr(A) = 0.6, \Pr(B) = 0.7$ and $\Pr(A \cap B) = 0.4$, determine the following.

   a. $\Pr(A \cup B)$ b. $\Pr(A \mid B)$ c. $\Pr(B \mid A')$ d. $\Pr(A' \mid B')$

9. **WE5** The table shows the results of a survey of 100 people aged between 16 and 29 who were asked whether they rode a bike and what drink they preferred.

	Drink containing caffeine (C)	**Caffeine-free drink (C')**	**Total**
Bike rider (B)	28	16	44
Non–bike rider (B')	36	20	56
Total	64	36	100

   One person is selected at random from those surveyed. Use the table to identify the following events and calculate the probabilities.

   a. $\Pr(B' \cap C')$ b. $\Pr(B' \mid C')$ c. $\Pr(C \mid B)$ d. $\Pr(B)$

10. Two six-sided dice are rolled. Using appropriate symbols, calculate the probability that:

    a. the sum of 8 is obtained
    b. a sum of 8 is obtained given the numbers are not the same
    c. the sum of 8 is obtained, but the numbers are not the same
    d. the numbers are not the same given the sum of 8 is obtained.

11. **WE6** a. If $\Pr(A') = 0.6$, $\Pr(B \mid A) = 0.3$ and $\Pr(B) = 0.5$, calculate $\Pr(A \cap B)$ and $\Pr(A \mid B)$.

    b. Three girls each select one ribbon at random, one after the other, from a bag containing 8 green ribbons and 4 red ribbons. Calculate the probability that all three girls select a green ribbon.

12. If $\Pr(A) = 0.61$, $\Pr(B) = 0.56$ and $\Pr(A \cup B) = 0.81$, calculate the following.

    a. $\Pr(A \mid B)$ b. $\Pr(A \mid A \cup B)$ c. $\Pr(A \mid A \cap B)$

13. **WE7** A box of jubes contains 5 green jubes and 7 red jubes. One jube is selected at random and once it is eaten, a second jube is chosen.
    Let G_i be the event a green jube is chosen on the ith selection and R_i the event that a red jube is chosen on the ith selection, $i = 1, 2$.

    a. Deduce the value of $\Pr(G_2 \mid R_1)$.
    b. Construct a probability tree diagram to illustrate the possible outcomes.
    c. Calculate the probability that the first jube is green and the second is red.
    d. Calculate the probability that either both jubes are green or both are red.

Technology active

14. Two cards are drawn randomly from a standard pack of 52 cards. Find the probability that:
    a. both cards are diamonds
    b. at least 1 card is a diamond
    c. both cards are diamonds, given that at least one card is a diamond
    d. both cards are diamonds, given that the first card drawn is a diamond.

15. A bag contains 5 red marbles and 7 green marbles. Two marbles are drawn from the bag, one at a time, without replacement. Find the probability that:
    a. both marbles are green
    b. at least 1 marble is green
    c. both marbles are green given that at least 1 is green
    d. the first marble drawn is green, given that the marbles are of different colours.

16. In an attempt to determine the efficacy of a test used to detect a particular disease, 100 subjects, of whom 27 had the disease, were tested. A positive result means the test detected the disease and a negative result means the test did not detect the disease. Only 23 of the 30 people who tested positive actually had the disease. Draw up a two-way table to show this information and hence find the probability that a subject selected at random:
    a. does not have the disease
    b. tested positive but did not have the disease
    c. had the disease, given that the subject tested positive
    d. did not have the disease, given that the subject tested negative

17. In a survey designed to check the number of male and female smokers in a population, it is found that there are 32 male smokers, 41 female smokers, 224 female non-smokers and 203 male non-smokers.
    A person is selected at random from this group of people. Find the probability that the person selected is:
    a. a non-smoker
    b. male
    c. female, given that the person is a non-smoker.

18. In a sample of 1000 people, it is found that:
    82 people are overweight and suffer from hypertension
    185 are overweight but do not suffer from hypertension
    175 are not overweight but suffer from hypertension
    558 are not overweight and do not suffer from hypertension.
    A person is selected at random from the sample. Find the probability that the person:
    a. is overweight
    b. suffers from hypertension
    c. suffers from hypertension given that the person is overweight
    d. is overweight given that the person does not suffer from hypertension.

19. A group of 400 people were tested for allergic reactions to two new medications. The results are shown in the table below:

	Allergic reaction	No allergic reaction
Medication A	25	143
Medication B	47	185

If a person is selected at random from the group, calculate the following.

a. The probability that the person suffers an allergic reaction
b. The probability that the person was administered medication A
c. Given there was an allergic reaction, the probability that medication B was administered
d. Given the person was administered medication A, the probability that the person did not have an allergic reaction

20. To get to school Rodney catches a bus and then walks the remaining distance. If the bus is on time, Rodney has a 98% chance of arriving at school on time. However, if the bus is late, Rodney's chance of arriving at school on time is only 56%. On average the bus is on time 90% of the time.

a. Draw a probability tree diagram to describe the given information, defining the symbols used.
b. Calculate the probability that Rodney will arrive at school on time.

21. When Incy Wincy Spider climbs up the waterspout, the chances of him falling are affected by whether or not it is raining. When it is raining, the probability that he will fall is 0.84. When it is not raining, the probability that he will fall is 0.02. On average it rains 1 day in 5 around Incy Wincy Spider's spout.
Draw a probability tree to show the sample space and hence find the probability that:

a. Incy Wincy will fall when it is raining
b. Incy Wincy falls, given that it is raining
c. it is raining given that Incy Wincy makes it to the top of the spout.

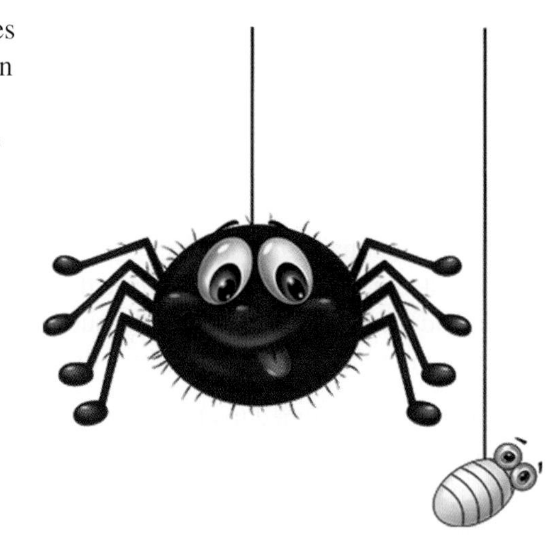

22. In tenpin bowling, a game is made up of 10 frames. Each frame represents one turn for the bowler. In each turn, a bowler is allowed up to 2 rolls of the ball to knock down the 10 pins. If the bowler knocks down all 10 pins with the first ball, this is called a strike. If it takes 2 rolls of the ball to knock the 10 pins, this is called a spare. Otherwise it is called an open frame. On average, Richard hits a strike 85% of the time. If he needs a second roll of the ball, on average, he will knock down the remaining pins 97% of the time. While training for the club championship, Richard plays a game.

Draw a tree diagram to represent the outcomes of the first 2 frames of his game. Hence, find the probability that:

a. both are strikes
b. all 10 pins are knocked down in both frames
c. the first frame is a strike, given that the second is a strike.

7.3 Exam questions

Question 1 (1 mark) TECH-ACTIVE

MC If $\Pr(A) = 0.7$, $\Pr(B) = 0.3$ and $\Pr(A \cup B) = 0.8$, then $\Pr(A \cap B)$ is

A. 1.0 **B.** 0.5 **C.** 0.3 **D.** 0.2 **E.** 0

Question 2 (1 mark) TECH-ACTIVE

MC If $\Pr(A) = 0.7$, $\Pr(B) = 0.3$ and $\Pr(A \cup B) = 0.8$, then $\Pr(B \mid A)$ is

A. $\frac{2}{7}$ **B.** $\frac{3}{7}$ **C.** $\frac{7}{8}$ **D.** $\frac{3}{8}$ **E.** $\frac{2}{3}$

Question 3 (1 mark) TECH-ACTIVE

MC Two classes completed a difficult Mathematics test and the results are shown in the table below. The probability that a randomly selected student was from class 1 and had passed the test was

	Passed	Failed
Class 1	15	10
Class 2	13	12

A. $\frac{2}{10}$ **B.** $\frac{3}{10}$ **C.** $\frac{3}{5}$ **D.** $\frac{2}{5}$ **E.** $\frac{5}{10}$

More exam questions are available online.

7.4 Independence

LEARNING INTENTION

At the end of this subtopic you should be able to:

- represent independent events in correct notation
- calculate independent events.

If a coin is tossed twice, the chance of obtaining a Head on the coin on its second toss is unaffected by the result of the first toss. The probability of a Head on the second toss given a Head is obtained on the first toss is still $\frac{1}{2}$.

Events that have no effect on each other are called **independent events**. For such events, $\Pr(A \mid B) = \Pr(A)$. The given information does not affect the chance of event A occurring.

Events that do affect each other are dependent events. For dependent events, $\Pr(A \mid B) \neq \Pr(A)$ and the conditional probability formula is used to evaluate $\Pr(A \mid B)$.

7.4.1 Test for mathematical independence

Although it may be obvious that the chance of obtaining a Head on a coin on its second toss is unaffected by the result of the first toss, in more complex situations it can be difficult to intuitively judge whether events are independent or dependent. For such situations there is a test for mathematical independence that will determine the matter.

The multiplication formula states $\Pr(A \cap B) = \Pr(A) \times \Pr(B \mid A)$.

If the events A and B are independent, then $\Pr(B \mid A) = \Pr(B)$.

Hence, for independent events the following applies.

Independent events

$$\Pr(A \cap B) = \Pr(A) \times \Pr(B)$$

This result is used to test whether events are mathematically independent or not.

WORKED EXAMPLE 8 Testing for independence

Consider the trial of tossing a coin twice.
Let A be the event of at least one Tail, B be the event of either two Heads or two Tails and C be the event that the first toss is a Head.

a. **List the sample space and the set of outcomes in each of A, B and C.**
b. **Test whether A and B are independent.**
c. **Test whether B and C are independent.**
d. **Use the addition formula to calculate $\Pr(B \cup C)$.**

THINK	WRITE
a. 1. List the elements of the sample space.	a. The sample space is the set of equiprobable outcomes {HH, HT, TH, TT}.
2. List the elements of A, B and C.	$A = \{\text{HT, TH, TT}\}$ $B = \{\text{HH, TT}\}$ $C = \{\text{HH, HT}\}$
b. 1. State the test for independence.	b. A and B are independent if $\Pr(A \cap B) = \Pr(A)\Pr(B)$.
2. Calculate the probabilities needed for the test for independence to be applied.	$\Pr(A) = \frac{3}{4}$ and $\Pr(B) = \frac{2}{4}$ Since $A \cap B = \{\text{TT}\}$, $\Pr(A \cap B) = \frac{1}{4}$.
3. Determine whether the events are independent.	Substitute values into the formula $\Pr(A \cap B) = \Pr(A)\Pr(B)$. $\text{LHS} = \frac{1}{4}$ $\text{RHS} = \frac{3}{4} \times \frac{2}{4}$ $= \frac{3}{8}$ Since LHS $\neq$ RHS, the events A and B are not independent.
c. 1. State the test for independence.	c. B and C are independent if $\Pr(B \cap C) = \Pr(B)\Pr(C)$.
2. Calculate the probabilities needed for the test for independence to be applied.	$\Pr(B) = \frac{2}{4}$ and $\Pr(C) = \frac{2}{4}$ Since $B \cap C = \{\text{HH}\}$, $\Pr(B \cap C) = \frac{1}{4}$.

3. Determine whether the events are independent.

Substitute values into $\Pr(B \cap C) = \Pr(B)\Pr(C)$.

$$\text{LHS} = \frac{1}{4}$$

$$\text{RHS} = \frac{2}{4} \times \frac{2}{4} = \frac{1}{4}$$

Since LHS = RHS, the events B and C are independent.

d. 1. State the addition formula for $\Pr(B \cup C)$.

d. $\Pr(B \cup C) = \Pr(B) + \Pr(C) - \Pr(B \cap C)$

2. Replace $\Pr(B \cap C)$.

Since B and C are independent,
$\Pr(B \cap C) = \Pr(B)\Pr(C)$.
$\therefore\ \Pr(B \cup C) = \Pr(B) + \Pr(C) - \Pr(B) \times \Pr(C)$

3. Complete the calculation.

$$\Pr(B \cup C) = \frac{1}{2} + \frac{1}{2} - \frac{1}{2} \times \frac{1}{2} = \frac{3}{4}$$

7.4.2 Independent trials

Consider choosing a ball from a bag containing 6 red and 4 green balls, noting its colour, returning the ball to the bag and then choosing a second ball. These trials are independent, as the chance of obtaining a red or green ball is unaltered for each draw. This is an example of **sampling with replacement**.

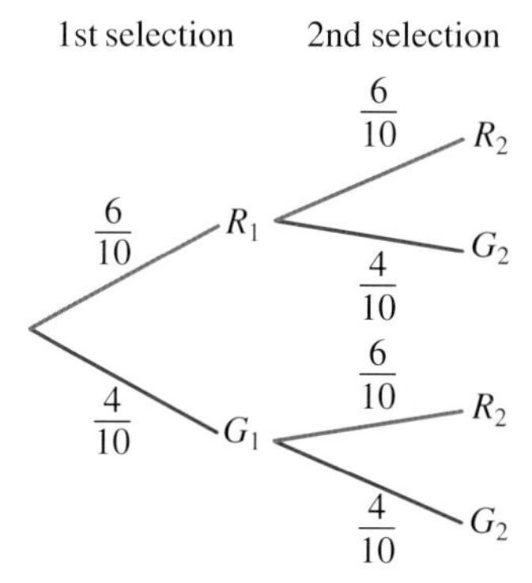

The probability tree diagram is as shown.

The second stage branch outcomes are not dependent on the results of the first stage.

The probability that both balls are red is $\Pr(R_1 R_2) = \frac{6}{10} \times \frac{6}{10}$.

As the events are independent:

$$\begin{aligned}\Pr(R_1 \cap R_2) &= \Pr(R_1) \times \Pr(R_2 \mid R_1)\\ &= \Pr(R_1) \times \Pr(R_2)\end{aligned}$$

However, if we are **sampling without replacement**, then $\Pr(R_1 \cap R_2) = \frac{6}{10} \times \frac{5}{9}$, since the events are not independent.

Sequences of independent events

If the events A, B, C, … are independent, then $\Pr(A \cap B \cap C \cap \ldots) = \Pr(A) \times \Pr(B) \times \Pr(C) \times \ldots$

WORKED EXAMPLE 9 Determining probabilities with independent events

Ari, Barry and Chris commence a meal program. The chances that each person sticks to the meal program are 0.6, 0.8 and 0.7 respectively, independent of each other. Calculate the probability that:
a. all three people stick to the meal program
b. only Chris sticks to the meal program
c. at least one of the three people sticks to the meal program.

THINK	WRITE
a. 1. Define the independent events.	a. Let A be the event that Ari sticks to the meal program, B be the event that Barry sticks to the meal program and C be the event that Chris sticks to the meal program. $\Pr(A) = 0.6, \Pr(B) = 0.8$ and $\Pr(C) = 0.7$
2. State an expression for the required probability.	The probability that all three people stick to the meal program is $\Pr(A \cap B \cap C)$.
3. Calculate the required probability.	Since the events are independent, $\Pr(A \cap B \cap C) = \Pr(A) \times \Pr(B) \times \Pr(C)$ $= 0.6 \times 0.8 \times 0.7$ $= 0.336$ The probability that all three stick to the meal program is 0.336.
b. 1. State an expression for the required probability.	b. If only Chris sticks to the meal program, then neither Ari nor Barry do. The probability is $\Pr(A' \cap B' \cap C)$.
2. Calculate the probability. *Note:* For complementary events, $\Pr(A') = 1 - \Pr(A)$.	$\Pr(A' \cap B' \cap C) = \Pr(A') \times \Pr(B') \times \Pr(C)$ $= (1 - 0.6) \times (1 - 0.8) \times 0.7$ $= 0.4 \times 0.2 \times 0.7$ $= 0.056$ The probability that only Chris sticks to the meal program is 0.056.
c. 1. Express the required event in terms of its complementary event.	c. The event that at least one of the three people sticks to the meal program is the complement of the event that no-one sticks to the meal program.
2. Calculate the required probability.	Pr(at least one sticks to the meal program) $= 1 -$ Pr(no-one sticks to the meal program) $= 1 - \Pr(A' \cap B' \cap C')$ $= 1 - 0.4 \times 0.2 \times 0.3$ $= 1 - 0.024$ $= 0.976$ The probability that at least one person sticks to the meal program is 0.976.

7.4 Exercise

Students, these questions are even better in jacPLUS

Receive immediate feedback and access sample responses

Access additional questions

Track your results and progress

Find all this and MORE in jacPLUS

Technology free

1. Two events A and B are such that $\Pr(A) = 0.7$, $\Pr(B) = 0.8$ and $\Pr(A \cup B) = 0.94$. Determine whether events A and B are independent.

2. Events A and B are such that $\Pr(A) = 0.75$, $\Pr(B) = 0.64$ and $\Pr(A \cup B) = 0.91$. Determine whether events A and B are independent.

3. a. Given $\Pr(A) = 0.3$, $\Pr(B) = p$ and $\Pr(A \cap B) = 0.12$, determine the value of p if the events A and B are independent.
   b. Given $\Pr(A) = p$, $\Pr(B) = 2p$ and $\Pr(A \cap B) = 0.72$, determine the value of p if the events A and B are independent.
   c. Given $\Pr(A) = q$, $\Pr(B) = \frac{1}{5}$ and $\Pr(A \cup B) = \frac{4}{5}$, determine the value of q if the events A and B are independent.

4. **WE8** Consider the experiment of tossing a coin twice. Let A be the event the first toss is a Tail, B the event of one Head and one Tail, and C be the event of no more than one Tail.
   a. List the sample space and the set of outcomes in each of A, B and C.
   b. Test whether A and B are independent.
   c. Test whether B and C are independent.
   d. Use the addition formula to calculate $\Pr(B \cup A)$.

5. A student estimates her chance of passing a History test is $\frac{8}{10}$ and passing a Chemistry test is $\frac{7}{10}$. Assuming independence, calculate the probability that the student:
   a. passes both tests
   b. passes the Chemistry test but not the History test
   c. does not pass the tests in either subject
   d. passes at least one of the tests.

6. Clinical trials of a drug indicate that patients with a particular disease who are treated with the drug have a cure rate of 4 in 5. Three patients are given the drug. Calculate the probability that:
   a. all three patients are cured
   b. none of the patients are cured
   c. the first patient is cured, the second patient is not cured, and the third patient is cured.

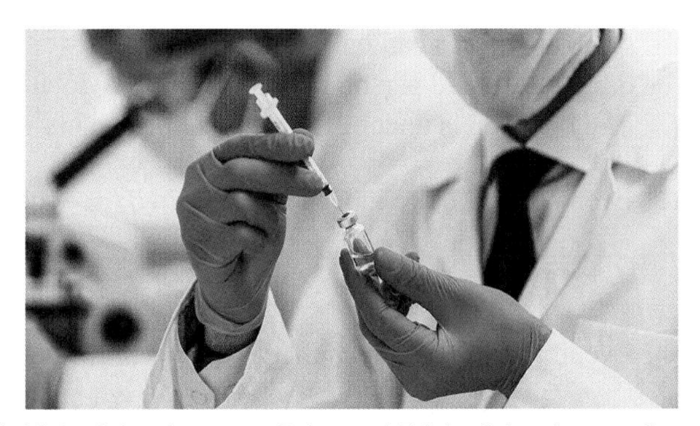

7. A family owns two cars, car A and car B. Car A is used 65% of the time, car B is used 74% of the time and at least one of the cars is used 97% of the time. Determine whether the two cars are used independently.

8. Two unbiased six-sided dice are rolled. Let A be the event the same number is obtained on each die and B be the event the sum of the numbers on each die exceeds 8.
   a. State whether events A and B mutually exclusive.
   b. State whether events A and B independent.
   Justify your answers.
   c. If C is the event the sum of the two numbers equals 8, determine whether B and C are:
      i. mutually exclusive
      ii. independent.

9. **WE9** Bree, Seiko and Hadiyah commence a balanced diet. The chances that each person sticks to the diet are 0.4, 0.9 and 0.6 respectively, independent of each other. Calculate the probability that:
   a. all three people stick to the diet
   b. only Bree and Hadiyah stick to the diet
   c. at least one of the three people does not stick to the diet.

10. A box of toy blocks contains 10 red blocks and 5 yellow blocks. A child draws out two blocks at random.
   a. Draw the tree diagram if the sampling is with replacement and calculate the probability that one block of each colour is obtained.
   b. Draw the tree diagram if the sampling is without replacement and calculate the probability that one block of each colour is obtained.
   c. If the child was to draw out three blocks, rather than two, calculate the probability of obtaining three blocks of the same colour if the sampling is with replacement.

Technology active

11. Sheena and Kala sit a Biology exam. The probability that Sheena passes the exam is 0.9 and the probability that Kala passes the exam is 0.8. Find the probability that:
   a. both Sheena and Kala pass the exam
   b. at least one of the two girls passes the exam
   c. only 1 girl passes the exam, given that Sheena passes.

12. A survey of 200 people was carried out to determine the number of traffic violations committed by different age groups. The results are shown in the table below.
   If one person is selected at random from the group, find the probability that:

	Number of violations		
Age group	**0**	**1**	**2**
Under 25	8	30	7
25–45	47	15	2
45–65	45	18	3
65+	20	5	0

   a. the person belongs to the under-25 age group
   b. the person has had at least one traffic violation
   c. given that the person has had at least 1 traffic violation, he or she has had only 1 violation
   d. a 38-year-old person has had no traffic violations
   e. the person is under 25, given that he/she has had 2 traffic violations.

13. Events A and B are independent. If $\Pr(B) = \frac{2}{3}$ and $\Pr(A \mid B) = \frac{4}{5}$, find:
   a. $\Pr(A)$
   b. $\Pr(B \mid A)$
   c. $\Pr(A \cap B)$
   d. $\Pr(A \cup B)$.

14. Events A and B are such that $\Pr(B) = \frac{3}{5}$, $\Pr(A \mid B) = \frac{1}{3}$ and $\Pr(A \cup B) = \frac{23}{30}$.
   a. Find $\Pr(A \cap B)$.
   b. Find $\Pr(A)$.
   c. Determine whether events A and B are independent.

15. Events A and B are independent such that $\Pr(A \cup B) = 0.8$ and $\Pr(A \mid B') = 0.6$. Find $\Pr(B)$.

16. Six hundred people were surveyed about whether they watched movies on their TV or on their devices. They were classified according to age, with the following results.

	Age 15 to 30	Age 30 to 70	
TV	95	175	270
Device	195	135	330
	290	310	600

Based on these findings, determine whether the device used to watch movies is independent of a person's age.

17. A popular fast-food restaurant has studied the customer service provided by a sample of 100 of its employees across Australia. They wanted to know if employees who were with the company longer received more positive feedback from customers than newer employees.

	Positive customer feedback (P)	Negative customer feedback (P'')	
Employed 2 years or more (T)	34	16	50
Employed fewer than 2 years (T'')	22	26	50
	58	42	100

State whether the events P and T are independent. Justify your answer.

18. Roll two fair dice and record the number uppermost on each. Let A be the event of rolling a 6 on one die, B be the event of rolling a 3 on the other, and C be the event of the product of the numbers on the dice being at least 20.
    a. State whether the events A and B independent. Explain your answer.
    b. State whether the events B and C independent. Explain your answer.
    c. State whether the events A and C independent. Explain your answer.

7.4 Exam questions

Question 1 (1 mark) TECH-ACTIVE

MC If A and B are independent events and $\Pr(A) = 0.4$ and $\Pr(B) = 0.5$, then $\Pr(A \cup B)$ is

A. 1 **B.** 0.2 **C.** 0.9 **D.** 0.7 **E.** 0

Question 2 (1 mark) TECH-ACTIVE

MC Two independent events, A and B, are such that $\Pr(A) = 0.7$, $\Pr(B) = 0.8$ and $\Pr(A \cup B) = 0.94$. $\Pr(A \cap B)$ is equal to

A. 0 **B.** 0.7 **C.** 0.56 **D.** 0.14 **E.** 0.24

Question 3 (3 marks) TECH-FREE

The probabilities that three swimmers — Sue, Bill and Fred — can swim 100 metres in less than 1 minute are 0.7, 0.8 and 0.4 respectively.

a. Calculate the probability that all the swimmers will swim 100 metres in less than 1 minute, assuming they are not in the same race and their times are independent of each other. **(1 mark)**

b. Calculate the probability that at least one swimmer will break 1 minute. **(2 marks)**

More exam questions are available online.

7.5 Counting techniques

LEARNING INTENTION

At the end of this subtopic you should be able to:
- represent counting techniques in correct notation
- apply counting techniques to probability problems.

When calculating the probability of an event A from the fundamental rule $\Pr(A) = \dfrac{n(A)}{n(\xi)}$, the number of elements in both A and the sample space need to be able to be counted. Here we shall consider two counting techniques, one where order is important and one where order is not important. Respectively, these are called **arrangements** and **selections** or, alternatively, **permutations** and **combinations**.

7.5.1 Arrangements or permutations

The arrangement AB is a different arrangement to BA.

If two of three people designated by A, B and C are to be placed in a line, the possible arrangements are AB, BA, AC, CA, BC, CB. There are 6 possible arrangements or permutations.

Rather than list the possible arrangements, the number of possibilities can be calculated as follows using a **box table**.

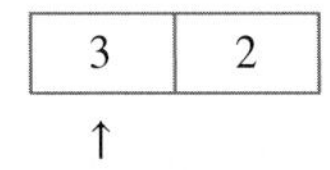

There are 3 people who can occupy the left position.

Once that position is filled, this leaves 2 people who can occupy the remaining position.

Multiplying these figures together gives the total number of $3 \times 2 = 6$ arrangements. This is an illustration of the **multiplication principle**.

Multiplication principle

If there are m ways of doing the first procedure and for each one of these there are n ways of doing the second procedure, then there are $m \times n$ ways of doing the first **and** the second procedures. This can be extended.

Suppose either two or three of four people A, B, C, D are to be arranged in a line. The possible arrangements can be calculated as follows:

Arrange two of the four people:

4	3

This gives $4 \times 3 = 12$ possible arrangements using the multiplication principle.

Arrange three of the four people:

4	3	2

This gives $4 \times 3 \times 2 = 24$ possible arrangements using the multiplication principle.

The total number of arrangements of either two or three from the four people is $12 + 24 = 36$. This illustrates the **addition principle**.

Addition principle for mutually exclusive events

For mutually exclusive procedures, if there are m ways of doing one procedure and n ways of doing another procedure, then there are $m+n$ ways of doing one **or** the other procedure.

AND $\times$ (Multiplication principle)

OR $+$ (Addition principle)

WORKED EXAMPLE 10 Counting the number of arrangements

Consider the set of five digits $\{2,6,7,8,9\}$.
Assume no repetition of digits in any one number can occur.

a. **Determine how many three-digit numbers can be formed from this set.**
b. **Determine how many numbers with at least four digits can be formed.**
c. **Determine how many five-digit odd numbers can be formed.**
d. **One of the five-digit numbers is chosen at random. Calculate the probability that it will be an odd number.**

THINK

WRITE

a. 1. Draw a box table with three divisions.

a. There are five choices for the first digit, leaving four choices for the second digit and then three choices for the third digit.

5	4	3

2. Calculate the answer.

Using the multiplication principle, there are $5\times4\times3=60$ possible three-digit numbers that could be formed.

b. 1. Interpret the event described.

b. At least four digits means either four-digit or five-digit numbers are to be counted.

2. Draw the appropriate box tables.

For four-digit numbers:

5	4	3	2

For five-digit numbers:

5	4	3	2	1

3. Calculate the answer.

There are $5\times4\times3\times2=120$ four-digit numbers and $5\times4\times3\times2\times1=120$ five-digit numbers. Using the addition principle there are $120+120=240$ possible four- or five-digit numbers.

c. 1. Draw the box table showing the requirement imposed on the number.

c. For the number to be odd its last digit must be odd, so the number must end in either 7 or 9. This means there are two choices for the last digit.

				2

2. Complete the box table.

Once the last digit has been formed, there are four choices for the first digit, then three choices for the second digit, two choices for the third digit and one choice for the fourth digit.

4	3	2	1	2

3. Calculate the answer.

Using the multiplication principle, there are $4 \times 3 \times 2 \times 1 \times 2 = 48$ odd five-digit numbers possible.

d. 1. Define the sample space and state $n(\xi)$.

d. The sample space is the set of five-digit numbers. From part **b**, $n(\xi) = 120$.

2. State the number of elements in the required event.

Let A be the event the five-digit number is odd. From part **c**, $n(A) = 48$.

3. Calculate the probability.
Note: The last digit must be odd. Of the five possible last digits, two are odd. Hence, the probability that the number is odd is $\frac{2}{5}$.

$$\Pr(A) = \frac{n(A)}{n(\xi)} = \frac{48}{120} = \frac{2}{5}$$

The probability that the five-digit number is odd is $\frac{2}{5}$.

Resources

Interactivity Counting techniques (int-6293)

Factorial notation

The number of ways that four people can be arranged in a row can be calculated using the box table shown.

4	3	2	1

Using the multiplication principle, the total number of arrangements is $4 \times 3 \times 2 \times 1 = 24$. This can be expressed using factorial notation as 4!.

Factorial notation

In general, the number of ways of arranging n objects in a row is:

$$n! = n \times (n-1) \times (n-2) \times \ldots \times 2 \times 1.$$

7.5.2 Arrangements in a circle

Consider arranging the letters A, B and C.

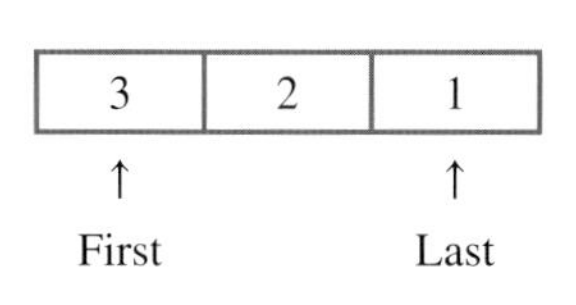

If the arrangements are in a row, then there are $3! = 6$ arrangements.

In a row arrangement there is a first and a last position.

In a circular arrangement there is no first or last position; order is only created clockwise or anticlockwise from one letter once this letter is placed. This means ABC and BCA are the same circular arrangement, as the letters have the same anticlockwise order relative to A.

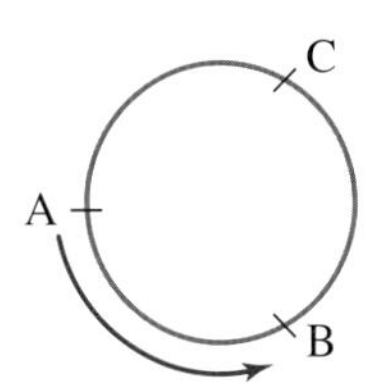

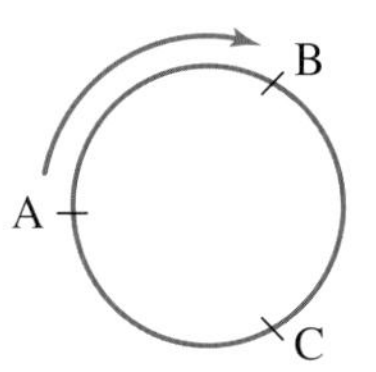

There are only two distinct circular arrangements of three letters: ABC or ACB as shown.

Three objects A, B and C can be arranged in a row in 3! ways; three objects can be arranged in a circle in $(3-1)! = 2!$ ways.

Types of arrangements

***n* objects can be arranged:**

$$n! \text{ ways in a row}$$

$$(n-1)! \text{ ways in a circle.}$$

WORKED EXAMPLE 11 Counting the number of circular arrangements

A group of 7 students queue in a straight line at a canteen to buy a drink.

a. Calculate the number of ways in which the queue can be formed.

b. The students carry their drinks to a circular table. Calculate the number of different arrangements in which the students can sit around the table.

c. This group of students has been shortlisted for the Mathematics, History and Art prizes. Calculate the probability that one person in the group receives all three prizes.

THINK	WRITE
a. 1. Use factorial notation to describe the number of arrangements.	**a.** Seven people can be arranged in a straight line in 7! ways.
2. Calculate the answer. *Note:* A calculator could be used to evaluate the factorial.	Since $7! = 7 \times 6 \times 5!$ and $5! = 120$, $7! = 7 \times 6 \times 120$ $= 7 \times 720$ $= 5040$ There are 5040 ways in which the students can form the queue.
b. 1. State the rule for circular arrangements.	**b.** For circular arrangements, 7 people can be arranged in $(7-1)! = 6!$ ways.
2. State the answer.	Since $6! = 720$, there are 720 different arrangements in which the 7 students may be seated.

THINK	WRITE
c. 1. Form the number of elements in the sample space.	**c.** There are three prizes. Each prize can be awarded to any one of the 7 students. | 7 | 7 | 7 | The total number of ways the prizes can be awarded is $7 \times 7 \times 7$. $\therefore n(\xi) = 7 \times 7 \times 7$
2. Form the number of elements in the event under consideration.	Let A be the event that the same student receives all three prizes. There are seven choices for that student. $\therefore n(A) = 7$
3. Calculate the required probability.	$\Pr(A) = \dfrac{n(A)}{n(\xi)} = \dfrac{7}{7 \times 7 \times 7} = \dfrac{1}{49}$ The probability that one student receives all three prizes is $\dfrac{1}{49}$.

TI | THINK	DISPLAY/WRITE
a. 1. On a Calculator page, complete the entry line as: 7! then press ENTER. *Note:* The factorial symbol can be found by pressing the ?▶ button	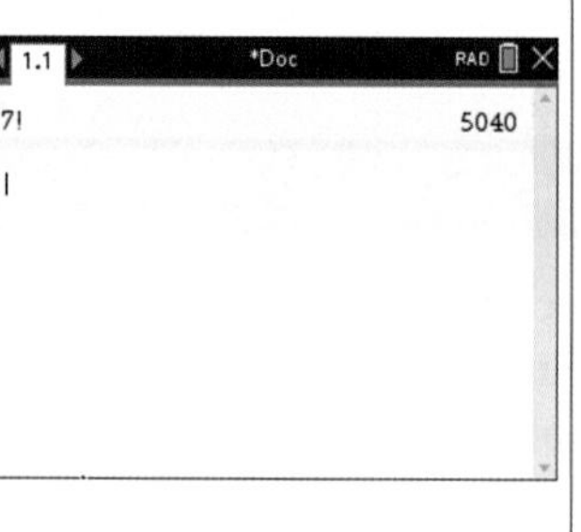
2. The answer appears on the screen.	5040

CASIO | THINK	DISPLAY/WRITE
a. 1. On a Main screen, complete the entry line as: 7! then press EXE. *Note:* The factorial symbol can be found in the Advance tab in the Keyboard menu.	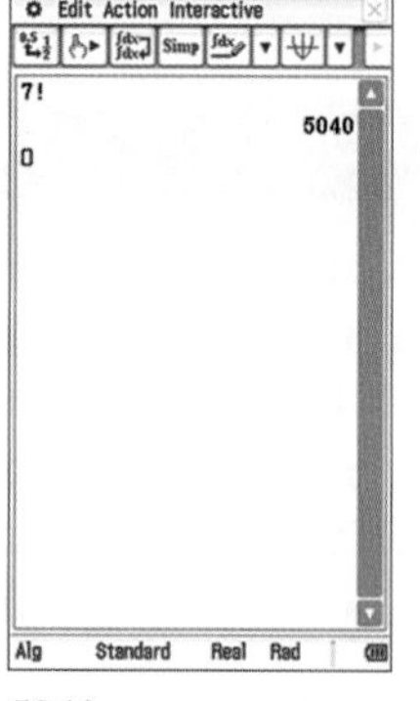
2. The answer appears on the screen.	5040

7.5.3 Arrangements with objects grouped together

Where a group of objects are to be together, treat them as one unit in order to calculate the number of arrangements. Having done this, then allow for the number of internal rearrangements within the objects grouped together and apply the multiplication principle.

For example, consider arranging the letters A, B, C and D with the restriction that ABC must be together.

Treating ABC as one unit would mean there are two objects to arrange: D and the unit (ABC).

Two objects arrange in 2! ways.

For each of these arrangements, the unit (ABC) can be internally arranged in 3! ways.

The multiplication principle then gives the total number of possible arrangements that satisfy the restriction, which is $2! \times 3!$ or $2 \times 6 = 12$ arrangements.

Arrangements where some objects may be identical

The word SUM has three distinct letters. These letters can be arranged in 3! ways. The six arrangements can be listed as 3 pairs — SUM and MUS, USM and UMS, SMU and MSU — formed when the S and the M are interchanged.

Now consider the word MUM. Although this word also has three letters, two of the letters are identical. Interchanging the two M's will not create a new arrangement. SUM and MUS where the S and the M are interchanged are different, but MUM and MUM are the same.

There are only three arrangements: MUM, UMM and MMU.

The number of arrangements is $\frac{3!}{2!} = 3$.

This can be considered as cancelling out the rearrangements of the 2 identical letters from the total number of arrangements of 3 letters.

Identical arrangements

For n objects, if p is of one type and q is of another type, number of arrangements $= \frac{n!}{p!\,q!\,\ldots}$.

WORKED EXAMPLE 12 Counting with groups with identical objects

Consider the two words 'PARALLEL' and 'LINES'.
a. Calculate how many arrangements of the letters of the word LINES have the vowels grouped together.
b. Calculate how many arrangements of the letters of the word LINES have the vowels separated.
c. Calculate how many arrangements of the letters of the word PARALLEL are possible.
d. Calculate the probability that in a randomly chosen arrangement of the word PARALLEL, the letters A are together.

THINK	WRITE
a. 1. Group the required letters together.	a. There are two vowels in the word LINES. Treat these letters, I and E, as one unit.
2. Arrange the unit of letters together with the remaining letters.	Now there are four groups to arrange: (IE), L, N, S. These arrange in 4! ways.
3. Use the multiplication principle to allow for any internal rearrangements.	The unit (IE) can internally rearrange in 2! ways. Hence, the total number of arrangements is: $4! \times 2! = 24 \times 2$ $= 48$
b. 1. State the method of approach to the problem.	b. The number of arrangements with the vowels separated is equal to the total number of arrangements minus the number of arrangements with the vowels together.
2. State the total number of arrangements.	The five letters of the word LINES can be arranged in $5! = 120$ ways.
3. Calculate the answer.	From part **a**, there are 48 arrangements with the two vowels together. Therefore, there are $120 - 48 = 72$ arrangements in which the two vowels are separated.

c. 1. Count the letters, stating any identical letters.

c. The word PARALLEL contains 8 letters, of which there are 2 A's and 3 L's.

2. Using the rule number of arrangements $= \frac{n!}{p!\ q!\ ...}$, state the number of distinct arrangements.

There are $\frac{8!}{2! \times 3!}$ arrangements of the word PARALLEL.

3. Calculate the answer.

$$\frac{8!}{2! \times 3!} = \frac{8 \times 7 \times 6 \times 5 \times \cancel{4}^2 \times \cancel{3!}}{\cancel{2} \times \cancel{3!}}$$

$$= 3360$$

There are 3360 arrangements.

d. 1. State the number of elements in the sample space.

d. There are 3360 total arrangements of the word PARALLEL, so $n(\xi) = 3360$ or $\frac{8!}{2! \times 3!}$.

2. Group the required letters together.

For the letters A to be together, treat these two letters as one unit. This creates seven groups (AA), P, R, L, L, E, L, of which three are identical L's.

3. Calculate the number of elements in the event.

The seven groups arrange in $\frac{7!}{3!}$ ways. As the unit (AA) contains two identical letters, there are no distinct internal rearrangements of this unit that need to be taken into account.

Hence, $\frac{7!}{3!}$ is the number of elements in the event.

4. Calculate the required probability.
Note: It helps to use factorial notation in the calculations.

The probability that the A's are together

$$= \frac{\text{number of arrangements with the As together}}{\text{total number of arrangements}}$$

$$= \frac{7!}{3!} \div \frac{8!}{2! \times 3!}$$

$$= \frac{7!}{3!} \times \frac{2! \times 3!}{8 \times 7!}$$

$$= \frac{2}{8}$$

$$= \frac{1}{4}$$

The formula for permutations

The number of arrangements of n objects taken r at a time is shown in the box table.

n	$n-1$	$n-2$	… … … … … … … … … … … …	$n-r+1$
↑	↑	↑		↑
1st	2nd	3rd	… … … … … … … … … … … …	r th object

The number of arrangements equals $n(n-1)(n-2)...(n-r+1)$.

This can be expressed using factorial notation as:

$$\begin{aligned}
&n(n-1)(n-2)\ldots(n-r+1)\\
&=n(n-1)(n-2)\ldots(n-r+1)\times\frac{(n-r)(n-r-1)\times\ldots 2\times 1}{(n-r)(n-r-1)\times\ldots 2\times 1}\\
&=\frac{n!}{(n-r)!}
\end{aligned}$$

The formula for the number of permutations or arrangements of n objects taken r at a time is expressed as follows.

Permutations

$$^{n}\mathrm{P}_{r}=\frac{n!}{(n-r)!},$$

$0 \leq r \leq n$ where r and n are non-negative integers.

Although we have preferred to use a box table, it is possible to count arrangements using this formula.

7.5.4 Combinations or selections

Now we shall consider the counting technique for situations where order is unimportant. This is the situation where the selection AB is the same as the selection BA. For example, the entry Alan and Bev is no different to the entry Bev and Alan as a pair of mixed doubles players in a tennis match: they are the same entry.

The number of combinations of r objects from a total group of n distinct objects is calculated by counting the number of arrangements of the objects r at a time and then dividing that by the number of ways each group of these r objects can rearrange between themselves. This is done in order to cancel out counting these as different selections.

The number of combinations is therefore $\frac{^{n}\mathrm{P}_{r}}{r!}=\frac{n!}{(n-r)!\,r!}$.

The symbol for the number of ways of choosing r objects from a total of n objects is $^{n}\mathrm{C}_{r}$ or $\binom{n}{r}$.

The number of combinations of r objects from a total of n objects is expressed as follows.

Combinations

$$^{n}\mathrm{C}_{r}=\frac{n!}{r!\,(n-r)!}=\binom{n}{r},$$

$0 \leq r \leq n$ where r and n are non-negative integers.

The formula for $^{n}\mathrm{C}_{r}$ is exactly that for the binomial coefficients used in the binomial theorem. These values are the combinatoric terms in Pascal's triangle as encountered in an earlier topic.

Drawing on that knowledge, we have the following results:

- $^{n}\mathrm{C}_{0}=1$, there being only one way to choose none or all of the n objects.
- $^{n}\mathrm{C}_{1}=n$, there being n ways to choose one object from a group of n objects.
- $^{n}\mathrm{C}_{r}={}^{n}C_{n-r}$, since choosing r objects must leave behind a group of $(n-r)$ objects and vice versa.

Calculations

The formula is always used for calculations in selection problems. Most calculators have a ${}^{n}C_r$ key to assist with the evaluation when the figures become large.

Both the multiplication and addition principles apply and are used in the same way as for arrangements.

The calculation of probabilities from the rule $\Pr(A) = \dfrac{n(A)}{n(\xi)}$ requires that the same counting technique used for the numerator is also used for the denominator. We have seen for arrangements that it can assist calculations to express the numerator and denominator in terms of factorials and then simplify. Similarly for selections, express the numerator and denominator in terms of the appropriate combinatoric coefficients and then carry out the calculations.

WORKED EXAMPLE 13 Counting combinations

A committee of 5 students is to be chosen from 7 boys and 4 girls.

a. Calculate how many committees can be formed.

b. Calculate how many of the committees contain exactly 2 boys and 3 girls.

c. Calculate how many committees have at least 3 girls.

d. Calculate the probability of the oldest and youngest students both being on the committee.

THINK	WRITE
a. 1. As there is no restriction, choose the committee from the total number of students.	a. There are 11 students in total, from whom 5 students are to be chosen. This can be done in ${}^{11}C_5$ ways.
2. Use the formula ${}^{n}C_r = \dfrac{n!}{r! \times (n-r)!}$ to calculate the answer.	${}^{11}C_5 = \dfrac{11!}{5! \times (11-5)!}$ $= \dfrac{11!}{5! \times 6!}$ $= \dfrac{11 \times 10 \times 9 \times 8 \times 7 \times 6!}{5! \times 6!}$ $= 462$ There are 462 possible committees.
b. 1. Select the committee to satisfy the given restriction.	b. The 2 boys can be chosen from the 7 boys available in ${}^{7}C_2$ ways. The 3 girls can be chosen from the 4 girls available in ${}^{4}C_3$ ways.
2. Use the multiplication principle to form the total number of committees. *Note:* The upper numbers on the combinatoric coefficients sum to the total available, $7 + 4 = 11$, and the lower numbers sum to the number that must be on the committee, $2 + 3 = 5$.	The total number of committees that contain two boys and three girls is ${}^{7}C_2 \times {}^{4}C_3$.

3. Calculate the answer.

$$^7C_2 \times {}^4C_3 = \frac{7!}{2! \times 5!} \times 4$$
$$= \frac{7 \times 6}{2!} \times 4$$
$$= 21 \times 4$$
$$= 84$$

There are 84 committees possible with the given restriction.

c. 1. List the possible committees that satisfy the given restriction.

c. As there are 4 girls available, at least 3 girls means either 3 or 4 girls.
The committees of 5 students that satisfy this restriction have either 3 girls and 2 boys or 4 girls and 1 boy.

2. Write the number of committees in terms of combinatoric coefficients.

3 girls and 2 boys are chosen in $^4C_3 \times {}^7C_2$ ways. 4 girls and 1 boy are chosen in $^4C_4 \times {}^7C_1$ ways.

3. Use the addition principle to state the total number of committees.

The number of committees with at least three girls is $^4C_3 \times {}^7C_2 + {}^4C_4 \times {}^7C_1$.

4. Calculate the answer.

$$^4C_3 \times {}^7C_2 + {}^4C_4 \times {}^7C_1 = 84 + 1 \times 7$$
$$= 91$$

There are 91 committees with at least 3 girls.

d. 1. State the number in the sample space.

d. The total number of committees of 5 students is $^{11}C_5 = 462$ from part **a**.

2. Form the number of ways the given event can occur.

Each committee must have 5 students. If the oldest and youngest students are placed on the committee, then 3 more students need to be selected from the remaining 9 students to form the committee of 5. This can be done in $^9C_3$ ways.

3. State the probability in terms of combinatoric coefficients.

Let A be the event the oldest and the youngest students are on the committee.

$$\Pr(A) = \frac{n(A)}{n(\xi)}$$
$$= \frac{^9C_3}{^{11}C_5}$$

4. Calculate the answer.

$$\Pr(A) = \frac{9!}{3! \times 6!} \div \frac{11!}{5! \times 6!}$$
$$= \frac{9!}{3! \times 6!} \times \frac{5! \times 6!}{11!}$$
$$= \frac{1}{3!} \times \frac{5!}{11 \times 10}$$
$$= \frac{5 \times 4}{110}$$
$$= \frac{2}{11}$$

The probability of the committee containing the youngest and the oldest students is $\frac{2}{11}$.

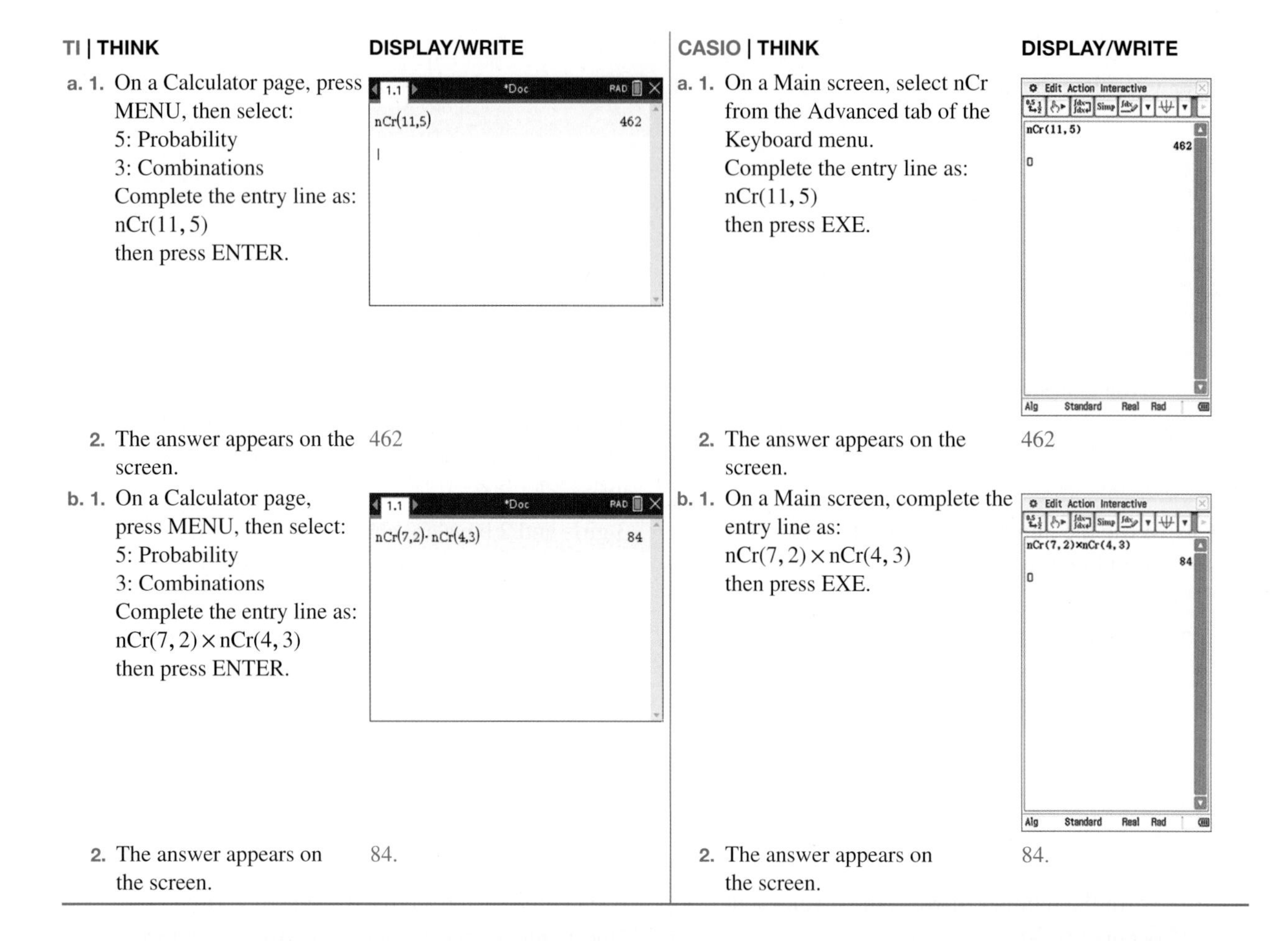

TI | THINK	DISPLAY/WRITE	CASIO | THINK	DISPLAY/WRITE
a. 1. On a Calculator page, press MENU, then select: 5: Probability 3: Combinations Complete the entry line as: nCr(11, 5) then press ENTER.	nCr(11,5) 462	**a. 1.** On a Main screen, select nCr from the Advanced tab of the Keyboard menu. Complete the entry line as: nCr(11, 5) then press EXE.	nCr(11, 5) 462
2. The answer appears on the screen.	462	**2.** The answer appears on the screen.	462
b. 1. On a Calculator page, press MENU, then select: 5: Probability 3: Combinations Complete the entry line as: nCr(7, 2) × nCr(4, 3) then press ENTER.	nCr(7,2)·nCr(4,3) 84	**b. 1.** On a Main screen, complete the entry line as: nCr(7, 2) × nCr(4, 3) then press EXE.	nCr(7, 2)×nCr(4, 3) 84
2. The answer appears on the screen.	84.	**2.** The answer appears on the screen.	84.

7.5 Exercise

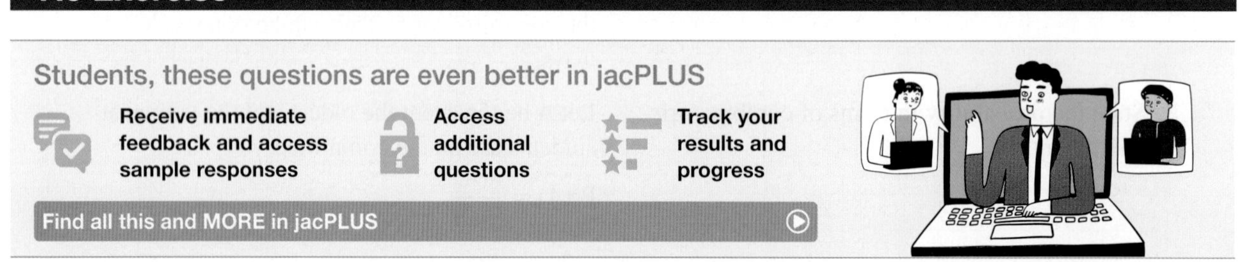

Technology free

1. Calculate the number of ways in which three students can be seated in a row.
2. Calculate how many even two-digit numbers can be formed from the digits 4, 5, 6, 7, 8 if each digit can be used only once.
3. a. Calculate how many arrangements of the letters of the word SAGE are possible.
   b. Calculate how many arrangements of the letters of the word THYME are possible if the letter T must always be first.
4. Calculate how many two- or three-digit numbers can be formed from the set of digits $\{9, 8, 5, 3\}$ if no digit can be repeated more than once in any number.

5. A person wanting to travel between Bendigo and Maldon gathers the following information: You can drive from Bendigo to Maldon, via Castlemaine, or via Newbridge or via Lockwood South. Alternatively you can take the train from Bendigo to Castlemaine and then the bus from Castlemaine to Maldon.
Using this information, determine the number of ways the person can travel from Bendigo to Maldon and back to Bendigo.

6. a. Baby Amelie has 10 different bibs and 12 different body suits. Calculate how many different combinations of bib and body suit they can wear.
b. A teacher, Christine, has the option of taking the motorway or the highway to work each morning. She then must travel through some suburban streets and has the option of three different routes through suburbia. If Christine wishes to take a different route to work each day, calculate how many days she will be able to take a different route before she must use a route already travelled.
c. When selecting his new car, Abdul has the option of a manual or an automatic. He is also offered a choice of 5 exterior colours, 3 interior colours, leather or vinyl seats, and the options of individual seat heating and self-parking. Calculate how many different combinations of new car Abdul can choose from.

d. Sarah has a choice of 3 hats, 2 pairs of sunglasses, 7 T-shirts and 5 pairs of shorts. When going out in the sun, she chooses one of each of these items to wear. Calculate how many different combinations are possible.
e. In order to start a particular game, each player must roll an unbiased die, then select a card from a standard pack of 52. Calculate how many different starting combinations are possible.
f. On a recent bushwalking trip, a group of friends had a choice of travelling by car, bus or train to the Blue Mountains. They decided to walk to one of the waterfalls, then to a mountaintop scenic view. Calculate how many different trips were possible if there are six different waterfalls and 12 mountaintop scenic views

Technology active

7. WE10 Consider the set of five digits $\{3, 5, 6, 7, 9\}$. Assume no repetition of digits in any one number can occur.
a. Determine how many four-digit numbers can be formed from this set.
b. Determine how many numbers with at least three digits can be formed.
c. Determine how many five-digit even numbers can be formed.
d. One of the five-digit numbers is chosen at random. Calculate the probability that it will be an even number.

8. a. Registration plates on a vehicle consist of 2 letters of the English alphabet followed by 2 digits followed by another letter. Calculate how many different number plates are possible if repetitions are allowed.
b. Calculate how many five-letter words can be formed using the letters B, C, D, E, G, I and M if repetitions are allowed.
c. A die is rolled three times. Calculate the number of outcomes.
d. Calculate how many three-digit numbers can be formed using the digits $2, 3, 4, 5, 6, 7$ if repetitions are allowed.
e. Three friends on holidays decide to stay at a hotel that has 4 rooms available. Calculate the number of ways in which the rooms can be allocated if there are no restrictions and each person has their own room.

9. Eight people, consisting of 4 boys and 4 girls, are to be arranged in a row. Find the number of ways this can be done if:

   a. there are no restrictions
   b. the boys and girls are to alternate
   c. the end seats must be occupied by a girl
   d. the brother and sister must not sit together
   e. the girls must sit together.

10. Calculate how many three-digit numbers can be formed from the digits 0, 1, 2, 3, 4, 5, 6, 7, 8, 9 given that no repetitions are allowed, the number cannot start with 0, and the following conditions hold.

   a. There are no other restrictions.
   b. The number must be even.
   c. The number must be less than 400.
   d. The number is made up of odd digits only.

11. Six girls are in a row.

   a. Calculate the total number of ways in which the girls can stand.
   b. Calculate how many of the arrangements have the two girls Agnes and Betty together.
   c. Use the results of parts **a** and **b** to calculate the probability of Agnes and Betty being next to each other.

12. From a group of 8 books on a shelf, 3 are hardbacks and 5 are paperbacks.

   a. Calculate the number of ways in which 2 hardbacks and 3 paperbacks can be chosen from this group of books.
   b. Calculate the number of ways in which any set of 5 books can be selected from this group of books.
   c. Use the results of parts **a** and **b** to calculate the probability of choosing 2 hardbacks and 3 paperbacks when selecting 5 books from this shelf.

13. A car's number plate consists of two letters of the English alphabet followed by three of the digits 0 to 9, followed by one single letter. Repetition of letters and digits is allowed.

   a. Calculate how many such number plates are possible.
   b. Calculate how many of the number plates use the letter X exactly once.
   c. Calculate the probability that the first two letters are identical and all three numbers are the same, but the single letter differs from the other two.

14. **WE11** A group of six students queue in a straight line to take out books from the library.

   a. Calculate the number of ways in which the queue can be formed.
   b. The students carry their books to a circular table. Calculate the number of different arrangements in which the students can sit around the table.
   c. This group of students have been shortlisted for the Mathematics, History and Art prizes. Calculate the probability that one person in the group receives all three prizes.

15. a. In how many ways can 7 men be selected from a group of 15 men.
   b. Calculate how many five-card hands can be dealt from a standard pack of 52 cards.
   c. Calculate how many five-card hands that contain all 4 aces, can be dealt from a standard pack of 52 cards.
   d. Calculate the number of ways in which 3 prime numbers can be selected from the set containing the first 10 prime numbers.

16. A panel of 8 is to be selected from a group of 8 men and 10 women. Find how many panels can be formed if:

   a. there are no restriction.
   b. there are 5 men and 3 women on the panel
   c. there are at least 6 men on the panel
   d. two particular men cannot both be included
   e. a particular man and woman must both be included.

17. A group of four boys seat themselves in a circle.
   a. Calculate how many different seating arrangements of the boys are possible.
   b. Four girls join the boys and seat themselves around the circle so that each girl is sat between two boys. Calculate how many ways the girls can be seated.
18. **WE12** Consider the words SIMULTANEOUS and EQUATIONS.
   a. Calculate how many arrangements of the letters of the word EQUATIONS have the letters Q and U grouped together.
   b. Calculate how many arrangements of the letters of the word EQUATIONS have the letters Q and U separated.
   c. Calculate how many arrangements of the letters of the word SIMULTANEOUS are possible.
   d. Calculate the probability that in a randomly chosen arrangement of the word SIMULTANEOUS, both the letters U are together.
19. **WE13** A committee of 5 students is to be chosen from 6 boys and 8 girls.
   a. Calculate how many committees can be formed.
   b. Calculate how many of the committees contain exactly 2 boys and 3 girls.
   c. Calculate how many committees have at least 4 boys.
   d. Calculate the probability of neither the oldest nor the youngest student being on the committee.
20. Calculate how many words can be formed from the letters of the word BANANAS given the following conditions.
   a. All the letters are used.
   b. A four-letter word is to be used including at least one A.
   c. A four-letter word using all different letters is to be used.
21. a. A set of 12 mugs that are identical except for colour are to be placed on a shelf. Calculate the number of ways in which this can be done if 4 of the mugs are blue, 3 are orange and 5 are green.
   b. Calculate the number of ways in which the 12 mugs from part **a** can be arranged in 2 rows of 6 if the green ones must be in the front row.

22. Calculate the number of ways in which 6 men and 3 women can be arranged at a circular table if:
   a. there are no restrictions
   b. the men can only be seated in pairs.
23. Consider the universal set $S = \{2, 3, 4, 5, 7, 8, 10, 11, 13, 14, 15, 17\}$.
   a. Calculate how many subsets of S there are.
   b. Determine the number of subsets whose elements are all even numbers.
   c. Find the probability that a subset selected at random will contain only even numbers.
   d. Find the probability that a subset selected at random will contain at least 3 elements.
   e. Find the probability that a subset selected at random will contain exactly 3 numbers, all of which are prime numbers.
24. Calculate the number of ways in which the thirteen letters PARALLEL LINES can be arranged:
   a. in a row
   b. in a circle
   c. in a row with the vowels together.
25. A cricket team of eleven players is to be selected from a list of 3 wicketkeepers, 6 bowlers and 8 batsmen. Calculate the probability that the team chosen consists of one wicketkeeper, four bowlers and six batsmen.

26. A representative sport committee consisting of 7 members is to be formed from 21 tennis players, 17 squash players and 18 badminton players. Find the probability that the committee will contain:
   a. 7 squash players
   b. at least 5 tennis players
   c. at least one representative from each sport
   d. exactly 3 badminton players, given that it contains at least 1 badminton player.

27. One three-digit number is selected at random from all the possible three-digit numbers. Find the probability that:
   a. the digits are all primes
   b. the number has just a single repeated digit
   c. the digits are perfect squares
   d. there are no repeated digits
   e. the number lies between 300 and 400 inclusive, given that the number is greater than 200.

7.5 Exam questions

Question 1 (1 mark) TECH-ACTIVE

MC Five students have made an appointment to see their teacher. The number of ways the appointments can be arranged if Sam, a struggling student, must go first is

A. 120 **B.** 60 **C.** 24 **D.** 16 **E.** 12

Question 2 (1 mark) TECH-ACTIVE

MC The number of arrangements of the letters in the word 'FINDER' where the vowels are together is

A. 720 **B.** 360 **C.** 240 **D.** 120 **E.** 60

Question 3 (4 marks) TECH-ACTIVE

A review panel is to be established. It is to consist of the principal and four other teachers selected from the 7 male and 9 female staff members.

a. Calculate how many different panels can be formed. **(1 mark)**

b. Calculate how many panels can be formed that consist of at least 3 female teachers. **(1 mark)**

c. Calculate the probability that only 1 male is on the panel. **(2 marks)**

More exam questions are available online.

7.6 Binomial coefficients and Pascal's triangle

LEARNING INTENTION

At the end of this subtopic you should be able to:
- write binomial expansions using Pascal's triangle
- calculate binomial coefficients.

In this section, the link between counting techniques and the **binomial coefficients** will be explored.

7.6.1 Binomial coefficients

Consider the following example.

A coin is biased in such a way that the probability of tossing a Head is 0.6.
The coin is tossed three times. The outcomes and their probabilities are shown below.

$\Pr(TTT) = (0.4)^3$ or $\Pr(0H) = (0.4)^3$

$$\left.\begin{array}{l}\Pr(HTT) = (0.6)\,(0.4)^2\\ \Pr(THT) = (0.6)\,(0.4)^2\\ \Pr(TTH) = (0.6)\,(0.4)^2\end{array}\right\} \rightarrow \Pr(1H) = \binom{3}{1}(0.6)\,(0.4)^2 \text{ or } \Pr(1H) = 3(0.6)(0.4)^2$$

$$\left.\begin{array}{l}\Pr(HHT) = (0.6)^2(0.4)\\ \Pr(HTH) = (0.6)^2(0.4)\\ \Pr(THH) = (0.6)^2(0.4)\end{array}\right\} \rightarrow \Pr(2H) = \binom{3}{2}(0.6)^2(0.4) \text{ or } \Pr(2H) = 3(0.6)^2(0.4)$$

$\Pr(HHH) = 0.6^3 \rightarrow \Pr(3H) = 0.6^3$

Since this represents the sample space, the sum of these probabilities is 1.

So we have $\Pr(0H) + \Pr(1H) + \Pr(2H) + \Pr(3H) = 1$.

$$\binom{3}{0}(0.4)^3 + \binom{3}{1}(0.4)^2(0.6) + \binom{3}{2}(0.4)(0.6)^2 + \binom{3}{3}0.6^3 = 1$$

$$\rightarrow 1 = (0.4)^3 + 3(0.4)^2(0.6) + 3(0.4)(0.6)^2 + 0.6^3$$

Note that $\binom{3}{2}$, which is the number of ways in which we can get 2 Heads from 3 tosses, is also the number of ordered selections of 3 objects where 2 are alike of one type (Heads) and 1 of another (Tails), as was noted previously.

Now compare the following expansion:

$$(q+p)^3 = q^3 + 3q^2p + 3qp^2 + p^3$$

We see that if we let $p = 0.6$ (the probability of getting a Head on our biased die) and $q = 0.4$ (the probability of not getting a Head), we have identical expressions. Hence:

$$\begin{aligned}(0.4+0.6)^3 &= (0.4)^3 + 3(0.4)^2(0.6) + 3(0.4)\,(0.6)^2 + 0.6^3\\ &= \binom{3}{0}(0.4)^3 + \binom{3}{1}(0.4)^2(0.6) + \binom{3}{2}(0.4)\,(0.6)^2 + \binom{3}{3}0.6^3\end{aligned}$$

So the coefficients in the binomial expansion are equal to the number of ordered selections of 3 objects of just 2 types.

If we were to toss the coin n times, with p being the probability of a Head and q being the probability of not Head (Tail), we have the following.

The probability of 0 Heads (n Tails) in n tosses is $\binom{n}{0}q^n$.

The probability of 1 Head ($n-1$ Tails) in n tosses is $\binom{n}{1}q^{n-1}p$.

The probability of 2 Heads ($n-2$ Tails) in n tosses is $\binom{n}{2} q^{n-2}p^2$.

These probabilities are known as **binomial probabilities**.

In general, the probability of getting r favourable and $n-r$ non-favourable outcomes in n repetitions of an experiment is $\binom{n}{r} q^{n-r}p^r$, where $\binom{n}{r}$ is the number of ways of getting r favourable outcomes.

So again, equating the sum of all the probabilities in the sample space to the binomial expansion of $(q+p)^n$, we have:

$$(q+p)^n = \binom{n}{0} q^n + \binom{n}{1} q^{n-1}p + \binom{n}{2} q^{n-2}p^2 + \ldots + \binom{n}{r} q^{n-r}p^r$$
$$+ \ldots \binom{n}{n-1} qp^{n-1} + \binom{n}{n} p^n$$

Generalising, in the binomial expansion of $(a+b)^n$, the coefficient of $a^{n-r}b^r$ is $\binom{n}{r}$, where $\binom{n}{r}$ is the number of ordered selections of n objects, in which r are alike of one type and $n-r$ are alike of another type.

Hence, we have:

$$(a+b)^n = \binom{n}{0} a^n + \binom{n}{1} a^{n-1}b + \binom{n}{2} a^{n-2}b^2 + \ldots + \binom{n}{r} a^{n-r}b^r + \ldots$$
$$+ \binom{n}{n-1} ab^{n-1} + \binom{n}{n} b^n$$

We now have a simple formula for calculating binomial coefficients.

This can be written using sigma notation as follows.

Binomial coefficients

$$(a+b)^n = \sum_{r=0}^{n} \binom{n}{r} a^{n-r}b^r$$

Important features to note in the expansion are:
- The powers of a decrease as the powers of b increase.
- The sum of the powers of a and b for each term in the expansion is equal to n.
- As we would expect, $\binom{n}{0} = \binom{n}{n} = 1$

  since there is only 1 way of selecting no objects or of selecting all n objects.
- If we look at the special case of

  $(1+x)^n = \binom{n}{0} + \binom{n}{1} x + \binom{n}{2} x^2 + \binom{n}{3} x^3 + \ldots + \binom{n}{r} x^r + \ldots + \binom{n}{n-1} x^{n-1} + \binom{n}{n} x^n$

  $\binom{n}{r}$, or ${}^n\mathrm{C}_r$, is the coefficient of x^r.

WORKED EXAMPLE 14 Binomial expansions

Write down the expansion of each of the following.

a. $(p+q)^4$ **b.** $(x-y)^3$ **c.** $(2m+5)^3$

THINK	WRITE
a. 1. Determine the values of n, a and b in the expansion of $(a+b)^n$.	**a.** $n=4;\ a=p;\ b=q$
2. Use the formula to write the expansion and simplify.	$(p+q)^4 = \binom{4}{0}p^4 + \binom{4}{1}p^3q + \binom{4}{2}p^2q^2 + \binom{4}{3}pq^3 + \binom{4}{4}q^4$ $= p^4 + 4p^3q + 6p^2q^2 + 4pq^3 + q^4$
b. 1. Determine the values of n, a and b in the expansion of $(a+b)^n$.	**b.** $n=3;\ a=x;\ b=-y$
2. Use the formula to write the expansion and simplify.	$(x-y)^3 = \binom{3}{0}x^3 + \binom{3}{1}x^2(-y) + \binom{3}{2}x(-y)^2 + \binom{3}{3}(-y)^3$ $= x^3 - 3x^2y + 3xy^2 - y^3$
c. 1. Determine the values of n, a and b in the expansion of $(a+b)^n$.	**c.** $n=3;\ a=2m;\ b=5$
2. Use the formula to write the expansion and simplify.	$(2m+5)^3 = \binom{3}{0}(2m)^3 + \binom{3}{1}(2m)^2(5) + \binom{3}{2}(2m)(5)^2 + \binom{3}{3}(5)^3$ $= 8m^3 + 60m^2 + 150m + 125$

TI | THINK	DISPLAY/WRITE	CASIO | THINK	DISPLAY/WRITE
a. 1. On a Calculator page, press MENU, then select: 3: Algebra 3: Expand Complete the entry line as: expand $((p+q)^4)$ then press ENTER.	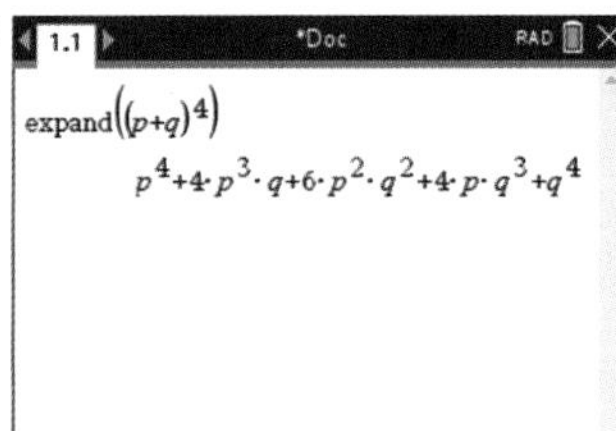	**a. 1.** On a Main screen, select: • Interactive • Transformation • expand Complete the entry line as: $(p+q)^4$ then select OK.	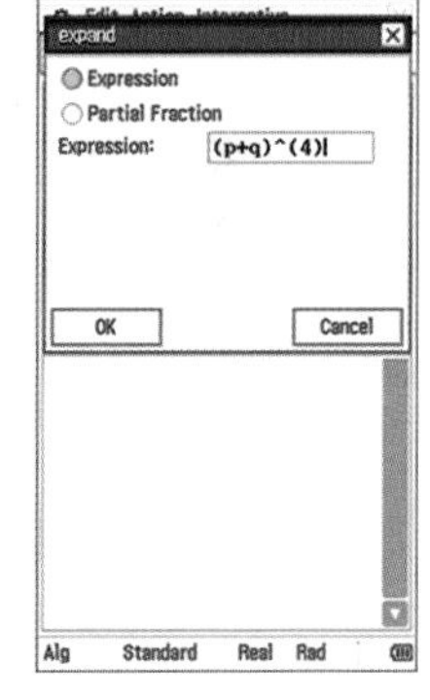 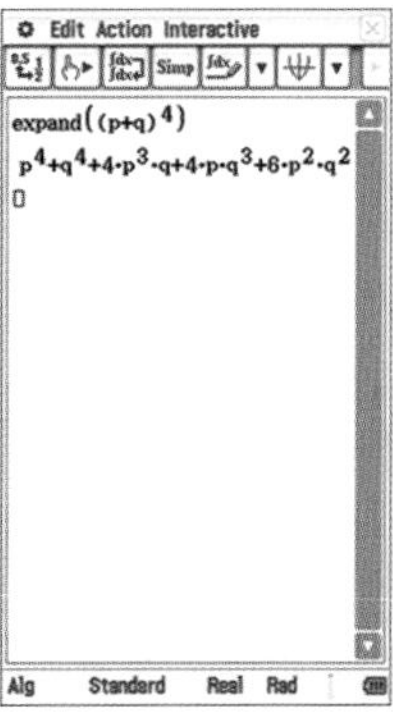
2. The answer appears on the screen.	$(p+q)^4 = p^4 + 4p^3q + 6p^2q^2 + 4pq^3 + q^4$	**2.** The answer appears on the screen.	$(p+q)^4 = p^4 + q^4 + 4p^3q + 4pq^3 + 6p^2q^2$

Sigma notation

The expansion of $(1+x)^n$ can also be expressed in sigma notation as $(1+x)^n = \sum_{r=0}^{n} \binom{n}{r} x^r$.

WORKED EXAMPLE 15 Determining a term in binomial expansions

Find each of the following in the expansion of $\left(3x^2 - \frac{1}{x}\right)^6$.

a. **The term independent of x**

b. **The term in x^{-6}**

THINK	WRITE
a. 1. Express the expansion in sigma notation.	a. $\left(3x^2 - \frac{1}{x}\right)^6 = \sum_{r=0}^{6} \binom{6}{r} (3x^2)^{6-r} \left(-\frac{1}{x}\right)^r$
2. Simplify by collecting powers of x.	$\left(3x^2 - \frac{1}{x}\right)^6 = \sum_{r=0}^{6} (-1)^r \binom{6}{r} 3^{6-r} x^{12-2r} x^{-r}$ $= \sum_{r=0}^{6} (-1)^r \binom{6}{r} 3^{6-r} x^{12-3r}$
3. For the term independent of x, we need the power of x to be 0.	$12 - 3r = 0$ $r = 4$
4. Substitute the value of r in the expression.	$(-1)^r \binom{6}{r} 3^{6-r} x^{12-3r} = (-1)^4 \binom{6}{4} 3^2$ $= 135$
5. Answer the question.	The 5th term, 135, is independent of x.
b. 1. For the term in x^{-6}, we need to make the power of x equal to -6.	b. $12 - 3r = -6$ $r = 6$
2. Substitute the value of r in the expression.	$(-1)^r \binom{6}{r} 3^{6-r} x^{12-3r} = (-1)^6 \binom{6}{6} 3^0 x^{-6}$ $= x^{-6}$
3. Answer the question.	The 7th term is x^{-6}.

7.6.2 Pascal's triangle

We saw earlier that the coefficient of $p^{n-r}q^r$ in the binomial expansion of $(p+q)^n$ is $\binom{n}{r}$, the number of ordered selections of n objects, in which r are alike of one type and $n-r$ are alike of another type. These coefficients reappear in Pascal's triangle, shown in the diagram.

1
1 1
1 2 1
1 3 3 1
1 4 6 4 1
1 5 10 10 5 1
1 6 15 20 15 6 1
1 7 21 35 35 21 7 1

Compare the following expansions.

$(p+q)^0 = 1$	Coefficient: 1	$\binom{0}{0}$
$(p+q)^1 = p+q$	Coefficients: 1 1	$\binom{1}{0}, \binom{1}{1}$
$(p+q)^2 = p^2 + 2pq + q^2$	Coefficients: 1 2 1	$\binom{2}{0}, \binom{2}{1}, \binom{2}{2}$
$(p+q)^3 = p^3 + 3p^2q + 3pq^2 + q^3$	Coefficients: 1 3 3 1	$\binom{3}{0}, \binom{3}{1}, \binom{3}{2}, \binom{3}{3}$
$(p+q)^4 = p^4 + 4p^3q + 6p^2q^2 + 4pq^3 + q^4$	Coefficients: 1 4 6 4 1	$\binom{4}{0}, \binom{4}{1}, \binom{4}{2}, \binom{4}{3}, \binom{4}{4}$

The coefficients in the binomial expansion are equal to the numbers in Pascal's triangle.

Hence, the triangle can be written using the $\binom{n}{r}$ or ${}^nC_r$ notation as follows.

$n = 0$:	${}^0C_0$	1
$n = 1$:	${}^1C_0$ ${}^1C_1$	1 1
$n = 2$:	${}^2C_0$ ${}^2C_1$ ${}^2C_2$	1 2 1
$n = 3$:	${}^3C_0$ ${}^3C_1$ ${}^3C_2$ ${}^3C_3$	1 3 3 1
$n = 4$:	${}^4C_0$ ${}^4C_1$ ${}^4C_2$ ${}^4C_3$ ${}^4C_4$	1 4 6 4 1
$n = 5$:	${}^5C_0$ ${}^5C_1$ ${}^5C_2$ ${}^5C_3$ ${}^5C_4$ ${}^5C_5$	1 5 10 10 5 1

Note that the first and last number in each row is always 1.

Each coefficient is obtained by adding the two coefficients immediately above it.

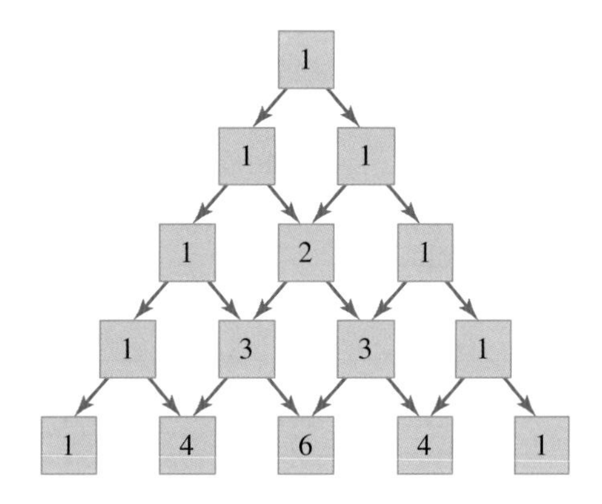

Pascal's identity

$$^{n}C_r = {}^{n-1}C_{r-1} + {}^{n-1}C_r$$

for $0 < r < n$.

WORKED EXAMPLE 16 Calculating binomial coefficients

a. Write down the notation for the 8th coefficient in the 13th row of Pascal's triangle.

b. Calculate the value of $\binom{7}{5}$.

c. A coin is biased in such a way that the probability of tossing a Head is 0.6.
Form an expression for the probability of obtaining 5 Heads in 7 tosses of the coin.

THINK	WRITE
a. 1. Write down the values of n and r. Remember, numbering starts at 0.	a. $n = 13$ and $r = 7$
2. Substitute into the notation.	The 8th coefficient in the 13th row is $^{13}C_7$.
b. 1. Use the formula.	b. $\binom{7}{5} = \frac{7!}{(7-5)!\,5!} = \frac{7 \times 6}{2!} = 21$
2. Verify using a calculator.	$^{7}C_5 = 21$
c. 1. State the two types of outcomes in the event of obtaining 5 Heads in 7 tosses of the coin.	c. 5 Heads in 7 tosses of the coin means 5 Heads and 2 Tails were obtained.
2. Write down the symbol for the number of ordered selections of n objects in which r are alike of one type and $n - r$ are alike of another.	The binomial coefficient $\binom{n}{r}$ gives the number of ordered selections of n objects in which r are alike of one type and $n - r$ are alike of another.
3. State the values of n, r and $n - r$, and write down the binomial coefficient.	For 5 Heads in 7 tosses of the coin, $n = 7$, $r = 5$ and $n - r = 2$. The number of ordered selections of 5 Heads and 2 Tails is $\binom{7}{5}$.
4. Write down the probabilities of a Head and a Tail for a single trial.	In each toss of the coin, $\Pr(H) = 0.6$ and $\Pr(T) = 0.4$.
5. Form the expression for the probability of obtaining 5 Heads in 7 tosses of the coin.	The probability of 5 Heads in 7 tosses of the coin is $\binom{7}{5}(0.6)^5(0.4)^2$.

on Resources

 **Interactivity** Pascal's triangle and binomial coefficients (int-2554)

7.6 Exercise

Technology free

1. A coin is biased in such a way that the probability of tossing a Head is p and the probability of a Tail is q. The coin is tossed twice.
   Let T be the event of obtaining a Tail in a single toss of the coin and let H be the event of obtaining a Head in a single toss of the coin.

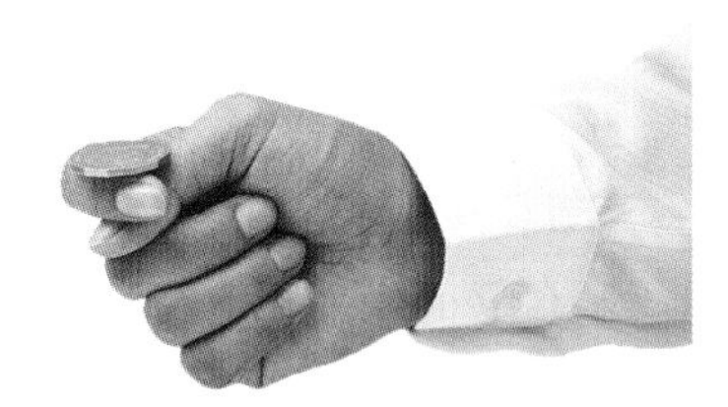

   a. Determine whether the events 'H on the first toss' and 'T on the second toss' are independent or dependent.
   b. Express the following probabilities in terms of p and q.
      i. $\Pr(TT)$ ii. $\Pr(HH)$ iii. $\Pr(TH)$ iv. $\Pr(TH \text{ or } HT)$
   c. Write down the expansion of $(p+q)^2$.
   d. Identify which term in the expansion of $(p+q)^2$ gives the probability of
      i. two Heads ii. two Tails
      iii. one Head and one Tail.

2. A drug for pain relief is administered to three patients in a hospital. The outcome for each patient is independent of the others. It is known that the probability this drug will be effective is 0.7.

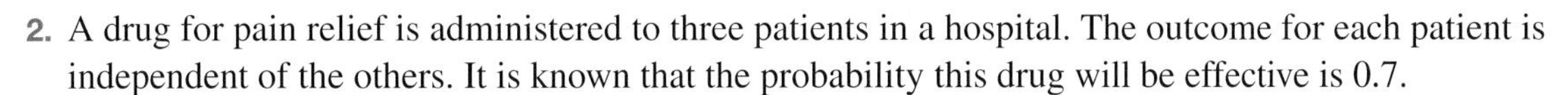

   a. Calculate the probability that the drug will not be effective for pain relief.
   b. Consider the binomial expansion of $(0.7+0.3)^3 = 0.7^3 + \binom{3}{1}(0.7)^2(0.3) + \binom{3}{2}(0.7)(0.3)^2 + 0.3^3$.
      i. Identify which term in the expansion gives the probability that the drug does not give pain relief to exactly one of the patients.
      ii. Identify the probability given by each of the other three terms in the expansion.

3. a. Calculate the number of ways in which the letters of the word *egg* can be arranged.
   b. i. Write down the expansion of $(e+g)^3$.
      ii. State the coefficient of the term g^2e in the expansion.
   c. Explain why the answers to part **a** and part **b ii** are the same.

4. a. Calculate the number of ways in which the letters of the word *abba* can be arranged.
   b. In the binomial expansion of $(a+b)^4$, show that the coefficient of the a^2b^2 term is the same as the answer given in part **a**.

5. An unbiased six sided die is rolled four times.
   Let S be the event of obtaining a six in a single roll of the die and let F be the event of not obtaining a six in a single roll of the die.

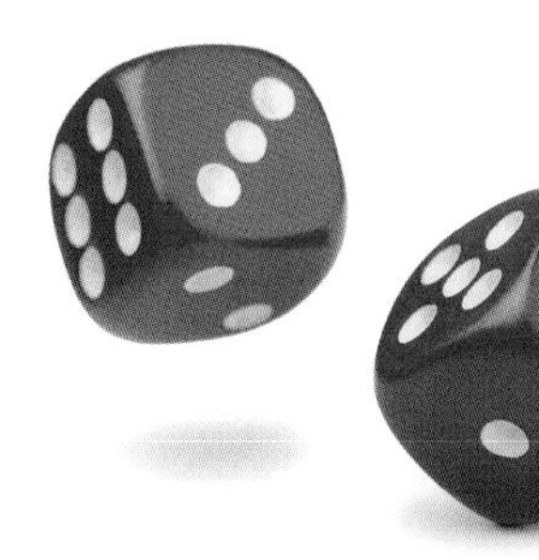

   a. State the values of $\Pr(S)$ and $\Pr(F)$.
   b. i. Describe the outcome 'SSFS'.
      ii. Write down all the possible arrangements of 'SSFS'.
      iii. Calculate the probability of obtaining exactly three sixes in four rolls of the die.
   c. i. Expand $(q+p)^4$.
      ii. With $p = \Pr(S)$ and $q = \Pr(F)$, given in part **a**, identify the term in the expansion that gives the probability of obtaining exactly three sixes in four rolls of the die.

6. Explain why $0.5^4 + {}^4C_1(0.5)^3(0.5) + {}^4C_2(0.5)^2(0.5)^2 + {}^4C_3(0.5)(0.5)^3 + 0.5^4$ must equal 1.

7. **WE14** Write down the expansion of each of the following.
   a. $(a+b)^4$
   b. $(2+x)^4$
   c. $(t-2)^3$

8. Expand and simplify each of the following.
   a. $(x+y)^3$
   b. $(a+2)^4$
   c. $(m-3)^4$
   d. $(2-x)^5$

9. Write down the expansion of each of the following.
   a. $(m+3b)^2$
   b. $(2d-x)^4$
   c. $\left(h+\frac{2}{h}\right)^3$

10. a. A family has 4 children. Assume there is an equal probability of each child being a boy or a girl. Without evaluating the term, write down a term from the expansion given in question 6 that gives the probability that:
      i. all 4 children are girls
      ii. all 4 children are boys
      iii. there are 2 boys and 2 girls.
      iv. there is only one boy.
    b. Calculate the probability that at least one of the children is a girl.

11. Consider $(p+q)^n, n \in N$.
    a. Determine what relationship must exist between p and q for the terms in the expansion of $(p+q)^n, n \in N$ to represent the sample space probabilities of a set of independent trials.
    b. Calculate how many outcomes there are for each trial.
    c. Explain what n represents.
    d. Calculate how many elements there are in the sample space.

12. There are 5 chocolate biscuits and 3 jam-centred biscuits remaining in a storage container. Consider the two cases:
    Case A: Three biscuits are chosen with replacement.
    Case B: Three biscuits are chosen without replacement.
    a. For Case A, identify the values of p, q and n so that the terms in the expansion of $(p+q)^n$ represent the sample space probabilities for the number of chocolate biscuits chosen.
    b. Explain why it is not possible to do the same for Case B.

13. The expansion of $(p+q)^6$ in sigma notation can be written as $\sum_{r=0}^{6} \binom{6}{r} p^{6-r} q^r$.
    A die is rolled six times. Write down the term that gives the probability that exactly 2 fives are obtained.

Technology active

14. For each of the following, find the term specified in the expansion.
    a. The term in w^3 in $(2w-3)^5$
    b. The term in b^{-4} in $\left(3-\frac{1}{b}\right)^7$
    c. The constant term in $\left(y-\frac{3}{y}\right)^4$

15. **WE15** a. Find the term independent of x in the expansion of $\left(x+\frac{1}{2x}\right)^4$.
    b. Find the term in m^2 in the expansion of $\left(2m-\frac{1}{3m}\right)^6$.

16. For each of the following, find the term specified in the expansion.

a. The term in d^5 in $(2b+3d)^5$

b. The coefficient of the term x^2y^3 in $(3x-5y)^5$

17. WE16 a. Write down the notation for the 4th coefficient in the 7th row of Pascal's triangle.

b. Calculate the value of $\binom{9}{6}$.

c. A coin is biased in such a way that the probability of tossing a Head is 0.55.
Form an expression for the probability of obtaining 12 Heads in 18 tosses of the coin.

18. a. Verify that $\binom{22}{8}=\binom{22}{14}$.

b. Show that $\binom{15}{7}+\binom{15}{6}=\binom{16}{7}$.

c. Show that ${}^nC_r + {}^nC_{r+1} = {}^{n+1}C_{r+1}$.

19. Find integers a and b such that $\left(2-\sqrt{5}\right)^4 = a+b\sqrt{5}$.

20. Simplify $1-6m+15m^2-20m^3+15m^4-6m^5+m^6$.

21. Simplify $(1-x)^4-4(1-x)^3+6(1-x)^2-4(1-x)+1$.

22. The first 3 terms in the expansion of $(1+kx)^n$ are 1, $2x$ and $\frac{3}{2}x^2$. Find the values of k and n.

23. In the binomial expansion of $\left(1+\frac{x}{2}\right)^n$, the coefficient of x^3 is 70. Find the value of n.

24. Three consecutive terms of Pascal's triangle are in the ratio 13 : 8 : 4. Find the three terms.

7.6 Exam questions

Question 1 (3 marks) TECH-ACTIVE

Find the term independent of x in the expansion of $\left(2x+\frac{3}{x^3}\right)^4$.

Question 2 (1 mark) TECH-ACTIVE

MC The value of $\binom{7}{5}$ is

A. 42
B. 210
C. 21
D. 2530
E. 35

Question 3 (1 mark) TECH-ACTIVE

MC The notation for the ninth coefficient in the 14th row of Pascal's triangle is

A. ${}^{15}C_8$
B. ${}^{14}C_8$
C. ${}^{15}C_9$
D. ${}^{14}C_9$
E. ${}^{9}C_{14}$

More exam questions are available online.

7.7 Review

7.7.1 Summary

doc-37023

Hey students! Now that it's time to revise this topic, go online to:

 Access the topic summary | Review your results |  Watch teacher-led videos | Practise exam questions

Find all this and MORE in jacPLUS

7.7 Exercise

Technology free: short answer

1. Given $\Pr(A) = 0.4$, $\Pr(A \cup B) = 0.58$ and $\Pr(B \mid A) = 0.3$:
   a. calculate $\Pr(B)$
   b. determine whether A and B are independent events.

2. From a bag that contains 6 red and 7 green balls, 2 balls are drawn without replacement. Calculate the probability that:
   a. one ball of each colour is chosen
   b. at least one green ball is chosen
   c. one ball of each colour is chosen given at least one green ball is chosen.

3. a. Determine the coefficient of x^5 in the expansion of $\left(1 + \frac{x}{2}\right)^7$.
   b. Express $\left(3 - \sqrt{2}\right)^3$ in the form $m + n\sqrt{2}$.
   c. Expand $\left(x^2 - \frac{3}{x}\right)^4$ in decreasing powers of x.
   d. Determine the term independent of y in the expansion of $\left(\frac{3}{y^2} - y\right)^{12}$. (Do not attempt to fully evaluate this term.)

4. At a teacher's college, 70% of students are female. On average, 75% of female and 85% of male students graduate. A student who graduates is selected at random. Determine the probability that the student is male.

5. A new lie detector machine is being tested to determine its degree of accuracy. One hundred people, 20 of whom were known liars, were tested on the machine. Some of the results are shown in the following table.

	Liars (L)	**Honest people (H)**	**Totals**
Correctly tested (C)	18		85
Incorrectly tested (I)			
Totals			100

   a. Complete the table.

   Hence, find the probability that a person selected at random:

   b. was incorrectly tested
   c. was honest and correctly tested
   d. was correctly tested, given that they were honest.

6. a. Calculate the number of ways in which a group of 2 girls and 3 boys can:
   i. be arranged in a row
   ii. be arranged in a row with the girls occupying the end positions
   iii. be arranged in a row with the two girls together.

   b. Calculate the number of ways in which 2 girls and 3 boys can be chosen from a group of 6 girls and 5 boys.

   c. On a particular day, 5 babies are born in a maternity hospital. Assuming the probability of a girl is 0.52, form an expression for the probability that 2 of the newborn babies are girls. (Do not attempt to evaluate this expression.)

Technology active: multiple choice

7. **MC** Let the universal set ξ be the set of integers from 1 to 50 inclusive. Three subsets, A, B and C, are defined as follows:
   A is the set of multiples of 3 less than 50.
   B is the set of prime numbers less than 50.
   C is the set of multiples of 10 less than or equal to 50.
   Identify which of the following statements is correct.

   A. $B \subseteq C$
   B. $A \cap B = \varnothing$
   C. A and B are independent.
   D. B and C are mutually exclusive.
   E. $n(A) = 15$

8. **MC** Hockey players Ash and Ben believe their independent chances of scoring a goal in a match are $\frac{2}{3}$ and $\frac{2}{5}$ respectively. Calculate the probability that in the next match they play, neither scores a goal.

   A. $\frac{1}{5}$
   B. $\frac{8}{15}$
   C. $\frac{11}{15}$
   D. $\frac{4}{5}$
   E. $\frac{14}{15}$

9. **MC** At the carnival, the mystery house offers a choice of different adventures to those daring to enter. Adventurers can enter the house through any one of 3 doors. Behind each door there are 4 mystery envelopes containing maps leading to different adventures. Calculate how many choices of adventures are possible.

   A. 3
   B. 7
   C. 12
   D. 24
   E. 36

10. **MC** One bag contains 2 green and 6 blue counters. A second identical bag contains 4 green counters. A bag is selected at random and a counter is drawn. Calculate the probability that the counter is green.

   A. $\frac{5}{8}$
   B. $\frac{1}{2}$
   C. $\frac{1}{4}$
   D. $\frac{1}{8}$
   E. $\frac{3}{16}$

11. **MC** The number of children per family for the 40 students in a class was recorded in the table below.

Number of children	Frequency
1	7
2	12
3	13
4	5
5	2
6	1

   The probability that a family selected at random has at least 3 children is:

   A. $\frac{21}{40}$
   B. $\frac{3}{7}$
   C. $\frac{1}{2}$
   D. $\frac{13}{40}$
   E. $\frac{1}{5}$

12. **MC** At the end of a tennis tournament, a television station has selected the 8 best points from the matches played. Calculate the number of ways in which the best shots can be ranked.

A. 1 B. 8 C. 56 D. 70 E. 40 320

13. **MC** The results of a survey concluded that 62% of people were in favour of a particular proposal. If 7 people are selected at random, the probability that exactly 5 of them will be in favour of the proposal is:

A. $(0.62)^5$ B. ${}^7C_5(0.62)^5(0.38)^2$ C. $(0.62)^5(0.38)^2$

D. $\binom{7}{5}(0.38)^5(0.62)^2$ E. $1-(0.38)^2$

14. **MC** There are 14 teams competing in a rugby championship. The number of different ways in which the first, second and third positions can be filled is:

A. ${}^{14}P_3$ B. ${}^{14}C_3$ C. 14^3 D. 3^{14} E. $\frac{12!}{3!}$

15. **MC** Calculate how many different 9-letter arrangements can be formed from the letters of the word EMERGENCY.

A. 181 440 B. 36 C. 60 480 D. 30 240 E. 57 960

16. **MC** Calculate how many committees consisting of 3 men and 4 women are possible if there are 7 men and 6 women available for selection to the committee.

A. 8 648 640 B. 1716 C. 700 D. 525 E. 350

Technology active: extended response

17. Jo, who owns a corner grocery, imports tins of chickpeas and lentils. When unpacking the tins, Jo finds that one box contains 10 tins that have lost their labels. The tins are identical, but after looking through his invoices, Jo has calculated that 7 of the tins contain chickpeas and 3 contain lentils.
He decides to take them home since he is unable to sell them without a label. He wants to use the chickpeas to make some hummus, so he opens the tins at random until he opens a tin of chickpeas.

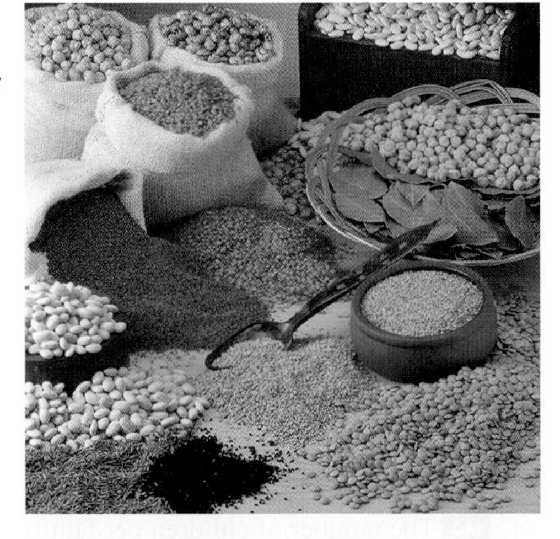

a. List the outcomes for this sample space.
b. Find the minimum number of tins Jo must open to ensure he opens a tin of chickpeas.
c. Find the probability that the chickpeas are in the third tin he opens.

In fact, Jo found the chickpeas in the second tin. After making the hummus, Jo decides he will open one tin each day and use whatever it contains.

d. Draw up a probability tree to show the outcomes for the next 3 days.
e. Calculate the probability that he opens at least one tin of each over the next 3 nights.
f. Given that the first tin opened contains chickpeas, calculate the probability that the third tin contains lentils.

18. There are 7 girls and 6 boys in the school's chess club. Two boys and two girls are to be chosen to represent the school at the High School Chess Tournament.

   a. Calculate the number of different ways in which the 4 representatives can be selected.
   b. Calculate the number of ways in which the team can be selected if both the captain, Sarah, and the oldest student, Kristi, must be included.
   c. Find the probability that both Sarah and Kristi are selected.
   d. Given that the youngest member of the team, Emma, is selected, calculate the probability that the other girl selected is either Sarah or Kristi.

19. Twenty-two items are to be sold at a small charity auction. The amount, in dollars, paid for each article, is to be decided by the throw of a dart at a dartboard. The rules are as follows.
   - The price of any item is the number scored multiplied by \$100.
   - If more than one person wants to bid for an item, the person with the highest score wins.
   - If a buyer misses the board the buyer can throw the dart again until he/she hits the board, unless there are others who want to buy the same item, in which case the buyer does not get a second chance.
   - Anyone hitting the inner bull will automatically win and will be sold the item for the modest price of \$5000.
   - Anyone hitting the outer bull will automatically win over any score other than an inner bull, and will be sold the item for \$2500.

   Aino and Bryan attend the auction. The associated probabilities are as follows.

   For Aino:
   The probability of hitting any one of the numbers from 1 to 20 is 0.045; the probability of hitting the inner bull is 0.005; and the probability of hitting the outer bull is 0.006.

   For Bryan:
   The probability of hitting any one of the numbers from 1 to 20 is 0.046; the probability of hitting the inner bull is 0.004; and the probability of hitting the outer bull is 0.003.

   a. Find the probability that Aino misses the board.

   Aino is the only buyer interested in a particular item. Calculate the probability that:

   b. she needs 2 attempts to buy her item
   c. she buys the item for \$900 with her first throw of the dart
   d. she only needs the one throw but her item costs more than \$1900.

   Aino then decides to buy a second item in which Bryan is also interested. They each throw a dart.

   e. Calculate the probability that Bryan wins with his first dart.
   f. Given that Bryan has already scored a 15 on his first throw, calculate the probability that Aino will beat Bryan's score on her first throw.

20. In a table tennis competition, each team must play every other team twice.

a. Calculate how many games must be played if there are 5 teams in the competition.

b. Calculate how many games must be played if there are n teams in the competition.

A regional competition consists of 16 teams, labelled A, B, C, …, N, O, P.

c. Calculate how many games must each team play.

d. Calculate the total number of games played.

7.7 Exam questions

Question 1 (1 mark) TECH-ACTIVE

MC A and B are mutually exclusive events. If $\Pr(A) = 0.6$ and $\Pr(B) = 0.3$, then $\Pr(A \cup B)$ is

A. 0
B. 0.18
C. 0.9
D. 0.3
E. cannot be determined

Question 2 (3 marks) TECH-FREE

A local shop sells mixed bags containing a dozen small muffins. Within each bag, five muffins have a chocolate surprise inside and the others don't. Sue randomly selects a muffin and then Margaret randomly selects a muffin.

Draw a probability tree diagram and work out the probability that Sue's muffin doesn't contain a chocolate surprise while Margaret's does contain a chocolate surprise.

Question 3 (1 mark) TECH-ACTIVE

MC A box contained 6 red balls, 9 blue balls and 5 orange balls. One ball was chosen randomly, the colour noted then replaced, and another ball was chosen. The probability that a blue and a red ball were selected is

A. 0.45
B. 0.27
C. 0.41
D. 0.03
E. 0.66

Question 4 (4 marks) TECH-ACTIVE

Emma has a special collection of bottles. Eight are red, five are blue and seven are green. She likes to arrange them in threes on a shelf so that one of each colour is displayed with green in the middle.

a. If the bottles are all different in shape and size, determine how many arrangements she can make. **(2 marks)**

b. If she randomly placed three bottles together ignoring the colour, calculate the probability that they will all be green. **(2 marks)**

Question 5 (1 mark) TECH-ACTIVE

MC The number of different three-digit numbers that can be formed using {3, 4, 5, 7} if each digit can be used only once is

A. 30
B. 24
C. 12
D. 10
E. 8

More exam questions are available online.

Answers

Topic 7 Probability

7.2 Probability review

7.2 Exercise

1. a. $\frac{5}{23}$ b. $\frac{12}{23}$ c. $\frac{15}{23}$ d. 0

2. a. $\frac{1}{6}$ b. $\frac{5}{6}$ c. $\frac{1}{2}$
 d. $\frac{2}{3}$ e. $\frac{1}{3}$ f. 0

3. a. $\frac{1}{26}$ b. $\frac{5}{26}$ c. $\frac{3}{26}$
 d. $\frac{25}{26}$ e. 1 f. $\frac{9}{26}$

4. a. $\frac{1}{200}$
 b. $\frac{199}{39\,980}$
 c. $\frac{1}{11\,094\,450} = \frac{9}{25 \times 1999 \times 1998}$

5. a. 1 b. $\frac{1}{2}$ c. $\frac{1}{4}$
 d. $\frac{7}{13}$ e. $\frac{12}{13}$

6. a. $\frac{1}{16}$ b. $\frac{1}{4}, \frac{3}{4}$

7. a. $\frac{1}{4}$ b. $\frac{1}{8}$ c. $\frac{9}{16}$

8. a. i. $\frac{3}{4}$ ii. $\frac{7}{10}$ iii. $\frac{1}{4}$
 b. 8 additional red balls

9. a.

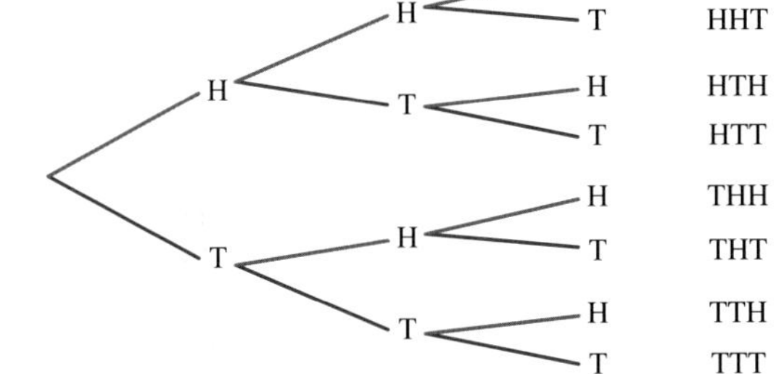

 b. $\frac{7}{8}$
 c. $\frac{3}{4}$

10. a. $\frac{3}{8}$ b. $\frac{1}{4}$ c. $\frac{1}{2}$
 d. $\frac{7}{8}$ e. $\frac{1}{2}$

11. a. $\frac{4}{9}$ b. $\frac{7}{27}$ c. $\frac{2}{3}$ d. $\frac{8}{27}$

12. a. $\frac{89}{400}$ b. $\frac{53}{800}$ c. $\frac{21}{25}$ d. $\frac{439}{800}$

13. a. $\frac{23}{100}$ b. $\frac{81}{200}$

14. a. 5
 b. i. $\frac{1}{4}$ ii. $\frac{1}{5}$ iii. $\frac{11}{20}$
 c. $\frac{9}{10}$

15. a. ξ (42)

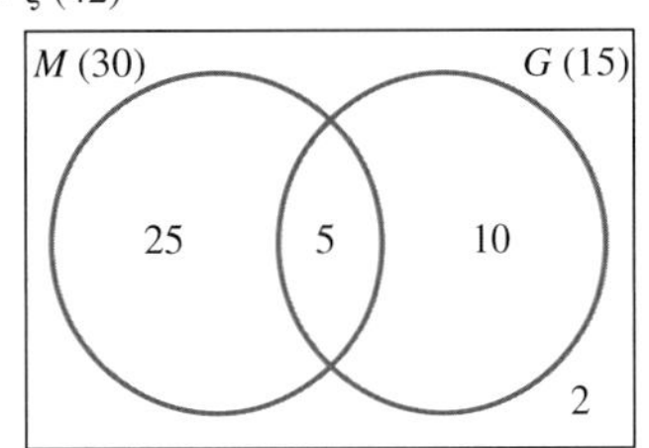

 b. $\frac{25}{42}$
 c. $\frac{1}{21}$
 d. $\frac{35}{42} = \frac{5}{6}$

16. a. $\xi = \{1, 2, 3, \ldots, 18\}$, $A = \{3, 6, 9, 12, 15, 18\}$, $B = \{4, 8, 12, 16\}$ and $C = \{5, 10, 15\}$
 ξ (18)

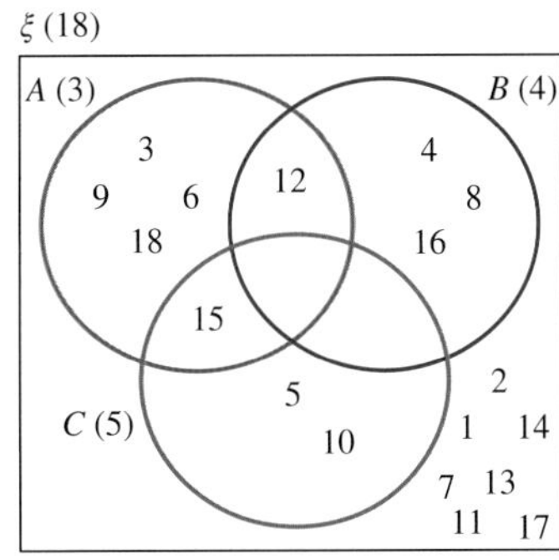

 b. B and C
 c. $\frac{1}{3}$
 d. i. $\frac{4}{9}$ ii. $\frac{5}{18}$ iii. $\frac{7}{18}$

17. a.

	B	B'	
A	0.35	0.3	0.65
A'	0.15	0.2	0.35
	0.5	0.5	1

 b. 0.85

18. a.

	B	B'	
A	0.38	0.2	0.58
A'	0.17	0.25	0.42
	0.55	0.45	1

b. 0.25

c. $\Pr(A \cup B) = 0.75, \Pr(A \cup B)' = 0.25$
$\Pr(A' \cap B') = 0.25 \Rightarrow \Pr(A \cup B)'$
$= \Pr(A' \cap B')$

d. From the probability table, $\Pr(A \cap B) = 0.38$.
$$1 - \Pr(A' \cup B') = 1 - [\Pr(A') + \Pr(B') - \Pr(A' \cap B')]$$
$$= 1 - 0.42 - 0.45 + 0.25$$
$$= 0.38$$

So, $\Pr(A \cap B) = 1 - \Pr(A' \cup B')$

e. $\Pr(\xi) = 1$

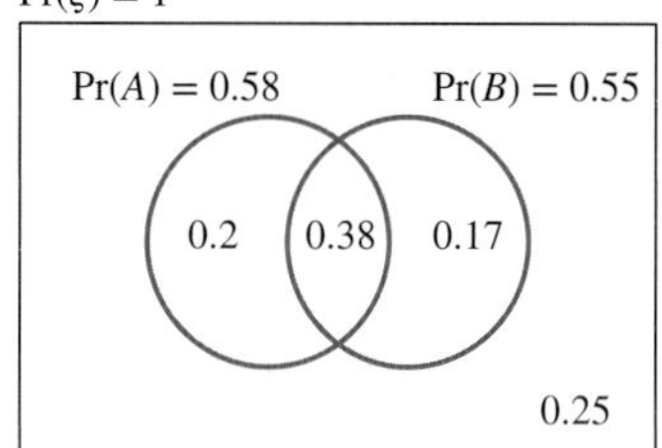

19. a. $\frac{2}{9}$ b. $\frac{1}{36}$ c. $\frac{7}{18}$
d. 0 e. $\frac{1}{12}$ f. 1

20. a. i.

ii. $\frac{1}{12}$

iii. $\frac{1}{4}$

b. $\frac{7}{13}$

21. a. $\frac{7}{20}$ b. $\frac{2}{5}$

22. a. Use int(6 * rand( ) + 1) to generate random numbers between 1 and 6.

b. Results from the simulation will vary. Sample responses can be found in the worked solutions in the online resources.

c. Increase the number of trials, since probabilities are long term proportions.

23. a. $\frac{18}{25}$ b. $\frac{6}{25}$ c. 288 students
d. 0.0576 e. 0.5184 f. 0.0784

7.2 Exam questions

Note: Mark allocations are available with the fully worked solutions online.

1. a.

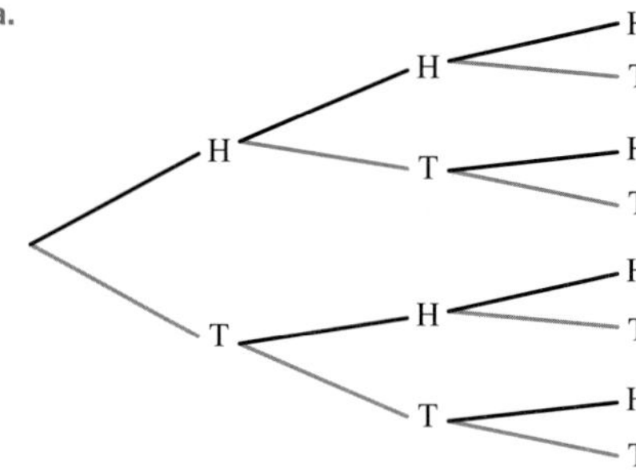

b. $\frac{1}{8}$

c. $\frac{1}{4}$

2. C

3. a. $\xi = 55$

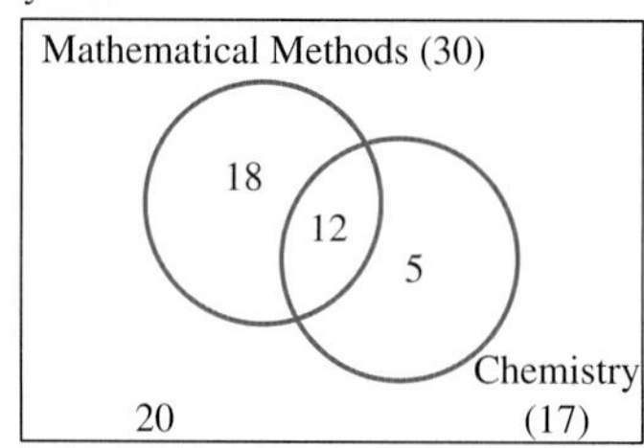

b. $\frac{18}{55}$

c. $\frac{4}{11}$

7.3 Conditional probability

7.3 Exercise

1. a. $\Pr(R_2 \mid Y_1) = \frac{3}{8}$ b. $\Pr(Y_2 \mid Y_1) = \frac{1}{8}$
c. $\Pr(P_2 \mid R_1) = \frac{1}{2}$ d. $\Pr(R_2 \mid R_1') = \frac{3}{8}$

2. a. T b. F c. F d. T
e. T f. F

3. a. $\frac{3}{20}$ b. 0.5

4. $\frac{2}{9}$

5. $\frac{1}{2}$

6. a. 0.2 b. $\frac{2}{3}$ c. $\frac{2}{7}$ d. $\frac{5}{7}$

7. a. 0.3 b. $\frac{3}{5}$ c. $\frac{1}{2}$ d. $\frac{3}{5}$

8. a. 0.9 b. $\frac{4}{7}$ c. $\frac{3}{4}$ d. $\frac{1}{3}$

9. a. $\frac{1}{5}$ b. $\frac{5}{9}$ c. $\frac{7}{11}$ d. $\frac{11}{25}$

10. a. $\frac{5}{36}$ b. $\frac{2}{15}$ c. $\frac{1}{9}$ d. $\frac{4}{5}$

11. a. $\Pr(A \cap B) = 0.12$; $\Pr(A \mid B) = \frac{6}{25}$

b. $\frac{14}{55}$

12. a. $\frac{9}{14}$ b. $\frac{61}{81}$ c. 1

13. a. $\frac{5}{11}$

b.

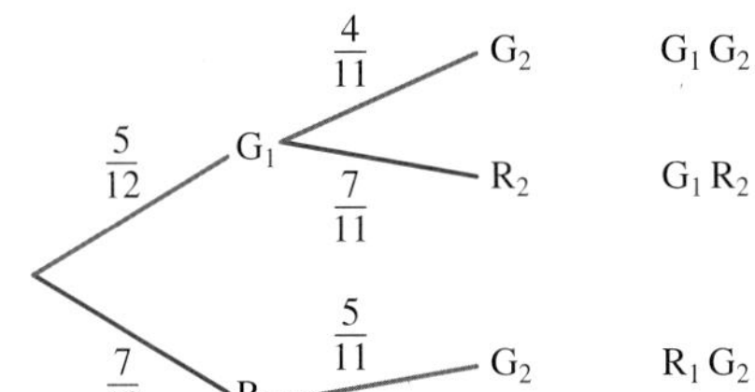

c. $\frac{35}{132}$

d. $\frac{31}{66}$

14. a. $\frac{1}{17}$ b. $\frac{15}{34}$ c. $\frac{2}{15}$ d. $\frac{4}{17}$

15. a. $\frac{7}{22}$ b. $\frac{28}{33}$ c. $\frac{3}{8}$ d. $\frac{1}{2}$

16. a. $\frac{73}{100}$ b. $\frac{7}{100}$ c. $\frac{23}{30}$ d. $\frac{66}{70}$

17. a. $\frac{427}{500}$ b. $\frac{47}{100}$ c. $\frac{32}{61}$

18. a. $\frac{267}{1000}$ b. $\frac{257}{1000}$ c. $\frac{82}{267}$ d. $\frac{185}{743}$

19. a. $\frac{9}{50}$ b. $\frac{21}{50}$ c. $\frac{47}{72}$ d. $\frac{143}{168}$

20. a. Let T = the bus being on time and T′ = the bus being late.
Let S = Rodney gets to school on time and S′ = Rodney gets to school late.

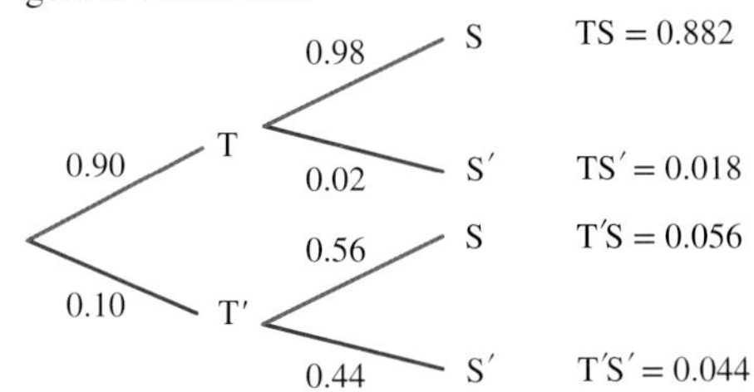

b. Pr(Rodney will arrive at school on time)
$= 0.882 + 0.056$
$= 0.938$

21.

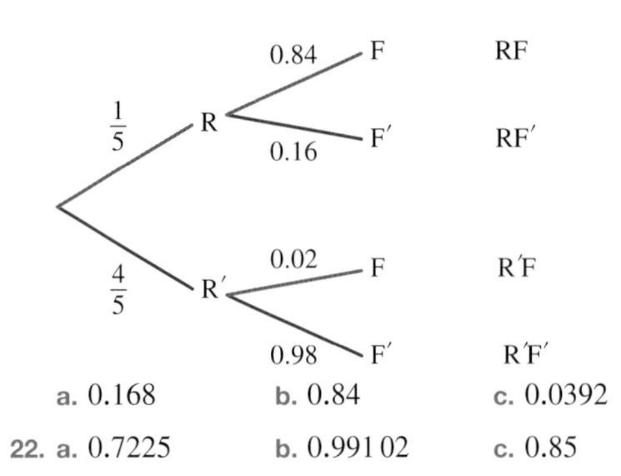

a. 0.168 b. 0.84 c. 0.0392

22. a. 0.7225 b. 0.991 02 c. 0.85

7.3 Exam questions

Note: Mark allocations are available with the fully worked solutions online.

1. D 2. A 3. B

7.4 Independence

7.4 Exercise

1. Yes
2. Yes
3. a. $p = 0.4$ b. $p = 0.6$ c. $q = \frac{3}{4}$
4. a. $\xi = \{HH, HT, TH, TT\}$; $A = \{TH, TT\}$; $B = \{HT, TH\}$; $C = \{HH, HT, TH\}$
   b. A and B are independent.
   c. B and C are not independent.
   d. $\frac{3}{4}$
5. a. 0.56 b. 0.14 c. 0.06 d. 0.94
6. a. $\frac{64}{125}$ b. $\frac{1}{125}$ c. $\frac{16}{125}$
7. No
8. a. No
   b. No
   c. i. Yes ii. No
9. a. 0.216 b. 0.024 c. 0.784
10. a. $\frac{4}{9}$

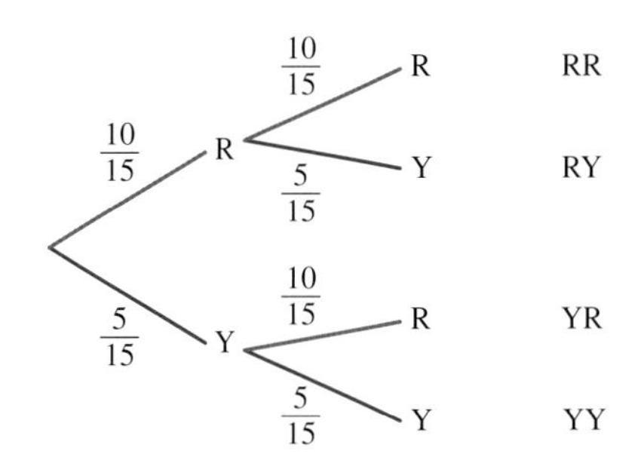

b. $\frac{10}{21}$

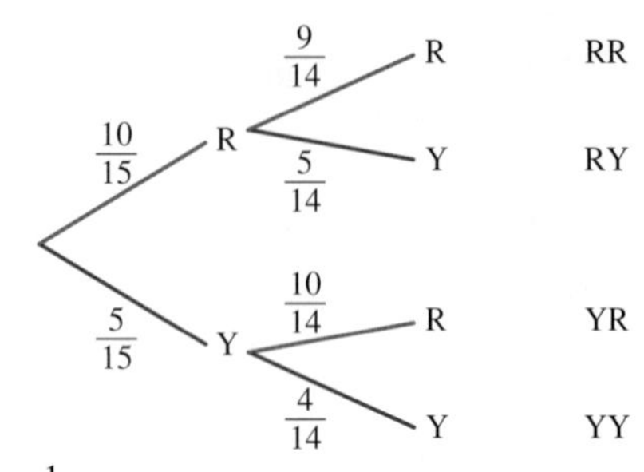

c. $\frac{1}{3}$

11. a. 0.72 b. 0.98 c. 0.2

12. a. $\frac{9}{40}$ b. $\frac{2}{5}$ c. $\frac{17}{20}$ d. $\frac{47}{200}$ e. $\frac{7}{12}$

13. a. $\frac{4}{5}$ b. $\frac{2}{3}$ c. $\frac{8}{15}$ d. $\frac{14}{15}$

14. a. $\frac{1}{5}$ b. $\frac{11}{30}$ c. No

15. $\Pr(B) = 0.5$

16. No; sample responses can be found in the worked solutions in the online resources.

17. No; sample responses can be found in the worked solutions in the online resources.

18. a–c. No; sample responses can be found in the worked solutions in the online resources.

7.4 Exam questions

Note: Mark allocations are available with the fully worked solutions online.

1. D
2. C
3. a. 0.224 b. 0.964

7.5 Counting techniques

7.5 Exercise

1. 6
2. 12
3. a. 24 b. 24
4. 36
5. 16
6. a. 120 b. 6 days c. 240 d. 210 e. 312 f. 216
7. a. 120 b. 300 c. 24 d. $\frac{1}{5}$
8. a. 1 757 600 b. 16 807 c. 216 d. 216 e. 24
9. a. 40 320 b. 1152 c. 8640 d. 30 240 e. 2880
10. a. 648 b. 328 c. 216 d. 60
11. a. 720 b. 240 c. $\frac{1}{3}$
12. a. 30 b. 56 c. $\frac{15}{28}$
13. a. 17 576 000 b. 1 875 000 c. $\frac{1}{2704}$
14. a. 720 b. 120 c. $\frac{1}{36}$
15. a. 6435 b. 2 598 960 c. 48 d. 120
16. a. 43 758 b. 6720 c. 1341 d. 35 750 e. 8008
17. a. 6 b. 24
18. a. 80 640 b. 282 240 c. 119 750 400 d. $\frac{1}{6}$
19. a. 2002 b. 840 c. 126 d. $\frac{36}{91}$
20. a. 420 b. 102 c. 24
21. a. 27 720 b. 210
22. a. 40 320 b. 14 400
23. a. 4096 b. 31 c. $\frac{31}{4096}$ d. $\frac{4017}{4096}$ e. $\frac{35}{4096}$
24. a. 64 864 800 b. 4 989 600 c. 453 600
25. $\frac{45}{442}$
26. a. $\frac{1}{11\,925}$ b. $\frac{19}{312}$ c. 0.8495 d. $\frac{24\,877}{73\,099\,048} \approx 0.2747$
27. a. $\frac{16}{225}$ b. $\frac{1}{100}$ c. $\frac{4}{75}$ d. $\frac{18}{25}$ e. $\frac{101}{799}$

7.5 Exam questions

Note: Mark allocations are available with the fully worked solutions online.

1. C
2. C
3. a. 1820 b. 714 c. 0.3231

7.6 Binomial coefficients and Pascal's triangle

7.6 Exercise

1. a. Independent, $\Pr(T\,|\,H) = \Pr(T)$
   b. i. q^2 ii. p^2 iii. qp iv. $2pq$
   c. $(p+q)^2 = p^2 + 2pq + q^2$
   d. i. First term ii. Third term iii. Second term

2. a. 0.3

b. i. Second term, $\binom{3}{1}(0.7)^2(0.3)$

ii. The first term is the probability that the drug is effective for all three patients; the second term is the probability that the drug is effective for only two of the three patients; the third term is the probability that the drug is effective for only one of the three patients; the fourth term is the probability that the drug is not effective for any of the three patients.

3. a. $\frac{3!}{2!} = 3$

b. i. $(e+g)^3 = e^3 + 3e^2g + 3eg^2 + g^3$

ii. 3

c. The coefficient of the third term is the number of arrangements of $eg^2 = egg$, calculated to be 3 in part a.

4. a. 6 b. $\binom{4}{2} = 6$

5. a. $\Pr(S) = \frac{1}{6}$, $\Pr(F) = \frac{5}{6}$

b. i. Six on the first roll, six on the second roll, not a six on the third roll and a six on the fourth roll of the die

ii. SSSF, SSFS, SFSS, FSSS

iii. $\frac{5}{324}$

c. i. $(q+p)^4 = q^4 + \binom{4}{1}q^3p + \binom{4}{2}q^2p^2 + \binom{4}{3}qp^3 + p^4$

ii. $p = \frac{1}{6}, q = \frac{5}{6}, \binom{4}{3}qp^3$

6. $(0.5+0.5)^4 = 1.$

7. a. $a^4 + 4a^3b + 6a^2b^2 + 4ab^3 + b^4$

b. $16 + 32x + 24x^2 + 8x^3 + x^4$

c. $t^3 - 6t^2 + 12t - 8$

8. a. $x^3 + 3x^2y + 3xy^2 + y^3$

b. $a^4 + 8a^3 + 24a^2 + 32a + 16$

c. $m^4 - 12m^3 + 54m^2 - 108m + 81$

d. $32 - 80x + 80x^2 - 40x^3 + 10x^4 - x^5$

9. a. $m^2 + 6bm + 9b^2$

b. $16d^4 - 32d^3x + 24d^2x^2 - 8dx^3 + x^4$

c. $h^3 + 6h + \frac{12}{h} + \frac{8}{h^3}$

10. a. i. 0.5^4 ii. 0.5^4

iii. ${}^4C_2(0.5)^2(0.5)^2$ iv. ${}^4C_1(0.5)(0.5)^3$

b. $\frac{15}{16}$

11. a. $p + q = 1, 0 \le p \le 1$ and $0 \le q \le 1$.

b. Two

c. The number of trials

d. $n + 1$

12. a. $p = \frac{5}{8}, q = \frac{3}{8}, n = 3.$

b. The trials are not independent.

13. $\binom{6}{4}p^2q^4$, $15\left(\frac{1}{6}\right)^2\left(\frac{5}{6}\right)^4$

14. a. $720w^3$ b. $945b^{-4}$ c. 54

15. a. $\frac{3}{2}$ b. $\frac{80}{3}m^2$

16. a. $243d^5$ b. -11250

17. a. ${}^7C_3$

b. 84

c. $\binom{18}{12}(0.55)^{12}(0.45)^6$

18. a–c. Sample responses can be found in the worked solutions in the online resources.

19. $a = 161$ and $b = -72$

20. $(1-m)^6$

21. x^4

22. $n = 4$; $k = \frac{1}{2}$

23. $n = 16$

24. 125 970, 77 570, 38 760 $\left(\text{or } {}^{20}C_{12}, {}^{20}C_{13}, {}^{20}C_{14}\right)$

7.6 Exam questions

Note: Mark allocations are available with the fully worked solutions online.

1. 96
2. C
3. B

7.7 Review

7.7 Exercise

Technology free: short answer

1. a. 0.3 b. Independent

2. a. $\frac{7}{13}$ b. $\frac{21}{26}$ c. $\frac{2}{3}$

3. a. $\frac{21}{32}$

b. $45 - 29\sqrt{2}$

c. $x^8 - 12x^5 + 54x^2 - 108x^{-1} + 81x^{-4}$

d. 40 095

4. $\frac{17}{52}$

5. a. Sample responses can be found in the worked solutions in the online resources.

b. $\frac{3}{20}$

c. $\frac{67}{100}$

d. $\frac{67}{80}$

6. a. i. 120 ii. 12 iii. 48

b. 150

c. $\binom{5}{2}(0.52)^2(0.48)^3$

Technology active: multiple choice

7. D
8. A
9. C
10. A
11. A
12. E
13. B
14. A
15. C
16. D

Technology active: extended response

17. a. {C, LC, LLC, LLLC}
 b. 4
 c. $\frac{7}{120}$
 d.

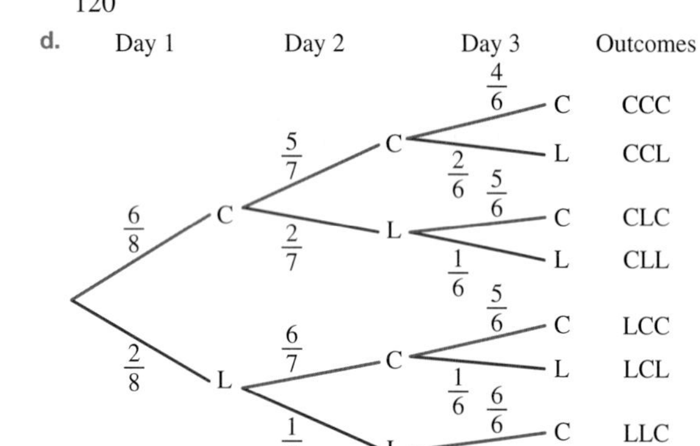

 e. $\frac{9}{14}$
 f. $\frac{2}{7}$

18. a. 315 b. 15 c. $\frac{1}{21}$ d. $\frac{1}{3}$

19. a. 0.089 b. 0.081 079 c. 0.045
 d. 0.056 e. 0.482 165 f. 0.236

20. a. 20 games b. $n(n-1)$
 c. 30 games d. 240 games

7.7 Exam questions

Note: Mark allocations are available with the fully worked solutions online.

1. C
2. $\frac{35}{132}$

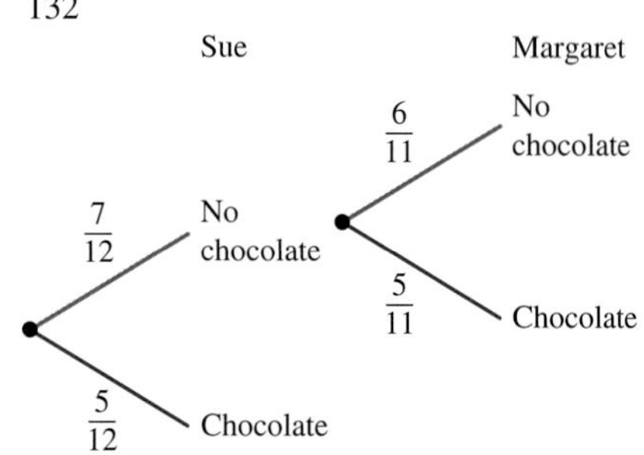

3. B
4. a. 560 b. 0.0307
5. B

8 Trigonometric functions

LEARNING SEQUENCE

Fully worked solutions for this topic are available online.

8.1 Overview

Hey students! Bring these pages to life online

Watch videos

Engage with interactivities

Answer questions and check results

Find all this and MORE in jacPLUS

8.1.1 Introduction

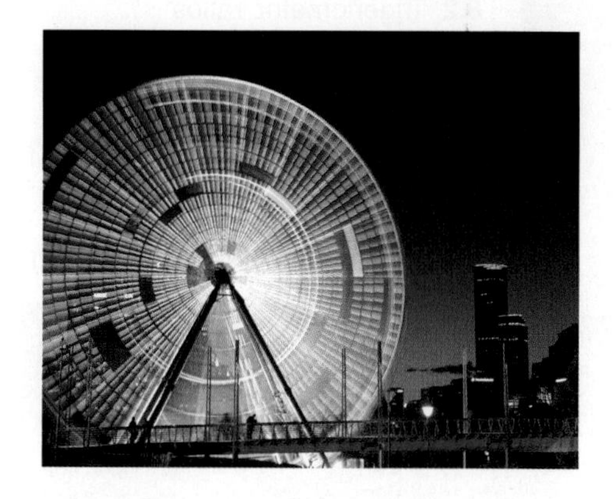

Periodicity forms a natural part of our lives. Our heart rates, blood pressure and other vital statistics fluctuate over a 24-hour period, our long-term sleep and wakefulness patterns are periodic, and our mood-affecting biorhythms and ovulation cycles are periodic. Even in the world of fashion, such measures as dress length or width of trouser leg can be cyclical. What is old-fashioned today will, in the fullness of time, often regain its popularity.

The length of time for one repetition of a cycle is known as the period. The rotation of the Earth creates a day–night cycle with a period of 24 hours; the revolution of the Earth about the sun creates seasonal cycles with a period of 12 months.

The trigonometric sine and cosine functions are the most important examples of periodic functions. These form models for many periodic phenomena, smoothing out random fluctuations to show the overall oscillatory nature of the phenomena about an equilibrium position. The functions model the wave form of alternating current in most electrical power circuits, they model the depth of water at a pier from low to high tide levels, they model sound and light waves, weather patterns and temperature fluctuations, and they model the height above ground of a person on a Ferris wheel; all of these are examples of their applicability.

The simple harmonic motion of a pendulum swinging under gravity is modelled by a sine function. Such pendulums have provided a means of timekeeping ever since their invention in 1656. London's Big Ben is one such example. Despite quartz clocks being more accurate, grandfather clocks are still found today in private homes and antique shops.

Radian measure makes trigonometric functions mathematically simpler. The concept of a radian was first recognised in 1714 by the English mathematician Roger Cotes, a colleague of Isaac Newton. The actual term 'radian', however, was introduced in 1873 by James Thomson, mathematics professor at Queens College, Belfast. Cotes died from illness at a young age, with Newton quoted as saying, 'If he had lived we would have known something'.

KEY CONCEPTS

This topic covers the following key concepts from the VCE Mathematics Study Design:

- the unit circle, radians, arc length and sine, cosine and tangent as functions of a real variable
- the relationships $\sin(x) \approx x$ for small values of x, $\sin^2(x) + \cos^2(x) = 1$ and $\tan(x) = \dfrac{\sin(x)}{\cos(x)}$
- exact values for sine, cosine and tangent of $\dfrac{n\pi}{6}$ and $\dfrac{n\pi}{4}$, $n \in Z$
- symmetry properties, complementary relations and periodicity properties for sine, cosine and tangent functions.

Note: *Concepts shown in grey are covered in other topics.*

8.2 Trigonometric ratios

LEARNING INTENTION

At the end of this subtopic you should be able to:
- calculate sides, angles and areas in right-angled triangles
- determine exact values for trigonometric ratios of 30°, 45° and 60°.

The process of calculating all side lengths and all angle magnitudes of a triangle is called **solving the triangle**. Here we review the use of trigonometry to solve right-angled triangles.

8.2.1 Right-angled triangles

The hypotenuse is the longest side of a right-angled triangle and it lies opposite the 90° angle, the largest angle in the triangle. The other two sides are labelled relative to one of the other angles in the triangle, an example of which is shown in the diagram.

It is likely that the trigonometric ratios of sine, cosine and tangent, possibly together with Pythagoras' theorem ($a^2 + b^2 = c^2$), will be required to solve a right-angled triangle.

Trigonometric ratios

$$\sin(\theta) = \frac{\text{opposite}}{\text{hypotenuse}}$$

$$\cos(\theta) = \frac{\text{adjacent}}{\text{hypotenuse}}$$

$$\tan(\theta) = \frac{\text{opposite}}{\text{adjacent}}$$

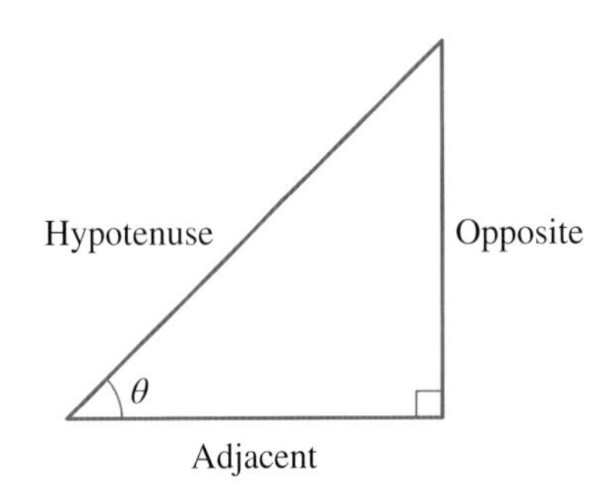

The trigonometric ratios, usually remembered as SOH, CAH, TOA, cannot be applied to triangles that do not have a right angle. However, isosceles and equilateral triangles can easily be divided into two right-angled triangles by dropping a perpendicular from the vertex between a pair of equal sides to the midpoint of the opposite side.

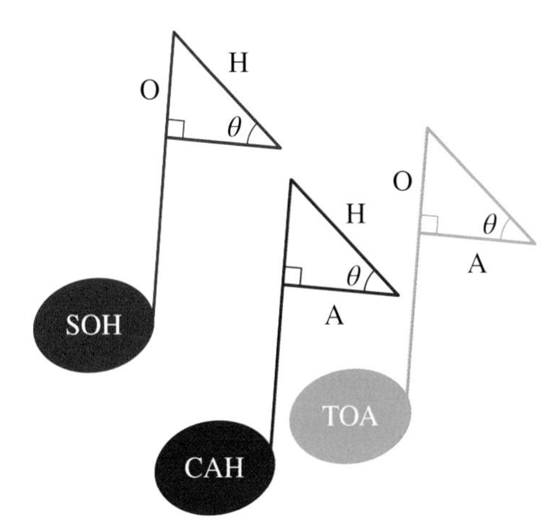

WORKED EXAMPLE 1 Calculating sides and angles in right-angled triangles

Calculate, to 2 decimal places, the value of the pronumeral shown in each diagram.

a.

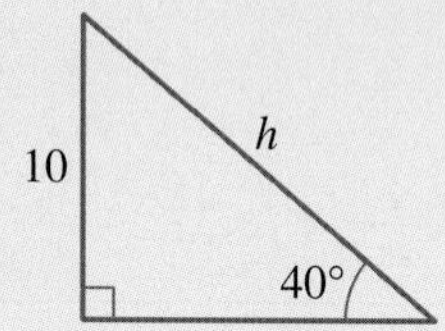

b.

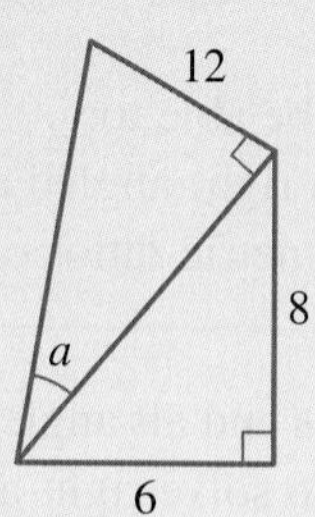

THINK	WRITE
a. 1. Choose the appropriate trigonometric ratio.	a. Relative to the angle, the sides marked are the opposite and the hypotenuse. $\sin(40°) = \frac{10}{h}$
2. Rearrange to make the required side the subject and evaluate, checking the calculator is in degree mode.	$h = \frac{10}{\sin(40°)}$ $= 15.56$ (to 2 decimal places)
b. 1. Obtain the hypotenuse length of the lower triangle.	b. From Pythagoras' theorem, the sides 6, 8, 10 form a Pythagorean triple, so the hypotenuse is 10.
2. In the upper triangle choose the appropriate trigonometric ratio.	The opposite and adjacent sides to the angle $a°$ are now known. $\tan(a) = \frac{12}{10}$
3. Rearrange to make the required angle the subject and evaluate.	$\tan(a) = 1.2$ $\therefore a = \tan^{-1}(1.2)$ $= 50.19°$ (to 2 decimal places)

TI | THINK	DISPLAY/WRITE
a. 1. Put the calculator in DEGREE mode by pressing the top right-hand corner until it displays DEG. On a Calculator page, press MENU then select: 3: Algebra 1: Solve Complete the entry line as: solve $\left(\sin(40) = \frac{10}{h}, h\right)$ then press ENTER. To put in decimal form, press MENU, then select: 2: Number 1: Convert to Decimal	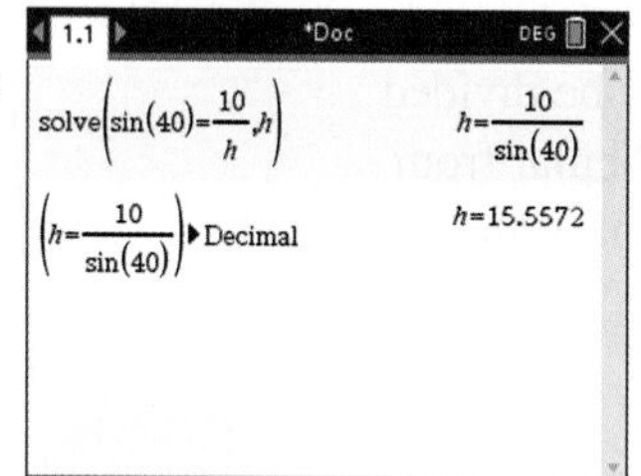
2. The answer appears on the screen.	$h = 15.56$ (to 2 decimal places)

CASIO | THINK	DISPLAY/WRITE
a. 1. Put the calculator in DEGREE mode. On a Main screen, complete the entry line as: solve $\left(\sin(40) = \frac{10}{h}, h\right)$ then press EXE.	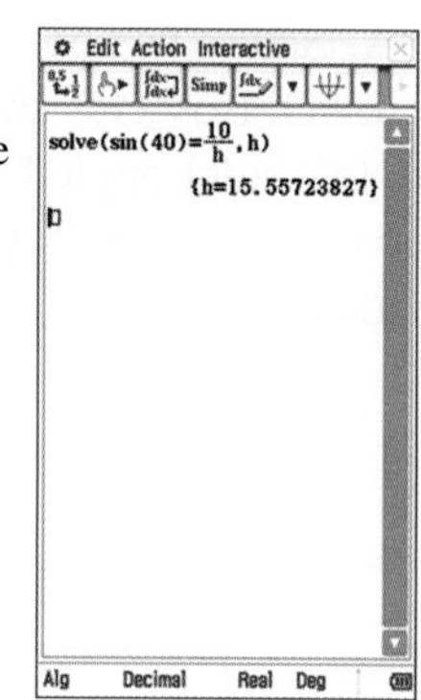
2. The answer appears on the screen.	$h = 15.56$ (to 2 decimal places)

b. 1. Put the calculator in DEGREE mode. On a Calculator page, press MENU, then select: 3: Algebra 1: Solve Complete the entry line as: $\text{solve}\left(\tan(a)=\dfrac{12}{10},a\right)$ $\|0<a>90$ then press ctrl ENTER. *Note:* Using ctrl ENTER gives you an approximate (decimal) value.	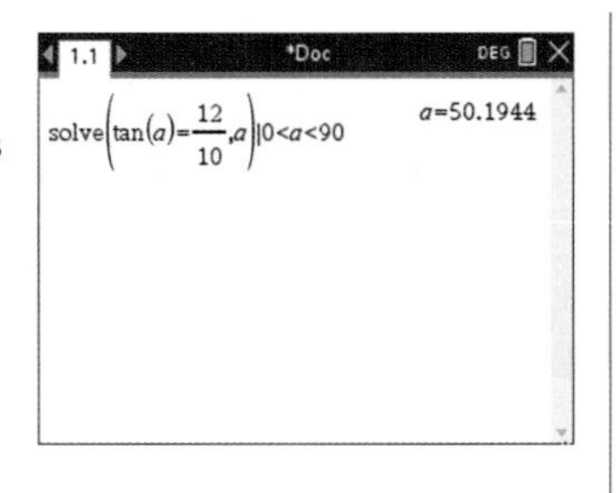 	**b. 1.** Put the calculator in DEGREE mode. On a Main screen, complete the entry line as: solve $\left(\tan(a)=\dfrac{12}{10},a\right)$ $\|0<a>90$ then press EXE.	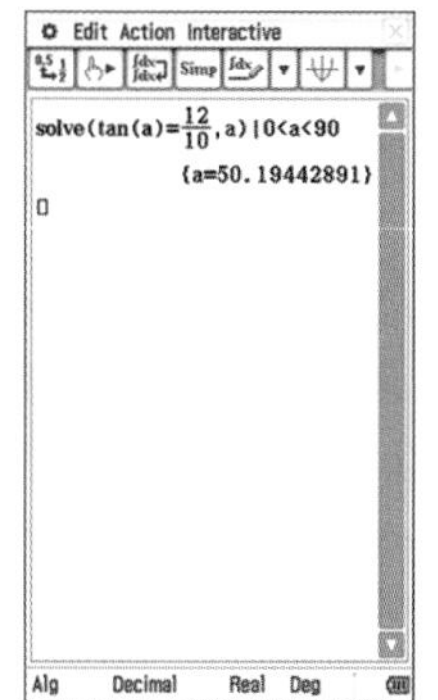
The answer appears on the screen.	$a=50.19°$ (to 2 decimal places)	The answer appears on the screen.	$a=50.19°$ (to 2 decimal places)

8.2.2 Exact values for trigonometric ratios of 30°, 45° and 60°

By considering the isosceles right-angled triangle with equal sides of 1 unit, the trigonometric ratios for 45° can be obtained. Using Pythagoras' theorem, the hypotenuse of this triangle will be $\sqrt{2}$ units.

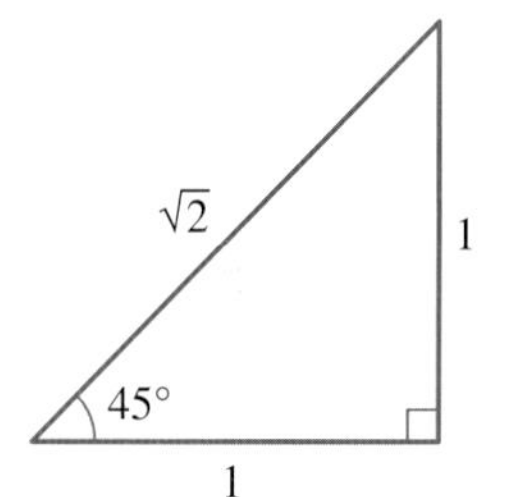

The equilateral triangle with the side length of 2 units can be divided in half to form a right-angled triangle containing 60° and 30°. The right-angled triangle has a hypotenuse of 2 units and the side divided in half has length 1 unit. Using Pythagoras' theorem, the third side will be $\sqrt{3}$ units.

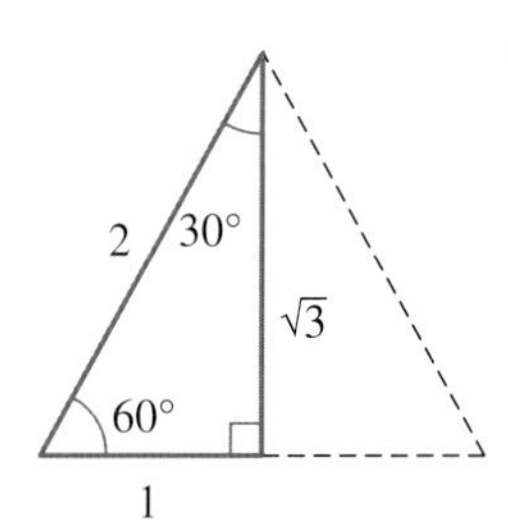

The exact values for trigonometric ratios of 30°, 45° and 60° can be calculated from these triangles using SOH, CAH, TOA. Alternatively, these values can be displayed in a table and committed to memory.

θ	30°	45°	60°
sin(θ)	$\dfrac{1}{2}$	$\dfrac{1}{\sqrt{2}}=\dfrac{\sqrt{2}}{2}$	$\dfrac{\sqrt{3}}{2}$
cos(θ)	$\dfrac{\sqrt{3}}{2}$	$\dfrac{1}{\sqrt{2}}=\dfrac{\sqrt{2}}{2}$	$\dfrac{1}{2}$
tan(θ)	$\dfrac{1}{\sqrt{3}}=\dfrac{\sqrt{3}}{3}$	1	$\sqrt{3}$

As a memory aid, notice the sine values in the table are in the order $\dfrac{\sqrt{1}}{2}, \dfrac{\sqrt{2}}{2}, \dfrac{\sqrt{3}}{2}$. The cosine values reverse this order, and the tangent values are the sine values divided by the cosine values.

For other angles, a calculator or other technology is required. It is essential to set the calculator mode to degrees in order to evaluate a trigonometric ratio involving angles in degree measure.

on Resources

Interactivity Trigonometric ratios (int-2577)

WORKED EXAMPLE 2 Calculating exact lengths

A ladder of length 4 metres leans against a fence. If the ladder is inclined at 30° to the ground, calculate how far exactly the foot of the ladder is from the fence.

THINK	WRITE 
1. Draw a diagram showing the given information.	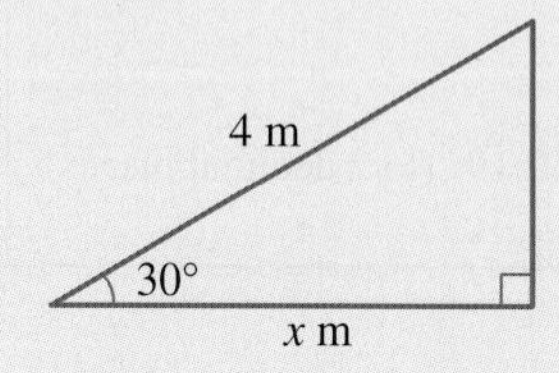 Let the distance of the ladder from the fence be x m.
2. Choose the appropriate trigonometric ratio.	Relative to the angle, the sides marked are the adjacent and the hypotenuse. $\cos(30°) = \frac{x}{4}$
3. Calculate the required length using the exact value for the trigonometric ratio.	$x = 4\cos(30°) = 4 \times \frac{\sqrt{3}}{2} = 2\sqrt{3}$
4. State the answer.	The foot of the ladder is $2\sqrt{3}$ metres from the fence.

8.2.3 Deducing one trigonometric ratio from another

Given the sine, cosine or tangent value of some unspecified angle, it is possible to obtain the exact value of the other trigonometric ratios of that angle using Pythagoras' theorem.

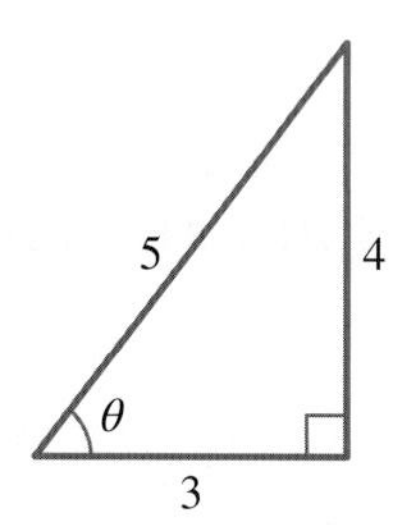

One common example is that given $\tan(\theta) = \frac{4}{3}$, it is possible to deduce that $\sin(\theta) = \frac{4}{5}$ and $\cos(\theta) = \frac{3}{5}$ without evaluating θ. The reason for this is that $\tan(\theta) = \frac{4}{3}$ means that the sides opposite and adjacent to the angle θ in a right-angled triangle are in the ratio 4 : 3.

Labelling these sides 4 and 3 respectively and using Pythagoras' theorem (or recognising the Pythagorean triad '3, 4, 5') leads to the hypotenuse being 5 and hence the ratios $\sin(\theta) = \frac{4}{5}$ and $\cos(\theta) = \frac{3}{5}$ are obtained.

WORKED EXAMPLE 3 Determining exact trigonometric ratios

A line segment AB is inclined at a degrees to the horizontal, where $\tan(a) = \frac{1}{3}$.

a. Deduce the exact value of $\sin(a)$.

b. Calculate the vertical height of B above the horizontal through A if the length of AB is $\sqrt{5}$ cm.

THINK	WRITE
a. 1. Draw a right-angled triangle with two sides in the given ratio and calculate the third side.	**a.** $\tan(a) = \frac{1}{3} \Rightarrow$ sides opposite and adjacent to angle a are in the ratio 1 : 3. 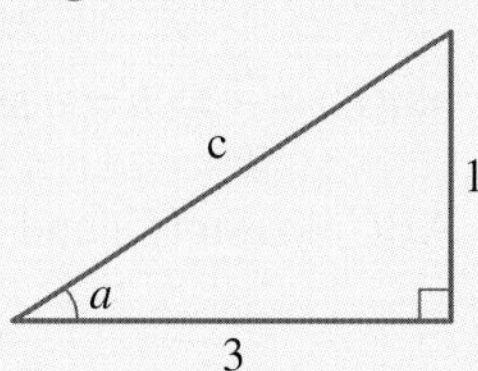 Using Pythagoras' theorem: $c^2 = 1^2 + 3^2$ $\therefore \; c = \sqrt{10}$
2. State the required trigonometric ratio.	$\sin(a) = \frac{1}{\sqrt{10}}$
b. 1. Draw the diagram showing the given information.	**b.** Let the vertical height be y cm. 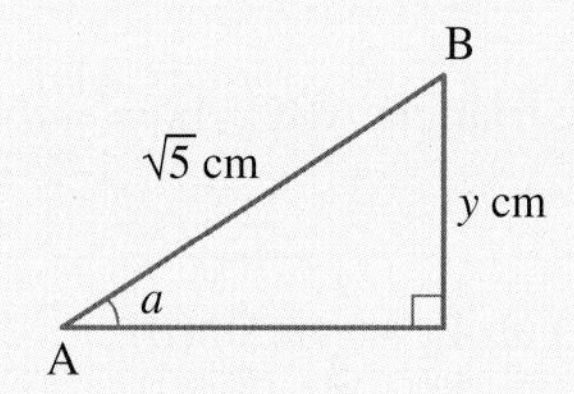
2. Choose the appropriate trigonometric ratio and calculate the required length.	$\sin(a) = \frac{y}{\sqrt{5}}$ $y = \sqrt{5}\sin(a)$ $= \sqrt{5} \times \frac{1}{\sqrt{10}}$ as $\sin(a) = \frac{1}{\sqrt{10}}$ $= \frac{1}{\sqrt{2}}$ or $\frac{\sqrt{2}}{2}$
3. State the answer.	The vertical height of B above the horizontal through A is $\frac{\sqrt{2}}{2}$ cm.

8.2.4 Area of a triangle

The formula for calculating the area of a triangle given its base and perpendicular height is shown below.

Area of a triangle

$$\textbf{Area} = \frac{1}{2}\textbf{(base)} \times \textbf{(perpendicular height)}$$

If two sides and the angle included between these two sides are known, it is also possible to calculate the area of the triangle from that information.

Consider the triangle ABC shown, where the convention of labelling the sides opposite the angles A, B and C with lower case letters a, b and c respectively has been adopted in the diagram.

In triangle ABC, construct the perpendicular height, h, from B to a point D on AC. As this is not necessarily an isosceles triangle, D is not the midpoint of AC.

In the right-angled triangle BCD, $\sin(C) = \frac{h}{a} \Rightarrow h = a\sin(C)$.

This means the height of triangle ABC is $a\sin(C)$ and its base is b.

The area of the triangle ABC can now be calculated.

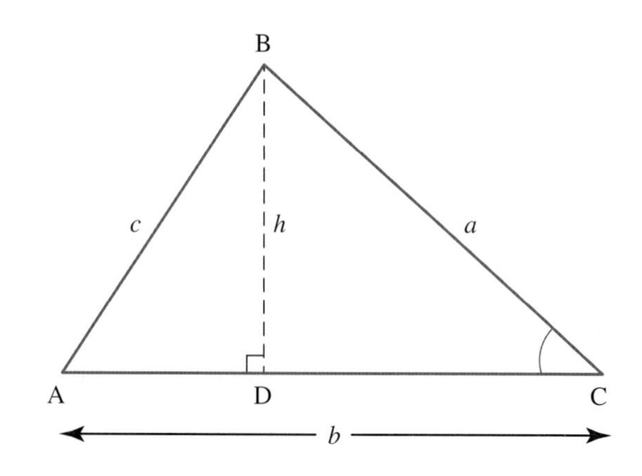

$$\begin{aligned}\text{Area} &= \frac{1}{2}(\text{base}) \times (\text{height})\\ &= \frac{1}{2}b \times a\sin(C)\\ &= \frac{1}{2}ab\sin(C)\end{aligned}$$

The formula for the area of the triangle ABC, $A_\Delta = \frac{1}{2}ab\sin(C)$, is expressed in terms of two of its sides and the angle included between them.

Alternatively, $A_\Delta = \frac{1}{2}bc\ \sin(A)$ or $A_\Delta = \frac{1}{2}ac\sin(B)$.

Hence, the area of a triangle is:

$$\frac{1}{2} \times (\text{product of two sides}) \times (\text{sine of the angle included between the two given sides}).$$

Area of a triangle with an angle given

$$\textbf{Area: } A_\Delta = \frac{1}{2}ab\sin(C)$$

WORKED EXAMPLE 4 Calculating the area of a triangle

Calculate the exact area of the triangle ABC for which $a = \sqrt{62}, b = 5\sqrt{2}, c = 6\sqrt{2}$ cm, and $A = 60°$.

THINK	WRITE
1. Draw a diagram showing the given information. *Note:* The naming convention for labelling the angles and the sides opposite them with upper- and lower-case letters is commonly used.	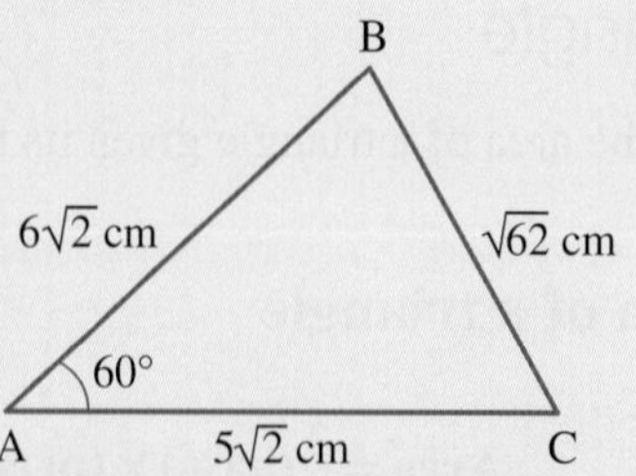
2. State the two sides and the angle included between them.	The given angle A is included between the sides b and c.

3. State the appropriate area formula and substitute the known values.

The area formula is:

$$A_{\Delta} = \frac{1}{2}\, bc \sin(A),$$

$$b = 5\sqrt{2}, c = 6\sqrt{2}, A = 60°$$

$$\therefore\ A = \frac{1}{2} \times 5\sqrt{2} \times 6\sqrt{2} \times \sin(60°)$$

4. Evaluate, using the exact value for the trigonometric ratio.

$$\therefore\ A = \frac{1}{2} \times 5\sqrt{2} \times 6\sqrt{2} \times \frac{\sqrt{3}}{2}$$

$$= \frac{1}{2} \times 30 \times 2 \times \frac{\sqrt{3}}{2}$$

$$= 15\sqrt{3}$$

5. State the answer.

The area of the triangle is $15\sqrt{3}$ cm^2.

8.2 Exercise

Students, these questions are even better in jacPLUS

Receive immediate feedback and access sample responses

Access additional questions

Track your results and progress

Find all this and MORE in jacPLUS

Technology free

Note: You may use a scientific calculator for Questions 1–6.

1. WE1 Calculate, to 2 decimal places, the value of the pronumeral shown in each diagram.

a.

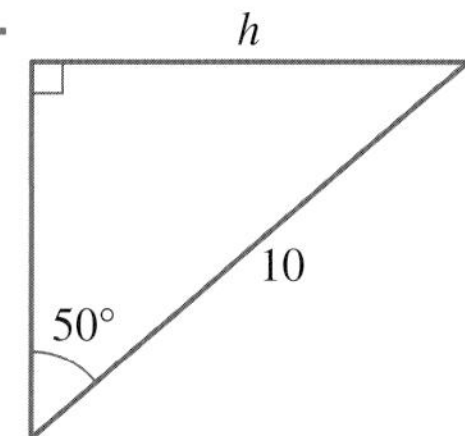

b.

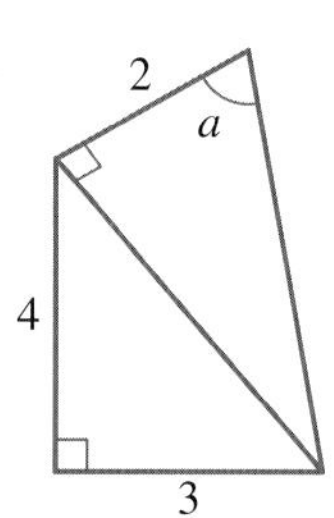

2. Calculate the values of the unknown marked sides correct to 2 decimal places.

a.

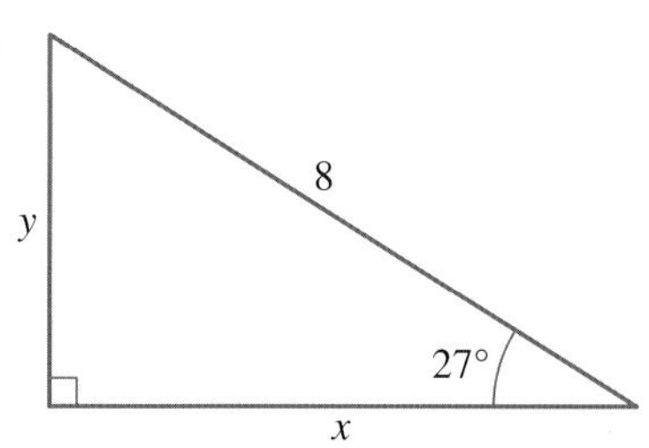

b.

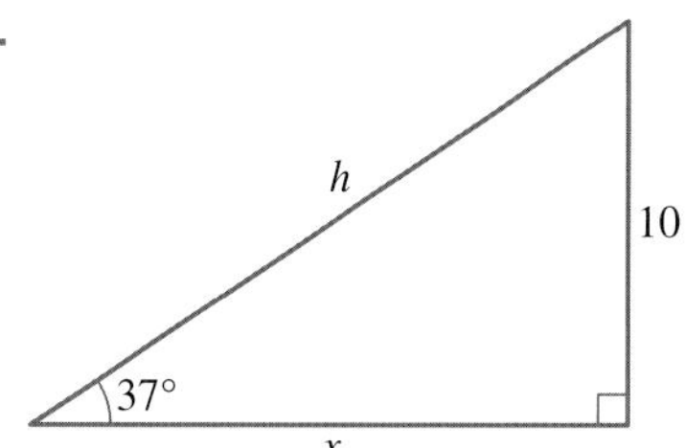

3. A ladder 6 m long rests against a vertical wall and forms an angle of 40° to the horizontal ground. Calculate how high up the wall the ladder reaches, correct to 2 decimal places.

4. The angle of depression of a boat from a cliff 60 m high is 10°. Calculate how far, to the nearest metre, the boat is from the base of the cliff.

5. Calculate the value of angle θ, correct to 2 decimal places.

a.

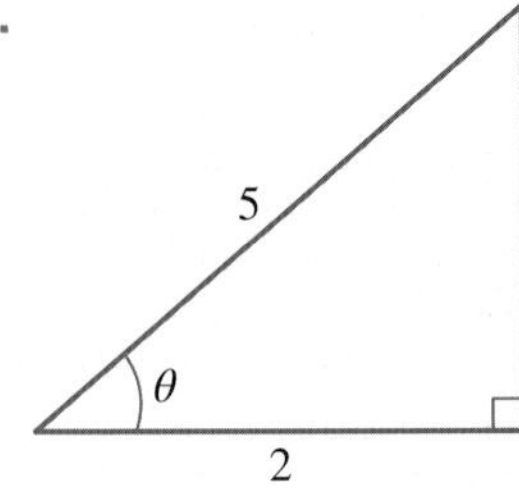

b.

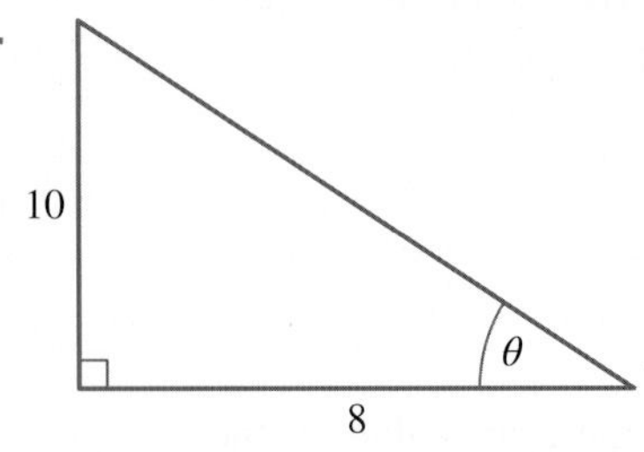

6. An 800 metre long taut chairlift cable enables the chairlift to rise 300 metres vertically. Calculate the angle of elevation (to the nearest degree) of the cable.

7. Find the exact values of the following.

a. $\sin(45°)$

b. $\tan(30°)$

c. $\cos(60°)$

d. $\tan(45°) + \cos(30°) - \sin(60°)$

8. A 3-metre ladder propped against a wall makes an angle of 60° with the ground. Calculate exactly how far up the wall the ladder reaches.

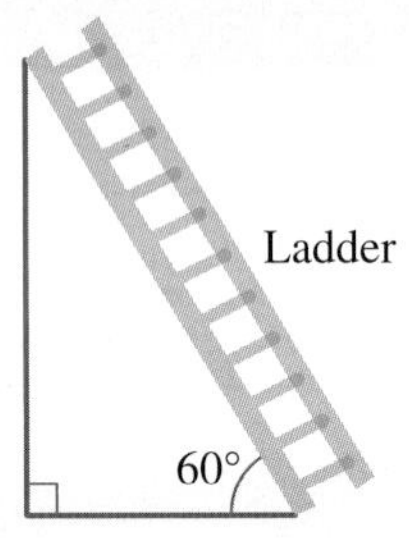

9. A 1-metre long broom leaning against a wall makes an angle of 30° with the wall. Calculate exactly how far up the wall the broom reaches.

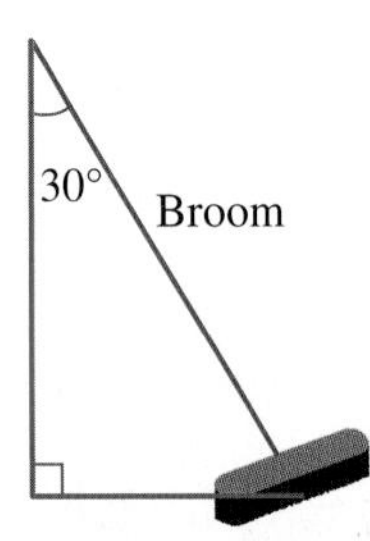

10. **WE2** A ladder of length 4 metres leans against a fence. If the ladder is inclined at 45° to the horizontal ground, calculate exactly how far exactly the foot of the ladder is from the fence.

11. Evaluate $\dfrac{\cos(30°)\ \sin(45°)}{\tan(45°) + \tan(60°)}$, expressing the answer in exact form with a rational denominator.

12. Evaluate $\dfrac{\sin(30°)\ \cos(45°)}{\tan(60°)}$, expressing the answer in exact form with a rational denominator.

13. For an acute angle θ, obtain the following trigonometric ratios without evaluating θ.

a. Given $\tan(\theta) = \dfrac{\sqrt{3}}{2}$, form the exact value of $\sin(\theta)$.

b. Given $\cos(\theta) = \dfrac{5}{6}$, form the exact value of $\tan(\theta)$.

c. Given $\sin(\theta) = \dfrac{\sqrt{5}}{3}$, form the exact value of $\cos(\theta)$.

14. For an acute angle θ, $\cos(\theta) = \frac{3\sqrt{5}}{7}$. Calculate the exact values of $\sin(\theta)$ and $\tan(\theta)$.

15. **WE3** A line segment AB is inclined at a degrees to the horizontal, where $\tan(a) = \frac{2}{3}$.

    a. Deduce the exact value of $\cos(a)$.
    b. Calculate the run of AB along the horizontal through A if the length of AB is 26 cm.

16. A right-angled triangle contains an angle θ where $\sin(\theta) = \frac{3}{5}$. If the longest side of the triangle is 60 cm, calculate the exact length of the shortest side.

17. An A-frame house has an angle at the roof apex of 30° and two sides each 7 m as shown. Calculate the exact area of the front of the house.

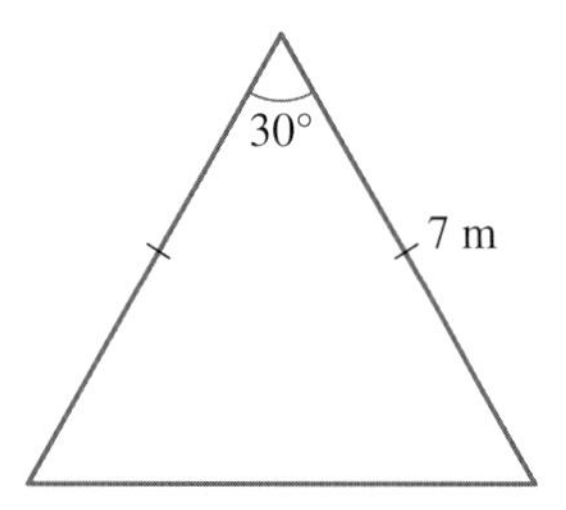

18. Calculate the exact area of the triangle ABC where:

    a. $b = 11,\ c = 8,\ A = 60°$
    b. $a = 4\sqrt{2},\ c = 5,\ B = 45°$

19. **WE4** Calculate the exact area of the triangle ABC for which $a = 10, b = 6\sqrt{2}, c = 2\sqrt{13}$ cm and $C = 45°$.

Technology active

20. In order to check the electricity supply, a technician uses a ladder to reach the top of an electricity pole. The ladder reaches 5 metres up the pole and its inclination to the horizontal ground is 54°.

    a. Calculate the length of the ladder to 2 decimal places.
    b. If the foot of the ladder is moved 0.5 metres closer to the pole, calculate its new inclination to the ground and the new vertical height it reaches up the electricity pole, both to 1 decimal place.

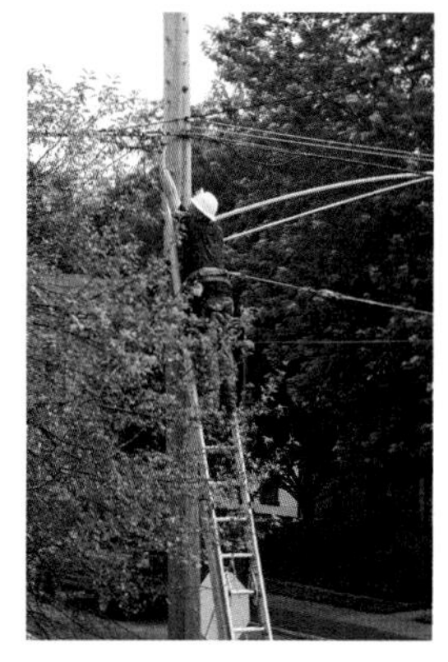

21. The two legs of a builder's ladder are each of length 2 metres. The ladder is placed on horizontal ground so that the distance between its 2 feet is 0.75 metres. Calculate the magnitude of the angle between the legs of the ladder.

22. Triangle ACB is an isosceles triangle with equal sides CA and CB. If the third side AB has length of 10 cm and the angle CAB is 72°, solve this triangle by calculating the length of the equal sides and the magnitudes of the other two angles.

23. Triangle ABC has angles such that $\angle CAB = 60°$ and $\angle ABC = 45°$. The perpendicular distance from C to AB is 18 cm. Calculate the exact lengths of each of its sides.

24. An isosceles triangle ABC has sides BC and AC of equal length 5 cm. If the angle enclosed between the equal sides is 20°, calculate:

    a. the area of the triangle to 3 decimal places
    b. the length of the third side AB to 3 decimal places.

25. An equilateral triangle has a vertical height of 10 cm. Calculate the exact perimeter and area of the triangle.

8.2 Exam questions

Question 1 (3 marks) TECH-ACTIVE

For the diagram shown

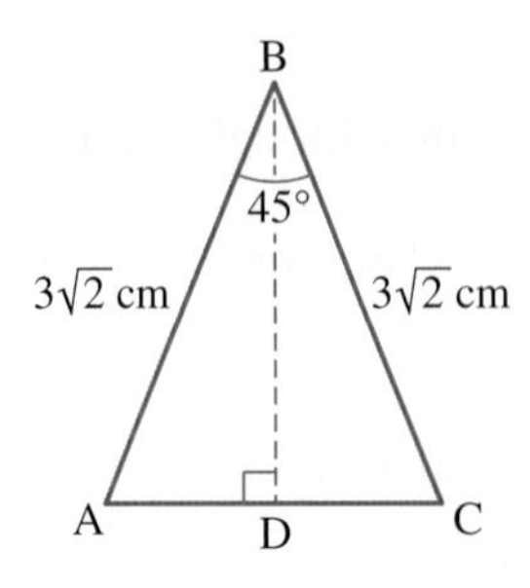

a. calculate the perpendicular height of triangle ABC (correct to 2 decimal places) **(1 mark)**

b. determine the area of ΔABC in exact form. **(2 marks)**

Question 2 (1 mark) TECH-ACTIVE

MC For the right-angled triangle shown, the exact value of x would be

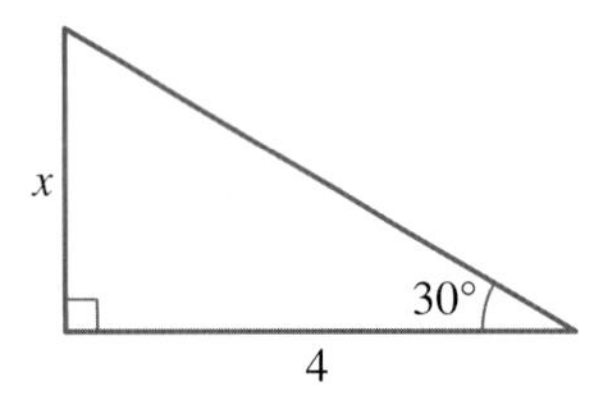

A. $2\sqrt{3}$ **B.** $4\sqrt{3}$ **C.** $\frac{4\sqrt{3}}{3}$ **D.** 4 **E.** 2

Question 3 (1 mark) TECH-ACTIVE

MC $4\sin(60°)\cos(60°)$ is equal to

A. 1 **B.** $\sqrt{3}$ **C.** $\sqrt{2}$ **D.** $\frac{1}{2}$ **E.** 0

More exam questions are available online.

8.3 Circular measure

LEARNING INTENTION

At the end of this subtopic you should be able to:

- convert between radians and degrees
- calculate arc lengths.

Measurements of angles up to now have been given in degree measure. An alternative to degree measure is **radian measure**. This alternative can be more efficient for certain calculations that involve circles, and it is essential for the study of trigonometric functions.

8.3.1 Definition of radian measure

Radian measure is defined in relation to the length of an **arc** of a circle. An arc is a part of the circumference of a circle.

One radian is the measure of the angle subtended at the centre of a circle by an arc equal in length to the radius of the circle.

In particular, an arc of 1 unit subtends an angle of one radian at the centre of a **unit circle**, a circle with radius 1 unit and, conventionally, a centre at the origin.

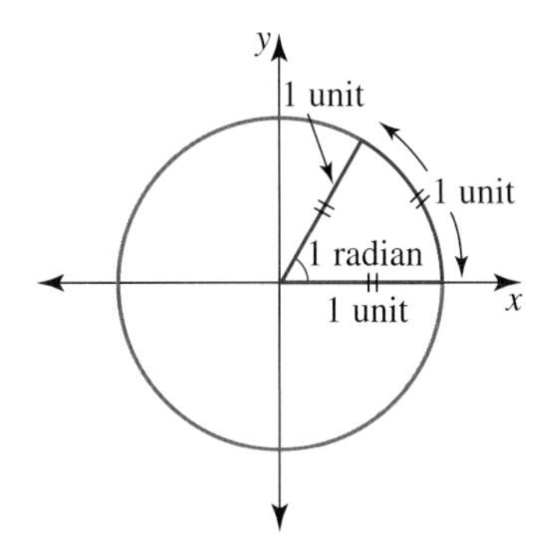

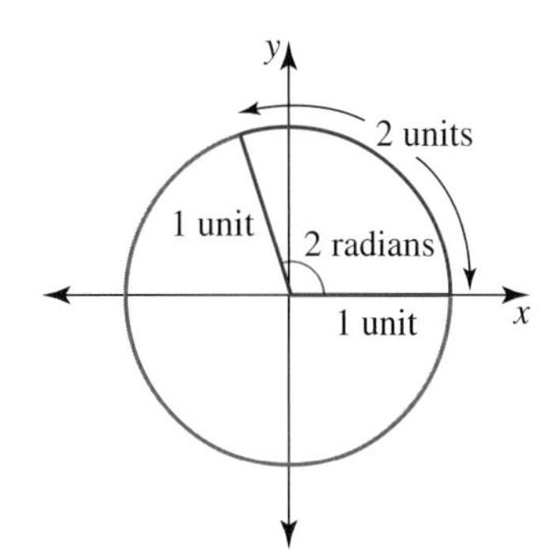

Doubling the arc length to 2 units doubles the angle to 2 radians. This illustrates the direct proportionality between the arc length and the angle in a circle of fixed radius.

The diagram suggests that an angle of one radian will be a little less than 60°, since the sector containing the angle has one 'edge' curved and therefore is not a true equilateral triangle.

The degree equivalent for 1 radian can be found by considering the angle subtended by an arc which is half the circumference. The circumference of a circle is given by $2\pi r$, so the circumference of a unit circle is 2π.

In a unit circle, an arc of π units subtends an angle of π radians at the centre. But we know this angle to be 180°.

This gives the relationship between radian and degree measure.

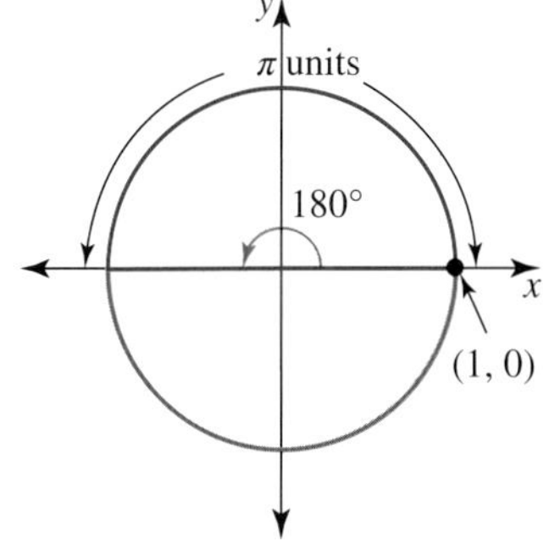

Radians to degrees

$$\pi \text{ radians} = 180°$$

Hence, 1 radian equals $\frac{180°}{\pi}$, which is approximately 57.3°; 1° equals $\frac{\pi}{180}$ radians, which is approximately 0.0175 radians.

From these relationships it is possible to convert from radians to degrees and vice versa.

Converting radians to degrees

To convert radians to degrees, multiply by $\frac{180}{\pi}$.

To convert degrees to radians, multiply by $\frac{\pi}{180}$.

Radians are often expressed in terms of π, perhaps not surprisingly, since a radian is a circular measure and π is so closely related to the circle.

Notation

π radians can be written as π^c, where c stands for circular measure. However, linking radian measure with the length of an arc, a real number, has such importance that the symbol c is usually omitted. Instead, the onus is on degree measure to always include the degree sign in order not to be mistaken for radian measure.

WORKED EXAMPLE 5 Converting between degrees and radians

a. Convert 30° to radian measure.

b. Convert $\frac{4\pi^c}{3}$ to degree measure.

c. Convert $\frac{\pi}{4}$ to degree measure and hence state the value of $\sin\left(\frac{\pi}{4}\right)$.

THINK	WRITE
a. Convert degrees to radians.	**a.** To convert degrees to radians, multiply by $\frac{\pi}{180}$. $30° = 30 \times \frac{\pi}{180}$ $= \cancel{30} \times \frac{\pi}{\cancel{180}_6}$ $= \frac{\pi}{6}$
b. Convert radians to degrees. *Note*: The degree sign must be used.	**b.** To convert radians to degrees, multiply by $\frac{180}{\pi}$. $\frac{4\pi^c}{3} = \left(\frac{4\pi}{3} \times \frac{180}{\pi}\right)^{\circ}$ $= \left(\frac{4\cancel{\pi}}{\cancel{3}} \times \frac{\cancel{180}^{60}}{\cancel{\pi}}\right)^{\circ}$ $= 240°$
c. 1. Convert radians to degrees.	**c.** $\frac{\pi}{4} = \frac{\cancel{\pi}}{4} \times \frac{180°}{\cancel{\pi}}$ $= 45°$
2. Calculate the trigonometric value.	$\sin\left(\frac{\pi}{4}\right) = \sin(45°)$ $= \frac{\sqrt{2}}{2}$

TI | THINK	DISPLAY/WRITE	CASIO | THINK	DISPLAY/WRITE
a. 1. On a Calculator page, type 30°. *Note:* Press the π▸ button to find the degree symbol. Press the [catalog] button and select ▶ Rad, then press the ENTER button twice.	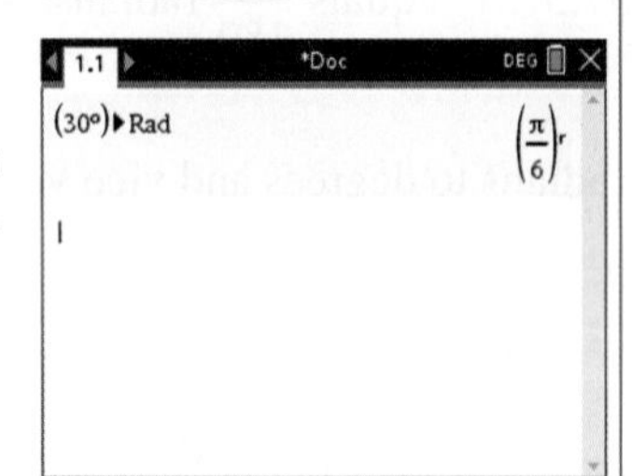	**a. 1.** Put the calculator in RADIAN mode. On a Main screen, type 30°, then press EXE. *Note:* The degree symbol can be found in the Trig tab of the keyboard menu.	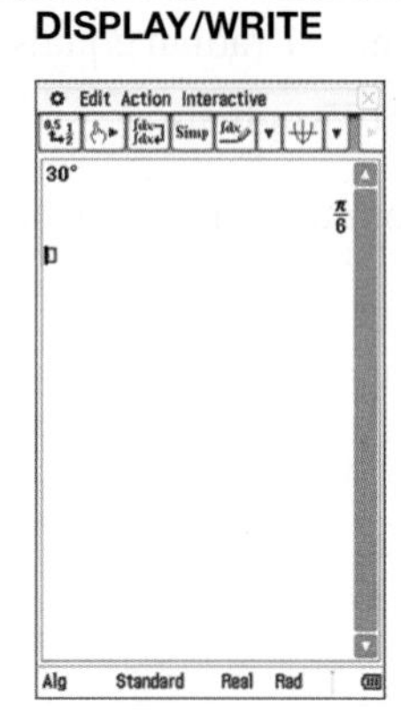
2. The answer appears on the screen.	$30° = \frac{\pi}{6}^c$	**2.** The answer appears on the screen.	$30° = \frac{\pi}{6}^c$

b. 1. On a Calculator page, type $\frac{4\pi^r}{3}$.
Note: Press the [π▸] button to find the radian symbol. Press the [book] button and select ▶DD, then press the ENTER button twice.

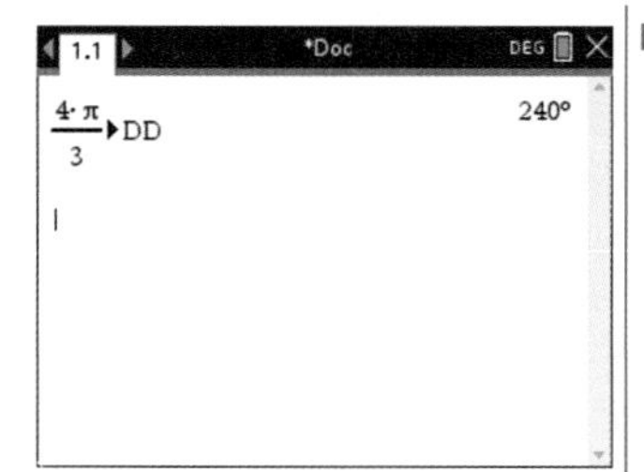

2. The answer appears on the screen.

$\frac{4\pi^c}{3} = 240°$

b. 1. Put the calculator in DEGREE mode.
On a Main screen, type $\frac{4\pi^r}{3}$, then press EXE.
Note: The radian symbol can be found in the Trig tab of the keyboard menu.

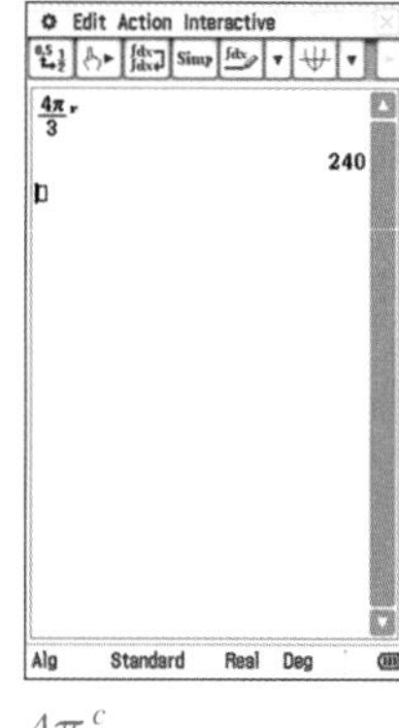

2. The answer appears on the screen.

$\frac{4\pi^c}{3} = 240°$

Angle measure and direction

By continuing to rotate around the circumference of the unit circle, larger angles are formed from arcs that are multiples of the circumference. For instance, an angle of 3π radians is formed from an arc of length 3π units created by one and a half revolutions of the unit circle: $3\pi = 2\pi + \pi$. This angle, in degrees, equals $360° + 180° = 540°$, and its endpoint on the circumference of the circle is in the same position as that of $180°$ or π^c; this is the case with any other angle that is a multiple of 2π added to π^c.

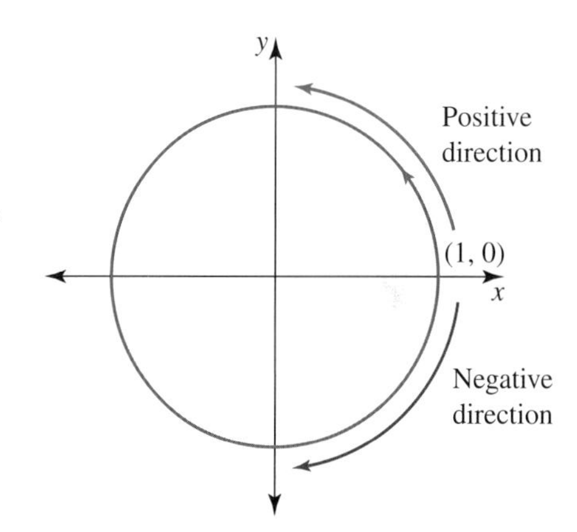

By convention, positive angles move **anticlockwise** around the unit circle from the point $(1, 0)$. Negative angles move **clockwise** around the unit circle from the point $(1, 0)$.

WORKED EXAMPLE 6 Converting radians to degrees

a. Convert -3^c to degree measure.
b. Draw a unit circle diagram to show angle -3^c.

THINK	WRITE
a. Convert radians to degrees.	a. $-3^c = -\left(3 \times \frac{180}{\pi}\right)^\circ$ As the π can't be cancelled, a calculator is used to evaluate. $-3^c \approx -171.9°$
b. Draw the unit circle diagram. Since the angle is negative, move clockwise around the unit circle from (1, 0).	b. 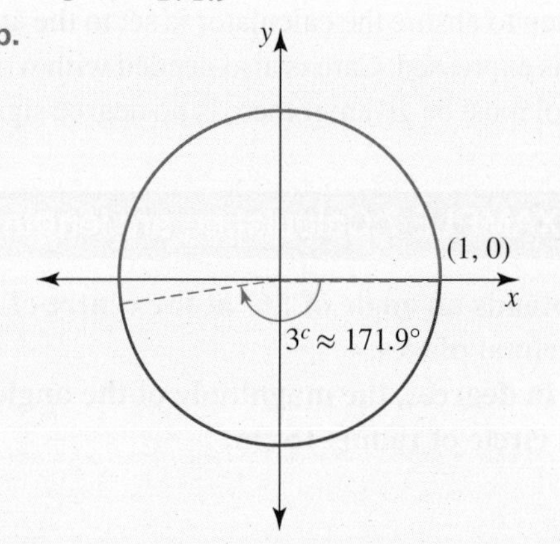

8.3.2 Using radians in calculations

From the definition of a radian, for any circle of radius r, an angle of 1^c is subtended at the centre of the circle by an arc of length r. So, if the angle at the centre of this circle is θ^c, then the length of the arc subtending this angle must be $\theta \times r$.

This gives a formula for calculating the length of an arc.

Length of an arc

$$l = r\theta$$

In the formula l is the arc length and θ is the angle, in radians, subtended by the arc at the centre of the circle of radius r.

Any angles given in degree measure will need to be converted to radian measure to use this arc length formula.

Some calculations may require recall of the geometry properties of the angles in a circle, such as the angle at the centre of a circle is twice the angle at the circumference subtended by the same arc.

Major and minor arcs

For a minor arc, $\theta < \pi$, and for a major arc, $\theta > \pi$, with the sum of the minor and major arc angles totalling 2π if the major and minor arcs have their endpoints on the same chord.

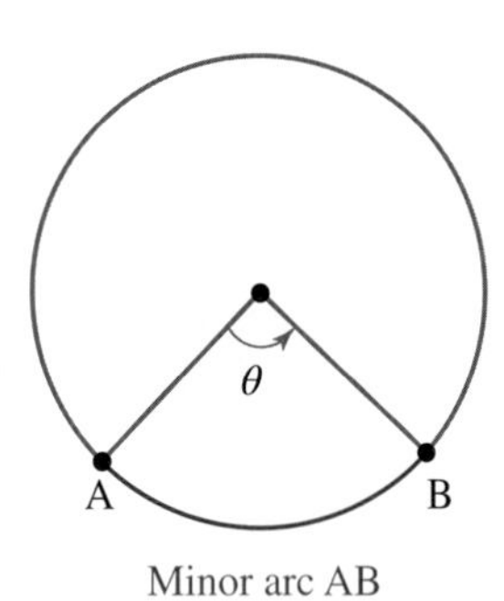

Minor arc AB

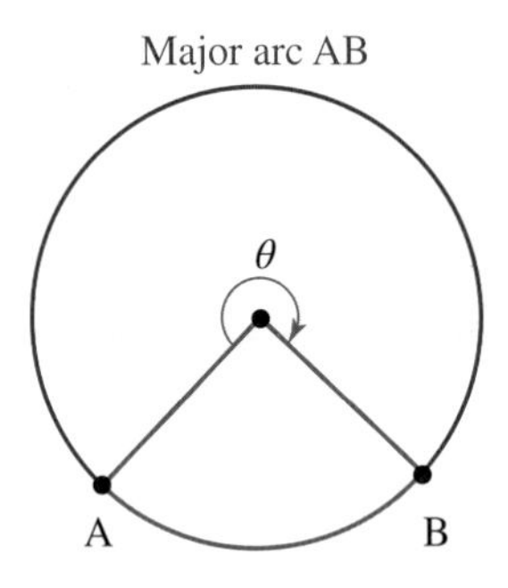

Major arc AB

Trigonometric ratios of angles expressed in radians

Problems in trigonometry may be encountered where angles are given in radian mode and their sine, cosine or tangent value is required to solve the problem. A calculator or other technology can be set on radian or 'rad' mode and the required trigonometric ratio evaluated directly without the need to convert the angle to degrees.

Care must be taken to ensure the calculator is set to the appropriate degree or radian mode to match the measure in which the angle is expressed. Care is also needed with written presentation: if the angle is measured in degrees, the degree symbol must be given; if there is no degree sign then it is assumed the measurement is in radians.

WORKED EXAMPLE 7 Calculating arc lengths

a. **An arc subtends an angle of 56° at the centre of a circle of radius 10 cm. Calculate the length of the arc to 2 decimal places.**

b. **Calculate, in degrees, the magnitude of the angle that an arc of length 20π cm subtends at the centre of a circle of radius 15 cm.**

THINK	WRITE
a. 1. The angle is given in degrees, so convert it to radian measure.	**a.** $\theta^{\circ} = 56^{\circ}$ $\theta^{c} = 56 \times \frac{\pi}{180}$ $= \frac{14\pi}{45}$
2. Calculate the arc length.	$l = r\theta, r = 10, \theta = \frac{14\pi}{45}$ $l = 10 \times \frac{14\pi}{45}$ $= \frac{28\pi}{9}$ ≈ 9.77 The arc length is 9.77 cm (to 2 decimal places).
b. 1. Calculate the angle at the centre of the circle subtended by the arc.	**b.** $l = r\theta, r = 15, l = 20\pi$ $15\theta = 20\pi$ $\therefore \theta = \frac{20\pi}{15}$ $= \frac{4\pi}{3}$ The angle is $\frac{4\pi}{3}$ radians.
2. Convert the angle from radians to degrees.	In degree measure: $\theta^{\circ} = \frac{4\pi}{3} \times \frac{180^{\circ}}{\pi}$ $= 240^{\circ}$ The magnitude of the angle is 240°.

TI | THINK	DISPLAY/WRITE
a. 1. Put the calculator in RADIAN mode. On a Calculator page, type 56°. *Note:* Press the [π▸] button to find the degree symbol. Press the [📖] button and select ▸Rad, then press the ENTER button twice.	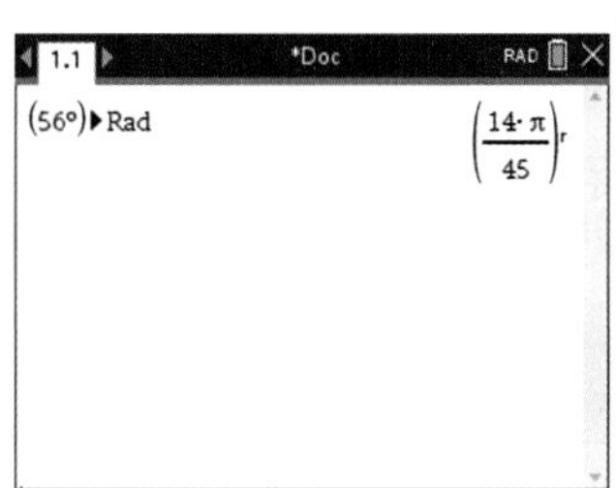
2. Complete the next entry line as: 10 × ans then press ENTER. *Note*: 'ans' can be found by pressing [ctrl] [(–)]. This will paste the previous answer onto the entry line. Press [ctrl] ENTER to get the decimal value.	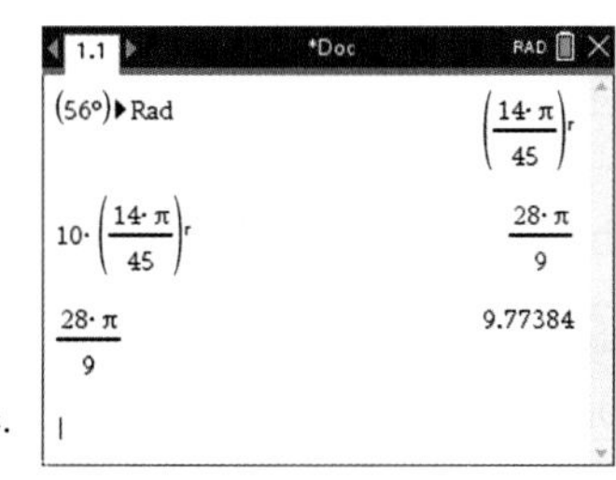

CASIO | THINK	DISPLAY/WRITE
a. 1. Put the calculator in RADIAN mode. On a Main screen, type 56°, then press EXE. *Note:* The degree symbol can be found in the Trig tab of the keyboard menu.	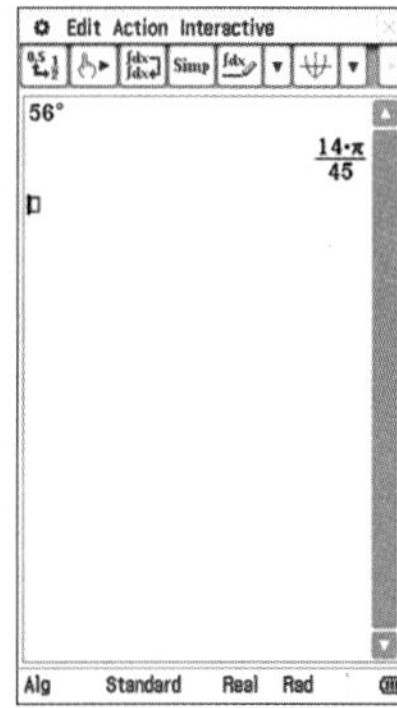
2. Complete the next entry line as: 10 × ans then press EXE. *Note:* 'ans' can be found in the Number tab in the Keyboard menu. This will paste the previous answer onto the entry line.	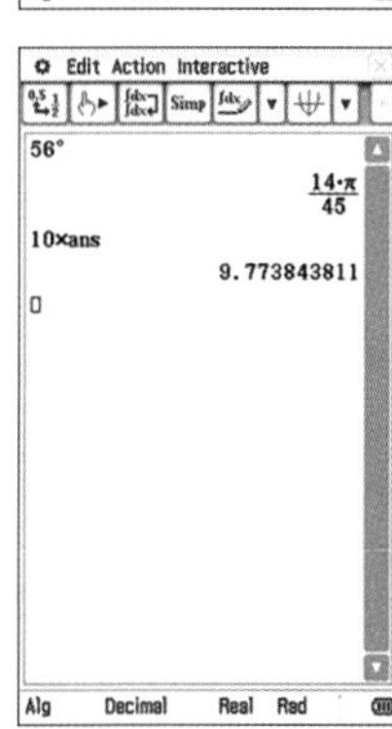

3. The answer appears on the screen.

The arc length is 9.77 cm (to 2 decimal places).

b. 1. On a Calculator page, press MENU, then select:
3: Algebra
1: Solve
Complete the entry line as:
solve $(15\theta = 20\pi, \theta)$
then press ENTER.
Note: The θ symbol can be found by pressing the $\boxed{\pi\blacktriangleright}$ button.

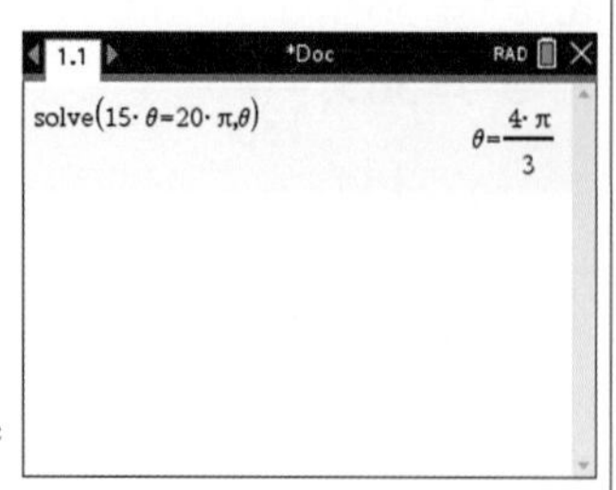

2. Press *ans*, then press [book icon] and select ▶ DD. Press ENTER twice.
Note: Press the $\boxed{\pi\blacktriangleright}$ button to find the radian symbol.

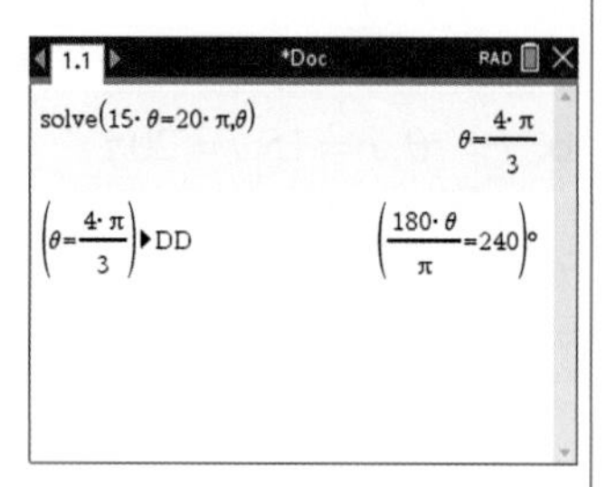

3. The answer appears on the screen.

240°

3. The answer appears on the screen.

The arc length is 9.77 cm (to 2 decimal places).

b. 1. On a Main screen, complete the entry line as:
solve $(15\theta = 20\pi, \theta)$
then press EXE.
Note: The θ symbol can be found in the Trig tab of the Keyboard menu.

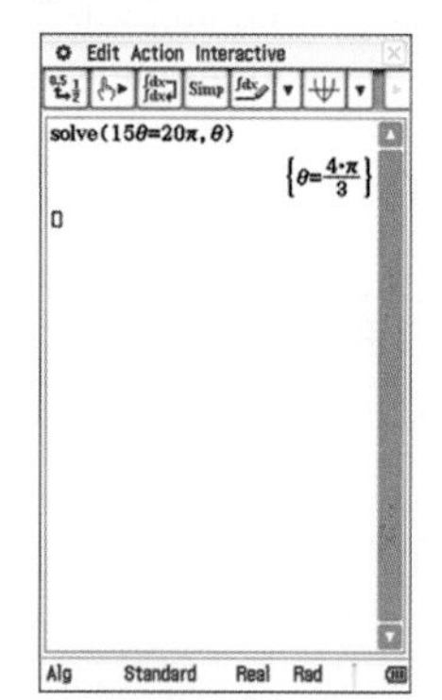

2. Put the calculator in DEGREE mode.
On a Main screen, type $\frac{4\pi}{3}^{r}$,
then press EXE.
Note: The radian symbol can be found in the Trig tab of the keyboard menu.

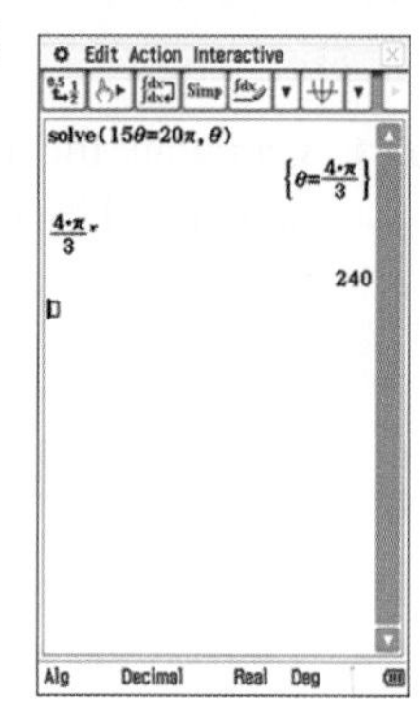

3. The answer appears on the screen.

240°

8.3 Exercise

Technology free

1. a. Copy, complete and learn the following table by heart.

Degrees	30°	45°	60°
Radians			

b. Copy, complete and learn the following table by heart.

Degrees	0°	90°	180°	270°	360°
Radians					

2. Convert the following to degrees.

a. $\dfrac{\pi^{c}}{5}$ **b.** $\dfrac{2\pi^{c}}{3}$ **c.** $\dfrac{5\pi}{12}$ **d.** $\dfrac{11\pi}{6}$ **e.** $\dfrac{7\pi}{9}$ **f.** $\dfrac{9\pi}{2}$

3. Convert the following to radian measure.

a. 40°   b. 150°   c. 225°   d. 300°   e. 315°   f. 720°

4. a. **WE5** Convert 60° to radian measure.

b. Convert $\frac{3\pi^c}{4}$ to degree measure.

c. Convert $\frac{\pi}{6}$ to degree measure and hence state the value of $\tan\left(\frac{\pi}{6}\right)$.

5. An arc subtends an angle of 2^c at the centre of a circle with radius 5 cm. Calculate the length of this arc.

6. An arc of length 6 mm subtends an angle of 0.5^c at the centre of a circle. Calculate the radius of the circle.

7. a. Express 36° in radian measure.

b. Hence, calculate the length of the arc that subtends an angle of 36° at the centre of the circle with radius 7 cm.

8. A circle has a radius of 6 cm. An arc of length $\frac{3\pi}{4}$ cm forms part of the circumference of this circle. Find the angle the arc subtends at the centre of the circle:

a. in radian measure

b. in degree measure.

9. Calculate the exact lengths of the following arcs.

a. The arc that subtends an angle of 150° at the centre of a circle of radius 12 cm

b. The major arc with endpoints on a chord that subtends an angle of 60° at the centre of a circle of radius 3 cm.

Technology active

10. a. Express the following in radian measure to 3 decimal places.

i. 3°   ii. 112°15′   iii. 215.36°

b. Express the following in degree measure to 3 decimal places.

i. 3^c ii. 2.3π

c. Rewrite $\left\{1.5^c, 50°, \frac{\pi^c}{7}\right\}$ with the magnitudes of the angles ordered from smallest to largest.

11. Express 145°12′ in radian measure, correct to 2 decimal places.

12. a. **WE6** Convert 1.8^c to degree measure.

b. Draw a unit circle diagram to show the angle 1.8^c.

13. For each of the following, draw a unit circle diagram to show the position of the angle and the arc that subtends the angle.

a. An angle of 2 radians   b. An angle of −2 radians   c. An angle of $-\frac{\pi}{2}$

14. a. **WE7** An arc subtends an angle of 75° at the centre of a circle of radius 8 cm. Calculate the length of the arc correct to 2 decimal places.

b. Calculate, in degrees, the magnitude of the angle that an arc of length 12π cm subtends at the centre of a circle of radius 10 cm.

15. A ball on the end of a rope of length 2.5 metres swings through an arc of 75 cm. Calculate the angle through which the ball swings, correct to the nearest tenth of a degree.

16. A fixed point on the rim of a wheel of radius 3 metres rolls along horizontal ground at a speed of 2 m/s. After 5 seconds, calculate the angle the point has rotated through and express the answer in both degrees and radians.

17. An analogue wristwatch has a minute hand of length 11 mm.
Calculate, to 2 decimal places, the arc length the minute hand traverses between 9.45 am and 9.50 am.

18. Evaluate the following to 3 decimal places.

a. $\tan(1.2)$
b. $\tan(1.2°)$

19. a. Calculate the following to 3 decimal places.

i. $\tan(1^c)$ ii. $\cos\left(\frac{2\pi}{7}\right)$ iii. $\sin(1.46°)$

b. Complete the following table with exact values.

θ	$\frac{\pi}{6}$	$\frac{\pi}{4}$	$\frac{\pi}{3}$
$\sin(\theta)$			
$\cos(\theta)$			
$\tan(\theta)$			

20. Calculate the exact value of x in the following diagram.

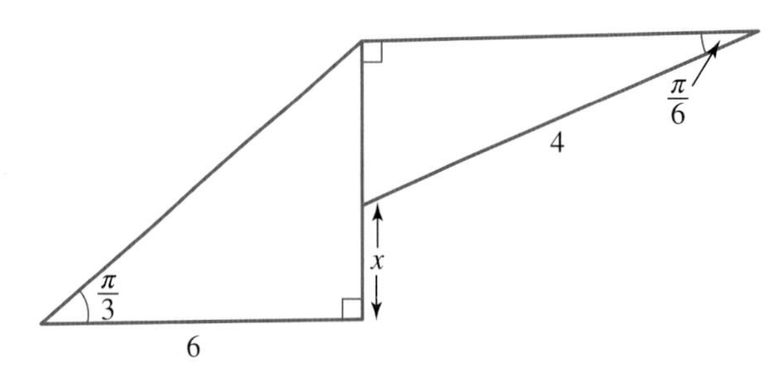

8.3 Exam questions

Question 1 (1 mark) TECH-ACTIVE

MC The angle 300° is equivalent to

A. $\frac{9\pi}{4}$ **B.** $\frac{7\pi}{4}$ **C.** 3π **D.** $\frac{3\pi}{2}$ **E.** $\frac{5\pi}{3}$

Question 2 (1 mark) TECH-ACTIVE

MC The angle that is equivalent to $\frac{7\pi}{6}$ is

A. 120° **B.** 210° **C.** 100° **D.** 150° **E.** 240°

Question 3 (1 mark) TECH-ACTIVE

MC The angle (to the nearest degree) subtended by an arc of length 3.5 cm at the centre of a circle with radius 5 cm is

A. 82° **B.** 68° **C.** 50° **D.** 40° **E.** 35°

More exam questions are available online.

8.4 Unit circle definitions

LEARNING INTENTION

At the end of this subtopic you should be able to:
- calculate and illustrate the location of trigonometric points on a unit circle
- understand the unit circle definitions for sine, cosine and tangent functions.

With the introduction of radian measure, we encountered positive and negative angles of any size. We now consider the conventions for angle rotations and the positions of the endpoints on the unit circle of these rotations.

8.4.1 Trigonometric points

The unit circle has centre $(0, 0)$ and radius 1 unit. Its Cartesian equation is $x^2 + y^2 = 1$.

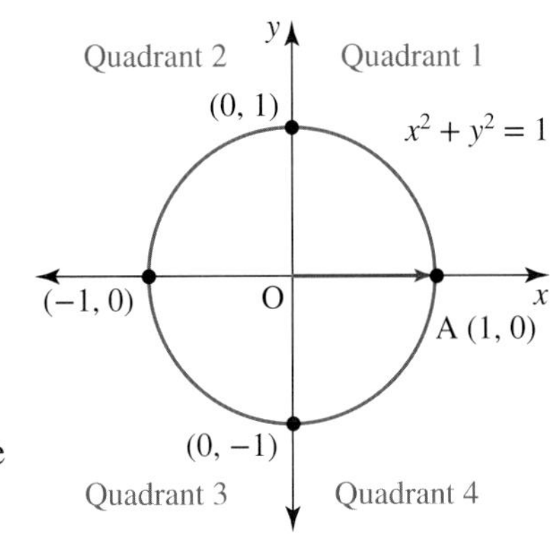

The coordinate axes divide the Cartesian plane into four quadrants. The points on the circle that lie on the boundaries between the quadrants are the endpoints of the horizontal and vertical diameters. These **boundary points** have coordinates $(-1, 0), (1, 0)$ on the horizontal axis and $(0, -1), (0, 1)$ on the vertical axis.

A rotation starts with an initial ray OA, where A is the point $(1, 0)$ and O $(0, 0)$. Angles are created by rotating the initial ray anticlockwise for positive angles and clockwise for negative angles. If the point on the circumference the ray reaches after a rotation of θ is P, then $\angle AOP = \theta$ and P is called the **trigonometric point** $[\theta]$. The angle of rotation θ may be measured in radian or degree measure. In radian measure, the value of θ corresponds to the length of the arc AP of the unit circle the rotation cuts off.

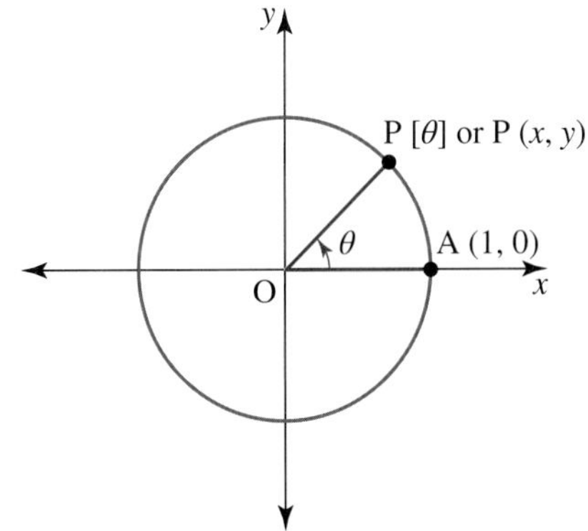

The point P $[\theta]$ has Cartesian coordinates (x, y) where:
- $x > 0, y > 0$ if P is in quadrant $1, 0 < \theta < \frac{\pi}{2}$
- $x < 0, y > 0$ if P is in quadrant $2, \frac{\pi}{2} < \theta < \pi$
- $x < 0, y < 0$ if P is in quadrant $3, \pi < \theta < \frac{3\pi}{2}$
- $x > 0, y < 0$ if P is in quadrant $4, \frac{3\pi}{2} < \theta < 2\pi$.

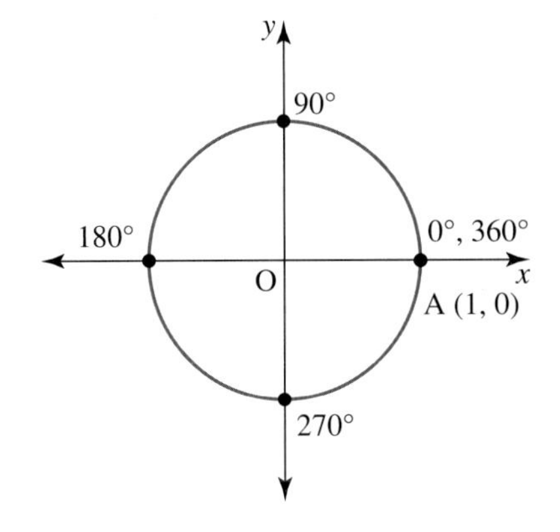

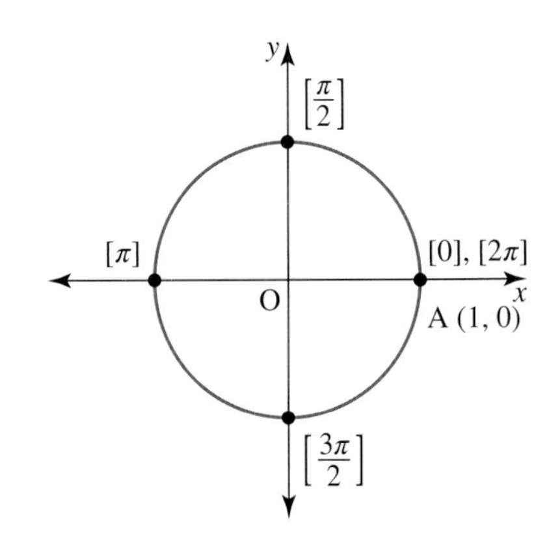

Continued rotation anticlockwise or clockwise can be used to form other values for θ greater than 2π or values less than 0 respectively. No trigonometric point has a unique θ value.

The angle θ is said to lie in the quadrant in which its endpoint P lies.

WORKED EXAMPLE 8 Identifying trigonometric points

a. Give a trigonometric value, using radian measure, of the point P on the unit circle which lies on the boundary between the quadrants 2 and 3.

b. Identify the quadrants the following angles would lie in: 250°, 400°, $\frac{2\pi^c}{3}$, $-\frac{\pi^c}{6}$.

c. Give two other trigonometric points, Q and R, one with a negative angle and one with a positive angle respectively, which would have the same position as the point P [250°].

THINK	WRITE
a. 1. State the Cartesian coordinates of the required point.	a. The point $(-1, 0)$ lies on the boundary of quadrants 2 and 3.
2. Give a trigonometric value of this point. *Note:* Other values are possible.	An anticlockwise rotation of 180° or π^c from the point $(1, 0)$ would have its endpoint at $(-1, 0)$. The point P has the trigonometric value $[\pi]$.
b. 1. Explain how the quadrant is determined.	b. For positive angles, rotate anticlockwise from $(1, 0)$; for negative angles, rotate clockwise from $(1, 0)$. The position of the endpoint of the rotation determines the quadrant.
2. Identify the quadrant the endpoint of the rotation would lie in for each of the given angles.	Rotating anticlockwise 250° from $(1, 0)$ ends in quadrant 3. Rotating anticlockwise from $(1, 0)$ through 400° would end in quadrant 1. Rotating anticlockwise from $(1, 0)$ by $\frac{2\pi}{3}$ would end in quadrant 2. Rotating clockwise from $(1, 0)$ by $\frac{\pi}{6}$ would end in quadrant 4.

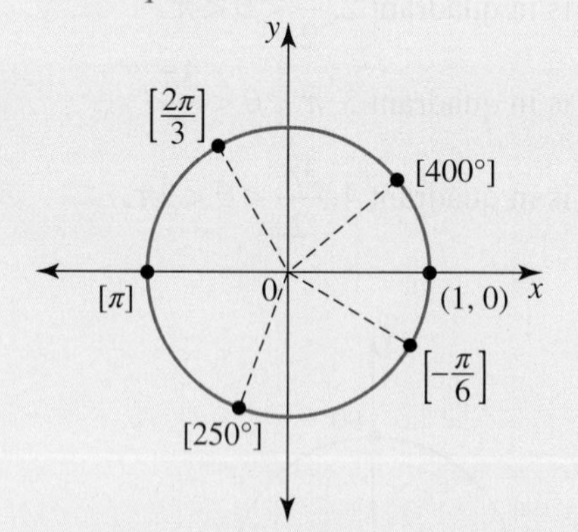

3. State the answer.	The angle 250° lies in quadrant 3, 400° in quadrant 1, $\frac{2\pi^c}{3}$ in quadrant 2, and $-\frac{\pi^c}{6}$ in quadrant 4.

c. 1. Identify a possible trigonometric point Q.	**c.** A rotation of 110° in the clockwise direction from $(1, 0)$ would end in the same position as P [250°]. Therefore, the trigonometric point could be Q [−110°].
2. Identify a possible trigonometric point R.	A full anticlockwise revolution of 360° plus another anticlockwise rotation of 250° would end in the same position as P [250°]. Therefore, the trigonometric point could be R [610°].

Resources

Interactivity The unit circle (int-2582)

8.4.2 Unit circle definitions of the sine and cosine functions

Consider the unit circle and trigonometric point P $[\theta]$ with Cartesian coordinates (x, y) on its circumference. In the triangle ONP, $\angle \text{NOP} = \theta = \angle \text{AOP}$, ON $= x$ and NP $= y$.

As the triangle ONP is right-angled, $\cos(\theta) = \frac{x}{1} = x$ and $\sin(\theta) = \frac{y}{1} = y$. This enables the following definitions to be given for a rotation from the point $(1, 0)$ of any angle θ with endpoint P $[\theta]$ on the unit circle.

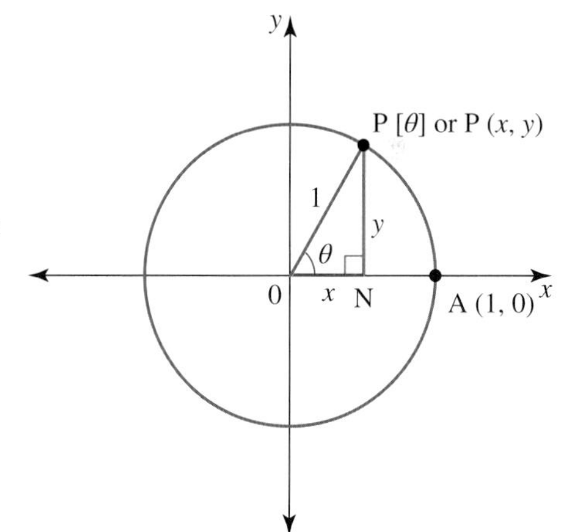

Coordinates of a trigonometric point

$\cos(\theta)$ is the x-coordinate of the trigonometric point P $[\theta]$.

$\sin(\theta)$ is the y-coordinate of the trigonometric point P $[\theta]$.

The importance of these definitions is that they enable sine and cosine functions to be defined for any real number θ. With θ measured in radians, the trigonometric point $[\theta]$ also marks the position of θ on the circumference of the unit circle, with zero placed at the point $(1, 0)$. This relationship enables the sine or cosine of a real number θ to be evaluated as the sine or cosine of the angle of rotation of θ radians in a unit circle: $\sin(\theta) = \sin(\theta^c)$ and $\cos(\theta) = \cos(\theta^c)$.

The sine and cosine functions are $f: R \to R, f(\theta) = \sin(\theta)$ and $f: R \to R, f(\theta) = \cos(\theta)$.

They are **trigonometric functions**, also referred to as **circular functions**. The use of parentheses in writing $\sin(\theta)$ or $\cos(\theta)$ emphasises their functionality.

The functions have a many-to-one correspondence, as many values of θ are mapped to the one trigonometric point. The functions have a **period** of 2π, since rotations of θ and of $2\pi + \theta$ have the same endpoint on the circumference of the unit circle. The cosine and sine values repeat after each complete revolution around the unit circle.

For $f(\theta) = \sin(\theta)$, the image of a number such as 4 is $f(4) = \sin(4) = \sin(4^c)$. This is evaluated as the y-coordinate of the trigonometric point [4] on the unit circle.

The values of a function for which $f(t) = \cos(t)$, where t is a real number, can be evaluated through the relation $\cos(t) = \cos(t^c)$, as t will be the trigonometric point $[t]$ on the unit circle.

The sine and cosine functions are periodic functions that have applications in contexts that may have nothing to do with angles, as we shall study in later topics.

WORKED EXAMPLE 9 Evaluating and drawing trigonometric points

a. Calculate the Cartesian coordinates of the trigonometric point P $\left[\frac{\pi}{3}\right]$ and show the position of this point on a unit circle diagram.

b. Illustrate cos(330°) and sin(2) on a unit circle diagram.

c. Use the Cartesian coordinates of the trigonometric point $[\pi]$ to obtain the values of $\sin(\pi)$ and $\cos(\pi)$.

d. If $f(\theta) = \cos(\theta)$, evaluate $f(0)$.

THINK	WRITE
a. 1. State the value of θ.	a. P $\left[\frac{\pi}{3}\right]$ This is the trigonometric point with $\theta = \frac{\pi}{3}$.
2. Calculate the exact Cartesian coordinates. *Note*: The exact values for sine and cosine of $\frac{\pi^c}{3}$, or 60°, need to be known.	The Cartesian coordinates are: $x = \cos(\theta)$ $y = \sin(\theta)$ $= \cos\left(\frac{\pi}{3}\right)$ $= \sin\left(\frac{\pi}{3}\right)$ $= \cos\left(\frac{\pi}{3}\right)^c$ $= \sin\left(\frac{\pi}{3}\right)^c$ $= \cos(60°)$ $= \sin(60°)$ $= \frac{1}{2}$ $= \frac{\sqrt{3}}{2}$ Therefore, P has coordinates $\left(\frac{1}{2}, \frac{\sqrt{3}}{2}\right)$.
3. Show the position of the given point on a unit circle diagram.	P $\left[\frac{\pi}{3}\right]$ or P $\left(\frac{1}{2}, \frac{\sqrt{3}}{2}\right)$ lies in quadrant 1 on the circumference of the unit circle.

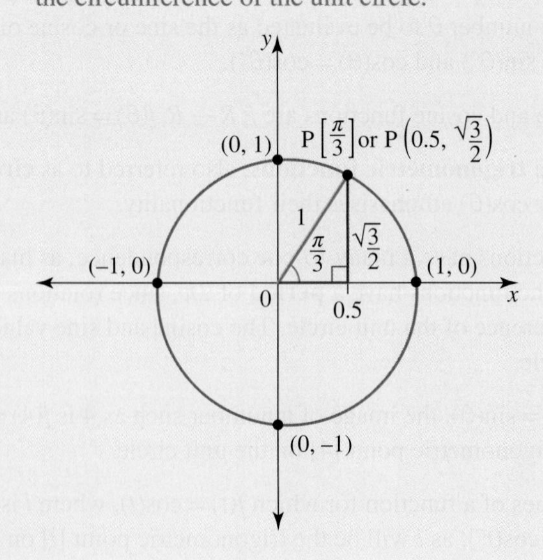

b. 1. Identify the trigonometric point and which of its Cartesian coordinates gives the first value.

b. $\cos(330°)$:
The value of $\cos(330°)$ is given by the x-coordinate of the trigonometric point [330°].

2. State the quadrant in which the trigonometric point lies.

The trigonometric point [330°] lies in quadrant 4.

3. Identify the trigonometric point and which of its Cartesian coordinates gives the second value.

$\sin(2)$:
The value of $\sin(2)$ is given by the y-coordinate of the trigonometric point [2].

4. State the quadrant in which the trigonometric point lies.

As $\frac{\pi}{2} \approx 1.57 < 2 < \pi \approx 3.14$, the trigonometric point [2] lies in quadrant 2.

5. Draw a unit circle showing the two trigonometric points and construct the line segments that illustrate the x- and y-coordinates of each point.

For each of the points on the unit circle diagram, the horizontal line segment gives the x-coordinate and the vertical line segment gives the y-coordinate.

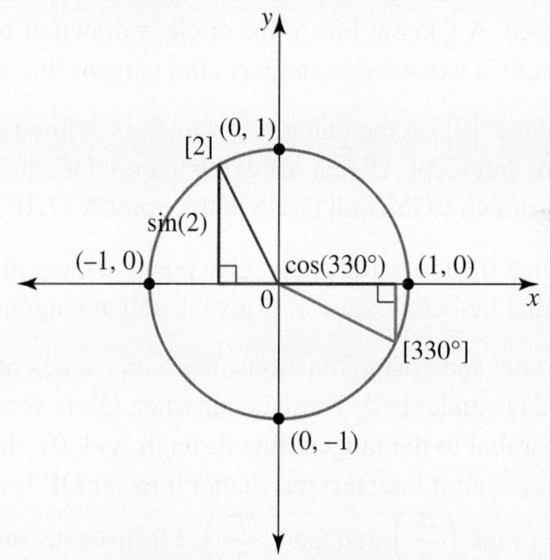

6. Label the line segments that represent the appropriate coordinate for each point.

The value of $\cos(330°)$ is the length measure of the horizontal line segment.
The value of $\sin(2)$ is the length measure of the vertical line segment.
The line segments illustrating these values are highlighted in orange on the diagram.

c. 1. State the Cartesian coordinates of the given point.

c. An anticlockwise rotation of π from $(1, 0)$ gives the endpoint $(-1, 0)$.
The trigonometric point $[\pi]$ is the Cartesian point $(-1, 0)$.

2. State the required values.

The point $(-1, 0)$ has $x = -1, y = 0$.
Since $x = \cos(\theta)$,

$$\begin{aligned}\cos(\pi) &= x \\ &= -1\end{aligned}$$

Since $y = \sin(\theta)$,

$$\begin{aligned}\sin(\pi) &= y \\ &= 0\end{aligned}$$

d. 1. Substitute the given value in the function rule. **d.** $f(\theta) = \cos(\theta)$

$\therefore f(0) = \cos(0)$

2. Identify the trigonometric point and state its Cartesian coordinates. The trigonometric point [0] has Cartesian coordinates $(1, 0)$.

3. Evaluate the required value of the function. The value of $\cos(0)$ is given by the x-coordinate of the point $(1, 0)$.

$\therefore \cos(0) = 1$

$\therefore f(0) = 1$

8.4.3 Unit circle definition of the tangent function

Consider again the unit circle with centre O $(0, 0)$ containing the points A $(1, 0)$ and the trigonometric point P $[\theta]$ on its circumference. A tangent line to the circle is drawn at point A. The radius OP is extended to intersect the tangent line at point T.

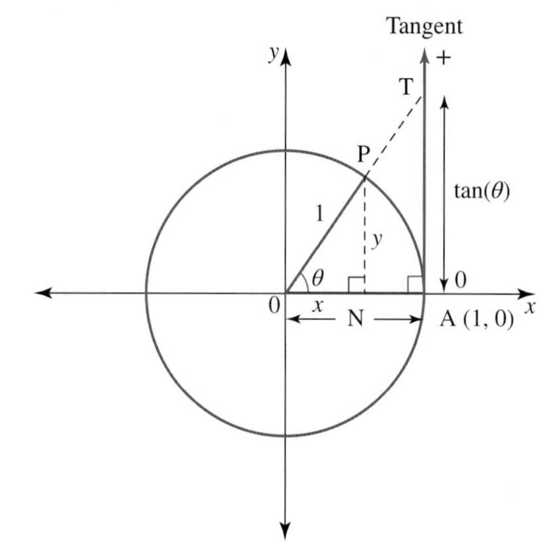

For any point P $[\theta]$ on the unit circle, $\tan(\theta)$ is defined as the length of the intercept AT that the extended ray OP cuts off on the tangent drawn to the unit circle at the point A $(1, 0)$.

Intercepts that lie above the x-axis give positive tangent values; intercepts that lie below the x-axis give negative tangent values.

Unlike the sine and cosine functions, there are values of θ for which $\tan(\theta)$ is undefined. These occur when OP is vertical and therefore parallel to the tangent line through A $(1, 0)$; these two vertical lines cannot intersect, no matter how far OP is extended. The values of $\tan\left(\frac{\pi}{2}\right)$ and $\tan\left(\frac{3\pi}{2}\right)$, for instance, are not defined.

The value of $\tan(\theta)$ can be calculated from the coordinates (x, y) of the point P $[\theta]$, provided the x-coordinate is not zero.

Using the ratio of sides of the similar triangles ONP and OAT:

$$\frac{\text{AT}}{\text{OA}} = \frac{\text{NP}}{\text{ON}}$$

$$\frac{\tan(\theta)}{1} = \frac{y}{x}$$

Hence:

$\tan(\theta) = \frac{y}{x}, x \neq 0$, where (x, y) are the coordinates of the trigonometric point P $[\theta]$.

Since $x = \cos(\theta), y = \sin(\theta)$, this can be expressed as the following relationship.

Unit circle definition of the tangent function

$$\tan(\theta) = \frac{\sin(\theta)}{\cos(\theta)}$$

WORKED EXAMPLE 10 Evaluating $\tan(\theta)$

a. Illustrate tan(130°) on a unit circle diagram and use a calculator to evaluate tan(130°) to 3 decimal places.

b. Use the Cartesian coordinates of the trigonometric point $P[\pi]$ to obtain the value of $\tan(\pi)$.

THINK	WRITE
a. 1. State the quadrant in which the angle lies.	a. 130° lies in the second quadrant.
2. Draw the unit circle with the tangent at the point A (1, 0). *Note*: The tangent line is always drawn at the point (1, 0).	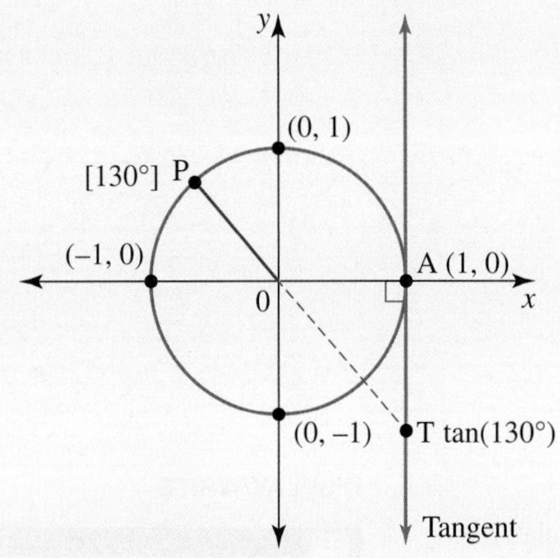
3. Extend PO until it reaches the tangent line.	Let T be the point where the extended radius PO intersects the tangent drawn at A. The intercept AT is tan(130°).
4. State whether the required value is positive, zero or negative.	The intercept lies below the x-axis, which shows that tan(130°) is negative.
5. Calculate the required value.	The value of $\tan(130°) = -1.192$, correct to 3 decimal places.
b. 1. Identify the trigonometric point and state its Cartesian coordinates.	b. The trigonometric point $P[\pi]$ is the endpoint of a rotation of π^c or 180°. It is the Cartesian point $P(-1, 0)$.
2. Calculate the required value.	The point $(-1, 0)$ has $x = -1, y = 0$. Since $\tan(\theta) = \frac{y}{x}$, $\tan(\pi) = \frac{0}{-1}$ $= 0$

3. Check the answer using the unit circle diagram.	Check: PO is horizontal and runs along the x-axis. Extending PO, it intersects the tangent at the point A. This means the intercept is 0, which means $\tan(\pi) = 0$.

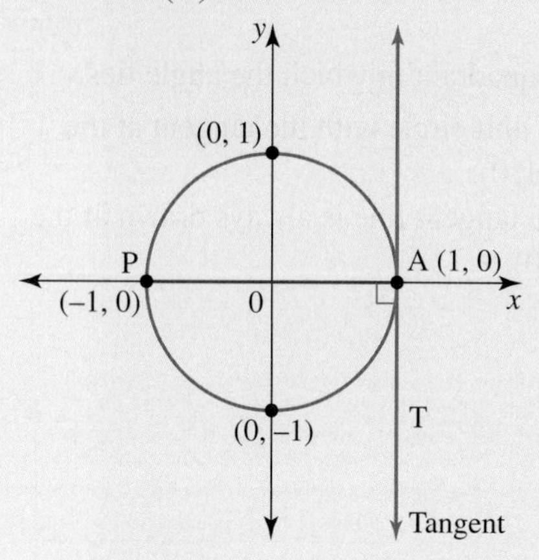

TI | THINK	DISPLAY/WRITE	CASIO | THINK	DISPLAY/WRITE
a. 1. Put the calculator in DEGREE mode. On the Calculator page, complete the entry line as: tan(130) then press ctrl ENTER.	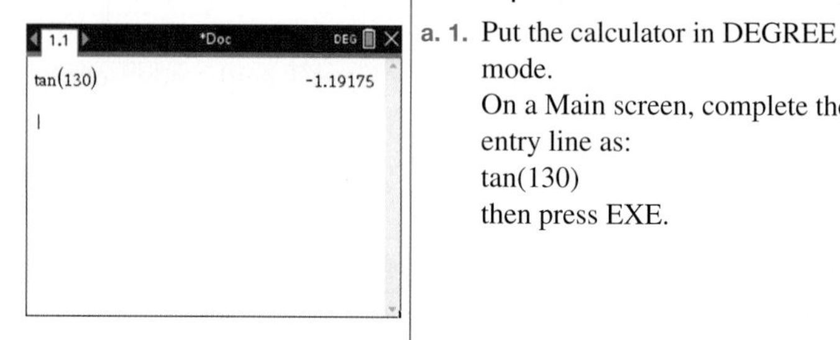	a. 1. Put the calculator in DEGREE mode. On a Main screen, complete the entry line as: tan(130) then press EXE.	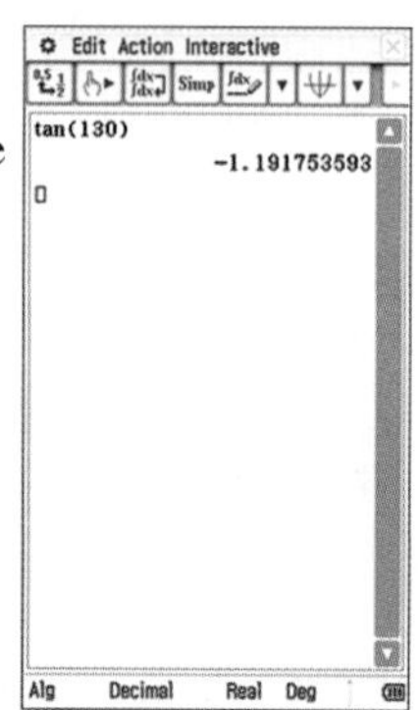
2. The answer appears on the screen.	$\tan(130°) = -1.192$ (to 3 decimal places)	2. The answer appears on the screen.	$\tan(130°) = -1.192$ (to 3 decimal places)

8.4 Exercise

Technology free

1. State the quadrant in which each of the following angles lies.

a. 24°
b. 240°
c. 123°
d. 365°
e. −50°
f. −120°

2. Consider the set $\left\{\frac{\pi}{3}, \frac{3\pi}{4}, \frac{7\pi}{6}, -\frac{\pi}{3}, -\frac{5\pi}{4}, -\frac{11\pi}{6}\right\}$ where the angles are expressed in radian measure.

a. Select the two angles that lie in the first quadrant.
b. Select the two angles that lie in the second quadrant.
c. State the angle that lies in the third quadrant.
d. State the angle that lies in the fourth quadrant.

3. a. The trigonometric point θ lies on the boundary between quadrants 1 and 2.

i. Calculate the Cartesian coordinates of the trigonometric point θ.
ii. State the value of $\sin(\theta)$.

b. The trigonometric point α lies on the boundary between quadrants 2 and 3.

i. Calculate the Cartesian coordinates of the trigonometric point α.
ii. State the value of $\cos(\alpha)$.

c. The trigonometric point β lies on the boundary between quadrants 3 and 4.

i. Calculate the Cartesian coordinates of the trigonometric point β.
ii. State the value of $\tan(\beta)$.

d. The trigonometric point ν lies on the boundary between quadrants 4 and 1.

i. Calculate the Cartesian coordinates of the trigonometric point ν.
ii. State the value of $\sin(\nu)$, $\cos(\nu)$ and $\tan(\nu)$.

4. Identify the quadrant in which each of the following lies.

a. $585°$
b. $\frac{11\pi}{12}$
c. -18π
d. $\frac{7\pi}{4}$

5. a. WE8 Give a trigonometric value, using radian measure, of the point P on the unit circle which lies on the boundary between the quadrants 1 and 2.

b. Identify the quadrants the following angles would lie in: $120°, -400°, \frac{4\pi^c}{3}, \frac{\pi^c}{4}$.

c. Give two other trigonometric points, Q with a negative angle and R with a positive angle, that would have the same position as the point $P[120°]$.

6. Using a positive and a negative radian measure, state trigonometric values of the point on the unit circle that lies on the boundary between quadrants 3 and 4.

7. a. WE9 Calculate the Cartesian coordinates of the trigonometric point $P\left[\frac{\pi}{6}\right]$ and show the position of this point on a unit circle diagram.

b. Illustrate $\sin(225°)$ and $\cos(1)$ on a unit circle diagram.

c. Use the Cartesian coordinates of the trigonometric point $\left[-\frac{\pi}{2}\right]$ to obtain the values of $\sin\left(-\frac{\pi}{2}\right)$ and $\cos\left(-\frac{\pi}{2}\right)$.

d. If $f(\theta) = \sin(\theta)$, evaluate $f(0)$.

8. a. For the function $f(t) = \sin(t)$, evaluate $f\left(\frac{3\pi}{2}\right)$.

b. For the function $g(t) = \cos(t)$, evaluate $g(4\pi)$.
c. For the function $h(t) = \tan(t)$, evaluate $h(-\pi)$.
d. For the function $k(t) = \sin(t) + \cos(t)$, evaluate $k(6.5\pi)$.

9. Identify the quadrants where:

a. $\sin(\theta)$ is always positive

b. $\cos(\theta)$ is always positive.

10. a. Calculate the Cartesian coordinates of the trigonometric point $P\left[\frac{\pi}{4}\right]$.

b. Express the Cartesian point $P(0, -1)$ as two different trigonometric points, one with a positive value for θ and one with a negative value for θ.

11. Illustrate the following on a unit circle diagram.

a. $\cos(40°)$

b. $\sin(165°)$

c. $\cos(-60°)$

d. $\sin(-90°)$

12. Illustrate the following on a unit circle diagram.

a. $\sin\left(\frac{5\pi}{3}\right)$

b. $\cos\left(\frac{3\pi}{5}\right)$

c. $\cos(5\pi)$

d. $\sin\left(-\frac{2\pi}{3}\right)$

13. Illustrate the following on a unit circle diagram.

a. $\tan(315°)$

b. $\tan\left(\frac{5\pi}{6}\right)$

c. $\tan\left(\frac{4\pi}{3}\right)$

d. $\tan(-300°)$

14. a. The trigonometric point $P[\theta]$ has Cartesian coordinates $(-0.8, 0.6)$. State the quadrant in which P lies and the values of $\sin(\theta), \cos(\theta)$ and $\tan(\theta)$.

b. The trigonometric point $Q[\theta]$ has Cartesian coordinates $\left(\frac{\sqrt{2}}{2}, -\frac{\sqrt{2}}{2}\right)$. State the quadrant in which Q lies and the values of $\sin(\theta), \cos(\theta)$ and $\tan(\theta)$.

c. For the trigonometric point $R[\theta]$ with Cartesian coordinates $\left(\frac{2}{\sqrt{5}}, \frac{1}{\sqrt{5}}\right)$, state the quadrant in which R lies and the values of $\sin(\theta), \cos(\theta)$ and $\tan(\theta)$.

d. The Cartesian coordinates of the trigonometric point $S[\theta]$ are $(0, 1)$. Describe the position of S and state the values of $\sin(\theta), \cos(\theta)$ and $\tan(\theta)$.

15. By locating the appropriate trigonometric point and its corresponding Cartesian coordinates, obtain the exact values of the following.

a. $\cos(0)$

b. $\sin\left(\frac{\pi}{2}\right)$

c. $\tan(\pi)$

d. $\cos\left(\frac{3\pi}{2}\right)$

e. $\sin(2\pi)$

f. $\tan(-11\pi)$

Technology active

16. a. **WE10** Illustrate $\tan(230°)$ on a unit circle diagram and use a calculator to evaluate $\tan(230°)$ to 3 decimal places.

b. Use the Cartesian coordinates of the trigonometric point $P[2\pi]$ to obtain the value of $\tan(2\pi)$.

17. Consider $\left\{\tan(-3\pi), \tan\left(\frac{5\pi}{2}\right), \tan(-90°), \tan\left(\frac{3\pi}{4}\right), \tan(780°)\right\}$.

a. State which elements in the set are not defined.

b. State which elements have negative values.

18. Consider O, the centre of the unit circle, and the trigonometric points $P\left[\frac{3\pi}{10}\right]$ and $Q\left[\frac{2\pi}{5}\right]$ on its circumference.

a. Sketch the unit circle showing these points.
b. Calculate how many radians are contained in the angle $\angle QOP$.
c. Express each of the trigonometric points P and Q with a negative θ value.
d. Express each of the trigonometric points P and Q with a larger positive value for θ than the given values $P\left[\frac{3\pi}{10}\right]$ and $Q\left[\frac{2\pi}{5}\right]$.

19. a. On a unit circle diagram, show the trigonometric point $P[2]$ and the line segments $\sin(2)$, $\cos(2)$ and $\tan(2)$. Label them with their length measures expressed to 2 decimal places.
b. State the Cartesian coordinates of P to 2 decimal places.

20. Use CAS technology to calculate the exact values of the following.

a. $\cos^2\left(\frac{7\pi}{6}\right)+\sin^2\left(\frac{7\pi}{6}\right)$
b. $\cos\left(\frac{7\pi}{6}\right)+\sin\left(\frac{7\pi}{6}\right)$
c. $\sin^2\left(\frac{7}{6}\right)+\cos^2\left(\frac{7}{6}\right)$
d. $\sin^2(76°)+\cos^2(76°)$
e. $\sin^2(t)+\cos^2(t)$; explain the result with reference to the unit circle.

8.4 Exam questions

Question 1 (1 mark) TECH-ACTIVE

MC The radian measure $\frac{5\pi}{4}$ is

A. in the first quadrant.
B. in the second quadrant.
C. in the third quadrant.
D. in the fourth quadrant.
E. exactly between the first and second quadrants.

Question 2 (1 mark) TECH-ACTIVE

MC The angle $-\frac{11\pi}{2}$ is

A. exactly between the first and second quadrants.
B. exactly between the third and fourth quadrants.
C. exactly between the third and fourth quadrants.
D. in the fourth quadrant.
E. in the second quadrant.

Question 3 (1 mark) TECH-ACTIVE

MC If $0<x<2\pi$, $\sin(x)<0$ and $\cos(x)<0$, state which of the following is true.

A. $0<x<\frac{\pi}{2}$
B. $\frac{\pi}{2}<x<\pi$
C. $\pi<x<\frac{3\pi}{2}$
D. $\frac{3\pi}{2}<x<2\pi$
E. $0<x<\pi$

More exam questions are available online.

8.5 Symmetry properties

LEARNING INTENTION

At the end of this subtopic you should be able to:
- identify the signs of trigonometric values in the four quadrants
- calculate exact trigonometric values
- use symmetry properties to calculate values of trigonometric functions.

There are relationships between the coordinates and associated trigonometric values of trigonometric points placed in symmetric positions in each of the four quadrants. These will now be investigated.

8.5.1 The signs of the sine, cosine and tangent values in the four quadrants

The definitions $\cos(\theta) = x$, $\sin(\theta) = y$ and $\tan(\theta) = \frac{y}{x}$, where (x, y) are the Cartesian coordinates of the trigonometric point $[\theta]$, have been established.

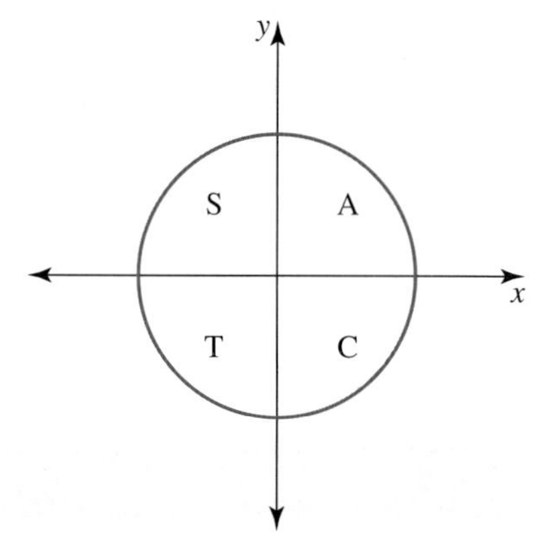

If θ lies in the first quadrant, **A**ll of the trigonometric values will be positive, since $x > 0, y > 0$.

If θ lies in the second quadrant, only the **S**ine value will be positive, since $x < 0, y > 0$.

If θ lies in the third quadrant, only the **T**angent value will be positive, since $x < 0, y < 0$.

If θ lies in the fourth quadrant, only the **C**osine value will be positive, since $x > 0, y < 0$.

This is illustrated in the diagram shown.

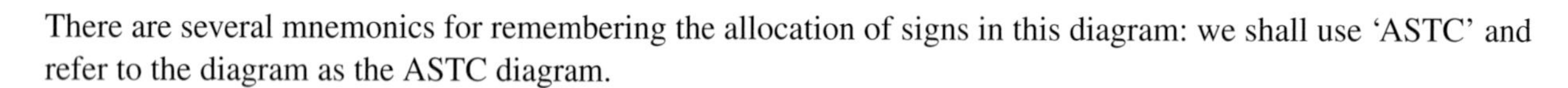

There are several mnemonics for remembering the allocation of signs in this diagram: we shall use 'ASTC' and refer to the diagram as the ASTC diagram.

A common saying is '**A**ll **S**tations **T**o **C**entral'.

The sine, cosine and tangent values at the boundaries of the quadrants

The points that do not lie within a quadrant are the four coordinate axis intercepts of the unit circle. These are called the boundary points. Since the Cartesian coordinates and the trigonometric positions of these points are known, the boundary values can be summarised by the following table.

Boundary point	**(1, 0)**	**(0, 1)**	**(−1, 0)**	**(0, −1)**	**(1, 0)**
θ radians	0	$\frac{\pi}{2}$	π	$\frac{3\pi}{2}$	2π
θ degrees	0°	90°	180°	270°	360°
$\sin(\theta)$	0	1	0	−1	0
$\cos(\theta)$	1	0	−1	0	1
$\tan(\theta)$	0	undefined	0	undefined	0

Other values of θ could be used for the boundary points, including negative values.

WORKED EXAMPLE 11 Identifying quadrants

a. Identify the quadrant(s) where both $\cos(\theta)$ and $\sin(\theta)$ are negative.
b. If $f(\theta) = \cos(\theta)$, evaluate $f(-6\pi)$.

THINK	WRITE
a. 1. Refer to the ASTC diagram.	a. $\cos(\theta) = x, \sin(\theta) = y$ The quadrant where both x and y are negative is quadrant 3.
b. 1. Substitute the given value in the function rule.	b. $f(\theta) = \cos(\theta)$ $\therefore\ f(-6\pi) = \cos(-6\pi)$
2. Identify the Cartesian coordinates of the trigonometric point.	A clockwise rotation of 6π from $(1, 0)$ shows that the trigonometric point $[-6\pi]$ is the boundary point with coordinates $(1, 0)$.
3. Evaluate the required value of the function.	The x-coordinate of the boundary point gives the cosine value. $\cos(-6\pi) = 1$ $\therefore\ f(-6\pi) = 1$

Exact trigonometric values of $\frac{\pi}{6}$, $\frac{\pi}{4}$ and $\frac{\pi}{3}$

As the exact trigonometric ratios are known for angles of 30°, 45° and 60°, these give the trigonometric ratios for $\frac{\pi}{6}$, $\frac{\pi}{4}$ and $\frac{\pi}{3}$, respectively. A summary of these is given with the angles in each triangle expressed in radian measure. The values should be memorised.

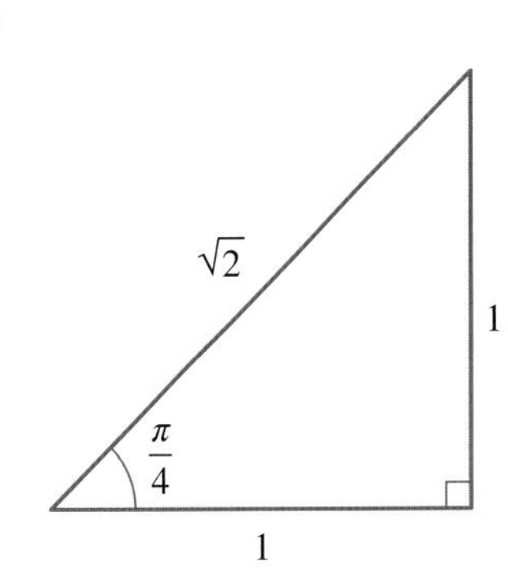

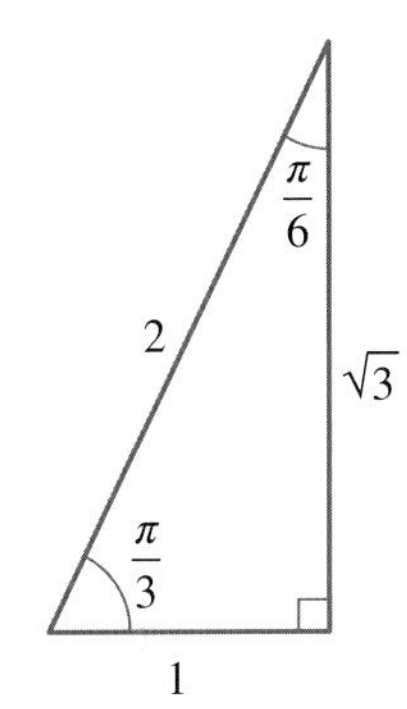

θ	$\frac{\pi}{6}$ or 30°	$\frac{\pi}{4}$ or 45°	$\frac{\pi}{3}$ or 60°
$\sin(\theta)$	$\frac{1}{2}$	$\frac{1}{\sqrt{2}} = \frac{\sqrt{2}}{2}$	$\frac{\sqrt{3}}{2}$
$\cos(\theta)$	$\frac{\sqrt{3}}{2}$	$\frac{1}{\sqrt{2}} = \frac{\sqrt{2}}{2}$	$\frac{1}{2}$
$\tan(\theta)$	$\frac{1}{\sqrt{3}} = \frac{\sqrt{3}}{3}$	1	$\sqrt{3}$

These values can be used to calculate the exact trigonometric values for other angles that lie in positions symmetric to these first-quadrant angles.

Resources

Interactivity The 'All Sin Tan Cos' rule (int-2583)

8.5.2 Trigonometric points symmetric to $[\theta]$ where $\theta \in \left\{30°, 45°, 60°, \frac{\pi}{6}, \frac{\pi}{4}, \frac{\pi}{3}\right\}$

The symmetrical points to [45°] are shown in the diagram.

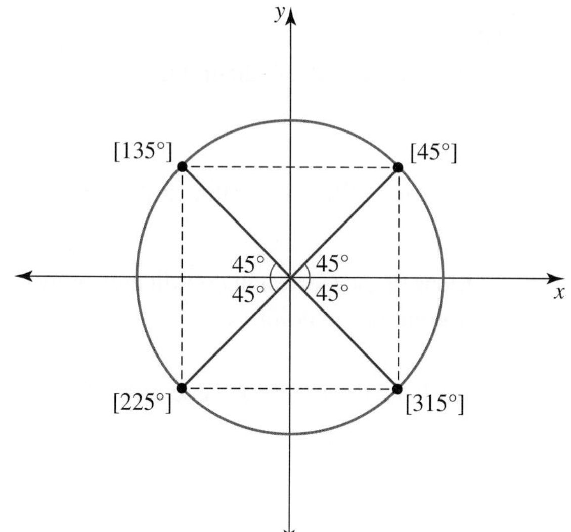

Each radius of the circle drawn to each of the points makes an acute angle of 45° with either the positive or the negative x-axis. The symmetric points to [45°] are the end points of a rotation that is 45° short of, or 45° beyond, the horizontal x-axis. The calculations 180° − 45°, 180° + 45° and 360° − 45° give the symmetric trigonometric points [135°], [225°] and [315°] respectively.

Comparisons between the coordinates of these trigonometric points with those of the first quadrant point [45°] enable the trigonometric values of these non-acute angles to be calculated from those of the acute angle 45°.

Consider the y-coordinate of each point.

As the y-coordinates of the trigonometric points [135°] and [45°] are the same, $\sin(135°) = \sin(45°)$. Similarly, the y-coordinates of the trigonometric points [225°] and [315°] are the same, but both are the negative of the y-coordinate of [45°].

Hence, $\sin(225°) = \sin(315°) = -\sin(45°)$. This gives the following exact sine values:

$$\sin(45°) = \frac{\sqrt{2}}{2};\ \sin(135°) = \frac{\sqrt{2}}{2};\ \sin(225°) = -\frac{\sqrt{2}}{2};\ \sin(315°) = -\frac{\sqrt{2}}{2}$$

Now consider the x-coordinate of each point.

As the x-coordinates of the trigonometric points [315°] and [45°] are the same, $\cos(315°) = \cos(45°)$. Similarly, the x-coordinates of the trigonometric points [135°] and [225°] are the same, but both are the negative of the x-coordinate of [45°]. Hence, $\cos(135°) = \cos(225°) = -\cos(45°)$. This gives the following exact cosine values:

$$\cos(45°) = \frac{\sqrt{2}}{2};\ \cos(135°) = -\frac{\sqrt{2}}{2};\ \cos(225°) = -\frac{\sqrt{2}}{2};\ \cos(315°) = \frac{\sqrt{2}}{2}$$

By using $\tan(\theta) = \frac{y}{x} = \frac{\sin(\theta)}{\cos(\theta)}$, you will find that the corresponding relationships for the four points are $\tan(225°) = \tan(45°)$ and $\tan(135°) = \tan(315°) = -\tan(45°)$. Hence, the exact tangent values are:

$$\tan(45°) = 1;\ \tan(135°) = -1;\ \tan(225°) = 1;\ \tan(315°) = -1$$

The relationships between the Cartesian coordinates of [45°] and each of [135°], [225°] and [315°] enable the trigonometric values of 135°, 225° and 315° to be calculated from those of 45°.

If, instead of degree measure, the radian measure of $\frac{\pi}{4}$ is used, the symmetric points to $\left[\frac{\pi}{4}\right]$ are the end points of rotations that lie $\frac{\pi}{4}$ short of, or $\frac{\pi}{4}$ beyond, the horizontal x-axis. The positions of the symmetric points are calculated as $\pi - \frac{\pi}{4}, \pi + \frac{\pi}{4}, 2\pi - \frac{\pi}{4}$, giving the symmetric trigonometric points $\left[\frac{3\pi}{4}\right], \left[\frac{5\pi}{4}\right], \left[\frac{7\pi}{4}\right]$ respectively.

By comparing the Cartesian coordinates of the symmetric points with those of the first quadrant point $\left[\frac{\pi}{4}\right]$, it is possible to obtain results such as the following selection.

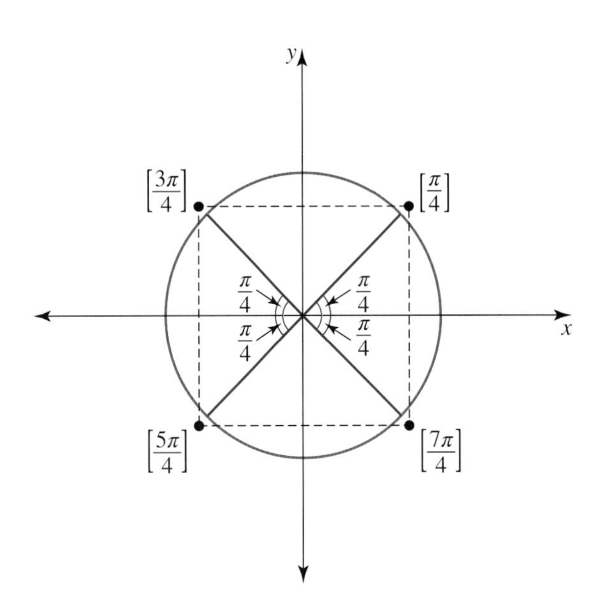

Second quadrant	Third quadrant	Fourth quadrant
$\cos\left(\frac{3\pi}{4}\right) = -\cos\left(\frac{\pi}{4}\right)$ $= -\frac{\sqrt{2}}{2}$	$\tan\left(\frac{5\pi}{4}\right) = \tan\left(\frac{\pi}{4}\right)$ $= 1$	$\sin\left(\frac{7\pi}{4}\right) = -\sin\left(\frac{\pi}{4}\right)$ $= -\frac{\sqrt{2}}{2}$

A similar approach is used to generate symmetric points to the first quadrant points $\left[\frac{\pi}{6}\right]$ and $\left[\frac{\pi}{3}\right]$.

WORKED EXAMPLE 12 Calculating exact values

Calculate the exact values of the following.

a. $\cos\left(\frac{5\pi}{3}\right)$ **b.** $\sin\left(\frac{7\pi}{6}\right)$ **c.** $\tan(-30°)$

THINK	WRITE
a. 1. State the quadrant in which the trigonometric point lies.	**a.** $\cos\left(\frac{5\pi}{3}\right)$ As $\frac{5\pi}{3} = \frac{5}{3}\pi = 1\frac{2}{3}\pi$, the point $\left[\frac{5\pi}{3}\right]$ lies in quadrant 4.
2. Identify the first-quadrant symmetric point.	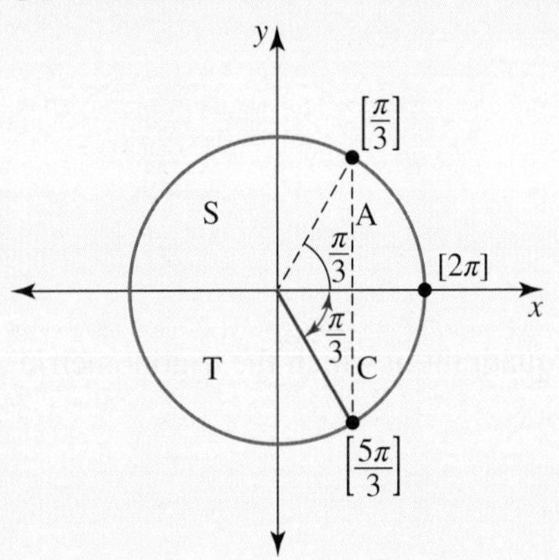

Since $\frac{5\pi}{3} = 2\pi - \frac{\pi}{3}$, the rotation of $\frac{5\pi}{3}$ stops short of the x-axis by $\frac{\pi}{3}$.

The points $\left[\frac{\pi}{3}\right]$ and $\left[\frac{5\pi}{3}\right]$ are symmetric.

3. Compare the coordinates of the symmetric points and obtain the required value.
Note: Check the +/− sign follows the ASTC diagram rule.

The x-coordinates of the symmetric points are equal.

$$\cos\left(\frac{5\pi}{3}\right) = +\cos\left(\frac{\pi}{3}\right)$$
$$= \frac{1}{2}$$

Check: cosine is positive in quadrant 4.

b. 1. State the quadrant in which the trigonometric point lies.

b. $\sin\left(\frac{7\pi}{6}\right)$

$\frac{7\pi}{6} = 1\frac{1}{6}\pi$

The point lies in quadrant 3.

2. Identify the first-quadrant symmetric point.

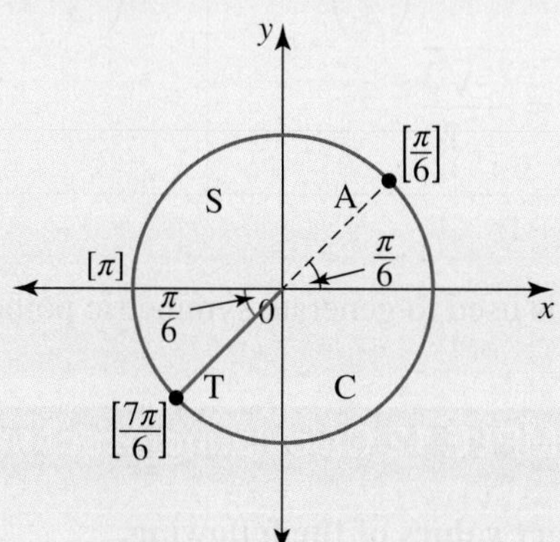

As $\frac{7\pi}{6} = \pi + \frac{\pi}{6}$, the rotation of $\frac{7\pi}{6}$ goes beyond the x-axis by $\frac{\pi}{6}$. The points $\left[\frac{\pi}{6}\right]$ and $\left[\frac{7\pi}{6}\right]$ are symmetric.

3. Compare the coordinates of the symmetric points and obtain the required value.

The y-coordinate of $\left[\frac{7\pi}{6}\right]$ is the negative of that of $\left[\frac{\pi}{6}\right]$ in the first quadrant.

$$\sin\left(\frac{7\pi}{6}\right) = -\sin\left(\frac{\pi}{6}\right)$$
$$= -\frac{1}{2}$$

Check: sine is negative in quadrant 3.

c. 1. State the quadrant in which the trigonometric point lies.

c. $\tan(-30°)$

$[-30°]$ lies in quadrant 4.

2. Identify the first-quadrant symmetric point.

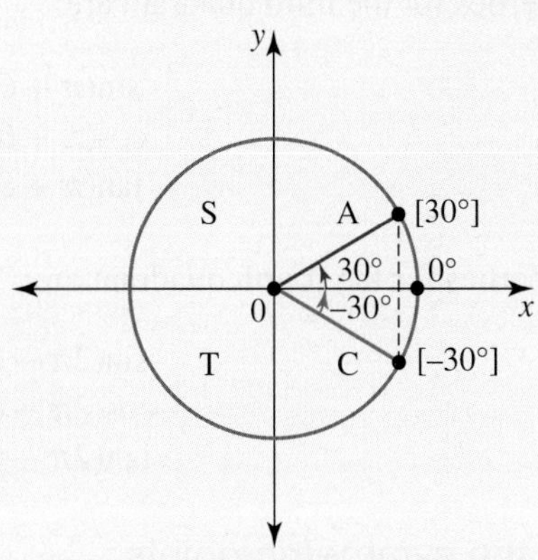

$-30°$ is a clockwise rotation of $30°$ from the horizontal so the symmetric point in the first quadrant is $[30°]$.

3. Compare the coordinates of the symmetric points and obtain the required value.

The points $[30°]$ and $[-30°]$ have the same x-coordinates but opposite y-coordinates. The tangent value is negative in quadrant 4.

$$\tan(-30°) = -\tan(30°)$$
$$= -\frac{\sqrt{3}}{3}$$

8.5.3 Symmetry properties

The **symmetry properties** give the relationships between the trigonometric values in quadrants 2, 3, 4 and that of the first quadrant value, called the base, with which they are symmetric. The symmetry properties are simply a generalisation of what was covered for the bases $\frac{\pi}{6}, \frac{\pi}{4}, \frac{\pi}{3}$.

For any real number θ where $0 < \theta < \frac{\pi}{2}$, the trigonometric point $[\theta]$ lies in the first quadrant. The other quadrant values can be expressed in terms of the base θ, since the symmetric values will either be θ short of or θ beyond the horizontal x-axis.

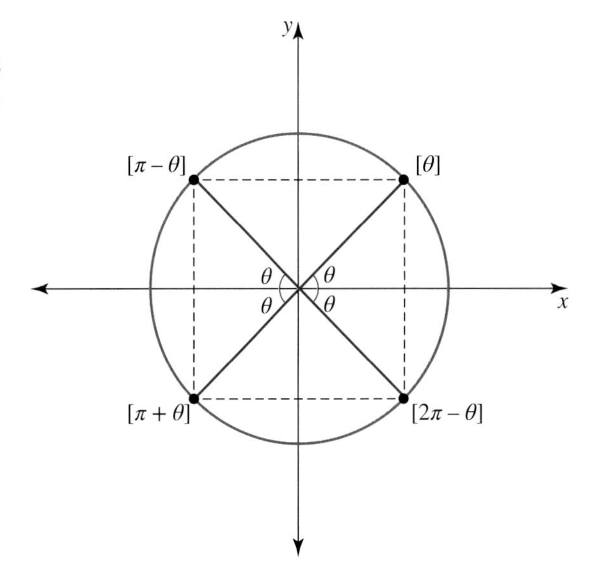

The symmetric points to $[\theta]$ are:

- second quadrant $[\pi - \theta]$
- third quadrant $[\pi + \theta]$
- fourth quadrant $[2\pi - \theta]$.

Comparing the Cartesian coordinates with those of the first-quadrant base leads to the following general statements.

The symmetry properties for the second quadrant are:

$$\sin(\pi - \theta) = \sin(\theta)$$
$$\cos(\pi - \theta) = -\cos(\theta)$$
$$\tan(\pi - \theta) = -\tan(\theta)$$

The symmetry properties for the third quadrant are:

$$\sin(\pi+\theta) = -\sin(\theta)$$
$$\cos(\pi+\theta) = -\cos(\theta)$$
$$\tan(\pi+\theta) = \tan(\theta)$$

The symmetry properties for the fourth quadrant are:

$$\sin(2\pi-\theta) = -\sin(\theta)$$
$$\cos(2\pi-\theta) = \cos(\theta)$$
$$\tan(2\pi-\theta) = -\tan(\theta)$$

Other forms for the symmetric points

The rotation assigned to a point is not unique. With clockwise rotations or repeated revolutions, other values are always possible. However, the symmetry properties apply no matter how the points are described.

The trigonometric point $[2\pi+\theta]$ would lie in the first quadrant where all ratios are positive. Hence:

$$\sin(2\pi+\theta) = \sin(\theta)$$
$$\cos(2\pi+\theta) = \cos(\theta)$$
$$\tan(2\pi+\theta) = \tan(\theta)$$

The trigonometric point $[-\theta]$ would lie in the fourth quadrant where only cosine is positive. Hence:

$$\sin(-\theta) = -\sin(\theta)$$
$$\cos(-\theta) = \cos(\theta)$$
$$\tan(-\theta) = -\tan(\theta)$$

For negative rotations, the points symmetric to $[\theta]$ could be given as:
- fourth quadrant $[-\theta]$
- third quadrant $[-\pi+\theta]$
- second quadrant $[-\pi-\theta]$
- first quadrant $[-2\pi+\theta]$.

Using symmetry properties to calculate values of trigonometric functions

Trigonometric values are either the same as or the negative of the associated trigonometric values of the first-quadrant base; the sign is determined by the ASTC diagram.

The base involved is identified by noting the rotation needed to reach the *x*-axis and determining how far short of or how far beyond this the symmetric point is. It is important to emphasise that for the points to be symmetric, this is always measured from the horizontal and not the vertical axis.

To calculate a value of a trigonometric function, follow these steps.
- Locate the quadrant in which the trigonometric point lies
- Identify the first-quadrant base with which the trigonometric point is symmetric
- Compare the coordinates of the trigonometric point with the coordinates of the base point or use the ASTC diagram rule to form the sign in the first instance
- Evaluate the required value exactly if there is a known exact value involving the base.

With practice, the symmetry properties allow us to recognise, for example, that $\sin\left(\frac{8\pi}{7}\right) = -\sin\left(\frac{\pi}{7}\right)$ because $\frac{8\pi}{7} = \pi + \frac{\pi}{7}$ and sine is negative in the third quadrant. Recognition of the symmetry properties is very important, and we should aim to be able to apply these quickly. For example, to evaluate $\cos\left(\frac{3\pi}{4}\right)$, think, 'Second quadrant; cosine is negative; base is $\frac{\pi}{4}$', and write:

$$\begin{aligned}\cos\left(\frac{3\pi}{4}\right) &= -\cos\left(\frac{\pi}{4}\right)\\ &= -\frac{\sqrt{2}}{2}\end{aligned}$$

WORKED EXAMPLE 13 Using symmetry properties to calculate exact values

a. Identify the symmetric points to [20°]. Determine at which of these points the tangent value is the same as tan(20°).

b. Express $\sin\left(\frac{6\pi}{5}\right)$ in terms of a first-quadrant value.

c. If $\cos(\theta) = 0.6$, give the values of $\cos(\pi - \theta)$ and $\cos(2\pi - \theta)$.

d. Calculate the exact values of the following.

i. $\tan\left(\frac{7\pi}{6}\right)$ ii. $\sin\left(\frac{11\pi}{3}\right)$

THINK	WRITE
a. 1. Calculate the symmetric points to the given point.	a. Symmetric points to [20°] will be ±20° from the x-axis. The points are: second quadrant [180° − 20°] = [160°] third quadrant [180° + 20°] = [200°] fourth quadrant [360° − 20°] = [340°]
2. Identify the quadrant.	The point [20°] is in the first quadrant, so tan(20°) is positive. As tangent is also positive in the third quadrant, tan(200°) = tan(20°).
3. State the required point.	The tangent value at the trigonometric point [200°] has the same value as tan(20°).
b. 1. Express the trigonometric value in the appropriate quadrant form.	b. $\frac{6\pi}{5}$ is in the third quadrant. $\sin\left(\frac{6\pi}{5}\right) = \sin\left(\pi + \frac{\pi}{5}\right)$
2. Apply the symmetry property for that quadrant.	$\therefore \sin\left(\frac{6\pi}{5}\right) = -\sin\left(\frac{\pi}{5}\right)$
c. 1. Use the symmetry property for the appropriate quadrant.	c. $(\pi - \theta)$ is the second quadrant form. $\therefore \cos(\pi - \theta) = -\cos(\theta)$
2. State the answer.	Since $\cos(\theta) = 0.6$, $\cos(\pi - \theta) = -0.6$.
3. Use the symmetry property for the appropriate quadrant.	$2\pi - \theta$ is the fourth quadrant form. $\therefore \cos(2\pi - \theta) = \cos(\theta)$
4. State the answer.	Since $\cos(\theta) = 0.6$, $\cos(2\pi - \theta) = 0.6$.

THINK	WRITE
d. i. 1. Express the trigonometric value in an appropriate quadrant form.	d. i. $\tan\left(\frac{7\pi}{6}\right) = \tan\left(\pi + \frac{\pi}{6}\right)$
2. Apply the symmetry property and evaluate.	$= \tan\left(\frac{\pi}{6}\right)$ $= \frac{\sqrt{3}}{3}$ $\therefore \tan\left(\frac{7\pi}{6}\right) = \frac{\sqrt{3}}{3}$
ii. 1. Express the trigonometric value in an appropriate quadrant form.	ii. $\frac{11\pi}{3}$ is in quadrant 4.
2. Apply the symmetry property and evaluate.	$\sin\left(\frac{11\pi}{3}\right) = \sin\left(4\pi - \frac{\pi}{3}\right)$ $= \sin\left(2\pi - \frac{\pi}{3}\right)$ $= -\sin\left(\frac{\pi}{3}\right)$ $= -\frac{\sqrt{3}}{2}$ $\therefore \sin\left(\frac{11\pi}{3}\right) = -\frac{\sqrt{3}}{2}$

Resources

Interactivity Symmetry points and quadrants (int-2584)

8.5 Exercise

Students, these questions are even better in jacPLUS

Receive immediate feedback and access sample responses

Access additional questions

Track your results and progress

Find all this and MORE in jacPLUS

Technology free

1. State the exact values of the following.
   a. $\sin(120°)$ b. $\tan(210°)$ c. $\cos(135°)$ d. $\cos(300°)$ e. $\tan(150°)$ f. $\sin(315°)$
2. Calculate the exact values of the following.
   a. $\cos(150°)$ b. $\sin(240°)$ c. $\tan(330°)$ d. $\tan(-45°)$ e. $\sin(-30°)$ f. $\cos(-60°)$
3. Calculate the exact values of the following.
   a. $\tan(420°)$ b. $\sin(405°)$ c. $\cos(480°)$ d. $\sin(765°)$ e. $\cos(-510°)$ f. $\tan(-585°)$

4. Calculate the exact values of the following.

a. $\cos(120°)$ b. $\tan(225°)$ c. $\sin(330°)$ d. $\tan(-60°)$ e. $\cos(-315°)$ f. $\sin(510°)$

5. State the exact values of the following.

a. $\tan\left(\frac{\pi}{3}\right)$ b. $\tan\left(\frac{5\pi}{3}\right)$ c. $\sin\left(\frac{\pi}{6}\right)$ d. $\sin\left(\frac{5\pi}{6}\right)$ e. $\cos\left(\frac{\pi}{4}\right)$ f. $\cos\left(\frac{5\pi}{4}\right)$

6. Calculate the exact values of the following.

a. $\cos\left(-\frac{5\pi}{6}\right)$ b. $\sin\left(-\frac{4\pi}{3}\right)$ c. $\tan\left(-\frac{3\pi}{4}\right)$ d. $\tan\left(\frac{13\pi}{6}\right)$ e. $\sin\left(\frac{17\pi}{4}\right)$ f. $\cos\left(\frac{8\pi}{3}\right)$

7. Calculate the exact values of the following.

a. $\sin\left(\frac{3\pi}{4}\right)$ b. $\tan\left(\frac{2\pi}{3}\right)$ c. $\cos\left(\frac{5\pi}{6}\right)$ d. $\cos\left(\frac{4\pi}{3}\right)$ e. $\tan\left(\frac{7\pi}{6}\right)$ f. $\sin\left(\frac{11\pi}{6}\right)$

8. Calculate the exact values of the following.

a. $\cos\left(-\frac{\pi}{4}\right)$ b. $\sin\left(-\frac{\pi}{3}\right)$ c. $\tan\left(-\frac{5\pi}{6}\right)$ d. $\sin\left(\frac{8\pi}{3}\right)$ e. $\cos\left(\frac{9\pi}{4}\right)$ f. $\tan\left(\frac{23\pi}{6}\right)$

9. Show the boundary points on a diagram and then state the value of the following.

a. $\cos(4\pi)$ b. $\tan(9\pi)$ c. $\sin(7\pi)$ d. $\sin\left(\frac{13\pi}{2}\right)$ e. $\cos\left(-\frac{9\pi}{2}\right)$ f. $\tan(-20\pi)$

10. Identify the quadrant(s) or boundaries for which the following apply.

a. $\cos(\theta) > 0, \sin(\theta) < 0$
b. $\tan(\theta) > 0, \cos(\theta) > 0$
c. $\sin(\theta) > 0, \cos(\theta) < 0$
d. $\cos(\theta) = 0$
e. $\cos(\theta) = 0, \sin(\theta) > 0$
f. $\sin(\theta) = 0, \cos(\theta) < 0$

11. **WE11** a. Identify the quadrant(s) where $\cos(\theta)$ is negative and $\tan(\theta)$ is positive.

b. If $f(\theta) = \tan(\theta)$, evaluate $f(4\pi)$.

12. **WE12** Calculate the exact values of the following.

a. $\sin\left(\frac{4\pi}{3}\right)$ b. $\tan\left(\frac{5\pi}{6}\right)$ c. $\cos(-30°)$

13. Calculate the exact values of $\sin\left(-\frac{5\pi}{4}\right)$, $\cos\left(-\frac{5\pi}{4}\right)$ and $\tan\left(-\frac{5\pi}{4}\right)$.

14. If $\cos(\theta) = 0.2$, use the symmetry properties to write down the value of the following.

a. $\cos(\pi - \theta)$ b. $\cos(\pi + \theta)$ c. $\cos(-\theta)$ d. $\cos(2\pi + \theta)$

15. If $\sin(t) = 0.9$ and $\tan(x) = 4$, calculate the value of the following.

a. $\tan(-x)$
b. $\sin(\pi - t)$
c. $\tan(2\pi - x)$
d. $\sin(-t) + \tan(\pi + x)$

16. Given $\cos(\theta) = 0.91, \sin(t) = 0.43$ and $\tan(x) = 0.47$, use the symmetry properties to obtain the values of the following.

a. $\cos(\pi + \theta)$ b. $\sin(\pi - t)$ c. $\tan(2\pi - x)$ d. $\cos(-\theta)$ e. $\sin(-t)$ f. $\tan(2\pi + x)$

17. If $\sin(\theta) = p$, express the following in terms of p.

a. $\sin(2\pi - \theta)$ b. $\sin(3\pi - \theta)$ c. $\sin(-\pi + \theta)$ d. $\sin(\theta + 4\pi)$

18. WE13 a. Identify the symmetric points to [75°]. Determine at which of these points the cosine value is the same as $\cos(75°)$.

b. Express $\tan\left(\frac{6\pi}{7}\right)$ in terms of a first quadrant value.

c. If $\sin(\theta) = 0.8$, give the values of $\sin(\pi - \theta)$ and $\sin(2\pi - \theta)$.

d. Calculate the exact values of the following.

i. $\cos\left(\frac{5\pi}{4}\right)$

ii. $\sin\left(\frac{25\pi}{6}\right)$

Technology active

19. a. Verify that $\sin^2\left(\frac{5\pi}{4}\right) + \cos^2\left(\frac{5\pi}{4}\right) = 1$.

b. Explain, with the aid of a unit circle diagram, why $\cos(-\theta) = \cos(\theta)$ is true for $\theta = \frac{5\pi}{6}$.

c. The point $[\phi]$ lies in the second quadrant and has Cartesian coordinates $(-0.5, 0.87)$. Show this on a diagram and give the values of $\sin(\pi + \phi)$, $\cos(\pi + \phi)$ and $\tan(\pi + \phi)$.

d. Simplify $\sin(-\pi + t) + \sin(-3\pi - t) + \sin(t + 6\pi)$.

e. Use the unit circle to give two values of an angle A for which $\sin(A) = \sin(144°)$.

f. With the aid of the unit circle, give three values of B for which $\sin(B) = -\sin\left(\frac{2\pi}{11}\right)$.

20. Consider the point Q $[\theta]$, $\tan(\theta) = 5$.

a. Identify the two quadrants in which could Q lie.

b. Determine, to 4 decimal places, the value of θ for each of the two points.

c. Calculate the exact sine and cosine values for each θ and state the exact Cartesian coordinates of each of the two points.

8.5 Exam questions

Question 1 (1 mark) TECH-ACTIVE

MC The value of $\cos\left(\frac{3\pi}{4}\right)$ is

A. $\frac{1}{2}$

B. $-\frac{\sqrt{2}}{2}$

C. $\frac{1}{\sqrt{2}}$

D. $\frac{\sqrt{3}}{2}$

E. $\frac{2}{\sqrt{2}}$

Question 2 (1 mark) TECH-ACTIVE

MC If $\sin(\theta) = -\frac{1}{\sqrt{2}}$ and $0° \le \theta \le 360°$, then θ equals

A. 225°

B. 135°, 225°

C. 135°

D. 225°, 315°

E. 45°

Question 3 (2 marks) TECH-FREE

Calculate the exact value of $4\sin\left(\frac{5\pi}{6}\right) - 2\cos\left(\frac{\pi}{3}\right) + \tan\left(\frac{5\pi}{4}\right) + \sin\left(\frac{7\pi}{3}\right)$.

More exam questions are available online.

8.6 Graphs of the sine and cosine functions

LEARNING INTENTION

At the end of this subtopic you should be able to:
- graph sine and cosine functions
- identify key features and properties of sine and cosine functions
- approximate for small values of x in trigonometric functions.

As the two functions sine and cosine are closely related, we shall initially consider their graphs together.

8.6.1 The graphs of $y = \sin(x)$ and $y = \cos(x)$

The functions sine and cosine are both periodic and have many-to-one correspondences, which means values repeat after every revolution around the unit circle. This means both functions have a period of 2π, since $\sin(x + 2\pi) = \sin(x)$ and $\cos(x + 2\pi) = \cos(x)$.

The graphs of $y = \sin(x)$ and $y = \cos(x)$ can be plotted using the boundary values from continued rotations clockwise and anticlockwise around the unit circle.

x	0	$\frac{\pi}{2}$	π	$\frac{3\pi}{2}$	2π
$\sin(x)$	0	1	0	−1	0
$\cos(x)$	1	0	−1	0	1

The diagram shows four cycles of the graphs drawn on the domain $[-4\pi, 4\pi]$. The graphs continue to repeat their wavelike pattern over their maximal domain R; the interval, or **period**, of each repetition is 2π.

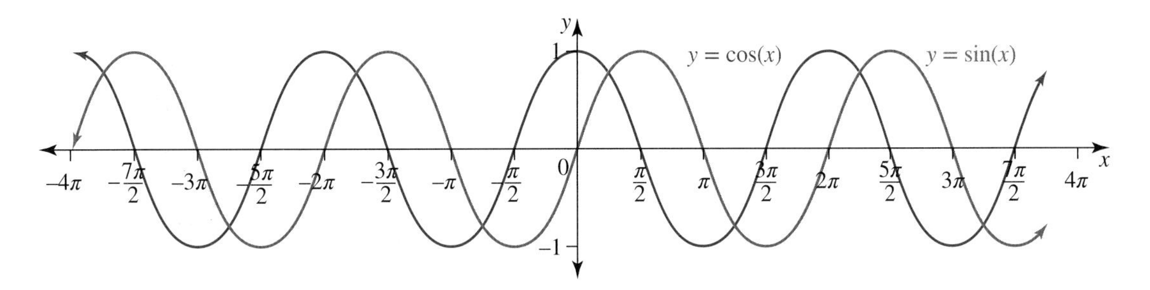

The first observation that strikes us about these graphs is how remarkably similar they are: a horizontal translation of $\frac{\pi}{2}$ to the right will transform the graph of $y = \cos(x)$ into the graph of $y = \sin(x)$, and a horizontal translation of $\frac{\pi}{2}$ to the left transforms the graph of $y = \sin(x)$ into the graph of $y = \cos(x)$.

Recalling our knowledge of transformations of graphs, this observation can be expressed as:

$$\cos\left(x - \frac{\pi}{2}\right) = \sin(x)$$

$$\sin\left(x + \frac{\pi}{2}\right) = \cos(x)$$

The two functions are said to be '**out of phase**' by $\frac{\pi}{2}$ or to have a **phase difference** or **phase shift** of $\frac{\pi}{2}$.

Both graphs oscillate up and down 1 unit from the x-axis. The x-axis is the **equilibrium** or **mean position**, and the distance the graphs oscillate up and down from this mean position to a maximum or minimum point is called the **amplitude**.

The graphs keep repeating this cycle of oscillations up and down from the equilibrium position, with the amplitude measuring half the vertical distance between maximum and minimum points and the period measuring the horizontal distance between successive maximum points or between successive minimum points.

 Resources

 **Interactivities** Sine and cosine graphs (int-2976)
The unit circle, sine and cosine graphs (int-6551)

One cycle of the graph of $y = \sin(x)$

The basic graph of $y = \sin(x)$ has the domain $[0, 2\pi]$, which restricts the graph to one cycle.

The graph of the function $f: [0, 2\pi] \to R$, $f(x) = \sin(x)$ is shown.

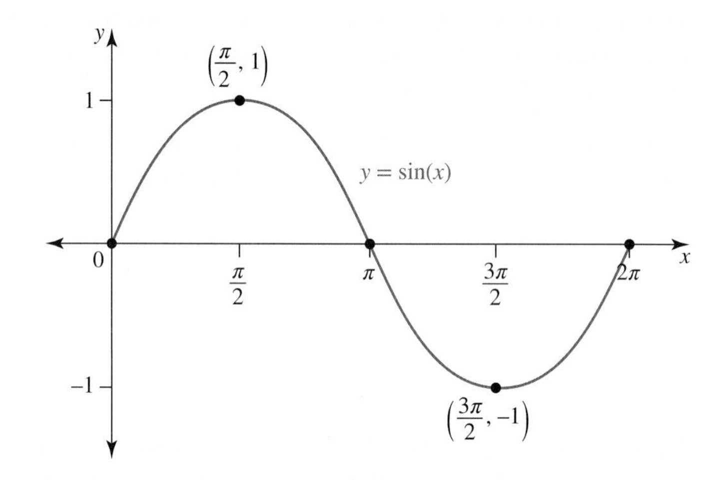

Key features of the graph of $y = \sin(x)$:
- The equilibrium position is the x-axis, the line with equation $y = 0$.
- The amplitude is 1 unit.
- The period is 2π units.
- The domain is $[0, 2\pi]$.
- The range is $[-1, 1]$.
- The x-intercepts occur at $x = 0, \pi, 2\pi$.
- The type of correspondence is many-to-one.

The graph lies above the x-axis for $x \in (0, \pi)$ and below for $x \in (\pi, 2\pi)$, which matches the quadrant signs of sine given in the ASTC diagram.

The symmetry properties of sine are displayed in its graph as $\sin(\pi - \theta) = \sin(\theta)$ and $\sin(\pi + \theta) = \sin(2\pi - \theta) = -\sin(\theta)$.

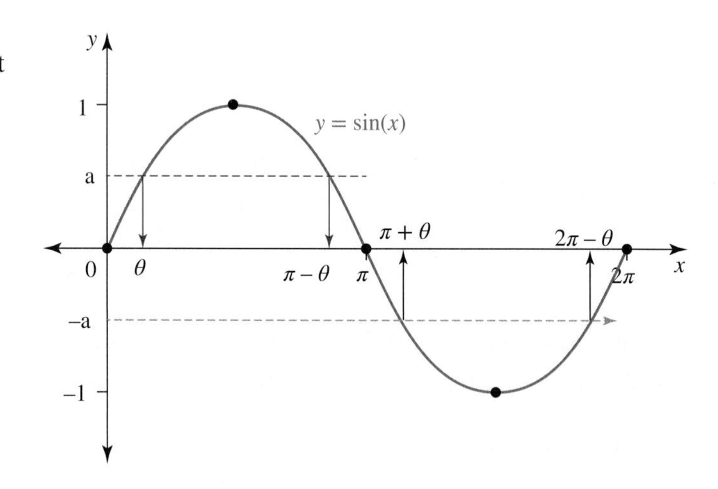

One cycle of the graph of $y = \cos(x)$

The basic graph of $y = \cos(x)$ has the domain $[0, 2\pi]$, which restricts the graph to one cycle.

The graph of the function $f: [0, 2\pi] \to R, f(x) = \cos(x)$ is shown.

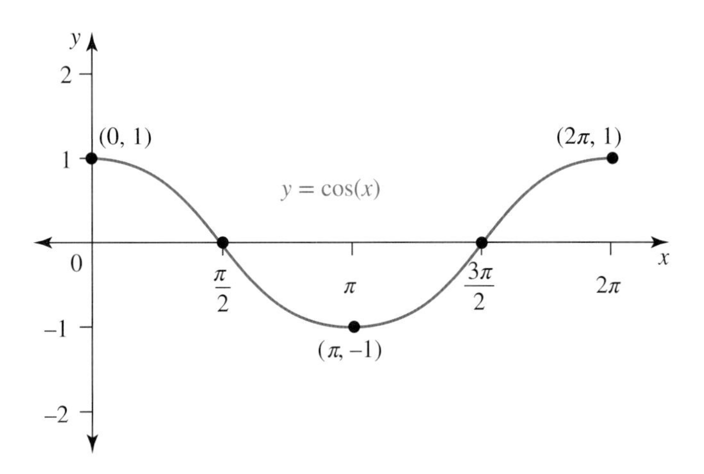

Key features of the graph of $y = \cos(x)$:

- The equilibrium position is the x-axis, the line with equation $y = 0$.
- The amplitude is 1 unit.
- The period is 2π units.
- The domain is $[0, 2\pi]$.
- The range is $[-1, 1]$.
- The x-intercepts occur at $x = \frac{\pi}{2}, \frac{3\pi}{2}$.
- The type of correspondence is many-to-one.

The graph of $y = \cos(x)$ has the same amplitude, period, equilibrium (or mean) position, domain, range and type of correspondence as the graph of $y = \sin(x)$.

Guide to sketching the graphs on extended domains

There is a pattern of 5 points to the shape of the basic sine and cosine graphs created by the division of the period into four equal intervals.

For $y = \sin(x)$, the first point starts at the equilibrium; the second point at $\frac{1}{4}$ of the period is up one amplitude to the maximum point; the third point, at $\frac{1}{2}$ of the period, is back at equilibrium; the fourth point at $\frac{3}{4}$ of the period is down one amplitude to the minimum point; and the fifth point at the end of the period interval returns back to equilibrium.

In other words:

equilibrium → range maximum → equilibrium → range minimum → equilibrium.

For $y = \cos(x)$, the pattern for one cycle is summarised as:

range maximum → equilibrium → range minimum → equilibrium → range maximum.

This pattern only needs to be continued in order to sketch the graph of $y = \sin(x)$ or $y = \cos(x)$ on a domain other than $[0, 2\pi]$.

WORKED EXAMPLE 14 Sketching the sine graph for a given domain

Sketch the graph of $y = \sin(x)$ over the domain $[-2\pi, 4\pi]$ and state the number of cycles of the sine function drawn.

THINK	WRITE
1. Draw the graph of the function over $[0, 2\pi]$.	The basic graph of $y = \sin(x)$ over the domain $[0, 2\pi]$ is drawn.

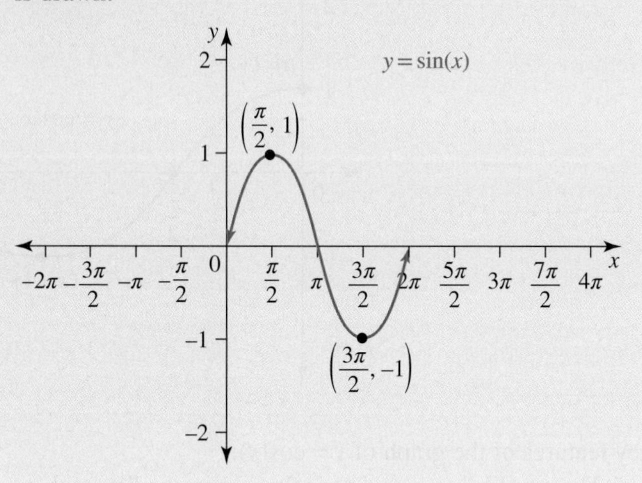

THINK	WRITE
2. Extend the pattern to cover the domain specified.	The pattern is extended for one cycle in the negative direction and one further cycle in the positive direction to cover the domain $[-2\pi, 4\pi]$.

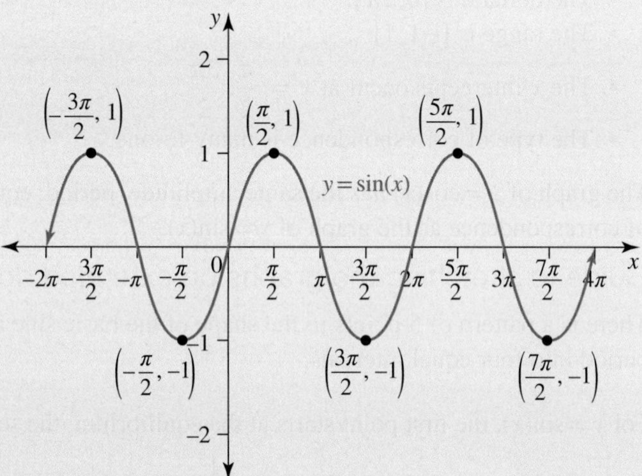

THINK	WRITE
3. State the number of cycles of the function that are shown in the graph.	Altogether, 3 cycles of the sine function have been drawn.

TI | THINK	DISPLAY/WRITE	CASIO | THINK	DISPLAY/WRITE
a. 1. On a Graph page, complete the entry line for function 1 as: $f1(x) = \sin(x)\| - 2\pi \le x \le 4\pi$ then press ENTER. Press MENU, then select: 9: Settings... Alternate between radians and degrees by pressing on RAD/DEG in the top right corner of the screen. Press MENU, then select: 4: Window/Zoom 1: Window Settings... to change the view shown on the screen.	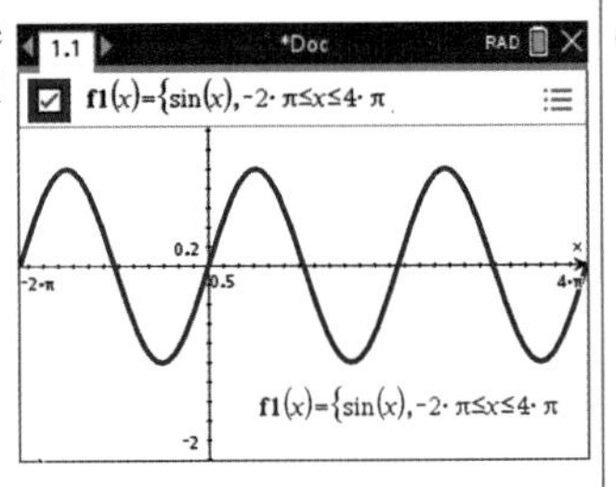	**a. 1.** Put the calculator in RADIAN mode. On a Graph & Table screen, complete the entry line for y1 as: $y1 = \sin(x)\| - 2\pi \le x \le 4\pi$ Click the tick box, then select the graph icon. Click the window icon to change the view on the screen.	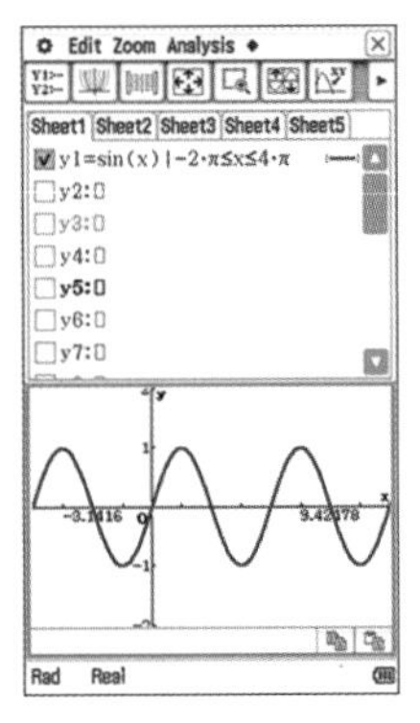
2. Answer the question.	Three cycles have been drawn.	**2.** Answer the question.	Three cycles have been drawn.

8.6.2 Graphical and numerical solutions to equations

The sine and cosine functions are **transcendental functions**, meaning they cannot be expressed as algebraic expressions in powers of x. There is no algebraic method of solution for an equation such as $\sin(x) = 1 - 2x$, because it contains a transcendental function and a linear polynomial function. However, whether solutions exist or not can usually be determined by graphing both functions to see if or in how many places they intersect. If a solution exists, an interval in which the root lies can be specified and the bisection method can refine this interval.

When sketching, care is needed with the scaling of the x-axis. The polynomial function is normally graphed using an integer scale, whereas the trigonometric function normally uses multiples of π. It can be helpful to remember that $\pi \approx 3.14$, so $\frac{\pi}{2} \approx 1.57$ and so on.

WORKED EXAMPLE 15 Solving equations graphically

Consider the equation $\sin(x) = 1 - 2x$.

a. Sketch the graphs of $y = \sin(x)$ and $y = 1 - 2x$ on the same set of axes and explain why the equation $\sin(x) = 1 - 2x$ has only one root.

b. Use the graph to give an interval in which the root of the equation $\sin(x) = 1 - 2x$ lies.

c. Use the bisection method to create two narrower intervals for the root and hence give an estimate of its value.

THINK	WRITE
a. 1. Calculate the points needed to sketch the two graphs.	**a.** $y = \sin(x)$ One cycle of this graph has the domain $[0, 2\pi]$. The axis intercepts are $(0, 0)$, $(\pi, 0)$ and $(2\pi, 0)$. $y = 1 - 2x$ has axis intercepts at $(0, 1)$ and $(0.5, 0)$.

2. Sketch the graphs on the same set of axes.	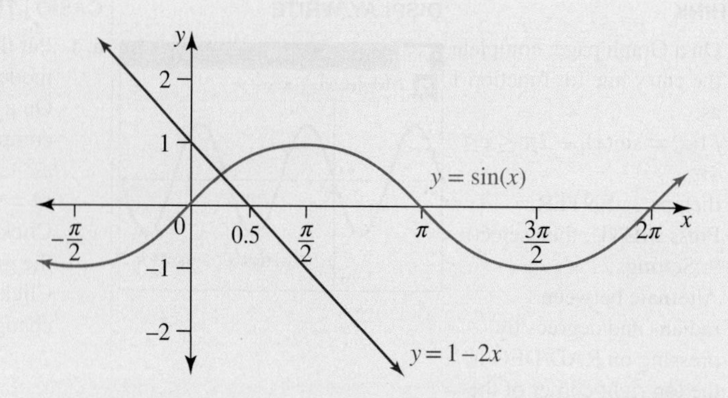
3. Give an explanation about the number of roots to the equation.	The two graphs intersect at one point only, so the equation $\sin(x) = 1 - 2x$ has only one root.
b. State an interval in which the root of the equation lies.	From the graph it can be seen that the root lies between the origin and the x-intercept of the line. The interval in which the root lies is therefore $[0, 0.5]$.
c. 1. Define the function whose sign is to be tested in the bisection method procedure.	**c.** $\sin(x) = 1 - 2x$ $\therefore \sin(x) - 1 + 2x = 0$ Let $f(x) = \sin(x) - 1 + 2x$.
2. Test the sign at the endpoints of the interval in which the root has been placed.	At $x = 0$, the sine graph lies below the line, so $f(0) < 0$. Check: $f(0) = \sin(0) - 1 + 2(0)$ $= -1$ < 0 At $x = 0.5$, the sine graph lies above the line, so $f(0.5) > 0$. Check: $f(0.5) = \sin(0.5) - 1 + 2(0.5)$ $= \sin(0.5^c) - 1 + 1$ $= 0.48\ldots$ > 0
3. Create the first of the narrower intervals.	The midpoint of $[0, 0.5]$ is $x = 0.25$. $f(0.25) = \sin(0.25) - 1 + 2(0.25)$ $= -0.25\ldots$ < 0 The root lies in the interval $[0.25, 0.5]$.
4. Create the second interval.	The midpoint of $[0.25, 0.5]$ is $x = 0.375$. $f(0.375) = \sin(0.375) - 1 + 2(0.375)$ $= 0.116\ldots$ > 0 The root lies in the interval $[0.25, 0.375]$.
5. State an estimated value of the root of the equation.	The midpoint of $[0.25, 0.375]$ is an estimate of the root. An estimate is $x = 0.3125$.

8.6.3 Approximations to the sine and cosine functions for values close to zero

Despite the sine and cosine functions having no algebraic form, for small values of x it is possible to approximate them by simple polynomial functions for a domain close to zero.

Comparing the graphs of $y = \sin(x)$ and $y = x$, we can see the two graphs resemble each other for a domain around $x = 0$.

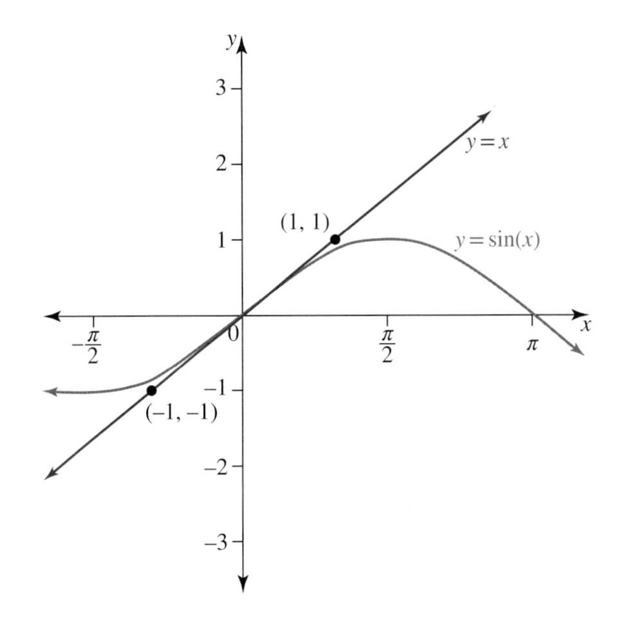

> **Small values of sin(x)**
>
> **For small values of x,**
>
> $$\sin(x) \approx x.$$

This offers another way to obtain an estimate of the root of the equation $\sin(x) = 1 - 2x$, as the graphs drawn in Worked example 15 placed the root in the small interval $[0, 0.5]$.

Replacing $\sin(x)$ by x, the equation becomes $x = 1 - 2x$, the solution to which is $x = \frac{1}{3}$ or 0.333 ... To 1-decimal-place accuracy, this value agrees with that obtained with greater effort using the bisection method.

Returning to the unit circle definitions of the sine and cosine functions, the line segments whose lengths give the values of sine and cosine are shown in the unit circle diagram for a small value of θ, the angle of rotation.

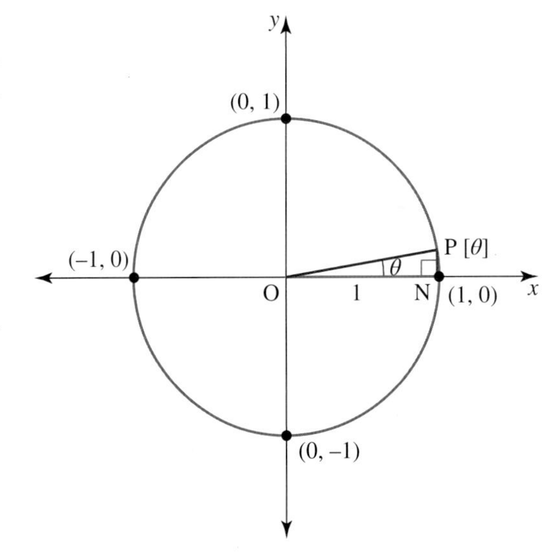

The length $PN = \sin(\theta)$, and for small θ this length is approximately the same as the length of the arc that subtends the angle θ at the centre of the unit circle. The arc length is $r\theta = 1\theta = \theta$. Hence, for small θ, $\sin(\theta) \approx \theta$.

The length $ON = \cos(\theta)$ and for small θ, $ON \approx 1$, the length of the radius of the circle. Hence, for small θ, $\cos(\theta) \approx 1$.

> **Small values of cos(x)**
>
> **For small values of x,**
>
> $$\cos(x) \approx 1.$$

WORKED EXAMPLE 16 Approximate $\sin(x)$ and $\cos(x)$ for small values of x

a. **Use the linear approximation for $\sin(x)$ to evaluate $\sin(3°)$ and compare the accuracy of the approximation with the calculator value for $\sin(3°)$.**
b. **Show there is a root to the equation $\cos(x) - 12x^2 = 0$ for which $0 \le x \le 0.4$.**
c. **Use an approximation for $\cos(x)$ to obtain an estimate of the root in part b, expressed to 2 decimal places.**

THINK	WRITE
a. 1. Express the angle in radian mode.	**a.** $3° = 3 \times \frac{\pi}{180}$ $\therefore 3° = \frac{\pi^c}{60}$
2. Use the approximation for the trigonometric function to estimate the value of the trigonometric ratio.	$\sin(x) \approx x$ for small values of x. As $\frac{\pi}{60}$ is small, $\sin\left(\frac{\pi^c}{60}\right) \approx \frac{\pi}{60}$ $\therefore \sin(3°) \approx \frac{\pi}{60}$
3. Compare the approximate value with that given by a calculator for the value of the trigonometric ratio.	From a calculator, $\sin(3°) = 0.052\,34$ and $\frac{\pi}{60} = 0.052\,36$ to 5 decimal places. The two values would be the same when expressed correct to 3 decimal places.
b. Show there is a root to the equation in the given interval.	**b.** $\cos(x) - 12x^2 = 0$ Let $f(x) = \cos(x) - 12x^2$ $f(0) = \cos(0) - 12(0)$ $= 1$ > 0 $f(0.4) = \cos(0.4) - 12(0.4)^2$ $= -0.9989...$ < 0 Therefore, there is a root of the equation for which $0 \le x \le 0.4$.
c. Use the approximation to obtain an estimate of the root in part **b**.	**c.** Values of x in the interval $0 \le x \le 0.4$ are small and for small x, $\cos(x) \approx 1$. Let $\cos(x) = 1$. The equation $\cos(x) - 12x^2 = 0$ becomes: $0 = 1 - 12x^2$ $1 = 12x^2$ $x^2 = \frac{1}{12}$ $x = \pm\frac{\sqrt{2}}{5}$ $x = \pm\frac{1}{2\sqrt{3}} = \pm\frac{\sqrt{3}}{6}$ $= \pm 0.2887$ The root for which x is positive is 0.29 to 2 decimal places.

8.6 Exercise

Students, these questions are even better in jacPLUS

Receive immediate feedback and access sample responses

Access additional questions

Track your results and progress

Find all this and MORE in jacPLUS

Technology free

1. Consider the graph of the function $y = f(x)$ shown.
   a. State the domain and range of the graph.
   b. Select the appropriate equation for the function from $y = \sin(x)$ or $y = \cos(x)$.
   c. Write down the coordinates of each of the four turning points.
   d. Identify the period and the amplitude of the graph.
   e. Give the equation of the mean (or equilibrium) position.
   f. Identify the values of x for which $f(x) > 0$.

2. Consider the graph of the function $y = g(x)$ shown.
   a. State the domain and range of the graph.
   b. Select the appropriate equation for the function from: $y = \sin(x)$ or $y = \cos(x)$.
   c. Write down the coordinates of each of the minimum turning points.
   d. Identify the period and amplitude of the graph and state the equation of its mean (or equilibrium) position.
   e. Write down the coordinates of the x intercepts of the graph.
   f. Identify the values of x for which $g(x) < 0$.

3. Sketch the following graphs over the given domain intervals.
   a. $y = \sin(x), 0 \le x \le 6\pi$
   b. $y = \cos(x), -4\pi \le x \le 2\pi$
   c. $y = \cos(x), -\dfrac{\pi}{2} \le x \le \dfrac{3\pi}{2}$
   d. $y = \sin(x), -\dfrac{3\pi}{2} \le x \le \dfrac{\pi}{2}$.

4. **WE14** Sketch the graph of $y = \cos(x)$ over the domain $[-2\pi, 4\pi]$ and state the number of cycles of the cosine function drawn.

5. a. State the number of maximum turning points on the graph of the function $f: [-4\pi, 0] \to R, f(x) = \sin(x)$.
   b. State the number of minimum turning points of the graph of the function $f: [0, 14\pi] \to R, f(x) = \cos(x)$.

6. State the number of intersections that the graphs of the following make with the x-axis.
   a. $y = \cos(x), 0 \le x \le \dfrac{7\pi}{2}$
   b. $y = \sin(x), -2\pi \le x \le 4\pi$
   c. $y = \sin(x), 0 \le x \le 20\pi$
   d. $y = \cos(x), \pi \le x \le 4\pi$

7. On the same set of axes, sketch the graphs of $y = \cos(x)$ and $y = \sin(x)$ over the domain $[0, 2\pi]$ and shade the region $\{(x, y) : \sin(x) \geq \cos(x), x \in [0, 2\pi]\}$.

8. a. Draw one cycle of the cosine graph over $[0, 2\pi]$ and give the values of x in this interval for which $\cos(x) < 0$.
   b. Explain how the graph in part a illustrates what the ASTC diagram says about the sign of the cosine function.

9. The graph of the function $f : [0, a] \to R, f(x) = \cos(x)$ has 10 intersections with the x-axis. Determine the smallest value possible for a.

10. Use the linear approximation $\sin(x) \approx x$ to estimate the values of the following.

    a. $\sin\left(\dfrac{\pi}{12}\right)$

    b. $\sin(2.5°)$

11. Consider the equation $\sin(x) - 11x + 4 = 0$.
    a. The number of solutions to this equation can be determined by using the graph of $y = \sin(x)$ and the graph of a straight line. Identify the straight line.
    b. The graphs of $y = \sin(x)$ and the straight line are shown in the diagram.

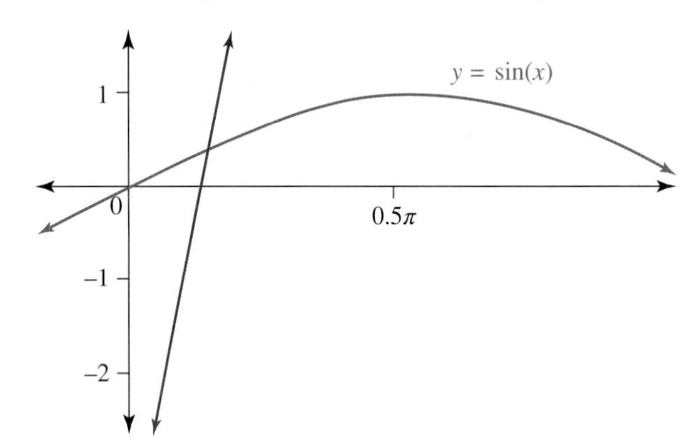

    Justifying your answer, state the number of solutions to the equation $\sin(x) - 11x + 4 = 0$.
    c. Use the linear approximation for $\sin(x)$ to express $\sin(x) - 11x + 4 = 0$ as an algebraic equation and, hence obtain an estimate of any solutions.

12. Use the linear approximation for $\sin(x)$ to calculate the following.
    a. $\sin(1°)$
    b. $\sin\left(\dfrac{\pi}{9}\right)$
    c. $\sin(-2°)$
    d. $\sin\left(\dfrac{\pi}{6}\right)$; comment on the reason for the discrepancy with its exact value.

13. The solution to the equation $\cos(x) + 5x - 2 = 0$ is small. Use an approximation for $\cos(x)$ to obtain an estimate of the solution.

Technology active

14. **WE15** Consider the equation $\sin(x) = x - 2$.
    a. Sketch the graphs of $y = \sin(x)$ and $y = x - 2$ on the same set of axes and explain why the equation $\sin(x) = x - 2$ has only one root.
    b. Use the graph to give an interval in which the root of the equation $\sin(x) = x - 2$ lies.
    c. Use the bisection method to create two narrower intervals for the root and hence give an estimate of its value.

15. Consider the equation $\cos(x) - x^2 = 0$.

a. Give the equations of the two graphs whose intersection determines the number of solutions to the equation.
b. Sketch the two graphs and hence determine the number of roots of the equation $\cos(x) - x^2 = 0$.
c. Determine the value of k for which the equation $\cos(x) = x^2 + k$ has exactly one solution.

16. WE16 a. Use the linear approximation for $\sin(x)$ to evaluate $\sin(1.8°)$ and compare the accuracy of the approximation with the calculator value for $\sin(1.8°)$.

b. Show there is a root to the equation $\cos(x) - 10.5x^2 = 0$ for which $0 \le x \le 0.4$.
c. Use an approximation for $\cos(x)$ to obtain an estimate of the root in part b, expressed to 2 decimal places.

17. a. State an approximate value of $\cos(0.05)$ and compare the value with that given by a calculator.
b. Explain why the quadratic approximation is not applicable for calculating the value of $\cos(5)$.

18. Consider the equation $4x\ \sin(x) - 1 = 0$.

a. Show the equation has a solution for which $0 \le x \le 0.6$ and use the linear polynomial approximation for $\sin(x)$ to estimate this solution.
b. Specify the function whose intersection with $y = \sin(x)$ determines the number of solutions to the equation and hence explain why there would be other positive solutions to the equation.
c. Explain why these other solutions were not obtained using the linear approximation for $\sin(x)$.
d. Analyse the behaviour of the function specified in part b to determine how many solutions of $4x\ \sin(x) - 1 = 0$ lie in the interval $[-4\pi, 4\pi]$.

19. a. i. Use a CAS technology to obtain the two solutions to the equation $\sin(x) = 1 - x^2$.
ii. For the solution closer to zero, compare its value with that obtained using the linear approximation for $\sin(x)$.
b. Investigate over what interval the approximation $\sin(x) = x$ is reasonable.

20. Sketch the graph of $y = \cos^2(x)$ for $x \in [0, 4\pi]$ and state its period.

8.6 Exam questions

Question 1 (1 mark) TECH-ACTIVE

MC State which of the following is not true of both $f(x) = \sin(x)$ and $g(x) = \cos(x)$.

A. The period is 2π.
B. The domain is all real numbers.
C. The range is $-1 \le y \le 1$.
D. They pass through the point $(0, 0)$.
E. The amplitude is 1.

Question 2 (1 mark) TECH-ACTIVE

MC For the domain $-3\pi \le x \le 3\pi$, the values of x for which $\sin(x) = 0$ are

A. $0,\ \pm\pi,\ \pm 2\pi,\ \pm 3\pi$
B. $0,\ \pm\pi,\ \pm 2\pi$
C. $0,\ \pm\frac{\pi}{2},\ \pm\frac{3\pi}{2},\ \pm\frac{5\pi}{2}$
D. $\pm\frac{\pi}{2},\ \pm\frac{3\pi}{2},\ \pm\frac{5\pi}{2}$
E. $0,\ \pm\frac{\pi}{4},\ \pm\frac{3\pi}{4},\ \pm\frac{5\pi}{4}\ \pm\frac{7\pi}{4}$

Question 3 (1 mark) TECH-ACTIVE

MC For the domain $-2\pi \le x \le 3\pi$, the values of x for which $\cos(x) = -1$ are

A. $0,\ \pm 2\pi,$
B. $\pm\frac{3\pi}{2},\ \pm\frac{3\pi}{2},\ \pm\frac{5\pi}{2}$
C. $\pm\frac{\pi}{2},\ \pm\frac{3\pi}{2}$
D. $\pm\pi,\ \pm 3\pi$
E. $\pm\pi,\ 3\pi$

More exam questions are available online.

8.7 Review

 8.7.1 Summary

doc-37024

Hey students! Now that it's time to revise this topic, go online to:

Access the topic summary

Review your results

Watch teacher-led videos

Practise exam questions

Find all this and MORE in jacPLUS

8.7 Exercise

Technology free: short answer

1. Express the following in degree measure.

   a. $\dfrac{11\pi^c}{9}$
   b. $-3.5\pi^c$

2. Evaluate $\cos\left(\dfrac{11\pi}{6}\right) - \tan\left(\dfrac{11\pi}{3}\right) + \sin\left(-\dfrac{11\pi}{4}\right)$ exactly.

3. If $\cos(t) = 0.6$, evaluate the following.

   a. $\cos(-t)$
   b. $\cos(\pi + t)$
   c. $\cos(3\pi - t)$
   d. $\cos(-2\pi + t)$

4. Sketch the following graphs over the given domain.

   a. $y = \sin(x), 0 \le x \le \dfrac{7\pi}{2}$
   b. $y = \cos(x), -\pi \le x \le 3\pi$

5. Use a graphical method to determine the number of roots of the following equations.

   a. $\cos(x) - x = 0$
   b. $\sin(x) - \sqrt{1 - x} = 0$

6. A window ledge 4 metres above the ground can just be reached by a 10-metre ladder. The foot of the ladder makes an angle θ with the ground (assumed horizontal).

   a. Write down the value of $\sin(\theta)$.
   b. Calculate exactly how far up the ladder a person of height 1.8 metres needs to climb in order for the top of the person's head to be level with the window ledge.

Technology active: multiple choice

7. **MC** In a rectangle ABCD, the angle CAD is 27° and the side AD is 3 cm. The length of the diagonal AC in cm is closest to:

   A. 6.61
   B. 5.89
   C. 3.37
   D. 2.67
   E. 1.53

8. **MC** The exact value of $\sin(45°) + \tan(30°) \times \cos(60°)$ is:

   A. $\dfrac{3}{2}$
   B. $\dfrac{\sqrt{2}+1}{2}$
   C. $\dfrac{3\sqrt{2}+\sqrt{3}}{6}$
   D. $\dfrac{3\sqrt{2}+2\sqrt{3}}{12}$
   E. $\dfrac{3\sqrt{6}+6}{12}$

9. **MC** An angle of 100° has a radian equivalent of:

   A. $\dfrac{5\pi}{9}$
   B. $\dfrac{7\pi}{9}$
   C. $\dfrac{2\pi}{3}$
   D. 0.573
   E. 1.8

10. **MC** An arc subtends an angle of 30° at the centre of a circle of radius 3 cm. The length of the arc, in cm, is:

    A. $\dfrac{\pi}{2}$
    B. 90
    C. π
    D. 180
    E. $\dfrac{3\pi}{4}$

Questions 11 and 12 refer to the given unit circle diagram.

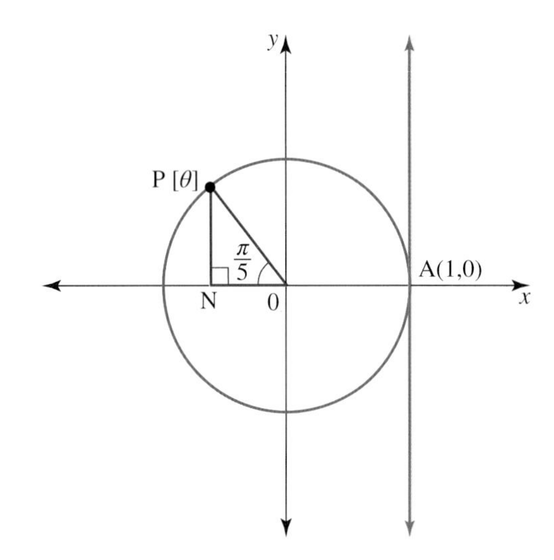

11. **MC** A possible value of θ for the trigonometric point P $[\theta]$ is:

    A. $\frac{\pi}{5}$
    B. $\frac{4\pi}{5}$
    C. $\frac{6\pi}{5}$
    D. $\frac{7\pi}{10}$
    E. $-\pi$

12. **MC** The value of $\sin(\theta)$ is given by the length of the line segment:

    A. OP
    B. PA
    C. ON
    D. NP
    E. AT, where T is the point where PO extended meets the vertical line through A.

13. **MC** Identify the quadrant(s) in which $\sin(\theta) < 0$ and $\cos(\theta) > 0$.

    A. First quadrant
    B. Second quadrant
    C. Third quadrant
    D. Fourth quadrant
    E. Both the second and fourth quadrants

14. **MC** The exact value of $\tan(330°)$ is:

    A. 1
    B. -1
    C. $\sqrt{3}$
    D. $-\sqrt{3}$
    E. $-\frac{\sqrt{3}}{3}$

15. **MC** The exact value of $\cos(-5\pi)$ is:

    A. 1
    B. -1
    C. 0
    D. 5
    E. -5

16. **MC** The value of $\sin(4.5°)$ is approximately equal to:

    A. $\frac{\sqrt{2}}{20}$
    B. 4.5
    C. $\frac{\pi}{40}$
    D. $\frac{\pi}{400}$
    E. 1

Technology active: extended response

17. Two circular pulleys, with radii 3 cm and 8 cm respectively, have their centres 13 cm apart.

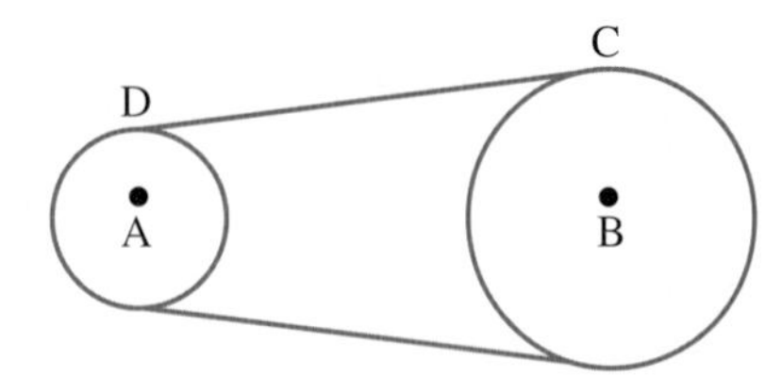

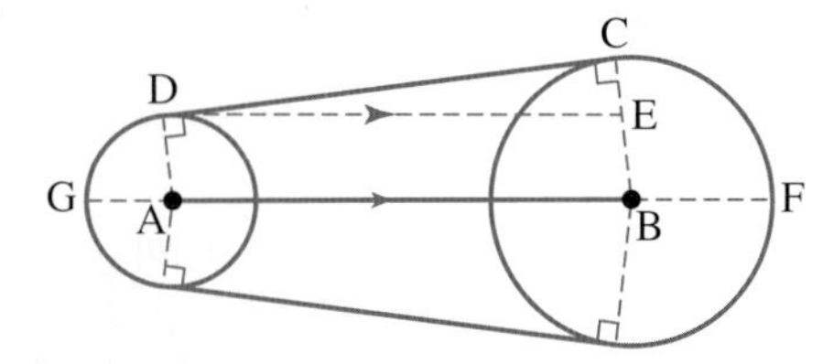

Calculate:

a. the length of DC
b. the angle CED in radians
c. the angle EBF in radians
d. the angle GAD in radians
e. the length of the belt required to pass tightly around the pulleys, giving the answer to 1 decimal place.

18. a. Show on a diagram of the unit circle the positions of -2 and $\frac{3\pi}{4}$.

b. Let P be the trigonometric point $\left[\frac{3\pi}{4}\right]$. Give the exact Cartesian coordinates of point P.

c. Let Q be the trigonometric point $[-2]$. Give the Cartesian coordinates of Q to 2 decimal places.

d. R is the trigonometric point $[\theta]$ where $0 < \theta < \frac{\pi}{2}$. The points P and R are symmetric points. Determine the exact value of θ.

e. i. Determine the exact value for the angle POQ where O is the centre of the unit circle.
ii. Express the angle POQ in degrees, correct to 2 decimal places.

f. Give another angle that would be mapped to the same position as each of -2 and $\frac{3\pi}{4}$ on the circumference of a unit circle.

19. A real estate agent has three land sites for sale. All three sites are triangular in shape.

a. The first site is in the shape of an equilateral triangle of side length $\sqrt{12}$ km. Calculate the exact area of this site.

b. The second site is a triangle ABC where $A = 40°$, $a = 50$ m, $B = 25°$ and $b = 78$ m using the naming convention. Calculate the area of this site, correct to the nearest square metre.

c. The third site is known to be in the shape of an isosceles triangle. The unequal side is 4 km in length and the equal angles are β where $\cos(\beta) = \frac{\sqrt{5}}{3}$.

i. Calculate the exact value of $\tan(\beta)$.
ii. Hence, calculate the exact area of this site.
iii. Express the third angle of the triangle in terms of β.
iv. Use the area measure to calculate the exact value of $\sin(2\beta)$.
v. Use your results to verify that $\sin(2\beta) = 2\sin(\beta)\cos(\beta)$.

20. The diagram shows a sketch of the function for which $y = 2\sin(x) + x - 2$. The graph intersects the x-axis exactly once.

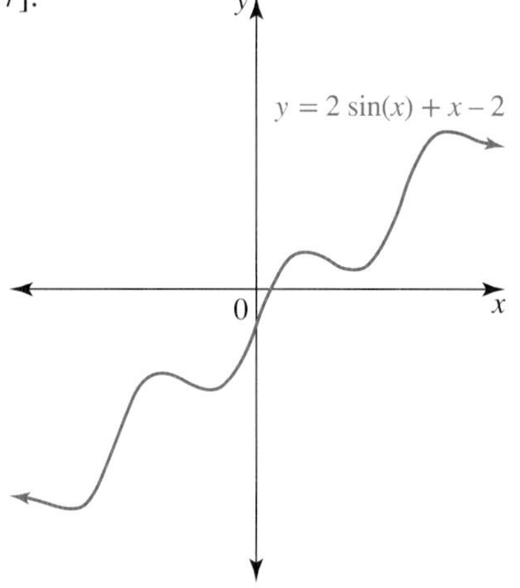

a. Show that the x-intercept of the graph does not lie in the interval $[0, 0.7]$.
b. Determine the value of a, $0 < a < 1$ for which the graph will cross the x-axis in the interval $[a, a + 0.1]$ and state this interval.
c. Given $x = \phi$ is the root of the equation $2\sin(x) + x - 2 = 0$, explain why ϕ lies in the interval $[a, a + 0.1]$.
d. Apply three iterations of the method of bisection to estimate the value of ϕ.
e. Compare the accuracy of the estimated value of ϕ using the method of bisection with that obtained using the linear approximation for $\sin(x)$.
f. The equation $2\sin(x) + x - 2 = 0$ can be solved by obtaining the point of intersection of two graphs, one of which is $y = \sin(x)$.
   i. Determine the equation of the other graph.
   ii. Express the coordinates of the point of intersection of the two graphs as algebraic expressions in terms of ϕ.

8.7 Exam questions

Question 1 (4 marks) TECH-FREE

Evaluate $\dfrac{\tan(30°)\sin(60°)}{\cos(60°)\sin(30°) - \tan(45°)} + \dfrac{\sin(45°)\cos(45°)}{\tan(45°)}$.

Question 2 (1 mark) TECH-ACTIVE

MC Identify which of the following statements is false.

A. $\tan(60°) = \dfrac{\sin(60°)}{\cos(60°)}$
B. $\sin^2(45°) + \cos^2(45°) = 1$
C. $\cos^2(60°) = 1 - \sin^2(60°)$
D. $\tan(90°) = \tan(45°) + \tan(45°)$
E. $\sin(60°) = \cos(30°)$

Question 3 (1 mark) TECH-ACTIVE

MC $\sin(\pi + \theta)$ is equal to

A. $\sin(\theta)$
B. $-\sin(\theta)$
C. $\cos(\theta)$
D. $\tan(\theta)$
E. $\cos(\pi + \theta)$

Question 4 (4 marks) TECH-FREE

Given $\sin(\theta) = 0.61$, $\cos(t) = 0.48$, $\tan(x) = 1.6$ and θ, x, t are all acute, use the symmetry properties to obtain the values of $\sin(\pi + \theta)$, $\cos(\pi - t)$, $\tan(2\pi + x)$ and $\cos(-t)$.

Question 5 (4 marks) TECH-ACTIVE

Sketch the graphs of $y = \cos(x)$ and $y = 3 - x$ on the same set of axes. Explain why the equation $\cos(x) = 3 - x$ has only one root, and give an interval in which the root lies.

More exam questions are available online.

Answers

Topic 8 Trigonometric functions

8.2 Trigonometric ratios

8.2 Exercise

1. a. $h = 7.66$ b. $a \approx 68.20$
2. a. $x = 7.13, y = 3.63$ b. $x = 13.27, h = 16.62$
3. 3.86 m
4. 340 m
5. a. 66.42° b. 51.34°
6. 22°
7. a. $\frac{\sqrt{2}}{2}$ b. $\frac{\sqrt{3}}{3}$ c. $\frac{1}{2}$ d. 1
8. $\frac{3\sqrt{3}}{2}$ m
9. $\frac{\sqrt{3}}{2}$ m
10. $2\sqrt{2}$ metres
11. $\frac{3\sqrt{2} - \sqrt{6}}{8}$
12. $\frac{\sqrt{6}}{12}$
13. a. $\frac{\sqrt{21}}{7}$ b. $\frac{\sqrt{11}}{5}$ c. $\frac{2}{3}$
14. $\sin(\theta) = \frac{2}{7}; \tan(\theta) = \frac{2\sqrt{5}}{15}$
15. a. $\cos(a) = \frac{3}{\sqrt{13}}$ b. $6\sqrt{13}$ cm
16. 36 cm
17. 12.25 m^2
18. a. $22\sqrt{3}$ sq units b. 10 sq units
19. 30 cm^2
20. a. 6.18 metres b. 59.5°; 5.3 metres
21. 21.6°
22. CA = CB = 16.18 cm; ∠CBA = ∠CAB = 72°; ∠ACB = 36°
23. $AC = 12\sqrt{3}$ cm; $BC = 18\sqrt{2}$ cm; $AB = 18 + 6\sqrt{3}$ cm
24. a. 4.275 cm^2
    b. 1.736 cm
25. $20\sqrt{3}$ cm; $\frac{100\sqrt{3}}{3}$ cm^2

8.2 Exam questions

Note: Mark allocations are available with the fully worked solutions online.

1. a. 3.92 cm b. $\frac{9\sqrt{2}}{2}$ cm^2
2. C
3. B

8.3 Circular measure

8.3 Exercise

1. a.

Degrees	30°	45°	60°
Radians	$\frac{\pi}{6}$	$\frac{\pi}{4}$	$\frac{\pi}{3}$

b.

Degrees	0°	90°	180°	270°	360°
Radians	0	$\frac{\pi}{2}$	π	$\frac{3\pi}{2}$	2π

2. a. 36° b. 120° c. 75°
   d. 330° e. 140° f. 810°
3. a. $\frac{2\pi}{9}$ b. $\frac{5\pi}{6}$ c. $\frac{5\pi}{4}$
   d. $\frac{5\pi}{3}$ e. $\frac{7\pi}{4}$ f. 4π
4. a. $\frac{\pi}{3}$ b. 135° c. 30°; $\frac{\sqrt{3}}{3}$
5. 10 cm
6. 12 mm
7. a. $\frac{\pi^c}{5}$ b. $\frac{7\pi}{5}$ cm
8. a. $\frac{\pi}{8}$ b. 22.5°
9. a. 10π cm b. 5π cm
10. a. i. 0.052 ii. 1.959 iii. 3.759
    b. i. 171.887° ii. 414°
    c. $\left\{\frac{\pi^c}{7}, 50°, 1.5^c\right\}$
11. 2.53
12. a. 103.1°
    b.

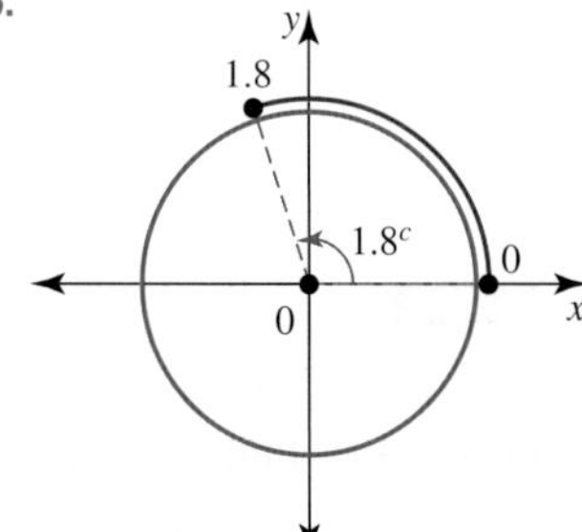

13. a.

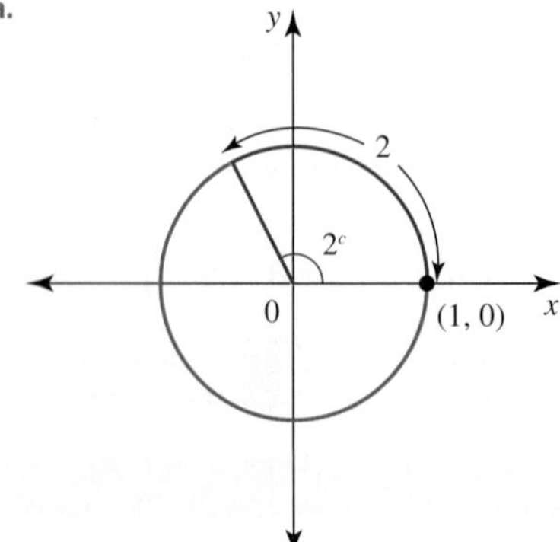

b.

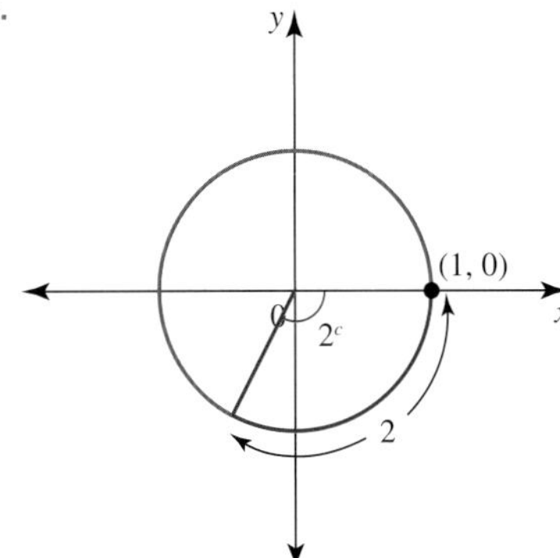

c.

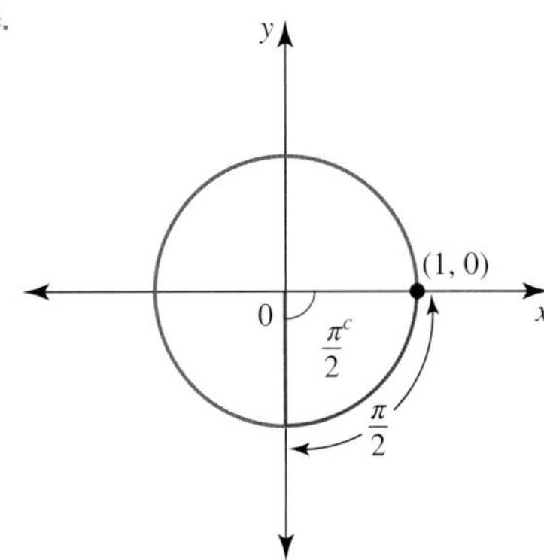

14. a. 10.47 cm b. 216°

15. 17.2°

16. 191° or $\frac{10^c}{3}$

17. 5.76 mm

18. a. 2.572 b. 0.021

19. a. i. 1.557 ii. 0.623 iii. 0.025

b.

θ	$\frac{\pi}{6}$	$\frac{\pi}{4}$	$\frac{\pi}{3}$
$\sin(\theta)$	$\frac{1}{2}$	$\frac{\sqrt{2}}{2}$	$\frac{\sqrt{3}}{2}$
$\cos(\theta)$	$\frac{\sqrt{3}}{2}$	$\frac{\sqrt{2}}{2}$	$\frac{1}{2}$
$\tan(\theta)$	$\frac{\sqrt{3}}{3}$	1	$\sqrt{3}$

20. $6\sqrt{3}-2$

8.3 Exam questions

Note: Mark allocations are available with the fully worked solutions online.

1. E
2. B
3. D

8.4 Unit circle definitions

8.4 Exercise

1. a. Quadrant 1 b. Quadrant 3
   c. Quadrant 2 d. Quadrant 1
   e. Quadrant 4 f. Quadrant 3

2. a. $\frac{\pi}{3}, -\frac{11\pi}{6}$ b. $\frac{3\pi}{4}, -\frac{5\pi}{4}$
   c. $\frac{7\pi}{6}$ d. $-\frac{\pi}{3}$

3. a. i. (0, 1) ii. $\sin(\theta) = 1$
   b. i. $(-1, 0)$ ii. $\cos(\alpha) = -1$
   c. i. $(0, -1)$ ii. $\tan(\beta)$ is undefined.
   d. i. (1, 0)
   ii. $\sin(\nu) = 0$, $\cos(\nu) = 1$, $\tan(\nu) = 0$

4. a. Quadrant 3
   b. Quadrant 2
   c. Boundary of quadrants 1 and 4
   d. Quadrant 4

5. a. $P\left[\frac{\pi}{2}\right]$

b.

120°	$-400° = -360° - 40°$	$\frac{4\pi}{3} = \pi + \frac{1}{3}\pi$	$\frac{\pi}{4}$
Quadrant 2	Quadrant 4	Quadrant 3	Quadrant 1

c. Q[−240°]; R[480°]

6. $P\left[\frac{3\pi}{2}\right]$ or $P\left[-\frac{\pi}{2}\right]$

7. a. $\left(\frac{\sqrt{3}}{2}, \frac{1}{2}\right)$

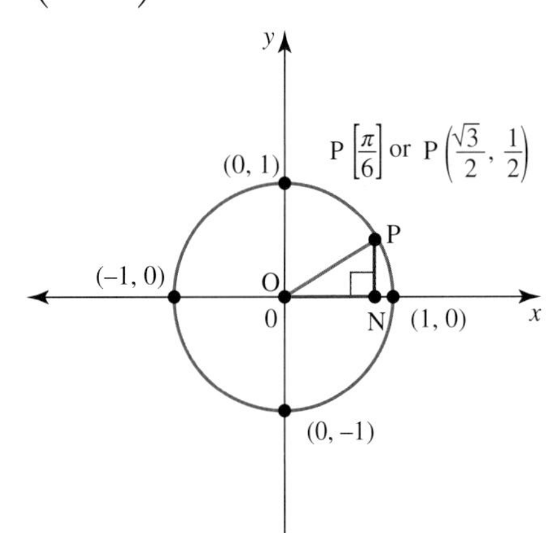

b.

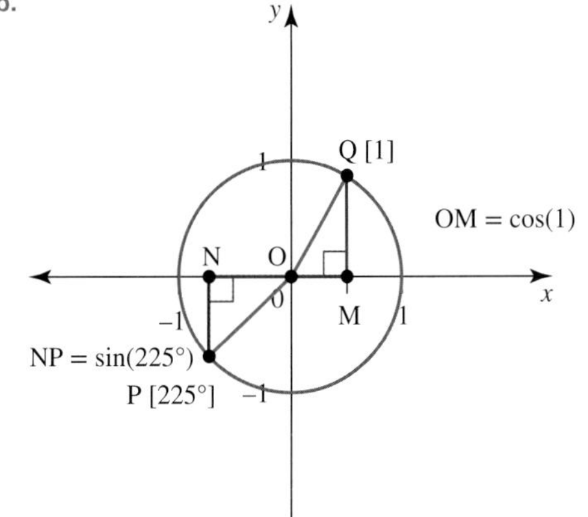

c. $\cos\left(-\frac{\pi}{2}\right) = 0; \sin\left(-\frac{\pi}{2}\right) = -1$

d. $f(0) = 0$

8. a. $f\left(\frac{3\pi}{2}\right) = \sin\left(\frac{3\pi}{2}\right) = -1$

b. $g(4\pi) = \cos(4\pi) = 1$

c. $h(-\pi) = \tan(-\pi) = 0$

d. $k(6.5\pi) = \sin\left(\frac{13\pi}{2}\right) + \cos\left(\frac{13\pi}{2}\right) = 1$

9. a. First and second quadrants

b. First and fourth quadrants

10. a. $\left(\frac{\sqrt{2}}{2}, \frac{\sqrt{2}}{2}\right)$ **b.** $\left[\frac{3\pi}{2}\right]$ or $\left[-\frac{\pi}{2}\right]$

11.

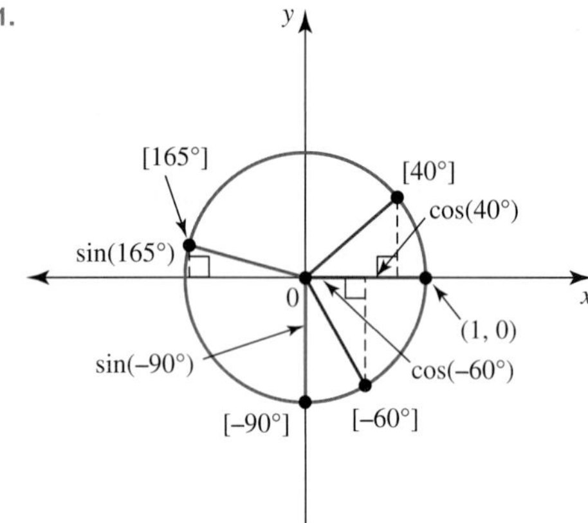

12.

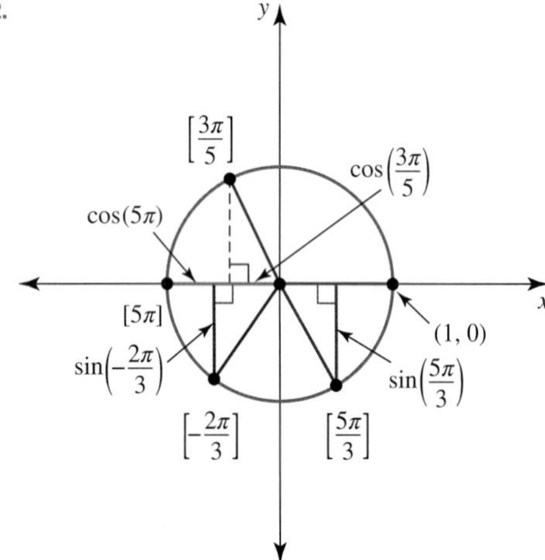

13.

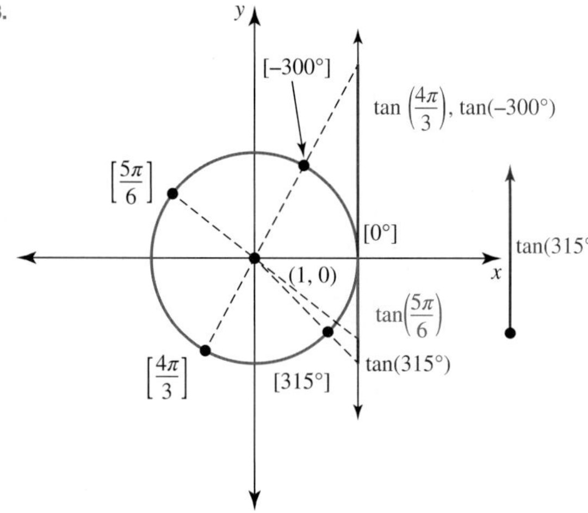

14. a. Quadrant 2; $\sin(\theta) = 0.6, \cos(\theta) = -0.8,$ $\tan(\theta) = -0.75$

b. Quadrant 4; $\sin(\theta) = -\frac{\sqrt{2}}{2}, \cos(\theta) = \frac{\sqrt{2}}{2}, \tan(\theta) = -1$

c. Quadrant 1; $\sin(\theta) = \frac{1}{\sqrt{5}}, \cos(\theta) = \frac{2}{\sqrt{5}}, \tan(\theta) = \frac{1}{2}$

d. Boundary between quadrants 1 and 2; $\sin(\theta) = 1, \cos(\theta) = 0, \tan(\theta)$ undefined

15. a. 1 **b.** 1 **c.** 0

d. 0 **e.** 0 **f.** 0

16. a. $\tan(230°) \approx 1.192$

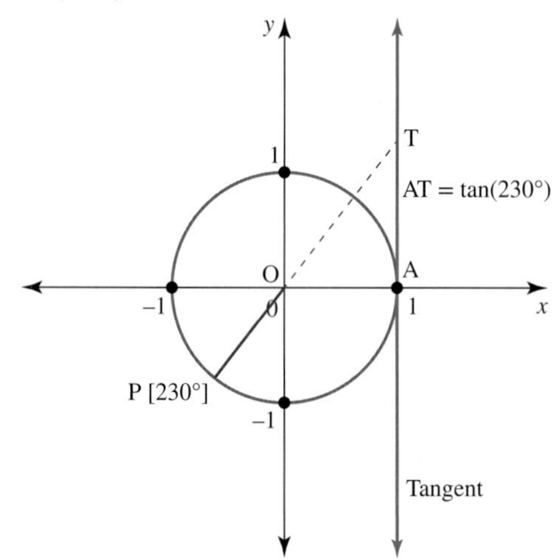

b. $\tan(2\pi) = 0$

17. a. $\tan\left(\frac{5\pi}{2}\right)$; $\tan(-90°)$ **b.** $\tan\left(\frac{3\pi}{4}\right)$

18. a.

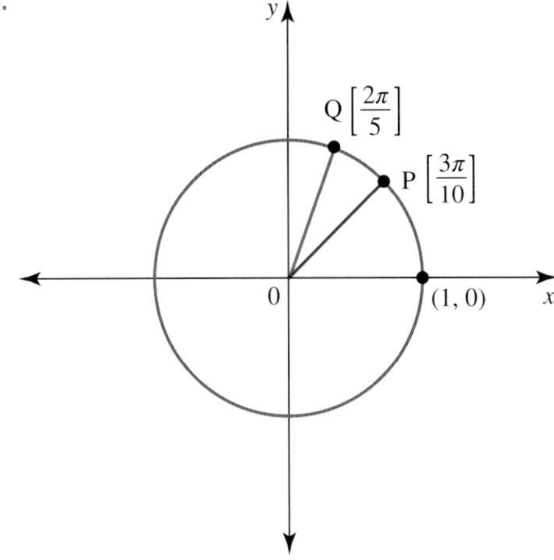

b. $\frac{\pi}{10}$

c. P $\left[-\frac{17\pi}{10}\right]$; Q $\left[-\frac{8\pi}{5}\right]$ (other answers are possible)

d. P $\left[\frac{23\pi}{10}\right]$; Q $\left[\frac{12\pi}{5}\right]$ (other answers are possible)

19. a.

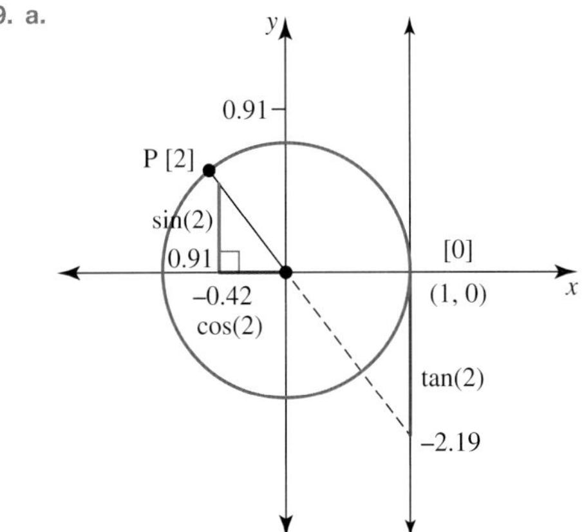

b. $(-0.42, 0.91)$

20. a. 1

b. $-\frac{\sqrt{3}}{2} - \frac{1}{2}$

c. 1

d. 1

e. 1; sample responses can be found in the worked solutions in the online resources.

8.4 Exam questions

Note: Mark allocations are available with the fully worked solutions online.

1. C
2. A
3. C

8.5 Symmetry properties

8.5 Exercise

1. a. $\frac{\sqrt{3}}{2}$ b. $\frac{\sqrt{3}}{3}$ c. $-\frac{\sqrt{2}}{2}$
   d. $\frac{1}{2}$ e. $-\frac{\sqrt{3}}{3}$ f. $-\frac{\sqrt{2}}{2}$
2. a. $-\frac{\sqrt{3}}{2}$ b. $-\frac{\sqrt{3}}{2}$ c. $-\frac{\sqrt{3}}{3}$
   d. -1 e. $-\frac{1}{2}$ f. $\frac{1}{2}$
3. a. $\sqrt{3}$ b. $\frac{\sqrt{2}}{2}$ c. $-\frac{1}{2}$
   d. $\frac{\sqrt{2}}{2}$ e. $-\frac{\sqrt{3}}{2}$ f. -1
4. a. $-\frac{1}{2}$ b. 1 c. $-\frac{1}{2}$
   d. $-\sqrt{3}$ e. $\frac{\sqrt{2}}{2}$ f. $\frac{1}{2}$
5. a. $\sqrt{3}$ b. $-\sqrt{3}$ c. $\frac{1}{2}$
   d. $\frac{1}{2}$ e. $\frac{\sqrt{2}}{2}$ f. $-\frac{\sqrt{2}}{2}$
6. a. $-\frac{\sqrt{3}}{2}$ b. $\frac{\sqrt{3}}{2}$ c. 1
   d. $\frac{\sqrt{3}}{3}$ e. $\frac{\sqrt{2}}{2}$ f. $-\frac{1}{2}$
7. a. $\frac{\sqrt{2}}{2}$ b. $-\sqrt{3}$ c. $-\frac{\sqrt{3}}{2}$
   d. $-\frac{1}{2}$ e. $\frac{\sqrt{3}}{3}$ f. $-\frac{1}{2}$
8. a. $\frac{\sqrt{2}}{2}$ b. $-\frac{\sqrt{3}}{2}$ c. $\frac{\sqrt{3}}{3}$
   d. $\frac{\sqrt{3}}{2}$ e. $\frac{\sqrt{2}}{2}$ f. $-\frac{\sqrt{3}}{3}$
9. a. 1 b. 0 c. 0
   d. 1 e. 0 f. 0
10. a. Fourth
    b. First
    c. Second
    d. Boundary between quadrants 1 and 2, and boundary between quadrants 3 and 4
    e. Boundary between quadrants 1 and 2
    f. Boundary between quadrants 2 and 3
11. a. Third quadrant
    b. 0
12. a. $-\frac{\sqrt{3}}{2}$ b. $-\frac{\sqrt{3}}{3}$ c. $\frac{\sqrt{3}}{2}$

13. $\sin\left(-\frac{5\pi}{4}\right) = \frac{\sqrt{2}}{2}$; $\cos\left(-\frac{5\pi}{4}\right) = -\frac{\sqrt{2}}{2}$;
$\tan\left(-\frac{5\pi}{4}\right) = -1$

14. a. -0.2 b. -0.2 c. 0.2 d. 0.2

15. a. -4 b. 0.9 c. -4 d. 3.1

16. a. -0.91 b. 0.43 c. -0.47
d. 0.91 e. -0.43 f. 0.47

17. a. $-p$ b. p c. $-p$ d. p

18. a. $[105°]$; $[255°]$; $[285°]$;
$\cos(285°) = \cos(75°)$; $[285°]$

b. $-\tan\left(\frac{\pi}{7}\right)$

c. $\sin(\pi - \theta) = 0.8, \sin(2\pi - \theta) = -0.8$

d. i. $\cos\left(\frac{5\pi}{4}\right) = -\frac{\sqrt{2}}{2}$

ii. $\sin\left(\frac{25\pi}{6}\right) = \frac{1}{2}$

19. a. Sample responses can be found in the worked solutions in the online resources.

b. Same x-coordinates

c. $\sin(\pi + \phi) = -0.87, \cos(\pi + \phi) = -0.5$,
$\tan(\pi + \phi) = -1.74$

d. $\sin(t)$

e. $A = 36°$ or $-216°$ (other answers are possible)

20. a. Quadrants 1 or 3

b. Quadrant 1, $\theta = 1.3734$; quadrant 3, $\theta = 4.5150$

c. Quadrant 1, $\cos(\theta) = \frac{\sqrt{26}}{26}$, $\sin(\theta) = \frac{5\sqrt{26}}{26}$,
$\left(\frac{\sqrt{26}}{26}, \frac{5\sqrt{26}}{26}\right)$; or quadrant 3, $\cos(\theta) = -\frac{\sqrt{26}}{26}$,
$\sin(\theta) = -\frac{5\sqrt{26}}{26}$, $\left(-\frac{\sqrt{26}}{26}, -\frac{5\sqrt{26}}{26}\right)$

8.5 Exam questions

Note: Mark allocations are available with the fully worked solutions online.

1. B

2. D

3. $\frac{4+\sqrt{3}}{2}$

8.6 Graphs of the sine and cosine functions

8.6 Exercise

1. a. Domain $[0, 4\pi]$, range $[-1, 1]$

b. $y = \sin(x)$

c. $\left(\frac{\pi}{2}, 1\right), \left(\frac{3\pi}{2}, -1\right), \left(\frac{5\pi}{2}, 1\right), \left(\frac{7\pi}{2}, -1\right)$.

d. Period 2π, amplitude 1

e. $y = 0$ (the x-axis)

f. $x \in (0, \pi) \cup (2\pi, 3\pi)$.

2. a. Domain $[-2\pi, 2\pi]$, range $[-1, 1]$

b. $y = \cos(x)$

c. $(-\pi, -1), (\pi, -1)$

d. Period 2π, amplitude 1, mean position $y = 0$ (the x-axis)

e. $\left(-\frac{3\pi}{2}, 0\right), \left(-\frac{\pi}{2}, 0\right), \left(\frac{\pi}{2}, 0\right), \left(\frac{3\pi}{2}, 0\right)$

f. $x \in \left(-\frac{3\pi}{2}, -\frac{\pi}{2}\right) \cup \left(\frac{\pi}{2}, \frac{3\pi}{2}\right)$.

3. a.

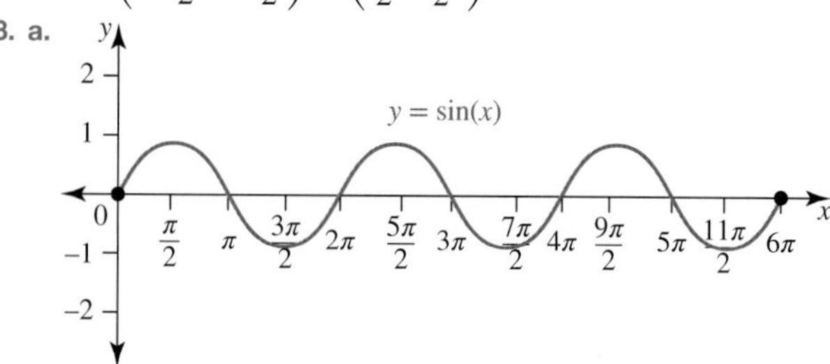

b.

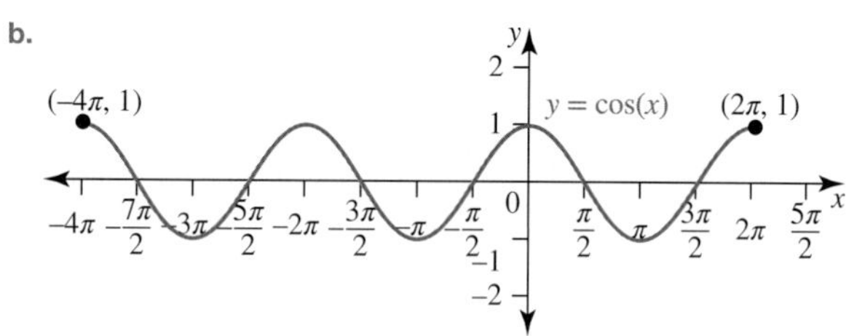

c.

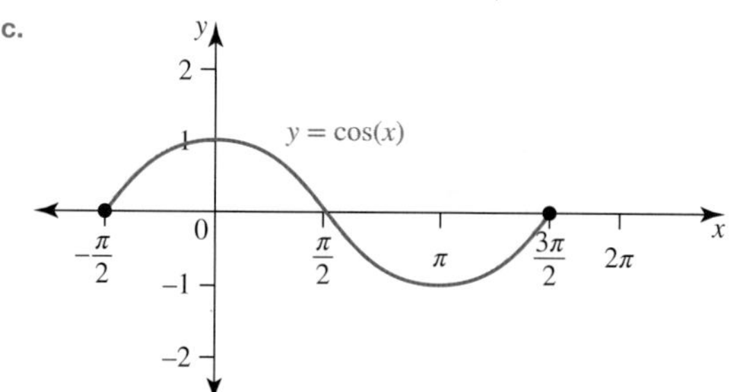

d.

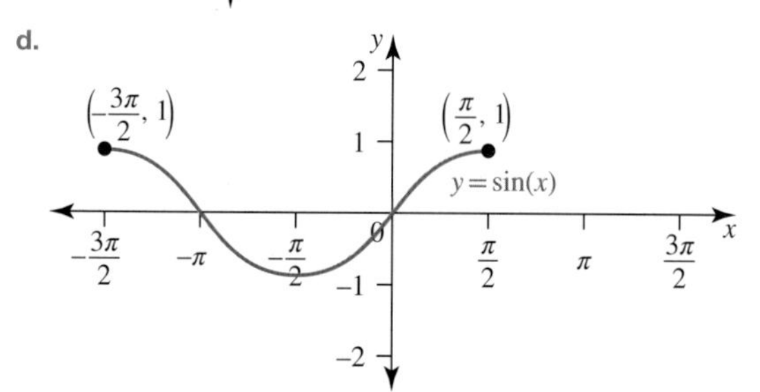

4. Three cycles

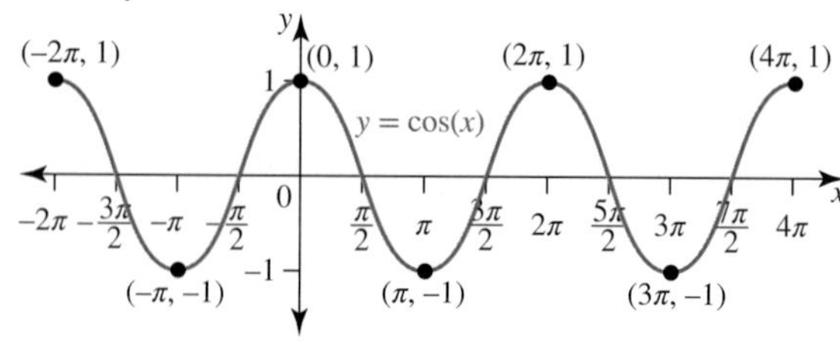

5. a. 2 maximum turning points

b. 7 minimum turning points

6. a. 4 b. 7 c. 21
d. 3

7.

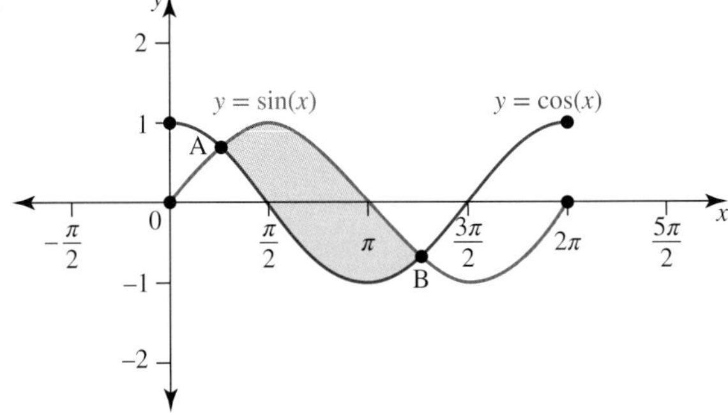

The region required lies below the sine graph and above the cosine graph between their points of intersection.

8. a. $\frac{\pi}{2} < x < \frac{3\pi}{2}$

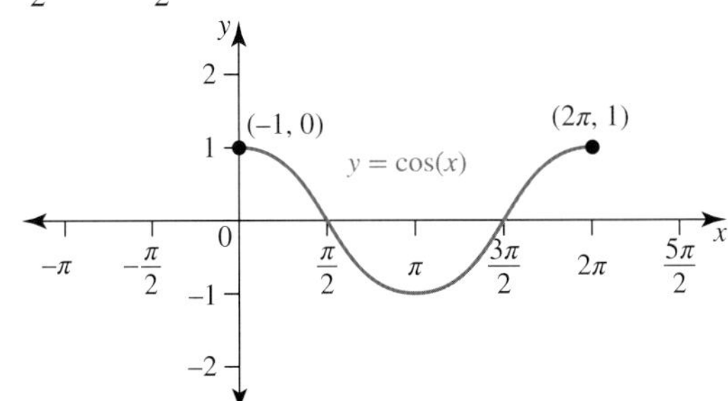

b. Sample responses can be found in the worked solutions in the online resources.

9. $a = \frac{19\pi}{2}$

10. a. $\frac{\pi}{12}$ b. $\frac{\pi}{72}$

11. a. $y = 11x - 4$ b. One solution c. $x \approx 0.4$

12. a. $\frac{\pi}{180}$

b. $\frac{\pi}{9}$

c. $-\frac{\pi}{90}$

d. $\frac{\pi}{6}$; the approximation becomes less accurate as the size of x increases.

13. a. $x = 0.2$

14. a.

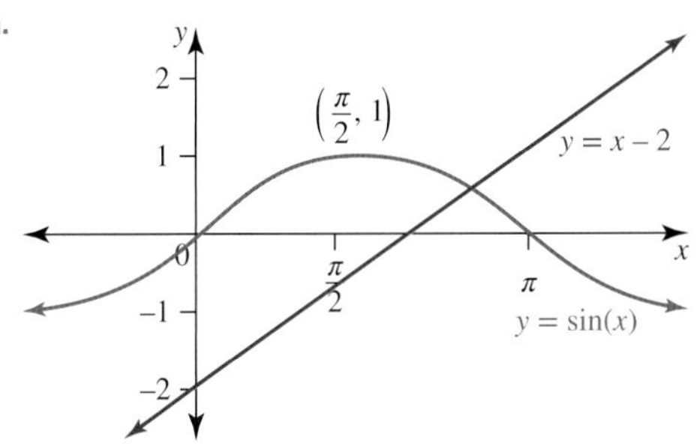

Only one point of intersection of the graphs

b. Between $x = 2$ and $x = 3$

c. $[2.5, 3]$; $[2.5, 2.75]$; $x = 2.625$

15. a. $y = \cos(x)$ and $y = x^2$

b. Two solutions

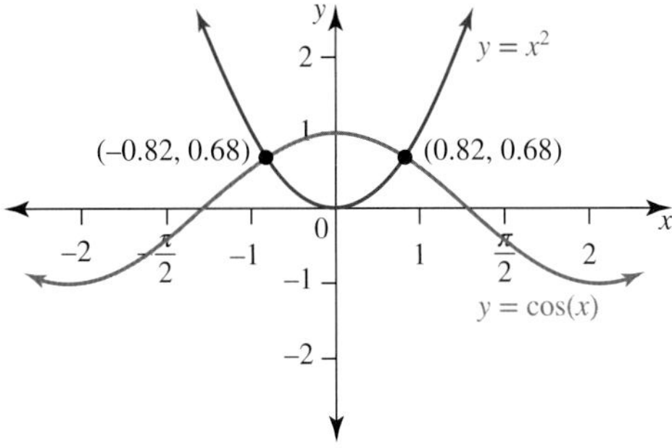

c. $k = 1$

16. a. $\sin(1.8°) \approx 0.01\pi$, accurate to 4 decimal places.

b. Sample responses can be found in the worked solutions in the online resources.

17. a. $\cos(0.05) \approx 1$; calculator gives 0.9987.

b. The approximation is only applicable for small numbers close to zero.

18. a. Sample responses can be found in the worked solutions in the online resources; $x = 0.5$.

b. $y = \frac{1}{4x}$; sample responses can be found in the worked solutions in the online resources.

c. They did not have small solution values for x.

d. 8

19. a. i. $x = -1.4096$ or $x = 0.6367$

ii. $x = 0.618$

b. $-0.6 \le x \le 0.6$

20. π

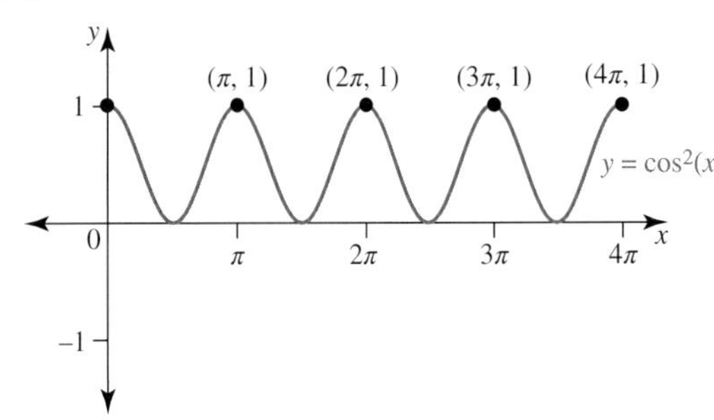

8.6 Exam questions

Note: Mark allocations are available with the fully worked solutions online.

1. D 2. A 3. E

8.7 Review

8.7 Exercise

Technology free: short answer

1. a. 220° b. −630°

2. $\frac{3\sqrt{3} - \sqrt{2}}{2}$

3. a. 0.6 b. −0.6 c. −0.6 d. 0.6

4. a.

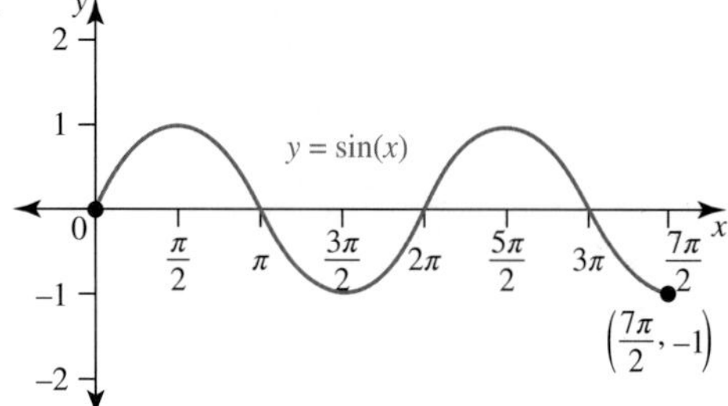

b.

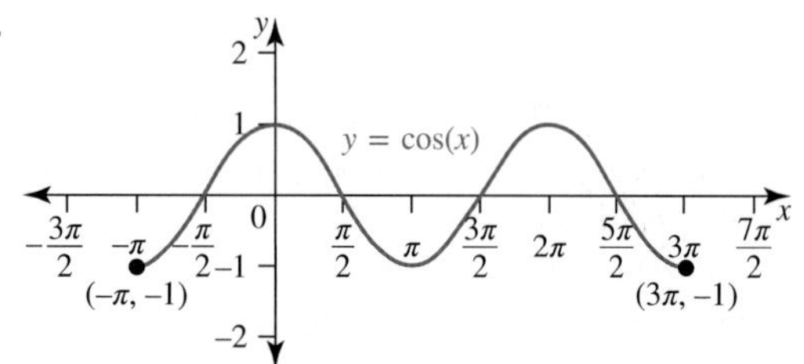

5. a.

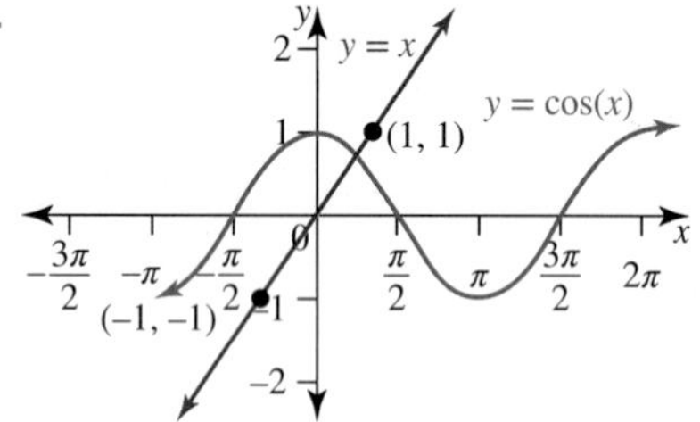

One solution

b.

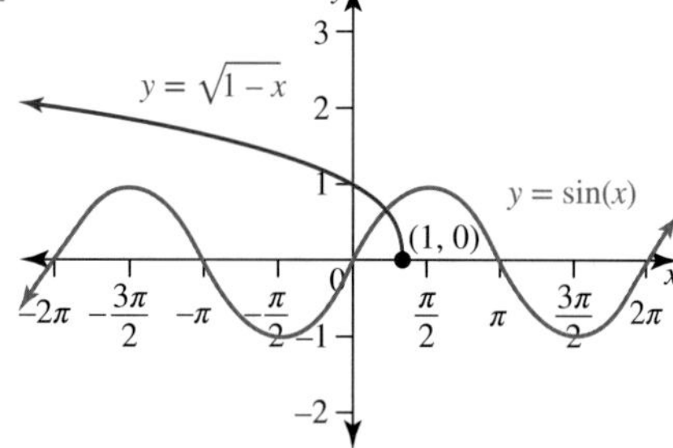

One solution

6. a. 0.4 b. 5.5 metres

Technology active: multiple choice

7. C 8. C 9. A 10. A 11. B
12. D 13. D 14. E 15. B 16. C

Technology active: extended response

17. a. 12 cm
b. 1.176^c
c. 1.966^c
d. 1.176^c
e. 62.5 cm

18. a.

b. $\left(-\frac{\sqrt{2}}{2}, \frac{\sqrt{2}}{2}\right)$

c. $(-0.42, -0.91)$

d. $\theta = \frac{\pi}{4}$

e. i. $\left(\frac{5\pi}{4} - 2\right)$ radians

ii. 110.41°

f. $-2\pi - 2, \frac{11\pi}{4}$ (other answers are possible)

19. a. $3\sqrt{3}$ km^2

b. 1767 m^2

c. i. $\tan(\beta) = \frac{2\sqrt{5}}{5}$

ii. $\frac{8\sqrt{5}}{5}$ km^2

iii. $180^\circ - 2\beta$ or $(\pi - 2\beta)$ radians

iv. $\sin(2\beta) = \frac{4\sqrt{5}}{9}$

v. Referring to the triangle drawn in part i:

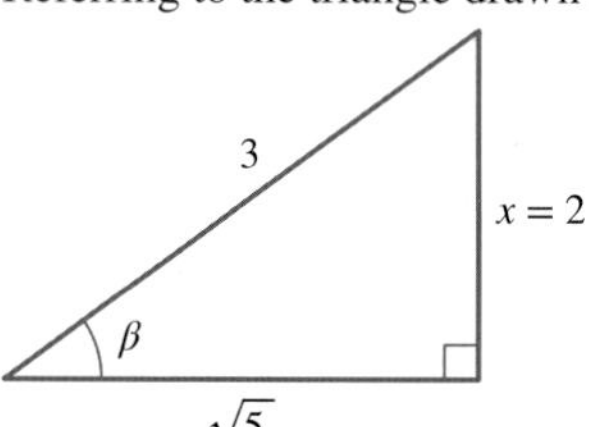

$\cos(\beta) = \frac{\sqrt{5}}{3}$ and $\sin(\beta) = \frac{2}{3}$

From part iv, $\sin(2\beta) = \frac{4\sqrt{5}}{9}$.

Substitute these values in each side of $\sin(2\beta) = 2\sin(\beta)\cos(\beta)$.

$$\begin{aligned}\text{RHS} &= 2 \times \frac{2}{3} \times \frac{\sqrt{5}}{3} \\ &= \frac{4\sqrt{5}}{9} \\ &= \text{LHS}\end{aligned}$$

The result is verified.

20. a. Let $f(x) = 2\sin(x) + x - 2$.

$$\begin{aligned}f(0) &= 2\sin(0) + (0) - 2 \\ &= -2 \\ &< 0 \\ f(0.7) &= 2\sin(0.7) + (0.7) - 2 \\ &= -0.01156 \\ &< 0\end{aligned}$$

$f(x)$ has not changed its sign. Since the sign is negative, the graph is still below the x-axis between $x = 0$ and $x = 0.7$.

b. $a = 0.7$, $[0.7, 0.8]$

c. When the graph crosses the x-axis, its y-coordinate is zero.
Also, when $y = 0$, $0 = 2\sin(x) + x - 2$.
Given $x = \phi$ is the root of this equation, $x = \phi$ is where the graph crosses the x-axis.
Therefore, ϕ lies in the interval $[0.7, 0.8]$.

d. $\phi = 0.7125$

e. $\phi = \frac{2}{3}$, which does not lie in $[0.7, 0.8]$.

f. i. $y = -\frac{1}{2}x + 1$

ii. $\left(\phi, -\frac{1}{2}\phi + 1\right)$

8.7 Exam questions

Note: Mark allocations are available with the fully worked solutions online.

1. $-\frac{1}{6}$
2. D
3. B
4. $\sin(\pi + \theta) = -0.61$
$\cos(\pi - t) = -0.48$
$\tan(2\pi + x) = 1.6$
$\cos(-t) = 0.48$

5.

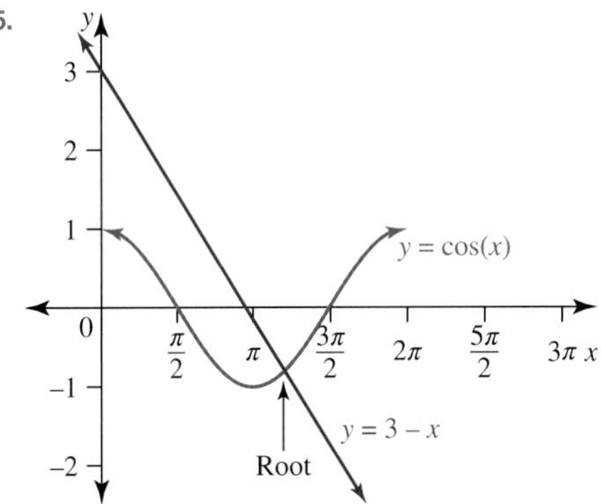

$\left[\pi, \frac{3\pi}{2}\right]$

9 Trigonometric functions and applications

LEARNING SEQUENCE

Fully worked solutions for this topic are available online.

9.1 Overview

Hey students! Bring these pages to life online

Watch videos

Engage with interactivities

Answer questions and check results

Find all this and MORE in jacPLUS

9.1.1 Introduction

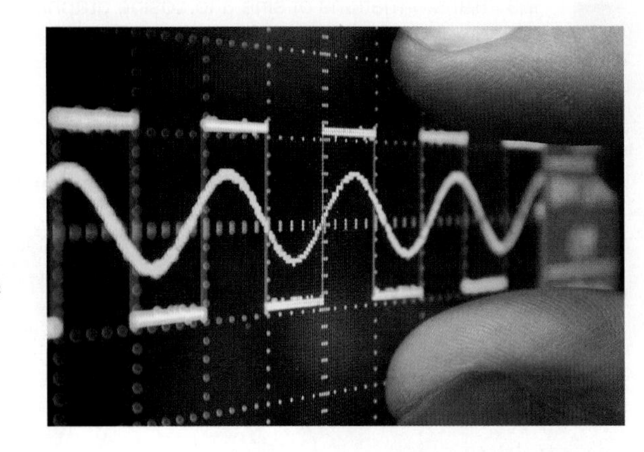

Trigonometry was once thought of as an independent branch of mathematics, the one associated with the solution of triangles. From possibly the beginning of the 17th century, the analytic functionality of the trigonometric or circular functions has become central to mainstream mathematics. The trigonometric functions sine, cosine and tangent, together with their reciprocal functions and their inverse functions, have a critical place in higher-level mathematics.

There are many and various identities and relationships that exist between the trigonometric functions. The French mathematician François Viète discovered many of these relationships and was among the first to recognise their algebraic potential. He applied trigonometric substitutions to the solution of cubic equations, for example, and he was aware of the relationships and links with logarithms, exponentials and possibly complex numbers that Euler and others developed further.

Viète ushered in the beginning of what is referred to as modern mathematics. He improved notation and recognised mathematics as a form of reasoning and generality, rather than a set of clever tricks. His work paved the way for the later application of trigonometric functions in mathematical analysis and in the calculus of Newton and Leibniz.

KEY CONCEPTS

This topic covers the following key concepts from the VCE Mathematics Study Design:

- the relationships $\sin(x) \approx x$ for small values of x, $\sin^2(x) + \cos^2(x) = 1$ and $\tan(x) = \dfrac{\sin(x)}{\cos(x)}$
- exact values for sine, cosine and tangent of $\dfrac{n\pi}{6}$ and $\dfrac{n\pi}{4}$, $n \in Z$
- symmetry properties, complementary relations and periodicity properties for sine, cosine and tangent functions
- circular functions of the form $y = Af(nx) + c$ and their graphs, where f is the sine, cosine or tangent function, and $A, n, c \in R$ with $A, n \neq 0$
- simple applications of sine and cosine functions of the above form, with examples from various modelling contexts, the interpretation of period, amplitude and mean value in these contexts and their relationship to the parameters A, n and c
- use of inverse functions and transformations to solve equations of the form $Af(nx) + c = k$, where $A, n, c, k \in R$ for $A, n \neq 0$, and f is sine, cosine, tangent or a^x, using exact or approximate values on a given domain.

Note: *Concepts shown in grey are covered in other topics.*

9.2 Trigonometric equations

LEARNING INTENTION

At the end of this subtopic you should be able to:
- solve simple trigonometric equations over a specified interval.

The symmetry properties of trigonometric functions can be used to obtain solutions to equations of the form $f(x) = a$ where f is sine, cosine or tangent. If f is sine or cosine, then $-1 \leq a \leq 1$. If f is tangent, then $a \in R$.

Once the appropriate base value of the first quadrant is known, symmetric points in any other quadrant can be obtained. However, there are many values generated by both positive and negative rotations that can form these symmetric quadrant points. Consequently, the solution of a trigonometric equation such as $\sin(x) = a, x \in R$ would have infinite solutions. We shall consider trigonometric equations in which a subset of R is specified as the domain in order to have a finite number of solutions.

9.2.1 Solving trigonometric equations on finite domains

To solve the basic type of equation $\sin(x) = a, 0 \leq x \leq 2\pi$, perform the following steps:
- Identify the quadrants in which solutions lie from the sign of a.
  - If $a > 0$, x must lie in quadrants 1 and 2 where sine is positive.
  - If $a < 0$, x must be in quadrants 3 and 4 where sine is negative.
- Obtain the base value, or first-quadrant value, by solving $\sin(x) = a$ if $a > 0$ or ignoring the negative sign if $a < 0$ (to ensure the first-quadrant value is obtained).
  - This may require recognition of an exact value ratio or it may require the use of a calculator.
- Once obtained, use the base value to generate the values for the quadrants required from their symmetric forms.

The basic equations $\cos(x) = a$ or $\tan(x) = a$, $0 \leq x \leq 2\pi$ are solved in a similar manner, with the sign of a determining the quadrants in which solutions lie.
- For $\cos(x) = a$:
  - if $a > 0$, x must lie in quadrants 1 and 4 where cosine is positive;
  - if $a < 0$, x must lie in quadrants 2 and 3 where cosine is negative.
- For $\tan(x) = a$:
  - if $a > 0$, x must lie in quadrants 1 and 3 where tangent is positive;
  - if $a < 0$, x must lie in quadrants 2 and 4 where tangent is negative.

Symmetric forms

For one positive and one negative rotation, the symmetric points to the first-quadrant base are shown in the diagrams.

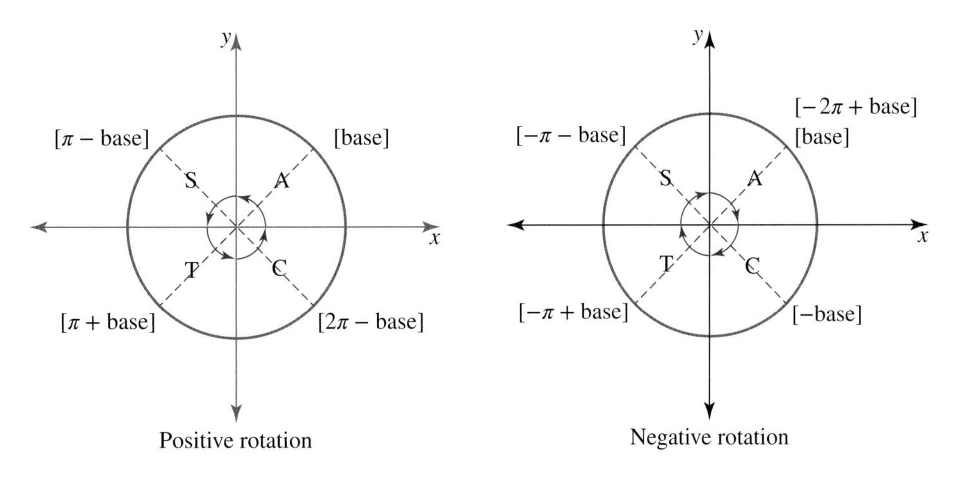

WORKED EXAMPLE 1 Solving equations with exact values

Solve the following equations to obtain exact values for x.

a. $\sin(x) = \dfrac{\sqrt{3}}{2}, 0 \le x \le 2\pi$

b. $\sqrt{2}\cos(x) + 1 = 0, 0 \le x \le 2\pi$

c. $\sqrt{3} - 3\tan(x) = 0, -2\pi \le x \le 2\pi$

THINK	WRITE
a. 1. Identify the quadrants in which the solutions lie.	**a.** $\sin(x) = \dfrac{\sqrt{3}}{2}, 0 \le x \le 2\pi$ Sine is positive in quadrants 1 and 2.
	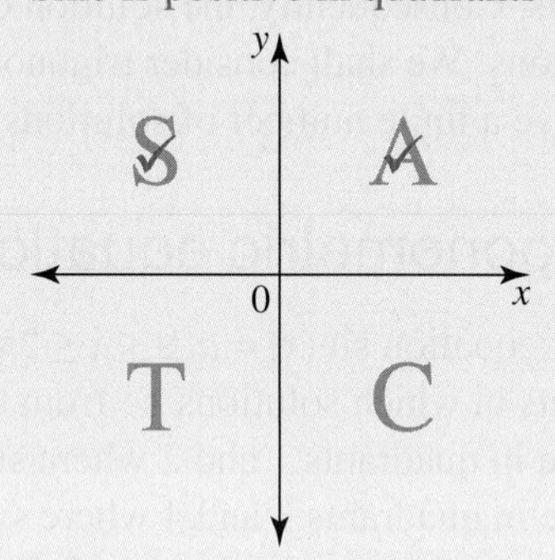
2. Use knowledge of exact values to state the first-quadrant base.	The base is $\dfrac{\pi}{3}$ since $\sin\left(\dfrac{\pi}{3}\right) = \dfrac{\sqrt{3}}{2}$.
3. Generate the solutions using the appropriate quadrant forms.	Since $x \in [0, 2\pi]$, there will be two positive solutions, one from quadrant 1 and one from quadrant 2. $\therefore x = \dfrac{\pi}{3}$ or $x = \pi - \dfrac{\pi}{3}$
4. Calculate the solutions from their quadrant forms.	$\therefore x = \dfrac{\pi}{3}$ or $\dfrac{2\pi}{3}$
b. 1. Rearrange the equation so the trigonometric function is isolated on one side.	**b.** $\sqrt{2}\cos(x) + 1 = 0, 0 \le x \le 2\pi$ $\sqrt{2}\cos(x) = -1$ $\cos(x) = -\dfrac{1}{\sqrt{2}}$
2. Identify the quadrants in which the solutions lie.	Cosine is negative in quadrants 2 and 3.
	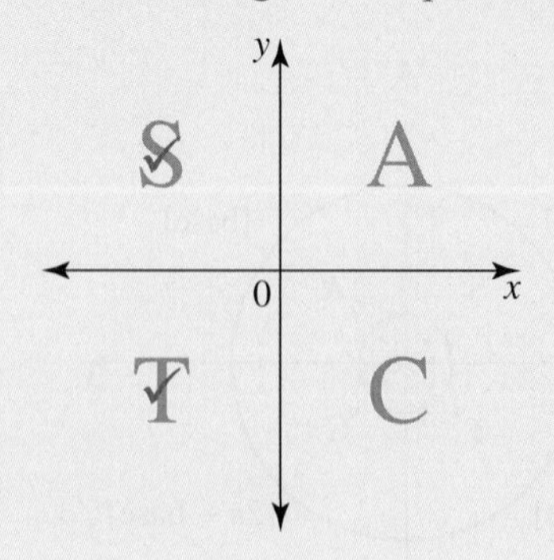

3. Identify the base.
Note: The negative sign is ignored in identifying the base since the base is the first-quadrant value.

Since $\cos\left(\frac{\pi}{4}\right) = \frac{1}{\sqrt{2}}$, the base is $\frac{\pi}{4}$.

4. Generate the solutions using the appropriate quadrant forms.

Since $x \in [0, 2\pi]$, there will be two positive solutions.
$\therefore x = \pi - \frac{\pi}{4}$ or $x = \pi + \frac{\pi}{4}$

5. Calculate the solutions from their quadrant forms.

$\therefore x = \frac{3\pi}{4}$ or $\frac{5\pi}{4}$

c. 1. Rearrange the equation so the trigonometric function is isolated on one side.

c. $\sqrt{3} - 3\tan(x) = 0, -2\pi \le x \le 2\pi$
$\therefore \tan(x) = \frac{\sqrt{3}}{3}$

2. Identify the quadrants in which the solutions lie.

Tangent is positive in quadrants 1 and 3.

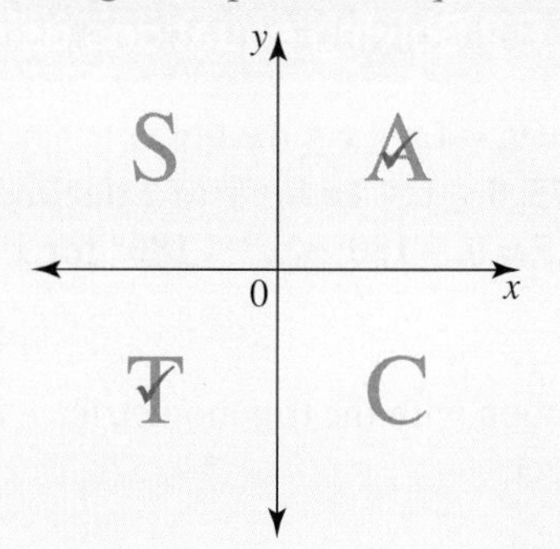

3. Identify the base.

The base is $\frac{\pi}{6}$ since $\tan\left(\frac{\pi}{6}\right) = \frac{\sqrt{3}}{3}$.

4. Generate the solutions using the appropriate quadrant forms.

Since $-2\pi \le x \le 2\pi$, there will be four solutions: two from a positive rotation and two from a negative rotation.

$x = \frac{\pi}{6}, \pi + \frac{\pi}{6}$ or $x = -\pi + \frac{\pi}{6}, -2\pi + \frac{\pi}{6}$

5. Calculate the solutions from their quadrant forms.

$\therefore x = \frac{\pi}{6}, \frac{7\pi}{6}, -\frac{5\pi}{6}, -\frac{11\pi}{6}$

TI | THINK — **DISPLAY/WRITE**

a. 1. Put the calculator in RADIAN mode.
On a Calculator page, press MENU, then select:
3: Algebra
1: Solve
Complete the entry line as:
solve
$\left(\sin(x) = \frac{\sqrt{3}}{2}, x\right) \mid$
$0 \le x \le 2\pi$
Then press ENTER.

```
solve(sin(x)=√3/2, x)|0≤x≤2·π
        x=π/3 or x=2·π/3
```

2. The answer appears on the screen.

$x = \frac{\pi}{3}$ or $x = \frac{2\pi}{3}$

CASIO | THINK — **DISPLAY/WRITE**

a. 1. Put the calculator in RADIAN mode.
On a Main screen, complete the entry line as:
solve
$\left(\sin(x) = \frac{\sqrt{3}}{2} \mid 0 \le x \le 2\pi\right)$
Then press EXE.

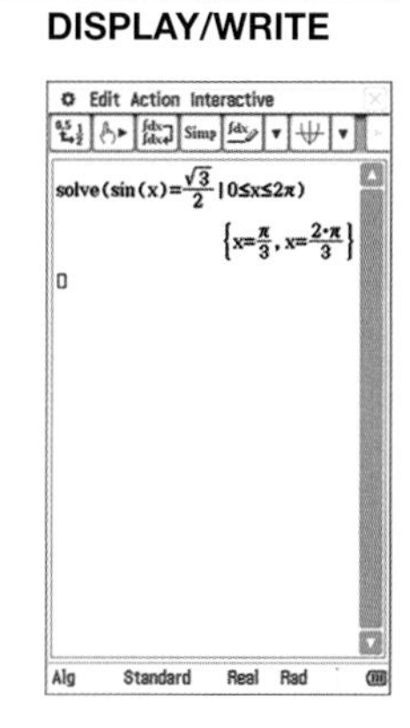

2. The answer appears on the screen.

$x = \frac{\pi}{3}$ or $x = \frac{2\pi}{3}$

9.2.2 Trigonometric equations with boundary value solutions

Recognition of exact trigonometric values allows us to identify the base for solving trigonometric equations to obtain exact solutions. However, there are also exact trigonometric values for boundary points. These need to be recognised should they appear in an equation. The simplest strategy to solve trigonometric equations involving boundary values is to use a unit circle diagram to generate the solutions. The domain for the equation determines the number of rotations required around the unit circle. It is not appropriate to consider quadrant forms to generate solutions, since boundary points lie between two quadrants.

Using technology

When bases are not recognisable from exact values, calculators are needed to identify the base. Whether the calculator or other technology is set on radian mode or degree mode is determined by the given equation. For example, if $\sin(x) = -0.7, 0 \le x \le 2\pi$, the base is calculated as $\sin^{-1}(0.7)$ in radian mode. However for $\sin(x) = 0.7, 0° \le x \le 360°$, degree mode is used when calculating the base as $\sin^{-1}(0.7)$. The degree of accuracy required for the answer is usually specified in the question; if not, express answers rounded to 2 decimal places.

WORKED EXAMPLE 2 Solving equations using technology

a. Solve $3\cos(x) + 3 = 0, -4\pi \le x \le 4\pi$ for x.
b. Solve $\sin(x) = -0.75, 0 \le x \le 4\pi$ for x to 2 decimal places.
c. Solve $\tan(x°) + 12.5 = 0, -180° \le x° \le 180°$ for x to 1 decimal place.

THINK	WRITE
a. 1. Express the equation with the trigonometric function as subject.	a. $3\cos(x) + 3 = 0 \quad -4\pi \le x \le 4\pi$, $\therefore \cos(x) = -1$
2. Identify any boundary points.	-1 is a boundary value since $\cos(\pi) = -1$. The boundary point $[\pi]$ has Cartesian coordinates $(-1, 0)$.
3. Use a unit circle to generate the solutions.	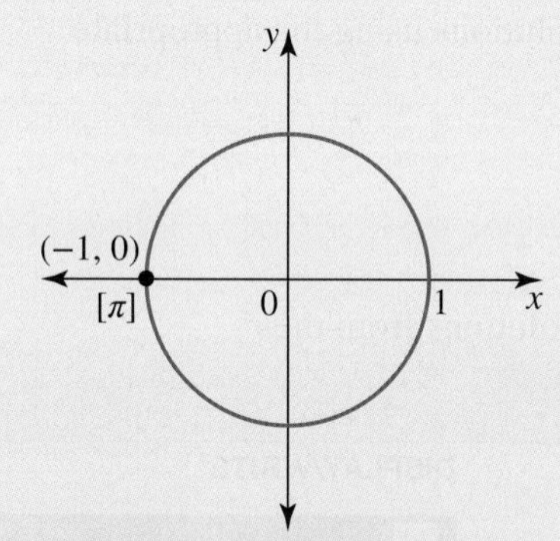 As $-4\pi \le x \le 4\pi$, this means 2 anticlockwise revolutions and 2 clockwise revolutions around the circle are required, with each revolution generating one solution. The solutions are: $x = \pi, 3\pi$ and $x = -\pi, -3\pi$ $\therefore x = \pm\pi, \pm 3\pi$
b. 1. Identify the quadrants in which the solutions lie.	b. $\sin(x) = -0.75, 0 \le x \le 4\pi$ Sine is negative in quadrants 3 and 4.

2. Calculate the base.	The base is $\sin^{-1}(0.75)$. Using radian mode, $\sin^{-1}(0.75) = 0.848$ to 3 decimal places.
3. Generate the solutions using the appropriate quadrant forms.	Since $x \in [0, 4\pi]$, there will be four solutions from two anticlockwise rotations. $x = \pi + 0.848, 2\pi - 0.848$ or $x = 3\pi + 0.848, 4\pi - 0.848$
4. Calculate the solutions to the required accuracy. *Note:* If the base is left as $\sin^{-1}(0.75)$, then the solutions such as $x = \pi + \sin^{-1}(0.75)$ could be calculated in radian mode in one step.	$\therefore x = 3.99, 5.44, 10.27, 11.72$ (correct to 2 decimal places)
c. 1. Identify the quadrants in which the solutions lie.	**c.** $\tan(x°) + 12.5 = 0, -180° \leq x° \leq 180°$ $\tan(x°) = -12.5$ Tangent is negative in quadrants 2 and 4.
2. Calculate the base.	The base is $\tan^{-1}(12.5)$. Using degree mode, $\tan^{-1}(12.5) = 85.43°$ to 2 decimal places.
3. Generate the solutions using the appropriate quadrant forms.	Since $-180° \leq x° \leq 180°$, a clockwise rotation of 180° gives one negative solution in quadrant 4 and an anticlockwise rotation of 180° gives one positive solution in quadrant 2. $x° = -85.43°$ or $x° = 180° - 85.43°$
4. Calculate the solutions to the required accuracy.	$\therefore x = -85.4, 94.6$ (correct to 1 decimal place)

9.2.3 Further types of trigonometric equations

Trigonometric equations may require algebraic techniques or the use of relationships between the functions before they can be reduced to the basic form $f(x) = a$, where f is either sin, cos or tan.

- Equations of the form $\sin(x) = a\cos(x)$ can be converted to $\tan(x) = a$ by dividing both sides of the equation by $\cos(x)$.
- Equations of the form $\sin^2(x) = a$ can be converted to $\sin(x) = \pm\sqrt{a}$ by taking the square roots of both sides of the equation.
- Equations of the form $\sin^2(x) + b\sin(x) + c = 0$ can be converted to standard quadratic equations by using the substitution $u = \sin(x)$.

Since $-1 \leq \sin(x) \leq 1$ and $-1 \leq \cos(x) \leq 1$, neither $\sin(x)$ nor $\cos(x)$ can have values greater than 1 or less than -1. This may have implications requiring the rejection of some steps when working with sine or cosine trigonometric equations. As $\tan(x) \in R$, there is no restriction on the values the tangent function can take.

WORKED EXAMPLE 3 Solving equations by substitution

Solve the following for x, where $0 \le x \le 2\pi$.

a. $\sqrt{3}\sin(x) = \cos(x)$

b. $\cos^2(x) + \cos(x) - 2 = 0$

THINK	WRITE
a. 1. Reduce the equation to one trigonometric function.	a. $\sqrt{3}\sin(x) = \cos(x),\ 0 \le x \le 2\pi$ Divide both sides by $\cos(x)$. $\frac{\sqrt{3}\sin(x)}{\cos(x)} = 1$ $\therefore \sqrt{3}\tan(x) = 1$ $\therefore \tan(x) = \frac{1}{\sqrt{3}}$
2. Calculate the solutions.	Tangent is positive in quadrants 1 and 3. The base is $\frac{\pi}{6}$. $x = \frac{\pi}{6}, \pi + \frac{\pi}{6}$ $= \frac{\pi}{6}, \frac{7\pi}{6}$
b. 1. Use substitution to form a quadratic equation.	b. $\cos^2(x) + \cos(x) - 2 = 0,\ 0 \le x \le 2\pi$ Let $a = \cos(x)$. The equation becomes $a^2 + a - 2 = 0$.
2. Solve the quadratic equation.	$(a+2)(a-1) = 0$ $\therefore a = -2$ or $a = 1$
3. Substitute back for the trigonometric function.	Since $a = \cos(x)$, $\cos(x) = -2$ or $\cos(x) = 1$. Reject $\cos(x) = -2$ since $-1 \le \cos(x) \le 1$. $\therefore \cos(x) = 1$
4. Solve the remaining trigonometric equation.	$\cos(x) = 1, 0 \le x \le 2\pi$ Boundary value since $\cos(0) = 1$ $\therefore x = 0, 2\pi$

9.2.4 Solving trigonometric equations that require a change of domain

Equations such as $\sin(2x) = 1, 0 \le x \le 2\pi$ can be expressed in the basic form by the substitution of $\theta = 2x$. The accompanying domain must be changed to be the domain for θ. This requires the end points of the domain for x to be multiplied by 2. Hence, $0 \le x \le 2\pi \Rightarrow 2 \times 0 \le 2x \le 2 \times 2\pi$ gives the domain requirement for θ as $0 \le \theta \le 4\pi$.

This allows the equation to be written as $\sin(\theta) = 1, 0 \le \theta \le 4\pi$.

This equation can then be solved to give $\theta = \frac{\pi}{2}, \frac{5\pi}{2}$.

Substituting back for x gives $2x = \frac{\pi}{2}, \frac{5\pi}{2} \Rightarrow x = \frac{\pi}{4}, \frac{5\pi}{4}$. The solutions are in the domain specified for x.

The change of domain ensures all possible solutions are obtained.

However, in practice, it is quite common not to formally introduce the pronumeral substitution for equations such as $\sin(2x) = 1, 0 \le x \le 2\pi$.

With the domain change, the equation can be written as $\sin(2x) = 1, 0 \le 2x \le 4\pi$ and the equation solved for x as follows:

$$\sin(2x) = 1, 0 \le 2x \le 4\pi$$
$$\therefore 2x = \frac{\pi}{2}, \frac{5\pi}{2}$$
$$\therefore x = \frac{\pi}{4}, \frac{5\pi}{4}$$

WORKED EXAMPLE 4 Solving equations with changed domains

a. Solve $\cos(3x) = -\frac{1}{2}$ for x, $0 \le x \le 2\pi$.

b. Use substitution to solve the equation $\tan\left(2x - \frac{\pi}{4}\right) = -1, 0 \le x \le \pi$.

THINK	WRITE
a. 1. Change the domain to be that for the given multiple of the variable.	a. $\cos(3x) = -\frac{1}{2}, 0 \le x \le 2\pi$ Multiply the end points of the domain of x by 3 $\therefore \cos(3x) = -\frac{1}{2}, 0 \le 3x \le 6\pi$ 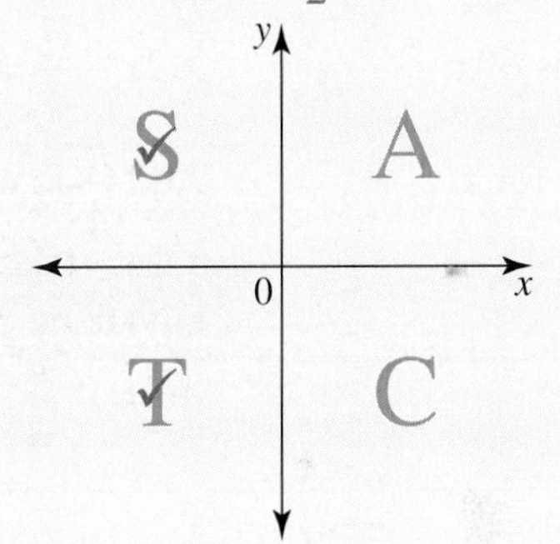
2. Solve the equation for $3x$. *Note:* Alternatively, substitute $\theta = 3x$ and solve for θ.	Cosine is negative in quadrants 2 and 3. The base is $\frac{\pi}{3}$. As $3x \in [0, 6\pi]$, each of the three revolutions will generate 2 solutions, giving a total of 6 values for $3x$. $3x = \pi - \frac{\pi}{3}, \pi + \frac{\pi}{3}, 3\pi - \frac{\pi}{3}, 3\pi + \frac{\pi}{3}, 5\pi - \frac{\pi}{3}, 5\pi + \frac{\pi}{3}$ $3x = \frac{2\pi}{3}, \frac{4\pi}{3}, \frac{8\pi}{3}, \frac{10\pi}{3}, \frac{14\pi}{3}, \frac{16\pi}{3}$
3. Calculate the solutions for x.	Divide each of the 6 values by 3 to obtain the solutions for x. $x = \frac{2\pi}{9}, \frac{4\pi}{9}, \frac{8\pi}{9}, \frac{10\pi}{9}, \frac{14\pi}{9}, \frac{16\pi}{9}$

b. 1. State the substitution required to express the equation in basic form.	**b.** $\tan\left(2x-\frac{\pi}{4}\right)=-1, 0\le x\le\pi$ Let $\theta=2x-\frac{\pi}{4}$.
2. Change the domain of the equation to that of the new variable.	For the domain change: $0\le x\le\pi$ $\therefore 0\le 2x\le 2\pi$ $\therefore -\frac{\pi}{4}\le 2x-\frac{\pi}{4}\le 2\pi-\frac{\pi}{4}$ $\therefore -\frac{\pi}{4}\le\theta\le\frac{7\pi}{4}$
3. State the equation in terms of θ.	The equation becomes $\tan(\theta)=-1, -\frac{\pi}{4}\le\theta\le\frac{7\pi}{4}$.
4. Solve the equation for θ.	Tangent is negative in quadrants 2 and 4. The base is $\frac{\pi}{4}$. $\theta=-\frac{\pi}{4}, \pi-\frac{\pi}{4}, 2\pi-\frac{\pi}{4}$ $=-\frac{\pi}{4}, \frac{3\pi}{4}, \frac{7\pi}{4}$
5. Substitute back in terms of x.	$\therefore 2x-\frac{\pi}{4}=-\frac{\pi}{4}, \frac{3\pi}{4}, \frac{7\pi}{4}$
6. Calculate the solutions for x.	Add $\frac{\pi}{4}$ to each value. $\therefore 2x=0, \pi, 2\pi$ Divide by 2. $\therefore x=0, \frac{\pi}{2}, \pi$

9.2 Exercise

Technology free

1. Solve the following for x, given $0\le x\le 2\pi$.

a. $\cos(x)=\frac{1}{\sqrt{2}}$

b. $\sin(x)=-\frac{1}{\sqrt{2}}$

c. $\tan(x)=-\frac{1}{\sqrt{3}}$

d. $2\sqrt{3}\cos(x)+3=0$

e. $4-8\sin(x)=0$

f. $2\sqrt{2}\tan(x)=\sqrt{24}$

2. **WE1** Solve the following equations to obtain exact values for x.

a. $\sin(x) = \frac{1}{2}, 0 \leq x \leq 2\pi$

b. $\sqrt{3} - 2\cos(x) = 0, 0 \leq x \leq 2\pi$

c. $4 + 4\tan(x) = 0, -2\pi \leq x \leq 2\pi$

3. Determine the exact solutions for $\theta \in [-2\pi, 2\pi]$ for which:

a. $\tan(\theta) = 1$

b. $\cos(\theta) = -0.5$

c. $1 + 2\sin(\theta) = 0.$

4. a. Determine the solutions to the equation $3\tan(x) + 3\sqrt{3} = 0$ over the domain $x \in [0, 3\pi]$.

b. For $0 \leq t \leq 4\pi$, find the exact solutions to $10\sin(t) - 3 = 2$.

c. Calculate the exact values of v that satisfy $4\sqrt{2}\cos(v) = \sqrt{2}\cos(v) + 3, -\pi \leq v \leq 5\pi$.

5. Determine the exact solutions, in degrees, to each of the following equations for the domain $\theta \in [0°, 360°]$.

a. $\cos(\theta) = \frac{\sqrt{2}}{2}$

b. $\sin(\theta) = -\frac{1}{2}$

c. $\tan(\theta) = \sqrt{3}$

d. $6\tan(\theta) = -6$

e. $6\sin(\theta) - 3\sqrt{2} = 0$

f. $-\sqrt{3} - 2\cos(\theta) = 0$

6. Calculate the exact solutions for $\alpha \in [-360°, 360°]$ for which:

a. $\tan(\alpha) = 1$

b. $\cos(\alpha) = 0$

c. $5\sin(\alpha) + 5 = 0$

d. $2\tan(\alpha) + 7 = 7.$

Technology active

7. **WE2** a. Solve $1 - \sin(x) = 0, -4\pi \leq x \leq 4\pi$ for x.

b. Solve $\tan(x) = 0.75, 0 \leq x \leq 4\pi$ for x to 2 decimal places.

c. Solve $4\cos(x°) + 1 = 0, -180° \leq x° \leq 180°$ for x to 1 decimal place.

8. Consider the equation $\cos(\theta) = -\frac{1}{2}, -180° \leq \theta \leq 180°$.

a. Determine how many solutions for θ the equation has.

b. Calculate the solutions of the equation.

9. Calculate the values of θ, correct to 2 decimal places, that satisfy the following conditions.

a. $2 + 3\cos(\theta) = 0, 0 \leq \theta \leq 2\pi$

b. $\tan(\theta) = \frac{1}{\sqrt{2}}, -2\pi \leq \theta \leq 3\pi$

c. $5\sin(\theta°) + 4 = 0, -270° \leq \theta° \leq 270°$

d. $\cos^2(\theta°) = 0.04, 0° \leq \theta° \leq 360°$

10. Obtain all values for t, $t \in [-\pi, 4\pi]$, for which:

a. $\tan(t) = 0$

b. $\cos(t) = 0$

c. $\sin(t) = -1$

d. $\cos(t) = 1$

e. $\sin(t) = 1$

f. $\tan(t) = 1.$

11. Consider the function $f: [0, 2] \to R, f(x) = \cos(\pi x)$.

a. Calculate $f(0)$.

b. Obtain $\{x : f(x) = 0\}$.

12. **WE3** Solve the following for x, given $0 \leq x \leq 2\pi$.

a. $\sqrt{3}\sin(x) = 3\cos(x)$

b. $\sin^2(x) - 5\sin(x) + 4 = 0$

13. Solve $\cos^2(x) = \frac{3}{4}, 0 \leq x \leq 2\pi$ for x.

14. Solve the following for x, where $0 \leq x \leq 2\pi$.

a. $\sin(x) = \sqrt{3}\cos(x)$

b. $\sin(x) = -\frac{\cos(x)}{\sqrt{3}}$

c. $\sin(2x) + \cos(2x) = 0$

d. $\frac{3\sin(x)}{8} = \frac{\cos(x)}{2}$

e. $\sin^2(x) = \cos^2(x)$

f. $\cos(x)(\cos(x) - \sin(x)) = 0$

15. Solve the following for x, where $0 \le x \le 2\pi$.

a. $\sin^2(x) = \frac{1}{2}$

b. $2\cos^2(x) + 3\cos(x) = 0$

c. $2\sin^2(x) - \sin(x) - 1 = 0$

d. $\tan^2(x) - 2\tan(x) + 1 = 0$

e. $\sin^2(x) + 2\sin(x) + 1 = 0$

f. $\cos^2(x) - 9 = 0$

16. **WE4** a. Solve $\sin(2x) = \frac{1}{\sqrt{2}}$, $0 \le x \le 2\pi$ for x.

b. Use substitution to solve the equation $\cos\left(2x + \frac{\pi}{6}\right) = 0, 0 \le x \le \frac{3\pi}{2}$.

17. Solve $\sin\left(\frac{x}{2}\right) = \sqrt{3}\cos\left(\frac{x}{2}\right)$, $0 \le x \le 2\pi$ for x.

18. Solve the following for θ, given $0 \le \theta \le 2\pi$.

a. $\sqrt{3}\tan(3\theta) + 1 = 0$

b. $2\sqrt{3}\sin\left(\frac{3\theta}{2}\right) - 3 = 0$

c. $4\cos^2(-\theta) = 2$

d. $\sin\left(2\theta + \frac{\pi}{4}\right) = 0$

19. Consider the function $f: [0, 2\pi] \to R, f(x) = a\sin(x)$.

a. If $f\left(\frac{\pi}{6}\right) = 4$, calculate the value of a.

b. Use the answer to part **a** to find, where possible, any values of x, to 2 decimal places, for which the following apply.

i. $f(x) = 3$

ii. $f(x) = 8$

iii. $f(x) = 10$

20. a. Give the exact value of $\sin(75°)$ using CAS technology.

b. Hence, solve the equation $\sin(A°) = \frac{\sqrt{6} + \sqrt{2}}{4}$, $-270 \le A \le 270$ for $A°$.

9.2 Exam questions

Question 1 (1 mark) TECH-ACTIVE

MC The equation $\sin(x) = \frac{1}{2}, 0 \le x \le 2\pi$ has two solutions in quadrants

A. 1 and 2.
B. 1 and 3.
C. 2 and 3.
D. 2 and 4.
E. 3 and 4.

Question 2 (1 mark) TECH-ACTIVE

MC Consider the equation $\sin(\theta) = -\frac{1}{2}, -180° \le \theta \le 720°$.

The number of solutions for $\theta°$ is

A. 2
B. 3
C. 4
D. 5
E. 6

Question 3 (3 marks) TECH-FREE

Solve the equation $2\cos(x) + \sqrt{3} = 0, 0 \le x \le 2\pi$, and obtain exact values for x.

More exam questions are available online.

9.3 Transformations of sine and cosine graphs

LEARNING INTENTION

At the end of this subtopic you should be able to:
- describe and apply the effect of transformations of sine and cosine graphs
- determine the amplitude, period and equilibrium position of sine and cosine graphs
- form the equation of a given sine or cosine graph.

The shapes of the graphs of the two functions, sine and cosine, are now familiar. Each graph has a wavelike pattern of period 2π and the pair are 'out of phase' with each other by $\frac{\pi}{2}$. We now consider the effect of transformations on the shape and on the properties of these basic sine and cosine graphs.

9.3.1 Transformations of the sine and cosine graphs

The following transformations are applied to the graph of $y = f(x) \rightarrow y = af(nx) + c$:
- dilation of factor a (assuming a is positive) from the x-axis parallel to the y-axis
- reflection in the x-axis if $a < 0$
- dilation of factor $\frac{1}{n}, n > 0$ from the y-axis parallel to the x-axis
- vertical translation of c units up if $c > 0$ or down if $c < 0$.

We can therefore infer that the graph of $y = a\sin(nx) + c$ is the image of the basic $y = \sin(x)$ graph after the same set of transformations are applied.

However, the period, amplitude and equilibrium, or mean, position of a trigonometric graph are such key features that we shall consider each transformation in order to interpret its effect on each of these features.

Amplitude changes

Consider the graphs of $y = \sin(x)$ and $y = 2\sin(x)$.

Comparison of the graph of $y = 2\sin(x)$ with the graph of $y = \sin(x)$ shows the dilation of factor 2 from the x-axis affects the amplitude, but neither the period nor the equilibrium position is altered.

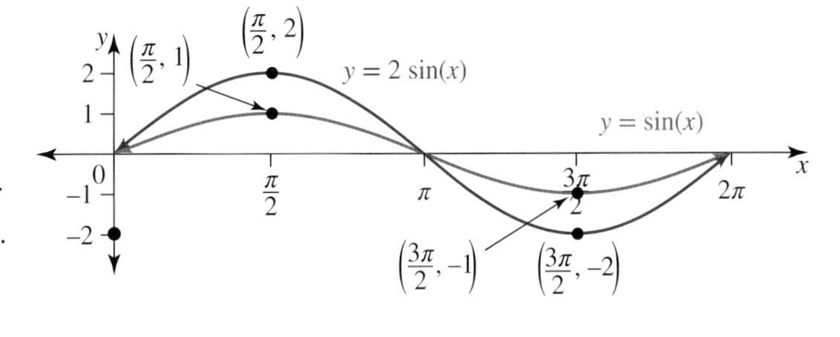

Consider the graphs of $y = \cos(x)$ and $y = -\frac{1}{2}\cos(x)$.

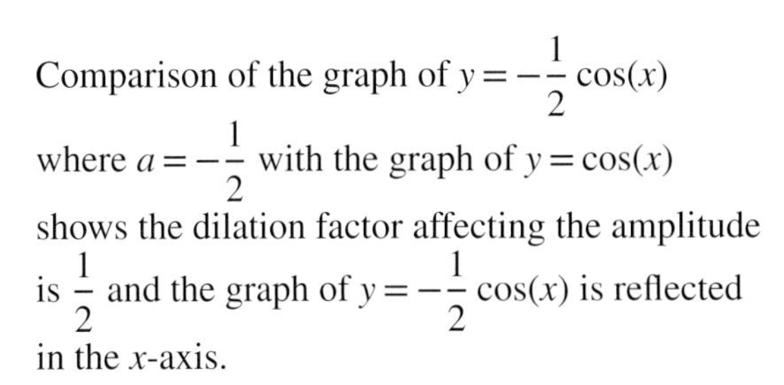

Comparison of the graph of $y = -\frac{1}{2}\cos(x)$ where $a = -\frac{1}{2}$ with the graph of $y = \cos(x)$ shows the dilation factor affecting the amplitude is $\frac{1}{2}$ and the graph of $y = -\frac{1}{2}\cos(x)$ is reflected in the x-axis.

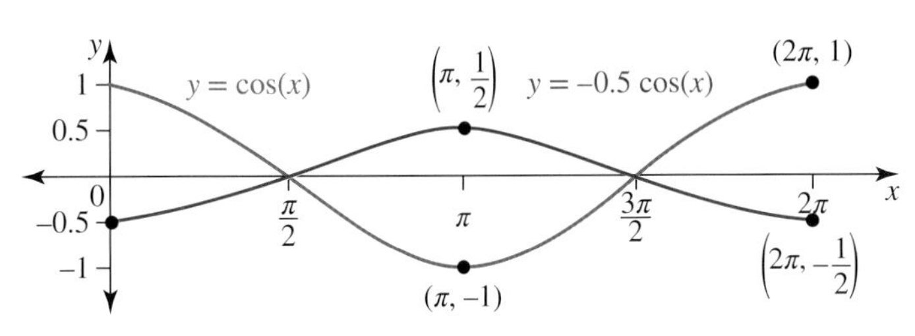

- The graphs of $y = a\sin(x)$ and $y = a\cos(x)$ have amplitude a, if $a > 0$.
- If $a < 0$, the graph is reflected in the x-axis (inverted) and the amplitude is the positive part of a (or $|a|$).

Period changes

Comparison of the graph of $y = \cos(2x)$ with the graph of $y = \cos(x)$ shows the dilation factor of $\frac{1}{2}$ from the y-axis affects the period: it halves the period. The period of $y = \cos(x)$ is 2π, and the period of $y = \cos(2x)$ is $\frac{1}{2}$ of 2π; that is, $y = \cos(2x)$ has a period of $\frac{2\pi}{2} = \pi$. Neither the amplitude nor the equilibrium position has been altered.

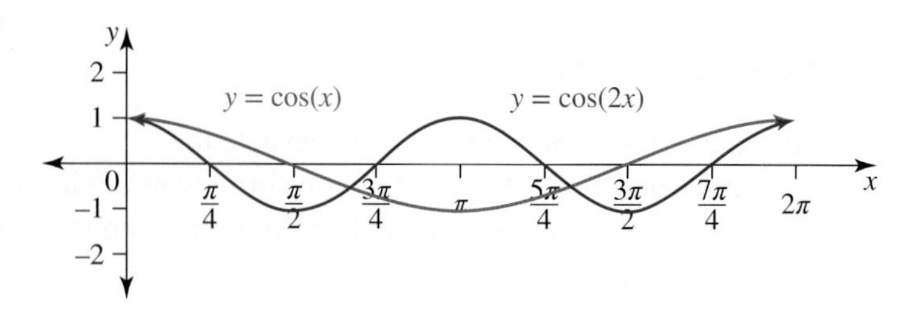

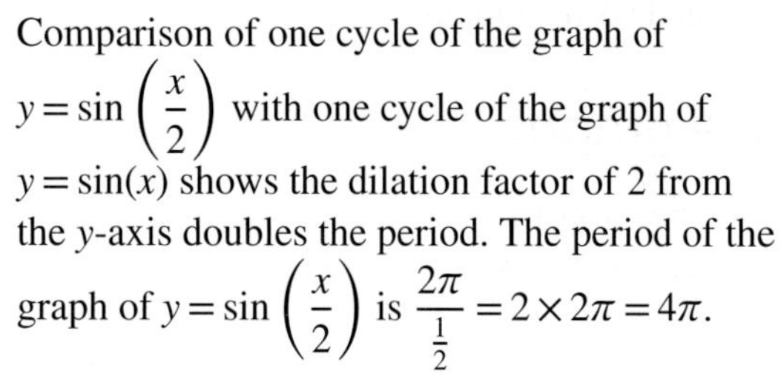

Comparison of one cycle of the graph of $y = \sin\left(\frac{x}{2}\right)$ with one cycle of the graph of $y = \sin(x)$ shows the dilation factor of 2 from the y-axis doubles the period. The period of the graph of $y = \sin\left(\frac{x}{2}\right)$ is $\frac{2\pi}{\frac{1}{2}} = 2 \times 2\pi = 4\pi$.

- The graphs of $y = \sin(nx)$ and $y = \cos(nx)$ have period $\frac{2\pi}{n}$, $n > 0$.

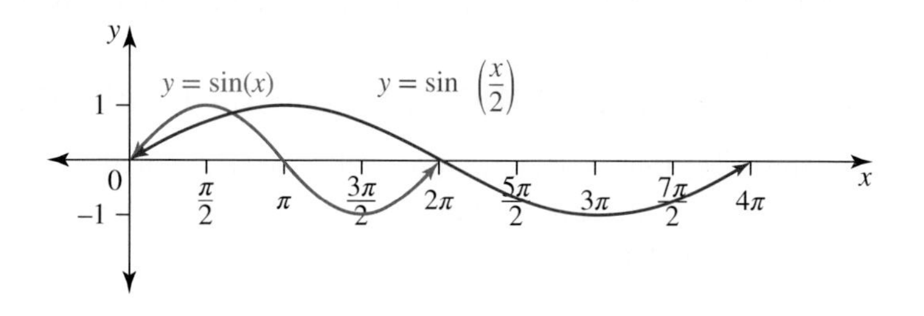

Equilibrium (or mean) position changes

Comparison of the graph of $y = \sin(x) - 2$ with the graph of $y = \sin(x)$ shows that vertical translation affects the equilibrium position. The graph of $y = \sin(x) - 2$ oscillates about the line $y = -2$, so its range is $[-3, -1]$. Neither the period nor the amplitude is affected.

- The graphs of $y = \sin(x) + c$ and $y = \cos(x) + c$ both oscillate about the equilibrium (or mean) position $y = c$.
- The range of both graphs is $[c - 1, c + 1]$ since the amplitude is 1.

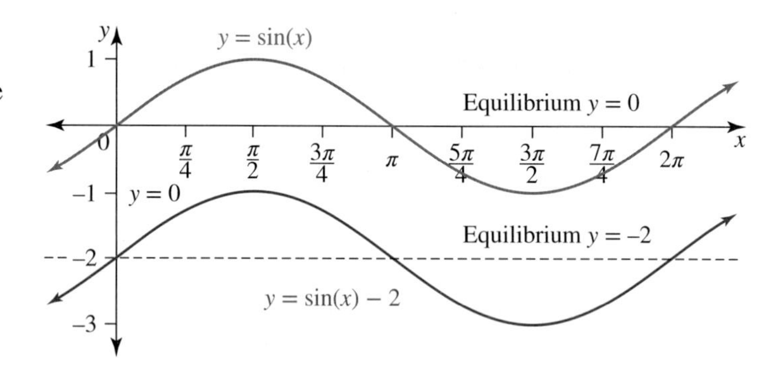

Summary of amplitude, period and equilibrium changes

Key features of $y = a\sin(nx) + c$ and $y = a\cos(nx) + c$

The graphs of $y = a\sin(nx) + c$ and $y = a\cos(nx) + c$ have:

- **amplitude a for $a > 0$; the graphs are reflected in the x-axis (inverted) if $a < 0$**
- **period $\frac{2\pi}{n}$ (for $n > 0$)**
- **equilibrium or mean position $y = c$**
- **range $[c - a, c + a]$.**

The oscillation about the equilibrium position of the graph of $y = a\sin(nx) + c$ always starts at the equilibrium with the pattern, for each period divided into quarters, of:

- equilibrium → range maximum → equilibrium → range minimum → equilibrium if $a > 0$, or:
  equilibrium → range minimum → equilibrium → range maximum → equilibrium if $a < 0$.

The oscillation about the equilibrium position of the graph of $y = a\cos(nx) + c$ either starts from its maximum or minimum point with the pattern:

- range maximum → equilibrium → range minimum → equilibrium → range maximum if $a > 0$, or:
  range minimum → equilibrium → range maximum → equilibrium → range minimum if $a < 0$.

When sketching the graphs, any intercepts with the x-axis are usually obtained by solving the trigonometric equation $a\sin(nx) + c = 0$ or $a\cos(nx) + c = 0$.

WORKED EXAMPLE 5 Sketching trigonometric graphs with transformations

Sketch the graphs of the following functions.

a. $y = 2\cos(x) - 1, 0 \le x \le 2\pi$

b. $y = 4 - 2\sin(3x), -\frac{\pi}{2} \le x \le \frac{3\pi}{2}$

THINK	WRITE
a. 1. State the period, amplitude and equilibrium position by comparing the equation with $y = a\cos(nx) + c$.	**a.** $y = 2\cos(x) - 1, 0 \le x \le 2\pi$ $a = 2, n = 1, c = -1$ Amplitude 2, period 2π, equilibrium position $y = -1$
2. Determine the range and whether there will be x-intercepts.	The graph oscillates between $y = -1 - 2 = -3$ and $y = -1 + 2 = 1$, so it has range $[-3, 1]$. It will have x-intercepts.
3. Calculate the x-intercepts.	x-intercepts: let $y = 0$. $2\cos(x) - 1 = 0$ $\therefore \cos(x) = \frac{1}{2}$ Base $\frac{\pi}{3}$, quadrants 1 and 4 $x = \frac{\pi}{3}, 2\pi - \frac{\pi}{3}$ $= \frac{\pi}{3}, \frac{5\pi}{3}$ The x-intercepts are $\left(\frac{\pi}{3}, 0\right), \left(\frac{5\pi}{3}, 0\right)$.
4. Scale the axes by marking $\frac{1}{4}$-period intervals on the x-axis. Mark the equilibrium position and end points of the range on the y-axis. Then plot the graph using its pattern.	The period is 2π, so the scale on the x-axis is in multiples of $\frac{\pi}{2}$. Since $a > 0$, the graph starts at range maximum at its y-intercept $(0, 1)$. 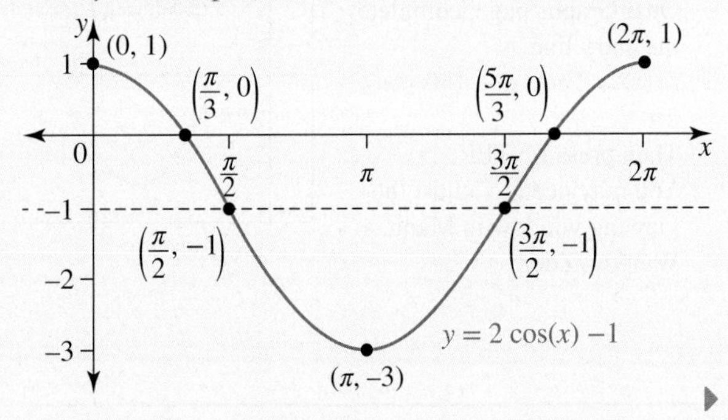

5. Label all key features of the graph including the maximum and minimum points.

The maximum points are $(0, 1)$ and $(2\pi, 1)$.
The minimum point is $(\pi, -3)$.

b. 1. State the information the equation provides by comparing the equation with $y = a\sin(nx) + c$.
Note: The amplitude is always a positive value.

b. $y = 4 - 2\sin(3x), -\frac{\pi}{2} \le x \le \frac{3\pi}{2}$
$y = -2\sin(3x) + 4$, domain $\left[-\frac{\pi}{2}, \frac{3\pi}{2}\right]$
$a = -2, n = 3, c = 4$
Amplitude 2; graph is inverted; period $\frac{2\pi}{3}$; equilibrium $y = 4$

2. Determine the range and whether there will be x-intercepts.

The graph oscillates between $y = 4 - 2 = 2$ and $y = 4 + 2 = 6$, so its range is $[2, 6]$.
There are no x-intercepts.

3. Scale the axes and extend the $\frac{1}{4}$-period intervals on the x-axis to cover the domain. Mark the equilibrium position and end points of the range on the y-axis. Then plot the graph using its pattern and continue the pattern over the given domain.

Dividing the period of $\frac{2\pi}{3}$ into four gives a horizontal scale of $\frac{\pi}{6}$. The first cycle of the graph starts at its equilibrium position at its y-intercept $(0, 4)$ and decreases as $a < 0$.

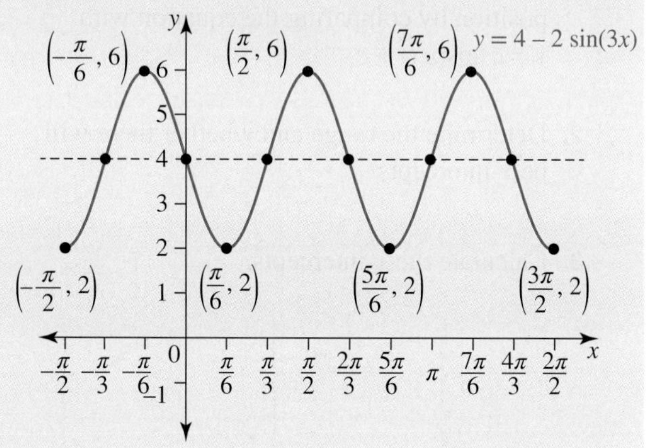

4. Label all key features of the graph including the maximum and minimum points.
Note: Successive maximum points are one period apart, as are the successive minimum points.

Maximum points are $\left(-\frac{\pi}{6}, 6\right)$, $\left(\frac{\pi}{2}, 6\right)$ and $\left(\frac{7\pi}{6}, 6\right)$.
Minimum points are
$\left(-\frac{\pi}{2}, 2\right)$, $\left(\frac{\pi}{6}, 2\right)$, $\left(\frac{5\pi}{6}, 2\right)$ and $\left(\frac{3\pi}{2}, 2\right)$.

TI | THINK

DISPLAY/WRITE

a. 1. Put the calculator in RADIAN mode.
On a Graphs page, complete the entry line as:
$f1(x) = 2\cos(x) - 1 | 0 \le x \le 2\pi$
Then press ENTER.
You may need to adjust the viewing window in Menu, 4: Window/Zoom

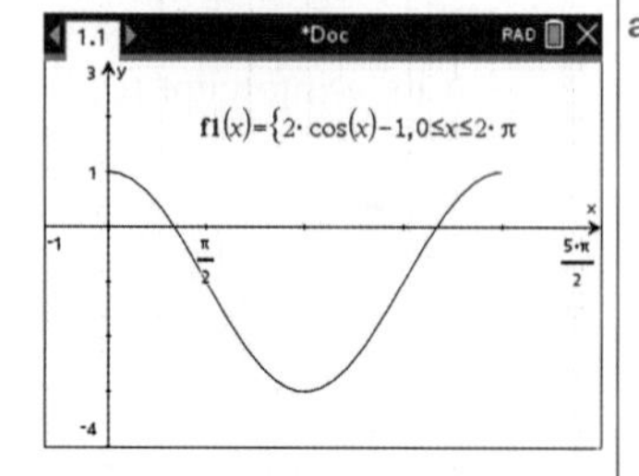

CASIO | THINK

DISPLAY/WRITE

a. 1. Put the calculator in RADIAN mode.
On a Graphs & Table screen, complete the entry line as:
$y1 = 2\cos(x) - 1 | 0 \le x \le 2\pi$
Then press EXE.
You may need to adjust the viewing window.

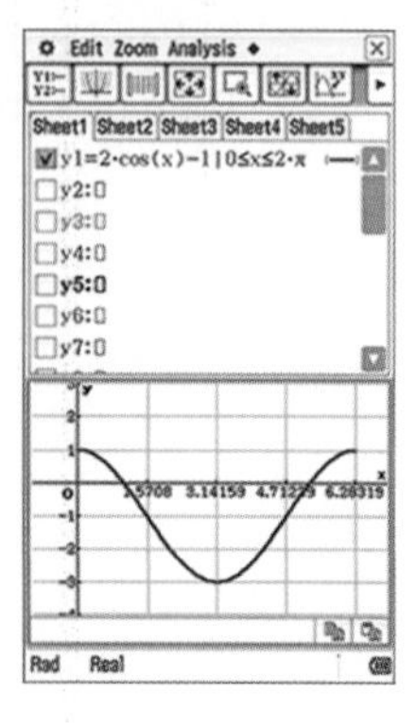

9.3.2 Forming the equation of a sine or cosine graph

The sine and cosine graphs have the same shape but different y-intercepts. To determine the equation of a graph, identify:

- the shape, $y = a\sin(nx) + c$ or $y = a\cos(nx) + c$
- the amplitude
- the period
- the equilibrium position, or vertical translation.

WORKED EXAMPLE 6 Determining the equation of a sine or cosine graph

Determine an equation for the following graph.

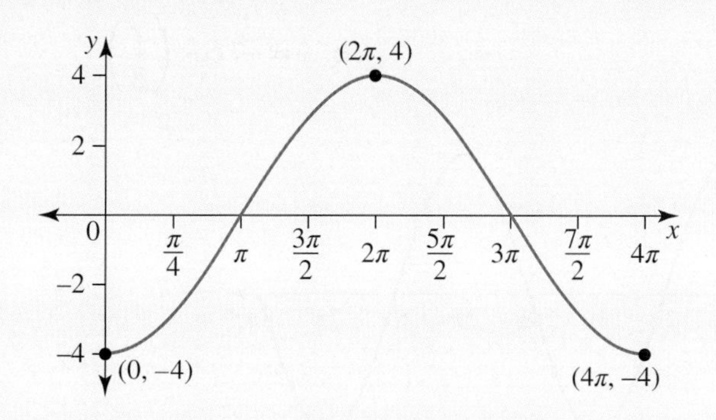

THINK	WRITE
1. Identify the key features of the given graph.	The graph has a period of 4π, amplitude 4, and the equilibrium position is $y = 0$.
2. Form the equation for the graph.	The graph is an inverted cosine graph. A cosine equation for the graph is $y = a\cos(nx) + c$ with $a = -4$ and $c = 0$. $\therefore y = -4\cos(nx)$ The period is $\frac{2\pi}{n}$. From the diagram the period is 4π. $\frac{2\pi}{n} = 4\pi$ $\frac{2\pi}{4\pi} = n$ $n = \frac{1}{2}$ Therefore, the equation is $y = -4\cos\left(\frac{1}{2}x\right)$.

9.3 Exercise

Students, these questions are even better in jacPLUS

Receive immediate feedback and access sample responses

Access additional questions

Track your results and progress

Find all this and MORE in jacPLUS

Technology free

1. State the period and amplitude of the following.

a. $y=6\cos(2x)$

b. $y=-7\cos\left(\frac{x}{2}\right)$

c.

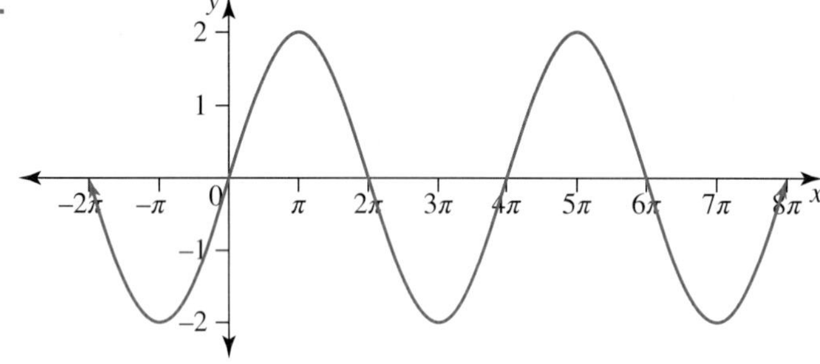

2. State the amplitude, range and period, then sketch the graph of one complete period of:

a. $y=\frac{1}{4}\sin(4x)$

b. $y=\frac{1}{2}\sin\left(\frac{1}{4}x\right)$

c. $y=-\sin(3x)$

d. $y=3\cos(2x)$

e. $y=-6\cos\left(\frac{2x}{5}\right)$

f. $y=\frac{1}{3}\cos(5\pi x)$.

3. Sketch the following graphs over the given domains.

a. $y=3\cos(2x), 0\le x\le 2\pi$

b. $y=2\sin\left(\frac{1}{2}x\right), 0\le x\le 4\pi$

c. $y=-5\sin(4x), 0\le x\le 2\pi$

d. $y=-\cos(\pi x), 0\le x\le 4$

4. **WE5** Sketch the graphs of the following functions.

a. $y=2\sin(x)+1, 0\le x\le 2\pi$

b. $y=4-3\cos(2x), -\pi\le x\le 2\pi$

5. Sketch each of the following over the domain specified and state the range.

a. $y=\sin(x)+3, 0\le x\le 2\pi$

b. $y=\cos(x)-1, 0\le x\le 2\pi$

c. $y=\cos(x)+2, -\pi\le x\le \pi$

d. $y=4-\sin(x), -\pi\le x\le 2\pi$

6. Sketch the graph of $y=f(x)$ for the function $f:[0,12]\to R, f(x)=\sin\left(\frac{\pi x}{6}\right)$.

7. State the period, mean position, amplitude and range, then sketch the graph showing one complete cycle of each of the following.

a. $y=4\sin(2x)+5$

b. $y=-2\sin(3x)+2$

c. $y=\frac{3}{2}-\frac{1}{2}\cos\left(\frac{x}{2}\right)$

d. $y=2\cos(\pi x)-\sqrt{3}$

8. Sketch the following graphs over the given domains and state the ranges of each.

a. $y = 2\cos(2x) - 2, 0 \leq x \leq 2\pi$

b. $y = 2\sin(x) + \sqrt{3}, 0 \leq x \leq 2\pi$

c. $y = 3\sin\left(\dfrac{x}{2}\right) + 5, -2\pi \leq x \leq 2\pi$

d. $y = -4 - \cos(3x), 0 \leq x \leq 2\pi$

e. $y = 1 - 2\sin(2x), -\pi \leq x \leq 2\pi$

f. $y = 2[1 - 3\cos(x)], 0° \leq x \leq 360°$

9. a. Give the range of $f: R \to R, f(x) = 3 + 2\sin(5x)$.

b. Determine the minimum value of the function $f: [0, 2\pi] \to R, f(x) = 10\cos(2x) - 4$.

c. Determine the maximum value of the function $f: [0, 2\pi] \to R, f(x) = 56 - 12\sin(x)$ and the value of x for which the maximum occurs.

d. Describe the sequence of transformations that must be applied for the following.

i. $\sin(x) \to 3 + 2\sin(5x)$

ii. $\cos(x) \to 10\cos(2x) - 4$

iii. $\sin(x) \to 56 - 12\sin(x)$

10. The equation of the graph shown is of the form $y = a\sin(nx)$. Determine the values of a and n and hence state the equation of the graph.

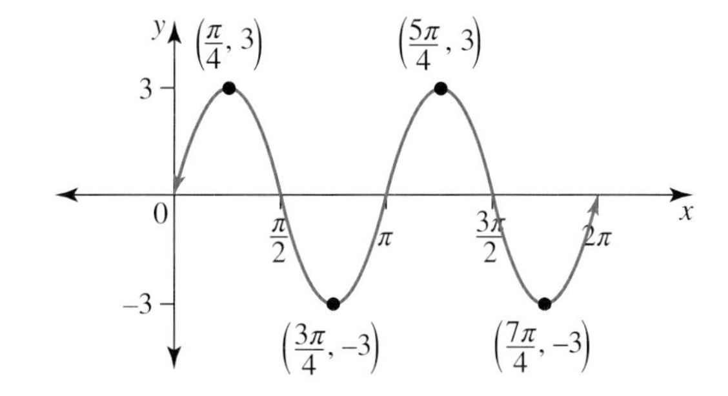

11. The equation of the graph shown is of the form $y = a\cos(nx)$. Determine the values of a and n and hence state the equation of the graph.

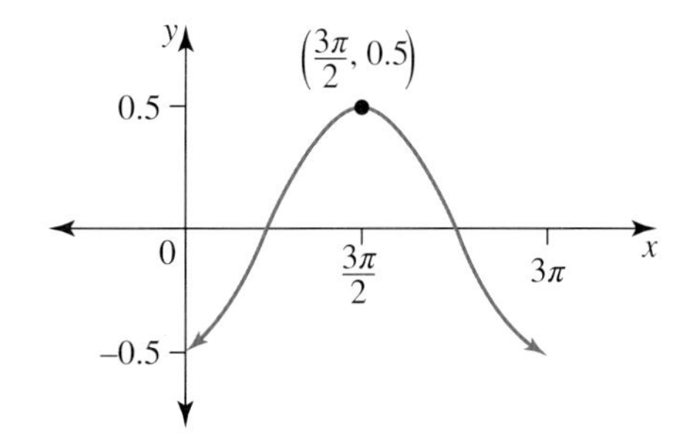

12. Determine the equation of the graph shown in the form $y = a\cos(nx)$.

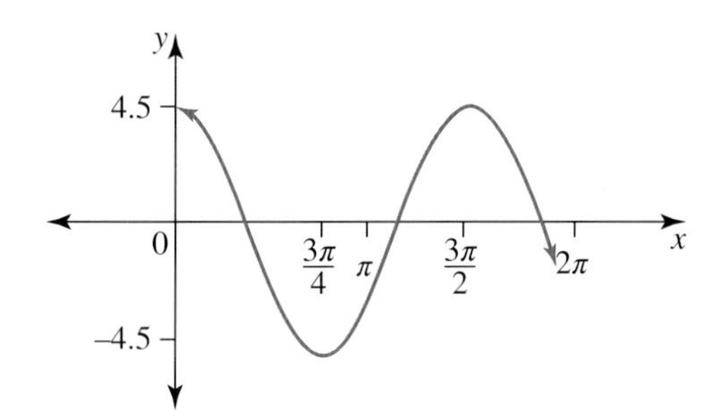

13. Determine the equation of each of the following graphs, given that the equation of each is either of the form $y = a\cos(nx)$ or $y = a\sin(nx)$.

a.

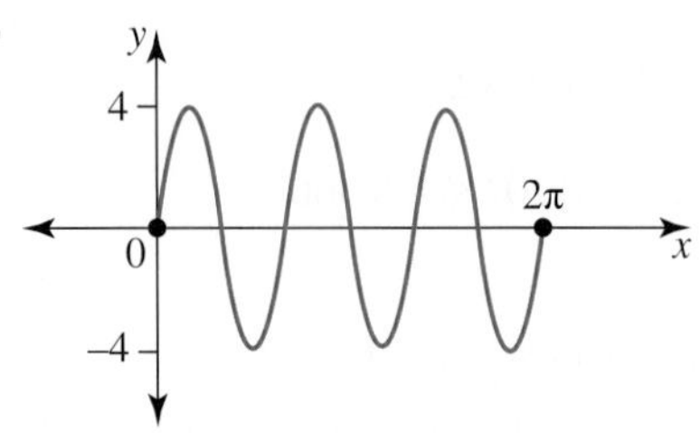

b.

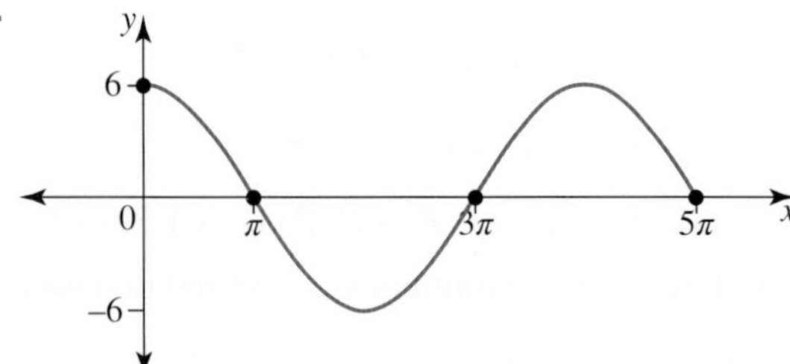

c.

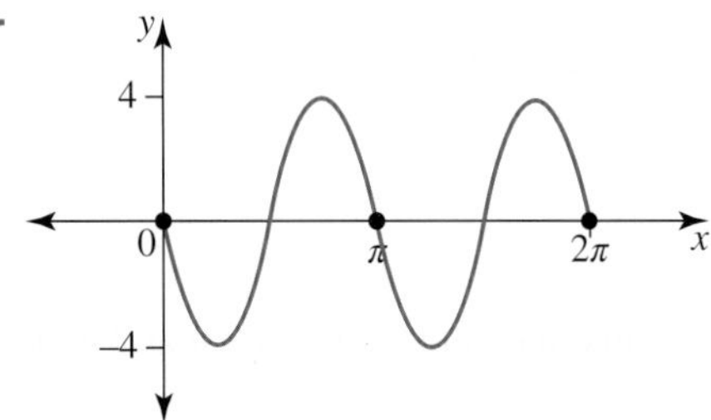

14. The following graphs are either of the form $y = a\cos(nx) + c$ or the form $y = a\sin(nx) + c$. Determine the appropriate equation for each graph.

a.

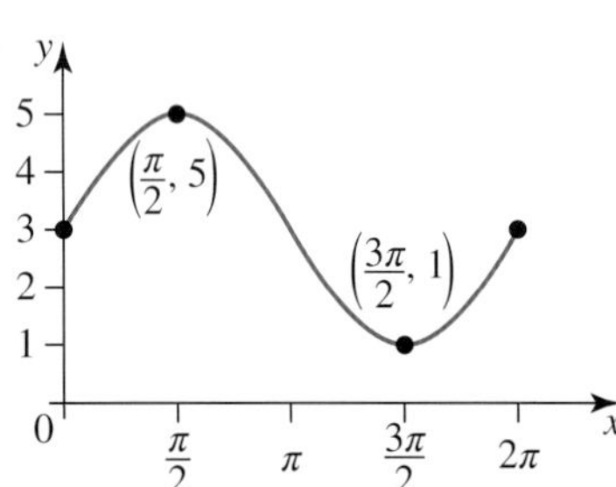

b.

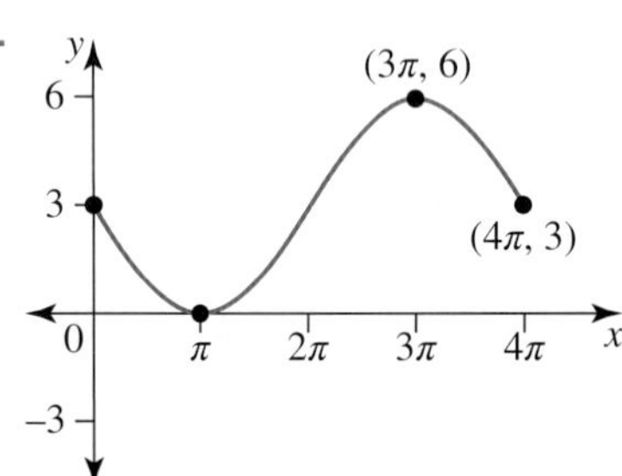

c.

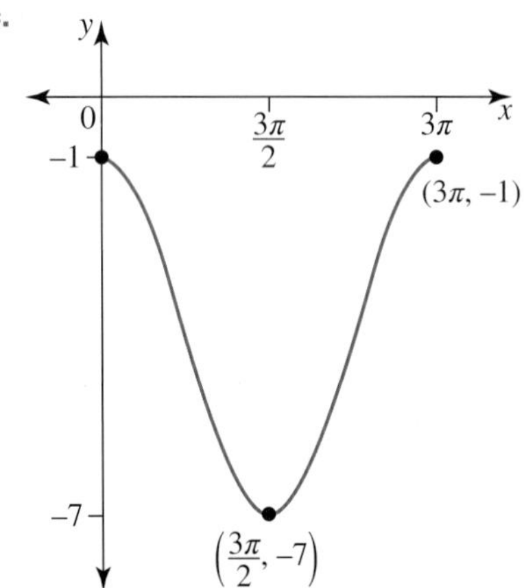

d.

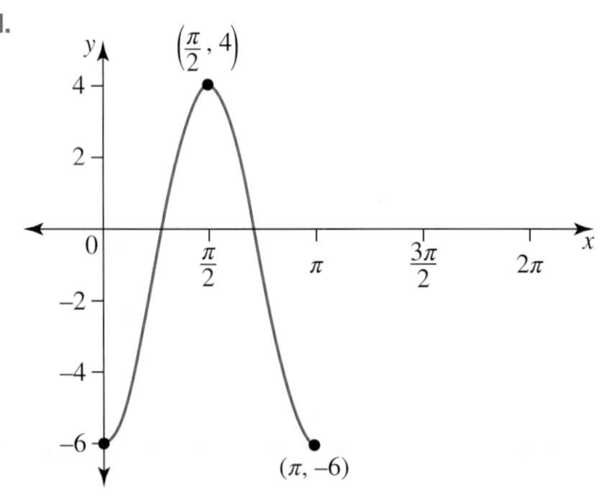

15. Determine a possible equation for the following graph.

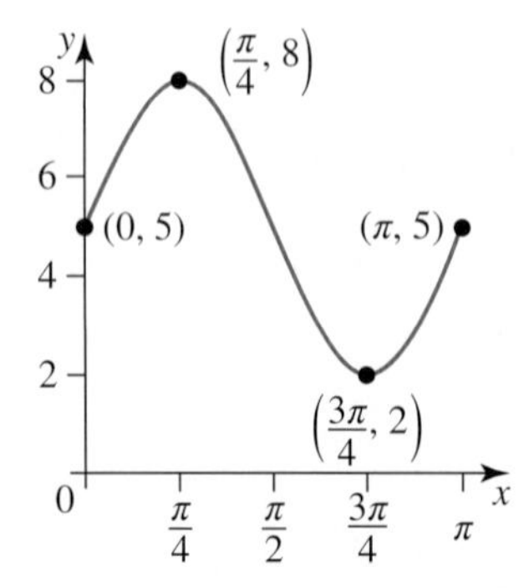

16. Obtain a possible equation for each of the following graphs.

a. 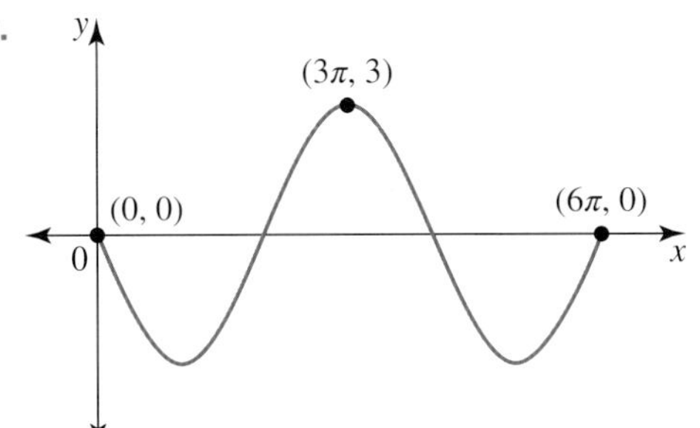

b. 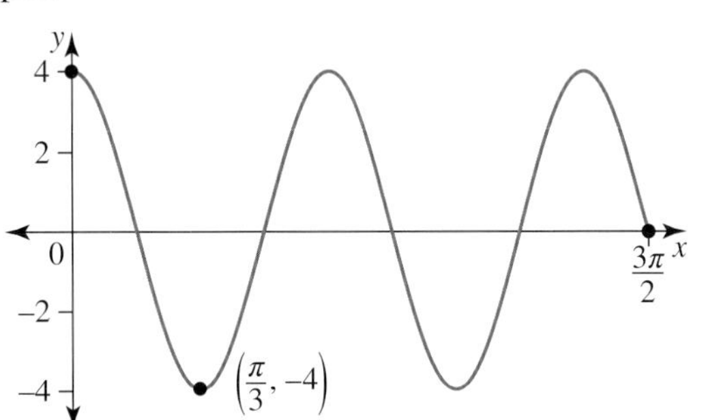

c. 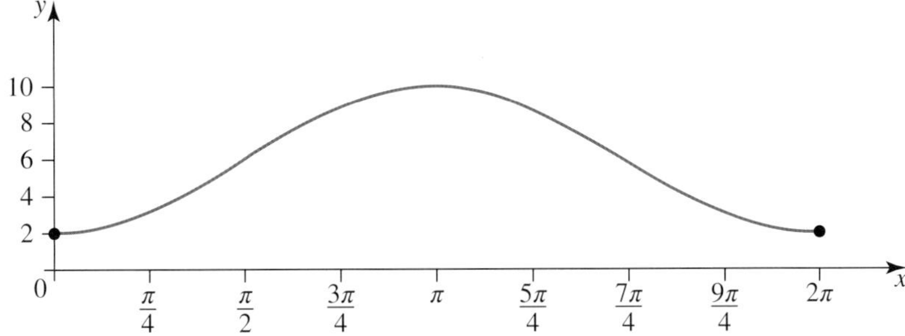

Technology active

17. WE6 Determine the equation for the following graph.

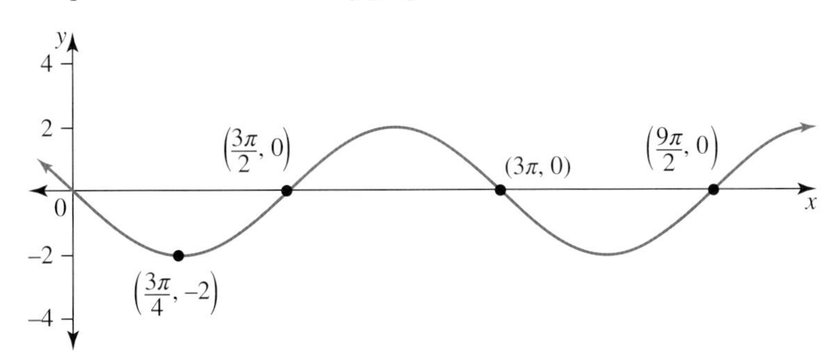

18. A function has the rule $f(x) = a\sin(bx) + c$, $a > 0$. It has a domain, D, of $\left[0, \frac{2\pi}{3}\right]$, a range of $[5, 9]$ and a maximum at $\left(\frac{\pi}{6}, 9\right)$.

a. State the period of the function.
b. Obtain values for the constants a, b and c.
c. Sketch the graph $y = f(x)$ for the given domain, D.
d. A second function has the rule $g(x) = a\cos(bx) + c$, where a, b and c have the same values as those of $y = f(x)$. Sketch the graph of $y = g(x)$, $x \in D$ on the same axes as the graph of $y = f(x)$.
e. Obtain the coordinates of any points of intersection of the graphs $y = f(x)$ and $y = g(x)$.
f. Give the values of x that are solutions to the inequation $f(x) \geq g(x)$, $x \in D$.

19. The graph of $y=\sin(x)$ is vertically translated upwards 3 units and is then reflected in the x-axis; this is followed by a dilation of factor 3 from the y-axis. Give the equation of its final image and determine if the graph of the image would intersect the graph of $y=\sin(x)$.

20. Under the function $f:[0,4]\rightarrow R$, $f(x)=a-20\sin\left(\frac{\pi x}{3}\right)$, the image of 4 is $10\left(\sqrt{3}+1\right)$.

a. Determine the value of a.
b. Determine the value(s) of x for which $f(x)=0$.
c. Sketch the graph of $y=f(x)$, stating its range.

9.3 Exam questions

Question 1 (1 mark) TECH-ACTIVE

MC The amplitude of the graph $y=4-2\sin(3x)$ is

A. 3 **B.** 4 **C.** -3 **D.** 2 **E.** -2

Question 2 (1 mark) TECH-ACTIVE

MC The maximum and minimum points of the graph $y=2\sin(3x)-1$ are

A. -1 and 1. **B.** -2 and 0. **C.** -3 and 1. **D.** 2 and 3. **E.** -2 and 3.

Question 3 (1 mark) TECH-ACTIVE

MC The period of the graph $y=\frac{2}{3}\sin\left(\frac{3}{2}x\right)-\frac{1}{4}$ is

A. $\frac{3\pi}{2}$ **B.** $\frac{4\pi}{3}$ **C.** $\frac{2\pi}{3}$ **D.** 2π **E.** 4π

More exam questions are available online.

9.4 Applications of sine and cosine functions

LEARNING INTENTION

At the end of this subtopic you should be able to:

- determine maximum and minimum values of sine and cosine functions
- use sine and cosine functions to solve real-world problems.

Phenomena that are cyclical in nature can often be modelled by a sine or cosine function.

Examples of periodic phenomena include sound waves, ocean tides and ovulation cycles. Trigonometric models may be able to approximate things like the movement in the value of the All Ordinaries Index of the stock market, fluctuations in temperature or the vibrations of violin strings about a mean position.

9.4.1 Maximum and minimum values

As $-1\leq\sin(x)\leq 1$ and $-1\leq\cos(x)\leq 1$, the maximum value of both $\sin(x)$ and $\cos(x)$ is 1 and the minimum value of both functions is -1. This can be used to calculate, for example, the maximum value of $y=2\sin(x)+4$ by substituting 1 for $\sin(x)$:

$$y_{\max}=2\times 1+4\Rightarrow y_{\max}=6$$

The minimum value can be calculated as:

$$y_{min} = 2 \times (-1) + 4 \Rightarrow y_{min} = 2$$

To calculate the maximum value of $y = 5 - 3\cos(2x)$, the largest negative value of $\cos(2x)$ would be substituted for $\cos(2x)$. Thus:

$$y_{max} = 5 - 3 \times (-1) \Rightarrow y_{max} = 8$$

The minimum value can be calculated by substituting the largest positive value of $\cos(2x)$:

$$y_{min} = 5 - 3 \times 1 \Rightarrow y_{min} = 2$$

Alternatively, identifying the equilibrium position and amplitude enables the range to be calculated. For $y = 5 - 3\cos(2x)$, with amplitude 3 and equilibrium at $y = 5$, the range is calculated from $y = 5 - 3 = 2$ to $y = 5 + 3 = 8$, giving a range of $[2, 8]$. This also shows the maximum and minimum values.

WORKED EXAMPLE 7 Determining maximum and minimum values

The temperature, T °C, during 1 day in April is given by $T = 17 - 4\sin\left(\frac{\pi}{12}t\right)$, where t is the time in hours after midnight.

a. **Calculate the temperature at midnight.**
b. **Determine the minimum temperature during the day and at what time it occurred.**
c. **Calculate the interval over which the temperature varied that day.**
d. **State the period and sketch the graph of the temperature for $t \in [0, 24]$.**
e. **If the temperature was below k degrees for 2.4 hours, obtain the value of k to 1 decimal place.**

THINK	WRITE
a. State the value of t and use it to calculate the required temperature.	a. At midnight, $t = 0$. Substitute $t = 0$ into $T = 17 - 4\sin\left(\frac{\pi}{12}t\right)$. $T = 17 - 4\sin(0)$ $= 17$ The temperature at midnight was 17°.
b. 1. State the condition on the value of the trigonometric function for the minimum value of T to occur.	b. $T = 17 - 4\sin\left(\frac{\pi}{12}t\right)$ The minimum value occurs when $\sin\left(\frac{\pi}{12}t\right) = 1$.
2. Calculate the minimum temperature.	$T_{min} = 17 - 4 \times 1$ $= 13$ The minimum temperature was 13°.
3. Calculate the time when the minimum temperature occurred.	The minimum value occurs when $\sin\left(\frac{\pi}{12}t\right) = 1$ Solving this equation, $\frac{\pi}{12}t = \frac{\pi}{2}$ $t = 6$ The minimum temperature occurred at 6 am.

c. Use the equilibrium position and amplitude to calculate the range of temperatures.

c. $T = 17 - 4\sin\left(\dfrac{\pi}{12}t\right)$

Amplitude 4, equilibrium $T = 17$

The range is $[17-4, 17+4] = [13, 21]$.

Therefore, the temperature varied between 13 °C and 21 °C.

d. 1. Calculate the period.

d. $2\pi \div \dfrac{\pi}{12} = 2\pi \times \dfrac{12}{\pi}$

$= 24$

The period is 24 hours.

2. Sketch the graph.

Dividing the period into quarters gives a horizontal scale of 6.

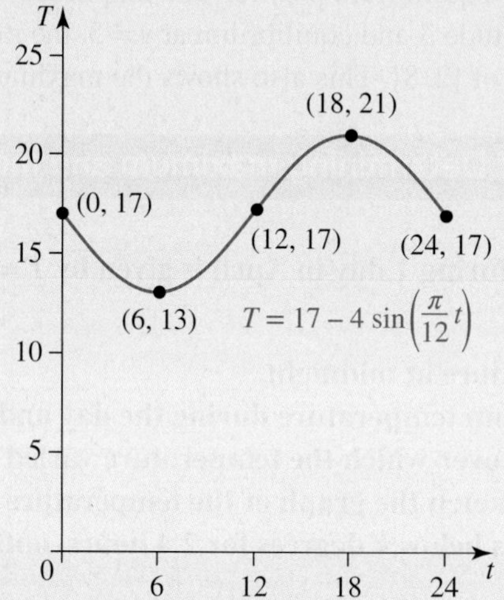

e. 1. Use the symmetry of the curve to deduce the end points of the interval involved.

e. $T < k$ for an interval of 2.4 hours.

For T to be less than the value for a time interval of 2.4 hours, the 2.4-hour interval is symmetric about the minimum point.

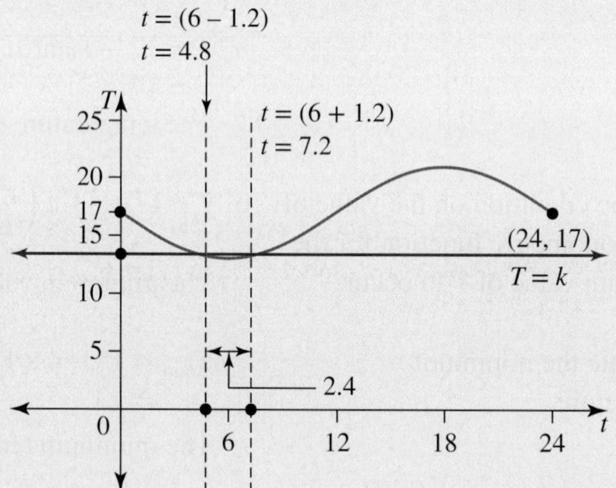

The end points of the t interval must each be $\dfrac{1}{2} \times 2.4$ from the minimum point where $t = 6$. The end points of the interval occur at $t = 6 \pm 1.2$.

$\therefore T = k$ when $t = 4.8$ or 7.2.

2. Calculate the required value.

Substituting either end point into the temperature model will give the value of k.

If $t = 4.8$:

$$T = 17 - 4\sin\left(\frac{\pi}{12} \times 4.8\right)$$
$$= 17 - 4\sin(0.4\pi)$$
$$= 13.2$$

Therefore, $k = 13.2$ and the temperature is below 13.2° for 2.4 hours.

Resources

Interactivity Oscillation (int-2977)

9.4 Exercise

Students, these questions are even better in jacPLUS

Receive immediate feedback and access sample responses

Access additional questions

Track your results and progress

Find all this and MORE in jacPLUS

Technology active

1. A child plays with a yo-yo attached to the end of an elastic string. The yo-yo rises and falls about its rest position so that its vertical distance, y cm, above its rest position at time t seconds is given by $y = -40\cos(t)$.
   a. Sketch the graph of $y = -40\cos(t)$, showing two complete cycles.
   b. Determine the greatest distance the yo-yo falls below its rest position.
   c. Determine the times at which the yo-yo returns to its rest position during the two cycles.
   d. Determine how many seconds the yo-yo takes to first reach a height of 20 cm above its rest position.

2. The temperature from 8 am to 10 pm on a day in February is shown.
   If T is the temperature in degrees Celsius t hours from 8 am, form the equation of the temperature model and use this to calculate the times during the day when the temperature exceeded 30 degrees.

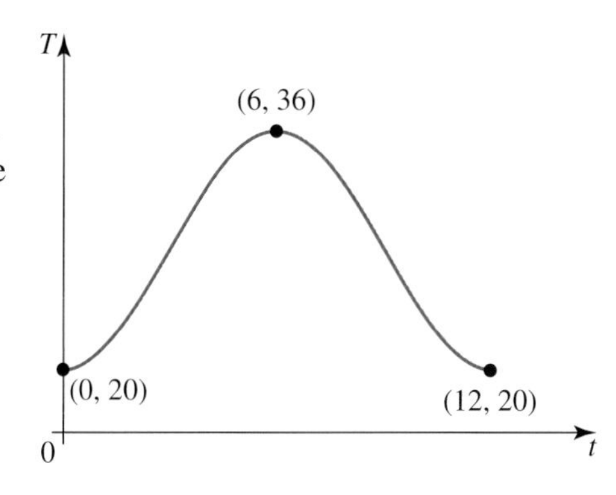

3. Emotional ups and downs are measured by a wellbeing index that ranges from 0 to 10 in increasing levels of happiness. A graph of this index over a 4-week cycle is shown.

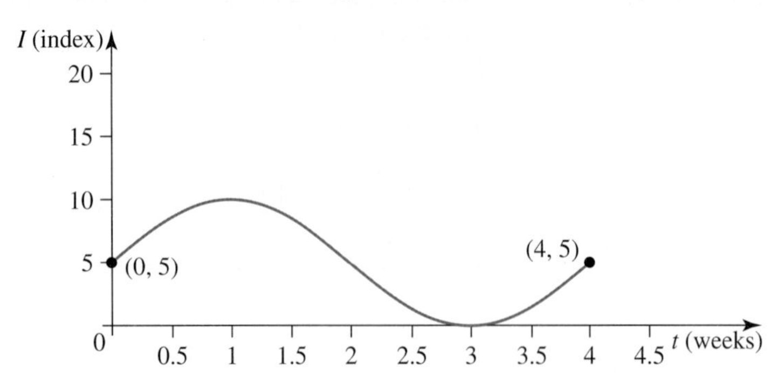

a. Express the relationship between the wellbeing index I and the time t in terms of a trigonometric equation.

b. A person with a wellbeing index of 6 or higher is considered to experience a high level of happiness. Determine the percentage of the 4-week cycle for which the model predicts this feeling would last.

4. **WE7** During one day in October the temperature T°C is given by $T = 19 - 3\sin\left(\frac{\pi}{12}t\right)$, where t is the time in hours after midnight.

a. Calculate the temperature at midnight.

b. Determine the maximum temperature during the day and at what time it occurred.

c. Calculate the interval over which the temperature varied that day.

d. State the period and sketch the graph of the temperature for $t \in [0, 24]$.

e. If the temperature was below k degrees for 3 hours, obtain the value of k to 1 decimal place.

5. The diagram below shows the cross-section of a sticky tape holder with a circular hole through its side. It is bounded by a curve with the equation $h = a\sin\left(\frac{\pi}{5}x\right) + b$, where h cm is the height of the curve above the base of the holder at distance x cm along the base.

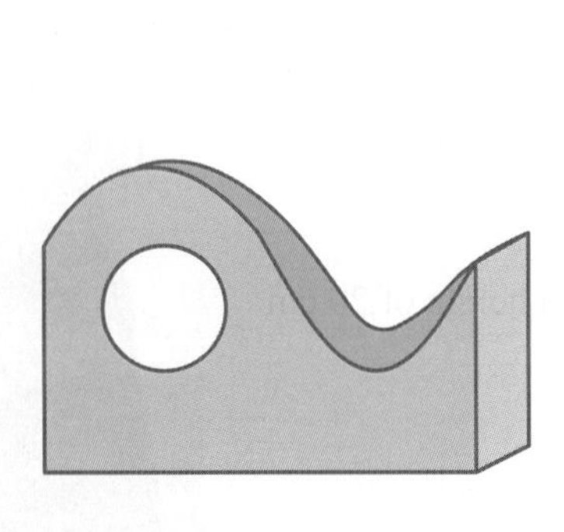

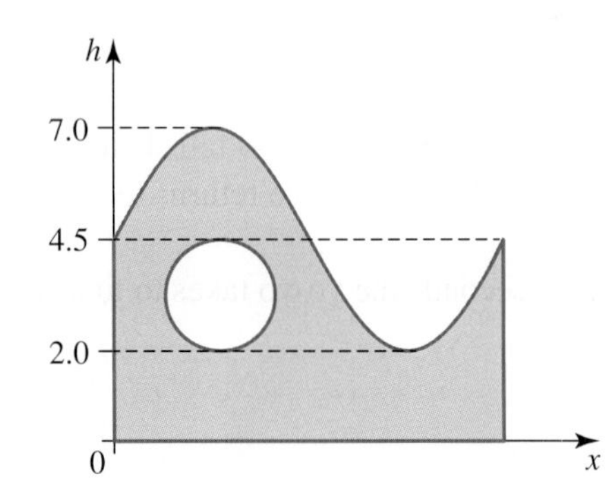

a. Use the measurements given on the diagram to state the values of a and b and hence write down the equation of the bounding curve.

b. Calculate the length of the base of the holder.

c. If the centre of the circular hole lies directly below the highest point of the curve, determine the coordinates of the centre.

d. Using the symmetry of the curve, calculate the cross-sectional area shaded in the diagram, to 1 decimal place.

6. In a laboratory, an organism is being grown in a test tube. To help increase the rate at which the organism grows, the test tube is placed in an incubator where the temperature T°C varies according to the rule $T = 30 - \cos\left(\frac{\pi}{12}t\right)$, where t is the time in minutes since the test tube has been placed in the incubator.

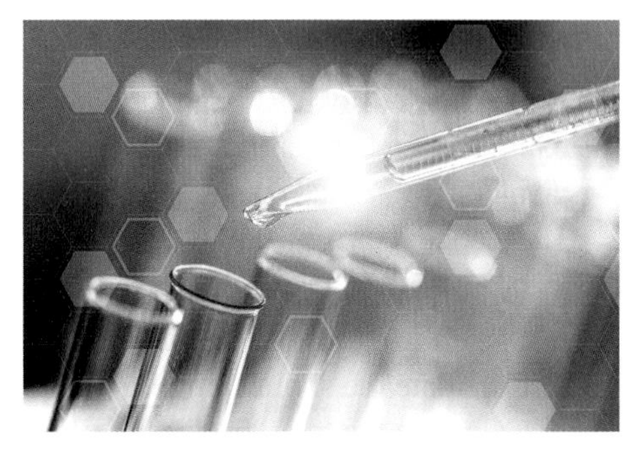

a. State the range of temperature in the incubator.
b. Determine when the temperature reaches its greatest value.
c. Sketch the graph of the temperature against time for the first half hour.
d. Determine how many cycles are completed in 1 hour.
e. If the organism is kept in the incubator for 2.5 hours, calculate the temperature at the end of this time.
f. Express the rule for the temperature in terms of a sine function.

7. During a particular day in a Mediterranean city, the temperature inside an office building between 10 am to 7.30 pm fluctuates so that t hours after 10 am, the temperature T°C is given by $T = 19 + 6\sin\left(\frac{\pi t}{6}\right)$.

a. i. State the maximum temperature and the time it occurs.
ii. State the minimum temperature and the time it occurs.
b. i. Calculate the temperature in the building at 11.30 am. Answer to 1 decimal place.
ii. Calculate the temperature in the building at 7.30 pm. Answer to 1 decimal place.
c. Sketch the graph of the temperature against time from 10 am to 7.30 pm.
d. When the temperature reaches 24 °C, an air conditioner in the boardroom is switched on and it is switched off when the temperature in the rest of the building falls below 24 °C. Determine the amount of time that the air conditioner is on in the boardroom.

8. The height above ground level of a particular carriage on a Ferris wheel is given by $h = 10 - 8.5\cos\left(\frac{\pi}{60}t\right)$ where h is the height in metres above ground level after t seconds.

a. Calculate how far above the ground the carriage is initially.
b. Calculate how high the carriage will be after 1 minute.
c. Determine how many revolutions the Ferris wheel will complete in a 4-minute time interval.
d. Sketch the graph of h against t for the first 4 minutes.
e. Calculate the time, to the nearest second, in one revolution, that the carriage is higher than 12 metres above the ground.
f. The carriage is attached by strong wire radial spokes to the centre of the circular wheel. Calculate the length of a radial spoke.

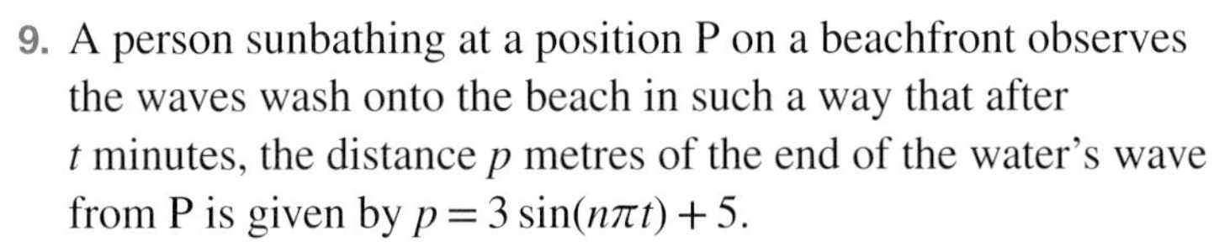

9. A person sunbathing at a position P on a beachfront observes the waves wash onto the beach in such a way that after t minutes, the distance p metres of the end of the water's wave from P is given by $p = 3\sin(n\pi t) + 5$.

a. Calculate the closest distance the water reaches to the sunbather at P.
b. Over a 1-hour interval, the sunbather counts 40 complete waves that have washed onto the beach. Calculate the value of n.

c. At some time later in the day, the distance p metres of the end of the water's wave from P is given by $p = a\sin(4\pi t) + 5$. If the water just reaches the sunbather who is still at P, deduce the value of a and determine how many times in half an hour the water reaches the sunbather at P.

d. Determine in which of the two models of the wave motion, $p = 3\sin(n\pi t) + 5$ or $p = a\sin(4\pi t) + 5$, the number of waves per minute is greater.

10. A discarded drink can floating in the waters of a creek oscillates vertically up and down 20 cm about its equilibrium position. Its vertical displacement, x metres, from its equilibrium position after t seconds is given by $x = a\sin(bt)$. Initially the can moved vertically downwards from its mean position and it took 1.5 seconds to rise from its lowest point to its highest point.

    a. Determine the values of a and b and state the equation for the vertical displacement.
    b. Sketch one cycle of the motion.
    c. Calculate the shortest time, T seconds, for the can's displacement to be one half the value of its amplitude.
    d. Calculate the total distance the can moved in one cycle of the motion.

11. The intensity, I, of sound emitted by a device is given by $I = 4\sin(t) - 3\cos(t)$, where t is the number of hours the device has been operating.

    a. Use the I–t graph to obtain the maximum intensity the device produces.
    b. State, to 2 decimal places, the first value of t for which $I = 0$.
    c. Express the equation of the I–t graph in the form $I = a\sin(t + b)$.
    d. Express the equation of the I–t graph in the form $I = a\cos(t + b)$.

12. The teeth of a tree saw can be approximated by the function $y = x + 4 + 4\cos(6x)$, $0 \le x \le 4\pi$, where y cm is the vertical height of the teeth at a horizontal distance x cm from the end of the saw.

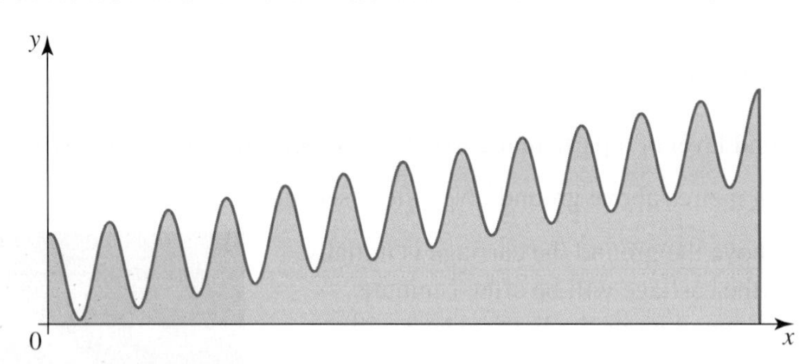

Note: This diagram is representative only.
Each peak of the graph represents one of the teeth.

    a. Determine how many teeth the saw has.
    b. Calculate exactly how far apart the successive peaks of the teeth are.
    c. The horizontal length of the saw is 4π cm. Determine the greatest width measurement of this saw.
    d. Give the equation of the linear function which will touch each of the teeth of the saw.

9.4 Exam questions

Question 1 (1 mark) TECH-ACTIVE

MC The minimum and maximum values for $y = 7 - 4\cos(3x)$ are

A. 0 and 7
B. 3 and 11
C. 4 and 7
D. 4 and 14
E. -3 and 5

Question 2 (1 mark) TECH-ACTIVE

MC The temperature, T degrees Celsius, during one day in November is given by $T = 20 - 4\sin\left(\frac{\pi}{12}t\right)$, where t is the time in hours after midnight. The minimum temperature during the day and the time at which it occurred were

- **A.** 20 °C and midnight.
- **B.** 16 °C and 6 am.
- **C.** 14 °C and 8 am.
- **D.** 12 °C and 10 am.
- **E.** 14 °C and 4 am.

Question 3 (4 marks) TECH-ACTIVE

An ecologist studying the population of kangaroos in a national park calculated that the population roughly followed the function $p = 1000 + 500\sin\left(\frac{\pi t}{4}\right)$, where t is the time in months and p is the number of kangaroos in the population.

Determine:

- **a.** the initial population **(1 mark)**
- **b.** the highest and lowest population sizes in the first 2 years **(2 marks)**
- **c.** the population at the end of 2 years. **(1 mark)**

More exam questions are available online.

9.5 The tangent function

LEARNING INTENTION

At the end of this subtopic you should be able to:

- sketch the tangent function and identify any vertical asymptotes
- describe and apply the effect of transformations on the tangent function.

The graph of $y = \tan(x)$ has a distinct shape quite different to that of the wave shape of the sine and cosine graphs. As values such as $\tan\left(\frac{\pi}{2}\right)$ and $\tan\left(\frac{3\pi}{2}\right)$ are undefined, a key feature of the graph of $y = \tan(x)$ is the presence of vertical asymptotes at odd multiples of $\frac{\pi}{2}$.

Trigonometric relationships

The relationship $\tan(x) = \frac{\sin(x)}{\cos(x)}$ shows that:

- **$\tan(x)$ will be undefined whenever $\cos(x) = 0$**
- **$\tan(x) = 0$ whenever $\sin(x) = 0$.**

on Resources

Interactivity The tangent function (int-2878)

9.5.1 The graph of $y = \tan(x)$

The diagram shows the graph of $y = \tan(x)$ over the domain $[0, 2\pi]$.

The key features of the graph of $y = \tan(x)$ are as follows:

- Vertical asymptotes exist at $x = \frac{\pi}{2}, x = \frac{3\pi}{2}$ for the domain $[0, 2\pi]$. For extended domains this pattern continues with asymptotes at $x = (2n+1)\frac{\pi}{2}, n \in Z$.
- The period is π. Two cycles are completed over the domain $[0, 2\pi]$.
- The range is R.
- x-intercepts occur at $x = 0, \pi, 2\pi$ for the domain $[0, 2\pi]$. For extended domains, this pattern continues with x-intercepts at $x = n\pi, n \in Z$.
- The mean position is $y = 0$.
- The asymptotes are one period apart.
- The x-intercepts are one period apart.

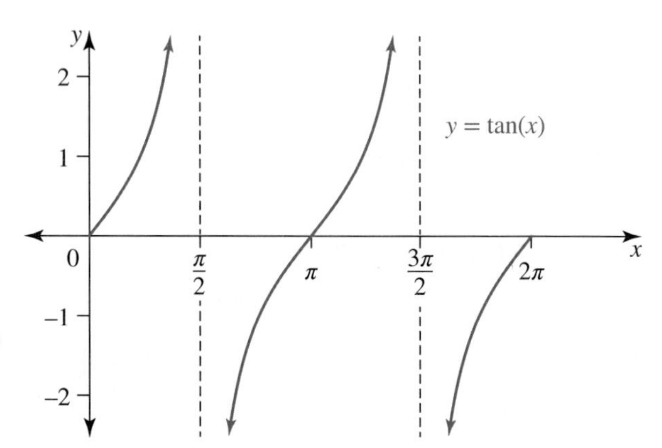

Unlike the sine and cosine graphs, 'amplitude' has no meaning for the tangent graph. As for any graph, the x-intercepts of the tangent graph are the solutions to the equation formed when $y = 0$.

WORKED EXAMPLE 8 Sketching $y = \tan(x)$

Sketch the graph of $y = \tan(x)$ for $x \in \left(-\frac{3\pi}{2}, \frac{\pi}{2}\right)$.

THINK	WRITE
1. State the equations of the asymptotes.	Period: the period of $y = \tan(x)$ is π. Asymptotes: the graph has an asymptote at $x = \frac{\pi}{2}$. As the asymptotes are a period apart, for the domain $\left(-\frac{3\pi}{2}, \frac{\pi}{2}\right)$ there is an asymptote at $x = \frac{\pi}{2} - \pi = -\frac{\pi}{2}$ and another at $x = -\frac{\pi}{2} - \pi = -\frac{3\pi}{2}$.
2. State where the graph cuts the x-axis. *Note:* The x-intercepts could be found by solving the equation $\tan(x) = 0$, $x \in \left(-\frac{3\pi}{2}, \frac{\pi}{2}\right)$.	The x-intercepts occur midway between the asymptotes. $(-\pi, 0)$ and $(0, 0)$ are the x-intercepts.

3. Sketch the graph.

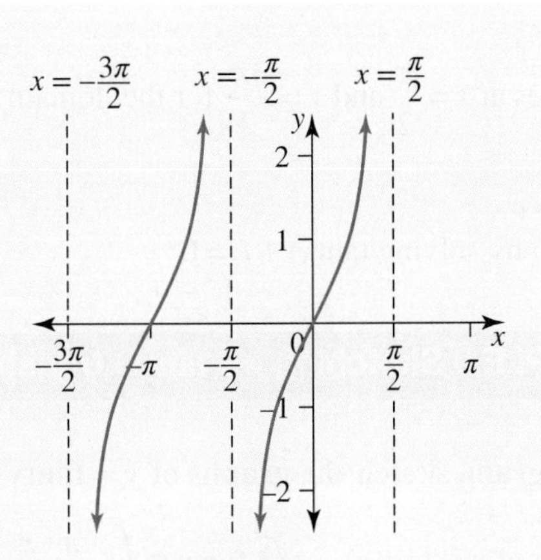

TI | THINK	DISPLAY/WRITE	CASIO | THINK	DISPLAY/WRITE
1. Put the calculator in RADIAN mode. On a Graphs page, complete the entry line as: $f1(x) = \tan(x) \mid -\frac{3\pi}{2} < x < \frac{\pi}{2}$ Then press ENTER. You may need to adjust the viewing window in Menu by selecting: 4: Window/Zoom 1: Window Settings	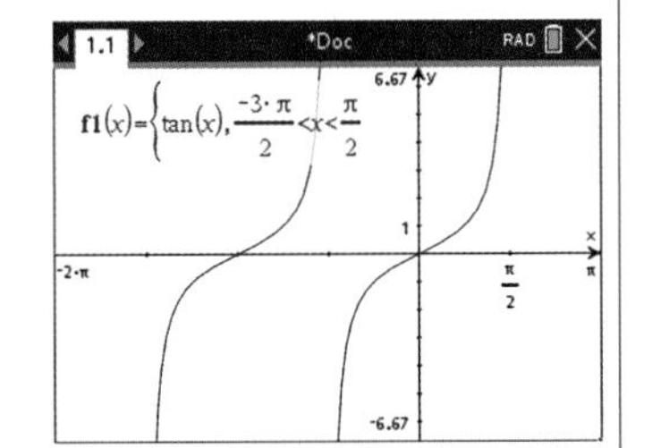	1. Put the calculator in RADIAN mode. On a Graphs & Table screen, complete the entry line as: $y1 = \tan(x) \mid -\frac{3\pi}{2} < x < \frac{\pi}{2}$ Then press EXE. You may need to adjust the viewing window.	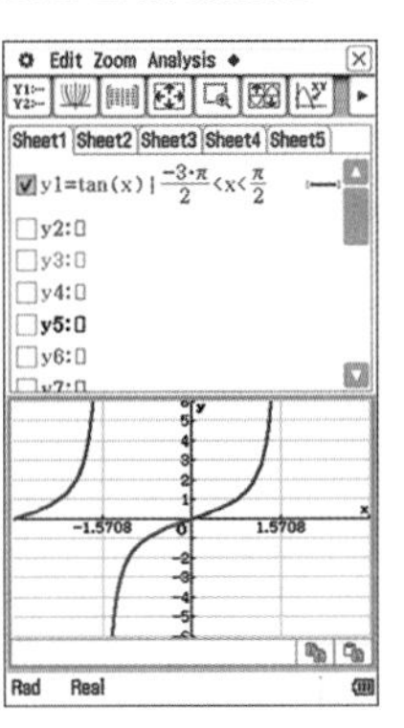

9.5.2 Transformations of the graph of $y = \tan(x)$

The vertical asymptotes are major features of the graph of any tangent function. Their positions may be affected by dilations from the y-axis. This is not the case if the graph is reflected in either axis, dilated from the x-axis or vertically translated.

The graph of $y = a\ \tan(x)$

The key features of the graph of $y = a\ \tan(x)$ are:

- dilation of factor a from the x-axis
  - if $a > 1$, the graph is narrower than $y = \tan(x)$
  - if $0 < a < 1$, the graph is wider than $y = \tan(x)$
  - if $a < 0$, the graph is reflected in the x-axis
- period π
- vertical asymptotes at $x = \frac{\pi}{2}$ and $x = \frac{3\pi}{2}$ for the domain $[0, 2\pi]$, the same as $y = \tan(x)$
- range R
- mean position $y = 0$
- x-intercepts midway between successive pairs of asymptotes.

The graph of $y = \tan(x) + c$

The key features of the graph of $y = \tan(x) + c$ are:

- vertical translation of $y = \tan(x)$ by c units
  - if $c > 0$, the graph is translated upwards
  - if $c < 0$, the graph is translated downwards

- period π
- vertical asymptotes at $x = \frac{\pi}{2}$ and $x = \frac{3\pi}{2}$ for the domain $[0, 2\pi]$, the same as $y = \tan(x)$
- range R
- mean position $y = c$
- x-intercepts found by solving $\tan(x) + c = 0$.

WORKED EXAMPLE 9 Sketching $y = \tan(x)$ transformations

a. On the same diagram, sketch the graphs of $y = \tan(x)$ and $y = -2\tan(x)$ for $x \in \left(-\frac{\pi}{2}, \frac{\pi}{2}\right)$.

b. Sketch the graph of $y = \tan(x) - \sqrt{3}$ for $x \in \left(-\frac{\pi}{2}, \frac{\pi}{2}\right)$.

THINK	WRITE
a. 1. State the key features of the graph of $y = \tan(x)$ for the given interval.	**a.** $y = \tan(x)$ The period is π. Asymptotes: $x = -\frac{\pi}{2}, x = \frac{\pi}{2}$ x-intercept: the x-intercept occurs midway between the asymptotes. Therefore, $(0, 0)$ is the x-intercept and the y-intercept.
2. Describe the transformations applied to $y = \tan(x)$ to obtain the second function.	$y = -2\tan(x)$ Dilation from the x-axis by a factor of 2 and reflection in the x-axis
3. State the key features of the graph of the second function.	The transformations do not alter the period, x-intercept or asymptotes. The period is π, the graph contains the point $(0, 0)$, and the asymptotes have equations $x = \pm\frac{\pi}{2}$.
4. To illustrate the relative positions of the graphs, calculate the coordinates of another point on each graph.	$y = \tan(x)$ Let $x = \frac{\pi}{4}$: $y = \tan\left(\frac{\pi}{4}\right)$ $= 1$ The point $\left(\frac{\pi}{4}, 1\right)$ lies on the graph of $y = \tan(x)$. $y = -2\tan(x)$ Let $x = \frac{\pi}{4}$: $y = -2\tan\left(\frac{\pi}{4}\right)$ $= -2$ The point $\left(\frac{\pi}{4}, -2\right)$ lies on the graph of $y = -2\tan(x)$.

5. Sketch the two graphs.

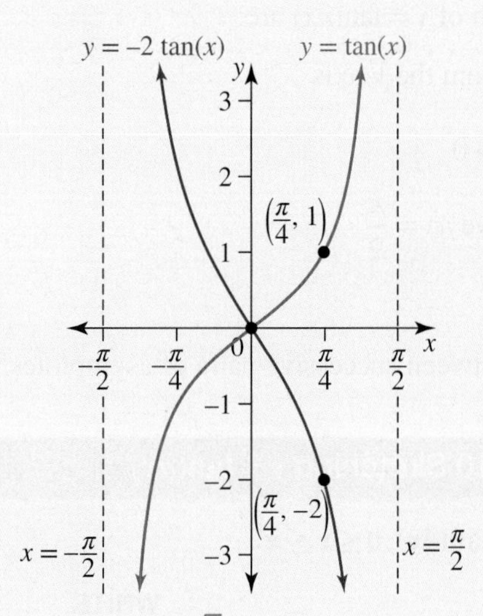

b. 1. Describe the transformation.

b. $y = \tan(x) - \sqrt{3}$

There is a vertical translation down of $\sqrt{3}$ units.

2. Calculate any x-intercepts.

x-intercepts: let $y = 0$.

$\tan(x) = \sqrt{3}, -\frac{\pi}{2} < x < \frac{\pi}{2}$

$\therefore x = \frac{\pi}{3}$

$\left(\frac{\pi}{3}, 0\right)$ is the x-intercept.

3. State any other features of the graph.

The transformation does not alter the period or asymptotes. The period is π and the asymptotes are $x = \pm\frac{\pi}{2}$.

The mean position is $y = -\sqrt{3}$.

The y-intercept is $\left(0, -\sqrt{3}\right)$.

4. Sketch the graph.

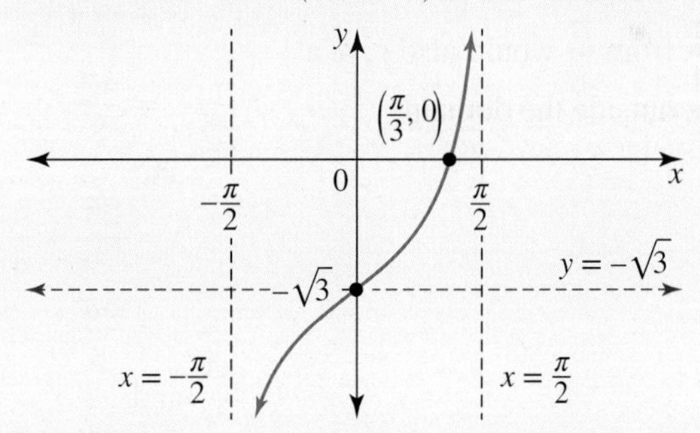

9.5.3 Period changes: the graph of $y = \tan(nx)$

Altering the period will alter the position of the asymptotes, although they will still remain one period apart. Since the graph of $y = \tan(x)$ has an asymptote at $x = \frac{\pi}{2}$, one way to obtain the equations of the asymptotes is to solve $nx = \frac{\pi}{2}$ to obtain one asymptote. Other asymptotes can then be generated by adding or subtracting multiples of the period.

The key features of the graph of $y = \tan(nx)$ are:

- a dilation factor of $\frac{1}{n}$ from the y-axis
- period $\frac{\pi}{n}$, assuming $n > 0$
- vertical asymptote: solve $nx = \frac{\pi}{2}$
- range R
- mean position $y = 0$
- x-intercepts midway between successive pairs of asymptotes, or found by solving $\tan(nx) = 0$.

WORKED EXAMPLE 10 Sketching $y = \tan(nx)$

Sketch the graph of $y = \tan(4x), 0 \le x \le \pi$.

THINK	WRITE
1. State the period by comparing the equation to $y = \tan(nx)$.	For $y = \tan(4x)$, $n = 4$. The period is $\frac{\pi}{n}$. Therefore, the period is $\frac{\pi}{4}$.
2. Obtain the equation of an asymptote.	An asymptote occurs when: $4x = \frac{\pi}{2}$ $\therefore x = \frac{\pi}{8}$
3. Calculate the equations of all the asymptotes in the given domain. *Note:* Adding $\frac{\pi}{4}$ to $\frac{7\pi}{8}$ would give a value that exceeds the domain end point of π; subtracting $\frac{\pi}{4}$ from $\frac{\pi}{8}$ would also give a value that lies outside the domain.	The other asymptotes in the domain are formed by adding multiples of the period to $x = \frac{\pi}{8}$. The other asymptotes in the domain $[0, \pi]$ occur at: $x = \frac{\pi}{8} + \frac{\pi}{4}$ $= \frac{3\pi}{8}$ and: $x = \frac{3\pi}{8} + \frac{\pi}{4}$ $= \frac{5\pi}{8}$ and: $x = \frac{5\pi}{8} + \frac{\pi}{4}$ $= \frac{7\pi}{8}$ The equations of the asymptotes are $x = \frac{\pi}{8}, \frac{3\pi}{8}, \frac{5\pi}{8}, \frac{7\pi}{8}$.

4. Calculate the x-intercepts.
Note: The x-intercepts lie midway between each pair of asymptotes, one period apart.

x-intercepts: let $y = 0$.

$$\tan(4x) = 0, \quad 0 \le x \le \pi$$
$$\tan(4x) = 0, \quad 0 \le 4x \le 4\pi$$
$$4x = 0, \pi, 2\pi, 3\pi, 4\pi$$
$$x = 0, \frac{\pi}{4}, \frac{2\pi}{4}, \frac{3\pi}{4}, \frac{4\pi}{4}$$
$$= 0, \frac{\pi}{4}, \frac{\pi}{2}, \frac{3\pi}{4}, \pi$$

The x-intercepts are
$(0,0)$, $\left(\frac{\pi}{4}, 0\right)$, $\left(\frac{\pi}{2}, 0\right)$, $\left(\frac{3\pi}{4}, 0\right)$, $(\pi, 0)$.

5. Sketch the graph.

The period is $\frac{\pi}{4}$, so the horizontal axis is scaled in multiples of $\frac{1}{2}$ of $\frac{\pi}{4} = \frac{\pi}{8}$.

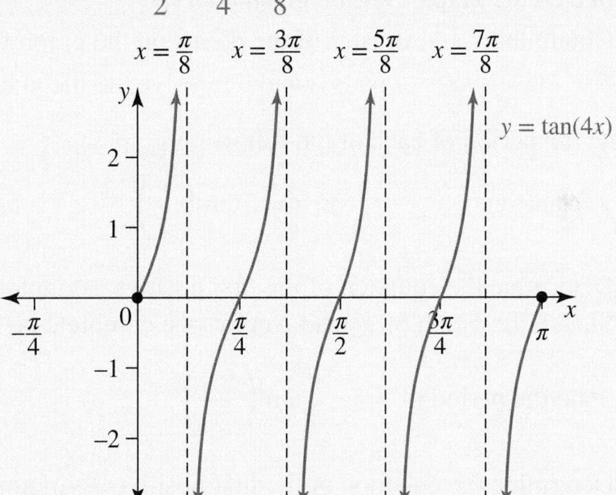

9.5 Exercise

Students, these questions are even better in jacPLUS

Receive immediate feedback and access sample responses

Access additional questions

Track your results and progress

Find all this and MORE in jacPLUS

Technology free

1. **WE8** Sketch the graph of $y = \tan(x)$ for $x \in \left(-\frac{\pi}{2}, \frac{3\pi}{2}\right)$.

2. Sketch each of the following over the given interval.
   a. $y = \tan(x), x \in \left(\frac{\pi}{2}, \frac{3\pi}{2}\right)$
   b. $y = \tan(x), x \in [-\pi, 0]$
   c. $y = \tan(x), x \in \left(0, \frac{5\pi}{2}\right)$

3. Sketch the graph of $y = \tan(x)$ over the interval for which $x \in [-3\pi, 3\pi]$ and state the domain and range of the graph.

4. **WE9** a. On the same diagram, sketch the graphs of $y = -\tan(x)$ and $y = 3\tan(x)$ for $x \in \left(-\frac{\pi}{2}, \frac{\pi}{2}\right)$.

   b. Sketch the graph of $y = \tan(x) + \sqrt{3}$ for $x \in \left(-\frac{\pi}{2}, \frac{\pi}{2}\right)$.

5. On the same diagram, sketch the graphs of $y = \tan(x)$ and $y = \tan(x) + 1$ for $x \in [0, 2\pi]$, showing any intercepts with the coordinate axes.

6. Sketch the following two graphs over the interval $x \in [0, 2\pi]$.

   a. $y = 4\tan(x)$

   b. $y = -0.5\tan(x)$

   c. On the same set of axes, sketch the graphs of $y = -\tan(x)$, $y = 1 - \tan(x)$ and $y = -1 - \tan(x)$ for $-\frac{\pi}{2} < x < \frac{\pi}{2}$.

7. Consider the function defined by $y = \tan(x), -\pi \leq x \leq \pi$.

   a. Sketch the graph over the given interval.

   b. Obtain the x-coordinates of the points on the graph for which the y-coordinate is $-\sqrt{3}$.

   c. Use the answers to parts **a** and **b** to solve the inequation $\tan(x) + \sqrt{3} < 0$ for $x \in [-\pi, \pi]$.

8. State the period of each of the following graphs.

   a. $y = \tan(6x)$

   b. $y = 5\tan\left(\frac{x}{4}\right)$

   c. $y = -2\tan\left(\frac{3x}{2}\right) + 5$

   d. $y = \tan(\pi x)$

9. a. Determine the equation of the first positive asymptote for the graph of $y = \tan(5x)$.

   b. Sketch the graph of $y = \tan(5x)$ for one complete period.

10. a. State the period of $y = -\frac{1}{4}\tan\left(\frac{\pi x}{2}\right)$.

    b. Determine the equation of the first positive asymptote for $y = -\frac{1}{4}\tan\left(\frac{\pi x}{2}\right)$.

    c. Sketch the graph of $y = -\frac{1}{4}\tan\left(\frac{\pi x}{2}\right)$ for two complete periods.

11. **WE10** Sketch the graph of $y = \tan(3x), 0 \leq x \leq \pi$.

12. Sketch the graph of $y = -2\tan\left(\frac{x}{2}\right), 0 \leq x \leq 2\pi$.

13. For $x \in [0, \pi]$, sketch the graphs of the following functions.

    a. $y = \tan(4x)$

    b. $y = 2\tan(2x)$

    c. $y = -\tan\left(\frac{x}{3}\right)$

14. A trigonometric graph has the general equation $y = a\tan(nx)$. The equation of the first positive asymptote is $x = \frac{\pi}{8}$. The graph passes through the point $\left(\frac{\pi}{3}, \frac{\sqrt{3}}{2}\right)$.

    Determine the values of a and n.

15. A trigonometric graph has the general equation $y = a\tan(nx)$. The equation of the first negative asymptote is $x = -\frac{3\pi}{4}$. The graph passes through the point $\left(-\frac{\pi}{4}, 3\right)$.

    a. Determine the values of a and n.

    b. State the period of the graph.

16. The graph of $y = \tan(bx), 0 \le x \le \frac{3\pi}{2}$ is shown.

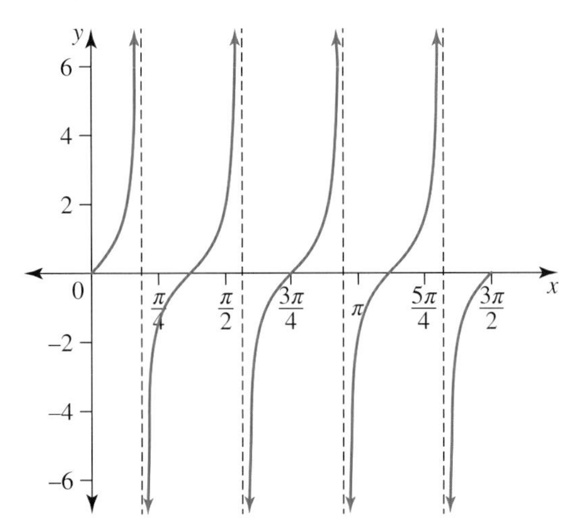

a. State the number of cycles shown.
b. State the period.
c. Obtain the value of the positive constant b.
d. Give the equations of the four vertical asymptotes.

17. Given that the first negative value for which $f(x) = a - \tan(cx)$ is undefined is $x = -\frac{6}{5}$ and that $f(-12) = -2$, calculate the values of a and c.

Technology active

18. For $x \in [0, \pi]$, sketch the graphs of the following functions.

a. $y = 3\tan(x) + 2$ b. $y = 3(1 - \tan(x))$ c. $y = \tan(3x) - 1$

19. Consider the function $f: [-3, 6] \setminus D \to R, f(x) = \tan\left(\frac{\pi x}{3}\right)$.

a. Evaluate $f(0)$.
b. Obtain the values of x for which the function is not defined and hence state the set D excluded from the domain.
c. Sketch the graph of $y = f(x)$.
d. State the values of x for which $f(x) = 1$.

20. An underground earthquake triggers an ocean tsunami event. The height, h metres, of the tsunami wave above mean sea level after t minutes is modelled by $h = 12\tan\left(\frac{\pi}{144}t\right), 0 \le t \le 48$.

a. Calculate the height of the wave after the first 24 minutes.
b. Determine how many minutes it took for the height of the tsunami wave to reach 12 metres.
c. The severity of the tsunami started to ease after 48 minutes. Calculate the peak height that the wave reached. Describe how this compares with the highest tsunami wave of 40.5 metres in the Japanese tsunami of 2011.
d. Sketch the height against time for $0 \le t \le 48$.

9.5 Exam questions

Question 1 (1 mark) TECH-ACTIVE

MC Select the statement that is *not* true about the graph $y = \tan(x)$ for the domain $[0, 2\pi]$.

- **A.** The asymptotes are one period apart.
- **B.** The x-intercepts occur at $x = 0, \pi, 2\pi$.
- **C.** Two cycles are completed.
- **D.** The range is $[-1, 1]$.
- **E.** The mean position is $y = 0$.

Question 2 (1 mark) TECH-ACTIVE

MC The graph $y = \tan(4x), 0 \le x \le \frac{\pi}{2}$ has vertical asymptotes at

- **A.** $x = \frac{\pi}{8}, \frac{3\pi}{8}$
- **B.** $x = \frac{\pi}{6}, \frac{3\pi}{6}$
- **C.** $x = \frac{\pi}{4}, \frac{\pi}{2}$
- **D.** $x = 0, \frac{\pi}{2}$
- **E.** $x = \frac{2\pi}{9}, \frac{4\pi}{9}$

Question 3 (5 marks) TECH-FREE

Sketch the graph of $y = 2\tan(x),\ 0 \le x \le 2\pi$.

More exam questions are available online.

9.6 Trigonometric identities and properties

LEARNING INTENTION

At the end of this subtopic you should be able to:

- simplify trigonometric expressions using the Pythagorean identity, complementary and symmetry properties
- deduce a trigonometric ratio from another given ratio.

There are many properties and relationships that exist between the trigonometric functions. Two of these that have previously been established are the symmetry properties and the relationship $\tan(x) = \frac{\sin(x)}{\cos(x)}$. Some others will now be considered.

9.6.1 Pythagorean identity

Consider again the definitions of the sine and cosine functions. The trigonometric point P $[\theta]$ lies on the circumference of the unit circle with equation $x^2 + y^2 = 1$. By definition, the Cartesian coordinates of P $[\theta]$ are $x = \cos(\theta), y = \sin(\theta)$. Substituting these coordinate values into the equation of the circle gives:

$$\begin{aligned} x^2 + y^2 &= 1 \\ (\cos(\theta))^2 + (\sin(\theta))^2 &= 1 \\ \cos^2(\theta) + \sin^2(\theta) &= 1 \end{aligned}$$

This relationship is known as the **Pythagorean identity**. It is a true statement for any value of θ and can be rearranged. These identities should be committed to memory. Together with the application of standard algebraic techniques, they can enable other relationships to be verified.

Pythagorean identity

$$\cos^2(\theta) + \sin^2(\theta) = 1$$
$$\sin^2(\theta) = 1 - \cos^2(\theta)$$
$$\cos^2(\theta) = 1 - \sin^2(\theta)$$

WORKED EXAMPLE 11 Simplifying trigonometric expressions

a. **Simplify $2\sin^2(\theta) + 2\cos^2(\theta)$.**

b. **Show that $\dfrac{3 - 3\cos^2(A)}{\sin(A)\cos(A)} = 3\tan(A)$.**

THINK	WRITE
a. 1. Factorise the given expression. *Note:* $\sin^2(\theta) + \cos^2(\theta) = 1$	a. $2\sin^2(\theta) + 2\cos^2(\theta) = 2(\sin^2(\theta) + \cos^2(\theta))$
2. Simplify by using the Pythagorean identity.	$= 2 \times 1$ $= 2$
b. 1. Simplify the expression on the left-hand side.	b. $\dfrac{3 - 3\cos^2(A)}{\sin(A)\cos(A)} = 3\tan(A)$ There is a common factor in the numerator of the LHS. $\text{LHS} = \dfrac{3 - 3\cos^2(A)}{\sin(A)\cos(A)}$ $= \dfrac{3(1 - \cos^2(A))}{\sin(A)\cos(A)}$
2. Apply a form of the Pythagorean identity.	Since $1 - \cos^2(A) = \sin^2(A)$, $\text{LHS} = \dfrac{3\sin^2(A)}{\sin(A)\cos(A)}$
3. Complete the simplification to show the two sides are equal.	Cancelling $\sin(A)$ from the numerator and denominator, $\text{LHS} = \dfrac{3\sin(A)}{\cos(A)}$ $= 3\tan(A)$ $= \text{RHS}$ $\therefore \dfrac{3 - 3\cos^2(A)}{\sin(A)\cos(A)} = 3\tan(A)$

9.6.2 Deducing trigonometric ratios for any real number

Given $\sin(\theta) = \frac{3}{5}$, where θ is acute, we have used the Pythagorean triple '3, 4, 5' to create a right-angled triangle from which to deduce $\cos(\theta) = \frac{4}{5}$ and $\tan(\theta) = \frac{3}{4}$ without actually evaluating θ. This same method can be employed to find trigonometric ratios for any real-valued θ if one ratio is known. The additional step needed is to ensure the quadrant in which the triangle lies is taken into consideration and the $\pm$ signs allocated as appropriate to the trigonometric values for that quadrant.

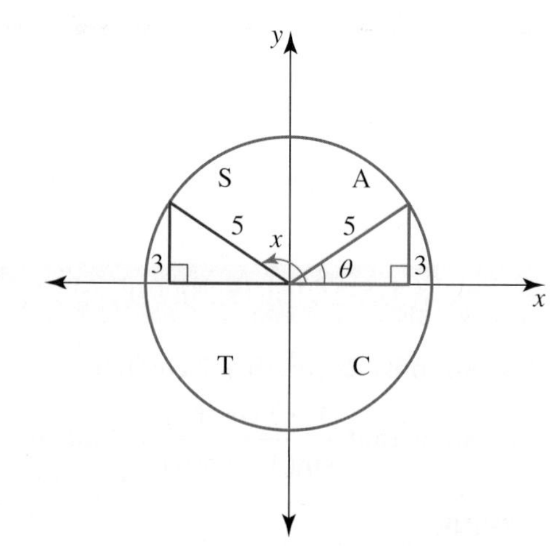

For example, if $\sin(x) = \frac{3}{5}, \frac{\pi}{2} \leq x \leq \pi$, the triangle would lie in the second quadrant, giving $\cos(x) = -\frac{4}{5}$ and $\tan(x) = -\frac{3}{4}$.

In the diagram shown, the circle in which the triangles are drawn has a radius of 5 units. In practice, just the first quadrant triangle needs to be drawn as it, together with the appropriate sign from the ASTC diagram, will determine the values for the other quadrants.

An alternative approach

Alternatively, the Pythagorean identity could be used so that given $\sin(x) = \frac{3}{5}, \frac{\pi}{2} \leq x \leq \pi$, we can write $\cos^2(x) = 1 - \sin^2(x)$ and substitute the value of $\sin(x)$. This gives:

$$\cos^2(x) = 1 - \left(\frac{3}{5}\right)^2 = 1 - \frac{9}{25} = \frac{16}{25}$$

$$\therefore \cos(x) = \pm\sqrt{\frac{16}{25}}$$

As $\cos(x) < 0$ for $\frac{\pi}{2} < x < \pi$, the negative square root is required.

$$\cos(x) = -\sqrt{\frac{16}{25}}$$
$$= -\frac{4}{5}$$

The value of $\tan(x)$ can then be calculated from the identity $\tan(x) = \frac{\sin(x)}{\cos(x)}$.

$$\tan(x) = \frac{3}{5} \div -\frac{4}{5}$$
$$= -\frac{3}{4}$$

Although both methods are appropriate and both are based on Pythagoras' theorem, the triangle method is often the quicker to apply.

WORKED EXAMPLE 12 Evaluating exact values

Given $\cos(x) = -\frac{1}{3}, \pi \leq x \leq \frac{3\pi}{2}$, deduce the exact values of $\sin(x)$ and $\tan(x)$.

THINK	WRITE
1. Identify the quadrant required.	$\cos(x) = -\frac{1}{3}, \pi \leq x \leq \frac{3\pi}{2}$ The sign of $\cos(x)$ indicates either the second or the third quadrant. The condition that $\pi \leq x \leq \frac{3\pi}{2}$ makes it the third, not the second quadrant. Quadrant 3 is required.
2. Draw a right-angled triangle as if for the first quadrant, label its sides in the ratio given and calculate the third side.	First quadrant, $\cos(x) = \frac{1}{3}$ The ratio adjacent : hypotenuse is 1 : 3. 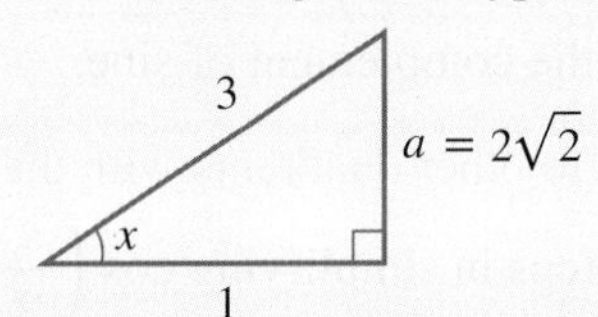 Using Pythagoras' theorem: $a^2 + 1^2 = 3^2$ $a = \sqrt{8}$ (positive root required) $a = 2\sqrt{2}$
3. Calculate the required values. *Note:* The alternative approach is to use the Pythagorean identity to obtain $\sin(x)$ and then to use $\tan(x) = \frac{\sin(x)}{\cos(x)}$ to obtain $\tan(x)$.	From the triangle, the ratio opposite : hypotenuse is $2\sqrt{2} : 3$ and the ratio opposite : adjacent is $2\sqrt{2} : 1$. In quadrant 3, sine is negative and tangent is positive. Allocating these signs gives $\sin(x) = -\frac{2\sqrt{2}}{3}$ and $\tan(x) = +\frac{2\sqrt{2}}{1}$. For the third quadrant where $\cos(x) = -\frac{1}{3}$, $\sin(x) = -\frac{2\sqrt{2}}{3}$, $\tan(x) = 2\sqrt{2}$.

9.6.3 Complementary properties for sine and cosine

In geometry, complementary angles add to 90° or $\frac{\pi}{2}$ in radian measure, so θ and $\frac{\pi}{2} - \theta$ are called a complementary pair.

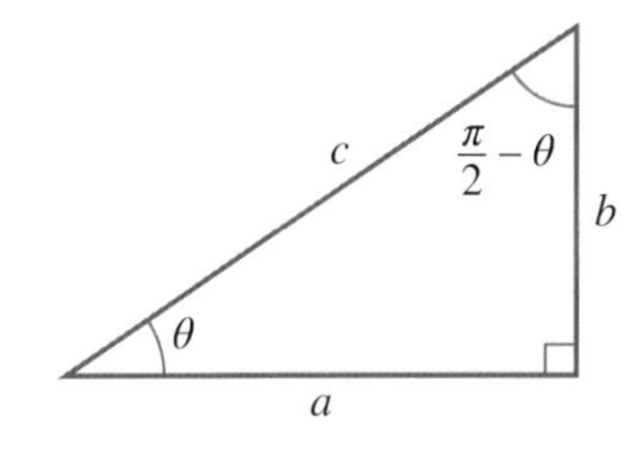

In a right-angled triangle, the trigonometric ratios give $\sin(\theta) = \frac{b}{c} = \cos\left(\frac{\pi}{2} - \theta\right)$ and $\cos(\theta) = \frac{a}{c} = \sin\left(\frac{\pi}{2} - \theta\right)$.

For example, $\sin(60°) = \cos(90° - 60°)$

$$= \cos(30°)$$

and $\cos\left(\frac{\pi}{3}\right) = \sin\left(\frac{\pi}{2} - \frac{\pi}{3}\right)$

and $\cos\left(\frac{\pi}{3}\right) = \sin\left(\frac{\pi}{6}\right)$.

Complementary properties

$$\sin\left(\frac{\pi}{2} - \theta\right) = \cos(\theta)$$

$$\cos\left(\frac{\pi}{2} - \theta\right) = \sin(\theta)$$

The name cosine is derived from the complement of sine.

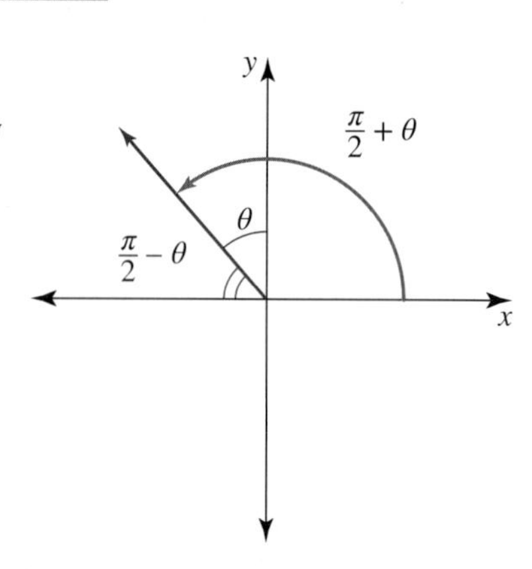

These properties can be extended to other quadrants with the aid of symmetry properties. To illustrate this, the steps in simplifying $\cos\left(\frac{\pi}{2} + \theta\right)$ would be as follows.

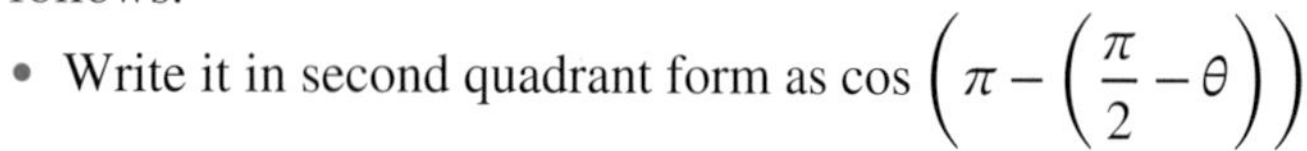

- Write it in second quadrant form as $\cos\left(\pi - \left(\frac{\pi}{2} - \theta\right)\right)$

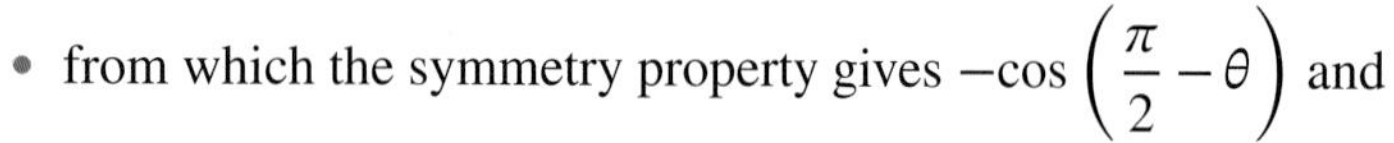

- from which the symmetry property gives $-\cos\left(\frac{\pi}{2} - \theta\right)$ and
- from which the complementary property gives $-\sin(\theta)$.

  Therefore, $\cos\left(\frac{\pi}{2} + \theta\right) = -\sin(\theta)$.

Distinction between complementary and symmetry properties

Note that for complementary properties θ is measured from the vertical axis whereas the base needed for symmetry properties is always measured from the horizontal axis. Hence, the three trigonometric points $\left[\frac{\pi}{2} + \theta\right], \left[\frac{3\pi}{2} - \theta\right], \left[\frac{3\pi}{2} + \theta\right]$ will each have the same base $\frac{\pi}{2} - \theta$, whereas the three trigonometric points $[\pi - \theta], [\pi + \theta], [2\pi - \theta]$ each have the base θ.

$\cos\left(\frac{\pi}{2} + \theta\right) = -\cos\left(\frac{\pi}{2} - \theta\right)$ using the second quadrant symmetry property

$= -\sin(\theta)$ using the complementary property.

With practice, seeing $\cos\left(\frac{\pi}{2} + \theta\right)$ should trigger the reaction that this is a complementary $\sin(\theta)$ with the second-quadrant negative sign, since the original cosine expression is negative in the second quadrant. This is much the same as $\cos(\pi + \theta)$ triggering the reaction this is a symmetry $\cos(\theta)$ with the third-quadrant negative sign.

$\cos\left(\frac{\pi}{2} + \theta\right) = -\sin(\theta)$ (complementary property)

$\cos(\pi + \theta) = -\cos(\theta)$ (symmetry property)

WORKED EXAMPLE 13 Using complementary and symmetry properties

a. Simplify $\sin\left(\frac{3\pi}{2}+\theta\right)$.

b. Give a value for t if $\cos\left(\frac{2\pi}{5}\right)=\sin(t)$.

THINK	WRITE
a. 1. Write the given expression in the appropriate quadrant form.	a. $\sin\left(\frac{3\pi}{2}+\theta\right)$ Quadrant 4 $=\sin\left(2\pi-\left(\frac{\pi}{2}-\theta\right)\right)$
2. Apply the symmetry property.	$=-\sin\left(\frac{\pi}{2}-\theta\right)$ using the symmetry property
3. Apply the complementary property.	$=-\cos(\theta)$ using the complementary property
4. State the answer. *Note:* Since θ is inclined to the vertical axis, this implies the complementary trigonometric function is required together with the appropriate sign from the ASTC diagram and the answer could be written in one step.	$\therefore \sin\left(\frac{3\pi}{2}+\theta\right)=-\cos(\theta)$
b. 1. Apply a complementary property to one of the given expressions.	b. $\cos\left(\frac{2\pi}{5}\right)$ Quadrant 1 Using the complementary property $\cos(\theta)=\sin\left(\frac{\pi}{2}-\theta\right)$, $\cos\left(\frac{2\pi}{5}\right)=\sin\left(\frac{\pi}{2}-\frac{2\pi}{5}\right)$ $=\sin\left(\frac{\pi}{10}\right)$
2. State an answer. *Note:* Other answers are possible using the symmetry properties of the sine function.	Since $\cos\left(\frac{2\pi}{5}\right)=\sin(t)$, $\sin\left(\frac{\pi}{10}\right)=\sin(t)$ $\therefore t=\frac{\pi}{10}$

on Resources

Interactivity Complementary properties for sin and cos (int-2979)

9.6 Exercise

Students, these questions are even better in jacPLUS

Receive immediate feedback and access sample responses

Access additional questions

Track your results and progress

Find all this and MORE in jacPLUS

Technology free

1. Evaluate each of the following.

   a. $\sin^2(45°)$
   b. $\cos^2\left(\frac{\pi}{3}\right)$
   c. $\tan^2\left(\frac{\pi}{6}\right)$
   d. $\cos^2(\pi)$

2. Pair each of the given values $2, \sqrt{3}, 1, 0$ with the matching expression from **a** to **d**.

   a. $\sin^2(2) + \cos^2(2)$
   b. $\dfrac{\sin(60°)}{\cos(60°)}$
   c. $1 - \sin^2\left(\frac{\pi}{2}\right)$
   d. $2\cos^2(D) + 2\sin^2(D)$

3. **WE11** a. Simplify $2 - 2\sin^2(\theta)$.

   b. Show that $\dfrac{\sin^3(A) + \sin(A)\ \cos^2(A)}{\cos^3(A) + \cos(A)\ \sin^2(A)} = \tan(A)$.

4. Simplify $\tan(u) + \dfrac{1}{\tan(u)}$.

5. Simplify each of the following expressions.

   a. $4 - 4\cos^2(\theta)$
   b. $\dfrac{2\sin(\alpha)}{\cos(\alpha)}$
   c. $8\cos^2(\beta) + 8\sin^2(\beta)$
   d. $(1 - \sin(A))(1 + \sin(A))$

6. a. Show that $\tan^2(\theta) + 1 = \dfrac{1}{\cos^2(\theta)}$.

   b. Show that $\cos^3(\theta) + \cos(\theta)\ \sin^2(\theta) = \cos(\theta)$.

   c. Show that $\dfrac{1}{1 - \cos(\theta)} + \dfrac{1}{1 + \cos(\theta)} = \dfrac{2}{\sin^2(\theta)}$.

   d. Show that $(\sin(\theta) + \cos(\theta))^2 + (\sin(\theta) - \cos(\theta))^2 = 2$.

7. **WE12** Given $\cos(x) = -\frac{2}{7}, \frac{\pi}{2} \le x \le \pi$, deduce the exact values of $\sin(x)$ and $\tan(x)$.

8. Given $\tan(x) = -3, \pi \le x \le 2\pi$, deduce the exact values of $\sin(x)$ and $\cos(x)$.

9. a. Given $\tan(x) = -\frac{4}{5}, \frac{3\pi}{2} \le x \le 2\pi$, obtain the exact values of $\sin(x)$ and $\cos(x)$.

   b. Given $\sin(x) = \frac{\sqrt{3}}{2}, \frac{\pi}{2} \le x \le \frac{3\pi}{2}$, obtain the exact values of:

      i. $\cos(x)$
      ii. $1 + \tan^2(x)$.

   c. Given $\cos(x) = -0.2, \pi \le x \le \frac{3\pi}{2}$, obtain the exact values of:

      i. $\tan(x)$
      ii. $1 - \sin^2(x)$
      iii. $\sin(x)\tan(x)$.

10. a. Given $\cos(x) = \dfrac{2\sqrt{3}}{5}$, use the identity $\sin^2(x) + \cos^2(x) = 1$ to obtain $\sin(x)$ if:

   i. x lies in the first quadrant
   ii. x lies in the fourth quadrant.

   b. Calculate $\tan(x)$ for each of parts **a i** and **ii** using $\tan(x) = \dfrac{\sin(x)}{\cos(x)}$.

   c. Check your answer to part **b** using the triangle method to obtain the values of $\sin(x)$ and $\tan(x)$ if x lies in the first quadrant.

11. Given $\sin(\theta) = 0.8$, $0 \le \theta \le \dfrac{\pi}{2}$, evaluate the following.

   a. $\sin(\pi + \theta)$
   b. $\cos\left(\dfrac{\pi}{2} - \theta\right)$
   c. $\cos(\theta)$
   d. $\tan(\pi + \theta)$

12. a. Verify that $\sin\left(\dfrac{\pi}{2} + \theta\right) = \cos(\theta)$ for $\theta = \dfrac{\pi}{6}$.

   b. Verify that $\cos\left(\dfrac{\pi}{2} + \theta\right) = -\sin(\theta)$ for $\theta = \dfrac{\pi}{6}$.

   c. Verify that $\sin\left(\dfrac{3\pi}{2} - \theta\right) = -\cos(\theta)$ for $\theta = \dfrac{\pi}{3}$.

   d. Verify that $\cos\left(\dfrac{3\pi}{2} + \theta\right) = \sin(\theta)$ for $\theta = \dfrac{\pi}{4}$.

13. **WE13** a. Simplify $\cos\left(\dfrac{3\pi}{2} - \theta\right)$.

   b. Give a value for t, if $\sin\left(\dfrac{\pi}{12}\right) = \cos(t)$.

14. Simplify $\sin\left(\dfrac{\pi}{2} + \theta\right) + \sin(\pi + \theta)$.

15. Simplify the following expressions.

   a. $\cos\left(\dfrac{3\pi}{2} + \theta\right)$
   b. $\cos\left(\dfrac{3\pi}{2} - \theta\right)$
   c. $\sin\left(\dfrac{\pi}{2} + \theta\right)$
   d. $\sin\left(\dfrac{5\pi}{2} - \theta\right)$

Technology active

16. Simplify the following expressions.

   a. $\dfrac{\cos(90° - a)}{\cos(a)}$

   b. $1 - \sin^2\left(\dfrac{\pi}{2} - \theta\right)$

   c. $\cos(\pi - x) + \sin\left(x + \dfrac{\pi}{2}\right)$

   d. $\dfrac{1 - \cos^2\left(\frac{\pi}{2} - \theta\right)}{\cos^2(\theta)}$

17. a. Give a value for a for which $\cos(a) = \sin(27°)$.
  b. Give a value for b for which $\sin(b) = \cos\left(\frac{\pi}{9}\right)$.
  c. Give a value, to 2 decimal places, for c so that $\sin(c) = \cos(0.35)$.
  d. Identify two values between 0 and 2π for d so that $\cos(d) = \sin\left(\frac{\pi}{4}\right)$.

18. Solve the following equations for $x, 0 \le x \le 2\pi$.
  a. $\cos^2(x) + 3\sin(x) - 1 = 0$
  b. $3\cos^2(x) + 6\sin(x) + 3\sin^2(x) = 0$
  c. $\frac{1-\cos(x)}{1+\sin(x)} \times \frac{1+\cos(x)}{1-\sin(x)} = 3$
  d. $\sin^2(x) = \cos(x) + 1$

19. Show that the following are true statements.
  a. $1 - \cos^2(40°) = \cos^2(50°)$
  b. $\frac{\cos^3(x) - \sin^3(x)}{\cos(x) - \sin(x)} = 1 + \sin(x)\cos(x)$
  c. $\frac{\sin^3(\theta) + \cos\left(\frac{\pi}{2} - \theta\right)\cos^2(\theta)}{\cos(\theta)} = \tan(\theta)$
  d. $\cos(2\pi - A)\cos(2\pi + A) - \cos\left(\frac{3\pi}{2} - A\right)\cos\left(\frac{3\pi}{2} + A\right) = 1$

20. Use CAS technology to simplify the following.
  a. $\sin(5.5\pi - x) \div \cos\left(x - \frac{3\pi}{2}\right)$
  b. $\sin(3x)\cos(2x) - 2\cos^2(x)\sin(3x)$

9.6 Exam questions

Question 1 (1 mark) TECH-ACTIVE

MC State which of the following is true.

A. $\tan(x) = \frac{\cos(x)}{\sin(x)}$
B. $\sin^2(x) = 1 + \cos^2(x)$
C. $\sin(x) + \cos(x) = 1$
D. $\cos(x) = \sqrt{1 - \sin^2(x)}$
E. $5\cos^2(x) = 5\sin^2(x) + 5$

Question 2 (1 mark) TECH-ACTIVE

MC Given $\sin(\theta) = \frac{3}{5}, 0 < \theta < \frac{\pi}{2}$, $\cos(\theta)$ is

A. $\frac{4}{5}$
B. $\frac{5}{8}$
C. $\frac{3}{4}$
D. $\frac{2}{5}$
E. $\frac{3}{8}$

Question 3 (4 marks) TECH-FREE

Given $\tan(x) = -\frac{2}{5}, 0 \le x \le \pi$, deduce the exact values of $\sin(x)$ and $\cos(x)$.

More exam questions are available online.

9.7 Review

9.7.1 Summary

doc-37025

Hey students! Now that it's time to revise this topic, go online to:

Access the topic summary

Review your results

Watch teacher-led videos

Practise exam questions

Find all this and MORE in jacPLUS

9.7 Exercise

Technology free: short answer

1. Solve the following equations for x.
   a. $\sqrt{6}\cos(x) = -\sqrt{3}, 0 \le x \le 2\pi$
   b. $2 - 2\cos(x) = 0, 0 \le x \le 4\pi$
   c. $2\sin(x) = \sqrt{3}, -2\pi \le x \le 2\pi$
   d. $\sqrt{5}\sin(x) = \sqrt{5}\cos(x), 0 \le x \le 2\pi$
2. Solve the following equations for x.
   a. $8\sin(3x) + 4\sqrt{2} = 0, -\pi \le x \le \pi$
   b. $\tan(2x°) = -2, 0 \le x \le 270$, given $\tan^{-1}(2) = 63.43°$
   c. $2\sin^2(x) - \sin(x) - 3 = 0, 0 \le x \le 2\pi$
   d. $2\sin^2(x) - 3\cos(x) - 3 = 0, 0 \le x \le 2\pi$
3. If $\cos(\theta) = p$, express the following in terms of p.
   a. $\sin\left(\frac{\pi}{2} - \theta\right)$
   b. $\sin\left(\frac{3\pi}{2} + \theta\right)$
   c. $\sin^2(\theta)$
   d. $\cos(5\pi - \theta)$
4. a. Given $\tan(x) = \frac{5}{2}, x \in \left(\pi, \frac{3\pi}{2}\right)$, obtain the exact values for $\cos(x)$ and $\sin(x)$.
   b. Given $\sin(y) = -\frac{\sqrt{5}}{3}, y \in \left(\frac{3\pi}{2}, 2\pi\right)$, obtain the exact values for $\cos(y)$ and $\tan(y)$.
5. Sketch the following graphs, stating the range and identifying the key features.
   a. $y = 5 - 3\cos(x), 0 \le x \le 2\pi$
   b. $y = 5\sin(3x), 0 \le x \le 2\pi$
   c. $y = -4\tan(x), 0 \le x \le 2\pi$
   d. $y = 1 + 2\cos\left(\frac{x}{2}\right), -2\pi \le x \le 3\pi$
6. a. Simplify $\sin^2(4x) + \cos^2(4x)$.
   b. Show that $\cos^2(A)(1 + 2\tan(A)) + \sin^2(A) = (\cos(A) + \sin(a))^2$.
   c. Simplify $(1 + \cos(\theta))\left(1 - \sin\left(\theta + \frac{\pi}{2}\right)\right)$.

Technology active: multiple choice

7. **MC** The trigonometric equation $f(x) = 5$ will have solutions if:
   A. f is the sine function
   B. f is the cosine function
   C. f is the tangent function
   D. f is either of the sine or cosine function
   E. f is any of the functions sine, cosine or tangent

8. **MC** The number of solutions to the equation $\cos(x)=0, -\frac{\pi}{2} \le x \le \frac{5\pi}{2}$ is:

A. 1
B. 2
C. 3
D. 4
E. 5

9. **MC** The solutions to the equation $\sin(x)=\frac{1}{2}, 0 \le x \le 2\pi$ are:

A. $x=\frac{\pi}{6}, \frac{5\pi}{6}$
B. $x=\frac{2\pi}{3}, \frac{4\pi}{3}$
C. $x=\frac{5\pi}{6}, \frac{7\pi}{6}$
D. $x=\frac{4\pi}{3}, \frac{5\pi}{3}$
E. $x=\frac{7\pi}{6}, \frac{11\pi}{6}$

10. **MC** The period and amplitude for $y=-4\cos\left(\frac{x}{4}\right)$ are:

A. $2\pi, 4$
B. $\frac{\pi}{2}, 4$
C. $8\pi, -4$
D. $\frac{\pi}{2}, -4$
E. $8\pi, 4$

11. **MC** A possible equation for the given graph could be:

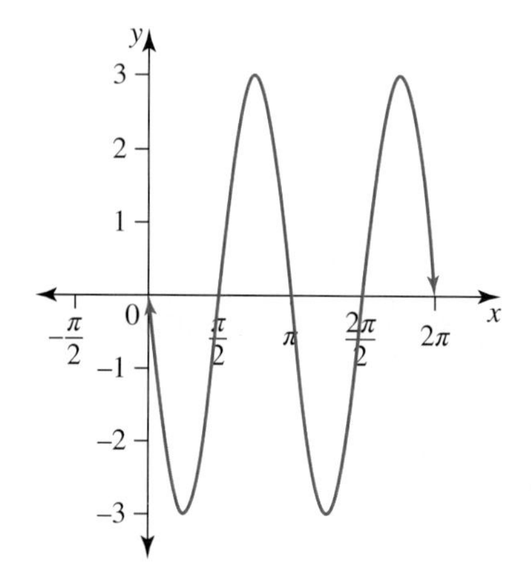

A. $y=-3\cos(2x)$
B. $y=3\cos\left(\frac{x}{2}\right)$
C. $y=-3\sin(2x)$
D. $y=-3\sin\left(\frac{x}{2}\right)$
E. $y=3\sin\left(\frac{x}{2}\right)$

12. **MC** The range of the graph of $y=2+4\cos\left(\frac{3x}{2}\right)$ is:

A. R
B. $[2,6]$
C. $[-2,6]$
D. $[-6,2]$
E. $\left[0, \frac{4\pi}{3}\right]$

13. **MC** The period of the graph of $y=2\tan(3x)$ is:

A. $\frac{\pi}{3}$
B. $\frac{\pi}{2}$
C. π
D. $\frac{2\pi}{3}$
E. 3π

14. **MC** The number of vertical asymptotes for the graph of $y = \tan(x)$ over the interval $x \in (-\pi, 2\pi)$ is:

A. 1
B. 2
C. 3
D. 4
E. 5

15. **MC** The expression $\sqrt{9 - 9\sin^2(\theta)}, 0 \le \theta \le \frac{\pi}{2}$, is equal to:

A. $\sin(\theta)$
B. $3 - 3\sin(\theta)$
C. $9\cos(\theta)$
D. $3\cos(\theta)$
E. 0

16. **MC** $\cos\left(\frac{\pi}{2} + x\right)$ is equal to:

A. $-\cos(x)$
B. $-\sin(x)$
C. $\cos(x)$
D. $\sin(x)$
E. $\sqrt{1 + \sin^2\left(\frac{\pi}{2} + x\right)}$

Technology active: extended response

17. Consider the function $f: [-12, 18] \to R, f(x) = 10\sin\left(\frac{\pi x}{6}\right) + 3$.

a. i. Calculate the maximum and minimum values of the function.
   ii. Obtain the first positive values of x for which a maximum and a minimum occur.
b. Evaluate $f(-12)$ and $f(18)$.
c. Sketch the graph of $y = f(x)$ without calculating the x-intercepts.
d. Determine how many solutions there are to the equation $f(x) = 0$, Obtain the values of these solutions to 1 decimal place.

18. a. The solutions to the equation $\cos(x) = \sin(x) + 1$ can be found graphically by sketching the graphs of $y = \cos(x)$ and $y = \sin(x) + 1$. Use this method to solve the equation for $x \in [0, 2\pi]$.
b. i. Use a graphical method to determine the number of roots of the equation $\cos(x) = \tan(x), x \in [0, 2\pi]$ and show that one of these roots lies in the interval for which $0.6 \le x \le 0.8$.
   ii. Use an algebraic method to show that the sine value of the root of the equation $\cos(x) = \tan(x)$ for which $0 < x < \frac{\pi}{2}$ is given by $\sin(x) = \frac{1}{2}\left(\sqrt{5} - 1\right)$. Calculate this root to 3 decimal places.

19. Sound waves emitted from an amplifier during a band practice session are illustrated over a 5-second interval, with the vertical axis measuring the sound in decibels.

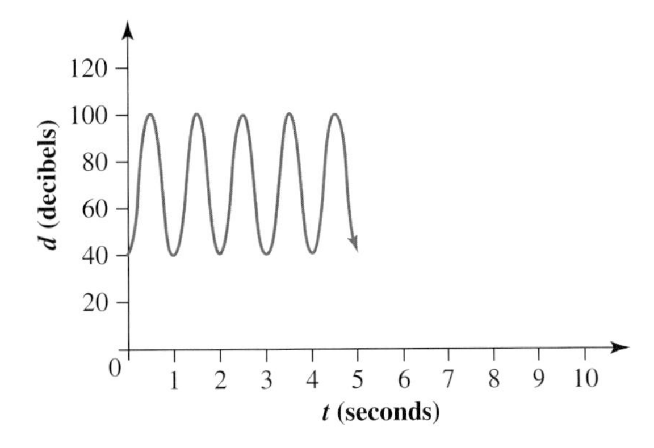

a. State the period of the sound waves.
b. State the amplitude of the sound waves.
c. Give the values of a, b and c for which the equation of the graph can be expressed in the form $d = a\cos(bt) + c$ where d is the number of decibels of the sound after t seconds.
d. Normal speech range is 50 to 60 decibels. Calculate the percentage of the first second for which the sound level fit this category.
e. Determine how many times the sound was at its greatest level during an interval of 10 minutes of the band practice session.
f. Sound under 80 decibels is considered to be a safe level. Calculate the percentage of time during the band practice when the level of the sound was potentially damaging to anyone present with unprotected hearing.

20. Part of a tree was brought down by gale force winds one evening. The following day a tree lopper tidied up the damage and reduced the tree to just a stump, the cross-section of which is shown in the diagram.

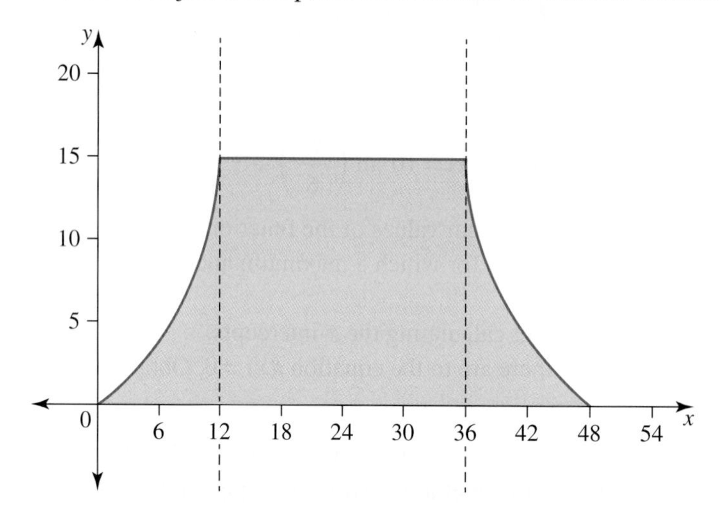

The full width of the base is 48 cm, the horizontal top of the stump has a width of 24 cm, and the vertical height of the stump is 15 cm.
The tree stump cross-section can be modelled by a hybrid function composed of three parts, with y cm as its height at width x cm.

a. The part of the tree stump for which $x \in [0, 12]$ is modelled by the function $y = a\tan(bx)$ over an interval of a quarter of its period. Obtain the values of a and b.
b. State the rule for the part of the tree stump for which $x \in (12, 36)$.
c. Given the two end parts of the tree stump are symmetric, form the equation for the tangent function, a quarter of a period of which is the model of the section where $x \in [36, 48]$. Express the rule in the form $y = c\tan(dx)$.
d. Use the answers to parts **a–c** to write the model of the tree stump as a hybrid function.
e. The two tangent curves in the model can be approximated by line segments joining the end points of each curve.
    i. For $x \in [0, 12]$, sketch $y = a\tan(bx)$ together with the line segment joining the end points of this graph. State the equation of the line segment.
    ii. State the equation of the line segment at the other end.
f. Using these line segments from part **e**, estimate the cross-sectional area of the tree stump. State whether this estimate of the area is an underestimate or an overestimate of the actual area.

9.7 Exam questions

Question 1 (4 marks) TECH-FREE

Solve the following equation for x.

$\sin(2x) = \frac{1}{2}, 0 \le x \le 2\pi$

Question 2 (1 mark) TECH-ACTIVE

MC For $\sin(x) = \sqrt{3}\cos(x)$, the solutions of x, where $0 \le x \le 2\pi$, are

A. $\frac{\pi}{6}, \frac{4\pi}{6}$

B. $\frac{\pi}{\sqrt{3}}, \frac{4\pi}{\sqrt{3}}$

C. $\frac{\pi}{4}, \frac{3\pi}{4}$

D. $\frac{\pi}{3}, \frac{4\pi}{3}$

E. $\frac{\pi}{2}, \frac{5\pi}{2}$

Question 3 (3 marks) TECH-FREE

Find the period, amplitude and range for the graph of $y = -3\cos(5x) + 7$.

Question 4 (1 mark) TECH-ACTIVE

MC For the equation $y = r\cos(\pi x) + c$, the maximum value of y is 19 and the minimum value is 8. The values of r and c are

A. 5.5 and 13.5

B. 8.0 and 11.0

C. 4.0 and 11.5

D. 3.5 and 10.5

E. 8.0 and 19.0

Question 5 (3 marks) TECH-ACTIVE

A windmill manufacturer attaches a flashing light on the far end of the rotor blade of a particular model to help scare birds from flying into the blades. The height of the light is given by $h = 80 - 25\cos\left(\frac{\pi}{2}t\right)$, where h is the height in metres above ground level after t seconds.

a. Determine how far above the ground the light is initially. **(1 mark)**

b. Calculate how many revolutions the rotor blade will complete in a 2-minute time interval. **(1 mark)**

c. Calculate the area that the windmill blade sweeps over (correct to 2 decimal places). **(1 mark)**

More exam questions are available online.

Answers

Topic 9 Trigonometric functions and applications

9.2 Trigonometric equations

9.2 Exercise

1. a. $\frac{\pi}{4}, \frac{7\pi}{4}$ b. $\frac{5\pi}{4}, \frac{7\pi}{4}$ c. $\frac{5\pi}{6}, \frac{11\pi}{6}$
   d. $\frac{5\pi}{6}, \frac{7\pi}{6}$ e. $\frac{\pi}{6}, \frac{5\pi}{6}$ f. $\frac{\pi}{3}, \frac{4\pi}{3}$
2. a. $\frac{\pi}{6}, \frac{5\pi}{6}$ b. $\frac{\pi}{6}, \frac{11\pi}{6}$
   c. $-\frac{5\pi}{4}, -\frac{\pi}{4}, \frac{3\pi}{4}, \frac{7\pi}{4}$
3. a. $\theta = -\frac{7\pi}{4}, -\frac{3\pi}{4}, \frac{\pi}{4}, \frac{5\pi}{4}$
   b. $\theta = -\frac{4\pi}{3}, -\frac{2\pi}{3}, \frac{2\pi}{3}, \frac{4\pi}{3}$
   c. $\theta = -\frac{5\pi}{6}, -\frac{\pi}{6}, \frac{7\pi}{6}, \frac{11\pi}{6}$
4. a. $x = \frac{2\pi}{3}, \frac{5\pi}{3}, \frac{8\pi}{3}$
   b. $t = \frac{\pi}{6}, \frac{5\pi}{6}, \frac{13\pi}{6}, \frac{17\pi}{6}$
   c. $v = -\frac{\pi}{4}, \frac{\pi}{4}, \frac{7\pi}{4}, \frac{9\pi}{4}, \frac{15\pi}{4}, \frac{17\pi}{4}$
5. a. $45°, 315°$ b. $210°, 330°$ c. $60°, 240°$
   d. $135°, 315°$ e. $45°, 135°$ f. $150°, 210°$
6. a. $-315°, -135°, 45°, 225°$
   b. $-270°, -90°, 90°, 270°$
   c. $-90°, 270°$
   d. $-360°, -180°, 0°, 180°, 360°$
7. a. $-\frac{7\pi}{2}, -\frac{3\pi}{2}, \frac{\pi}{2}, \frac{5\pi}{2}$
   b. $0.64, 3.79, 6.93, 10.07$
   c. $x° = \pm 104.5°$, so $x = \pm 104.5$.
8. a. 2 solutions b. $\pm 120°$
9. a. $2.30, 3.98$
   b. $-5.67, -2.53, 0.62, 3.76, 6.90$
   c. $-53.13, -126.87, 233.13$
   d. $78.46, 101.54, 258.46, 281.54$
10. a. $-\pi, 0, \pi, 2\pi, 3\pi, 4\pi$ b. $-\frac{\pi}{2}, \frac{\pi}{2}, \frac{3\pi}{2}, \frac{5\pi}{2}, \frac{7\pi}{2}$
    c. $-\frac{\pi}{2}, \frac{3\pi}{2}, \frac{7\pi}{2}$ d. $0, 2\pi, 4\pi$
    e. $\frac{\pi}{2}, \frac{5\pi}{2}$ f. $-\frac{3\pi}{4}, \frac{\pi}{4}, \frac{5\pi}{4}, \frac{9\pi}{4}, \frac{13\pi}{4}$
11. a. 1 b. $\left\{\frac{1}{2}, \frac{3}{2}\right\}$
12. a. $\frac{\pi}{3}, \frac{4\pi}{3}$ b. $\frac{\pi}{2}$
13. $\frac{\pi}{6}, \frac{5\pi}{6}, \frac{7\pi}{6}, \frac{11\pi}{6}$
14. a. $\frac{\pi}{3}, \frac{4\pi}{3}$
    b. $\frac{5\pi}{6}, \frac{11\pi}{6}$
    c. $\frac{3\pi}{8}, \frac{7\pi}{8}, \frac{11\pi}{8}, \frac{15\pi}{8}$
    d. $\tan^{-1}\left(\frac{4}{3}\right), \ \pi + \tan^{-1}\left(\frac{4}{3}\right)$
    e. $\frac{\pi}{4}, \frac{3\pi}{4}, \frac{5\pi}{4}, \frac{7\pi}{4}$
    f. $\frac{\pi}{4}, \frac{5\pi}{4}, \frac{\pi}{2}, \frac{3\pi}{2}$
15. a. $\frac{\pi}{4}, \frac{3\pi}{4}, \frac{5\pi}{4}, \frac{7\pi}{4}$ b. $\frac{\pi}{2}, \frac{3\pi}{2}$
    c. $\frac{\pi}{2}, \frac{7\pi}{6}, \frac{11\pi}{6}$ d. $\frac{\pi}{4}, \frac{5\pi}{4}$
    e. $\frac{3\pi}{2}$ f. No solution
16. a. $x = \frac{\pi}{8}, \frac{3\pi}{8}, \frac{9\pi}{8}, \frac{11\pi}{8}$ b. $x = \frac{\pi}{6}, \frac{2\pi}{3}, \frac{7\pi}{6}$
17. $x = \frac{2\pi}{3}$
18. a. $\frac{5\pi}{18}, \frac{11\pi}{18}, \frac{17\pi}{18}, \frac{23\pi}{18}, \frac{29\pi}{18}, \frac{35\pi}{18}$
    b. $\frac{2\pi}{9}, \frac{4\pi}{9}, \frac{14\pi}{9}, \frac{16\pi}{9}$
    c. $\frac{\pi}{4}, \frac{3\pi}{4}, \frac{5\pi}{4}, \frac{7\pi}{4}$
    d. $\frac{3\pi}{8}, \frac{7\pi}{8}, \frac{11\pi}{8}, \frac{15\pi}{8}$
19. a. $a = 8$
    b. i. $0.38, 2.76$
       ii. 1.57
       iii. No solution
20. a. $\frac{\sqrt{2}\left(\sqrt{3}+1\right)}{4}$
    b. $-255°, 75°, 105°$

9.2 Exam questions

Note: Mark allocations are available with the fully worked solutions online.

1. A
2. E
3. $x = \frac{5\pi}{6}$ or $\frac{7\pi}{6}$

9.3 Transformations of sine and cosine graphs

9.3 Exercise

1.	Amplitude	Period
a.	6	π
b.	7	4π
c.	2	4π

2. a. Amplitude $\frac{1}{4}$, range $\left[-\frac{1}{4}, \frac{1}{4}\right]$, period $\frac{\pi}{2}$

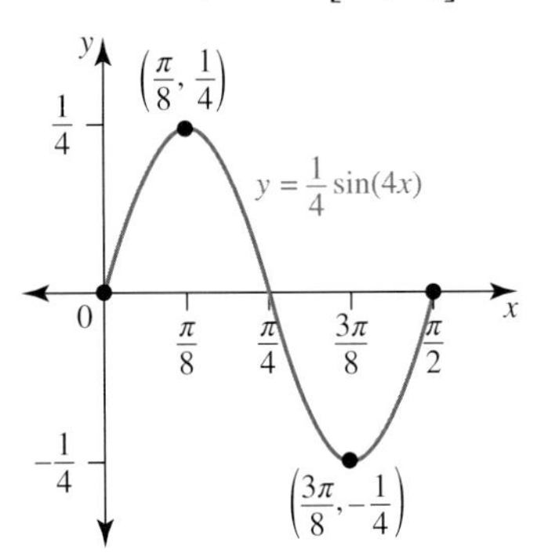

b. Amplitude $\frac{1}{2}$, range $\left[-\frac{1}{2}, \frac{1}{2}\right]$, period 8π

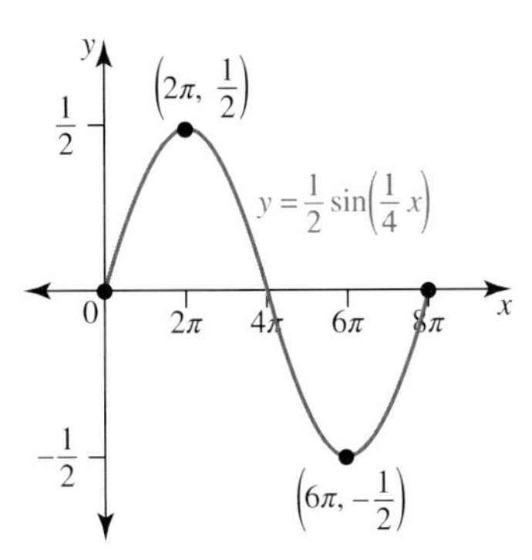

c. Amplitude 1, range $[-1, 1]$, period $\frac{2\pi}{3}$

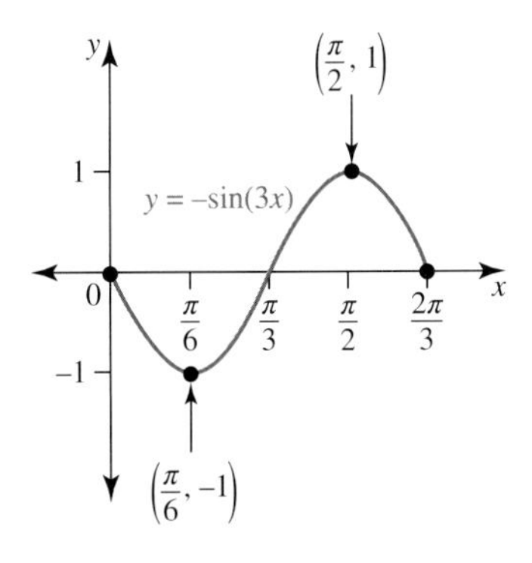

d. Amplitude 3, range $[-3, 3]$, period π

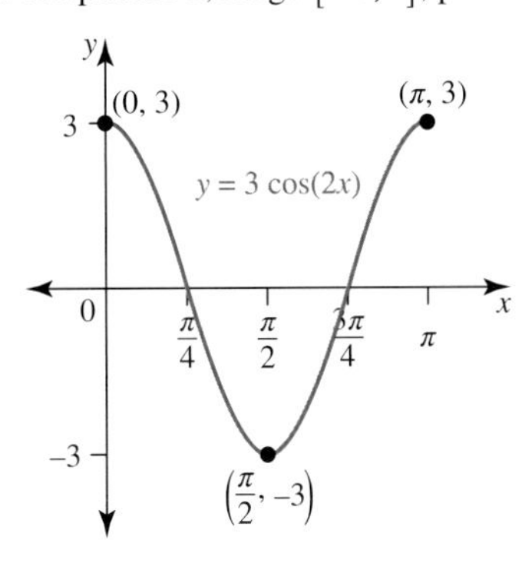

e. Amplitude 6, range $[-6, 6]$, period 5π

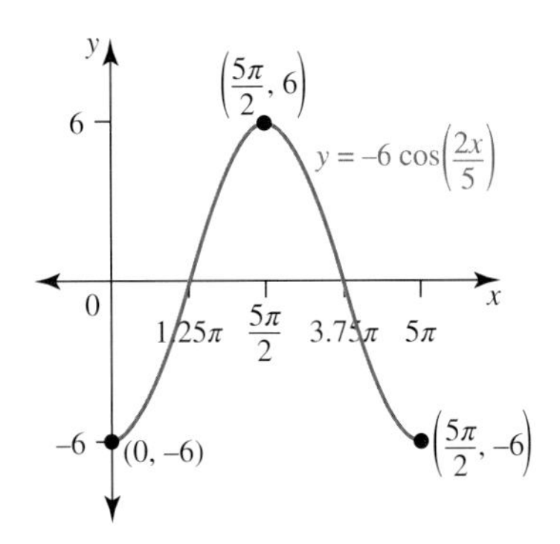

f. Amplitude $\frac{1}{3}$, range $\left[-\frac{1}{3}, \frac{1}{3}\right]$, period $\frac{2}{5}$

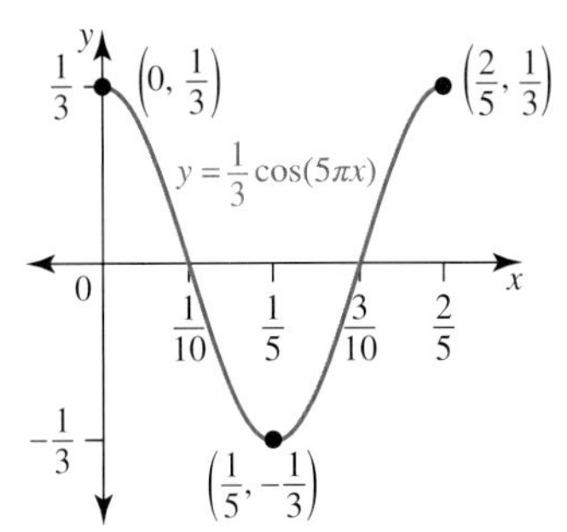

3. a.

b.

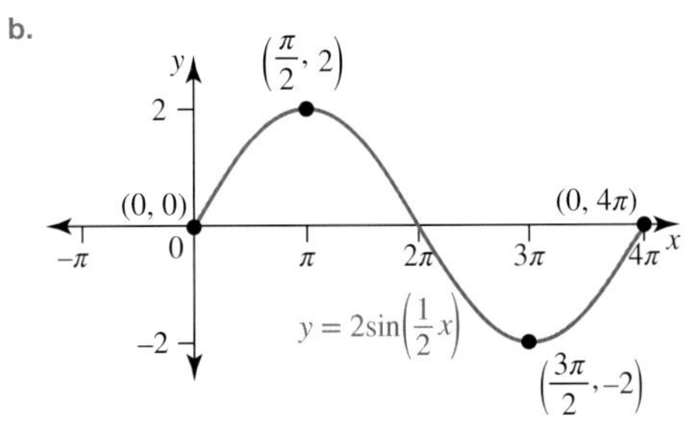

c.

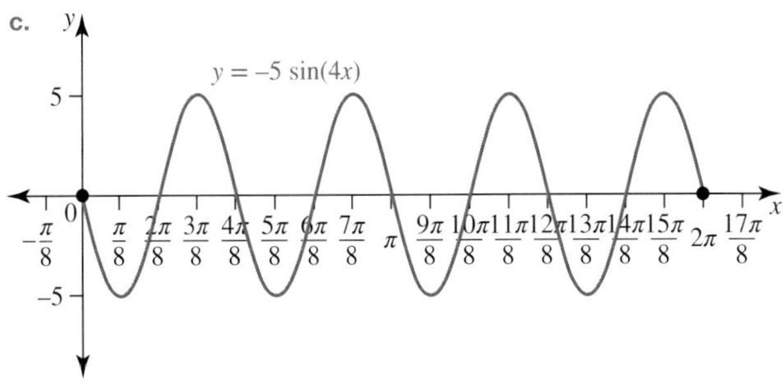

d.

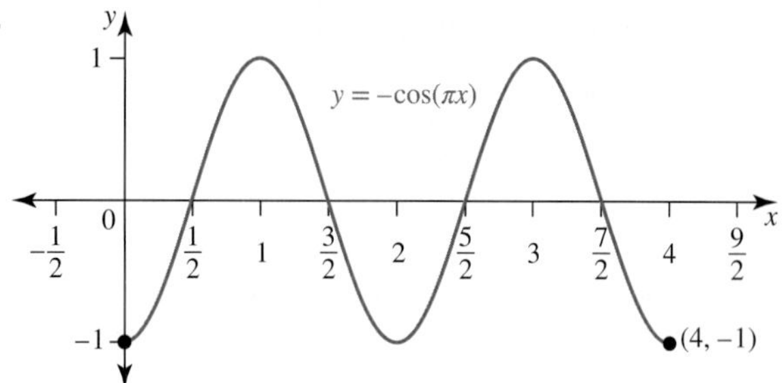

4. a.

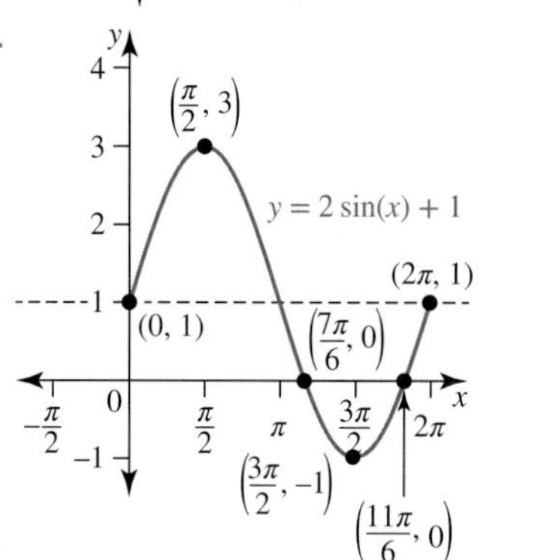

Period 2π; amplitude 2; equilibrium position $y = 1$; range $[-1, 3]$; x-intercepts at $x = \dfrac{7\pi}{6}, \dfrac{11\pi}{6}$

b.

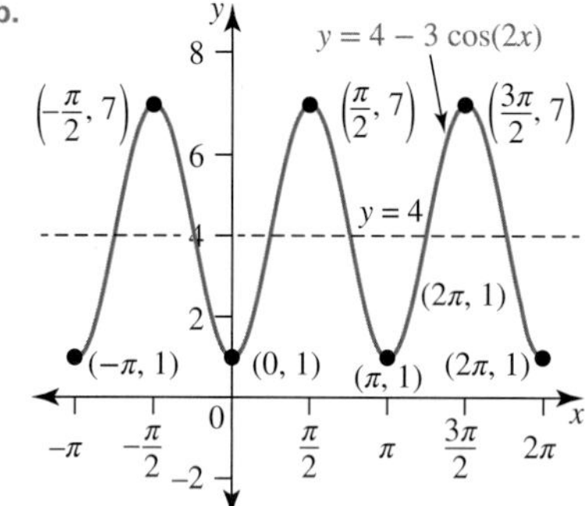

Period π; amplitude 3; inverted; equilibrium $y = 4$; range $[1, 7]$

5. a. $[2, 4]$

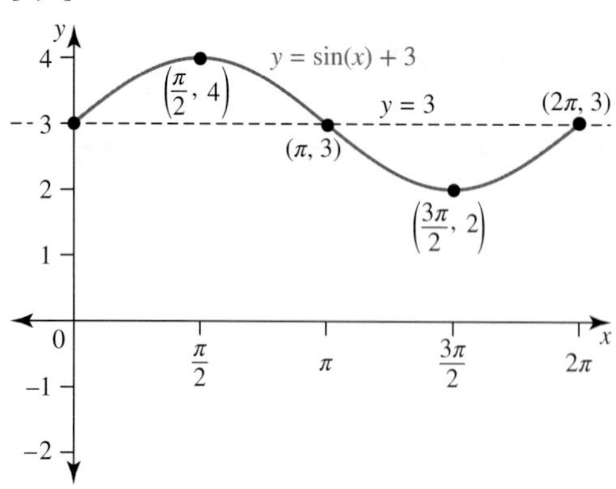

b. $[-2, 0]$

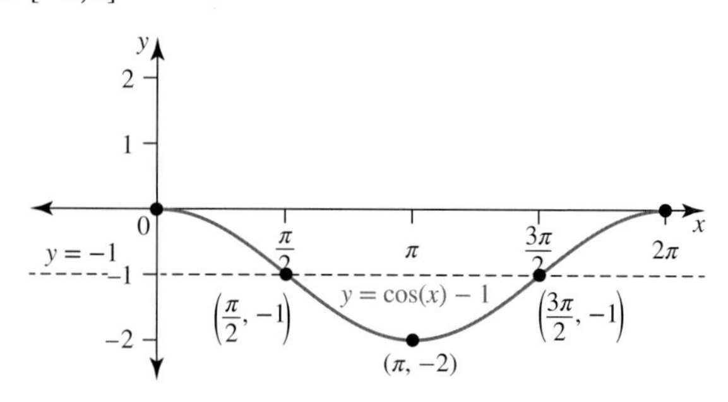

c. $[1, 3]$

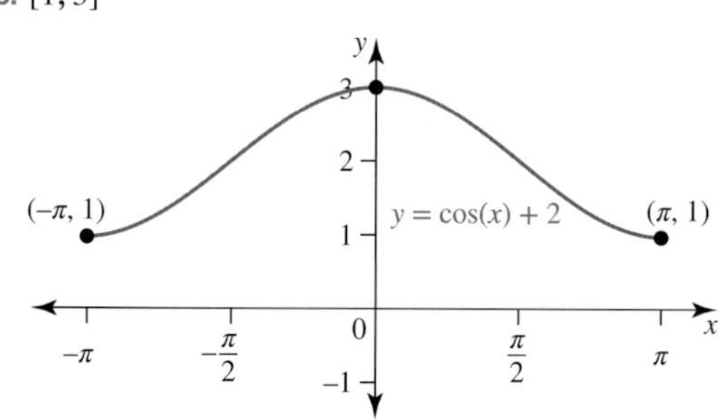

d. $[3, 5]$

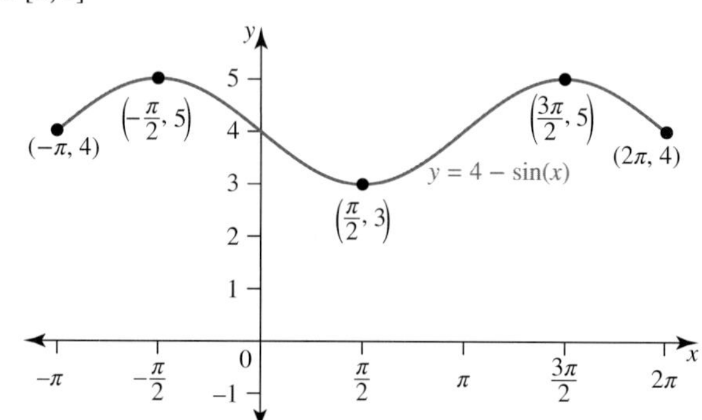

6.

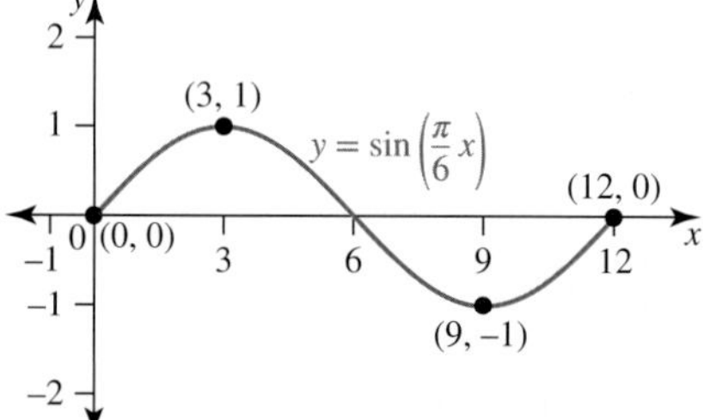

Period 12; amplitude 1; equilibrium $y = 0$; range $[-1, 1]$; domain $[0, 12]$

7. a. Period π, mean $y = 5$, amplitude 4, range $[1, 9]$

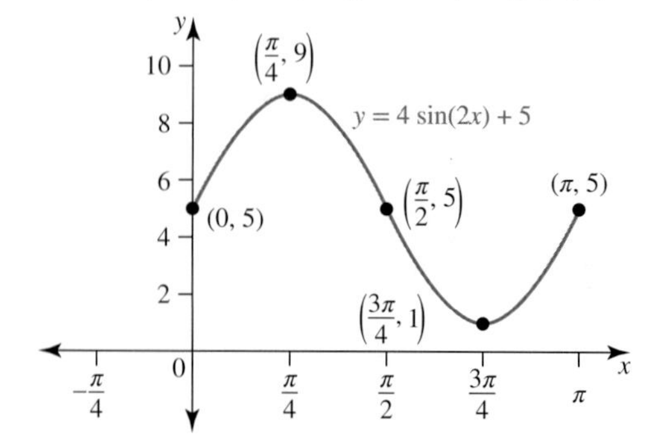

b. Period $\frac{2\pi}{3}$, mean $y = 2$, amplitude 2, range $[0, 4]$

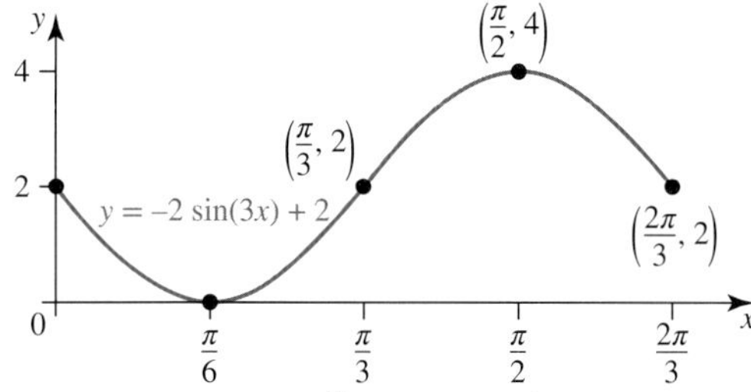

c. Period 4π, mean $y = \frac{3}{2}$, amplitude $\frac{1}{2}$, range $[1, 2]$

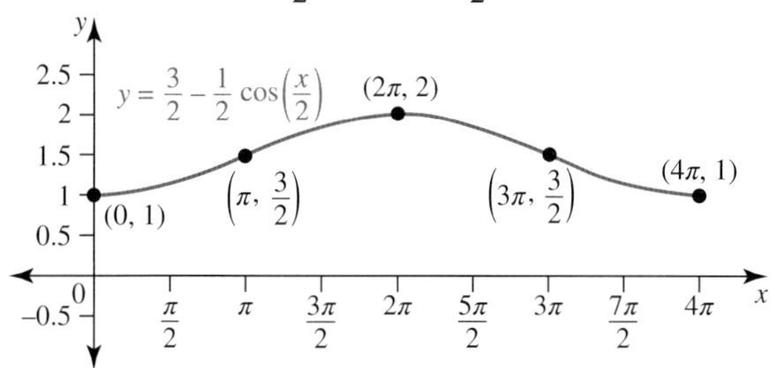

d. Period 2, mean $y = -\sqrt{3}$, amplitude 2, range $\left[-\sqrt{3} - 2, -\sqrt{3} + 2\right]$

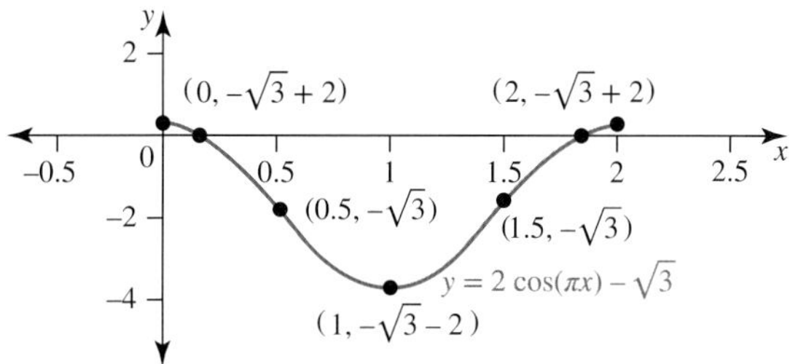

8. a. Range $[-4, 0]$

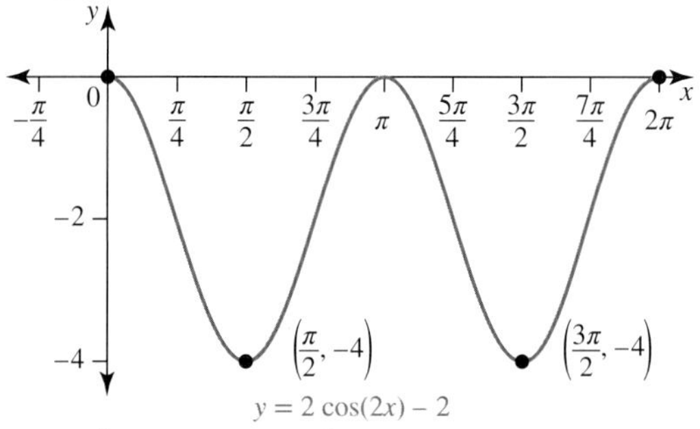

b. Range $\left[\sqrt{3} - 2, \sqrt{3} + 2\right]$

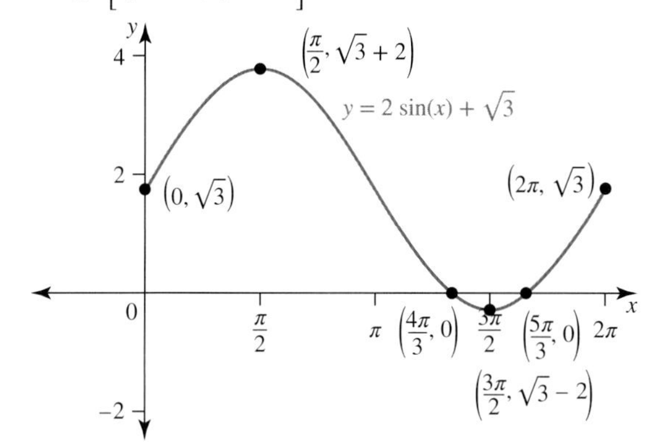

c. Range $[2, 8]$

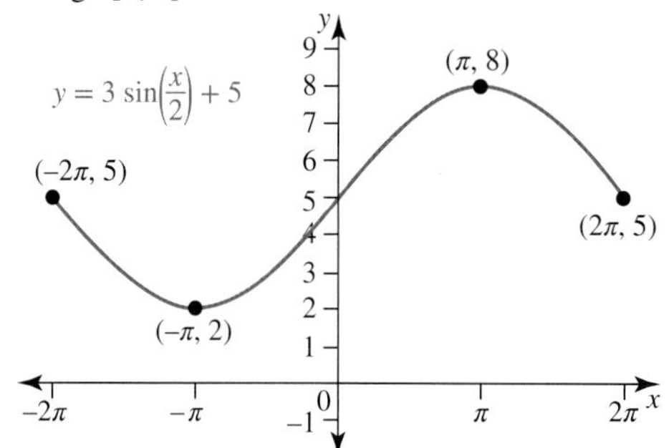

d. Range $[-5, -3]$

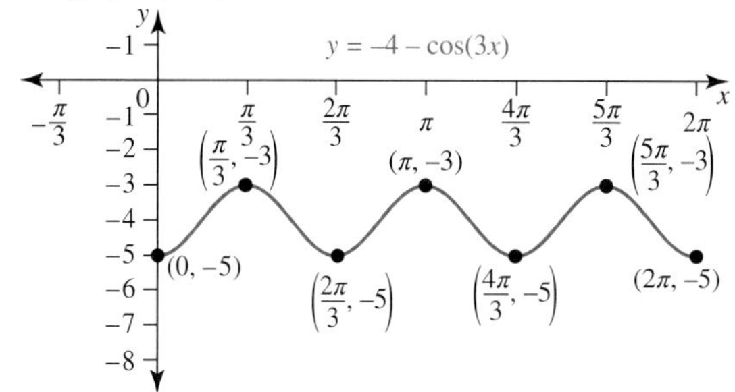

e. Range $[-1, 3]$

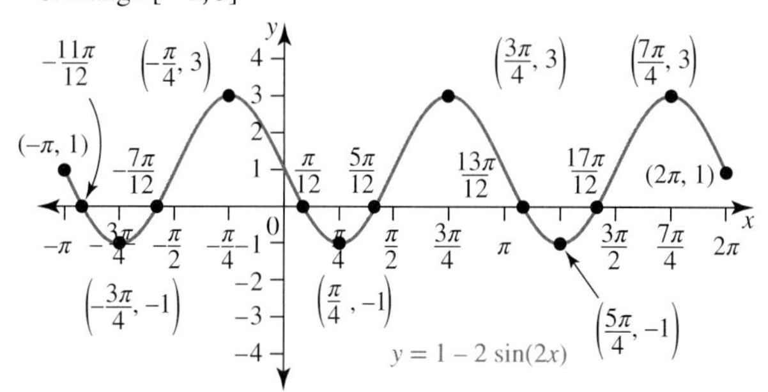

f. Range $[-4, 8]$

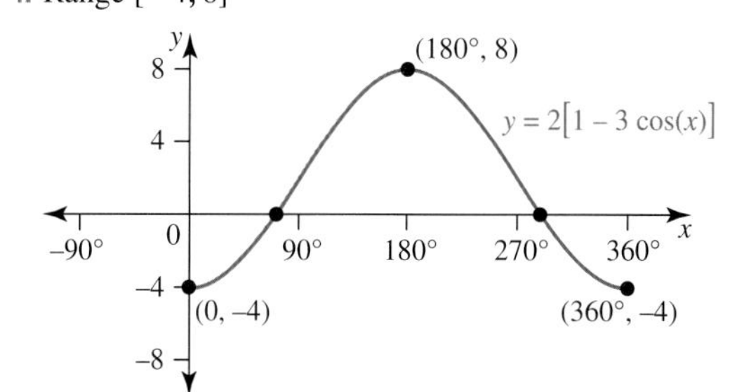

9. a. $[1, 5]$

b. -14

c. 68; $x = \frac{3\pi}{2}$

d. i. Dilation factor $\frac{1}{5}$ from y-axis; dilation factor 2 from x-axis; vertical translation up 3

ii. Dilation factor $\frac{1}{2}$ from y-axis; dilation factor 10 from x-axis; vertical translation down 4

iii. Dilation factor 12 from x-axis; reflection in x-axis; vertical translation up 56

10. $a = 3, n = 2, y = 3\sin(2x)$

11. $a = -0.5, n = \frac{2}{3}, y = -0.5\cos\left(\frac{2x}{3}\right)$

12. $a = 4.5, n = \frac{4}{3}, y = 4.5\cos\left(\frac{4}{3}x\right)$

13. a. $y = 4\sin(3x)$
b. $y = 6\cos\left(\frac{x}{2}\right)$
c. $y = -4\sin(2x)$

14. a. $y = 2\sin(x) + 3$
b. $y = -3\sin\left(\frac{x}{2}\right) + 3$
c. $y = 3\cos\left(\frac{2x}{3}\right) - 4$
d. $y = -5\cos(2x) - 1$

15. $y = 3\ \sin(2x) + 5$

16. a. $y = -3\sin\left(\frac{x}{2}\right)$
b. $y = 4\cos(3x)$
c. $y = -4\cos(x) + 6$

17. $y = -2\sin\left(\frac{2x}{3}\right)$

18. a. $\frac{2\pi}{3}$
b. $a = 2, b = 3, c = 7$
c. $D = \left[0, \frac{2\pi}{3}\right]$

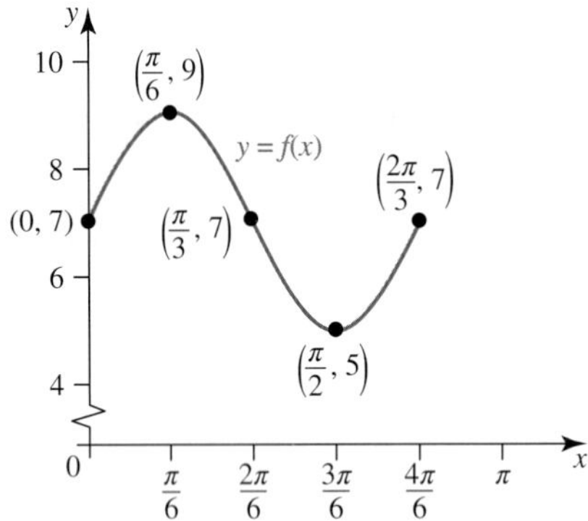

d.

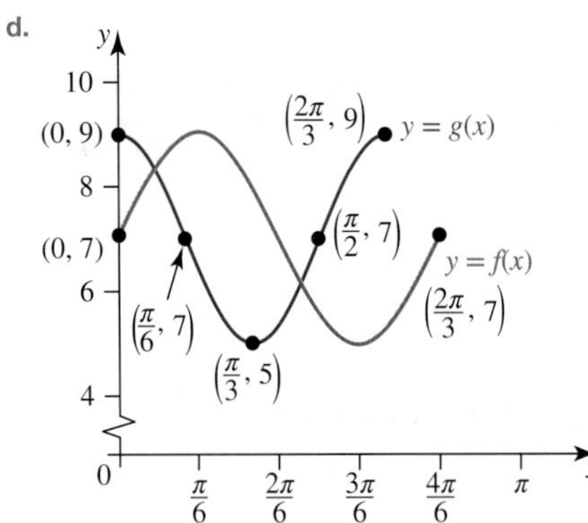

e. $\left(\frac{\pi}{12}, 7 + \sqrt{2}\right), \left(\frac{5\pi}{12}, 7 - \sqrt{2}\right)$

f. $\frac{\pi}{12} \leq x \leq \frac{5\pi}{12}$

19. $y = -\sin\left(\frac{x}{3}\right) - 3$; no intersections

20. a. $a = 10$
b. $x = 0.5, 2.5$
c. Range $[-10, 10\sqrt{3} + 10]$

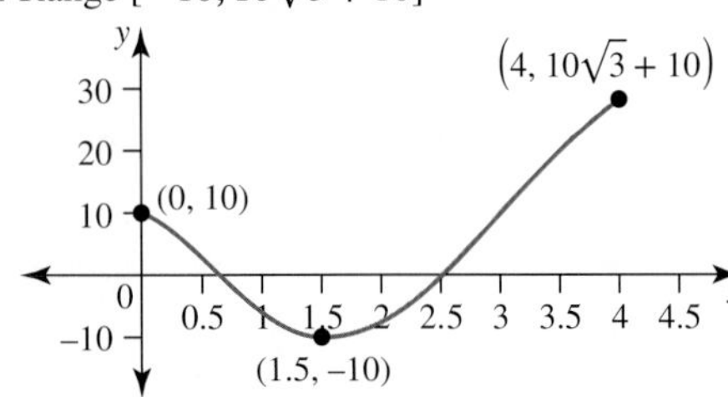

9.3 Exam questions

Note: Mark allocations are available with the fully worked solutions online.

1. D
2. C
3. B

9.4 Applications of sine and cosine functions

9.4 Exercise

1. a.

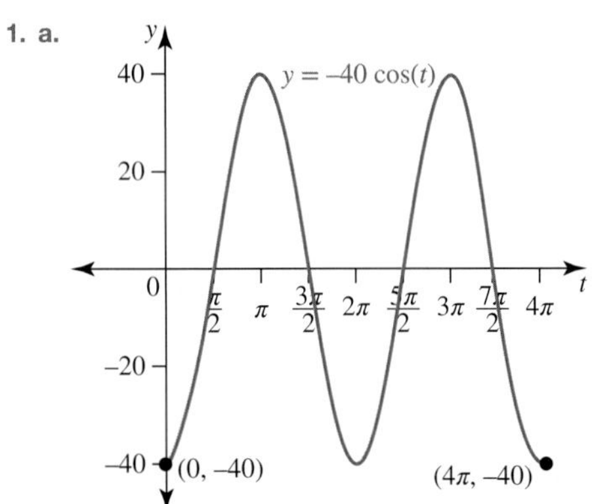

b. 40 cm
c. $t = \frac{\pi}{2}, \frac{3\pi}{2}, \frac{5\pi}{2}, \frac{7\pi}{2}$
d. $\frac{2\pi}{3}$ seconds ≈ 2.1 seconds

2. $T = 28 - 8\cos\left(\frac{\pi}{6}t\right)$; between 11.29 am and 4.31 pm

3. a. $I = 5\sin\left(\frac{\pi}{2}t\right) + 5$
b. 44%

4. a. 19°
b. 22° at 6 pm
c. Between 16° and 22°
d. Period is 24 hours.

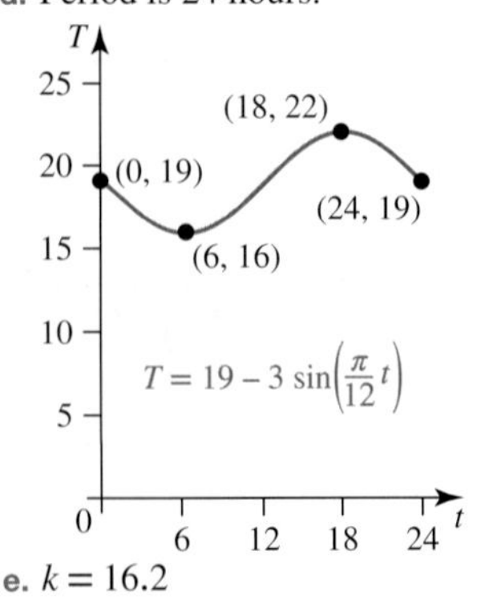

e. $k = 16.2$

5. a. $a = 2.5; b = 4.5; h = 2.5 \sin\left(\frac{\pi}{5}x\right) + 4.5$

b. 10 cm

c. $(2.5, 3.25)$

d. 40.1 cm^2

6. a. Between 29° and 31°

b. 12 minutes

c.

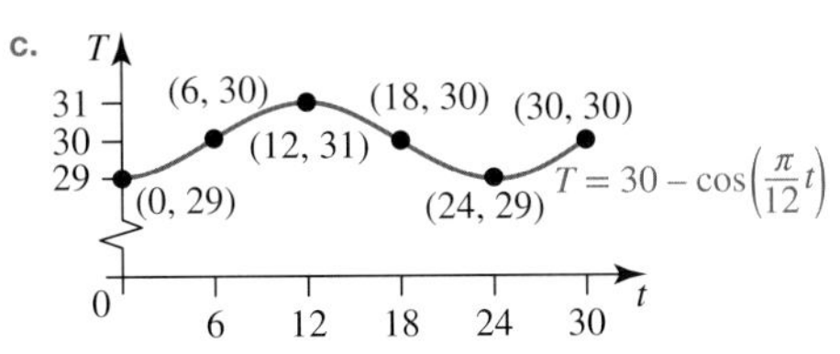

d. 2.5 cycles

e. 30 °C

f. $T = \sin\left(\frac{\pi}{12}(t-6)\right) + 30$

7. a. i. 25° at 1 pm   ii. 13° at 7 pm

b. i. 23.2 °C   ii. 13.2 °C

c.

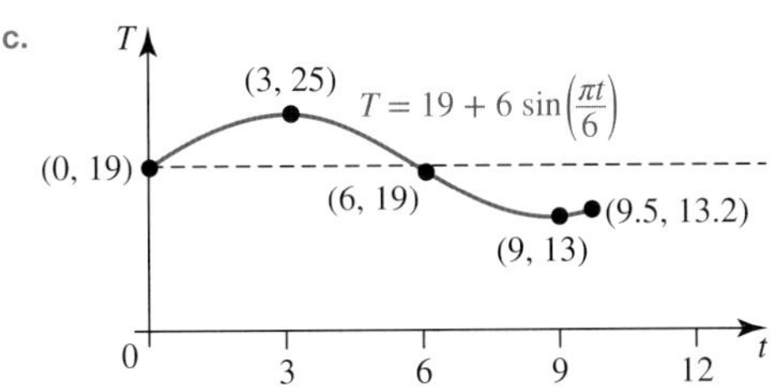

d. 2.24 hours

8. a. 1.5 metres

b. 18.5 metres

c. 2

d.

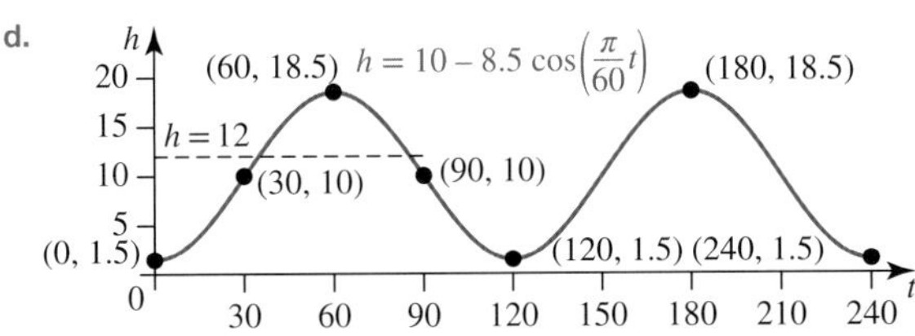

e. 51 seconds

f. 8.5 metres

9. a. 2 metres   b. $n = \frac{4}{3}$

c. $a = 5$; 60 times   d. $p = a\sin(4\pi t) + 5$

10. a. $a = -20, b = \frac{2\pi}{3}$

b.

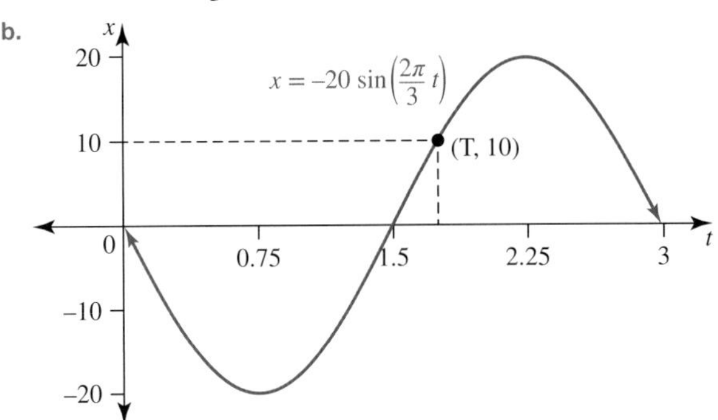

c. $T = 1.75$

d. 80 cm

11. a. 5 units

b. $t = 0.64$

c. $I = 5\sin(t - 0.64)$

d. $I = 5\cos(t - 2.21)$

12. a. 13 (includes the single-sided teeth at the ends)

b. $\frac{\pi}{3}$ cm

c. $(4\pi + 8)$ cm

d. $y = x + 8$

9.4 Exam questions

Note: Mark allocations are available with the fully worked solutions online.

1. B

2. B

3. a. 1000

b. Highest population = 1500; lowest population = 500

c. 1000

9.5 The tangent function

9.5 Exercise

1.

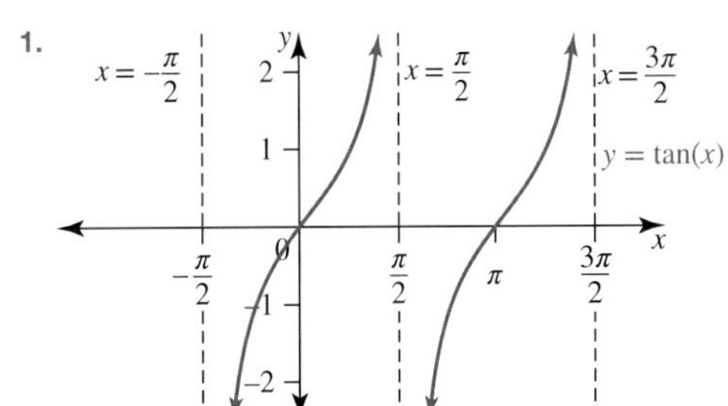

2. a.

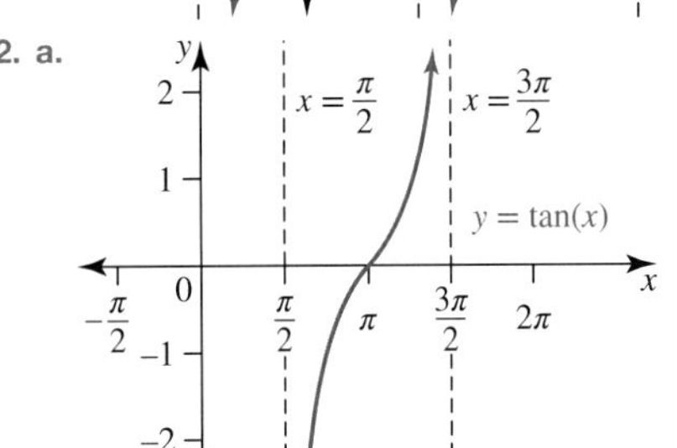

b.

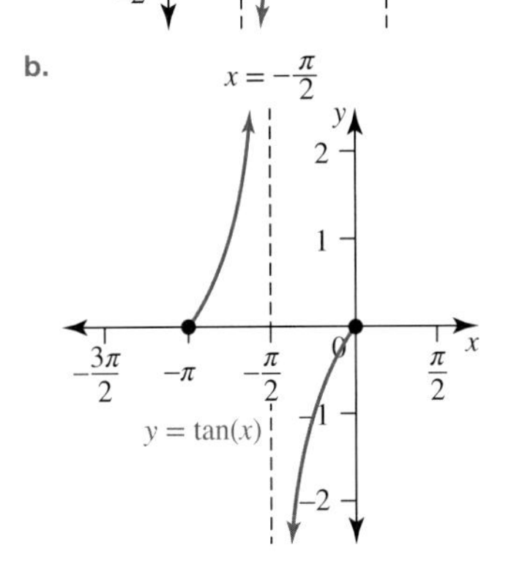

c.

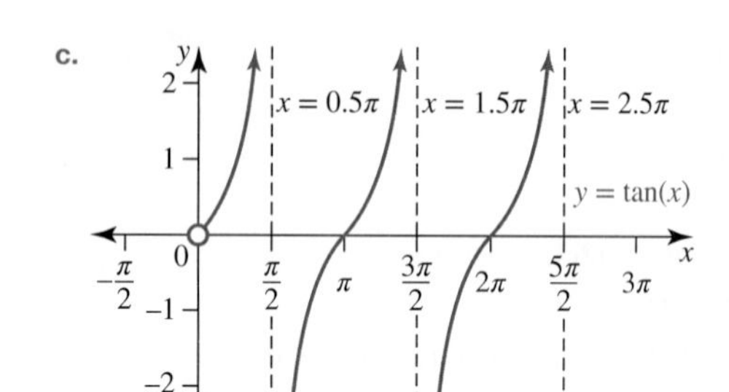

3. Domain $[-3\pi, 3\pi] \backslash \left\{\pm \frac{\pi}{2}, \pm \frac{3\pi}{2}, \pm \frac{5\pi}{2}\right\}$; range R

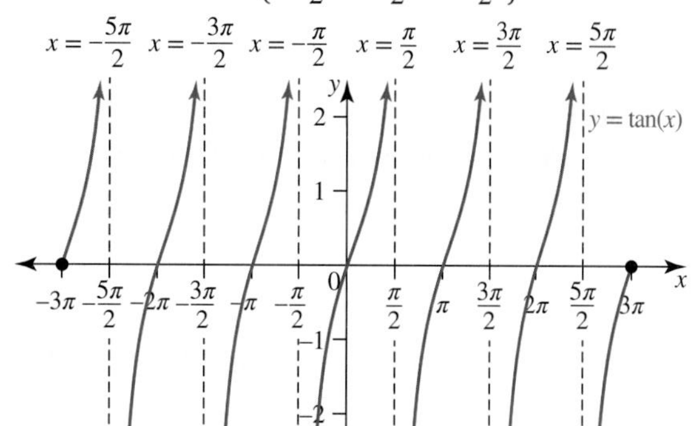

4. a.

$x = -\frac{\pi}{2}$

$y = -\tan(x)$

$\left(\frac{\pi}{4}, 3\right)$

$x = \frac{\pi}{2}$

y, 3, 2, 1, 0, −1, −2, −3

$-\frac{\pi}{2}$, $-\frac{\pi}{4}$, $\frac{\pi}{4}$, $\frac{\pi}{2}$, x

$y = 3\tan(x)$

$\left(\frac{\pi}{4}, -1\right)$

b.

$x = -\frac{\pi}{2}$

$x = \frac{\pi}{2}$

y, 4, 3, 2, 1, 0, −1, −2

$\sqrt{3}$

$y = \tan(x) + \sqrt{3}$

$-\frac{\pi}{2}$, $-\frac{\pi}{3}$, $\frac{\pi}{2}$, x

5.

$y = \tan(x) + 1$

$x = \frac{\pi}{2}$

$x = \frac{3\pi}{2}$

$y = \tan(x)$

$(\pi, 1)$

$(2\pi, 1)$

$(0, 1)$

$(2\pi, 0)$

$(0, 0)$

y, 2, 0, −2

$\frac{\pi}{4}$, $\frac{\pi}{2}$, $\frac{3\pi}{4}$, π, $\frac{5\pi}{4}$, $\frac{3\pi}{2}$, $\frac{7\pi}{4}$, 2π, x

6. a.

$\left(\frac{\pi}{4}, 4\right)$

$x = \frac{\pi}{2}$

$x = \frac{3\pi}{2}$

$y = 4\tan(x)$

y, 4, 2, 0, −2, −4

$-\frac{\pi}{2}$, $\frac{\pi}{2}$, π, $\frac{3\pi}{2}$, 2π, x

b.

$x = \frac{\pi}{2}$

$x = \frac{3\pi}{2}$

y, 4, 2, 0, −2, −4

$\left(\frac{\pi}{4}, -0.5\right)$

$\frac{\pi}{2}$, π, $\frac{3\pi}{2}$, 2π, x

$y = 0.5\tan(x)$

c.

$x = -\frac{\pi}{2}$

$x = \frac{\pi}{2}$

y, 3, 2, 1, 0, −1, −2, −3

$y = -\tan(x)$

$y = 1 - \tan(x)$

$y = -1 - \tan(x)$

$-\frac{\pi}{2}$, $-\frac{\pi}{4}$, $\frac{\pi}{4}$, $\frac{\pi}{2}$, x

7. a.

$x = -\frac{\pi}{2}$

$x = \frac{\pi}{2}$

y, 3, 2, 1, 0, −1, −2, −3

$-\pi$, $-\frac{\pi}{2}$, $\frac{\pi}{2}$, π, x

$y = -\sqrt{3}$

b. $x = -\frac{\pi}{3}, \frac{2\pi}{3}$

c. $x \in \left(-\frac{\pi}{2}, -\frac{\pi}{3}\right) \cup \left(\frac{\pi}{2}, \frac{2\pi}{3}\right)$

8. a. $\frac{\pi}{6}$

b. 4π

c. $\frac{2\pi}{3}$

d. 1

9. a. $x = \frac{\pi}{10}$

b.

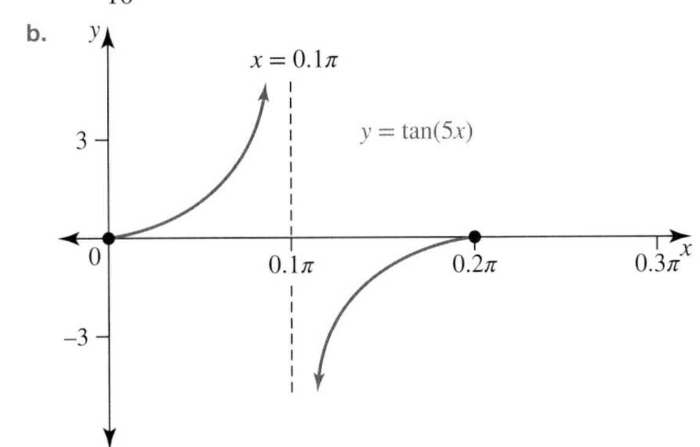

10. a. 2

b. $x = 1$

c.

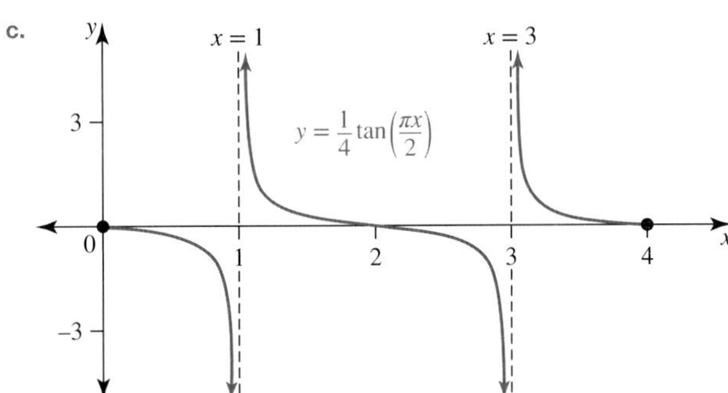

11. Period $\frac{\pi}{3}$; asymptotes $x = \frac{\pi}{6}, \frac{\pi}{2}, \frac{5\pi}{6}$; x-intercepts $x = 0, \frac{\pi}{3}, \frac{2\pi}{3}$

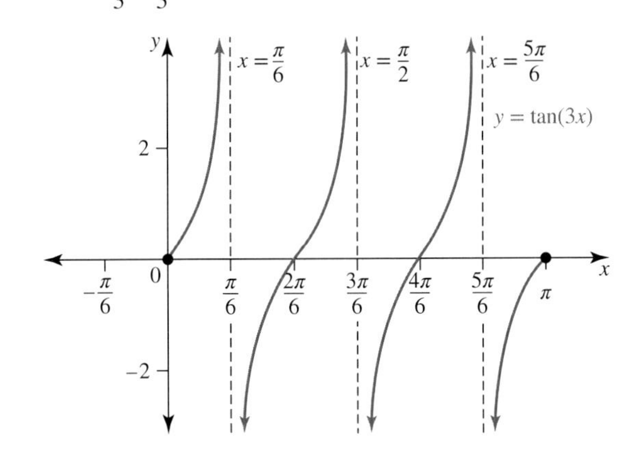

12. Period 2π; asymptote $x = \pi$; x-intercepts $x = 0, 2\pi$; point $\left(\frac{\pi}{2}, -2\right)$

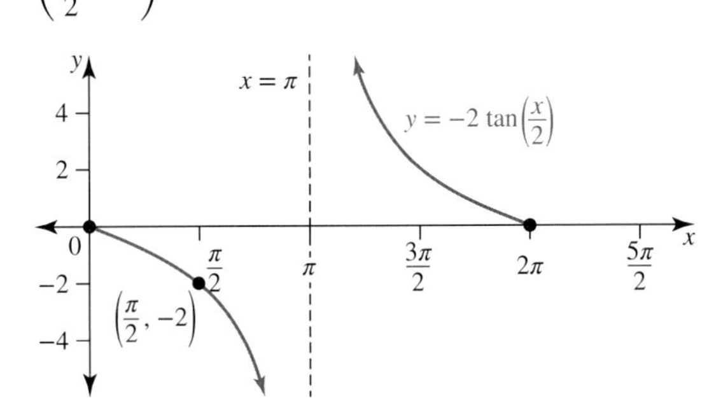

13. a. Period $\frac{\pi}{4}$; asymptotes $x = \frac{\pi}{8}, \frac{3\pi}{8}, \frac{5\pi}{8}, \frac{7\pi}{8}$; x-intercepts $x = 0, \frac{\pi}{4}, \frac{\pi}{2}, \frac{3\pi}{4}, \pi$

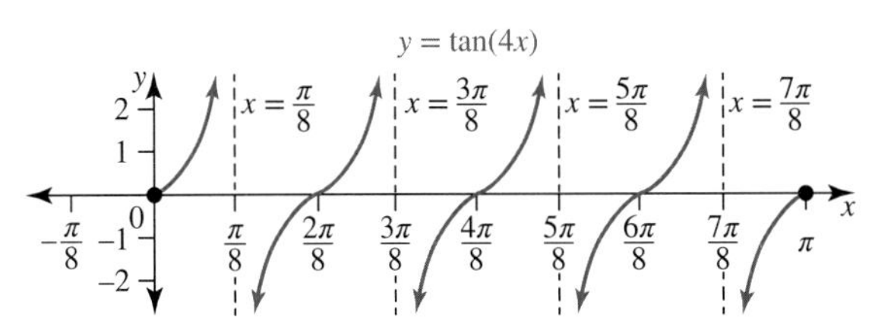

b. Period $\frac{\pi}{2}$; asymptotes $x = \frac{\pi}{4}, \frac{3\pi}{4}$; x-intercepts $x = 0, \frac{\pi}{2}, \pi$

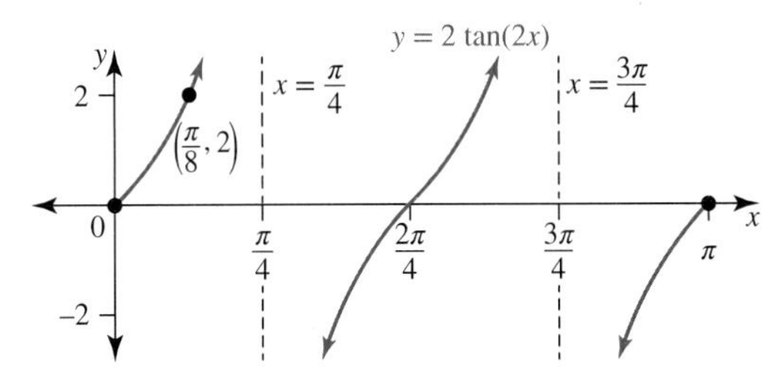

c. Period 3π; no asymptotes in the domain; x-intercept $x = 0$

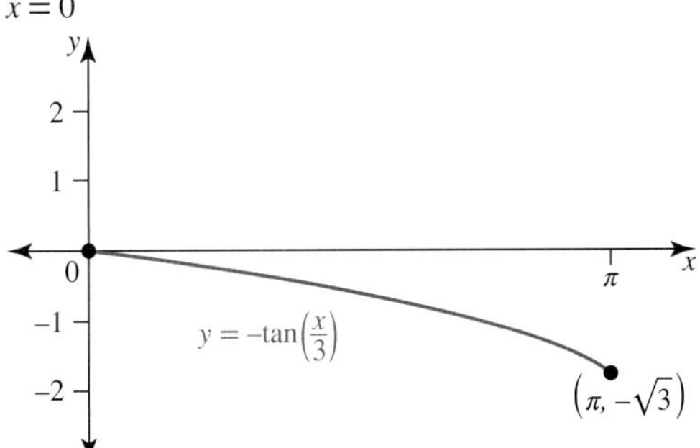

14. $n = 4, a = \frac{1}{2}$

15. a. $n = \frac{2}{3}, a = -3\sqrt{3}$ b. $\frac{3\pi}{2}$

16. a. 4

b. $\frac{3\pi}{8}$

c. $b = \frac{8}{3}$

d. $x = \frac{3\pi}{16}, \frac{9\pi}{16}, \frac{15\pi}{16}, \frac{21\pi}{16}$

17. $a = -2, c = \frac{5\pi}{12}$

18. a. Period π; asymptote $x = \dfrac{\pi}{2}$; x-intercept $x \approx 2.55$

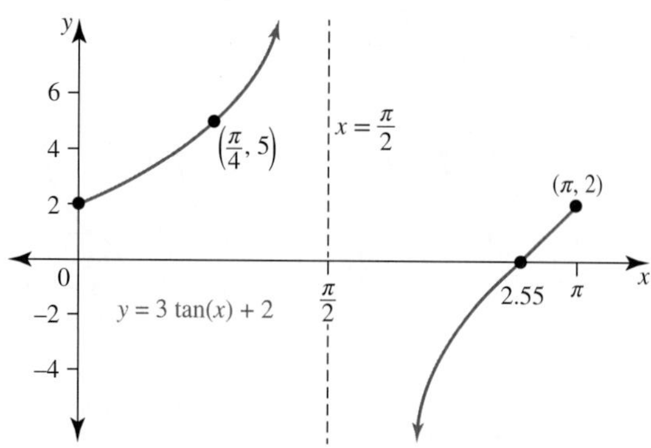

b. Period π; asymptote $x = \dfrac{\pi}{2}$; x-intercept $x = \dfrac{\pi}{4}$

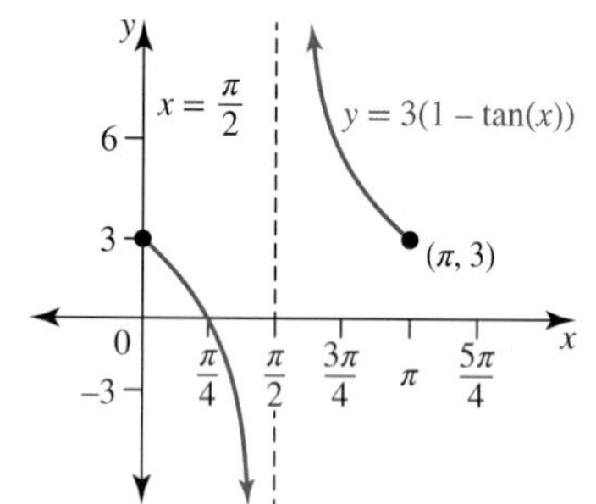

c. Period $\dfrac{\pi}{3}$; asymptotes $x = \dfrac{\pi}{6}, \dfrac{\pi}{2}, \dfrac{5\pi}{6}$; x-intercepts

$x = \dfrac{\pi}{12}, \dfrac{5\pi}{12}, \dfrac{3\pi}{4}$

19. a. 0

b. $D = \left\{-\dfrac{3}{2}, \dfrac{3}{2}, \dfrac{9}{2}\right\}$

c. Period 3; asymptotes $x = -\dfrac{3}{2}, \dfrac{3}{2}, \dfrac{9}{2}$; x-intercepts

$x = -3, 0, 3, 6$

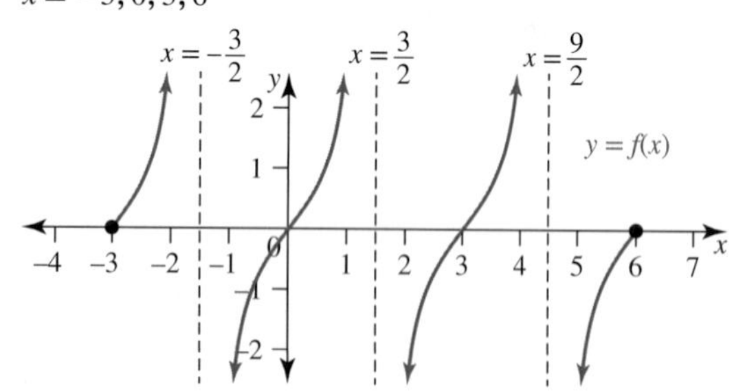

d. $x = -\dfrac{9}{4}, \dfrac{3}{4}, \dfrac{15}{4}$

20. a. $4\sqrt{3} \approx 6.9$ metres

b. 36 minutes

c. $12\sqrt{3} \approx 20.8$ metres, 19.7 metres lower than the Japanese tsunami

d.

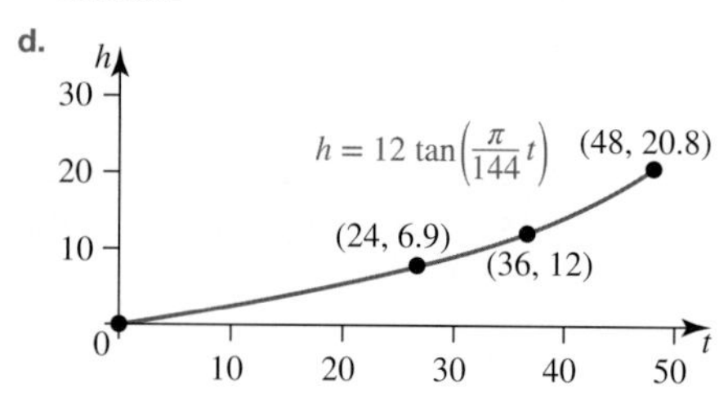

9.5 Exam questions

Note: Mark allocations are available with the fully worked solutions online.

1. D

2. A

3.

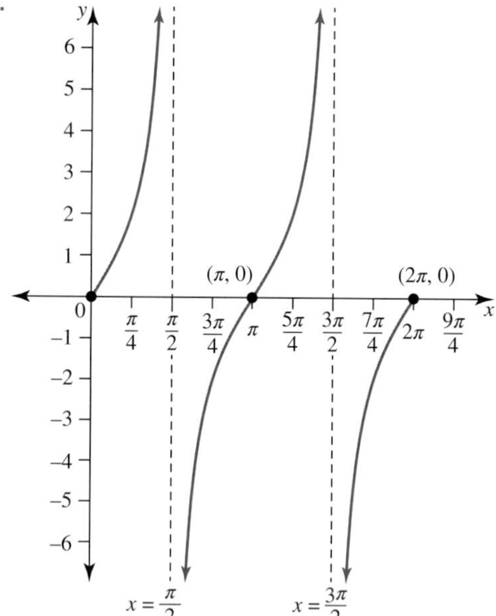

9.6 Trigonometric identities and properties

9.6 Exercise

1. a. $\dfrac{1}{2}$ b. $\dfrac{1}{4}$ c. $\dfrac{1}{3}$ d. 1

2. a. 1 b. $\sqrt{3}$ c. 0 d. 2

3. a. $2\cos^2(\theta)$

b. Sample responses can be found in the worked solutions in the online resources.

4. $\dfrac{1}{\cos(u)\sin(u)}$

5. a. $4\sin^2(\theta)$ b. $2\tan(\alpha)$

c. 8 d. $\cos^2(A)$

6. Sample responses can be found in the worked solutions in the online resources.

7. $\sin(x) = \dfrac{3\sqrt{5}}{7}$; $\tan(x) = -\dfrac{3\sqrt{5}}{2}$

8. $\sin(x) = -\frac{3}{\sqrt{10}}; \cos(x) = \frac{1}{\sqrt{10}}$

9. a. $\sin(x) = -\frac{4}{\sqrt{41}}; \cos(x) = \frac{5}{\sqrt{41}}$

b. i. $-\frac{1}{2}$ ii. 4

c. i. $2\sqrt{6}$ ii. $\frac{1}{25}$ iii. $-\frac{24}{5}$

10. a. i. $\sin(x) = \frac{\sqrt{13}}{5}$ ii. $\sin(x) = -\frac{\sqrt{13}}{5}$

b. $\frac{\sqrt{39}}{6}; -\frac{\sqrt{39}}{6}$

c. Answers should agree.

11. a. -0.8 b. 0.8 c. $\frac{3}{5}$ d. $\frac{4}{3}$

12. a. LHS = RHS = $\frac{\sqrt{3}}{2}$ b. LHS = RHS = $-\frac{1}{2}$

c. LHS = RHS = $-\frac{1}{2}$ d. LHS = RHS = $\frac{\sqrt{2}}{2}$

13. a. $-\sin(\theta)$ b. $t = \frac{5\pi}{12}$

14. $\cos(\theta) - \sin(\theta)$

15. a. $\sin(\theta)$ b. $-\sin(\theta)$

c. $\cos(\theta)$ d. $\cos(\theta)$

16. a. $\tan(a)$ b. $\sin^2(\theta)$ c. 0 d. 1

17. a. $a = 63°$ b. $b = \frac{7\pi}{18}$

c. $c = 1.22$ d. $d = \frac{\pi}{4}, \frac{7\pi}{4}$

18. a. $x = 0, \pi, 2\pi$ b. $x = \frac{7\pi}{6}, \frac{11\pi}{6}$

c. $x = \frac{\pi}{3}, \frac{2\pi}{3}, \frac{4\pi}{3}, \frac{5\pi}{3}$ d. $x = \frac{\pi}{2}, \pi, \frac{3\pi}{2}$

19. Sample responses can be found in the worked solutions in the online resources.

20. a. $\frac{1}{\tan(x)}$ b. $-\sin(3x)$

9.6 Exam questions

Note: Mark allocations are available with the fully worked solutions online.

1. D

2. A

3. $\sin(x) = +\frac{2}{\sqrt{29}}$ and $\cos(x) = -\frac{5}{\sqrt{29}}$

9.7 Review

9.7 Exercise

Technology free: short answer

1. a. $\frac{3\pi}{4}, \frac{5\pi}{4}$ b. $0, 2\pi, 4\pi$

c. $\frac{-5\pi}{3}, \frac{-4\pi}{3}, \frac{\pi}{3}, \frac{2\pi}{3}$ d. $\frac{\pi}{4}, \frac{5\pi}{4}$

2. a. $-\frac{11\pi}{12}, -\frac{3\pi}{4}, -\frac{\pi}{4}, -\frac{\pi}{12}, \frac{5\pi}{12}, \frac{7\pi}{12}$

b. 58.28, 148.28, 238.28

c. $\frac{3\pi}{2}$

d. $\frac{2\pi}{3}, \pi, \frac{4\pi}{3}$

3. a. p b. $-p$ c. $1 - p^2$ d. $-p$

4. a. $\cos(x) = -\frac{2}{\sqrt{29}}; \sin(x) = -\frac{5}{\sqrt{29}}$

b. $\cos(y) = \frac{2}{3}; \tan(y) = -\frac{\sqrt{5}}{2}$

5. a. Period 2π; amplitude 3; range [2, 8]

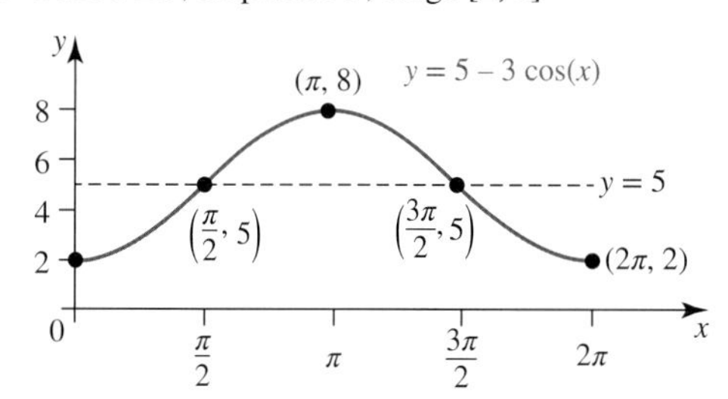

b. Period $\frac{2\pi}{3}$; amplitude 5; range $[-5, 5]$

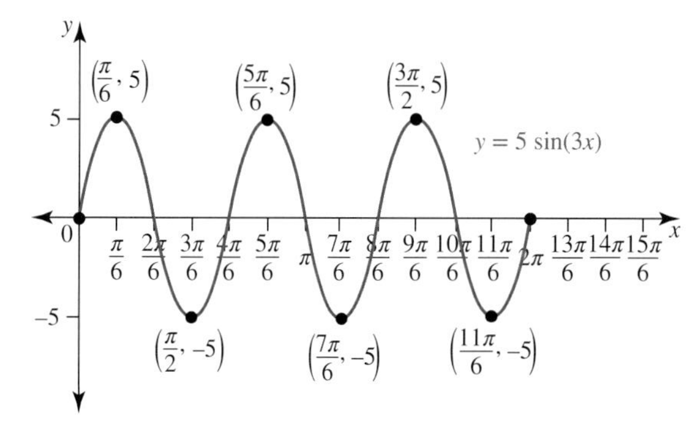

c. Period π; asymptotes $x = \frac{\pi}{2}, \frac{3\pi}{2}$; range R

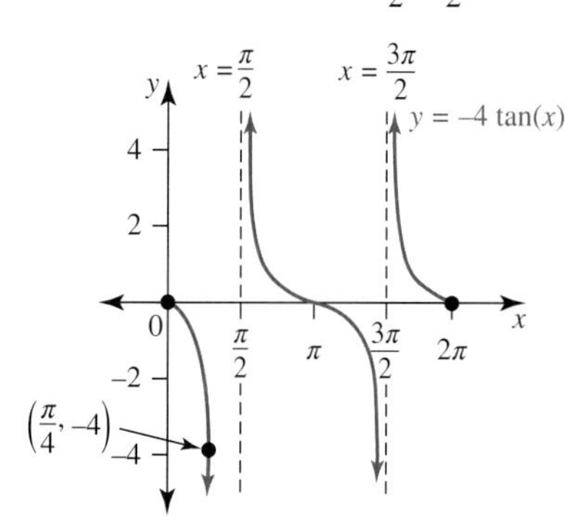

d. Period 4π; amplitude 2; range $[-1, 3]$

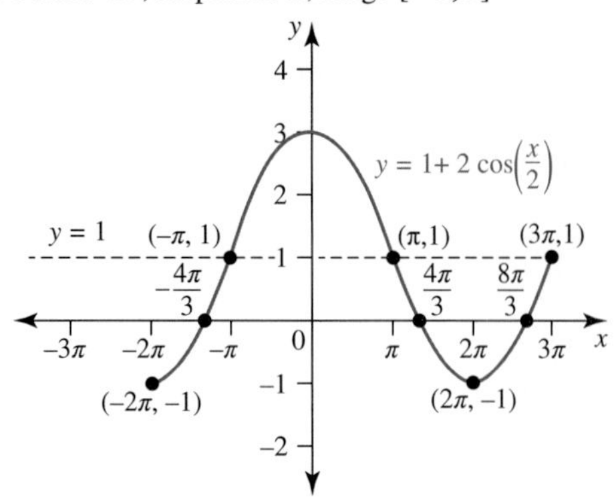

6. a. 1

b. $\cos^2(A)(1 + 2\tan(A)) + \sin^2(A) = (\cos(A) + \sin(A))^2$

$$\begin{aligned}\text{LHS} &= \cos^2(A) + \cos^2(A) \times 2\tan(A) + \sin^2(A)\\ &= \cos^2(A) + \cos^2(A) \times \frac{2\sin(A)}{\cos(A)} + \sin^2(A)\\ &= \cos^2(A) + \cos(A) \times 2\sin(A) + \sin^2(A)\\ &= \cos^2(A) + 2\cos(A)\sin(A) + \sin^2(A)\\ &= (\cos(A) + \sin(A))^2\\ &= \text{RHS}\end{aligned}$$

c. $\sin^2(\theta)$

Technology active: multiple choice

7. C

8. D

9. A

10. E

11. C

12. C

13. A

14. C

15. D

16. B

Technology active: extended response

17. a. i. The maximum is 13; the minimum is −7.

ii. The maximum occurs when $x = 3$; the minimum occurs when $x = 9$.

b. $f(-12) = 3; f(18) = 3$

c. Period 12; amplitude 10; range $[-7, 13]$

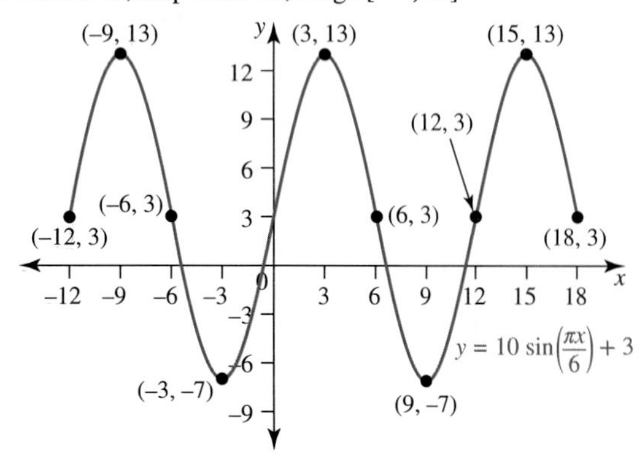

d. 4 solutions: $x = -5.4, -0.6, 6.6, 11.4$

18. a.

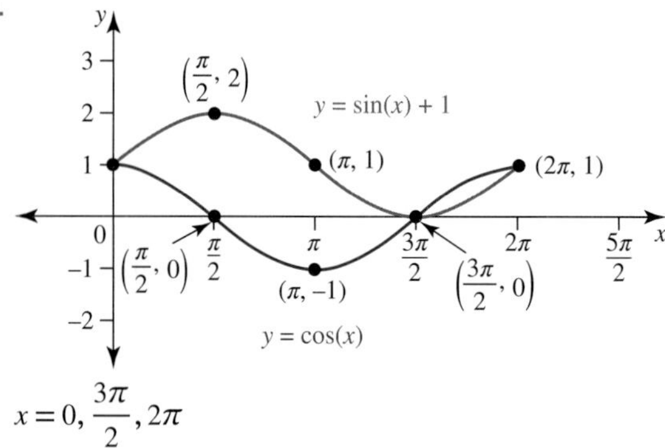

$x = 0, \frac{3\pi}{2}, 2\pi$

b. i. Two solutions

To show one root lies in the interval $0.6 \le x \le 0.8$:

When $x = 0.6$, $\cos(0.6) \approx 0.825$ and $\tan(0.6) \approx 0.684$. The graph of $y = \cos(x)$ lies above the graph of $y = \tan(x)$.

When $x = 0.8$, $\cos(0.8) \approx 0.697$ and $\tan(0.8) \approx 1.03$. The graph of $y = \cos(x)$ lies below the graph of $y = \tan(x)$.

Therefore, at some value of x between 0.6 and 0.8, the graphs have crossed each other.

There is a root that lies between 0.6 and 0.8.

ii. $\cos(x) = \tan(x)$

$\therefore \cos(x) = \frac{\sin(x)}{\cos(x)}$

$\therefore \cos^2(x) = \sin(x)$

$\therefore 1 - \sin^2(x) = \sin(x)$

$\therefore \sin^2(x) + \sin(x) - 1 = 0$

Using the formula for quadratic equations,

$\sin(x) = \frac{-b \pm \sqrt{b^2 - 4ac}}{2a}$, $a = 1, b = 1, c = -1$

$\therefore \sin(x) = \frac{-1 \pm \sqrt{5}}{2}$

As $0 < x < \frac{\pi}{2}$, $\sin(x) > 0$.

$\therefore \sin(x) = \frac{-1 + \sqrt{5}}{2}$

$\therefore \sin(x) = \frac{1}{2}\left(\sqrt{5} - 1\right)$

To 3 decimal places, $x = 0.618$.

19. a. 1 second

b. 30 decibels

c. $a = -30, b = 2\pi, c = 70, d = 70 - 30\cos(2\pi t)$

d. 12.4%

e. 600

f. 39.2%

20. a. $a = 15, b = \frac{\pi}{48}$

b. $y = 15$

c. One answer is $y = -15\tan\left(\frac{\pi}{48}x\right)$ or

$y = 15\tan\left(-\frac{\pi}{48}x\right)$.

$15\tan\left(\frac{\pi}{48}x\right), \quad 0 \le x \le 12$

d. $y = \begin{cases} 15\tan\left(\frac{\pi}{48}x\right), & 0 \le x \le 12 \\ 15, & 12 < x < 36 \\ -15\tan\left(\frac{\pi}{48}x\right), & 36 \le x \le 48 \end{cases}$

e. i. $y = \frac{5}{4}x, 0 \le x \le 12$

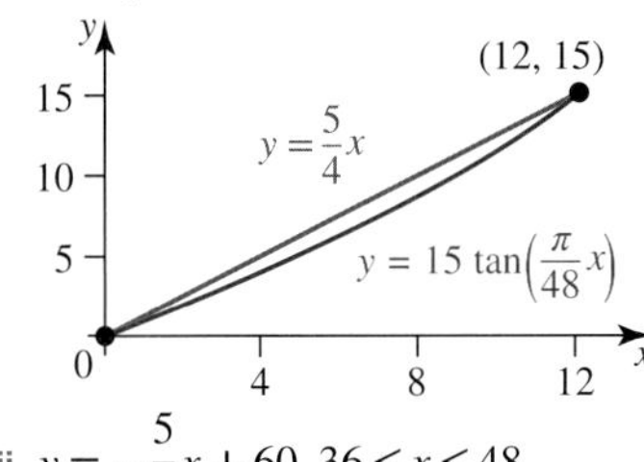

ii. $y = -\frac{5}{4}x + 60, 36 \le x \le 48$

f. Approximately 540 cm^2; overestimate

9.7 Exam questions

Note: Mark allocations are available with the fully worked solutions online.

1. $x = \frac{\pi}{12}, \frac{5\pi}{12}, \frac{13\pi}{12}, \frac{17\pi}{12}$
2. D
3. Period: $\frac{2\pi}{n} = \frac{2\pi}{5}$

   Amplitude: $a = 3$
   Range: [4, 10]
4. A
5. a. 55 m

   b. 30

   c. 1963.50 m^2

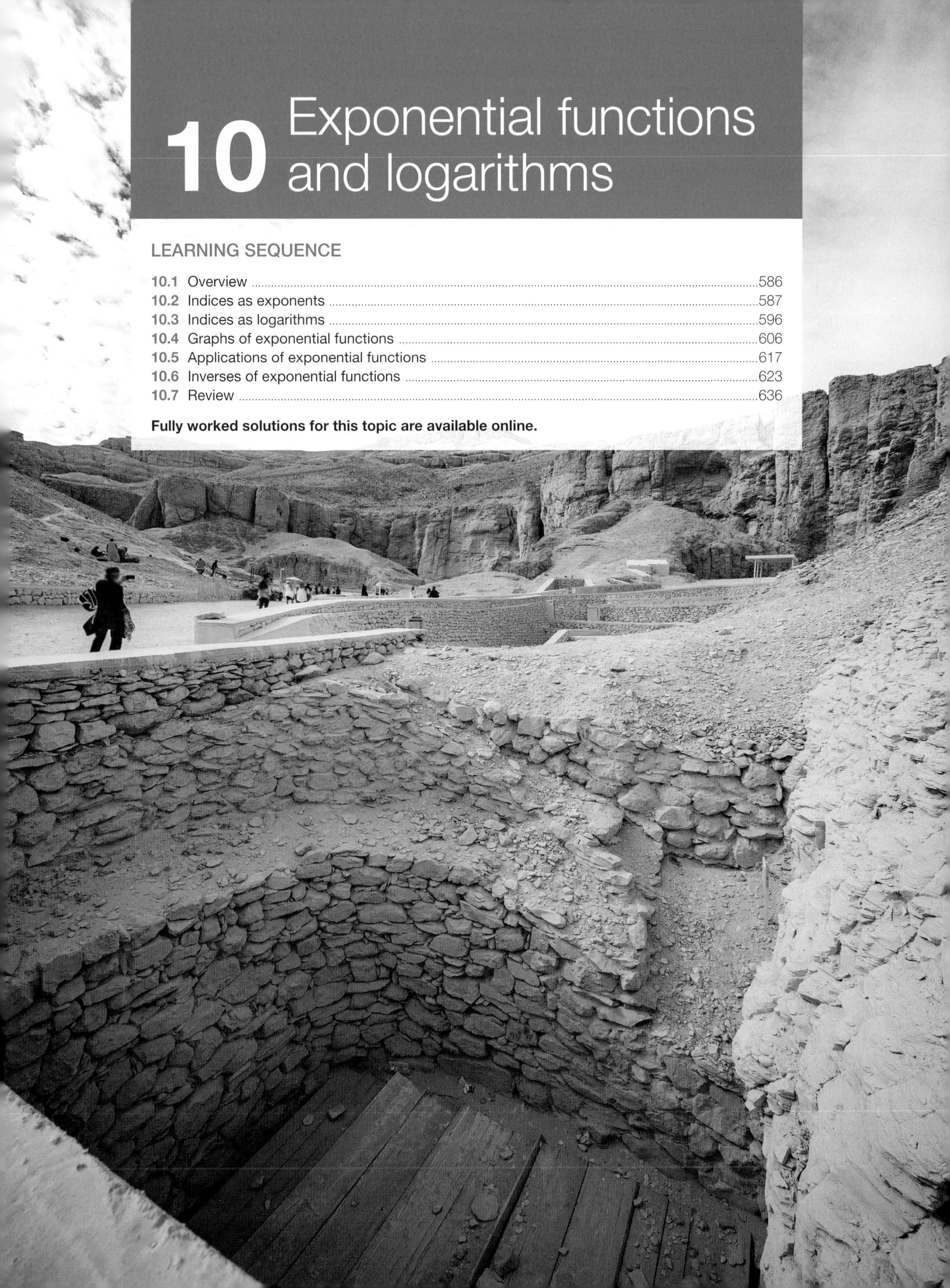

10 Exponential functions and logarithms

LEARNING SEQUENCE

Fully worked solutions for this topic are available online.

10.1 Overview

Hey students! Bring these pages to life online

 Watch videos

 Engage with interactivities

 Answer questions and check results

Find all this and MORE in jacPLUS

10.1.1 Introduction

In 1949 Willard Libby and his team at the University of Chicago made a highly significant discovery. By measuring the amount of radiation emitted by decaying carbon particles, Libby devised a method for determining the age of organic materials. The method, known as carbon dating, is based on the mathematics of exponential functions, because the radioactive isotope carbon-14 undergoes exponential decay.

Carbon is present in the tissues of living organisms and is replenished through natural means throughout the life of the organism. However, this replenishment ceases at death and the carbon starts to decay.

A mass spectrometer can measure the radiation emitted by the decaying carbon remaining in the dead organism. Once this data is substituted into the exponential model, the date of death can be estimated for samples up to 60 000 years old. As the half-life of carbon-14 is 5730 years, carbon dating is less reliable over very long time periods.

The ability to date carbon was revolutionary and resulted in changing several historical dates and theories. For example, the discovery of fossil remains in 1969 and 1974 at Lake Mungo, a World Heritage site in New South Wales, radically altered the estimates of the length of time Indigenous Australians have inhabited this country. The remains are known as Mungo Lady (1969) and Mungo Man (1974). Through carbon dating it was estimated Mungo Lady lived over 25 000 years ago; other methods have dated Mungo Man's death to 62 000 years ago. In 2015 Mungo Man's remains were formally returned to the traditional owners of the land around Lake Mungo, where he was subsequently reburied.

KEY CONCEPTS

This topic covers the following key concepts from the VCE Mathematics Study Design:

- exponential functions of the form $f: R \to R, f(x) = A\,a^{kx} + c$ and their graphs, where $a \in R^+, A, k, c \in R, A \neq 0$
- logarithmic functions of the form $f: R^+ \to R,\ f(x) = \log_a(x)$, where $a > 1$, and their graphs, and as the inverse function of $y = a^x$, including the relationships $a^{\log_a(x)} = x$ and $\log_a(a^x) = x$
- simple applications of exponential functions of the above form, with examples from various modelling contexts, and the interpretation of initial value, rate of growth or decay, half-life, doubling time and long run value in these contexts and their relationship to the parameters A, k and c
- use of inverse functions and transformations to solve equations of the form $Af(nx) + c = k$, where $A, n, c, k \in R$ for $A, n \neq 0$, and f is sine, cosine, tangent or a^x, using exact or approximate values on
- exponent laws and logarithm laws, including their application to the solution of simple exponential equations.

Note: Concepts shown in grey are covered in other topics.

10.2 Indices as exponents

LEARNING INTENTION

At the end of this subtopic you should be able to:
- simplify expressions with index laws
- solve equations with indices
- convert to and from scientific notation.

10.2.1 Index or exponential form

When the number 8 is expressed as a power of 2, it is written as $8 = 2^3$. In this form, the base is 2 and the power (also known as index or **exponent**) is 3. The form of 8 expressed as 2^3 is known both as its index form with index 3 and base 2, and its **exponential form** with exponent 3 and base 2. The words 'index' and 'exponent' are interchangeable.

For any positive number n where $n = a^x$, the statement $n = a^x$ is called an index or exponential statement. For our study:
- the base is a where $a \in R^+ \backslash \{1\}$
- the exponent, or index, is x where $x \in R$
- the number n is positive, so $a^x \in R^+$.

Index laws control the simplification of expressions that have the same base.

Review of index laws

The index laws are summarised below.

Index laws

$$a^m \times a^n = a^{m+n}$$

$$a^m \div a^n = a^{m-n}$$

$$(a^m)^n = a^{mn}$$

$$a^0 = 1$$

$$a^{-n} = \frac{1}{a^n}$$

$$(ab)^n = a^n b^n$$

$$\left(\frac{a}{b}\right)^n = \frac{a^n}{b^n}$$

$$\left(\frac{a}{b}\right)^{-n} = \left(\frac{b}{a}\right)^n$$

Fractional indices

Since $\left(a^{\frac{1}{2}}\right)^2 = a$ and $\left(\sqrt{a}\right)^2 = a$, $a^{\frac{1}{2}} = \sqrt{a}$. Thus, surds such as $\sqrt{3}$ can be written in index form $\sqrt{3} = 3^{\frac{1}{2}}$.

The symbols $\sqrt{\ }, \sqrt[3]{\ }, \ldots \sqrt[n]{\ }$ are radical signs. Any radical can be converted to and from a fractional index using the index law.

Fractional indices

$$a^{\frac{1}{n}} = \sqrt[n]{a}$$

A combination of index laws allows $a^{\frac{m}{n}}$ to be expressed as $a^{\frac{m}{n}} = \left(a^{\frac{1}{n}}\right)^m = \left(\sqrt[n]{a}\right)^m$. It can also be expressed as $a^{\frac{m}{n}} = (a^m)^{\frac{1}{n}} = \sqrt[n]{a^m}$.

Fractional indices

$$a^{\frac{m}{n}} = \left(\sqrt[n]{a}\right)^m \text{ or } a^{\frac{m}{n}} = \sqrt[n]{a^m}$$

WORKED EXAMPLE 1 Simplifying expressions with indices

a. **Express $\dfrac{3^{2n} \times 9^{1-n}}{81^{n-1}}$ as a power of 3.**

b. **Simplify $\left(3a^2b^{\frac{5}{2}}\right)^2 \times 2\left(a^{-1}b^2\right)^{-2}$.**

c. **Evaluate $8^{\frac{2}{3}} + \left(\dfrac{4}{9}\right)^{-\frac{1}{2}}$.**

THINK

a. 1. Express each term with the same base.

2. Apply an appropriate index law.
Note: To raise a power to a power, multiply the indices.

3. Apply an appropriate index law.
Note: To multiply numbers with the same base, add the indices.

4. Apply an appropriate index law and state the answer.
Note: To divide numbers with the same base, subtract the indices.

WRITE

a. $\dfrac{3^{2n} \times 9^{1-n}}{81^{n-1}} = \dfrac{3^{2n} \times \left(3^2\right)^{1-n}}{\left(3^4\right)^{n-1}}$

$= \dfrac{3^{2n} \times 3^{2(1-n)}}{3^{4(n-1)}}$

$= \dfrac{3^{2n} \times 3^{2-2n}}{3^{4n-4}}$

$= \dfrac{3^{2n+2-2n}}{3^{4n-4}}$

$= \dfrac{3^2}{3^{4n-4}}$

$= 3^{2-(4n-4)}$

$= 3^{6-4n}$

$\therefore \dfrac{3^{2n} \times 9^{1-n}}{81^{n-1}} = 3^{6-4n}$

b. 1. Use an index law to remove the brackets.	**b.** $\left(3a^2b^{\frac{5}{2}}\right)^2 \times 2\left(a^{-1}b^2\right)^{-2}$ $= 3^2a^4b^5 \times 2 \times a^2b^{-4}$ $= 9a^4b^5 \times 2a^2b^{-4}$
2. Apply the index laws to terms with the same base to simplify the expression.	$= 18a^6b^1$ $= 18a^6b$
c. 1. Express each term with a positive index. *Note:* There is no index law for the addition of numbers.	**c.** $8^{\frac{2}{3}} + \left(\frac{4}{9}\right)^{-\frac{1}{2}} = 8^{\frac{2}{3}} + \left(\frac{9}{4}\right)^{\frac{1}{2}}$
2. Apply the index law for fractional indices. *Note:* The fractional indices could be interpreted in other ways including, for example as $8^{\frac{2}{3}} = (2^3)^{\frac{2}{3}}$.	$= \left(\sqrt[3]{8}\right)^2 + \frac{\sqrt{9}}{\sqrt{4}}$
3. Evaluate each term separately and calculate the answer required.	$= (2)^2 + \frac{3}{2}$ $= 4 + 1\frac{1}{2}$ $= 5\frac{1}{2}$

TI | THINK	DISPLAY/WRITE	CASIO | THINK	DISPLAY/WRITE
b. 1. On a Calculator page, complete the entry line as: $\left(3a^2 \times b^{\frac{5}{2}}\right)^2 \times 2\left(a^{-1} \times b^2\right)^{-2}$ Then press ENTER.	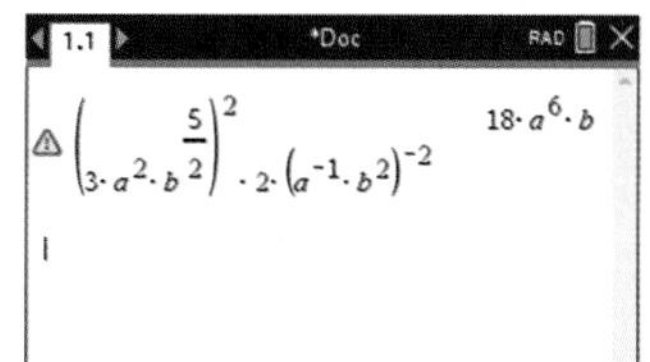	**b. 1.** On a Main screen, complete the entry line as: $\left(3a^2 \times b^{\frac{5}{2}}\right)^2 \times 2\left(a^{-1} \times b^2\right)^{-2}$ Then press EXE.	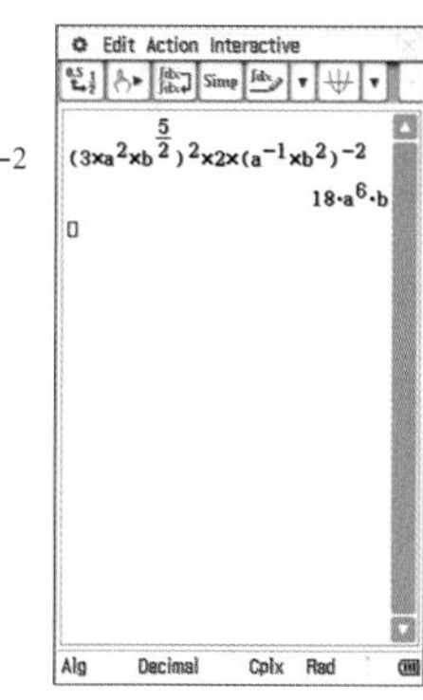
2. The answer appears on the screen.	$18a^6b$	**2.** The answer appears on the screen.	$18a^6b$

10.2.2 Indicial equations

An **indicial equation** has the unknown variable as an exponent. In this section we shall consider indicial equations that have rational solutions.

Method of equating indices

If index laws can be used to express both sides of an equation as single powers of the same base, then this allows indices to be equated. For example, if an equation can be simplified to the form $2^{3x} = 2^6$, then for the equality to hold, $3x = 6$. Solving this linear equation gives the solution to the indicial equation as $x = 2$.

To solve for the exponent x in equations of the form $a^x = n$, perform the following steps:

- Express both sides as powers of the same base.
- Equate the indices and solve the equation formed to obtain the solution to the indicial equation.

Inequations are solved in a similar manner. If $2^{3x} < 2^6$, for example, then $3x < 6$. Solving this linear inequation gives the solution to the indicial inequation as $x < 2$. However, you need to ensure the base a is greater than 1 before expressing the indices with the corresponding order sign between them.

For example, you first need to write $\left(\frac{1}{2}\right)^x < \left(\frac{1}{2}\right)^3$ as $2^{-x} < 2^{-3}$ and then solve the linear inequation $-x < -3$ to obtain the solution $x > 3$.

WORKED EXAMPLE 2 Solving equations with indices

Solve $5^{3x} \times 25^{4-2x} = \frac{1}{125}$ for x.

THINK	WRITE
1. Use the index laws to express the left-hand side of the equation as a power of a single base.	$5^{3x} \times 25^{4-2x} = \frac{1}{125}$ $5^{3x} \times 5^{2(4-2x)} = \frac{1}{125}$ $5^{3x+8-4x} = \frac{1}{125}$ $5^{8-x} = \frac{1}{125}$
2. Express the right-hand side as a power of the same base.	$5^{8-x} = \frac{1}{5^3}$ $5^{8-x} = 5^{-3}$
3. Equate indices and calculate the required value of x.	Equating indices, $8 - x = -3$ $x = 11$

TI | THINK	DISPLAY/WRITE	CASIO | THINK	DISPLAY/WRITE
b. 1. On a Calculator page, press MENU followed by: 3: Algebra 1: Solve Complete the entry line as: solve $\left(5^{3x} \times 25^{4-2x} = \frac{1}{125}, x\right)$ Then press ENTER.	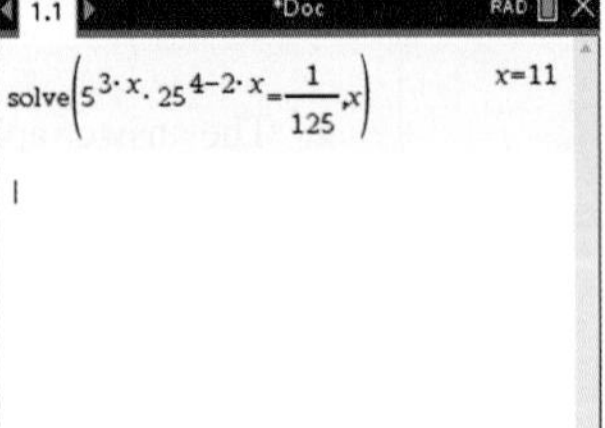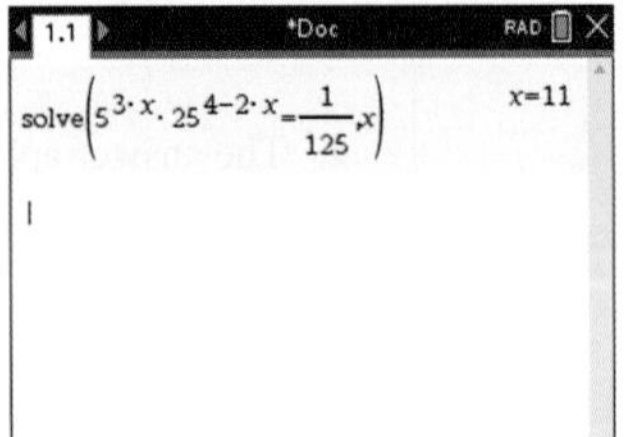	b. 1. On a Main screen, complete the entry line as: solve $\left(5^{3x} \times 25^{4-2x} = \frac{1}{125}, x\right)$ Then press EXE.	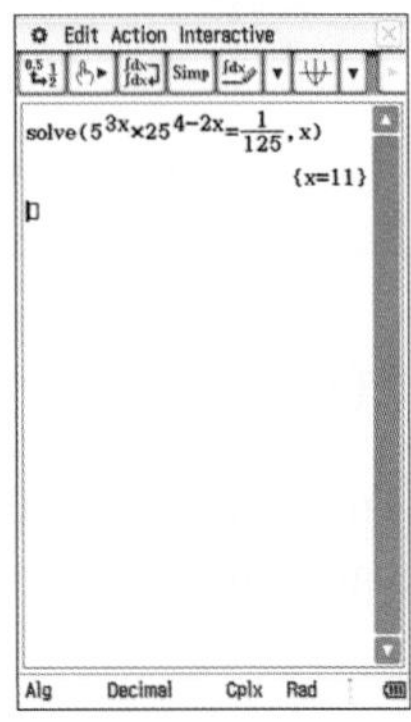
2. The answer appears on the screen.	$x = 11$	2. The answer appears on the screen.	$x = 11$

10.2.3 Indicial equations that reduce to quadratic form

The technique of substitution to form a quadratic equation may be applicable to indicial equations.

To solve equations of the form $p \times a^{2x} + q \times a^x + r = 0$, perform the following steps:

- Note that $a^{2x} = (a^x)^2$.
- Reduce the indicial equation to quadratic form by using a substitution for a^x.
- Solve the quadratic and then substitute back for a^x.
- Since a^x must always be positive, solutions for x can only be obtained for $a^x > 0$; reject any negative or zero values for a^x.

WORKED EXAMPLE 3 Solving indicial equations in quadratic form

Solve $3^{2x} - 6 \times 3^x - 27 = 0$ for x.

THINK	WRITE
1. Use a substitution technique to reduce the indicial equation to quadratic form. *Note:* The subtraction signs prevent the use of index laws to express the left-hand side as a power of a single base.	$3^{2x} - 6 \times 3^x - 27 = 0$ Let $a = 3^x$. $\therefore a^2 - 6a - 27 = 0$
2. Solve the quadratic equation.	$(a - 9)(a + 3) = 0$ $a = 9, a = -3$
3. Substitute back and solve for x.	Replace a with 3^x. $\therefore 3^x = 9$ or $3^x = -3$ (reject negative value) $3^x = 9$ $\therefore 3^x = 3^2$ $\therefore x = 2$

10.2.4 Scientific notation (standard form)

Index notation provides a convenient way to express numbers that are either very large or very small. Writing a number as $a \times 10^b$ (the product of a number a where $1 \leq a < 10$ and a power of 10) is known as writing the number in **scientific notation** (or **standard form**). The age of the earth since the Big Bang is estimated to be 4.54×10^9 years, and the mass of a carbon atom is approximately 1.994×10^{-23} grams. These numbers are written in scientific notation.

To convert scientific notation back to a basic numeral:

- move the decimal point b places to the **right** if the power of 10 has a positive index to obtain the large number $a \times 10^b$ represents
- move the decimal point b places to the **left** if the power of 10 has a negative index to obtain the small number $a \times 10^{-b}$ represents. Remember multiplying by 10^{-b} is equivalent to dividing by 10^b.

Significant figures

When a number is expressed in scientific notation as either $a \times 10^b$ or $a \times 10^{-b}$, the number of digits in a determines the number of **significant figures** in the basic numeral. The age of the Earth is 4.54×10^9 years in scientific notation or 4 540 000 000 years to 3 significant figures. To 1 significant figure, the age would be 5 000 000 000 years.

WORKED EXAMPLE 4 Expressing numerals in scientific notation

a. Express each of the following numerals in scientific notation and state the number of significant figures each numeral contains.

i. 3 266 400

ii. 0.009 876 03

b. Express each of the following as a basic numeral.

i. 4.54×10^9

ii. 1.037×10^{-5}

THINK	WRITE
a. i. 1. Write the given number as a value between 1 and 10 multiplied by a power of 10. *Note:* The number is large, so the power of 10 should be positive and is the number of places the decimal point has moved to the left.	a. i. In scientific notation, $3\,266\,400 = 3.2664 \times 10^6$.
2. Count the number of digits in the number a in the scientific notation form $a \times 10^b$ and state the number of significant figures.	There are 5 significant figures in the number 3 266 400.
ii. 1. Write the given number as a value between 1 and 10 multiplied by a power of 10. *Note:* The number is small so the power of 10 should be negative and is the number of places the decimal point has moved to the right.	ii. $0.009\,876\,03 = 9.876\,03 \times 10^{-3}$
2. Count the number of digits in the number a in the scientific notation form and state the number of significant figures.	0.009 876 03 has 6 significant figures.
b. i. 1. Perform the multiplication. *Note:* The power of 10 is positive, so a large number should be obtained.	b. i. 4.54×10^9 Move the decimal point 9 places to the right. $\therefore 4.54 \times 10^9 = 4\,540\,000\,000$
ii. 2. Perform the multiplication. *Note:* The power of 10 is negative, so a small number should be obtained.	ii. 1.037×10^{-5} Move the decimal point 5 places to the left. $\therefore 1.037 \times 10^{-5} = 0.000\,010\,37$

10.2 Exercise

Students, these questions are even better in jacPLUS

Receive immediate feedback and access sample responses

Access additional questions

Track your results and progress

Find all this and MORE in jacPLUS

Technology free

1. a. Express the following in index (exponent) form.

i. $\sqrt{a^3b^4}$ ii. $\sqrt{\dfrac{a^5}{b^{-4}}}$ iii. $\sqrt[3]{a^2b}$

b. Express the following in surd form.

i. $a^{\frac{1}{2}} \div b^{\frac{3}{2}}$ ii. $2^{\frac{5}{2}}$ iii. $3^{-\frac{2}{5}}$

2. Evaluate the following without a calculator.

a. $4^{\frac{3}{2}}$

b. $3^{-1} + 5^0 - 2^2 \times 9^{-\frac{1}{2}}$

c. $2^3 \times \left(\dfrac{4}{9}\right)^{-\frac{1}{2}}$

d. $\dfrac{15 \times 5^{\frac{3}{2}}}{125^{\frac{1}{2}}}$

3. Simplify the following.

a. $a^{3-3n} \times \left(a^2\right)^{2n} \div \left(a^3\right)^{n+2}$

b. $\dfrac{\left(x^4y\right)^3 \times \left(9xy^2\right)^2}{(2y)^0 \times \left(3x^2\right)^3}$

c. $\dfrac{\left(a^3b^2\right)^2 \times \left(4ab^3\right)^3}{8\left(a^3b^9\right)^2}$

d. $\dfrac{2\left(x^2y\right)^4 \times \left(2x^{\frac{1}{3}}y^2\right)^3}{y^3\left(2x^{-1}\right)^5}$

e. $\dfrac{81^2 \times 27^{-\frac{2}{3}} \times \sqrt{3}}{243^{\frac{3}{5}}}$

f. $\dfrac{\left(a^nb^{3n}\right)^{-2} \times \left(a^{3n}b^{\frac{3}{2}}\right)^{\frac{1}{3}}}{\left(a^{-4n}b\right)^{-\frac{3}{2}}}$

4. **WE1** a. Express $\dfrac{2^{1-n} \times 8^{1+2n}}{16^{1-n}}$ as a power of 2.

b. Simplify $\left(9a^3b^{-4}\right)^{\frac{1}{2}} \times 2\left(a^{\frac{1}{2}}b^{-2}\right)^{-2}$.

c. Evaluate $27^{-\frac{2}{3}} + \left(\dfrac{49}{81}\right)^{\frac{1}{2}}$.

5. Simplify $\dfrac{20p^5}{m^3q^{-2}} \div \dfrac{5\left(p^2q^{-3}\right)^2}{-4m^{-1}}$.

6. Simplify and express the following with positive indices.

a. $\dfrac{3\left(x^2y^{-2}\right)^3}{\left(3x^4y^2\right)^{-1}}$

b. $\dfrac{2a^{\frac{2}{3}}b^{-3}}{3a^{\frac{1}{3}}b^{-1}} \times \dfrac{3^2 \times 2 \times (ab)^2}{\left(-8a^2\right)^2b^2}$

c. $\dfrac{\left(2mn^{-2}\right)^{-2}}{m^{-1}n} \div \dfrac{10n^4m^{-1}}{3\left(m^2n\right)^{\frac{3}{2}}}$

d. $\dfrac{4m^2n^{-2} \times -2\left(m^2n^{\frac{3}{2}}\right)^2}{\left(-3m^3n^{-2}\right)^2}$

e. $\dfrac{m^{-1} - n^{-1}}{m^2 - n^2}$

f. $\sqrt{4x-1} - 2x(4x-1)^{-\frac{1}{2}}$

7. a. Express $\dfrac{32 \times 4^{3x}}{16^x}$ as a power of 2.

b. Express $\dfrac{3^{1+n} \times 81^{n-2}}{243^n}$ as a power of 3.

c. Express $0.001 \times \sqrt[3]{10} \times 100^{\frac{5}{2}} \times (0.1)^{-\frac{2}{3}}$ as a power of 10.

d. Express $\dfrac{5^{n+1} - 5^n}{4}$ as a power of 5.

8. Solve the following for x.

a. $2^{3x} = 16$

b. $7^{3x} = 49^{1+x}$

c. $3^{1-5x} = 1$

d. $2^{x-4} = 8^{4-x}$

e. $25^{3-x} = \dfrac{125^{2x}}{5^{x+2}}$

f. $3^{x+5} - \left(\dfrac{1}{3}\right)^{1-2x} = 0$

9. **WE2** Solve $\dfrac{2^{5x-3} \times 8^{9-2x}}{4^x} = 1$ for x.

10. Solve the following inequations.

a. $2 \times 5^x + 5^x < 75$

b. $\left(\dfrac{1}{9}\right)^{2x-3} > \left(\dfrac{1}{9}\right)^{7-x}$

11. Solve the following for x.

a. $2^{2x} \times 8^{2-x} \times 16^{-\frac{3x}{2}} = \dfrac{2}{4^x}$

b. $25^{3x-3} \leq 125^{4+x}$

c. $9^x \div 27^{1-x} = \sqrt{3}$

d. $\left(\dfrac{2}{3}\right)^{3-2x} > \left(\dfrac{27}{8}\right)^{-\frac{1}{3}} \times \dfrac{1}{\sqrt{2\frac{1}{4}}}$

e. $4^{5x} + 4^{5x} = \dfrac{8}{2^{4x-5}}$

f. $5^{\frac{2x}{3}} \times 5^{\frac{3x}{2}} = 25^{x+4}$

12. Determine the values of x that are solutions to the following equations.

a. $2^{2x} - 9 \times 2^x + 8 = 0$

b. $3^{2x} + 6 \times 3^x - 27 = 0$

c. $4^{2x} + 20 \times 4^x + 64 = 0$

d. $5^{2x} - 2 \times 5^x + 1 = 0$

13. **WE3** Solve $30 \times 10^{2x} + 17 \times 10^x - 2 = 0$ for x.

14. Solve $2^x - 48 \times 2^{-x} = 13$ for x.

15. Use a suitable substitution to solve the following equations.

a. $3^{2x} - 10 \times 3^x + 9 = 0$

b. $24 \times 2^{2x} + 61 \times 2^x = 2^3$

c. $25^x + 5^{2+x} - 150 = 0$

d. $(2^x + 2^{-x})^2 = 4$

e. $10^x - 10^{2-x} = 99$

f. $2^{3x} + 3 \times 2^{2x-1} - 2^x = 0$

16. **WE4** a. Express each of the following numbers in scientific notation and state the number of significant figures each number contains.

i. 1 409 000

ii. 0.000 130 6

b. Express each of the following as a basic numeral.

i. 3.04×10^5

ii. 5.803×10^{-2}

17. Calculate $\left(4 \times 10^6\right)^2 \times \left(5 \times 10^{-3}\right)$ without using a calculator, expressing the answer in standard form.

18. a. Express the following in scientific notation.

i. $-0.000\,000\,050\,6$
ii. The diameter of the Earth, given its radius is 6370 km
iii. $3.2 \times 10^4 \times 5 \times 10^{-2}$
iv. The distance between Roland Garros and Kooyong tennis stadiums of 16 878.7 km

b. Express the following as basic numerals.

i. $6.3 \times 10^{-4} + 6.3 \times 10^4$ ii. $\left(1.44 \times 10^6\right)^{\frac{1}{2}}$

c. Express the following to 2 significant figures.

i. 60 589 people attended a football match.
ii. The probability of winning a competition is 1.994×10^{-2}.
iii. The solution to an equation is $x = -0.006\,34$.
iv. The distance flown per year by the Royal Flying Doctor Service is 26 597 696 km.

19. a. Solve the pair of simultaneous equations for x and y.

$$5^{2x-y} = \frac{1}{125}$$
$$10^{2y-6x} = 0.01$$

b. Solve the pair of simultaneous equations for a and k.

$$a \times 2^{k-1} = 40$$
$$a \times 2^{2k-2} = 10$$

Technology active

20. Evaluate:

a. $5^{-4.3}$ and express the answer in scientific notation to 4 significant figures
b. $22.9 \div 1.3\text{E}2$ to 4 significant figures and explain what the 1.3E2 notation means
c. $5.04 \times 10^{-6} \div \left(3 \times 10^9\right)$, expressing the answer in standard form
d. $5.04 \times 10^{-6} \div \left(3.2 \times 10^{4.2}\right)$, expressing the answer in standard form to 4 significant figures.

10.2 Exam questions

Question 1 (1 mark) TECH-ACTIVE

MC The value of $3^{-4} \times \left(\frac{8}{27}\right)^{-\frac{1}{3}}$ is

A. $\frac{1}{8}$ B. $\frac{1}{12}$ C. $\frac{3}{32}$ D. $\frac{1}{54}$ E. $\frac{8}{81}$

Question 2 (1 mark) TECH-ACTIVE

MC If $27^{3-2x} = 729^x$, then x is equal to

A. 6 B. $\frac{3}{4}$ C. $\frac{1}{3}$ D. 1 E. $\frac{7}{9}$

Question 3 (2 marks) TECH-FREE

Solve for x.

$27^x \div 9^{1-x} = 3\sqrt{3}$

More exam questions are available online.

10.3 Indices as logarithms

LEARNING INTENTION

At the end of this subtopic you should be able to:
- recognise the relationship between an exponential function and a logarithmic function
- simplify expressions with logarithmic laws
- solve equations with logarithmic laws.

Not all solutions to indicial equations are rational. In order to obtain the solution to an equation such as $2^x = 5$, we need to learn about logarithms.

10.3.1 Index–logarithm forms

A **logarithm** is also another name for an index.

The index statement, $n = a^x$, with base a and index x, can be expressed with the index as the subject. This is called the logarithm statement and is written as $x = \log_a(n)$. The statement is read as 'x equals the log to base a of n' (adopting the abbreviation of 'log' for logarithm).

Logarithm form

The statements $n = a^x$ and $x = \log_a(n)$ are equivalent.

$$n = a^x \Leftrightarrow x = \log_a(n)$$

where

the base $a \in R^+\backslash\{1\}$,

the number $n \in R^+$ and

the logarithm, or index, $x \in R$.

Consider again the equation $2^x = 5$. The solution is obtained by converting this index statement to its logarithm statement.

$$2^x = 5$$
$$\therefore\ x = \log_2(5)$$

The number $\log_2(5)$ is irrational: the power of 2 that gives the number 5 is not rational. A decimal approximation for this logarithm can be obtained using a calculator. The exact solution to the indicial equation $2^x = 5$ is $x = \log_2(5)$; an approximate solution is $x \approx 2.3219$ to 5 significant figures.

Not all expressions containing logarithms are irrational. Solving the equation $2^x = 8$ by converting to logarithm form gives:

$$2^x = 8$$
$$\therefore\ x = \log_2(8)$$

In this case, the expression $\log_2(8)$ can be simplified. As the power of 2 that gives the number 8 is 3, the solution to the equation is $x = 3$; that is, $\log_2(8) = 3$.

Use of a calculator

Scientific calculators have two inbuilt logarithmic functions.

Base 10 logarithms are obtained from the LOG key. Thus, $\log_{10}(2)$ is evaluated as log (2), giving the value of 0.3010 to 4 decimal places. Base 10 logarithms are called common logarithms.

Base e logarithms are obtained from the LN key. Thus, $\log_e(2)$ is evaluated as ln (2), giving the value 0.6931 to 4 decimal places. Base e logarithms are called natural or Naperian logarithms. These logarithms occur extensively in calculus. The number e itself is known as Euler's number and, like π, it is a transcendental irrational number that has great importance in higher mathematical studies, as you will begin to discover in Units 3 and 4 of Mathematical Methods.

CAS technology enables logarithms to bases other than 10 or e to be evaluated.

WORKED EXAMPLE 5 Expressing equations as logarithmic or exponential equivalents

a. Express $3^4 = 81$ as a logarithm statement.

b. Express $\log_7\left(\frac{1}{49}\right) = -2$ as an index statement.

c. Solve the equation $10^x = 12.8$, expressing the exponent x to 2 significant figures.

d. Solve the equation $\log_5(x) = 2$ for x.

THINK	WRITE
a. 1. Identify the given form.	**a.** $3^4 = 81$ The index form is given with base 3, index or logarithm 4 and number 81.
2. Convert to the equivalent form.	Since $n = a^x \Leftrightarrow x = \log_a(n)$, $81 = 3^4 \Rightarrow 4 = \log_3(81)$. The logarithm statement is $4 = \log_3(81)$ or $\log_3(81) = 4$.
b. 1. Identify the given form.	**b.** $\log_7\left(\frac{1}{49}\right) = -2$ The logarithm form is given with base 7, number $\frac{1}{49}$ and logarithm, or index, of -2.
2. Convert to the equivalent form.	Since $x = \log_a(n) \Leftrightarrow n = a^x$, $-2 = \log_7\left(\frac{1}{49}\right) \Rightarrow \frac{1}{49} = 7^{-2}$ The index statement is $\frac{1}{49} = 7^{-2}$ or $7^{-2} = \frac{1}{49}$.
c. 1. Convert to the equivalent form.	**c.** $10^x = 12.8$ $\therefore x = \log_{10}(12.8)$
2. Evaluate using a calculator and state the answer to the required accuracy. *Note:* The base is 10, so use the LOG key on the calculator.	≈ 1.1 to 2 significant figures
d. 1. Convert to the equivalent form.	**d.** $\log_5(x) = 2$
2. Calculate the answer.	$\therefore x = 5^2$ $\therefore x = 25$

10.3.2 Logarithm laws

Since logarithms are indices, unsurprisingly there is a set of laws that control simplification of logarithmic expressions with the same base.

Logarithm laws

For any a, m, $n > 0$, $a \neq 1$, the laws are:
1. $\log_a(1) = 0$
2. $\log_a(a) = 1$
3. $\log_a(m) + \log_a(n) = \log_a(mn)$
4. $\log_a(m) - \log_a(n) = \log_a\left(\frac{m}{n}\right)$
5. $\log_a(m^p) = p\log_a(m)$.

Note that there is no logarithm law for either the product or quotient of logarithms or for expressions such as $\log_a(m \pm n)$.

Proofs of the logarithm laws

1. Consider the index statement.

$$a^0 = 1$$
$$\therefore \log_a(1) = 0$$

2. Consider the index statement.

$$a^1 = a$$
$$\therefore \log_a(a) = 1$$

3. Let $x = \log_a(m)$ and $y = \log_a(n)$.

$$\therefore m = a^x \text{ and } n = a^y$$
$$mn = a^x \times a^y$$
$$= a^{x+y}$$

Convert to logarithm form:

$$x + y = \log_a(mn)$$

Substitute back for x and y:

$$\log_a(m) + \log_a(n) = \log_a(mn)$$

4. With x and y as given in law 3:

$$\frac{m}{n} = \frac{a^x}{a^y}$$
$$= a^{x-y}$$

Converting to logarithm form and then substituting back for x and y gives:

$$x - y = \log_a\left(\frac{m}{n}\right)$$

Substitute back for x and y:

$$\therefore \log_a(m) - \log_a(n) = \log_a\left(\frac{m}{n}\right)$$

5. With x as given in law 3:

$$x = \log_a(m)$$
$$\therefore m = a^x$$

Raise both sides to the power p:

$$\therefore (m)^p = (a^x)^p$$
$$\therefore m^p = a^{px}$$

Express as a logarithm statement with base a:

$$px = \log_a(m^p)$$

Substitute back for x:

$$\therefore p\log_a(m) = \log_a(m^p)$$

WORKED EXAMPLE 6 Simplifying using logarithm laws

Simplify the following using the logarithm laws.

a. $\log_{10}(5) + \log_{10}(2)$

b. $\log_2(80) - \log_2(5)$

c. $\log_3\left(2a^4\right) + 2\log_3\left(\sqrt{\frac{a}{2}}\right)$

d. $\dfrac{\log_a\left(\frac{1}{4}\right)}{\log_a(2)}$

THINK	WRITE
a. 1. Apply the appropriate logarithm law.	**a.** $\log_{10}(5) + \log_{10}(2) = \log_{10}(5 \times 2)$
	$= \log_{10}(10)$
2. Simplify and state the answer.	$= 1 (\text{since } \log_a(a) = 1)$
b. 1. Apply the appropriate logarithm law.	**b.** $\log_2(80) - \log_2(5) = \log_2\left(\frac{80}{5}\right)$
	$= \log_2(16)$
2. Further simplify the logarithmic expression.	$= \log_2\left(2^4\right)$
	$= 4\log_2(2)$
3. Calculate the answer.	$= 4 \times 1$
	$= 4$

c. 1. Apply a logarithm law to the second term.
Note: As with indices, there is often more than one way to approach the simplification of logarithms.

c. $\log_3(2a^4) + 2\log_3\left(\sqrt{\frac{a}{2}}\right)$

$= \log_3(2a^4) + 2\log_3\left(\frac{a}{2}\right)^{\frac{1}{2}}$

$= \log_3(2a^4) + 2 \times \frac{1}{2}\log_3\left(\frac{a}{2}\right)$

$= \log_3(2a^4) + \log_3\left(\frac{a}{2}\right)$

2. Apply an appropriate logarithm law to combine the two terms as one logarithmic expression.

$= \log_3\left(2a^4 \times \frac{a}{2}\right)$

$= \log_3(a^5)$

3. State the answer.

$= 5\log_3(a)$

d. 1. Express the numbers in the logarithm terms in index form.
Note: There is no law for division of logarithms.

d. $\frac{\log_a\left(\frac{1}{4}\right)}{\log_a(2)} = \frac{\log_a(2^{-2})}{\log_a(2)}$

2. Simplify the numerator and denominator separately.

$= \frac{-2\log_a(2)}{\log_a(2)}$

3. Cancel the common factor in the numerator and denominator.

$= \frac{-2\cancel{\log_a(2)}}{\cancel{\log_a(2)}}$

$= -2$

TI | THINK	DISPLAY/WRITE
d. 1. On a Calculator page, complete the entry line as: $\frac{\log_a\left(\frac{1}{4}\right)}{\log_a(2)}$ Then press ENTER.	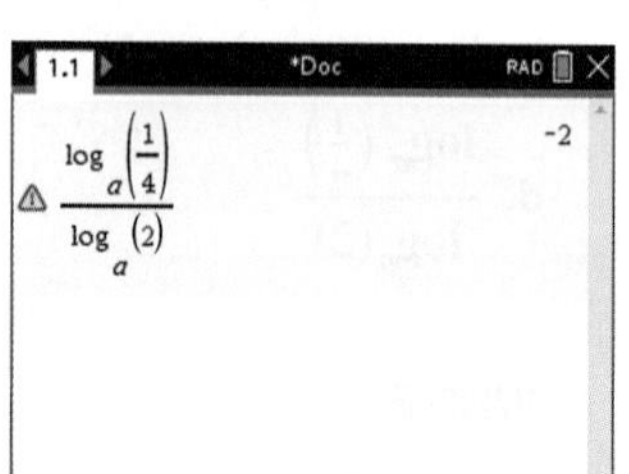
2. The answer appears on the screen.	$\frac{\log_a\left(\frac{1}{4}\right)}{\log_a(2)} = -2$

CASIO | THINK	DISPLAY/WRITE
d. 1. On a Main screen, complete the entry line as: $\frac{\log_a\left(\frac{1}{4}\right)}{\log_a(2)}$ Then press EXE.	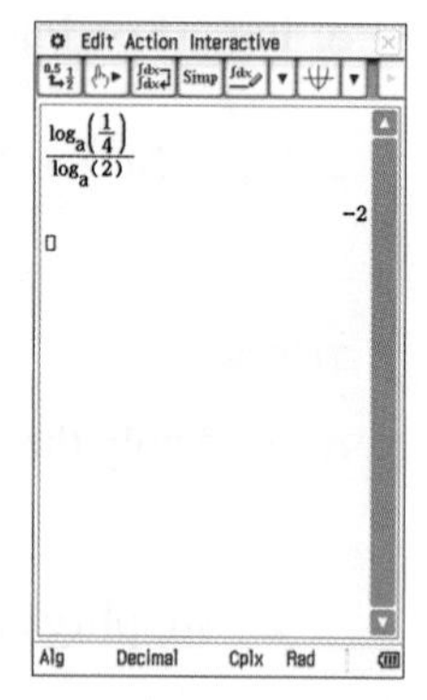
2. The answer appears on the screen.	$\frac{\log_a\left(\frac{1}{4}\right)}{\log_a(2)} = -2$

10.3.3 Logarithms as operators

Just as both sides of an equation may be raised to a power and the equality still holds, taking logarithms of both sides of an equation maintains the equality.

If $m = n$, then it is true that $\log_a(m) = \log_a(n)$ and vice versa, provided the same base is used for the logarithms of each side.

This application of logarithms can provide an important tool when solving indicial equations.

Consider again the equation $2^x = 5$, where the solution was given as $x = \log_2(5)$.

Take base 10 logarithms of both sides of this equation.

$$2^x = 5$$
$$\log_{10}(2^x) = \log_{10}(5)$$

Using one of the logarithm laws, this becomes $x\log_{10}(2) = \log_{10}(5)$, from which the solution to the indicial equation is obtained as $x = \frac{\log_{10}(5)}{\log_{10}(2)}$. This form of the solution can be evaluated on a scientific calculator and is the prime reason for choosing base 10 logarithms in solving the indicial equation.

It also demonstrates that $\log_2(5) = \frac{\log_{10}(5)}{\log_{10}(2)}$, which is a particular example of another logarithm law called the *change of base law*.

Change of base law for calculator use

The equation $a^x = p$ for which $x = \log_a(p)$ could be solved in a similar way to $2^x = 5$, giving the solution as $x = \frac{\log_{10}(p)}{\log_{10}(a)}$. Thus, $\log_a(p) = \frac{\log_{10}(p)}{\log_{10}(a)}$. This form enables decimal approximations to logarithms to be calculated on scientific calculators.

Change of base law

$$\log_a(p) = \frac{\log_{10}(p)}{\log_{10}(a)}$$

The change of base law is the more general statement allowing base a logarithms to be expressed in terms of any other base b as $\log_a(p) = \frac{\log_b(p)}{\log_b(a)}$. This more general form will be explored further in Units 3 and 4.

Convention

There is a convention that if the base of a logarithm is not stated, this implies it is base 10. As it is on a calculator, $\log(n)$ represents $\log_{10}(n)$. When working with base 10 logarithms it can be convenient to adopt this convention.

WORKED EXAMPLE 7 Solving equations using logarithms

a. **State the exact solution to $5^x = 8$ and calculate its value to 3 decimal places.**
b. **Calculate the exact value and the value to 3 decimal places of the solution to the equation $2^{1-x} = 6^x$.**

THINK	WRITE
a. 1. Convert to the equivalent form and state the exact solution.	a. $5^x = 8$ $\therefore x = \log_5(8)$ The exact solution is $x = \log_5(8)$.
2. Use the change of base law to express the answer in terms of base 10 logarithms.	Since $\log_a(p) = \dfrac{\log_{10}(p)}{\log_{10}(a)}$, $\log_5(8) = \dfrac{\log_{10}(8)}{\log_{10}(5)}$ $\therefore x = \dfrac{\log_{10}(8)}{\log_{10}(5)}$
3. Calculate the approximate value.	$\therefore x \approx 1.292$ to 3 decimal places.
b. 1. Take base 10 logarithms of both sides. *Note:* The convention is not to write the base 10.	b. $2^{1-x} = 6^x$ Take logarithms to base 10 of both sides: $\log\left(2^{1-x}\right) = \log\left(6^x\right)$
2. Apply the logarithm law so that x terms are no longer exponents.	$(1-x)\log(2) = x\log(6)$
3. Solve the linear equation in x. *Note:* This is no different to solving any other linear equation of the form $a - bx = cx$ except the constants a, b, c are expressed as logarithms.	Expand: $\log(2) - x\log(2) = x\log(6)$ Collect x terms together: $\log(2) = x\log(6) + x\log(2)$ $= x(\log(6) + \log(2))$ $x = \dfrac{\log(2)}{\log(6) + \log(2)}$ This is the exact solution.
4. Calculate the approximate value. *Note:* Remember to place brackets around the denominator for the division when entering it into a calculator.	$x \approx 0.279$ to 3 decimal places.

10.3.4 Equations containing logarithms

Although the emphasis in this topic is on exponential (indicial) relations for which some knowledge of logarithms is essential, it is important to know that logarithms contribute substantially to mathematics. As such, some equations involving logarithms are included, allowing further consolidation of the laws which logarithms must satisfy.

Remembering the requirement that x must be positive for $\log_a(x)$ to be real, it is advisable to check any solution to an equation involving logarithms. Any value of x that when substituted back into the original equation creates a '$\log_a$ (negative number)' term must be rejected as a solution. Otherwise, normal algebraic approaches together with logarithm laws are the techniques for solving such equations.

WORKED EXAMPLE 8 Solving equations containing logarithms

Solve the equation $\log_6(x) + \log_6(x-1) = 1$ for x.

THINK	WRITE
1. Apply the logarithm law that reduces the equation to one logarithm term.	$\log_6(x) + \log_6(x-1) = 1$ $\therefore \log_6(x(x-1)) = 1$ $\therefore \log_6(x^2 - x) = 1$
2. Convert the logarithm form to its equivalent form. *Note:* An alternative method is to write $\log_6(x^2 - x) = \log_6(6)$, from which $x^2 - x = 6$ is obtained.	Converting from logarithm form to index form gives: $x^2 - x = 6^1$ $\therefore x^2 - x = 6$
3. Solve the quadratic equation.	$x^2 - x - 6 = 0$ $(x-3)(x+2) = 0$ $\therefore x = 3, x = -2$
4. Check the validity of both solutions in the original equation.	Check in $\log_6(x) + \log_6(x-1) = 1$. If $x = 3$, $\text{LHS} = \log_6(3) + \log_6(2)$ $= \log_6(6)$ $= 1$ $= \text{RHS}$ If $x = -2$, $\text{LHS} = \log_6(-2) + \log_6(-3)$, which is not admissible. Therefore, reject $x = -2$.
5. State the answer.	The solution is $x = 3$.

TI | THINK	DISPLAY/WRITE	CASIO | THINK	DISPLAY/WRITE
1. On a Calculator page, complete the entry line as: solve $(\log_6(x) + \log_6(x-1) = 1, x)$ Then press ENTER.	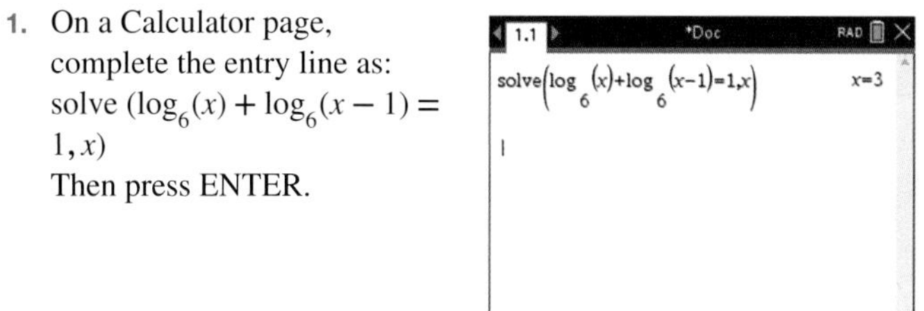	1. On a Main screen, complete the entry line as: solve $(\log_6(x) + \log_6(x-1) = 1, x)$ Then press EXE.	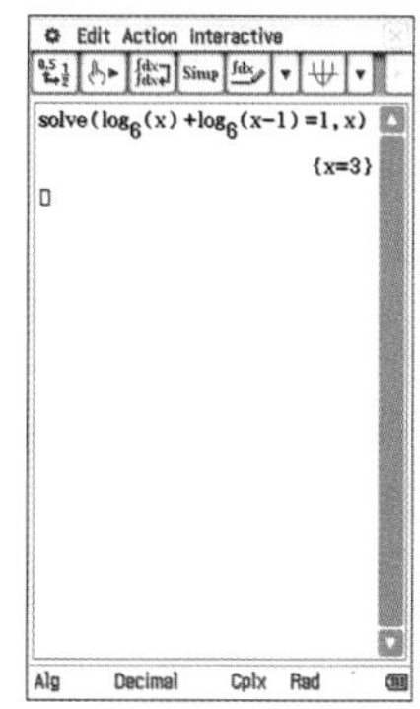
2. The answer appears on the screen.	$x = 3$	2. The answer appears on the screen.	$x = 3$

10.3 Exercise

Students, these questions are even better in jacPLUS

Receive immediate feedback and access sample responses

Access additional questions

Track your results and progress

Find all this and MORE in jacPLUS

Technology active

1. WE5 a. Express $5^4 = 625$ as a logarithm statement.
   b. Express $\log_{36}(6) = \frac{1}{2}$ as an index statement.
   c. Solve the equation $10^x = 8.52$, expressing the exponent x to 2 significant figures.
   d. Solve the equation $\log_3(x) = -1$ for x.
2. a. Evaluate $\log_e(5)$ to 4 significant figures and write the equivalent index statement.
   b. Evaluate $10^{3.5}$ to 4 significant figures and write the equivalent logarithm statement.
3. WE6 Simplify the following using the logarithm laws.
   a. $\log_{12}(3) + \log_{12}(4)$
   b. $\log_2(192) - \log_2(12)$
   c. $\log_3(3a^3) - 2\log_3\left(a^{\frac{3}{2}}\right)$
   d. $\frac{\log_a(8)}{\log_a(4)}$
4. Given $\log_a(2) = 0.3$ and $\log_a(5) = 0.7$, evaluate the following.
   a. $\log_a(0.5)$
   b. $\log_a(2.5)$
   c. $\log_a(20)$
5. Use the logarithm laws to evaluate the following.
   a. $\log_5(5 \div 5)$
   b. $\log_{10}(5) + \log_{10}(2)$
   c. $\log_3\left(\frac{1}{3}\right)$
   d. $\log_2(32)$
   e. $\log_4\left(\frac{7}{32}\right) - \log_4(14)$
   f. $\log_6(9) + \log_6(8) - \log_6(2)$
6. Use the logarithm laws to simplify the following.
   a. $\log_3(x^3) - \log_3(x^2)$
   b. $\log_a(2x^5) + \log_a\left(\frac{x}{2}\right)$
   c. $-\frac{1}{2}\log_{10}(a) + \log_{10}\left(\sqrt{a}\right)$
   d. $\frac{1}{2}\log_b(16a^4) - \frac{1}{3}\log_b(8a^3)$
   e. $3\log_{10}(x) - 2\log_{10}(x^3) + \frac{1}{2}\log_{10}(x^5)$
   f. $\frac{\log_3(x^6)}{\log_3(x^2)}$
7. Use the logarithm laws to evaluate the following.
   a. $\log_9(3) + \log_9(27)$
   b. $\log_9(3) - \log_9(27)$
   c. $2\log_2(4) + \log_2(6) - \log_2(12)$
   d. $\log_5(\log_3(3))$
   e. $\log_{11}\left(\frac{7}{3}\right) + 2\log_{11}\left(\frac{1}{3}\right) - \log_{11}\left(\frac{11}{3}\right) - \log_{11}\left(\frac{7}{9}\right)$
   f. $\log_2\left(\frac{\sqrt{x} \times x^{\frac{3}{2}}}{x^2}\right)$

8. a. Express the following as logarithm statements with the index as the subject.

i. $2^5 = 32$ ii. $4^{\frac{3}{2}} = 8$ iii. $10^{-3} = 0.001$

b. Express the following as index statements.

i. $\log_2(16) = 4$ ii. $\log_9(3) = \frac{1}{2}$ iii. $\log_{10}(0.1) = -1$

9. Express the following as single logarithms.

a. $\log_{10}(2) + \log_{10}(7)$ b. $\log_5(4) + \log_5(11)$ c. $\log_3(20) - \log_3(2)$
d. $\log_{10}(32) - \log_{10}(4)$ e. $2\log_2(5)$ f. $-3\log_5(2)$

10. Simplify the following.

a. $\log_7(7)$ b. $\log_6(3 \div 3)$ c. $\log_2(8)$
d. $\log_{10}\left(1 + 3^2\right)$ e. $\log_9(3)$ f. $2\log_2\left(\frac{1}{4}\right)$

11. Rewrite each of the following in the equivalent index or logarithm form and hence calculate the value of x.

a. $x = \log_2\left(\frac{1}{8}\right)$ b. $\log_{25}(x) = -0.5$ c. $10^{(2x)} = 4.$
d. $3 = e^{-x}.$ e. $\log_x(125) = 3$ f. $\log_x(25) = -2$

12. **WE7** a. State the exact solution to $7^x = 15$ and calculate its value to 3 decimal places.

b. Calculate the exact value and the value to 3 decimal places of the solution to the equation $3^{2x+5} = 4^x$.

13. Solve the following for x.

a. $\log_5(x - 1) = 2$ b. $\log_2(2x + 1) = -1$ c. $\log_x\left(\frac{1}{49}\right) = -2$ d. $\log_x(36) - \log_x(4) = 2$

14. Solve the following for x.

a. $\log_{10}(x + 5) = \log_{10}(2) + 3\log_{10}(3)$ b. $\log_4(2x) + \log_4(5) = 3$
c. $\log_{10}(x + 1) = \log_{10}(x) + 1$ d. $2\log_6(3x) + 3\log_6(4) - 2\log_6(12) = 2$

15. If $\log_2(3) - \log_2(2) = \log_2(x) + \log_2(5)$, solve for x, using logarithm laws.

16. **WE8** Solve the equation $\log_3(x) + \log_3(2x + 1) = 1$ for x.

17. Solve the equation $\log_6(x) - \log_6(x - 1) = 2$ for x.

18. Solve the following for x.

a. $\log_2(2x + 1) + \log_2(2x - 1) = 3\log_2(3)$ b. $\log_3(2x) + \log_3(4) = \log_3(x + 12) - \log_3(2)$
c. $\log_2(2x + 12) - \log_2(3x) = 4$ d. $\log_2(x) + \log_2(2 - 2x) = -1$
e. $\left(\log_{10}(x) + 3\right)\left(2\log_4(x) - 3\right) = 0$ f. $2\log_3(x) - 1 = \log_3(2x - 3)$

19. Solve the following indicial equations, expressing values for x in exact form.

a. $3^{x+2} = 7$ b. $2 \times 6^{3x-5} = 10$ c. $2^{3-x} = 5^x$

20. Given $\log_a(3) = p$ and $\log_a(5) = q$, express the following in terms of p and q.

a. $\log_a(15)$ b. $\log_a(125)$ c. $\log_a(45)$
d. $\log_a(0.6)$ e. $\log_a\left(\frac{25}{81}\right)$ f. $\log_a\left(\sqrt{5}\right) \times \log_a\left(\sqrt{27}\right)$

21. Solve the following equations.

a. $2^{\log_5(x)} = 8$ b. $2^{\log_2(x)} = 7$

22. Express y in terms of x.

a. $\log_{10}(y) = \log_{10}(x) + 2$

b. $x = 10^{y-2}$

c. $\log_{10}\left(10^{3xy}\right) = 3$

d. $10^{3\log_{10}(y)} = xy$

23. Solve the following equations, giving exact solutions.

a. $2^{2x} - 14 \times 2^x + 45 = 0$

b. $5^{-x} - 5^x = 4$

c. $9^{2x} - 3^{1+2x} + 2 = 0$

d. $\log_a\left(x^3\right) + \log_a\left(x^2\right) - 4\log_a(2) = \log_a(x)$

24. State the exact solution and then give the approximate solution to 4 significant figures for each of the following indicial equations.

a. $11^x = 18$

b. $5^{-x} = 8$

c. $7^{2x} = 3$

25. Obtain the approximate solution to 4 significant figures for each of the following inequations.

a. $3^x \leq 10$

b. $5^{-x} > 0.4$

10.3 Exam questions

Question 1 (1 mark) TECH-ACTIVE

MC $\log_6(42) - \log_6(7)$ equals

A. $\log_{12}(35)$

B. $\log_6(35)$

C. -1

D. 1

E. 6

Question 2 (1 mark) TECH-ACTIVE

MC $y = a^x$ in logarithmic form is

A. $\log_y(a) = x$

B. $\log_x(y) = a$

C. $\log_a(y) = x$

D. $\log_y(x) = a$

E. $\log_a(x) = y$

Question 3 (3 marks) TECH-FREE

Solve the equation $\log_{12}(x) + \log_{12}(x+1) = 1$ for x and check the validity of the solutions.

More exam questions are available online.

10.4 Graphs of exponential functions

LEARNING INTENTION

At the end of this subtopic you should be able to:

- determine the equation of an exponential function from a graph
- sketch graphs of exponential functions, showing all key features
- identify and graph transformations of exponential functions, showing all key features.

Exponential functions are functions of the form $f: R \to R,\ f(x) = a^x,\ a \in R^+ \backslash \{1\}$. They provide mathematical models of exponential growth and exponential decay situations such as population increase and radioactive decay.

10.4.1 Graphs of $y = a^x$

The graph of $y = a^x$ where $a > 1$

Before sketching such a graph, consider the table of values for the function with rule $y = 2^x$.

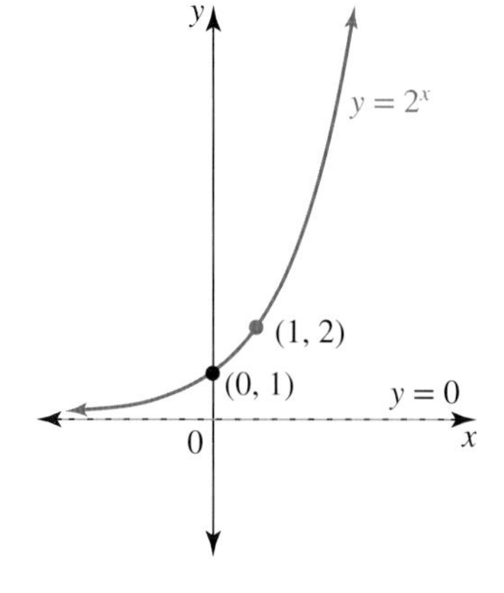

x	-3	-2	-1	0	1	2	3
$y = 2^x$	$\frac{1}{8}$	$\frac{1}{4}$	$\frac{1}{2}$	1	2	4	8

From the table it is evident that $2^x > 0$ for all values of x and that as $x \to -\infty$, $2^x \to 0$. This means that the graph will have a horizontal asymptote with equation $y = 0$.

It is also evident that as $x \to \infty$, $2^x \to \infty$ with the values increasing rapidly.

Since these observations are true for any function $y = a^x$ where $a > 1$, the graph of $y = 2^x$ will be typical of the basic graph of any exponential with base larger than 1.

The graph of $y = a^x$ where $a > 1$

The key features of the graph $y = a^x$ where $a > 1$ are:
- **a horizontal asymptote with equation $y = 0$**
- **y-intercept $(0, 1)$**
- **'exponential growth' shape**
- **domain R**
- **range R^+**
- **one-to-one correspondence.**

For $y = 2^x$, the graph contains the point $(1, 2)$; for the graph of $y = a^x, a > 1$, the graph contains the point $(1, a)$, showing that as the base increases, the graph becomes steeper more quickly for values $x > 0$. This is illustrated by the graphs of $y = 2^x$ and $y = 10^x$, with the larger base giving the steeper graph.

on Resources

Interactivity Exponential functions (int-5959)

The graph of $y = a^x$ where $0 < a < 1$

An example of a function whose rule is in the form $y = a^x$ where $0 < a < 1$ is $y = \left(\frac{1}{2}\right)^x$. Since $\left(\frac{1}{2}\right)^x = 2^{-x}$, the rule for the graph of this exponential function $y = \left(\frac{1}{2}\right)^x$ where the base lies between 0 and 1 is identical to the rule $y = 2^{-x}$ where the base is greater than 1.

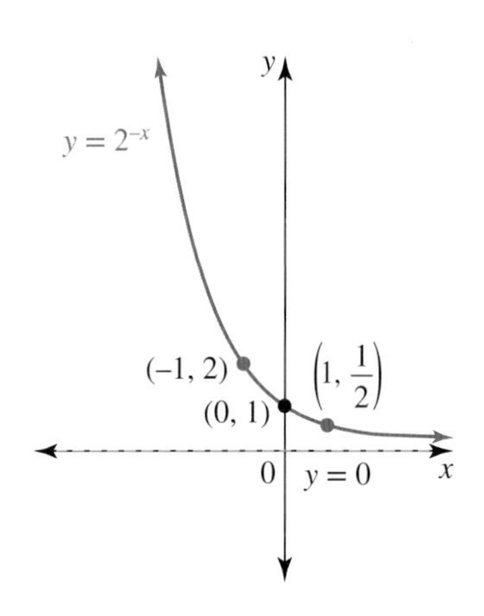

The graph of $y = 2^{-x}$ shown is typical of the graph of $y = a^{-x}$ where $a > 1$ and of the graph of $y = a^x$ where $0 < a < 1$.

The graph of $y = a^x$ where $0 < a < 1$ or $y = a^{-x}$ where $a > 1$

The key features of the graph of $y = a^x$ where $0 < a < 1$ or $y = a^{-x}$ where $a > 1$ are:

- **a horizontal asymptote with equation $y = 0$**
- **y-intercept $(0, 1)$**
- **'exponential decay' shape**
- **domain R**
- **range R^+**
- **one-to-one correspondence**
- **the graph is a reflection of $y = 2^x$ in the y-axis.**

The basic shape of an exponential function is one of either 'growth' or 'decay'.

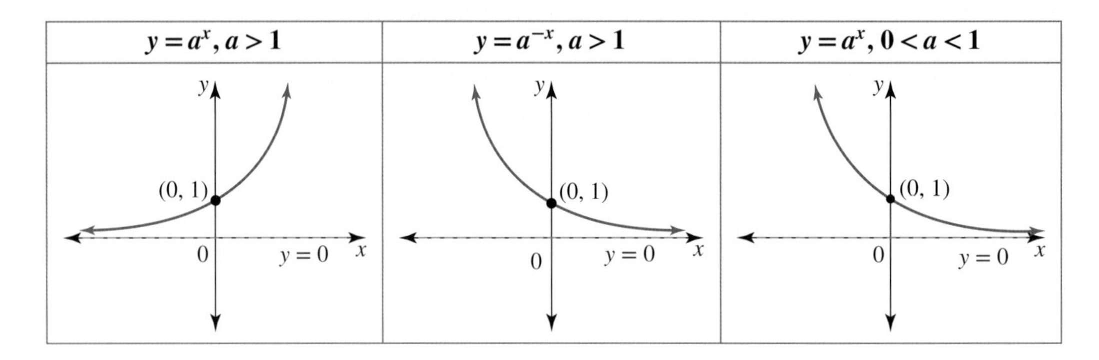

As with other functions, the graph of $y = -a^x$ will be inverted (reflected in the x-axis).

WORKED EXAMPLE 9 Sketching exponential graphs

a. On the same set of axes, sketch the graphs of $y = 5^x$ and $y = -5^x$, stating their ranges.

b. Give a possible equation for the graph shown.

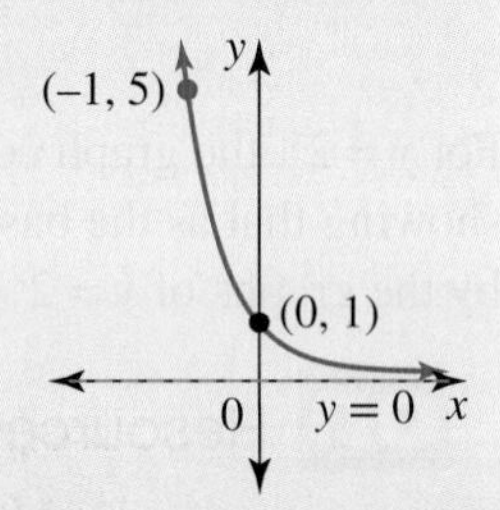

THINK	WRITE
a. 1. Identify the asymptote of the first function.	a. $y = 5^x$ The asymptote is the line with equation $y = 0$.
2. Find the y-intercept.	y-intercept: when $x = 0$, $y = 1 \Rightarrow (0, 1)$.
3. Calculate the coordinates of a second point.	Let $x = 1$. $y = 5^1$ $= 5$ $\Rightarrow (1, 5)$
4. Use the relationship between the two functions to deduce the key features of the second function.	$y = -5^x$ This is the reflection of $y = 5^x$ in the x-axis. The graph of $y = -5^x$ has the same asymptote as that of $y = 5^x$. The equation of its asymptote is $y = 0$. Its y-intercept is $(0, -1)$. The point $(1, -5)$ lies on the graph.

5. Sketch and label each graph.

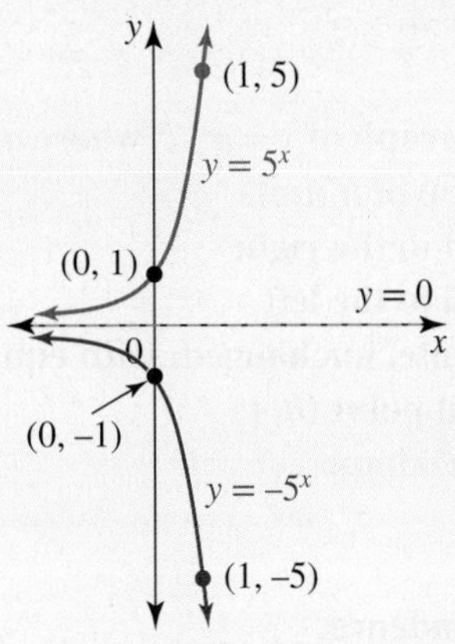

6. State the range of each graph.

The range of $y = 5^x$ is R^+ and the range of $y = -5^x$ is R^-.

b. 1. Use the shape of the graph to suggest a possible form for the rule.

b. The graph has a 'decay' shape. Let the equation be $y = a^{-x}$.

2. Use a given point on the graph to calculate a.

The point $(-1, 5) \Rightarrow 5 = a^1$

$\therefore a = 5$

3. State the equation of the graph.

The equation of the graph could be $y = 5^{-x}$. The equation could also be expressed as $y = \left(\frac{1}{5}\right)^x$ or $y = 0.2^x$.

10.4.2 Translations of exponential graphs

Once the basic exponential growth or exponential decay shapes are known, the graphs of exponential functions can be translated in similar ways to graphs of any other functions previously studied.

The graph of $y = a^x + k$

The key features of the graph of $y = a^x + k$ where $a > 1$ are:

- **vertical translation of k units**
  - **if $k > 0$, translated vertically upwards**
  - **if $k < 0$, translated vertically downwards**
- **a horizontal asymptote with equation $y = k$**
- **y-intercept $(0, 1 + k)$**
- **x-intercept, if $k < 0$, found by solving the exponential equation $a^x + k = 0$**
- **'exponential growth' shape**
- **domain R**
- **range (k, ∞)**
- **one-to-one correspondence.**

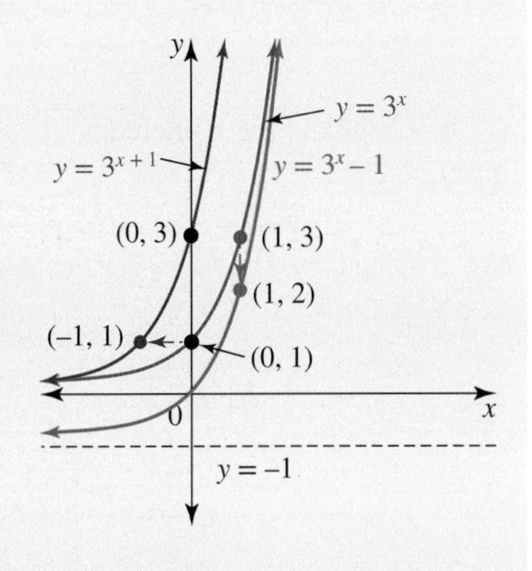

A vertical translation of the graph of $y = 3^x$ is illustrated by the graph of $y = 3^x - 1$; under the vertical translation of 1 unit down, the point $(1, 3) \rightarrow (1, 2)$.

The graph of $y = a^{x-h}$

The key features of the graph of $y = a^{x-h}$ where $a > 1$ are:

- **horizontal translation of h units**
  - **if $h > 0$, translated to the right**
  - **if $h < 0$, translated to the left**
- **a horizontal asymptote, unchanged, with equation $y = 0$**
- **an additional helpful point $(h, 1)$**
- **'exponential growth' shape**
- **domain R**
- **range $(0, \infty)$ or R^+**
- **one-to-one correspondence.**

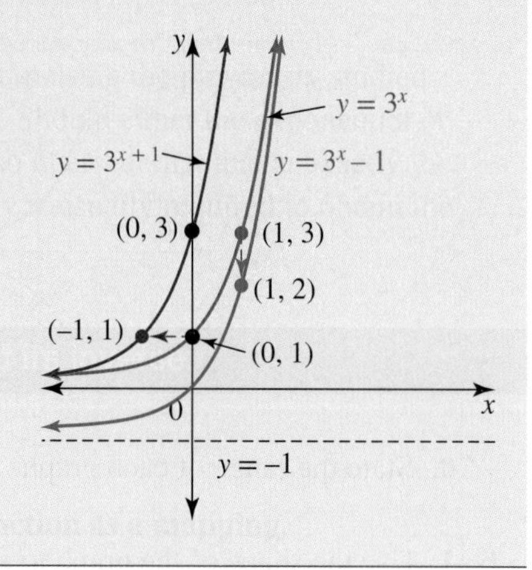

A horizontal translation of the graph of $y = 3^x$ is illustrated by the graph of $y = 3^{x+1}$. Under the horizontal translation of 1 unit to the left, the point $(0, 1) \to (-1, 1)$.

WORKED EXAMPLE 10 Sketching transformations of exponential graphs

Sketch the graphs of each of the following and state the range of each.

a. $y = 2^x - 4$ **b.** $y = 10^{-(x+1)}$

THINK	WRITE
a. 1. State the equation of the asymptote.	**a.** $y = 2^x - 4$ The vertical translation 4 units down affects the asymptote. The asymptote has the equation $y = -4$.
2. Calculate the y-intercept.	y-intercept: let $x = 0$. $y = 2^0 - 4$ $= 1 - 4$ $= -3$ The y-intercept is $(0, -3)$.
3. Calculate the x-intercept.	x-intercept: let $y = 0$. $2^x - 4 = 0$ $2^x = 4$ $2^x = 2^2$ $x = 2$ The x-intercept is $(2, 0)$.

4. Sketch the graph and state the range.

A 'growth' shape is expected since the coefficient of x is positive.

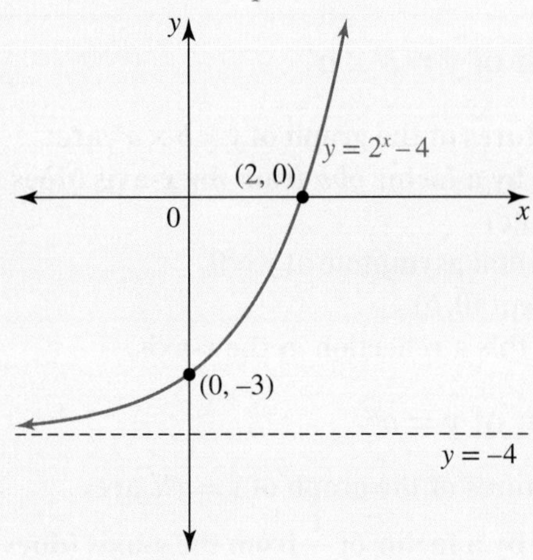

The range is $(-4, \infty)$.

b. 1. Identify the key features from the given equation.

b. $y = 10^{-(x+1)}$

Reflection in the y-axis, horizontal translation 1 unit to the left

The asymptote will not be affected.

Asymptote: $y = 0$

There is no x-intercept.

y-intercept: let $x = 0$.

$y = 10^{-1}$

$= \dfrac{1}{10}$

The y-intercept is $(0, 0.1)$.

2. Calculate the coordinates of a second point on the graph.

Let $x = -1$.

$y = 10^0$

$= 1$

The point $(-1, 1)$ lies on the graph.

3. Sketch the graph and state the range.

A 'decay' shape is expected since the coefficient of x is negative.

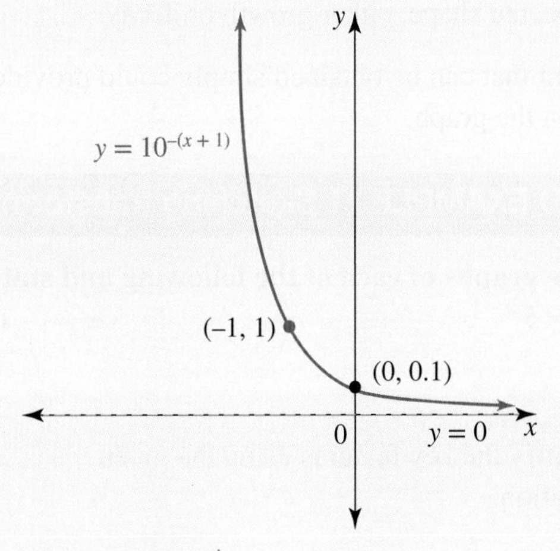

The range is R^+.

Dilations

Dilations from the x- and y-axes transform exponential graphs as follows.

The graph of $y = b \times a^x$

The key features of the graph of $y = b \times a^x$ are:
- **dilation by a factor of b from the x-axis (does not affect the asymptote)**
- **a horizontal asymptote at $y = 0$**
- **y-intercept $(0, b)$**
- **if $b < 0$, this a reflection in the x-axis.**

The graph of $y = a^{nx}$

The key features of the graph of $y = a^{nx}$ are:
- **dilation by a factor of $\frac{1}{n}$ from the y-axis (does not affect the asymptote or y-intercept)**
- **a horizontal asymptote at $y = 0$**
- **y-intercept $(0, 1)$**
- **if $n < 0$, this gives a reflection in the y-axis.**

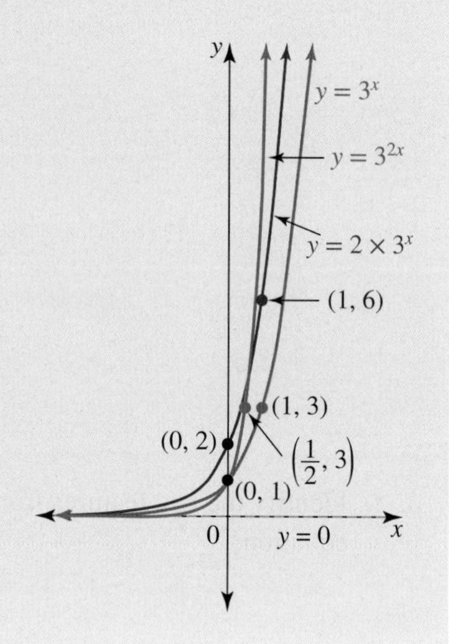

A dilation from the x-axis of factor 2 is illustrated by the graph $y = 2 \times 3^x$; a dilation from the y-axis of factor $\frac{1}{2}$ is illustrated by the graph of $y = 3^{2x}$. Under the dilation from the x-axis of factor 2, the point $(1, 3) \rightarrow (1, 6)$; under the dilation from the y-axis of factor $\frac{1}{2}$, the point $(1, 3) \rightarrow \left(\frac{1}{2}, 3\right)$.

Combinations of transformations

Exponential functions with equations of the form $y = b \times a^{n(x-h)} + k$ are derived from the basic graph of $y = a^x$ by applying a combination of transformations. The key features to identify in order to sketch the graphs of such exponential functions are:
- the asymptote
- the y-intercept
- the x-intercept, if there is one
- the expected shape, either growth or decay.

Another point that can be obtained simply could provide assurance about the shape. Always aim to show at least two points on the graph.

WORKED EXAMPLE 11 Sketching transformations of exponential graphs

Sketch the graphs of each of the following and state the range of each.

a. $y = 10 \times 5^{2x-1}$ **b.** $y = 1 - 3 \times 2^{-x}$

THINK	WRITE
a. 1. Identify the key features using the given equation.	a. $y = 10 \times 5^{2x-1}$ $\therefore y = 10 \times 5^{2\left(x-\frac{1}{2}\right)}$ Asymptote: $y = 0$ No x-intercept y-intercept: let $x = 0$.

$y = 10 \times 5^{-1}$
$= 10 \times \frac{1}{5}$
$= 2$
The y-intercept is $(0, 2)$.

2. Calculate the coordinates of a second point.

Since the horizontal translation is $\frac{1}{2}$ to the right, let $x = \frac{1}{2}$.

$y = 10 \times 5^{2 \times \frac{1}{2} - 1}$
$= 10 \times 5^0$
$= 10 \times 1$
$= 10$

The point $\left(\frac{1}{2}, 10\right)$ lies on the graph.

3. Sketch the graph and state the range.

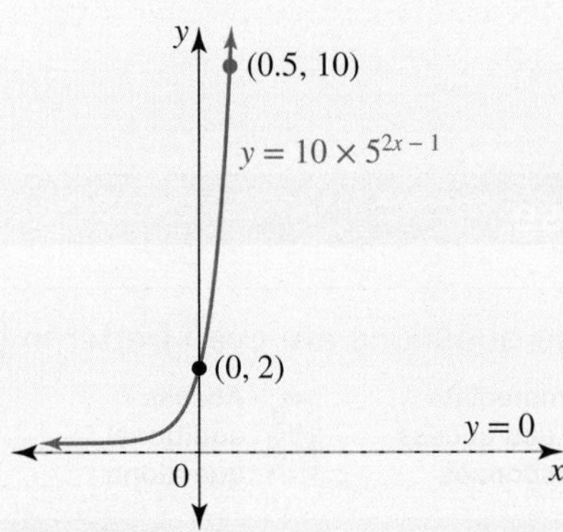

The range is R^+.

b. 1. Write the equation in the form $y = b \times a^{n(x-h)} + k$ and state the asymptote.

b. $y = 1 - 3 \times 2^{-x}$
$\therefore y = -3 \times 2^{-x} + 1$
Asymptote: $y = 1$

2. Calculate the y-intercept.

y-intercept: let $x = 0$.
$y = -3 \times 2^0 + 1$
$= -3 \times 1 + 1$
$= -2$
The y-intercept is $(0, -2)$.

3. Calculate the x-intercept.
Note: As the point $(0, -2)$ lies below the asymptote and the graph must approach the asymptote, there will be an x-intercept.

x-intercept: let $y = 0$.
$0 = 1 - 3 \times 2^{-x}$
$2^{-x} = \frac{1}{3}$
In logarithm form,
$-x = \log_2\left(\frac{1}{3}\right)$
$= \log_2\left(3^{-1}\right)$
$= -\log_2(3)$
$\therefore x = \log_2(3)$
The exact x-intercept is $\left(\log_2(3), 0\right)$.

4. Calculate an approximate value for the x-intercept to help determine its position on the graph.

$x = \log_2(3)$
$= \dfrac{\log_{10}(3)}{\log_{10}(2)}$
≈ 1.58
The x-intercept is approximately $(1.58, 0)$.

5. Sketch the graph and state the range.
Note: Label the x-intercept with its exact coordinates once the graph is drawn.

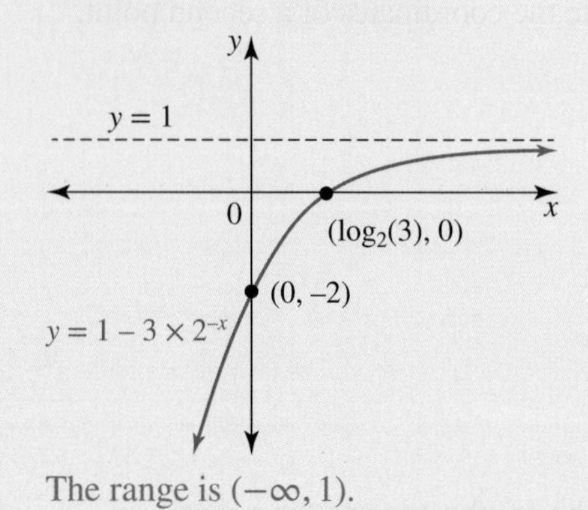

The range is $(-\infty, 1)$.

10.4 Exercise

Students, these questions are even better in jacPLUS

Receive immediate feedback and access sample responses

Access additional questions

Track your results and progress

Find all this and MORE in jacPLUS

Technology free

1. WE9 a. On the same set of axes, sketch the graphs of $y = 3^x$ and $y = -3^x$, stating their ranges.

   b. Give a possible equation for the graph shown.

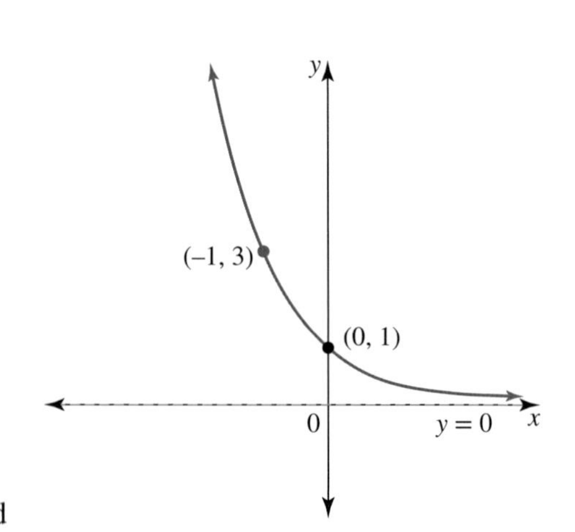

2. a. i. Sketch, on the same set of axes, the graphs of $y = 4^x, y = 6^x$ and $y = 8^x$.
   ii. Describe the effect produced by increasing the base.

   b. i. Sketch, on the same set of axes, the graphs of $y = \left(\dfrac{1}{4}\right)^x$, $y = \left(\dfrac{1}{6}\right)^x$ and $y = \left(\dfrac{1}{8}\right)^x$.
   ii. Express each rule in part **b i** in a different form.

3. Sketch, on the same set of axes, the graphs of $y = 5^{-x}, y = 7^{-x}$ and $y = 9^{-x}$, and describe the effect produced by increasing the base.

4. WE10 Sketch the graphs of each of the following and state the range of each.

   a. $y = 4^x - 2$

   b. $y = 3^{-(x+2)}$

5. Sketch the graph of $y = 4^{x-2} + 1$ and state its range.

6. For each of the following exponential functions, state:
   i. the equation of its asymptote
   ii. the coordinates of its y-intercept
   iii. the range of the function.

   a. $y = 3^{2x} - 1$
   b. $y = 2^{-x} + 3$
   c. $y = 2 - 5^{1-x}$
   d. $y = 10 \times 2^{3x}$

7. Determine the exact coordinates of any intercepts with the coordinate axes made by each of the following.
   a. $y = 2^{\frac{x}{2}} - 128$
   b. $y = 5^{-2x} - 10$

8. Sketch each of the following graphs, showing the asymptote and labelling any intersections with the coordinate axes with their exact coordinates.
   a. $y = 5^{-x} + 1$
   b. $y = 1 - 4^x$
   c. $y = 10^x - 2$
   d. $y = 6.25 - (2.5)^{-x}$

9. Sketch the graphs of:
   a. $y = 2^{x-2}$
   b. $y = -3^{x+2}$
   c. $y = 4^{x-0.5}$
   d. $y = 7^{1-x}$

10. **WE11** Sketch the graphs of each of the following and state the range of each.
   a. $y = \frac{1}{2} \times 10^{1-2x}$
   b. $y = 5 - 4 \times 3^{-x}$

11. Sketch the graphs of:
   a. $y = 3 \times 2^x$
   b. $y = 2^{\frac{3x}{4}}$
   c. $y = -3 \times 2^{-3x}$
   d. $y = 1.5 \times 10^{-\frac{x}{2}}$

12. Sketch the graphs of $y = (1.5)^x$ and $y = \left(\frac{2}{3}\right)^x$ on the same set of axes.

13. a. Sketch the graphs of $y = 3^{2x}$ and $y = 9^x$ and explain the result.
   b. i. Use index laws to obtain another form of the rule for $y = 2 \times 4^{0.5x}$.
      ii. Hence or otherwise, sketch the graph of $y = 2 \times 4^{0.5x}$.

14. Determine the rule, stating its domain and range, for each of the following.
   a. An exponential function with equation $y = 2^x + k$ and a y-intercept at $y = 7$
   b. The exponential function $y = 5^{ax} + b$ that passes through the points $(0, -4)$ and $(2, 0)$
   c. The exponential function $y = b - a \times 3^{-x}$ with asymptote $y = 10$ and passing through the origin

15. The graph shown has the equation $y = a \times 3^x + b$. Determine the values of a and b.

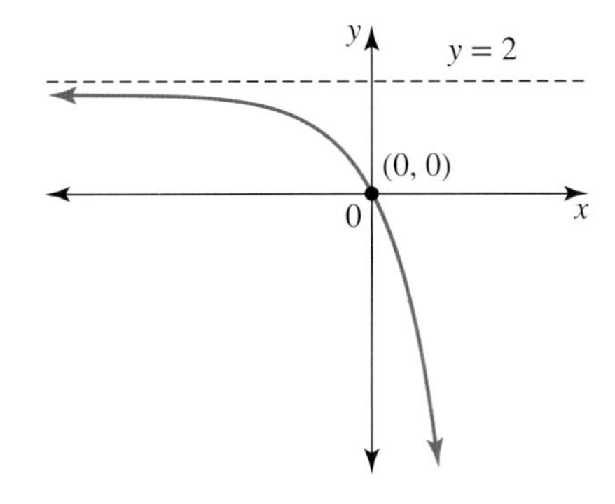

16. **a.** Determine a possible rule for the given graph in the form $y = a \times 10^x + b$.

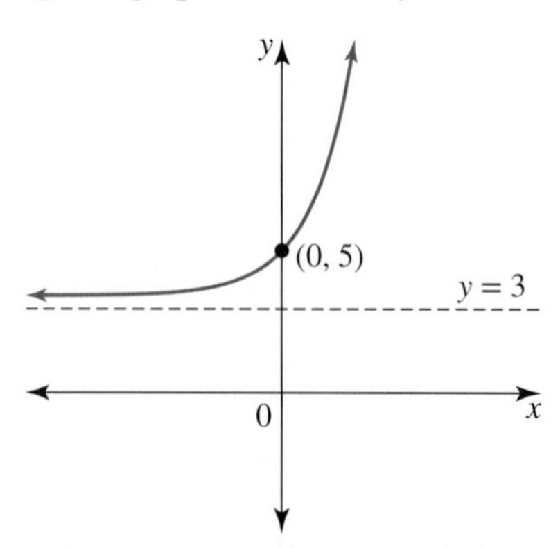

b. The graph of an exponential function of the form $y = a \times 3^{kx}$ contains the points $(1, 36)$ and $(0, 4)$. Determine its rule and state the equation of its asymptote.

c. For the graph shown, determine a possible rule in the form $y = a - 2 \times 3^{b-x}$.

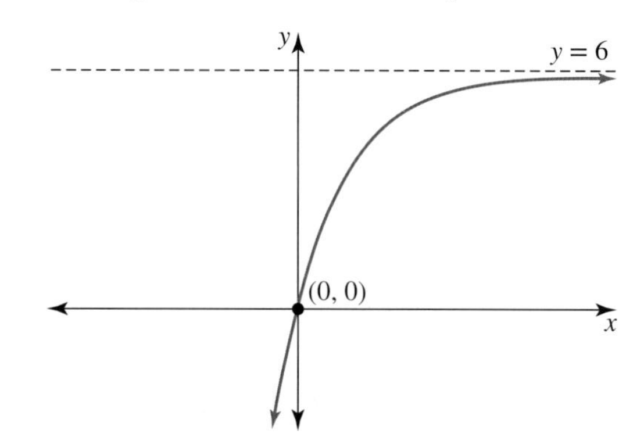

d. Express the equation given in part **c** in another form not involving a horizontal translation.

Technology active

17. Sketch, on the same set of axes, the graphs of $y = (0.8)^x$, $y = (1.25)^x$ and $y = (0.8)^{-x}$, and describe the relationships between the three graphs.

18. Sketch the graphs of the following exponential functions and state their ranges. Where appropriate, any intersections with the coordinate axes should be given to 1 decimal place.

a. $y = 2 \times 10^{2x} - 20$

b. $y = 5 \times 2^{1-x} - 1$

c. $y = 3 - 2\left(\dfrac{2}{3}\right)^x$

d. $y = 2\,(3.5)^{x+1} - 7$

e. $y = 8 - 4 \times 5^{2x-1}$

f. $y = -2 \times 10^{3x-1} - 4$

19. Consider the function $f: R \to R, f(x) = 3 - 6 \times 2^{\frac{x-1}{2}}$.

a. Evaluate:

i. $f(1)$

ii. $f(0)$, expressing the answer in simplest surd form.

b. Determine for what values of x, if any, the following hold.

i. $f(x) = -9$

ii. $f(x) = 0$

iii. $f(x) = 9$

c. Sketch the graph of $y = f(x)$ and state its range.

d. Solve the inequation $f(x) \geq -1$ correct to 2 significant figures.

20. Use a graphical means to determine the number of intersections between:

a. $y = 2^x$ and $y = -x$, specifying an interval in which the x-coordinate of any point of intersection lies
b. $y = 2^x$ and $y = x^2$
c. $y = e^x$ and $y = 2^x$
d. $y = 2^{-x} + 1$ and $y = \sin(x)$
e. $y = 3 \times 2^x$ and $y = 6^x$, determining the coordinates of any points of intersection algebraically
f. $y = 2^{2x-1}$ and $y = \frac{1}{2} \times 16^{\frac{x}{2}}$, giving the coordinates of any points of intersection.

10.4 Exam questions

Question 1 (1 mark) TECH-ACTIVE

MC State which one of the following is *not* a key feature of the graph $y = 2^x$ or any such function $y = a^x$ where $a > 1$.

A. Horizontal asymptote with equation $y = 0$
B. y-intercept (0, 1)
C. Shape of 'exponential growth'
D. Domain R
E. Range R

Question 2 (1 mark) TECH-ACTIVE

MC A possible equation for the graph shown is

A. $y = 10^x$
B. $y = 10^{-x}$
C. $y = 2^x$
D. $y = 2^{2x}$
E. $y = -2^x$

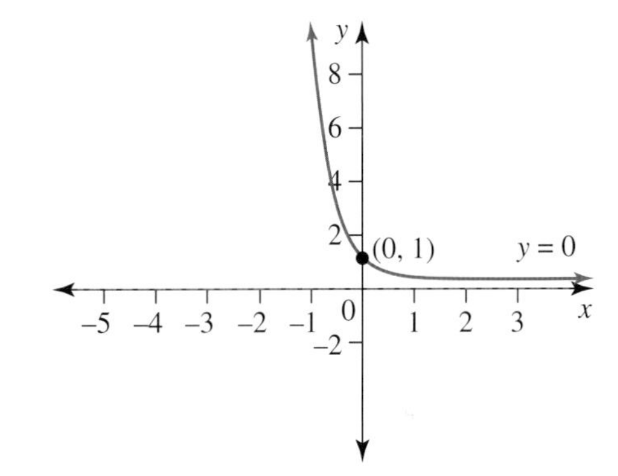

Question 3 (4 marks) TECH-FREE

Sketch the graph of $y = 9 \times 3^{3x-2}$ and state its range.

More exam questions are available online.

10.5 Applications of exponential functions

LEARNING INTENTION

At the end of this subtopic you should be able to:
- using exponential functions to solve real-world problems such as problems involving growth and decay.

The importance of exponential functions lies in the frequency with which they occur in models of phenomena involving growth and decay situations, in chemical and physical laws of nature, and in higher-level mathematical analysis.

10.5.1 Exponential growth and decay models

For time t, the exponential function defined by $y = b \times a^{nt}$ where $a > 1$ represents exponential growth over time if $n > 0$ and exponential decay over time if $n < 0$. The domain of this function is restricted according to the way the independent time variable t is defined. The rule $y = b \times a^{nt}$ may also be written as $y = b \cdot a^{nt}$.

In some mathematical models such as population growth, the initial population may be represented by a symbol such as N_0. For an exponential decay model, the time it takes for 50% of the initial amount of the substance to decay is called its half-life.

WORKED EXAMPLE 12 Evaluating exponential decay models

The decay of a radioactive substance is modelled by $Q(t) = Q_0 \times 2.7^{-kt}$, where Q kg is the amount of the substance present at time t years and Q_0 and k are positive constants.

a. Show that the constant Q_0 represents the initial amount of the substance.

b. If the half-life of the radioactive substance is 100 years, calculate k to 1 significant figure.

c. If initially there was 25 kg of the radioactive substance, calculate how many kilograms would decay in 10 years. Use the value of k from part b in your calculations.

THINK	WRITE
a. 1. Calculate the initial amount.	a. $Q(t) = Q_0 \times 2.7^{-kt}$ The initial amount is the value of Q when $t = 0$. Let $t = 0$: $Q(0) = Q_0 \times 2.7^0$ $= Q_0$ Therefore, Q_0 represents the initial amount of the substance.
b. 1. Form an equation in k from the given information. *Note:* It does not matter that the value of Q_0 is unknown, since the Q_0 terms cancel.	b. The half-life is the time it takes for 50% of the initial amount of the substance to decay. Since the half-life is 100 years, when $t = 100$, $Q(100) = 50\%$ of Q_0. $Q(100) = 0.50Q_0$ [1] From the equation, $Q(t) = Q_0 \times 2.7^{-kt}$. When $t = 100$, $Q(100) = Q_0 \times 2.7^{-k(100)}$. $\therefore Q(100) = Q_0 \times 2.7^{-100k}$ [2] Equate equations [1] and [2]: $0.50Q_0 = Q_0 \times 2.7^{-100k}$ Cancel Q_0 from each side: $0.50 = 2.7^{-100k}$
2. Solve the exponential equation to obtain k to the required accuracy.	Convert to the equivalent logarithm form. $-100k = \log_{2.7}(0.5)$ $k = -\frac{1}{100}\log_{2.7}(0.5)$ $= -\frac{1}{100} \times \frac{\log_{10}(0.5)}{\log_{10}(2.7)}$ ≈ 0.007
c. 1. Use the values of the constants to state the actual rule for the exponential decay model.	c. $Q_0 = 25$, $k = 0.007$ $\therefore Q(t) = 25 \times 2.7^{-0.007t}$
2. Calculate the amount of the substance present at the time given.	When $t = 10$, $Q(10) = 25 \times 2.7^{-0.07}$ ≈ 23.32

3. Calculate the amount that has decayed. *Note:* Using a greater accuracy for the value of k would give a slightly different answer for the amount decayed.	Since $25 - 23.32 = 1.68$, in 10 years approximately 1.68 kg will have decayed.

10.5 Exercise

Students, these questions are even better in jacPLUS

Receive immediate feedback and access sample responses

Access additional questions

Track your results and progress

Find all this and MORE in jacPLUS

Technology active

1. The value V of a new car depreciates so that its value after t years is given by $V = V_0 \times 2^{-kt}$.
   a. If 50% of the purchase value is lost in 5 years, calculate k.
   b. Calculate how long it takes for the car to lose 75% of its purchase value.

2. The manager of a small business is concerned about the amount of time she spends dealing with the growing number of emails she receives. The manager starts keeping records and finds the average number of emails received per day can be modelled by $D = 42 \times 2^{\frac{t}{16}}$, where D is the average number of emails received per day, t weeks from the start of the records.
   a. Calculate how many daily emails on average the manager was receiving when she commenced her records.
   b. Calculate how many weeks the model predicts it will take for the average number of emails received per day to double.
3. The number of drosophilae (fruit flies), N, in a colony after t days of observation is modelled by $N = 30 \times 2^{0.072t}$. Give whole number answers to the following.
   a. Calculate how many drosophilae were present when the colony was initially observed.
   b. Calculate how many of the insects were present after 5 days.
   c. Calculate how many days it takes for the population number to double from its initial value.
   d. Sketch a graph of N versus t to show how the population changes.
   e. Calculate how many days it will take for the population to first exceed 100.
4. **WE12** The decay of a radioactive substance is modelled by $Q(t) = Q_0 \times 1.7^{-kt}$, where Q is the amount of the substance present at time t years and Q_0 and k are positive constants.
   a. Show that the constant Q_0 represents the initial amount of the substance.
   b. If the half-life of the radioactive substance is 300 years, calculate k to 1 significant figure.
   c. If initially there was 250 kg of the radioactive substance, calculate how many kilograms would decay in 10 years. Use the value of k from part **b** in the calculations.

5. The value of an investment that earns compound interest can be calculated from the formula $A = P\left(1 + \frac{r}{n}\right)^{nt}$, where P is the initial investment, r the interest rate per annum (yearly) expressed as a decimal, n the number of times per year the interest is compounded and t the number of years of the investment.

An investor deposits \$2000 in an account where interest is compounded monthly.

a. If the interest rate is 3% per annum:

i. show that the formula giving the value of the investment is $A = 2000\,(1.0025)^{12t}$

ii. calculate how much the investment is worth after a 6-month period

iii. calculate the time period that would be needed for the value of the investment to reach \$2500.

b. The investor would like the \$2000 to grow to \$2500 in a shorter time period. calculate the interest rate, still compounded monthly, that would be required for the goal to be achieved in 4 years.

6. A cup of coffee is left to cool on a kitchen table inside a Brisbane home. The temperature of the coffee, T (°C), after t minutes is thought to be given by $T = 85 \times 3^{-0.008t}$.

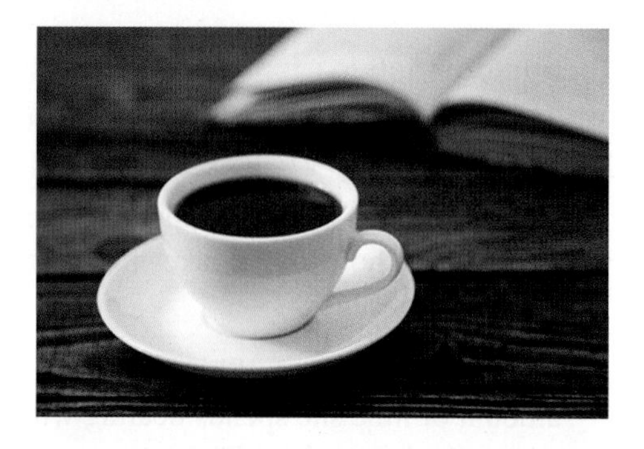

a. Calculate how many degrees the coffee cools by in 10 minutes.

b. Calculate how long it takes for the coffee to cool to 65 °C .

c. Sketch a graph of the temperature of the coffee for $t \in [0, 40]$.

d. By considering the temperature the model predicts the coffee will eventually cool to, explain why the model is not realistic in the long term.

7. The contents of a meat pie immediately after being heated in a microwave have a temperature of 95 °C. The pie is removed from the microwave and left to cool. A model for the temperature of the pie as it cools is given by $T = a \times 3^{-0.13t} + 25$, where T is the temperature after t minutes of cooling.

a. Calculate the value of a.

b. Calculate the temperature of the contents of the pie after being left to cool for 2 minutes.

c. Determine how long, to the nearest minute, it will take for the contents of the meat pie to cool to 65 °C.

d. Sketch the graph showing the temperature over time and state the temperature to which this model predicts the contents of the pie will eventually cool if left unattended.

8. The barometric pressure P, measured in kilopascals, at height h above sea level, measured in kilometres, is given by $P = P_o \times 10^{-kh}$ where P_o and k are positive constants. The pressure at the top of Mount Everest is approximately one-third that of the pressure at sea level.

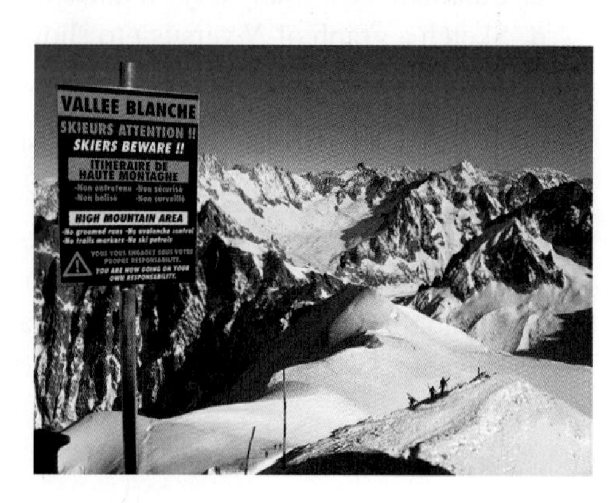

a. Given the height of Mount Everest is approximately 8848 metres, calculate the value of k to 2 significant figures.

Use the value obtained for k for the remainder of question 8.

b. Mount Kilimanjaro has a height of approximately 5895 metres. If the atmospheric pressure at its summit is approximately 48.68 kilopascals, calculate the value of P_0 to 3 decimal places.

c. Use the model to estimate the atmospheric pressure to 2 decimal places at the summit of Mont Blanc, 4810 metres, and of Mount Kosciuszko, 2228 metres in height.

d. Draw a graph of the atmospheric pressure against height showing the readings for the four mountains from the above information.

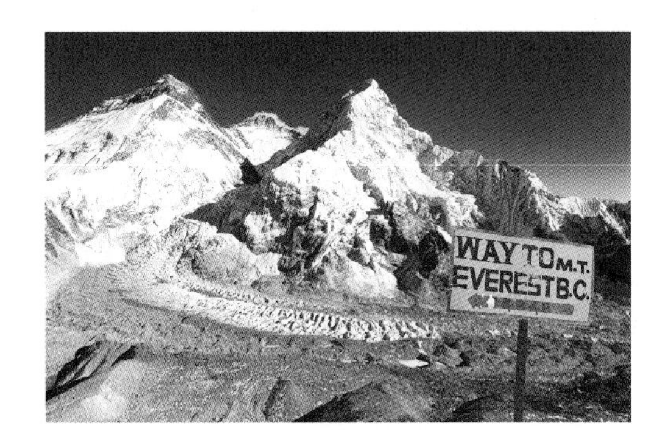

9. The common Indian mynah bird was introduced into Australia in order to control insects affecting market gardens in Melbourne. It is now considered to be Australia's most important pest problem. In 1976, the species was introduced to an urban area in New South Wales. By 1991 the area averaged 15 birds per square kilometre, and by 1994 the density reached an average of 75 birds per square kilometre. A model for the increasing density of the mynah bird population is thought to be $D = D_0 \times 10^{kt}$, where D is the average density of the bird per square kilometre t years after 1976 and D_0 and k are constants.

a. Use the given information to set up a pair of simultaneous equations in D and t.

b. Solve these equations to show that $k = \frac{1}{3}\log(5)$ and $D_0 = 3 \times 5^{-4}$, and hence that $k \approx 0.233$ and $D_0 \approx 0.005$.

c. A project was started in 1996 to curb the growth in numbers of these birds. Calculate what the model predicts was the average density of the mynah bird population at the time the project began in the year 1996. Use $k \approx 0.233$ and $D_0 \approx 0.005$ and round the answer to the nearest whole number.

d. Sometime after the project is successfully implemented, a different model for the average density of the bird population becomes applicable. This model is given by $D = 30 \times 10^{-\frac{t}{3}} + b$. Four years later, the average density is reduced to 40 birds per square kilometre. Calculate how much the average density could be expected to be reduced by.

10. Carbon dating enables estimates of the age of fossils of once living organisms to be ascertained by comparing the amount of the radioactive isotope carbon-14 remaining in the fossil with the normal amount present in the living entity, which can be assumed to remain constant during the organism's life. It is known that carbon-14 decays with a half-life of approximately 5730 years according to an exponential model of the form $C = C_0 \times \left(\frac{1}{2}\right)^{kt}$, where C is the amount of the isotope remaining in the fossil t years after death and C_0 is the normal amount of the isotope that would have been present when the organism was alive.

a. Calculate the exact value of the positive constant k.

b. The bones of an animal are unearthed during digging explorations by a mining company. The bones are found to contain 83% of the normal amount of the isotope carbon-14. Estimate how old the bones are.

11. The data shown in the table gives the population of Australia, in millions, in years since 1960.

	1975	1990
t (years since 1960)	15	30
p (population in millions)	13.9	17.1

A model for increasing population is thought to be $P = P_0 \times 10^{kt}$, where P is the population (in millions) t years after 1960 and P_0 and k are constant.

a. Use this information to set up a pair of simultaneous equations in P and t.
b. Show that $k \approx 0.006$ and $P_0 \approx 10.3$.
c. Calculate how many years it took for the population to reach double the 1960 population.
d. It is said that the population of Australia is likely to exceed 28 million by the year 2030. Explain whether this model supports this claim.

12. Following a fall from their bike, Stephan is feeling some shock but not, initially, a great deal of pain. However, their doctor gives them an injection for relief from the pain that they will start to feel once the shock of the accident wears off. The amount of pain Stephan feels over the next 10 minutes is modelled by the function $P(t) = (200t + 16) \times 2.7^{-t}$, where P is the measure of pain on a scale from 0 to 100 that Stephan feels t minutes after receiving the injection.

a. Give the measure of pain Stephan is feeling:
   i. at the time the injection is administered
   ii. 15 seconds later when their shock is wearing off but the injection has not reached its full effect.
b. Use technology to draw the graph showing Stephan's pain level over the 10-minute interval and hence give, to 2 decimal places:
   i. the maximum measure of pain Stephan feels
   ii. the number of seconds it takes for the injection to start lowering pain level
   iii. their pain levels after 5 minutes and after 10 minutes have elapsed.
c. At the end of the 10 minutes, Stephan receives a second injection modelled by $P(t) = (100(t - 10) + a) \times 2.7^{-(t-10)}, 10 \le t \le 20$.
   i. Determine the value of a.
   ii. Sketch the pain measure over the time interval $t \in [0, 20]$ and label the maximum points with their coordinates.

10.5 Exam questions

Question 1 (1 mark) TECH-ACTIVE

MC At a secondhand car dealer, the depreciation of the value of a car is calculated using $V = V_0 \times 3^{-0.1t}$, where t is the age of the car (in years). The number of years it would take to lose two-thirds of the purchase value is

A. 3 **B.** 5 **C.** 10 **D.** 15 **E.** 20

Question 2 (1 mark) TECH-ACTIVE

MC The decay of a radioactive substance is modelled by the equation $Q(t) = 15 \times 5^{-0.0025t}$, where Q kg is the amount of the substance present at time t years. The initial amount of the substance was

A. 5 **B.** 15 **C.** 25 **D.** 3 **E.** 0.0025

Question 3 (1 mark) TECH-ACTIVE

MC Sue calculates that a cup of tea cools at a rate of $T = 90 \times 3^{-0.01t}$, where T (°C) is the temperature of the tea after t minutes. After 30 minutes, the tea cools by

A. 34.21 °C **B.** 25.27 °C **C.** 22.32 °C **D.** 14.32 °C **E.** 64.73 °C

More exam questions are available online.

10.6 Inverses of exponential functions

LEARNING INTENTION

At the end of this subtopic you should be able to:
- identify and graph the inverses of exponential functions
- simplify expressions using inverse relationships
- sketch graphs of logarithmic functions, showing all key features
- identify and graph transformations of logarithmic functions, showing all key features.

10.6.1 The inverse of $y = a^x, a \in R^+ \backslash \{1\}$

The exponential function has a one-to-one correspondence, so its inverse must also be a function. To form the inverse of $y = a^x$, interchange the x-and y-coordinates.

The inverse of $y = ax$

Function: $y = a^x$ domain R, range R^+

Inverse function: $x = a^y$ domain R^+, range R

$$\therefore y = \log_a(x)$$

Therefore, the inverse of an exponential function is a logarithmic function: $y = \log_a(x)$ and $y = a^x$ are the rules for a pair of inverse functions. These are transcendental functions, not algebraic functions. However, they can be treated similarly to the inverse pairs of algebraic functions previously encountered. This means the graph of $y = \log_a(x)$ can be obtained by reflecting the graph of $y = a^x$ in the line $y = x$.

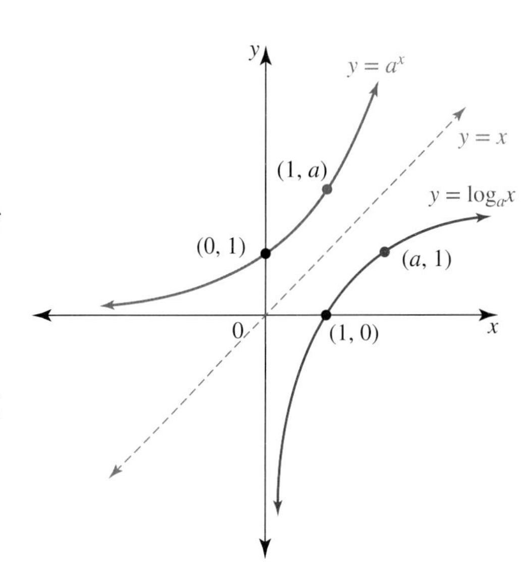

The graph of $y = \log_a(x)$, for $a > 1$

The shape of the basic logarithmic graph with rule $y = \log_a(x)$, $a > 1$ is shown as the reflection in the line $y = x$ of the exponential graph with rule $y = a^x$, $a > 1$.

The key features of the graph of $y = \log_a(x)$ can be deduced from those of the exponential graph.

$y = a^x$	$y = \log_a(x)$
Horizontal asymptote with equation $y = 0$	Vertical asymptote with equation $x = 0$
y-intercept $(0, 1)$	x-intercept $(1, 0)$
The point $(1, a)$ lies on the graph.	The point $(a, 1)$ lies on the graph.
Range R^+	Domain R^+
Domain R	Range R
One-to-one correspondence	One-to-one correspondence

Note that logarithmic growth is much slower than exponential growth. Also note that unlike a^x, which is always positive, $\log_a(x)\begin{cases} >0, & \text{if } x>1 \\ =0, & \text{if } x=1 \\ <0, & \text{if } 0<x<1 \end{cases}$.

The logarithmic function is formally written as $f: R^+ \to R,\ f(x) = \log_a(x)$.

WORKED EXAMPLE 13 Deducing the logarithmic graph from the exponential graph

a. Form the exponential rule for the inverse of $y = \log_2(x)$ and hence deduce the graph of $y = \log_2(x)$ from the graph of the exponential.

b. Given the points $(1, 2)$, $(2, 4)$ and $(3, 8)$ lie on the exponential graph in part a, explain how these points can be used to illustrate the logarithm law $\log_2(m) + \log_2(n) = \log_2(mn)$.

THINK	WRITE
a. 1. Form the rule for the inverse by interchanging coordinates and then make y the subject of the rule.	a. $y = \log_2(x)$ Inverse: $x = \log_2(y)$ $\therefore y = 2^x$
2. Sketch the exponential function.	$y = 2^x$ Asymptote: $y = 0$ y-intercept: $(0, 1)$ Second point: let $x = 1$. $\therefore y = 2$ The point $(1, 2)$ is on the graph. 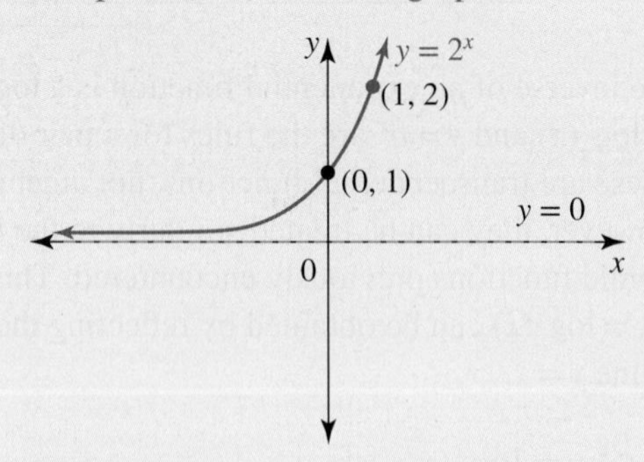
3. Reflect the exponential graph in the line $y = x$ to form the required graph.	$y = \log_2(x)$ has: asymptote: $x = 0$ x-intercept: $(1, 0)$ second point: $(2, 1)$

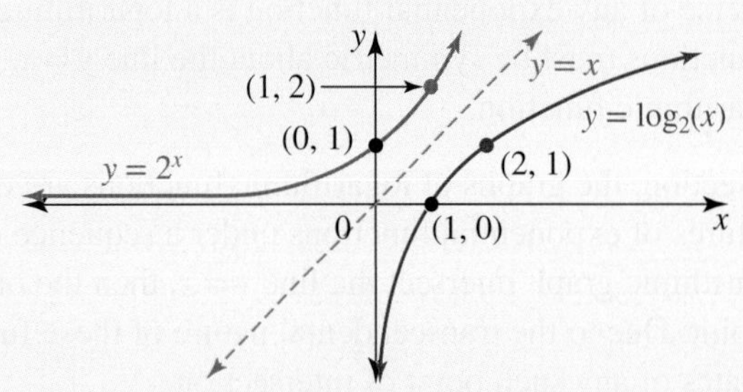

b. 1. State the coordinates of the corresponding points on the logarithm graph.

b. Given the points $(1, 2)$, $(2, 4)$ and $(3, 8)$ lie on the exponential graph, the points $(2, 1)$, $(4, 2)$ and $(8, 3)$ lie on the graph of $y = \log_2(x)$.

2. State the x- and y-values for each of the points on the logarithmic graph.

$y = \log_2(x)$
Point $(2, 1)$: when $x = 2, y = 1$.
Point $(4, 2)$: when $x = 4, y = 2$.
Point $(8, 3)$: when $x = 8, y = 3$.

3. Use the relationship between the y-coordinates to illustrate the logarithm law.

The sum of the y-coordinates of the points on $y = \log_2(x)$ when $x = 2$ and $x = 4$ equals the y-coordinate of the point on $y = \log_2(x)$ when $x = 8$, as $1 + 2 = 3$.
$\log_2(2) + \log_2(4) = \log_2(8)$
$\log_2(2) + \log_2(4) = \log_2(2 \times 4)$
This illustrates the logarithm law $\log_2(m) + \log_2(n) = \log_2(mn)$ with $m = 2$ and $n = 4$.

10.6.2 The inverse of exponential functions of the form $y = b \times a^{n(x-h)} + k$

The rule for the inverse of $y = b \times a^{n(x-h)} + k$ is calculated in the usual way by interchanging x- and y-coordinates to give $x = b \times a^{n(y-h)} + k$.

Expressing this equation as an index statement:

$$x - k = b \times a^{n(y-h)}$$
$$\frac{x-k}{b} = a^{n(y-h)}$$

Converting to the equivalent logarithm form:

$$\log_a\left(\frac{x-k}{b}\right) = n(y-h)$$

Rearranging to make y the subject:

$$\frac{1}{n}\log_a\left(\frac{x-k}{b}\right) = y - h$$

$$y = \frac{1}{n}\log_a\left(\frac{x-k}{b}\right) + h$$

$$f^{-1}(x) = \frac{1}{n}\log_a\left(\frac{x-k}{b}\right) + h$$

The inverse of any exponential function is a logarithmic function. Since the corresponding pair of graphs of these functions must be symmetric about the line $y = x$, this provides one approach for sketching the graph of any logarithmic function.

In this section, the graphs of logarithmic functions are obtained by deduction using the previously studied key features of exponential functions under a sequence of transformations. Should either the exponential or the logarithmic graph intersect the line $y = x$, then the other graph must also intersect that line at exactly the same point. Due to the transcendental nature of these functions, technology is usually required to obtain the coordinates of any such point of intersection.

WORKED EXAMPLE 14 Sketching an exponential graph and its inverse

Consider the function $f: R \to R$, $f(x) = 5 \times 2^{-x} + 3$.

a. Determine the domain of its inverse.

b. Form the rule for the inverse function and express the inverse function as a mapping.

c. Sketch $y = f(x)$ and $y = f^{-1}(x)$ on the same set of axes.

THINK	WRITE
a. 1. Determine the range of the given function.	a. $f(x) = 5 \times 2^{-x} + 3$ Asymptote: $y = 3$ y-intercept: $(0, 8)$ This point lies above the asymptote, so the range of f is $(3, \infty)$.
2. State the domain of the inverse.	The domain of the inverse function is the range of the given function. The domain of the inverse function is $(3, \infty)$.
b. 1. Form the rule for the inverse function by interchanging x-and y-coordinates and rearranging the equation obtained. *Note:* Remember to express the rule for the inverse function f^{-1} with $f^{-1}(x)$ in place of y as its subject.	b. Let $f(x) = y$. Function: $y = 5 \times 2^{-x} + 3$ Inverse: $x = 5 \times 2^{-y} + 3$ $\therefore x - 3 = 5 \times 2^{-y}$ $\therefore \frac{x-3}{5} = 2^{-y}$ Converting to logarithm form: $-y = \log_2\left(\frac{x-3}{5}\right)$ $\therefore y = -\log_2\left(\frac{x-3}{5}\right)$ The inverse function has the rule $f^{-1}(x) = -\log_2\left(\frac{x-3}{5}\right)$.
2. Write the inverse function as a mapping.	The inverse function has domain $(3, \infty)$ and its rule is $f^{-1}(x) = -\log_2\left(\frac{x-3}{5}\right)$. Hence, as a mapping: $f^{-1}: (3, \infty) \to R$, $f^{-1}(x) = -\log_2\left(\frac{x-3}{5}\right)$

c. 1. Sketch the graph of the exponential function f and use this to deduce the graph of the inverse function f^{-1}.

c. $f(x) = 5 \times 2^{-x} + 3, f^{-1}(x) = -\log_2\left(\frac{x-3}{5}\right)$

The key features of f determine the key features of f^{-1}.

$y = f(x)$	$y = f^{-1}(x)$
Asymptote: $y = 3$	Asymptote: $x = 3$
$y-$intercept$(0, 8)$	$x-$intercept$(8, 0)$

second point on $y = f(x)$:

Let $x = -1$.

$$\begin{aligned} y &= 5 \times 2^1 + 3 \\ &= 13 \end{aligned}$$

The point $(-1, 13)$ is on $y = f(x)$ and the point $(13, -1)$ is on $y = f^{-1}(x)$.

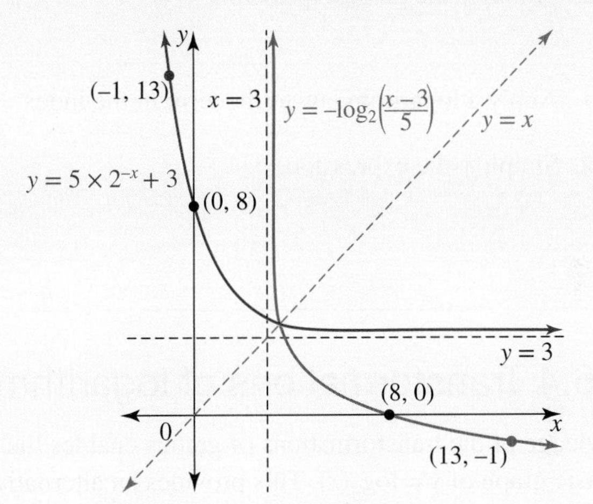

The graphs intersect on the line $y = x$.

10.6.3 Relationships between the inverse pairs

As the exponential and logarithmic functions are a pair of inverses, each 'undoes' the effect of the other. Thus, we can state the following.

Inverse pairs

$$\log_a\left(a^x\right) = x \text{ and } a^{\log_a(x)} = x$$

The first of these statements could also be explained using logarithm laws:

$$\begin{aligned} \log_a(a^x) &= x\log_a(a) \\ &= x \times 1 \\ &= x \end{aligned}$$

The second statement can also be explained from the index–logarithm definition that $a^n = x \Leftrightarrow n = \log_a(x)$. Replacing n by its logarithm form in the definition gives:

$$\begin{aligned} a^n &= x \\ a^{\log_a(x)} &= x \end{aligned}$$

WORKED EXAMPLE 15 Simplifying expressions using inverse relationships

a. Simplify $\log_{12}\left(2^{2x} \times 3^x\right)$ using the inverse relationship between exponentials and logarithms.
b. Evaluate $10^{2\log_{10}(5)}$.

THINK	WRITE
a. 1. Use index laws to simplify the product of powers given in the logarithm expression.	a. $\log_{12}\left(2^{2x} \times 3^x\right)$ Consider the product $2^{2x} \times 3^x$. $2^{2x} \times 3^x = \left(2^2\right)^x \times 3^x$ $= 4^x \times 3^x$ $= (4 \times 3)^x$ $= 12^x$
2. Simplify the given expression.	$\log_{12}\left(2^{2x} \times 3^x\right) = \log_{12}\left(12^x\right)$ $= x$
b. 1. Apply a logarithm law to the term in the index.	b. $10^{2\log_{10}(5)} = 10^{\log_{10}(5)^2}$
2. Simplify the expression.	$= 10^{\log_{10}(25)}$ $= 25$ Therefore, $10^{2\log_{10}(5)} = 25$.

10.6.4 Transformations of logarithmic graphs

Knowledge of the transformations of graphs enables the graph of any logarithmic function to be obtained from the basic graph of $y = \log_a(x)$. This provides an alternative to sketching the graph as the inverse of that of an exponential function. Further, given the logarithmic graph, the exponential graph could be obtained as the inverse of the logarithmic graph.

The logarithmic graph under a combination of transformations will be studied in Units 3 and 4. In this section we shall consider the effect simple transformations have on the key features of the graph of $y = \log_a(x)$.

Dilation from the x-axis: the graph of $y = n\log_a(x)$

The key features of the graph of $y = n\log_a(x)$ are:

- **dilation by a factor of n from the x-axis**
- **a vertical asymptote at $x = 0$**
- **x-intercept $(1, 0)$**
- **if $n < 0$, this also gives a reflection in the x-axis.**

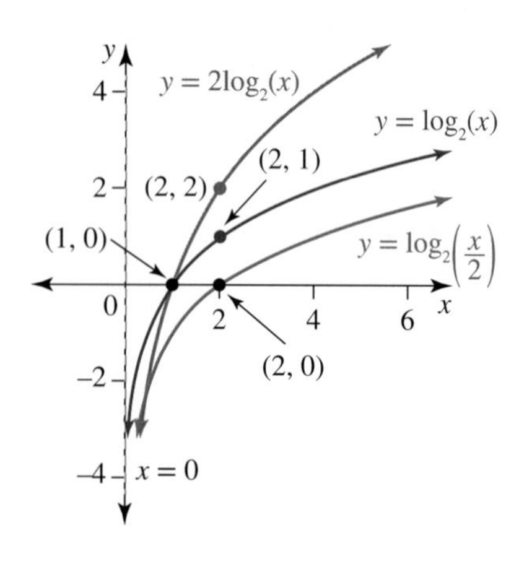

Dilation from the y-axis: the graph of $y = \log_a(nx)$

The key features of the graph of $y = \log_a(nx)$ are:

- **dilation by a factor of $\frac{1}{n}$ from the y-axis**
- **a vertical asymptote at $x = 0$**
- **x-intercept $\left(\frac{1}{n}, 0\right)$**
- **if $n < 0$, this also gives a reflection in the y-axis.**

For example, $y = 2\log_2(x)$ and $y = \log_2\left(\dfrac{x}{2}\right)$ give the images when $y = \log_2(x)$ undergoes a dilation of factor 2 from the x-axis and from the y-axis, respectively. The asymptote at $x = 0$ is not affected by either dilation. The position of the x-intercept is affected by the dilation from the y-axis as $(1, 0) \to (2, 0)$. The dilation from the x-axis does not affect the x-intercept.

Horizontal translation: the graph of $y = \log_a(x - h)$

The key features of the graph of $y = \log_a(x - h)$ are:

- **horizontal translation of h units**
  - **if $h > 0$, translated to the right**
  - **if $h < 0$, translated to the left**
- **a vertical asymptote with equation $x = h$**
- **x-intercept $(1 + h, 0)$**
- **domain $x \in (h, \infty)$.**

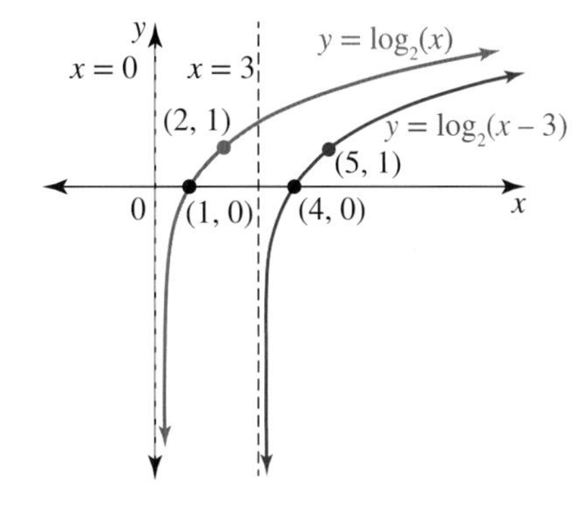

These changes are illustrated in the diagram by the graph of $y = \log_2(x)$ and its image, $y = \log_2(x - 3)$, after a horizontal translation of 3 units to the right.

The diagram shows that the domain of $y = \log_2(x - 3)$ is $(3, \infty)$. Its range is unaffected by the horizontal translation and remains R.

It is important to realise that the domain and the asymptote position can be calculated algebraically, since we only take logarithms of positive numbers. For example, the domain of $y = \log_2(x - 3)$ can be calculated by solving the inequation $x - 3 > 0 \Rightarrow x > 3$. This means that the domain is $(3, \infty)$ as the diagram shows. The equation of the asymptote of $y = \log_2(x - 3)$ can be calculated from the equation $x - 3 = 0 \Rightarrow x = 3$.

The function defined by $y = \log_a(nx + c)$ would have a vertical asymptote when $nx + c = 0$, and its domain can be calculated by solving $nx + c > 0$.

Vertical translation: the graph of $y = \log_a(x) + k$

The key features of the graph of $y = \log_a(x) + k$ are:

- **vertical translation of k units**
  - **if $k > 0$, translated upwards**
  - **if $k < 0$, translated downwards**
- **vertical asymptote with equation $x = 0$, unchanged from $y = \log_a(x)$**
- **an x-intercept found by solving the equation $\log_a(x) + k = 0$**
- **an additional helpful point $(1, k)$**
- **domain $x \in R^+$, unchanged from $y = \log_a(x)$**
- **range R.**

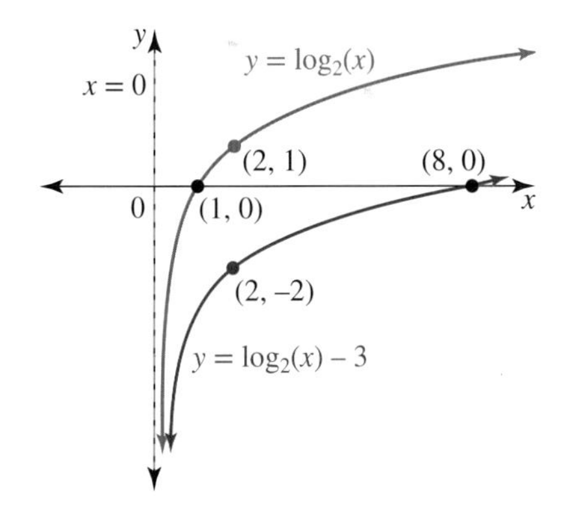

The graph of $y = \log_2(x) - 3$ is a vertical translation down by 3 units of the graph of $y = \log_2(x)$. Solving $\log_2(x) - 3 = 0$ gives $x = 2^3$, so the graph cuts the x-axis at $x = 8$, as illustrated.

Reflections: the graphs of $y = -\log_a(x)$ and $y = \log_a(-x)$

The graph of $y = -\log_a(x)$ is obtained by inverting the graph of $y = \log_a(x)$; that is, by reflecting it in the x-axis.

The graph of $y = \log_a(-x)$ is obtained by reflecting the graph of $y = \log_a(x)$ in the y-axis. For $\log_a(-x)$ to be defined, $-x > 0$, so the graph has domain $\{x : x < 0\}$.

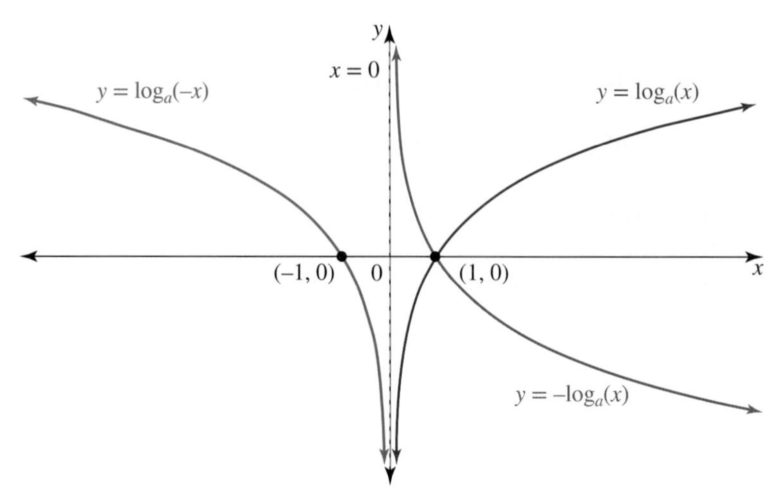

The relative positions of the graphs of $y = \log_a(x)$, $y = -\log_a(x)$ and $y = \log_a(-x)$ are illustrated in the diagram. The vertical asymptote at $x = 0$ is unaffected by either reflection.

WORKED EXAMPLE 16 Sketching logarithmic graphs

a. Sketch the graph of $y = \log_2(x + 2)$ and state its domain.
b. Sketch the graph of $y = \log_{10}(x) + 1$ and state its domain.
c. The graph of the function for which $f(x) = \log_2(b - x)$ is shown.
i. Determine the value of b.
ii. State the domain and range of the inverse function and form its rule.
iii. Sketch the graph of $y = f^{-1}(x)$.

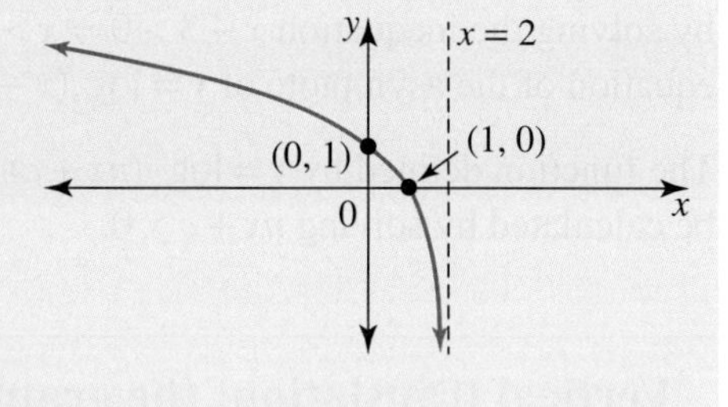

THINK	WRITE
a. 1. Identify the transformation involved.	**a.** $y = \log_2(x + 2)$ Horizontal translation 2 units to the left
2. Use the transformation to state the equation of the asymptote and the domain.	The vertical line $x = 0 \to$ the vertical line $x = -2$ under the horizontal translation. The domain is $\{x : x > -2\}$.
3. Calculate any intercepts with the coordinate axes. *Note:* The domain indicates there will be an intercept with the y-axis as well as the x-axis.	y-intercept: when $x = 0$, $y = \log_2(2)$ $= 1$ y-intercept $(0, 1)$ x-intercept: when $y = 0$, $\log_2(x + 2) = 0$ $x + 2 = 2^0$ $x + 2 = 1$ $x = -1$ x-intercept $(-1, 0)$ Check: the point $(1, 0) \to (-1, 0)$ under the horizontal translation.

4. Sketch the graph.

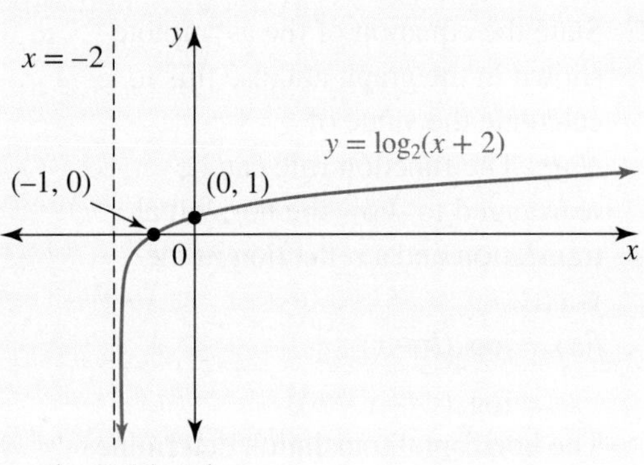

b. 1. Identify the transformation involved.

b. $y = \log_{10}(x) + 1$
Vertical translation of 1 unit upwards

2. State the equation of the asymptote and the domain.

The vertical transformation does not affect either the position of the asymptote or the domain.
Hence, the equation of the asymptote is $x = 0$. The domain is R^+.

3. Obtain any intercept with the coordinate axes.

Since the domain is R^+, there is no y-intercept.
x-intercept: when $y = 0$,

$$\log_{10}(x) + 1 = 0$$
$$\log_{10}(x) = -1$$
$$x = 10^{-1}$$
$$= \frac{1}{10} \text{ or } 0.1$$

The x-intercept is $(0.1, 0)$.

4. Calculate the coordinates of a second point on the graph.

Point: let $x = 1$.

$$y = \log_{10}(1) + 1$$
$$= 0 + 1$$
$$= 1$$

The point $(1, 1)$ lies on the graph.
Check: the point $(1, 0) \rightarrow (1, 1)$ under the vertical translation.

5. Sketch the graph.

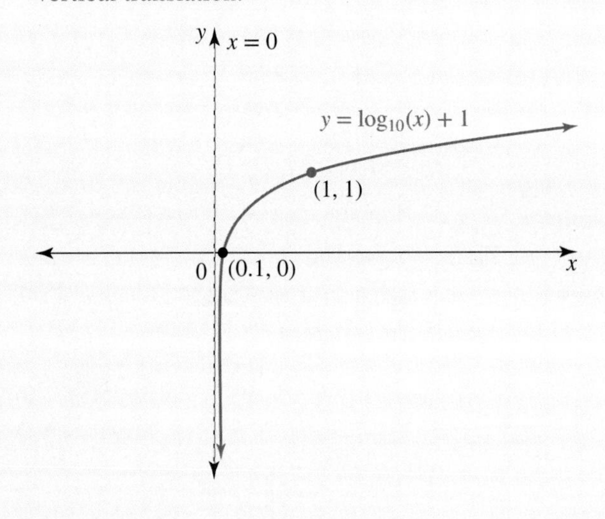

c. i. 1. State the equation of the asymptote shown in the graph and use this to calculate the value of b.
Note: The function rule can be rearranged to show the horizontal translation and a reflection in the y-axis.
$f(x) = \log_2(b - x)$
$\quad = \log_2(-(-x - b))$
The horizontal translation determines the position of the asymptote.

c. i. $f(x) = \log_2(b - x)$
From the diagram, the asymptote of the graph is $x = 2$.
From the function rule, the asymptote occurs when:
$b - x = 0$
$\quad x = b$
Hence, $b = 2$.

ii. 1. Give the domain and range of the inverse function.

ii. The given function has domain $(-\infty, 2)$ and range R. Therefore, the inverse function has domain R and range $(-\infty, 2)$.

2. Form the rule for the inverse function by interchanging x-and y-coordinates and rearranging the equation obtained.

Function f: $y = \log_2(2 - x)$
Inverse f^{-1}: $x = \log_2(2 - y)$
$2^x = 2 - y$
$\therefore y = 2 - 2^x$
$\therefore f^{-1}(x) = 2 - 2^x$

iii. 1. Use the features of the logarithm graph to deduce the features of the exponential graph.

iii. The key features of f give those for f^{-1}.

$y = f(x)$	$y = f^{-1}(x)$
Asymptote: $x = 2$	Asymptote: $y = 2$
x-intercept $(1, 0)$	y-intercept $(0, 1)$
y-intercept $(0, 1)$	x-intercept $(1, 0)$

2. Sketch the graph of $y = f^{-1}(x)$.

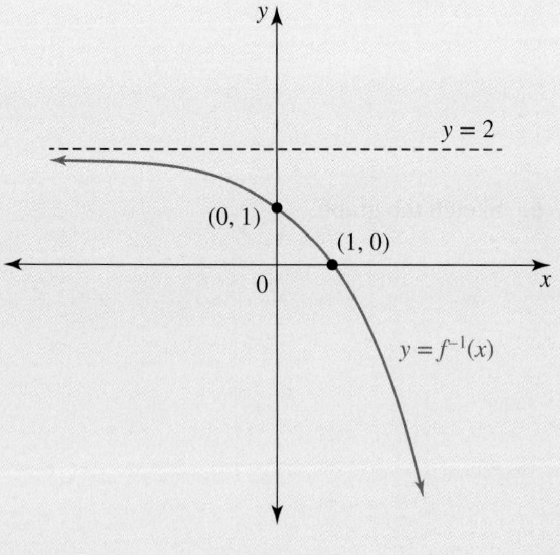

10.6 Exercise

Students, these questions are even better in jacPLUS

Receive immediate feedback and access sample responses

Access additional questions

Track your results and progress

Find all this and MORE in jacPLUS

Technology free

1. WE13 a. Form the exponential rule for the inverse of $y = \log_{10}(x)$ and hence deduce the graph of $y = \log_{10}(x)$ from the graph of the exponential.
   b. Given the points (1, 10), (2, 100) and (3, 1000) lie on the exponential graph in part a, explain how these points can be used to illustrate the logarithm law $\log_{10}(m) - \log_{10}(n) = \log_{10}\left(\frac{m}{n}\right)$.
2. a. On the same axes, sketch the graphs of $y = 3^x$ and $y = 5^x$ together with their inverses.
   b. State the rules for the inverses graphed in part a as logarithmic functions.
   c. Describe the effect of increasing the base on a logarithm graph.
3. a. On the same set of axes, sketch $y = 3^x + 1$ and draw its inverse.
   b. Give the equation of the inverse of $y = 3^x + 1$.
4. Sketch the graphs of $y = 5^{x+1}$ and its inverse on the same set of axes and give the rule for the inverse.
5. Sketch the graphs of $y = 2^{-x}$ and its inverse and form the rule for this inverse.
6. WE14 Consider the function $f: R \to R,\ f(x) = 4 - 2^{3x}$.
   a. Determine the domain of its inverse.
   b. Form the rule for the inverse function and express the inverse function as a mapping.
   c. Sketch $y = f(x)$ and $y = f^{-1}(x)$ on the same set of axes.
7. Consider the function $f: R \to R,\ f(x) = 8 - 2 \times 3^{2x}$:
   a. Determine the domain and the range of the inverse function f^{-1}.
   b. Evaluate $f(0)$.
   c. Identify the point at which the graph of f^{-1} cuts the x-axis.
   d. Obtain the rule for f^{-1} and express f^{-1} as a mapping.
8. WE15 a. Simplify $\log_6\left(2^{2x} \times 9^x\right)$.
   b. Evaluate $2^{-3\log_2(10)}$ using the inverse relationship between exponentials and logarithms.
9. WE16 a. Sketch the graph of $y = \log_{10}(x - 1)$ and state its domain.
   b. Sketch the graph of $y = \log_5(x) - 1$ and state its domain.
   c. The graph of the function for which $f(x) = -\log_2(x + b)$ is shown.
      i. Determine the value of b.
      ii. State the domain and range of, and form the rule for, the inverse function.
      iii. Sketch the graph of $y = f^{-1}(x)$.

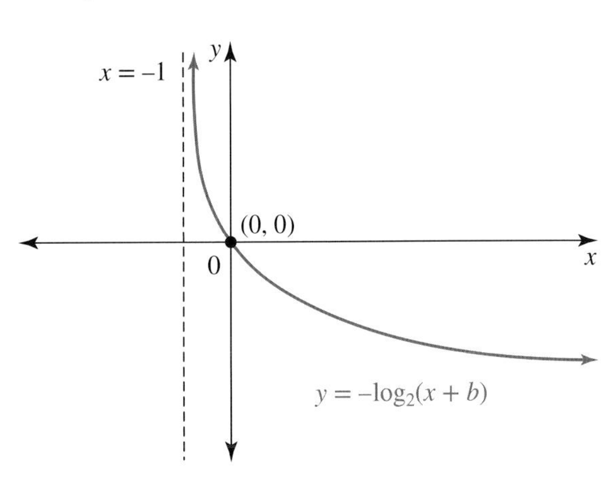

10. State the maximal domain and the equation of the vertical asymptote of each of the following logarithmic functions.

a. $y = \log_{10}(x+6)$
b. $y = \log_{10}(x-6)$
c. $y = \log_{10}(6-x)$
d. $y = \log_{10}(6x)$
e. $y = 3\log_5(-x)$
f. $y = -\log_4(2x-3)+1$

11. Consider the function with rule $y = \log_3(x+9)$.

a. State the domain.
b. State the equation of the vertical asymptote.
c. Calculate the coordinates of the y-intercept.
d. Calculate the coordinates of the x-intercept.
e. Sketch the graph of $y = \log_3(x+9)$.
f. State the range of the graph.

12. Consider the function with rule $y = \log_{10}(x-5)$.

a. State the domain.
b. State the equation of the vertical asymptote.
c. Calculate the coordinates of any intercepts with the coordinate axes.
d. Sketch the graph, stating its range.
e. On the same diagram, sketch the graph of $y = -\log_{10}(x-5)$.

13. Sketch the graphs of the following transformations of the graph of $y = \log_a(x)$, stating the domain and range, equation of the asymptote and any points of intersection with the coordinate axes.

a. $y = \log_5(x) - 2$
b. $y = \log_5(x-2)$
c. $y = \log_{10}(x) + 1$
d. $y = \log_3(x+1)$
e. $y = \log_3(4-x)$
f. $y = -\log_2(x+4)$

14. Use a logarithm law to describe the vertical translation that maps $y = \log_2(x) \to y = \log_2(2x)$.

15. Simplify $5^{x\log_5(2) - \log_5(3)}$.

16. a. Evaluate the following.

i. $3^{\log_3(8)}$
ii. $10^{\log_{10}(2)+\log_{10}(3)}$
iii. $5^{-\log_5(2)}$
iv. $6^{\frac{1}{2}\log_6(25)}$

b. Simplify the following.

i. $3^{\log_3(x)}$
ii. $2^{3\log_2(x)}$
iii. $\log_2(2^x) + \log_3(9^x)$
iv. $\log_6\left(\dfrac{6^{x+1} - 6^x}{5}\right)$

Technology active

17. a. The graph of the function with equation $y = a\log_7(bx)$ contains the points $(2, 0)$ and $(14, 14)$. Determine its equation.

b. The graph of the function with equation $y = a\log_3(x) + b$ contains the points $\left(\dfrac{1}{3}, 8\right)$ and $(1, 4)$.

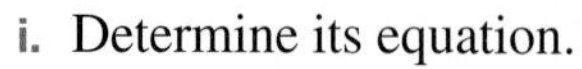

i. Determine its equation.
ii. Obtain the coordinates of the point where the graph of the inverse function would cut the y-axis.

c. i. For the graph illustrated in the diagram, determine a possible equation in the form $y = a\log_2(x-b)+c$.
ii. Use the diagram to sketch the graph of the inverse and form the rule for the inverse.

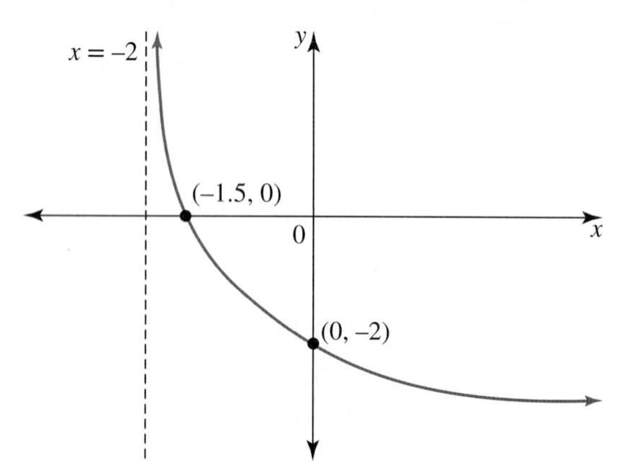

18. Consider the functions f and g for which $f(x) = \log_3(4x + 9)$ and $g(x) = \log_4(2 - 0.1x)$.

a. Determine the maximal domain of each function.
b. State the equations of the asymptotes of the graphs of $y = f(x)$ and $y = g(x)$.
c. Calculate the coordinates of the points of intersection of each of the graphs with the coordinate axes.
d. Sketch the graphs of $y = f(x)$ and $y = g(x)$ on separate diagrams.

19. Sketch the graphs of the following functions and their inverses, and form the rule for each inverse.

a. $y = 2^{-\frac{x}{3}}$
b. $y = \frac{1}{2} \times 8^x - 1$
c. $y = 2 - 4^x$
d. $y = 3^{x+1} + 3$
e. $y = -10^{-2x}$
f. $y = 2^{1-x}$

20. The diagram shows the graph of the exponential function $y = 2^{ax+b} + c$. The graph intersects the line $y = x$ twice and cuts the x-axis at $\left(\frac{1}{2}, 0\right)$ and the y-axis at $(0, -2)$.

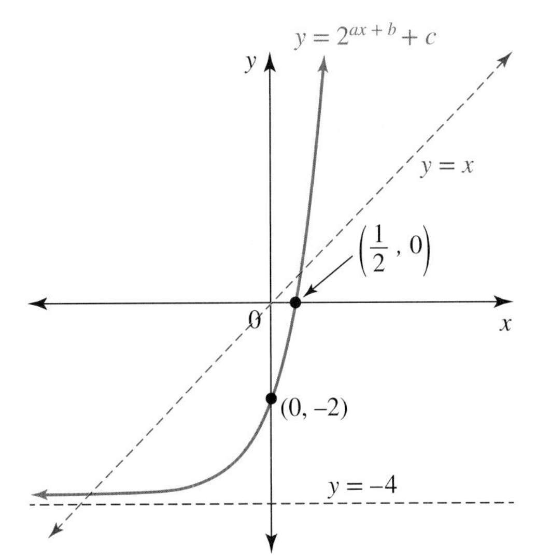

a. Form the rule for the exponential function.
b. Form the rule for the inverse function.
c. For the inverse, state the equation of its asymptote and the coordinates of the points where its graph would cut the x- and y-axes.
d. Copy the diagram and sketch the graph of the inverse on the same diagram. Determine how many points of intersection of the inverse and the exponential graphs there are.
e. The point $(\log_2(3), k)$ lies on the exponential graph. Calculate the exact value of k.
f. Using the equation for the inverse function, verify that the point $(k, \log_2(3))$ lies on the inverse graph.

10.6 Exam questions

Question 1 (1 mark) TECH-ACTIVE

MC State which of the following is *not* true about the inverse of $y = a^x, a > 1$.

A. It has a vertical asymptote with equation $x = 0$.
B. The point $(a, 1)$ lies on the graph.
C. The y-intercept is $(0, 1)$.
D. The domain is R^+.
E. The range is R.

Question 2 (1 mark) TECH-ACTIVE

MC Consider the function $f: R \to R, f(x) = 4 \times 2^{-x} + 5$. The domain of its inverse function is

A. $(4, \infty)$
B. $(5, \infty)$
C. $(\infty, 4)$
D. $(\infty, 5)$
E. R

Question 3 (4 marks) TECH-FREE

Sketch the graph of $y = \log_2(x + 4)$ and state its domain.

More exam questions are available online.

10.7 Review

10.7.1 Summary

doc-37026

Hey students! Now that it's time to revise this topic, go online to:

Access the topic summary

Review your results

Watch teacher-led videos

Practise exam questions

Find all this and MORE in jacPLUS

10.7 Exercise

Technology free: short answer

1. Simplify the following, expressing answers with positive indices where appropriate.

a. $\dfrac{\left(3a^2b^{\frac{1}{2}}\right)^{-2} \times 2\left(a^{-3}b\right)^{-1}}{\left(4a^{-4}\right)^{\frac{1}{2}}}$

b. $\dfrac{2^{n+3}-2^n}{14}$

c. $\dfrac{a-a^{-1}}{a+1}$

d. $18^{\frac{3}{4}} \times \left(\dfrac{2}{9}\right)^{-2} \div \sqrt{3\times 2}$

2. Evaluate the following.

a. $\log_6(9) - \log_6\left(\dfrac{1}{4}\right)$

b. $2\log_a(4) + 0.5\log_a(16) - 6\log_a(2)$

c. $\dfrac{\log_a(27)}{\log_a(3)}$

d. $-\log_{10}(5) \times \log_9(3) - \log_{10}\left(\sqrt{2}\right)$

3. Solve the following equations for x.

a. $3^{1-7x} = 81^{x-2} \times 9^{2x}$

b. $2^{2x} - 6 \times 2^x - 16 = 0$

c. $\log_5(x+2) + \log_5(x-2) = 1$

d. $2\log_{10}(x) - \log_{10}(101x - 10) = -1$

4. Obtain the value(s) of x, in exact form, for each of the following.

a. $3^{1-x} = 7$

b. $5^{x-1} = 2^{x+1}$

c. $0.2^x < 3$

d. $10^x = 12 \times 10^{-x} + 4$

5. Sketch the graphs of the following, stating the equation of the asymptote, the coordinates of any points of intersection with the axes and the domain and range.

a. $y = 2^{3x} - 1$

b. $y = -2 \times 3^{(x-1)}$

c. $y = 5 - 5^{-x}$

d. $y = 3 \times \left(\dfrac{2}{5}\right)^{x+1}$

6. a. Describe the transformations that map $y = \log_3(x) \to y = -\log_3(x+3)$ and hence state the equation of the asymptote of $y = -\log_3(x+3)$.

b. Give the rule for the inverse of $y = -\log_3(x+3)$, stating its domain and range, and sketch the graphs of $y = -\log_3(x+3)$ and its inverse on the same set of axes.

c. Describe the transformations that map $y = 7^x \to y = 5 \times 7^{1-x}$ and state the equation of the asymptote of $y = 5 \times 7^{1-x}$.

d. Given $f: R \to R$, $f(x) = 5 \times 7^{1-x}$, form the inverse function f^{-1}.

Technology active: multiple choice

7. **MC** $\dfrac{(2^n)^2 \times 2^{n-2}}{4}$ equals:

A. 2^{2n-2} B. 2^{3n-4} C. 2^{n^2+n-2} D. 2^{3n} E. $3n-2$

8. **MC** If $5^{2x+3} = 125^x$, x equals:

A. -3 B. 0 C. 1 D. 3 E. 5

9. **MC** In scientific notation, $(3.2 \times 10^{-2}) \times (5 \times 10^5)$ would equal:

A. 2.56×10^8 B. 16×10^{-10} C. 1.6×10^{-9}
D. 1.6×10^2 E. 1.6×10^4

10. **MC** The population of Melbourne on census night in August 2016 was reported to be 4 485 211. If this figure was expected to grow by 942 000 by the year 2020, then the estimated population of Melbourne in 2020, to 2 significant figures, would have been:

A. $54\,272 \times 10^2$ B. 5.42 million C. 5 400 000
D. 5 500 000 E. 5 427 000

11. **MC** The statement $3^5 = 243$ expressed in logarithm form would be:

A. $\log_3(5) = 243$ B. $\log_5(3) = 243$ C. $\log_5(243) = 3$
D. $\log_3(243) = 5$ E. $\log_{243}(5) = 3$

12. **MC** If $\log_{64}(x) = -\dfrac{2}{3}$, then x is equal to:

A. 2 B. -2 C. $\dfrac{1}{16}$ D. $-\dfrac{1}{16}$ E. -512

13. **MC** If $\log_5(35) = p$, then $\log_5(7)$ is equal to:

A. $p-1$ B. $\dfrac{p}{5}$ C. $5-p$
D. $p - \log_5(28)$ E. $\log_5(42) - p$

14. **MC** The expression $\sqrt[3]{5} + \sqrt[3]{625}$ is equal to:

A. $\sqrt[3]{630}$ B. $6 \times \sqrt[3]{5}$ C. $26 \times \sqrt[3]{5}$ D. $5^{\frac{13}{12}}$ E. $10^{\frac{5}{3}}$

15. **MC** A possible equation for the given graph could be:

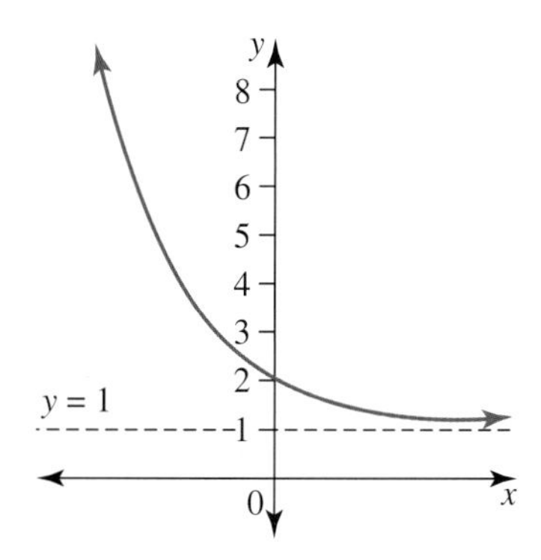

A. $y = 2 \times 3^{-x}$ B. $y = 2 \times 3^{x+1}$ C. $y = -3^x + 1$
D. $y = 6 \times 3^{x-1} + 1$ E. $y = 3^{-x} + 1$

16. **MC** $2^{-3\log_2(x)}$ simplifies to:

A. $-3x$ B. $-x^3$ C. x^{-3} D. $\dfrac{1}{8}\log_2(x)$ E. $-6\log_2(x)$

Technology active: extended response

17. The graph of the function defined by $f: R \rightarrow R$, $f(x) = 2^{x+b} + c$ is shown. The graph has an asymptote at $y = 2$ and contains the point $(1, 3)$.

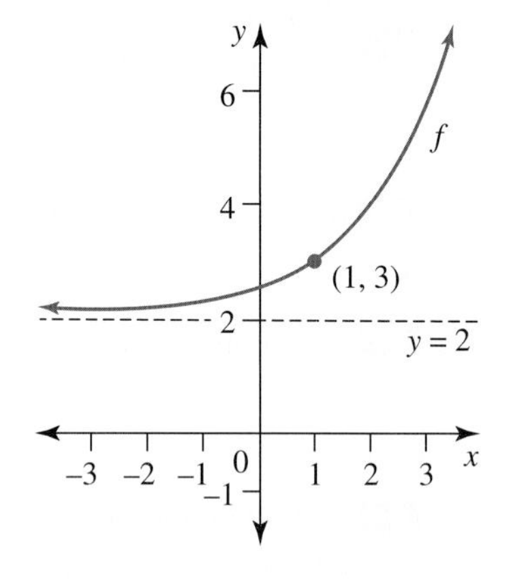

a. Determine the values of b and c.
b. Show that the rule for the function could also be expressed in the form $f(x) = a \times 2^x + c$.
c. Copy the diagram and draw the graph of the inverse function on the same axes, stating the domain of the inverse function.
d. Form the rule for the inverse function from each of the two rules for the exponential function and use logarithm laws to show that the two forms are equivalent.
e. Obtain the coordinates of:

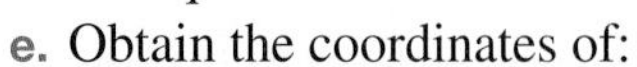

i. the point A on $y = f(x)$ for which $f(x) = 6$
ii. the point B on $y = f^{-1}(x)$ for which $f^{-1}(x) = 6$
iii. the point P on $y = f^{-1}(x)$ for which $f^{-1}(x) = 1$.

f. Calculate the area of the triangle ABP.

18. Polly fills a kettle with water, planning to make a pot of tea. They switch the kettle on and the water heats to boiling point of 100 °C at 10 am, when the kettle automatically switches off. However, Polly is distracted by their phone and forgets they have put the kettle on. The water in the kettle begins to cool in such a way that the temperature, T °C, can be modelled by $T = a \times \left(\frac{16}{5}\right)^{-kt}$, where t is the number of minutes since 10 am and a and k are constants.

a. Obtain the value of the constant a.
b. If the temperature of the water in the kettle was 75 °C at 10.12 am, obtain the value of k correct to 2 decimal places.
c. By the time Polly remembers they had put the kettle on, it is 10.30 am. Using the value for k obtained in part b, calculate, to the nearest degree, the temperature of the water at this time.
d. At 10.30 am Polly switches the kettle back on and the water is reheated. If the temperature of the water t minutes after 10.30 am is described by $T = 50 \times 2^{\frac{t}{9}}$, determine the exact time at which the water will re-reach its boiling point.
e. Sketch a graph to show the cooling and heating of the water from 10 am to the time the water re-reaches boiling point.
f. Determine the length of time, to the nearest minute, for which the temperature of the water in the kettle was below 75 °C.

19. Giana purchased a town house in an outer suburb of a large city. As the population of the suburb has grown, the time it takes Giana to drive to work has increased exponentially. A model for the amount of time in minutes it takes Giana to get to work is $T(n) = a \times 2^{kn}$ where T is the time in minutes and the number of drivers is n thousand; a and k are constants.

When Giana moved to the suburb, there were 1.2 thousand drivers. As this number doubled, so did the time for Giana's drive to work.

a. Show that $k = \frac{5}{6}$.
b. Initially it took Giana 20 minutes to drive to work. Determine how long it takes her now that the population of drivers has trebled.
c. Giana is not only concerned about the number of pollutants being released into the atmosphere from car emissions, but also about the fact that, as the population grows, so does the amount of litter. This is starting to result in an increase in pollutants getting into the water supply. The amount Q grams per litre of pollutant t years after Giana moved to the suburb is given by $Q = 0.5 - 0.27 \times 3^{-0.2t}$.

i. Calculate how much the amount of pollutant per litre increased by during the first 5 years Giana has lived in the suburb.
ii. Sketch the graph of Q against t.
iii. Calculate how many years it will take for the amount of pollutant per litre to reach double the value when Giana first moved to the suburb.
iv. Calculate the pollutant level that the model predicts the water supply will approach.

d. Adding to her woes, Giana's new neighbours are very noisy. It seems to Giana that the neighbours practise playing their electric guitars most evenings until quite late at night. The measure of loudness in decibels is given by $L = 10\log_{10}\left(I \times 10^{12}\right)$, where L is the number of decibels and I is the intensity of the sound measured in watts per square metre. Given the sound level produced by a guitar is 70 decibels, answer the following questions.

i. Calculate, in standard form, the intensity of the sound each guitar produces.
ii. Calculate the decibel reading when both guitars are played together.

20. a. Factorise:

i. $(x+2)^{\frac{3}{2}} - (x+2)^{\frac{1}{2}}$
ii. $4x(4x^2+1)^{-\frac{3}{2}}(2x-1)^{-1} - 2(2x-1)^{-2}(4x^2+1)^{-\frac{1}{2}}$, and hence obtain any values of x for which $4x(4x^2+1)^{-\frac{3}{2}}(2x-1)^{-1} - 2(2x-1)^{-2}(4x^2+1)^{-\frac{1}{2}} = 0$.

b. Simplify:

i. $\dfrac{a^{-3}-b^{-3}}{a^{-2}-b^{-2}} \div \dfrac{1}{a^{-1}+b^{-1}}$
ii. $\left[(m+n)^{\frac{1}{2}} + m^{\frac{1}{2}} - n^{\frac{1}{2}}\right]\left[(m+n)^{\frac{1}{2}} - m^{\frac{1}{2}} + n^{\frac{1}{2}}\right]$, and hence evaluate the expression when $m = \log_a(b)$ and $n = \log_b(a)$.

c. i. Show that $\log(n!) = \log(2) + \log(3) + \ldots + \log(n)$ for $n \in N$.
ii. Hence, evaluate $\log(10!) - \log(9!)$.

d. For any integer $x > 1$, it was established in the late 19th century that the number of prime numbers less than or equal to x approaches the ratio $\dfrac{x}{\log_e(x)}$ as x becomes large.
Let the function $p(x) = \dfrac{x}{\log_e(x)}$ be an estimate of the number of prime numbers less than or equal to x.
Obtain $p(10)$ and $p(30)$, and compare the estimates with the actual number of primes in each case.

10.7 Exam questions

Question 1 (1 mark) TECH-ACTIVE

MC $52.2 \times 10^{-4} \div \left(5 \times 10^{8}\right)$ expressed in standard form is

A. 262
B. 10.44
C. 2.62×10^{-32}
D. 1.044×10^{-11}
E. 2.62×10^{-12}

Question 2 (1 mark) TECH-ACTIVE

MC If $\log_{10}(2x-3) = 0$, then x is equal to

A. $\frac{1}{2}$ **B.** $\frac{3}{2}$ **C.** 2 **D.** 1 **E.** 0

Question 3 (1 mark) TECH-ACTIVE

MC The graph of $y = -4 \times 5^x$ is best represented by

A.

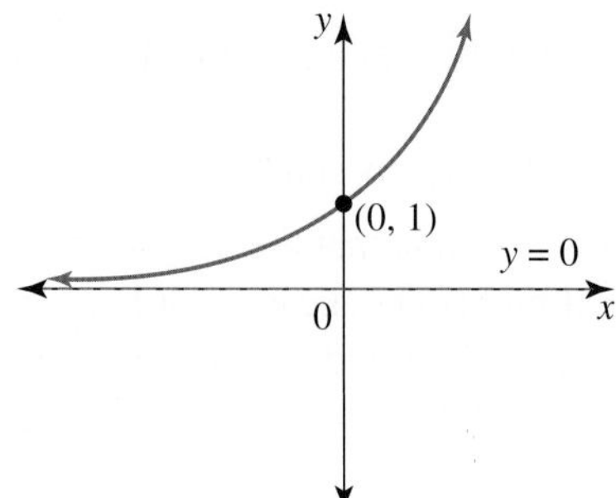

B.

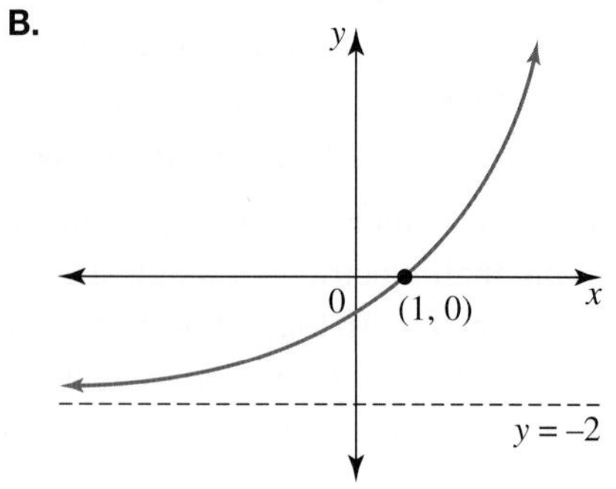

C.

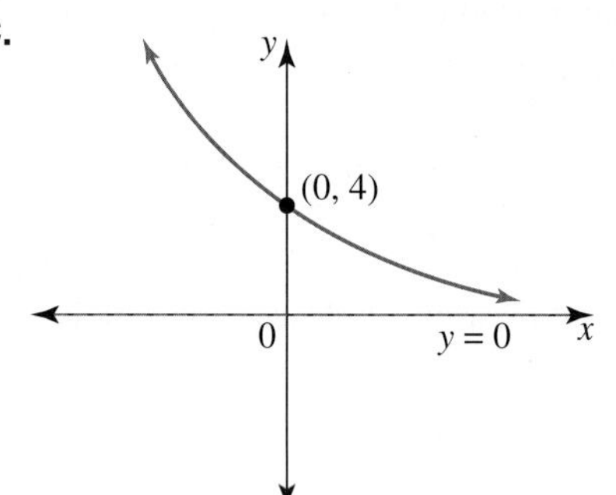

D.

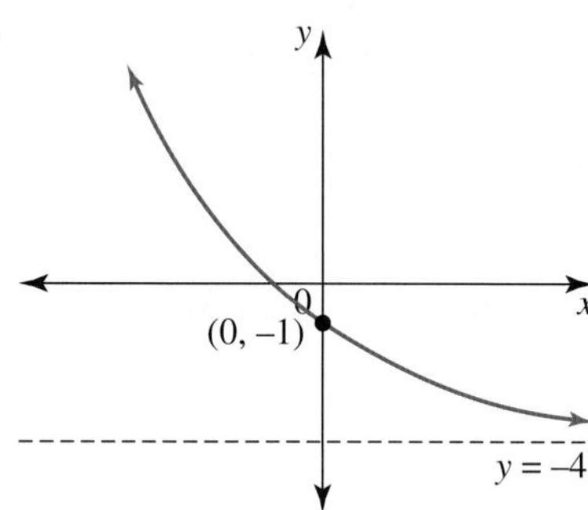

E.

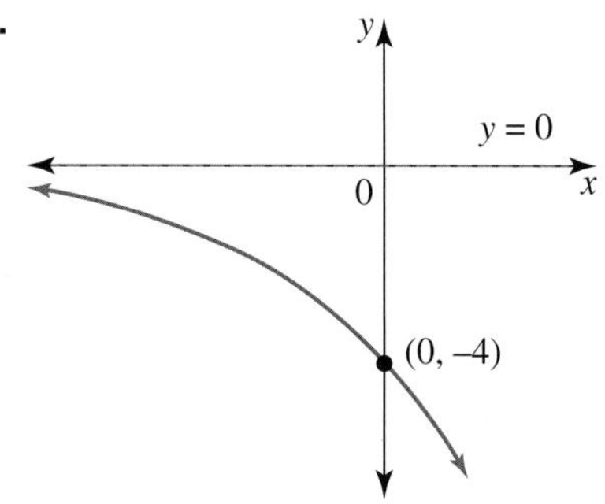

Question 4 (4 marks) TECH-FREE

Find the values of a, b and the equation if $y = a\ \log_9(bx)$ contains the points $(3, 0)$ and $(27, 4)$.

Question 5 (1 mark) TECH-ACTIVE

MC Simplify $\log_{20}\left(16^x \times 5^{2x}\right)$ using the inverse relationship between exponentials and logarithms.

A. 20 **B.** x **C.** $2x$ **D.** $16x$ **E.** $5x$

More exam questions are available online.

Answers

Topic 10 Exponential functions

10.2 Indices as exponents

10.2 Exercise

1. a. i. $a^{\frac{3}{2}}b^2$ ii. $a^{\frac{5}{2}}b^2$ iii. $a^{\frac{2}{3}}b^{\frac{1}{3}}$
   b. i. $\sqrt{\frac{a}{b^3}}$ ii. $\sqrt{32}$ iii. $\sqrt[5]{\frac{1}{9}}$ or $\frac{1}{\sqrt[5]{9}}$
2. a. 8 b. 0 c. 12 d. 15
3. a. $\frac{1}{a^{2n+3}}$ b. $3x^8y^7$ c. $\frac{8a^3}{b^5}$
   d. $\frac{x^{14}y^7}{2}$ e. $3^{\frac{7}{2}}$ f. $\frac{1}{a^{7n}b^{6n-2}}$
4. a. 2^{9n} b. $6a^{\frac{1}{2}}b^2$ c. $\frac{8}{9}$
5. $-\frac{16pq^8}{m^4}$
6. a. $\frac{9x^{10}}{y^4}$ b. $\frac{3}{16a^{\frac{5}{3}}b^2}$ c. $\frac{3m^3n^{\frac{1}{2}}}{40}$
   d. $-\frac{8n^5}{9}$ e. $-\frac{1}{mn(m+n)}$ f. $\frac{2x-1}{(4x-1)^{\frac{1}{2}}}$
7. a. 2^{5+2x} b. 3^{-7} c. 10^3 d. 5^n
8. a. $x=\frac{4}{3}$ b. $x=2$ c. $x=\frac{1}{5}$
   d. $x=4$ e. $x=\frac{8}{7}$ f. $x=6$
9. $x=8$
10. a. $x<2$ b. $x<\frac{10}{3}$
11. a. $x=1$ b. $x\leq 6$ c. $x=\frac{7}{10}$
    d. $x>\frac{1}{2}$ e. $x=\frac{1}{2}$ f. $x=48$
12. a. $x=0, x=3$ b. $x=1$
    c. No solution d. $x=0$
13. $x=-1$
14. $x=4$
15. a. $x=0,2$ b. $x=-3$ c. $x=1$
    d. $x=0$ e. $x=2$ f. $x=-1$
16. a. i. 1.409×10^6; 4 significant figures
    ii. 1.306×10^{-4}; 4 significant figures
    b. i. 304 000 ii. 0.058 03
17. 8×10^{10}
18. a. i. -5.06×10^{-8} ii. 1.274×10^4 km
    iii. 1.6×10^3 iv. 1.68787×10^4 km
    b. i. 63 000.000 63 ii. 1200
    c. i. 61 000 ii. 0.020
    iii. $x=-0.0063$ iv. 27 million km
19. a. $x=4, y=11$ b. $a=160, k=-1$
20. a. 9.873×10^{-4} b. 0.1762; $1.3\text{E}2=1.3\times10^2$
    c. 1.68×10^{-15} d. 9.938×10^{-11}

10.2 Exam questions

Note: Mark allocations are available with the fully worked solutions online.

1. D
2. B
3. $x=\frac{7}{10}$

10.3 Indices as logarithms

10.3 Exercise

1. a. $4=\log_5(625)$ b. $6=36^{\frac{1}{2}}$
   c. $x\approx0.93$ d. $x=\frac{1}{3}$
2. a. $\log_e(5)\approx1.609$; $5=e^{1.609}$
   b. $10^{3.5}\approx3162$; $3.5=\log_{10}(3162)$
3. a. 1 b. 4 c. 1 d. $\frac{3}{2}$
4. a. -0.3 b. 0.4 c. 1.3
5. a. 0 b. 1 c. -1
   d. 5 e. -3 f. 2
6. a. $\log_3(x)$ b. $6\log_a(x)$ c. 0
   d. $\log_b(2a)$ e. $-\frac{1}{2}\log_{10}(x)$ f. 3
7. a. 2 b. -1 c. 3
   d. 0 e. -1 f. 0
8. a. i. $5=\log_2(32)$
   ii. $\frac{3}{2}=\log_4(8)$
   iii. $-3=\log_{10}(0.001)$
   b. i. $2^4=16$
   ii. $9^{\frac{1}{2}}=3$
   iii. $10^{-1}=0.1$
9. a. $\log_{10}(14)$ b. $\log_5(44)$ c. $\log_3(10)$
   d. $\log_{10}(8)$ e. $\log_{10}(25)$ f. $\log_5\left(\frac{1}{8}\right)$
10. a. 1 b. 0 c. 3
    d. 1 e. $\frac{1}{2}$ f. -4
11. a. $2^x=\frac{1}{8}; x=-3$
    b. $25^{-0.5}=x; x=\frac{1}{5}$
    c. $2x=\log_{10}(4); x=\log_{10}(2)$
    d. $-x=\log_e(3); x=-\log_e(3)$
    e. $x^3=125; x=5$
    f. $x^{-2}=25; x=\frac{1}{5}$

12. a. $x = \log_7(15);\ x \approx 1.392$

b. $x = \dfrac{5\log(3)}{\log(4) - 2\log(3)};\ x = -6.774$

13. a. $x = 26$ b. $x = -\dfrac{1}{4}$

c. $x = 7$ d. $x = 3$

14. a. $x = 49$ b. $x = 6.4$

c. $x = \dfrac{1}{9}$ d. $x = 3$

15. $x = \dfrac{3}{10}$

16. $x = 1$

17. $x = \dfrac{36}{35}$

18. a. $\sqrt{7}$ b. $\dfrac{4}{5}$

c. $\dfrac{6}{23}$ d. $\dfrac{1}{2}$

e. $0.001, 8$ f. 3

19. a. $x = \log_3(7) - 2$

b. $x = \dfrac{1}{3}\log_6(5) + \dfrac{5}{3}$

c. $x = 3\log(2)$

20. a. $p + q$ b. $3q$

c. $2p + q$ d. $p - q$

e. $2q - 4p$ f. $\dfrac{3}{4}pq$

21. a. 125 b. 7

22. a. $y = 100x$ b. $y = \log_{10}(x) + 2$

c. $y = \dfrac{1}{x}$ d. $y = \sqrt{x}, x > 0$

23. a. $\log_2(5), \log_2(9)$

b. $\log_5\left(\sqrt{5} - 2\right)$

c. $0, \log_9(2)$ $\left(\text{alternatively}, x = 0, \dfrac{1}{2}\log_3(2)\right)$

d. 2

24. a. $\log_{11}(18) \approx 1.205$

b. $-\log_5(8) \approx -1.292$

c. $\dfrac{1}{2}\log_7(3) \approx 0.2823$

25. a. $x \le 2.096$ b. $x < 0.5693$

10.3 Exam questions

Note: Mark allocations are available with the fully worked solutions online.

1. D
2. C
3. $x = 3$

10.4 Graphs of exponential functions

10.4 Exercise

1. a. For $y = 3^x$, the range is R^+, and for $y = -3^x$, the range is R^-. The asymptote is $y = 0$ for both graphs.

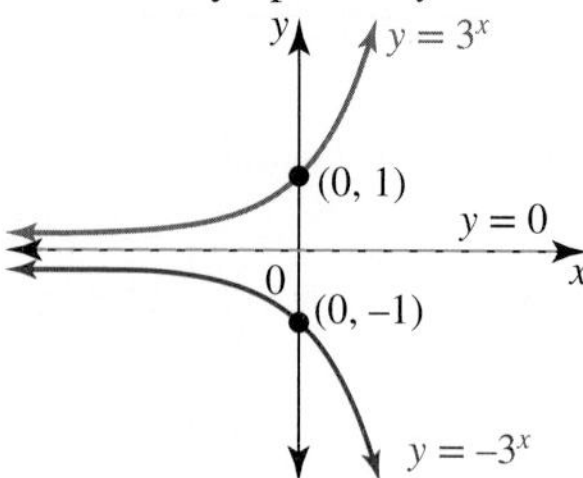

b. $y = 3^{-x}$

2. a. i.

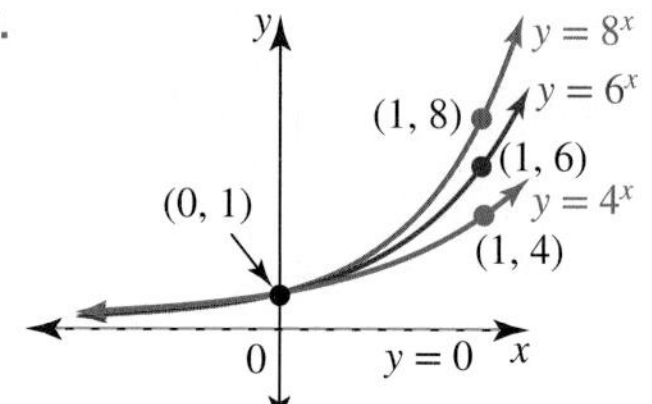

ii. As the base increases, the graphs become steeper (for $x > 0$).

b. i.

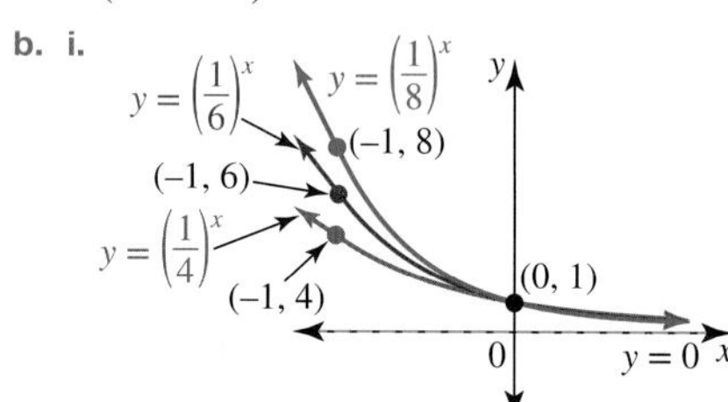

ii. $y = 4^{-x}; y = 6^{-x}; y = 8^{-x}$

3.

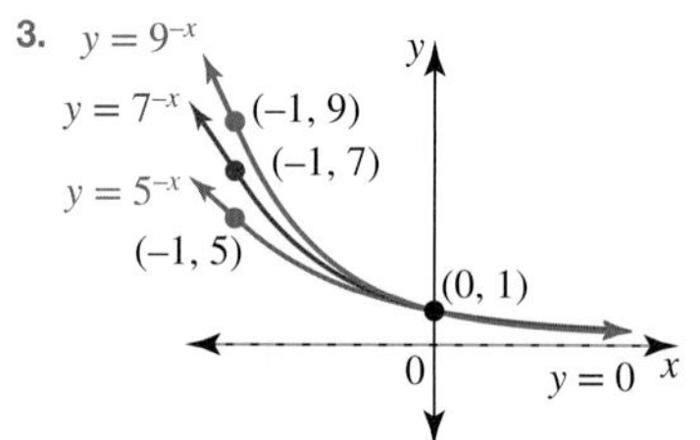

As the base increases, the graphs decrease more steeply (for $x < 0$).

4. a.

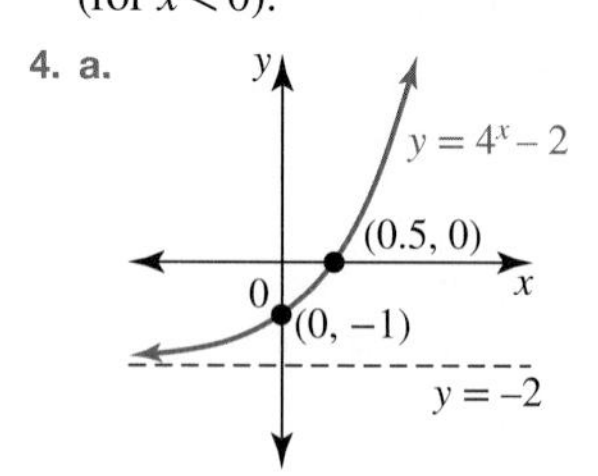

Range $(-2, \infty)$

b.

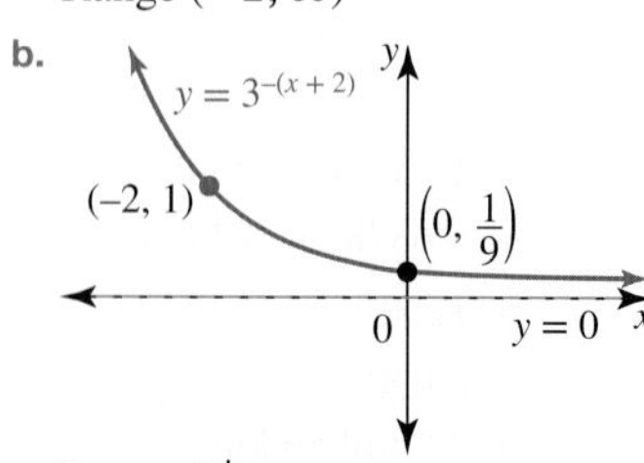

Range R^+

5.

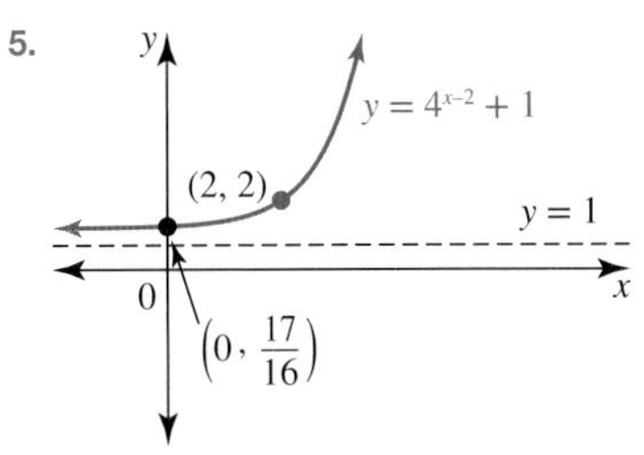

Range $(1, \infty)$

6. a. i. $y = -1$ ii. $(0, 0)$ iii. $(-1, \infty)$

b. i. $y = 3$ ii. $(0, 4)$ iii. $(3, \infty)$

c. i. $y = 2$ ii. $(0, -3)$ iii. $(-\infty, 2)$

d. i. $y = 0$ ii. $(0, 10)$ iii. $(0, \infty)$

7. a. $(0, -127)$ and $(14, 0)$

b. $(0, -9)$ and $\left(-\frac{1}{2}\log_5(10), 0\right)$

8.

	Asymptote	y-intercept	x-intercept	Point
a.	$y = 1$	$(0, 2)$	none	$(-1, 6)$
b.	$y = 1$	$(0, 0)$	$(0, 0)$	$(1, -3)$
c.	$y = -2$	$(0, -1)$	$(\log_{10}(2), 0)$	$(1, 8)$
d.	$y = 6.25$	$(0, 5.25)$	$(-2, 0)$	$(-1, 3.75)$

a.

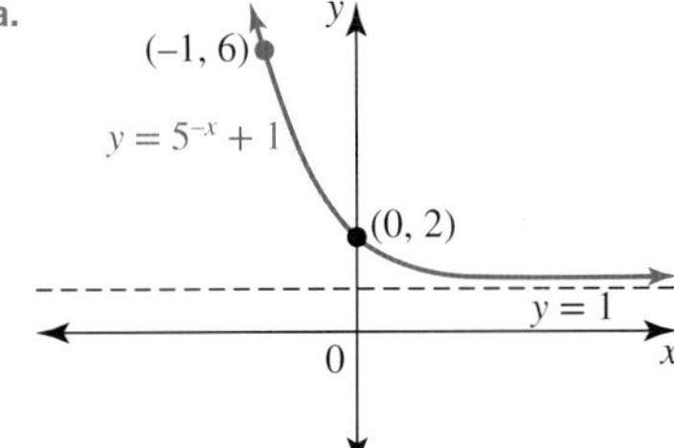

b.

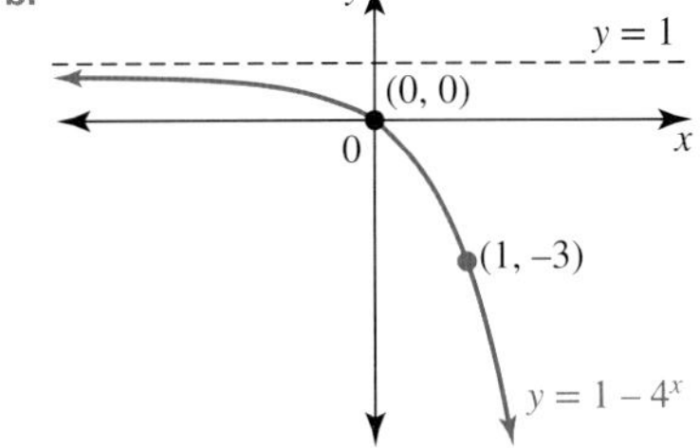

c.

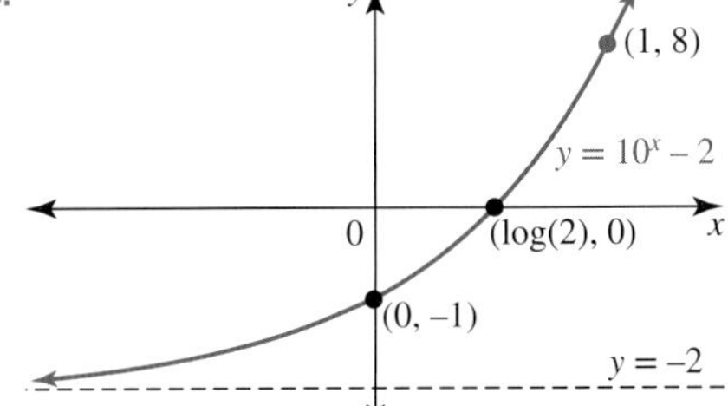

d.

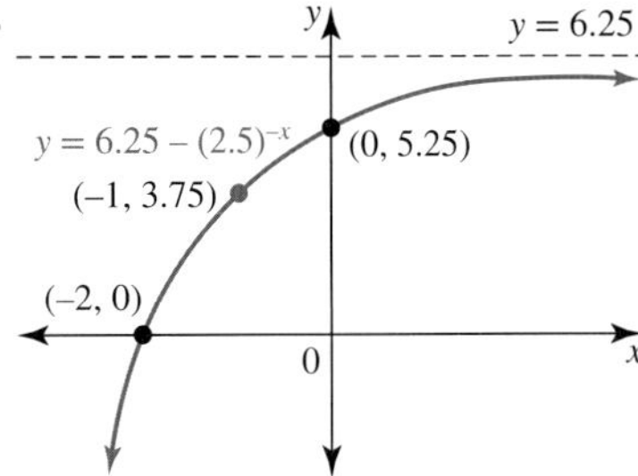

9. Each graph has an asymptote at $y = 0$.

a. $\left(0, \frac{1}{4}\right)$, $(2, 1)$

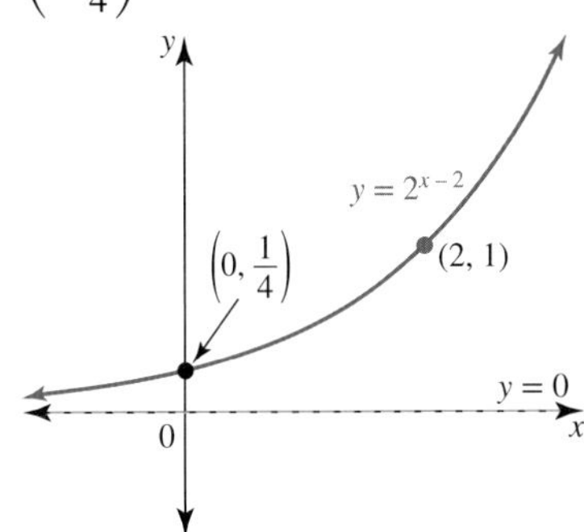

b. $(0, -9)$, $(-2, -1)$

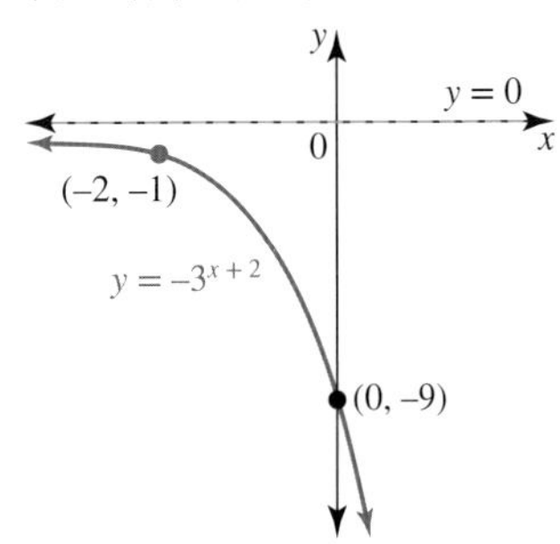

c. $\left(0, \frac{1}{2}\right)$, $\left(\frac{1}{2}, 1\right)$

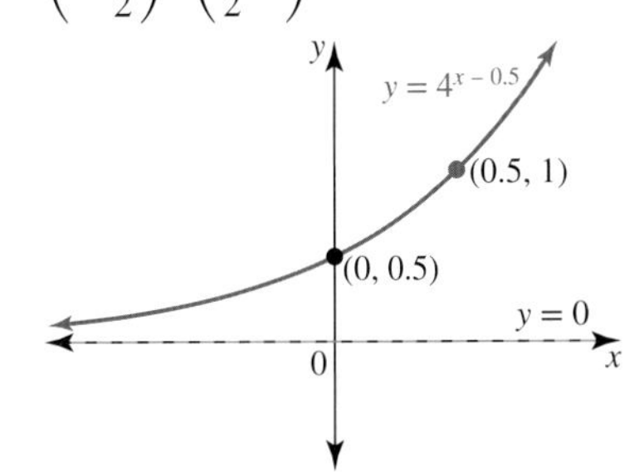

d. $(0, 7)$, $(1, 1)$

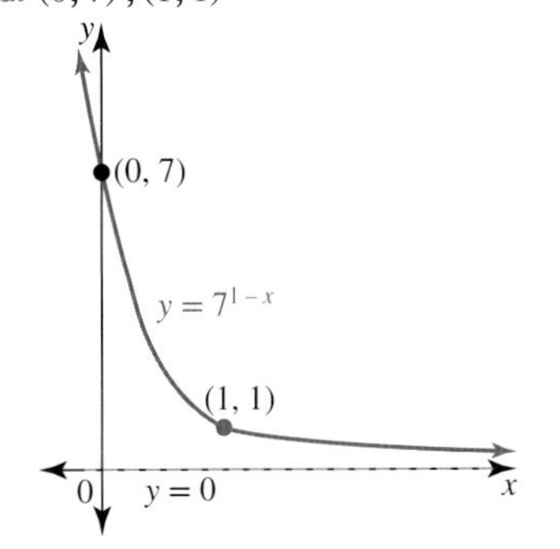

10. a.

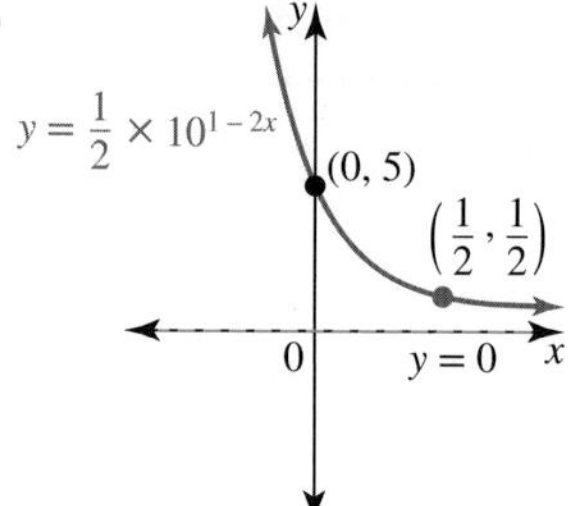

Range R^+

b.

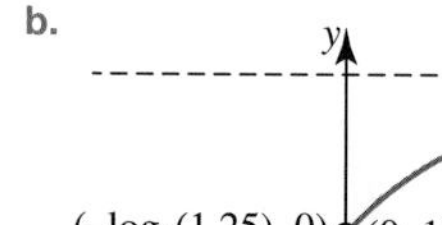

Range $(-\infty, 5)$

11. Each graph has an asymptote at $y = 0$.

a. $(0, 3), (1, 6)$

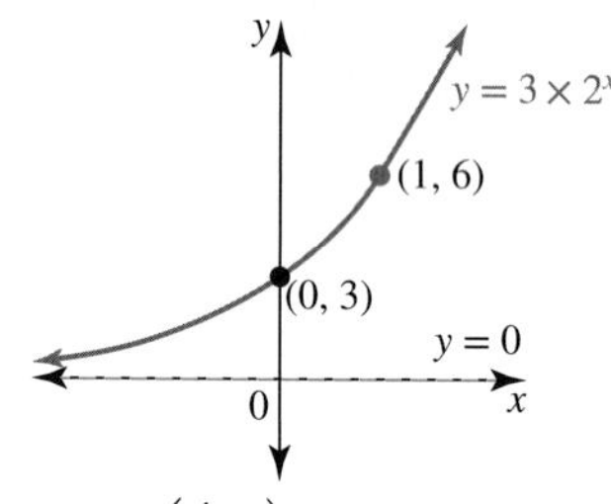

b. $(0, 1), \left(\frac{4}{3}, 2\right)$

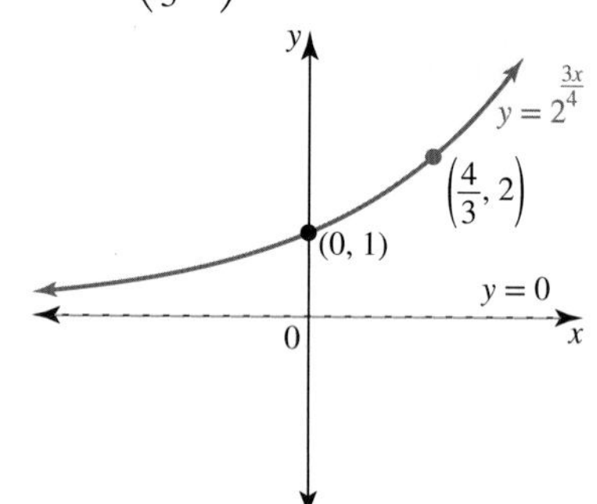

c. $(0, -3), \left(-\frac{1}{3}, -6\right)$

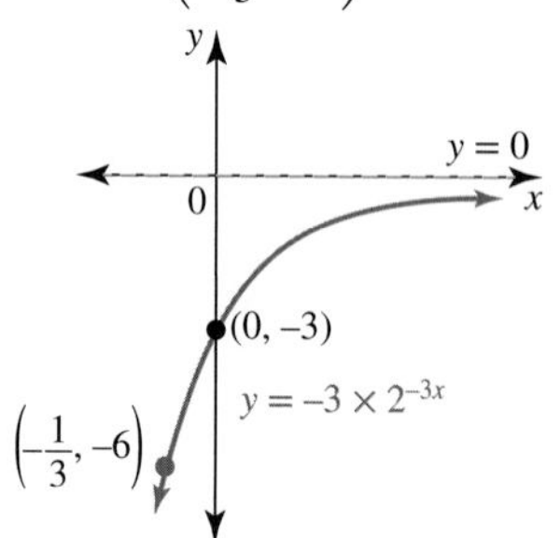

d. $(0, 1.5), (-2, 15)$

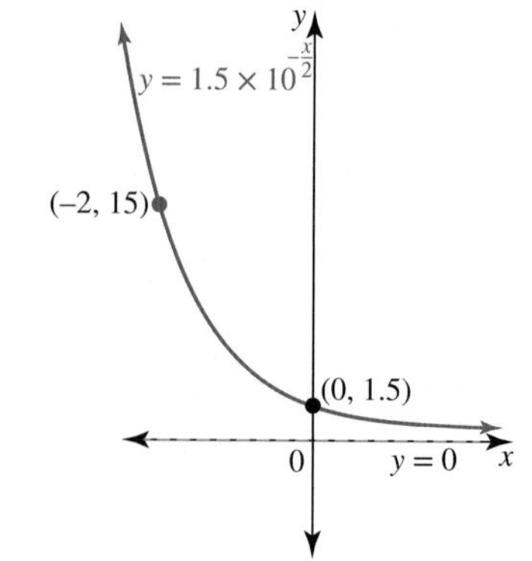

12.

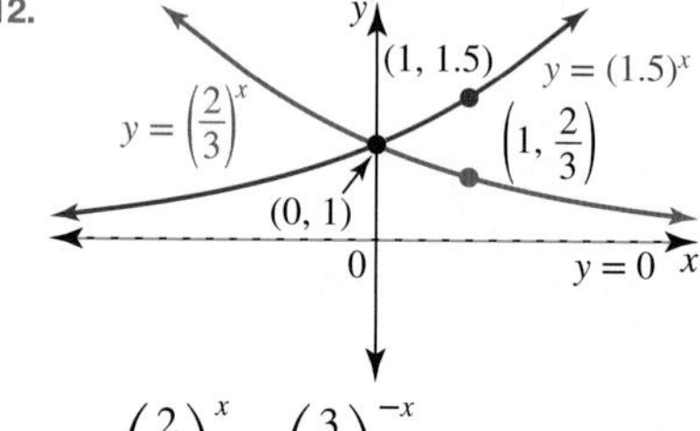

$$y = \left(\frac{2}{3}\right)^x = \left(\frac{3}{2}\right)^{-x} \text{ and } y = (1.5)^x = \left(\frac{3}{2}\right)^x$$

13. a. The graphs are identical with asymptote $y = 0$; y-intercept $(0, 1)$; point $(1, 9)$

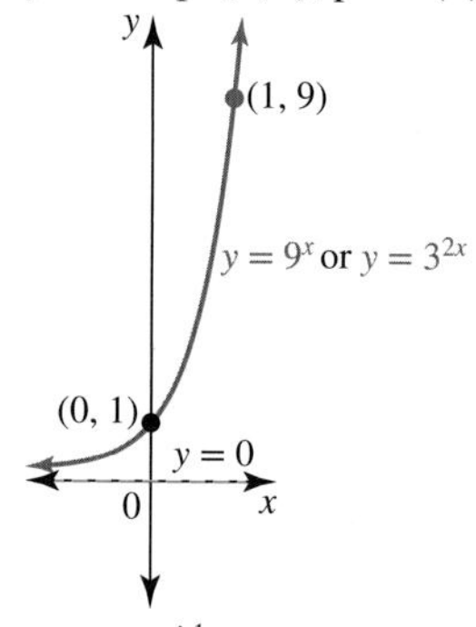

b. i. $y = 2^{x+1}$

ii.

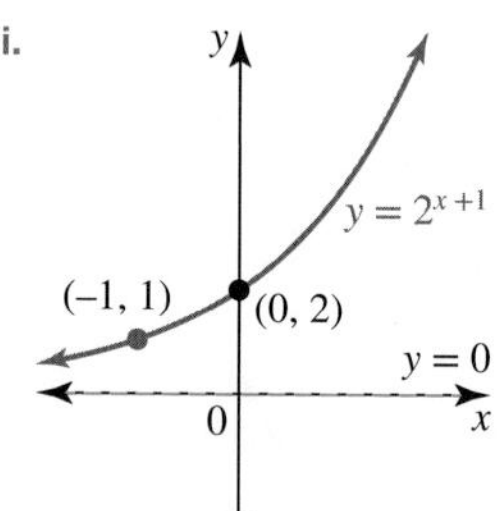

Asymptote $y = 0$; y-intercept $(0, 2)$; point $(-1, 1)$

14. a. $y = 2^x + 6$, domain R, range $(6, \infty)$

b. $y = 5^{\frac{1}{2}x} - 5$, domain R, range $(-5, \infty)$

c. $y = 10 - 10 \times 3^{-x}$, domain R, range $(-\infty, 10)$

15. $a = -2, b = 2$

16. a. $y = 2 \times 10^x + 3$

b. $y = 4 \times 3^{2x}$; asymptote $y = 0$

c. $y = 6 - 2 \times 3^{1-x}$

d. $y = 6 - 6 \times 3^{-x}$

17.

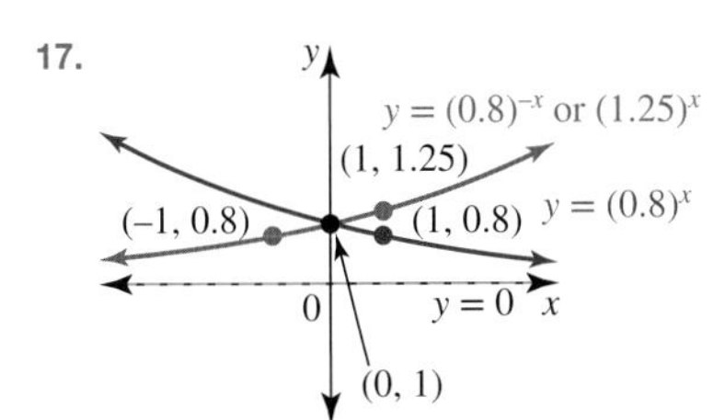

$y = (1.25)^x$ is the same as $y = 0.8^{-x}$. Reflecting these in the y-axis gives $y = 0.8^x$.

18.

	Asymptote	y-intercept	x-intercept	Range
a.	$y=-20$	$(0,-18)$	$(0.5,0)$	$(-20,\infty)$
b.	$y=-1$	$(0,9)$	$(3.3,0)$	$(-1,\infty)$
c.	$y=3$	$(0,1)$	$(-1,0)$	$(-\infty,3)$
d.	$y=-7$	$(0,0)$	$(0,0)$	$(-7,\infty)$
e.	$y=8$	$(0,7.2)$	$(0.7,0)$	$(-\infty,8)$
f.	$y=-4$	$(0,-4.2)$	none	$(-\infty,-4)$

a.

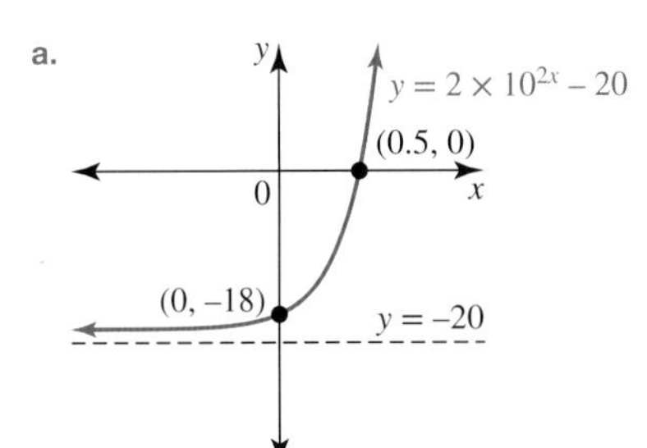

b.

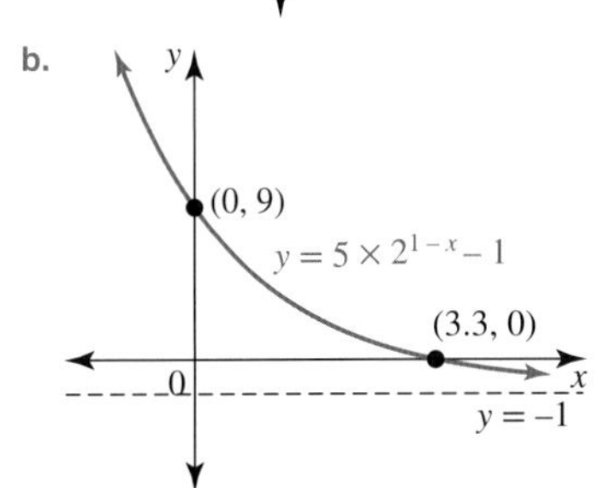

c.

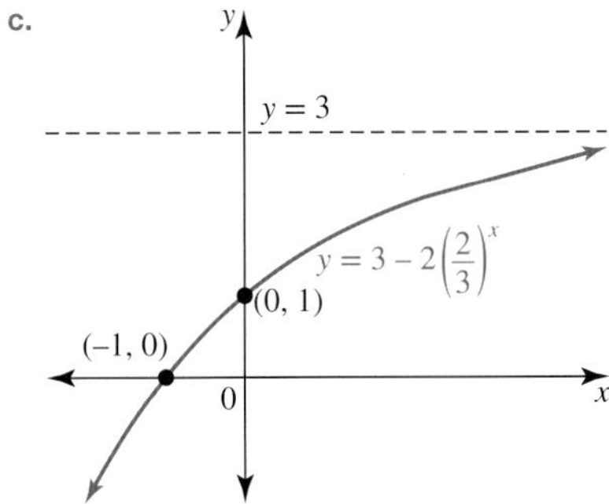

d.

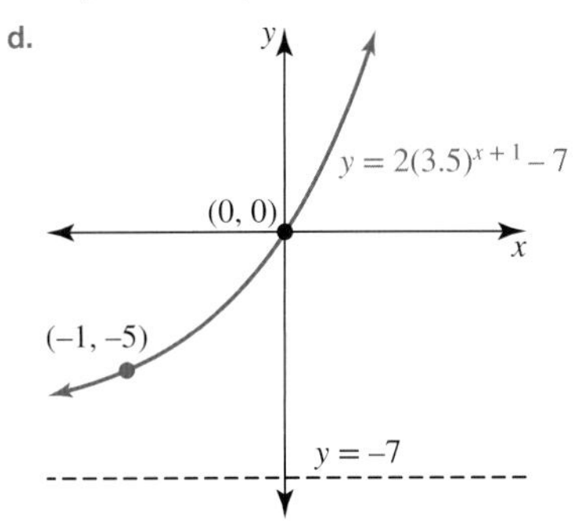

e.

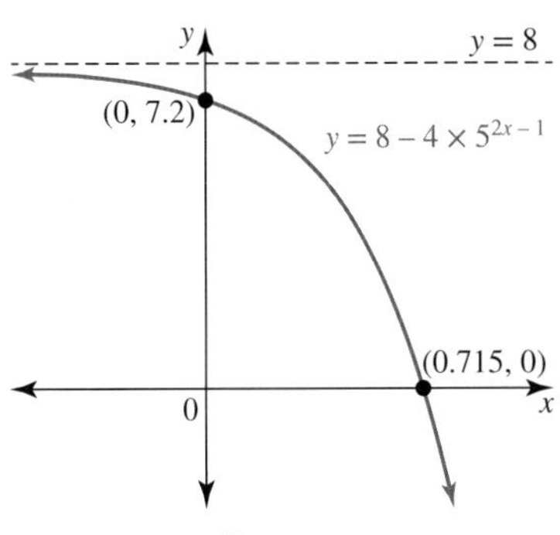

f.

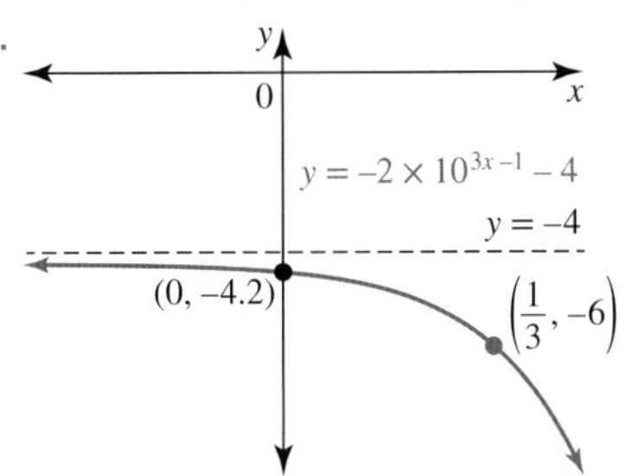

19. a. i. -3 ii. $3-3\sqrt{2}$

b. i. $x=3$ ii. $x=-1$ iii. Not possible

c.

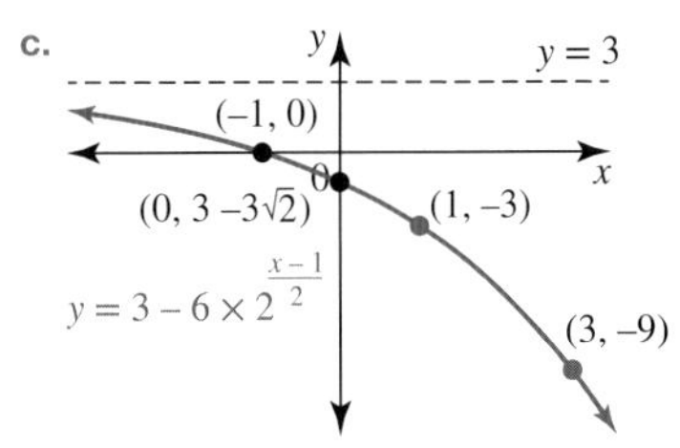

Asymptote $y=3$; range $(-\infty,3)$

d. $x \le -0.17$

20. a. One intersection for which $x \in (-1,-0.5)$

b. Three

c. One

d. None

e. One point of intersection $(1,6)$

f. Infinite points of intersection $\left(t, 2^{2t-1}\right), t \in R$

10.4 Exam questions

Note: Mark allocations are available with the fully worked solutions online.

1. E
2. B
3. Range R^+

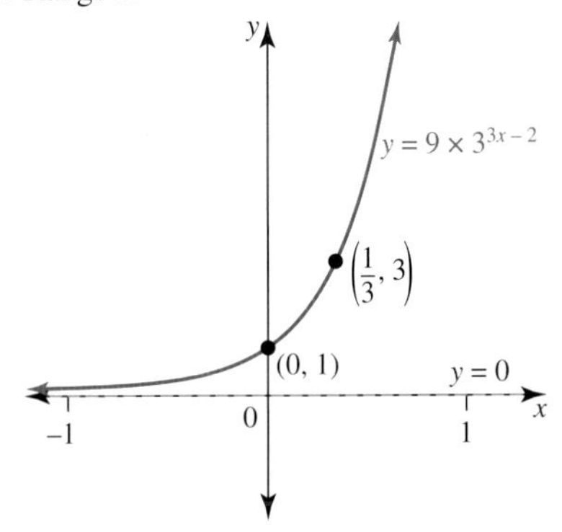

10.5 Applications of exponential functions

10.5 Exercise

1. a. $k = \frac{1}{5}$

   b. 10 years

2. a. 42 emails per day on average

   b. 16 weeks

3. a. 30 drosophilae

   b. 39 insects

   c. 14 days

   d.

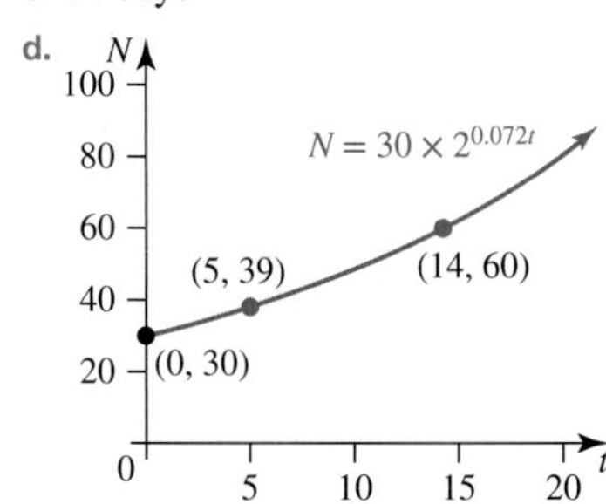

   e. After 25 days

4. a. Sample responses can be found in the worked solutions in the online resources.

   b. $k \approx 0.004$

   c. 5.3 kg

5. a. i. Sample responses can be found in the worked solutions in the online resources.

   ii. \$2030.19

   iii. 7.45 years

   b. 5.6%

6. a. 7 °C

   b. Approximately $\frac{1}{2}$ hour

   c.

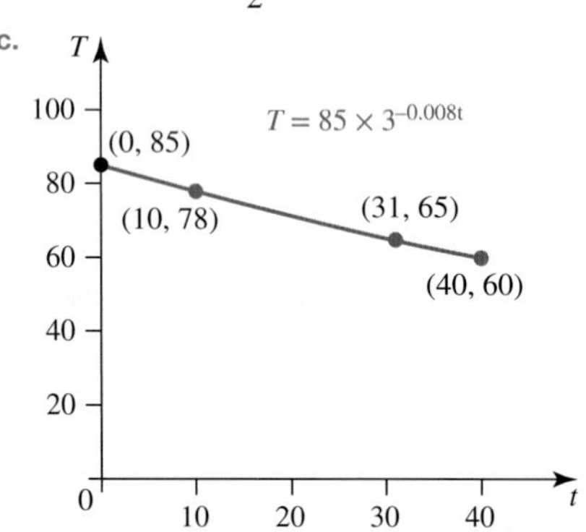

   d. It is unrealistic for the temperature to approach 0 °C.

7. a. $a = 70$

   b. 77.6 °C

   c. 4 minutes

   d.

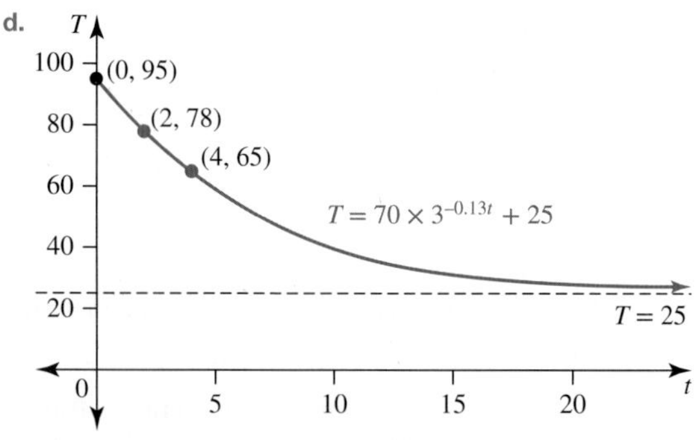

   The temperature approaches 25 °C.

8. a. $k \approx 0.054$

   b. $P_0 \approx 101.317$

   c. 55.71 kilopascals; 76.80 kilopascals

   d.

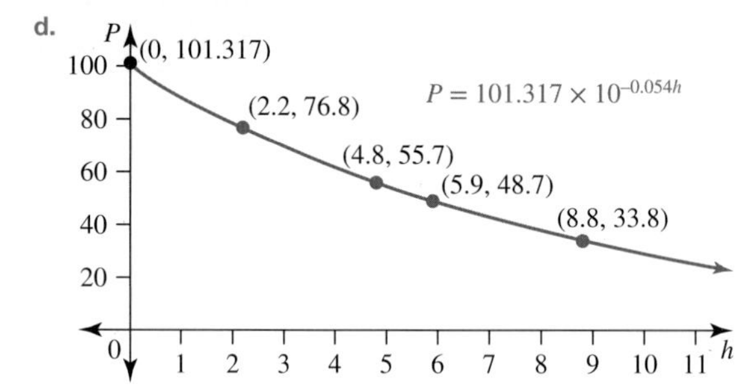

9. a. $D_0 \times 10^{15k} = 15$ and $D_0 \times 10^{18k} = 75$

   b. Sample responses can be found in the worked solutions in the online resources.

   c. 229 birds per square kilometre

   d. Not lower than 39 birds per square kilometre

10. a. $k = \frac{1}{5730}$ b. 1540 years old

11. a. Sample responses can be found in the worked solutions in the online resources.

    b. Sample responses can be found in the worked solutions in the online resources.

    c. Approximately 50 years

    d. The model supports the claim.

12. a. i. 16 ii. 51.5

    b. i. 80.20

    ii. 55.61 seconds

    iii. 7.08; 0.10

    c. i. $a = 0.10$

    ii.

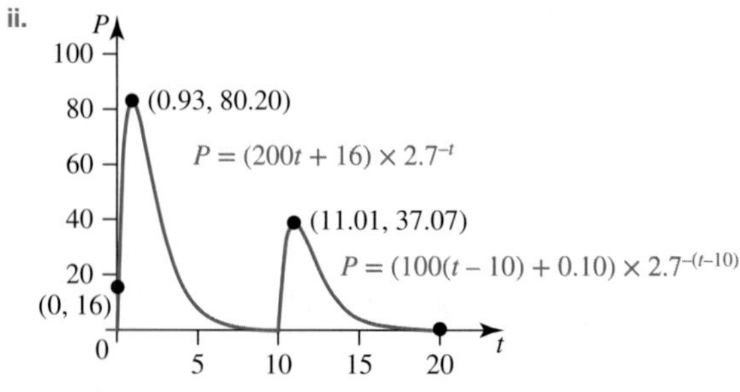

    Maximum turning points at $(0.93, 80.2), (11.01, 37.07)$

10.5 Exam questions

Note: Mark allocations are available with the fully worked solutions online.

1. C
2. B
3. B

10.6 Inverses of exponential functions

10.6 Exercise

1. a. $y = 10^x$

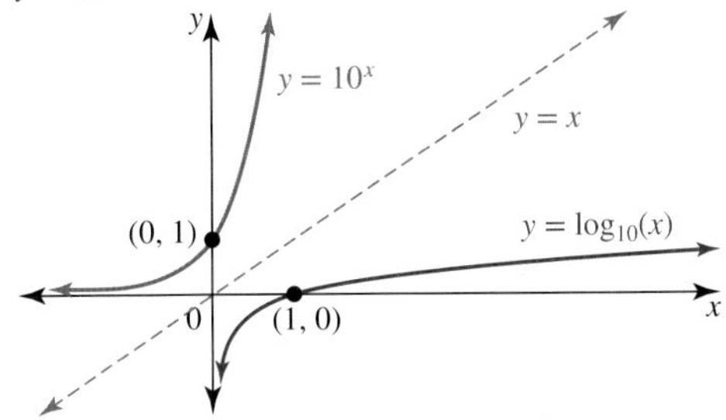

b. $\log_{10}(1000) - \log_{10}(100) = \log_{10}(10)$ and $3 - 2 = 1$

2. a.

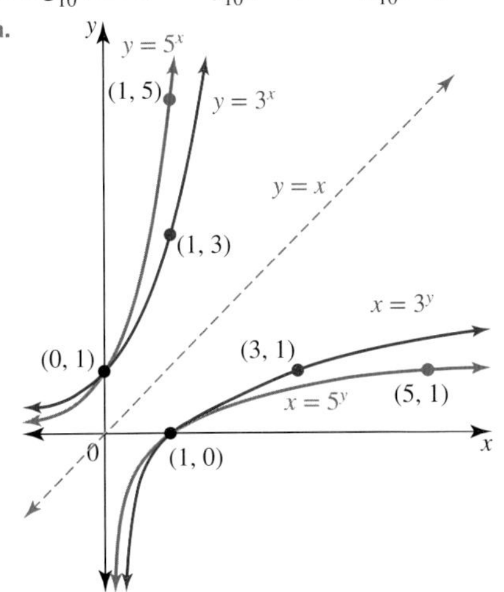

b. $y = \log_3(x), y = \log_5(x)$

c. The larger the base, the more slowly the graph increases for $x > 1$.

3. a.

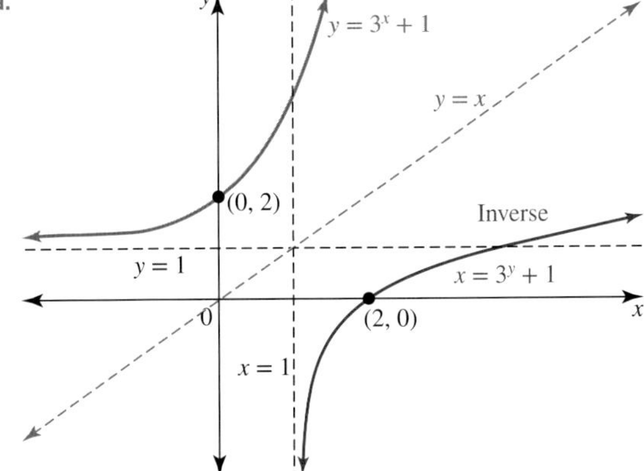

b. $y = \log_3(x - 1)$

4.

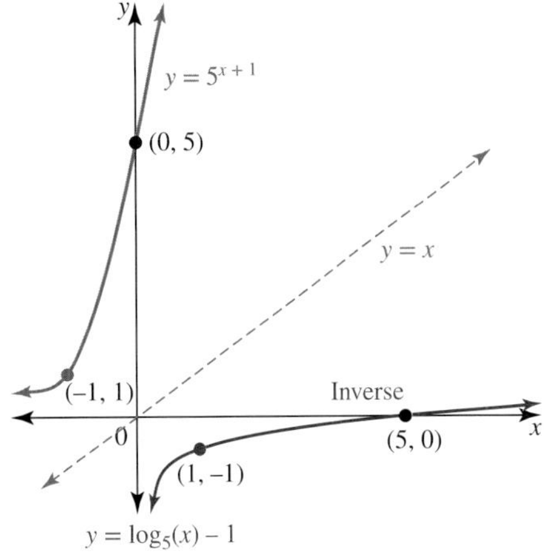

5. Inverse $y = -\log_2(x)$

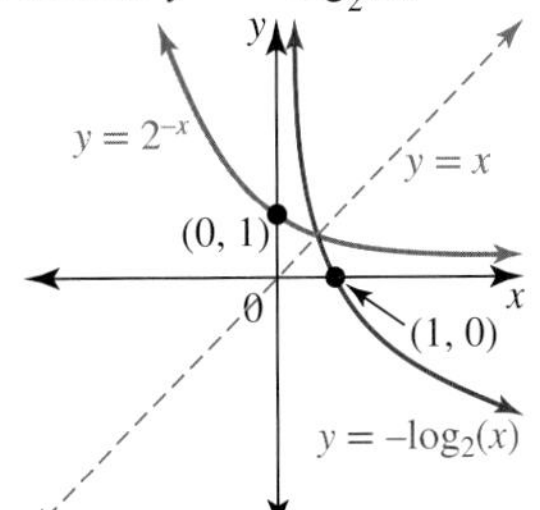

6. a. $(-\infty, 4)$

b. The rule is $f^{-1}(x) = \dfrac{1}{3}\log_2(4 - x)$; the mapping is

$$f^{-1}: (-\infty, 4) \rightarrow R,\ f^{-1}(x) = \frac{1}{3}\log_2(4 - x).$$

c. y, y = 4, (0, 3), y = x, f(x), x = 4, $\left(0, \frac{2}{3}\right)$, $f^{-1}(x)$, 0, $\left(\frac{2}{3}, 0\right)$, (3, 0), x

7. a. Domain $(-\infty, 8)$; range R

b. 6

c. $(6, 0)$

d. The rule is $f^{-1}(x) = \dfrac{1}{2}\log_3\left(\dfrac{8 - x}{2}\right)$; the mapping is

$$f^{-1}: (-\infty, 8) \rightarrow R,\ f^{-1}(x) = \frac{1}{2}\log_3\left(\frac{8 - x}{2}\right).$$

8. a. $2x$

b. 0.001

9. a. Asymptote $x = 1$; domain $(1, \infty)$; x-intercept $(2, 0)$

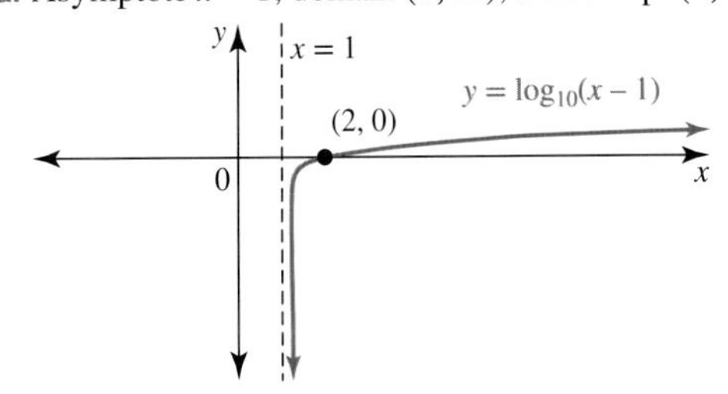

b. Asymptote $x = 0$; domain R^+; x-intercept $(5, 0)$

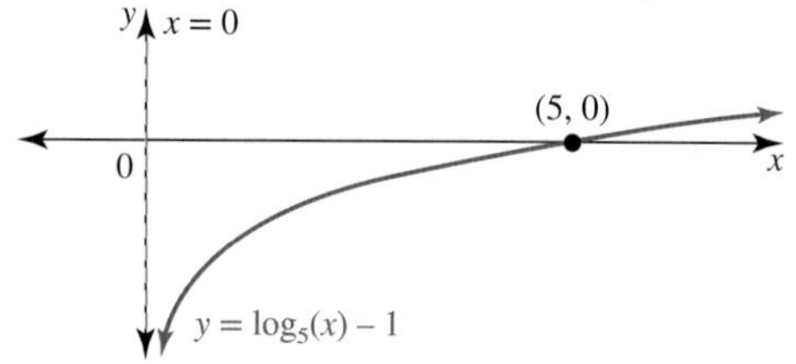

c. i. $b = 1$

ii. Domain R; range $(-1, \infty)$, $f^{-1}(x) = 2^{-x} - 1$

iii. Asymptote $y = -1$, contains the origin

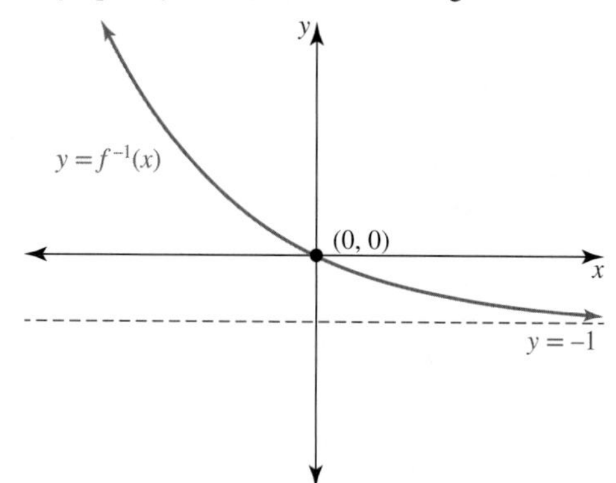

10. a. $(-6, \infty)$, $x = -6$
b. $(6, \infty)$, $x = 6$
c. $(-\infty, 6)$, $x = 6$
d. $(0, \infty)$, $x = 0$
e. $(-\infty, 0)$, $x = 0$
f. $(0, \infty)$, $x = 0$

11. a. $(-9, \infty)$
b. $x = -9$
c. $(0, 2)$
d. $(-8, 0)$
e.

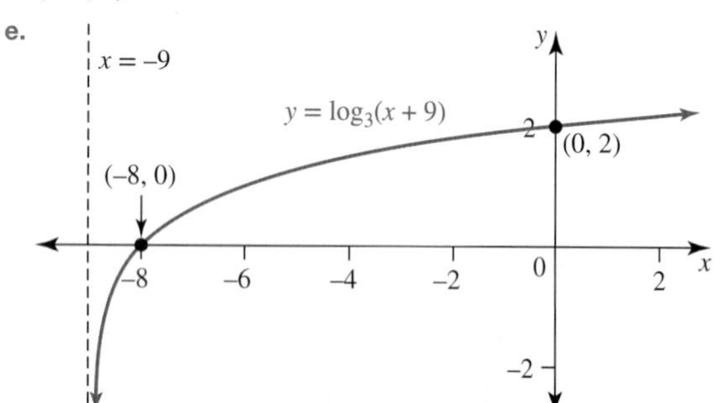

f. R

12. a. $(5, \infty)$
b. $x = 5$
c. $(6, 0)$, no y-intercept
d. and e.

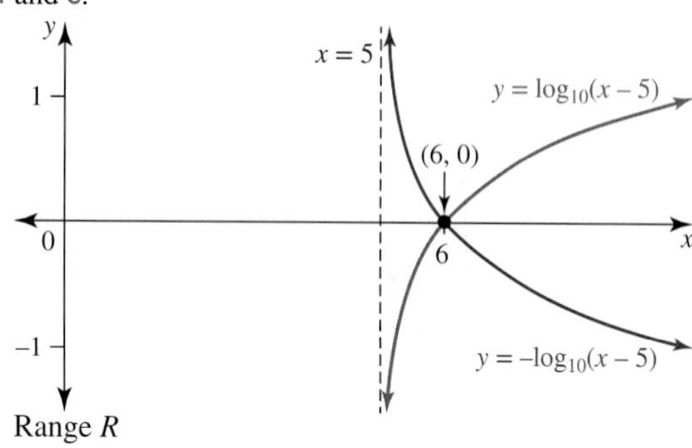

Range R

13. The range is R for each graph.

	Domain	Asymptote equation	Axis intercepts
a.	R^+	$x = 0$	$(25, 0)$
b.	$(2, \infty)$	$x = 2$	$(3, 0)$
c.	R^+	$x = 0$	$(0.1, 0)$
d.	$(-1, \infty)$	$x = -1$	$(0, 0)$
e.	$(-\infty, 4)$	$x = 4$	$(3, 0), (0, \log_3(4))$
f.	$(-4, \infty)$	$x = -4$	$(-3, 0), (0, -2)$

a.

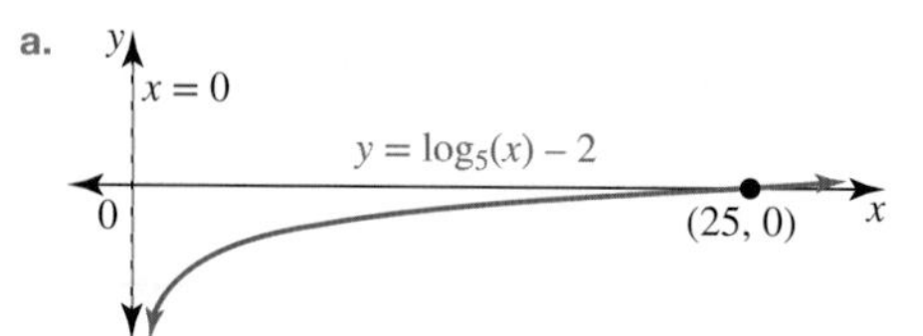

b.

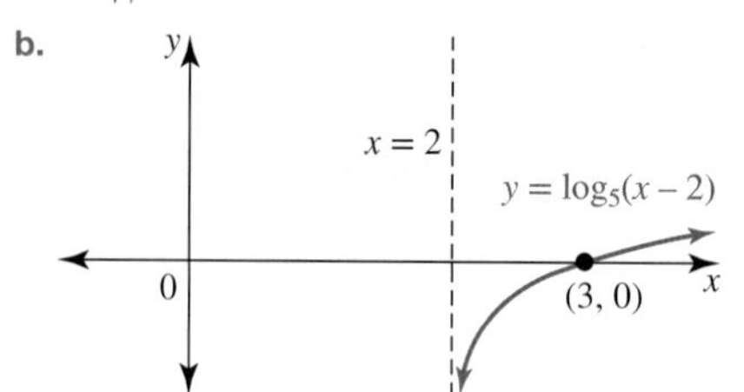

c.

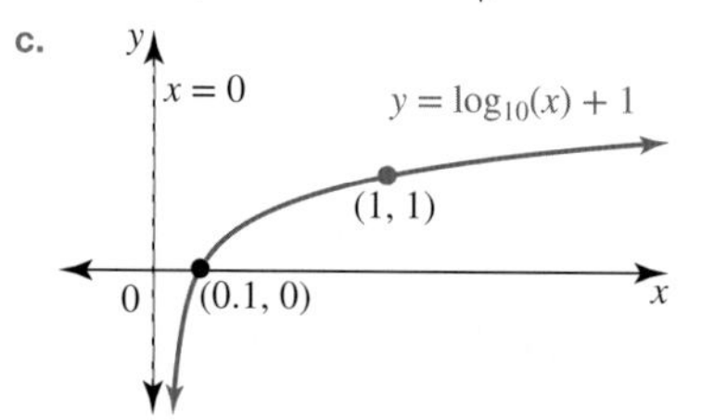

d.

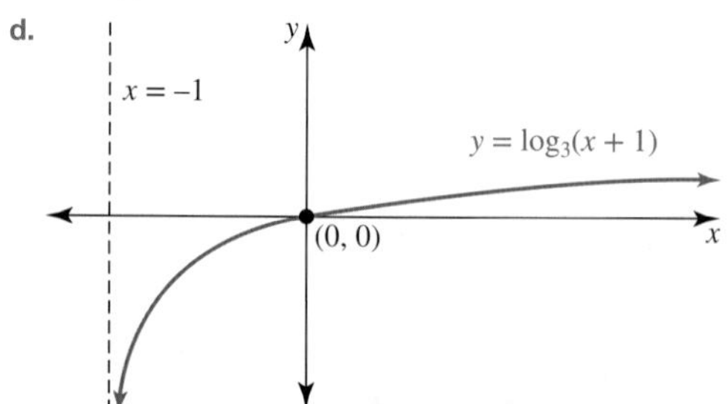

e.

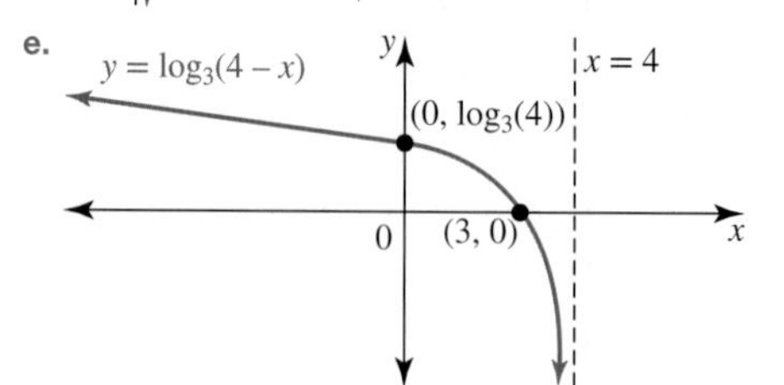

f.

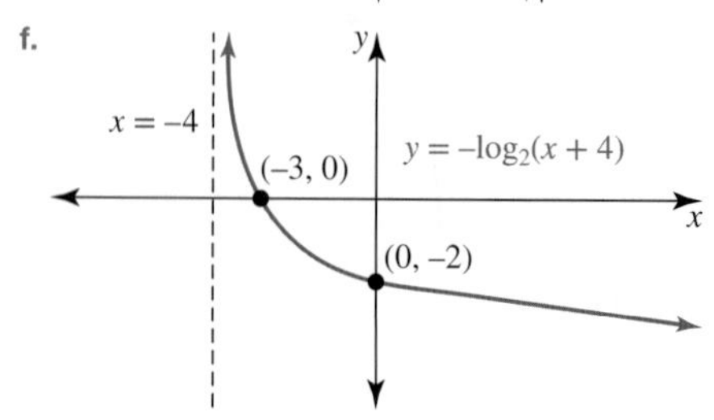

14. Vertical translation of 1 unit up

15. $\frac{1}{3} \times 2^x$

16. a. i. 8 ii. 6 iii. 0.5 iv. 5

b. i. x ii. x^3 iii. $3x$ iv. x

17. a. $y = 14\log_7\left(\frac{x}{2}\right)$

b. i. $y = -4\log_3(x) + 4$

ii. $(0, 3)$

c. i. $y = -\log_2(x + 2) - 1$

ii. $y = 2^{-(x+1)} - 2$

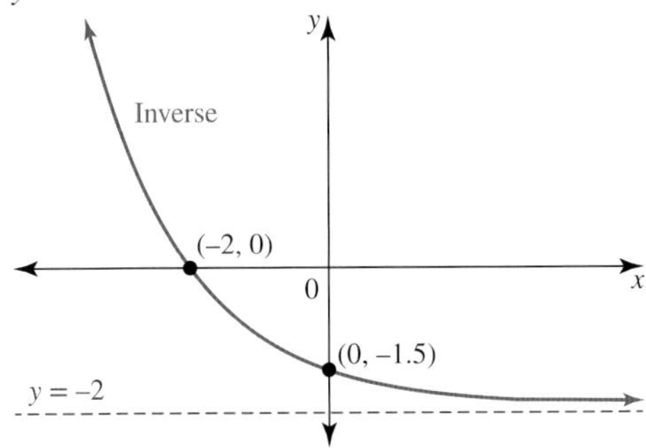

18. a. $d_f = \left(-\frac{9}{4}, \infty\right)$; $d_g = (-\infty, 20)$

b. $x = -\frac{9}{4}$, $x = 20$

c. f: $(-2, 0), (0, 2)$; g: $(10, 0), \left(0, \frac{1}{2}\right)$

d. Sketch using CAS technology.

19. a. $y = -3\log_2(x)$

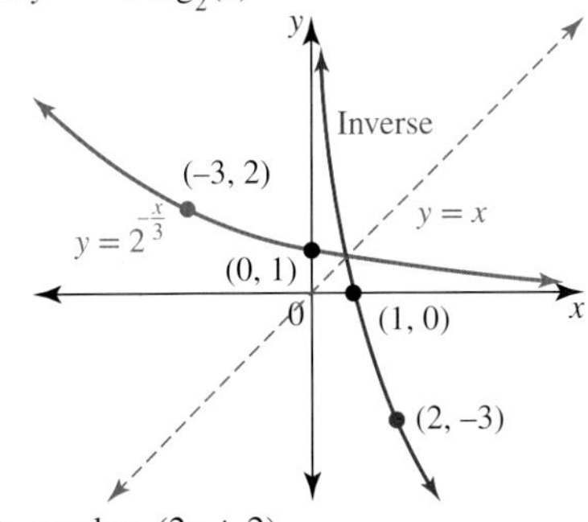

b. $y = \log_8(2x + 2)$

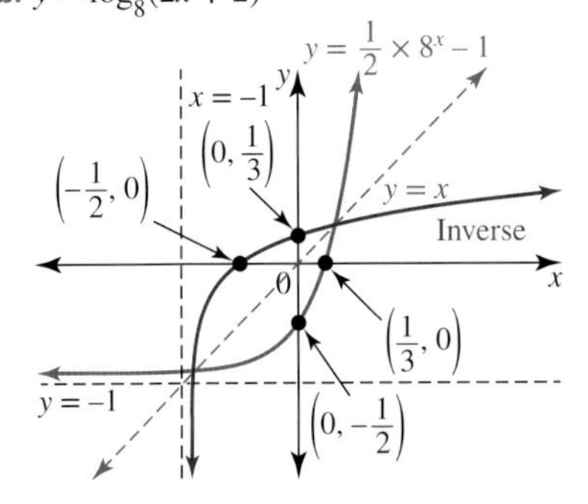

c. $y = \log_4(2 - x)$

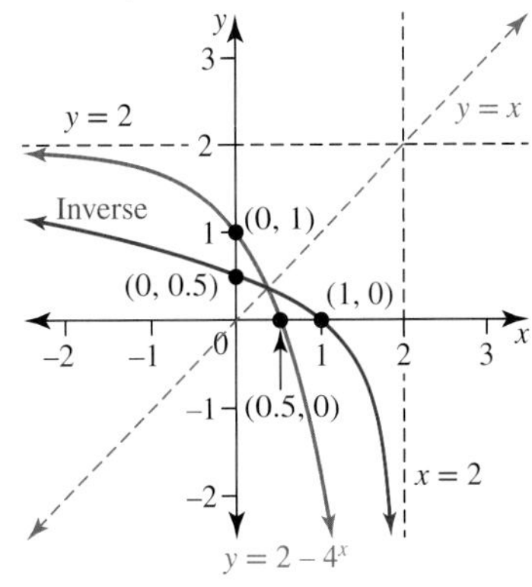

d. $y = \log_3(x - 3) - 1$

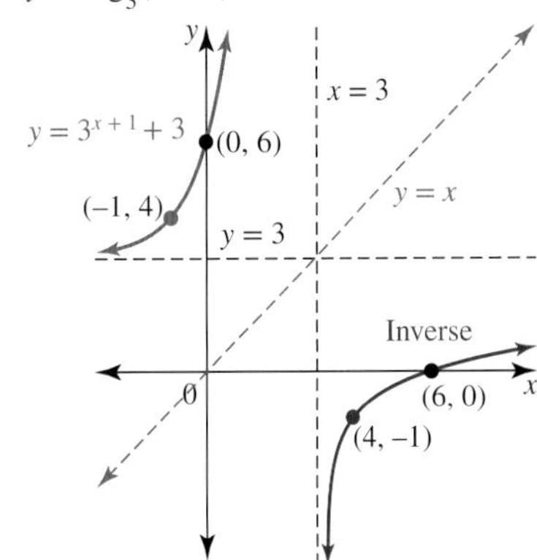

e. $y = -\frac{1}{2}\log_{10}(-x)$

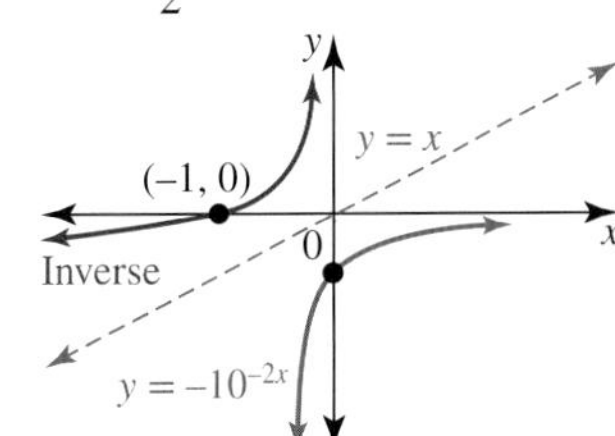

f. $y = 1 - \log_2(x)$

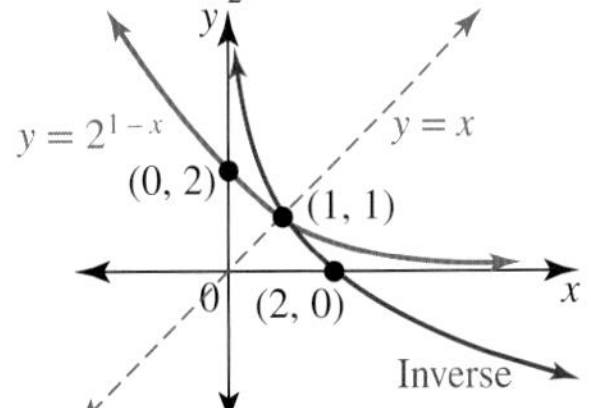

20. a. $y = 2^{2x+1} - 4$

b. $y = \frac{1}{2}\log_2(x + 4) - \frac{1}{2}$

c. Asymptote: $x = -4$; intercepts: $(-2, 0), (0, 0.5)$

d. Two points of intersection

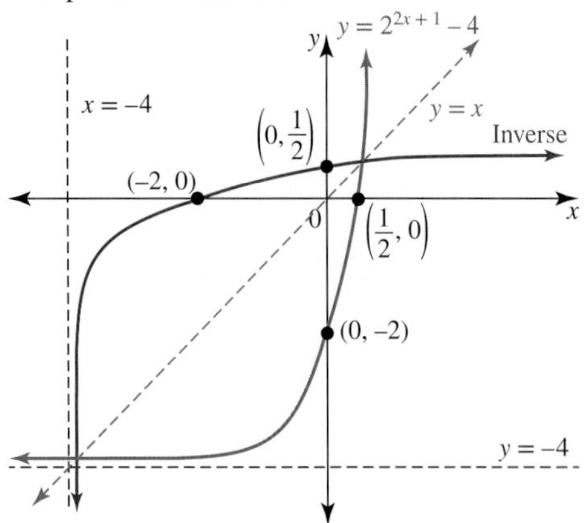

e. $k = 14$

f. Sample responses can be found in the worked solutions in the online resources.

10.6 Exam questions

Note: Mark allocations are available with the fully worked solutions online.

1. C
2. B
3.

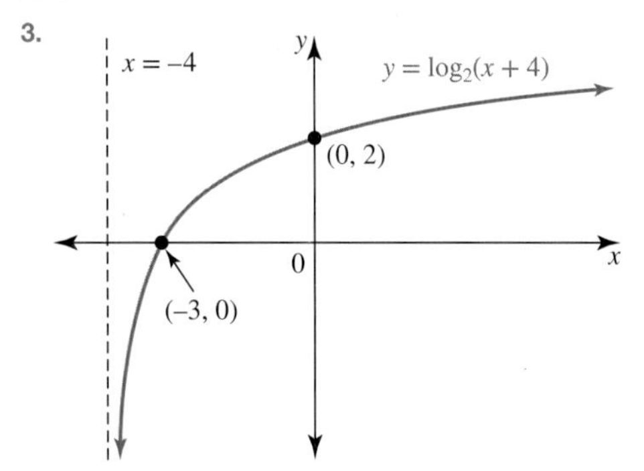

Domain $(-4, \infty)$

10.7 Review

10.7 Exercise

Technology free: short answer

1. a. $\dfrac{a}{9b^2}$ b. 2^{n-1} c. $\dfrac{a-1}{a}$ d. $\dfrac{3^5}{2^{\frac{7}{4}}}$

2. a. 2 b. 0 c. 3 d. $-\dfrac{1}{2}$

3. a. $x = \dfrac{3}{5}$ b. $x = 3$
   c. $x = 3$ d. $x = 0.1, 10$

4. a. $x = 1 - \log_3(7)$
   b. $x = \dfrac{\log(5) + \log(2)}{\log(5) - \log(2)} = \dfrac{1}{\log(2.5)}$
   c. $x > \dfrac{\log(3)}{\log(0.2)}$
   d. $x = \log_{10}(6)$

5.

	Asymptote	Intercepts with axes and other points	Domain	Range
a.	$y = -1$	$(0, 0), \left(\frac{1}{3}, 1\right)$	R	$(-1, \infty)$
b.	$y = 0$	$\left(0, -\frac{2}{3}\right), (1, -2)$	R	R^-
c.	$y = 5$	$(0, 4), (-1, 0)$	R	$(-\infty, 5)$
d.	$y = 0$	$\left(0, \frac{6}{5}\right), (-1, 3)$	R	R^+

a.

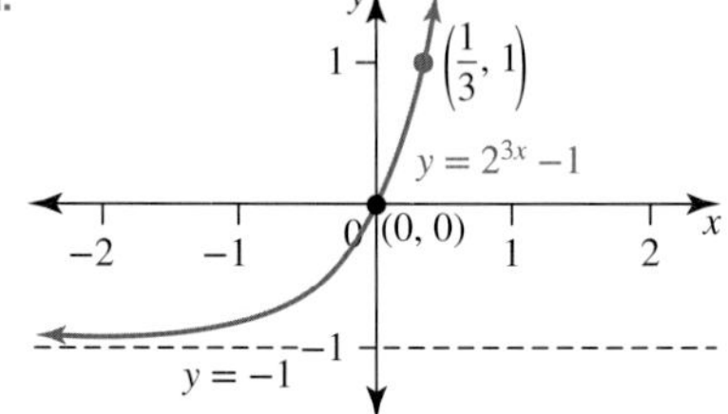

b.

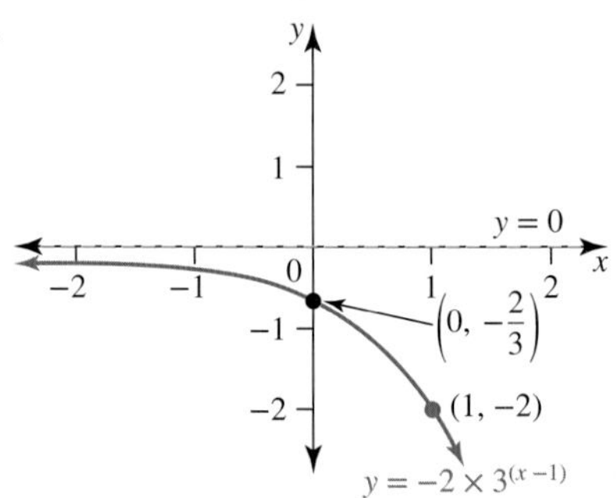

c.

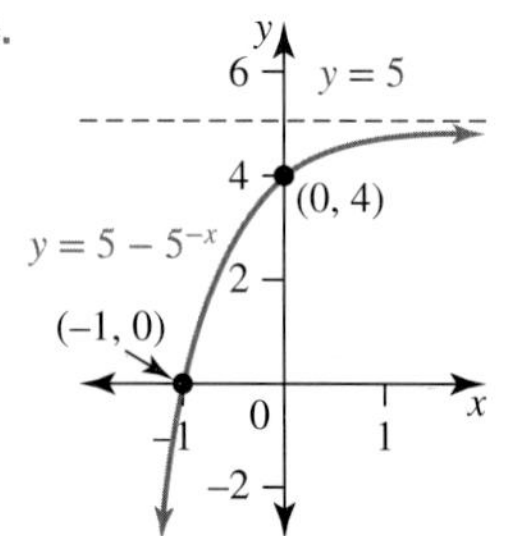

d.

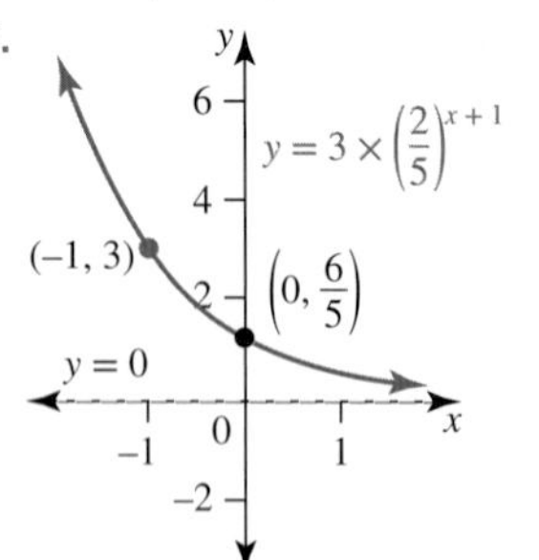

6. a. Reflection in x-axis; translation 3 units to the left; asymptote $x = -3$

b. $y = 3^{-x} - 3$; domain R; range $(-3, \infty)$

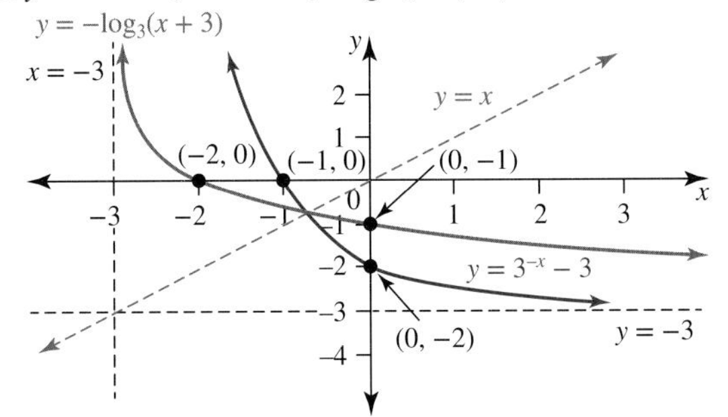

c. Dilation factor 5 from x-axis; reflection in y-axis; translation 1 unit to right; asymptote $y = 0$

d. $f^{-1}: R^+ \rightarrow R,\ f^{-1}(x) = 1 - \log_7\left(\frac{x}{5}\right)$

Technology active: multiple choice

7. B 8. D 9. E 10. C 11. D
12. C 13. A 14. B 15. E 16. C

Technology active: extended response

17. a. $b = -1, c = 2$

b. $a = \frac{1}{2}, c = 2$

c.

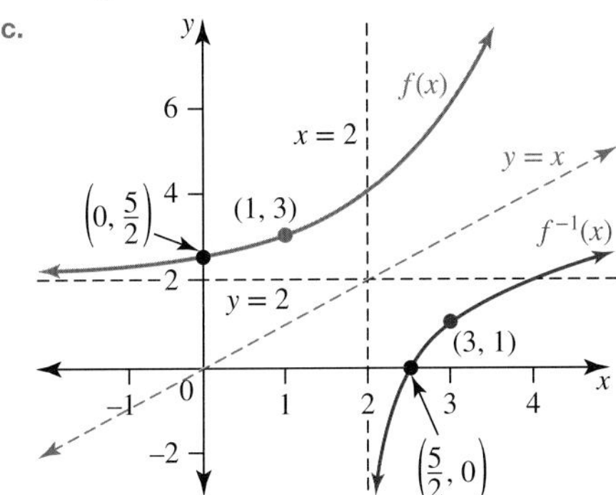

Domain $(2, \infty)$

d. $f^{-1}(x) = \log_2(x - 2) + 1$ or $f^{-1}(x) = \log_2(2x - 4)$

e. i. $A(3, 6)$ ii. $B(34, 6)$ iii. $P(3, 1)$

f. 77.5 square units

18. a. $a = 100$

b. $k \approx 0.02$

c. 50 °C

d. 10.39 am

e.

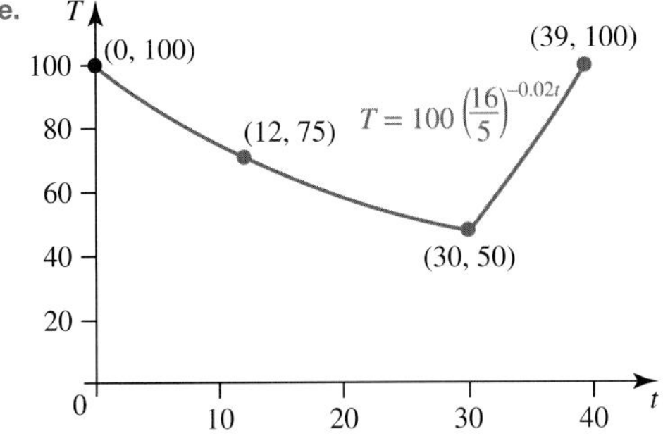

f. 23 minutes

19. a. Given $T(2.4) = 2T(1.2)$

$$a \times 2^{2.4k} = 2 \times a \times 2^{1.2k}$$
$$2^{2.4k} = 2 \times 2^{1.2k}$$
$$2^{2.4k} = 2^{1.2k+1}$$
$$2.4k = 1.2k + 1$$
$$1.2k = 1$$
$$\frac{6}{5}k = 1$$
$$k = \frac{5}{6}$$

b. 80 minutes

c. i. 0.18 grams/litre

ii.

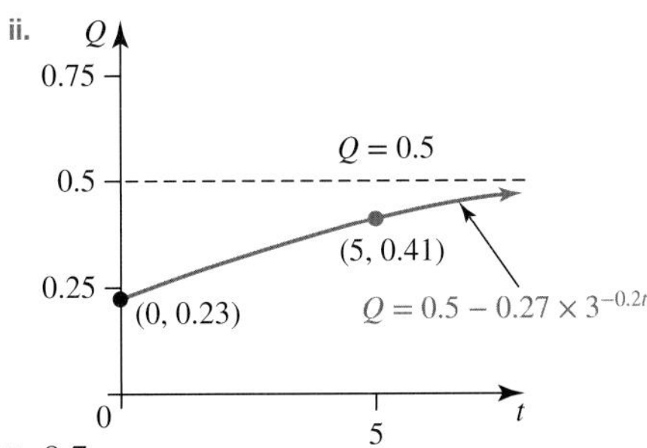

iii. 8.7 years

iv. 0.5 grams/litre

d. i. 1×10^{-5} watts/square metre

ii. 73 dB

20. a. i. $(x + 2)^{\frac{1}{2}}(x + 1)$

ii. $-2(4x^2 + 1)^{-\frac{3}{2}}(2x - 1)^{-2}(2x + 1);\ x = -\frac{1}{2}$

b. i. $\frac{a^2 + ab + b^2}{a^2b^2}$

ii. $2\sqrt{mn}$; 2

c. i. Show that $\log(n!) = \log(2) + \log(3) + ... + \log(n)$, $n \in N$.

$$\begin{aligned} \text{LHS} &= \log(n!) \\ &= \log(n \times (n-1) \times (n-2) \times ... \times 3 \times 2 \times 1) \\ &= \log(n) + \log(n-1) + \log(n-2) + ... \\ &\quad + \log(3) + \log(2) + \log(1) \\ &= \log(n) + \log(n-1) + \log(n-2) + ... \\ &\quad + \log(3) + \log(2) + 0 \\ &= \log(2) + \log(3) + ... + \log(n) \\ &= \text{RHS} \end{aligned}$$

ii. 1

d. $p(10) \approx 4$; four primes smaller than 10; $p(30) \approx 9$, but there are 10 primes smaller than 30

10.7 Exam questions

Note: Mark allocations are available with the fully worked solutions online.

1. D
2. C
3. E
4. $a = 4, b = \frac{1}{3}, y = 4\log_9\left(\frac{1}{3}x\right)$
5. C

11 Introduction to differential calculus

LEARNING SEQUENCE

Fully worked solutions for this topic are available online.

11.1 Overview

Hey students! Bring these pages to life online

Watch videos

Engage with interactivities

Answer questions and check results

Find all this and MORE in jacPLUS

11.1.1 Introduction

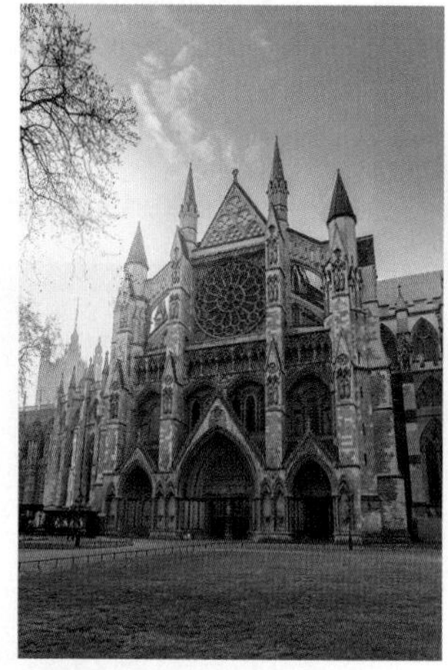

'If I have seen further [than others] it is by standing on the shoulders of giants.' So said Sir Isaac Newton, acknowledging the work of those who came before him and who paved the way for him to invent calculus and much else. Only in the 20th century did Einstein come close to rivalling Newton's genius in physics and mathematics.

Newton was educated at Cambridge University, rising to be appointed in 1669 to the Lucasian Chair of Mathematics, a prestigious position also held more recently by the late Stephen Hawking. Earlier, the bubonic plague of 1665–66 shut the university down, forcing Newton to return home. This was a blessing in disguise, as forced seclusion drove Newton to form his theory of gravitation and his method of fluxions, now called differential calculus. Newton's *Principia Mathematica* (1687) contains all his greatest work and remains the most admired and respected scientific publication of all time. An image of a page from the *Principia* was carried on the 1977 Voyager space mission as an illustration of the intellectual capacity of the human race. The State Library of Victoria possesses a first edition of the work.

Independently, unaware of what Newton had already accomplished, Gottfried Wilhelm Leibniz also invented calculus. As he was the first to publish his work, this led to controversy between the two men over who first invented it. Leibniz had huge admiration for Newton's achievements, so the dispute should have been settled amicably; instead it became political and descended to become acrimonious and unedifying. The subsequent lack of collaboration between British and European mathematicians lasted long after Newton's death. Newton died in 1727 and is buried in Westminster Abbey.

KEY CONCEPTS

This topic covers the following key concepts from the VCE Mathematics Study Design:

- average and instantaneous rates of change in a variety of practical contexts and informal treatment of instantaneous rate of change as a limiting case of the average rate of change
- use of gradient of a tangent at a point on the graph of a function to describe and measure instantaneous rate of change of the function, including consideration of where the rate of change is positive, negative or zero, and the relationship of the gradient function to features of the graph of the original function
- informal treatment of the gradient of the tangent to a curve at a point as a limit, and the limit definition of the derivative of a function $f'(x) = \lim_{h\to 0} \dfrac{f(x) - f(x-h)}{h} = \lim_{h\to 0} \dfrac{f(x+h) - f(x)}{h}$
- the central difference approximation $f'(x) \approx \dfrac{f(x+h) - f(x-h)}{2h}$ and its graphical interpretation
- the derivative as the gradient of the graph of a function at a point and its representation by a gradient function, and as a rate of change
- differentiation of polynomial functions by rule.

11.2 Rates of change

LEARNING INTENTION

At the end of this subtopic you should be able to:
- calculate average and instantaneous rates of change
- estimate the gradient of a curve.

A rate of change measures the change in one variable relative to another. An oil spill may grow at the rate of 20 square metres per day; the number of subscribers to a newspaper may decrease at the rate of 50 people per month. In practical situations, the rate of change is often expressed with respect to time.

11.2.1 Gradients as rates of change

A linear function has a straight-line graph whose gradient, or slope, is $m = \frac{y_2 - y_1}{x_2 - x_1}$. This gradient measures the rate at which the y-values change with respect to the change in the x-values. The gradient measures the rate of change of the function.

The rate of change of a linear function is a constant: either the function increases steadily if $m > 0$ or it decreases steadily if $m < 0$. The function neither increases nor decreases if $m = 0$.

The rate of change of any function is measured by the gradient or slope of its graph. However, the gradient of a non-linear function is not constant; it depends at what point on the graph, or at what instant, it is to be measured.

We know how to find the gradient of a linear function. Learning how to measure the gradient of a non-linear function is the aim of this topic.

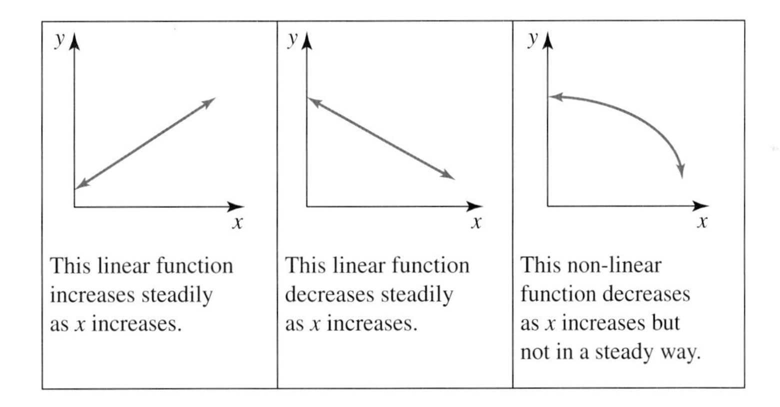

Average rate of change

A car journey of 240 km completed in 4 hours may not have been undertaken with a steady driving speed of 60 km/h. The speed of 60 km/h is the average speed for the journey found from the calculation: $\text{average speed} = \frac{\text{distance travelled}}{\text{time taken}}$. It gives the average rate of change of the distance with respect to the time.

The diagram shows a graph of the distance travelled by a motorist over a period of 4 hours.

A possible scenario for this journey is that slow progress was made for the first hour (O–A) due to city traffic congestion. Once on a freeway, good progress was made, with the car travelling at the constant speed limit on cruise control (A–B); then after a short break for the driver to rest (B–C), the journey was completed with the car's speed no longer constant (C–D).

The average speed of the journey is the gradient of the line segment joining the origin $(0, 0)$ and D $(4, 240)$, the end points for the start and finish of the journey.

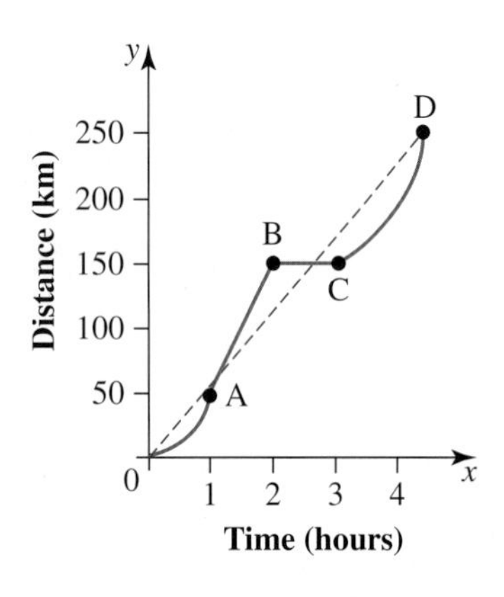

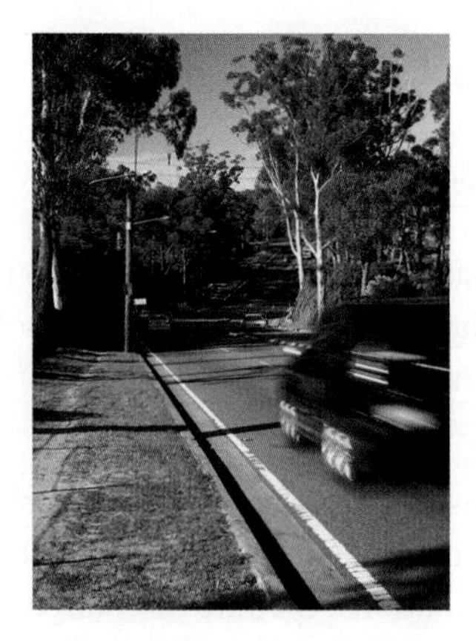

For any function $y = f(x)$, the **average rate of change** of the function over the interval $x \in [a,\ b]$ is calculated as $\dfrac{f(b) - f(a)}{b - a}$. This is the gradient of the line joining the end points of the interval.

Average rate of change

For the function $y = f(x)$,

$$\textbf{average rate of change} = \frac{f(b) - f(a)}{b - a} \textbf{ for } x \in [a, b].$$

WORKED EXAMPLE 1 Calculating the average rate of change

a. A water tank contains 1000 litres of water. The volume decreases steadily and after 4 weeks there are 900 litres of water remaining in the tank. Show this information on a diagram and calculate the rate at which the volume is changing with respect to time.

b. The profit made by a company in its first 4 years of operation can be modelled by the function p for which $p(t) = \sqrt{t}$, where the profit is $p \times \$10\,000$ after t years. Sketch the graph of the function over the time interval and calculate the average rate of change of the profit over the first 4 years.

THINK	WRITE
a. 1. Define the variables and state the information given about their values.	a. Let volume be V litres at time t weeks. Given: when $t = 0$, $V = 1000$ and when $t = 4$, $V = 900$. So the points (0, 1000) and (4, 900) are the end points of the volume–time graph.

2. Interpret the keyword 'steadily' and sketch the given information on a graph.

As the volume decreases steadily, the graph of V against t is linear and the rate of change is constant.

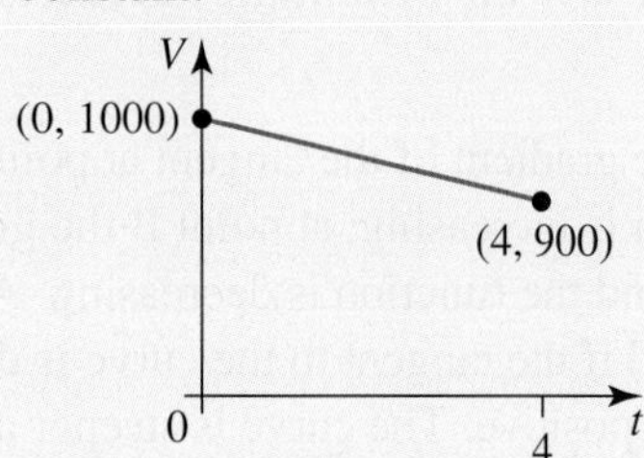

3. Calculate the rate of change of the volume with respect to time.

The rate of change of the volume is the gradient of the graph. It is measured by $\frac{\text{change in volume}}{\text{change in time}}$, so:

$$\begin{aligned}\frac{\text{change in volume}}{\text{change in time}} &= \frac{V(4) - V(0)}{4 - 0} \\ &= \frac{900 - 1000}{4} \\ &= -25\end{aligned}$$

4. Write the answer in context using appropriate units.
Note: The negative value for the rate represents the fact the volume is decreasing. The units for the rate are the volume unit per time unit.

The volume of water is decreasing at a rate of 25 litres/week.

b. 1. Sketch the graph of the function over the given time interval.
Note: The graph is not linear, so the rate of change is not constant.

b. $p(t) = \sqrt{t}$
When $t = 0$, $p = 0$.
When $t = 4$, $p = 2$.
(0, 0) and (4, 2) are the end points of the square root function graph.

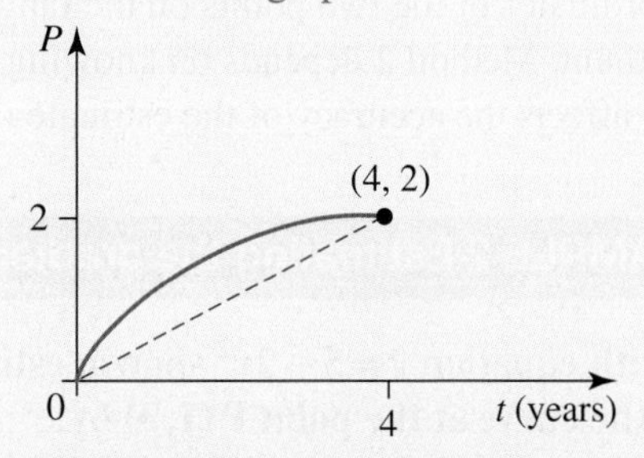

2. Calculate the average rate of the function over the time interval.

The average rate is measured by the gradient of the line joining the end points.

$$\begin{aligned}\text{Average rate of change of } p &= \frac{p(4) - p(0)}{4 - 0} \\ &= \frac{2 - 0}{4} \\ &= 0.5\end{aligned}$$

3. Calculate the answer.

Since the profit is $p \times \$10\,000$, the profit is increasing at an average rate of \$5000 per year over this time period.

11.2.2 Instantaneous rate of change

The gradient at any point on the graph of a function is given by the gradient of the tangent to the curve at that point. This gradient measures the **instantaneous rate of change**, or rate of change, of the function at that point or at that instant.

For the curve shown, the gradient of the tangent at point A is positive and the function is increasing; at point B the gradient of the tangent is negative and the function is decreasing. At point C, the curve is rising and if the tangent to the curve is drawn at C, its gradient would be positive. The curve is steeper at point A than at point C as the tangent at point A has a more positive gradient than the tangent at point C. The instantaneous rate of change of the function at A is greater than the instantaneous rate of change at C.

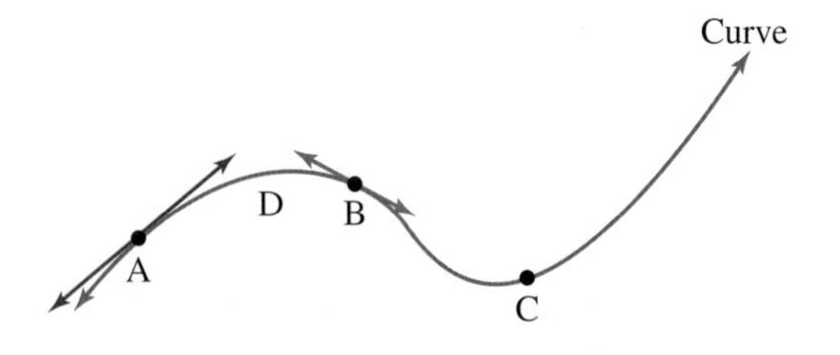

At the turning point D, the function is neither increasing nor decreasing and the tangent is horizontal, with a zero gradient. The function is said to be stationary at that instant.

- The rate of change of a function is measured by the gradient of the tangent to its curve
- For non-linear functions, the rate of change is not constant and its value depends on the point, or instant, at which it is evaluated.

Estimating the gradient of a curve

Two approximation methods that could be used to obtain an estimate of the gradient of a curve at a point P are as follows:

Method 1: Draw the tangent line at P and obtain the coordinates of two points on this line in order to calculate its gradient.

Method 2: Calculate the average rate of change between the point P and another point on the curve that is close to the point P.

The accuracy of the estimate of the gradient obtained using either of these methods may have limitations. In method 1 the accuracy of the construction of the tangent line, especially when drawn 'by eye', and the accuracy with which the coordinates of the two points on the tangent can be obtained will affect the validity of the estimate of the gradient. Method 2 depends on knowing the equation of the curve for validity. The closeness of the two points also affects the accuracy of the estimate obtained using this method.

WORKED EXAMPLE 2 Estimating the gradient at a point

For the curve with equation $y = 5 - 2x^2$ shown, estimate the gradient of the curve at the point $P(1, 3)$ by:

a. constructing a tangent at P and calculating its gradient

b. choosing a point on the curve close to P and calculating the average rate of change between P and this point.

THINK	WRITE
a. 1. Draw a tangent line that just touches the curve at P.	a. The tangent at P is shown. 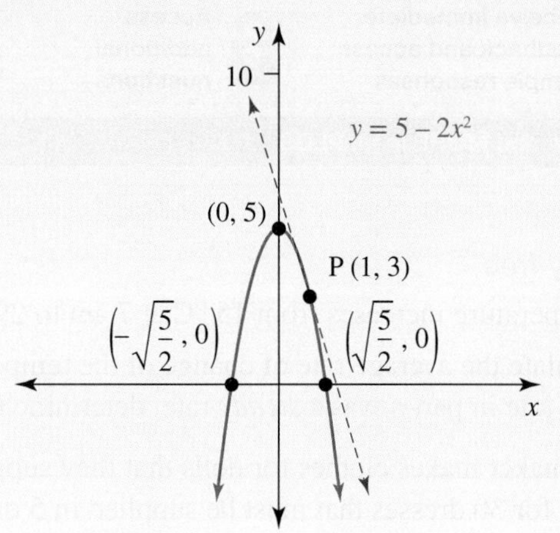
2. Determine the coordinates of two points that lie on the tangent. *Note:* The second point may vary and its coordinates may lack accuracy.	The tangent contains point P (1, 3). A second point is estimated to be (0, 7), the y-intercept.
3. Calculate the gradient using the two points.	$m = \frac{7-3}{0-1}$ $= -4$
4. State the answer. *Note:* Answers may vary.	The gradient of the curve at point P is approximately equal to −4.
b. 1. Choose a value for x close to the x-value at point P and use the equation of the curve to calculate the corresponding y-coordinate. *Note:* Other points could be chosen.	b. A point close to P (1, 3) could be the point for which $x = 0.9$. Equation of curve: $y = 5 - 2x^2$ If $x = 0.9$, $y = 5 - 2(0.9)^2$ $= 5 - 2 \times 0.81$ $= 3.38$ The point (0.9, 3.38) lies on the curve close to point P.
2. Calculate the average rate of change between the two points.	Average rate of change $= \frac{3.38-3}{0.9-1}$ $= -3.8$
3. State the answer. *Note:* Answers may vary.	The gradient of the curve at point P is approximately equal to −3.8.

on Resources

Interactivity Rates of change (int-5960)

11.2 Exercise

Students, these questions are even better in jacPLUS

Receive immediate feedback and access sample responses

Access additional questions

Track your results and progress

Find all this and MORE in jacPLUS

Technology free

1. The temperature increases from 15 °C at 7 am to 29 °C by 2 pm.
   a. Calculate the average rate of change of the temperature, in degrees Celsius per hour, over this time.
   b. If the rate in part a was a *steady* rate, determine the temperature at 9 am.
2. A dressmaker makes clothes for dolls that they supply to a retail outlet. The manager of the outlet gives them an order for 30 dresses that must be supplied in 5 days' time.
   a. Calculate the average rate of dresses per day that the dressmaker must maintain.
   b. If the dressmaker works each day from 9 am to 12 noon, calculate the average rate of dresses per hour that the dressmaker must maintain.
   c. On completion of the task, the dressmaker delivers the order by car. They drive a distance of 25 km in 30 minutes. Calculate their average speed of travel in km/h.
3. Two points A and B have coordinates $(2, 6)$ and $(-7, 12)$ respectively.
   a. Calculate the average rate of change between these two points.
   b. Calculate the gradient of the straight line that passes through both points A and B.
   c. If a function has a constant rate of change, state what type of function must it be.
   d. The point A also lies on the function with rule $y = 12 - 3x$. Determine the instantaneous rate of change of this function at the point A.
4. The diagram shows part of the graph of a function f passing through the points P, Q and R.

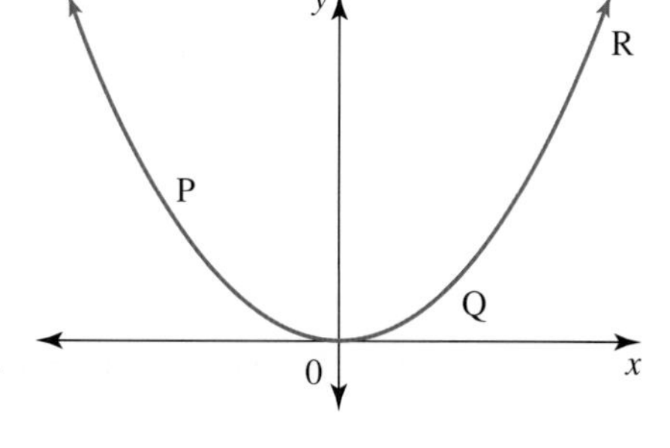

   a. Explain what the average rate of change between points P and R measures.
   b. State the pair of points between which the average rate of change is negative.
   c. The point S also lies on the function's graph. If the average rate of change between the points R and S is zero, describe the position of the point S.
   d. Explain how the instantaneous rate of change of the function at a point is measured.
   e. State at which of the points P, Q and R the instantaneous rate of change is negative.
   f. State at which of the points P, Q and R the instantaneous rate of change is the greatest.
5. A householder needs to purchase petrol for his lawnmower since both of his petrol containers are empty. One container is cylindrical in shape and it is filled to the top; the other container is more of a jerry can shape that is filled to just below the handle section as the diagram shows.

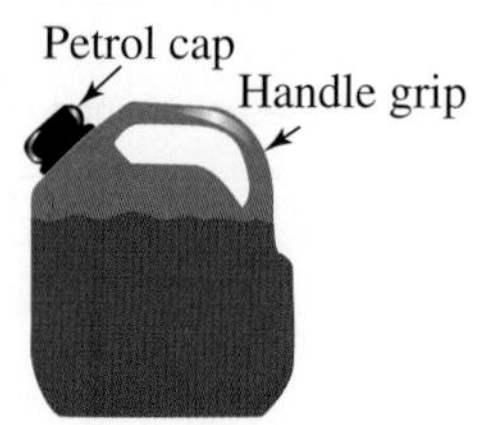

For each container, draw the graph of h versus t showing the rise in the depth h of petrol over time t as the man steadily fills each container.

6. Calculate the average rate of change of the function $f(x) = x^2 + 3$ over the interval for which $x \in [1, 4]$.

7. Calculate the constant speed of a car if in 1 hour 20 minutes it has driven a distance of 48 km.

8. **WE1** a. A Petri dish contains 40 bacterial cells that reproduce steadily so that after half an hour there are 100 cells in the dish. Show this information on a graph and calculate the rate at which the cells are reproducing with respect to time.

   b. After a review of its operations, part of a company's strategic plan is to improve efficiency in one of its administrative departments. The operating costs for this section of the company are modelled by the function c for which $c(t) = \frac{4000}{t+2}$, where t months after the strategic plan came into effect, the operating cost is \$$c$. Sketch the graph of the operating costs over time and calculate the average rate of change of the operating costs over the first 2 months.

9. The sketch graphs I–IV show the level of interest of four students in Maths Methods during semester 1.

   a. identify which student had a zero rate of change in interest level for the entire semester.
   b. Identify which student had a constant positive rate of change in interest level for the entire semester.
   c. Describe the other two students' interest levels in Maths Methods during semester 1.
   d. Sketch a graph showing your own interest level in Maths Methods during semester 1.

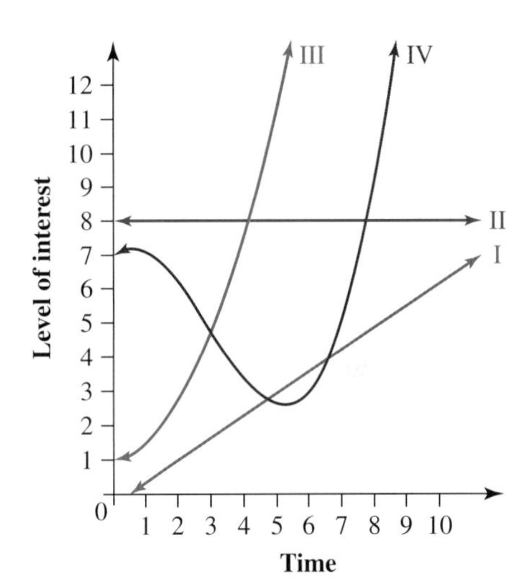

10. Calculate the average rate of change of each of the following functions over the given interval.

   a. $f(x) = 2x - x^2,\ x \in [-2, 6]$
   b. $f(x) = 2 + 3x,\ x \in [12, 16]$
   c. $f(x) = t^2 + 3t - 1,\ t \in [1, 3]$
   d. $f(t) = t^3 - t,\ t \in [-1, 1]$

11. Calculate the hourly rate of pay for each of the following people based on one event that may be atypical.

   a. A plumber who receives \$200 for a task that takes 20 minutes to complete
   b. A surgeon who is paid \$180 for a consultation lasting 15 minutes
   c. A teacher who receives \$1820 for working a 52-hour week

Technology active

12. a. One Australian dollar was worth 0.67 euros in November; by August of the following year one Australian dollar was worth 0.83 euros. Calculate the average rate of change per month in the exchange value of the Australian dollar with respect to the euro over this time period.

   b. \$1000 is invested in a term deposit for 3 years at a fixed rate of r% per annum interest. If the value of the investment is \$1150 at the end of the 3 years, calculate the value of r.

13. The distance travelled by a particle moving in a straight line can be expressed as a function of time. If x metres is the distance travelled after t seconds, and $x = t^3 + 4t^2 + 3t$.

   a. Calculate the average rate of change of the distance x over the time interval $t \in [0, 2]$ and over the time interval $t \in [0.9, 1.1]$.
   b. Explain which of the average speeds calculated in part **a** would be the better estimate for the instantaneous speed at $t = 1$.

14. WE2 For the curve with equation $y = 5 - 2x^2$ shown in Worked example 2, estimate the gradient of the curve at the point Q$(-1.5, 0.5)$ by:

a. constructing a tangent at Q and calculating its gradient

b. choosing a point on the curve close to Q and calculating the average rate of change between Q and this point.

15. a. Estimate the gradient at the point on the curve where $x = 2$.

b. Estimate the gradient at the points on the curve where $x = 2$ and $x = 4$.

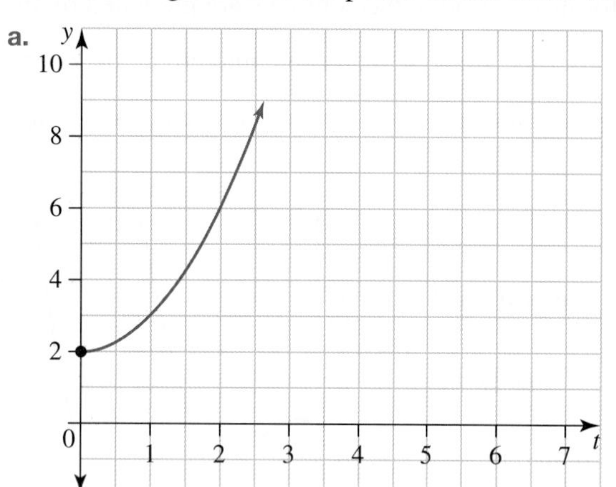

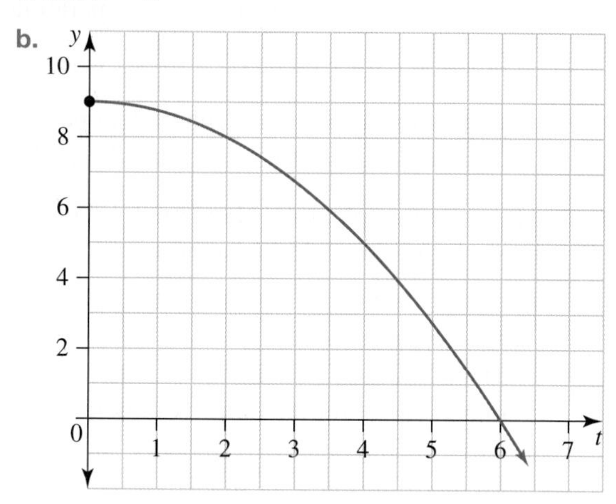

c. For the curve in part b, calculate the average rate of change over the interval $x \in [4, 6]$.

d. For the curve in part b, find an estimate, to the nearest degree, for the magnitude of the angle at which the curve cuts the x-axis (construct the tangent at the x intercept).

16. The temperature T °C over the 12-hour time interval (t) from midday to midnight is shown in the diagram.

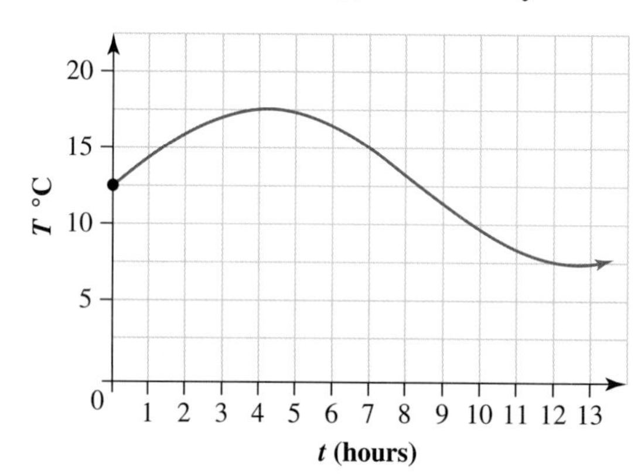

a. Use the graph and the tangent line method to estimate the rate at which the temperature is changing at:

i. 2 pm ii. 4 pm iii. 7 pm iv. 10 pm.

b. State the time at which the temperature appears to be decreasing most rapidly.

c. Calculate the average rate of change of temperature from 1 pm to 9.30 pm.

17. a. Draw the distance–time graph for a cyclist riding along a straight road, returning to point O after 9 hours, that fits the following description:

A cyclist travels from O to A in 2 hours at a constant speed of 10 km/h; she then has a rest break for 0.5 hours, after which she rides at constant speed to reach point B, 45 km from O. After a lunch break of 1 hour, the cyclist takes 2.5 hours to return to O at constant speed.

b. Calculate the time between the cyclist leaving A and arriving at B and hence calculate the constant speed with which the cyclist rode from A to B.
c. Calculate the constant speed with which the cyclist rode from B to return to O.
d. Calculate the cyclist's average speed over the entire 9-hour journey.

18. The top T of the stem of a small plant is 24 mm directly above its base B. From the base B a small shoot grows at a rate of 2 mm/week at an angle of 60° to the stem BT. The stem of the plant does not grow further.

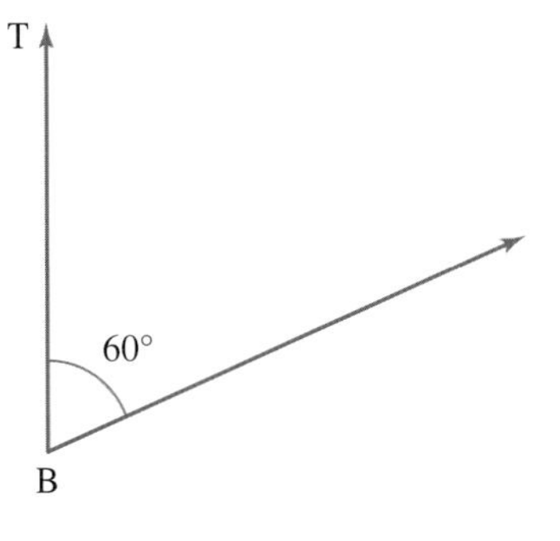

a. After 3 weeks, find how far vertically the tip of the small shoot is below T.
b. The shoot continues to grow at the constant rate of 2 mm/week. Calculate how far vertically its tip is below T after t weeks?
c. Determine how many weeks it will take for the tip of the shoot to be at the same height as T.

19. A small boat sets out to sea away from a jetty. Its distance in metres from the jetty after t hours is given by $d = \frac{200t}{t+1}$, $t \geq 0$.
a. Calculate the average speed of the boat over the first 4 hours.
b. Sketch the graph of the distance from the jetty against time and give a short description of the motion of the boat.
c. Draw a tangent to the curve at the point where $t = 4$ and hence find an approximate value for the instantaneous speed of the boat 4 hours after it leaves the jetty.
d. Use an interval close to $t = 4$ to obtain another estimate of this speed.

20. Use CAS technology to sketch the curve $y = 4 - 3x^2$ and its tangent at the point where $x = -2$ and to state the equation of the tangent. Hence state the gradient of $y = 4 - 3x^2$ at $x = -2$.

11.2 Exam questions

Question 1 (1 mark) TECH-ACTIVE

MC A gardener observed that, during spring, their tomato bush grew from 5 cm to 40 cm over a 2-week period. They estimated that the average daily growth of the tomato bush was

A. 2 cm/day
B. 35 cm/day
C. 2.5 cm/day
D. 4.3 cm/day
E. 20 cm/day

Question 2 (6 marks) TECH-ACTIVE

A small business calculated that its yearly profit followed the equation Profit $p(t) = 200\left(10t - t^2\right)$, where t is in years.

a. Sketch the graph of profit against time for the first 10 years. **(3 marks)**
b. Calculate the average rate of change of the profit over the first 3 years and compare it with the rate of change of profit from years 6 to 9. Comment on the different rates of change. **(3 marks)**

Question 3 (1 mark) TECH-ACTIVE

MC The average rate of change of the function $f(x) = 3 - x^2$ between $x = 1$ and $x = 5$ is

A. 3
B. −3
C. 6
D. −6
E. 5

More exam questions are available online.

11.3 Gradients of secants

LEARNING INTENTION

At the end of this subtopic you should be able to:
- determine the gradient of a secant or tangent to a curve at a given point graphically, numerical and algebraically.

The gradient of a curve can be estimated by finding the average rate of change between two nearby or near-neighbouring points. As the interval between these two points is reduced, the curve becomes 'straighter', which is why the average rate starts to become a good estimate for the actual gradient. This method of zooming in on two close points is called the **near neighbour** method or the **straightening the curve** method.

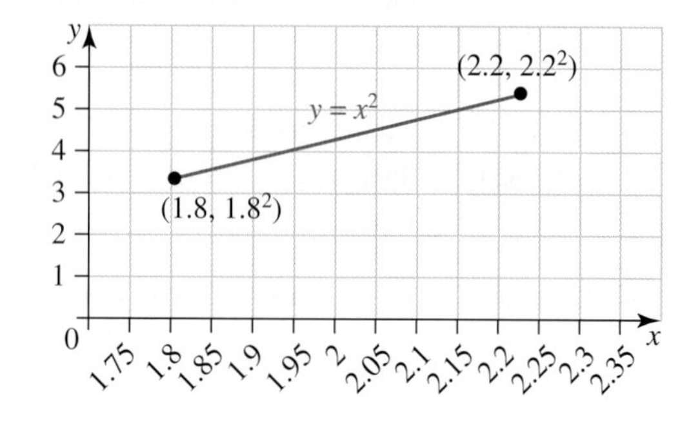

The graph of the curve $y = x^2$ appears quite straight if viewed on the domain [1.8, 2.2] and straighter still for intervals with end points closer together.

If this section of the curve is treated as straight, its gradient is $\dfrac{(2.2)^2 - (1.8)^2}{2.2 - 1.8} = 4$. This value should be a very good estimate of the gradient of the curve at the midpoint of the interval, $x = 2$. Zooming in closer would either boost confidence in this estimate or produce an even better estimate.

11.3.1 Gradient of a secant

Straightening the curve gives a good approximation to the gradient of a curve. However, it is still an approximation, since the curve is not absolutely straight. The completely straight line through the end points of the interval formed by the two neighbouring points is called a **secant**, or a chord if just the line segment joining the end points is formed. The gradient of a secant, the line passing through two points on a curve, can be used to find the gradient of the tangent, the line that just touches the curve at one point.

To illustrate this, consider finding the gradient of the curve $y = x^2$ at the point A where $x = 3$ by forming a secant line through A and a nearby close-neighbour point B.

As A has an x-coordinate of 3, let B have an x-coordinate of $3 + h$, where h represents the small difference between the x-coordinates of points A and B. It does not matter whether h is positive or negative. What matters is that it is small, so that B is close to A.

The y-coordinate of A will be $y = (3)^2 = 9$.

The y-coordinate of B will be $y = (3 + h)^2$.

The gradient of the secant through the points A $(3, 9)$ and B $(3 + h, \ (3 + h)^2)$ is:

$$\begin{aligned}
m_{\text{secant}} &= \frac{(3+h)^2 - 9}{(3+h) - 3} \\
&= \frac{9 + 6h + h^2 - 9}{3 + h - 3} \\
&= \frac{6h + h^2}{h} \\
&= \frac{h(6+h)}{h} \\
&= 6 + h, h \neq 0
\end{aligned}$$

Thus, the gradient of the secant is $6 + h$.

By choosing small values for h, the gradient of the secant can be calculated to any desired accuracy.

h	0.1	0.01	0.001	0.000 1	0.000 01
Gradient $6 + h$	6.1	6.01	6.001	6.000 1	6.000 01

As h becomes smaller, the gradient appears to approach the value of 6.

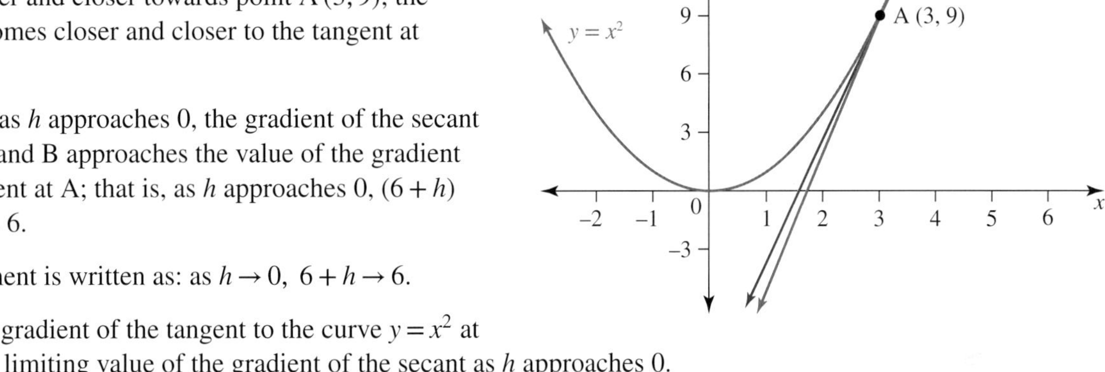

Also, as h becomes smaller, the point B $(3 + h, (3 + h)^2)$ moves closer and closer towards point A $(3, 9)$; the secant becomes closer and closer to the tangent at point A.

Therefore, as h approaches 0, the gradient of the secant through A and B approaches the value of the gradient of the tangent at A; that is, as h approaches 0, $(6 + h)$ approaches 6.

This statement is written as: as $h \to 0$, $6 + h \to 6$.

Hence, the gradient of the tangent to the curve $y = x^2$ at $x = 3$ is the limiting value of the gradient of the secant as h approaches 0.

This statement is written as: the gradient of the tangent is $\lim\limits_{h \to 0} (6 + h) = 6$.

Evaluating the limit expression

For two neighbouring points whose x-coordinates differ by h, the gradient of the secant will be an expression involving h. Algebraic techniques such as expansion and factorisation are used to simplify the expression, allowing any common factor of h in the numerator and denominator to be cancelled, for $h \neq 0$. Once this simplified expression for the gradient of the secant has been obtained, then the limit expression as $h \to 0$ can be calculated simply by replacing h by 0. It is essential that the expression is simplified first before replacing h by 0.

WORKED EXAMPLE 3 Determining the gradient of a secant or tangent at a point

Consider the curve with equation $y = 3x - x^2$.

a. **Express the gradient of the secant through the points on the curve where $x = 2$ and $x = 2 + h$ in terms of h.**

b. **Use $h = 0.01$ to obtain an estimate of the gradient of the tangent to the curve at $x = 2$.**

c. **Deduce the gradient of the tangent to the curve at the point where $x = 2$.**

THINK	WRITE
a. 1. Form the coordinates of the two points on the secant.	a. $y = 3x - x^2$ When $x = 2$, $y = 6 - 4$ $= 2$ Point $(2, 2)$ When $x = 2 + h$,

$y = 3(2+h) - (2+h)^2$
$= 6 + 3h - (4 + 4h + h^2)$
$= 6 + 3h - 4 - 4h - h^2$
$= 2 - h - h^2$

Point $(2+h, 2-h-h^2)$

The two neighbouring points are $(2, 2)$ and $(2+h, 2-h-h^2)$.

2. Calculate the gradient of the secant line through the two points and simplify the expression obtained.

$$m_{\text{secant}} = \frac{(2-h-h^2)-(2)}{(2+h)-2}$$
$$= \frac{-h-h^2}{h}$$
$$= \frac{-h(1+h)}{h}$$
$$= -(1+h), h \neq 0$$

Therefore, the gradient of the secant is $-(1+h) = -h-1$.

b. Calculate an estimate of the gradient of the tangent.

b. Since h is small, the secant's gradient with $h = 0.01$ should be an estimate of the gradient of the tangent.

If $h = 0.01$, $m_{\text{secant}} = -0.01 - 1 = -1.01$.

An estimate of the gradient of the tangent at $x = 2$ is -1.01.

c. Calculate the limiting value of the secant's gradient as h approaches 0.

c. As $h \to 0$, $(-h-1) \to -1$

$\therefore \lim\limits_{h \to 0}(-h-1) = -1$

$$m_{\text{tangent}} = \lim_{h \to 0}(-h-1)$$
$$= -1$$

The gradient of the tangent at the point $(2, 2)$ is -1.

11.3 Exercise

Technology free

1. Explain the difference between a secant, a tangent and a chord.

2. a. Calculate the gradient of the secant through the points $(1, 1)$ and $(1.5, 2.25)$.
   b. Calculate the gradient of the chord with end points $(1, 1)$ and $(1.5, 2.25)$.
   c. Explain how the gradients of the secant and the chord compare.

3. Calculate the y-coordinate of each of the following points.
   a. The point on the curve $y = x^2$ with an x-coordinate of $1 + h$
   b. The point on the curve $y = x^2 + 5x$ with an x-coordinate of $-2 + h$
   c. The point on the curve $y = 4 - 3x^2$ with an x-coordinate of $3 + h$
   d. The point on the curve $y = -2x^2 + 3x + 11$ with an x-coordinate of $-4 + h$
4. a. Sketch the curve $y = x^2$ and draw the secant through each of the following pairs of points: $(-1, 1)$ and $(3, 9)$; $(0, 0)$ and $(3, 9)$; $(2, 4)$ and $(3, 9)$.
   b. State which of the three secants in part a is the best approximation to the tangent at $(3, 9)$. Explain how this approximation could be improved.
5. For each of the following, calculate the gradient of the secant through P and Q.
   a. P and Q lie on $y = x^2$ with x-coordinates $x = 1$ and $x = 1 + h$, respectively.
   b. P and Q lie on $y = x^2 + 3x$ with x-coordinates $x = 2$ and $x = 2 + h$, respectively.
   c. P and Q lie on $y = 2x^2 - 6$ with x-coordinates $x = 3$ and $x = 3 + h$, respectively.
   d. P and Q lie on $y = x^2 - 10x + 20$ with x-coordinates $x = 4$ and $x = 4 + h$, respectively.
6. **WE3** Consider the curve with equation $y = 2x^2 - x$.
   a. Express the gradient of the secant through the points on the curve where $x = -1$ and $x = -1 + h$ in terms of h.
   b. Use $h = 0.01$ to obtain an estimate of the gradient of the tangent to the curve at $x = -1$.
   c. Deduce the gradient of the tangent to the curve at the point where $x = -1$.
7. a. Calculate the coordinates of the point A on the parabola $y = 4 - x^2$ for which $x = 2 + h$.
   b. Express the gradient of the chord joining the point A to the point C $(2, 0)$ in terms of h.
   c. If the gradient of the chord AC is -5, identify the value of h and state the coordinates of A.
   d. If A is the point $(2.1, -0.41)$, identify the value of h and evaluate the gradient of the chord AC.
   e. The point B for which $x = 2 - h$ also lies on the parabola $y = 4 - x^2$. Obtain an expression for the gradient of the chord BC and evaluate this gradient for $h = 0.1$.
   f. Calculate the gradient of the secant which passes through the points A and B.
8. The point D lies on the curve with equation $y = x^3 + x$.
   a. Obtain an expression for the y-coordinate of D given its x-coordinate is $3 + h$.
   b. Determine the gradient of the secant passing through the point D and the point on the curve $y = x^3 + x$ where $x = 3$.
   c. Evaluate this gradient if $h = 0.001$.
   d. Suggest the value for the gradient of the tangent to the curve at $x = 3$.
9. a. Obtain the coordinates of the points M and N on the hyperbola $y = \frac{1}{x}$ where $x = 1$ and $x = 1 + h$ respectively.
   b. Determine, in terms of h, the gradient of the secant passing through the points M and N.
   c. Determine the value that this gradient approaches as $h \to 0$.
   d. Deduce the gradient of the tangent to $y = \frac{1}{x}$ at the point M.
10. Consider the graph of $y = 1 + 3x - x^2$.
    a. Calculate the gradient of the secant passing through the points on the graph with x-coordinates 1 and $1 - h$.
    b. Write a limit expression for the gradient of the tangent to curve at the point where $x = 1$.
    c. Hence, state the gradient of the tangent to the curve at the point for which $x = 1$.

Technology active

11. Consider the curve with equation $y = \frac{1}{3}x^3 - x^2 + x + 5$.
    a. Calculate the gradient of the chord joining the points on the curve where $x = 1$ and $x = 1 - h$.
    b. Hence, deduce the gradient of the tangent at the point where $x = 1$.

12. a. Calculate the gradient of the chord joining the points where $x = 3.9$ and $x = 4$ on the curve with equation $y = \frac{x^4}{4}$.
b. Hence, state a whole-number estimate for the gradient of the tangent to the curve at the point where $x = 4$.
c. Explain how could the estimate be improved.

13. Consider the graph of $y = (x - 2)(x + 1)(x - 3)$.
a. Show that the gradient of the secant passing through the points on the graph with x-coordinates 0 and h is $h^2 - 4h + 1$ for $h \neq 0$.
b. Hence, write down a limit expression for the gradient of the tangent to curve at $x = 0$.
c. State the gradient of the tangent to the curve at the point for which $x = 0$.

14. Simplify the given secant gradient expression where possible for $h \neq 0$ and then evaluate the given limit.
a. $m_{\text{secant}} = 3 + h; \lim_{h \to 0} (3 + h)$
b. $m_{\text{secant}} = \frac{8h - 2h^2}{h}; \lim_{h \to 0} \left(\frac{8h - 2h^2}{h}\right)$
c. $m_{\text{secant}} = \frac{(2 + h)^3 - 8}{h}; \lim_{h \to 0} \left(\frac{(2 + h)^3 - 8}{h}\right)$
d. $m_{\text{secant}} = \frac{(1 + h)^4 - 1}{h}; \lim_{h \to 0} \left(\frac{(1 + h)^4 - 1}{h}\right)$

15. Simplify the given secant gradient expression where possible for $h \neq 0$ and then evaluate the given limit.
a. $m_{\text{secant}} = \frac{\frac{1}{(4+h)} - \frac{1}{4}}{h}; \lim_{h \to 0} \left(\frac{-1}{4(4 + h)}\right)$
b. $m_{\text{secant}} = \frac{h}{\sqrt{1 + h} - 1}; \lim_{h \to 0} \left(\frac{h}{\sqrt{1 + h} - 1}\right)$

16. Consider the function with the rule $f(x) = x^2$.
a. Draw a diagram showing the curve and the secant through the points A and B on the curve that have x-coordinates a and b respectively with $b > a > 0$.
b. Calculate the gradient of the secant through the points A and B in terms of a and b.
c. Hence, deduce the gradient of the tangent to curve at:
i. the point A
ii. the point B.

17. Evaluate the following using CAS technology.
a. $\lim_{h \to 0} \left(\frac{(h + 2)^2(h + 1) - 4}{h}\right)$
b. $\lim_{x \to \infty} \left(\frac{1}{x} + 3\right)$
c. $\lim_{x \to \infty} \frac{2x + 1}{3x - 2}$

18. Consider the graph of the function defined by $f(x) = \sqrt{x}$.
a. Calculate the gradient of the secant that passes through the points on the graph that have x-coordinates 9 and $9 + h$.
b. Hence, state the limit expression for the gradient of the tangent to the graph at the point for which $x = 9$.
c. Calculate this limit using CAS technology to obtain the gradient of the graph at the point where $x = 9$.

11.3 Exam questions

Question 1 (1 mark) TECH-ACTIVE

MC A secant is a line that
A. touches a curve at a point.
B. passes through two points on a curve.
C. is a tangent to the curve.
D. is parallel to the curve.
E. is at right angles to the curve.

Question 2 (4 marks) TECH-ACTIVE

a. For the curve with the equation $y = x^2 - 4x$, express the gradient of the secant through the points on the curve where $x = 6$ and $x = 6 + h$ in terms of h. **(2 marks)**
b. Use $h = 0.01$ to obtain an estimate of the gradient of the tangent to the curve at $x = 6$. **(1 mark)**
c. Deduce the gradient of the tangent to the curve at the point where $x = 6$. **(1 mark)**

Question 3 (1 mark) TECH-ACTIVE

MC The gradient of the secant passing through the points on the graph $y = (x-3)(x+1)(x-4)$ with x-coordinates 0 and h, where $h \neq 0$, is

A. $h^2 - 6h + 5$
B. $h^2 + 6h + 5$
C. $h^2 - 5h + 6$
D. $h^2 + 5h + 6$
E. $2h^2 - 5h + 6$

More exam questions are available online.

11.4 The derivative function

LEARNING INTENTION

At the end of this subtopic you should be able to:
- determine the gradient function
- determine the limit definition of the derivative of a function
- approximate the gradient using the central difference method.

Although the work of the seventeenth-century mathematicians Sir Isaac Newton and Gottfried Wilhelm Leibniz created a bitter rivalry between them, it is now widely accepted that they each independently invented calculus. Their geniuses bequeathed to the world the most profound of all mathematical inventions, one that underpins most advanced mathematics even today. In essence, it all started from the problem of finding the rate of change of a function at a particular instant, that is, of finding the gradient of a curve at any point.

The approach both Newton and Leibniz used is based on the concept of a limit: that if two points on a curve are sufficiently close, the tangent's gradient is the limiting value of the secant's gradient. The only difference in their approach was the form of notation each used.

Both mathematicians essentially found the gradient of a curve $y = f(x)$ at some general point $(x, f(x))$ by:
- taking a neighbouring point $(x+h, f(x+h))$, where h represents a small change in the x-coordinates
- forming the **difference quotient** $\frac{f(x+h) - f(x)}{h}$, which is the gradient of the secant or the average rate of change of the function between the two points
- calculating the gradient of the curve as the limiting value of the difference quotient as $h \to 0$.

This method created a new function called the **gradient function**, the **derived function** or the **derivative function**, which is frequently just referred to as the derivative. For the function f, the symbol for the gradient function is f', which is read as 'f dashed'. The rule for the derivative is $f'(x)$.

11.4.1 Definition of the gradient or derivative function

For the function $y = f(x)$, its gradient or derivative function, $y = f'(x)$, is defined as the limiting value of the difference quotient as $h \to 0$.

Gradient or derivative function

$$f'(x) = \lim_{h \to 0} \frac{f(x+h) - f(x)}{h}$$

Once the derivative function is obtained, the gradient at a given point on the curve $y = f(x)$ is evaluated by substituting the x-coordinate of the given point in place of x in the equation of the derivative function. The value of $f'(a)$ represents the rate of change of the function at the instant when $x = a$; it gives the gradient of the tangent to the curve $y = f(x)$ at the point on the curve where $x = a$.

The limit expression must be simplified in order to cancel the h term from its denominator before the limit can be evaluated by substituting $h = 0$. As a limit is concerned with the value approached as $h \to 0$, there is no need to add the proviso $h \neq 0$ when cancelling the h term from the denominator of the limit expression.

WORKED EXAMPLE 4 Determining the gradient function

Consider the function for which $f(x) = x^2 - 5x + 4$.

a. Use the difference quotient definition to form the rule $f'(x)$ for the gradient function.

b. Hence, obtain the gradient of the graph of the function at the point (1, 0).

c. Evaluate the instantaneous rate of change of the function when $x = 4$.

THINK	WRITE
a. 1. Form an expression for $f(x+h)$.	a. $f(x) = x^2 - 5x + 4$ Replacing every x in the function rule with $(x+h)$, we get $f(x+h) = (x+h)^2 - 5(x+h) + 4$.
2. State the difference quotient definition.	The difference quotient is $\dfrac{f(x+h) - f(x)}{h}$.
3. Substitute the expressions for the functions into the difference quotient.	$\therefore \dfrac{f(x+h) - f(x)}{h}$ $= \dfrac{\left[(x+h)^2 - 5(x+h) + 4\right] - \left[x^2 - 5x + 4\right]}{h}$
4. Expand the numerator and simplify. *Note:* The terms in the numerator not containing h have cancelled out.	$= \dfrac{x^2 + 2xh + h^2 - 5x - 5h + 4 - x^2 + 5x - 4}{h}$ $= \dfrac{2xh + h^2 - 5h}{h}$
5. State the definition of the gradient function.	By definition, $f'(x) = \lim\limits_{h \to 0} \dfrac{f(x+h) - f(x)}{h}$
6. Substitute the expression for the difference equation and simplify. *Note:* As the limit is yet to be evaluated, we must write $\lim\limits_{h \to 0}$ at each line of the simplification steps.	$f'(x) = \lim\limits_{h \to 0} \dfrac{2xh + h^2 - 5h}{h}$ $= \lim\limits_{h \to 0} \dfrac{h(2x + h - 5)}{h}$ $= \lim\limits_{h \to 0} (2x + h - 5)$
7. Evaluate the limit to calculate the rule for the gradient function.	Substitute $h = 0$ to calculate the limit. $f'(x) = 2x + 0 - 5$ $= 2x - 5$ The rule for the gradient function is $f'(x) = 2x - 5$.

b. Calculate the value of the gradient function at the given point.

b. For the point $(1, 0)$, $x = 1$.
The gradient at this point is the value of $f'(1)$.
$f'(x) = 2x - 5$
$f'(1) = 2(1) - 5$
$f'(1) = -3$
Therefore, the gradient at the point $(1, 0)$ is -3.

c. Evaluate the instantaneous rate of change at the given point.

c. The instantaneous rate of change at $x = 4$ is $f'(4)$.
$f'(x) = 2x - 5$
$f'(4) = 2(4) - 5$
$= 3$
The rate of change of the function at $x = 4$ is 3.

Other forms of notation

Other forms of notation include replacing h for the **small increment** or small change in x by one of the symbols Δx or δx, where Δ and δ are upper and lower case respectively of the Greek letter 'delta'. The gradient function can be expressed as $f'(x) = \lim\limits_{\Delta x \to 0} \dfrac{f(x + \Delta x) - f(x)}{\Delta x}$ or as $f'(x) = \lim\limits_{\delta x \to 0} \dfrac{f(x + \delta x) - f(x)}{\delta x}$.

Leibniz used the symbol δx ('delta x') for a small change or small increment in x and the symbol δy for a small change or small increment in y. These are used as single symbols, not as products of two symbols.

The coordinates of the two neighbouring points are (x, y) and$(x + \delta x, y + \delta y)$.

The gradient of the secant, or difference quotient, is $\dfrac{(y + \delta y) - y}{(x + \delta x) - x} = \dfrac{\delta y}{\delta x}$.

The gradient of the tangent at the point (x, y) is therefore $\lim\limits_{\delta x \to 0} \dfrac{\delta y}{\delta x}$.

Leibniz used the symbol $\dfrac{dy}{dx}$ for the rule of this gradient or derivative function.

This is different notation but exactly the same thinking to that used previously in the topic.

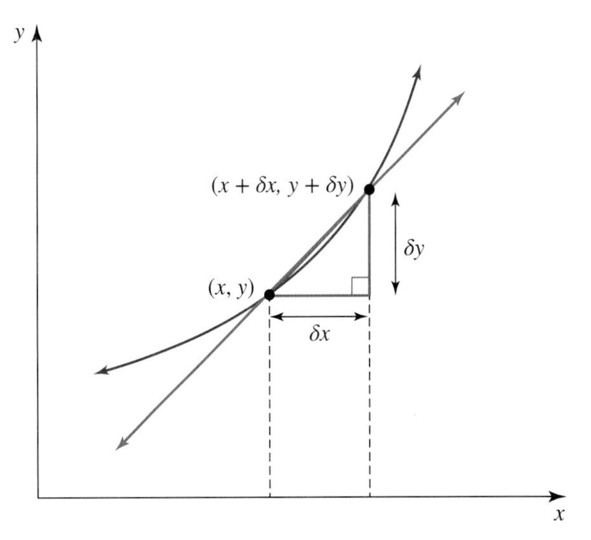

Note that $\dfrac{dy}{dx}$ (pronounced 'dee y by dee x') is a single symbol for the derivative; it is not a fraction. It represents the derivative of the dependent variable y with respect to the independent variable x.

Importantly, both symbols $\dfrac{dy}{dx}$ and $f'(x)$ are used for the derivative of $y = f(x)$.

11.4.2 The limit definition of the derivative

The process of obtaining the gradient or derivative function is called **differentiation**.

The derivative is calculated from forms such as $f'(x)=\lim_{h\to 0}\frac{f(x+h)-f(x)}{h}$.

The derivative with respect to x of the function $y=f(x)$ is $\frac{dy}{dx}=f'(x)$.

Once the derivative is known, the gradient of the tangent to the curve $y=f(x)$ at a point where $x=a$ is calculated as $\left.\frac{dy}{dx}\right|_{x=a}$ or $f'(a)$. Alternatively, this gradient value could be calculated directly from the limit definition using $f'(a)=\lim_{h\to 0}\frac{f(a+h)-f(a)}{h}$.

WORKED EXAMPLE 5 Using the limit definition to determine the derived function

Differentiate $3x-2x^3$ with respect to x using the limit definition.

THINK	WRITE
1. Define $f(x)$ and form the expression for $f(x+h)$. *Note:* Leibniz's notation could be used instead of function notation.	Let $f(x)=3x-2x^3$. $f(x+h)=3(x+h)-2(x+h)^3$
2. State the limit definition of the derivative.	$f'(x)=\lim_{h\to 0}\frac{f(x+h)-f(x)}{h}$
3. Substitute the expressions for $f(x)$ and $f(x+h)$.	$\therefore f'(x)=\lim_{h\to 0}\frac{\left[3(x+h)-2(x+h)^3\right]-\left[3x-2x^3\right]}{h}$
4. Expand and simplify the numerator. *Note:* Either the binomial theorem or Pascal's triangle could be used to expand $(x+h)^4$.	$f'(x)=\lim_{h\to 0}\frac{3(x+h)-2(x^3+3x^2h+3xh^2+h^3)-3x+2x^3}{h}$ $=\lim_{h\to 0}\frac{\cancel{3x}+3h-\cancel{2x^3}-6x^2h-6xh^2-2h^3-\cancel{3x}+\cancel{2x^3}}{h}$ $=\lim_{h\to 0}\frac{3h-6x^2h-6xh^2-2h^3}{h}$
5. Factorise to allow h to be cancelled from the denominator.	$f'(x)=\lim_{h\to 0}\frac{h\left[3-6x^2-6xh-2h^2\right]}{h}$ $=\lim_{h\to 0}\left(3-6x^2-6xh-2h^2\right)$
6. Evaluate the limit by substituting 0 for h.	$f'(x)=3-6x^2-6x(0)-2(0)^2$ $=3-6x^2$
7. State the answer.	The derivative of $3x-2x^3$ with respect to x is $3-6x^2$.

11.4.3 The central difference approximation of $f'(x)$

The limit definition for the derivative of a function $y = f(x)$ takes the gradient of the secant joining the point $(x, f(x))$ to a neighbouring point $(x+h, f(x+h))$ where h represents a small change in the x-coordinate. The derivative is then:

$$f'(x) = \lim_{h \to 0} \frac{f(x+h) - f(x)}{h}$$

This means that:

$$f'(x) \approx \frac{f(x+h) - f(x)}{h}$$

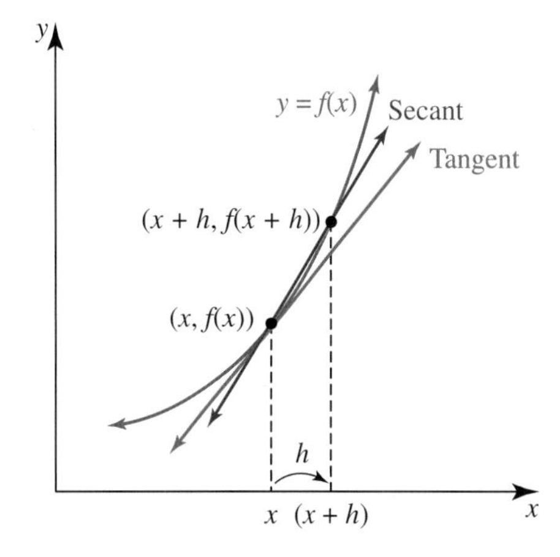

The central difference approximation also uses the gradient of the secant joining two points. The points are a small change, h, either side of the x-coordinate, giving the two points $(x-h, f(x-h))$ and $(x+h, f(x+h))$ with the x-coordinate now halfway between, or central to, the two points. Using the gradient of the secant:

$$m_{\text{secant}} = \frac{f(x+h) - f(x-h)}{(x+h) - (x-h)}$$

$$= \frac{f(x+h) - f(x-h)}{2h}$$

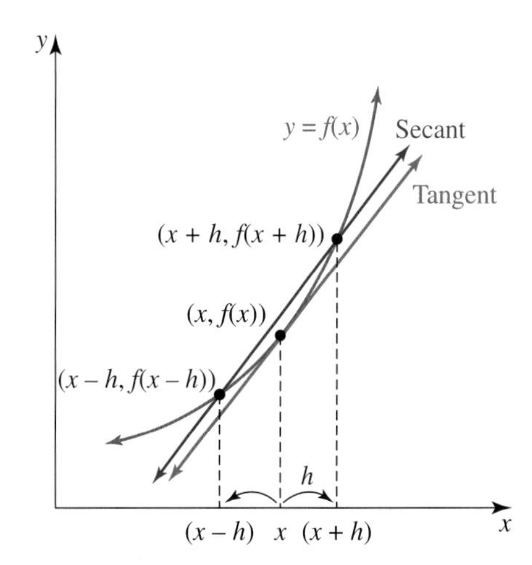

The central difference approximation for the derivative of the function $y = f(x)$ is given by the following formula.

Central difference approximation

$$f'(x) \approx \frac{f(x+h) - f(x-h)}{2h}$$

where h is a small change in the x-coordinate.

The two diagrams illustrate graphically these two methods to approximate the gradient of the tangent or the derivative function. Both have been 'zoomed' in around the point $(x, f(x))$. It is considered that the central difference method gives a better approximation to the gradient of the function $y = f(x)$ at the point $(x, f(x))$. This is supported by comparing the two diagrams.

WORKED EXAMPLE 6 Approximating $f'(x)$ by the central difference method

Consider the function $f(x) = 3x^2 - 4x$.

a. **Determine the central difference approximation of $f'(x)$.**

b. **Hence, obtain the gradient of the graph of the function at the point (2, 4).**

c. **Evaluate the instantaneous rate of change of the function when $x = -2$.**

THINK	WRITE
a. 1. Form an expression for $f(x+h)$.	$f(x) = 3x^2 - 4x$ Replacing every x with $(x+h)$: $f(x+h) = 3(x+h)^2 - 4(x+h)$
2. Form an expression for $f(x-h)$.	Replacing every x with $(x-h)$: $f(x-h) = 3(x-h)^2 - 4(x-h)$
3. Substitute the expressions into the central difference approximation formula, expand and simplify. *Note:* Care must be taken with the negative signs.	$f'(x) \approx \dfrac{f(x+h) - f(x-h)}{2h}$ $\approx \dfrac{[3(x+h)^2 - 4(x+h)] - [3(x-h)^2 - 4(x-h)]}{2h}$ $\approx \dfrac{[3x^2 + 6xh + 3h^2 - 4x - 4h] - [3x^2 - 6xh + 3h^2 - 4x + 4h]}{2h}$ $\approx \dfrac{3x^2 + 6xh + 3h^2 - 4x - 4h - 3x^2 + 6xh - 3h^2 + 4x - 4h}{2h}$ $\approx \dfrac{12xh - 8h}{2h}$ $\approx \dfrac{4h(3x-2)}{2h}$ $\approx 2(3x-2)$
4. State the approximation.	$f'(x) \approx 6x - 4$
b. Calculate the gradient at the given point.	For the point (2, 4), $x = 2$. The gradient at this point is the value of $f'(2)$. $f'(2) = 6(2) - 4$ $f'(2) = 8$ Therefore, the gradient at the point (2, 4) is 8.
c. Evaluate the instantaneous rate of change of the given point.	The instantaneous rate of change at $x = -2$ is $f'(-2)$. $f'(-2) = 6(-2) - 4$ $f'(-2) = -16$ Therefore, the instantaneous rate of change at $x = -2$ is -16.

11.4 Exercise

Students, these questions are even better in jacPLUS

Receive immediate feedback and access sample responses

Access additional questions

Track your results and progress

Find all this and MORE in jacPLUS

Technology free

1. a. If $y = f(x)$, write down a symbol for the derivative of this function and state its limit definition.
   b. If the point where $x = a$ lies on a function $y = f(x)$, identify the feature of the function that $f'(a)$ measures.

2. a. Form an expression for $f(x+h)$ if:
      i. $f(x) = x^2 + 2x$
      ii. $f(x) = 3 - 2x^2$
   b. Form an expression for $f(x+h) - f(x)$ if:
      i. $f(x) = 3x^2 - 7x$
      ii. $f(x) = 5x^2 - 3x + 2$.

3. Consider the function defined by $f(x) = x^2$.
   a. Form the difference quotient $\dfrac{f(x+h) - f(x)}{h}$.
   b. Use the difference quotient to find $f'(x)$, the rule for the gradient function.
   c. Hence, state the gradient of the tangent at the point $(3, 9)$.
   d. Show that the gradient at the point $(3, 9)$ could be calculated by evaluating $\lim\limits_{h \to 0} \dfrac{f(3+h) - f(3)}{h}$.

4. Use the difference quotient definition to form the derivative of $f(x) = x^3$.

5. **WE4** Consider the function $f(x) = 2x^2 + x + 1$.
   a. Use the difference quotient definition to form the rule $f'(x)$ for the gradient function.
   b. Hence obtain the gradient of the function at the point $(-1, 2)$.
   c. Evaluate the instantaneous rate of change of the function when $x = 0$.

6. Explain the geometric meanings of the following.
   a. $\dfrac{f(x+h) - f(x)}{h}$
   b. $\lim\limits_{h \to 0} \dfrac{f(x+h) - f(x)}{h}$
   c. $\lim\limits_{h \to 0} \dfrac{f(3+h) - f(3)}{h}$
   d. $\dfrac{f(3) - f(3-h)}{h}$

7. Consider $f(x) = 3x - 2x^2$.
   a. Determine $f'(x)$ using the limit definition.
   b. Calculate the gradient of the tangent to the curve at $(0, 0)$.
   c. Sketch the curve $y = f(x)$ showing the tangent at $(0, 0)$.

8. The points $\mathrm{P}(1, 5)$ and $\mathrm{Q}(1 + hx, 3(1 + hx)^2 + 2)$ lie on the curve $y = 3x^2 + 2$.
   a. Express the average rate of change of $y = 3x^2 + 2$ between P and Q in terms of hx.
   b. Calculate the gradient of the tangent to the curve at P.
   c. S and T are also points on the curve $y = 3x^2 + 2$, where $x = 2$ and $x = 2 + hx$ respectively.
      i. Give the coordinates of the points S and T.
      ii. Form an expression in terms of hx for the gradient of the secant through S and T.
      iii. Calculate the gradient of the tangent to the curve at S.

9. **WE5** Differentiate $5x+3x^4$ with respect to x using the limit definition.

10. Using the limit definition, differentiate the following with respect to x.

a. $f(x)=8x^2+2$

b. $f(x)=\frac{1}{2}x^2-4x-1$

c. $f(x)=6-2x$

d. $f(x)=x^3-6x^2+2x$

11. Given $f(x)=(x+5)^2$:

a. determine $f'(x)$ using the limit definition

b. calculate $f'(-5)$ and explain its geometric meaning

c. calculate the gradient of the tangent to the curve $y=f(x)$ at its y-intercept

d. calculate the instantaneous rate of change of the function $y=f(x)$ at $(-2, 9)$.

12. **WE6** Consider the function $f(x)=5x^2-2x-6$.

a. Determine the central difference approximation of $f'(x)$.

b. Hence, obtain the gradient of the graph of the function at the point $(2, 10)$.

c. Evaluate the instantaneous rate of change of the function when $x=-3$.

13. Determine $f'(x)$ using the central difference approximation for the following functions.

a. $f(x)=8-3x^2$

b. $f(x)=2x^3+5x$

c. $f(x)=x^3+7x^2-4x$

d. $f(x)=(x+3)(x-7)$

14. Given $f(x)=2(x-4)(x+2)$:

a. sketch the graph of $y=f(x)$ showing its key points

b. determine $f'(x)$ using the central difference approximation

c. calculate $f'(1)$ and hence draw the tangent at $x=1$ on the graph

d. calculate the gradient of the tangent to the curve at each of its x-intercepts and draw these tangents on the graph.

Technology active

15. Consider the function $f: R\to R,\ f(x)=ax^2+bx+c$ where a, b and c are constants.

a. Use the central difference approximation to find $f'(x)$.

b. Form the derivative function of f and express it as a mapping.

c. Hence, write down the derivative of $f(x)=3x^2+4x+2$.

d. Compare the degree of the function f with that of the function f'.

16. Consider the function $g: R\to R,\ g(x)=x^3$.

a. Factorise $(2+h)^3-8$ as a difference of two cubes.

b. Write down a limit expression for $g'(2)$.

c. Use the answers to parts **a** and **b** to evaluate $g'(2)$.

17. Define the function $f(x)=x^2$ and use CAS technology to calculate $\lim\limits_{h\to 0}\frac{f(x+h)-f(x)}{h}$.

18. a. Define the function $f(x)=ax^3+bx^2+cx+d$ and use CAS technology to calculate $\lim\limits_{h\to 0}\frac{f(x+h)-f(x)}{h}$.

b. Compare the degree of $f(x)=ax^3+bx^2+cx+d$ with that of its derivative.

11.4 Exam questions

Question 1 (1 mark) TECH-ACTIVE

MC The gradient of the tangent to the curve $y = f(x)$ at the point where $x = 3$ is

A. $\lim\limits_{h \to 0} \dfrac{f(x+h) - f(x)}{h}$

B. $\dfrac{f(x+h) - f(x)}{h}$

C. $\lim\limits_{h \to 0} \dfrac{f(3+h) - f(3)}{3}$

D. $\lim\limits_{h \to 0} \dfrac{f(3+h) - f(3)}{h}$

E. $\dfrac{f(3+h) - f(3)}{h}$

Question 2 (4 marks) TECH-FREE

Use the limit definition the derivative of a function to differentiate $3x^3 - 2x$ with respect to x.

Question 3 (3 marks) TECH-FREE

Find $\dfrac{dy}{dx}$ using the central difference approximation for the function $y = 2 - x^2$.

More exam questions are available online.

11.5 Differentiation of polynomials by rule

LEARNING INTENTION

At the end of this subtopic you should be able to:

- determine the derivative function for a polynomial function of low degree by rule
- graph the derivative of a polynomial function
- determine the instantaneous rate of change of a function at a point.

Differentiation using the limit definition is an important procedure, yet it can be quite lengthy and for non-polynomial functions it can be quite difficult. Examining the results obtained from differentiation of polynomials by using the limit definition, some simple patterns appear. These allow some basic rules for differentiation to be established.

11.5.1 Differentiation of polynomial functions

The results obtained from some questions in the previous exercise are compiled into the following table.

$f(x)$	$f'(x)$
$x^2 - 5x + 4$	$2x - 5$
$2x^2 + x + 1$	$4x + 1$
x^3	$3x^2$
$x^2 + 1$	$2x$
$x^2 + x$	$2x + 1$
$3x - 2x^4$	$3 - 8x^3$
$5x + 3x^4$	$5 + 12x^3$

Derivative of x^n, $n \in N$

Close examination of these results suggests that the derivative of the polynomial function, $f(x) = x^n$, $n \in N$ is $f'(x) = nx^{n-1}$, a polynomial of one degree less. This observation can be proven to be correct by differentiating x^n using the limit definition.

Derivative of $f(x) = x^n$

If $f(x) = x^n$, $n \in N$,
then $f'(x) = nx^{n-1}$.

It also follows that any coefficient of x^n will be unaffected by the differentiation operation.

Derivative of $f(x) = ax^n$

If $f(x) = ax^n$, $n \in N, a \in R$,
then $f'(x) = anx^{n-1}$.

However, if a function is itself a constant, then its graph is a horizontal line with zero gradient. The derivative with respect to x of such a function must be zero.

Derivative of a constant

If $f(x) = c$ where c is a constant,
then $f'(x)=0$.

Where the polynomial function consists of the sum or difference of a number of terms of different powers of x, the derivative is calculated as the sum or difference of the derivative of each of these terms.

Derivative of polynomial functions

If $f(x) = a_n x^n + a_{n-1}x^{n-1} + \ldots + a_1 x + a_0$,
then $f'(x) = a_n n x^{n-1} + a_{n-1}(n-1)x^{n-2} + \ldots + a_1$.

The above results illustrate what are known as the **linearity properties** of the derivative:

- If $f(x) = g(x) \pm h(x)$, then $f'(x) = g'(x) \pm h'(x)$.
- If $f(x) = ag(x)$ where a is a constant, then $f'(x) = ag'(x)$.

Calculating the derivative

The linearity properties of the derivative enable the derivative of $f(x) = 3x^4$ to be calculated as $f'(x) = 3 \times 4x^3 \Rightarrow f'(x) = 12x^3$ and the derivative of $y = 3x^4 + x^2$ to be calculated as $\frac{dy}{dx} = 12x^3 + 2x$. This is differentiation by rule. Differentiation becomes an operation that can be applied to a polynomial function with relative ease. Although the limit definition defines the derivative function, from now on we will only use this definition to calculate the derivative if specific instructions to use this method are given.

The notation for the derivative has several forms including $f'(x)$, $\frac{dy}{dx}$, $\frac{d}{dx}(f(x))$, $\frac{df}{dx}$ or $D_x(f)$.

This means that the derivative of $3x^4$ equals $12x^3$, could be written as $f'(x) = 12x^3$ or $\frac{dy}{dx} = 12x^3$ or $\frac{d}{dx}(3x^4) = 12x^3$ or $\frac{df}{dx} = 12x^3$ or $D_x(f) = 12x^3$.

For functions of variables other than x, such as $p = f(t)$, the derivative would be $\frac{dp}{dt} = f'(t)$.

WORKED EXAMPLE 7 Differentiating a function

a. Calculate $\frac{d}{dx}(5x^3 - 8x^2)$.

b. Differentiate $\frac{1}{15}x^5 - 7x + 9$ with respect to x.

c. If $p(t) = (2t+1)^2$, calculate $p'(t)$.

d. If $y = \frac{x - 4x^5}{x}$, $x \neq 0$, calculate $\frac{dy}{dx}$.

THINK	WRITE
a. Apply the rules for differentiation of polynomials.	a. This is a difference of two functions. $\frac{d}{dx}(5x^3 - 8x^2) = \frac{d}{dx}(5x^3) - \frac{d}{dx}(8x^2)$ $= 5 \times 3x^{3-1} - 8 \times 2x^{2-1}$ $= 15x^2 - 16x$
b. Choose which notation to use and then differentiate.	b. Let $f(x) = \frac{1}{15}x^5 - 7x + 9$. $f'(x) = \frac{1}{15} \times 5x^{5-1} - 7 \times 1x^{1-1} + 0$ $= \frac{1}{3}x^4 - 7$
c. 1. Express the function in expanded polynomial form.	c. $p(t) = (2t+1)^2$ $= 4t^2 + 4t + 1$
2. Differentiate the function.	$\therefore p'(t) = 8t + 4$
d. 1. Express the function in partial fraction form and simplify. *Note:* An alternative method is to factorise the numerator as $y = \frac{x(1 - 4x^4)}{x}$ and cancel to obtain $y = 1 - 4x^4$. Another alternative is to use an index law to write $y = x^{-1}(x - 4x^5)$ and expand using index laws to obtain $y = 1 - 4x^4$.	d. $y = \frac{x - 4x^5}{x}$, $x \neq 0$ $= \frac{x}{x} - \frac{4x^5}{x}$ $= 1 - 4x^4$
2. Calculate the derivative.	$y = 1 - 4x^4$ $\frac{dy}{dx} = -16x^3$

TI | THINK

b. 1. On a Calculator page, press MENU and select:
4. Calculus
1. Derivative
Complete the entry line as:
$\frac{d}{dx}\left(\frac{1}{15}x^5 - 7x + 9\right)$
Then press ENTER.

DISPLAY/WRITE

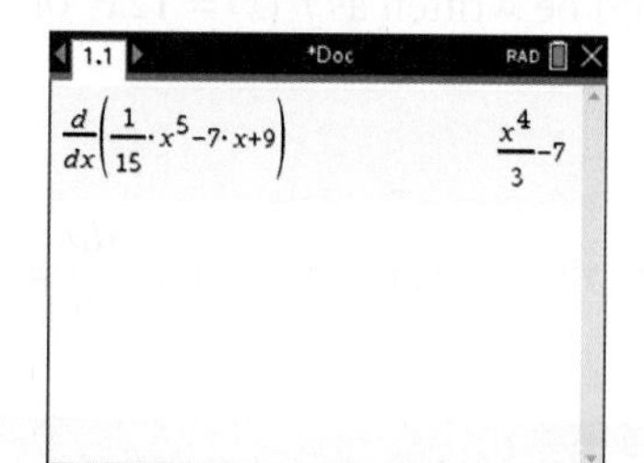

2. The answer appears on the screen.

$\frac{x^4}{3} - 7$

CASIO | THINK

b. 1. On a Main screen, complete the entry line as:
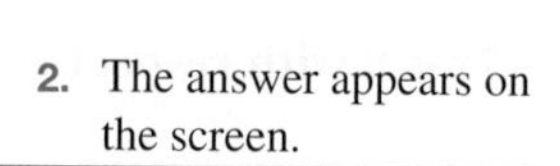
$\frac{d}{dx}\left(\frac{1}{15}x^5 - 7x + 9\right)$
Then press EXE.
Note: The template for the derivative is in the Math 2 Keyboard menu.

DISPLAY/WRITE

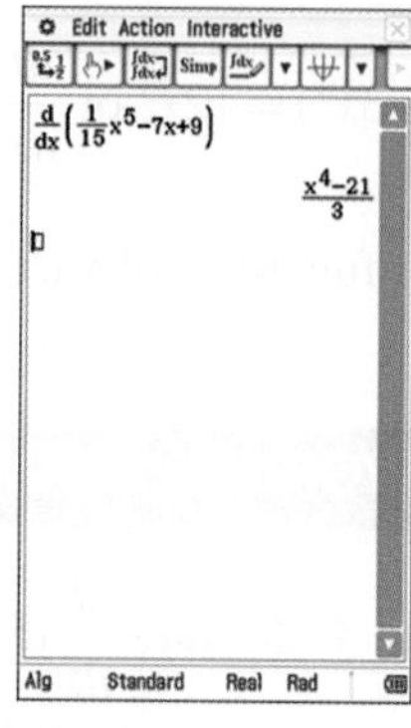

2. The answer appears on the screen.

$\frac{x^4 - 21}{3}$

11.5.2 Using the derivative

Differential calculus is concerned with the analysis of rates of change. At the start of this topic we explored ways to estimate the instantaneous rate of change of a function. Now, at least for polynomial functions, we can calculate these rates of change using calculus. The instantaneous rate of change is obtained by evaluating the derivative at a particular point. An average rate, however, is the gradient of the chord joining the end points of an interval and its calculation does not involve the use of calculus.

WORKED EXAMPLE 8 Calculating rates of change

Consider the polynomial function with equation $y = \frac{x^3}{3} - \frac{2x^2}{3} + 7$.

a. Calculate the rate of change of the function at the point (3, 10).

b. Calculate the average rate of change of the function over the interval $x \in [0, 3]$.

c. Obtain the coordinates of the point(s) on the graph of the function where the gradient is $-\frac{4}{9}$.

THINK

a. 1. Differentiate the function.

WRITE

a. $y = \frac{x^3}{3} - \frac{2x^2}{3} + 7$

$\frac{dy}{dx} = \frac{3x^2}{3} - \frac{4x}{3}$

$= x^2 - \frac{4x}{3}$

2. Calculate the rate of change at the given point.

At the point (3, 10), $x = 3$.

The rate of change is the value of $\frac{dy}{dx}$ when $x = 3$.

$\frac{dy}{dx} = (3)^2 - \frac{4(3)}{3}$

$= 5$

The rate of change of the function at (3, 10) is 5.

b. 1. Calculate the end points of the interval.

b. $y = \frac{x^3}{3} - \frac{2x^2}{3} + 7$ over $[0, 3]$

When $x = 0, y = 7 \Rightarrow (0, 7)$

When $x = 3, y = 10 \Rightarrow (3, 10)$ (given)

2. Calculate the average rate of change over the given interval.

The gradient of the line joining the end points of the interval is:

$$m = \frac{10 - 7}{3 - 0}$$
$$= 1$$

Therefore, the average rate of change is 1.

c. 1. Form an equation from the given information.

c. When the gradient is $-\frac{4}{9}$, $\frac{dy}{dx} = -\frac{4}{9}$.

$$\therefore x^2 - \frac{4x}{3} = -\frac{4}{9}$$

2. Solve the equation.

Multiply both sides by 9.

$$9x^2 - 12x = -4$$
$$9x^2 - 12x + 4 = 0$$
$$(3x - 2)^2 = 0$$
$$3x - 2 = 0$$
$$x = \frac{2}{3}$$

3. Calculate the y-coordinate of the point.

$$y = \frac{1}{3}x^3 - \frac{2}{3}x^2 + 7$$

When $x = \frac{2}{3}$,

$$y = \frac{1}{3} \times \left(\frac{2}{3}\right)^3 - \frac{2}{3} \times \left(\frac{2}{3}\right)^2 + 7$$
$$= \frac{8}{81} - \frac{8}{27} + 7$$
$$= -\frac{16}{81} + 7$$
$$= 6\frac{65}{81}$$

At the point $\left(\frac{2}{3}, 6\frac{85}{81}\right)$, the gradient is $-\frac{4}{9}$.

TI | THINK	DISPLAY/WRITE
a. 1. On a Calculator page, press MENU and select: 1. Actions 1. Define Complete the entry line as: Define $y = \frac{x^3}{3} - \frac{2x^2}{3} + 7$ Then press ENTER.	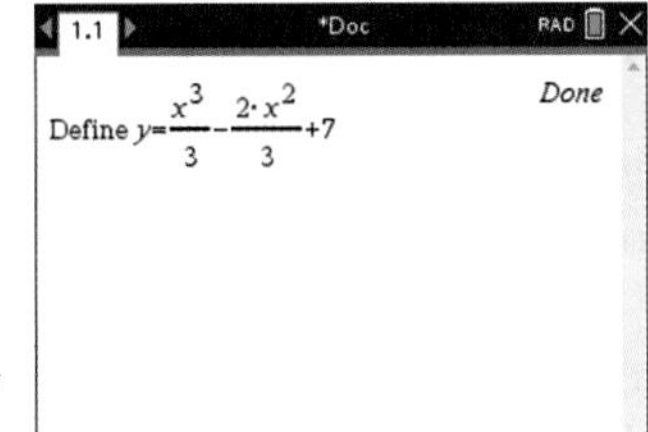

CASIO | THINK	DISPLAY/WRITE
a. 1. On a Main screen, define the equation by completing the entry line as: $y := \frac{x^3}{3} - \frac{2x^2}{3} + 7$ Then press EXE.	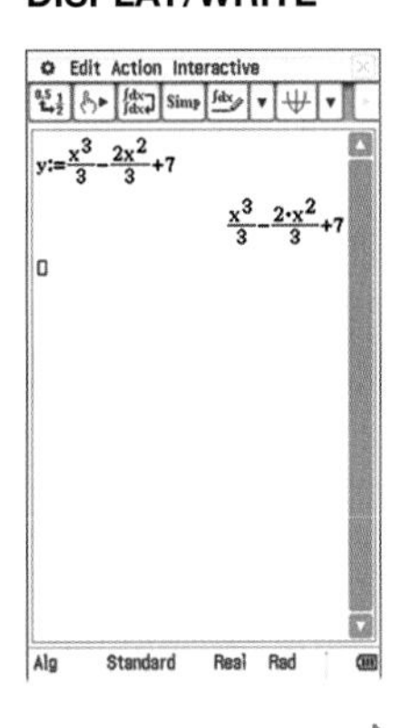

2. Press MENU and select:
   4: Calculus
   2: Derivative at a Point
   Variable: x
   Value: 3
   Derivative: 1st Derivative
   Complete the entry line as:
   $\frac{d}{dx}(y)|x = 3$
   Then press ENTER.

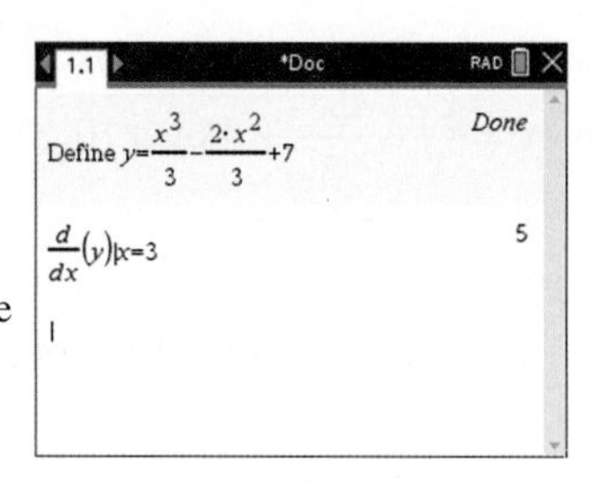

3. The answer appears on the screen.

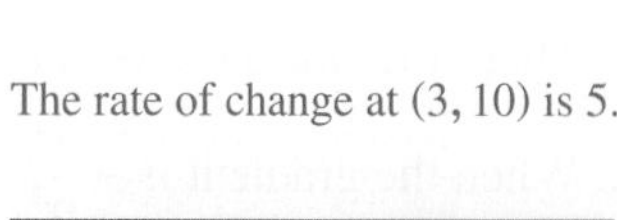

The rate of change at (3, 10) is 5.

C. 1. Press MENU and select:
   3: Algebra
   1: Solve
   Complete the entry line as:
   $\text{solve}\left(\frac{d}{dx}(y) = -\frac{4}{9}, x\right)$
   Then press ENTER.
   Note: To show $\frac{d}{dx}(y)$, press MENU, then select:
   4: Calculus
   1: Derivative

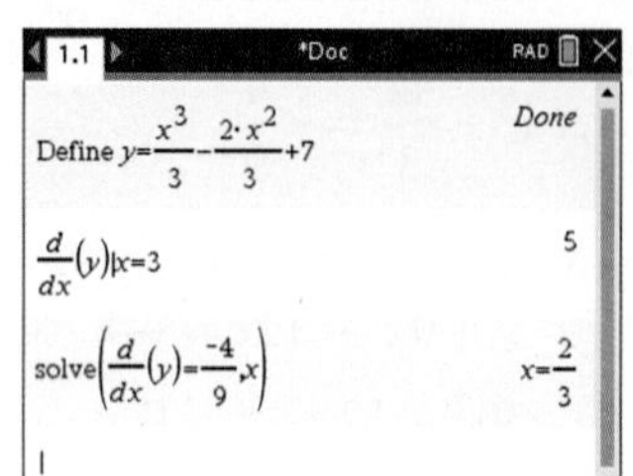

2. The answer appears on the screen.

$x = \frac{2}{3}$

2. Complete the entry line as:
   $\frac{d}{dx}(y)|x = 3$
   Then press EXE.

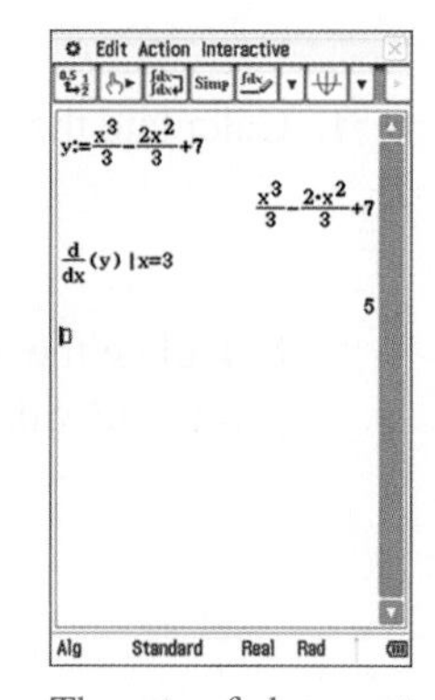

3. The answer appears on the screen.

The rate of change at (3, 10) is 5.

C. 1. Complete the entry line as:
   $\text{solve}\left(\frac{d}{dx}(y) = -\frac{4}{9}, x\right)$
   Then press EXE.

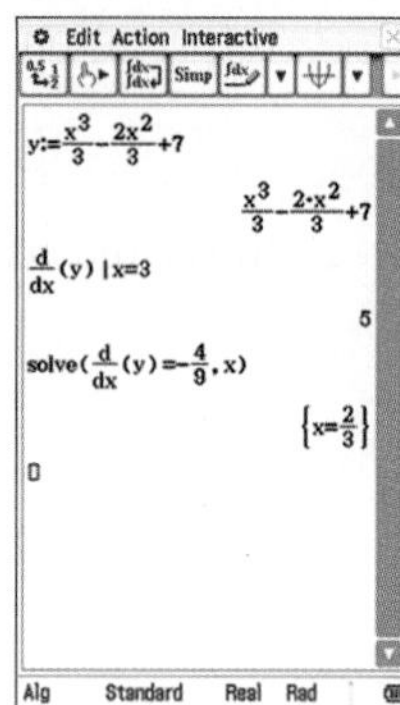

2. The answer appears on the screen.

$x = \frac{2}{3}$

11.5.3 The graph of the derivative function

The derivative of a polynomial function in x is also a polynomial function, but with one degree less than the original polynomial.

As a mapping, for $n \in N$, if $f: R \to R,\ f(x) = x^n$,
then $f': R \to R,\ f'(x) = nx^{n-1}$.

The graphs of a function and its derivative function are interrelated. To illustrate this, consider the graphs of $y = f(x)$ and $y = f'(x)$ for the functions defined as $f: R \to R,\ f(x) = x^2 + 1$
and $f': R \to R,\ f'(x) = 2x$.

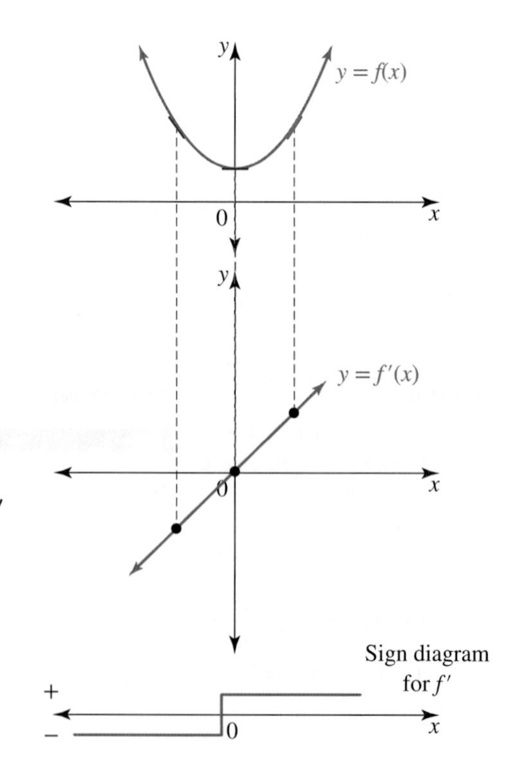

A comparison of the graphs of $y = f(x)$ and $y = f'(x)$ shows the following:

- The graph of f is quadratic (degree 2) and the graph of f' is linear (degree 1).
- The x-coordinate of the turning point of f is the x-intercept of f' (this is because the tangent at the turning point is horizontal so its gradient, $f'(x)$, is zero).
- Over the section of the domain where any tangent to the graph of f would have a negative gradient, the graph of f' lies below the x-axis.
- Over the section of the domain where any tangent to the graph of f would have a positive gradient, the graph of f' lies above the x-axis.

- The turning point on the graph of f is a minimum and the graph of f' cuts the x-axis from below to above; that is, from negative to positive f' values.

These connections enable the graph of the derivative function to be deduced from the graph of a function.

WORKED EXAMPLE 9 Graphing the derivative function

The graph of a cubic polynomial function $y = f(x)$ is given. The turning points occur where $x = \pm\ a$. Sketch a possible graph for its gradient function.

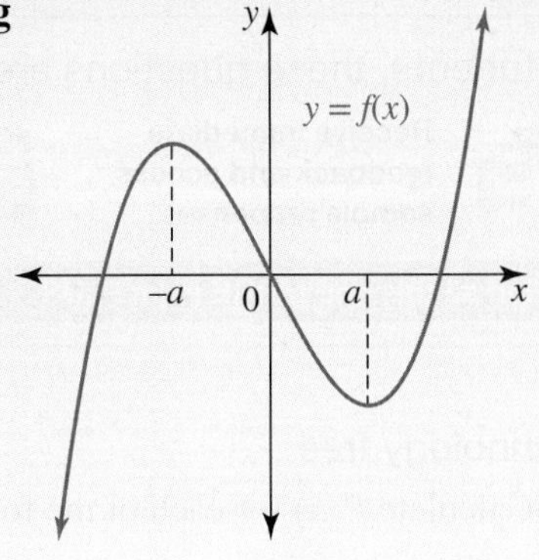

THINK	WRITE
1. Identify the values of the x-intercepts of the gradient function's graph.	The x-coordinates of the turning points of $y = f(x)$ give the x-intercepts of the gradient graph. Turning points of $y = f(x)$ occur at $x = \pm a$ $\therefore f'(\pm a) = 0$ $\Rightarrow$ the gradient graph has x-intercepts at $x = \pm a$.
2. By considering the gradient of any tangent drawn to the given curve at sections on either side of the turning points, determine whether the corresponding gradient graph lies above or below its x-axis.	For $x < -a$, the gradient of the curve is positive $\Rightarrow$ the gradient graph lies above the x-axis. For $-a < x < a$, the gradient of the curve is negative $\Rightarrow$ the gradient graph lies below the x-axis. For $x > a$, the gradient of the curve is positive $\Rightarrow$ the gradient graph lies above the x-axis.
3. State the degree of the gradient function.	The given function has degree 3, so the gradient function has degree 2.
4. Identify any other key features of the gradient graph. *Note:* It is often not possible to locate the y-coordinate of the turning point of the gradient graph without further information.	The gradient function is a quadratic. Its axis of symmetry is the vertical line halfway between $x = \pm a$. $\Rightarrow$ the gradient graph is symmetric about $x = 0$ (y-axis). The x-coordinate of its turning point is $x = 0$.
5. Sketch the shape of the gradient function. *Note:* The y-axis scale is not known.	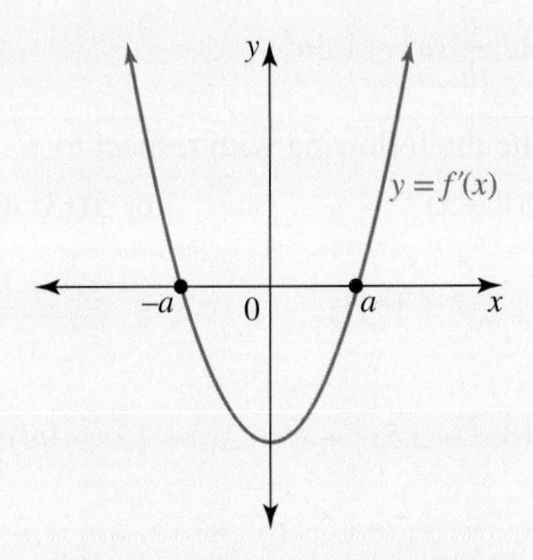

Resources

Interactivity Graph of a derivative function (int-5961)

11.5 Exercise

Students, these questions are even better in jacPLUS

Receive immediate feedback and access sample responses

Access additional questions

Track your results and progress

Find all this and MORE in jacPLUS

Technology free

1. Calculate $f'(x)$ for each of the following polynomial functions.

a. $f(x) = 16$

b. $f(x) = 21x + 9$

c. $f(x) = 3x^2 - 12x + 2$

d. $f(x) = 0.3x^4$

e. $f(x) = -\frac{2}{3}x^6$

f. $f(x) = 10x^4 - 2x^3 + 8x^2 + 7x + 1$

2. Express each function in expanded polynomial form and then calculate $\frac{dy}{dx}$.

a. $y = -2x(5x - 4)$

b. $y = (6x + 5)(4x + 1)$

c. $y = \left(\frac{x^2}{2} + 1\right)\left(\frac{x^2}{2} - 1\right)$

d. $y = (2x - 3)(4x^2 + x + 9)$

e. $y = (5x + 4)^2$

f. $y = 0.5x^3(2 - x)^2$

3. **WE7** a. Calculate $\frac{d}{dx}\left(5x^8 + \frac{1}{2}x^{12}\right)$.

b. Differentiate $2t^3 + 4t^2 - 7t + 12$ with respect to t.

c. If $f(x) = (2x + 1)(3x - 4)$, calculate $f'(x)$.

d. If $y = \frac{4x^3 - x^5}{2x^2}$, calculate $\frac{dy}{dx}$.

4. a. If $f(x) = x^7$, determine $f'(x)$.

b. If $y = 3 - 2x^3$, determine $\frac{dy}{dx}$.

c. Calculate $\frac{d}{dx}(8x^2 + 6x - 4)$.

d. If $f(x) = \frac{1}{6}x^3 - \frac{1}{2}x^2 + x - \frac{3}{4}$, determine $D_x(f)$.

e. Calculate $\frac{d}{du}(u^3 - 1.5u^2)$.

f. If $z = 4(1 + t - 3t^4)$, determine $\frac{dz}{dt}$.

5. Differentiate the following with respect to x.

a. $(2x + 7)(8 - x)$

b. $5x(3x + 4)^2$

c. $(x - 2)^4$

d. $250(3x + 5x^2 - 17x^3)$

e. $\frac{3x^2 + 10x}{x}, x \neq 0$

f. $\frac{4x^3 - 3x^2 + 10x}{2x}, x \neq 0$

6. If $z = \frac{1}{420}(3x^5 - 3.5x^4 + x^3 - 2x^2 + 12x - 99)$, calculate $\frac{dz}{dx}$.

7. Calculate each of the following.

a. $f'(-1)$ given $f(x) = (3x^4)^2$

b. $g'(2)$ given $g(x) = \dfrac{2x^3 + x^2}{x}$, $x \neq 0$

c. The gradient of the tangent drawn to the curve $y = -\dfrac{1}{3}x^3 + 4x^2 + 8$ at the point $(6, 80)$

d. The coordinates of the point on the curve $h = 10 + 20t - 5t^2$ where $\dfrac{dh}{dt} = 0$

8. a. Consider the function $f(x) = 3x^2 - 2x$.

i. Calculate the instantaneous rate of change of the function at the point where $x = -1$.
ii. Calculate the average rate of change of the function over the interval $x \in [-1, 0]$.

b. Consider the function $g(x) = x^2(2 + x - x^2)$.

i. Calculate the rate of change of the function at the point $(1, 2)$.
ii. Calculate the average rate of change of the function over the interval $x \in [0, 2]$.

Technology active

9. **WE8** Consider the polynomial function with equation $y = \dfrac{2x^3}{3} - x^2 + 3x - 1$.

a. Calculate the rate of change of the function at the point $(6, 125)$.
b. Calculate the average rate of change of the function over the interval $x \in [0, 6]$.
c. Obtain the coordinates of the point(s) on the graph of the function where the gradient is 3.

10. a. Calculate the gradient of the tangent to the curve $y = 3x^2 - x + 4$ at the point where $x = \dfrac{1}{2}$.

b. Calculate the gradient of the tangent to the curve $f(x) = \dfrac{1}{6}x^3 + 2x^2 - 4x + 9$ at the point $(0, 9)$.

c. Show that the gradient of the tangent to the curve $y = (x + 6)^2 - 2$ at its turning point is 0.
d. Show that the gradient of the tangent to the curve $y = 2 - (x - 3)^3$ at its point of inflection is 0.

11. For the curve with equation $f(x) = 5x - \dfrac{3}{4}x^2$, calculate:

a. $f'(-6)$
b. $f'(0)$
c. $\{x : f'(x) = 0\}$
d. $\{x : f'(x) > 0\}$
e. the gradient of the curve at the points where it cuts the x-axis
f. the coordinates of the point where the tangent has a gradient of 11.

12. Determine the coordinates of the point(s) on the curve $f(x) = x^2 - 2x - 3$ where:

a. the gradient is zero
b. the tangent is parallel to the line $y = 5 - 4x$
c. the tangent is perpendicular to the line $x + y = 7$
d. the slope of the tangent is half that of the slope of the tangent at point $(5, 12)$.

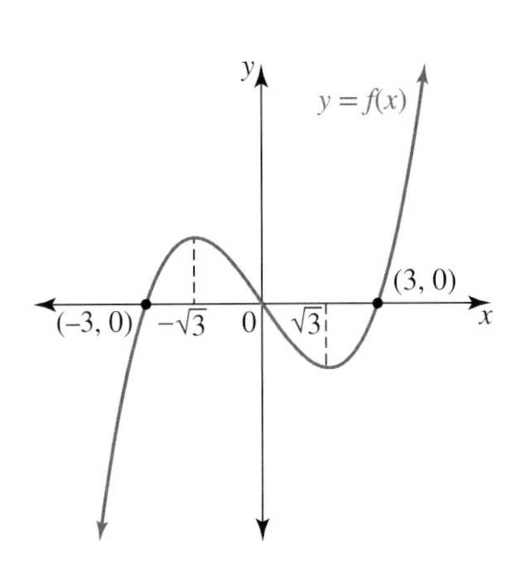

13. a. For the graph of $y = f(x)$ shown, state:

i. $\{x : f'(x) = 0\}$
ii. $\{x : f'(x) < 0\}$
iii. $\{x : f'(x) > 0\}$.

b. Sketch a possible graph for $y = f'(x)$.

14. For the curve defined by $f(x) = (x - 1)(x + 2)$, calculate the coordinates of the point where the tangent is parallel to the line with equation $3x + 3y = 4$.

15. WE9 The graph of a cubic polynomial function $y = f(x)$ is given. The turning points occur when $x = -a$ and $x = b$. Sketch a possible graph for its gradient function.

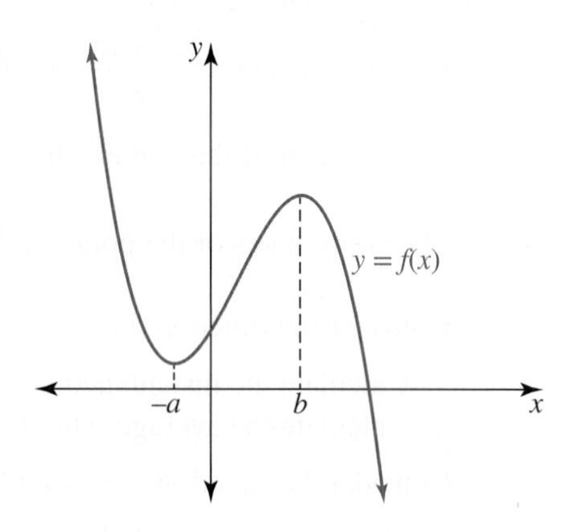

16. Sketch the graph of $y = (x + 1)^3$ and hence sketch the graph of its gradient $\frac{dy}{dx}$ versus x.

17. Sketch a possible graph of $\frac{dy}{dx}$ versus x for each of the following curves.

a.

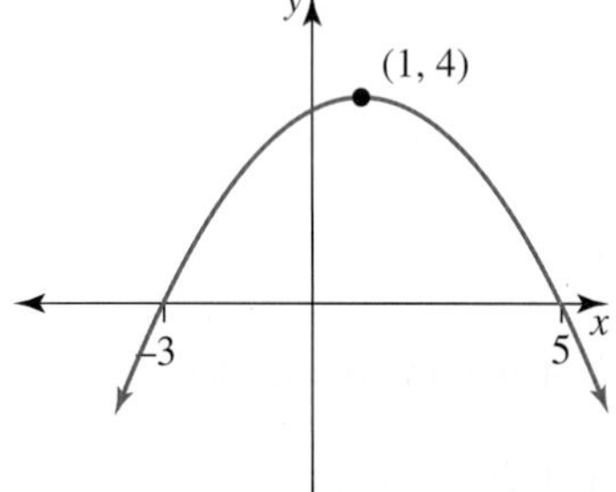

b.

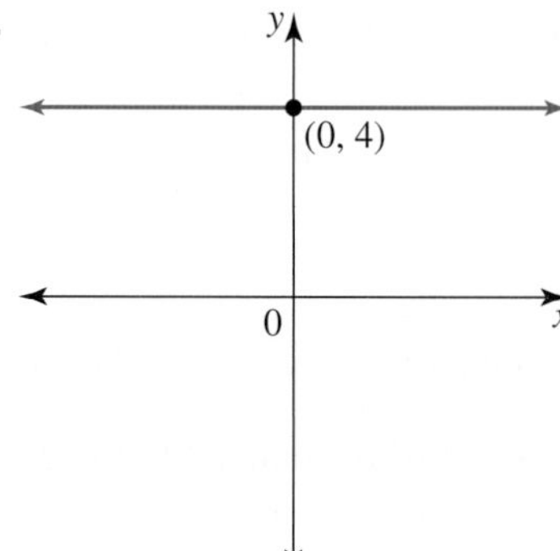

c.

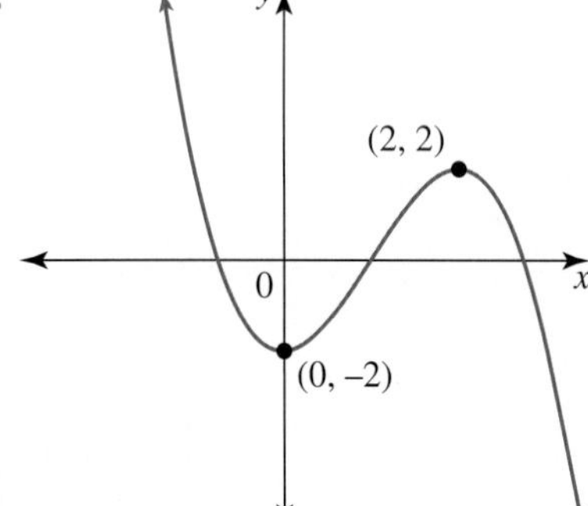

d.

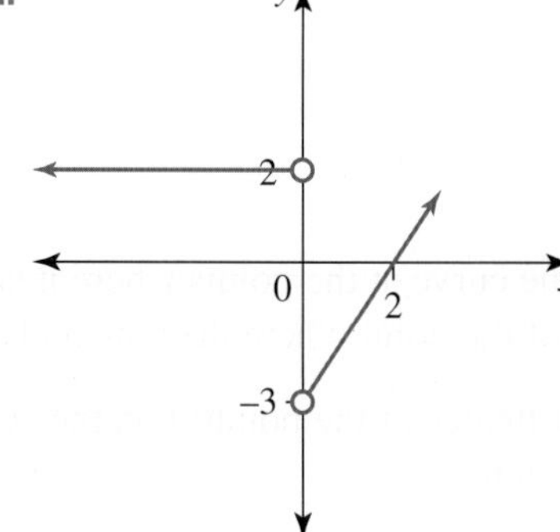

e.

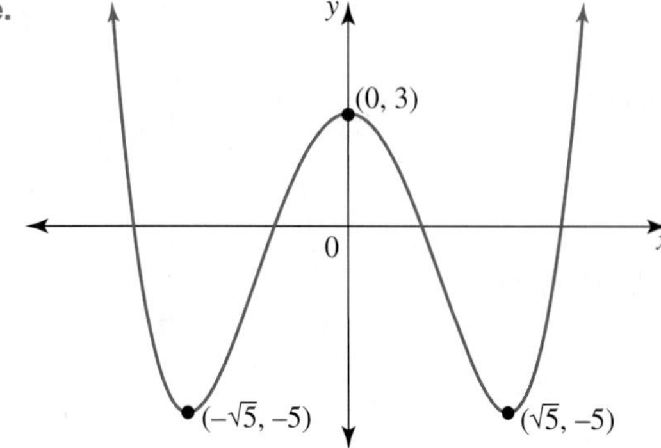

f.

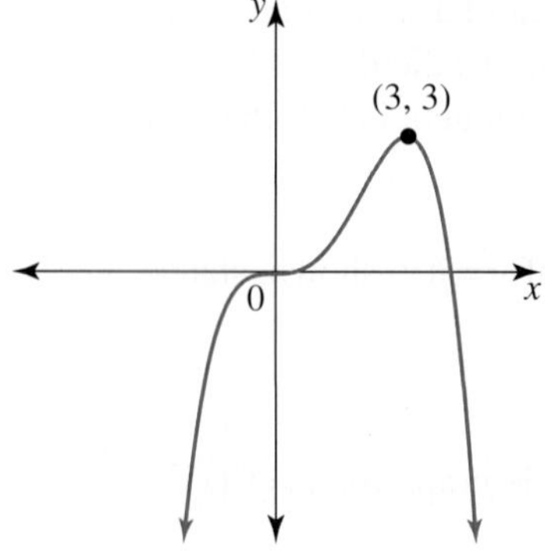

18. The size of a population of ants in a kitchen cupboard at time t hours after a pot of honey is spilt is given by $N = 0.5(t^2 + 1)^2 + 2.5$.
    a. Calculate the rate at which the population is growing at times $t = 1$ and $t = 2$.
    b. Calculate the average rate of growth over the first 2 hours.
    c. Calculate the time at which the population of ants is increasing at 60 ants/hour.

19. a. If the curve $y = 4x^2 + kx - 5$ has a gradient of 4 at the point where $x = -2$, determine the value of k.
    b. Use calculus to obtain the x-coordinate of the turning point of $y = ax^2 + bx + c$.
    c. i. Show that the gradient of the curve $f(x) = x^3 + 9x^2 + 30x + c$, $c \in R$ is always positive.
       ii. Hence, sketch the graph of the gradient function $y = f'(x)$ using the rule obtained for $f'(x)$.

    d. The radius, r metres, of the circular area covered by an oil spill after t days is $r = at^2 + bt$. After the first day the radius is growing at 6 metres/day and after 3 days the rate is 14 metres/day. Calculate a and b.

20. On the same set of axes, sketch the graphs of $y = f(x)$ and $y = f'(x)$, given:
    a. $f(x) = -x^2 - 1$
    b. $f(x) = x^3 - x^2$
    c. $f(x) = 6 - 3x$
    d. $f(x) = 1 - x^3$

11.5 Exam questions

Question 1 (1 mark) TECH-ACTIVE

MC If $f(x) = 2x^2 - 3x + 5$, then $f'(2)$ is

A. -3 **B.** 2 **C.** 5 **D.** 4 **E.** -4

Question 2 (1 mark) TECH-ACTIVE

MC The coordinates of the point on the curve $f(x) = 3x^2 - 12x + 8$ where the gradient is 0 are

A. $(3, 8)$ **B.** $(2, -4)$ **C.** $(3, -12)$ **D.** $(6, 12)$ **E.** $(0, 0)$

Question 3 (2 marks) TECH-FREE

Sam inflates a spherical balloon. If the volume of a sphere is $\frac{4\pi r^3}{3}$ where r is the radius, find the rate the volume is increasing when the radius is 15 cm.

More exam questions are available online.

11.6 Review

11.6.1 Summary

doc-37027

Hey students! Now that it's time to revise this topic, go online to:

Access the topic summary | Review your results | Watch teacher-led videos | Practise exam questions

Find all this and MORE in jacPLUS

11.6 Exercise

Technology free: short answer

1. a. Water leaks from a cylindrical container at a steady rate of 4 litres/hour. Draw a graph showing the depth of water remaining in the container at any time and describe how the depth changes.
   b. Rainwater is collected in a container in the shape of a right inverted cone. If the rainwater flows in at a steady rate of 4 litres/hour, draw a graph showing the depth of water collected in the initially empty container at any time and describe how the depth changes.
2. a. Obtain an expression in terms of h for the gradient of the chord joining the points $(2, 1)$ and $(2+h, f(2+h))$ on the curve $f(x) = \frac{x^2}{4}$ and evaluate this gradient if $h = 0.01$.
   b. Deduce the gradient of $f(x) = \frac{x^2}{4}$ at the point $(2, 1)$.
3. a. For $f(x) = 2x^2 - 3x + 4$, use the central difference approximation to determine $f'(x)$.
   b. For $f(x) = x(1 - x^3)$, use the limit definition to determine $f'(x)$.
4. a. If $y = \frac{1}{3}x^2(6+x)$, calculate $\frac{dy}{dx}$.
   b. If $f(x) = 2x^3 - \frac{3x(2x+1)}{2} - 7$, calculate $f'(x)$.
   c. Determine $D_x(f)$ for $f(x) = (4x+1)^3 - (1-2x)^2$.
   d. Differentiate $\frac{2x^3 + 4x^2 - 3x}{4x}$, $x \neq 0$ with respect to x.
5. Copy the graph and, on the same axes, sketch a possible graph of the derivative function.

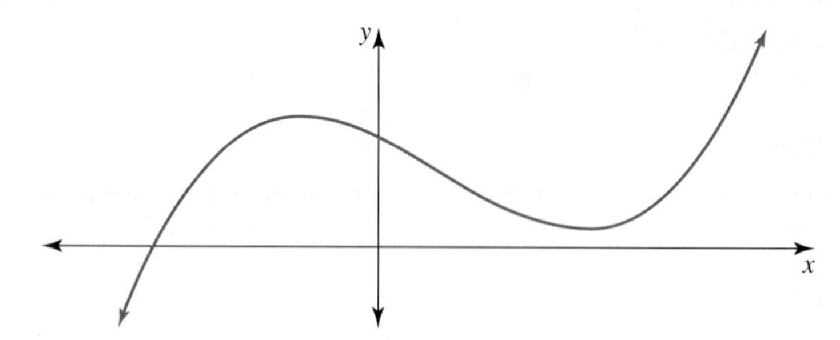

6. For the function defined by $f(x) = 3x^3 - 4x^2 + 2x + 5$, calculate:
   a. $f'(2)$
   b. the values of x for which $f'(x) = 3$.

Technology active: multiple choice

7. **MC** In order to get fit, Joan does daily walks as part of their exercise routine. The graph shows the distance Joan has walked on 1 day.
State which of the following is a correct statement about the graph.

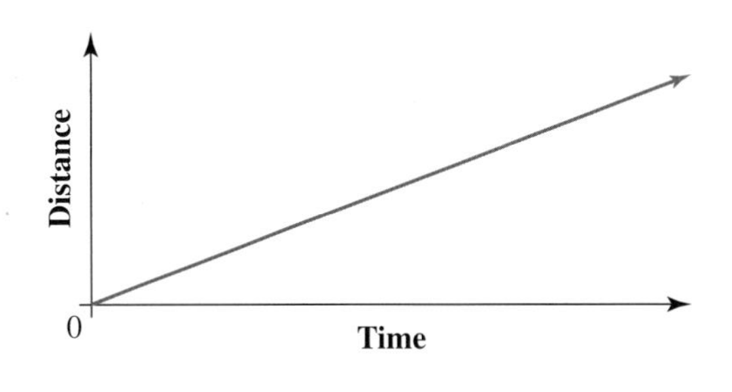

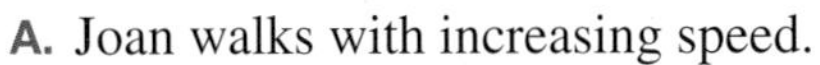

A. Joan walks with increasing speed.
B. Joan walks in a straight line with constant acceleration.
C. Joan walks with constant speed.
D. Joan is walking uphill with increasing speed.
E. Joan is walking uphill so their speed is decreasing.

Questions 8 and 9 relate to the following graph.

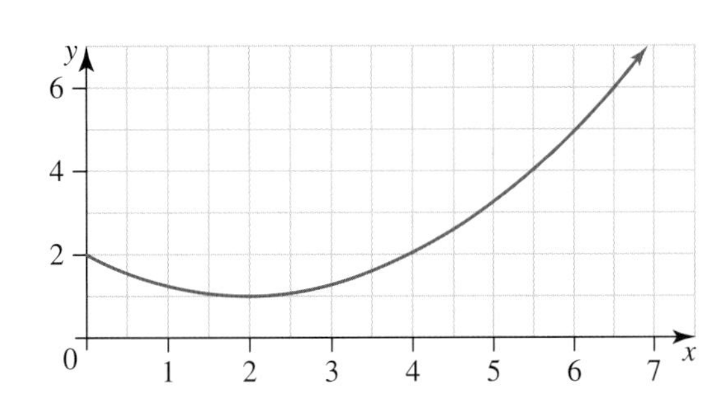

8. **MC** The average rate of change of the function for $x \in [0,\ 6]$ is closest to:

A. $\frac{1}{2}$ B. $\frac{2}{3}$ C. 1 D. $-\frac{2}{3}$ E. $-\frac{1}{2}$

9. **MC** The instantaneous rate of change of the function when $x = 4$ is best approximated by:

A. -2 B. 0 C. 1 D. 3 E. 5

10. **MC** The gradient of the secant passing through the points on $y = x^3$ where $x = 2$ and $x = 2.5$ is:

A. 6.75 B. 12 C. 13.5 D. 15.25 E. 15.375

11. **MC** The derivative of $f(x) = x^2 + 2x$ can be found by evaluating:

A. $\lim\limits_{h \to 0} \dfrac{(x^2 + 2x) - (h^2 + 2h)}{h}$

B. $\lim\limits_{h \to 0} \dfrac{(x^2 + h^2 + 2x) - (x^2 + 2x)}{h}$

C. $\lim\limits_{h \to 0} \dfrac{(x+h)^2 + 2(x+h) - (h^2 + 2h)}{h}$

D. $\lim\limits_{h \to 0} \dfrac{(x+h)^2 + 2(x+h) - (x^2 + 2h)}{x}$

E. $\lim\limits_{h \to 0} \dfrac{(x+h)^2 + 2(x+h) - (x^2 + 2x)}{h}$

12. **MC** If $f(x) = 10 - 7x^2 - 3x^3$, then $f'(-1)$ equals:

A. 6 B. 5 C. 2 D. 0 E. -8

13. **MC** If $y = (2x^2 - 5)(2x^2 + 5)$, then $\dfrac{dy}{dx}$ equals:

A. $8x$ B. $6x$ C. $2x$ D. $4x^2$ E. $16x^3$

14. **MC** $\dfrac{d}{dx}(2(x^3 - 5x^2 + 1))$ equals:

A. $2x^3 - 10x^2$ B. $6x^3 - 20x^2$ C. $2(3x - 5)$ D. $2x(3x - 10)$ E. 0

15. **MC** The tangent to the curve $y = \frac{1}{250}(1.5x^2 - 6x)$ is parallel to the x-axis at the point(s) where:

A. $x = 1$
B. $x = 2$
C. $x = 3$
D. $x = 4$
E. $x = 0, \ x = 4$

16. **MC** If the points (x, y) and $(x + h, \ y + k)$ are neighbouring points on the polynomial function defined by $y = 4 - x + x^2$, k would equal:

A. $(h)^2 + 2xh - h$
B. $(h)^2 + h$
C. $4 - h + (h)^2$
D. $-1 + 2x + h$
E. $-1 + 2x$

Technology active: extended response

17. Consider the function $f: R \to R, \ f(x) = x^3 - 4x^2 + 5x$.
   a. Write down the rule for $f'(x)$ and express f' as a mapping.
   b. Show that the graph of the function $y = f(x)$ has only one x-intercept and calculate the gradient of the curve at this x-intercept.
   c. Calculate the x-coordinates, in simplest surd form, of the points on the curve where the tangent is perpendicular to the line $3y + x = 6$.
   d. $f'(x)$ can be expressed in the form $(x - 1)(ax + b)$. Obtain the values of a and b.
   e. i. One of the points on the curve $y = f(x)$ where the gradient is zero has coordinates $\left(c, \frac{50}{27}\right)$. Calculate the value of c.
      ii. State the coordinates of the second point where the gradient is zero.
   f. Use the above information to draw a sketch of the graph of $y = f(x)$.

18. Consider the curve with equation $y = x(x + 3)$. The points P and Q lie on this curve where $x = 3$ and $x = 3 + h$ respectively.
   a. Determine the coordinates of P and Q.
   b. Calculate the gradient of the chord PQ.
   c. Hence, find the gradient of the curve at point P.
   d. Check your answer to part **c** by finding the derivative by rule.
   e. For $f(x) = x(x + 3)$, show that $\lim\limits_{a \to 3} \frac{f(3) - f(a)}{3 - a}$ also gives the gradient of the curve at P and explain with a diagram why this is so.

19. The population of Town N is thought to be modelled by $P_N = 2\left(1 + 4n + \frac{n^2}{4}\right)$, where P_N is the population in thousands n years after 2010.
   a. i. Calculate the average rate of growth of the population from 2010 to 2015.
      ii. Calculate the rate at which the population is predicted to be growing in 2020.

   The model for the population of neighbouring Town W is $P_W = 2\left(1 + n - \left(\frac{n}{4}\right)^2\right)$, where P_W is the population in thousands n years after 2010.

   b. i. Show that both towns N and W had the same population in the year 2010.
      ii. Calculate the rate at which the population of Town W is predicted to be changing in 2020.
   c. Determine the year in which the population of Town N was growing at 12 times the rate of Town W.
   d. Determine the year in which the population of Town W ceased to grow.
   e. Determine the year in which the population of Town W is predicted to fall back to only 2000 people.

f. Towns N and W lie on either side of a state's border. The state governments decide to combine the two towns into the one larger town to be renamed NEW Town, with this to come in to effect in the year 2020. Calculate:

i. the initial population of NEW Town in 2020
ii. the initial rate at which the population of NEW Town will grow in 2020.

20. Consider the function defined by $y = x^4 + 2x^2$

a. Sketch, using CAS technology, the graphs of $y = x^4 + 2x^2$ and its derivative.
b. Explain whether or not the derivative graph has a stationary point of inflection at the origin.
c. Obtain, to 4 decimal places, any non-zero values of x where $\frac{dy}{dx} = y$.
d. Form an equation where the solution(s) to the equation would give the x-values when the graphs of $y = x^4 + 2x^2$ and its derivative are parallel. Give the solution(s) correct to 2 decimal places where necessary.

11.6 Exam questions

Question 1 (1 mark) TECH-ACTIVE

MC The point D lies on the curve with equation $y = x^3 - x$. An expression for the y-coordinate of D, given that its x-coordinate is $3 + h$, is

A. $h^3 - h$
B. $h^3 + 9h^2 + 28h + 30$
C. $h^3 + 9h^2 + 26h + 24$
D. $h^3 + 9h^2 + 26h$
E. $h^3 + 9h^2 + 24h$

Question 2 (4 marks) TECH-ACTIVE

Calculate the gradient of the chord joining the points $x = 2.9$ and $x = 3$ on the curve with equation $y = \frac{x^3}{3} + 1$ and estimate the gradient of the tangent to the curve at the point where $x = 3$. Explain how you could improve your estimate.

Question 3 (1 mark) TECH-ACTIVE

MC The gradient of a secant for the function $f(x) = 2x^2 - 4$ is given by

A. $2h^2 - 4$
B. $2(x + h)^2 - 4$
C. $2h^2 - x^2$
D. $2(2x + h)$
E. $4(x + h) - 2$

Question 4 (1 mark) TECH-ACTIVE

MC The derivative of $\frac{4x^3 - 3x + 15}{3}$ is

A. $4x^2 - 1$
B. $\frac{4}{3}x^3 - x + 5$
C. $12x - 3$
D. $-3x + 15$
E. $3x^2 - 3$

Question 5 (4 marks) TECH-FREE

Sketch, on the same set of axes, the graphs of $y = f(x)$ and $y = f'(x)$ given that $f(x) = 2 - x^2$.

More exam questions are available online.

Answers

Topic 11 Introduction to differential calculus

11.2 Rates of change

11.2 Exercise

1. a. 2 °C/h b. 19 °C
2. a. 6 dresses per day
   b. 2 dresses per hour
   c. 50 km/h
3. a. $-\frac{2}{3}$ b. $-\frac{2}{3}$
   c. A linear function d. -3
4. a. The gradient of the line segment joining the two points
   b. P and Q
   c. S is the reflection in the y-axis of point R.
   d. The gradient of the tangent to the curve drawn at the point
   e. P
   f. R
5. 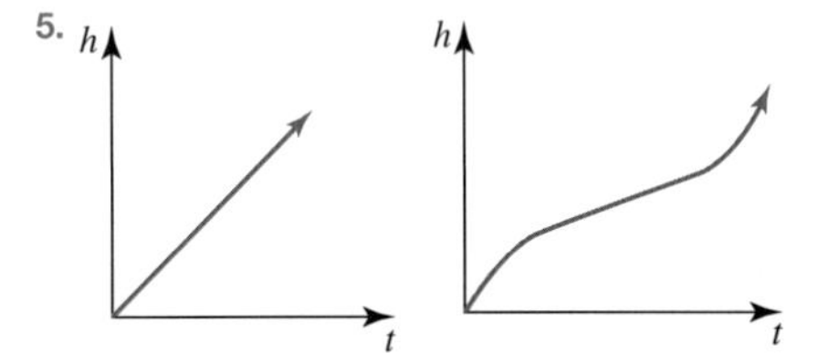

6. The average rate of change is 5.
7. 36 km/h
8. a. 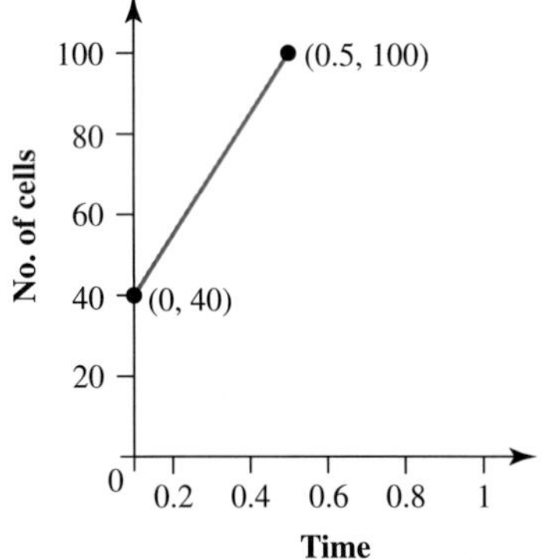

   Linear graph with positive gradient; the rate of growth is 120 bacterial cells per hour.

   b. 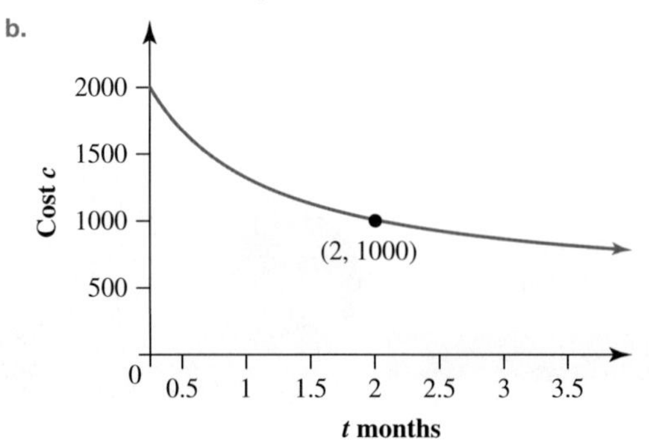

   Hyperbola shape for $t \geq 0$; costs decreased at an average rate of \$500 per month over the first 2 months.
9. a. Student II
   b. Student I
   c. Student III's level of interest increased slowly initially, but grew to a high level of interest; after a brief growth in interest, student IV started to lose interest, but eventually this slowed and the student gained a high level of interest quickly afterwards.
   d. Answers will vary.
10. a. -2 b. 3 c. 7 d. 0
11. a. \$600 per hour
    b. \$720 per hour
    c. \$35 per hour
12. a. Growing at approximately 0.018 euros per month
    b. $r = 5$ (5% per annum)
13. a. Average speed over the interval $t \in [0, 2]$ is 15 m/s; average speed over the interval $t \in [0.9, 1.1]$ is 14.01 m/s.
    b. 14.01 m/s
14. a. Approximately 6
    b. Approximately 5.8
15. a. Answers will vary; approximately 4
    b. Answers will vary; approximately -1 and -2
    c. -2.5
    d. Answers will vary; approximately 108°
16. Answers will vary. Approximately:
    a. i. 1.4 °C/hour
       ii. 0
       iii. -1.8 °C/hour
       iv. -1.4 °C/hour
    b. 8 pm
    c. -0.59 °C/hour
17. a. 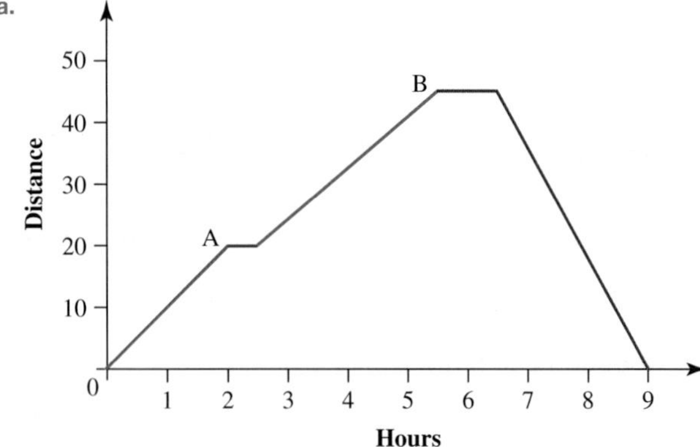

    b. 3 hours; $8\frac{1}{3}$ km/h
    c. 18 km/h
    d. 10 km/h
18. a. 21 mm
    b. $(24 - t)$ mm
    c. 24 weeks

19. a. 40 m/h

b. 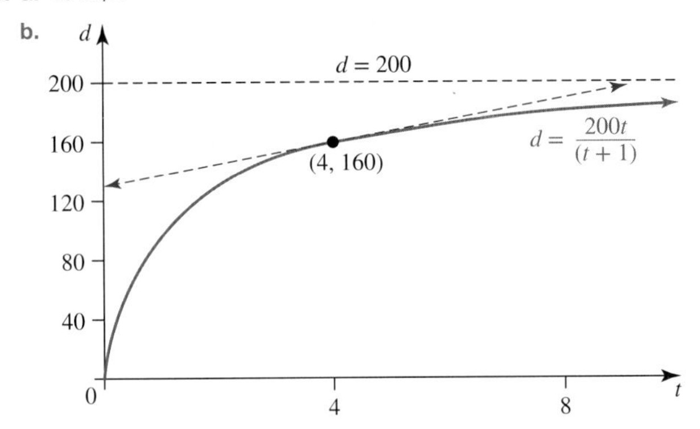

The boat initially travels away from the jetty quickly but slows almost to a stop as it nears the distance of 200 metres from the jetty.

c. Answers will vary; approximately 8 m/h

d. Answers will vary; for $t \in [4, 4.1]$ approximately 7.84 m/h

20. $y = 12x + 16$; $m = 12$

11.2 Exam questions

Note: Mark allocations are available with the fully worked solutions online.

1. C

2. a. 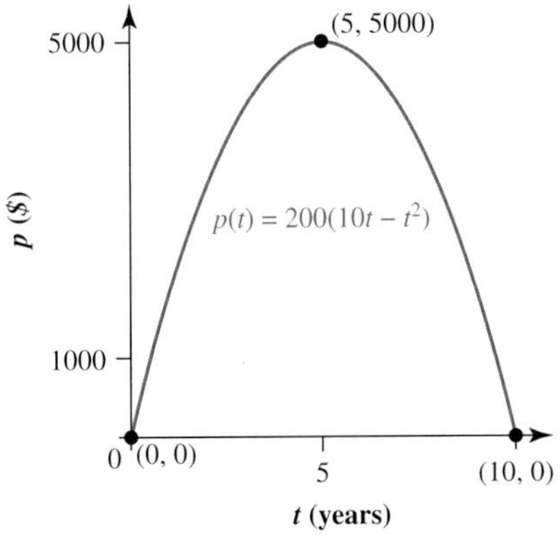

b. In the first 3 years, the rate of change of profit was positive (\$1400 per year) but, during years 6 to 9, the rate of change of profit was in decline at −\$1000 per year.

3. D

11.3 Gradients of secants

11.3 Exercise

1. A secant cuts a curve at two points; a tangent touches a curve at a single point; a chord is the line segment joining two points on a curve as its end points.

2. a. 2.5

b. 2.5

c. The same, the chord is a part of the secant.

3. a. $y = 1 + 2h + h^2$

b. $y = h^2 + h - 6$

c. $y = -3h^2 - 18h - 23$

d. $y = -2h^2 + 19h - 33$

4. a. 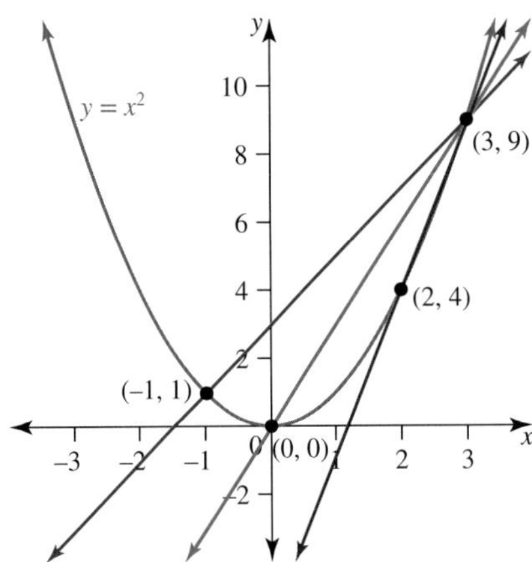

b. The best approximation is the secant through $(2, 4)$ and $(3, 9)$; choose a closer point.

5. a. $2 + h,\ h \neq 0$

b. $h + 7,\ h \neq 0$

c. $2(h + 6),\ h \neq 0$

d. $h - 2,\ h \neq 0$

6. a. $2h - 5,\ h \neq 0$

b. -4.98

c. -5

7. a. $(2 + h,\ -4h - h^2)$

b. $-4 - h,\ h \neq 0$

c. $h = 1$; $(3,\ -5)$

d. $h = 0.1$; -4.1

e. $h - 4,\ h \neq 0, -3.9$

f. -4

8. a. The y-coordinate is $h^3 + 9h^2 + 28h + 30$.

b. $h^2 + 9h + 28,\ h \neq 0$

c. 28.009 001

d. 28

9. a. $M(1, 1)$; $N\left(1 + h, \dfrac{1}{1 + h}\right)$

b. $-\dfrac{1}{1 + h},\ h \neq 0$

c. -1

d. -1

10. a. $1 + h, h \neq 0$

b. $\lim\limits_{h \to 0} (1 + h)$

c. 1

11. a. $\dfrac{h^2}{3},\ h \neq 0$

b. 0

12. a. 61.639 75

b. 62

c. Use closer points to form the chord.

13. a. Sample responses can be found in the worked solutions in the online resources.

b. $\lim\limits_{h \to 0} (h^2 - 4h + 1)$

c. 1

14. a. Cannot be simplified; 3
   b. $8 - 2h,\ h \neq 0;\ 8$
   c. $12 + 6h + h^2,\ h \neq 0;\ 12$
   d. $4 + 6h + 4h^2 + h^3,\ h \neq 0;\ 4$

15. a. $-\dfrac{1}{4(4+h)},\ h \neq 0;\ -\dfrac{1}{16}$
   b. $\sqrt{1+h} + 1,\ h \neq 0;\ 2$

16. a.

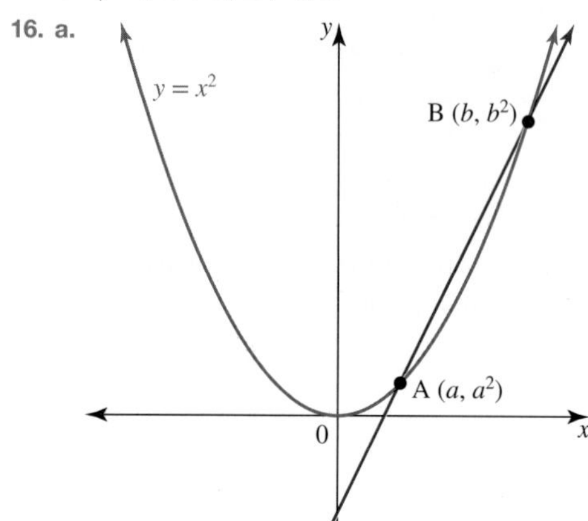

   b. $a + b, a \neq b$
   c. i. $2a$
      ii. $2b$

17. a. 8
   b. 3
   c. $\dfrac{2}{3}$

18. a. $\dfrac{\sqrt{9+h} - 3}{h}$
   b. $\lim\limits_{h \to 0} \dfrac{\sqrt{9+h} - 3}{h}$
   c. $\dfrac{1}{6}$

11.3 Exam questions

Note: Mark allocations are available with the fully worked solutions online.

1. B
2. a. $h + 8,\ h \neq 0$
   b. 8.01
   c. The gradient of the tangent to the curve at the point where $x = 6$ is 8.
3. A

11.4 The derivative function

11.4 Exercise

1. a. $f'(x), f'(x) = \lim\limits_{h \to 0} \dfrac{f(x+h) - f(x)}{h}$ or
   $\dfrac{dy}{dx}, \dfrac{dy}{dx} = \lim\limits_{\delta x \to 0} \dfrac{\delta y}{\delta x}$
   b. The rate of change of the function at $x = a$ or the gradient of the tangent at $x = a$

2. a. i. $x^2 + 2xh + h^2 + 2x + 2h$
      ii. $3 - 2x^2 - 4xh - 2h^2$
   b. i. $6xh + 3h^2 - 7h$
      ii. $10xh + 5h^2 - 3h$

3. a. $2x + h,\ h \neq 0$
   b. $f'(x) = 2x$
   c. 6
   d. 6

4. $f'(x) = 3x^2$

5. a. $f'(x) = 4x + 1$
   b. -3
   c. 1

6. a. Gradient of the secant through $(x, f(x))$ and $(x + h, f(x + h))$
   b. Gradient of the tangent at $(x, f(x))$
   c. Gradient of the tangent at $(3, f(3))$
   d. Gradient of the secant through $(3 - h, f(3 - h))$ and $(3, f(3))$

7. a. $f'(x) = 3 - 4x$
   b. 3
   c.

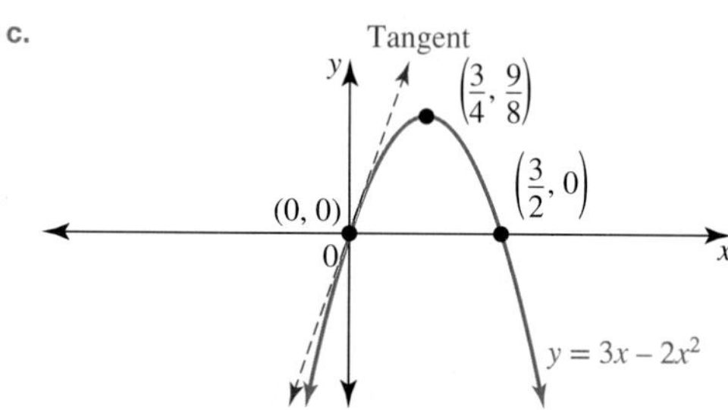

8. a. $6 + 3h,\ h \neq 0$
   b. 6
   c. i. $S(2, 14);\ T(2 + h, 3(2 + h)^2 + 2)$
      ii. $12 + 3h,\ h \neq 0$
      iii. 12

9. $5 + 12x^3$

10. a. $f'(x) = 16x$
    b. $f'(x) = x - 4$
    c. $f'(x) = -2$
    d. $f'(x) = 3x^2 - 12x + 2$

11. a. $f'(x) = 2x + 10$
    b. 0; the tangent at $x = -5$ has zero gradient.
    c. 10
    d. 6

12. a. $f'(x) \approx 10x - 2$
    b. 18
    c. -32

13. a. $f'(x) \approx -6x$
    b. $f'(x) \approx 6x^2 + 5 + h^2$
    c. $f'(x) \approx 3x^2 + 14x - 4 + h^2$
    d. $f'(x) \approx 2x - 4$

14. a. 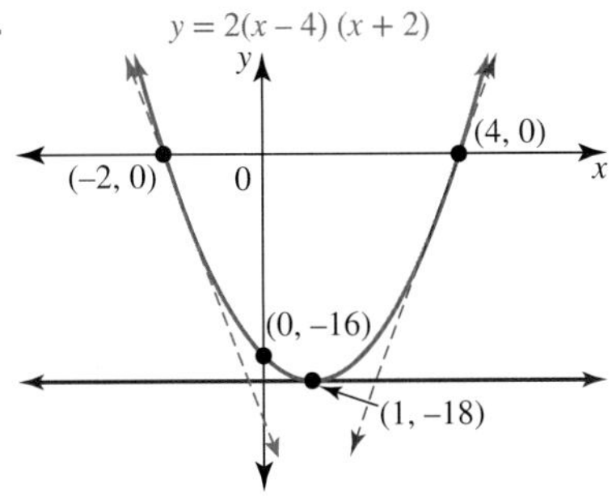

x-intercepts at $x = -2$, $x = 4$; y-intercept at $y = -16$; minimum turning point $(1, -18)$

b. $f'(x) = 4x - 4$

c. 0; the tangent at the turning point is horizontal.

d. -12; 12

15. a. $f'(x) = 2ax + b$

b. $f': R \to R$, $f'(x) = 2ax + b$

c. $f'(x) = 6x + 4$

d. f has degree 2; f' has degree 1.

16. a. $h\left(h^2 + 6h + 12\right)$

b. $g'(2) = \lim_{h \to 0} \dfrac{(2+h)^3 - 8}{h}$

c. 12

17. $2x$

18. a. $3ax^2 + 2bx + c$

b. f has degree 3; f' has degree 2.

11.4 Exam questions

Note: Mark allocations are available with the fully worked solutions online.

1. D
2. $f'(x) = 9x^2 - 2$
3. $\dfrac{dy}{dx} = -2x$

11.5 Differentiation of polynomials by rule

11.5 Exercise

1. a. 0

b. 21

c. $6x - 12$

d. $1.2x^3$

e. $-4x^5$

f. $40x^3 - 6x^2 + 16x + 7$

2. a. $-20x + 8$

b. $48x + 26$

c. x^3

d. $24x^2 - 20x + 15$

e. $50x + 40$

f. $6x^2 - 8x^3 + 2.5x^4$

3. a. $40x^7 + 6x^{11}$

b. $6t^2 + 8t - 7$

c. $12x - 5$

d. $2 - \dfrac{3}{2}x^2$

4. a. $f'(x) = 7x^6$

b. $\dfrac{dy}{dx} = -6x^2$

c. $16x + 6$

d. $D_x(f) = \dfrac{1}{2}x^2 - x + 1$

e. $3u^2 - 3u$

f. $\dfrac{dz}{dt} = 4\left(1 - 12t^3\right)$

5. a. $-4x + 9$

b. $135x^2 + 240x + 80$

c. $4x^3 - 24x^2 + 48x - 32$

d. $250\left(3 + 10x - 51x^2\right)$

e. 3

f. $4x - \dfrac{3}{2}$

6. $\dfrac{1}{420}(15x^4 - 14x^3 + 3x^2 - 4x + 12)$

7. a. -72

b. 9

c. 12

d. $(2, 30)$

8. a. i. -8

ii. -5

b. i. 3

ii. 0

9. a. 63

b. 21

c. $(0, -1)$, $\left(1, \dfrac{5}{3}\right)$

10. a. 2

b. -4

c. Sample responses can be found in the worked solutions in the online resources.

d. Sample responses can be found in the worked solutions in the online resources.

11. a. 14

b. 5

c. $\left\{\dfrac{10}{3}\right\}$

d. $\left\{x : x < \dfrac{10}{3}\right\}$

e. 5 and -5

f. $(-4, -32)$

12. a. $(1, -4)$

b. $(-1, 0)$

c. $\left(\dfrac{3}{2}, -\dfrac{15}{4}\right)$

d. $(3, 0)$

13. a. i. $\left\{\pm\sqrt{3}\right\}$

ii. $\{x: -\sqrt{3} < x < \sqrt{3}\}$

iii. $\{x: x < -\sqrt{3}\} \cup \{x : x > \sqrt{3}\}$

b. 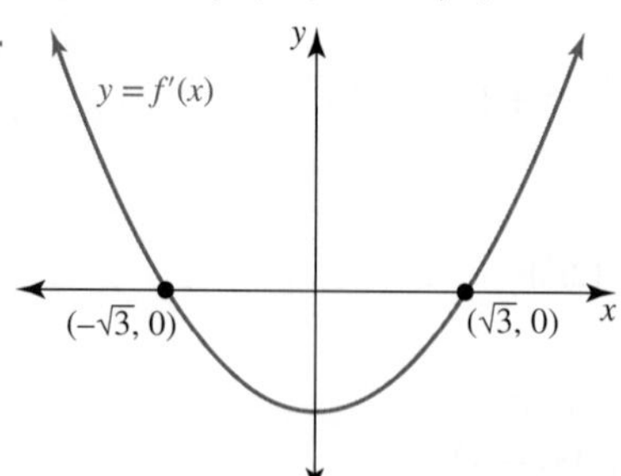

Concave up parabola with x-intercepts at $x = \pm\sqrt{3}$

14. $(-1, -2)$

15. 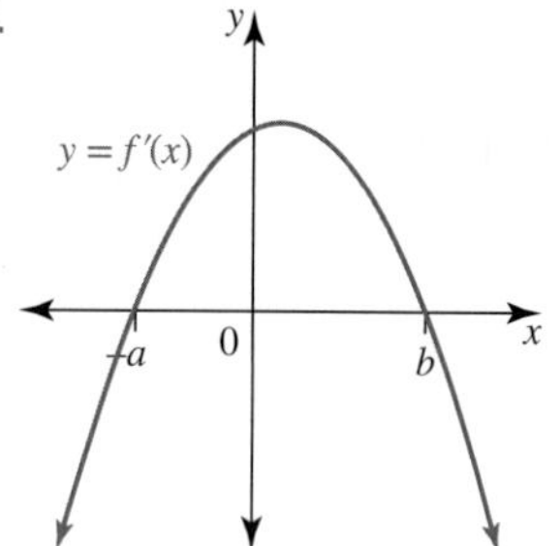

Concave down parabola with x-intercepts $x = -a$ and $x = b$

16. a. 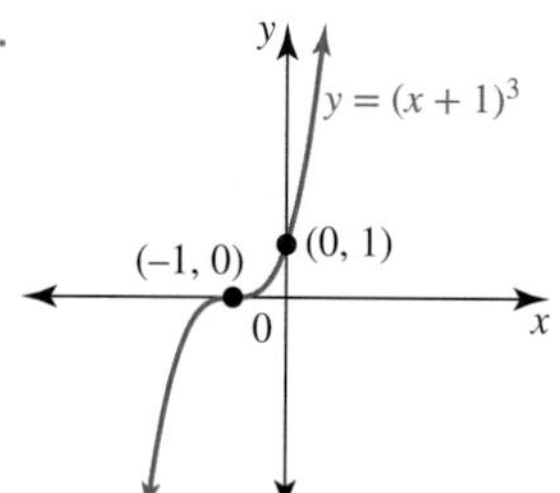

b. 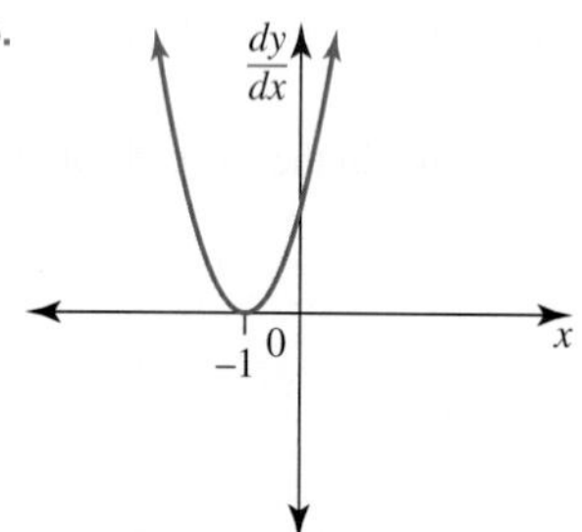

$y = (x+1)^3$ has a stationary point of inflection at $(-1, 0)$, y-intercept $(0, 1)$; the gradient graph is a concave up parabola with turning point at its x-intercept at $(-1, 0)$

17. a. 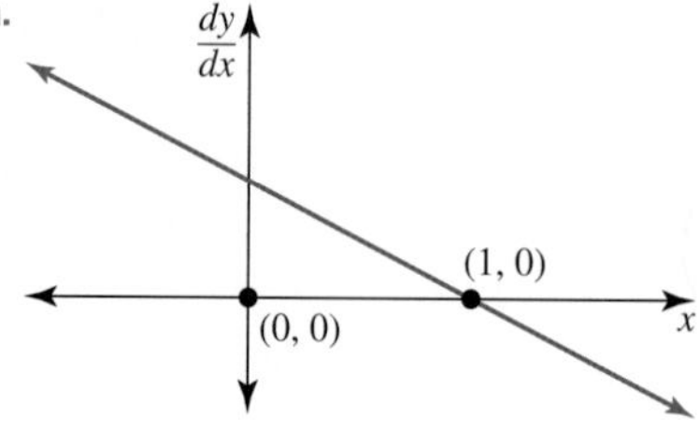

b. 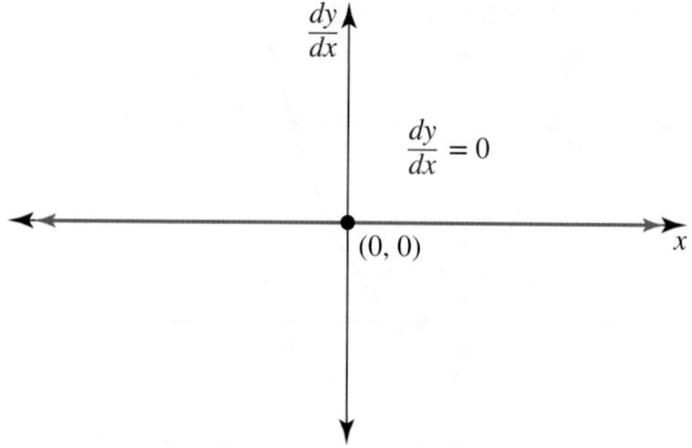

c. 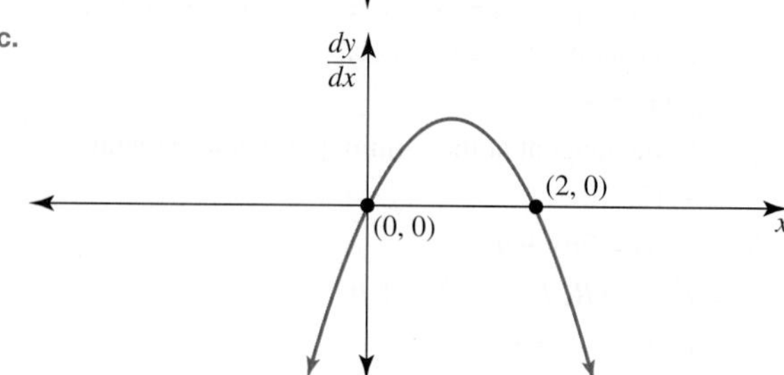

d. 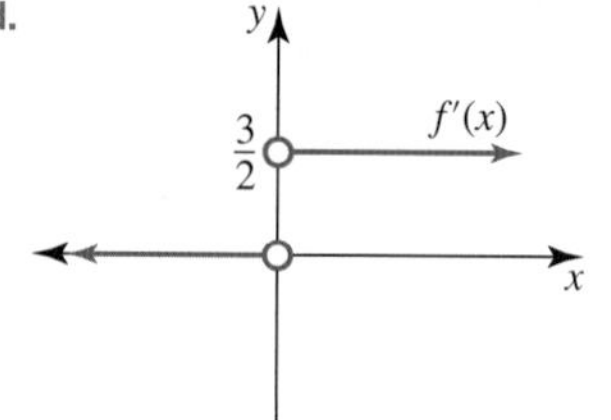

e. 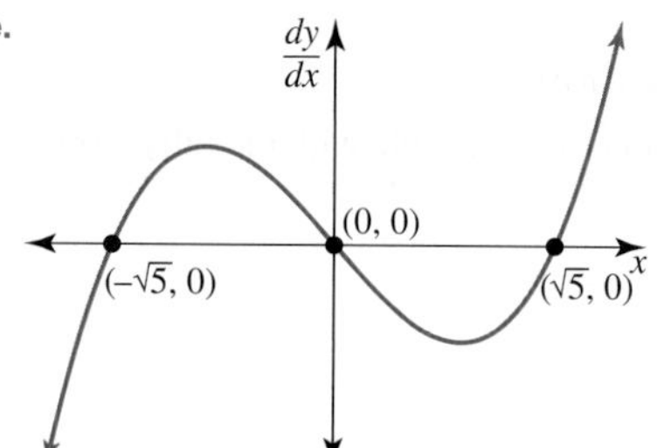

f. 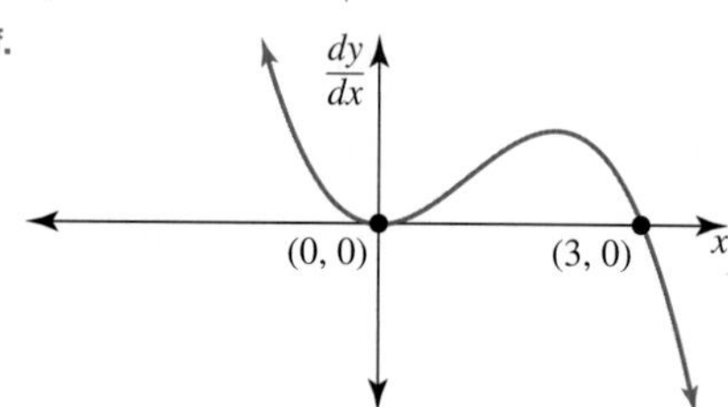

18. a. 4 ants/hour when $t = 1$ and 20 ants/hour when $t = 2$

b. 6 ants/hour

c. 3 hours

19. a. $k = 20$

b. $x = -\dfrac{b}{2a}$

c. i. Sample responses can be found in the worked solutions in the online resources.

ii.

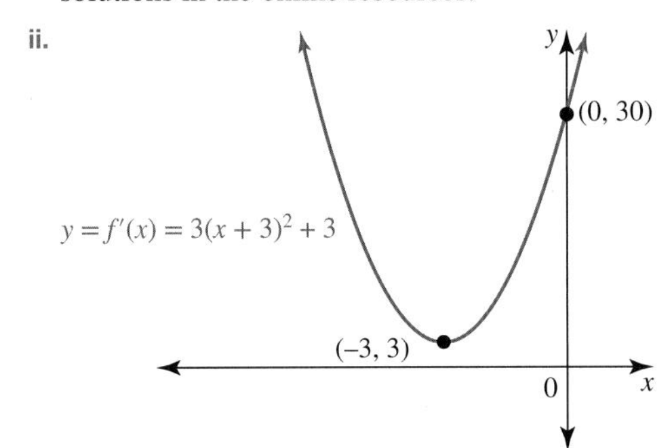

d. $a = 2; b = 2$

20. a. $f'(x) = -2x$

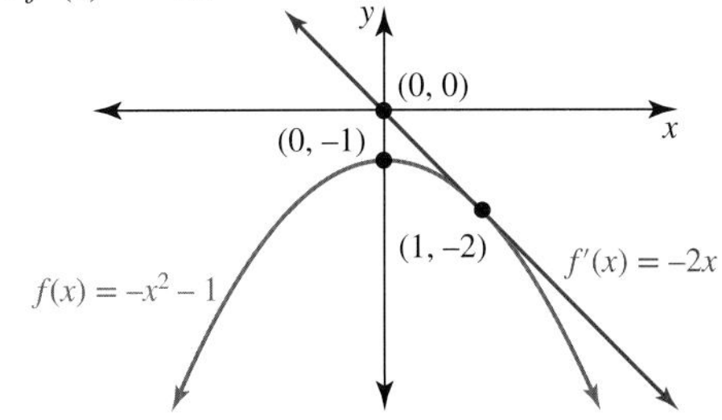

b. $f'(x) = 3x^2 - 2x$

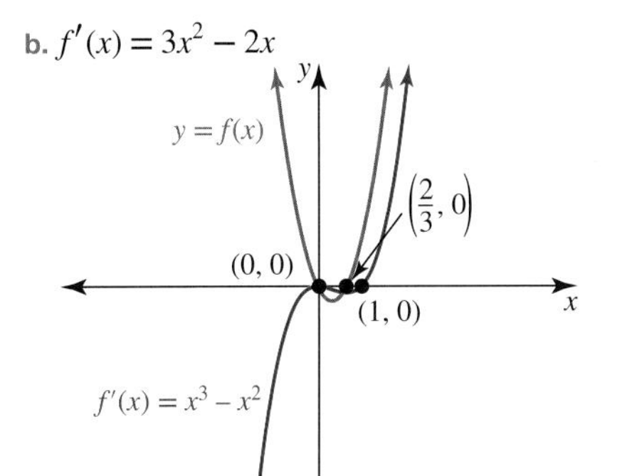

c. $f'(x) = -3$

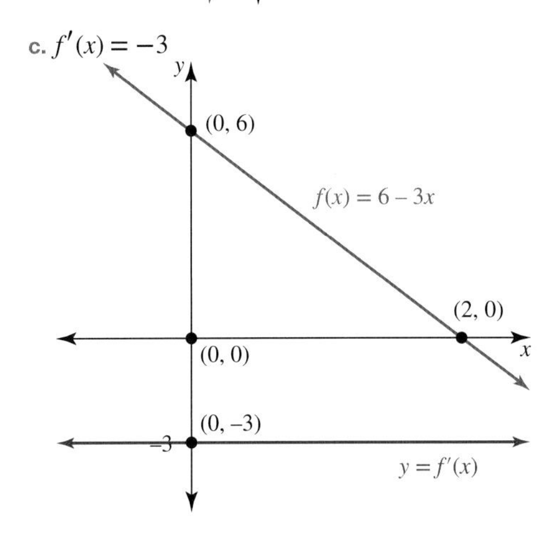

d. $f'(x) = -3x^2$

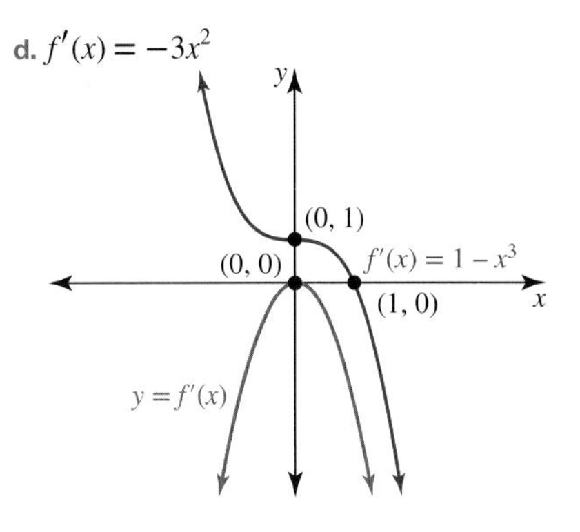

11.5 Exam questions

Note: Mark allocations are available with the fully worked solutions online.

1. C
2. B
3. $\dfrac{dV}{dr} = 900\pi \text{ cm}^3/\text{cm}$

11.6 Review

11.6 Exercise

Technology free: short answer

1. a.

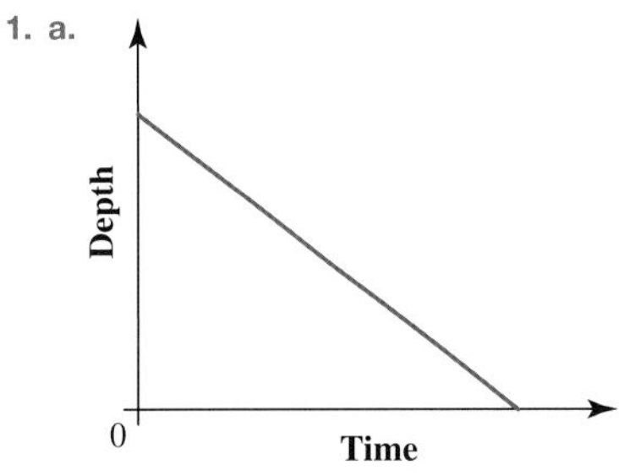

The depth will decrease steadily to zero. The graph is linear with a negative gradient, $m = -4$.

b.

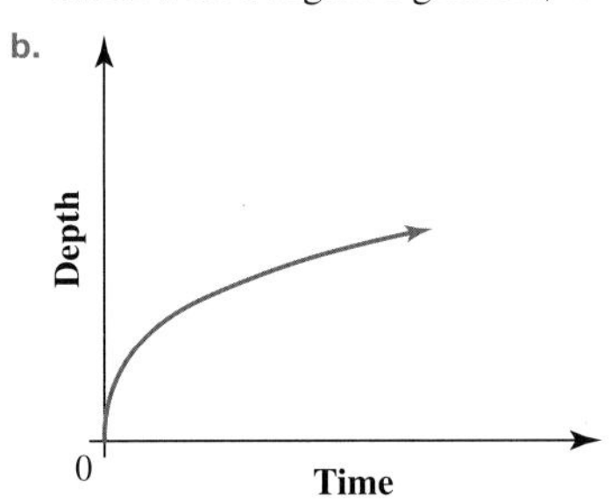

The depth will increase quickly at first, then more slowly. The graph is not linear.

2. a. $1 + \dfrac{h}{4}, \ h \neq 0, 1.0025$

   b. 1

3. a. $f'(x) \approx 4x - 3$

   b. $f'(x) = 1 - 4x^3$

4. a. $4x + x^2$

   b. $6x^2 - 6x - \dfrac{3}{2}$

   c. $192x^3 + 88x + 16$

   d. $x + 1$

5. 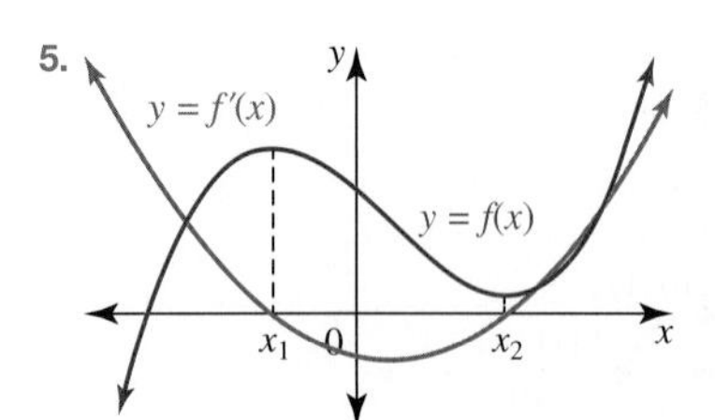

Concave up parabola with x-intercepts the same as x-coordinates of the turning points

6. a. 22

b. $-\frac{1}{9}, 1$

Technology active: multiple choice

7. C	8. A	9. C	10. D	11. E
12. B	13. E	14. D	15. B	16. A

Technology active: extended response

17. a. $f'(x) = 3x^2 - 8x + 5, f': R \to R, f'(x) = 3x^2 - 8x + 5$

b. Let $f(x) = 0$.

$x^3 - 4x^2 + 5x = 0$

$x\left(x^2 - 4x + 5\right) = 0$

$\therefore \; x = 0$ or $x^2 - 4x + 5 = 0$

Consider the discriminant of $x^2 - 4x + 5 = 0$.

$\Delta = (-4)^2 - 4(1)(5)$
$= 16 - 20$
$= -4$
< 0

There are no real solutions to the quadratic equation.

This means there is only one x-intercept at $x = 0$.

Since $f'(0) = 5$, the gradient of the curve at $(0, 0)$ is 5.

c. $\dfrac{4 \pm \sqrt{10}}{3}$

d. $a = 3, \; b = -5$

e. i. $c = \dfrac{5}{3}$

ii. $(1, 2)$

f. 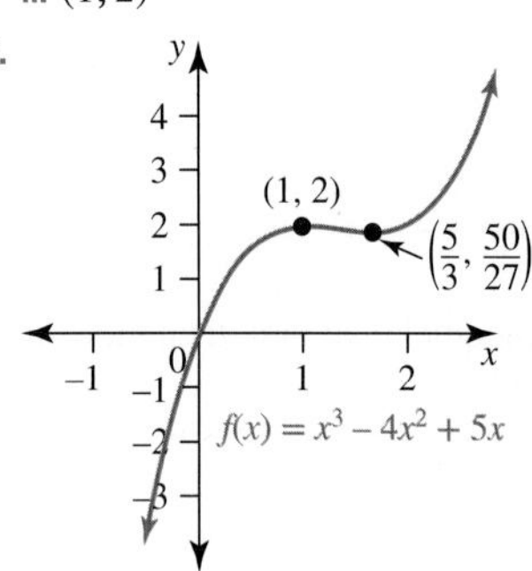

$(1, 2)$ and $\left(\dfrac{5}{3}, \dfrac{50}{27}\right)$ are maximum and minimum turning points respectively.

18. a. $P(3, 18)$, $Q\left(3 + h, \; 18 + 9h + h^2\right)$

b. $9 + h, \; h \neq 0$

c. 9

d. $y = x^2 + 3x$

$\therefore \dfrac{dy}{dx} = 2x + 3$

At P, $x = 3$.

When $x = 3$, $\dfrac{dy}{dx} = 2 \times 3 + 3 = 9$.

e. Let $f(x) = x(x + 3) = x^2 + 3x$.

$$\begin{aligned}\lim_{a \to 3} \frac{f(3) - f(a)}{3 - a} &= \lim_{a \to 3} \frac{18 - (a^2 + 3a)}{3 - a} \\ &= \lim_{a \to 3} \frac{18 - a^2 - 3a}{3 - a} \\ &= \lim_{a \to 3} \frac{18 - 3a - a^2}{3 - a} \\ &= \lim_{a \to 3} \frac{(6 + a)(3 - a)}{3 - a} \\ &= \lim_{a \to 3} (6 + a) \\ &= 9\end{aligned}$$

Let A be the point $(a, \; f(a))$ on the curve $y = f(x) = x^2 + 3x$. $P(3, 18)$ lies on this curve, so its coordinates could be expressed as $(3, \; f(3))$.

The line through A and P is a secant with gradient $m_{\text{secant}} = \dfrac{f(3) - f(a)}{3 - a}$.

As the point A is moved closer to P, $a \to 3$ and the gradient of the secant approaches the value of the gradient of the tangent at P.

$\therefore \lim_{a \to 3} \dfrac{f(3) - f(a)}{3 - a} = f'(3)$

19. a. i. 10 500 people per year

ii. 18 000 people per year

b. i. Both towns had a population of 20 000 in 2010.

ii. Decreasing at a rate of 500 people per year

c. 2014

d. 2018

e. 2026

f. i. 141 500

ii. 17 500 people per year

20. a. 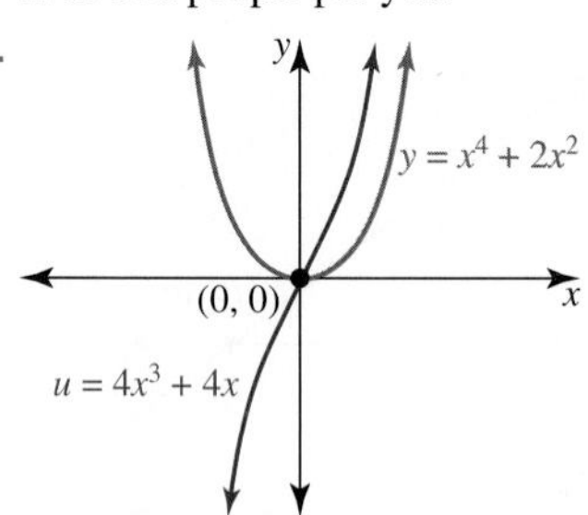

b. The gradient of the derivative graph at the origin is not 0, so it is not a stationary point of inflection.

c. $x = 3.7511$

d. $4x^3 - 12x^2 + 4x - 4 = 0, \; x = 2.77$

11.6 Exam questions

Note: Mark allocations are available with the fully worked solutions online.

1. C
2. Gradient of chord = 8.703
   Estimated gradient of the tangent to the curve = 9
   Improve the estimate by using a point closer to 3, such as $x = 2.99$.
3. D
4. A
5. 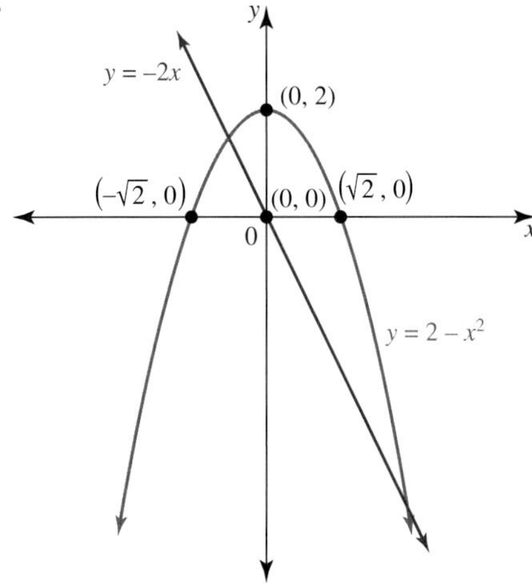

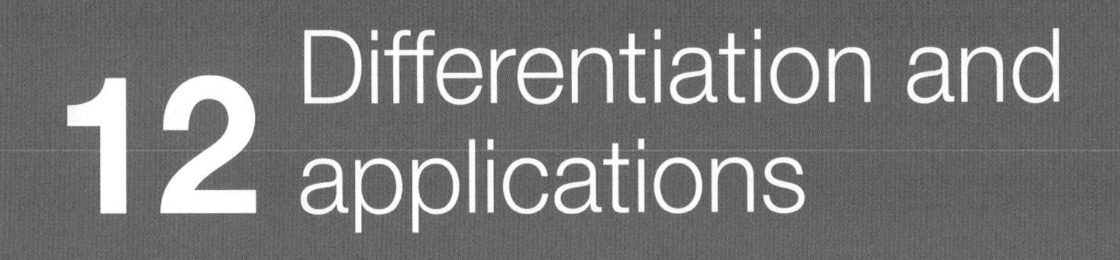

12 Differentiation and applications

LEARNING SEQUENCE

Fully worked solutions for this topic are available online.

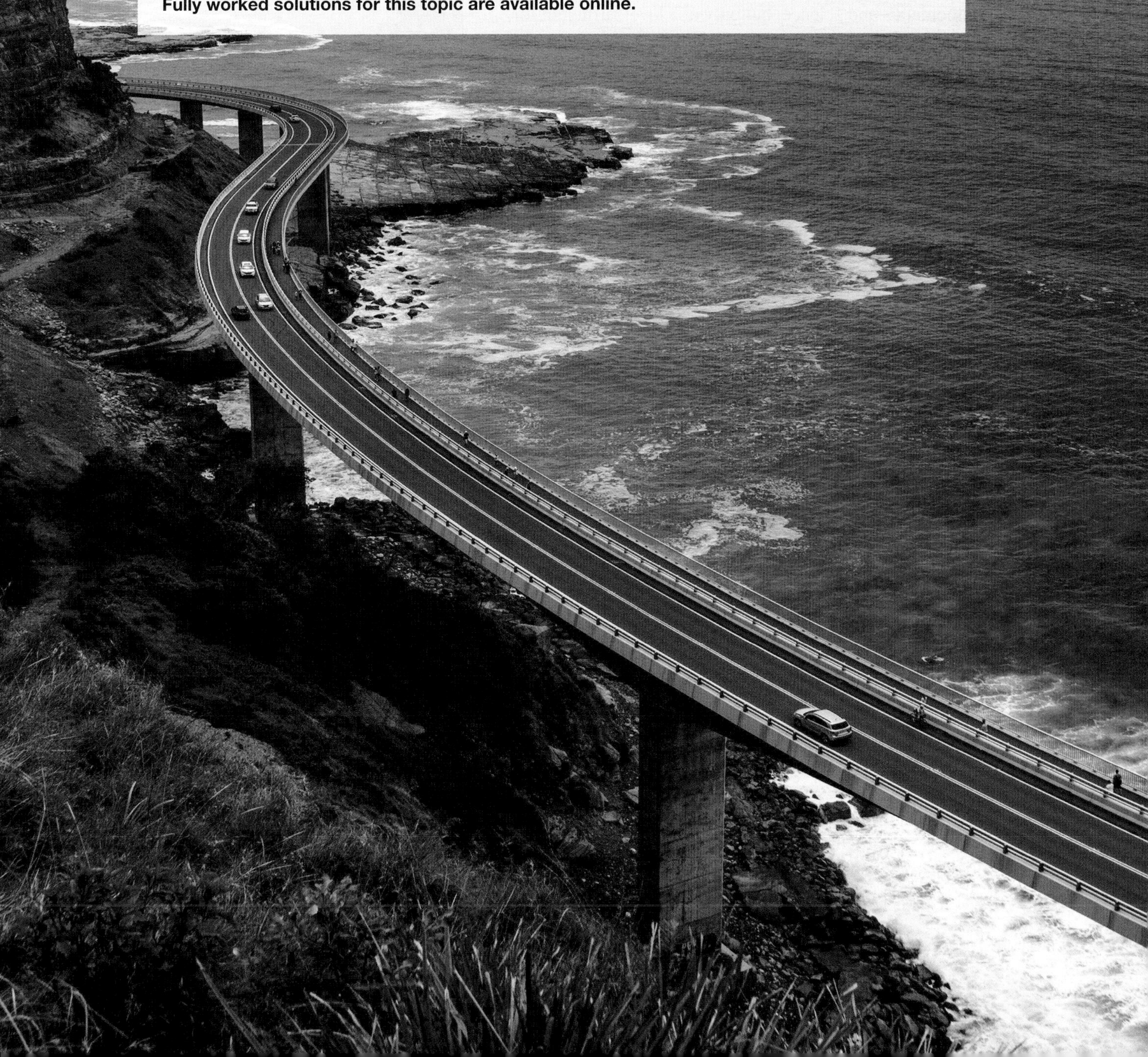

12.1 Overview

Hey students! Bring these pages to life online

Watch videos

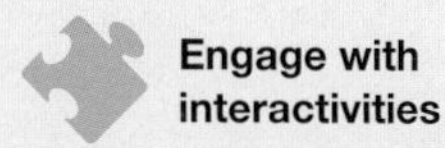

Engage with interactivities

Answer questions and check results

Find all this and MORE in jacPLUS

12.1.1 Introduction

Before the 20th century, education in Mathematics and Science was considered to be unsuitable for women. Despite this, there were some female mathematicians who made significant contributions to the collective knowledge of their times.

Being born into the minor French aristocracy, Émilie du Châtelet (1706–49) had the advantage of an education that included some mathematics and science, but only at a level sufficient for participation in polite conversation. She was presented to the court of Versailles aged 16; marriage followed and she became Madame la Marquise du Chastellet or Châtelet. This enabled her to employ tutors to properly teach her Mathematics. She became known and recognised for her ability, yet when she attempted to join a meeting of leading mathematicians she was turned away because she was female. She came to the next meeting dressed as a man, not as a disguise but to illustrate the stupidity of the ruling and this time she was admitted.

Émilie was one of the few people at the time who really understood calculus. Together with her lover and companion Voltaire, she recognised the greatness of the work of Newton. This was not at the time a popular viewpoint in France, due to the aftermath of the fued between Newton and Leibniz.

The mathematical accomplishment for which du Châtelet is most lauded is her translation into French, with commentary, of Newton's *Principia Mathematica*. This was an enormous undertaking and could only have been done by someone with a thorough understanding of both mathematics and physics. It remains the standard French translation of *Principia*.

KEY CONCEPTS

This topic covers the following key concepts from the VCE Mathematics Study Design:

- numerical approximation of roots of cubic polynomial functions using the Newton's method algorithm
- interpretation of graphs of empirical data with respect to rate of change such as temperature or pollution levels over time, motion graphs and the height of water in containers of different shapes that are being filled at a constant rate, with informal consideration of continuity and smoothness
- the derivative as the gradient of the graph of a function at a point and its representation by a gradient function, and as a rate of change
- differentiation of polynomial functions by rule
- applications of differentiation, including finding instantaneous rates of change, stationary values of functions, local maxima or minima, points of inflection, analysing graphs of functions including motion graphs, and solving maximum and minimum problems with consideration of modelling domain and local and global maxima and minima.

12.2 Limits, continuity and differentiability

LEARNING INTENTION

At the end of this subtopic you should be able to:
- evaluate the limit and continuity of a function
- test if a function is differentiable.

Differential calculus is founded on the concept of limits. To develop an understanding of when a function can and cannot be differentiated, we need to have a closer look at how limits can be calculated.

12.2.1 Limits

For any function $y = f(x)$, the statement $\lim\limits_{x \to a} f(x) = L$ means that as the x-values approach the value a, the y-values approach the value L; that is, as $x \to a$, $f(x) \to L$.

A **limit** is only concerned with the behaviour of a function as it approaches a particular point, and not with its behaviour at the point. What happens to the function at $x = a$ is immaterial; indeed, $f(a)$ may not even exist.

It is not only in calculus that limits occur. We have encountered limits when we considered the long-term behaviour of a hyperbolic function as $x \to \infty$, thereby identifying its limiting behaviour. The graph of the hyperbola with the rule
$y = 80 - \dfrac{60}{x+1}$ illustrates that as $x \to \infty$, $y \to 80$.

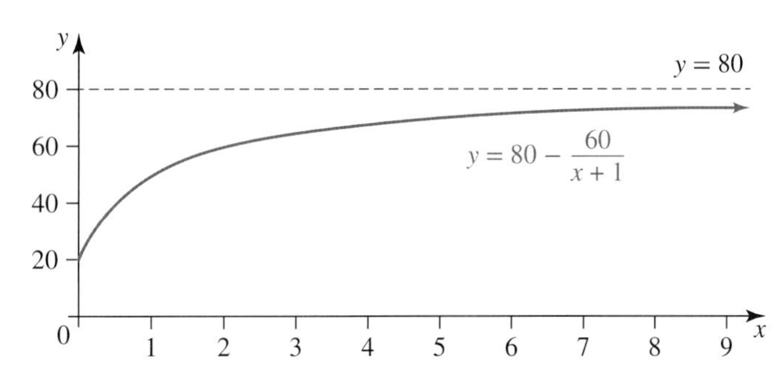

Although the graph never reaches the horizontal asymptote at $y = 80$, as the x-values increase indefinitely, the y-values come closer and closer to 80.

This is written as $\lim\limits_{x \to \infty} \left(80 - \dfrac{60}{x+1}\right) = 80$.

Calculating limits algebraically

In the limit definition of the derivative, we encountered limits such as $\lim\limits_{h \to 0} \dfrac{2xh + h^2}{h}$. The value $h = 0$ could not be substituted immediately, because the expression $\dfrac{2xh + h^2}{h}$ is not a *well-behaved* function at $h = 0$: substitution of $h = 0$ is not possible, since division by zero is not possible. Factorisation, $\dfrac{\not{h}(2x + h)}{\not{h}}$, was used to create an equivalent expression, $2x + h$, which was well-behaved at $h = 0$ and therefore allowed for $h = 0$ to be substituted.

To evaluate the limit of a function as $x \to a$:
- if the function is well-behaved, substitute $x = a$
- if the function is not well-behaved at $x = a$, simplify it to obtain a well-behaved function, then substitute $x = a$
- if it is not possible to obtain a well-behaved function, then the limit does not exist.

WORKED EXAMPLE 1 Calculating limits

Calculate, where possible, the following limits.

a. $\lim_{x \to 2}(4x+3)$ **b.** $\lim_{x \to -1}\left(\frac{x^2-1}{x+1}\right)$ **c.** $\lim_{x \to 1}\left(\frac{1}{x-1}\right)$

THINK	WRITE
a. 1. State whether or not the function is well-behaved at the value of x.	**a.** $\lim_{x \to 2}(4x+3)$ The function $f(x)=4x+3$ is well-behaved at $x=2$, since it is possible to substitute $x=2$ into the equation.
2. Calculate the limit.	Substitute $x=2$: $\lim_{x \to 2}(4x+3) = 4(2)+3$ $= 11$
b. 1. State whether or not the function is well-behaved at the value of x.	**b.** $\lim_{x \to -1}\left(\frac{x^2-1}{x+1}\right)$ The function $f(x)=\frac{x^2-1}{x+1}$ is not well-behaved at $x=-1$, since its denominator would be zero.
2. Use algebra to simplify the function to create an equivalent limit statement.	$\lim_{x \to -1}\left(\frac{x^2-1}{x+1}\right) = \lim_{x \to -1}\left(\frac{(x-1)\cancel{(x+1)}}{\cancel{x+1}}\right)$ $= \lim_{x \to -1}(x-1)$
3. Calculate the limit.	$= -1-1$ $= -2$ $\therefore \lim_{x \to -1}\left(\frac{x^2-1}{x+1}\right) = -2$
c. Determine whether the limit exists.	**c.** $\lim_{x \to 1}\left(\frac{1}{x-1}\right)$ The function $f(x)=\frac{1}{x-1}$ is not well-behaved at $x=1$, but it cannot be simplified any further. Therefore, $\lim_{x \to 1}\left(\frac{1}{x-1}\right)$ does not exist.

TI | THINK	DISPLAY/WRITE	CASIO | THINK	DISPLAY/WRITE
b. 1. On a Calculator page, press MENU and select: 4. Calculus 4. Limit Complete the entry line as: $\lim_{x \to -1}\left(\frac{x^2-1}{x+1}\right)$ Then press ENTER.	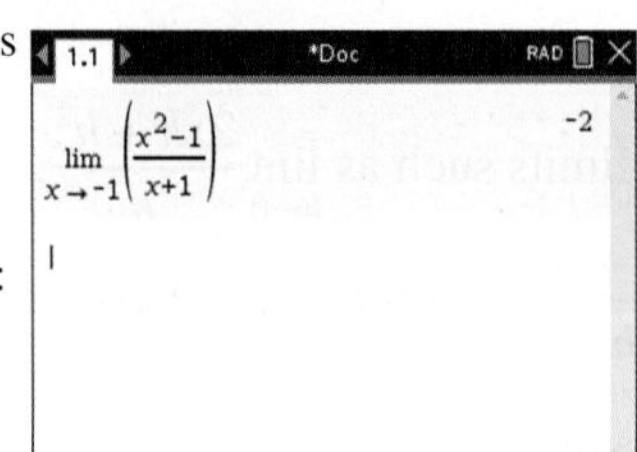	**b. 1.** On a Main screen, complete the entry line as: $\lim_{x \to -1}\left(\frac{x^2-1}{x+1}\right)$ Then press EXE. *Note:* The template for the limit is in the Math 2 Keyboard menu.	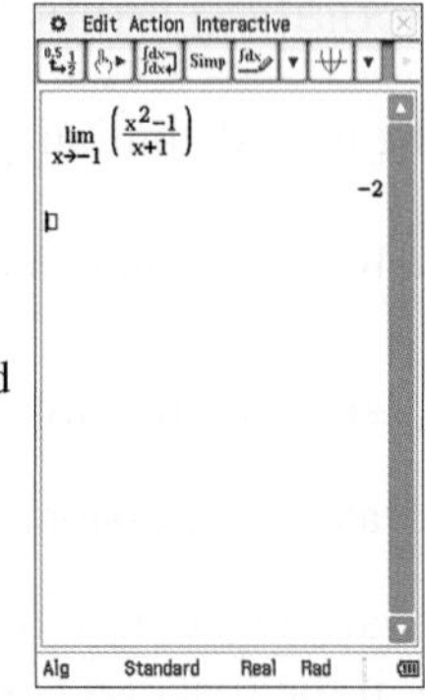
2. The answer appears on the screen.	-2	**2.** The answer appears on the screen.	-2

12.2.2 Left and right limits

For $\lim\limits_{x \to a} f(x)$ to exist, the limit of the function as x approaches a from values smaller than a should be the same as its limit as x approaches a from values larger than a. These are called left and right limits and are written as follows.

> **Left and right limits**
>
> $$L^- = \lim_{x \to a^-} f(x) \text{ and } L^+ = \lim_{x \to a^+} f(x).$$

The direction as x approaches a must not matter if the limit is to exist.

- If $L^- = L^+$, then $\lim\limits_{x \to a} f(x)$ exists and $\lim\limits_{x \to a} f(x) = L^- = L^+$.
- If $L^- \neq L^+$, then $\lim\limits_{x \to a} f(x)$ does not exist.

For the graph of the hybrid function shown, the limiting value of the function as $x \to 2$ depends on which of the two branches are considered.

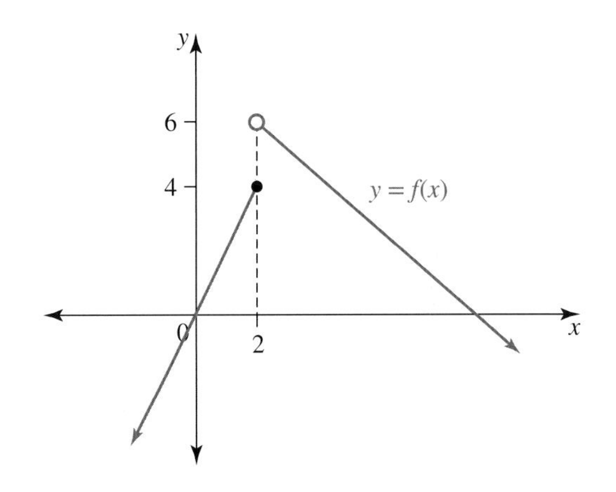

The limit as $x \to 2$ from the left of $x = 2$ is:

$$L^- = \lim_{x \to 2^-} f(x) = 4$$

The limit as $x \to 2$ from the right of $x = 2$ is:

$$L^+ = \lim_{x \to 2^+} f(x) = 6$$

Since $L^- \neq L^+$, the limit of the function as $x \to 2$ does not exist.

Limits and continuity

For a function to be **continuous** at $x = a$, there should be no break in its graph at $x = a$, so both its behaviour as it approaches $x = a$ and its behaviour at that point are important.

> **Continuous functions**
>
> **If a function is continuous at a point where $x = a$, then:**
> - **$f(a)$ exists**
> - **$\lim\limits_{x \to a} f(x)$ exists**
> - **$\lim\limits_{x \to a} f(x) = f(a)$.**

For the graph of the hybrid function previously considered, the two branches do not join. There is a break in the graph at $x = 2$, so the function is not continuous at $x = 2$.

However, to illustrate the definition of continuity, we can reason that although $f(2) = 4$, so the function is defined at $x = 2$, $\lim\limits_{x \to 2} f(x)$ does not exist since $L^- \neq L^+$ and therefore the function is not continuous at $x = 2$.

From a graph, places where a function is not continuous (the points of **discontinuity**) are readily identified. The graph may contain:

- a vertical asymptote
- branches that do not join
- a gap at some point.

Without a graph, we may need to test the definition to decide if a function is continuous or discontinuous at any point.

WORKED EXAMPLE 2 Determining continuity

a. A function is defined as $f(x) = \begin{cases} x, & x < 1 \\ 1, & x = 1 \\ x^2, & x > 1 \end{cases}$.

i. Calculate $\lim\limits_{x \to 1} f(x)$.

ii. Determine whether the function is continuous at $x = 1$ and sketch the graph of $y = f(x)$ to illustrate the answer.

b. Consider the function with rule $g(x) = \dfrac{x^2 + x}{x + 1}$.

i. State its maximal domain.

ii. Calculate $\lim\limits_{x \to -1} g(x)$.

iii. Explain why the function is not continuous at $x = -1$.

iv. Sketch the graph of $y = g(x)$.

THINK	WRITE
a. i. 1. Calculate the left and right limits of the function at the value of x.	**a. i.** $f(x) = \begin{cases} x, & x < 1 \\ 1, & x = 1 \\ x^2, & x > 1 \end{cases}$ Limit from the left of $x = 1$: $L^- = \lim\limits_{x \to 1^-} f(x)$ $= \lim\limits_{x \to 1}(x)$ $= 1$ Limit from the right of $x = 1$: $L^+ = \lim\limits_{x \to 1^+} f(x)$ $= \lim\limits_{x \to 1}(x^2)$ $= 1$
2. State the answer.	Since $L^- = L^+$, $\lim\limits_{x \to 1} f(x) = 1$.
ii. 1. Identify the value of the function at the x-value under consideration.	**ii.** From the given rule, $f(1) = 1$.
2. Explain whether or not the function is continuous at the given x-value.	Both $f(1)$ and $\lim\limits_{x \to 1} f(x)$ exist, and since $f(1) = \lim\limits_{x \to 1} f(x)$, the function is continuous at $x = 1$.
3. Sketch the hybrid function.	The branch of the graph for $x < 1$ is the straight line $y = x$. Two points to plot are $(0, 0)$ and $(1, 1)$. The branch of the graph for $x > 1$ is the parabola $y = x^2$. Two points to plot are $(1, 1)$ and $(2, 4)$. The two branches join at $(1, 1)$, so there is no break in the graph.

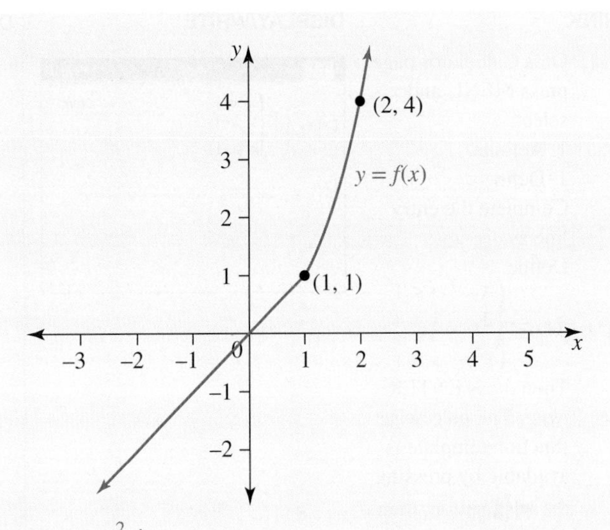

b. i. Determine the maximal domain. The function has a denominator $x+1$, which would become zero if $x=-1$. This value must be excluded from the domain of the function.

b. i. $g(x)=\dfrac{x^2+x}{x+1}$

The maximal domain is $R\setminus\{-1\}$.

ii. Calculate the required limit of the function.

ii.
$$\begin{aligned}\lim_{x\to-1} g(x) &= \lim_{x\to-1}\frac{x^2+x}{x+1}\\ &= \lim_{x\to-1}\frac{x(x+1)}{x+1}\\ &= \lim_{x\to-1}(x)\\ &= -1\end{aligned}$$
$$\therefore \lim_{x\to-1} g(x) = -1$$

iii. Explain why the function is not continuous at the given point.

iii. The domain of the function is $R\setminus\{-1\}$, so $g(-1)$ does not exist. Therefore, the function is not continuous at $x=-1$.

iv. 1. Identify the key features of the graph.

iv. $g(x)=\dfrac{x(x+1)}{x+1}$
$= x,\ x\neq-1$

Therefore, the graph of $y=g(x)$ is the graph of $y=x$, $x\neq-1$.

The straight-line graph has a hole, or point of discontinuity, at $(-1,-1)$.

2. Sketch the graph.

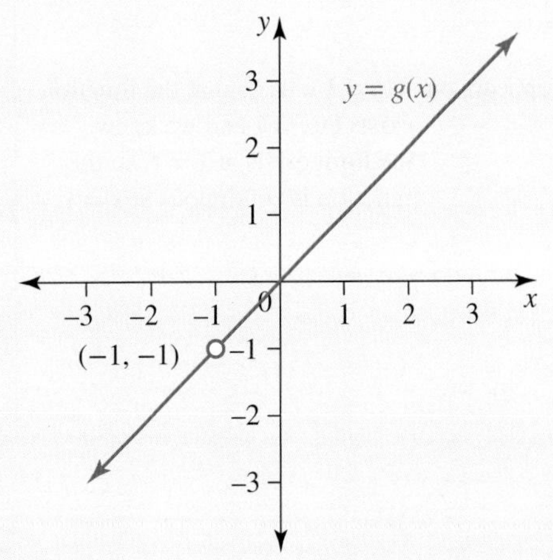

TI | THINK — **DISPLAY/WRITE**

a. i. 1. On a Calculator page, press MENU and select:
1. Actions
1. Define
Complete the entry line as:
Define
$f(x) = \begin{cases} x, & x < 1 \\ 1, & x = 1 \\ x^2, & x > 1 \end{cases}$
Then press ENTER.
Note: The piecewise function template is available by pressing the button, then .

2. Press MENU and select:
4: Calculus
4: Limit
Complete the entry line as:
$\lim\limits_{x \to 1} \left(f(x)\right)$
Then press ENTER.

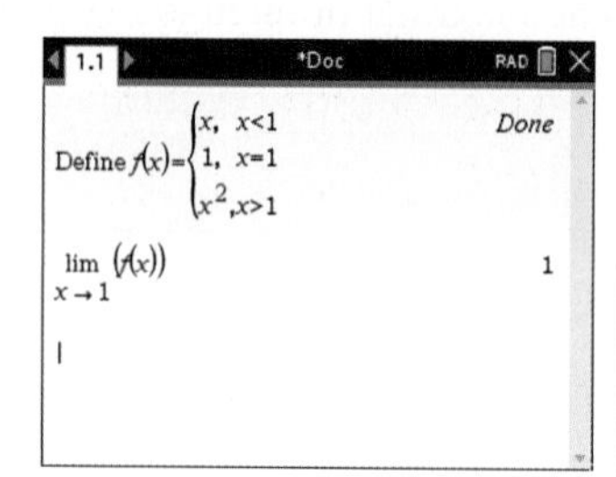

3. The answer appears on the screen.

$\lim\limits_{x \to 1} \left(f(x)\right) = 1$

a. ii. 1. Complete the entry line as:
$f(1)$
Then press ENTER.

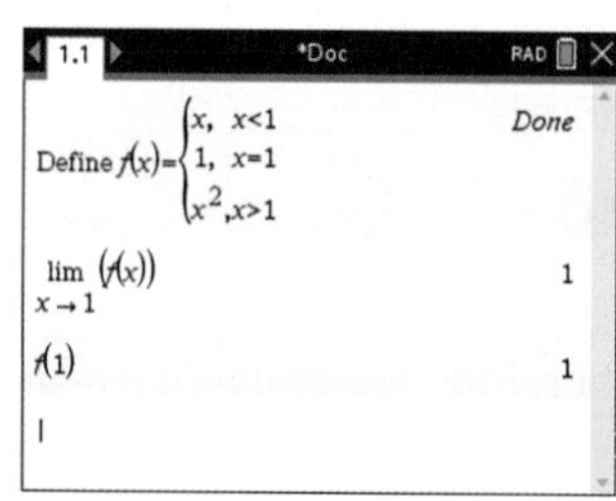

2. The answer appears on the screen.

$f(1) = 1$ tells us that the function exists at $x = 1$ and we know the limit exists at $x = 1$, so the function is continuous at $x = 1$.

CASIO | THINK — **DISPLAY/WRITE**

a. i. 1. On a Main screen, define the function by completing the entry line as:
Define
$f(x) = \begin{cases} x, & x < 1 \\ 1, & x = 1 \\ x^2, & x > 1 \end{cases}$
Then press EXE.
Note: The piecewise function template is located in the Math 3 Keyboard menu. Tap twice to create a third row.

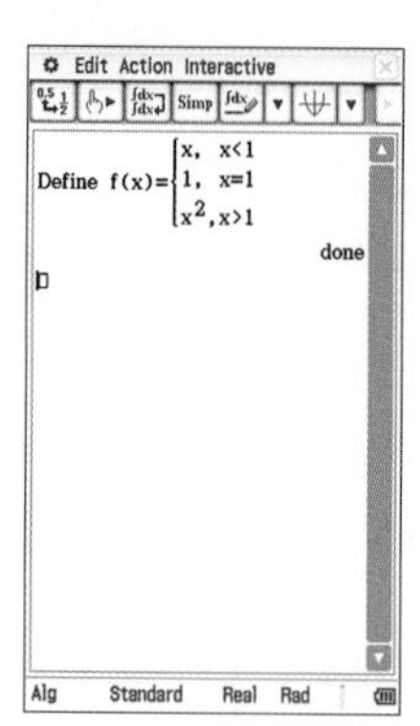

2. Complete the entry line as:
$\lim\limits_{x \to 1} \left(f(x)\right)$
Then press EXE.

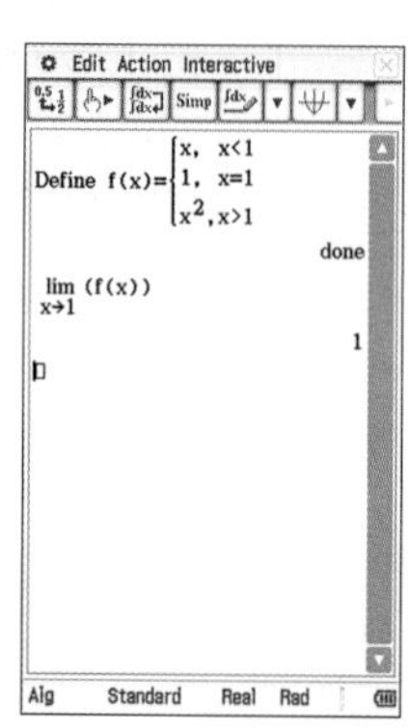

3. The answer appears on the screen.

$\lim\limits_{x \to 1} \left(f(x)\right) = 1$

a. ii. 1. Complete the entry line as:
$f(1)$
Then press EXE.

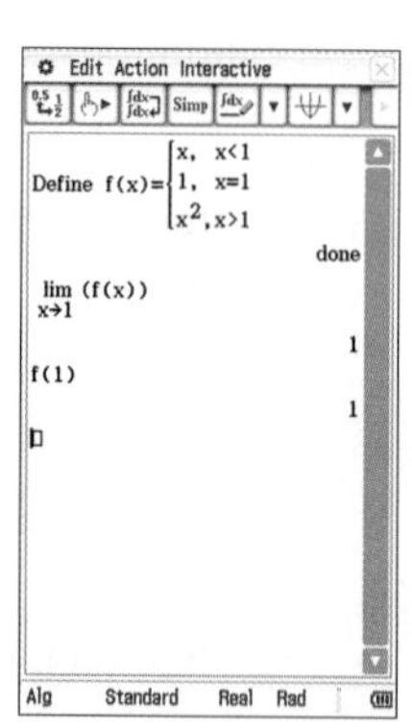

2. The answer appears on the screen.

$f(1) = 1$ tells us that the function exists at $x = 1$ and we know the limit exists at $x = 1$, so the function is continuous at $x = 1$.

3. To confirm, open a Graphs page.
Complete the entry line as:
$f1(x) = f(x)$
Then press ENTER.

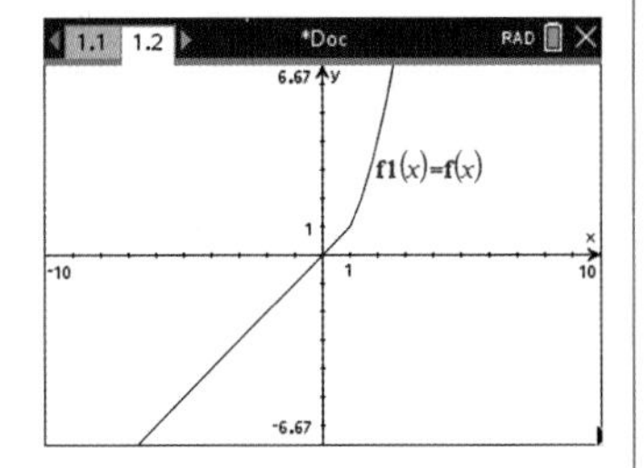

3. To confirm, open a Graphs & Table screen.
Complete the entry line as:
$y1 = f(x)$
Then press ENTER.

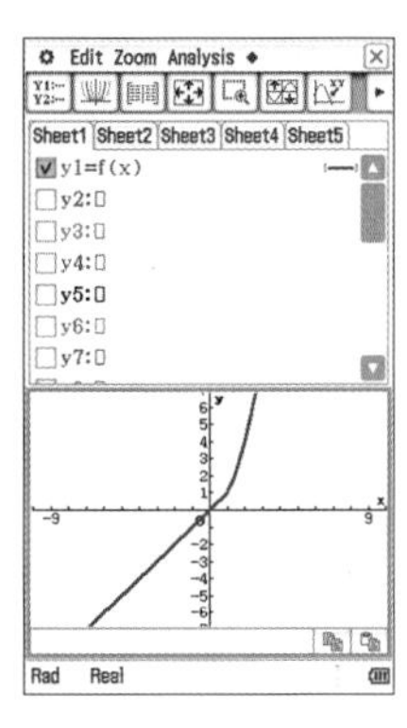

12.2.3 Differentiability

Although a function must be continuous at a point in order to find the gradient of its tangent, that condition is necessary but not sufficient. The derivative function has a limit definition, and as a limit exists only if the limit from the left equals the limit from the right, to test whether a continuous function is **differentiable** at a point $x = a$ in its domain, the derivative from the left of $x = a$ must equal the derivative from the right of $x = a$.

> **Differentiability**
>
> **A function f is differentiable at $x = a$ if:**
> - **f is continuous at $x = a$, and**
> - **$f'(a^-) = f'(a^+)$.**

This means that only **smoothly continuous functions** are differentiable at $x = a$: their graph contains no 'sharp' points at $x = a$. Polynomials are smoothly continuous functions, but a hybrid function, for example, may not always join its ends smoothly. An example of a hybrid function that is continuous but not differentiable at $x = a$ is shown in the diagram.

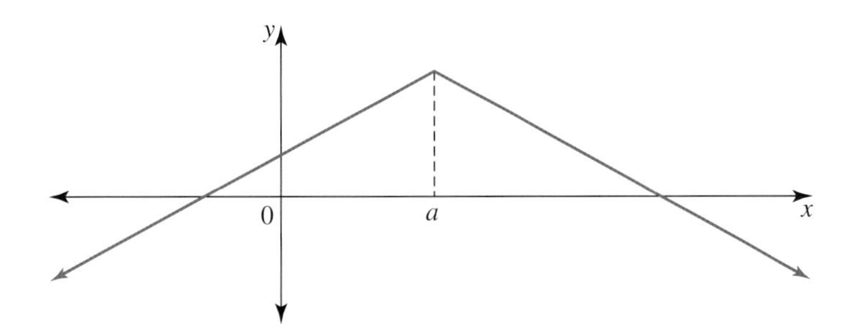

For this hybrid function, the line on the left of a has a positive gradient, but the line on the right of a has a negative gradient; $\left.\frac{dy}{dx}\right|_{x=a^-} \neq \left.\frac{dy}{dx}\right|_{x=a^+}$. The 'sharp' point at $x = a$ is clearly visible. However, we cannot always rely on the graph to determine if a continuous function is differentiable because visually it may not always be possible to judge whether the join is smooth or not.

A function with a restricted domain is not differentiable at the end points.

For a function f with domain $x \in [a, \infty)$, $f(a)$ exists. However, $f'(a^-)$ does not exist, so $f'(a^-) \neq f'(a^+)$ and the function is not differentiable at $x = a$, the end point.

WORKED EXAMPLE 3 Testing for differentiability

The function defined as $f(x) = \begin{cases} x, & x \le 1 \\ x^2, & x > 1 \end{cases}$ **is continuous at** $x = 1$.

a. **Test whether the function is differentiable at** $x = 1$.

b. **Give the rule for** $f'(x)$**, stating its domain, and sketch the graph of** $y = f'(x)$.

THINK	WRITE
a. 1. Calculate the derivative from the left and the derivative from the right at the given value of x.	a. $f(x) = \begin{cases} x, & x \le 1 \\ x^2, & x > 1 \end{cases}$ Derivative from the left of $x = 1$: $f(x) = x$ $\therefore f'(x) = 1$ $\therefore f'(1^-) = 1$ Derivative from the right of $x = 1$: $f(x) = x^2$ $f'(x) = 2x$ $\therefore f'(1^+) = 2(1)$ $= 2$
2. State whether the function is differentiable at the given value of x.	Since the derivative from the left does not equal the derivative from the right, the function is not differentiable at $x = 1$.
b. 1. State the rule for the derivative in the form of a hybrid function.	b. Each branch of this function is a polynomial, so the derivative of the function is $f'(x) = \begin{cases} 1, & x < 1 \\ 2x, & x > 1 \end{cases}$ with domain $R\backslash\{1\}$.
2. Sketch the graph.	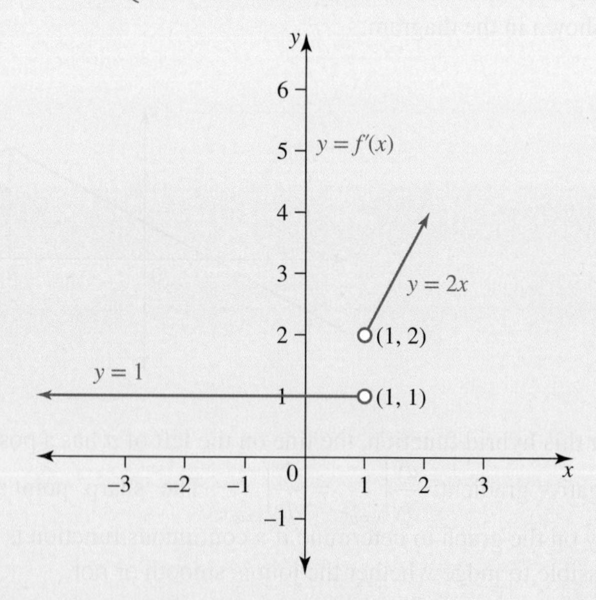

12.2 Exercise

Students, these questions are even better in jacPLUS

Receive immediate feedback and access sample responses

Access additional questions

Track your results and progress

Find all this and MORE in jacPLUS

Technology free

1. Calculate the value of the following.

a. $\lim\limits_{x\to 3}(x^2+1)$ b. $\lim\limits_{x\to 1}(x+1)(7-2x)$ c. $\lim\limits_{x\to 0}(5x^3+x^2+3x+8)$ d. $\lim\limits_{x\to -1}\dfrac{x+6}{x-4}$

2. Simplify the given expression and then calculate the value of the following limits.

a. $\lim\limits_{x\to 0}\dfrac{13x}{x}$ b. $\lim\limits_{x\to 0}\dfrac{2x(x+1)}{x}$

c. $\lim\limits_{x\to 9}\dfrac{(x-9)(x+11)}{x-9}$ d. $\lim\limits_{x\to -8}\dfrac{(4x+1)(x+8)}{x+8}$

3. Calculate the following.

a. $\lim\limits_{x\to -10}\dfrac{x^2-100}{x+10}$ b. $\lim\limits_{x\to 4}\dfrac{5x-x^2}{x}$ c. $\lim\limits_{x\to \frac{1}{3}}\dfrac{6x^2-5x+1}{3x-1}$ d. $\lim\limits_{x\to -1}\dfrac{x^3+1}{x^2-1}$

4. **WE1** Calculate, where possible, the following limits.

a. $\lim\limits_{x\to -5}(8-3x)$ b. $\lim\limits_{x\to 3}\left(\dfrac{x^2-9}{x-3}\right)$ c. $\lim\limits_{x\to -3}\left(\dfrac{1}{x+3}\right)$

5. Calculate the following limits.

a. $\lim\limits_{x\to 2}\left(\dfrac{x^3-8}{x-2}\right)$ b. $\lim\limits_{x\to 2}\left(\dfrac{1}{x+3}\right)$

6. Evaluate the following limits.

a. $\lim\limits_{x\to 3}(6x-1)$ b. $\lim\limits_{x\to 3}\dfrac{2x^2-6x}{x-3}$ c. $\lim\limits_{x\to 1}\dfrac{2x^2+3x-5}{x^2-1}$

d. $\lim\limits_{x\to 0}\dfrac{3x-5}{2x-1}$ e. $\lim\limits_{x\to -4}\dfrac{64+x^3}{x+4}$ f. $\lim\limits_{x\to \infty}\dfrac{x+1}{x}$

7. **WE2** a. A function is defined as $f(x)=\begin{cases} x^2, & x<1 \\ 1, & x=1 \\ 2x-1, & x>1 \end{cases}$.

i. Calculate $\lim\limits_{x\to 1}f(x)$.

ii. Determine whether the function is continuous at $x=1$ and sketch the graph of $y=f(x)$ to illustrate the answer.

b. Consider the function with rule $g(x)=\dfrac{x^2-x}{x}$.

i. State its maximal domain.

ii. Calculate $\lim\limits_{x\to 0}g(x)$.

iii. Explain why the function is not continuous at $x=0$.

iv. Sketch the graph of $y=g(x)$.

8. For each of the following functions:

   i. calculate $\lim_{x \to 2} f(x)$ if it exists

   ii. explain whether the function is continuous at $x = 2$.

   a. $f(x) = \begin{cases} x^2, & x \le 2 \\ -2x, & x > 2 \end{cases}$

   b. $f(x) = \begin{cases} (x-2)^2, & x < 2 \\ x-2, & x \ge 2 \end{cases}$

   c. $f(x) = \begin{cases} -x, & x < 2 \\ -2, & x = 2 \\ x-4, & x > 2 \end{cases}$

   d. $f(x) = \dfrac{x^2-4}{x-2}, \ x \ne 2$

9. For the function defined by

$$f(x) = \begin{cases} x^2 - 4, & x < 0 \\ 4 - x^2, & x \ge 0 \end{cases}$$

   calculate, where possible, $f(0)$ and $\lim_{x \to 0} f(x)$.
   Hence, determine the domain over which the function is continuous.

10. Identify which of the following functions are continuous over R.

   a. $f(x) = x^2 + 5x + 2$

   b. $g(x) = \dfrac{4}{x+2}$

   c. $h(x) = \begin{cases} x^2 + 5x + 2, & x < 0 \\ 5x + 2, & x > 0 \end{cases}$

   d. $k(x) = \begin{cases} x^2 + 5x + 2, & x < 0 \\ \dfrac{4}{x+2}, & x \ge 0 \end{cases}$

11. Determine the value of a so that the following is a continuous function.

$$y = \begin{cases} x + a, & x < 1 \\ 4 - x, & x \ge 1 \end{cases}$$

12. Determine the values of a and b so that the following is a continuous function.

$$y = \begin{cases} ax + b, & x < -1 \\ 5, & -1 \le x \le 2 \\ 2bx + a, & x > 2 \end{cases}$$

13. **WE3** The function defined as below is continuous at $x = 1$.

$$f(x) = \begin{cases} x^2, & x \le 1 \\ 2x - 1, & x > 1 \end{cases}$$

   a. Test whether the function is differentiable at $x = 1$.

   b. Give the rule for $f'(x)$, stating its domain, and sketch the graph of $y = f'(x)$.

14. The graph shown consists of a set of straight-line segments and rays.

   a. State any values of x for which the function is not continuous.

   b. State any values of x for which the function is not differentiable.

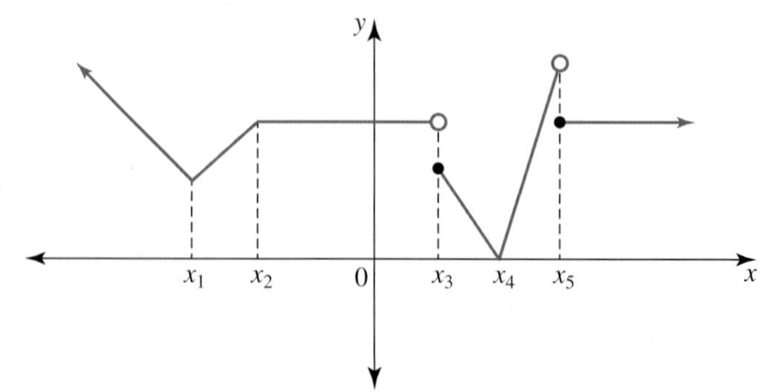

Technology active

15. Consider the function defined by the rule $f(x)=\begin{cases}3-2x, & x<0\\ x^2+3, & x\geq 0\end{cases}$.

a. Determine whether the function is differentiable at $x=0$.
b. Sketch the graph of $y=f(x)$. State whether the two branches join smoothly at $x=0$.
c. Form the rule for $f'(x)$ and state its domain.
d. Calculate $f'(3)$.
e. Sketch the graph of $y=f'(x)$.

16. Determine the values of a and b so that the function below is smoothly continuous at $x=2$.

$$f(x)=\begin{cases}ax^2, & x\leq 2\\ 4x+b, & x>2\end{cases}$$

17. a. Explain why the function below is not differentiable at $x=1$.

$$f(x)=\begin{cases}4x^2-5x+2, & x\leq 1\\ -x^3+3x^2, & x>1\end{cases}$$

b. Form the rule for $f'(x)$ for $x\in R\backslash\{1\}$.
c. Calculate the values of x for which $f'(x)=0$.
d. Sketch the graph of $y=f'(x)$, showing all key features.
e. Calculate the coordinates of the points on the graph of $y=f(x)$ where the gradient of the tangent to the curve is zero.
f. Sketch the graph of $y=f(x)$, locating any turning points and intercepts with the coordinate axes.

18. Determine the values of a, b, c and d so that the function below is a differentiable function for $x\in R$.

$$y=\begin{cases}ax^2+b, & x\leq 1\\ 4x, & 1<x<2\\ cx^2+d, & x\geq 2\end{cases}$$

19. For the function $f(x)=\begin{cases}x^2, & x<2\\ 2^x, & x\geq 2\end{cases}$:

a. show the function is continuous at $x=2$
b. write down the limit definition for the derivative from the left of $x=2$ using a nearby point with $x=2-h$
c. write down the limit definition for the derivative from the right of $x=2$ using a nearby point with $x=2+h$
d. evaluate these limits and hence determine whether the function is differentiable at $x=2$.

12.2 Exam questions

Question 1 (1 mark) TECH-ACTIVE

MC Evaluate $\lim_{x\to 3} f(x)$ for the function $f(x)=\dfrac{x^2-9}{x-3}$, $x\neq 3$.

A. 3 **B.** 0 **C.** 6 **D.** 9 **E.** Does not exist

Question 2 (1 mark) TECH-ACTIVE

MC The value of v such that $y=\begin{cases}x+2, & x<1\\ v-x, & x\geq 1\end{cases}$ is a continuous function is

A. 4 **B.** 0 **C.** 1 **D.** 2 **E.** undefined.

Question 3 (3 marks) TECH-ACTIVE

Determine the values of a and b such that $f(x) = \begin{cases} ax^2 + 4, & x \leq 1 \\ 6x + b, & x > 1 \end{cases}$ is smooth and continuous at $x = 1$.

More exam questions are available online.

12.3 Coordinate geometry applications of differentiation

LEARNING INTENTION

At the end of this subtopic you should be able to:

- determine the equation of a tangent and the angle it makes with the horizontal axis
- determine intervals where a function is increasing or decreasing
- use Newton's method to find an approximation to a root of a cubic polynomial.

The derivative measures the gradient of the tangent to a curve at any point. This has several applications in coordinate geometry.

12.3.1 Equations of tangents

A **tangent** is a straight line, so, to form its equation, its gradient and a point on the tangent are needed. The equation can then be formed using $y - y_1 = m(x - x_1)$.

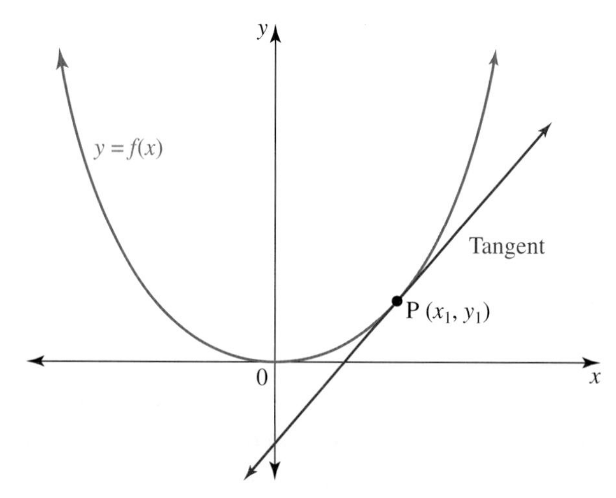

For the tangent to a curve $y = f(x)$ at a point P, the gradient m is found by evaluating the curve's derivative $f'(x)$ at P, the point of contact or point of tangency. The coordinates of P provide the point (x_1, y_1) on the line.

WORKED EXAMPLE 4 Determining the equation of a tangent

Form the equation of the tangent to the curve $y = 2x^3 + x^2$ at the point on the curve where $x = -2$.

THINK	WRITE
1. Obtain the coordinates of the point of contact of the tangent with the curve by substituting $x = -2$ into the equation of the curve.	$y = 2x^3 + x^2$ When $x = -2$, $y = 2(-2)^3 + (-2)^2$ $= -12$ The point of contact is $(-2, -12)$.

2. Calculate the gradient of the tangent at the point of contact by finding the derivative of the function at the point $x = -2$.

$$y = 2x^3 + x^2$$
$$\therefore \frac{dy}{dx} = 6x^2 + 2x$$
When $x = -2$,
$$\frac{dy}{dx} = 6(-2)^2 + 2(-2)$$
$$= 20$$
The gradient of the tangent is 20.

3. Form the equation of the tangent line.

Equation of the tangent:
$$y - y_1 = m(x - x_1),\ m = 20,\ (x_1, y_1) = (-2, -12)$$
$$\therefore\ y + 12 = 20(x + 2)$$
$$\therefore\ y = 20x + 28$$
The equation of the tangent is $y = 20x + 28$.

TI | THINK	DISPLAY/WRITE
1. On a Calculator page, complete the entry line as: Define $y = 2x^3 + x^2$ Then press ENTER. Press MENU and select: 4. Calculus 9. TangentLine Complete the entry line as: tangentLine $(y, x = -2)$ Then press ENTER.	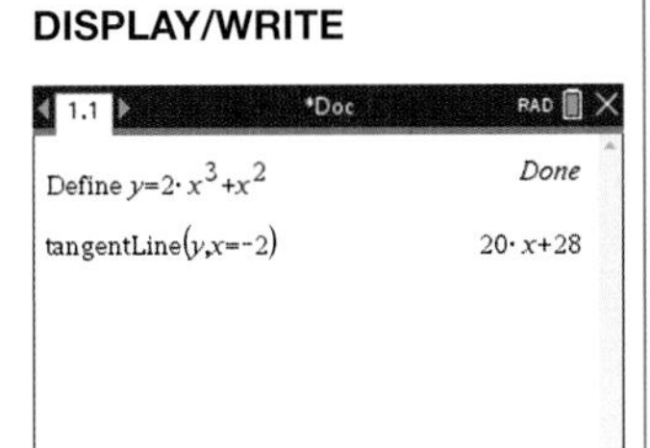
2. The answer appears on the screen.	$y = 20x + 28$

CASIO | THINK	DISPLAY/WRITE
1. On a Main screen, complete the entry line as: tanLine $(2x^3 + x^2, x, -2)$ Then press EXE. Alternatively, type the equation on the Main screen, highlight it and tap: • Interactive • Calculation • line • tanLine Insert the value $x = -2$ and press EXE.	
2. The answer appears on the screen.	$y = 20x + 28$

Resources

Interactivity Equations of tangents (int-5962)

12.3.2 Coordinate geometry

Since the tangent is a straight line, all the results obtained for the coordinate geometry of a straight line apply to the tangent line. These include the following:

- The angle of inclination of the tangent to the horizontal can be calculated using $m = \tan(\theta)$.
- Tangents that are parallel have the same gradient.
- The gradient of a line perpendicular to the tangent is found using $m_1 m_2 = -1$.
- The gradient of a horizontal tangent is zero.
- The gradient of a vertical tangent is undefined.
- Coordinates of points of intersections of tangents with other lines and curves can be found using simultaneous equations.

WORKED EXAMPLE 5 Using the tangent to determine angles

At the point (2, 1) on the curve $y = \frac{1}{4}x^2$, a line is drawn perpendicular to the tangent to the curve at that point. This line meets the curve $y = \frac{1}{4}x^2$ again at the point Q.

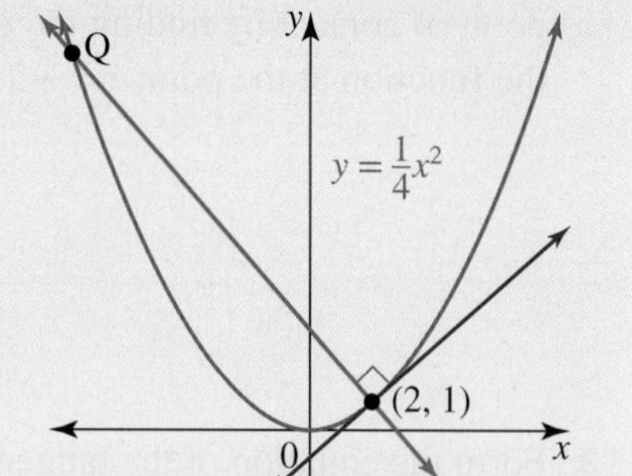

a. **Calculate the coordinates of the point Q.**
b. **Calculate the magnitude of the angle that the line passing through Q and the point (2, 1) makes with the positive direction of the x-axis.**

THINK	WRITE
a. 1. Calculate the gradient of the tangent at the given point.	a. $y = \frac{1}{4}x^2$ $\frac{dy}{dx} = \frac{1}{2}x$ At the point (2, 1), $\frac{dy}{dx} = \frac{1}{2}(2) = 1$ The gradient of the tangent at the point (2, 1) is 1.
2. Calculate the gradient of the line perpendicular to the tangent.	For perpendicular lines, $m_1 m_2 = -1$. Since the gradient of the tangent is 1, the gradient of the line perpendicular to the tangent is -1.
3. Form the equation of the perpendicular line.	Equation of the line perpendicular to the tangent: $y - y_1 = m(x - x_1),\ m = -1,\ (x_1, y_1) = (2, 1)$ $y - 1 = -(x - 2)$ $\therefore y = -x + 3$
4. Use simultaneous equations to calculate the coordinates of Q.	Point Q lies on the line $y = -x + 3$ and the curve $y = \frac{1}{4}x^2$. At Q, $\frac{1}{4}x^2 = -x + 3$ $x^2 = -4x + 12$ $x^2 + 4x - 12 = 0$ $(x + 6)(x - 2) = 0$ $\therefore x = -6, x = 2$ $x = 2$ is the x-coordinate of the given point. Therefore, the x-coordinate of Q is $x = -6$. Substitute $x = -6$ into $y = -x + 3$ $y = -(-6) + 3$ $= 9$ Point Q has the coordinates $(-6, 9)$.

b. Calculate the angle of inclination required.

b. For the angle of inclination, $m = \tan\theta$.
As the gradient of the line passing through Q and the point (2, 1) is −1, $\tan\theta = -1$.
Since the gradient is negative, the required angle is obtuse.
The second quadrant solution is

$$\begin{aligned}\theta &= 180° - \tan^{-1}(1)\\ &= 180° - 45°\\ &= 135°\end{aligned}$$

The angle made with the positive direction of the x-axis is 135°.

12.3.3 Increasing and decreasing functions

If the tangents to a function's curve have a positive gradient over some domain interval, then the function is increasing over that domain; if the gradients of the tangents are negative, then the function is decreasing. From this it follows that for a function $y = f(x)$ which is differentiable over an interval:

- if $f'(x) > 0$, then the function increases as x increases over the interval
- if $f'(x) < 0$, then the function decreases as x increases over the interval.

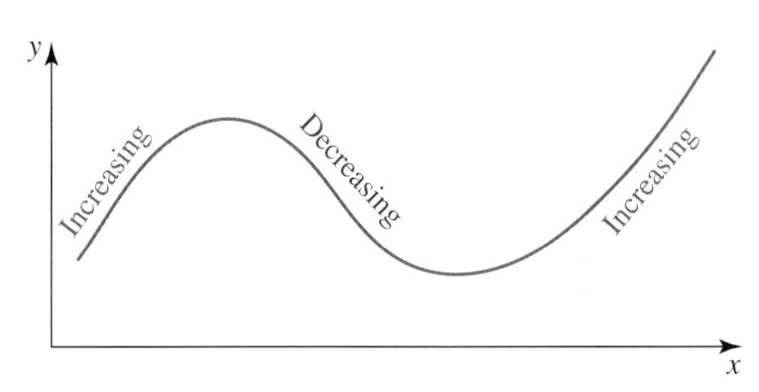

WORKED EXAMPLE 6 Determining domain intervals

a. Determine the domain interval over which the function $f(x) = \frac{1}{3}x^3 + \frac{7}{2}x^2 + 6x - 5$ is decreasing.

b. i. Form the equation of the tangent to the curve $y = x^2 + ax + 4$ at the point where $x = -1$.
ii. Hence, find the values of a for which the function $y = x^2 + ax + 4$ is increasing at $x = -1$.

THINK	WRITE
a. 1. Apply the condition for a function to be decreasing to the given function.	**a.** For a decreasing function, $f'(x) < 0$. $f(x) = \frac{1}{3}x^3 + \frac{7}{2}x^2 + 6x - 5$ $f'(x) = x^2 + 7x + 6$ $\therefore\ x^2 + 7x + 6 < 0.$
2. Solve the inequation using a sign diagram of $f'(x)$. *Note:* Alternatively, sketch the gradient function.	$\therefore\ (x+6)(x+1) < 0$ The zeros are $x = -6, x = -1$; $f'(x)$ is concave up. 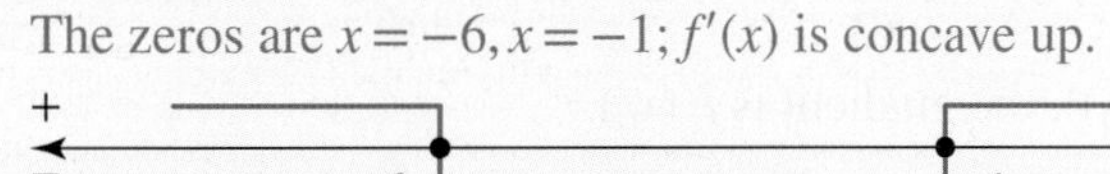 $\therefore\ -6 < x < -1$
3. State the answer.	The function is decreasing over the interval $x \in (-6, -1)$.

b. i. 1. Express the coordinates of the point of tangency in terms of a.	**b. i.** $y = x^2 + ax + 4$ When $x = -1$, $y = (-1)^2 + a(-1) + 4$ $= 5 - a$ The point is $(-1,\ 5 - a)$.
2. Obtain the gradient of the tangent in terms of a.	Gradient: $\frac{dy}{dx} = 2x + a$ At the point $(-1, 5 - a)$, $\frac{dy}{dx} = -2 + a$. Therefore, the gradient is $-2 + a$.
3. Form the equation of the tangent.	Equation of tangent: $y - y_1 = m(x - x_1),\ m = -2 + a,\ (x_1, y_1) = (-1,\ 5 - a)$ $y - (5 - a) = (-2 + a)(x + 1)$ $y = (a - 2)x - 2 + a + 5 - a$ $y = (a - 2)x + 3$
ii. Apply the condition on the tangent for a function to be increasing, and calculate the values of a.	If a function is increasing, the tangent to its curve must have a positive gradient. The gradient of the tangent at $x = -1$ is $a - 2$. Therefore, the function $y = x^2 + ax + 4$ is increasing at $x = -1$ when: $a - 2 > 0$ $\therefore a > 2$

12.3.4 Newton's method

Using tangents to obtain approximate solutions to equations

Isaac Newton and his assistant Joseph Raphson devised an iterative method for obtaining numerical approximations to the roots of equations of the form $f(x) = 0$. This method is known as **Newton's method** or the **Newton–Raphson method**.

Consider a curve $y = f(x)$ in the region near its intersection with the x-axis. An increasing curve has been chosen. The solution to, or root of, the equation $f(x) = 0$ is given by the x-intercept of the graph of $y = f(x)$.

Let x_0 be a first estimate of the x-intercept, that is the first estimate of the root of the equation $f(x) = 0$.

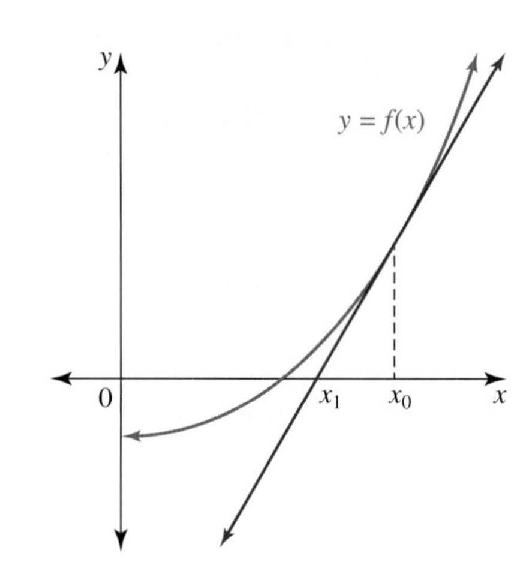

The tangent to the curve $y = f(x)$ is drawn at the point on the curve where $x = x_0$. To obtain the equation of this tangent, its gradient and the coordinates of the point on the curve are needed.

The gradient of the tangent is $\frac{dy}{dx} = f'(x)$.

At the point $(x_0,\ f(x_0))$, the gradient is $f'(x_0)$.

The equation of the tangent is:

$$y - f(x_0) = f'(x_0)(x - x_0)$$
$$\therefore y = f(x_0) + f'(x_0)(x - x_0)$$

The intercept, x_1, that this tangent makes with the x-axis is a better estimate of the required root. Its value is obtained from the equation of the tangent.

Let $y = 0$.

$$\begin{aligned} 0 &= f(x_0) + f'(x_0)(x - x_0) \\ f'(x_0)(x - x_0) &= -f(x_0) \\ (x - x_0) &= -\frac{f(x_0)}{f'(x_0)} \\ x &= x_0 - \frac{f(x_0)}{f'(x_0)} \end{aligned}$$

Hence, the improved estimate is $x_1 = x_0 - \dfrac{f(x_0)}{f'(x_0)}$.

The tangent at the point on the curve where $x = x_1$ is then constructed and its x-intercept x_2 calculated from $x_2 = x_1 - \dfrac{f(x_1)}{f'(x_1)}$.

The procedure continues to be repeated until the desired degree of accuracy to the root of the equation is obtained.

The formula for the iterative relation is generalised as follows.

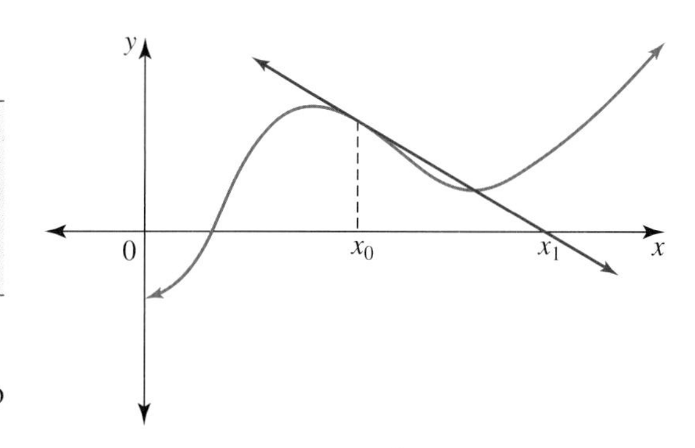

Newton's method

$$x_{n+1} = x_n - \frac{f(x_n)}{f'(x_n)}, \quad n = 0, 1, 2...$$

The first estimate needs to be reasonably close to the desired root for the procedure to require only a few iterations. It is also possible that a less appropriate choice of first estimate could lead to subsequent estimates going further away from the true value.

WORKED EXAMPLE 7 Applying Newton's method to approximate a root

Use Newton's method to calculate the root of the equation $x^3 - 2x - 2 = 0$ that lies near $x = 2$. Express the answer correct to 4 decimal places.

THINK	WRITE
1. Define $f(x)$ and state $f'(x)$.	Let $f(x) = x^3 - 2x - 2$. Then $f'(x) = 3x^2 - 2$.
2. State Newton's formula for calculating x_1 from x_0.	$x_{n+1} = x_n - \dfrac{f(x_n)}{f'(x_n)}$ $\therefore x_1 = x_0 - \dfrac{f(x_0)}{f'(x_0)}$

3. Use the value of x_0 to calculate the value of x_1.

Let $x_0 = 2$.

$$f(2) = (2)^3 - 2(2) - 2$$
$$= 2$$
$$f'(2) = 3(2)^2 - 2$$
$$= 10$$
$$x_1 = 2 - \frac{2}{10} = 1.8$$

4. Use the value of x_1 to calculate the value of x_2.

$$f(1.8) = (1.8)^3 - 2(1.8) - 2$$
$$= 0.232$$
$$f'(1.8) = 3(1.8)^2 - 2$$
$$= 7.72$$
$$x_2 = x_1 - \frac{f(x_1)}{f'(x_1)}$$
$$= 1.8 - \frac{0.232}{7.72}$$
$$\therefore x_2 \approx 1.769\,948\,187$$

5. Use the *Ans* key to carry the value of x_2 in a calculator, and calculate the value of x_3.
Note: In practice, carrying the previous value as *Ans* can be commenced from the beginning.

With the value of x_2 as *Ans*:

$$x_3 = x_2 - \frac{f(x_2)}{f'(x_2)}$$
$$= Ans - \frac{f(Ans)}{f'(Ans)}$$
$$= Ans - \frac{(Ans^3 - 2 \times Ans - 2)}{(3 \times Ans^2 - 2)}$$
$$\therefore x_3 = 1.769\,292\,663$$

6. Continue the procedure until the 4th decimal place remains unchanged.

$$x_4 = Ans - \frac{f(Ans)}{f'(Ans)}$$
$$= 1.769\ 292\ 354$$

Both x_3 and x_4 are the same value for the desired degree of accuracy.

7. State the solution.

To 4 decimal places, the solution to $x^3 - 2x - 2 = 0$ is $x = 1.7693$.

12.3 Exercise

Students, these questions are even better in jacPLUS

Receive immediate feedback and access sample responses

Access additional questions

Track your results and progress

Find all this and MORE in jacPLUS

Technology free

1. WE4 Form the equation of the tangent to the curve $y = 5x - \frac{1}{3}x^3$ at the point on the curve where $x = 3$.

2. Form the equation of the tangent to the curve at the given point.
   a. $y = 2x^2 - 7x + 3; (0, 3)$
   b. $y = 5 - 8x - 3x^2; (-1, 10)$
   c. $y = \frac{1}{2}x^3; (2, 4)$
   d. $y = \frac{1}{3}x^3 - 2x^2 + 3x + 5; (3, 5)$

3. a. Determine the equation of the tangent to the parabola $y = x^2$ at the point $(3, 9)$.
   b. Determine the equation of the tangent to the parabola $y = -3x^2$ at the point $(1, -3)$.
   c. A tangent is drawn to the parabola $y = x^2 + 2x + 5$ at the point where the parabola cuts the y-axis. Form the equation of this tangent.
   d. Form the equation of the tangent to $y = 7 - 4x^2$ at the point where $x = -\frac{1}{2}$.
   e. Form the equation of the tangent to $y = \frac{4}{3}x^3$ at the point where $x = -\frac{3}{2}$.
   f. A tangent is drawn to the curve $y = -x^3 + 8$ at the point where the curve cuts the x-axis. Form the equation of this tangent.

4. a. i. Determine the coordinates of the point on the curve $y = x^2 - 3x + 9$ where the tangent has a gradient of 3.
   ii. Hence, form the equation of the tangent.
   b. i. Determine the coordinates of the point on the curve $y = 6x(x + 2)$ where the gradient of the tangent is zero.
   ii. Hence, state the equation of the tangent.
   c. The tangent to the curve $y = x^2 + x$ is parallel to the line $y = 5x$. Form the equation of the tangent.
   d. The tangent to the curve $y = x^3 - 4x$ is parallel to the line $y = -4x + 21$. Form the equation of the tangent.

5. Form the equation of the tangent to the curve $y = 4x^2 + 3$ at the point where the tangent is parallel to the line $y = -8x$.

6. Form the equation of the tangent to $y = x^2 - 6x + 3$ that is:
   a. parallel to the line $y = 4x - 2$
   b. parallel to the x-axis
   c. perpendicular to the line $6y + 3x - 1 = 0$.

7. Consider the curve $y = x^2 - 2x + 5$
   a. Obtain the equation of the tangent at the point where $x = 2$.
   b. Calculate the equation of the line which is perpendicular to the tangent at the point where $x = 2$.
   c. A tangent to the curve is inclined at 135° to the horizontal. Determine its equation.

8. a. Calculate the coordinates of the point of intersection of the tangents drawn to the x-intercepts of the curve $y = x(4 - x)$.
   b. Consider the tangent drawn to A, the positive x-intercept of the curve $y = x(4 - x)$. A line perpendicular to this tangent is drawn through point A. Determine the coordinates of the point on the curve $y = x(4 - x)$ where this line intersects this curve.

9. Show that the tangents drawn to the x-intercepts of the parabola $y = (x - a)(x - b)$ intersect on the parabola's axis of symmetry.

10. The diagram shows the tangent drawn to a point P on the curve $y = x^2 + 1$. The coordinates of P are $\left(t, t^2 + 1\right)$, $t > 0$ and the tangent passes through the origin, O.

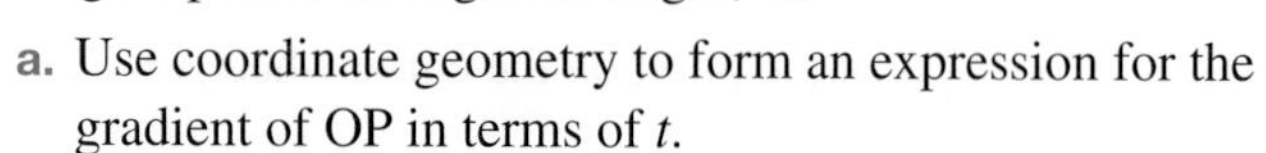

   a. Use coordinate geometry to form an expression for the gradient of OP in terms of t.
   b. Use calculus to form an expression for the gradient of the tangent at P in terms of t.
   c. Hence, determine the value of t and state the coordinates of point P.
   d. Hence, form the equation of the tangent.

11. **WE5** At the point $(-1, 1)$ on the curve $y = x^2$, a line is drawn perpendicular to the tangent to the curve at that point. This line meets the curve $y = x^2$ again at the point Q.

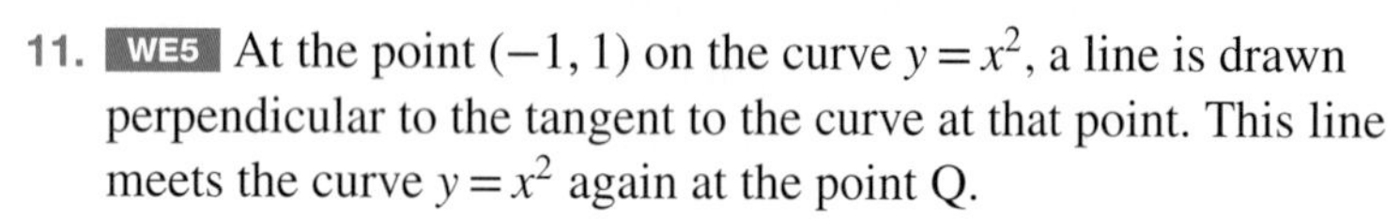

   a. Calculate the coordinates of the point Q.
   b. Calculate the magnitude of the angle that the line passing through Q and the point $(-1, 1)$ makes with the positive direction of the x-axis, correct to 1 decimal place.

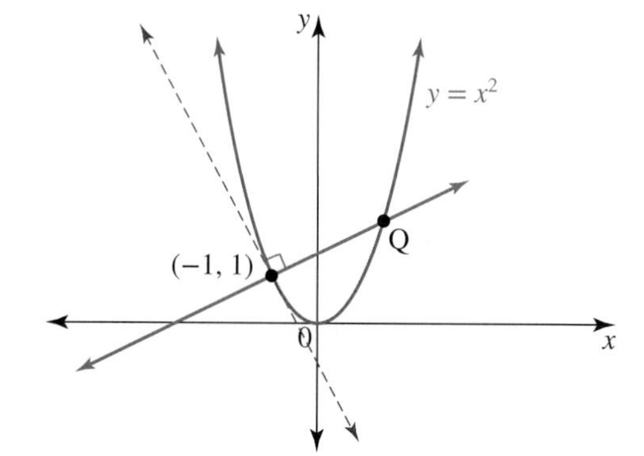

12. The tangent to the curve $y = x^3 + 3x^2$ at point A where $x = 1$ meets the curve again point B.

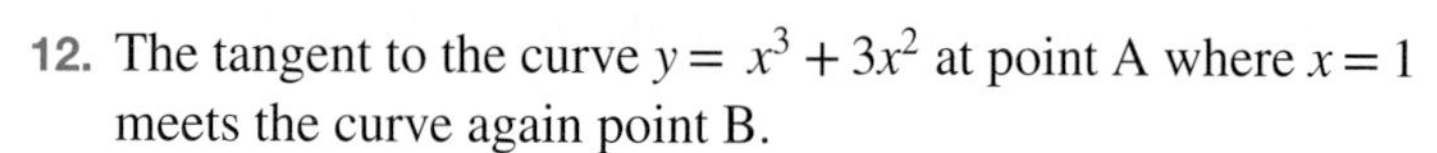

   a. Calculate the coordinates of point B.
   b. Calculate the coordinates of the midpoint of the line segment AB.

13. **WE6** a. Determine the domain interval over which the function $f(x) = \frac{1}{3}x^3 + x^2 - 8x + 6$ is increasing.
   b. i. Form the equation of the tangent to the curve $y = ax^2 + 4x + 5$ at the point where $x = 1$.
      ii. Hence, find the values of a for which the function $y = ax^2 + 4x + 5$ is decreasing at $x = 1$.

14. a. Determine the interval over which the function $f(x) = 3 - 7x + 4x^2$ is decreasing.
   b. Determine the interval over which the function $y = -12x + x^3$ is increasing.
   c. i. Show that any tangent to the curve $y = x^3 + 3x + 5$ will have a positive gradient.
      ii. Form the equation of the tangent at the point on the curve where $x = -1$.

15. For the function $f(x) = x^3 + 3$:
   a. determine the interval where the function is increasing
   b. sketch the function for $x \in R$.

Technology active

16. **WE7** Use Newton's method to calculate the root of the equation $\frac{1}{3}x^3 + 7x - 3 = 0$ that lies near $x = 2$. Express the answer correct to 4 decimal places.

17. Consider the equation $-x^3 + x^2 - 3x + 5 = 0$.
   a. Show that there is a root that lies between $x = 1$ and $x = 2$.
   b. Use Newton's method to calculate the root to an accuracy of 4 decimal places.

18. a. Use Newton's method to calculate the root of the equation $x^3 + x - 4 = 0$ near $x = 1$. Express the answer to an accuracy of 4 decimal places.
b. The equation $x^3 - 8x - 10 = 0$ has only one solution.
i. Identify the two integers between which this solution lies.
ii. Use Newton's method to obtain the solution to an accuracy of 4 decimal places.
c. Obtain the root of the equation $x^3 + 5x^2 = 10$ near $x = -2$ using Newton's method, to an accuracy of 4 decimal places.

19. a. Use a suitable cubic equation and Newton's method to calculate $\sqrt[3]{16}$ correct to 4 decimal places.
b. Calculate the exact roots of the quadratic equation $x^2 - 8x + 9 = 0$ and hence, using Newton's method to obtain one of the roots, calculate the value of $\sqrt{7}$ to 4 decimal places.

20. Consider the quartic function defined by $y = x^4 - 2x^3 - 5x^2 + 9x$.
a. Show the function is increasing when $x = 0$ and decreasing when $x = 1$.
b. Use Newton's method to obtain the value of x where the quartic function changes from increasing to decreasing in the interval $[0, 1]$. Express the value to an accuracy of 3 decimal places.

21. Consider the curve with equation $y = \frac{1}{3}x(x+4)(x-4)$.
a. Sketch the curve and draw a tangent to the curve at the point where $x = 3$.
b. Form the equation of this tangent.
c. i. The tangent meets the curve again at a point P. Show that the x-coordinate of the point P satisfies the equation $x^3 - 27x + 54 = 0$.
ii. Explain why $(x-3)^2$ must be a factor of this equation and hence calculate the coordinates of P.
d. Show that the tangents to the curve at the points where $x = \pm 4$ are parallel.
e. i. For $a \in R\backslash\{0\}$, show that the tangents to the curve $y = x(x+a)(x-a)$ at the points where $x = \pm a$ are parallel.
ii. Calculate the coordinates, in terms of a, of the points of intersection of the tangent at $x = 0$ with each of the tangents at $x = -a$ and $x = a$.

22. Consider the family of curves $C = \{(x, y): y = x^2 + ax + 3,\ a \in R\}$.
a. Express the equation of the tangent to the curve $y = x^2 + ax + 3$ at the point where $x = -a$ in terms of a.
b. If the tangent at $x = -a$ cuts the y-axis at $y = -6$ and the curve $y = x^2 + ax + 3$ is increasing at $x = -a$, calculate the value of a.
c. Determine the domain interval over which all the curves in the family C are decreasing functions.
d. Show that for $a \neq 0$, each line through $x = -a$ perpendicular to the tangents to the curves in C at $x = -a$ will pass through the point $(0, 4)$. If $a = 0$, explain whether or not this would still hold.

12.3 Exam questions

Question 1 (1 mark) TECH-ACTIVE

MC The gradient of the line perpendicular to the tangent $(2, 3)$ on the curve $y = 3 + 7x - 2x^2$ is

A. 1 **B.** 2 **C.** 3 **D.** -1 **E.** -2

Question 2 (4 marks) TECH-ACTIVE

Calculate the coordinates of the point of intersection of the tangents drawn to the x-intercepts of the curve $y = x(6 - x)$.

Question 3 (3 marks) TECH-ACTIVE

Determine the domain interval over which the function $f(x) = \frac{1}{3}x^3 + 5x^2 + 24x - 16$ is decreasing.

More exam questions are available online.

12.4 Curve sketching

LEARNING INTENTION

At the end of this subtopic you should be able to:
- determine the sign of the gradient at and near a point
- use derivatives to identify stationary points, determine their nature and sketch simple functions
- determine global or absolute maximum and minimum values of a function for a restricted domain.

At the points where a differentiable function is neither increasing nor decreasing, the function is stationary and its gradient is zero. Identifying such stationary points provides information which assists curve sketching.

12.4.1 Stationary points

At a **stationary point** on a curve $y = f(x)$, $f'(x) = 0$.

There are three types of stationary points:
- **(local) minimum** turning point
- **(local) maximum** turning point
- **stationary point of inflection**.

The word 'local' means the point is a minimum or a maximum in a particular locality or neighbourhood. Beyond this section of the graph there could be other points on the graph that are lower than the local minimum or higher than the local maximum. Our purpose for the time being is simply to identify the turning points and their nature, so we shall continue to refer to them just as minimum or maximum turning points.

Determining the nature of a stationary point

At each of the three types of stationary points, $f'(x) = 0$. This means that the tangents to the curve at these points are horizontal. By examining the slope of the tangent to the curve immediately before and immediately after the stationary point, the nature or type of stationary point can be determined.

	Minimum turning point	Maximum turning point	Stationary point of inflection
Stationary point			 or 
Slope of tangent	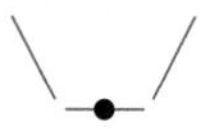	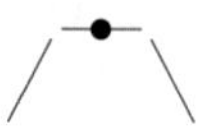	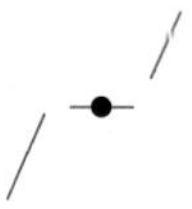 or 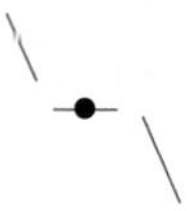
Sign diagram of gradient function $f'(x)$	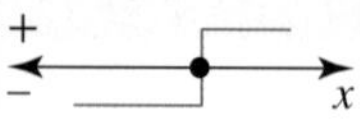 + − x	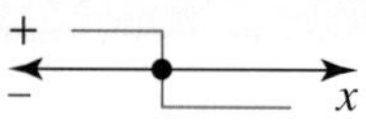 + − x	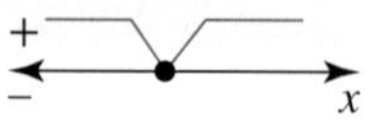 + − x or + − x 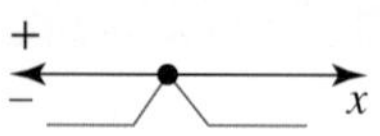

For a minimum turning point, the behaviour of the function changes from decreasing just before the point, to stationary at the point, to increasing just after the point; the slope of the tangent changes from negative to zero to positive.

For a maximum turning point, the behaviour of the function changes from increasing just before the point, to stationary at the point, to decreasing just after the point; the slope of the tangent changes from positive to zero to negative.

For a stationary point of inflection, the behaviour of the function remains either increasing or decreasing before and after the point, and stationary at the point; the slope of the tangent is zero at the point but it does not change sign either side of the point.

To identify stationary points and their nature:

- establish where $f'(x) = 0$
- determine the nature by drawing the sign diagram of $f'(x)$ around the stationary point or, alternatively, by testing the slope of the tangent at selected points either side of, and in the neighbourhood of, the stationary point.

For polynomial functions, the sign diagram is usually the preferred method for determining the nature, or for justifying the anticipated nature, of the turning point, as the gradient function is also a polynomial. If the function is not a polynomial, then testing the slope of the tangent is usually the preferred method for determining the nature of a stationary point.

WORKED EXAMPLE 8 Determining stationary points and their nature

a. Determine the stationary points of $f(x) = 2 + 4x - 2x^2 - x^3$ and justify their nature.
b. The curve $y = ax^2 + bx - 24$ has a stationary point at $(-1, -25)$. Calculate the values of a and b.

THINK	WRITE
a. 1. Calculate the x-coordinates of the stationary points. *Note:* Always include the reason why $f'(x) = 0$.	**a.** $f(x) = 2 + 4x - 2x^2 - x^3$ $f'(x) = 4 - 4x - 3x^2$ At stationary points, $f'(x) = 0$, so: $4 - 4x - 3x^2 = 0$ $(2 - 3x)(2 + x) = 0$ $x = \frac{2}{3}$ or $x = -2$
2. Calculate the corresponding y-coordinates.	When $x = \frac{2}{3}$, $f(x) = 2 + 4\left(\frac{2}{3}\right) - 2\left(\frac{2}{3}\right)^2 - \left(\frac{2}{3}\right)^3$ $= \frac{94}{27}$ When $x = -2$, $f(x) = 2 + 4(-2) - 2(-2)^2 - (-2)^3$ $= -6$ The stationary points are $\left(\frac{2}{3}, \frac{94}{27}\right)$, $(-2, -6)$.
3. To justify the nature of the stationary points, draw the sign diagram of $f'(x)$. *Note:* The shape of the cubic graph would suggest the nature of the stationary points.	Since $f'(x) = 4 - 4x - 3x^2$ $= (2 - 3x)(2 + x)$ the sign diagram of $f'(x)$ is that of a concave down parabola with zeros at $x = -2$ and $x = \frac{2}{3}$. 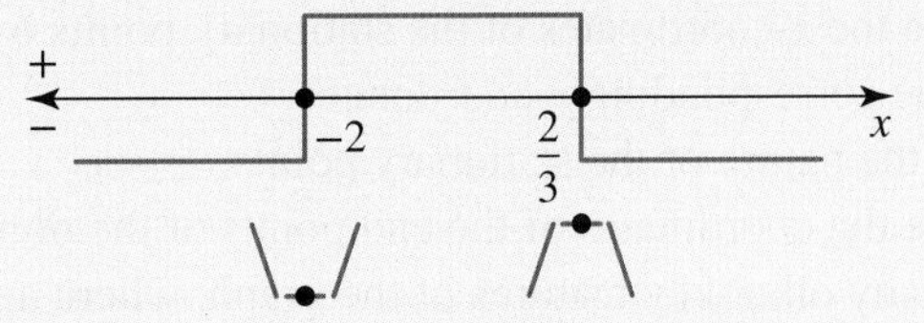
4. Identify the nature of each stationary point by examining the sign of the gradient before and after each point.	At $x = -2$, the gradient changes from negative to positive, so $(-2, -6)$ is a minimum turning point. At $x = \frac{2}{3}$, the gradient changes from positive to negative, so $\left(\frac{2}{3}, \frac{94}{27}\right)$ is a maximum turning point.

b. 1. Use the coordinates of the given point to form an equation. *Note:* As there are two unknowns to determine, two pieces of information are needed to form two equations in the two unknowns.	**b.** $y = ax^2 + bx - 24$ The point $(-1, -25)$ lies on the curve. $-25 = a(-1)^2 + b(-1) - 24$ $\therefore a - b = -1 \quad [1]$
2. Use the other information given about the point to form a second equation.	The point $(-1, -25)$ is a stationary point, so $\frac{dy}{dx} = 0$ at this point. $y = ax^2 + bx - 24$ $\frac{dy}{dx} = 2ax + b$ At $(-1 - 24)$, $\frac{dy}{dx} = 2a(-1) + b$ $= -2a + b$ $\therefore -2a + b = 0 \quad [2]$
3. Solve the simultaneous equations and state the answer.	$a - b = -1 \quad [1]$ $-2a + b = 0 \quad [2]$ Adding the equations, $-a = -1$ $\therefore a = 1$ Substitute $a = 1$ into equation [2]: $-2 + b = 0$ $\therefore b = 2$ The values are $a = 1$ and $b = 2$.

on Resources

 **Interactivity** Stationary points (int-5963)

12.4.2 Curve sketching

To sketch the graph of any function $y = f(x)$, perform the following steps:

- Obtain the y-intercept by evaluating $f(0)$.
- Obtain any x-intercepts by solving, if possible, $f(x) = 0$. This may require the use of factorisation techniques including the factor theorem.
- Calculate the x-coordinates of the stationary points by solving $f'(x) = 0$ and use the equation of the curve to obtain the corresponding y-coordinates.
- Identify the nature of the stationary points.
- Calculate the coordinates of the end points of the domain, where appropriate.
- Identify any other key features of the graph, where appropriate.

WORKED EXAMPLE 9 Sketching a cubic polynomial

Sketch the function $y = \frac{1}{2}x^3 - 3x^2 + 6x - 8$. Locate any intercepts with the coordinate axes and any stationary points, and justify their nature.

THINK

1. State the y-intercept.

WRITE

$y = \frac{1}{2}x^3 - 3x^2 + 6x - 8$

y-intercept: $(0, -8)$

2. Calculate any x-intercepts.

x-intercepts: when $y = 0$,

$\frac{1}{2}x^3 - 3x^2 + 6x - 8 = 0$

$x^3 - 6x^2 + 12x - 16 = 0$

Let $p(x) = x^3 - 6x^2 + 12x - 16$.

$p(4) = 64 - 96 + 48 - 16$

$\quad = 0$

$\therefore (x - 4)$ is a factor.

$x^3 - 6x^2 + 12x - 16 = (x - 4)(x^2 - 2x + 4)$

$\therefore x = 4$ or $x^2 - 2x + 4 = 0$

The discriminant of $x^2 - 2x + 4$ is $\Delta = 4 - 16 < 0$.

There is only one x-intercept: $(4, 0)$.

3. Obtain the derivative in order to locate any stationary points.

Stationary points:

$y = \frac{1}{2}x^3 - 3x^2 + 6x - 8$

$\frac{dy}{dx} = \frac{3}{2}x^2 - 6x + 6$

At stationary points, $\frac{dy}{dx} = 0$, so:

$\frac{3}{2}x^2 - 6x + 6 = 0$

$\frac{3}{2}(x^2 - 4x + 4) = 0$

$\frac{3}{2}(x - 2)^2 = 0$

$x = 2$

Substitute $x = 2$ into the function's equation:

$y = \frac{1}{2}(2)^3 - 3(2)^3 + 6(2) - 8$

$\quad = -4$

The stationary point is $(2, -4)$.

4. Identify the type of stationary point using the sign diagram of the gradient function.

Sign diagram of $\frac{dy}{dx} = \frac{3}{2}(x - 2)^2$:

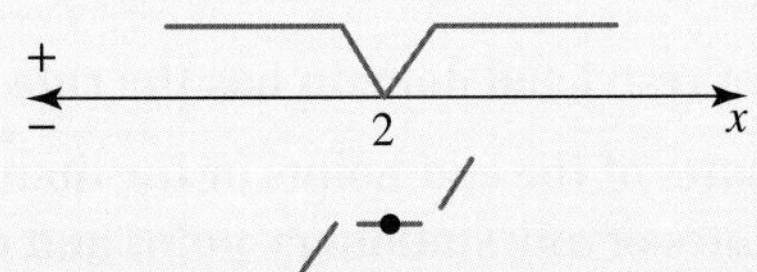

The point $(2, -4)$ is a stationary point of inflection.

5. Sketch the curve showing the intercepts with the axes and the stationary point.

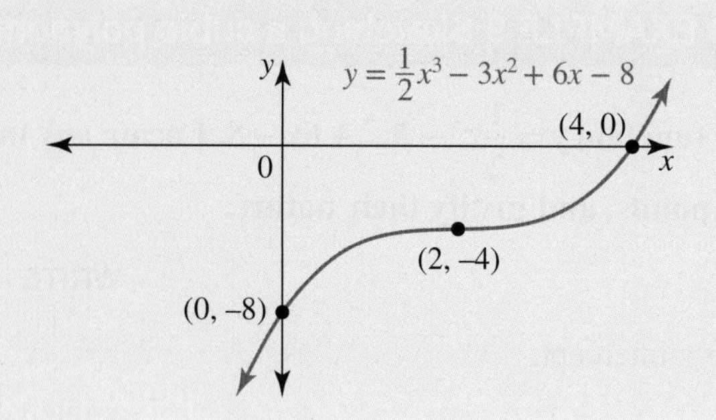

12.4.3 Local and global maxima and minima

The diagram shows the graph of a function sketched over a domain with end points D and E.

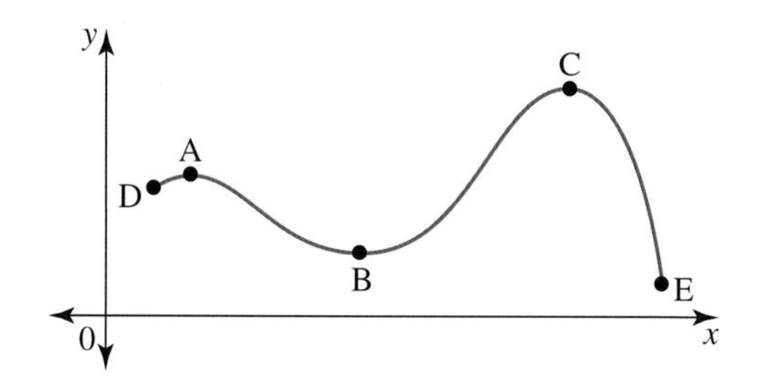

There are three turning points: A and C are maximum turning points, and B is a minimum turning point.

The y-coordinate of point A is greater than those of its neighbours, so A is a local maximum point. At point C, not only is its y-coordinate greater than those of its neighbours, it is greater than that of any other point on the graph. For this reason, C is called the **global** or **absolute maximum** point.

The **global** or **absolute minimum** point is the point whose y-coordinate is smaller than any others on the graph. For this function, point E, an end point of the domain, is the global or absolute minimum point. Point B is a local minimum point; it is not the global minimum point.

Global maximums and global minimums may not exist for all functions. For example, a cubic function on its maximal domain may have one local maximum turning point and one local minimum turning point, but there is neither a global maximum nor a global minimum point since as $x \to \pm\infty$, $y \to \pm\infty$ (assuming a positive coefficient of x^3).

If a differentiable function has a global maximum or a global minimum value, then this will either occur at a turning point or at an end point of the domain. The y-coordinate of such a point gives the value of the global maximum or the global minimum.

Maxima and minima

- **A function $y = f(x)$ has a global maximum $f(a)$ if $f(a) \geq f(x)$ for all x-values in its domain.**
- **A function $y = f(x)$ has a global minimum $f(a)$ if $f(a) \leq f(x)$ for all x-values in its domain.**
- **A function $y = f(x)$ has a local maximum $f(x_0)$ if $f(x_0) \geq f(x)$ for all x-values in the neighbourhood of x_0.**
- **A function $y = f(x)$ has a local minimum $f(x_0)$ if $f(x_0) \leq f(x)$ for all x-values in the neighbourhood of x_0.**

WORKED EXAMPLE 10 Determining absolute maximum and minimum values

A function defined on a restricted domain has the rule $y = \frac{3}{8}x(x-3)^2$, $x \in [-1, 4]$.

a. **Specify the coordinates of the end points of the domain.**
b. **Obtain the coordinates of any stationary point and determine its nature.**
c. **Sketch the graph of the function.**
d. **State the global, or absolute, maximum and global minimum values of the function.**

THINK	WRITE
a. 1. Use the given domain to calculate the coordinates of the end points.	$y = \frac{3}{8}x(x-3)^2$
2. Substitute each of the end values of the domain in the function's rule.	For the domain $-1 \leq x \leq 4$: Left end point: when $x = -1$, $y = \frac{3}{8}(-1)(-1-3)^2$ $= -\frac{3}{8} \times 16$ $= -6$ Right end point: when $x = 4$, $y = \frac{3}{8}(4)(4-3)^2$ $= \frac{3}{2} \times 1$ $= \frac{3}{2}$
3. Answer the question.	The end points are $(-1, -6)$, $\left(4, \frac{3}{2}\right)$.
b. 1. Expand the function to find the derivative.	$y = \frac{3}{8}x(x-3)^2$ $= \frac{3}{8}x(x^2 - 6x + 9)$ $= \frac{3}{8}(x^3 - 6x^2 + 9x)$
2. Calculate the derivative of the function.	$\frac{dy}{dx} = \frac{3}{8}\left(3x^2 - 12x + 9\right)$
3. Calculate the coordinates of any stationary points.	At a stationary point, $\frac{dy}{dx} = 0$, so: $\frac{3}{8}\left(3x^2 - 12x + 9\right) = 0$ $3x^2 - 12x + 9 = 0$ $3(x^2 - 4x + 3) = 0$ $3(x-1)(x-3) = 0$ $x = 1, 3$ When $x = 1$, $y = \frac{3}{8}(1)(1-3)^2$ $= \frac{3}{8} \times 4$ $= \frac{3}{2}$ When $x = 3$, $y = \frac{3}{8}(3)(3-3)^2$ $= 0$

4. Test the gradient either side of the stationary points to determine their nature.

Stationary points: $\left(1, \frac{3}{2}\right)$, $(3, 0)$

x	0	1	2	3	4
$\frac{dy}{dx}$	$\frac{3}{8}\times(9)>0$	0	$\frac{3}{8}\times(12-24+9)<0$	0	$\frac{3}{8}\times(48-48+9)>0$
Slope of the tangent	/	—	\	—	/

5. State the nature of the stationary points

The point $\left(1, \frac{3}{2}\right)$ is a maximum turning point and the point $(3, 0)$ is a minimum turning point.

c. 1. Calculate any intercepts with the coordinate axes.

For the y-intercept, $x=0$: point $(0,0)$

For the x-intercepts, $y=0$:

$$\frac{3}{8}x(x-3)^2=0$$

$\therefore x=0$ or $x=3$

Axis intercepts: $(0,0)$ and $(3,0)$

2. Sketch the graph using the end points, axis intercepts and stationary points.

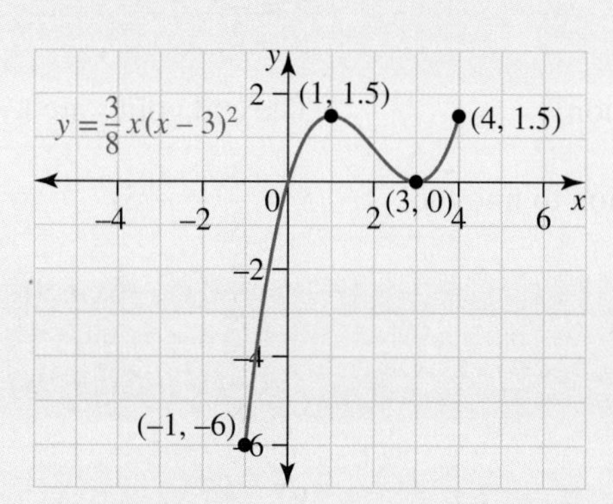

d. Examine the graph and the y-coordinates to identify the global, or absolute extreme values.

The function has a global maximum of $\frac{3}{2}$ at the maximum turning point and the right end point, and a global minimum of -6 at the left end point.

12.4 Exercise

Students, these questions are even better in jacPLUS

Receive immediate feedback and access sample responses

Access additional questions

Track your results and progress

Find all this and MORE in jacPLUS

Technology free

1. Use calculus to obtain the coordinates of any stationary points on each of the following curves.

a. $y=4x-x^2$
b. $y=12x^2-24x+5$
c. $y=-x(6+x)$
d. $y=x^3+1$
e. $y=x^3-3x$
f. $y=2-x^4$

2. **MC** A continuous polynomial function $y = f(x)$, has a local maximum turning point at the point where $x = 7$. Select the statement that describes how the gradient $f'(x)$ changes in the neighbourhood of $x = 7$.

A. $f'(x) > 0$ just before the turning point, $f'(7) = 0$ and $f'(x) > 0$ just after the turning point
B. $f'(x) > 0$ just before the turning point, $f'(7) = 0$ and $f'(x) < 0$ just after the turning point
C. $f'(x) < 0$ just before the turning point, $f'(7) = 0$ and $f'(x) > 0$ just after the turning point
D. $f'(x) < 0$ just before the turning point, $f'(7) = 0$ and $f'(x) < 0$ just after the turning point
E. None of the above

3. Consider $f(x) = 2x^3 - 3x^2 - 12x$.
   a. Show that $f'(2) = 0$.
   b. Calculate the values of $f'(1)$ and $f'(3)$.
   c. Hence, state the nature of the stationary point at $x = 2$ and find its coordinates.
   d. Determine the coordinates of the other stationary point.
   e. Draw a sketch of the graph of the cubic function $f(x) = 2x^3 - 3x^2 - 12x$ without identifying any non-zero x-intercepts.

4. **WE8** a. Determine the stationary points of $f(x) = x^3 + x^2 - x + 4$ and justify their nature.
   b. The curve $y = ax^2 + bx + c$ contains the point $(0, 5)$ and has a stationary point at $(2, -14)$. Calculate the values of a, b and c.

5. a. Use calculus to identify the coordinates of the turning points of the following parabolas.
      i. $y = x^2 - 8x + 10$
      ii. $y = -5x^2 + 6x - 12$
   b. The point $(4, -8)$ is a stationary point of the curve $y = ax^2 + bx$.
      i. Calculate the values of a and b.
      ii. Sketch the curve.

6. Consider the function defined by $f(x) = x^3 + 3x^2 + 8$.
   a. Show that $(-2, 12)$ is a stationary point of the function.
   b. Use a slope diagram to determine the nature of this stationary point.
   c. Give the coordinates of the other stationary point.
   d. Use a sign diagram of the gradient function to justify the nature of the second stationary point.

7. Obtain any stationary points of the following curves and justify their nature using the sign of the derivative.
   a. $y = \frac{1}{3}x^3 + x^2 - 3x - 1$
   b. $y = -x^3 + 6x^2 - 12x + 18$
   c. $y = \frac{23}{6}x(x - 3)(x + 3)$
   d. $y = 4x^3 + 5x^2 + 7x - 10$

8. a. There is a stationary point on the curve $y = x^3 + x^2 + ax - 5$ at the point where $x = 1$. Determine the value of a.
   b. There is a stationary point on the curve $y = -x^4 + bx^2 + 2$ at the point where $x = -2$. Determine the value of b.
   c. The curve $y = ax^2 - 4x + c$ has a stationary point at $(1, -1)$. Calculate the values of a and c.

9. The curve $y = x^3 + ax^2 + bx - 11$ has stationary points when $x = 2$ and $x = 4$.
   a. Calculate a and b.
   b. Determine the coordinates of the stationary points and their nature.

10. **WE9** Sketch the function $y = 2x^2 - x^3$. Locate any intercepts with the coordinate axes and any stationary points, and justify their nature.

11. Sketch a possible graph of the function $y = f(x)$ for which:

a. $f'(-4) = 0, f'(6) = 0, f'(x) < 0$ for $x < -4, f'(x) > 0$ for $-4 < x < 6$ and $f'(x) < 0$ for $x > 6$
b. $f'(1) = 0, f'(x) > 0$ for $x \in R \setminus \{1\}$ and $f(1) = -3$.

12. Sketch the graphs of each of the following functions and label any axis intercepts and any stationary points with their coordinates. Justify the nature of the stationary points.

a. $f: R \to R,\ f(x) = 2x^3 + 6x^2$
b. $g: R \to R,\ g(x) = -x^3 + 4x^2 + 3x - 12$
c. $p: [-1,\ 1] \to R,\ p(x) = x^3 + 2x$
d. $\{(x, y): y = x^4 - 6x^2 + 8\}$

13. a. State the greatest and least number of turning points a cubic function can have.
b. Show that $y = 3x^3 + 6x^2 + 4x + 6$ has one stationary point and determine its nature.
c. Determine the values of k so the graph of $y = 3x^3 + 6x^2 + kx + 6$ will have no stationary points.
d. If a cubic function has exactly one stationary point, explain why it is not possible for that stationary point to be a maximum turning point. State the type of stationary point it must be.
e. State the degree of the gradient function of a cubic function and use this to explain whether it is possible for the graph of a cubic function to have two stationary points: one a stationary point of inflection and the other a maximum turning point.

14. **WE10** A function defined on a restricted domain has the rule $y = 16x(x+1)^3,\ x \in [-2, 1]$.

a. Specify the coordinates of the end points of the domain.
b. Obtain the coordinates of any stationary point and determine its nature.
c. Sketch the graph of the function.
d. State the global, or absolute, maximum and global minimum values of the function.

15. Determine, if possible, the global maximum and minimum values of the function $f(x) = 4x^3 - 12x$ over the domain $\{x: x \le \sqrt{3}\}$.

16. Sketch the graphs and state any local and global maximum or minimum values over the given domain for each of the following functions.

a. $y = 4x^2 - 2x + 3,\ -1 \le x \le 1$
b. $y = x^3 + 2x^2, -3 \le x \le 3$
c. $y = 3 - 2x^3,\ x \le 1$
d. $f: [0,\ \infty) \to R,\ f(x) = x^3 + 6x^2 + 3x - 10$

17. The point $(2, -54)$ is a stationary point of the curve $y = x^3 + bx^2 + cx - 26$.

a. Find the values of b and c.
b. Obtain the coordinates of the other stationary point.
c. Identify where the curve intersects each of the coordinate axes.
d. Sketch the curve and label all key points with their coordinates.

Technology active

18. Sketch the function $y = x^4 + 2x^3 - 2x - 1$. Locate any intercepts with the coordinate axes and any stationary points, and justify their nature.

19. Show that the line through the turning points of the cubic function $y = xa^2 - x^3$ must pass through the origin for any real positive constant a.

20. Consider the function $y = ax^3 + bx^2 + cx + d$. The graph of this function passes through the origin at an angle of 135° with the positive direction of the x-axis.

a. Obtain the values of c and d.
b. Calculate the values of a and b given that the point $(2, -2)$ is a stationary point.
c. Find the coordinates of the second stationary point.
d. Calculate the coordinates of the point where the tangent at $(2, -2)$ meets the curve again.

12.4 Exam questions

Question 1 (4 marks) TECH-FREE

The point $(2, -16)$ is a stationary point on the curve $y = ax^2 + bx$. Calculate the values of a and b, and sketch the curve.

Question 2 (3 marks) TECH-FREE

A polynomial with equation $y = f(x)$ has the following properties:

- $f(x)$ has a y-intercept at $(0, -1)$ and x-intercepts at $(-4, 0)$ and $(-1, 0)$.
- $f(x)$ is a decreasing function for $\{x : x < -3\} \cup \{x : x > -1\}$.
- $f'(-3) = 0$ and $f'(-1) = 0$.

Sketch a possible shape of the graph $f(x)$.

Question 3 (5 marks) TECH-FREE

Consider the function $f(x) = -x^3 + 3x^2 - 4$ for $-2 \le x \le 3$.

a. Find the exact coordinates of any stationary points. **(3 marks)**

b. Sketch the graph of $y = -x^3 + 3x^2 - 4$ for $x \in [-2, 3]$ given that an x-intercept occurs at $(-1, 0)$. **(1 mark)**

c. State the global minimum and global maximum values. **(1 mark)**

More exam questions are available online.

12.5 Optimisation problems

LEARNING INTENTION

At the end of this subtopic you should be able to:

- using derivatives to solve simple maximum and minimum optimisation problems.

Optimisation problems involve determining the greatest possible or least possible value of some quantity, subject to certain conditions. These types of problems provide an important practical application of differential calculus and global extrema.

12.5.1 Solving optimisation problems

If the mathematical model of the quantity to be optimised is a function of more than one variable, then it is necessary to reduce it to a function of one variable before its derivative can be calculated. The given conditions may enable one variable to be expressed in terms of another. Techniques that may be required include the use of Pythagoras' theorem, trigonometry, similar triangles or standard mensuration formulas.

To solve optimisation problems:

- draw a diagram of the situation, where appropriate
- identify the quantity to be maximised or minimised and define the variables involved
- express the quantity to be optimised as a function of one variable
- find the stationary points of the function and determine their nature
- consider the domain of the function as there are likely to be restrictions in practical problems and, if appropriate, find the end points of the domain
- the maximum or minimum value of the function will occur either at a stationary point or at an end point of the domain.

WORKED EXAMPLE 11 Solving optimisation problems

The new owner of an apartment wants to install a window in the shape of a rectangle surmounted by a semicircle in order to allow more light into the apartment.

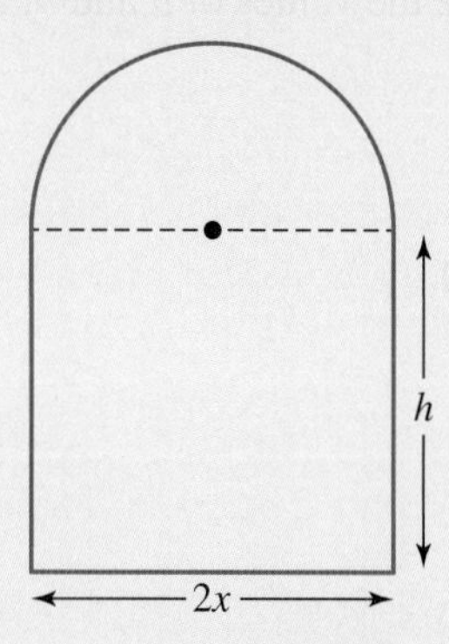

The owner has 336 cm of waterproofing tape to surround the window and wants to determine the dimensions, as shown in the diagram, that will allow as much light into the apartment as possible.

a. **Show that the area A, in cm^2, of the window is $A = 336x - \frac{1}{2}(4 + \pi)x^2$.**

b. **Hence, determine, to the nearest cm, the width and the height of the window for which the area is greatest.**

c. **Structural problems require that the width of the window should not exceed 84 cm. Determine what the new dimensions of the window should be for maximum area.**

THINK	WRITE
a. 1. Form an expression for the total area. *Note:* This expression involves more than one variable.	a. The total area is the sum of the areas of the rectangle and semicircle. Rectangle: length $2x$ cm, width h cm $\therefore A_{\text{rectangle}} = 2xh$ Semicircle: diameter $2x$ cm, radius x cm $\therefore A_{\text{semicircle}} = \frac{1}{2}\pi x^2$ The total area of the window is $A = 2xh + \frac{1}{2}\pi x^2$.
2. Use the given condition to form an expression connecting the two variables.	The perimeter of the window is 336 cm. The circumference of the semicircle is $\frac{1}{2}(2\pi x)$, so the perimeter of the shape is $h + 2x + h + \frac{1}{2}(2\pi x)$. Hence, $2h + 2x + \pi x = 336$.
3. Express one appropriately chosen variable in terms of the other.	The required expression for the area is in terms of x, so express h in terms of x. $2h = 336 - 2x - \pi x$ $h = \frac{1}{2}(336 - 2x - \pi x)$

4. Write the area as a function of one variable in the required form.	Substitute h into the area function: $A = x(2h) + \frac{1}{2}\pi x^2$ $= x(336 - 2x - \pi x) + \frac{1}{2}\pi x^2$ $= 336x - 2x^2 - \pi x^2 + \frac{1}{2}\pi x^2$ $= 336x - \left(2 + \frac{1}{2}\pi\right)x^2$ $\therefore A = 336x - \frac{1}{2}(4 + \pi)x^2$ as required.
b. 1. Determine where the stationary point occurs and justify its nature.	**b.** At the stationary point, $\frac{dA}{dx} = 0$. $\frac{dA}{dx} = 336 - (4 + \pi)x$ $336 - (4 + \pi)x = 0$ $x = \frac{336}{4 + \pi}$ ≈ 47.05 As the area function is a concave down quadratic, the stationary point at $x = \frac{336}{4 + \pi}$ is a maximum turning point.
2. State the values of both variables.	When $x = \frac{336}{4 + \pi}$, $2h = 336 - 2 \times \left(\frac{336}{4 + \pi}\right) - \pi \times \left(\frac{336}{4 + \pi}\right)$ $h \approx 47.05$
3. Calculate the required dimensions and state the answer.	The width of the window is $2x \approx 94$ cm. The total height of the window is $h + x \approx 94$ cm. Therefore, the area of the window will be greatest if its width is 94 cm and its height is 94 cm.
c. 1. Give the restricted domain of the area function.	**c.** If the width is not to exceed 84 cm, then: $2x \leq 84$ $x \leq 42$ With the restriction, the domain of the area function is $(0, 42]$.
2. Determine where the function is greatest.	As the stationary point occurs when $x = 47$, for the domain $(0, 42]$ there is no stationary point, so the greatest area must occur at an end point of the domain.

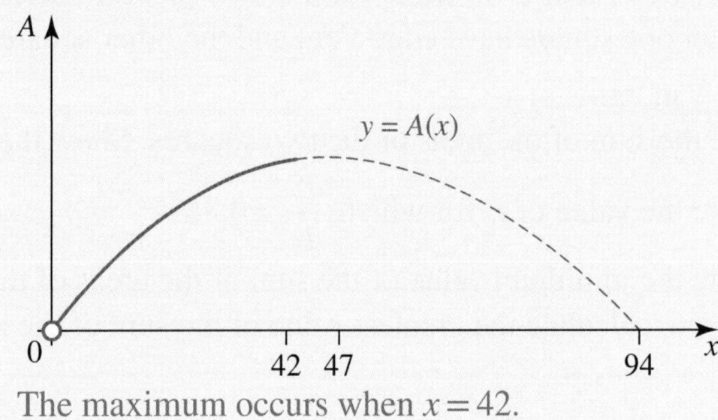

The maximum occurs when $x = 42$.

3. Calculate the required dimensions and state the answer.

When $x = 42$,

$$h = \frac{1}{2}(336 - 84 - 42\pi)$$
$$\approx 60$$

The width of the window is $2x = 84$ cm.
The height of the window is $h + x = 102$ cm.
With the restriction, the area of the window will be greatest if its width is 84 cm and its height is 102 cm.

12.5 Exercise

Students, these questions are even better in jacPLUS

Receive immediate feedback and access sample responses

Access additional questions

Track your results and progress

Find all this and MORE in jacPLUS

Technology free

1. Murray sells small pots containing tomato seedlings at local farmers' markets. Murray believes that the amount of money they can expect to make at one of these markets can be modelled by the function $M = 40x - 2x^2$, where M is the amount in dollars that Murray makes from selling x pots.
   a. Calculate $\frac{dM}{dx}$.
   b. Hence, find the value of x for which M is greatest.
   c. Determine the maximum amount of money that Murray expects to earn at a market.
2. Mai has 32 cm of tape to place around the perimeter of a rectangular area of material to create a placemat. The dimensions of the rectangle are x cm by y cm.
   a. Express y in terms of x.
   b. Show that the area, A cm^2, of the rectangular placemat is given by $A = 16x - x^2$ and state any restrictions on the domain of this area function.
   c. Use calculus to calculate the value of x for which the area is greatest.
   d. Determine the dimensions of the placemat with the greatest area. State the greatest area.
3. A piece of wire of length 40 cm is cut into two pieces. Each of these two pieces is then bent into the shape of a square. Let one square have edge x cm and the other square have edge y cm.
   a. Express y in terms of x.
   b. Let S be the sum of the areas of the two squares. Show that $S = 2x^2 - 20x + 100$.
   c. Calculate the value of x for which $\frac{dS}{dx} = 0$.
   d. Calculate the minimum value of the sum of the areas of the two squares.
   e. If $3 \le x \le 6$, calculate the greatest value of the sum of the areas of the two squares.

4. A rectangular vegetable garden patch uses part of a back fence as the length of one side. There are 40 metres of fencing available for enclosing the other three sides of the vegetable garden.
   a. Draw a diagram of the garden and express the area in terms of the width (the width being the length of the sides perpendicular to the back fence).
   b. Use calculus to obtain the dimensions of the garden for maximum area.
   c. State the maximum area.
   d. If the width is to be between 5 and 7 metres, calculate the greatest area of garden that can be enclosed with this restriction.

5. The cost in dollars of employing n people per hour in a small distribution centre is modelled by $C = n^3 - 10n^2 - 32n + 400,\ 5 \le n \le 10$. Calculate the number of people who should be employed in order to minimise the cost and justify your answer.

Technology active

6. The opening batter in a cricket match strikes the ball so that its height y metres above the ground after it has travelled a horizontal distance x metres is given by $y = 0.0001x^2(625 - x^2)$.
   a. Calculate, to 2 decimal places, the greatest height the ball reaches and justify the maximum nature.
   b. Determine how far the ball travels horizontally before it strikes the ground.

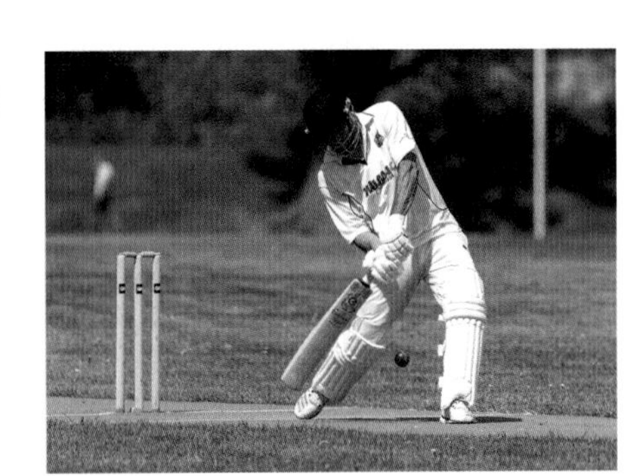

7. A rectangular box with an open top is to be constructed from a rectangular sheet of cardboard measuring 20 cm by 12 cm by cutting equal squares of side length x cm out of the four corners and folding the flaps up.
   a. Express the volume as a function of x, stating the domain restrictions.
   b. Determine the dimensions of the box with greatest volume and give this maximum volume to the nearest whole number.

8. **WE11** The owner of an apartment wants to create a stained glass feature in the shape of a rectangle surmounted by an isosceles triangle of height equal to half its base, next to the door opening on to a balcony.
   The owner has 150 cm of plastic edging to place around the perimeter of the feature and wants to determine the dimensions of the figure with the greatest area.
   a. Show that the area A, in cm^2, of the stained glass figure is $A = 150x - (2\sqrt{2} + 1)x^2$.
   b. Hence, determine, to 1 decimal place, the width and the height of the figure for which the area is greatest.
   c. Structural problems require that the width of the figure should not exceed 30 cm. Calculate the dimensions of the stained glass figure that has maximum area within this requirement.

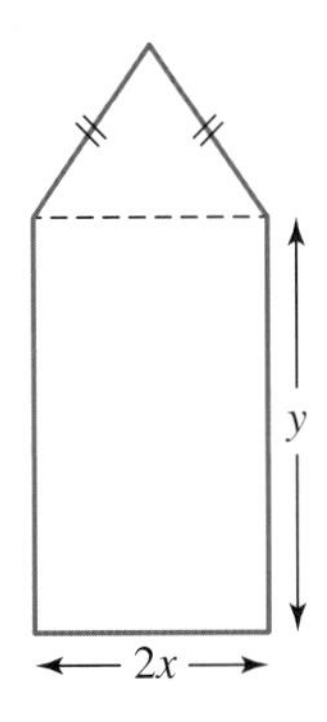

9. The total surface area of a closed cylinder is 200 cm^2. Let the base radius of the cylinder be r cm and let the height be h cm.
   a. Express h in terms of r.
   b. Show that the volume, V cm^3, is $V = 100r - \pi r^3$.
   c. Hence, show that for maximum volume the height must equal the diameter of the base.
   d. Calculate, to the nearest integer, the minimum volume if $2 \le r \le 4$.

10. A right circular cone is inscribed in a sphere of radius 7 cm. In the diagram shown, O is the centre of the sphere, C is the centre of the circular base of the cone, and V is the vertex of the cone.
The formula for the volume of a cone of height h cm and base radius r cm is $V_{\text{cone}} = \frac{1}{3}\pi r^2 h$.

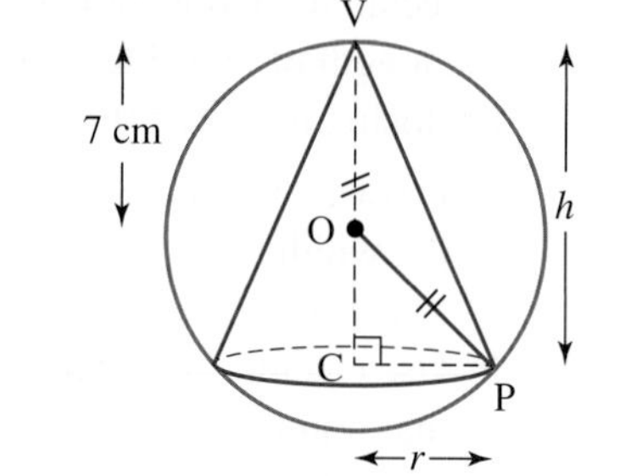

a. Show that the volume V cm^3 of the cone satisfies the relationship $V = \frac{1}{3}\pi(14h^2 - h^3)$.
b. Hence, obtain the exact values of r and h for which the volume is greatest, justifying your answer.

11. A shop offers a gift-wrapping service where purchases are wrapped in the shape of closed rectangular prisms.
For one gift-wrapped purchase, the rectangular prism has length $2w$ cm, width w cm, height h cm and a surface area of 300 cm^2.

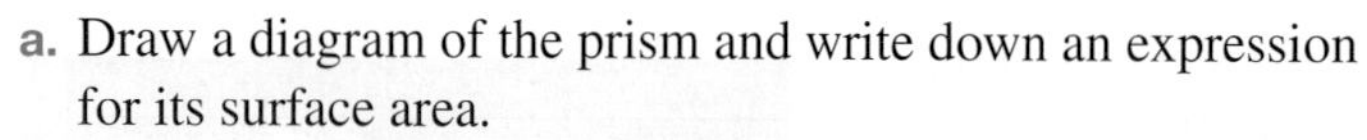

a. Draw a diagram of the prism and write down an expression for its surface area.
b. Express its volume as a function of w and determine the maximum value of the volume.
c. Give the exact dimensions of the prism with greatest volume.
d. If $2 \le w \le 6$, determine the range of values for the volume.

12. A piece of wire 240 cm long is used to reinforce the edges of a rectangular box where the length of the box is three times the width. Let the width of the box be x cm.
a. Show that the height, h cm, is $h = 60 - 4x$.
b. Show that the volume, V cm^3, is $V = 180x^2 - 12x^3$.
c. Determine the maximum volume of the box and state its dimensions.
d. Sketch the graph of $V = 180x^2 - 12x^3$, stating its domain.
e. If the length of the box needs to be restricted to 12 cm or less, determine the maximum volume of the box and state its dimensions.

13. An open rectangular storage bin has its edges reinforced with 100 metre of steel. The storage bin is to be constructed from different strength materials. The cost of the materials for its sides is \$10 per square metre and the material for the stronger base costs \$25 per square metre. If the dimensions of the base are x and y metres and the bin has a height of 1.5 metres, find with justification, the cost, to the nearest dollar, of the largest bin that can be formed under these conditions.

14. A section of a rose garden is enclosed by edging to form the shape of a sector ABC of radius r metres and arc length l metres.

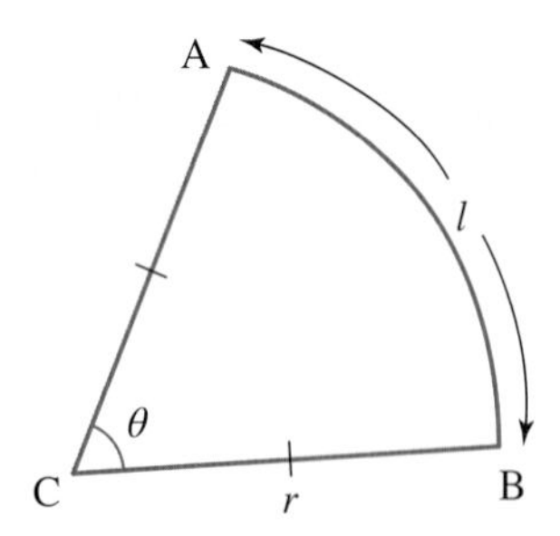

The perimeter of this section of the garden is 8 metres.
a. If θ is the angle, in radian measure, subtended by the arc at C, express θ in terms of r.
b. The formula for the area of a sector is $A_{\text{sector}} = \frac{1}{2}r^2\theta$. Express the area of the sector in terms of r.
c. Calculate the value of θ when the area is greatest.

12.5 Exam questions

Question 1 (1 mark) TECH-ACTIVE

MC The total surface area of a closed cylinder is $100\,\text{cm}^2$, where the base radius is r cm and the height is h cm. An expression for h in terms of r is

A. $h = \dfrac{50 - \pi r^2}{\pi r}$ **B.** $h = 100r$ **C.** $h = \dfrac{100 - \pi r^2}{\pi}$ **D.** $h = \dfrac{50r - \pi r^2}{\pi}$ **E.** $h = 100\pi r^2$

Question 2 (1 mark) TECH-ACTIVE

MC The shape shown has a perimeter of 200 cm, where h and x are lengths of the sides.

The area of this shape in terms of x is

A. $50x - x^2\left(\dfrac{1+\sqrt{2}}{5}\right)$ **B.** $x^2\left(\dfrac{1-\sqrt{2}}{2}\right)$ **C.** $100x - x^2\left(\dfrac{1+\sqrt{2}}{2}\right)$

D. $100x - x^2\sqrt{2}$ **E.** $50x + x^2\left(\dfrac{3+\sqrt{2}}{2}\right)$

Question 3 (6 marks) TECH-ACTIVE

A rectangular box with an open top is to be constructed from a rectangular sheet of cardboard measuring 10 cm by 10 cm. Equal squares of side length x cm are cut out of the four corners and the flaps are folded up to form the box.

a. Express the volume of the box as a function of x. **(2 marks)**

b. Determine the dimensions of the box with the greatest volume, and determine this maximum volume to the nearest whole number. **(4 marks)**

More exam questions are available online.

12.6 Rates of change and kinematics

LEARNING INTENTION

At the end of this subtopic you should be able to:

- use differentiation to determine rates of change
- interpret position velocity and acceleration as functions of time and their relationships to derivatives
- understand the difference between velocity, average velocity and average speed.

Calculus enables us to analyse the behaviour of a changing quantity. Many of the fields of interest in the biological, physical and social sciences involve the study of **rates of change**. In this section we shall consider the application of calculus to rates of change in general contexts and then as applied to the motion of a moving object.

12.6.1 Rates of change

The derivative of a function measures the instantaneous rate of change. For example, the derivative $\dfrac{dV}{dt}$ could be the rate of change of volume with respect to time, with possible units being litres per minute; the rate of change of volume with respect to height would be $\dfrac{dV}{dh}$, with possible units being litres per cm. To calculate these rates, V would need to be expressed as a function of one independent variable: either time or height. Similar methods to those encountered in optimisation problems may be required to connect variables together in order to obtain this function of one variable.

To solve rates of change problems:

- draw a diagram of the situation, where appropriate
- identify the rate of change required and define the variables involved
- express the quantity that is changing as a function of one independent variable, the variable the rate is measured with respect to
- calculate the derivative that measures the rate of change
- to obtain the rate at a given value or instant, substitute the given value into the derivative expression
- a negative value for the rate of change means the quantity is decreasing; a positive value for the rate of change means the quantity is increasing.

WORKED EXAMPLE 12 Determining rates of change

A container in the shape of an inverted right cone of radius 2 cm and depth 5 cm is being filled with water. When the depth of water is *h* cm, the radius of the water level is *r* cm.

a. Use similar triangles to express *r* in terms of *h*.

b. Express the volume of the water as a function of *h*.

c. Determine the rate, with respect to the depth of water, at which the volume of water is changing when its depth is 1 cm.

THINK

a. 1. Draw a diagram of the situation.

2. Obtain the required relationship between the variables.

b. 3. Express the function in the required form.

c. 1. Calculate the derivative of the function.

WRITE

a.

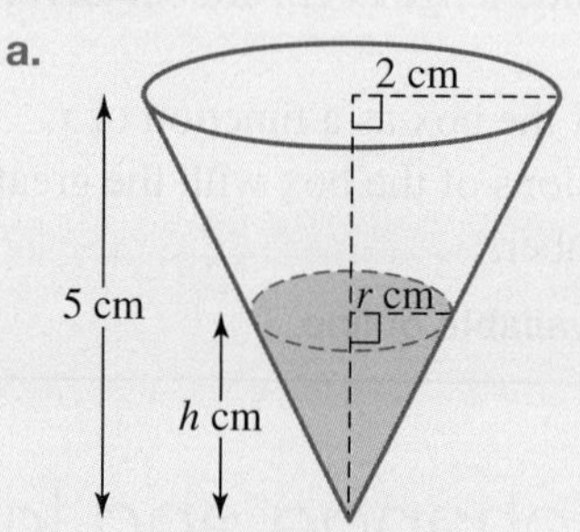

Using similar triangles,

$$\frac{r}{h} = \frac{2}{5}$$

$$\therefore r = \frac{2h}{5}$$

b. The volume of a cone is $V_{\text{cone}} = \frac{1}{3}\pi r^2 h$.

Therefore, the volume of water is $V = \frac{1}{3}\pi r^2 h$.

Substitute $r = \frac{2h}{5}$:

$$V = \frac{1}{3}\pi\left(\frac{2h}{5}\right)^2 h$$

$$= \frac{4\pi h^3}{75}$$

c. The derivative gives the rate of change at any depth.

$$\frac{dV}{dh} = \frac{4\pi}{75} \times 3h^2$$

$$= \frac{4\pi}{25}h^2$$

2. Evaluate the derivative at the given value.

When $h = 1$,

$$\frac{dV}{dh} = \frac{4\pi}{25} = 0.16\pi$$

3. Write the answer in context, with the appropriate units.

At the instant the depth is 1 cm, the volume of water is increasing at the rate of 0.16π cm^3 per cm.

 Resources

 **Interactivity** Kinematics (int-5964)

12.6.2 Kinematics

Many quantities change over time, so many rates measure that change with respect to time. Motion is one such quantity. The study of the motion of a particle without considering the causes of the motion is called **kinematics**. Analysing motion requires interpretation of **position**, **velocity** and **acceleration**, and this analysis depends on calculus. For the purpose of our course, only motion in a straight line, also called **rectilinear motion**, will be considered.

Position

The position x, gives the position of a particle by specifying both its distance and direction from a fixed origin.

Common units for position and distance are cm, m and km.

The commonly used conventions for motion along a horizontal straight line are:

- if $x > 0$, the particle is to the right of the origin
- if $x < 0$, the particle is to the left of the origin
- if $x = 0$, the particle is at the origin.

For example, if $x = -10$, this means the particle is 10 units to the left of the origin, O.

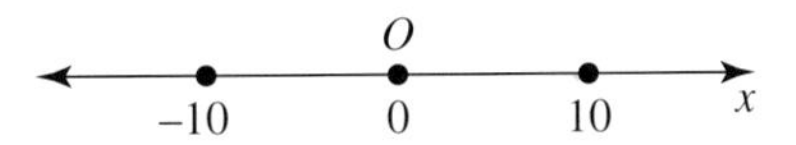

Note that its distance from the origin of 10 units is the same as the distance from the origin of a particle with a position $x = 10$.

Distance is not concerned with the direction of motion. This can have implications if there is a change of direction in a particle's motion. For example, suppose a particle initially 3 cm to the right of the origin travels 2 cm further to the right, then 2 cm to the left, to return to where it started.

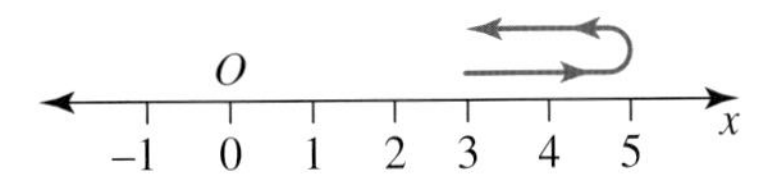

Its change in position is zero, but the distance it has travelled is 4 cm.

Velocity

Velocity, v, measures the rate of change of position so $v = \frac{dx}{dt}$.

> **Velocity**
>
> $$v = \frac{dx}{dt}$$
>
> **Velocity is the instantaneous rate of change of position.**

For a particle moving in a horizontal straight line, the sign of the velocity indicates the direction of motion:
- if $v > 0$, the particle is moving to the right
- if $v < 0$, the particle is moving to the left
- if $v = 0$, the particle is stationary (instantaneously at rest).

Common units for velocity and speed include m/s or km/h.

As for distance, speed is not concerned with the direction in which the particle travels, and it is never negative. A velocity of -10 m/s means the particle is travelling at 10 m/s to the left. Its speed, however, is 10 m/s regardless of whether the particle is moving to the left or to the right; that is, the speed is 10 m/s for $v = \pm 10$ m/s.

Since $v = \frac{dx}{dt}$, if the position x of a particle is plotted against time t to create the position–time graph $x = f(t)$, then the gradient of the tangent to its curve at any point represents the velocity of the particle at that point: $v = \frac{dx}{dt} = f'(t)$.

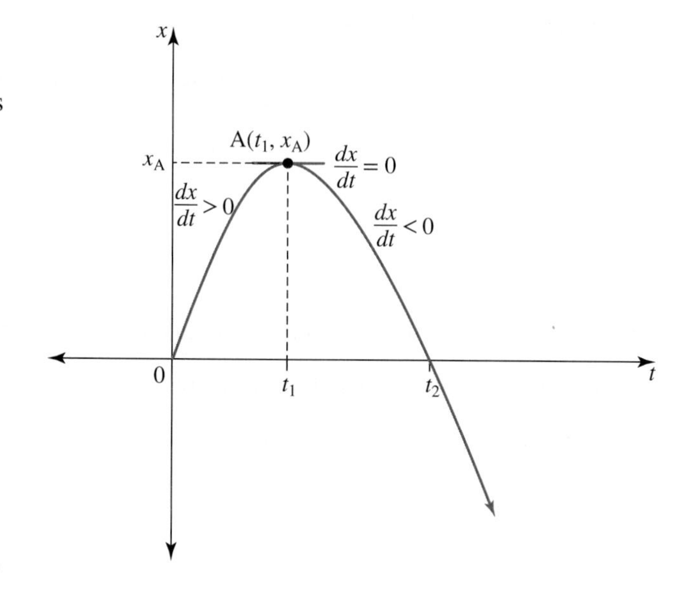

The **position–time graph** shows the position of a particle that starts at the origin and initially moves to the right as the gradient of the graph, the velocity, is positive.

At point A the tangent is horizontal and the velocity is zero, indicating the particle changes its direction of motion at that point.

The particle then starts to move to the left, which is indicated by the gradient of the graph, the velocity, having a negative sign. It returns to the origin and continues to move to the left, so its position becomes negative.

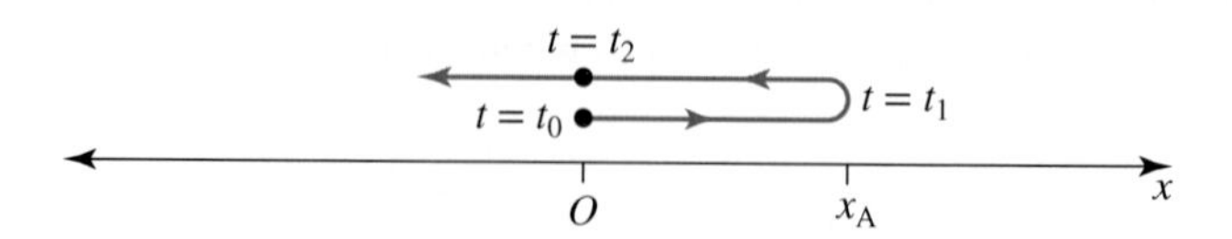

The same motion is also shown along the horizontal position line.

Average velocity

Average velocity is the average rate of change of the position. It is measured by the gradient of the chord joining two points on the position–time graph and requires coordinate geometry to evaluate it, not calculus.

Average velocity and speed

$$\textbf{Average velocity} = \frac{\textbf{change in position}}{\textbf{change in time}} = \frac{x_2 - x_1}{t_2 - t_1}$$

$$\textbf{Average speed} = \frac{\textbf{distance travelled}}{\textbf{time taken}}$$

Acceleration

The **acceleration,** a, measures the rate of change of velocity, so $a = \dfrac{dv}{dt}$.

Acceleration

$$a = \frac{dv}{dt}$$

Acceleration is the instantaneous rate of change of velocity.

A common unit for acceleration is metres per second per second, m/s^2.

Since acceleration is the instantaneous rate of change of velocity, if velocity is plotted against time t to create the **velocity–time graph** $v = f(t)$, then the gradient of the tangent to the curve at any point represents the acceleration of the particle at that point.

Average acceleration

Average acceleration is the average rate of change of the velocity. It is measured by the gradient of the chord joining two points on the velocity–time graph.

Average acceleration

$$\textbf{Average acceleration} = \frac{\textbf{change in velocity}}{\textbf{change in time}} = \frac{v_2 - v_1}{t_2 - t_1}$$

Position, velocity and acceleration are linked by calculus. Differentiation enables the velocity function to be obtained from the position function; it also enables the acceleration function to be obtained from the velocity function.

Position, velocity and acceleration

$$x \rightarrow v \rightarrow a$$

$$x \rightarrow \frac{dx}{dt} \rightarrow \frac{dv}{dt}$$

WORKED EXAMPLE 13 Determining the average speed and velocity of a particle

A particle moves in a straight line such that its position, x metres, from a fixed origin at time t seconds is modelled by $x = t^2 - 4t - 12$, $t \geq 0$.

a. **Identify its initial position.**
b. **Obtain its velocity and hence state its initial velocity and describe its initial motion.**
c. **Determine the time and position at which the particle is momentarily at rest.**
d. **Show the particle is at the origin when $t = 6$ and calculate the distance it has travelled to reach the origin.**
e. **Calculate the average speed over the first 6 seconds.**
f. **Calculate the average velocity over the first 6 seconds.**

THINK	WRITE
a. Calculate the value of x when $t = 0$.	a. $x = t^2 - 4t - 12,\ t \geq 0$ When $t = 0$, $x = -12$. Initially the particle is 12 metres to the left of the origin.
b. 1. Calculate the rate of change required.	b. $v = \dfrac{dx}{dt}$ $\therefore v = 2t - 4$
2. Calculate the value of v at the given instant.	When $t = 0$, $v = -4$. The initial velocity is -4 m/s.
3. Describe the initial motion.	Since the initial velocity is negative, the particle starts to move to the left with an initial speed of 4 m/s.
c. 1. Calculate when the particle is momentarily at rest. *Note:* This usually represents a change of direction of motion.	c. The particle is momentarily at rest when its velocity is zero. When $v = 0$, $2t - 4 = 0$ $\therefore t = 2$ The particle is at rest after 2 seconds.
2. Calculate where the particle is momentarily at rest.	The position of the particle when $t = 2$ is: $x = (2)^2 - 4(2) - 12$ $= -16$ Therefore, the particle is momentarily at rest after 2 seconds at the position 16 metres to the left of the origin.
d. 1. Calculate the position to show the particle is at the origin at the given time.	d. When $t = 6$, $x = 36 - 24 - 12$ $= 0$ The particle is at the origin when $t = 6$.
2. Track the motion on a horizontal line and calculate the required distance.	The motion of the particle for the first 6 seconds is shown.

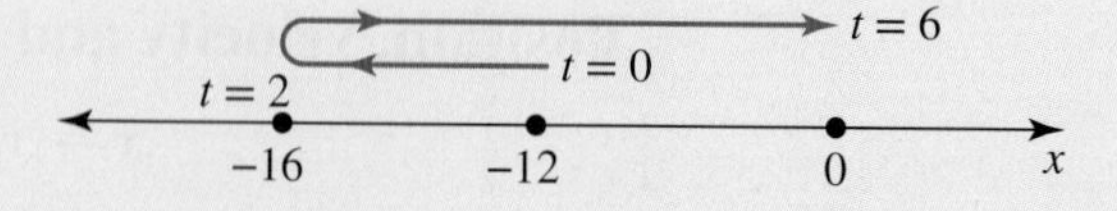

The distances travelled are 4 metres to the left then 16 metres to the right.
The total distance travelled is the sum of the distances in each direction.
The particle has travelled a total distance of 20 metres.

e. Calculate the value required.

e. $\text{Average speed} = \dfrac{\text{distance travelled}}{\text{time taken}}$

$= \dfrac{20}{6}$

$= 3\dfrac{1}{3}$

The average speed is $3\dfrac{1}{3}$ m/s.

f. Calculate the average rate of change required. *Note:* As there is a change of direction, the average velocity will not be the same as the average speed.

f. Average velocity is the average rate of change of position.

For the first 6 seconds,
$(t_1, x_1) = (0, -12)$, $(t_2, x_2) = (6, 0)$

$\text{Average velocity} = \dfrac{x_2 - x_1}{t_2 - t_1}$

$= \dfrac{0 - (-12)}{6 - 0}$

$= 2$

The average velocity is 2 m/s.

12.6 Exercise

Students, these questions are even better in jacPLUS

Receive immediate feedback and access sample responses

Access additional questions

Track your results and progress

Find all this and MORE in jacPLUS

Technology free

Where appropriate in this exercise, express answers as multiples of π.

1. **WE12** A container in the shape of an inverted right cone of radius 5 cm and depth 10 cm is being filled with water. When the depth of water is h cm, the radius of the water level is r cm.
   a. Use similar triangles to express r in terms of h.
   b. Express the volume of the water as a function of h.
   c. Determine the rate, with respect to the depth of water, at which the volume of water is changing when its depth is 3 cm.
2. A container in the shape of an inverted right circular cone is being filled with water. The cone has a height of 12 cm and a radius of 6 cm. Determine the rate at which the volume of water is changing with respect to the depth of water when the depth of water reaches half the height of the cone.

3. a. The formula for the circumference of a circle is given by $C = 2\pi r$. Write down the expression for the rate of change of C with respect to r.
   b. The formula for the area of a circle is given by $A = \pi r^2$.
      i. Determine the rate of change of A with respect to r.
      ii. When the radius of a circle is 3 mm, determine the rate at which its area is changing.
   c. The surface area, S, of a sphere of radius r is given by $S = 4\pi r^2$.
      i. Determine the rate of change of S with respect to r.
      ii. Determine the rate at which the surface area is changing when the radius of the sphere is 5 cm.
      iii. Determine the radius if the rate of change of the surface area is 16π cm^2/cm.
   d. The volume, V, of a sphere of radius r is given by $V = \frac{4}{3}\pi r^3$.
      i. Determine the rate of change of the volume of the sphere with respect to its radius.
      ii. Determine the rate at which the volume is changing when the radius of the sphere is 0.5 m.

4. a. Write down the formula for the volume, V, of a cube of side length x and hence calculate the rate at which the volume is changing when the side length is 2 cm.
   b. Write down the formula for the surface area, A, of a cube of side length l and hence calculate the rate at which the surface area is changing when the side length is 5 cm.
   c. Water is poured into a container in such a way that the volume of water, V cm^3, is given by $V = \frac{1}{3}h^3 + 2h$, where h cm is the depth of the water.

      i. Determine the rate of change of the volume of water with respect to the depth of water.
      ii. Calculate the volume of water when its depth is 3 cm and the rate at which the volume is changing at that instant.

5. a. As the area covered by an ink spill from a printer cartridge grows, it maintains a circular shape. Calculate the rate at which the area of the circle is changing with respect to its radius at the instant its radius is 0.2 metres.
   b. An ice cube melts in such a way that it maintains its shape as a cube. Calculate the rate at which its surface area is changing with respect to its side length at the instant the side length is 8 mm.

6. An equilateral triangle has side length x cm.
   a. Express its area in terms of x.
   b. Determine the rate, with respect to x, at which the area is changing when $x = 2$.
   c. Determine the rate, with respect to x, at which the area is changing when the area is $64\sqrt{3}$ cm^2.

7. a. A streetlight pole is 9 metres above the horizontal ground level. When a person of height 1.5 metres is x metres from the foot of the pole, the length of the person's shadow is s metres.
      i. Use similar triangles to express s in terms of x.

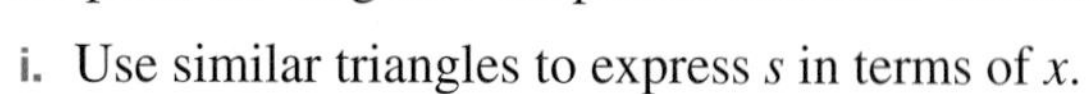

      ii. Calculate $\frac{ds}{dx}$ and explain what this measures.

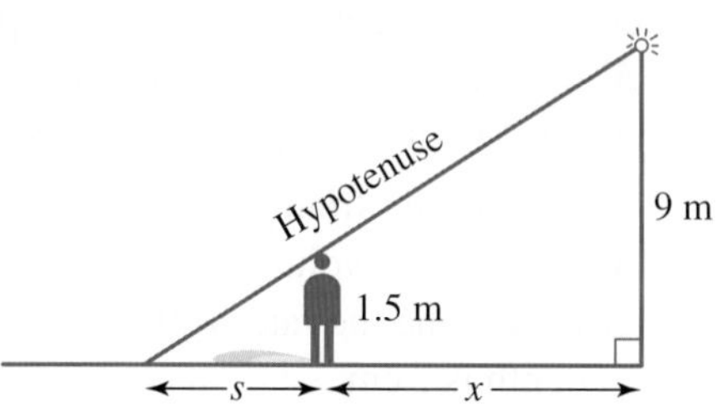

   b. Oil starts to leak from a small hole in the base of a cylindrical oil drum. The oil drum has a height of 0.75 metres and is initially full, with a volume of 0.25 m^3 of oil.
      i. Calculate the area of the circular base.
      ii. Determine the rate at which the volume is decreasing with respect to the depth of the oil in the oil drum.

8. **WE13** A particle moves in a straight line such that its position, x metres, from a fixed origin at time t seconds is $x = 3t^2 - 6t,\ t \geq 0$.

 a. Identify its initial position.
 b. Obtain its velocity and hence state its initial velocity and describe its initial motion.
 c. Determine the time and position at which the particle is momentarily at rest.
 d. Show that the particle is at the origin when $t = 2$ and calculate the distance it has travelled to reach the origin.
 e. Calculate the average speed over the first 2 seconds.
 f. Calculate the average velocity over the first 2 seconds.

9. A particle moves in a straight line so that at time t seconds, its position x metres from a fixed origin is given by $x = 12t + 9,\ t \geq 0$.

 a. Determine the rate of change of the position.
 b. Show the particle moves with a constant velocity.
 c. Calculate the distance that the particle travels in the first second.
 d. Calculate the time at which the particle is 45 metres to the right of the origin.

10. The position from a fixed origin, x metres, at time t seconds, of a particle moving in a straight line is given by $x = t^2 - 6t - 7,\ t \geq 0$.

 a. Describe the initial position of the particle.
 b. Calculate the velocity at any time t and hence obtain the particle's initial velocity.
 c. Determine the time at which the particle is momentarily at rest.
 d. Determine the time at which the particle returns to its initial position and calculate its velocity at that time.

11. A particle travels in a straight line with its position from a fixed origin, x metres, at time t seconds, given by $x = t^3 - 4t^2 - 3t + 12,\ t \geq 0$.

 a. Form the expressions for the velocity and the acceleration at any time t.
 b. Determine the time at which the velocity zero, and calculate its acceleration and position at that instant.
 c. Calculate the average velocity over the first 3 seconds.

12. The position, x cm, relative to a fixed origin of a particle moving in a straight line at time t seconds is $x = 5t - 10,\ t \geq 0$.

 a. Give its initial position and its position after 3 seconds.
 b. Calculate the distance travelled in the first 3 seconds.
 c. Show the particle is moving with a constant velocity.
 d. Sketch the $x - t$ and $v - t$ graphs and explain their relationship.

13. Relative to a fixed origin, the position of a particle moving in a straight line at time t seconds, $t \geq 0$, is given by $x = 6t - t^2$, where x is the position in metres.

 a. Write down expressions for its velocity and its acceleration at time t.
 b. Sketch the three motion graphs showing position velocity and acceleration versus time and describe their shapes.
 c. Use the graphs to determine the velocity is zero; find the value of x at that time.
 d. Use the graphs to determine when the particle is moving to the right and what the sign of the velocity is for that time interval.

14. The population, $P(t)$, of a colony of bacteria after t hours is given by $P(t) = 3t(200 - 2t)$, where $t \geq 0$.

 a. Determine the rate of change of the population of bacteria with respect to time.
 b. Calculate the rate at which the population is changing after 20 hours. State whether the population is increasing or decreasing.

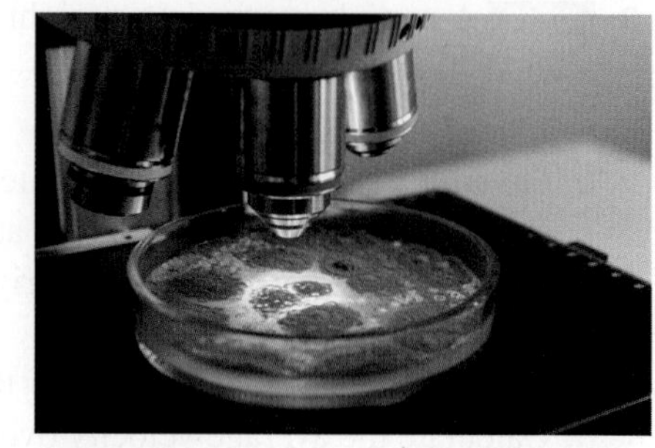

c. Calculate the rate at which the population is changing after 60 hours. State whether the population is increasing or decreasing.

d. Calculate the instant of time at which the population is:

i. increasing at 240 bacteria per hour
ii. decreasing at 240 bacteria per hour
iii. neither increasing nor decreasing.

Technology active

15. The position of a particle after t seconds is given by $x(t) = -\frac{1}{3}t^3 + t^2 + 8t + 1,\ t \geq 0$.

a. Find its initial position and initial velocity.
b. Calculate the distance travelled before it changes its direction of motion.
c. Determine its acceleration at the instant it changes direction.

16. The position x cm relative to a fixed origin of a particle moving in a straight line at time t seconds is $x = \frac{1}{3}t^3 - t^2,\ t \geq 0$.

a. Show the particle starts at the origin from rest.
b. Determine the time and position at which the particle is next at rest.
c. Determine when the particle returns to the origin.
d. Calculate the particle's speed and acceleration when it returns to the origin.
e. Sketch the three motion graphs, x–t, v–t and a–t, and comment on their behaviour at $t = 2$.
f. Describe the motion at $t = 1$.

17. A ball is thrown vertically upwards into the air so that after t seconds, its height h metres above the ground is $h = 40t - 5t^2$.

a. Calculate the rate at which its height is changing after 2 seconds.
b. Calculate its velocity when $t = 3$.
c. Determine the time at which its velocity is -10 m/s and the direction in which the ball is then travelling.
d. Determine when its velocity is zero.
e. Determine the greatest height the ball reaches.
f. Determine the time at which the ball strikes the ground and its speed at that time.

18. A cone has a slant height of 10 cm. The diameter of its circular base is increased in such a way that the cone maintains its slant height of 10 cm while its perpendicular height decreases. When the base radius is r cm, the perpendicular height of the cone is h cm.

a. Use Pythagoras' theorem to express r in terms of h.
b. Express the volume of the cone as a function of h.
c. Determine the rate of change of the volume with respect to the perpendicular height when the height is 6 cm.

19. A tent in the shape of a square-based right pyramid has perpendicular height h metres, side length of its square base x metres and volume $\frac{1}{3}Ah$, where A is the area of its base.

a. Express the length of the diagonal of the square base in terms of x. The slant height of the pyramid is 10 metres.
b. Show that $x^2 = 200 - 2h^2$ and hence express the volume of air inside the tent in terms of h.
c. Calculate the rate of change of the volume with respect to height when the height is $2\sqrt{3}$ metres.
d. Determine the value of h for which the rate of change of the volume equals zero. Explain what is significant about this value for h.

20. A particle moves in a straight line so that at time t seconds, its position, x metres, from a fixed origin O is given by $x(t) = 3t^2 - 24t - 27,\ t \geq 0$.
   a. Calculate the distance the particle is from O after 2 seconds.
   b. Calculate the speed at which the particle is travelling after 2 seconds.
   c. Calculate the average velocity of the particle over the first 2 seconds of motion.
   d. Determine the time at which the particle reaches O and its velocity at that time.
   e. Calculate the distance the particle travels in the first 6 seconds of motion.
   f. Calculate the average speed of the particle over the first 6 seconds of motion.

21. A particle P moving in a straight line has a position x metres, from a fixed origin O of $x_P(t) = t^3 - 12t^2 + 45t - 34$ for time t seconds.
   a. Determine the time(s) at which the particle is stationary.
   b. Determine the time interval over which the velocity is negative.
   c. Determine when its acceleration is negative.

   A second particle Q also travels in a straight line with its position from O at time t seconds given by $x_Q(t) = -12t^2 + 54t - 44$.
   d. Determine the time at which P and Q are travelling with the same velocity.
   e. Determine the times at which P and Q have the same position from O.

22. A particle moving in a straight line has position, x m, from a fixed origin O of $x = 0.25t^4 - t^3 + 1.5t^2 - t - 3.75$ for time t seconds.
   a. Calculate the distance it travels in the first 4 seconds of motion.
   b. Determine the time at which the particle is at O.
   c. Show that the acceleration is never negative.
   d. Sketch the three motion graphs showing position velocity and acceleration versus time, for $0 \leq t \leq 4$.
   e. Explain, using the velocity and acceleration graphs, why the particle will not return to O at a later time.

12.6 Exam questions

Question 1 (5 marks) TECH-FREE

A particle moves in a straight line such that its position, x metres, from a fixed origin at time t seconds is modelled by $x = 2t^2 - 8t + 9,\ t \geq 0$.

a. Find its initial velocity and the time and position that the particle is momentarily at rest. **(3 marks)**

b. Calculate the average velocity after the first 3 seconds. **(2 marks)**

Question 2 (1 mark) TECH-ACTIVE

MC A particle moves in a straight line such that its position, x metres, from a fixed origin at time t seconds is given by $x = 2t^2 - 15t + 23,\ t \geq 0$. Its speed after it has travelled for 3 seconds is

A. 6 m/s
B. −6 m/s
C. 3 m/s
D. −3 m/s
E. 0 m/s

Question 3 (1 mark) TECH-ACTIVE

MC Oil dropped into water spreads in a circular shape on the surface of the water. When the radius of the spill is 3 metres, the rate of change of the area with respect to the radius is

A. $\pi\ \text{cm}^2/\text{cm}$
B. $2\pi\ \text{cm}^2/\text{cm}$
C. $4\pi\ \text{cm}^2/\text{cm}$
D. $6\pi\ \text{cm}^2/\text{cm}$
E. $9\pi\ \text{cm}^2/\text{cm}$

More exam questions are available online.

12.7 Derivatives of power functions (extending)

***Note:* Derivatives of power functions are part of the Units 3 and 4 course, but you may be introduced to the concept in Units 1 and 2 to extend your understanding.**

A **power function** is of the form $y = x^n$ where n is a rational number. The hyperbola $y = \frac{1}{x} = x^{-1}$ and the square root function $y = \sqrt{x} = x^{\frac{1}{2}}$ are examples of power functions that we have already studied.

12.7.1 The derivative of $y = x^{-n},\ n \in N$

Let $f(x) = x^{-n},\ n \in N$.

Then $f(x) = \frac{1}{x^n}$.

Since this function is undefined and therefore not continuous at $x = 0$, it cannot be differentiated at $x = 0$.

For $x \neq 0$, the derivative can be calculated using the limit definition of differentiation.

The derivative of $y = x^{-n}$

For $x \neq 0$, if $f(x) = x^{-n}$ where $n \in N$, then

$$f'(x) = -nx^{-n-1}.$$

This rule means the derivative of the hyperbola $y = \frac{1}{x}$ can be obtained in a very similar way to differentiating a polynomial function.

$$\begin{aligned} y &= \frac{1}{x} = x^{-1} \\ \frac{dy}{dx} &= (-1)x^{-1-1} \\ &= -x^{-2} \\ &= -\frac{1}{x^2} \end{aligned}$$

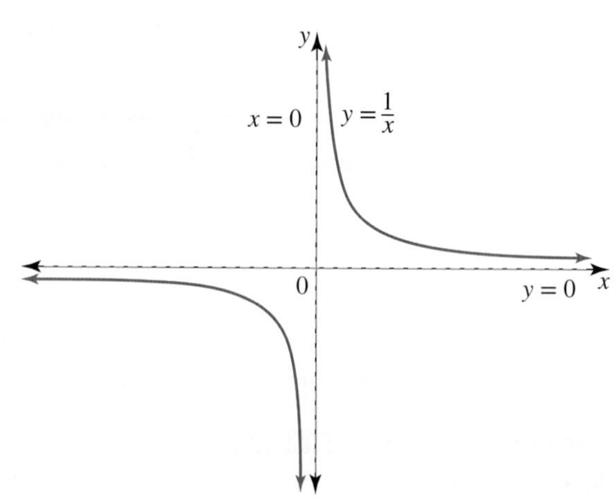

Since this derivative is always negative, it tells us that the hyperbola $y = \frac{1}{x}$ is always decreasing, as the gradient of its graph is always negative. This is consistent with what we already know about this hyperbola.

12.7.2 Derivative of $y = x^{\frac{1}{n}},\ n \in N$

Derivative of $y = x^{\frac{1}{n}}$

If $y = x^{\frac{1}{n}}, n \in N$, then

$$\frac{dy}{dx} = \frac{1}{n}x^{\frac{1}{n}-1}.$$

This rule means the derivative of the square root function $y = \sqrt{x}$ can also be obtained in a very similar way to differentiating a polynomial function.

In index form, $y = \sqrt{x}$ is written as $y = x^{\frac{1}{2}}$.

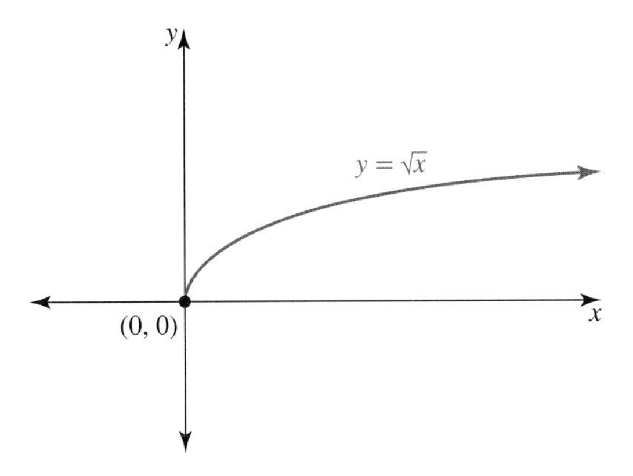

Its derivative is calculated as:

$$\begin{aligned}\frac{dy}{dx} &= \frac{1}{2}x^{\frac{1}{2}-1} \\ &= \frac{1}{2}x^{-\frac{1}{2}} \\ &= \frac{1}{2\sqrt{x}}\end{aligned}$$

This derivative is always positive, which tells us that the square root function $y = \sqrt{x}$ is an increasing function, as its gradient is always positive. This is consistent with our knowledge of the graph of $y = \sqrt{x}$.

Although the domain of the function $y = \sqrt{x}$ is $R^+ \cup \{0\}$, the domain of its derivative is R^+ as $\frac{1}{2\sqrt{x}}$ is undefined when $x = 0$.

12.7.3 The derivative of any power function

The two previous types of power functions considered lead us to generalise that any power function can be differentiated using the same basic rule that was applied to polynomial functions. In fact, $y = x^n$ for any real number n is differentiated using that same basic rule.

Differentiation of any power function

For $n \in R$, if $f(x) = x^n$, then

$$f'(x) = nx^{n-1}.$$

The linearity properties of the derivative also hold for rational as well as polynomial functions.

Linearity properties of the derivative

$$\frac{d(ky)}{dx} = k\frac{dy}{dx},\ k \in R$$

$$\frac{d(f \pm g)}{dx} = \frac{df}{dx} \pm \frac{dg}{dx}$$

If $n \in N$, the domain of the derivative of $f(x) = x^n$, a polynomial function, will be R. However, if the power n is a negative number, then $x = 0$, at least, must be excluded from the domains of both the function and its derivative function. This is illustrated by the hyperbola $y = \frac{1}{x}$ where the domain of the function and its derivative are both $R \setminus \{0\}$.

The derivative function will be undefined at any value of x for which the tangent to the graph of the function is undefined.

Other restrictions on the domain of the derivative may arise when considering, for example, a function that is a sum of rational terms, or a hybrid function. Any function must be smoothly continuous for it to be differentiable.

Before the derivative can be calculated, index laws may be required to express power functions as a sum of terms where each x term is placed in the numerator with a rational index. For example, $y = \frac{1}{x^p} + \sqrt[p]{x}$ would be expressed as $y = x^{-p} + x^{\frac{1}{p}}$, and then its derivative can be calculated using the rule for differentiation.

WORKED EXAMPLE 14 Differentiating power functions

a. If $y = \frac{2 + x^2 - 8x^5}{2x^3}$, calculate $\frac{dy}{dx}$ and state its domain.

b. If $f(x) = 2\sqrt{3x}$:

i. calculate $f'(x)$ and state its domain

ii. calculate the gradient of the tangent at the point where $x = 3$.

THINK	WRITE
a. 1. Express the rational function as the sum of its partial fractions.	a. $y = \frac{2 + x^2 - 8x^5}{2x^3}$ $= \frac{2}{2x^3} + \frac{x^2}{2x^3} - \frac{8x^5}{2x^3}$ $= \frac{1}{x^3} + \frac{1}{2x} - 4x^2$
2. Express each term as a power of x.	$\therefore\ y = x^{-3} + \frac{1}{2}x^{-1} - 4x^2$
3. Differentiate with respect to x.	$\frac{dy}{dx} = -3x^{-3-1} - \frac{1}{2}x^{-1-1} - 8x^{2-1}$ $= -3x^{-4} - \frac{1}{2}x^{-2} - 8x$
4. Express each term in x with a positive index.	$\therefore\ \frac{dy}{dx} = \frac{-3}{x^4} - \frac{1}{2x^2} - 8x$
5. State the domain of the derivative.	There are terms in x in the denominator, so $x \neq 0$. The domain of the derivative is $R \setminus \{0\}$.

b. i. 1. Express the term in x with a rational index.

b. i. $f(x) = 2\sqrt{3x}$

$f(x) = 2\sqrt{3}\sqrt{x}$

$= 2\sqrt{3}x^{\frac{1}{2}}$

2. Calculate the derivative.

$f'(x) = 2\sqrt{3} \times \frac{1}{2}x^{\frac{1}{2}-1}$

$= \sqrt{3}x^{-\frac{1}{2}}$

3. Express x with a positive index.

$f'(x) = \frac{\sqrt{3}}{\sqrt{x}}$

$= \sqrt{\frac{3}{x}}$

4. State the domain of the derivative.

Due to the square root term on the denominator, $x > 0$.
The domain of the derivative is R^+.

ii. Use the derivative to calculate the gradient at the given point.

ii. The gradient at $x = 3$ is $f'(3)$.

$f'(3) = \sqrt{\frac{3}{3}}$

$= \sqrt{1}$

$= 1$

Therefore, the gradient is 1.

12.7 Exercise (extending)

Students, these questions are even better in jacPLUS

Receive immediate feedback and access sample responses

Access additional questions

Track your results and progress

Find all this and MORE in jacPLUS

Technology free

1. Calculate $f'(x)$ for each of the following rational functions.

   a. $f(x) = x^{-3}$
   b. $f(x) = x^{-4}$
   c. $f(x) = x^{\frac{5}{2}}$
   d. $f(x) = x^{\frac{2}{3}}$
   e. $f(x) = 4x^{-2}$
   f. $f(x) = \frac{1}{2}x^{-\frac{6}{5}}$

2. Differentiate the following with respect to x.

   a. $y = 4x^{-1} + 5x^{-2}$
   b. $y = 4x^{\frac{1}{2}} - 3x^{\frac{2}{3}}$
   c. $y = 2 + 8x^{-\frac{1}{2}}$
   d. $y = 0.5x^{1.8} - 6x^{3.1}$

3. Calculate $\frac{dy}{dx}$ for each of the following.

a. $y = 1 + \frac{1}{x} + \frac{1}{x^2}$

b. $y = 5\sqrt{x}$

c. $y = 3x - \sqrt[3]{x}$

d. $y = x\sqrt{x}$

e. $y = \frac{3}{x^8}$

f. $y = \frac{4}{x} - 5x^3$

4. **WE14** a. If $y = \frac{4 - 3x + 7x^4}{x^4}$, calculate $\frac{dy}{dx}$ and state its domain.

b. If $f(x) = 4\sqrt{x} + \sqrt{2x}$:

i. calculate $f'(x)$ and state its domain

ii. calculate the gradient of the tangent at the point where $x = 1$.

5. a. Calculate the gradient of the tangent to the graph of $y = 3\sqrt{x}$ at the point where $x = 9$.

b. Calculate the gradient of the tangent to the graph of $y = \sqrt[3]{x} + 10$ at the point $(-8, 8)$.

c. Determine the gradient of the tangent to the curve $y = \frac{5}{x} - 1$ at its x-intercept.

d. Obtain the coordinates of the point on the curve $y = 6 - \frac{2}{x^2}$ where the gradient of the tangent is $\frac{1}{2}$.

e. Determine the points on the curve $y = \frac{4}{3x}$ at which the tangent to the curve has a gradient of -12.

f. Calculate the coordinates of the point on $y = 3 - 2\sqrt{x}$ where the tangent is parallel to the line $y = -\frac{2}{3}x$.

6. Consider the function $f: R \setminus \{0\} \rightarrow R,\ f(x) = x^2 + \frac{2}{x}$.

a. Evaluate $f'(2)$.

b. Determine the coordinates of the point for which $f'(x) = 0$.

c. Calculate the exact values of x for which $f'(x) = -4$.

7. The height of a magnolia tree, in metres, is modelled by $h = 0.5 + \sqrt{t}$ where h is the height t years after the tree was planted.

a. Calculate how tall the tree was when it was planted.

b. Determine the rate at which the tree is growing 4 years after it was planted.

c. Calculate when the tree will be 3 metres tall.

d. Determine the average rate of growth of the tree over the time period from planting to a height of 3 metres.

Technology active

8. On a warm day in a garden, water in a bird bath evaporates in such a way that the volume, V mL, at time t hours is given by

$$V=\frac{60t+2}{3t},\ t>0$$

a. Show that $\frac{dV}{dt}<0$.

b. Calculate the rate at which the water is evaporating after 2 hours.

c. Sketch the graph of $V=\frac{60t+2}{3t}$ for $t\in\left[\frac{1}{3},\ 2\right]$.

d. Calculate the gradient of the chord joining the end points of the graph for $t\in\left[\frac{1}{3},\ 2\right]$ and explain what the value of this gradient measures.

9. Give the gradient of the tangent to the hyperbola $y=4-\frac{1}{3x-2}$ at the point $\left(\frac{1}{3},\ \frac{9}{2}\right)$.

10. a. Sketch the graphs of $y=x^{\frac{2}{3}}$ and $y=x^{\frac{5}{3}}$ using CAS technology and find the coordinates of the points of intersection.

b. Compare the gradients of the tangents to each curve at the points of intersection.

12.7 Exam questions

Question 1 (1 mark) TECH-ACTIVE

MC If $y=4x\sqrt{x}$, $\frac{dy}{dx}$ is

A. $6\sqrt{x}$
B. $4\sqrt{x}$
C. $\frac{6}{\sqrt{x}}$
D. $\frac{4}{\sqrt{x}}$
E. $2x$

Question 2 (4 marks) TECH-ACTIVE

A function f is defined as $f:[0,\infty)\to R, f(x)=3-2\sqrt{x}$.

a. Define the derivative function f'. **(1 mark)**

b. Obtain the gradient of the graph of $y=f(x)$ as the graph cuts the x-axis. **(3 marks)**

Question 3 (4 marks) TECH-ACTIVE

Consider $f(x)=2x+\frac{1}{3x}$.

a. State the domain of the function. **(1 mark)**

b. Form the rule for its gradient function, stating its domain. **(1 mark)**

c. Find the coordinates of the points on the curve where the tangent has a gradient of -1. **(2 marks)**

More exam questions are available online.

12.8 Review

 12.8.1 Summary

doc-37028

Hey students! Now that it's time to revise this topic, go online to:

 Access the topic summary  Review your results  Watch teacher-led videos Practise exam questions

Find all this and MORE in jacPLUS

12.8 Exercise

Technology free: short answer

1. Evaluate the following where possible.

a. $\lim\limits_{x \to -6} \dfrac{x^3 - 36x}{x+6}$

b. $\lim\limits_{x \to 4} \dfrac{x+12}{x^2 - 1}$

c. $\lim\limits_{x \to 0} \dfrac{x+9}{9x}$

d. $\lim\limits_{x \to 1} f(x)$ if $f(x) = \begin{cases} \dfrac{3x+5}{2}, & x < 1 \\ 4 - x & x \geq 1 \end{cases}$

2. Consider the function $f(x) = x^3 - 5x^2 - 8x$.

a. Show that $f'(4) = 0$.

b. Calculate the values of $f'(3)$ and $f'(5)$.

c. Hence, state the nature of the stationary point at $x = 4$.

d. Determine the x-coordinate of the other stationary point.

3. Identify the subset of the domain of the function $f(x) = x^3 + 9x^2$ where it is:

a. decreasing
b. stationary
c. increasing.

4. Find the equation of the tangent drawn to the graph of $y = 2x^2 + 8x - 9$ at the point where the tangent:

a. is parallel to the line $y + 4x = 1$
b. is parallel to the x-axis.

5. The curve $y = x^4 - 4x^3 + 16x + 16$ has two stationary points, one of which occurs at $x = -1$.

a. Show that there is a stationary point at $x = 2$ and determine its nature using the slope of the tangent.

b. Determine the nature of the stationary point at $x = -1$.

c. State the coordinates of the two stationary points and the y-intercept.

d. Hence, draw a sketch of the curve $y = x^4 - 4x^3 + 16x + 16$.

6. Determine the rate, with respect to its radius, at which the volume of a sphere is changing when:

a. the radius is 5 cm

b. the volume is 4.5π cm^3.

(The formula for the volume of a sphere is $V_{\text{sphere}} = \dfrac{4}{3}\pi r^3$.)

Technology active: multiple choice

7. MC Given $f(x)=\begin{cases}2x, & x\le 1\\ 2x^2, & x>1\end{cases}$ at $x=1$, the function is:

A. continuous and differentiable.
B. not continuous but differentiable.
C. continuous but not differentiable.
D. not continuous and not differentiable.
E. undefined.

8. MC The gradient of the tangent at the point $(2, 3)$ on the curve $y=3x^2-9$ is:

A. 3 B. 12 C. 18 D. 9 E. 0

9. MC The equation of the tangent to the curve $y=4x-x^2$ at the point where $x=0$ is:

A. $y=4$ B. $y=4x$ C. $y=-4x$ D. $y=-\frac{1}{4}x$ E. $y=(4-2x)x$

10. MC The tangent to the curve $f(x)=x^2-8x+7$ has a negative slope when:

A. $-1<x<7$ B. $x<-1$ or $x>7$ C. $x>4$ D. $x<4$ E. $x<0$

11. MC Given that $f(2)=-1, f'(2)=0, f'(x)>0$ for $x<2$ and $f'(x)<0$ for $x>2$, select the correct statement from the following.

A. The point $(2,-1)$ is a local minimum turning point.
B. The point $(-1,2)$ is a local minimum turning point.
C. The point $(2,-1)$ is a local maximum turning point.
D. The point $(-1,2)$ is a local maximum turning point.
E. The point $(2,-1)$ is a stationary point of inflection.

12. MC The stationary points of the curve $y=4x(x^2-3)$ occur when:

A. $x=0$ B. $x=0, x=\pm\sqrt{3}$ C. $x=\pm\sqrt{3}$ D. $x=\pm 1$ E. $x=\pm\frac{1}{2}$

13. MC Water is poured into a jug and after t seconds, the volume V mL of water in the jug is given by $V=2t(8-t)$. The rate of change of the volume, in mL/s, at $t=2$ is:

A. 24 B. 12 C. 8 D. 4 E. 2

Questions* 14 *and* 15 *refer to the following information.

A particle moves in a straight line so that at time t seconds its position, x metres, from a fixed origin O is given by $x(t)=t^2-10t+24,\ t\ge 0$.

14. MC The velocity of the particle is zero when time t equals:

A. -10 s B. 0 s C. 5 s D. 10 s E. 24 s

15. MC The distance travelled in the first second of motion, in metres, is:

A. 15 B. -15 C. 9 D. -9 E. 2

16. MC Select the correct statement about the function $y=f(x)$, the graph of which is shown.

A. $\lim_{x\to 2} f(x)=2$
B. $\lim_{x\to 2} f(x)=1$
C. $f'(2)$ exists
D. $f(2)$ exists
E. The function is continuous at $x=2$.

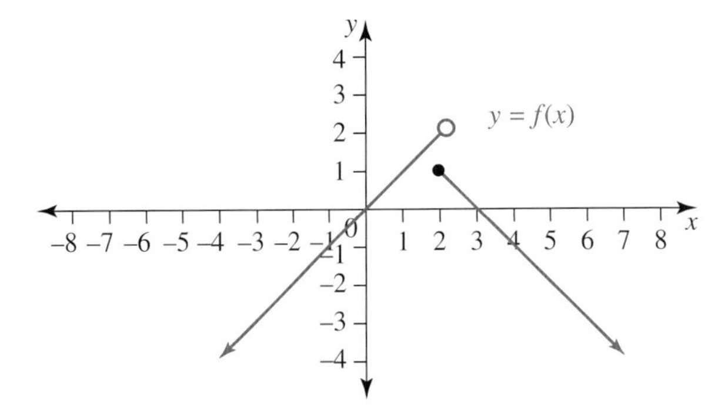

Technology active: extended response

17. A rectangular box with an open top is to be constructed from a rectangular sheet of cardboard measuring 25 cm by 18 cm by cutting equal squares of side length x cm out of the four corners and folding the flaps up.
    a. Express the volume V as a function of x and state the domain.
    b. Find, to 1 decimal place, the length of the side of the squares cut out for which the volume of the box is maximum, justifying the nature.
    c. State the maximum volume to the nearest whole number.
    d. To reduce wastage, restrictions are placed on the length of the square to be cut out. One idea proposed was that the sum of the areas of the four equal squares cut out from the cardboard should not exceed 36 cm^2. Determine the greatest volume possible for this proposal and calculate the dimensions of the box with greatest volume.
    e. Too small or too large a square cut-out results in a box with too small a volume. To ensure the volume of the box is adequate, the side length of each square cut-out is proposed to lie in the interval $2.5 \leq x \leq 6$. Find the greatest and the least volumes of the box that are possible with this proposed restriction.

18. a. Consider the curve $y = x^4 - 3x^3 + 16x + 16$.
        i. Write down the cubic equation from which the stationary points of $y = x^4 - 3x^3 + 16x + 16$ are obtained.
        ii. Given there is a stationary point near $x = -1$, use the Newton–Raphson method to obtain the x-coordinate of this stationary point correct to 4 decimal places.
        iii. Determine the nature of this stationary point.

    b. A function f is defined by $f(x) = ax^3 + bx^2 + cx + d$.
        i. The tangent to the curve at the origin is inclined at 45° to the positive direction of the x-axis. Calculate the values of c and d.
        ii. At the point $\left(1, \frac{5}{6}\right)$ on the curve, the tangent is perpendicular to the line $y + 2x = 0$. Calculate the values of a and b and hence show that $f(x) = x - \frac{x^3}{6}$.
        iii. Locate the stationary points of the function f and justify their nature using the sign of $f'(x)$.
        iv. Sketch the graph of $f: \left[-\sqrt{6}, \sqrt{6}\right] \to R,\ f(x) = x - \frac{x^3}{6}$ and state the local maximum and minimum values and any global maximum and minimum values.
        v. The tangent at the point $\left(1, \frac{5}{6}\right)$ meets the curve again at point P. Determine the coordinates of P.
        vi. The function f approximates a sine function over the domain $\left[-\sqrt{6}, \sqrt{6}\right]$. Form the equation of this sine function in the form $y = a\sin(nx)$.

19. A particle A moves in a straight line so that at time t seconds, its position, x metres, from a fixed origin O is given by $x = 5t^2 - 40t - 12,\ t \geq 0$.
    a. Describe the initial position of the particle and the direction in which it starts to move.
    b. Calculate how far the particle has travelled by the time its velocity becomes zero.
    c. Calculate the distance the particle travels in the first 10 seconds of motion and its average speed over this time.

    A second particle B moves on the same straight line so that at time t seconds, it has position s metres from the origin O, where $s = t^2 + 8t - 12,\ t \geq 0$.

    d. Show that the two particles start from the same initial position but move in opposite directions.
    e. Determine the time and position at which particle A overtakes particle B.
    f. Determine the times and positions at which the two particles are travelling with the same speed.

20. Sweet biscuits are to be packaged in closed cylindrical containers made from cardboard and sold as gifts at a local market.

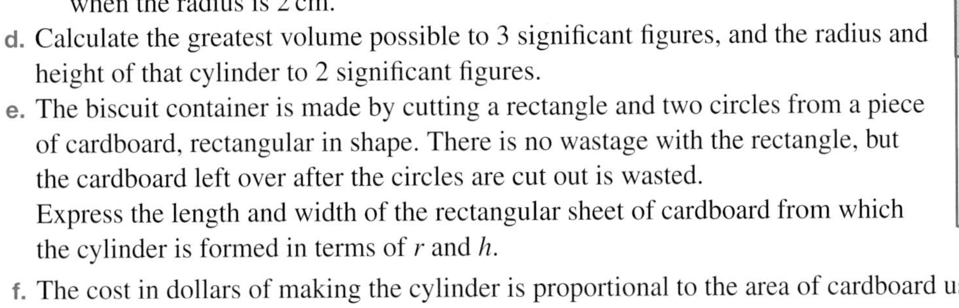

a. For a container with a total surface area of 308 square centimetres, express the height h in terms of the radius r.
b. Show that the volume V of the container is $V = 154r - \pi r^3$.
c. i. If the radius changes from 2 cm to 3 cm, find the average rate of change of the volume in the form $a + b\pi$ cm^3/cm.
   ii. Determine the rate at which the volume is changing with respect to its radius when the radius is 2 cm.
d. Calculate the greatest volume possible to 3 significant figures, and the radius and height of that cylinder to 2 significant figures.
e. The biscuit container is made by cutting a rectangle and two circles from a piece of cardboard, rectangular in shape. There is no wastage with the rectangle, but the cardboard left over after the circles are cut out is wasted.
   Express the length and width of the rectangular sheet of cardboard from which the cylinder is formed in terms of r and h.
f. The cost in dollars of making the cylinder is proportional to the area of cardboard used, including the wasted part, with the constant of proportionality of 0.01. Use the dimensions obtained in part **d** to calculate the cost of making the cylindrical container of greatest volume.

12.8 Exam questions

Question 1 (1 mark) TECH-ACTIVE

MC Given $f(x) = \begin{cases} x^2 - 2, & x \le 1 \\ 2x - 3, & x > 1 \end{cases}$ at $x = 1$, the function is

A. continuous and differentiable.
B. not continuous but differentiable.
C. continuous but not differentiable.
D. not continuous and not differentiable.
E. undefined.

Question 2 (1 mark) TECH-ACTIVE

MC For the graph shown, the gradient is positive for

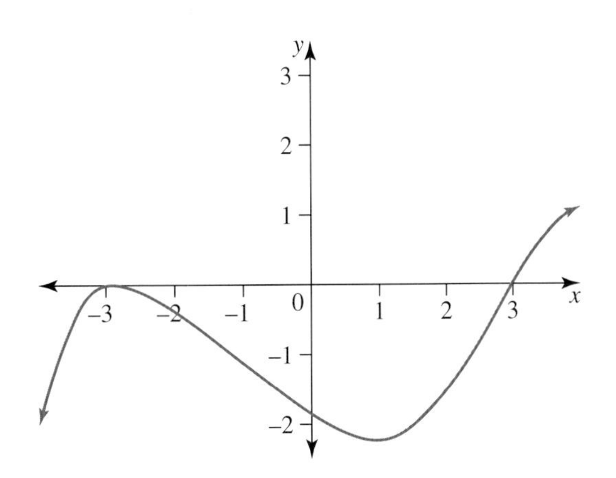

A. $x \in (-\infty, 3)$
B. $x \in (-3, 3)$
C. $x \in (-\infty, -3) \cup (1, \infty)$
D. $x \in (-\infty, -3) \cup (3, \infty)$
E. $x \in (-3, 1)$

Question 3 (1 mark) TECH-ACTIVE

MC If $f'(x) < 0$ where $x < 3, f'(x) = 0$ at $x = 3$ and $f'(x) < 0$ when $x > 3$, then at point $x = 3$, $f(x)$ has

A. a gradient of 3.
B. a negative gradient for all x.
C. a discontinuity at $x = 3$.
D. a point of inflection at $x = 3$.
E. a turning point at $x = 3$.

Question 4 (4 marks) TECH-ACTIVE

Morgan decides to build a rectangular vegetable garden bed using the corner formed by the back and side fences. There is 20 metres of fencing available for enclosing the other two sides of the garden bed.

Draw a diagram of the garden bed, and express the area in terms of x and y (the lengths of the two fenced sides). Use calculus to obtain the dimensions of the garden for maximum area, and state the maximum area.

Question 5 (3 marks) TECH-ACTIVE

A triangle has a base that is half the length of the other two equal sides (cm) as shown.

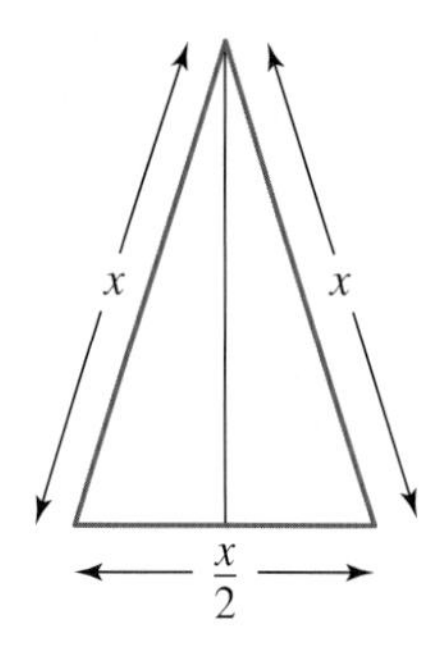

a. Express the area of the triangle in terms of x. **(2 marks)**
b. Determine the rate with respect to x at which the area is changing when $x = 8$ cm. **(1 mark)**

More exam questions are available online.

Answers

Topic 12 Differentiation and applications

12.2 Limits, continuity and differentiability

12.2 Exercise

1. a. 10 b. 10 c. 8 d. -1
2. a. 13 b. 2 c. 20 d. -31
3. a. -20 b. 1 c. $-\frac{1}{3}$ d. $-\frac{3}{2}$
4. a. 23 b. 6
   c. The limit does not exist.
5. a. 12 b. $\frac{1}{5}$
6. a. 17 b. 6 c. 3.5
   d. 5 e. 48 f. 1
7. a. i. 1
   ii. Continuous at $x = 1$ since both $f(1)$ and $\lim_{x\to1} f(x)$ exist, and $f(1) = 1 = \lim_{x\to1} f(x)$; graph required showing that the two branches join at $x = 1$.
   b. i. $R\setminus\{0\}$
   ii. -1
   iii. $g(0)$ is not defined.
   iv. The graph required is $y = x - 1$ with an open circle at $(0, -1)$.
8. i. a. Does not exist b. 0
   c. -2 d. 4
   ii. a–d. Only b and c are continuous at $x = 2$; explanations are required.
9. $f(0) = 4$, $\lim_{x\to0} f(x)$ does not exist; continuous for domain $R\setminus\{0\}$
10. a. Yes b. No c. No d. Yes
11. $a = 2$
12. $a = -3, b = 2$
13. a. Differentiable at $x = 1$
    b. $f'(x) = \begin{cases} 2x, & x \le 1 \\ 2, & x > 1 \end{cases}$; domain R; graph required showing that the two branches join.
14. a. $x = x_3, x_5$ b. $x = x_1, x_2, x_3, x_4, x_5$
15. a. Not differentiable at $x = 0$
    b.

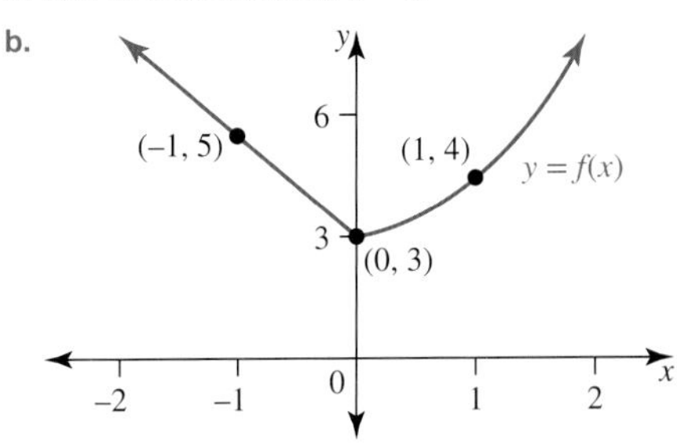

The branches do not join smoothly.

c. $f'(x) = \begin{cases} -2, & x < 0 \\ 2x, & x > 0 \end{cases}$, domain $R\setminus\{0\}$

d. 6

e.

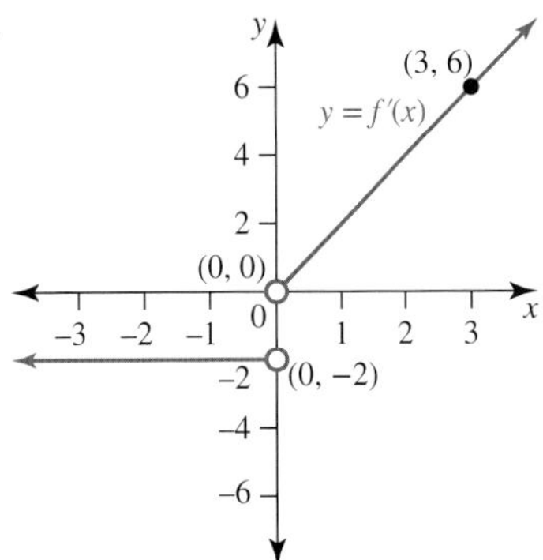

The graph is not continuous at $x = 0$.

16. $a = 1,\ b = -4$
17. a. Not continuous at $x = 1$
    b. $f'(x) = \begin{cases} 8x - 5, & x < 1 \\ -3x^2 + 6x, & x > 1 \end{cases}$
    c. $x = \frac{5}{8},\ x = 2$
    d.

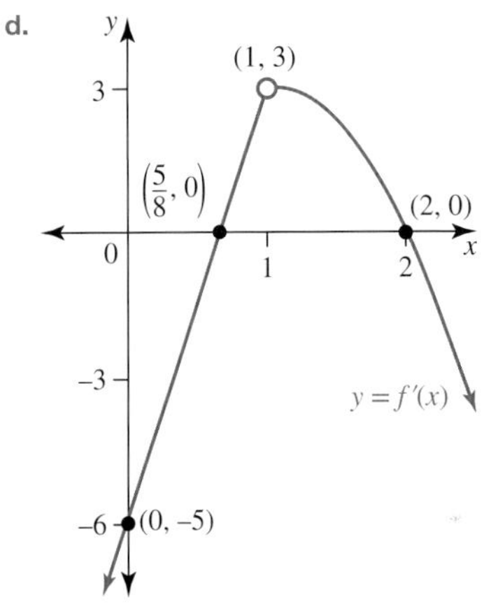

Point of discontinuity at $(1, 3)$; intercepts with axes at $(0, -5)$, $\left(\frac{5}{8}, 0\right)$, $(2, 0)$

e. $\left(\frac{5}{8}, \frac{7}{16}\right)$ and $(2, 4)$

f.

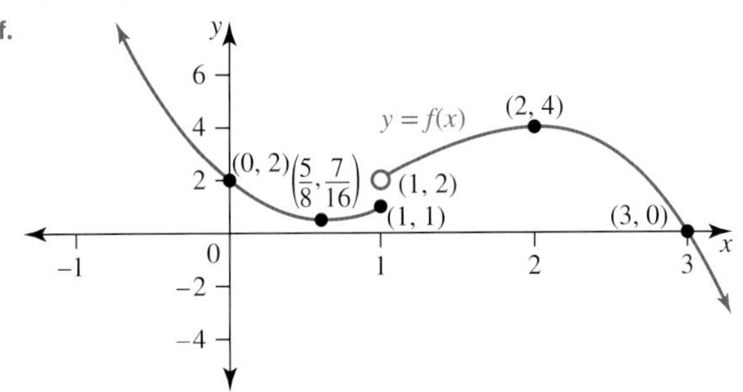

Minimum turning point at $\left(\frac{5}{8}, \frac{7}{16}\right)$, maximum turning point at $(2, 4)$; intercepts with axes $(0, 2)$, $(3, 0)$; end points $(1, 1)$ closed, $(1, 2)$ open

18. $a=2,\ b=2,\ c=1,\ d=4$

19. a. Sample responses can be found in the worked solutions in the online resources.

b. $\lim\limits_{h \to 0} \dfrac{4-(2-h)^2}{h}$

c. $\lim\limits_{h \to 0} \dfrac{2^{2+h}-4}{h}$

d. 4, $4\log_e(2)$ respectively; not differentiable at $x=2$

12.2 Exam questions

Note: Mark allocations are available with the fully worked solutions online.

1. C
2. A
3. $a=3,\ b=1$

12.3 Coordinate geometry applications of differentiation

12.3 Exercise

1. $y=-4x+18$

2. a. $y=-7x+3$ b. $y=-2x+8$ c. $y=6x-8$ d. $y=5$

3. a. $y=6x-9$ b. $y=-6x+3$ c. $y=2x+5$ d. $y=4x+8$ e. $y=9x+9$ f. $y=-12x+24$

4. a. i. $(3,9)$ ii. $y=3x$

b. i. $(-1,-6)$ ii. $y=-6$

c. $y=5x-4$

d. $y=-4x$

5. $y=-8x-1$

6. a. $y=4x-22$ b. $y=-6$ c. $y=2x-13$

7. a. $y=2x+1$ b. $y=-\dfrac{1}{2}x+6$ c. $y=-x+\dfrac{19}{4}$

8. a. $(2,8)$ b. $\left(-\dfrac{1}{4},-\dfrac{17}{16}\right)$

9. Sample responses can be found in the worked solutions in the online resources.

10. a. $\dfrac{t^2+1}{t}, t>0$ b. $2t$ c. $t=1$, P $(1,2)$ d. $y=2x$

11. a. $\left(\dfrac{3}{2},\dfrac{9}{4}\right)$ b. $26.6°$

12. a. $(-5,\ -50)$ b. $(-2,\ -23)$

13. a. $(-\infty,-4)\cup(2,\infty)$

b. i. $y=(2a+4)x+5-a$

ii. $a<-2$

14. a. $x\in\left(-\infty,\dfrac{7}{8}\right)$

b. $x\in(-\infty,-2)\cup(2,\infty)$

c. i. Sample responses can be found in the worked solutions in the online resources.

ii. $y=6x+7$

15. a. $x\in R\backslash\{0\}$

b. At $x=0$, the function is stationary at the point $(0,3)$.

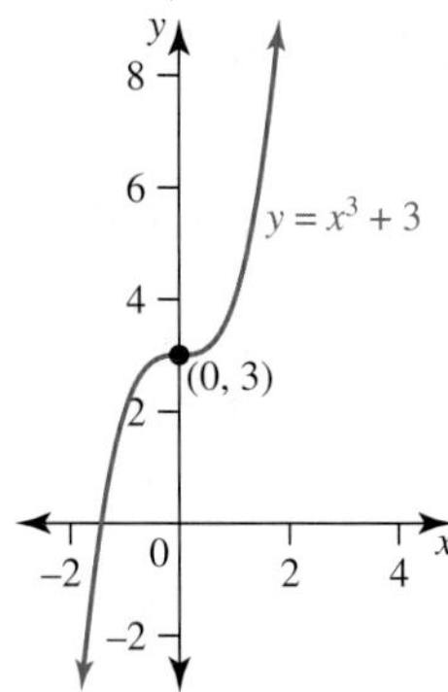

16. $x=0.4249$

17. a. The polynomial changes sign.

b. $x=1.4026$

18. a. $x=1.3788$

b. i. Between $x=3$ and $x=4$

ii. $x=3.3186$

c. $x=-1.7556$

19. a. $\sqrt[3]{16}=2.5198$

b. $x=4\pm\sqrt{7},\ \sqrt{7}\approx 2.6458$

20. a. The gradient is positive at $x=0$ and negative at $x=1$.

b. $x=0.735$

21. a.

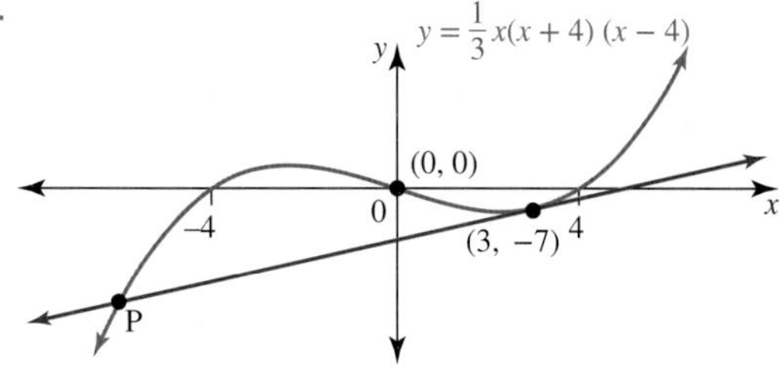

b. $y=\dfrac{11}{3}x-18$

c. i. Sample responses can be found in the worked solutions in the online resources.

ii. $(-6,-40)$

d. Same gradient

e. i. Same gradient

ii. $\left(-\dfrac{2a}{3},\dfrac{2a^3}{3}\right),\left(\dfrac{2a}{3},-\dfrac{2a^3}{3}\right)$

22. a. $y=-ax+3-a^2$

b. $a=-3$

c. $\left(-\infty,-\dfrac{a}{2}\right)$

d. Sample responses can be found in the worked solutions in the online resources; true also when $a=0$

12.3 Exam questions

Note: Mark allocations are available with the fully worked solutions online.

1. A
2. $(3,18)$

3. $x \in (-6, -4)$

+

– -6 -4 x

12.4 Curve sketching

12.4 Exercise

1. a. $(2, 4)$ b. $(1, -7)$
   c. $(-3, 9)$ d. $(0, 1)$
   e. $(1, -2)$ and $(-1, 2)$ f. $(0, 2)$

2. B

3. a. Sample responses can be found in the worked solutions in the online resources.
   b. $f'(1) = -12, f'(3) = 24$
   c. Local minimum turning point at $(2 - 20)$
   d. $(-1, 7)$
   e.

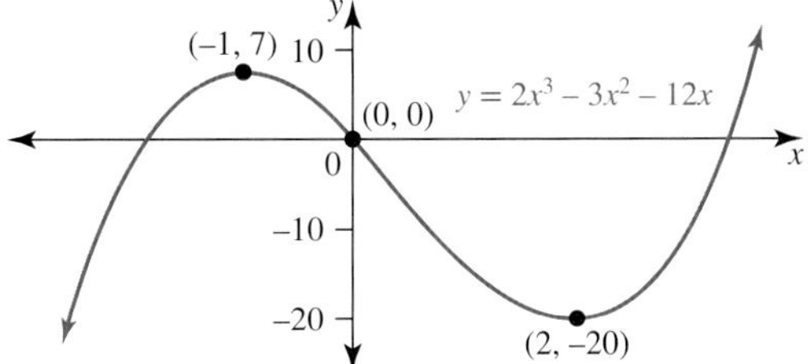

4. a. $(-1, 5)$ is a maximum turning point; $\left(\frac{1}{3}, \frac{103}{27}\right)$ is a minimum turning point.
   b. $a = \frac{19}{4}$, $b = -19$, $c = 5$

5. a. i. $(4, -6)$ ii. $\left(\frac{3}{5}, -\frac{51}{5}\right)$
   b. i. $a = \frac{1}{2}, b = -4$
   ii.

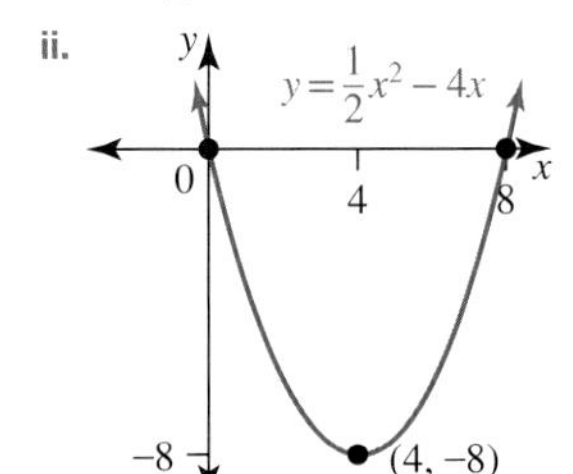

6. a. Sample responses can be found in the worked solutions in the online resources.
   b.

x	-3	-2	-1
$f'(x)$	9	0	-3
Slope	/	—	\

Maximum turning point

   c. $(0, 8)$
   d.

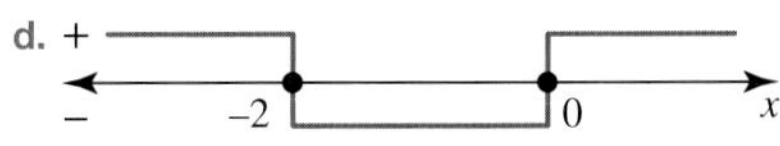

Minimum turning point

7. a. $(-3, 8)$ is a maximum turning point; $\left(1, -\frac{8}{3}\right)$ is a minimum turning point.
   b. $(2,\ 10)$ is a stationary point of inflection.
   c. $\left(-\sqrt{3}, 23\sqrt{3}\right)$ is a maximum turning point; $\left(\sqrt{3}, -23\sqrt{3}\right)$ is a minimum turning point.
   d. There are no stationary points.

8. a. $a = -5$ b. $b = 8$ c. $a = 2,\ c = 1$

9. a. $a = -9,\ b = 24$
   b. $(2, 9)$ is a maximum turning point; $(4, 5)$ is a minimum turning point.

10.

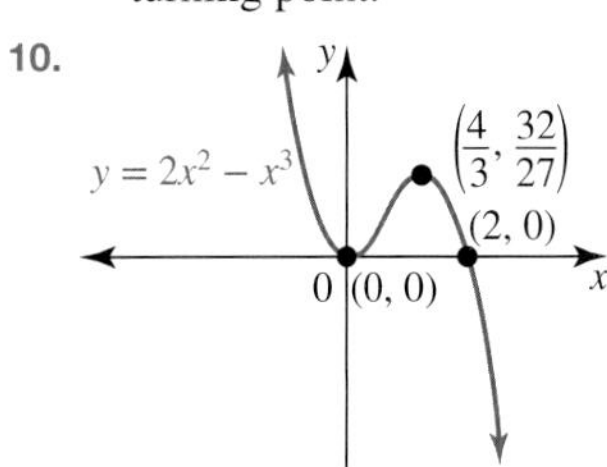

x-intercepts when $x = 2,\ x = 0$; $(0, 0)$ is a minimum turning point; $\left(\frac{4}{3}, \frac{32}{27}\right)$ is a maximum turning point.

11. a.

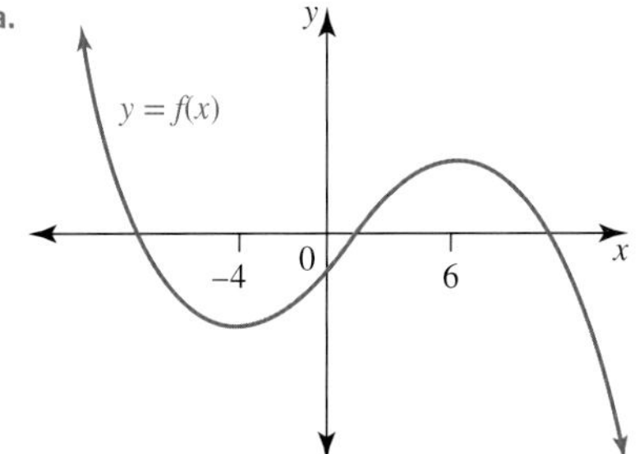

Negative cubic with minimum turning point where $x = -4$ and maximum turning point where $x = 6$

   b.

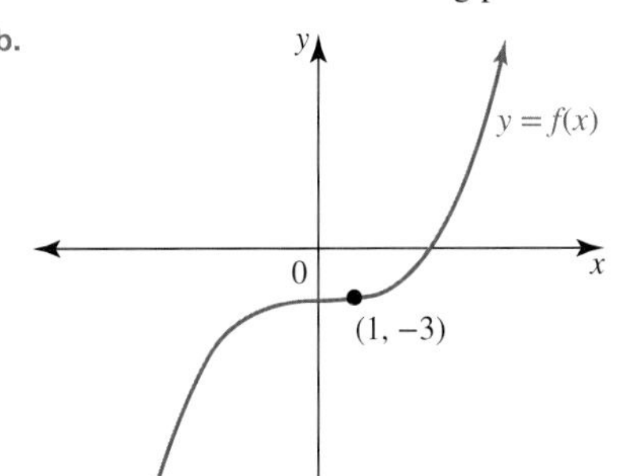

Positive cubic with stationary point of inflection at $(1, -3)$

12. a.

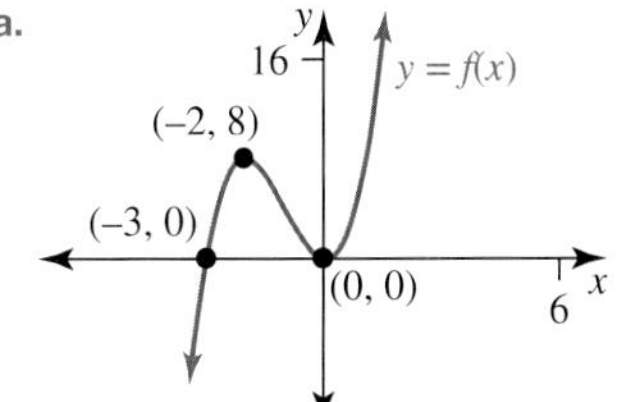

Minimum turning point $(0, 0)$; maximum turning point $(-2, 8)$, intercept $(-3, 0)$

b. 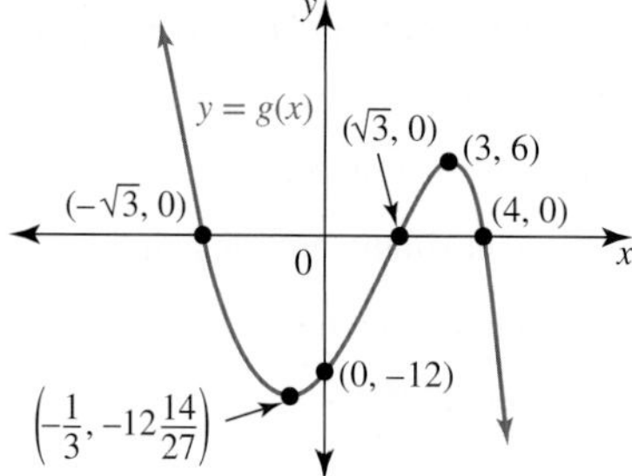

Minimum turning point $\left(-\frac{1}{3}, -12\frac{14}{27}\right)$; maximum turning point $(3, 6)$; intercepts $\left(\pm\sqrt{3}, 0\right), (4, 0), (0, -12)$

c. 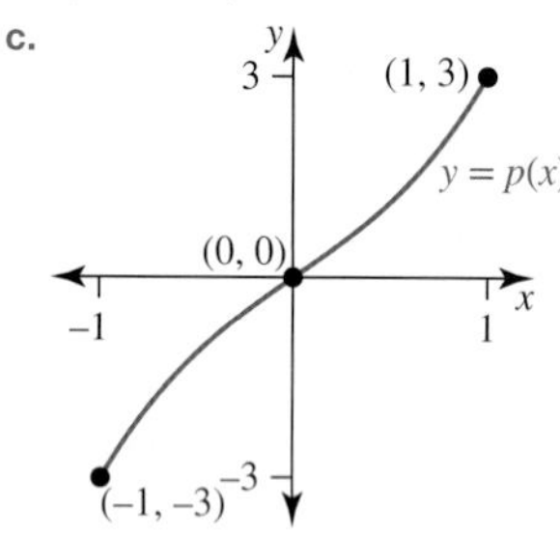

End points $(-1, -3), (1, 3)$; intercept $(0, 0)$; no stationary points

d. 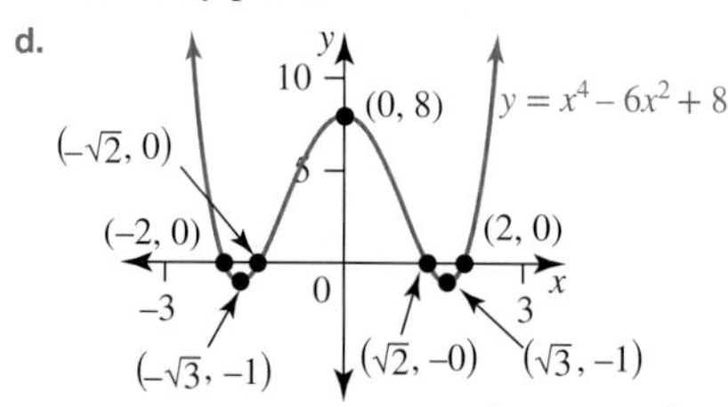

Minimum turning points $\left(\pm\sqrt{3}, -1\right)$; maximum turning point and y-intercept $(0, 8)$; x-intercepts $\left(\pm\sqrt{2}, 0\right), (\pm 2, 0)$

13. a. $2, 0$
 b. Sample responses can be found in the worked solutions in the online resources; stationary point of inflection.
 c. $k > 4$
 d. Explanation required; stationary point of inflection
 e. Degree 2; not possible; explanation required

14. a. $(-2, 32),\ (1, 128)$
 b. The point $(-1,\ 0)$ is a stationary point of inflection and the point $\left(-\frac{1}{4}, -\frac{27}{16}\right)$ is a minimum turning point.
 c. 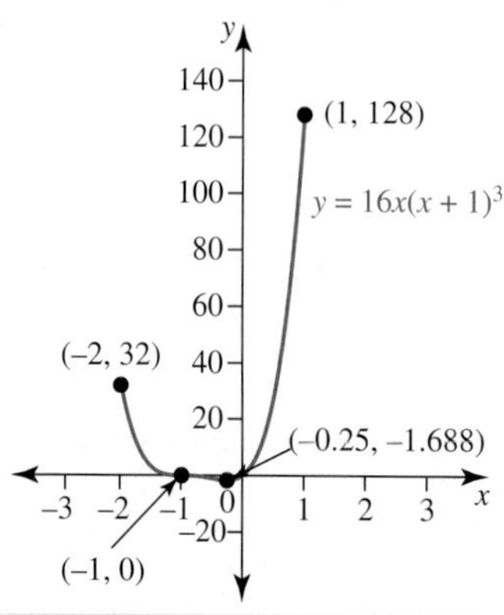

 d. Absolute maximum value: 128; absolute minimum value: $-\frac{27}{16}$

15. No global minimum; global maximum 8

16. a. 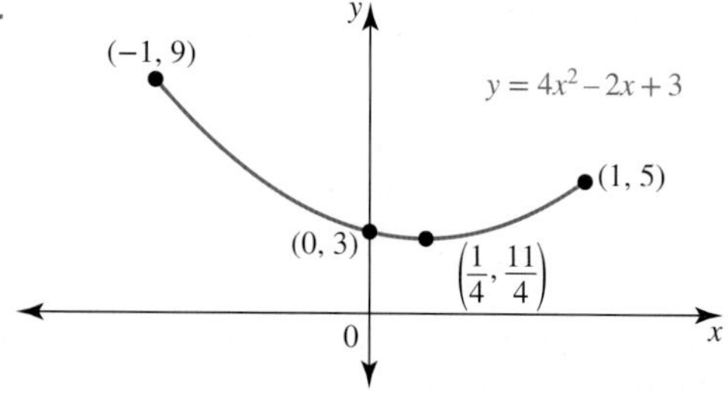

Local and global minimum $\frac{11}{4}$; no local maximum; global maximum 9

 b. 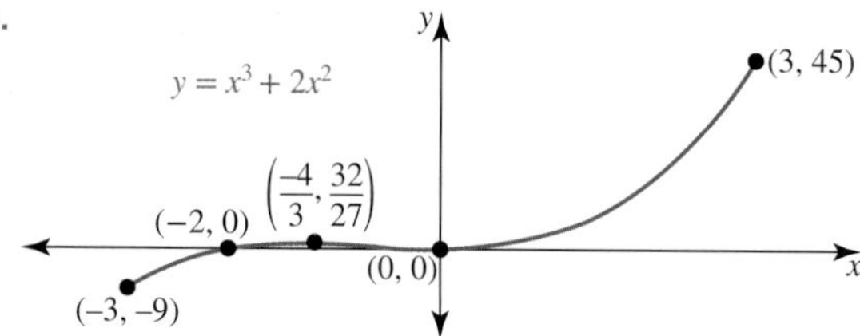

Local maximum $\frac{32}{27}$; local minimum 0; global maximum 45; global minimum -9

 c. 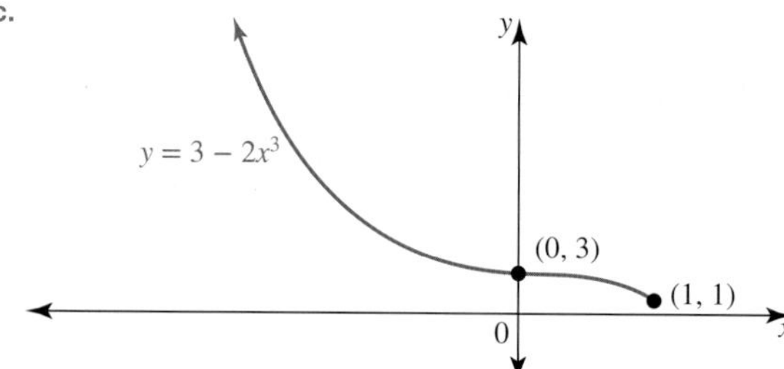

No local or global maximum; no local minimum; global minimum 1

 d. 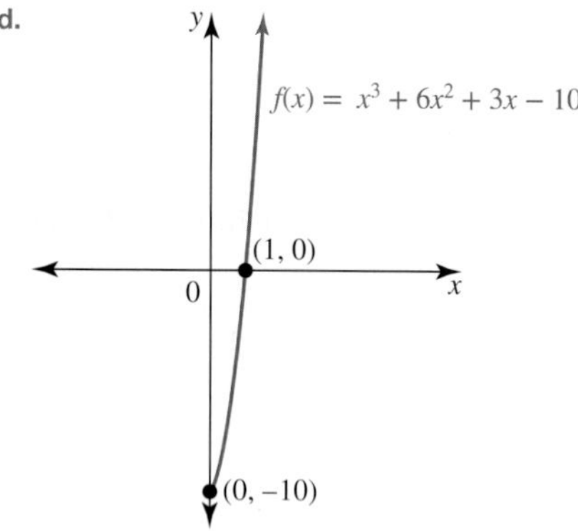

No local or global maximum; no local minimum; global minimum -10

17. a. $b = 3, c = -24$
 b. $(-4, 54)$
 c. $(0, -26), \left(-1 \pm 3\sqrt{3}, 0\right), (-1, 0)$

d. 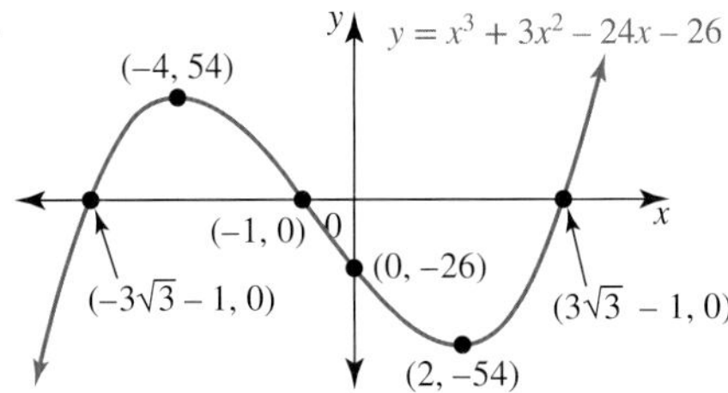

Key points are $(-1-3\sqrt{3}, 0)$, $(-4, 54)$, $(-1, 0)$, $(0, -26)$, $(0, -26)$, $(2, -54)$, $(-1+3\sqrt{3}, 0)$.

18. 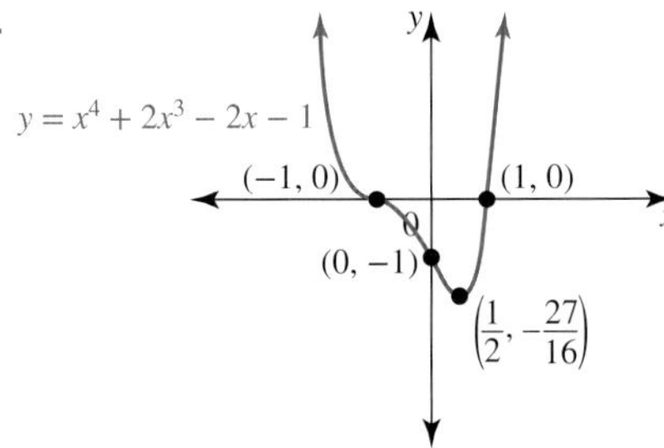

$(0, -1)$, $(\pm 1, 0)$ intercepts with axes; $(-1, 0)$ is a stationary point of inflection; $\left(\frac{1}{2}, -\frac{27}{16}\right)$ is a minimum turning point.

19. The origin lies on the line $y = \frac{2a^2x}{3}$.

20. a. $c = -1,\ d = 0$

b. $a = \frac{1}{4},\ b = -\frac{1}{2}$

c. $\left(-\frac{2}{3}, \frac{10}{27}\right)$

d. $(-2, -2)$

12.4 Exam questions

Note: Mark allocations are available with the fully worked solutions online.

1. $a = 4, b = -16$
   $(0, 0)$ and $(4, 0)$ are x-intercepts.

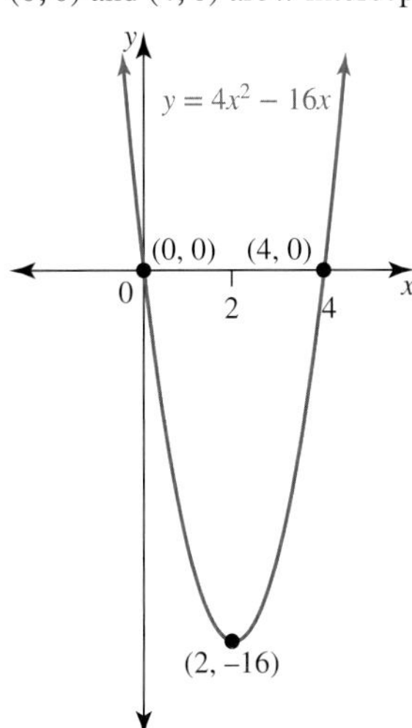

2. 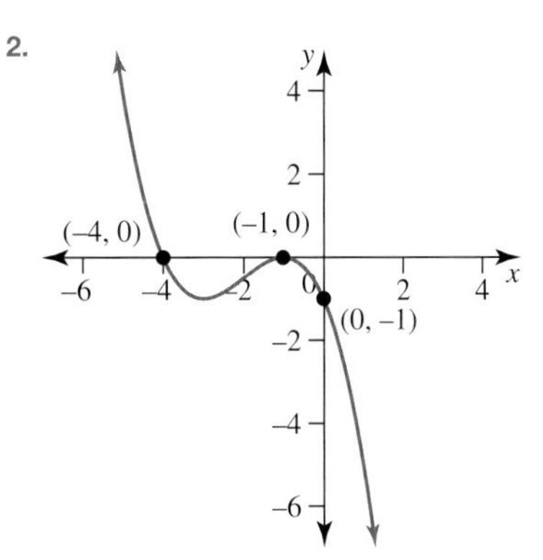

3. a. $(2, 0), (0, -4)$

b. 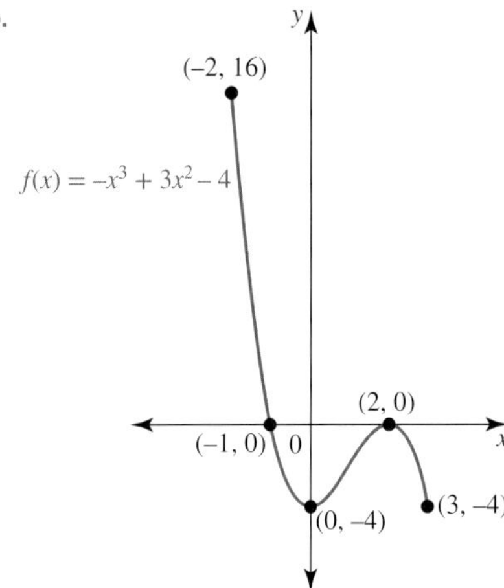

c. Global minimum $= -4$, global maximum $= 16$

12.5 Optimisation problems

12.5 Exercise

1. a. $\frac{dM}{dx} = 40 - 4x$ b. $x = 10$
   c. \$200
2. a. $y = 16 - x$
   b. Sample responses can be found in the worked solutions in the online resources. $0 < x < 16$
   c. $x = 8$
   d. 8 cm by 8 cm, 64 cm^2
3. a. $y = 10 - x$
   b. Sample responses can be found in the worked solutions in the online resources.
   c. $x = 5$
   d. 50 cm^2
   e. 58 cm^2
4. a. $A = 40x - 2x^2$
   b. Width 10 metres; length 20 metres
   c. 200 m^2
   d. 182 m^2
5. 8 people
6. a. 9.77 metres b. 25 metres

7. a. $V = 240x - 64x^2 + 4x^3, 0 < x < 6$
   b. Length 15.15 cm; width 7.15 cm; height 2.43 cm; volume 263 cm^3
8. a. Sample responses can be found in the worked solutions in the online resources.
   b. Width 39.2 cm and height 47.3 cm
   c. Width 30 cm and height 53.8 cm
9. a. $h = \frac{100 - \pi r^2}{\pi r}$
   b. Sample responses can be found in the worked solutions in the online resources.
   c. Sample responses can be found in the worked solutions in the online resources.
   d. 175 cm^3
10. a. Sample responses can be found in the worked solutions in the online resources.
   b. $r = \frac{14\sqrt{2}}{3}$, $h = \frac{28}{3}$
11. a.

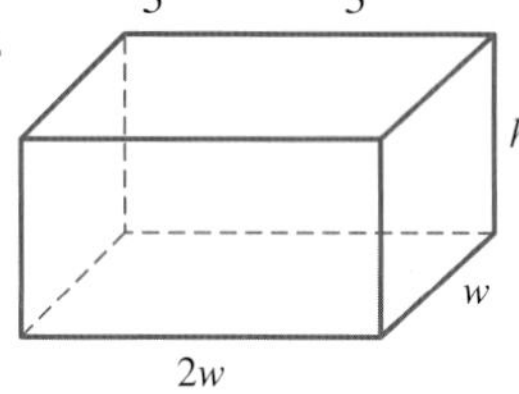

Surface area: $4w^2 + 6wh$

   b. $V = 100w - \frac{4}{3}w^3$; $\frac{1000}{3}$ cm^3
   c. 10 cm by 5 cm by $\frac{20}{3}$ cm
   d. $\left[\frac{568}{3}, \frac{1000}{3}\right]$
12. a. Sample response can be found in the worked solutions in the online resources.
   b. Sample response can be found in the worked solutions in the online resources.
   c. 6000 cm^3; 30 cm by 10 cm by 20 cm
   d. $x \in (0, 15)$

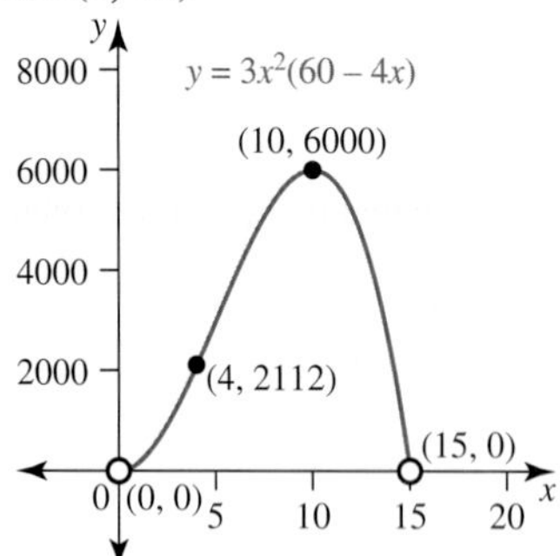

   e. 2112 cm^3; 12 cm by 4 cm by 44 cm
13. \$4157
14. a. $\theta = \frac{8 - 2r}{r}$
   b. $A = 4r - r^2$
   c. 2 radians

12.5 Exam questions

Note: Mark allocations are available with the fully worked solutions online.

1. A
2. C
3. a. $x(10 - 2x)^2$ or $100x - 40x^2 + 4x^3$
   b. Width and length $\frac{20}{3}$ cm, height $\frac{5}{3}$ cm; the maximum volume is 74 cm^3.

12.6 Rates of change and kinematics

12.6 Exercise

1. a. $r = \frac{1}{2}h$
   b. $V = \frac{1}{12}\pi h^3$
   c. $\frac{9\pi}{4}$ cm^3/cm
2. 9π cm^3/cm
3. a. $\frac{dC}{dr} = 2\pi$
   b. i. $\frac{dA}{dr} = 2\pi r$
      ii. 6π mm^2/mm
   c. i. $\frac{dS}{dr} = 8\pi r$
      ii. 40π cm^2/cm
      iii. 2 cm
   d. i. $\frac{dV}{dr} = 4\pi r^2$
      ii. π m^3/m
4. a. $V = x^3$, 12 cm^3/cm
   b. $A = 6l^2$, 60 cm^2/cm.
   c. i. $\frac{dV}{dh} = h^2 + 2$
      ii. 15 cm^3, 11 cm^3/cm
5. a. 0.4π m^2/m
   b. 96 mm^2/mm
6. a. $A = \frac{\sqrt{3}}{4}x^2$
   b. $\sqrt{3}$ cm^2/cm
   c. $8\sqrt{3}$ cm^2/cm
7. a. i. $s = \frac{x}{5}$
      ii. $\frac{ds}{dx} = \frac{1}{5}$, rate of change of shadow length with respect to distance of person from light pole is a constant value of $\frac{1}{5}$ m/m.
   b. i. $\frac{1}{3}$ m^2
      ii. $\frac{1}{3}$ m^3/m
8. a. At the origin
   b. $v = 6t - 6$; initial velocity -6 m/s; move to the left
   c. After 1 second and 3 metres to the left of the origin
   d. 6 metres
   e. 3 m/s
   f. 0 m/s

9. a. $\frac{dx}{dt} = 12$

b. The velocity is constant at 12 m/s.

c. 12 m

d. 3 seconds

10. a. 7 m to the left of the origin

b. $v = 2t - 6$, -6 m/s

c. 3 seconds

d. 6 s, 6 m/s

11. a. $v = \frac{dx}{dt} = 3t^2 - 8t - 3$, $a = \frac{dv}{dt} = 6t - 8$

b. 3 s, 10 m/s^2, 6 m to the left of origin.

c. -6 m/s

12. a. 10 cm to left of origin, 5 cm to right of origin

b. 15 cm

c. $v = 5$ cm/s

d. 

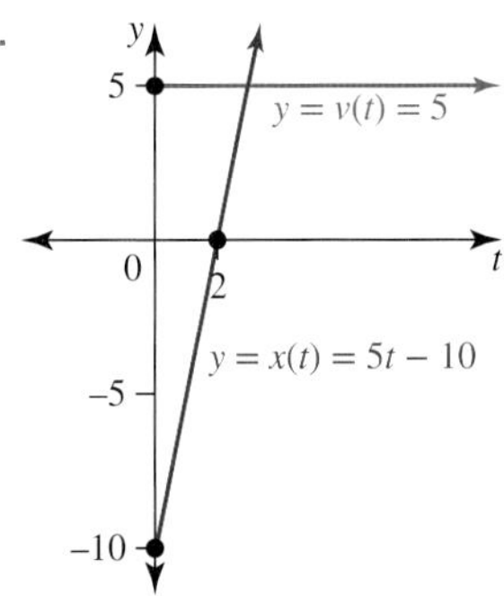

The velocity graph is the gradient graph of the position.

13. a. $v = 6 - 2t$, $a = -2$

b. The position graph is quadratic with maximum turning point when $t = 3$; velocity graph is linear with t-intercept at $t = 3$; acceleration graph is horizontal with constant value of -2.

c. $v = 0 \Rightarrow t = 3$, $t = 3 \Rightarrow x = 9$

d. Position increases for $0 < t < 3$ and $v > 0$.

14. a. $P'(t) = 600 - 12t$

b. Increasing at 360 bacteria per hour

c. Decreasing at 120 bacteria per hour

d. i. 30 hours ii. 70 hours iii. 50 hours

15. a. 1 metre to right of origin; 8 m/s

b. $26\frac{2}{3} = \frac{80}{3}$ metres

c. -6 m/s^2

16. a. $t = 0$, $x = 0$, $v = 0$

b. 2 seconds; $1\frac{1}{3}$ cm left of origin

c. 3 seconds

d. 3 cm/s, 4 cm/s^2

e. 

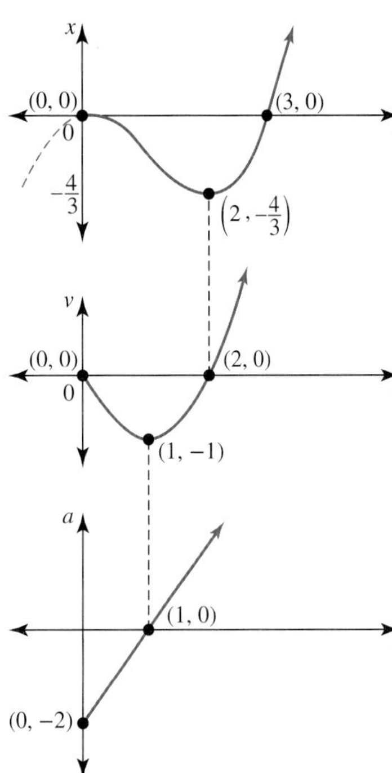

At $t = 2$, position is most negative, velocity is zero, acceleration is 2 cm/s^2

f. The particle is moving to the left with greatest velocity in that direction; acceleration is momentarily zero as it changes from negative to positive.

17. a. 20 m/s

b. 10 m/s

c. 5 seconds, travelling down towards the ground

d. 4 seconds

e. 80 metres

f. 8 seconds, 40 m/s

18. a. $r = \sqrt{100 - h^2}$

b. $V = \frac{1}{3}\pi\left(100h - h^3\right)$

c. The volume is decreasing at the rate of $\frac{8\pi}{3}$ cm^3/cm.

19. a. $\sqrt{2x}$ m

b. $V = \frac{1}{3}\left(200h - 2h^3\right)$

c. $42\frac{2}{3}$ m^3/m

d. $h = \frac{10\sqrt{3}}{3}$; gives the height for maximum volume

20. a. 63 m b. 12 m/s

c. -18 m/s d. 9 seconds; 30 m/s

e. 60 m f. 10 m/s

21. a. 3 seconds and 5 seconds

b. $t \in (3, 5)$

c. $t \in [0, 4)$

d. $\sqrt{3}$ seconds

e. 2 seconds and $(\sqrt{6} - 1)$ seconds

22. a. 20.5 metres

b. 3 seconds

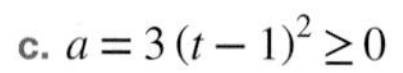

c. $a = 3(t-1)^2 \geq 0$

d.

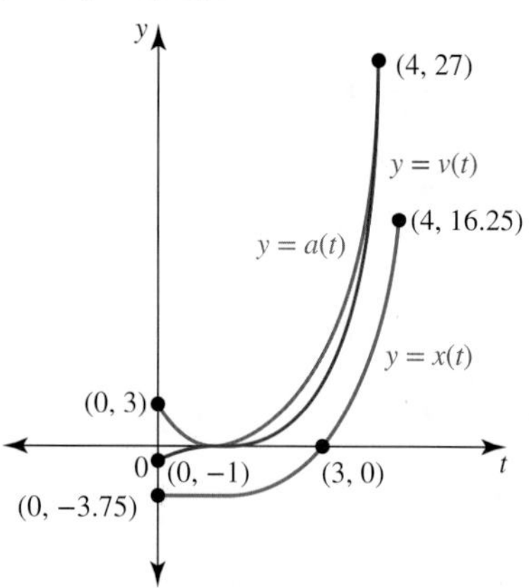

e. Velocity and acceleration are positive and increasing for all t values greater than 1. This causes the particle to forever move to the right for $t > 1$ passing through O at $t = 3$ and never returning.

12.6 Exam questions

Note: Mark allocations are available with the fully worked solutions online.

1. a. $v = -8$ m/s $t = 2$ seconds
   The particle is momentarily at rest after 2 seconds at the position 1 metre to the right of the origin.
   b. -2 m/s
2. C
3. D

12.7 Derivatives of power functions (extending)

12.7 Exercise

1. a. $f'(x) = -3x^{-4}$ b. $f'(x) = -4x^{-5}$
   c. $f'(x) = \frac{5}{2}x^{\frac{3}{2}}$ d. $f'(x) = \frac{2}{3}x^{-\frac{1}{3}}$
   e. $f'(x) = -8x^{-3}$ f. $f'(x) = -\frac{3}{5}x^{-\frac{11}{5}}$
2. a. $\frac{dy}{dx} = -4x^{-2} - 10x^{-3}$ b. $\frac{dy}{dx} = 2x^{-\frac{1}{2}} - 2x^{-\frac{1}{3}}$
   c. $\frac{dy}{dx} = -4x^{-\frac{3}{2}}$ d. $\frac{dy}{dx} = 0.9x^{0.8} - 18.6x^{2.1}$
3. a. $\frac{dy}{dx} = -\frac{1}{x^2} - \frac{2}{x^3}$ b. $\frac{dy}{dx} = \frac{5}{2\sqrt{x}}$
   c. $\frac{dy}{dx} = 3 - \frac{1}{3x^{\frac{2}{3}}}$ d. $\frac{dy}{dx} = \frac{3\sqrt{x}}{2}$
   e. $\frac{dy}{dx} = -\frac{24}{x^9}$ f. $\frac{dy}{dx} = -\frac{4}{x^2} - 15x^2$
4. a. $\frac{dy}{dx} = -\frac{16}{x^5} + \frac{9}{x^4}$; domain $R \setminus \{0\}$
   b. i. $f'(x) = \frac{2}{\sqrt{x}} + \frac{\sqrt{2}}{2\sqrt{x}}$; domain R^+
   ii. Gradient $= \frac{4 + \sqrt{2}}{2}$
5. a. $\frac{1}{2}$
   b. $\frac{1}{12}$
   c. $-\frac{1}{5}$
   d. $(2, 5.5)$
   e. $\left(\frac{1}{3}, 4\right)$ and $\left(-\frac{1}{3}, -4\right)$.
   f. $\left(\frac{9}{4}, 0\right)$
6. a. 3.5
   b. $(1, 3)$
   c. $x = -1,\ x = \frac{-1 \pm \sqrt{5}}{2}$
7. a. 0.5 metres b. 0.25 metres/year
   c. $6\frac{1}{4}$ years d. 0.4 metres/year
8. a. $\frac{dV}{dt} = -\frac{2}{3t^2}$, which is < 0.
   b. $\frac{1}{6}$ mL/h
   c.

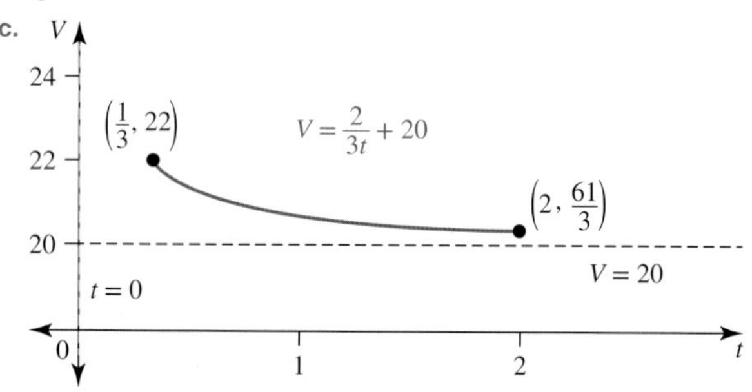

   Part of hyperbola with asymptotes at $V = 20,\ t = 0$; end points $\left(\frac{1}{3}, 22\right), \left(2, \frac{61}{3}\right)$
   d. -1, average rate of evaporation
9. 3
10. a.

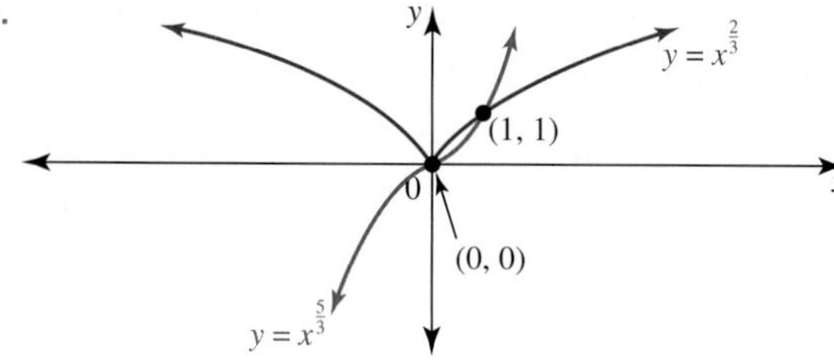

   $(0, 0)$ and $(1, 1)$
   b. At $(1, 1)$, the gradient of $y = x^{\frac{2}{3}}$ is $\frac{2}{3}$ and the gradient of $y = x^{\frac{5}{3}}$ is $\frac{5}{3}$, so $y = x^{\frac{5}{3}}$ is steeper; at $(0, 0)$, the gradient of $y = x^{\frac{2}{3}}$ is undefined and the gradient of $y = x^{\frac{5}{3}}$ is zero.

12.7 Exam questions

Note: Mark allocations are available with the fully worked solutions online.

1. A
2. a. $-\frac{1}{\sqrt{x}}$

   b. The gradient at $y = 0$ is $-\frac{2}{3}$.

3. a. $R\backslash(0)$

   b. $2 - \frac{1}{3x^2}$, domain $R\backslash(0)$

   c. $\left(\frac{1}{3}, \frac{5}{3}\right)$ and $\left(-\frac{1}{3}, -\frac{5}{3}\right)$.

12.8 Review

12.8 Exercise

Technology free: short answer

1. a. 72   b. $\frac{16}{15}$

   c. Does not exist   d. Does not exist

2. a. Sample responses can be found in the worked solutions in the online resources.

   b. $f'(3) = -11$ and $f'(5) = 17$

   c. Minimum stationary point

   d. $x = -\frac{2}{3}$

3. a. $(-6, 0)$ b. $\{-6, 0\}$

   c. $(-\infty, -6) \cup (0, \infty)$

4. a. $y = -4x - 27$ b. $y = -17$

5. a. Stationary point of inflection

   b. Minimum turning point

   c. Point of inflection $(2, 32)$, minimum turning point $(-1, 5)$, y-intercept $(0, 16)$.

   d.

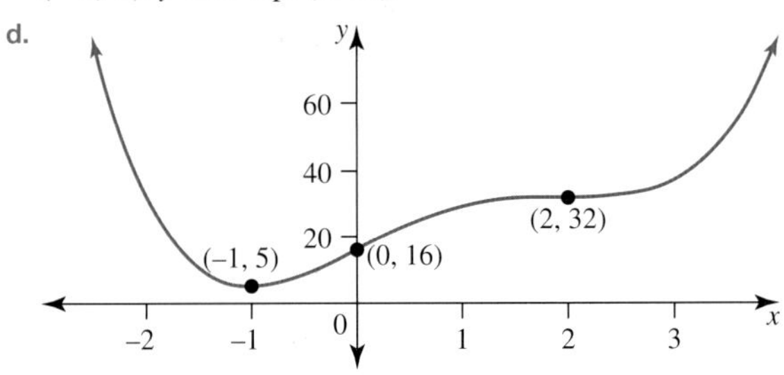

6. a. $100\pi \text{ cm}^3/\text{cm}$ b. $9\pi \text{ cm}^3/\text{cm}$

Technology active: multiple choice

7. C
8. B
9. B
10. D
11. C
12. D
13. C
14. C
15. C
16. D

Technology active: extended response

17. a. $V = 4x^3 - 86x^2 + 450x, x \in (0, 9)$

    b. 3.4 cm; sample responses can be found in the worked solutions in the online resources.

    c. 693 cm^3

    d. 19 cm by 12 cm by 3 cm, 684 cm^3

    e. 693 cm^3, 468 cm^3

18. a. i. $4x^3 - 9x^2 + 16 = 0$

    ii. -1.0937

    iii. Minimum turning point

    b. i. $c = 1,\ d = 0$

    ii. $a = -\frac{1}{6},\ b = 0$

    iii. $\left(-\sqrt{2}, -\frac{2\sqrt{2}}{3}\right)$ is a minimum turning point;

    $\left(\sqrt{2}, \frac{2\sqrt{2}}{3}\right)$ is a maximum turning point.

    iv. $\frac{2\sqrt{2}}{3}$ is a local and global maximum; $-\frac{2\sqrt{2}}{3}$ is a local and global minimum.

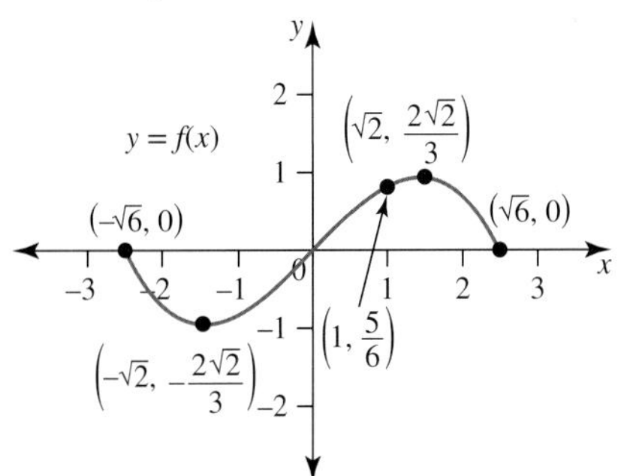

    v. $\left(-2, -\frac{2}{3}\right)$

    vi. $y = \frac{2\sqrt{2}}{3} \sin\left(\frac{\pi x}{\sqrt{6}}\right)$

19. a. 12 metres to left of O; moves to the left

    b. 80 metres

    c. 260 metres; 26 m/s

    d. For particle B, $s = t^2 + 8t - 12,\ t \geq 0$.
    Initial position: when $t = 0, s = -12$.
    Therefore, both particles A and B start from the same initial position 12 metres to the left of O.

$$\text{B's velocity: } v = \frac{ds}{dt} = 2t + 8$$

$$\text{When } t = 0,\ v = 8$$

Since particle B's initial velocity is positive, it starts to move to the right toward O. Hence, particles A and B start from the same initial position but they start to move in opposite directions.

e. 12 seconds; 228 metres to right of O

f. Time 6 seconds, positions $x = -72, s = 72$;
time $2\frac{2}{3}$ seconds, positions $x = -83\frac{1}{9}, s = 16\frac{4}{9}$

20. a. $h = \dfrac{154 - \pi r^2}{\pi r}$

b. The volume of a cylinder is $\pi r^2 h$.
Substituting the expression for h in terms of r from part a,

$$V = \pi r^2 \times \frac{154 - \pi r^2}{\pi r}$$
$$= r \times \left(154 - \pi r^2\right)$$
$$\therefore V = 154r - \pi r^3$$

c. i. $(154 - 19\pi)\,\text{cm}^3/\text{cm}$
ii. $(154 - 12\pi)\,\text{cm}^3/\text{cm}$

d. $415\,\text{cm}^3$; radius 4.0 cm; height 8.1 cm

e. $2\pi r$; $4r + h$

f. \$6.06

12.8 Exam questions

Note: Mark allocations are available with the fully worked solutions online.

1. A
2. C
3. D
4. The dimensions of the garden bed are 10×10 m and the area is $100\,\text{m}^2$.

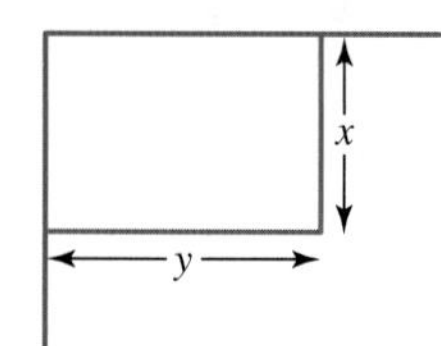

5. a. $A = \dfrac{\sqrt{15}}{16}x^2$ b. $\sqrt{15}\,\text{cm}^2/\text{cm}$

13 Anti-differentiation and introduction to integral calculus

LEARNING SEQUENCE

Fully worked solutions for this topic are available online.

13.1 Overview

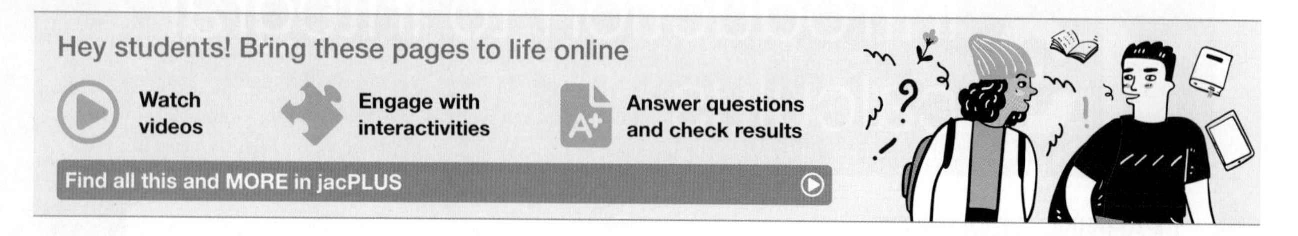

13.1.1 Introduction

Why is calculus so important? Newton showed us it governs the universe.

Nevertheless, the parents of the 18th-century French mathematician Sophie Germaine were alarmed when, as a young girl, she taught herself calculus from books in the family's library. They tried switching off the heating and removing her candles, but Sophie found a way to continue her studies. Eventually her parents relented and thereafter accepted her interest in this 'unfeminine' subject.

Higher mathematical study beckoned, but this was open only to men. Again, Sophie prevailed. Dressed as a man to impersonate a student who had withdrawn from his studies, she collected his lecture notes and submitted his assigned work using the pseudonym Monsieur LeBlanc. However, the submitted work was so outstanding that Lagrange, one of France's great mathematicians, insisted on meeting M. LeBlanc. Despite being stunned to find the student was female, Lagrange kept her secret, becoming Sophie's mentor and friend throughout her mathematical career.

Germaine claimed that her interest in mathematics was ignited by reading about Archimedes of Syracuse. Legend has it that Archimedes was kneeling in the sand so engrossed in a mathematical problem that he refused to comply with a Roman soldier's order until the solution was completed. Enraged, the soldier killed him with a spear. Germaine concluded that mathematics must be the most captivating of all subjects.

KEY CONCEPTS

This topic covers the following key concepts from the VCE Mathematics Study Design:

- anti-differentiation as the inverse process of differentiation and identification of families of curves with the same gradient function, and the use of a boundary condition to determine a specific anti-derivative of a given function.

Source: VCE Mathematics Study Design (2023–2027) extracts @VCAA; reproduced by permission.

13.2 Anti-derivatives

LEARNING INTENTION

At the end of this subtopic you should be able to:
- determine anti-derivatives and indefinite integrals.

Calculus is made up of two parts: differential calculus and integral calculus. Differential calculus arose from the need to measure an instantaneous rate of change, or gradient; integral calculus arose from the need to measure the area enclosed between curves. The fundamental theorem of calculus, which established the connection between the two branches of calculus, is one of the most important theorems in mathematics. Before we can gain some understanding of this theorem, we must first consider the reverse operation to differentiation.

13.2.1 Anti-derivatives of polynomial functions

Anti-differentiation is the reverse process (or 'undoing' process) to differentiation.

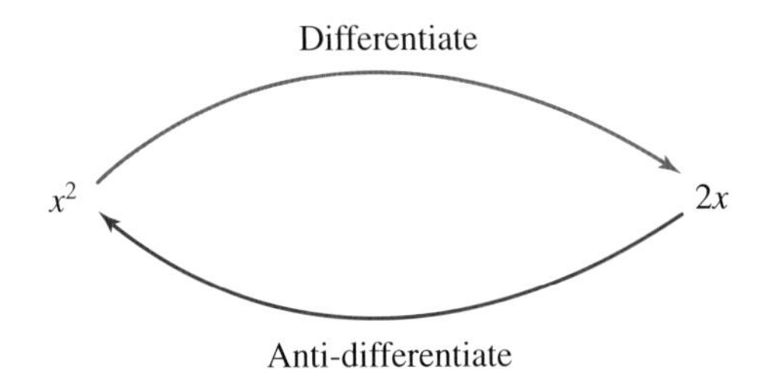

The derivative of x^2 is $2x$; hence, reversing the process, an anti-derivative of $2x$ is x^2.

The expression 'an' anti-derivative is of significance in the above example. For any constant c, $\frac{d}{dx}(x^2+c)=2x$, so 'the' anti-derivative of $2x$ is x^2+c. The constant c is referred to as an arbitrary constant.

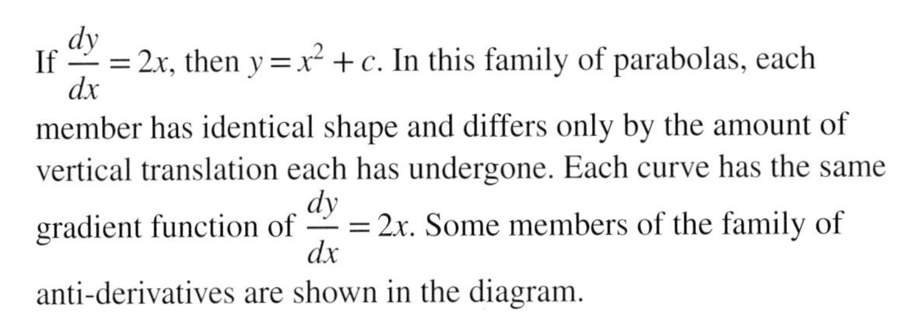

If $\frac{dy}{dx}=2x$, then $y=x^2+c$. In this family of parabolas, each member has identical shape and differs only by the amount of vertical translation each has undergone. Each curve has the same gradient function of $\frac{dy}{dx}=2x$. Some members of the family of anti-derivatives are shown in the diagram.

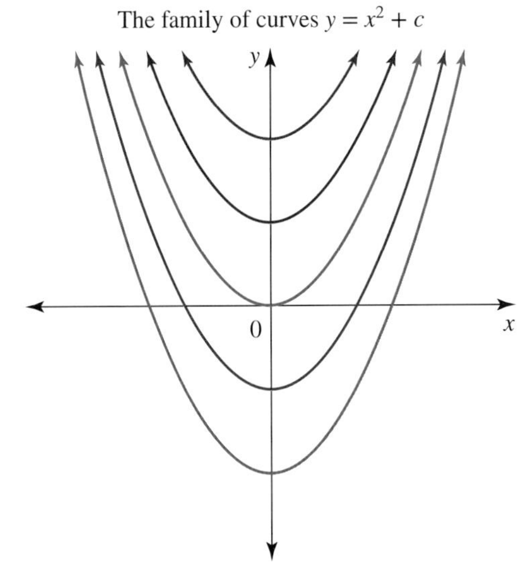

Given $\frac{dy}{dx}$ or $f'(x)$, the process of obtaining y or $f(x)$ is called anti-differentiation. The process is also referred to as finding the **primitive function**, given the gradient function.

The basic rule for anti-differentiation of polynomials

By studying the patterns in the following examples, the basic rule of anti-differentiation can be deduced. The values in the table can be verified by differentiating the polynomial y.

y	$5x+c$	x^2+c	x^3+c	$\frac{1}{3}x^3+c$	$\frac{1}{4}x^4+c$
$\frac{dy}{dx}$	5	$2x$	$3x^2$	x^2	x^3

The rule the table illustrates is:

Anti-differentiation rule

If $\frac{dy}{dx} = x^n$, $n \in N$,

then $y = \frac{1}{n+1}x^{n+1} + c$, where c is an arbitrary constant.

This rule can be verified by differentiating y with respect to x.

The linearity properties allow anti-derivatives of sums and differences of polynomial terms to be calculated. For example, if $f'(x) = a_1 + a_2x + a_3x^2 + ... + a_nx^{n-1}$, $n \in N$, then

$$f(x) = a_1x + a_2\frac{x^2}{2} + a_3\frac{x^3}{3} + ... + a_n\frac{x^n}{n} + c$$

$$= a_1x + \frac{1}{2}a_2x^2 + \frac{1}{3}a_3x^3 + ... + \frac{1}{n}a_nx^n + c$$

Note that the degree of the primitive function is one higher than the degree of the gradient function.

The rule for anti-differentiation of a polynomial shows that differentiation and anti-differentiation are inverse operations.

WORKED EXAMPLE 1 Determining anti-derivatives

a. **If $\frac{dy}{dx} = 6x^8$, determine y.**

b. **Calculate an anti-derivative of $3x^2 - 5x + 2$.**

c. **Given $f'(x) = x(x-1)(x+1)$, obtain the rule for the primitive function.**

THINK	WRITE
a. 1. State the reverse process required to find y.	a. $\frac{dy}{dx} = 6x^8$ The derivative of y is $6x^8$. Hence, the anti-derivative of $6x^8$ is y.
2. Use the rule to anti-differentiate the polynomial term. *Note:* It's a good idea to mentally differentiate the answer to check its derivative is $6x^8$.	$y = 6 \times \left(\frac{1}{8+1}x^{8+1}\right) + c$ $= 6 \times \frac{1}{9}x^9 + c$ $= \frac{2}{3}x^9 + c$
b. 1. Apply the anti-differentiation rule to each term of the polynomial.	b. $3x^2 - 5x + 2 = 3x^2 - 5x^1 + 2x^0$ The anti-derivative is: $3 \times \frac{1}{3}x^3 - 5 \times \frac{1}{2}x^2 + 2 \times \frac{1}{1}x^1 + c$ $= x^3 - \frac{5}{2}x^2 + 2x + c$

2. Choose any value for the constant c to obtain an anti-derivative.
Note: The conventional choice is $c = 0$.

When $c = 0$, an anti-derivative is $x^3 - \frac{5}{2}x^2 + 2x$.

c. 1. Express the rule for the derivative function in expanded polynomial form.

c. $f'(x) = x(x-1)(x+1)$
$= x(x^2 - 1)$
$= x^3 - x$

2. Calculate the anti-derivative and state the answer.

$f(x) = \frac{1}{4}x^4 - \frac{1}{2}x^2 + c$
The rule for the primitive function is
$f(x) = \frac{1}{4}x^4 - \frac{1}{2}x^2 + c$.

TI | THINK	DISPLAY/WRITE	CASIO | THINK	DISPLAY/WRITE
b. 1. On a Calculator page, press MENU and select: 4. Calculus 3. Integral Complete the entry line as: $\int (3x^2 - 5x + 2)\,dx$ Then press ENTER. *Note:* The concept of the integral will be introduced in the next section.	1.1 *Doc RAD $\int(3\cdot x^2-5\cdot x+2)dx$ $x^3-\frac{5\cdot x^2}{2}+2\cdot x$	b. 1. On a Main screen, complete the entry line as: $\int (3x^2 - 5x + 2)\,dx$ Then press EXE. *Note:* The template for the integral is in the Math 2 Keyboard menu. *Note:* The concept of the integral will be introduced in the next section.	Edit Action Interactive $\int 3\cdot x^2-5\cdot x+2dx$ $x^3-\frac{5\cdot x^2}{2}+2\cdot x$ Alg Standard Real Rad
2. The answer appears on the screen. Include the constant term.	$y = x^3 - \frac{5x^2}{2} + 2x + c$	2. The answer appears on the screen. Include the constant term.	$y = x^3 - \frac{5x^2}{2} + 2x + c$

13.2.2 The indefinite integral

Anti-differentiation is not only about undoing the effect of differentiation. It is an operation that can be applied to functions. There are different forms of notation for the anti-derivative, just as there are for the derivative.

It is customary to write 'the anti-derivative of $f(x)$' as $\int f(x)\,dx$. This is called the **indefinite integral**.

Using this symbol, we could write 'the anti-derivative of $2x$ with respect to x equals $x^2 + c$' as $\int 2x\,dx = x^2 + c$.

For any function, the indefinite integral is defined as follows.

The indefinite integral

$$\int f(x)\,dx = F(x) + c, \text{ where } F'(x) = f(x).$$

$F(x)$ is an anti-derivative or primitive of $f(x)$. It can also be said that $F(x)$ is the indefinite integral, or just the integral, of $f(x)$. The arbitrary constant c is called the constant of integration. Here we take integration to be the process of using the integral to obtain an anti-derivative. A little more will be said about integration later.

The rule for anti-differentiation of a polynomial term could be written in this notation as follows.

Anti-differentiation of a polynomial

$$\text{For } n \in N, \int x^n \, dx = \frac{x^{n+1}}{n+1} + c.$$

The linearity properties of anti-differentiation, or integration, could be expressed as:

$$\int \left(f(x) \pm g(x) \right) dx = \int f(x) \, dx \pm \int g(x) \, dx$$

$$\int kf(x) \, dx = k \int f(x) \, dx$$

WORKED EXAMPLE 2 Determining indefinite integrals

a. **Use the linearity properties to calculate** $\int (2x^3 + 4x - 3) \, dx$.

b. **Calculate** $\int \frac{7x^5 - 8x^7}{5x} \, dx$.

c. **Given** $f(x) = \frac{3}{2}x^2$, **form** $F(x)$, **where** $F'(x) = f(x)$.

d. **Obtain** $\int (3x - 2)^2 \, dx$.

THINK	WRITE
a. 1. Express the integral as the sum or difference of integrals of each term.	**a.** $\int (2x^3 + 4x - 3) \, dx = \int 2x^3 \, dx + \int 4x \, dx - \int 3 \, dx$ $= 2\int x^3 \, dx + 4\int x \, dx - 3\int 1 \, dx$
2. Anti-differentiate term by term. *Note:* In practice, the calculation usually omits the steps illustrating the linearity properties.	$= 2 \times \frac{x^4}{4} + 4 \times \frac{x^2}{2} - 3x + c$ $= \frac{x^4}{2} + 2x^2 - 3x + c$
b. 1. Express the term to be anti-differentiated in polynomial form.	**b.** $\int \frac{7x^5 - 8x^7}{5x} \, dx = \int \left(\frac{7x^5}{5x} - \frac{8x^7}{5x} \right) dx$ $= \int \left(\frac{7}{5}x^4 - \frac{8}{5}x^6 \right) dx$
2. Calculate the anti-derivative.	$= \frac{7}{5} \times \frac{x^5}{5} - \frac{8}{5} \times \frac{x^7}{7} + c$ $= \frac{7}{25}x^5 - \frac{8}{35}x^7 + c$

c. Use the rule to obtain the anti-derivative.

c. $f(x) = \frac{3}{2}x^2$

$F(x)$ is the anti-derivative function.

$F(x) = \frac{3}{2} \times \frac{x^3}{3} + c$

$\therefore F(x) = \frac{x^3}{2} + c$

d. 1. Prepare the expression to be anti-differentiated by expanding the perfect square and simplifying.

d. $\int (3x-2)^2\, dx = \int \left(9x^2 - 12x + 4\right) dx$

$= 9 \times \frac{x^3}{3} - 12 \times \frac{x^2}{2} + 4x + c$

2. Calculate the anti-derivative.

$= 3x^3 - 6x^2 + 4x + c$

13.2 Exercise

Students, these questions are even better in jacPLUS

Receive immediate feedback and access sample responses

Access additional questions

Track your results and progress

Find all this and MORE in jacPLUS

Technology free

1. Use anti-differentiation to obtain y in terms of x.

a. $\frac{dy}{dx} = x^3$
b. $\frac{dy}{dx} = x^6$
c. $\frac{dy}{dx} = x^8$
d. $\frac{dy}{dx} = 12x$
e. $\frac{dy}{dx} = -6x^2$
f. $\frac{dy}{dx} = \frac{1}{3}x^4$

2. Determine the following.

a. $\int x^5\, dx$
b. $\int x^7\, dx$
c. $\int 9x^8\, dx$
d. $\int 6x^{11}\, dx$
e. $\int (2x+3)\, dx$
f. $\int \left(10 - x^2\right) dx$

3. For the following functions, obtain $F(x)$, where $F'(x) = f(x)$.

a. $f(x) = 10x^9$
b. $f(x) = -x^{12}$
c. $f(x) = 6x^5 + 4x$
d. $f(x) = x^3 - 7x^2$
e. $f(x) = 4 + x + x^2$
f. $f(x) = x^3 + 9x^2 - 5x + 1$

4. **WE1** a. If $\frac{dy}{dx} = 12x^5$, determine y.

b. Calculate an anti-derivative of $4x^2 + 2x - 5$.

c. Given $f'(x) = (x-2)(3x+8)$, obtain the rule for the primitive function.

5. Given $\frac{dy}{dx}$, obtain y in terms of x for each of the following.

a. $\frac{dy}{dx} = 5x^9$

b. $\frac{dy}{dx} = -3 + 4x^7$

c. $\frac{dy}{dx} = 2(x^2 - 6x + 7)$

d. $\frac{dy}{dx} = (8 - x)(2x + 5)$

6. For each $f'(x)$ expression, obtain an expression for $f(x)$.

a. $f'(x) = \frac{1}{2}x^5 + \frac{7}{3}x^6$

b. $f'(x) = \frac{4x^6}{3} + 5$

c. $f'(x) = \frac{4x^4 - 6x^8}{x^2}, x \neq 0$

d. $f'(x) = (3 - 2x^2)^2$

7. Calculate the following indefinite integrals.

a. $\int \frac{3x^8}{5}\, dx$

b. $\int 2\, dx$

c. $4\int (20x - 5x^7)\, dx$

d. $\int \frac{1}{100}(9 + 6x^2 - 5.5x^{10})\, dx$

8. a. Calculate the primitive function of $2ax + b$ for $a, b \in R$.
b. Calculate the anti-derivative of $0.05x^{99}$.
c. Calculate an anti-derivative of $(2x + 1)^3$.
d. Calculate the primitive of $7 - x(5x^3 - 4x - 8)$.

9. Given the gradient function $\frac{dy}{dx} = \frac{2x^3 - 3x^2}{x}$, $x \neq 0$, obtain the primitive function.

10. Find $F(x)$, where $F'(x) = f(x)$ for the following.

a. $f(x) = \frac{3x^2 - 2}{4}$

b. $f(x) = \frac{3x}{4} + \frac{2(1 - x)}{3}$

c. $f(x) = 0.25(1 + 5x^{14})$

d. $f(x) = \frac{12(x^5)^2 - (4x)^2}{3x^2}$

11. WE2 a. Use the linearity properties to calculate $\int \left(-7x^4 + 3x^2 - 6x\right) dx$.

b. Calculate $\int \frac{5x^8 + 3x^3}{4x^2}\, dx$.

c. Given $f(x) = 2x^4$, form $F(x)$, where $F'(x) = f(x)$.

d. Obtain $\int (3x + 4)^2\, dx$.

12. Calculate each expression to show that $\int 2\left(t^2 + 2\right) dt$ is the same as $2\int \left(t^2 + 2\right) dt$.

13. Calculate the following indefinite integral.

$$\int \frac{2x^5 + 7x^3 - 5x^2}{x^2}\, dx$$

14. For the following, find $F(x)$, where $F'(x) = f(x)$, given $a, b \in R$:

$$f(x) = (2ax)^2 + b^3$$

15. a. Calculate the primitive of $(x-1)(x+4)(x+1)$.
b. Calculate the anti-derivative of $(x-4)(x+4)$.
c. Calculate an anti-derivative of $\dfrac{(2x)^3-4x^2}{2x^2}$.
d. Calculate $\dfrac{d}{dx}\left(\displaystyle\int (4x+7)\,dx\right)$.

Technology active

16. a. The calculator enables the anti-derivative to be calculated through a number of menus. Familiarise yourself with these by calculating the anti-derivative of x^3 using each of these methods. Remember to insert the constant yourself if the calculator does not.
b. Write down the answers to the following from the calculator.
i. Given $\dfrac{dy}{dx}=\left(x^4+1\right)^2$, obtain y.
ii. If $f(t)=100(t-5)$, obtain $F(t)$, where $F'(t)=f(t)$
iii. $\displaystyle\int \left(y^5-y^3\right)dy$
iv. $\displaystyle\int (4u+5)^2\,du$

13.2 Exam questions

Question 1 (1 mark) TECH-ACTIVE

MC Given that $\dfrac{dy}{dx}=6x^{11}$, then

A. $y=\dfrac{x^{10}}{5}+c$
B. $y=12x^{12}+c$
C. $y=5x^{10}+c$
D. $y=\dfrac{x^{12}}{2}+3x+c$
E. $y=\dfrac{x^{12}}{2}+c$

Question 2 (1 mark) TECH-ACTIVE

MC Given that $f'(x)=3x^5-3x^8$, an expression for $f(x)$ is

A. $\dfrac{1}{2}x^6-\dfrac{1}{3}x^9+c$
B. $3x^6-9x^9+c$
C. $\dfrac{1}{6}x^6-\dfrac{1}{9}x^9+c$
D. $15x^4-24x^7+c$
E. $2x^6-8x^9+c$

Question 3 (1 mark) TECH-ACTIVE

MC Calculate $\displaystyle\int \frac{7x^6-4x^5}{2x^3}\,dx$.

A. $7x^4-4x^3+c$
B. $8x^4-3x^3+c$
C. $\dfrac{8}{7}x^4-\dfrac{3}{2}x^3+c$
D. $\dfrac{7}{9}x^6-\dfrac{2}{5}x^5+c$
E. $\dfrac{7}{8}x^4-\dfrac{2}{3}x^3+c$

More exam questions are available online.

13.3 Anti-derivative functions and graphs

LEARNING INTENTION

At the end of this subtopic you should be able to:
- determine the equation of the anti-derivative function
- sketch the graph of the anti-derivative function.

If some characteristic information is known about the anti-derivative function, such as a point on its curve, then this information allows the constant of integration to be calculated.

13.3.1 Determining the constant

Once the constant of integration is known, we can determine more than just a family of anti-derivatives. A specific anti-derivative function can be found, since only one member of the family will possess the given characteristic information.

Although there is a family of anti-derivative functions $y = x^2 + c$ for which $\frac{dy}{dx} = 2x$, if it is known the anti-derivative must contain the point $(0, 0)$, substituting $(0, 0)$ in $y = x^2 + c$ gives $0 = 0 + c$ and therefore $c = 0$. Only the specific anti-derivative function $y = x^2$ satisfies this condition.

WORKED EXAMPLE 3 Determining equations of differentiated curves

a. The gradient of a curve is given by $\frac{dy}{dx} = a + 3x$, where a is a constant. Given the curve has a stationary point at $(2, 5)$, determine its equation.

b. If $f'(x) = x^2 - 4$ and $f(3) = 4$, obtain the value of $f(1)$.

THINK	WRITE
a. 1. Use the given information to determine the value of a.	a. $\frac{dy}{dx} = a + 3x$ At the stationary point $(2, 5)$, $\frac{dy}{dx} = 0$. $\therefore 0 = a + 3(2)$ $\therefore a = -6$
2. Calculate the anti-derivative.	$\frac{dy}{dx} = -6 + 3x$ $y = -6x + 3 \times \frac{x^2}{2} + c$ $= -6x + \frac{3x^2}{2} + c$
3. Use the given information to calculate the constant of integration.	Substitute the point $(2, 5)$: $5 = -6(2) + \frac{3(2)^2}{2} + c$ $5 = -12 + 6 + c$ $c = 11$
4. State the answer.	$y = -6x + \frac{3x^2}{2} + 11$ The equation of the curve is $y = \frac{3}{2}x^2 - 6x + 11$.

b. 1. Calculate the anti-derivative.	**b.** $f'(x) = x^2 - 4$ $f(x) = \frac{x^3}{3} - 4x + c$
2. Use the given information to calculate the constant of integration.	Substitute $f(3) = 4$ to calculate c. $4 = \frac{3^3}{3} - 4 \times 3 + c$ $4 = 9 - 12 + c$ $c = 7$
3. Calculate the required value.	$f(x) = \frac{x^3}{3} - 4x + 7$ $f(1) = \frac{1}{3} - 4 + 7$ $= \frac{10}{3}$

 Resources

 **Interactivity** Sketching the anti-derivative graph (int-5965)

13.3.2 Sketching the anti-derivative graph

Given the graph of a function $y = f(x)$ whose rule is not known, the shape of the graph of the anti-derivative function $y = F(x)$ may be able to be deduced from the relationship $F'(x) = f(x)$. This means interpreting the graph of $y = f(x)$ as the gradient graph of $y = F(x)$.

- The x-intercepts of the graph of $y = f(x)$ identify the x-coordinates of the stationary points of $y = F(x)$.
- The nature of any stationary point on $y = F(x)$ is determined by the way the sign of the graph of $y = f(x)$ changes about its x-intercepts.
- If $f(x)$ is a polynomial of degree n, then $F(x)$ will be a polynomial of degree $(n + 1)$.

If the rule for $y = f(x)$ can be formed from its graph, then the equation of possible anti-derivative functions can also be formed. Additional information to determine the specific function would be needed, as in the previous Worked example.

WORKED EXAMPLE 4 Determining the equation of a graph from a derivative

The gradient function of a curve is illustrated in the diagram shown.

a. Describe the position and nature of any stationary point of the curve $y = F(x)$ with this gradient function.

b. Draw three possible graphs of $y = F(x)$.

c. Obtain the rule for $f(x)$.

d. Given $F(0) = -2$, determine the rule for $F(x)$ and sketch the graph of $y = F(x)$.

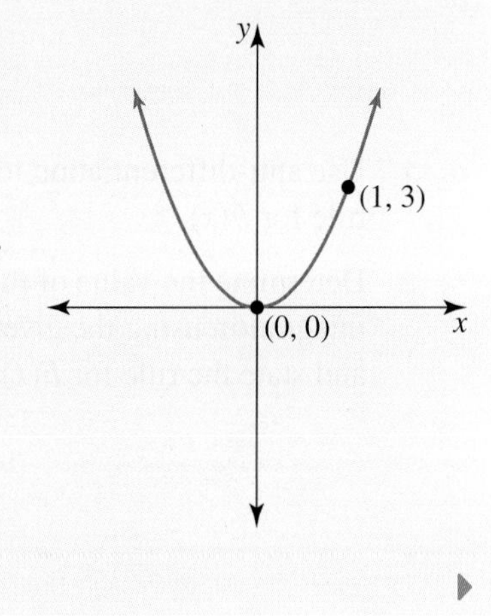

THINK	WRITE
a. 1. Identify the position of any stationary point on the graph of $y = f(x)$.	**a.** The x-intercept on the given graph identifies a stationary point on $y = f(x)$. Therefore, the graph of $y = f(x)$ has a stationary point at the point where $x = 0$.
2. Determine the nature of the stationary point.	As x increases, the sign of the graph of $y = f(x)$ changes from positive to zero to positive about its x-intercept. Therefore, there is a stationary point of inflection at the point where $x = 0$ on the graph of $y = f(x)$.
b. 1. State the degree of $f(x)$.	**b.** Since $f(x)$ is quadratic, its anti-derivative $F(x)$ will be a polynomial of degree 3.
2. Draw three possible graphs for the anti-derivative function.	Other than at $x = 0$, the graph of $y = f(x)$ is positive, so the gradient of the graph of $y = F(x)$ is positive other than at $x = 0$. The y-coordinate of the stationary point on $y = F(x)$ is not known. Three graphs of an increasing cubic function with a stationary point of inflection at the point where $x = 0$ are shown for $y = F(x)$.

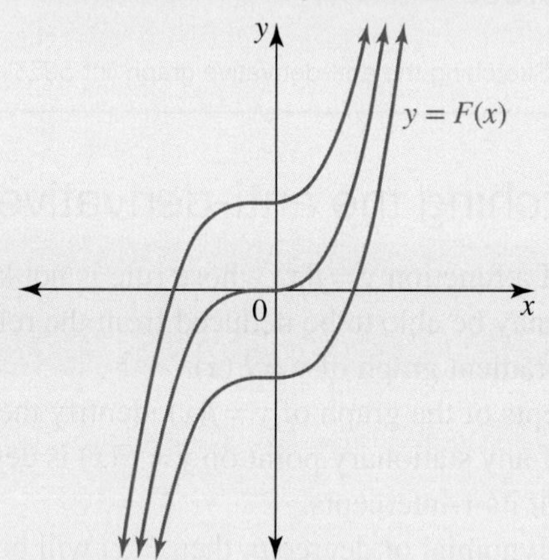

c. Determine the equation of the quadratic function using the information on its graph.	**c.** The graph of $y = f(x)$ has a minimum turning point at $(0, 0)$ and contains the point $(1, 3)$. Let $f(x) = a(x - h)^2 + k$. The turning point is $(0, 0)$. $\therefore f(x) = ax^2$ Substitute the point $(1, 3)$: $3 = a(1)^2$ $\therefore a = 3$ The rule for the quadratic function is $f(x) = 3x^2$.
d. 1. Use anti-differentiation to obtain the rule for $F(x)$.	**d.** $f(x) = 3x^2$ $\therefore F(x) = x^3 + c$
2. Determine the value of the constant of integration using the given information and state the rule for $F(x)$.	Since $F(0) = -2$, $-2 = (0)^3 + c$ $c = -2$ $\therefore F(x) = x^3 - 2$

3. Sketch the graph of $y = F(x)$.

The point $(0, -2)$ is a stationary point of inflection at the y-intercept.

x-intercept: let $y = 0$.

$$x^3 - 2 = 0$$
$$x^3 = 2$$
$$x = \sqrt[3]{2}$$

$\left(\sqrt[3]{2}, 0\right)$ is the x-intercept.

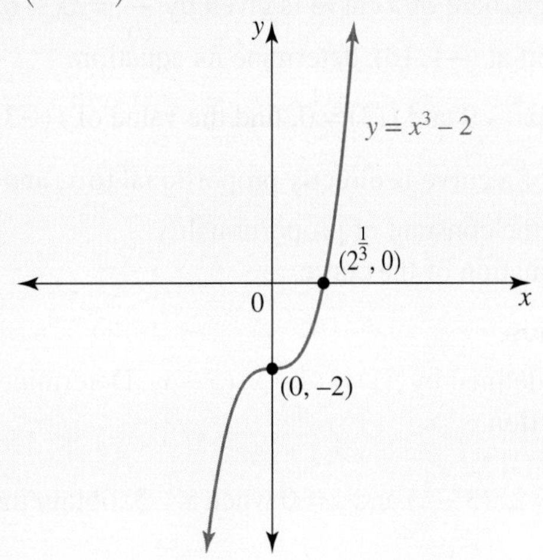

13.3 Exercise

Students, these questions are even better in jacPLUS

Receive immediate feedback and access sample responses

Access additional questions

Track your results and progress

Find all this and MORE in jacPLUS

Technology free

1. a. If $f'(x) = 16x$ and $f(1) = 3$, find an expression for $f(x)$.
   b. If $f'(x) = 3x^2 + 4x$ and $f(0) = 8$, find an expression for $f(x)$.
   c. If $f'(x) = x^2(9 - 8x)$ and $f(1) = -12$, find an expression for $f(x)$.
   d. If $f'(x) = (5x + 1)^2$ and $f(1) = 6$, find an expression for $f(x)$.
2. The gradient of a curve is given by $f'(x) = -3x^2 + 4$. Find the equation of the curve if it passes through the point $(-1, 2)$.
3. a. Determine the equation of the curve that contains the point $(2, 3)$ and for which $\frac{dy}{dx} = 4 - 5x$.
   b. Determine the equation of the curve for which the gradient is $\frac{dy}{dx} = (x + 1)(x - 2)$ at any point (x, y), and the point $(0, -5)$ lies on the curve.

4. The gradient of a curve is given by $\frac{dy}{dx} = \frac{2x}{5} - 3$. It is also known that the curve passes through the point $(5, 0)$.

   a. Determine the equation of the curve.
   b. Find its x-intercepts.

5. If $\frac{dy}{dx} = 2x$ and $y = 10$ when $x = 4$, find y when $x = 5$.

6. **WE3** a. The gradient of a curve is given by $\frac{dy}{dx} = ax - 6$, where a is a constant. Given the curve has a stationary point at $(-1, 10)$, determine its equation.

   b. If $F'(x) = 2x^2 - 9$ and $f(3) = 0$, find the value of $f(-3)$.

7. The gradient of a curve is directly proportional to x, and at the point $(2, 5)$ on the curve, the gradient is -3.

   a. Determine the constant of proportionality.
   b. Find the equation of the curve.

Technology active

8. A function is defined by $f(x) = (4 - x)(5 - x)$. Determine its primitive function if the point $(1, -1)$ lies on the primitive function.

9. a. Given $\frac{dy}{dx} = 2x(3 - x)$ and $y = 0$ when $x = 3$, obtain the value of y when $x = 0$.

   b. Given $\frac{dz}{dx} = (10 - x)^2$ and $z = 200$ when $x = 10$, obtain the value of z when $x = 4$.

10. At any point (x, y) on a curve, the gradient of the tangent to the curve is given by $\frac{dy}{dx} = a - x^3$. The curve has a stationary point at $(2, 9)$.

    a. Find the value of a.
    b. Determine the equation of the curve.
    c. Calculate the equation of the tangent to the curve at the point where $x = 1$.

11. The graph of the gradient, $\frac{dy}{dx}$, of a particular curve is shown.

    Given that $(-2, 3)$ lies on the curve with this gradient, determine the equation of the curve and deduce the coordinates of, and the nature of, any stationary point on the curve.

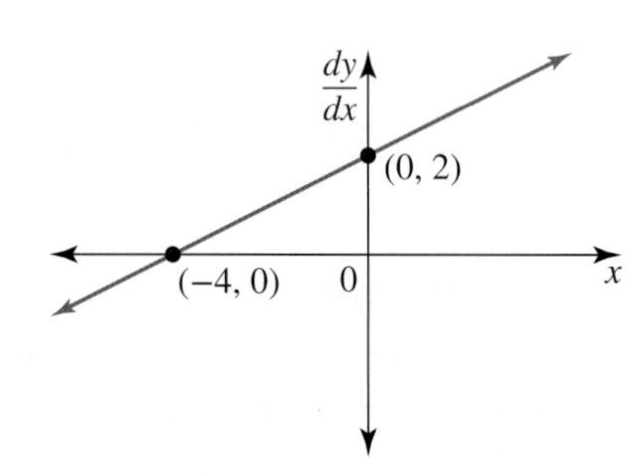

12. **WE4** The gradient function of a curve is illustrated in the diagram shown.

    a. Describe the position and nature of any stationary point of the curve $y = F(x)$ with this gradient function.
    b. Draw three possible graphs of $y = F(x)$.
    c. Obtain the rule for $f(x)$.
    d. Given $F(0) = 1$, determine the rule for $F(x)$ and sketch the graph of $y = F(x)$.

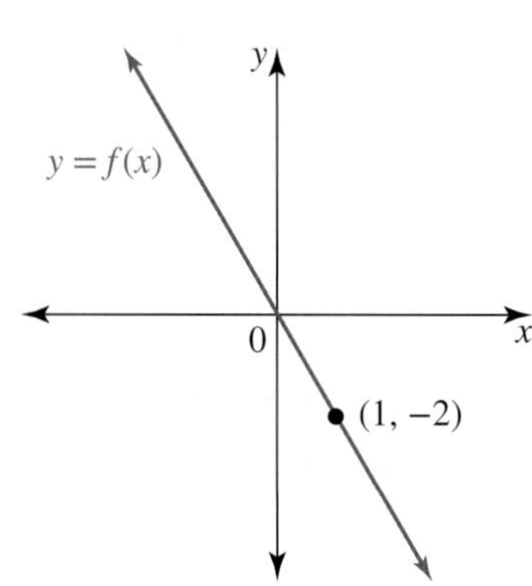

13. The diagram shows the graph of a function $y = f'(x)$.

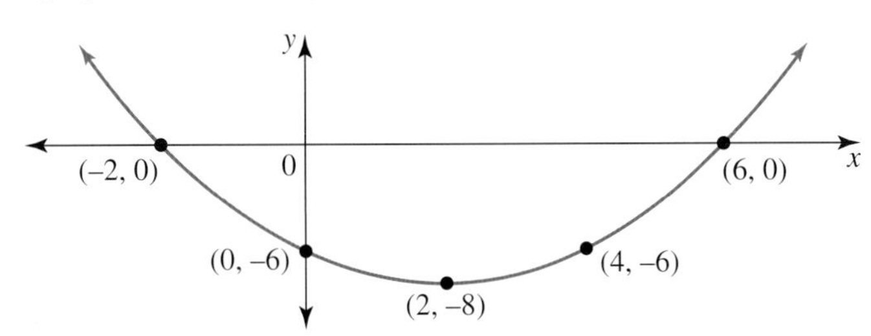

***The following information relates to parts* a *and* b.**

Select the correct statements about the behaviour of the anti-derivative function $y = f(x)$.

a. **MC** There is a minimum turning point on the graph of $y = f(x)$ when:

A. $x = -2$ B. $x = 0$ C. $x = 2$
D. $x = 4$ E. $x = 6$

b. **MC** There is a maximum turning point on the graph of $y = f(x)$ when:

A. $x = -2$ B. $x = 0$ C. $x = 2$
D. $x = 4$ E. $x = 6$

c. State which of the following statements about the gradient of the graph of $y = f(x)$ at the point where $x = 4$ are true (T) or false (F).

i. The gradient at $x = 4$ on the graph of $y = f(x)$ is positive
ii. The gradient at $x = 4$ on the graph of $y = f(x)$ is negative
iii. On the graph of $y = f(x)$, the gradient at $x = 4$ will be the same as the gradient at $x = 0$.

d. Draw a possible graph for $y = f(x)$.

14. The gradient function of a curve is illustrated in the diagram.

a. Describe the position and nature of the stationary points of the curve with this gradient function.
b. Sketch a possible curve that has this gradient function.
c. Determine the rule for the gradient graph and hence obtain the equation which describes the family of curves with this gradient function.
d. One of the curves belonging to the family of curves cuts the x-axis at $x = 3$. Obtain the equation of this particular curve and calculate the slope with which it cuts the x-axis at $x = 3$.

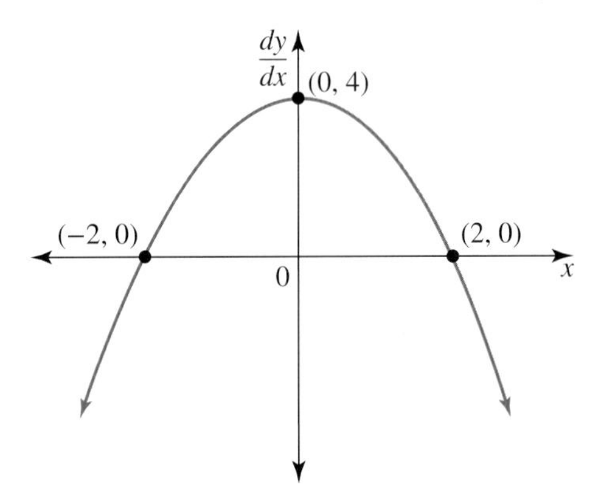

15. For each of the following graphs of $y = f(x)$, draw a sketch of a possible curve for $y = F(x)$ given $F'(x) = f(x)$.

a.

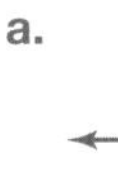

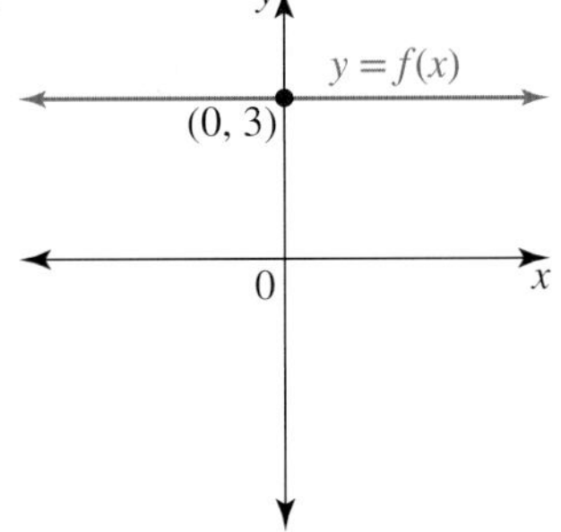

b.

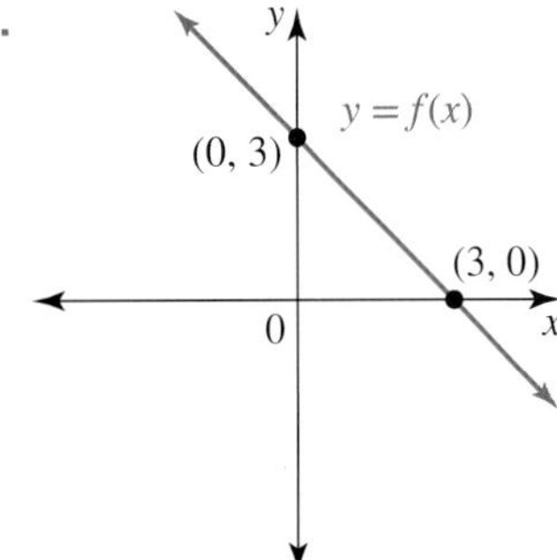

c.

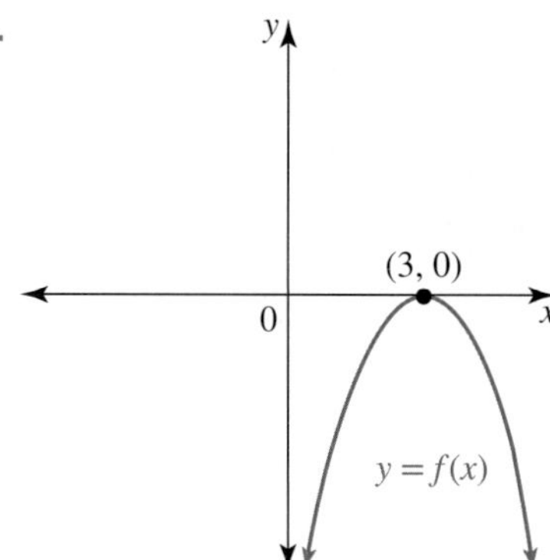

d.

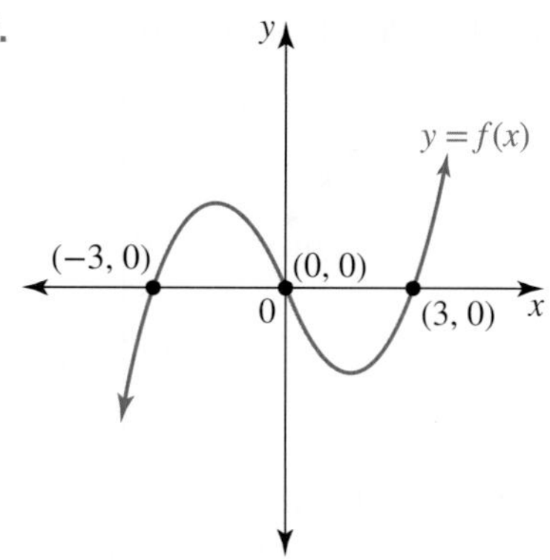

16. The diagram shows the graph of a cubic function $y = f(x)$ with a stationary point of inflection at $(0, 18)$ and passing through the point $(1, 9)$.

   a. Find $\int f(x)\,dx$.
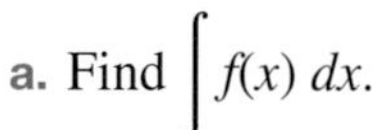

   b. A particular anti-derivative function of $y = f(x)$ passes through the origin. For this particular function, answer the following.
      i. Determine its equation.
      ii. Find the coordinates of its other x-intercept and determine at which of its x-intercepts the graph of this function is steeper.
      iii. Deduce the exact x-coordinate of its turning point and its nature.
      iv. Hence, show the y-coordinate of its turning point can be expressed as $2^p \times 3^q$, specifying the values of p and q.
      v. Draw a sketch graph of this particular anti-derivative function.

17. Using CAS technology, sketch the shape of the graphs of the anti-derivative functions given by the following equations.

   a. $y_1 = \int (x-2)(x+1)\,dx$
   b. $y_2 = \int (x-2)^2(x+1)\,dx$
   c. $y_3 = \int (x-2)^2(x+1)^2\,dx$
   d. Comment on the effect of changing the multiplicity of each factor.

18. Define $f(x) = x^2 - 6x$ and use the calculator to sketch on the same set of axes the graphs of $y = f(x)$, $y = F'(x)$ and $y = \int f(x)\,dx$. Comment on the connections between the three graphs.

13.3 Exam questions

Question 1 (1 mark) TECH-ACTIVE

MC The gradient function of a curve is given by $f'(x) = -6x^2 + 2$. If the curve passes through the point $(-1, 8)$, its equation is

A. $-3x^3 + 2x + 6$ **B.** $-2x^2 + x + 4$ **C.** $-2x^3 + 2x + 8$ **D.** $-x^3 + 2x + 8$ **E.** $-2x^3 + 2x - 1$

Question 2 (3 marks) TECH-FREE

The gradient of a curve is given by $\frac{dy}{dx} = 5x + b$, where b is a constant. Given that the curve has a stationary point at $(4, 6)$, determine its equation.

Question 3 (3 marks) TECH-FREE

The graph of the gradient function of a particular curve is shown. Given that $(8, 9)$ lies on the curve with this gradient, determine the equation of the curve.

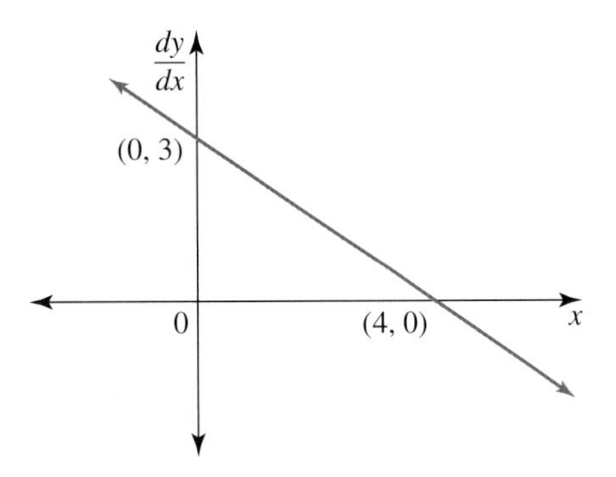

More exam questions are available online.

13.4 Applications of anti-differentiation

LEARNING INTENTION

At the end of this subtopic you should be able to:
- use anti-differentiation of functions to solve real-world problems.

In this section, anti-differentiation is applied in calculations involving functions defined by their rates of change.

13.4.1 The anti-derivative in kinematics

Recalling that velocity, v, is the rate of change of position, x, means that we can now interpret position as the anti-derivative of velocity.

Note: convention states that positive position corresponds to right of the origin and negative position corresponds to the left of the origin.

$$v = \frac{dx}{dt} \Leftrightarrow x = \int v\,dt$$

Further, since acceleration is the rate of change of velocity, it follows that velocity is the anti-derivative of acceleration.

$$a = \frac{dv}{dt} \Leftrightarrow v = \int a\,dt$$

The role of differentiation and anti-differentiation in kinematics is displayed in the diagram below.

Velocity is of particular interest for two reasons:
- The anti-derivative of velocity with respect to time gives position.
- The derivative of velocity with respect to time gives acceleration.

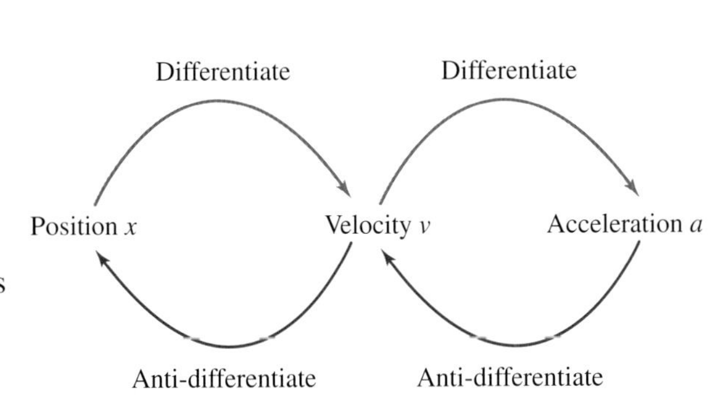

WORKED EXAMPLE 5 Applying anti-derivatives to kinematics

A particle moves in a straight line so that its velocity at time t seconds is given by $v = 3t^2 + 4t - 4,\ t \geq 0$. Initially the particle is 8 metres to the left of a fixed origin.

a. **Calculate the number of seconds it takes for the particle to reach the origin.**

b. **Calculate the particle's acceleration when its velocity is zero.**

THINK	WRITE
a. 1. Calculate the position function from the velocity function.	a. $v = 3t^2 + 4t - 4$ Anti-differentiate the velocity to obtain position. $x = \frac{3t^3}{3} + \frac{4t^2}{2} - 4t + c$ $= t^3 + 2t^2 - 4t + c$
2. Use the initial conditions to calculate the constant of integration.	When $t = 0$, $x = -8$. $\therefore -8 = c$ $\therefore x = t^3 + 2t^2 - 4t - 8$
3. Set up and solve the equation which gives the time the particle is at the required position.	When the particle is at the origin, $x = 0$. $t^3 + 2t^2 - 4t - 8 = 0$ $\therefore t^2(t+2) - 4(t+2) = 0$ $\therefore (t+2)\left(t^2 - 4\right) = 0$ $\therefore (t+2)^2(t-2) = 0$ $\therefore t = -2,\ t = 2$ As $t \geq 0$, reject $t = -2$. $\therefore t = 2$ The particle reaches the origin after 2 seconds.
b. 1. Calculate the acceleration function from the velocity function.	b. $v = 3t^2 + 4t - 4$ Acceleration, $a = \frac{dv}{dt}$. $\therefore a = 6t + 4$.
2. Calculate the time when the velocity is zero.	When velocity is zero, $3t^2 + 4t - 4 = 0$ $\therefore (3t-2)(t+2) = 0$ $\therefore t = \frac{2}{3},\ t = -2$ $t \geq 0$, so $t = \frac{2}{3}$.
3. Calculate the acceleration at the time required.	When $t = \frac{2}{3}$, $a = 6 \times \frac{2}{3} + 4$ $= 8$ The acceleration is 8 m/s^2 when the velocity is zero.

13.4.2 Acceleration and position

To obtain the position from an expression for acceleration will require anti-differentiating twice in order to proceed from a to v to x. The first anti-differentiation operation will give velocity and the second will obtain position from velocity.

This will introduce two constants of integration, one for each step. Where possible, evaluate the first constant using given v–t information before commencing the second anti-differentiation operation. To evaluate the second constant, x–t information will be needed. The notation used for each constant should distinguish between them. For example, the first constant could be written as c_1 and the second as c_2.

WORKED EXAMPLE 6 Calculating velocity and position from acceleration

A particle moves in a straight line so that its acceleration at time t seconds is given by $a = 4 + 6t,\ t \geq 0$. If $v = 30$ and $x = 2$ when $t = 0$, calculate the particle's velocity and position when $t = 1$.

THINK	WRITE
1. Calculate the velocity function from the acceleration function.	$a = 4 + 6t$ Anti-differentiate the acceleration to obtain the velocity. $v = 4t + 3t^2 + c_1$ $= 3t^2 + 4t + c_1$
2. Evaluate the first constant of integration using the given v–t information.	When $t = 0$, $v = 30$. $\Rightarrow 30 = c_1$ $\therefore v = 3t^2 + 4t + 30$
3. Calculate the position function from the velocity function.	Anti-differentiate the velocity to obtain the position. $x = t^3 + 2t^2 + 30t + c_2$
4. Evaluate the second constant of integration using the given x–t information.	When $t = 0$, $x = 2$. $\Rightarrow 2 = c_2$ $\therefore x = t^3 + 2t^2 + 30t + 2$
5. Calculate the velocity at the given time.	$v = 3t^2 + 4t + 30$ When $t = 1$, $v = 3 + 4 + 30$ $= 37$
6. Calculate the position at the given time.	$x = t^3 + 2t^2 + 30t + 2$ When $t = 1$, $x = 1 + 2 + 30 + 2$ $= 35$
7. State the answer.	The particle has a velocity of 37 m/s and its position is 35 metres to the right of the origin when $t = 1$.

13.4.3 Other rates of change

The process of anti-differentiation can be used to solve problems involving other rates of change, not only those involved in kinematics. For example, if the rate at which the area of an oil spill is changing with respect to t is given by $\frac{dA}{dt} = f(x)$, then the area of the oil spill at time t is given by $A = \int f(x)\, dt$, the anti-derivative with respect to t.

WORKED EXAMPLE 7 Determining volume given its rate of change

An ice block with initial volume 36π cm^3 starts to melt. The rate of change of its volume, V cm^3, after t seconds is given by $\frac{dV}{dt} = -0.2$.

a. Use anti-differentiation to express V in terms of t.

b. Calculate, to 1 decimal place, the number of minutes it takes for all of the ice block to melt.

THINK	WRITE
a. 1. Calculate the volume function from the given derivative function.	a. $\frac{dV}{dt} = -0.2$ Anti-differentiate with respect to t: $V = -0.2t + c$
2. Use the initial conditions to evaluate the constant of integration.	When $t = 0$, $V = 36\pi$. $\therefore 36\pi = c$ Therefore, $V = -0.2t + 36\pi$
b. 1. Identify the value of V and calculate the corresponding value of t.	b. When the ice block has melted, $V = 0$. $0 = -0.2t + 36\pi$ $\therefore t = \frac{36\pi}{0.2}$ $= 180\pi$
2. Express the time in the required units and state the answer to the required degree of accuracy. *Note:* The rate of decrease of the volume is constant, so this problem could have been solved without calculus.	The ice block melts after 180π seconds. Divide this by 60 to convert to minutes: 180π seconds is equal to 3π minutes. To 1 decimal place, the time for the ice block to melt is 9.4 minutes.

13.4 Exercise

Technology free

1. a. Starting from the origin, a particle moves in a straight line. Calculate its position at any time t if its velocity is given by $v = 2t + 5$.

   b. Starting from 2 metres to the left of the origin, a particle moves in a straight line. Calculate its position at any time t if its velocity is given by $v = 8t - 3t^2$.

   c. The velocity of a particle moving in a straight line is $v = 2(t - 3)^2$. Form an expression for:

      i. its acceleration

      ii. its position, given the particle is at the origin after 3 seconds.

2. A particle moves in a straight line in such a way that after t seconds its acceleration is given by $a = 4t$. If the particle starts from rest at a position 1 metre to the right of the origin, form the expression for:

   a. its velocity at any time t
   b. its position at any time t.

3. A particle moves in a straight line so that its velocity at time t seconds is given by $v = 6 - 2t, t \geq 0$.

   a. Form an expression for its position, $x(t)$, given that $x(1) = 2$.
   b. Determine the time at which the particle is 6 metres to the right of the origin.

4. A particle moves in a straight line so that its velocity at time t seconds is given by $v = 2\left(3t^2 + 1\right),\ t \geq 0$. When $t = 1$, the particle is 1 metre to the left of a fixed origin. Obtain expressions for the particle's position and acceleration after t seconds.

5. The velocity, v m/s, of a particle moving in a straight line at time t seconds is given by $v = 8t^2 - 20t - 12$, $t \geq 0$. Initially the particle is 54 metres to the right of a fixed origin.

   a. Obtain an expression for the particle's position at time t seconds.
   b. Determine how far from its initial position the particle is after the first second.
   c. Determine the position of the particle when its velocity is zero.

6. A particle starts from rest and moves in a straight line so that its velocity at time t is given by $v = -3t^3, t \geq 0$. When its position is 1 metre to the right of a fixed origin, its velocity is -24 m/s. Determine the initial position of the particle relative to the fixed origin.

7. A particle moving in a straight line has a velocity of 3 m/s and a position of 2 metres from a fixed origin after 1 second. If its acceleration after time t seconds is $a = 8 + 6t$, obtain expressions in terms of t for:

   a. its velocity
   b. its position.

Technology active

8. **WE5** A particle moves in a straight line so that its velocity at time t seconds is given by $v = 3t^2 - 10t - 8$, $t \geq 0$. Initially the particle is 40 metres to the right of a fixed origin.

   a. Calculate the number of seconds it takes for the particle to first reach the origin.
   b. Calculate the particle's acceleration when its velocity is zero.

9. **WE6** A particle moves in a straight line so that its acceleration at time t seconds is given by $a = 8 - 18t$, $t \geq 0$. If $v = 10$ and $x = -2$ when $t = 0$, calculate the particle's velocity and position when $t = 1$.

10. A particle moves in a straight line so that its acceleration, in m/s^2, at time t seconds is given by $a = 9.8$, $t \geq 0$. If $v = x = 0$ when $t = 0$, calculate the particle's position when $t = 5$.

11. The acceleration of a moving object is given by $a = -10$. Initially, the object was at the origin and its initial velocity was 20 m/s.

    a. Determine when and where its velocity is zero.
    b. Determine how many seconds it will take for the object to return to its starting point.

12. Starting from a point 9 metres to the right of a fixed origin, a particle moves in a straight line in such a way that its velocity after t seconds is $v = 6 - 6t,\ t \geq 0$.

    a. Show that the particle moves with constant acceleration.
    b. Determine when and where its velocity is zero.
    c. How far does the particle travel before it reaches the origin?
    d. Calculate the average speed of this particle over the first 3 seconds.
    e. Calculate the average velocity of this particle over the first 3 seconds.

13. When first purchased, the height of a small rubber plant was 50 cm. The rate of growth of its height over the first year after its planting is measured by $h'(t) = 0.2t,\ 0 \le t \le 12$, where h is its height in cm t months after being planted. Calculate its height at the end of the first year after its planting.

14. Pouring boiling water on weeds is a means of keeping unwanted weeds under control. Along the cracks in an asphalt driveway, a weed had grown to cover an area of 90 cm^2, so it was given the boiling water treatment. The rate of change of the area, A cm^2, of the weed t days after the boiling water is poured on it is given by $\frac{dA}{dt} = -18t$.

a. Express the area as a function of t.
b. Determine how many whole days it will take for the weed to be completely removed according to this model.
c. Determine the exact values of t for which the model $\frac{dA}{dt} = -18t$ is valid.

15. Rainwater collects in a puddle on part of the surface of an uneven path. For the time period for which the rain falls, the rate at which the area, A m^2, grows is modelled by $A'(t) = 4 - t$, where t is the number of days of rain.

a. Determine how many days it takes for the area of the puddle to stop increasing.

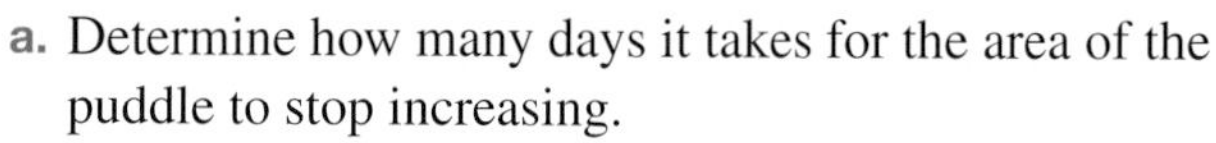

b. Express the area function $A(t)$ in terms of t and state its domain.
c. Determine the greatest area the puddle grows to.
d. On the same set of axes, sketch the graphs of $A'(t)$ versus t and $A(t)$ versus t over an appropriate domain.

16. After t seconds the velocity v m/s of a particle moving in a straight line is given by $v(t) = 3(t-2)(t-4)$, $t \ge 0$.
a. Calculate the particle's initial velocity and initial acceleration.
b. Obtain an expression for $x(t)$, the particle's position from a fixed origin after time t seconds, given that the particle was initially at this origin.
c. Calculate $x(5)$.
d. Determine how far the particle travels during the first 5 seconds.

17. WE7 An ice block with initial volume 4.5π cm^3 starts to melt. The rate of change of its volume, V cm^3, after t seconds is given by $\frac{dV}{dt} = -0.25$.
a. Use anti-differentiation to express V in terms of t.
b. Calculate, to 1 decimal place, the number of seconds it takes for all of the ice block to melt.

18. The diagram shows a velocity–time graph.

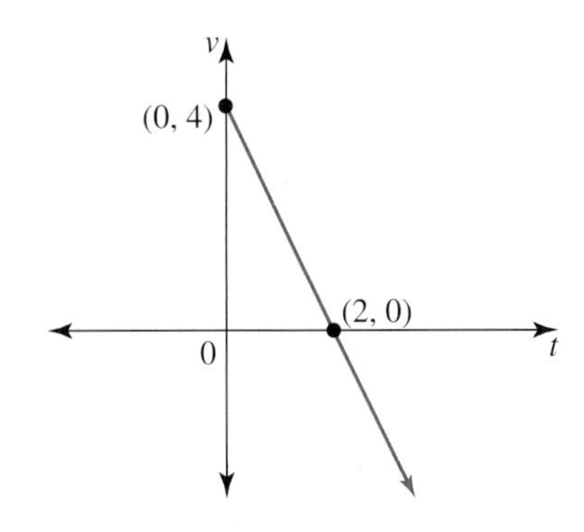

a. Draw a possible position–time graph.
b. Draw the acceleration–time graph.
c. If it is known that the initial position is −4, determine the rule for the position–time graph.
d. Sketch the position–time graph with the rule obtained in part **c**.

19. The rate of growth of a colony of microbes in a laboratory can be modelled by $\frac{dm}{dt} = kt^2$, where m is the number of microbes after t days. Initially there were 20 microbes and after 5 days the population was growing at 300 microbes per day.

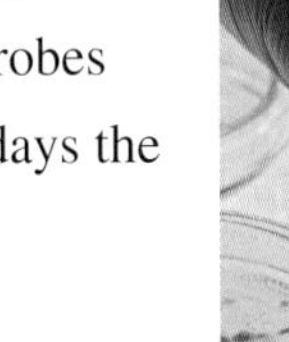

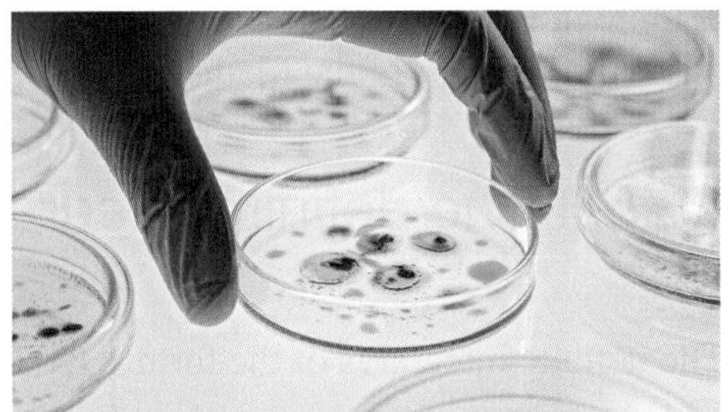

a. Determine the value of k.
b. Express m as a function of t.
c. Hence, calculate the number of days it takes for the population size to reach 6420.

20. The velocity of an object that moves in a straight line is $v = (2t + 1)^4$, $t \geq 0$.
Use CAS technology to:

a. state the acceleration
b. calculate the position, given that initially the object's position was 4.2 metres from a fixed origin
c. find the time and the velocity, to 2 decimal places, when the position is 8.4 metres.

13.4 Exam questions

Question 1 (1 mark) TECH-ACTIVE

MC A particle moves in a straight line so that its acceleration, in m/s^2, at time t seconds is given by $a = 2t + 1$, $t \geq 0$. If $v = 10$ m/s when $t = 0$, then the particle's velocity when $t = 2$ is

A. 0 m/s **B.** 5 m/s **C.** 10 m/s **D.** 16 m/s **E.** 20 m/s

Question 2 (3 marks) TECH-ACTIVE

A particle moves in a straight line so that its acceleration, in m/s^2, at time t seconds is given by $a = 16$, $t \geq 0$. If $t = 0$ when $v = 2$ m/s and $x = 2$ m, calculate the particle's position when $t = 3$.

Question 3 (3 marks) TECH-ACTIVE

While riding home from work, Angela discovers a puncture in a tyre on her bicycle. If the volume of her tyre was 2880 cm^3 when fully pumped, and the air is escaping at a rate of $0.4t$ cm^3/s, where t is the number of seconds since the puncture occurred, determine how many minutes it will take for the tyre to be fully flat (assuming volume goes to 0).

More exam questions are available online.

13.5 The definite integral

LEARNING INTENTION

At the end of this subtopic you should be able to:

- calculate definite integrals
- determine the area bound by a function
- solve problems involving definite integrals.

The indefinite integral has been used as a symbol for the anti-derivative. In this section we will look at the definite integral. We shall learn how to evaluate a definite integral, but only briefly explore its meaning, since this will form part of the Mathematical Methods Units 3 & 4 course.

13.5.1 The definite integral

The definite integral is of the form $\int_a^b f(x)\,dx$; it is quite similar to the indefinite integral, but has the numbers a and b placed on the integral symbol. These numbers are called terminals. They define the end points of the interval or the limits on the values of the variable x over which the integration takes place. Due to their relative positions, a is referred to as the **lower terminal** and b as the **upper terminal**.

For example, $\int_1^3 2x\,dx$ is a definite integral with lower terminal 1 and upper terminal 3, whereas $\int 2x\,dx$ without terminals is an indefinite integral.

In the definite integral the term $f(x)$ is called the **integrand**. It is the expression being integrated with respect to x.

Calculation of the definite integral

The definite integral results in a numerical value when it is evaluated.

It is evaluated by the calculation: $\int_a^b f(x)\,dx = F(b) - F(a)$, where $F(x)$ is an anti-derivative of $f(x)$.

The calculation takes two steps:

- Anti-differentiate $f(x)$ to obtain $F(x)$.
- Substitute the terminals in $F(x)$ and carry out the subtraction calculation.

This calculation is commonly written as follows.

Definite integrals

$$\begin{aligned}\int_a^b f(x)\,dx &= [F(x)]_a^b \\ &= F(b) - F(a)\end{aligned}$$

For example, to evaluate $\int_1^3 2x\,dx$, we write:

$$\begin{aligned}\int_1^3 2x\,dx &= \left[x^2\right]_1^3 \\ &= (3)^2 - (1)^2 \\ &= 8\end{aligned}$$

Note that only 'an' anti-derivative is needed in the calculation of the definite integral. Had the constant of integration been included in the calculation, these terms would cancel out, as illustrated by the following:

$$\begin{aligned}\int_1^3 2x\,dx &= \left[x^2 + c\right]_1^3 \\ &= \left((3)^2 + c\right) - \left((1)^2 + c\right) \\ &= 9 + \not{c} - 1 - \not{c} \\ &= 8\end{aligned}$$

The value of a definite integral may be a positive, zero or negative real number. It is not a family of functions as obtained from an indefinite integral.

WORKED EXAMPLE 8 Evaluating definite integrals

Evaluate $\int_{-1}^{2} (3x^2 + 1)\,dx$.

THINK	WRITE
1. Calculate an anti-derivative of the integrand.	$\int_{-1}^{2} (3x^2 + 1)\,dx = \left[x^3 + x\right]_{-1}^{2}$
2. Substitute the upper and then the lower terminal in place of x and subtract the two expressions.	$= \left(2^3 + 2\right) - \left((-1)^3 + (-1)\right)$
3. Evaluate the expression.	$= (8 + 2) - (-1 - 1)$ $= (10) - (-2)$ $= 12$
4. State the answer.	$\int_{-1}^{2} (3x^2 + 1)\,dx = 12$

TI | THINK	DISPLAY/WRITE	CASIO | THINK	DISPLAY/WRITE
d. 1. On a Calculator page, press MENU and select: 4. Calculus 3. Integral Complete the entry line as: $\int_{-1}^{2}(3x^2+1)\,dx$ Then press ENTER.	 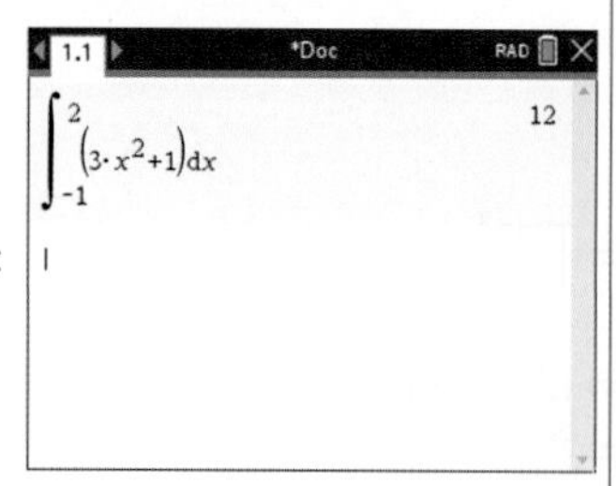	**d. 1.** On a Main screen, complete the entry line as: $\int_{-1}^{2}(3x^2+1)\,dx$ Then press EXE.	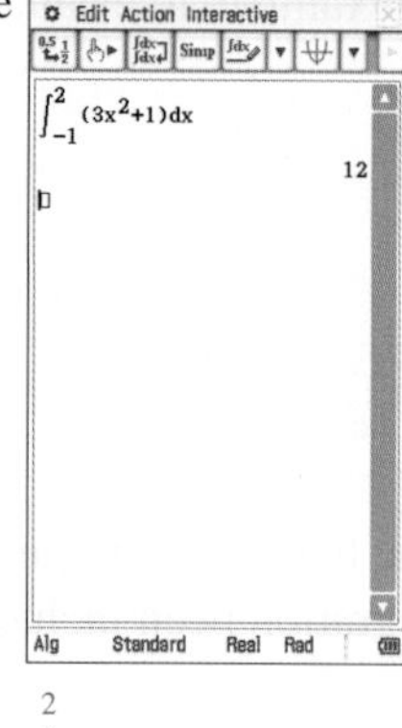
2. The answer appears on the screen.	$\int_{-1}^{2}(3x^2+1)\,dx=12$	**2.** The answer appears on the screen.	$\int_{-1}^{2}(3x^2+1)\,dx=12$

13.5.2 Integration and area

Areas of geometric shapes such as rectangles, triangles and trapeziums can be used to approximate areas enclosed by edges that are not straight. Leibniz invented the method for calculating the exact measure of an area enclosed by a curve, thereby establishing the branch of calculus known as integral calculus. His method bears some similarity to a method used in the geometry studies of Archimedes around 230 BC.

Leibniz's approach involved summing a large number of very small areas and then calculating the limiting sum, a process known as **integration**. Like differential calculus, integral calculus is based on the concept of a limit.

To calculate the area bounded by a curve $y=f(x)$ and the x-axis between $x=1$ and $x=3$, the area could be approximated by a set of rectangles as illustrated in the diagram. Here, 8 rectangles, each of width 0.25 units, have been constructed.

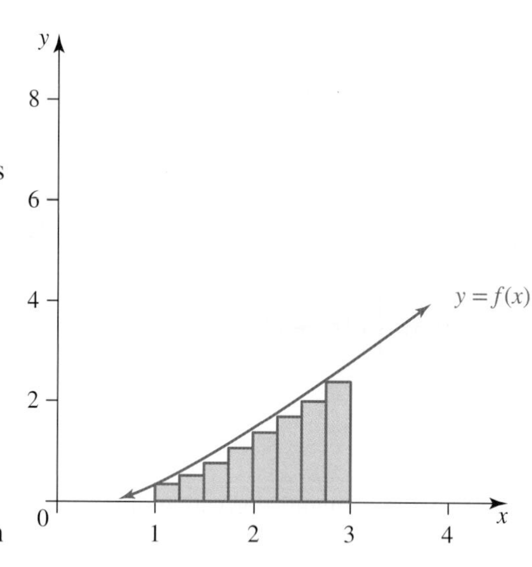

The larger the number of rectangles, the better the accuracy of the approximation to the actual area; also, the larger the number, the smaller the widths of the rectangles.

To calculate the area between $x=a$ and $x=b$, Leibniz partitioned (divided) the interval $[a, b]$ into a large number of strips of very small width. A typical strip would have width δx and area δA, with the total area approximated by the sum of the areas of such strips.

For δx, the approximate area of a typical strip is $\delta A \approx y\delta x$, so the total area A is approximated from the sum of these areas. This is written as $A \approx \sum_{x=a}^{b} y\,\delta x$. The approximation improves as $\delta x \to 0$ and the number of strips increases.

The actual area is calculated as a limit — the limiting sum as $\delta x \to 0$; therefore, $A = \lim_{\delta x \to 0} \sum_{x=a}^{b} y\,\delta x$.

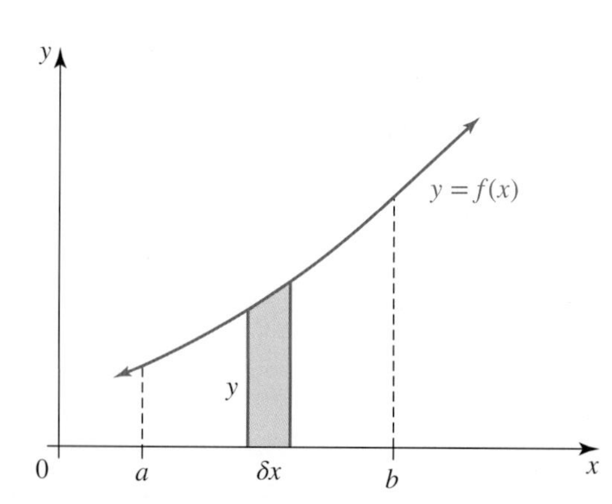

Leibniz chose to write the symbol for the limiting sum with the long capital 'S' used in his time. That symbol is the now familiar integral sign. This means that the area bounded by the curve, the x-axis and $x = a, x = b$ is $A = \int_a^b y\,dx$ or $A = \int_a^b f(x)\,dx$. The process of evaluating this integral or limiting sum expression is called integration.

The fundamental theorem of calculus links together the process of anti-differentiation with the process of integration. The full proof of this theorem is left to Units 3 and 4.

Signed area

As we have seen, the definite integral can result in a positive, negative or zero value. Area, however, can only be positive.

If $f(x) > 0$, the graph of $y = f(x)$ lies above the x-axis and $\int_a^b f(x)\,dx$ will be positive. Its value will be a measure of the area bounded by the curve $y = f(x)$ and the x-axis between $x = a$ and $x = b$.

However, if $f(x) < 0$, the graph of $y = f(x)$ is below the x-axis and $\int_a^b f(x)\,dx$ will be negative. Its value gives a **signed area**. By ignoring the negative sign, the value of the integral will still measure the actual area bounded by the curve $y = f(x)$ and the x-axis between $x = a$ and $x = b$.

If over the interval $[a, b]$ $f(x)$ is partly positive and partly negative, then $\int_a^b f(x)\,dx$ could be positive or negative, or even zero, depending on which of the positive or negative signed areas is the 'larger'.

This adds a complexity to using the signed area measure of the definite integral to calculate area. Perhaps you can already think of a way around this situation. However, for now, another treat awaiting you in Units 3 and 4 will be to explore signed areas.

WORKED EXAMPLE 9 Calculating the area under a function

The area bounded by the line $y = 2x$, the x-axis and $x = 0$, $x = 4$ is illustrated in the diagram.

a. **Calculate the area using the formula for the area of a triangle.**

b. **Write down the definite integral that represents the measure of this area.**

c. **Hence, calculate the area using calculus.**

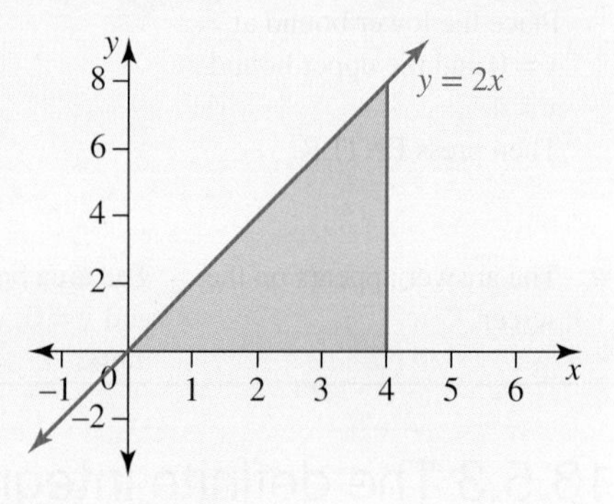

THINK	WRITE
a. Calculate the area using the formula for the area of a triangle.	**a.** The base is 4 units; the height is 8 units. Area of triangle: $A = \frac{1}{2}bh$ $= \frac{1}{2}(4)(8)$ $= 16$ The area is 16 square units.
b. State the definite integral that gives the area.	**b.** The definite integral $\int_a^b f(x)\,dx$ gives the area. For this area $f(x) = 2x$, $a = 0$ and $b = 4$. $\int_0^4 2x\,dx$ gives the area measure.
c. 1. Evaluate the definite integral. *Note:* There are no units, just a real number for the value of a definite integral.	**c.** $\int_0^4 2x\,dx = \left[x^2\right]_0^4$ $= 4^2 - 0^2$ $= 16$
2. State the area using appropriate units.	Therefore, the area is 16 square units.

TI | THINK	DISPLAY/WRITE	CASIO | THINK	DISPLAY/WRITE
1. On a Graphs page, complete the entry line as: $f1(x) = 2x$. Then press ENTER. Then press MENU and select: 6. Analyze Graph 7. Integral Place the lower bound at $x = 0$ and the upper bound at $x = 4$. Then press ENTER.	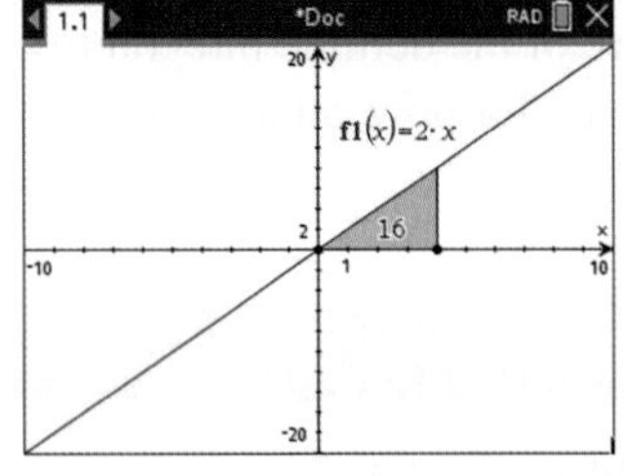	**1.** On a Graphs & Table screen, complete the entry line as: $y1 = 2x$ Then press: • Analysis • G-Solve • Integral • $\int dx$ Place the lower bound at $x = 0$ and the upper bound at $x = 4$. Then press EXE.	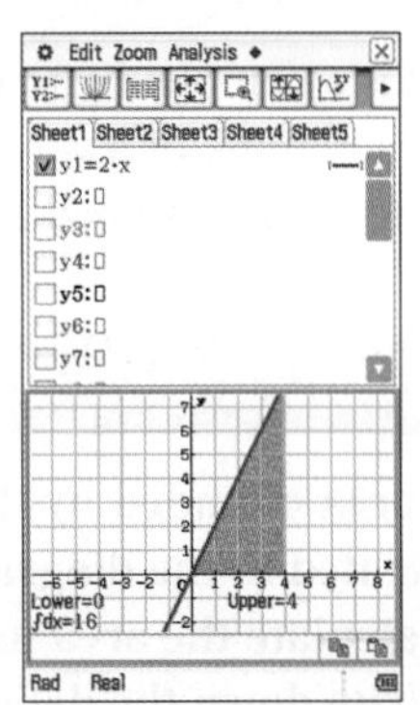
2. The answer appears on the screen.	The area bounded by the line and $x = 0$, $x = 4$ is 16 square units.	**2.** The answer appears on the screen.	The area bounded by the line and $x = 0$, $x = 4$ is 16 square units.

13.5.3 The definite integral in kinematics

A particle that travels at a constant velocity of 2 m/s will travel a distance of 10 metres in 5 seconds. For this motion, the velocity–time graph is a horizontal line. Looking at the rectangular area under the velocity–time graph bounded by the horizontal axis between $t = 0$ and $t = 5$ shows that its area measure is also equal to 10.

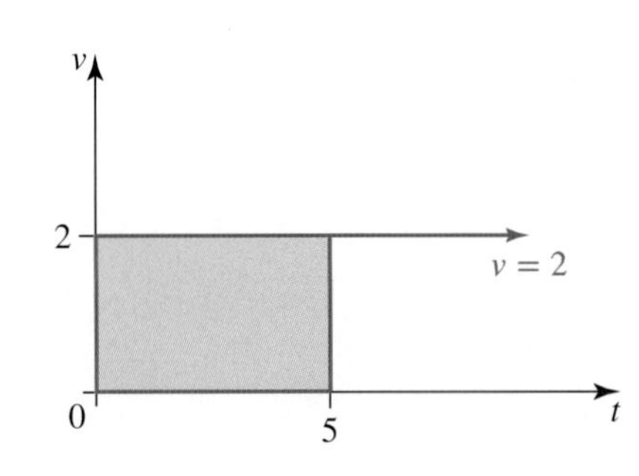

- The area under the velocity–time graph gives the measure of the distance travelled by a particle.
- For a positive signed area, the definite integral $\int_{t_1}^{t_2} v\,dt$ gives the distance travelled over the time interval $t \in [t_1, t_2]$.

The velocity–time graph is the most important of the motion graphs as it gives the velocity at any time.
- The gradient of a tangent to the graph gives the instantaneous acceleration at that time.
- The area under the graph gives the distance travelled.

WORKED EXAMPLE 10 Determining distance travelled

The velocity of a particle moving in a straight line is given by $v = 3t^2 + 1$, $t \geq 0$. Assume distance is in metres and time is in seconds.

a. **Give the particle's velocity and acceleration when $t = 1$.**

b. i. **Sketch the velocity–time graph and shade the area that represents the distance the particle travels over the interval $t \in [1, 3]$.**

ii. **Use a definite integral to calculate the distance.**

THINK	WRITE
a. 1. Calculate the velocity at the given time.	a. $v = 3t^2 + 1$ When $t = 1$, $v = 3(1)^2 + 1$ $= 4$ The velocity is 4 m/s.
2. Obtain the acceleration from the velocity function and evaluate it at the given time.	$a = \frac{dv}{dt}$ $= 6t$ When $t = 1$, $a = 6$. The acceleration is 6 m/s^2.
b. i. Sketch the v–t graph and shade the area required.	b. i. $v = 3t^2 + 1$, $t \geq 0$ The v–t graph is part of the parabola that has turning point (0, 1) and passes through the point (1, 4). The area under the graph bounded by the t-axis and $t = 1$, $t = 3$ represents the distance. 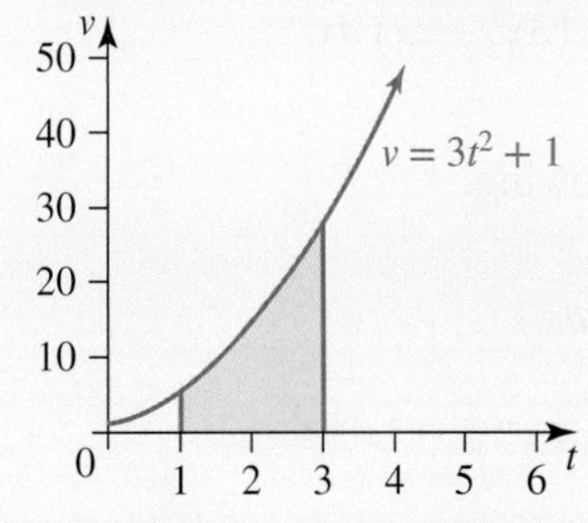
ii. 1. State the definite integral that gives the measure of the area shaded. *Note:* The integrand is a function of t and is to be integrated with respect to t.	ii. The area measure is given by $\int_1^3 (3t^2 + 1)dt$.

2. Evaluate the definite integral.	$\int_1^3 (3t^2+1)dt = [t^3+t]_1^3$ $= (3^3+3)-(1^3+1)$ $= 28$
3. State the distance travelled.	The distance travelled by the particle over the time interval [1, 3] is 28 metres.

Resources

Interactivity Kinematics (int-5964)

13.5 Exercise

Students, these questions are even better in jacPLUS

Receive immediate feedback and access sample responses

Access additional questions

Track your results and progress

Find all this and MORE in jacPLUS

Technology free

1. Evaluate the following integrals.

a. $\int_1^4 2x\,dx$

b. $\int_0^2 3x^2\,dx$

c. $\int_{-5}^{-2} 7\,dx$

d. $\int_{-2}^{2} (5x+4)\,dx$

e. $\int_5^{10} (6-4x)\,dx$

f. $\int_{-\frac{1}{2}}^{0} (1+5x+3x^2)\,dx$

2. **WE8** Evaluate $\int_0^3 \left(3x^2-2x\right)\,dx$.

3. Evaluate the following.

a. $2\int_0^1 (x(1-x))\,dx$

b. $\int_0^3 (3-x)\,dx - \int_3^4 (3-x)\,dx$

c. $\int_{-2}^{1} \left(1-y^3\right)dy$

d. $\int_{-2}^{-1} (t(3t+2))\,dt$

4. Evaluate the following.

a. $\int_{-2}^{2} (x-2)(x+2)\,dx$

b. $\int_{-1}^{1} x^3\,dx$

5. Evaluate each of the following.

a. $\int_{-2}^{2} 5x^4\, dx$

b. $\int_{0}^{2} (7-2x^3)\, dx$

c. $\int_{1}^{3} (6x^2+5x-1)\, dx$

d. $\int_{-3}^{0} 12x^2(x-1)\, dx$

e. $\int_{-4}^{-2} (x+4)^2\, dx$

f. $\int_{-\frac{1}{2}}^{\frac{1}{2}} (x+1)(x^2-x)\, dx$

6. a. Calculate the value of a so that $\int_{0}^{1} (ax-2)\, dx = 7$.

b. Calculate the value of b so that $\int_{-2}^{1} (b+8x)\, dx = 0$.

c. Calculate the value of k so that $\int_{3}^{k} 2\, dx = 14$.

d. Calculate the value of n so that $\int_{-n}^{n} 9x^2\, dx = 48$.

7. Show, by calculation, that the following statements are true.

a. $\int_{1}^{2} (20x+15)\, dx = 5\int_{1}^{2} (4x+3)\, dx$

b. $\int_{-1}^{2} (x^2+2)\, dx = \int_{-1}^{2} x^2\, dx + \int_{-1}^{2} 2\, dx$

c. $\int_{1}^{3} 3x^2\, dx = \int_{1}^{3} 3t^2 dt$

d. $\int_{a}^{a} 3x^2\, dx = 0$

e. $\int_{b}^{a} 3x^2\, dx = -\int_{a}^{b} 3x^2\, dx$

f. $\int_{-a}^{a} dx = 2a$

8. Determine the value of n so that $\int_{4}^{n} 3\, dx = 9$.

9. Express the measured areas in each of the following diagrams in terms of a definite integral and calculate the area using both integration and a known formula for the area of the geometric shape.

a. The triangular area bounded by the line $y = 3x + 6$, the x-axis and $x = -2$, $x = 2$ as illustrated in the diagram

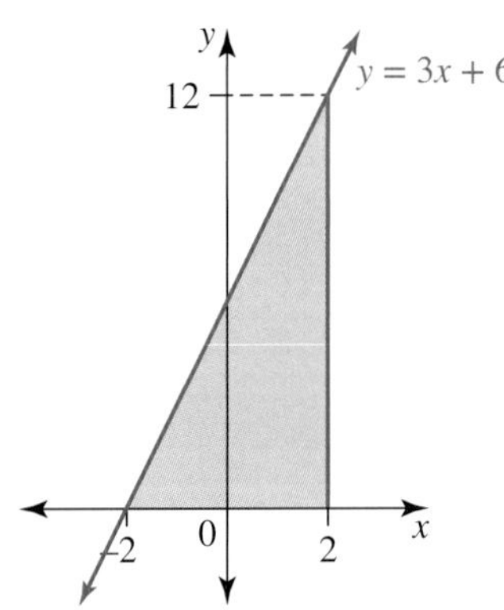

b. The rectangular area bounded by the line $y = 1$, the x-axis and $x = -6$, $x = -2$ as illustrated in the diagram

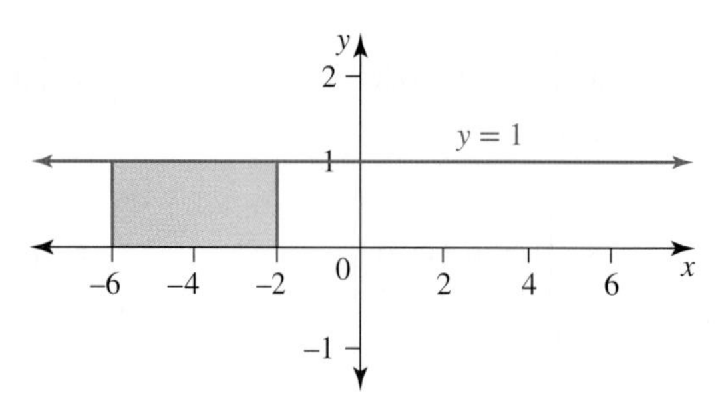

c. The trapezoidal area bounded by the line $y = 4 - 3x$, the coordinate axes and $x = -2$ as illustrated in the diagram

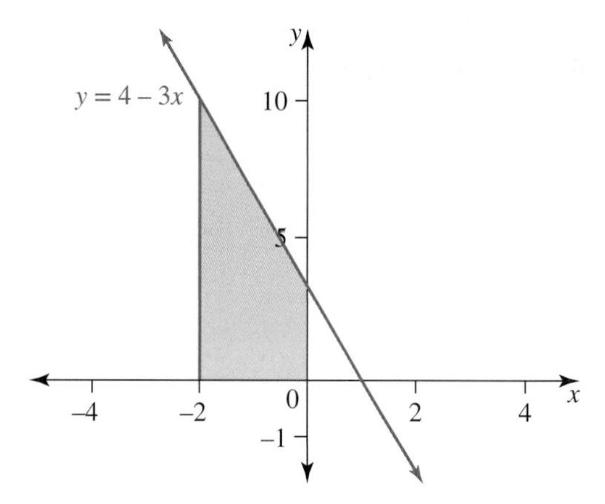

10. **WE9** The area bounded by the line $y = 0.75x$, the x-axis and $x = 0, x = 4$ is illustrated in the diagram.

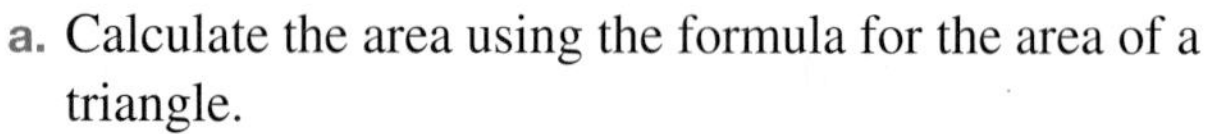

a. Calculate the area using the formula for the area of a triangle.
b. Write down the definite integral that represents the measure of this area.
c. Hence, calculate the area using calculus.

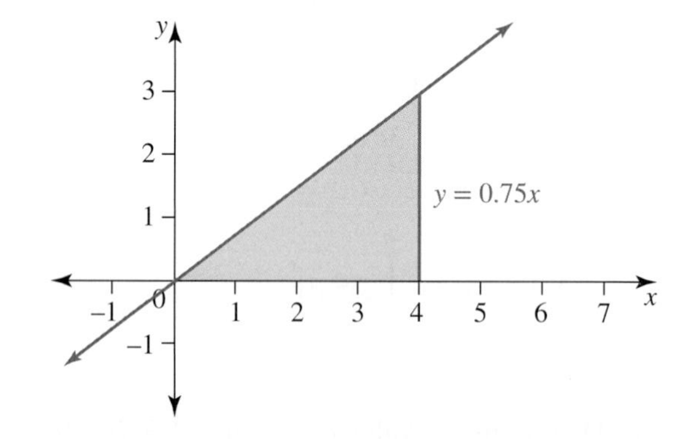

11. The area bounded by the curve $y = 16 - x^2$, the x-axis and $x = 1$, $x = 3$ is illustrated in the diagram.

a. Write down the definite integral that represents the measure of this area.
b. Hence, calculate the area.

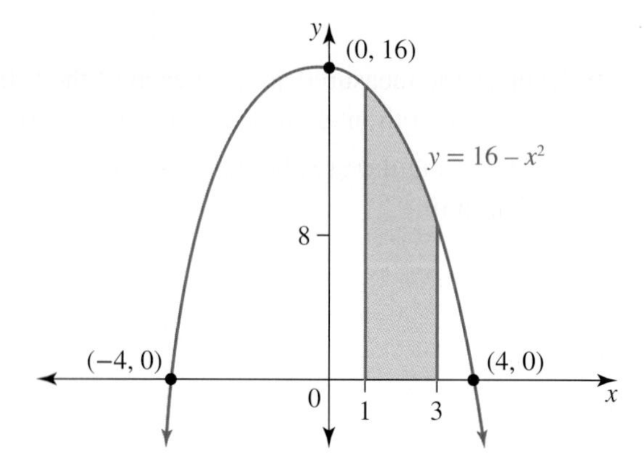

12. For each of the following areas:
    i. write a definite integral, the value of which gives the area measure
    ii. calculate the area.
    a. The area bounded by the curve $y = x^2 + 1$, the x-axis and $x = -1, x = 1$ as illustrated in the diagram

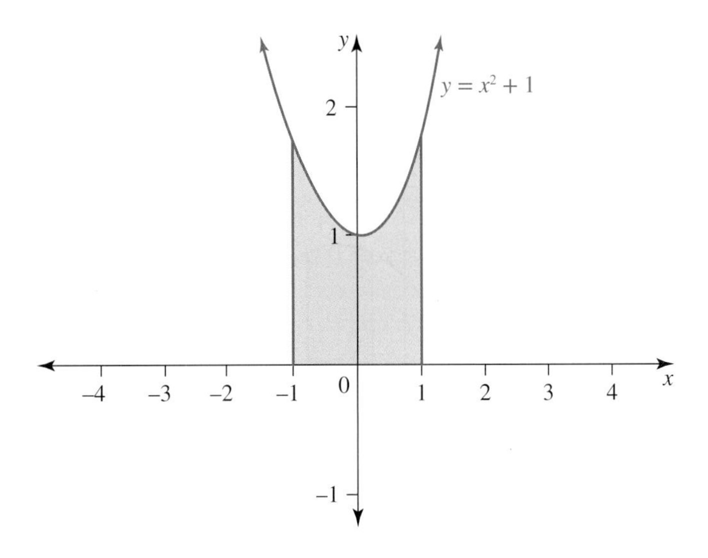

    b. The area bounded by the curve $y = 1 - x^3$, the x-axis and $x = -2$ as illustrated in the diagram

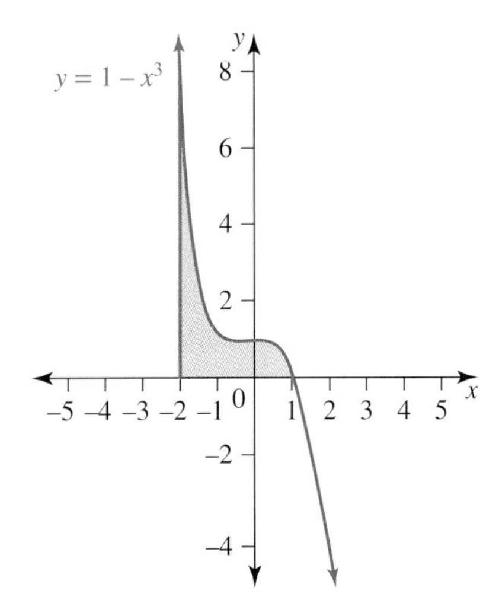

13. Sketch a graph to show each of the areas represented by the given definite integrals and then calculate the area.

    a. $\int_{0}^{3} 4x^2\,dx$ b. $\int_{-1}^{1} 1 - x^2\,dx$ c. $\int_{-2}^{4} dx$

Technology active

14. **WE10** The velocity of a particle moving in a straight line is given by $v = t^3 + 2,\ t \geq 0$. Assume distance is in metres and time is in seconds.
    a. Give the particle's velocity and acceleration when $t = 1$.
    b. i. Sketch the velocity–time graph and shade the area that represents the distance the particle travels over the interval $t \in [2, 4]$.
       ii. Use a definite integral to calculate the distance.

15. The velocity, v m/s, of a particle moving in a straight line after t seconds is given by $v = 3t^2 - 2t + 5, t \geq 0$.

a. Show that velocity is always positive.

b. Calculate the distance the particle travels in the first 2 seconds using:

i. a definite integral

ii. anti-differentiation.

16. Calculate the area bounded by the curve $y = (1 + x)(3 - x)$, the x-axis and $x = 0.5, x = 1.5$ as illustrated in the diagram.

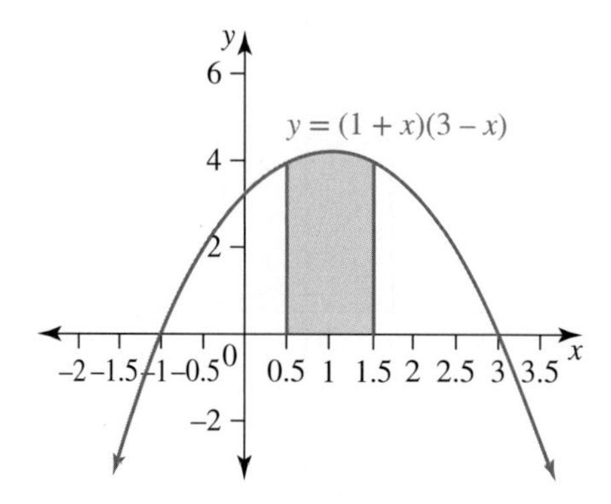

17. A particle starts from rest at the origin and moves in a straight line so that after t seconds its velocity is given by $v = 10t - 5t^2, \ t \geq 0$.

a. Calculate an expression for the position at time t.

b. Calculate how far the particle will have travelled from its starting point when it is next at rest.

c. Draw the velocity–time graph for $t \in [0, 2]$.

d. Use a definite integral to calculate the area enclosed by the velocity–time graph and the horizontal axis for $t \in [0, 2]$.

e. Explain what the area in part **d** measures.

18. An athlete training to compete in a marathon race runs along a straight road with velocity v km/h. The velocity–time graph over a 5-hour period for this athlete is shown.

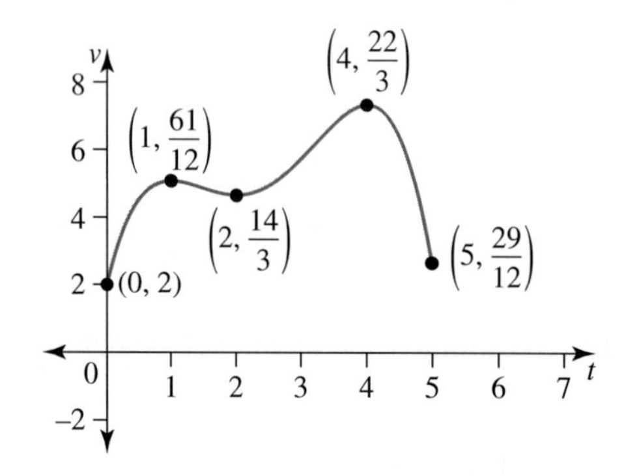

a. Calculate the athlete's highest and lowest speeds over the 5-hour period.

b. Determine the times at which the athlete's acceleration becomes zero.

c. Given the equation of the graph shown is $v = 2 + 8t - 7t^2 + \frac{7t^3}{3} - \frac{t^4}{4}$, use a definite integral to calculate the distance the athlete ran during this 5-hour training period.

19. Write down the values of:

a. $\int_{-3}^{4} (2-3x+x^2)\, dx$

b. $\int_{4}^{-3} (2-3x+x^2)\, dx$

c. Explain the relationship between parts **a** and **b**.

20. a. Use the calculator to obtain $\int_{-2}^{3} (x+2)(x-3)\, dx$.

b. Draw the graph of $y=(x+2)(x-3)$.

c. Calculate the area enclosed between the graph of $y=(x+2)(x-3)$ and the x-axis.

d. Explain why the answers to part **a** and part **c** differ. Write down a definite integral that does give the area.

13.5 Exam questions

Question 1 (4 marks) TECH-FREE

The velocity of a particle moving in a straight line is given by $v=t^3-2,\ t\geq 0$.

a. Find the particle's velocity and acceleration when $t=2$. **(2 marks)**

b. Sketch the velocity–time graph and shade the area that represents the distance the particle travels over the interval $t\in[2,3]$. **(1 mark)**

c. Use a definite integral to calculate the distance. **(1 mark)**

Assume units for distance are in metres and time in seconds.

Question 2 (2 marks) TECH-FREE

Sketch a graph to show the area represented by $\int_{-2}^{2} 3x^2\, dx$ and calculate the area.

Question 3 (1 mark) TECH-FREE

MC Evaluate $\int_{0}^{3} (x+2)^2\, dx$.

A. 4
B. 16
C. 24
D. 30
E. 39

More exam questions are available online.

13.6 Review

13.6.1 Summary

doc-37029

Hey students! Now that it's time to revise this topic, go online to:

 Access the topic summary

 Review your results

 Watch teacher-led videos

 Practise exam questions

Find all this and MORE in jacPLUS

13.6 Exercise

Technology free: short answer

1. Calculate the following.

a. $\int 2\left(1-5x-3x^4\right)\,dx$

b. $\int 2x(x+3)(x-3)\,dx$

c. $\int \dfrac{2x^3-5x^4}{8x}\,dx$

d. $\int \left(x^2+1\right)^3\,dx$

2. If $f'(x)=\dfrac{3x}{2}-4$ and $f(2)=10$, obtain the value of $f(4)$.

3. A curve contains the point (24, 12) and has a gradient at any point (x, y) that is equal to $1+\dfrac{1}{12}x$.

a. Determine the equation of the curve.

b. Calculate the equation of the tangent to the curve at its y-intercept.

4. Starting from rest, a particle moves in a straight line with its velocity, in m/s, at time t seconds given by $v=2t(t-3)$.

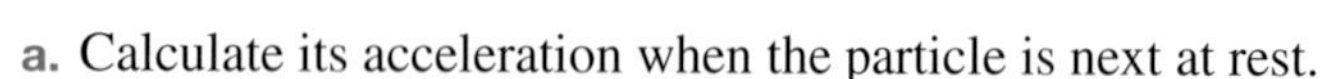

a. Calculate its acceleration when the particle is next at rest.

b. Form an expression for its position if initially the particle was 3 metres to the right of the origin.

c. Determine how far it is from its starting point when it next comes to rest.

5. Evaluate the following.

a. $\int_1^3 (3-2x)\,dx$

b. $\int_0^2 (4+x)(2-x)\,dx$

6. The graph of $y=f(x)$ is shown in the diagram. Sketch a possible graph of the anti-derivative function of this function.

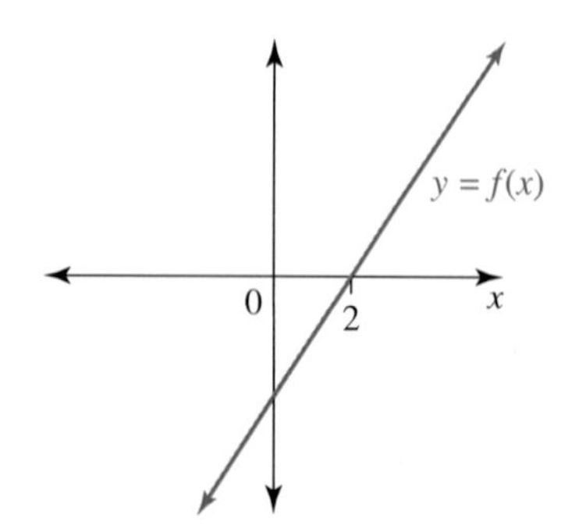

Technology active: multiple choice

7. **MC** If $f'(x) = x(2x + 5)$, then $f(x)$ equals:

A. $2x^2 + 5x$ B. $4x + 5$ C. $\frac{2x^3}{3} + \frac{5x^2}{2} + c$ D. $6x^3 + 10x^2 + c$ E. $\frac{x^2}{2}\left(x^2 + 5x\right) + c$

8. **MC** An anti-derivative of $2x^3$ with respect to x could be:

A. $6x^2 + c$ B. $6x^4 + c$ C. $\frac{2}{3}x^2 + c$ D. $\frac{2}{3}x^4 + c$ E. $\frac{1}{2}x^4 + c$

9. **MC** $\int (3x - 2)^2\, dx$ equals:

A. $\frac{3}{2}x^2 - 2x + c$ B. $\frac{(3x-2)^3}{3} + c$ C. $3x^3 + 4x + c$
D. $3x^3 - 4x + c$ E. $3x^3 - 6x^2 + 4x + c$

10. **MC** If $f(x) = \frac{3x^4 - 5x^3}{2x^2}$, $x \neq 0$, then $F(x)$, where $F'(x) = f(x)$, would equal:

A. $3x - \frac{5}{2} + c$ B. $\frac{3}{2}x^3 - \frac{5}{2}x^2 + c$ C. $\frac{x^3}{3} - \frac{5x^2}{4} + c$ D. $\frac{x^3}{2} - \frac{5x^2}{4} + c$ E. $3x^2 - \frac{15}{4}x + c$

11. **MC** Given $\frac{dy}{dx} = 4x^3$ and $y = 2$ when $x = 1$, then the value of y when $x = 2$ would be:

A. 15 B. 16 C. 17 D. 32 E. 38

12. **MC** A particle, initially 1 unit to the left of a fixed origin, moves in a straight line so that its velocity at time t is given by $v = 3t^2 + 2t - 1$, $t \geq 0$. Its position relative to the origin at time t is:

A. $x = t^3 + t^2 - 1$ B. $x = t^3 + t^2 - t - 1$ C. $x = t^3 + t^2 - t + 1$
D. $x = 6t + 1$ E. $x = 3t^2 + 2t$

13. **MC** The graph of $y = f(x)$ is shown in the diagram. Given $F'(x) = f(x)$, determine which of the following is a correct statement about the graph of $y = F(x)$.

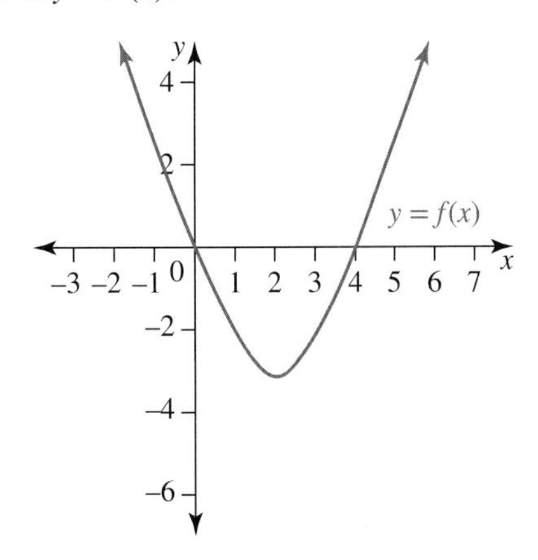

A. The graph of $y = F(x)$ has a stationary point when $x = 2$.
B. The graph of $y = F(x)$ always has an x-intercept at $(2, 0)$.
C. The graph of $y = F(x)$ has a local minimum turning point when $x = 0$ and a local maximum turning point when $x = 4$.
D. The graph of $y = F(x)$ has a local maximum turning point when $x = 0$ and a local minimum turning point when $x = 4$.
E. The graph of $y = F(x)$ has a linear shape with a positive gradient.

14. **MC** The value of $\int_{0}^{1} 4x^7\,dx$ is:

A. 0.5 B. 2 C. 4 D. 28 E. 32

15. **MC** The value of $\int_{1}^{2} \left(4 - 3x^2\right)dx$ is:

A. 9 B. 7 C. 3 D. −3 E. −7

16. **MC** State which of the following integrals represents the measure of the area enclosed by the graph of $y = (4 - x)(1 + x)$ and the x-axis.

A. $\int_{4}^{1} (4 - x)(1 + x)\,dx$ B. $\int_{1}^{4} (4 - x)(1 + x)\,dx$ C. $\int_{-1}^{4} (4 - x)(1 + x)\,dx$

D. $\int_{-4}^{1} (4 - x)(1 + x)\,dx$ E. $\int_{-4}^{-1} (4 - x)(1 + x)\,dx$

Technology active: extended response

17. a. A particle, P, moves in a straight line with a constant acceleration of 4 m/s^2. Obtain expressions for its velocity and position in terms of time t given that the particle starts from the position $x = -6$ with an initial velocity of 3 m/s.

b. A second particle Q starts from the origin at the same time as P sets off and moves on the same straight line with constant velocity of 7 m/s.

i. Determine how far apart P and Q are at the time when they are moving with the same velocity.

ii. Determine when and where P overtakes Q.

c. After Q is overtaken, its velocity reduces uniformly, reaching zero 6 seconds after P overtook it.

i. Draw a graph of the velocity–time graph of Q during its travel from $t = 0$ to when it comes to rest.

ii. Hence, or otherwise, calculate the total distance Q travelled.

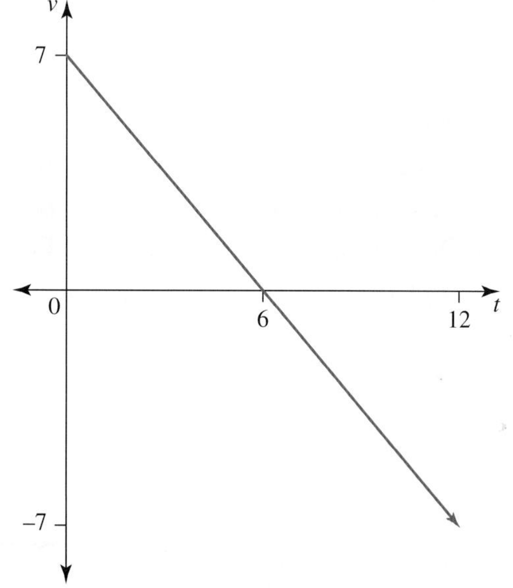

d. The velocity–time graph of a third particle R is shown.

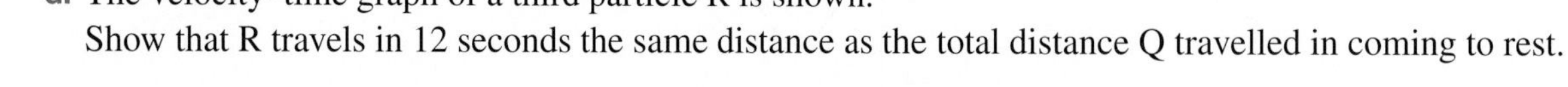
Show that R travels in 12 seconds the same distance as the total distance Q travelled in coming to rest.

18. A curve $y = f(x)$ has a turning point at $(-2, 6)$ and a y-intercept at $(0, 4)$. Its gradient at any point is given by $f'(x) = ax + b$.

a. Determine the values of a and b.

b. State the equation of the curve $y = f(x)$.

c. Calculate the exact x-coordinates of the stationary points on the curve $y = F(x)$ where $f(x) = F'(x)$.

d. If the curve $y = F(x)$ also has its y-intercept at $(0,\ 4)$, determine its equation.

19. a. If $\int f(x)\,dx = F(x) + c$, show that:

i. $\int_a^b f(x)\,dx = -\int_b^a f(x)\,dx$

ii. $\int_a^b f(x)\,dx + \int_b^c f(x)\,dx = \int_a^c f(x)\,dx$

iii. $\int_a^a f(x)\,dx = 0$

b. Given $\int_{-1}^{2} f(x)\,dx = 5$, use the results in part **a** and the linearity properties of integration to calculate the values of:

i. $\int_2^{-1} f(x)\,dx$

ii. $\int_{-1}^{0} f(x)\,dx + \int_0^2 f(x)\,dx$

iii. $\int_{-1}^{2} 3f(x)\,dx$

iv. $\int_{-1}^{2} (1 - f(x))\,dx$

20. The graph of $y = \frac{1}{2}x^2$ is shown. The shaded area is bounded by the curve, the x-axis and $x = -4$.

a. i. State a definite integral, the value of which gives the measure of the shaded area.

ii. If the shaded area is A square units, calculate the value of A.

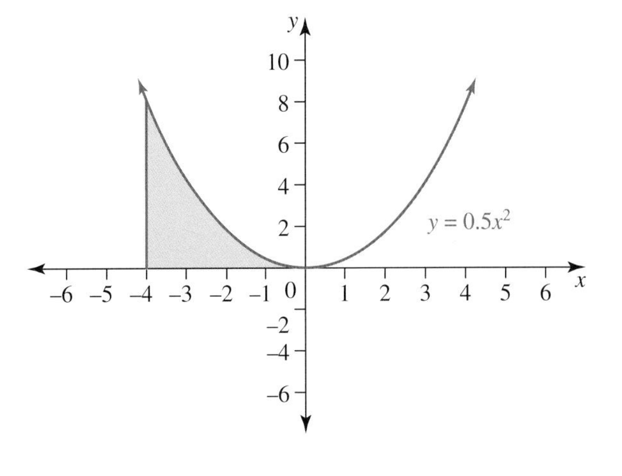

b. The area represents the side cross-section of a waterslide at a swimming pool, with measurements in metres and the water surface the positive x-axis. Determine the height at which the slide starts.

c. The enjoyment factor E is measured on a scale from 1 to 10 and changes as a person slides down the slide and into the water. The enjoyment factor of Sam, who feels apprehensive at the start, is initially 4. If the rate of change of Sam's enjoyment factor is $\frac{dE}{dt} = t$ and it takes her 3 seconds to enter the water:

i. calculate her enjoyment factor as she reached the end of the slide and entered the water

ii. determine what her initial enjoyment factor would need to be for E to reach 10 at the end of the slide.

d. For young children, the starting height of the slide is thought to be too high. A modification to the slide will allow the children to start at the point where $x = p$ and where the area enclosed by the curve $y = \frac{1}{2}x^2$, the x-axis, $x = -4$ and $x = p$ is $\frac{1}{2}A$ units^2, with A the value calculated in part **a**.

i. Draw a diagram to show a possible area and position for $x = p$ and determine the exact value of p.

ii. Express the slide's height at this starting point for children as a power of 2 and give its value to the nearest metre.

13.6 Exam questions

Question 1 (2 marks) TECH-FREE

Calculate the anti-derivative (primitive function) of $(x-2)(2x+5)(x+2)$.

Question 2 (1 mark) TECH-ACTIVE

MC The gradient of a curve is directly proportional to the square of x, and, at the point $(3, 9)$ on the curve, the gradient is -18. The constant of proportionality is

A. 9
B. -2
C. 1
D. 3
E. -9

Question 3 (1 mark) TECH-ACTIVE

MC State which of the following is **not** correct.

A. The anti-derivative of acceleration gives velocity.
B. Acceleration is the rate of change of velocity.
C. Velocity is the rate of change of distance.
D. The derivative of velocity gives acceleration.
E. The anti-derivative of velocity gives position.

Question 4 (1 mark) TECH-ACTIVE

MC A particle moves in a straight line so that its velocity at time t seconds is given by $v = 2, t \geq 0$. Initially, the particle is 6 metres to the left of a fixed origin. The time (in seconds) that it takes to reach the origin is

A. 0
B. 1
C. 2
D. 3
E. 4

Question 5 (1 mark) TECH-ACTIVE

MC Evaluate $\int_{-2}^{2} 3x^2 \, dx$.

A. 0
B. 6
C. 12
D. 16
E. 24

More exam questions are available online.

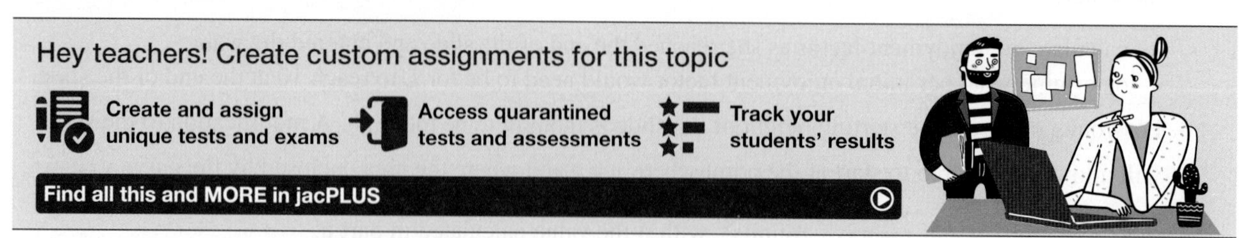

Answers

Topic 13 Anti-differentiation and introduction to integral calculus

13.2 Anti-derivatives

13.2 Exercise

1. a. $y = \frac{1}{4}x^4 + c$ b. $y = \frac{1}{7}x^7 + c$ c. $y = \frac{1}{9}x^9 + c$
   d. $y = 6x^2 + c$ e. $y = -2x^3 + c$ f. $y = \frac{1}{15}x^5 + c$
2. a. $\frac{1}{6}x^6 + c$ b. $\frac{1}{8}x^8 + c$ c. $x^9 + c$
   d. $\frac{1}{2}x^{12} + c$ e. $x^2 + 3x + c$ f. $10x - \frac{x^3}{3} + c$
3. a. $F(x) = x^{10} + c$
   b. $F(x) = -\frac{x^{13}}{13} + c$
   c. $F(x) = x^6 + 2x^2 + c$
   d. $F(x) = \frac{x^4}{4} - \frac{7x^3}{3} + c$
   e. $F(x) = 4x + \frac{x^2}{2} + \frac{x^3}{3} + c$
   f. $F(x) = \frac{x^4}{4} + 3x^3 - \frac{5x^2}{2} + x + c$
4. a. $y = 2x^6 + c$
   b. $\frac{4}{3}x^3 + x^2 - 5x$
   c. $f(x) = x^3 + x^2 - 16x + c$
5. a. $y = \frac{1}{2}x^{10} + c$
   b. $y = -3x + \frac{1}{2}x^8 + c$
   c. $y = \frac{2}{3}x^3 - 6x^2 + 14x + c$
   d. $y = -\frac{2}{3}x^3 + \frac{11}{2}x^2 + 40x + c$
6. a. $f(x) = \frac{1}{12}x^6 + \frac{1}{3}x^7 + c$
   b. $f(x) = \frac{4}{21}x^7 + 5x + c$
   c. $f(x) = \frac{4}{3}x^3 - \frac{6}{7}x^7 + c$
   d. $f(x) = 9x - 4x^3 + \frac{4x^5}{5} + c$
7. a. $\frac{1}{15}x^9 + c$
   b. $2x + c$
   c. $40x^2 - \frac{5x^8}{2} + c$
   d. $\frac{1}{100}\left[9x + 2x^3 - 0.5x^{11}\right] + c$
8. a. $ax^2 + bx + c$
   b. $0.0005x^{100} + c$
   c. $2x^4 + 4x^3 + 3x^2 + x$
   d. $7x - x^5 + \frac{4x^3}{3} + 4x^2 + c$
9. $y = \frac{2x^3}{3} - \frac{3x^2}{2} + c$
10. a. $F(x) = \frac{1}{4}(x^3 - 2x) + c$ b. $F(x) = \frac{1}{24}x^2 + \frac{2}{3}x + c$
    c. $F(x) = \frac{x}{4} + \frac{x^{15}}{12} + c$ d. $F(x) = \frac{4x^9}{9} - \frac{16x}{3} + c$
11. a. $-\frac{7x^5}{5} + x^3 - 3x^2 + c$ b. $\frac{5}{28}x^7 + \frac{3}{8}x^2 + c$
    c. $F(x) = \frac{2}{5}x^5 + c$ d. $3x^3 + 12x^2 + 16x + c$
12. $\int (2t^2 + 4)\, dt = \frac{2t^3}{3} + 4t + c = 2\int (t^2 + 2)\, dt$
13. $\frac{x^4}{2} + \frac{7x^2}{2} - 5x + c$
14. $F(x) = \frac{4a^2x^3}{3} + b^3x + c$
15. a. $\frac{1}{4}x^4 + \frac{4}{3}x^3 - \frac{1}{2}x^2 - 4x + c$
    b. $\frac{x^3}{3} - 16x + c$
    c. $2x^2 - 2x$
    d. $4x + 7$
16. a. $\frac{x^4}{4} + c$
    b. i. $\frac{x^9}{9} + \frac{2x^5}{5} + x$ ii. $50t(t - 10) = 50t^2 - 500t$
    iii. $\frac{y^6}{6} - \frac{y^4}{4}$ iv. $\frac{(4u + 5)^3}{12}$

13.2 Exam questions

Note: Mark allocations are available with the fully worked solutions online.

1. E
2. A
3. E

13.3 Anti-derivative functions and graphs

13.3 Exercise

1. a. $f(x) = 8x^2 - 5$
   b. $f(x) = x^3 + 2x^2 + 8$
   c. $f(x) = 3x^3 - 2x^2 - 13$
   d. $f(x) = \frac{25x^3}{3} + 5x^2 + x - \frac{25}{3}$
2. $f(x) = -x^3 + 4x + 5$

3. a. $y = -\frac{5}{2}x^2 + 4x + 5$

b. $y = \frac{1}{3}x^3 - \frac{1}{2}x^2 - 2x - 5$

4. a. $y = \frac{x^2}{5} - 3x + 10$ b. $(5, 0), (10, 0)$

5. $y = 19$

6. a. $y = -3x^2 - 6x + 7$ b. $f(-3) = 18$

7. a. $k = -\frac{3}{2}$ b. $y = -\frac{3x^2}{4} + 8$

8. $F(x) = \frac{x^3}{3} - \frac{9}{2}x^2 + 20x - \frac{101}{6}$

9. a. $y = -9$ b. $z = 128$

10. a. $a = 8$ b. $y = 8x - \frac{x^4}{4} - 3$

c. $y = 7x - \frac{9}{4}$

11. $y = \frac{1}{4}x^2 + 2x + 6$; minimum turning point at $(-4, 2)$

12. a. Maximum turning point at $x = 0$

b.

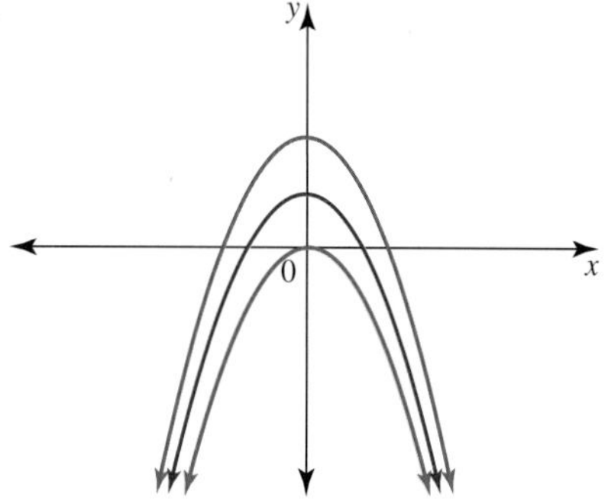

Family of curves $y = -x^2 + c$

c. $f(x) = -2x$

d. $f(x) = -x^2 + 1$

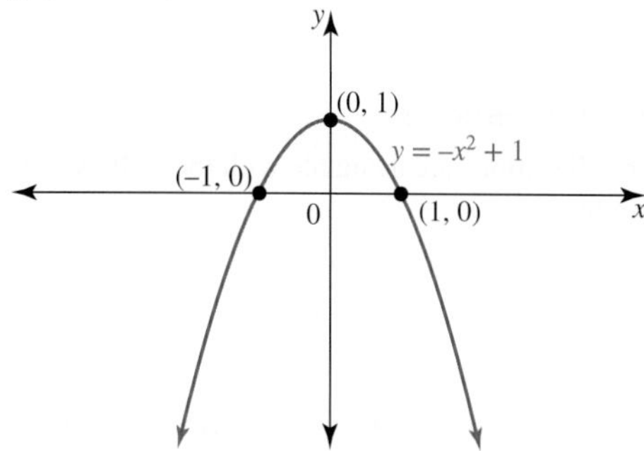

13. a. E

b. A

c. i. F ii. T iii. T

d.

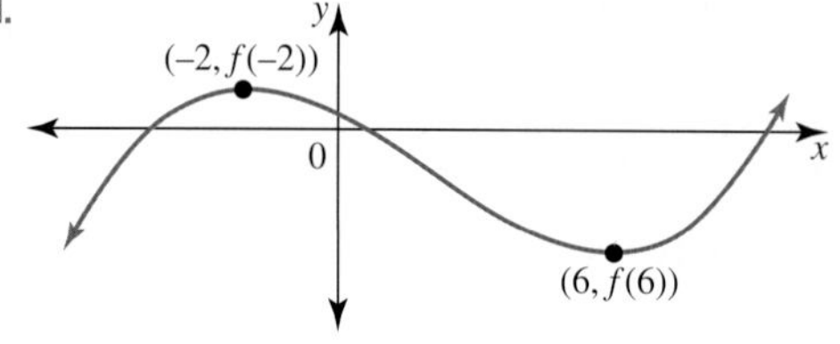

14. a. Minimum turning point at $x = -2$; maximum turning point at $x = 2$

b.

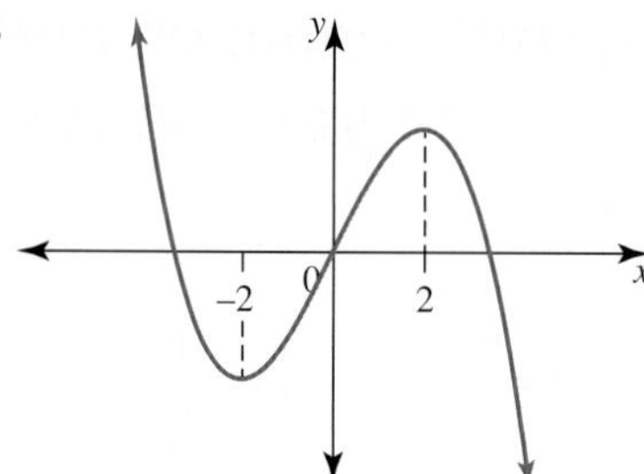

c. $\frac{dy}{dx} = 4 - x^2$; $y = 4x - \frac{x^3}{3} + c$

d. $y = -\frac{x^3}{3} + 4x - 3$; slope of -5

15. a. Any line with gradient of 3

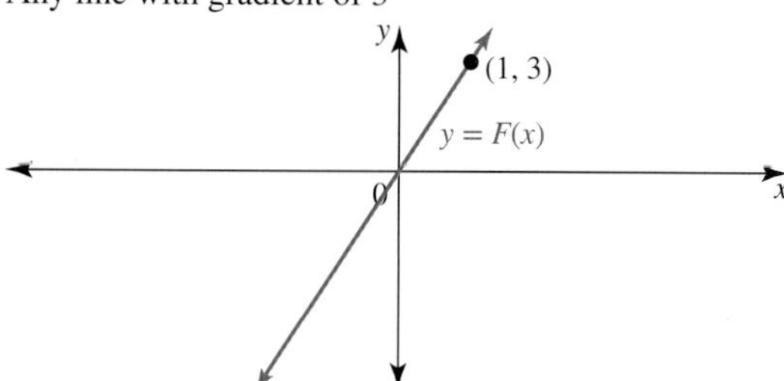

b. Any parabola with a maximum turning point when $x = 3$

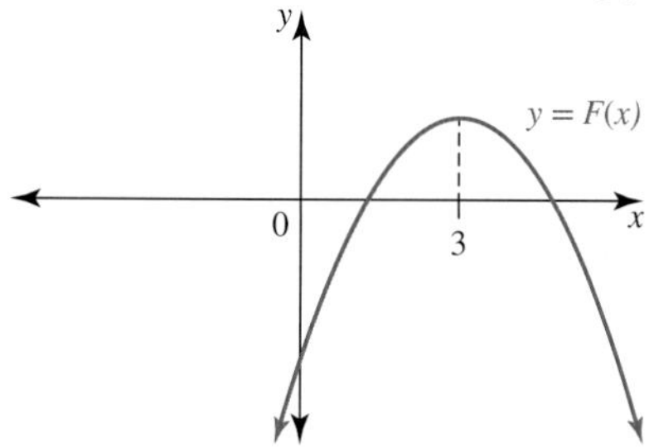

c. Any inverted cubic with a stationary point of inflection when $x = 3$

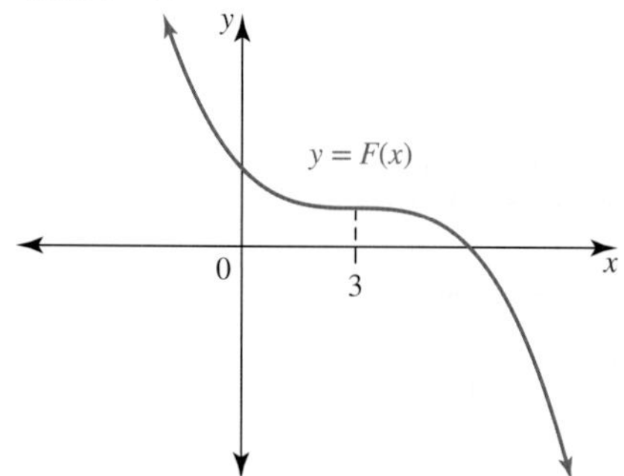

d. Any quartic with minimum turning points when $x = \pm 3$ and a maximum turning point when $x = 0$

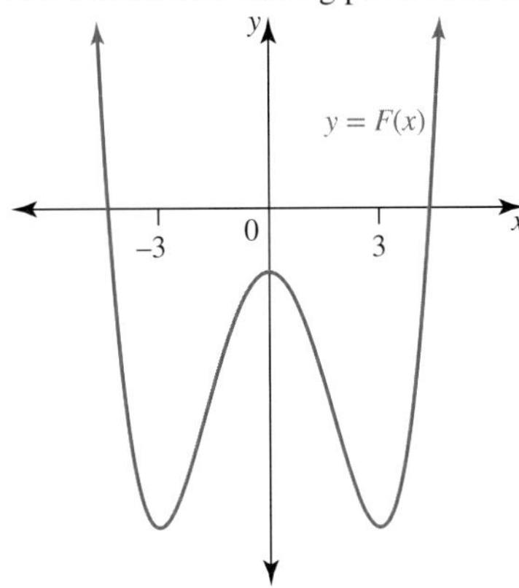

16. a. $-\dfrac{9x^4}{4} + 18x + c$

b. i. $y = -\dfrac{9x^4}{4} + 18x$

ii. $(2, 0)$; steeper at $(2, 0)$

iii. Maximum turning point when $x = \sqrt[3]{2}$

iv. $p = -\dfrac{2}{3}, q = 3$

v.

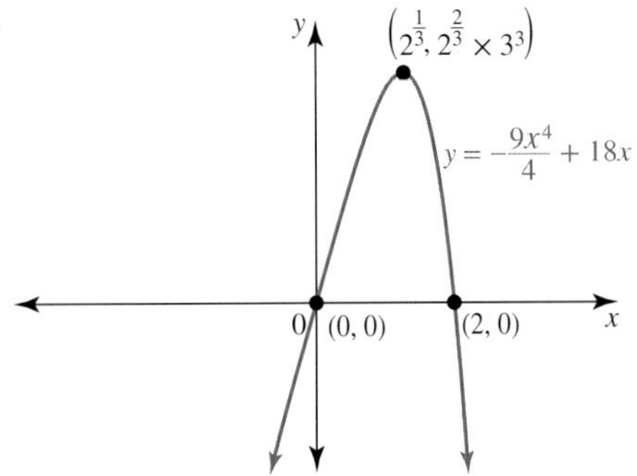

17. a–c. Use CAS technology to sketch the graphs.

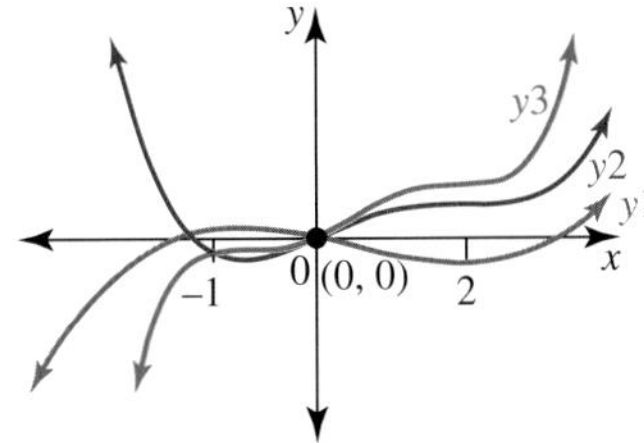

d. For factors of multiplicity 1, the anti-derivative graph has turning points at the zeros of each factor.
For factors of multiplicity 2, the anti-derivative graph has stationary points of inflection at the zeros of each factor.

18. Use your CAS technology to sketch the graphs.

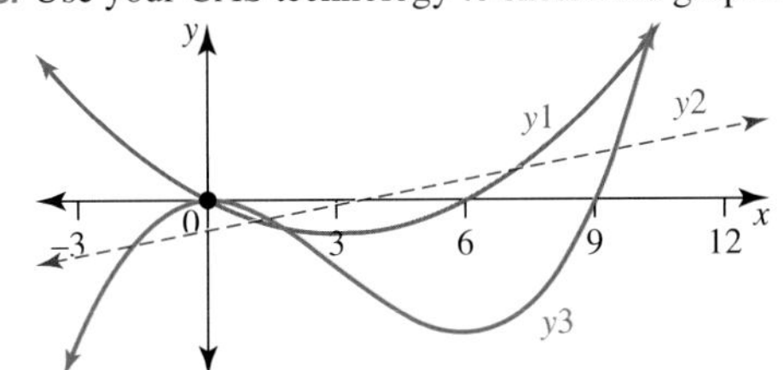

The x-intercept of $y = f'(x)$ is connected to the turning point of $y = f(x)$.

The turning points of $y = \int f(x)\,dx$ are connected to the x-intercepts of $y = f(x)$.

13.3 Exam questions

Note: Mark allocations are available with the fully worked solutions online.

1. C

2. $y = \dfrac{5x^2}{2} - 20x + 46$

3. $y = -\dfrac{3}{8}x^2 + 3x + 9$

13.4 Applications of anti-differentiation

13.4 Exercise

1. a. $x = t^2 + 5t$
   b. $x = -t^3 + 4t^2 - 2$
   c. i. $a = 4t - 12$ ii. $x = \dfrac{2t^3}{3} - 6t^2 + 18t - 18$
2. a. $v = 2t^2$ b. $x = \dfrac{2}{3}t^3 + 1$
3. a. $x = 6t - t^2 - 3$ b. 3 seconds
4. $x = 2t^3 + 2t - 5$; $a = 12t$
5. a. $x = \dfrac{8t^3}{3} - 10t^2 - 12t + 54$
   b. $19\frac{1}{3}$ m
   c. At the origin
6. 13 metres to the right of the origin
7. a. $v = 3t^2 + 8t - 8$ b. $x = t^3 + 4t^2 - 8t + 5$
8. a. $2\sqrt{2}$ seconds b. 14 m/s^2
9. Velocity 9 m/s; position 9 m to the right of the origin
10. 122.5 m
11. a. 2 seconds; 20 m to right of origin
    b. 4 seconds
12. a. $a = -6$
    b. 1 second; 12 m to the right of origin
    c. 15 m
    d. 5 m/s
    e. −3 m/s
13. 64.4 cm
14. a. $A = -9t^2 + 90$ b. 4 days
    c. $0 \le t \le \sqrt{10}$
15. a. 4 days
    b. $A(t) = 4t - \dfrac{t^2}{2}$; domain $[0, 8]$
    c. 8 m^2

d. 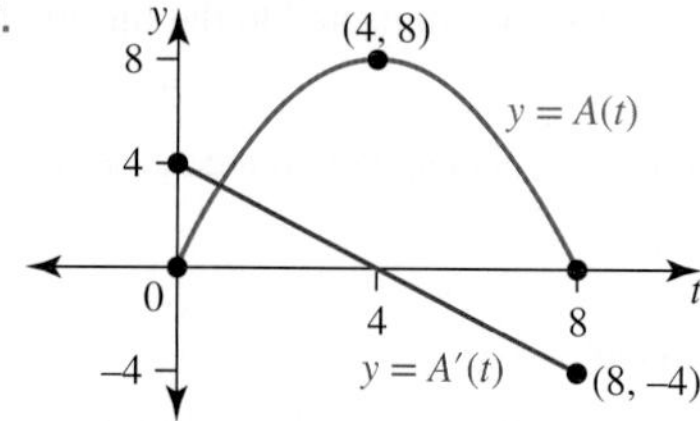

16. a. 24 m/s; -18 m/s^2 b. $x(t)=t^3-9t^2+24t$
c. 20 d. 28 m

17. a. $V=-0.25t+4.5\pi$ b. 56.5 seconds

18. a. Part of a parabola with a maximum turning point at $t=2$

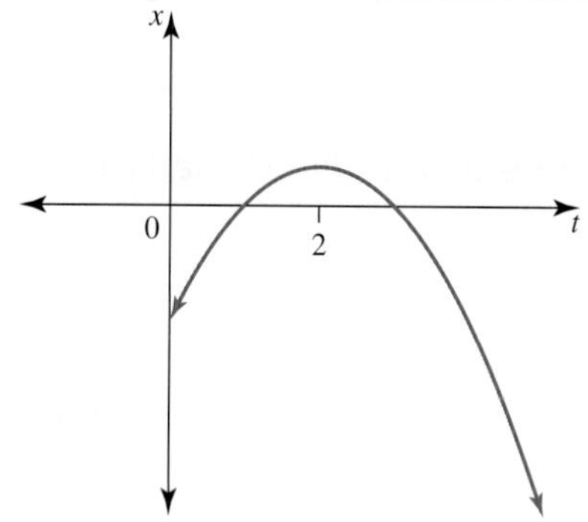

b. Part of a horizontal line below the t-axis through $(0,\ -2)$

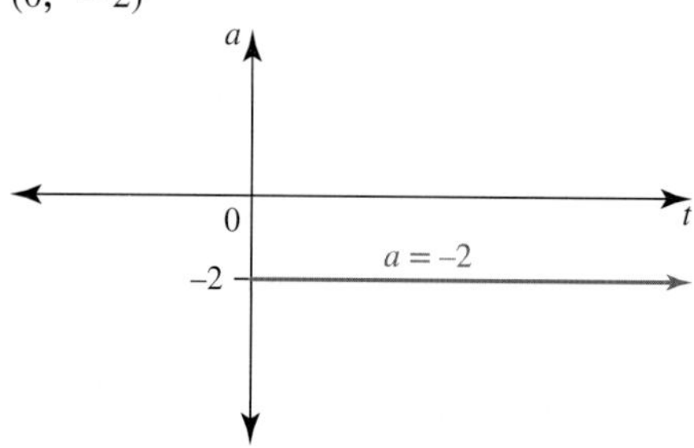

c. $x=-(t-2)^2,\ t\geq 0$

d. Part of a parabola with maximum turning point at $(2,\ 0)$

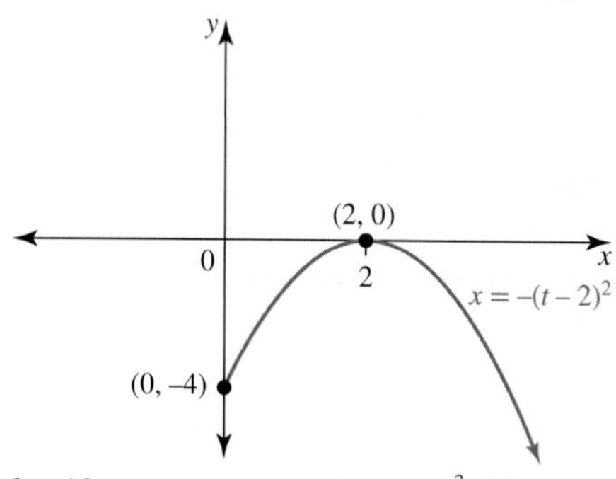

19. a. $k=12$ b. $m=4t^3+20$
c. 12 days

20. a. $a=8(2t+1)^3$ b. $x=\dfrac{(2t+1)^5}{10}+4.1$
c. 0.56 seconds; 20.27 m/s

13.4 Exam questions

Note: Mark allocations are available with the fully worked solutions online.

1. D
2. Position at $t=3$ is 80 m.
3. $t=2$ minutes

13.5 The definite integral

13.5 Exercise

1. a. 15 b. 8 c. 21
d. 16 e. -120 f. 0
2. 18
3. a. $\frac{1}{3}$ b. 5 c. $6\frac{3}{4}$ d. 4
4. a. $-\frac{32}{3}$ b. 0
5. a. 64 b. 6 c. 70
d. -351 e. $\frac{8}{3}$ f. 0
6. a. $a=18$ b. $b=4$
c. $k=10$ d. $n=2$
7. Sample responses can be found in the worked solutions in the online resources.
8. $n=7$
9. a. $\int_{-2}^{2}(3x+6)\,dx$; 24 square units
b. $\int_{-6}^{-2}1\,dx$; 4 square units
c. $\int_{-2}^{0}(4-3x)\,dx$; 14 square units
10. a. 6 square units b. $\int_{0}^{4}0.75x\,dx$ c. 6 square units
11. a. $\int_{1}^{3}(16-x^2)\,dx$ b. $23\frac{1}{3}$ square units
12. a. i. $\int_{-1}^{1}(x^2+1)\,dx$ ii. $2\frac{2}{3}$ square units
b. i. $\int_{-2}^{1}(1-x^3)\,dx$ ii. $6\frac{3}{4}$ square units

13. a. The area enclosed by $y = 4x^2$, the x-axis and $x = 0,\ x = 3$ is 36 square units.

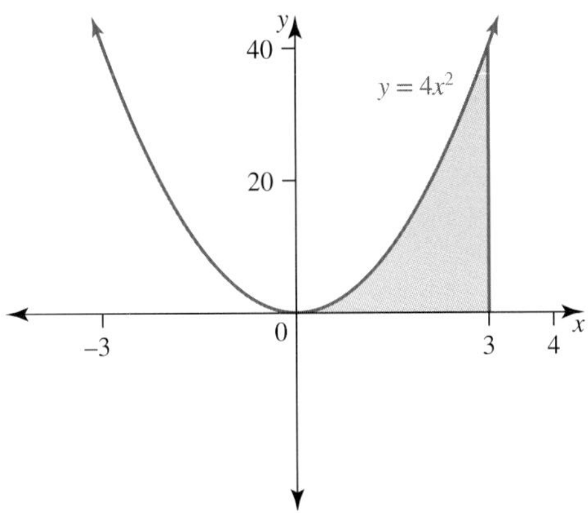

b. The area enclosed by $y = 1 - x^2$ and the x-axis is $\frac{4}{3}$ square units.

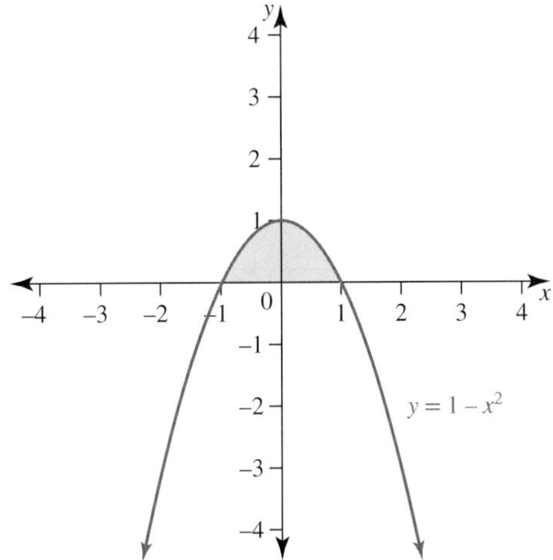

c. The rectangular area enclosed by the line $y = 1$, the x-axis and $x = -2,\ x = 4$ is 6 square units.

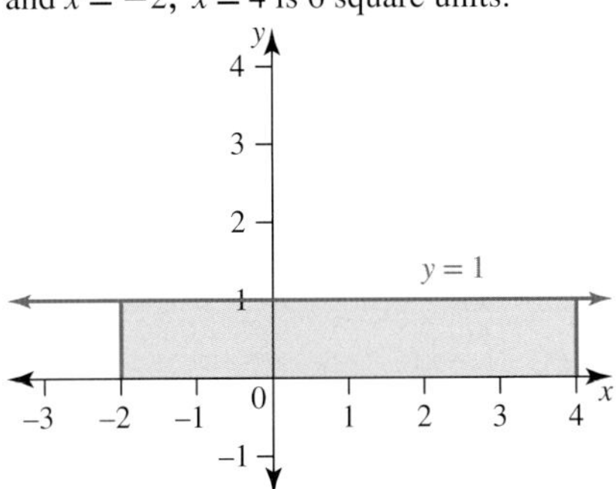

14. a. $v = 3$ m/s; $a = 3$ m/s^2

b. i.

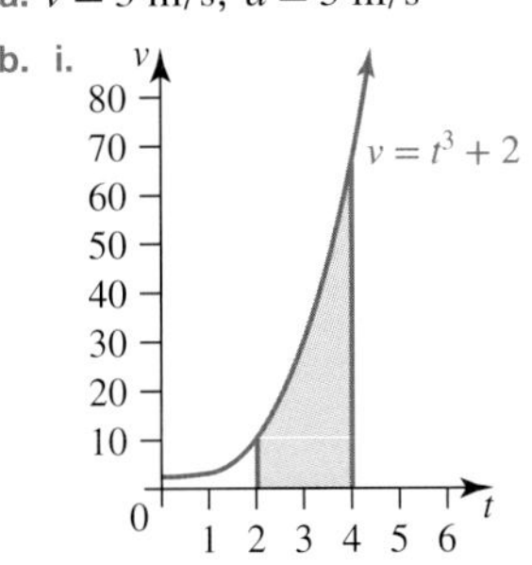

ii. $\int_2^4 (t^3 + 2)\, dt = 64$, so the distance is 64 metres.

15. a. Velocity is a positive definite quadratic function in t.

b. i. 14 m ii. 14 m

16. $\int_{0.5}^{1.5} (1 + x)(3 - x)\, dx,\ 3\frac{11}{12}$ square units

17. a. $x = 5t^2 - \frac{5t^3}{3}$

b. $6\frac{2}{3}$ m

c.

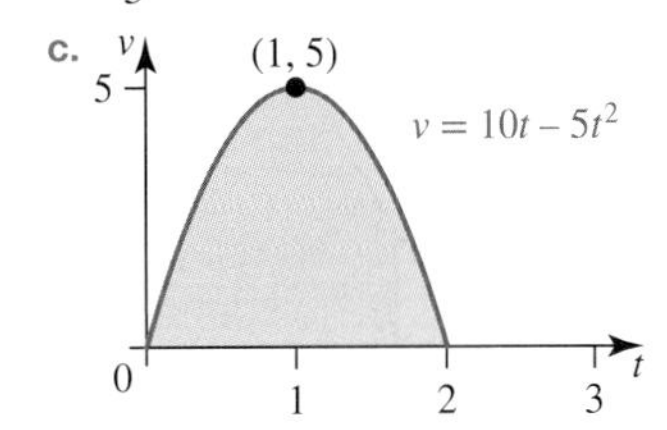

d. $6\frac{2}{3}$ m

e. Distance travelled in first 2 seconds.

18. a. $7\frac{1}{3}$ km/h; 2 km/h

b. 1 hour, 2 hours and 4 hours

c. $26\frac{2}{3}$ km

19. a. $\frac{203}{6}$

b. $-\frac{203}{6}$

c. Interchanging the terminals changes the sign of the integral.

20. a. $-\frac{125}{6}$

b.

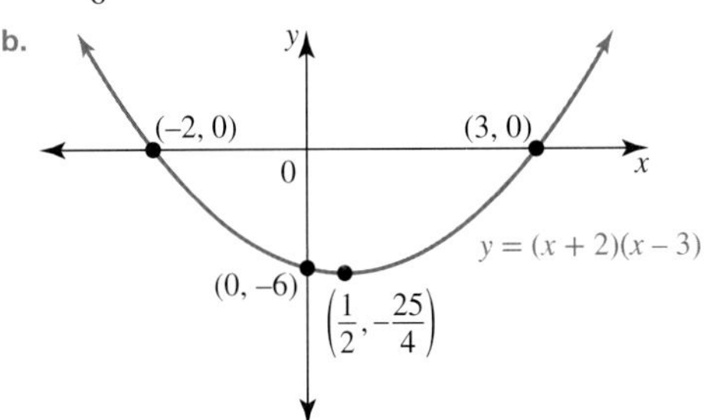

c. $\frac{125}{6}$ square units

d. The area lies below the x-axis, so the signed area is negative.

Possible integrals are $-\int_{-2}^{3} (x + 2)(x - 3)\, dx$ or $\int_{3}^{-2} (x + 2)(x - 3)\, dx$.

13.5 Exam questions

Note: Mark allocations are available with the fully worked solutions online.

1. a. 6 m/s; 12 m/s^2

b. 

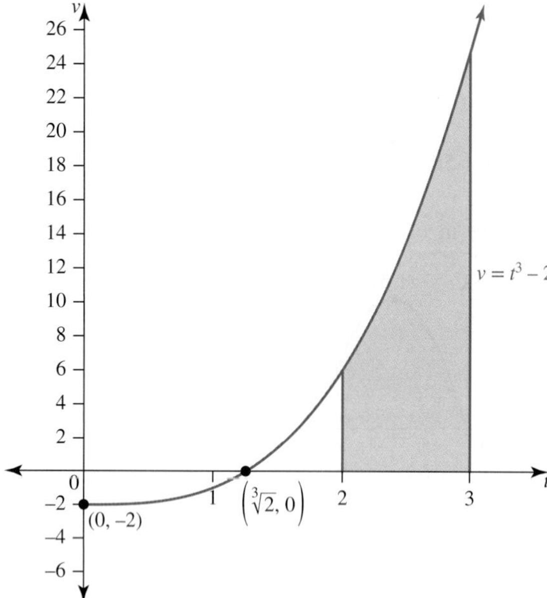

c. The distance travelled by the particle over the time interval [2, 3] is 14.25 metres.

2. 

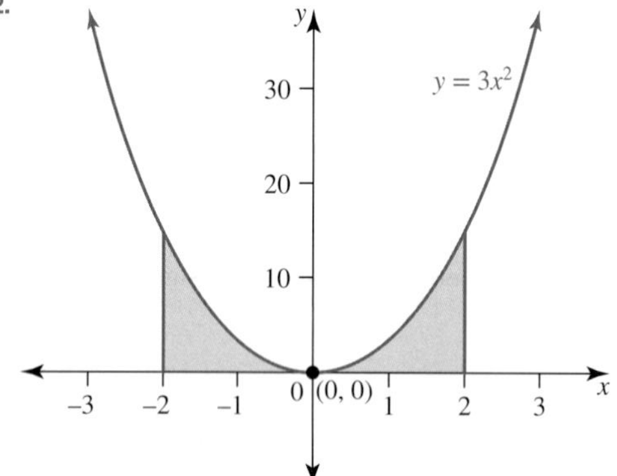

16 square units

3. E

4. a. 6 m/s^2

b. $x = \dfrac{2t^3}{3} - 3t^2 + 3$

c. 9 m

5. a. -2

b. $\dfrac{28}{3}$

6.

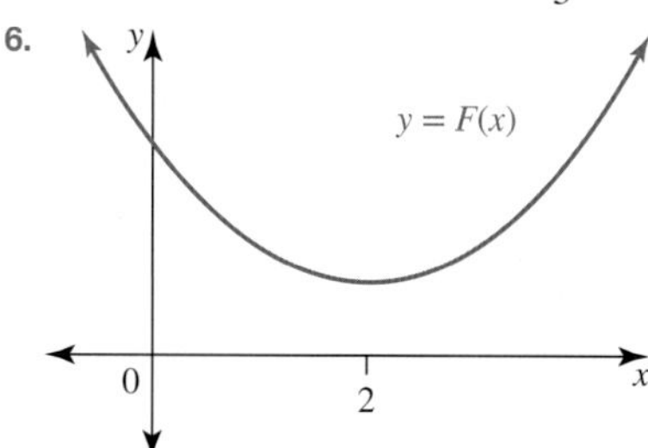

Technology active: multiple choice

7. C 8. E 9. E 10. D 11. C
12. B 13. D 14. A 15. D 16. C

Technology active: extended response

17. a. $v = 4t + 3,\ x = 2t^2 + 3t - 6$

b. i. 8 metres

ii. After 3 seconds, 21 metres to right of the origin

c. i.

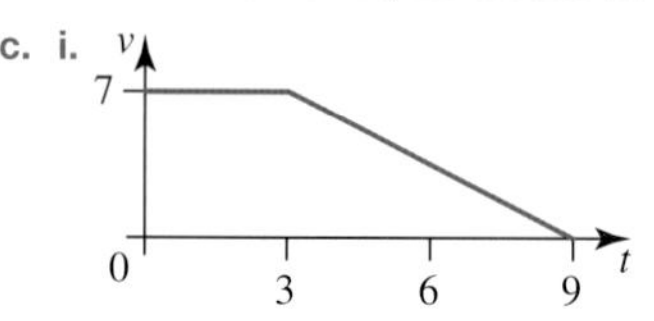

ii. 42 metres

d. 42 metres

18. a. $a = -1,\ b = -2$

b. $f(x) = -\dfrac{1}{2}x^2 - 2x + 4$

c. $x = -2 \pm 2\sqrt{3}$

d. $f(x) = -\dfrac{x^3}{6} - x^2 + 4x + 4$

19. a. Given $\int f(x)\,dx = f(x) + c$, $f(x)$ is an anti-derivative of $f(x)$.

i. $\displaystyle\int_a^b f(x)\,dx = -\int_b^a f(x)\,dx$

$$\text{LHS} = \int_a^b f(x)\,dx$$
$$= [F(x)]_a^b$$
$$= F(b) - F(a)$$

$$\text{RHS} = -\int_b^a f(x)\,dx$$
$$= -[F(x)]_b^a$$
$$= -(F(a) - F(b))$$
$$= -F(a) + F(b)$$
$$= F(b) - F(a)$$

$\therefore$ LHS = RHS

13.6 Review

13.6 Exercise

Technology free: short answer

1. a. $2x - 5x^2 - \dfrac{6x^5}{5} + c$

b. $\dfrac{x^4}{2} - 9x^2 + c$

c. $\dfrac{x^3}{12} - \dfrac{5x^4}{32} + c$

d. $\dfrac{1}{7}x^7 + \dfrac{3}{5}x^5 + x^3 + x + c$

2. 11

3. a. $y = \dfrac{x^2}{24} + x - 36$

b. $y = x - 36$

ii. $\int_a^b f(x)\,dx + \int_b^c f(x)\,dx = \int_a^c f(x)\,dx$

$$\begin{aligned}\text{LHS} &= \int_a^b f(x)\,dx + \int_b^c f(x)\,dx \\ &= [f(x)]_a^b + [f(x)]_b^c \\ &= F(b) - F(a) + F(c) - F(b) \\ &= F(c) - f(a)\end{aligned}$$

$$\begin{aligned}\text{RHS} &= \int_a^c f(x)\,dx \\ &= [f(x)]_a^c \\ &= F(c) - F(a)\end{aligned}$$

$\therefore$ LHS = RHS

iii. $\int_a^a f(x)\,dx = 0$

$$\begin{aligned}\text{LHS} &= \int_a^a f(x)\,dx \\ &= [f(x)]_a^a \\ &= F(a) - F(a) \\ &= 0 \\ &= \text{RHS}\end{aligned}$$

b. i. -5

ii. 5

iii. 15

iv. -2

20. a. i. $\int_{-4}^{0} \frac{1}{2}x^2\,dx$

ii. $A = \frac{32}{3}$ square units

b. 8 m

c. i. 8.5

ii. 5.5

d. i.

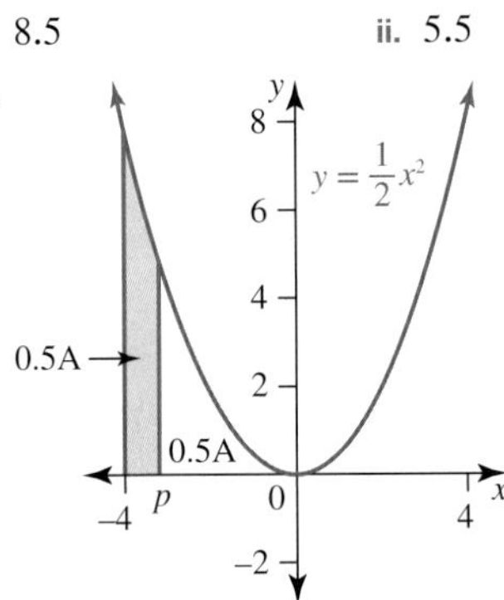

$p = -\sqrt[3]{32}$

ii. $2^{\frac{7}{3}} \approx 5$ m

13.6 Exam questions

Note: Mark allocations are available with the fully worked solutions online.

1. $\frac{1}{2}x^4 + \frac{5}{3}x^3 - 4x^2 - 20x + c$
2. B
3. C
4. D
5. D

GLOSSARY

absolute maximum *see* global maximum
absolute minimum *see* global minimum
acceleration the rate of change of velocity with respect to time
addition principle states that, for mutually exclusive procedures, if there are m ways of doing one procedure and n ways of doing another procedure, then there are $m+n$ ways of doing one or the other procedure
amplitude the distance a sine or cosine graph oscillates up and down from its equilibrium, or mean, position
anti-differentiation the reverse process to differentiation
anticlockwise moving around a circle in the opposite direction to the hands of a clock
arc a section of the circumference of a circle. A *minor arc* is shorter than the circumference of the semicircle; a *major arc* is longer than the circumference of the semicircle.
arrangement in probability, a counting technique in which order is important; also known as a permutation
asymptotes a line that a graph approaches but never reaches. A horizontal asymptote shows the long-term behaviour as, for example, $x \to \infty$; a vertical asymptote may occur where a function is undefined such as at $x=0$ for the hyperbola $y=\frac{1}{x}$.
average rate of change of a function f over an interval $[a, b]$ is measured by $\frac{f(b)-f(a)}{b-a}$. This is the gradient of the chord joining the end points of the interval on the curve $y=f(x)$.
average velocity average rate of change of position with respect to time or $\frac{\text{change in position}}{\text{change in time}}$
axis of symmetry a line about which a graph is symmetrical
binomial coefficients the coefficients of the terms in the binomial expansion of $(x+y)^n$, with $\binom{n}{r}$ or ${}^nC_r$ the coefficient of the $(r+1)$th term. For r and n non-negative whole numbers, $\binom{n}{r}=\frac{n!}{r!\,(n-r)!}$, $0 \le r \le n$.
binomial probability probability with two outcomes, favourable and non-favourable, the sum of which is 1
bisection method a procedure for improving the estimate for a root of an equation by halving the interval in which the root is known to lie
boundary points in trigonometry, the four points $(1, 0)$, $(0, 1)$, $(-1, 0)$, $(0, -1)$ on the circumference of the unit circle that lie on the boundaries between the four quadrants
box table a method of calculating the number of possible arrangements or permutations
circular functions sine, cosine, tangent. On a unit circle, $\cos(\theta)$ is the x-coordinate of the trigonometric point $[\theta]$; $\sin(\theta)$ is the y-coordinate of the trigonometric point $[\theta]$; $\tan(\theta)$ is the length of the intercept the line through the origin and the trigonometric point $[\theta]$ cuts off on the tangent drawn to the unit circle at $(1, 0)$.
clockwise moving around a circle in the same direction as the hands of a clock
co-domain the set of all y-values available for pairing with x-values to form a mapping according to a function rule $y=f(x)$
collinear describes points that lie on the same straight line
combination *see* selection
combinatoric coefficients *see* binomial coefficients
combined transformation a transformation made up of two or more transformations
complement the set A' containing all the elements of the sample space that are not in set A
completing the square technique used to express a quadratic as a difference of two squares. To apply this technique to a monic quadratic expression in x, add the square of half the coefficient of the term in x and then subtract this. This enables $x^2 \pm mx$ to be expressed as $\left(x \pm \frac{m}{2}\right)^2 - \left(\frac{m}{2}\right)^2$.
concave down describes parabolas that open downwards
concave up describes parabolas that open upwards
concurrent describes lines that intersect at a common point

conditional probability the probability of an event given that another event has occurred

conjugate surds pairs of surd expressions that, when multiplied, always result in a rational number, since $\left(\sqrt{a}+\sqrt{b}\right)\left(\sqrt{a}-\sqrt{b}\right)=a-b$; the expressions $\sqrt{a}+\sqrt{b}$ and $\sqrt{a}-\sqrt{b}$ are a pair of conjugate surds

constant term (of a polynomial or other expression) the term that does not contain the variable

continuous describes a graph that forms a curve with no breaks

correspondence Four types of correspondence between coordinates can be used to classify relations: one-to-one, many-to-one, one-to-many and many-to-many. Graphically, the type of correspondence is determined by the number of intersections of a horizontal line (one or many) to the number of intersections of a vertical line (one or many) with the graph.

cube root function the inverse of the cubic function; expressed as $f: R \to R, f(x) = \sqrt[3]{x}$

decreasing describes a function with a negative gradient; a function $y=f(x)$ is decreasing if, as x increases, y decreases

degree of a polynomial the highest power of a variable in a polynomial; for example, polynomials of degree 1 are linear, degree 2 are quadratic and degree 3 are cubic

determinant of a matrix a value associated with a square matrix, evaluated by multiplying the elements in the leading diagonal and subtracting the product of the elements in the other diagonal;

$\det(A)=\Delta=|A|=\begin{vmatrix} a & b \\ c & d \end{vmatrix}=ad-bc$

derived function or **derivative function** *see* gradient function

difference quotient $\dfrac{f(x+h)-f(x)}{h}$; measures the gradient of a secant or chord drawn through two points on the curve of a function f. The difference quotient represents the average rate of change of the function over the domain interval $[x, x+h]$.

differentiable describes a function that is continuous at $x=a$ and its derivative from the left of $x=a$ equals its derivative from the right of $x=a$: $f'(a^-)=f'(a^+)$. Smoothly continuous functions are differentiable.

differentiation the process of obtaining the gradient function or derivative

dilation a linear transformation that enlarges or reduces the size of a figure by a scale factor k parallel to either axis or both

dilation factor measures the amount of stretching or compression of a graph from an axis

dilation from the *x*-axis a linear transformation that enlarges or reduces the size of a figure by a scale factor k parallel to the y-axis where $(x, y) \to (x, ky)$

dilation from the *y*-axis a linear transformation that enlarges or reduces the size of a figure by a scale factor k parallel to the x-axis where $(x, y) \to (kx, y)$

discontinuity describes a point in a graph at which there is a break

discontinuous describes a graph that has a break at a given point

discriminant for the quadratic expression ax^2+bx+c, the discriminant is $\Delta=b^2-4ac$

disjoint describes sets that do not intersect; if $A \cap B = \varnothing$, the null set, then the sets A and B are disjoint

Distributive Law $a(b+c)=ab+ac$

division of polynomials can be performed using CAS technology, long division or the inspection method:

$\dfrac{\text{dividend}}{\text{divisor}}=\text{quotient}+\dfrac{\text{remainder}}{\text{divisor}}$

domain the set of all x-values of the ordered pairs (x, y) that make up a relation

element a member of a set; $a \in A$ means a is an element of, or belongs to, the set A. If a is not an element of the set A, this is written as $a \notin A$.

equating coefficients if $ax^2+bx+c \equiv 2x^2+5x+7$, then $a=2$, $b=5$ and $c=7$ for all values of x

equilibrium position or **mean position** the position about which a trigonometric graph oscillates

equiprobable having an equal likelihood of occurring

event a particular set of outcomes that is a subset of the sample space in an experiment

exclusion notation used to describe the set containing the elements in set A excluding any elements that are in set B; denoted as $A \backslash B$

expansion the process of writing an algebraic expression in extended form, removing brackets

exponent an index or power. For the number $n=a^p$, the base is a and the power, or index, or exponent is p.

exponential form also known as index form; a way of expressing a standard number n using a base a and an exponent, or index (power) x: $n = a^x$

exponential functions functions of the form $f: R \to R, f(x) = a^x, a \in R^+ \setminus \{1\}$

factor theorem states that for the polynomial $P(x)$, if $P(a) = 0$, then $(x - a)$ is a factor of $P(x)$

factors two or more numbers or algebraic expressions that divide a given quantity without a remainder

function a relation in which the set of ordered pairs (x, y) have each x-coordinate paired to a unique y-coordinate

global or **absolute maximum** a point on a curve where the y-coordinate is greater than that of any other point on the curve for the given domain

global or **absolute minimum** a point on a curve where the y-coordinate is smaller than that of any other point on the curve for the given domain

gradient measures the steepness of a line as the ratio $m = \frac{\text{rise}}{\text{run}}$. If (x_1, y_1) and (x_2, y_2) are two points on the line, $m = \frac{y_2 - y_1}{x_2 - x_1}$.

gradient function or the **derived** or the **derivative function** (of a function f) the function f' whose rule is defined as $f'(x) = \lim_{h \to 0} \frac{f(x+h) - f(x)}{h}$; for $y = f(x)$, $\frac{dy}{dx} = f'(x)$ where $\frac{dy}{dx} = \lim_{\delta x \to 0} \frac{\delta y}{\delta x}$

gradient–y-intercept form the equation of a line in the form $y = mx + c$, for a line with gradient m cutting the y-axis at c

horizontal parallel to the x-axis or with zero gradient

horizontal line test test that determines whether a function is one-to-one or many-to-one by considering the number of intersections a horizontal line makes with the graph of the function

horizontal translation a transformation that moves a figure a given distance in a horizontal direction where $(x, y) \to (x + a, y)$

hybrid function a function whose rule takes different forms for different sections of its domain

hyperbola a function with two branches of the form $y = \frac{a}{x - h} + k$

identically equal describes polynomials with the same coefficients of corresponding like terms. The $\equiv$ symbol is used to emphasise the fact that the equality must hold for all values of the variable.

image a figure after a transformation; for the function $x \to f(x)$, $f(x)$ is the image of x under the mapping f

implied *see* maximal

increasing describes a function with a positive gradient; a function $y = f(x)$ is increasing if, as x increases, y increases

indefinite integral the anti-derivative or primitive of $f(x)$, denoted as $F(x)$; $\int f(x)dx = F(x) + c$ where $f'(x) = f(x)$

independent events events that have no effect on each other

indeterminate not known or defined and unable to be known or defined

indicial equation equation in which the unknown variable is an exponent

inequation an expression that contains one of the order symbols $<$, $\leq$, $>$ or $\geq$ rather than $=$, the equality symbol

instantaneous rate of change the derivative of a function evaluated at a particular instant

integrand the expression that is to be integrated

integration the process of calculating the limiting sum of the area of a large number of strips of very small width in order to obtain the total area under a curve

intersections (of sets) the set containing the elements common to both A and B; denoted as $A \cap B$

interval notation a convenient alternative way of describing sets of numbers; $[a, b] = \{x: a \leq x \leq b\}$, $(a, b) = \{x: a < x < b\}$, $[a, b) = \{x: a \leq x < b\}$, $(a, b] = \{x: a < x \leq b\}$

invariant point or **fixed point** a point of the domain of a function which is mapped onto itself after a transformation

inverse the opposite; for example, the inverse of 2 is −2. Also, a pair of relations for which the rule of one can be formed from the rule of the other by interchanging x- and y-coordinates. The inverse of a one-to-one function f is given the symbol f^{-1}.

inverse proportion a relationship between two variables such that when one variable increases the other decreases in proportion so that the product is unchanged; if y is inversely proportional to x, then $y = \frac{k}{x}$, where k is the constant of proportionality

irrational describes a number that cannot be expressed as a ratio of two integers

irrational number a number that cannot be expressed as a ratio of two integers

irreducible describes a quadratic that cannot be factorised over R

iteration a method of approximation in which a value is estimated and then the process repeated using the values obtained from one stage to calculate the value of the next stage, and so on

Karnaugh maps *see* probability table

kinematics the study of motion

leading term (of a polynomial or other expression) the term containing the highest power of a variable

limit the behaviour of a function as it approaches a point, not its behaviour at that point. If $\lim_{x \to a} f(x) = L$, then the function approaches the value L as x approaches a. The limit must be independent of which direction $x \to a$ so $L^- = L^+ = L$ for the limit to exist.

line segment a section of a line that is of finite length, with two end points

linear equation an equation in which the highest power of any variable is 1

linearity properties The derivative of the sum of two functions is the sum of the derivatives of the two functions: $\frac{d}{dx}(f + g) = \frac{d(f)}{dx} + \frac{d(g)}{dx}$. A constant factor of a function is a factor of the derivative: if $y = af(x)$ then $\frac{dy}{dx} = af'(x)$ and a is a constant.

literal equations equations containing pronumerals rather than numerals as terms or coefficients

local maximum a stationary point on a curve where the gradient is zero and the y-coordinate is greater than those of neighbouring points

local minimum a stationary point on a curve where the gradient is zero and the y-coordinate is smaller than those of neighbouring points

logarithm an index or power. If $n = a^x$, then $x = \log_a(n)$ is an equivalent statement.

long division a method of performing division

long division algorithm a method to divide a polynomial, $P(x)$, by another polynomial, $A(x)$, to obtain the quotient, $Q(x)$, and the remainder, $R(x)$, where $P(x) = A(x) \times Q(x) + R(x)$ or $\frac{P(x)}{A(x)} = Q(x) + \frac{R(x)}{A(x)}$

long-term behaviour refers to the effect on the y-values of a graph as the x-values approach $\pm\infty$

lower terminal the number a placed in the lower position on the integral symbol, $\int_a^b f(x)\,dx$, defining one end point of the interval or the lower limit on the values of the variable x over which the integration takes place

many-to-many correspondence two or more elements in the domain map to two or more elements in the range

many-to-one correspondence two or more elements in the domain map to one element in the range

mapping a relation that pairs each element of a given set (the domain) with one or more elements of a second set (the range)

maximal or **implied** describes the domain over which a relation or function is defined

mean position *see* equilibrium position

midpoint the point on a line segment that sits halfway between the end points

monic a single-variable polynomial in which the leading term has a coefficient that is equal to 1

multiplication principle states that, if there are m ways of doing the first procedure and for each one of these there are n ways of doing the second procedure, then there are $m \times n$ ways of doing the first and the second procedures

multiplicity if $(x - a)^2$ is a factor of a polynomial then $x = a$ is a zero of multiplicity 2 or a repeated zero; if $(x - a)^3$ is a factor then $x = a$ is a zero of multiplicity 3

mutually exclusive describes events that cannot occur simultaneously

near neighbour a method for estimating the gradient of a curve at a point by calculating the average rate of change between two very close points on the curve

negative definite describes a quadratic expression $ax^2 + bx + c$ if $ax^2 + bx + c < 0$ for all $x \in R$

Newton's method or the **Newton–Raphson method** a method of solving an equation of the form $f(x) = 0$ by obtaining an initial approximation x_0 to the solution and improving this value by carrying out the iterative procedure $x_{n+1} = x_n - \dfrac{f(x_n)}{f'(x_n)}$, $n = 0, 1, 2, ...$

Null Factor Law mathematical law stating that if $ab = 0$ then either $a = 0$ or $b = 0$ or both a and b equal zero. This allows equations for which the product of two or more terms equals zero to be solved.

one-to-many correspondence one element in the domain maps to two or more elements in the range

one-to-one correspondence one element in the domain maps to one element in the range

outcome the result of an experiment

out of phase *see* phase difference

parabola (with a vertical axis of symmetry) the graph of a quadratic function; its equation may be expressed in turning point form $y = a(x - h)^2 + k$ with turning point (h, k), factored form $y = a(x - x_1)(x - x_2)$ with x-intercepts at $x = x_1, x = x_2$, or general form $y = ax^2 + bx + c$

parallel describes lines with the same gradient

parameter a varying constant in a common equation used to describe a family of polynomials

partial fractions fractions that sum together to form a single fraction

Pascal's identity ${}^nC_r = {}^{n-1}C_{r-1} + {}^{n-1}C_r$ for $0 < r < n$

Pascal's triangle a triangle formed by rows of the binomial coefficients of the terms in the expansion of $(a + b)^n$, with n as the row number

perfect cube $(a + b)^3 = a^3 + 3a^2b + 3ab^2 + b^3$

perfect square $(a + b)^2 = a^2 + 2ab + b^2$

period on a trigonometric graph, the length of the domain interval required to complete one full cycle. For the sine and cosine functions, the period is 2π since $\sin(x + 2\pi) = \sin(x)$ and $\cos(x + 2\pi) = \cos(x)$. The tangent function has a period of π since $\tan(x + \pi) = \tan(x)$.

permutation *see* arrangement

perpendicular bisector the line passing through the midpoint of, and at right angles to, a line segment

perpendicular lines a pair of lines with an angle between them of 90° the gradients have a product of -1, that is, $m_1m_2 = -1$

phase difference or **phase shift** the horizontal translation of a trigonometric graph. For example, $\sin(x - a)$ has a phase shift or a phase change of a or a horizontal translation of a from $\sin(x)$; and $\sin(x)$ and $\cos(x)$ are out of phase by $\frac{\pi}{2}$ or have a phase difference of $\frac{\pi}{2}$.

point–gradient form the equation of a line in the form $y - y_1 = m(x - x_1)$, for a line with gradient m and known point (x_1, y_1)

polynomial an algebraic expression in which each power of the variable is a positive whole number

position measures both distance and direction from a fixed origin

position–time graph a graph of the position of a particle plotted against time

positive definite describes a quadratic expression $ax^2 + bx + c$ if $ax^2 + bx + c > 0$ for all $x \in R$

power function function with rules of the form $f(x) = x^n, n \in Q$

pre-image the original figure before a transformation

primitive function given $f'(x)$, the primitive function is $f(x)$; the process of finding the primitive function is anti-differentiation

probability the long-term proportion, or relative frequency, of the occurrence of an event

probability table a format for presenting all known probabilities of events in rows and columns

probability tree diagram a graphic method used to represent a sample space

proper rational function form of rational function in which the degree of the numerator is smaller than that of the denominator

Pythagorean identity states that, for any θ, $\sin^2(\theta) + \cos^2(\theta) = 1$

quadrants the four sections into which the Cartesian plane is divided by the coordinate axes

quadratic formula the formula $x = \dfrac{-b \pm \sqrt{b^2 - 4ac}}{2a}$, used to find the solutions to the quadratic equation $ax^2 + bx + c = 0$

radian measure the measure of the angle subtended at the centre of a circle by an arc of length equal to the radius of the circle
radical (symbol) the symbol $\sqrt{}$ indicating the root of a number, for example $\sqrt{x}$ or $\sqrt[3]{8}$. For a square root of a positive real number, the radical indicates the positive root, for example $\sqrt{4}=2$.
range the set of y-values of the ordered pairs (x, y) that make up a relation
rate of change (instantaneous) the derivative of a function evaluated at a given value or instant
rational describes a number that can be expressed as a ratio of two integers
rational function function expressed as the quotient of two polynomials $\frac{P(x)}{Q(x)}, Q(x) \neq 0$
rational root theorem states that if a polynomial with integer coefficients has a rational zero $\frac{p}{q}$, then its constant term is divisible by p and its leading coefficient is divisible by q
rectangular hyperbola the function $f: R\backslash\{0\} \rightarrow R, f(x)=x^{-1}$
rectilinear motion motion in a straight line
reflection a transformation of a figure defined by the line of reflection where the image point is a mirror image of the pre-image point
reflection in the x-axis a transformation of a figure in the x-axis where $(x, y) \rightarrow (x, -y)$
reflection in the y-axis a transformation of a figure in the y-axis where$(x, y) \rightarrow (-x, y)$
relation any set of ordered pairs
remainder theorem states that if the polynomial $P(x)$ is divided by $(x-a)$, then the remainder is $P(a)$
restricted domain a subset of a function's maximal domain, often due to practical limitations on the independent variable in modelling situations
roots (of an equation) the solutions to an equation
sample space the set of all possible outcomes in an experiment, denoted as ξ
sampling with replacement a type of sampling whereby the object is replaced, resulting in an independent trial
sampling without replacement a type of sampling whereby the object is not replaced, resulting in a dependent trial
scientific notation or **standard form** expression of a number in the form $a \times 10^{b}$, the product of a number a, where $1 \leq a < 10$, and a power of 10
secant a line passing through two points on a curve
selection in probability, a counting technique in which order is not important; also known as a combination
set a collection of elements
set of integers the positive and negative whole numbers and the number zero; $Z=\{\ldots-2,-1,0,1,2,\ldots\}$
set of irrational numbers numbers that cannot be expressed in fraction form as the ratio of two integers, including numbers such as $\sqrt{2}$ and π; $Q'=R\backslash Q$
set of natural numbers the positive whole numbers (counting numbers); $N=\{1,2,3, \ldots\}$
set of rational numbers the numbers that can be expressed as quotients in the form $\frac{p}{q}$, including finite and recurring decimals as well as fractions and integers; Q is the set of quotients $\frac{p}{q}, q \neq 0$ with $p, q \in Z$
set of real numbers the union of the set of rational and irrational numbers; $R=Q \cup Q'$
sign diagram a diagram showing where the graph of a function lies above or below the x-axis and where the function is positive or negative
signed area the value of a definite integral, which can be positive, negative or zero depending on the position relative to the x-axis of the area it measures
significant figures the number of digits that would occur in a if the number was expressed in scientific notation as $a \times 10^{b}$ or as $a \times 10^{-b}$
slope the steepness of a line; *see* gradient
small increment a small change in a variable x, denoted by δx
smoothly continuous function a function for which the graph has no breaks or sharp points
solving the triangle the process of calculating all side lengths and all angle magnitudes of a triangle

square root function a function of the form $y = a\sqrt{x - h} + k$
standard form *see* scientific notation
stationary point a point where a function has a gradient of zero
stationary point of inflection a point on a curve where the tangent to the curve is horizontal and the curve changes concavity. For example, the graph of $y = x^3$ has a stationary point of inflection at $(0, 0)$.
straightening the curve a method for estimating the gradient of a curve at a point by calculating the average rate of change between two very close points on the curve
subset a set whose elements form part of a larger set; denoted as $A \subset B$ if every element of A is also an element of B. $A \not\subset B$ means A is not a subset of B.
surd irrational number containing a radical sign such as a square or cube root (but not all numbers with radical signs are surds)
symmetry properties indicate the relationships between trigonometric values of symmetric points in the four quadrants. If $[\theta]$ lies in the first quadrant, the symmetric points to it could be expressed as $[\pi - \theta]$, $[\pi + \theta]$, $[2\pi - \theta]$. This gives, for example, the symmetric property $\cos(\pi - \theta) = -\cos(\theta)$.
system of equations A 2×2 system of simultaneous equations has two equations in two unknowns, whereas a 3×3 system has three equations in three unknowns. The solutions for the values of the unknowns must satisfy each equation of the system: that is, they must satisfy each equation simultaneously.
tangent *see* tangent line
tangent line a line that touches a curve at a single point; for a circle, the tangent is perpendicular to the radius drawn to the point of contact
transcendental describes a number that cannot be expressed as the root of a polynomial equation, such as π
transcendental functions non-algebraic functions such as the trigonometric and exponential functions
transformations geometric operators that may change the shape and/or the position of a graph:

- under a dilation of factor a from the x-axis, or parallel to the y-axis, $y = f(x) \rightarrow y = af(x)$
- under a dilation of factor a from the y-axis, or parallel to the x-axis, $y = f(x) \rightarrow y = f\left(\frac{x}{a}\right)$
- under a reflection in the x-axis, $y = f(x) \rightarrow y = -f(x)$
- under a reflection in the y-axis, $y = f(x) \rightarrow y = f(-x)$
- under a translation of a parallel to the x-axis, $y = f(x) \rightarrow y = f(x - a)$
- under a translation of a parallel to the y-axis, $y = f(x) \rightarrow y = f(x) + a$

translation a transformation of a figure where each point in the plane is moved a given distance in a horizontal or vertical direction
trigonometric functions sine, cosine, tangent. On a unit circle, $\cos(\theta)$ is the x-coordinate of the trigonometric point $[\theta]$; $\sin(\theta)$ is the y-coordinate of the trigonometric point $[\theta]$; $\tan(\theta)$ is the length of the intercept the line through the origin and the trigonometric point $[\theta]$ cuts off on the tangent drawn to the unit circle at $(1, 0)$.
trigonometric point On a unit circle, the end point of an arc, drawn from the initial point $(1, 0)$ and of length θ, where θ is any real number, is the trigonometric point $[\theta]$. This point has Cartesian coordinates $(\cos(\theta), \sin(\theta))$.
truncus a function with two branches of the form $y = \frac{a}{(x - h)^2} + k$
union the set of all elements in any one set and in a combination of sets; $A \cup B$ contains the elements in A or in B or in both A and B
unit circle in trigonometry, the circle $x^2 + y^2 = 1$ with the centre at the origin and a radius of 1 unit
universal set the set of all possibilities in a given situation
upper terminal the number b placed in the upper position on the integral symbol, $\int_a^b f(x)dx$, defining one end point of the interval or the upper limit on the values of the variable x over which the integration takes place
velocity the instantaneous rate of change of position with respect to time
velocity–time graph a graph of the position of a particle plotted against time
Venn diagram a diagram that displays the union and intersection of sets

vertex (plural vertices) the turning point of a curve; the point where two lines meet to form an angle; the meeting point of two or more sides of a polygon or three or more edges of a solid
vertical parallel to the y-axis or with undefined gradient
vertical translation a transformation that moves a figure a given distance in a vertical direction where $(x, y) \rightarrow (x, y + a)$
zeros (of an equation) the values of x for which $f(x) = 0$

INDEX